AA002445

2018 31st International Conference on VLSI Design (VLSID 2018) and 2018 17th International Conference on Embedded Systems (ES 2018)

Pune, India
6-10 January 2018

IEEE Catalog Number: CFP18041-POD
ISBN: 978-1-5386-3693-0

Copyright © 2018 by the Institute of Electrical and Electronics Engineers, Inc.
All Rights Reserved

Copyright and Reprint Permissions: Abstracting is permitted with credit to the source. Libraries are permitted to photocopy beyond the limit of U.S. copyright law for private use of patrons those articles in this volume that carry a code at the bottom of the first page, provided the per-copy fee indicated in the code is paid through Copyright Clearance Center, 222 Rosewood Drive, Danvers, MA 01923.

For other copying, reprint or republication permission, write to IEEE Copyrights Manager, IEEE Service Center, 445 Hoes Lane, Piscataway, NJ 08854. All rights reserved.

*** This is a print representation of what appears in the IEEE Digital Library. Some format issues inherent in the e-media version may also appear in this print version.

IEEE Catalog Number:	CFP18041-POD
ISBN (Print-On-Demand):	978-1-5386-3693-0
ISBN (Online):	978-1-5386-3692-3
ISSN:	1063-9667

Additional Copies of This Publication Are Available From:

Curran Associates, Inc
57 Morehouse Lane
Red Hook, NY 12571 USA
Phone: (845) 758-0400
Fax: (845) 758-2633
E-mail: curran@proceedings.com
Web: www.proceedings.com

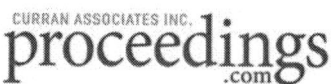

2018 31st International Conference on VLSI Design (VLSID 2018) and 2018 17th International Conference on Embedded Systems (ES 2018)

Pune, India
6-10 January 2018

IEEE Catalog Number: CFP18041-POD
ISBN: 978-1-5386-3693-0

Proceedings

31st International Conference on VLSI Design

Held concurrently with
17th International Conference on Embedded Systems

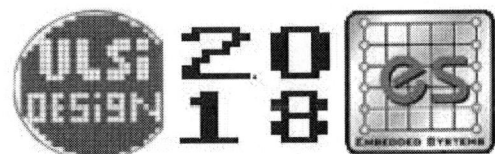

Pune, India   **6-10 January, 2018**

Technical Co-Sponsorship
Association for Computing Machinery (ACM)
IEEE Circuits and Systems Society
IEEE Computer Society
IEEE Council on Electronic Design Automation (CEDA)
Technical Committee on VLSI

Sponsored by
VLSI Society of India

Los Alamitos, California

Washington • Tokyo

Proceedings

31st International Conference on VLSI Design

Held concurrently with
17th International Conference on Embedded Systems

VLSID & ES 2018

Technical Sponsors

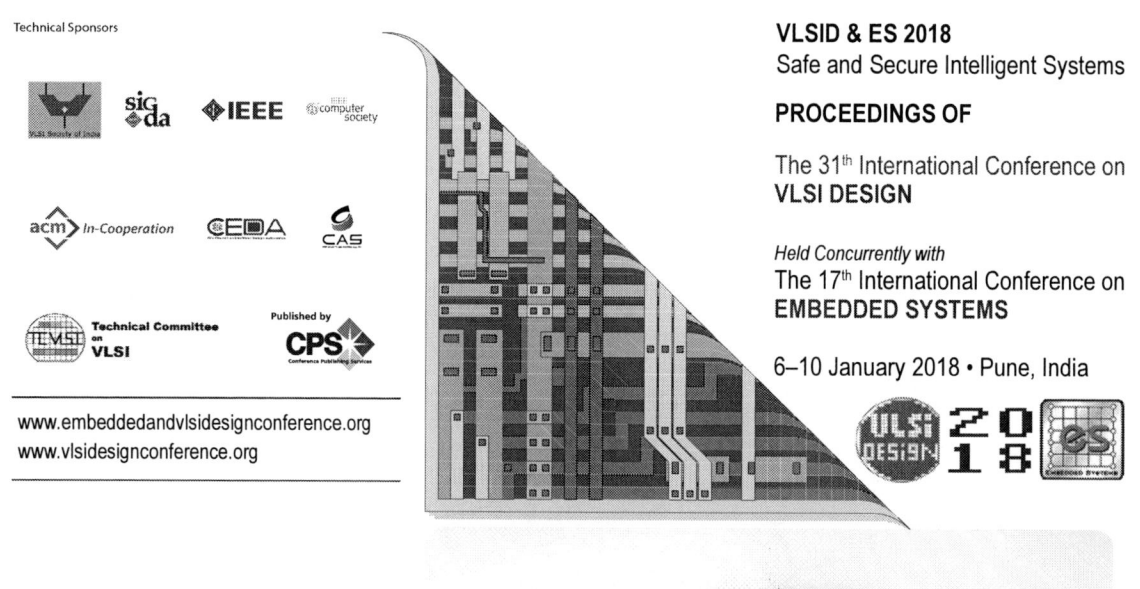

VLSID & ES 2018
Safe and Secure Intelligent Systems

PROCEEDINGS OF

The 31th International Conference on
VLSI DESIGN

Held Concurrently with
The 17th International Conference on
EMBEDDED SYSTEMS

6–10 January 2018 • Pune, India

www.embeddedandvlsidesignconference.org
www.vlsidesignconference.org

2018 31th International Conference on VLSI Design and 2018 17th International Conference on Embedded Systems

VLSID 2018

Table of Contents

Message from the General Chair...xiii

Message from the Technical Program Co-Chairs...xiv

Message from the Organizing Co-Chairs..xv

Message from the Tutorial Chair..xvi

Message from the Steering Committee Chair..xvii

Message from the President, VLSI Society of India...xviii

About the Cover from the Publication Chair..xix

VLSI Design 2017 Conference Steering Committee..xx

VLSI Design and ES 2018 Conference Committees...xxi

Technical Program Committee...xxiii

Reviewers..xxiv

VLSI Design Conference History...xxviii

Embedded Systems Design Conference History..xxx

Tutorial T1A: Assistive Technology for Visually Impaired:

Embedded & Vision Solutions..xxxi

Tutorial T1B: Emerging Computational Devices,

Architectures and Computational Models..xxxiii

Tutorial T1C: High-Speed Serial Links: Architectures

and Circuits for Clock and Data Recovery (CDR)...xxxv

Tutorial T2A: Safe Autonomous Systems: Real-Time Error

Detection and Correction in Safety-Critical Signal

Processing and Control Algorithms...xxxvii

Tutorial T2B: Hardware Intellectual Property (IP) Security

and Trust: Challenges and Solutions..xxxix

Tutorial T2C: Beyond von-Neumann Computing: Devices,

Circuits, and Applications...xli

Tutorial T2D: Privacy Assurances in the Internet of Things (IoT) World...xlii

Tutorial T2E: Pre-Silicon Verification and Post-Silicon Validation: Dramatic Improvements through Disruptive Innovations...xliii

Tutorial T2F: Silicon Nanophotonics for Future Manycore Chips: Opportunities and Challenges...xlv

Special Session: Design of Energy-Efficient and Reliable VLSI Systems: A Data-Driven Perspective...xlvii

Special Session: Advances in Power Management Integrated Circuits...li

---

Track 1A: Analog/RF - I

A Systematic Approach to Determining the Weights of the Capacitors in the DAC of a Non-binary Redundant SAR ADCs 1

    Uma Mukund Kulkarni, Chetan Parikh, and Subhajit Sen

Mitigating Aperture Error in Pipelined ADCs Without a Front-end Sample-and-Hold Amplfier 7

    Diego James, Abishek T. Kunnath, A. Purushothaman, and Bibhu Datta Sahoo

A 12.5Gbps Transmitter for Multi-standard SERDES in 40nm Low Leakage CMOS Process 13

    Biman Chattopadhyay, Sharath N. Bhat, Gopalkrishna Nayak, and Ravi Mehta

An Ultra Low Power, 10-Bit Two-Step Flash ADC for Signal Processing Applications 19

    Mahesh Kumar Adimulam, Amit Kapoor, Sreehari Veeramachaneni, and M.B. Srinivas

Track 1B: Power Management

Rescheduling of Power Gating Instructions for Reduction of In-rush Current 25

    Sumanta Pyne

Dynamic Thermal Management by Using Task Migration in Conjunction with Frequency Scaling for Chip Multiprocessors 31

    Alankar V. Umdekar, Arijit Nath, Shirshendu Das, and Hemangee K. Kapoor

Formal Methods for Coverage Analysis of Power Management Logic with Mixed-Signal Components 37

    Sudipa Mandal, Aritra Hazra, Pallab Dasgupta, and Rama Mohan Chunduri

DPFair Scheduling with Slowdown and Suspension 43

    Sanjay Moulik, Arnab Sarkar, and Hemangee K. Kapoor

Track 1C: FPGA - I

Fault-Tolerant Learning in Spiking Astrocyte-Neural Networks on FPGAs 49

Anju P. Johnson, Junxiu Liu, Alan G. Millard, Shvan Karim, Andy M. Tyrrell, Jim Harkin,
Jon Timmis, Liam McDaid, and David M. Halliday

FPGA Implementation of an Improved Watchdog Timer for Safety-Critical Applications 55

Ravi Krishnan Unni, Vijayanand P., and Y. Dilip

Image Compression Using 2D-Discrete Wavelet Transform on a Light Weight Reconfigurable
Hardware 61

Nupur Jain, Mansi Singh, and Biswajit Mishra

YaNoC: Yet Another Network-on-Chip Simulation Acceleration Engine Using FPGAs 67

Khyamling Parane, Basavaraj Talawar, and Prabhu Prasad B. M.

Track 2A: Security - I

An Energy-Efficient Trusted FSM Design Technique to Thwart Fault Injection and Trojan
Attacks 73

Vijaypal Singh Rathor, Bharat Garg, and G. K. Sharma

Novel Variability Aware Path Selection for Self-Referencing Based Hardware Trojan Detection 79

Vaikuntapu Ramakrishna, Lava Bhargava, and Vineet Sahula

A Secure Low-Cost Edge Device Authentication Scheme for the Internet of Things 85

Ujjwal Guin, Adit Singh, Mahabubul Alam, Janice Cañedo, and Anthony Skjellum

Hardware Trojan Detection Using ATPG and Model Checking 91

Jonathan Cruz, Farimah Farahmandi, Alif Ahmed, and Prabhat Mishra

Track 2B: Test

Towards Single Pin Scan for Extremely Low Pin Count Test 97

Mudasir S. Kawoosa, Rajesh K. Mittal, Maheedhar Jalasuthram, and Rubin A. Parekhji

Test-Time Reduction for Power-Aware 3D-SoC 103

Sabyasachee Banerjee, Subhashis Majumder, and Bhargab B. Bhattacharya

Identification of Faulty TSVs in 3D IC During Pre-Bond Testing 109

Dilip Kumar Maity, Surajit Kumar Roy, and Chandan Giri

Modeling & Analysis of Redundancy Based Fault Tolerance for Permanent Faults in Chip
Multiprocessor Cache 115

Avishek Choudhury and Biplab K. Sikdar

Track 2C: Devices and Emerging Technologies

Exact Synthesis of Biomolecular Protocols for Multiple Sample Pathways on Digital
Microfluidic Biochips 121

Oliver Keszocze, Mohamed Ibrahim, Robert Wille, Krishnendu Chakrabarty, and Rolf Drechsler

Design Optimization at the Fluid-Level Synthesis for Safe and Low-Cost Droplet-Based
Microfluidic Biochips 127

Arpan Chakraborty, Piyali Datta, and Rajat Kumar Pal

Hysteresis Free sub-60 mV/dec Subthreshold Swing in Junctionless MOSFETs 133

Manish Gupta and Abhinav Kranti

New Asymmetric Atomistic Model for the Analysis of Phase-Engineered MoS2-Gold Top
Contact 139

Richa Chakravarty, Dipankar Saha, and Santanu Mahapatra

Track 3A: Security - II

Power Side Channel Resistance of RNS Secure Logic 143

Ravikumar Selvam and Akhilesh Tyagi

Positive Feedback Symmetric Adiabatic Logic Against Differential Power Attack 149

Bhuvana B. P. and V. S. Kanchana Bhaaskaran

Online Detection and Reactive Countermeasure for Leakage from BPU Using TVLA 155

Sarani Bhattacharya, Shivam Bhasin, and Debdeep Mukhopadhyay

Secure Neural Circuits to Mitigate Correlation Power Analysis on SHA-3 Hash Function 161

James Thesing and Dhireesha Kudithipudi

Track 3B: Oscillators

A Novel Zero Blind Zone Phase Frequency Detector for Fast Acquisition in Phase Locked
Loops 167

Sucheth S. Kuncham, Manasa Gadiyar, Sushmitha Din K., Kiran Kumar Lad, and Tonse Laxminidhi

A 1.2 pJ/cycle KHz Timer Circuit for Heavily Duty-Cycled Systems 171

Manikandan R. R., Vipul Singhal, Rajat Chauhan, Vinod Menezes, and Mahesh Mehendale

A Quadrature-Phase Voltage Controlled Oscillator for Offset Phase and Frequency
Compensation 177

*Pragya Maheshwari, Pavan Kumar Sadhu, Mukesh Deharia, Nandakumar Nambath,
and Shalabh Gupta*

CMOS Oscillator Having Stable Frequency with Process, Temperature and Voltage Variation 181

Vikas Rana

Track 3C: FPGA - II

High Speed FPGA Fabric Aware CSD Recoding with Run-Time Support for Fault Localization 186

Ayan Palchaudhuri and Anindya Sundar Dhar

A Low-Power Circuit for Adaptive Dynamic Programming 192

Nan Zheng and Pinaki Mazumder

Low Power Configurable Readout Integrated Circuit for Infrared Detector 198

*Hari Shanker Gupta, Pranoy Datta, Maryam Shojaei Baghini, A. S. Kiran Kumar,
and Dinesh K. Sharma*

Track 4A: Analog/RF - II

Feedback Biasing Based Adjustable Gain Ultrasound Preamplifier for CMUTs in 45nm CMOS 204
   Linga Reddy Cenkeramaddi

A High Efficiency Body Injected Differential Power Amplifier at 2.4GHz for Low Power
Applications 208
   Mannem Naga Sasikanth and Tarun Kanti Bhattacharyya

A Novel Low Power G m-C Continuous-Time Analog Filter with Wide Tuning Range 214
   P. S. Veerendranath, Vasantha M.H., Nithin Kumar Y.B., and Edoardo Bonizzoni

0.36 nJ/bit MedRadio Band OOK Transmitter for Wearable Healthcare Applications 220
   Abhishek Srivastava and Maryam Shojaei Baghini

Track 4B: Special Session - Power Management Integrated Circuits

Time-Based PWM Controller for Fully Integrated High Speed Switching DC-DC Converters —
An Alternative to Conventional Analog and Digital Controllers 226
   Qadeer A. Khan, Seong-Joong Kim, and Pavan Kumar Hanumolu

Flipped Voltage Follower Based Low Dropout (LDO) Voltage Regulators: A Tutorial Overview 232
   Punith R. Surkanti, Annajirao Garimella, and Paul M. Furth

Track 4C: Special Session - Energy Efficient and Reliable VLSI Systems

Overcoming Energy and Reliability Challenges for IoT and Mobile Devices with Data Analytics 238
   Sudeep Pasricha

Data-Driven Resiliency Solutions for Boards and Systems 244
   Shi Jin and Krishnendu Chakrabarty

Track 5A: Reliability and SRAMs

Single-Error Hardened and Multiple-Error Tolerant Guarded Dual Modular Redundancy
Technique 250
   Sai Aparna Aketi, Joycee Mekie, and Hemal Shah

Impact of Device Aging on Early Mode Failures in Pulsed Latches 256
   Ankur Shukla, Rahul M. Rao, and James D. Warnock

A 0.6V Retention VMIN Ultra-Low Leakage High Density 6T SRAM in 40nm CMOS
Technology Using Adaptive Source Bias 261
   Ashish Kumar and G.S. Visweswaran

A 7-Nm Dual Port 8T SRAM with Duplicated Inter-Port Write Data to Mitigate Write
Disturbance 266
   *M. Sultan M. Siddiqui, Sumit Srivastav, Dattatray Ramrao Wanjul, Manankumar Suthar,
   and Sudhir Kumar*

Track 5B: VLSI Architecture

An Efficient VLSI Architecture for Convolution Based DWT Using MAC 271
Mohamed Asan Basiri M. and Noor Mahammad Sk

Hardware-Efficient and Wide-Band Frequency-Domain Energy Detector for Cognitive-Radio
Wireless Network 277
Mahesh S. Murty and Rahul Shrestha

Area and Power Efficient VLSI Architecture of Distributed Arithmetic Based LMS Adaptive
Filter 283
Mohd. Tasleem Khan and Shaik Rafi Ahamed

Novel High Speed Vedic Multiplier Proposal Incorporating Adder Based on Quaternary Signed
Digit Number System 289
Preyesh Dalmia, Vikas, Abhinav Parashar, Akshi Tomar, and Neeta Pandey

Track 5C: Design Automation

Design Space Exploration of an Execution-Driven Functional Simulation Methodology 295
Ipsita Biswas Mahapatra, Utkarsh Agarwal, Chandrashekhar Azad, and S. K. Nandy

A Practical Methodology to Compress Technology Libraries Using Recursive Polynomial
Representation 301
Sneh Saurabh and Priyanka Mittal

CLRFrame: An Analysis Framework for Designing Cross-Layer Reliability in Embedded
Systems 307
Siva Satyendra Sahoo, Bharadwaj Veeravalli, and Akash Kumar

Computing Fréchet Distance Metric Based L-Shape Tile Decomposition for E-Beam
Lithography 313
Arindam Sinharay, Pranab Roy, and Hafizur Rahaman

Track 6A: Analog/RF - II

2nd Order Sallen Key Switched Capacitor LPF with N-type Transistors 319
Kamal Chapagai, Pydi Bahubalindruni, and Nishtha

Pseudo-Continuous Output Switched-Capacitor Amplifier for Rail-to-Rail Current Sensing
Application 325
Anjali Gopinath, Ravi Kumar Adusumalli, Rohit Ranganathan, and Arya S.

Design Considerations of a Sub-50 Mu-W Receiver Front-end for Implantable Devices
in MedRadio Band 329
Gregory Chang, Shovan Maity, Baibhab Chatterjee, and Shreyas Sen

A 0.6 mW 1.6 dB Noise Figure Inductorless Shunt Feedback Wideband LNA With Gm
Enhancement and Current Reuse in 65 nm CMOS 335
Narendra Nath Ghosh, Prakash Kumar Lenka, SriHarsa Vardan G, and Ashudeb Dutta

Track 6B: Regulators

An NMOS Low Drop-out Voltage Regulator with -17dB Wide-Band Power Supply Rejection
for SerDes in 22FDX 341
    Nitin Bansal and Rahul Gupta

Single Inductor Dual Output Buck Converter for Low Power Applications and Its Stability
Analysis 347
    Sowmya Sankaranarayanan, Kulkarni Chaitali Vinod, Aswanth Sreekumar, Tonse Laxminidhi,
    Vipul Singhal, and Rajat Chauhan

A 0.29ps FOM Fast Transient any Cap Stable LVR in 28FDSOI 353
    Nitin Bansal, Saurabh Kumar Singh, Hemant Shukla, and Madhvi Sharma

A High Performance Gated Voltage Level Translator with Integrated Multiplexer 358
    Dharshak B. S. and Rahul M. Rao

Track 6C: Embedded Systems - I

Accelerating Hash Computations Through Efficient Instruction-Set Customisation 362
    Mayuran Sivanesan, Anupam Chattopadhyay, and Ronak Bajaj

Lightweight Forth Programmable NoCs 368
    Vinay B. Y. Kumar, Deval Shah, Mandar Datar, and Sachin B. Patkar

An Adaptive Deflection Router with Dual Injection and Ejection Units for Mesh NoCs 374
    John Jose and Abhijit Das

Towards Near Data Processing of Convolutional Neural Networks 380
    Palash Das, Shivam Lakhotia, Prabodh Shetty, and Hemangee K. Kapoor

Track 7A: Security - III

PPU: Privacy-Aware Purchasing Unit for Residential Customers in Smart Electric Grids 386
    Amrita Roy Chowdhury and Parameswaran Ramanathan

A High-Performance and Area-Efficient VLSI Architecture for the PRESENT Lightweight
Cipher 392
    Jai Gopal Pandey, Tarun Goel, and Abhijit Karmakar

Security Vulnerabilities of Unmanned Aerial Vehicles and Countermeasures: An Experimental
Study 398
    Vishal Dey, Vikramkumar Pudi, Anupam Chattopadhyay, and Yuval Elovici

Track 7B: Verification and Validation

AMS-Miner: Mining AMS Assertions Using Interval Arithmetic 404
    Antonio Anastasio Bruto da Costa, Shriya Dharade, Sudipa Mandal, and Pallab Dasgupta

ELURA: A Methodology for Post-Silicon Gate-Level Error Localization Using Regression
Analysis 410
    Ankit Jindal, Binod Kumar, Kanad Basu, and Masahiro Fujita

Combined Inference and Satisfiability Based Methods for Complete Signal Restoration
in Post-Silicon Validation 416
Xiaobang Liu and Ranga Vemuri

Track 7C: Memory

Emerging FETs for Low Power and High Speed Embedded Dynamic Random Access Memory 422
Md. Hasan Raza Ansari, Nupur Navlakha, Jyi-Tsong Lin, and Abhinav Kranti

Energy-Efficient Dynamic Data Encoding for Multi-level STT-MRAM 428
Mohammad Gh. Alfailakawi, Imtiaz Ahmad, and Sarah Elghandour

Switching-Time Dependent PUF Using STT-MRAM 434
Ashwani Kumar, Shubham Sahay, and Manan Suri

Floating Point Multiplication Mapping on ReRAM Based In-memory Computing Architecture 439
Tarun Vatwani, Arko Dutt, Debjyoti Bhattacharjee, and Anupam Chattopadhyay

Poster Papers

Parasitic Aware Automatic Analog CMOS Circuit Design Environment Using ABC Algorithm 445
Subhash J. Patel and Rajesh A. Thakker

A Temperature Compensated Read Assist for Low Vmin and High Performance High Density
6T SRAM in FinFET Technology 447
Vinay Kumar, Ravindra kumar Shrivatava, and Madhav Mansukh Padaliya

FPGA Implementation of Power Management Algorithm for Wind Energy Storage System
with Kalman Filter MPPT Technique 449
Vulisi Narendra Kumar and Gayadhar Panda

A Novel Tool for Synthesis by Direct Mapping of Asynchronous Circuits from Extended STG
Specifications 451
Felipe Mendes, Tiago Curtinhas, Duarte L. Oliveira, Higor A. Delsoto, and Lester A. Faria

Optimized Concurrent Testing of Digital Microfluidic Biochips 453
Sourav Ghosh, Hafizur Rahaman, and Chandan Giri

Self-Powered IoT Device for Indoor Applications 455
*Rolf Arne Kjellby, Thor Eirik Johnsrud, Svein Erik Loetveit, Linga Reddy Cenkeramaddi,
Mohamed Hamid, and Baltasar Beferull-Lozano*

Fault Tolerance in Network on Chip Using Bypass Path Establishing Packets 457
Sharma Priya, Sukarn Agarwal, and Hemangee K. Kapoor

Impact of Variations on Synchronizer Performance: An Experimental Study 459
Joycee Mekie, Prashansa Mukim, and Kimaya Kale

TileNET: Scalable Architecture for High-Throughput Ternary Convolution Neural Networks
Using FPGAs 461
Sahu Sai Vikram, Vibha Panty, Mihir Mody, and Madhura Purnaprajna

SHIRT (Self Healing Intelligent Real Time) Scheduling for Secure Embedded Task Processing 463
Krishnendu Guha, Sangeet Saha, and Amlan Chakrabarti

Exploration of Loop Unroll Factors in High Level Synthesis 465
*Preeti Ranjan Panda, Namita Sharma, Srikanth Kurra, Khushboo Anil Bhartia,
and Neeraj Kumar Singh*

Securing Module-Less Synthesis on Cyberphysical Digital Microfluidic Biochips
from Malicious Intrusions 467
Sarit Chakraborty, Chandan Das, and Susanta Chakraborty

On the ESD Reliability Issues in Carbon Electronics: Graphene and Carbon Nano Tubes 469
Nagothu Karmel Kranthi, Abhishek Mishra, Adil Meersha, and Mayank Shrivastava

Author Index 471

Message from the General Chair

Dear Patron,

Welcome to the 31st International Conference on VLSI Design and 17th International Conference on Embedded Systems, being held in the second week of January, 2018 in the City of Pune, India.

Pune, popularly known as the Oxford of the East, has a thriving VLSI and Embedded Ecosystem and is proud to host this prestigious international conference for the second time in the city after having a stellar debut in 2013 – an event that firmly marked Pune on the semiconductor and embedded map of the world. This joint conference in 2018 will serve as a forum for the entire technical community engaged in semiconductor design, EDA, embedded systems, IoT and academic and industrial research thereof. The conference has traditionally been enriched by a fruitful dialogue with industry participants and I expect this edition of the conference to carry on that mantle to greater heights.

Driven by the vision of the VLSI Steering Committee, ably guided by the active efforts of the VLSI Society of India(VSI) and a strong partnership with India Electronics and Semiconductor Association(IESA), this conference platform has grown more popular over the years. India's prominence in the electronics world has been steadily increasing over the years. With a resurgent economy, a large and growing middle class and a consumer base with a strong appetite for all forms of intelligent electronic systems and gadgets, conditions are ripening for a thriving and indigenous electronics market in the country. Advent of cloud, IoT and device to device connectivity in the "edge" of the cloud has shown tremendous growth. With edge of the cloud, electronics has become more pervasive in sectors where connected devices interact -- such as -- automotive, medical and industrial, to name a few sectors. All of these augur well for an electronics ecosystem in India that is growing in the cities such as Pune. In this new wave of automation, safety and security of these intelligent systems is one of the top priorities of researchers and industries alike worldwide and is also the aptly selected theme of this conference.

Hosting this conference in Pune after a period of 5 years reaffirms our objective of bringing a multitude of opportunities for every stakeholder, from talented individuals in the industry to academics and students. I envision the conference as a prolific platform for exchange of ideas, results and future challenges in VLSI and Embedded systems. I look forward to bringing to you an enriching experience at this conference that is both informative and invigorating.

Once again, a very warm welcome to you!!!

With Warm Regards,

Niranjan A. Pol

General Chair, VLSID 2018

VLSID & ES 2018

31st International Conference on VLSI Design and
17th International Conference on Embedded Systems

Message from the Technical Program Co-Chairs

Welcome to Pune! The conference returns to this beautiful city, popularly known as the "Oxford of the East" and the cultural capital of Maharashtra, India after a gap of five years. Pune is known for its many premier prestigious educational institutions attracting the best students from all over India. In recent years, Pune has also made significant economic contribution to the country's growth with a strong industrial presence in fields of IT, electronics, manufacturing, automotive, and bio tech.

Recent advances in VLSI design and semiconductor technology are widely seen as key enablers for breakthroughs in healthcare, energy, security and safety-critical embedded systems. As we attempt to translate these advances to societal benefits, we are confronted with significant challenges related to the inclusive adoption of intelligent technology in a safe and secure manner. In keeping with this trend, the theme of this conference is ***Safe and Secure Intelligent Systems***. The conference begins with a two-day tutorial program that showcases the latest trends in embedded systems, VLSI design, and emerging technologies. The three-day technical program begins after the tutorials, and features nine keynote addresses by visionaries from industry and academia, panel discussions and special sessions cutting across a wide range of topics of current and emerging interest.

The conference received 268 regular paper submissions spread over 15 subject categories. More than 1000 reviews were submitted by our team of 155 committee members, assisted by over 100 secondary reviewers, from all across the world, led by our 22 track chairs. After detailed discussion on each of the submissions, 73 high quality papers (27.2% acceptance rate) were selected for presentation at the conference. This year's programme also includes Interactive Presentation papers. In order to encourage authors of papers that are not yet mature for acceptance in the regular category, but contain interesting germinating ideas, we have selected 13 papers under this new category for a 5-minute presentation, an interactive poster presentation, and a 2-page abridged version in the proceedings. The final selection of all papers was done through meetings of the Technical Program Committee at Mumbai, India and with the coordination of program committee participants across the globe.

The conference has six parallel tracks running through three days. The regular papers will be presented in 21 sessions distributed over 3 parallel tracks. The fourth parallel track focuses on industry trends, and user/design practices. The fifth track is devoted to the PhD Forum and poster presentations from doctoral students, in addition to special topics related to Women in Engineering. The sixth track is a student only conference, where technical and soft-skill topics essential for industry readiness are highlighted. As a last event of the conference, the 9th International Workshop on Reliability Aware System Design and Test (RASDAT) 2018 will be held on Jan 10 and 11.

Welcome again to Pune! Enjoy the exciting program of the 31st International Conference on VLSI Design and the 17th International Conference on Embedded Systems!

With warm regards,

Sudeep Pasricha, Rahul Rao, and Virendra Singh

Technical Program Co-Chairs, VLSID 2018

xiv

Message from the Organizing Co-Chairs

Wishing you all a very Happy and Prosperous New Year 2018!

On behalf of the VLSID & ES 2018 organizing committee, we are honored and delighted to welcome you to the 31st International Conference on VLSI Design and 17th International Conference on Embedded System, being held from 6th through 11th of January 2018 at Hyatt Regency, Pune, India. VLSID & ES 2018 is a premier Embedded Systems and VLSI Design conference and oldest conference series happening in India.

VLSI & Embedded Systems has continued its remarkable growth and with the rising demand from end-use industries such as telecom, electrical, industrial, automotive, aerospace, military and healthcare and others, the timing for this event in Pune couldn't have been better. Artificial Intelligence, Machine Learning, Robotics, Embedded Systems and IoT are attracting wide attention and reshaping the tech landscape. Learn the latest on these technologies, strategies and techniques for building the safe and secure competitive products straight from industry experts during a range of engaging formats that can advance your expertise and inspire breakthroughs. These innovative technologies have potentials to build borderless communication society, a symbiotic society between humans and robots, and deliver safe and secured solutions. All of these solutions will work in collaboration to deliver high performance and more intelligent processing which makes VLSI and Embedded systems an important field.

We believe we have chosen a venue that guarantees a successful technical conference amid the culture and beautiful & smart Pune. VLSID & ES 2018 is featuring regular paper sessions, full day tutorials given by experts in the topics. Topics include all areas of Embedded Systems and VLSI Design covering the latest innovations. The conference channelizes the networking and connectivity bringing together students, designers, engineers, researchers, technology visionaries and corporate leaders representing government departments, professional and trade institutions from every corner of the world.

This annual conference aims to report recent advances in VLSI and Embedded Technologies and it's significant contribution in the advancement for academic and industrial growth. The overall program is packed with key note speeches, technical papers, panels, exhibits, tutorial and workshop, student and fellowship program, industry forum, design contest and hackathon on emerging research and state-of-the-art advances on VLSI and embedded systems which have been a premier features of this conference.

We highly appreciate all the organizing committee members and volunteers for their dedicated effort and a phenomenal contribution in making this conference exciting and enriching. We would also like to express our heartfelt gratitude to the VLSI society and its steering committee for their timely guidance and for giving us the opportunity to host this event in Pune this year. We wish to thank all the Authors, Speakers, Panelist for their contribution and time. We are sure that all the participants will have an enjoyable moment reading your articles and listen to you. We sincerely thank all sponsors, exhibitors for aptly recognizing the value that the conference will bring to them and bringing the spotlight back on Pune ecosystem.

With our conference theme of **Safe and Secure Intelligent Systems**, we are confident that the conference will set the tone for achieving new benchmarks in India.

We look forward to welcoming you and seek your active participation in making this endeavor successful.

With Best Wishes and Warm Regards,

Kiran H. Dahimiwal and Prof. P V S Shastry

Organizing Co-chairs, VLSID & ES 2018

VLSID & ES 2018
31st International Conference on VLSI Design and
17th International Conference on Embedded Systems

Message from the Tutorial Chair

Welcome to the 31st International Conference on VLSI Design (VLSID 2018), held concurrently with 17th International Conference on Embedded Systems (ES 2018), at the "Oxford of the East," Pune, India. The conference features both full-day and half-day tutorials, to be held on Jan 6th and 7th, 2018. We have received an overwhelming number of high quality submissions and selected two full-day tutorials and four half-day tutorials. We have also included one full-day invited tutorial and four half-day invited tutorials. These tutorials focus on both established and emerging research topics and applications.

The tutorials below make up the final program:

- Assistive Technology for Visually Impaired: Embedded & Vision Solutions, M Balakrishnan (IIT Delhi) and Chetan Arora (IIIT Delhi)
- Emerging Computational Devices, Architectures and Computational Models, Vijaykrishnan Narayanan (Penn State University), Arijit Raychowdhury (Georgia Institute of Technology) and Sumeet Kumar Gupta (Pennsylvania State University/Purdue University)
- High-Speed Serial Links: Architectures and Circuits for Clock and Data Recovery (CDR), Saurabh Saxena (IIT Madras) and Nagendra Krishnapura (IIT Madras)
- Safe Autonomous Systems: Real-Time Error Detection and Correction in Safety-Critical Signal Processing and Control Algorithms, Jacob Abraham (University of Texas, Austin) and Abhijit Chatterjee (Georgia Institute of Technology)
- Hardware Intellectual Property (IP) Security and Trust: Challenges and Solutions, Prabhat Mishra (University of Florida) and Rajat Subhra Chakraborty, (IIT Kharagpur)
- Beyond von-Neumann Computing: Devices, Circuits, and Applications, Kaushik Roy (Purdue University)
- Designing Reliable and Secure SSD Controllers For Cloud Storage, Erich F. Haratsch (Seagate Technology)
- Privacy Assurances in the Internet of Things (IoT) World, Parameswaran Ramanathan (University of Wisconsin, Madison)
- Pre-Silicon Verification and Post-Silicon Validation: Dramatic Improvements through Disruptive Innovations, Subhasish Mitra (Stanford University), and Srinivas Shashank Nuthakki (Stanford University), Eshan Singh (Stanford University)
- Silicon Nanophotonics for Future Manycore Chips: Opportunities and Challenges, Sudeep Pasricha (Colorado State University)
- Automotive Battery Management Solutions, Philippe Perruchoud (NXP)

For VLSID & ES 2018, we are looking for ways to become environmentally friendly by "going green." One important way is to obviate printing the tutorial slides and save plenty of paper. Password-protected slides will be delivered in PDF format to the attendees. We sincerely appreciate tutorial speakers' consent in going green and offer thanks for their contributions and making this a world-class program.

I sincerely hope that you will enjoy the tutorial program at VLSID & ES 2018!

Annaji Row Garimella

Tutorial Chair, VLSID 2018

VLSID & ES 2018 | 31st International Conference on VLSI Design and
17th International Conference on Embedded Systems

Message from the Steering Committee Chair

Just after five years, the twin conference, this time *31st International Conference on VLSI Design* and *17th International Conference on Embedded Systems,* is back in Pune, India.

The conference is the result of a yearlong effort of the leaders and volunteers from industry and academia. The General Chair, Niranjan Pol, leads the team of organizers. We are thankful to our sponsor, the VLSI Society of India (VSI), technical co-sponsors, IEEE (CS, TCVLSI, CAS and CEDA) and ACM SIGDA, and Industry supporters, Seagate, Qualcomm, John Deere, Cadence, Tessolve and others, for their contributions to the conference.

Emphasizing the theme, "Safe and Secure Intelligent Systems," the conference offers a variety of activities. Tutorials are on January 6 and 7 followed by regular paper presentations on January 8, 9 and 10. You will network with peers and industry leaders, sample recent research results at the PhD Forum, see innovative VLSI designs at the Design Contest displays, and learn about prevailing practices through the User/Designer Track and Industry Forum. Regardless of where you spend most of your time, you must not miss the Industry Exhibition.

An established feature in this conference has been the Fellowship Program that provides financial assistance to delegates who might otherwise be unable to attend. A more recent addition, Student Conference now exposes college students to the field of VLSI as they are choosing the direction for their future. Although the student conference and the regular program are concurrent, participants have opportunities to interact. If you are a student, do not hesitate to query the matured participants on the life in the field. I am sure older colleagues will be curious to find out how growing up has changed in the cyber age.

The city of Pune has made a name in the fields of education and industry. Besides, there is history to explore. Visit the Agha Khan Palace located in Pune where Mohandas Karamchand Gandhi, the architect of India's independence, and his wife Kasturba were detained after Gandhi declared the start of the Quit India Movement in 1942. Once you learn the facts, you may have your own opinion on why the movement failed and what happened next. The Palace has a museum that contains historical information and artifacts including a memorial for Mrs. Kasturba Gandhi who died during the detention.

Going forward, the *Thirty-second International Conference on VLSI Design* and *18th International Conference on Embedded Systems* are jointly scheduled for January 2019 in the NCR area near New Delhi.

With warm regards,

Vishwani D. Agrawal

Chair, VLSI Design Conference Steering Committee
October 2, 2017

VLSID & ES 2018 | **31st International Conference on VLSI Design and 17th International Conference on Embedded Systems**

Message from the President, VLSI Society of India

On behalf of the VLSI Society of India, it is my pleasure to welcome you to the 31st International Conference on VLSI Design and the 17th International Conference on Embedded Systems being held concurrently in Pune, India.

This is the second time this prestigious international conference is being hosted in Pune, a testament to the increasing significance of Pune as a key hub for Electronic System Design and Manufacturing (ESDM) especially focused on verticals like industrial and automotive.

The theme for this years' conference is "Safe and Secure Intelligent Systems". With the rapid proliferation of smart connected devices, the aspects of safety, security and privacy are becoming increasingly important. They take even greater significance in applications like transportation, medical, personal-health, smart-home, smart-city, industrial and the like where they touch our daily lives at home or work more directly. Tackling these challenges will require novel approaches at the system level which will have profound impact on architecture and design of both hardware and software. This is driving a transformation of how design has to be approached starting with the system first ushering in the era of System Design Enablement (SDE).

I look forward to paper presentations, invited talks and keynotes that provide insight into trends, opportunities and the latest technology research in this space. While the conference provides an excellent platform for interaction between stakeholders from industry, academia and government, I hope it also serves as a catalyst to facilitate much needed on-going collaboration between these stakeholders.

Further, like each year in the past, the VLSI Society of India is delighted to be able to continue our highly successful student and academic Fellowship Program that funds the participation of several deserving students and academicians from across the country to further our goal for excellence in education and research in VLSI & Embedded Design and related areas.

Lastly, my sincere thanks and congratulations to this years' conference team for hosting what I am sure will be an excellent conference. Hope you find it insightful, stimulating and thought provoking.

Warm Regards,

Jaswinder S. Ahuja

Corporate VP & MD, Cadence Design Systems, India
President, VLSI Society of India

VLSID & ES 2018

**31st International Conference on VLSI Design and
17th International Conference on Embedded Systems**

About the Cover from the Publication Chair

Pune, the original 'Punawadi', and later Punyanagari or the city of rightful deeds is also known as the 'Oxford of the East' for its long-lasting fame as educational hub of India. It is also the cultural capital of the state of Maharashtra especially in the fields of classical dance and classical music. It is home to litterateurs, artists, film makers and social reformers. Mahatma Gandhi spent a lot of time here in Aga Khan Palace, now a memorial. Another Heritage site close to Pune is Sinhgad Fort amongst many of the great Maratha Emperor 'Chhatrapati Shivaji Maharaj'. 'Ganesh Chaturthi' is a renowned festival, celebrated with devotion and fanfare every year in Pune started by freedom fighter Bal Gangadhar Tilak. The city is now a famous auto- manufacturing and IT hub. The city also boasts of many Institutes International repute like IUCCA, NCL and is home to the first indigenous supercomputer 'PARAM'.

The cover photo is of Shaniwarwada, built by Bajirao Peshwa, the rulers of Pune, in 1736. It was a seven storied mansion and served as a seat of political power during its time. It acclaimed tremendous importance and is the city's main attraction till date. When visiting the palace, one can still get a view of the leftover plinths, the fortification walls with five gateways, and nine bastions that enclosed the whole complex. The elaborate foundations of the original palace showcases the wooden pillars and lattice work add up to the beauty of the mansion.

Much like Shaniwarwada, the International Conference on VLSI Design has over the years turned into the most prominent VLSI focused Conference in India bridging the VLSI communities in academia, industry and the government.

Like every year, the VLSID 2018 proceedings will be available online through IEEE*X*plore and other scientific databases. The VLSID conference proceedings series is available on IEEEXplore at persistent link http://ieeexplore.ieee.org/servlet/opac?punumber=1000799

Welcome to Pune for the 31st International Conference on VLSI Design – 2018!

Prachi Mukherji

Publication Chair, VLSID 2018

Photo: Shaniwar wada, Pune, India

xix

VLSID & ES 2018 | 31st International Conference on VLSI Design and
17th International Conference on Embedded Systems

VLSI Design Conference Steering Committee (2017)

Vishwani D. Agrawal

Jaswinder Ahuja

Preeti Ranjan Panda

Srimat T. Chakradhar

Annajirao Garimella

Manoj S. Gaur

Gangadhar Gude

Satya Gupta

Anshul Kumar

Nagi Naganathan

Niranjan Pol

N. Ranganathan

Virendra Singh

Susmita Sur-Kolay

VLSID & ES 2018 | 31st International Conference on VLSI Design and
17th International Conference on Embedded Systems

VLSI Design and Embedded Systems 2018
Conference Committee

STEERING COMMITTEE CHAIR
Vishwani D. Agrawal

GENERAL CHAIR
Niranjan A Pol

TECHNICAL PROGRAM CO-CHAIRS
Rahul Rao
Sudeep Pasricha
Virendra Singh

ORGANIZING CO-CHAIRS
Kiran Dahimiwal
P V S Shastry

TUTORIAL CHAIR
Annajirao Garimella

PHD FORUM CHAIRS
Devesh Dwivedi
Prashant Bartakke

FELLOWSHIP CHAIR
Mukul Sutoane

REGISTRATION CHAIR
Nitin Palan

DESIGN CONTEST CHAIRS
Amol Kodag
Sunil Desai

USER / DESIGNER TRACK CHAIRS
Ashish Bhattad
Dinesh Pathak

SPONSORSHIP CHAIR
S. Uma Mahesh

EXHIBITIONS CHAIRS
Shrikrishna Mehetre
Jaweed Akthar

VLSID & ES 2018
31st International Conference on VLSI Design and
17th International Conference on Embedded Systems

INDUSTRY FORUM CHAIR
Prasad Joshi
Amol Khanolkar

PUBLICATION CHAIRS
Prachi Mukherji

PUBLICITY CHAIRS
Abhijit Athavale
Monali Bhalerao

IEEE/ACM LIAISON
Nagi Naganathan

VSI LIAISON
Jaswinder S. Ahuja

IESA LIAISON
Som Shubro Pal Choudhury

LOCAL IEEE LIAISON
Rajesh Ingale

START-UP CHAIR
Ashwani Bhat

FINANCE CHAIR
Ritesh Jain

STUDENTS CONFERENCE CO- CHAIRS
Sarang Shelke
Jitendra Kanitkar
Vaibhav Kale

WOMEN IN ENGINEERING TRACK CO-CHAIRS
Ashlesha Karandikar
Supriya Kelkar
Vijayalaxmi Khanolkar

CONFERENCE PROGRAM MANAGER
Kedar A Patankar

VLSID & ES 2018

31st International Conference on VLSI Design and
17th International Conference on Embedded Systems

VLSID & ES 2018 Technical Program Committee

PROGRAM COMMITTEE CHAIRS

Sudeep Pasricha, *Colorado State University*
Rahul Rao, *IBM*
Virendra Singh, *Indian Institute of Technology, Mumbai*

Track Chairs

Umit Ogras, *Arizona State University*
Anshul Kumar, *Indian Institute of Technology, Delhi*
Partha Pande, *Washington State University*
Sourav Roy, *NXP Corporation*
Suhrud Khare, *Intel Corporation*
Tsung-Yi Ho, *National Tsing Hua University*
Deukhyoun Heo, *Washington State University*
Rajesh Zele, *Indian Institute of Technology, Mumbai*
Preeti Ranjan Panda, *Indian Institute of Technology, Delhi*
Vijaykrishnan Narayanan, *Pennsylvania State University*
Bipin Rajendra, *New Jersey Institute of Technology*
Sudeb Dasgupta, *Indian Institute of Technology, Roorkee*

Mehdi Tahoori, *Karlshure Institute of Technology*
Siddharta Duttagupta, *Indian Institute of Technology, Mumbai*
Prabhat Mishra, *University of Florida*
Masahiro Fujita, *University of Tokyo*
Adit Singh, *Auburn University*
Rubin Parekhji, *Texas Instruments*
Srinivas Katkoori, *University of Southern Florida*
Sridhar Rangarajan, *IBM*
Swarup Bhunia, *University of Florida*
Debdeep Mukhopadhyay, *Indian Institute of Technology, Kharagpur*

VLSID & ES 2018 – List of Reviewers

Abhinav Kranti	Rajat Subhra Chakraborty	Gaurav Saini
Abhishek Dixit	Rajendra Bishnoi	gautam hazari
Alex Doboli	Rajendra Patrikar	Govardhan Rao
Amey Kulkarni	Rajesh Bhagwat	Gregory Chang
Amir Masoud Gharehbaghi	Rajiv Joshi	Heba Khdr
Amit Patra	Ramesh Karri	Honglan Jiang
Amlan Ganguly	Rangharajan Venkatesan	Hwisoo So
Amol Dharangutte	Robert Karam	Jack Tang
Andreas Veneris	Robert Wille	Jiang Guiyuan
Ankur Gupta	Rolf Drechsler	Jing Ye
Ansuman Banerjee	Sachin Patkar	Ketan Budhiya
Anupam Chattopadhyay	Sachin Sapatnekar	Laxmeesha S.
Arijit Mondal	Sandip Ray	Lokesh Siddhu
Arijit Raychowdhury	Santanu Mahapatra	Mahesh Balasubramanian
Aritra Hazra	Santosh Balasubramanian	Mahesh Kumashikar
Arnab Sarkar	Santosh Vishvakarma	Mandar Datar
Arun Joseph	Seetharam Narasimhan	Manikandan R R
Arya Rahimi	Sheikh Nijam Ali	Manish Rana
Aryabartta Sahu	Shigeru Yamashita	Marc Stöttinger
Aviral Shrivastava	Shivam Bhasin	Marcel Walter
Bai Nguyen	Shobha Vasudevan	Michael Shamouilian
Basant Dwivedi	Shreepad Karmalkar	Michele Lora
Bhargab Bhattacharya	Shreyas Sen	Mike Hayenga
Bibhu Datta Sahoo	Siddharth Garg	Mohammad Khayatian
Bijan Alizadeh	Siew Kei Lam	Mohammadreza Mehrabian
Bodhisatwa Mazumdar	Smruti R. Sarangi	Morgan Ledwon
Brajesh Kumar Kaushik	Somyendu Raha	Moslem Didehban
Chandan Kumar Sarkar	Soumya Pandit	Piyoosh Purushottam Nair

VLSID & ES 2018 | 31st International Conference on VLSI Design and 17th International Conference on Embedded Systems

Charles Augustine

Chester Rebeiro

Chethan Kumar Y.B.

Chip Hong Chang

Debashis Banerjee

Debashis Mandal

Debasri Saha

Deleep Nair

Devanathan VR

Dey Soumyajit

Dhireesha Kudithipudi

Dimin Niu

Duo Liu

Elena Ioana Vatajelu

Erfan Ghaderi

Gautam Hazari

Gopal Iyer

Goutam Paul

Graziano Pravadelli

Hao Zheng

Hemangee Kapoor

Hiroyuki Tomiyama

Huan Hu

Huawei Li

Hyung Gyu Lee

Ishan Thakkar

Jaehyun Park

Jaeyoung Jung

Jason Xue

Jaydeep Kulkarni

Jie Han

Joe Baylon

Sree Hari Rao Patri

Srinivasan Gopal

Subhanshu Gupta

Sudeep Pasricha

Sudip Kundu

Sujay Deb

Swagath Venkataramani

Swaroop Ghosh

Talal Bonny

Tiju Jacob

Turbo Majumder

Utpal Desai

Veeresh Deshpande

Venkatnarayan Hariharan

Venkatraman Ramakrishnan

Vineet Sahula

Vineet Singh

Vishal Khatri

Vishwani Agrawal

Vivek Chickermane

Vivek De

Wei Zhang

Wookpyo Hong

Wujie Wen

Xiaoqing Wen

Xinmin Yu

Yaojun Zhang

Yier Jin

Yiran Chen

Yogesh Chauhan

Younghyun Kim

Zebo Peng

Rajesh Devaraj

Sandeep Saini

Sangeet Saha

Satish Kumar

Mrigank Sharad

Muhammad Abdullah Hanif

Muhammad Hassan

Nasim Imtiaz Khan

Naveen Kumar M

Neelam Surana

Oleg Oleynikov

Parul K Sharma

Pavan Torvi

Pham Hung Thinh

Pradeep Paul

Prof. Samah Saeed

Rachmad Vidy Achmad

Rekha Govindaraj

S Majumdar

Sachin Turkewadikar

Saeed Ansari

Sandeep

Sandeep Chandran

Santosh Narayanan

Satpathy, Sudhir

Saurav Kumar Ghosh

Sayandeep Saha

Sayandeep Sanyal

Shail Dave

Shayan Mohammad

Siddhartha Sen

Sikhar Patranabis

VLSID & ES 2018 | 31st International Conference on VLSI Design and 17th International Conference on Embedded Systems

Joerg Henkel	Zhiyuan Zhou	Sitansusekhar Roymohapatra
Joycee Mekie	Abhrajit Sengupta	Siting Liu
Kanad Basu	Ajay D. Thakur	Sonu Jha
Kartik Mohanram	Ajay Kumar Singh	Sriram Kashyap
Kyoungwoo Lee	Alessandro Danese	Sumana Ghosh
Lava Bhargava	Alireza Rohani	Sumedh Dhabu
Lionel Torres	Alok Prakash	Suresh Mallala
Madhu Mutyam	Aniruddha Roy	T.V. Kalyan
Mainak Banga	Anirudh Iyengar	Ujjwal Guin
Manan Suri	Anjan Kumar	Urbi Chatterjee
Manu Awasthi	Antonio A. Bruto da Costa	Vaikuntapu Ramakrishna
Marco Ottavi	Anupam Chattopadhyay	Venkata Ramanan R
Mark Po-Hung Lin	Arindam Raychaudhuri	Victor van Santen
Maryam Shojaei Baghini	Arnab Kumar Biswas	Vidya Govindan
Michael Niemier	Ashis Maity	Vijay Sheshadri
Mingoo Seok	Ashish Sharma	Vinay Kumar B. Y.
Mingsong Chen	Bei Zhang	Vishwanath
Mohammad Al Faruque	Bernhard Jungk	Vladimir Herdt
Mostafa Taha	Bharat Srivastava	Wei He
Mrugesh Walimbe	Bhaskar Pal	Xiaotong Cui
Muhammad Shafique	Chandan Kumar Jha	Xueqing Li
Nagesh Tamarapalli	David Lemma	Yao Xiao
Nihar Mohapatra	Debayan Das	Yidong Liu
Nijwm Wary	Debdut Biswas	Ying Wang
Nikhil Tripathi	Dinesh Somasekhar	Young Moon Kim
Niraj Jha	Dirmanto Jap	Yuan Cao
Nithin Shastri	Dr. Amit M Joshi	Yuankun Xue
Nitin Agarwal	Dr. Chitrakant Sahu	Yue Zheng
Pallab Dasgupta	Enrico Fraccaroli	Yuhua Liang
Paul Bogdan	Faiq Khalid	
Pawan Agarwal	Far`ad Samie	
Peter Marwedel	Florenc Demrozi	

VLSID & ES 2018

**31st International Conference on VLSI Design and
17th International Conference on Embedded Systems**

Prashant Sohani

Florian Kriebel

Prokash Ghosh

Ganapati Bhat

VLSI Design Conference History

No.	Year	Location and Dates	Papers / Tutorials	Proceedings Pages	Approx. Attendance
1st	1985	Chennai, Dec 26-28	29/1	193	75
2nd	1988	Bangalore, Dec 15-18	26/4	496	150
3rd	1990	Bangalore, Jan 6-9	30/4	390	150
4th	1991	New Delhi, Jan 4-8	45/9	315	250
5th	1992	Bangalore, Jan 4-7	57/4	378	300
6th	1993	Mumbai, Jan 3-6	70/6	371	300
7th	1994	Kolkata, Jan 5-8	87/6	448	400
8th	1995	New Delhi, Jan 4-7	77/6	456	450
9th	1996	Bangalore, Jan 3-6	75/6	480	550
9th	1996	Bangalore, Jan 3-6	75/6	480	550
10th	1997	Hyderabad, Jan 4-7	84/6	608	550
11th	1998	Chennai, Jan 4-7	98/6	624	600
12th	1999	Goa, Jan 7-10	103/6	682	600
13th	2000	Kolkata, Jan 3-7	93/6	590	700
14th	2001	Bangalore, Jan 3-7	77/9	592	750
15th *	2002	Bangalore, Jan 7-11	109/8	834	1000
16th	2003	New Delhi, Jan 4-8	84/6	622	800
17th	2004	Mumbai, Jan 5-9	120/8	1132	800
18th	2005	Kolkata, Jan 3-7	113/9	922	850
19th	2006	Hyderabad, Jan 3-7	136/11	880	
20th	2007	Bangalore, Jan 6-10	147/15	990	
21st	2008	Hyderabad, Jan 4-8	108/10	780	
22nd	2009	New Delhi, Jan 5-9	79/9	632	
23rd	2010	Bangalore, Jan 3-7	79/8	461	
24th	2011	Chennai, Jan 2-7	66/8	391	450
25th	2012	Hyderabad, Jan 7-11	69/10	458	1500
26th	2013	Pune, Jan 5-10	66/12	396	500

27th	2014	Mumbai, Jan 5-9	97/12	584	
28th	2015	Bangalore, Jan 3-7	72/9	558	1500
29th	2016	Kolkata, Jan 4-8	86/14	610	
30th	2017	Hyderabad, Jan 7-10	136/11	454	1836

* Jointly held with ASP-DAC

Embedded Systems Design Conference History

No.	Year	Location and Dates	No. of Papers	Proceedings Pages
1st	2002	New Delhi, Jan 2-4	8	70
2nd	2003	New Delhi, Jan 4-8	84	622
3rd	2004	Mumbai, Jan 5-9	120	1132
4th	2005	Kolkata, Jan 3-7	113	922
5th	2006	Hyderabad, Jan 3-7	136	880
6th	2007	Bangalore, Jan 6-10	147	990
7th	2008	Hyderabad, Jan 4-8	108	780
8th	2009	New Delhi, Jan 5-9	79	632
9th	2010	Bangalore, Jan 3-7	79	461
10th	2011	Chennai, Jan 2-7	66	391
11th	2012	Hyderabad, Jan 7-11	69	458
12th	2013	Pune, Jan 5-10	66	396
13th	2014	Mumbai, Jan 5-9	97	584
14th	2015	Bangalore, Jan 3-7	72	558
15th	2016	Kolkata, Jan 4-8	86	610
16th	2017	Hyderabad, Jan 7-10	136	454

Tutorial T1A: Assistive Technology for Visually Impaired: Embedded & Vision Solutions

M Balakrishnan, IIT Delhi
Chetan Arora, IIIT Delhi

Abstract

Census 2011 classifies more that 5 million people as visually disabled In India. AssisTech (Assistive Technology) group at IIT Delhi aims to develop technological solutions to address their two key challenges; namely independent mobility and access to education. In this tutorial, we will firstly brief about the challenge of mobility and education for visually impaired people. Solutions addressing these challenges could be developed using embedded sensors or with embedded computer vision. Our existing products like SmartCaneTM and OnBoard are embedded system based solutions and are aimed towards facilitating independent mobility. There have been major advances in vision based techniques including possibility of their implementation on low cost embedded platforms. This has encouraged us to visualize a device named MAVI (Mobility Assistant for Visually Impaired). Objective of MAVI is to create a platform that would be able to provide mobility assistance to a visually impaired person in an unstructured infrastructure of developing countries like India. In this tutorial we will discuss specific challenges and various vision techniques used in MAVI to address the requirements of safety, navigation and social inclusion. Designing a complex system like MAVI presents the designer with numerous choices of algorithms as well as platform/hardware resources. We will discuss the challenges in handling system level design complexities and explain our approach towards design space exploration for such systems. Further, we also observe that existing metrics employed in the literature to assess various vision algorithms are inadequate when it comes to complex systems like MAVI. We also propose the need for system level metrics. Finally, we will demonstrate apart from a MAVI prototype, other devices and solutions developed by the AssisTech group (http://assistech.iitd.ac.in/).

Speaker Bios:

M. Balakrishnan is a Professor in the Department of Computer Science & Engineering and currently Deputy Director (Strategy & Planning) at IIT Delhi. He obtained his B.E.(Hons.) in Electronics & Electrical Engg. from BITS Pilani in 1977 and Ph.D. from EE Dept. IIT Delhi in 1985. He worked as a Scientist in CARE, IIT Delhi from 1977 to 1985 where he was involved in designing and implementing real-time DSP systems. For the last 29 years, he is involved in teaching and research in the areas of digital systems design, electronic design automation and embedded systems. He has supervised 14 Ph.D. students, 3 MSR students, 173 M.Tech/B.Tech projects and published nearly 112 conference and journal papers. Further, he has held visiting positions in universities in Canada, USA and Germany. At IIT Delhi, he has been the Philips Chair Professor, Head of the Department of Computer Science & Engineering, Dean of Post Graduate Studies & Research and Deputy Director (Faculty). He has been associated with a number of initiatives to promote research at IIT Delhi. Along with seven graduating students and four other faculty colleagues he founded KritiKal Solutions in 2002. This was the first student-faculty led start-up in the Technology Business Incubation Unit at IIT Delhi.
ASSISTECH, a laboratory and research group co-founded by him is involved in developing a number of assistive devices targeted towards mobility and education of the visually impaired. He has been a recipient of two National awards for his work in the disability space. SmartCaneTM is a mobility aid for visually impaired developed by his group and currently it is used by thousands of users in India and other countries.

Chetan Arora received Ph.D. in Computer Science and B.Tech in Electrical Engineering, both from IIT Delhi in 2012 and 1999 respectively. He did his Post-Doctoral Research with Prof. Shmuel Peleg from 2012-2014 in Hebrew University, Israel. Since 2014, he is teaching as an Assistant Professor in Computer Science Department at IIIT Delhi. Chetan has published more than 20 papers in top computer vision journals and peer reviewed conferences. Prior to returning to academics, Chetan has spent over 10 years in industry, where he co-founded 3 startups, along with his counterparts in Japan and Israel, all working on computer vision products coming out of latest research ideas. Chetan has served as an area chair at ICVGIP 2016 and program co-chair for NCVPRIPG 2017. He has been actively involved in the area of computer vision for persons with disabilities and has organized multiple workshops to promote the same including Workshop on Assistive Vision held with ACCV 2016.

VLSID & ES 2018

31st International Conference on VLSI Design and
17th International Conference on Embedded Systems

Tutorial T1B: Emerging Computational Devices, Architectures and Computational Models

Vijaykrishnan Narayanan, Penn State University, University Park, PA
Arijit Raychowdhury, Georgia Institute of Technology, Atlanta, GA
Sumeet Kumar Gupta, Pennsylvania State University/Purdue University

Abstract

As traditional CMOS scaling nears the end of physical scaling, the need for new computational devices, models and architectures has become imperative. At the application level, the systems are evolving from number-crunching compute modules to intelligent systems capable of cognitive thinking. This course will look at the synergies that are required across the stack from new devices to new computational models for designing future computing systems. This course will enable attendees to understand the inter-twined nature of design optimization that requires one to interact with experts in different domains. The students will be exposed to simulation tools and modelling techniques to help explore new circuits and architectures. As physical dimensional scaling alone has ceased to be the key factor driving the industry, many innovations have occurred in designing new types of logic switches including changes to their structure (such as three-dimensional FinFETs), their underlying physics (use of tunneling for steep-switching devices), the material systems (integration of ferro-electrics in gate stack for Negative Capacitance FETs). The first part of the lecture will introduce these devices, simulation models and accompanying circuit innovations. There has been a world of revolution in the memory devices with the emergence of many non-volatile memory technologies and their tight integration in cross-point architectures. These memory systems have enabled new styles of computing systems such as the non-volatile processor for internet of thing systems and neuromorphic computing systems for cognitive computing. The lectures will focus on the synergistic coordination in advances in devices to system design. Neuromorphic systems also leverage new advances in technology such as cross-point memory arrays to integrate computing and store. Another emerging novel computational model is based on the principle "let physics do the computation". This technique focuses on using the intrinsic operation mechanism of devices (such as nanoscale electronic coupled oscillators) to do the computation, instead of building complex circuits with standard transistors to carry out the same function. The primary objective is to train the next generation researchers and practitioners that can understand the synergy across the stack from devices to applications. This will prepare the next generation workforce for the beyond Moore era using post-CMOS devices and new computational paradigms beyond Von-Neumann computing models. This course will introduce the following topics: [1] Emerging logic and memory devices: What value do they add for circuit designers/architects? [2] Circuit/Architecture design using Emerging Logic and Memory devices [3] Neuromorphic and Brain-Inspired Computing using emerging devices [4] Computing Using Coupled Dynamical Systems.

Speaker Bios:

Vijaykrishnan Narayanan is a Distinguished Professor of Computer Science and Engineering and Electrical Engineering at The Pennsylvania State University. He has more than 400-refereed publications and has 17000+ citations to his work with an h-index of 68. He is a Fellow of IEEE and Association of Computing Machinery. He is the Editor-in-Chief of IEEE Transactions of Computer-Aided Design of Integrated Circuits and Systems and served as the founding Co-editor-in-chief of ACM Journal of Emerging Technologies in Computing Systems. He is the chair of ACM Special Interest Group on Design Automation. He has won several awards in recognition of his research including the IEEE Micro Top Picks for 2016, ASPDAC ten-year retrospective most influential paper award, IEEE Transactions on VLSI Best Paper Award, One of the most significant papers of FPL in the 25-year history of Conference. He is listed in the Hall of Fame of the top computer architecture conferences: ISCA and HPCA. He also owns multiple patents on architectures for emerging technologies and applications. He has provided invited technical briefing at the US Senate and invited demonstration at the Science Fair at the US Congress. He has also been an invited attendee at the White House Brain Conference. His work has been widely featured in technical press. Prof. Narayanan leads an internationally renowned National Science Foundations (NSF) expeditions-in-computing center and is an investigator of three other major multi-university centers: DARPA/SRC LEAST Center, NSF ERC ASSIST and NSF/SRC E2CDA centers.

Arijit Raychowdhury is currently an Associate Professor in the School of Electrical and Computer Engineering at the Georgia Institute of Technology where he currently holds the ON Semiconductor Junior Research Professorship. He received his Ph.D. degree in Electrical and Computer Engineering from Purdue University. He joined Georgia Tech in January, 2013. His industry experience includes five years as a Staff Scientist in the Circuits Research Lab, Intel Corporation and a year as an Analog Circuit Designer with Texas Instruments Inc. His research interests include digital and mixed-signal circuit design, design of on-chip sensors, memory, and device-circuit interactions. Dr. Raychowdhury holds more than 25 U.S. and international patents and has published over 150 articles in journals and refereed conferences. He is the winner of the NSF CRII Award, 2015; Intel Labs Technical Contribution Award, 2011; Dimitris N. Chorafas Award for outstanding doctoral research, 2007; the Best Thesis Award, College of Engineering, Purdue University, 2007; Best Paper Awards at the International Symposium on Low Power Electronic Design (ISLPED) 2012, 2006; IEEE Nanotechnology Conference, 2003; SRC Technical Excellence Award, 2005; Intel Foundation Fellowship 2006, NASA INAC Fellowship 2004, and the Meissner Fellowship 2002. Dr. Raychowdhury is a Senior Member of the IEEE.

Sumeet Kumar Gupta received the B. Tech. degree in Electrical Engineering from the Indian Institute of Technology, Delhi, India in 2006, and the M.S. and Ph.D. degrees in Electrical and Computer Engineering from Purdue University, West Lafayette IN in 2008 and 2012, respectively. Dr. Gupta is currently an Assistant Professor of Electrical Engineering at The Pennsylvania State University. Previously, he was a Senior Engineer at Qualcomm Inc. in San Diego CA, where he developed circuit design techniques and benchmarking methodologies of standard cells in deeply scaled technologies. He has also worked as an intern at National Semiconductor, Advanced Micro Devices Inc. and Intel Corporation in 2005, 2007 and 2010, respectively. His research interests include nano-electronics and spintronics, device-circuit-architecture co-design in post-CMOS technologies, low power variation aware VLSI circuit design and nano-scale device-circuit modeling and simulations. He has published over 70 articles in refereed journals and conferences and is a member of IEEE. Dr. Gupta was the recipient of 2016 DARPA Young Faculty Award, an Early Career Professorship by Penn State in 2014, the 6th TSMC Outstanding Student Research Bronze Award in 2012 and Intel Ph.D. Fellowship in 2009. He has also received Magoon Award from the School of Electrical and Computer Engineering, Purdue University, and the Outstanding Teaching Assistant Award from the Teaching Academy and the Office of the Provost, Purdue University, both in 2007. He was awarded a certificate of recognition for outstanding job during the summer internship by Intel Labs and certificates of merit for excellent academic performance at IIT Delhi.

VLSID & ES 2018 | 31st International Conference on VLSI Design and 17th International Conference on Embedded Systems

Tutorial T1C: High-Speed Serial Links: Architectures and Circuits for Clock and Data Recovery (CDR)

Saurabh Saxena, IIT Madras
Nagendra Krishnapura, IIT Madras

Abstract

Serial links behave as arteries of computing systems while transferring data from one point to the other and clock and data recovery (CDR) block has the responsibility of recovering the bits error-free at the other end of transmission. This tutorial will present the basic requirements of CDRs and relate their performance metrics to different architectures and loop components with tradeoffs for different implementations. Particularly, this tutorial will compare different CDR architectures: linear vs. bang-bang, full-rate vs. sub-rate architectures, analog vs. digital vs. hybrid loops, oscillator vs. phase interpolator vs. embedded phase-locked loops (PLLs), reference-less vs. reference-based CDRs, and CMOS vs. CML vs. charge-based receiver front-end of CDRs.

Introduction and basic architectures of CDRs: We will start with introduction to serial links and motivation for learning CDR in the serial data communication. While identifying the basic requirements of a CDR, we will discuss its performance metrics like jitter tolerance (JTOL), jitter transfer function (JTRAN), and jitter generation (JGEN) and consequences of specific demands on CDR's architecture and design. We will discuss, analyze, and compare basic design of full rate/sub-rate CDRs with their analog, digital, and hybrid implementations. Realization of CDR's front-end with CMOS, CML, and charge-based circuits will also be studied.

CDRs for multi-lane chip-to-chip links: We will discuss tradeoffs between power/area and performance for CDR's basic architectures in a multi-lane chip-to-chip application. Addressing the problems with conventional architectures, we will compare the following: wide-range ring oscillator based CDR vs. narrow-range LC oscillator based CDR, feedback loop with oscillator vs. phase interpolator or phase rotator based frequency/phase tracking in the feedback loop, and ring/LC oscillator vs. embedded phase-locked loop as a digitally controlled oscillator (DCO). We will examine circuit-level design of building blocks like phase interpolator, phase rotator, and PLL for CDRs.

CDRs for repeaters in long-haul communication: CDRs in repeaters require a wide JTOL bandwidth and low JTRAN bandwidth and traditional type-II architectures experience a direct tradeoff between JTOL and JTRAN bandwidth. So, we will examine architectures with decoupled JTOL and JTRAN bandwidth and discuss such architectures employing phase-rotating PLL (PRPLL), or phase interpolator, or delay lines.

Frequency detectors: Frequency detection of a random bit sequence is a critical part of CDR. Traditionally, a prior knowledge of data rate and a fixed reference for each receiver have been used for phase/frequency tracking of the incoming data. Since, an access to the data rate and a reference clock for each CDR is costlier, we will examine a couple of frequency detection techniques and their implementation without prior information or reference clock for the received data.

Speaker Bios:

Nagendra Krishnapura obtained his B.Tech. from the Indian Institute of Technology, Madras, India and his Ph.D. from Columbia University, New York. He has worked as an analog design engineer at Celight Inc., Multilink, and Vitesse semiconductor. He has taught analog circuit design courses at Columbia University as an adjunct faculty. He is currently an associate professor at the Indian Institute of Technology Madras. His interests are analog and RF circuit design and analog signal processing. He has been an associate editor of the IEEE Transactions on Circuits and Systems II: Express Briefs and is currently an associate editor of the IEEE Transactions on Circuits and Systems I: Regular Papers.

Saurabh Saxena (S'10-M'16) received the B.Tech. degree in electrical engineering, the M.Tech. degree in microelectronics and VLSI design from the Indian Institute of Technology Madras, Chennai, India, in 2009, and the Ph.D. degree in electrical and computer engineering from the University of Illinois, Urbana-Champaign, IL, USA in 2015. He is currently an Assistant Professor in the Department of Electrical Engineering at Indian Institute of Technology Madras, Chennai, India. He serves as a reviewer for the IEEE Journal of Solid-State Circuits, IEEE Transactions on Circuits and Systems I: Regular Papers, IEEE Transactions on Very Large Scale Integration Systems, and IEEE International Symposium on Circuits and Systems. His research interests include delta-sigma modulators, high speed I/O interfaces, and clocking circuits.

VLSID & ES 2018
31st International Conference on VLSI Design and
17th International Conference on Embedded Systems

Tutorial T2A: Safe Autonomous Systems: Real-Time Error Detection and Correction in Safety-Critical Signal Processing and Control Algorithms

Jacob Abraham, University of Texas, Austin, TX
Abhijit Chatterjee, Georgia Tech, Atlanta, GA

Abstract

While the last two decades have seen revolutions in computing and communications systems, the next few decades will see a revolution in the use of every-day robotics and artificial intelligence in broad societal applications. Examples of such systems include sensor networks, the smart power grid, self-driven cars and autonomous drones. Such systems are driven by signal processing, control and learning algorithms that process sensor data, actuate control functions and learn about the environment in which these systems operate. The trustworthiness and safety of such systems is of paramount importance and has significant impact on the commercial viability of the underlying technology. As a consequence, anomalies in system operation due to computation errors in on-board processors, degradation and failure of embedded sensors, actuators and electro-mechanical subsystems and unforeseen changes in their operation environment need to detected with minimum latency. Such anomalies also need to be mitigated in ways that ensure the safety of such systems under all possible failure scenarios. Many future systems will be self-learning in the field. It is necessary to ensure that such learning does not compromise the safety of all human personnel involved in the operation of such systems.

To enable safe operation of such systems, the underlying hardware needs to be tuned in the field to maximize performance, reliability and error-resilience while minimizing power consumption. To enable such dynamic adaptation, device operating conditions and the onset of soft errors are sensed using post-manufacture and real-time checking mechanisms. These mechanisms rely on the use of built-in sensors and/or low-overhead function encoding techniques to detect anomalies in system functions. A key capability is that of being able to deduce multiple performance parameters of the system-under-test using compact optimized stimulus using learning algorithms. The sensors and function encodings assess the loss in performance of the relevant systems due to workload uncertainties, manufacturing process imperfections, soft errors and hardware malfunction and failures induced by electro-mechanical degradation. These are then mitigated through the use of algorithm-through-circuit level compensation techniques based on pre-deployment simulation and post-deployment self-learning. These techniques continuously trade off performance vs. power of the individual software and hardware modules in such a way as to deliver the end-to-end desired application level Quality of Service (QoS), while minimizing energy/power consumption and maximizing reliability and safety. Applications to signal processing, and control algorithms for example autonomous systems will be discussed.

Speaker Bios:

Jacob A. Abraham is a Professor in the Department of Electrical and Computer Engineering at the University of Texas at Austin. He is also director of the Computer Engineering Research Center and holds a Cockrell Family Regents Chair in Engineering. He received the Bachelor's degree in Electrical Engineering from the University of Kerala, India, in 1970. His M.S. degree, in Electrical Engineering, and Ph.D., in Electrical Engineering and

Computer Science, were received from Stanford University, Stanford, California, in 1971 and 1974, respectively. From 1975 to 1988 he was on the faculty of the University of Illinois, Urbana, Illinois.

Professor Abraham's research interests include VLSI design and test, formal verification, and fault-tolerant computing. He is the principal investigator of several contracts and grants in these areas, and a consultant to industry and government on testing and fault-tolerant computing. He has over 400 publications, and has been included in a list of the most cited researchers in the world. He has supervised more than 80 Ph.D. dissertations. He is particularly proud of the accomplishments of his students, many of whom occupy senior positions in academia and industry. He has served as associate editor of several IEEE Transactions, and as chair of the IEEE Computer Society Technical Committee on Fault-Tolerant Computing. He has been elected Fellow of the IEEE as well as Fellow of the ACM, and is the recipient of the 2005 IEEE Emanuel R. Piore Award.

Abhijit Chatterjee is a professor in the School of Electrical and Computer Engineering at Georgia Tech and a Fellow of the IEEE. He received his Ph.D in electrical and computer engineering from the University of Illinois at Urbana-Champaign in 1990. Dr. Chatterjee received the NSF Research Initiation Award in 1993 and the NSF CAREER Award in 1995. He has received seven Best Paper Awards and three Best Paper Award nominations. His work on self-healing chips was featured as one of General Electric's key technical achievements in 1992 and was cited by the *Wall Street Journal*. In 1996, he received the Outstanding Faculty for Research Award from the Georgia Tech Packaging Research Center, and in 2000, he received the Outstanding Faculty for Technology Transfer Award, also given by the Packaging Research Center. In 2007, his group received the Margarida Jacome Award for work on VIZOR: Virtually Zero Margin Adaptive RF from the Berkeley Gigascale Research Center (GSRC).

Dr. Chatterjee has authored over 400 papers in refereed journals and meetings and has 20 patents. He is a co-founder of Ardext Technologies Inc., a mixed-signal test solutions company and served as chairman and chief scientist from 2000-2002. His research interests include error-resilient signal processing and control systems, mixed-signal/RF/multi-GHz design and test and adaptive real-time systems.

VLSID & ES 2018

31st International Conference on VLSI Design and
17th International Conference on Embedded Systems

Tutorial T2B: Hardware Intellectual Property (IP) Security and Trust: Challenges and Solutions

Prabhat Mishra, University of Florida, Gainesville, Florida, USA
Rajat Subhra Chakraborty, IIT Kharagpur

Abstract

Reusable hardware intellectual property (IP) based System-on-Chip (SoC) design has emerged as a pervasive design practice in the industry to dramatically reduce design/verification cost while meeting aggressive time-tomarket constraints. Growing reliance on reusable, functionally pre-verified hardware IPs and wide array of CAD tools during SoC design – often gathered from untrusted 3rd party vendors – severely affects the security and trustworthiness of SoC computing platforms. Major security issues in the hardware IPs at different stages of SoC life cycle include piracy during IP evaluation, reverse engineering, and cloning, counterfeiting, as well as malicious, hard-to-detect hardware modifications in the hardware IPs. The global electronic piracy market is growing rapidly and is now estimated to be over $1B/day [1], of which a significant part is related to hardware IPs. Due to evergrowing computing demands, modern SoCs tend to include many heterogeneous processing IP cores, together with reconfigurable cores e.g. embedded FPGA in order to incorporate logic that is likely to change as standards and requirements evolve. Such design practices greatly increase the number of untrusted components in the SoC design flow and make the overall system security a pressing concern. There is a critical need to analyze the SoC security issues and attack models due to involvement of multiple untrusted entities through the 3rd party IP (3-PIP) route – and develop low-cost effective countermeasures. These countermeasures would encompass hardware encryption and obfuscation, intelligent automatic test pattern generation (ATPG), hardware watermarking and fingerprinting, and certain analytic methods derived from the behavioral aspects of the hardware IPs to enable trusted operation with untrusted components.

In this tutorial, we plan to provide a comprehensive coverage of both fundamental concepts and recent advances in validation of security and trust of hardware IPs. It examines the state-of-the-art in research in this challenging area as well as industrial practice, and points to important gaps that need to be filled in order to develop a validation and debug flow to establish the necessary trust level of hardware IPs, eventually leading to secure SoC systems. The tutorial presenters with complementary areas of expertise and extensive experience of consulting for leading companies and R&D labs will provide a unique snapshot of the challenges, cutting-edge solutions and open problems in this area. The selection of topics covers a broad spectrum and will be of interest to a wide audience including design, validation, security, and debug engineers.

The proposed tutorial consists of four parts. The first part introduces security vulnerabilities and various challenges associated with trust validation for hardware IPs. Part II covers various demonstrated attacks and design modification based countermeasures such as hardware watermarking and obfuscation. Part III covers formal methods, simulation-based approaches as well as side channel analysis for security and trust validation in hardware IPs. Finally, Part V concludes this tutorial with discussion on emerging issues and future directions.

Speaker Bios:

Prabhat Mishra is a Professor in the Department of Computer and Information Science and Engineering at the University of Florida. His research interests include design automation of embedded systems, energy-aware computing, hardware security and trust, system validation and verification, reconfigurable architectures, and postsilicon debug. He received his Ph.D. in Computer Science and Engineering from the University of California,

Irvine. He has published five books and more than 150 research articles in premier international journals and conferences. His research has been recognized by several awards including the NSF CAREER Award, IBM Faculty Award, three best paper awards, and EDAA Outstanding Dissertation Award. Prof. Mishra currently serves as the Deputy Editor-in-Chief of IET Computers & Digital Techniques, and as an Associate Editor of ACM Transactions on Design Automation of Electronic Systems, IEEE Transactions on VLSI Systems, and Journal of Electronic Testing. He has served on many conference organizing committees and technical program committees of premier ACM and IEEE conferences. He is currently serving as an ACM Distinguished Speaker. Prof. Mishra is an ACM Distinguished Scientist and a Senior Member of IEEE.

Rajat Subhra Chakraborty is currently an Associate Professor at CSE Department of IIT Kharagpur. He received his Ph.D. from Case Western Reserve University (U.S.A.) and B.E. from Jadavpur University. He has professional experience of working at National Semiconductor, Bangalore, India and Advanced Micro Devices (AMD), Santa Clara, USA. His research interests include Hardware Security, VLSI Design and Design Automation and Digital Content Protection. He holds 2 Granted U.S. patents, 2 edited volumes, and has co-authored 3 books, 7 book chapters, and over 75 publications in international journals and conferences. His work has received close to 2000 citations till date, and a paper co-authored by him won the Best Paper Award at the IWDW'16 workshop. He has been the Program Chair of SPACE'14, SPACE'15 and AHSA-DSD'17, and regularly features in the program committee of top international conferences. He has received several prestigious international and national awards such as IEI Young Engineers Award (2016), IBM Shared University Research (SUR) Award (2015), Royal Academy of Engineering (U.K.) RECI Fellowship (2014), IBM Faculty Award (2012). Dr. Chakraborty is a Senior Member of IEEE and a Senior Member of ACM.

VLSID & ES 2018

31st International Conference on VLSI Design and
17th International Conference on Embedded Systems

Tutorial T2C: Beyond von-Neumann Computing: Devices, Circuits, and Applications

Kaushik Roy, Purdue University, West Lafayette, IN

Speaker Bio:

Kaushik Roy received B.Tech. Degree in electronics and electrical communications engineering from the Indian Institute of Technology, Kharagpur, India, and Ph.D. degree from the electrical and computer engineering department of the University of Illinois at Urbana-Champaign in 1990. He was with the Semiconductor Process and Design Center of Texas Instruments, Dallas, where he worked on FPGA architecture development and low-power circuit design. He joined the electrical and computer engineering faculty at Purdue University, West Lafayette, IN, in 1993, where he is currently Edward G. Tiedemann Jr. Distinguished Professor. His research interests include spintronics, device-circuit co-design for nano-scale Silicon and non-Silicon technologies, low-power electronics for portable computing and wireless communications, and new computing models enabled by emerging technologies. Dr. Roy has published more than 700 papers in refereed journals and conferences, holds 15 patents, supervised 75 PhD dissertations, and is co-author of two books on Low Power CMOS VLSI Design (John Wiley & McGraw Hill).

Dr. Roy received the National Science Foundation Career Development Award in 1995, IBM faculty partnership award, ATT/Lucent Foundation award, 2005 SRC Technical Excellence Award, SRC Inventors Award, Purdue College of Engineering Research Excellence Award, Humboldt Research Award in 2010, 2010 IEEE Circuits and Systems Society Technical Achievement Award (IEEE Charles Doeser award) , Distinguished Alumnus Award from Indian Institute of Technology (IIT), Kharagpur, Fulbright-Nehru Distinguished Chair, DoD Vannevar Bush Faculty Fellow Faculty Fellow (2014-2019), Semiconductor Research Corporation Aristotle award in 2015, and best paper awards at 1997 International Test Conference, IEEE 2000 International Symposium on Quality of IC Design, 2003 IEEE Latin American Test Workshop, 2003 IEEE Nano, 2004 IEEE International Conference on Computer Design, 2006 IEEE/ACM International Symposium on Low Power Electronics & Design, and 2005 IEEE Circuits and system society Outstanding Young Author Award (Chris Kim), 2006 IEEE Transactions on VLSI Systems best paper award, 2012 ACM/IEEE International Symposium on Low Power Electronics and Design best paper award, 2013 IEEE Transactions on VLSI Best paper award. Dr. Roy was a Purdue University Faculty Scholar (1998-2003). He was a Research Visionary Board Member of Motorola Labs (2002) and held the M. Gandhi Distinguished Visiting faculty at Indian Institute of Technology (Bombay) and Global Foundries visiting Chair at National University of Singapore. He has been in the editorial board of IEEE Design and Test, IEEE Transactions on Circuits and Systems, IEEE Transactions on VLSI Systems, and IEEE Transactions on Electron Devices. He was Guest Editor for Special Issue on Low-Power VLSI in the IEEE Design and Test (1994) and IEEE Transactions on VLSI Systems (June 2000), IEE Proceedings — Computers and Digital Techniques (July 2002), and IEEE Journal on Emerging and Selected Topics in Circuits and Systems (2011). Dr. Roy is a fellow of IEEE.

VLSID & ES 2018
31st International Conference on VLSI Design and
17th International Conference on Embedded Systems

Tutorial T2D: Privacy Assurances in the Internet of Things (IoT) World

Parameswaran Ramanathan, University of Wisconsin, Madison, WI

Abstract

Most Internet-of-Things (IoT) applications collect and aggregate data from a large number of sensors/users. The data often contains sensitive information related to the associated users and system. In recent years, there is extensive research on techniques to assure confidentiality of sensitive information in such systems. In this tutorial, we will introduce the privacy challenge, review the different approaches for privacy assurances in literature, and highlight open issues that need further investigation. The tutorial will use example from smart grids, smart homes, smart cities, mobile computing, and intelligent transportation systems to illustrate the solution approaches. This tutorial will be of interest to undergraduate and graduate students working in different IoT applications areas such as smart city projects, smart homes, and mobile computing. Industrial participants working in IoT technologies will also benefit from this tutorial. Lecturers in Indian engineering colleges who are teaching courses in security and privacy can also use this tutorial to not only get a broad overview but also to understand recent advances in the area. The tutorial will be self-contained.

Speaker Bio:

Parmesh Ramanathan received the B. Tech degree from the Indian Institute of Technology, Bombay, India, in 1984, and the M.S.E. and Ph.D. degrees from the University of Michigan, Ann Arbor, in 1986 and 1989, respectively. Since 1989, Dr. Ramanathan has been faculty member in the Department of Electrical and Computer Engineering, University of Wisconsin, Madison, where he is presently an Associate Dean for Graduate Education. He also served as a Chair of Department of Electrical and Computer Engineering from 2005-2009. He has served as consultant to AT&T Laboratories, Telcordia Technologies, and Hewlett Packard Laboratories. He was also a Visiting Professor at Kanwal Rekhi School of Information Technology, Indian Institute of Technology, Bombay, India in 2004 and Visiting Researcher at Microsoft Research in 2010. Dr. Ramanathan's research interests include real-time systems, wireless and wireline networking, faulttolerant computing, and distributed systems. He has served as an Associate Editor for IEEE Transactions on Mobile Computing, Associate Editor for IEEE Transactions on Parallel and Distributed Computing (1996–1999) and Elsevier AdHoc Networks Journal (2002–2005). He was General Chair of Mobicom (2011) and MASS (2013). In 2009, he was elevated to Fellow of IEEE for his contributions to real-time systems and networks.

VLSID & ES 2018
31st International Conference on VLSI Design and
17th International Conference on Embedded Systems

Tutorial T2E: Pre-Silicon Verification and Post-Silicon Validation: Dramatic Improvements through Disruptive Innovations

Subhasish Mitra, Stanford University, Stanford, CA
Srinivas Shashank Nuthakki, Stanford University, Stanford, CA
Eshan Singh, Stanford University, Stanford, CA

Abstract

You have all spent weeks or months of onerous manual effort, from writing assertions to running long simulations (with limited success for corner-case bugs) or debugging false positives. This tutorial will give you a unique hands-on experience on how to detect and localize difficult bugs automatically, in just a few hours, during pre-silicon verification and post-silicon validation.

We present the Quick Error Detection (QED) technique for post-silicon validation and debug. QED drastically reduces error detection latency, the time elapsed between the occurrence of an error caused by a bug and its manifestation as an observable failure. Symbolic QED combines QED principles with a formal engine for both pre- and post-silicon validation.

Results from several commercial designs demonstrate:

1. For billion transistor-scale designs, you can now detect and localize difficult logic design bugs automatically (without having to write design-specific assertions) in only a few (~3) hours during pre-silicon verification.
2. You can now drastically improve error detection latencies of post-silicon validation tests by up to 9 orders of magnitude for quick debug, from billions of clock cycles to very few clock cycles, and simultaneously improve bug coverage.
3. You can now automatically localize bugs in billion transistor-scale designs during post-silicon debug, e.g., narrow locations of electrical bugs to a handful of flip-flops (~18 for a design with ~1million flip-flops), in only a few (~9) hours.

QED and Symbolic QED are effective for logic design bugs and electrical bugs inside processor cores, hardware accelerators, and uncore components such as cache controllers, memory controllers, interconnection networks or power management units. QED techniques have been successfully used in industry.

Speaker Bios:

Subhasish Mitra is Professor of Electrical Engineering and of Computer Science at Stanford University, where he directs the Stanford Robust Systems Group and co-leads the Computation focus area of the Stanford SystemX Alliance. He is also a faculty member of the Stanford Neurosciences Institute. Prof. Mitra holds the Carnot Chair of Excellence in Nanosystems at CEA-LETI in Grenoble, France. Before joining the Stanford faculty, he was a Principal Engineer at Intel Corporation.

Prof. Mitra's research interests range broadly across robust computing, nanosystems, VLSI design, validation, test and electronic design automation, and neurosciences. He, jointly with his students and collaborators, demonstrated the first carbon nanotube computer and the first three-dimensional nanosystem with computation immersed in data storage. These demonstrations received wide-spread recognitions (cover of NATURE, Research Highlight to the United States Congress by the National Science Foundation, highlight as "important, scientific breakthrough" by the BBC, Economist, EE Times, IEEE Spectrum, MIT Technology Review, National Public Radio, New York Times, Scientific American, Time, Wall Street Journal, Washington Post and numerous others worldwide). His earlier work on X-Compact test compression has been key to cost-effective manufacturing and high-quality testing of almost all electronic systems. X-Compact and its derivatives have been implemented in widely-used commercial Electronic Design Automation tools.

Prof. Mitra's honors include the ACM SIGDA/IEEE CEDA A. Richard Newton Technical Impact Award in Electronic Design Automation (a test of time honor), the Semiconductor Research Corporation's Technical Excellence Award, the Intel Achievement Award (Intel's highest corporate honor), and the Presidential Early Career Award for Scientists and Engineers from the White House (the highest United States honor for early-career outstanding scientists and engineers). He and his students published several award-winning papers at major venues: ACM/IEEE Design Automation Conference, IEEE International Solid-State Circuits Conference, IEEE International Test Conference, IEEE Transactions on CAD, IEEE VLSI Test Symposium, and the Symposium on VLSI Technology. At Stanford, he has been honored several times by graduating seniors "for being important to them during their time at Stanford."

Prof. Mitra served on the Defense Advanced Research Projects Agency's (DARPA) Information Science and Technology Board as an invited member. He is a Fellow of the Association for Computing Machinery (ACM) and the Institute of Electrical and Electronics Engineers (IEEE).

Eshan Singh received an ScB in Electrical Engineering, along with an AB in Economics, from Brown University in 2009. After completing an MS in Electrical Engineering from Stanford in 2011, Eshan spent three years at Intel as a Component Design Engineer. Eshan returned to Stanford in 2014 and is currently a PhD candidate in the Stanford Robust Systems Group with interests in VLSI design, 3-D integrated circuits, computer architecture, validation and debug. His current research focuses on addressing challenges in validation and debug, specifically aiming to improve bug localization, increase automation and reduce debug time.

Srinivasa Shashank Nuthakki is a PhD student working in Prof. Subhasish Mitra's Robust Systems Group at Stanford University. He received the B.Tech. (Hons.) degree in Electronics and Electrical Communication Engineering and the M.Tech. degree in Microelectronics and VLSI Design from the Indian Institute of Technology, Kharagpur, in 2016. His current research interests include pre-siiicon verification, post-silicon validation, formal methods, hardware/software security and computer architecture.

VLSID & ES 2018
31st International Conference on VLSI Design and
17th International Conference on Embedded Systems

Tutorial T2F: Silicon Nanophotonics for Future Manycore Chips: Opportunities and Challenges

Sudeep Pasricha, Colorado State University, Fort Collins, CO

Abstract

The need for high performance and energy-efficient communication between processing cores has never been more critical. With levels of integration growing from tens of cores on a single chip today to hundreds of cores in emerging manycore chip architectures, there is immense pressure on the onchip communication fabric to support the many more quality of service (QoS) requirements and higher bandwidths for data transfers than ever before. An important consequence of this trend is that chip power and performance are now beginning to be dominated not by processor cores but by the components that facilitate transport of data between processors and memory. Unfortunately, traditional electrical wires that make up the backbone of communication fabrics of manycore chips today are facing unprecedented challenges in ultra-deep nanoscale CMOS fabrication technologies. These wires are becoming slower, more power hungry, and less reliable. We are thus at a critical juncture where the power, bandwidth, and latency costs of communication must scale favorably to meet the needs of manycore processing chips in the near future. A failure to adequately respond to this challenge will create a brick wall that will impede advances in all forms of computing.

Silicon nanophotonics has emerged in recent years as one of the most promising solutions to overcome the challenge of worsening communication performance with technology scaling. Recent breakthroughs in silicon nanophotonic device fabrication and CMOS integration have presented computer architects with an opportunity to explore both on-chip and chip-to-chip communications with optical networks that have significant advantages in bandwidth density, energy-efficiency, and propagation delay over traditional electrical solutions. Not surprisingly, the challenge of designing silicon nanophotonic communication fabrics is today actively being pursued by a number of researchers worldwide, and from a variety of different perspectives. New nanophotonic devices (e.g., modulators, photodetectors, waveguides) are emerging from academia and research labs, as are new nanophotonics based on-chip network architectures, and tools for rapid design and analysis. Many semiconductor companies (e.g., Intel, IBM) have begun investing heavily into silicon nanophotonics and are releasing functional prototypes. Silicon photonic foundries are also emerging to allow low-cost fabrication of nanophotonic components on silicon chips. However, silicon nanophotonic interconnects in manycore chips will not result in a one-for-one replacement of electrical interconnects. The lack of practical buffering and the fundamental circuit switched nature of optical data communications requires holistic and innovative approaches to designing system-wide photonic interconnection networks. New network devices, circuits, architectures, and protocols are required that incorporate the unique characteristics of the optical physical layer.

This tutorial aims to provide a comprehensive overview of silicon nanophotonics for future manycore chip architectures. The first part of the tutorial will educate the audience on the basics of silicon nanophotonics technology as well as the state-of-the-art in silicon nanophotonic fabrication and prototyping from industry and academia. Next, the challenges related to reliability, energy consumption, performance, and thermal stability of silicon photonics building blocks will be discussed. Subsequently, device-level, circuit-level, and architecture-level solutions will be comprehensively surveyed and presented, where the goal is to overcome the key challenges to the low cost integration of silicon nanophotonic interconnects at the chip-scale. Lastly, cross-layer solutions that combine techniques at the device, circuit, architecture, and/or system levels will be presented. Such solutions represent a very promising approach to achieving fault tolerance, energy efficiency, high performance, and thermal stability with much lower overheads than traditional single layer solutions. The tutorial will conclude with a discussion of open challenges and research problems in the area of silicon nanophotonics. The target audience for this tutorial is quite broad, from students of VLSI design who want to learn about this exciting area, to researchers and industry practitioners who want to get an updated view of recent developments in the area.

Speaker Bio:

Sudeep Pasricha received the B.E. in electronics and communication engineering from Delhi Institute of Technology, Delhi, India, in 2000, and his Ph.D. in computer science from the University of California, Irvine in 2008. Between 2000 and 2008 he also worked as a design engineer for several years at STMicroelectronics and Conexant. He joined Colorado State University (CSU) in 2008. He is currently a Monfort and Rockwell-Anderson Associate Professor in the Department of Electrical and Computer Engineering and the Department of Computer Science at CSU, where he is also the Chair of Computer Engineering and Director of the Embedded Systems and High Performance Computing (EPiC) Laboratory. Dr. Pasricha received the 2015 IEEE/TCSC Award for Excellence for a mid-career researcher, the 2014 George T. Abell Outstanding mid-career faculty award, and the 2013 AFOSR Young Investigator Award. His research on silicon photonics for 2D and 3D manycore computing has been recognized with three Best Paper Awards at the ACM SLIP 2016, ACM GLSVLSI 2015, and IEEE ISQED 2010 conferences, as well as a Best Paper Award Candidate selection at IEEE ISQED 2016. He is currently in the Editorial Board of ACM TECS, IEEE TCAD, IEEE TMSCS, IEEE D&T, IEEE CM, and JPDC. He is currently or has been an Organizing Committee Member of several IEEE/ACM conferences, including DAC, ESWEEK, GLSVLSI, NOCS, RTCSA, IGSC, VLSID, and ICESS.

VLSID & ES 2018 | **31st International Conference on VLSI Design and**
17th International Conference on Embedded Systems

Special Session: Design of Energy-Efficient and Reliable VLSI Systems: A Data-Driven Perspective

Janardhan Rao Doppa, Washington State University, USA
Sudeep Pasricha, Colorado State University, USA
Partha Pratim Pande, Washington State University, USA
Krishnendu Chakrabarty, Duke University, USA

Motivation for Special Session:

The amount of data generated and collected across computing platforms every day is not only enormous, but growing at an exponential rate. Advanced data analytics and machine learning techniques have become increasingly essential to analyze and extract meaning from such "Big Data". These techniques can be very useful to detect patterns and trends to improve the operational behavior of computing systems, but they also introduce a number of outstanding challenges: (1) How can we design and deploy data analytics mechanisms to improve energy-efficiency and reliability in IoT and mobile devices, without introducing significant software overheads? (2) How to leverage emerging technologies (e.g.,3D integration) to design energy-efficient and reliable manycore systems for big data computing? (3) How to use machine learning and data mining techniques for effective design space exploration of computing systems, and enable adaptive control to improve energy-efficiency? (4) How can data analytics detect anomalies and increase robustness in the network backbone of emerging large-scale networking systems? To address these outstanding challenges, out-of-the-box approaches need to be explored. In this special session, we will discuss these outstanding problems and describe far-reaching solutions applicable across the interconnected ecosystem of IoT and mobile devices, manycore chips, datacenters, and networks. The special session brings together speakers with unique insights on applying data analytics and machine learning to real-world problems to achieve the most sought after features on multi-scale computing platforms, viz. intelligent data mining, energy-efficiency, and robustness.

By integrating data analytics and machine learning algorithms, statistical modeling, embedded hardware and software design, and cloud computing content, this session will engage a broad section of Embedded and VLSI Design conference attendees. This special session is targeted towards university researchers/professors, students, industry professionals, and embedded/VLSI system designers. This session will attract newcomers who want to learn how to apply data analytics to solve problems in computing systems, as well as experienced researchers looking for exciting new directions in embedded systems, VLSI design, EDA algorithms, and multi-scale computing.

Organizers' Biographies

Jana Doppa is an Assistant Professor in the School of Electrical Engineering and Computer Science at Washington State University, Pullman. He earned his PhD working with the Artificial Intelligence group at Oregon State University (2014); and his MTech from Indian Institute of Technology (IIT), Kanpur, India (2006). His general research interests are in the broad field of Artificial Intelligence (AI), where he focuses on machine learning, and data-driven science and engineering with applications to electronic design automation, databases, sustainability, natural language processing, and bioinformatics. He received an Outstanding Paper Award for his structured prediction work at the AAAI (2013) conference, a Google Faculty Research Award (2015), and the Outstanding

xlvii

Innovation in Technology Award from Oregon State University (2015). He is an elected editorial board member of the Journal of Artificial Intelligence Research, and regularly serves on the Program Committee of top-tier conferences including AAAI, IJCAI, ICML, NIPS, AISTATS, ICAPS, and KDD. He taught a tutorial on "Structured Prediction" at IJCAI-2016 conference and will teach a tutorial on "Recent Advances in Structured Prediction" at AAAI-2018 conference. He has organized successful workshops on "Structured Prediction" at ICML-2013 conference,and "Human is More than a Labeler" at IJCAI-2016 conference.

Sudeep Pasricha is a Monfort Professor and Rockwell-Anderson Professor in the Department of Electrical and Computer Engineering at the Walter Scott Jr. College of Engineering in Colorado State University. He received his Ph.D. in Computer Science from the University of California, Irvine in 2008. His research interests are broadly in the areas of algorithms and architectures for embedded systems, mobile computing, and high performance computing, with an emphasis on energy-efficient, reliable, and secure design. His work in recent years has extensively applied principles of data analytics and machine learning to the design of efficient software, particularly middleware, distributed resource managers, and runtime engines, for mobile devices, servers, and datacenters/supercomputers. His research has been recognized with five Best Paper Awards and multiple Best Paper nominations. He is a recipient of the 2015 IEEE/TCSC Award for Excellence for a mid-career researcher and the 2013 AFOSR Young Investigator Award. He is in the Editorial Board for six journals, including *ACM TECS, IEEE TCAD, and IEEE TMSCS*. He is currently or has been an Organizing Committee Member of several conferences such as DAC, ESWEEK, GLSVLSI, VLSID, NOCS, IGSC, RTCSA, etc. He is a Senior Member of the IEEE and ACM.

Partha Pratim Pande is a Professor and holder of the Boeing Centennial Chair in computer engineering at the school of Electrical Engineering and Computer Science, Washington State University, Pullman, USA. His current research interests are novel interconnect architectures for manycore chips, on-chip wireless communication networks, and hardware accelerators for biocomputing. Dr. Pande currently serves as the Editor-in-Chief (EIC) of *IEEE Transactions on Multi-Scale Computing Systems* (TMSCS) and Associate Editor-in-Chief (A-EIC) of *IEEE Design and Test* (D&T). He is on the editorial boards of *IEEE Transactions on VLSI (TVLSI)* and *ACM Journal of Emerging Technologies in Computing Systems (JETC)*. He was the technical program committee chair of IEEE/ACM Network-on-Chip Symposium 2015. He also serves in the program committee of many reputed international conferences. He has won the NSF CAREER award in 2009. He is the winner of the Anjan Bose outstanding researcher award from the college of engineering, Washington State University in 2013.

Krishnendu Chakrabarty received the B. Tech. degree from the Indian Institute of Technology, Kharagpur, in 1990, and the M.S.E. and Ph.D. degrees from the University of Michigan, Ann Arbor, in 1992 and 1995, respectively. He is now the William H. Younger Distinguished Professor and Chair in the Department of Electrical and Computer Engineering and Professor of Computer Science at Duke University. Prof. Chakrabarty is a recipient of the National Science Foundation CAREER award, the Office of Naval Research Young Investigator award, the Humboldt Research Award from the Alexander von Humboldt Foundation, Germany, the IEEE Transactions on CAD Donald O. Pederson Best Paper Award (2015), the ACM Transactions on Design Automation of Electronic Systems Best Paper Award (2017), and over a dozen best paper awards at major conferences. He is also a recipient of the IEEE Computer Society Technical Achievement Award (2015), the IEEE Circuits and Systems Society Charles A. Desoer Technical Achievement Award (2017), and the Distinguished Alumnus Award from the Indian Institute of Technology, Kharagpur (2014). He is a Research Ambassador of the University of Bremen (Germany) and a Hans Fischer Senior Fellow (named after Nobel Laureate Prof. Hans Fischer) at the Institute for Advanced Study, Technical University of Munich, Germany. He has held Visiting Professor positions at University of Tokyo and the Nara Institute of Science and Technology (NAIST) in Japan, and Visiting Chair Professor positions at Tsinghua University (Beijing, China) and National Cheng Kung University (Tainan, Taiwan). He is a De Tao Master of Emerging Information Technologies, DeTao Masters Academy of the DeTao Group in China.

Prof. Chakrabarty's current research projects include: testing and design-for-testability of integrated circuits and systems; digital microfluidics, biochips, and cyberphysical systems; data analytics for fault diagnosis, failure prediction, anomaly detection, and hardware security; neuromorphic computing systems. He has authored 20 books on these topics (with one translated into Chinese), published over 630 papers in journals and refereed conference proceedings, and given over 270 invited, keynote, and plenary talks. He has also presented 60 tutorials at major international conferences. He is a Fellow of ACM, a Fellow of IEEE, and a Golden Core Member of the IEEE

Computer Society. He holds 10 US patents, with several patents pending. He was a 2009 Invitational Fellow of the Japan Society for the Promotion of Science (JSPS). He is a recipient of the 2008 Duke University Graduate School Dean's Award for excellence in mentoring, and the 2010 Capers and Marion McDonald Award for Excellence in Mentoring and Advising, Pratt School of Engineering, Duke University. He has served as a Distinguished Visitor of the IEEE Computer Society (2005-2007, 2010-2012), a Distinguished Lecturer of the IEEE Circuits and Systems Society (2006-2007, 2012-2013), and an ACM Distinguished Speaker (2008-2016).

Prof. Chakrabarty served as the Editor-in-Chief of *IEEE Design & Test of Computers* during 2010-2012 and *ACM Journal on Emerging Technologies in Computing Systems* during 2010-2015. Currently he serves as the Editor-in-Chief of *IEEE Transactions on VLSI Systems*. He is also an Associate Editor of *IEEE Transactions on Biomedical Circuits and Systems, IEEE Transactions on Multiscale Computing Systems*, and *ACM Transactions on Design Automation of Electronic Systems*.

Contributed Presentations in the Special Session

Overcoming Energy and Reliability Challenges for IoT and Mobile Devices with Data Analytics

Sudeep Pasricha
Department of Electrical and Computer Engineering
Colorado State University, USA

Abstract: A very large amount of data is produced by mobile and Internet-of-Thing (IoT) devices today. Increasing computational abilities and more sophisticated operating systems (OS) on these devices have allowed us to create applications that are able to leverage this data to deliver better services. But today's mobile and IoT solutions are heavily limited by low battery capacity and limited cooling capabilities. This motivates a search for new ways to optimize for energy-efficiency. Advanced data analytics and machine-learning techniques are becoming increasingly popular to analyze and extract meaning from Big Data. In this paper, we make the case for designing and deploying data analytics and learning mechanisms to improve energy-efficiency in IoT and mobile devices with minimal overheads. We focus on middleware for inserting energy-efficient data analytics-driven solutions and optimizations in a robust manner, without altering the OS or application code. We discuss several case studies of powerful and promising developments in deploying data analytics middleware for energy-efficient and robust execution of a variety of applications on commodity mobile devices.

Data-Driven Optimization Algorithms for Computing Systems Design

Sourav Das, Wonje Choi, Nitthilan Kannappan Jayakodi, Janardhan Rao Doppa, and Partha Pratim Pande
School of Electrical Engineering and Computer Science
Washington State University, Pullman, USA

Abstract: In the emerging data-driven science paradigm, computing systems ranging from IoT and mobile to manycores and datacenters play different roles, and they need to be optimized for appropriate objectives as per the needs of the applications. The design space of computing systems is very large in general. In this talk, the speaker will describe how machine learning techniques can be leveraged to improve the computational-efficiency of hardware design optimization. We will describe generic methodologies that are applicable for any design space and employ manycore system design as an example for illustration purposes. First, to perform static optimization with a known objective function (no need to perform simulations), we will provide a guided design space exploration approach, where a learned evaluation function that improves with exploration identifies promising parts of the design space. Second, to perform static optimization with an unknown objective function (need to perform simulations to estimate the quality of a design), we will describe a Bayesian sequential decision-making approach that selects the next candidate design to be evaluated to quickly direct the search towards (near-) optimal designs by trading-off exploration and exploitation at each step. Third, to perform dynamic optimization (e.g., power/memory management), we will describe direct policy search methods as a simple and better alternative for reinforcement learning (RL) techniques such as Q-learning, which are very popular in the EDA community. We will present comprehensive experimental results for application-specific manycore system design optimization to demonstrate the efficacy of these methods over traditional EDA approaches.

Energy-Efficient and Reliable 3D Network-on-Chip (NoC): A Machine Learning-Inspired Methodology
Sourav Das, Biresh Kumar Joardar, Janardhan Rao Doppa, and Partha Pratim Pande
School of Electrical Engineering and Computer Science
Washington State University, Pullman, USA

Abstract: Existing 3D NoC architectures predominantly follow straightforward extension of regular 2D NoCs. However, this does not exploit the advantages provided by the 3D integration technology appropriately. In this context, design of small-world network-based NoC architectures is a notable example. We will show in this presentation how this concept of small-worldness can be adopted in 3D NoCs. More specifically, the vertical links in 3D NoC enable the design of long-range shortcuts necessary for a small-world network. These vertical connections give rise to the concept of physically far but logically near links while implementing the shortcuts. We will explain how machine learning can be leveraged to intelligently explore the design space to optimize the placement of both planar and vertical communication links for energy efficiency.

Data-Driven Resiliency Solutions for Boards and Systems
Shi Jin and Krishnendu Chakrabarty
Department of Electrical and Computer Engineering
Duke University, Durham, NC, USA

Abstract: Data analytics and real-time monitoring can be used to ensure that boards and systems operate as intended. This paper first describes how machine learning, statistical techniques, and information-theoretic analysis can be used to close the gap between working silicon and a working system. Next, it describes how time-series analysis can be used to analyze health status and detect anomalies in complex core router systems. Traditional techniques fail to identify abnormal or suspect patterns when the monitored data involves temporal measurements and exhibits significantly different statistical characteristics for its constituent features. This paper thus not only describes a feature-categorization-based hybrid method and a changepoint-based method to detect anomalies in time-varying features with different statistical characteristics, but also proposes a symbol-based health analyzer to obtain a full picture of the health status of monitored core routers. A comprehensive set of experimental results is presented for data collected during 30 days of field operation from over 20 core routers deployed by customers of a major telecom company.

Special Session: Advances in Power Management Integrated Circuits

Qadeer Khan, Indian Institute of Technology Madras, Chennai, India
Seong-Joong Kim, University of Illinois at Urbana-Champaign, IL, USA
Pavan Kumar Hanumolu, University of Illinois at Urbana-Champaign, IL, USA
Punith Surkanti, Nokia, Inc., Boston, MA, USA
Annajirao Garimella
Paul M. Furth, New Mexico State University, Las Cruces, NM, USA

Motivation for Special Session:

Recent advances in integrated power management circuits are facilitating the possibility of seamless integration of LDOs, DC-DC converters, energy harvesters and references on the same die along with analog, mixed-signal and RF circuits. In this special session, we will introduce several new designs emerging in this rapidly developing field. This session is a great place for beginners and experts alike to gain an in-detail look at this emerging field.

Organizers' Biographies

Qadeer A. Khan is an assistant professor in the Integrated Circuits and System group of the department of Electrical Engineering, Indian Institute of Technology Madras. He received the Bachelor's degree in electronics and communication engineering from Jamia Millia Islamia University, New Delhi, India, in 1999 and the Ph.D. degree in electrical and computer engineering from Oregon State University, USA in 2012. His Ph.D. work was focused on developing novel control techniques for high performance switching dc-dc converters. He has worked with Qualcomm, San Diego, CA, Qualcomm, Bangalore, Motorola and Freescale Semiconductor, India in various positions.

Dr. Qadeer Khan holds 17 U.S. patents and authored/co-authored over 20 IEEE publications in the area of analog, mixed-signal and power management ICs. He serves as reviewer of the *IEEE Journal of Solid-State Circuits, IEEE Transactions on Very Large Scale Integration Systems,* and *IEEE Transaction on Power Electronics and IEEE Power Electronics Letters.*
His research interests involve high-performance linear regulators, LDOs, switching dc-dc converters and power management ICs for portable electronics and energy harvesting.

Seong-Joong Kim received the Bachelor and the Master degree in electrical and computer engineering from the Hanyang University, Korea, in 2001 and in 2003, respectively. He is currently pursuing the Ph.D.degree at the University of Illinois at Urbana-Champaign, IL, USA. From 2003 to 2008, he was a research engineer at the research laboratory of the LG Electronics, Korea, where he was involved in the development of advanced display systems such as PM/AM-OLEDs, flexible and transparent displays. Then, he moved to the System-IC business division of the LG Electronics where he worked as an analog IC design engineer for low-power & high-resolution display driver ICs. His research interests include mixed-signal design, power converter design, and display electronics.

Punith R. Surkanti (S'07) is a Ph.D. candidate in electrical engineering at New Mexico State University, Las Cruces, NM. He obtained his M.S. degree in electrical engineering from New Mexico State University, Las Cruces,

NM in 2011. He is the recipient of the Outstanding Graduate Teaching Assistantship Award from NMSU in 2011. He is currently with Nokia as Senior IC Design Engineer and in the past he worked at Eta Devices Inc. His areas of interest include Analog, Mixed-signal and Power Management IC Design.

Annajirao Garimella (S'99–M'10–SM'15) received the B.E. degree in electrical and electronics engineering from the University of Madras, Chennai, India, in 2000, and the M.S. and Ph.D. degrees in electrical engineering with a minor in management from New Mexico State University, Las Cruces, NM in 2009 and 2010, respectively. He has authored or coauthored more than 45 papers, published in IEEE conferences and journals. His research interests lie in the area of Circuits and Systems, with emphasis on analog, mixed-signal, power management IC design, SoC design, and testing. He is a recipient of the HENAAC (Hispanic Engineer National Achievement Award Corporation) AMD Scholarship in 2005 and the HENAAC DaimlerChrysler Scholarship in 2006.

Dr. Garimella received the Outstanding Ph.D. Graduate Alumni Association Award at New Mexico State University for the class of fall, 2010. He has co-organized special sessions on analog and power management circuit design and delivered embedded tutorials at IEEE conferences. He is a Guest Editor for Springer Analog Integrated Circuits and Signal Processing journal and is serving on the steering committee of the IEEE Midwest Symposium on Circuits and Systems Conference (MWSCAS) and on the steering committee of the International Conference on VLSI Design (VLSID).

Paul M. Furth (S'90–M'96–SM'11) received a B.A. from Grinnell College, Ginnell, IA in 1984, the B.S.E.E. degree from the California Institute of Technology, Pasadena, CA in 1985 and the M.S. and Ph.D. degrees in electrical engineering from the Johns Hopkins University, Baltimore, MD, in 1992 and 1996, respectively. In 1995, he joined the Klipsch School of Electrical and Computer Engineering, New Mexico State University, Las Cruces, where he is currently an Associate Professor. He has work experience at Sandia National Labs, Micron, and Motorola. His areas of interest include analog and mixed-signal VLSI design, power management circuits, and CMOS image sensors. Dr. Furth received the Bromilow Teaching Excellence Award in 2008. He has organized Special Sessions on Power Management circuit design at MWSCAS 2011 and MWSCAS 2012 and currently serves as a Steering Committee Member for MWSCAS.

Contributed Presentations in the Special Session

Time-Based PWM Controller for Fully Integrated High Speed Switching DC-DC Converters – An Alternative to Conventional Analog and Digital Controllers
Qadeer A. Khan, Seong-Joong Kim and Pavan Kumar Hanumolu

Abstract:
This tutorial discusses design of a highly integrated, low quiescent current continuous time PWM controller using time-based signal processing. By virtue of the continuous-time digital nature of the time-based PWM controller, it is capable of achieving very high resolution and speed without using any error amplifier and large compensation capacitor or any high resolution A/D converter and digital PWM while preserving all the benefits of both analog and digital PWM controllers. Using time as the processing variable, the controller operates with CMOS-level digital-like signals but without adding any quantization error. A voltage or current controlled ring oscillator is used as an integrator in place of conventional opamp-RC or Gm-C integrator while a voltage or current controlled delay line is used to perform voltage-to-time conversion. Starting with trade-offs with high speed design of conventional analog and digital PWM controllers, the concept of time-based proportional-integral-derivative (PID) controller with complete architecture of a time-based buck converter is presented. The technique was successfully demonstrated on silicon with implementation of 10-25MHz single phase and 30- 70MHz 4-phase buck converters in 180nm and 65nm CMOS processes, respectively.

Flipped Voltage Follower based Low Dropout (LDO) Voltage Regulators: A Tutorial Overview

Punith R. Surkanti, Annajirao Garimella and Paul M. Furth

Abstract*:* This tutorial introduces flipped voltage follower (FVF) based low dropout voltage regulators (LDOs) and their recent advances in the literature. Conventional off-chip output capacitor LDOs deploy an off-chip output capacitor in the range of hundreds of nanofarads to tens of microfarads with an inherent equivalent series resistance (ESR) in order to attain stability and low overshoot and undershoot during load transients. Because of the large value of the output capacitor, the bandwidth of the LDO is often limited and the transient response is slow. Advancements in System-on-Chips (SoCs) and bill of materials cost often restrict designers from using off-chip components or too many input output (IO) pins. Thus low value integrated on-chip capacitor LDOs are gaining importance. One of the major requirements of on-chip capacitor LDOs is to achieve high bandwidth. An FVF is an analog voltage buffer circuit with fast local feedback. It is capable of sourcing high currents, which makes it ideal candidate for LDOs. Design details such as pole-zero equations, stability, output resistance and transient response of basic and advanced FVF LDOs are discussed in this tutorial. The designs are analyzed with mathematical analysis and simulation results.

2018 31th International Conference on VLSI Design and 2018 17th International Conference on Embedded Systems

A Systematic Approach to Determining the Weights of the Capacitors in the DAC of a Non-binary Redundant SAR ADCs

Abstract—The use of non-binary weighted DACs with re-dundant DAC capacitor weights has been shown to provide significant speed improvements in SAR ADCs. The non-binary SAR algorithm has an in-built correction mechanism against bit decision errors caused by incomplete DAC settling by providing overlap ranges in the search space and allowing alternative bit decision paths for approximating a given input voltage. In this paper we propose to use overlap ranges not only for speed-improvement but also for providing immunity to noise in the least-significant bit decisions thereby improving ADC robustness in noisy environments. Furthermore, we propose a systematic approach to the design of the redundant SAR ADC by providing an algorithm for determining the weights of the capacitors in the DAC to obtain the required DAC settling time improvement as well as noise immunity in the last few bit decisions as compared to a binary weighted SAR ADC. A 10-bit SAR ADC designed using the above weight distribution algorithm shows a 4X improvement in bit cycle-time (200ps) and a 5.5 dB improvement in noise immunity over the binary SAR ADC using 180 nm CMOS technology.

I. INTRODUCTION

Dramatic speed improvements are achieved using different SAR ADC architectures such as non-binary redundant SAR algorithm [1], asynchronous SAR logic [5] and monotonic capacitor switching DAC [3]. The scheme used in [1] offers flexibility, but at the cost of complex hardware that uses a ROM. The architecture in the scheme of [4] combines redundancy, asynchronous SAR logic and a monotonic switching DAC to get further speed improvement. It uses fixed weights of capacitors used in the DAC and hence, requires no ROM, thus simplifying the SAR logic.

In this paper, we propose the utilization of redundancy to improve SNR in addition to improved conversion speed of the SAR ADC. Redundancy can be used to correct comparator errors caused by various noise sources. The overlap range in the last few conversion cycles limits the error correction capability. This paper establishes a relationship between redundancy, speed and SNR of a non-binary redundant SAR ADC. For given specifications of speed and performance, it is possible to design capacitor weights used in the reference DAC. These weights can generalize the scheme suggested by [4].The algorithm presented in this paper determines these weights. It estimates the DAC settling time in each conversion cycle. These estimates can be used to design synchronous SAR ADC by considering worse DAC settling time, which may not necessarily happen in the first conversion cycle.

Asynchronous SAR ADC [5] scheme suggests variable bit clock period to improve conversion speed. It can be enhanced further by optimizing budget for DAC settling time in each conversion cycle. Irrespective of the capacitor weights used, the overlap range in the last conversion cycle is always zero. We suggest utilization of noise filter gear shifting [6] to overcome this inherent limitation of the non-binary redundant SAR algorithm

The remaining part of this paper is arranged as follows. Section-II summarizes the concept of the overlap range. Section-III explains limitations of the in-built error correction mechanism provided by the non-binary redundant search algo-rithm. Section-IV analyses the dependency of the DAC settling time on the distribution of weights. Section-V draws a trade-off between speed and performance. Section-VI presents an al-gorithm to find the weights of capacitors in the DAC in a non-binary redundant SAR ADC. Section-VII shows simulation results for a 10-bit, 12-bit-cycle non-binary redundant ADC. Simulation results of 10-bit binary ADC are also provided for comparison between the two schemes.

II. REDUNDANCY AND OVERLAP RANGE

The non-binary search algorithm requires M cycles to complete the conversion for an N-bit resolution. There are 2^M comparison patterns that result in 2^N output codes [2]. Hence the overall redundancy offered is

$$Redundancy_{total} = 2^M - 2^N \qquad (1)$$

The overall redundancy is distributed across all bit cycles. If the input lies within the overlap range of a particular bit cycle, the error in the bit cycle is corrected in subsequent bit cycles [2].

Algorithm1 shows the non-binary redundant search method [4] used in this paper.

Let $p[i]$ denote the weight of the bit cycle i. It can be proven that the overlap range $q[i]$ in i^{th} bit cycle is given by

$$q[i] = -p[i] + \sum_{k=i+1}^{M} p[k] + 1 \qquad (2)$$

The overlap range in the last bit cycle is always zero, irre-spective of the weight distribution used. Fig.1 shows overlap range in all bit cycles of an example distribution.

978-1-5386-3693-0/18 $31.00 © 2018 IEEE

Algorithm 1 Calculate $DOUT$ such that $Vin = Dout * LSB$

Require: $M \geq N \vee N \neq 0$

Ensure: $Dout = floor(Vin/LSB)$

   $V_{ref[1]} \leftarrow V_{CM}$

   $i \leftarrow 1$

   while $i \leq M$ **do**

     if $Vin > V_{ref[i]}$ **then**

       $V_{ref[i+1]} \leftarrow V_{ref[i+1]} + LSB \times p[i]/2$

       $dout[i] = 1$

     else

       $V_{ref[i+1]} \leftarrow V_{ref[i+1]} - LSB \times p[i]/2$

       $dout[i] = 0$

     end if

     $i \leftarrow i + 1$

   end while

   $DOUT = \sum_{k=1}^{M} dout[k] \times p[k]$

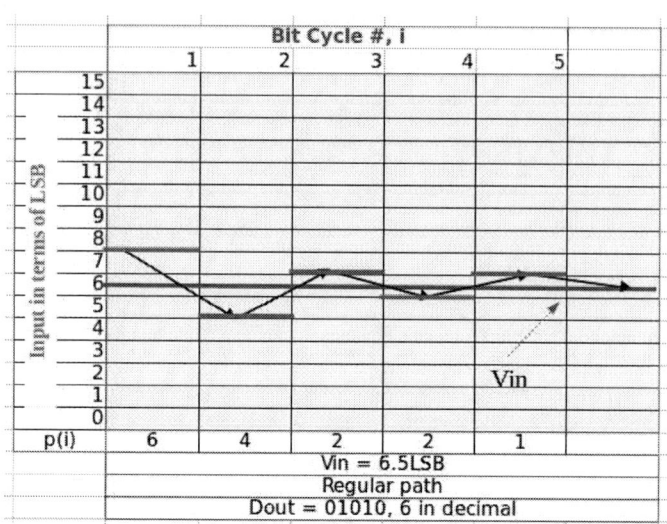

Fig. 2. Normal path to the input level $Vin = 6.5LSB$

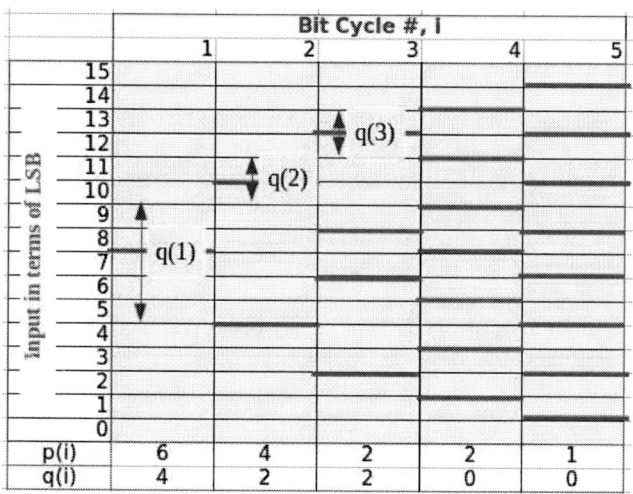

Fig. 1. Overlap range for p[1] = 6, p[2] = 4, p[3] = 2, p[4] = 2, p[5] = 1

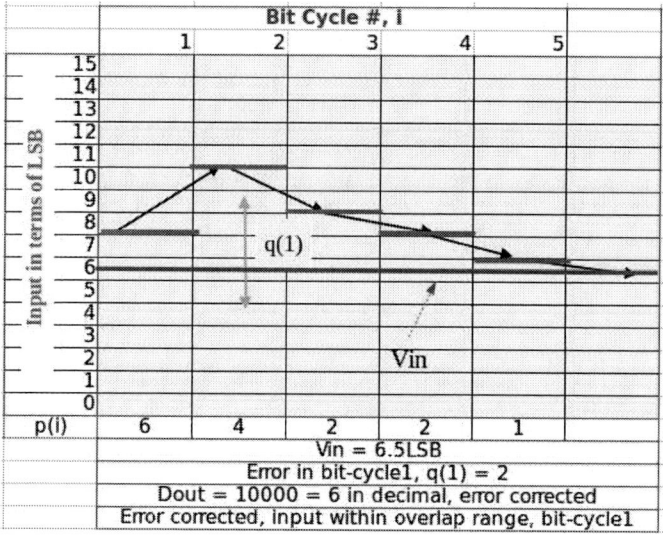

Fig. 3. Alternate path to the input level $Vin = 6.5LSB$

Fig.2 shows an example of the usual path taken to approximate to the input level of $6.5LSB$. Fig.3 shows an alternate path taken to approximate to the same input level when an error occurs in the bit-cycle1. Since the input lies withing the overlap range of the bit-cycle, there is no error in the output.

III. MINIMUM NON-ZERO OVERLAP RANGE

The non-binary redundant SAR ADC guarantees accurate conversion as long as the input lies within the overlap range of all bit cycles. From Eqn.(2), it can be shown that

$$q[i] > q[i+1] \tag{3}$$

Minimum non-zero overlap range limits the auto-correction mechanism obtained by the virtue of non-binary redundant search algorithm. It is, therefore, desirable to have the highest possible minimum non-zero overlap range.

To obtain the output in optimal number of bit-cycles, we propose to follow the conditions stated by Eqn.4 and Eqn.5

$$p[i] \geq p[i-1] \tag{4}$$

$$p[i] \leq 2 * p[i-1] \tag{5}$$

e.g. $p[i] = 4$, $p[i+1] = 3$, $p[i+2] = 1$ is unacceptable, but $p[i] = 4$, $p[i+1] = 3$, $p[i+2] = 2$ is acceptable. From Eqn.(2), the maximum $q[i]$ is obtained with the minimum $p[i]$ and the maximum sum of weights,p, in subsequent bit cycles. The maximum sum of weights in the subsequent bit cycles occurs when

$$p[k] = 2p[k+1]; k > i \tag{6}$$

However, this leads to

$$q[k] = 0; k > i \tag{7}$$

Consider a case where $Q_{min} = 4$, $N = 10$ and $M = 13$. The weight distribution to be used for the last 4 bit cycles would be

$p[10] = 4$, $p[11] = 4$, $p[12] = 2$, $p[13] = 1$.

This leads to

$q[10] = 4$, $q[11] = 0$, $q[12] = 0$, $q[13] = 0$.

In an attempt to maximize the overlap range of a bit cycle, the subsequent bit cycles get an overlap range of zero, thereby providing no correction for decision errors taking place in these bit cycles.

IV. OVERLAP RANGE AND INADEQUATE DAC SETTLING

The DAC settling time depends on the accuracy within which it needs to settle to the final value.

A. DAC settling time for binary SAR ADC

In case of a binary SAR ADC, the accuracy desired is 0.5LSB. The worst case DAC settling time for an N-bit binary SAR ADC is given by

$$t_{DACb} = \tau \ln(2^N) \tag{8}$$

where, τ is the time constant of the DAC.

B. DAC settling time for non-binary SAR ADC

- Bit-cycles with non-zero overlap range: The DAC can be allowed to settle so that the DAC voltage falls within half of the overlap range of a bit-cycle.

$$t_{DACnb1} = \tau \ln\left(\frac{2p}{q}\right) \tag{9}$$

  where p is the weight and q, the overlap range of the bit cycle.

- Bit-cycles with zero overlap range: The DAC has to settle within $0.5LSB$ of the final DAC voltage, since the DAC settling error in these bit-cycles cannot be corrected. If i is the number of the first bit-cycle with no overlap range, Eqn.10 represents the settling time required for this bit-cycle.

$$t_{DACnb2} = \tau \ln(2 * p(i)) \tag{10}$$

Thus, the worst case DAC settling time for a non-binary SAR ADC is given by Eqn.11

$$t_{DACnb} = \max(t_{DACnb1}, t_{DACnb2}) \tag{11}$$

It is possible to have a non-binary distribution p, such that

$$t_{DACnb} << t_{DACb} \tag{12}$$

Let R be the ratio of the DAC settling times such that

$$R = \frac{t_{DACb}}{t_{DACnb}} \tag{13}$$

While designing a non-binary SAR ADC, the desired improvement in DAC settling time can be expressed in terms of R. The maximum allowed $\frac{p}{q}$ ratio can be expressed as

$$R_m = 0.5e^{\frac{N*\ln(2)}{R}} \tag{14}$$

R_m can be used to determine the overlap range for each bit cycle of the non-binary redundant SAR ADC.

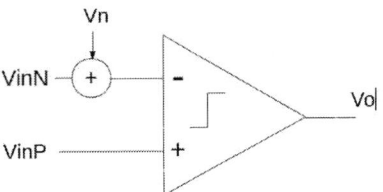

Fig. 4. Comparator noise model

V. SPEED VS. SNR - TRADEOFF

In the presence of noise, the comparator can be modelled as an ideal comparator with an additive noise source at the input as shown in Fig. 4. The comparator input referred noise is modeled as a Gaussian random variable with mean zero and variance σ^2 [8].

The bit-cycles that have an overlap range lesser than σ, need other mechanism to ensure the accuracy of the comparator decision. One such method is the noise gear shifting mechanism discussed in [6]. It allows more time for the pre-amplifier to settle during the desired bit-cycles. This averages out the effect of noise, thus making the comparator more immune to noise during these bit-cycles. The use of high noise - low power comparator in the bit-cycles with higher overlap range and low noise - high power comparator in the bit-cycles with lower overlap range can be an alternate solution, as explained in [7]. However, both of these methods increases power consumption. Moreover, the speed of conversion reduces due to stretching of the bit-cycles.

VI. A NOVEL APPROACH TO DECIDING THE DISTRIBUTION OF THE WEIGHTS IN A NON-BINARY SAR ADC

The methodologies of non-binary distribution of weights discussed in [1], [2] and [4] mainly focus on the improvement in the speed of the conversion, which is achieved by determining overlap range of the first bit-cycle. The overlap range of the bit-cycles towards the end of the conversion is not considered. The use of non-binary DAC to improve speed is mentioned in [6], which also proposes the use of noise gear shifting to protect the bit-cycles with lesser overlap range. However, the weights of the other bit-cycles are determined manually in all the above mentioned works. We propose a methodology to decide the weights of the capacitors in the DAC of a non-binary SAR ADC by taking into account the desired improvement in the DAC settling time required over the equivalent binary SAR ADC. Below are the findings of the analysis so far.

- Improvement in speed: The worst DAC settling time in the case of a non-binary sAR ADC does not necessarily be defined by the overlap range of the first bit-cycle. In fact, it is decided by the bit-cycle with the highest ratio $\frac{p}{q}$.
- Protection from noise: The last few bit-cycles get a very less or no overlap range. The variance σ of the Gaussian distributed noise determines the number of bit-cycles that

need protection from noise achieved by the noise gear shifting.

An algorithm to decide the distribution of weights in a non-binary SAR ADC is discussed below. The algorithm needs the following inputs.

- N, the resolution of the ADC.
- σ, the standard deviation of the gaussian distributed noise at the input of the comparator, expressed in terms of LSB of the ADC. It is used to decide the number of bit-cycles that need to be protected from noise as discussed in Section V.
- The DAC settling time improvement ratio, R. It is expressed as Eqn.15

$$R = \frac{t_{DACb}}{t_{DACnb}} \quad (15)$$

where t_{DACb} and t_{DACnb} are defined in Eqn. 8 and Eqn. 11.

As we know

$$R = N * \frac{\tau * \ln(2)}{\tau * ln(\frac{2*p}{q})} \quad (16)$$

We get the ratio $pbyq_{max}$ as Eqn.17

$$pbyq_{max} = 0.5 * e^{0.7 * \frac{N}{R}} \quad (17)$$

The distribution of weights is done from the last bit-cycle to the first bit-cycle, in the manner described below:

- The weights of the last two, M and $M-1$, bit cycles is always 1. Since the number of bit-cycles required to achieve the desired R is not yet known, an array p is populated from the first location. This leads to Eqn.18

$$p(1) = p(2) = 1 \quad (18)$$

- The remaining weights are decided using the ratio $pbyq_{max}$ as defined by Eqn.17, in the following manner.

  - Let i be the number of last bit-cycles that are already assigned weights. The remaining weight to be distributed is given by

$$remwt = 2^N - 1 - \sum_{k=1}^{i} p(k) \quad (19)$$

  - The weight temporarily assigned to p(i+1) is

$$p(i+1) = 2 * p(i) \quad (20)$$

The overlap range for this bit-cycle is calculated as

$$q(i+1) = -p(i+1) + \sum_{k=1}^{i} p(k) + 1 \quad (21)$$

  - The overlap range would be used for speed improvement and for correction of decision errors due to noise.

    * The overlap range that can be allocated for an incomplete DAC settling is $qtemp = q(i+1) - \sigma$. The ratio $\frac{p(i+1)}{qtemp}$ should be less than $pbyq_{max}$ to

have the desired DAC settling time in this bit-cycle. If $\frac{p(i+1)}{q(i+1)} > pbyq_{max}$ then some weight $w1$ needs to be removed from $p(i+1)$ as explained in Eqn.22

$$\frac{p(i+1) - w1}{q(i+1) + w1} = pbyq_{max} \quad (22)$$

    * An attempt is made to have the overlap range of the current bit-cycle to be greater than σ. The weight required to be removed from the current bit cycle for this purpose is denoted by $w2$
    * The final weight to be removed from the current bit-cycle is given by Eqn.23

$$w = \max(w1, w2) \quad (23)$$

The new value of p(i+1) would be $p(i+1) - w$

  - If the remaining weight $remwt$ is such that $remwt < p(i+1)$, then $p(i+1) = remwt$. When this condition is met, the distribution is complete, so that the total number of bit-cycles is $M = i+1$
  - It must be noted that during bit-cycles having overlap range of zero, the DAC must settle within $\pm LSB/2$ of the final value. By default, the last bit-cycle has zero overlap range. Hence, the maximum DAC settling time improvement ratio that can be achieved is limited by Eqn.24

$$R < \frac{N * \tau * \ln(2)}{\tau * ln(2)} \quad (24)$$

which sets the limit on R as $R < N$.

Algorithm2 is the detailed algorithm that achieves a non-binary weights distribution as explained above.

TableI shows an example output of the algorithm for $R = 3$ and $\sigma = 1$. TableII shows an example output of the algorithm

TABLE I
$N = 10, R = 3, \sigma = 1, M = 13, pbyq_{max} = 5.16$

i	p(i)	q(i)	$\frac{p(i)}{q(i)}$
1	423	178	2.4
2	273	55	5
3	149	30	5.1
4	81	17	5
5	44	10	4.9
6	24	6	4.8
7	13	4	4.3
8	7	3	3.5
9	4	2	4
10	2	2	2
11	1	2	1
12	1	1	1
13	1	0	NA

for $R = 4$ and $\sigma = 1$. The number of bit-cycles is $M = 14$ as compared to $M = 13$ in case of $R = 3$.

TableIII shows an example output of the algorithm for $R = 4$ and $\sigma = 2$. The number of bit-cycles is $M = 15$ as compared to $M = 14$ in case of $\sigma = 1$.

978-1-5386-3693-0/18 $31.00 © 2018 IEEE

Algorithm 2 Obtain a non-binary distribution, P

Require: $N, \sigma, R < N$

Ensure: $Non - binary\,Distribution\,P$

    $pbyq_{max} = 0.5 * e^{\frac{0.7*N}{R}}$

    while $(pbyq_{max} < 1 \,||\, pbyq_{max} == 1)$ **do**

        $R = R - 0.1$

        $pbyq_{max} = 0.5 * e^{\frac{0.7*N}{R}}$

    end while

    $p(1) = 1$

    $p(2) = 1$

    $i = 3$

    $distdone = 0$

    while $(distdone == 0)$ **do**

        $remwt = 2^N - 1 - \sum_{k=1}^{i-1} p_{asc}(k);$

        $p_{asc}(i) = 2 * p_{asc}(i-1);$

        $qtemp = -p_{asc}(i) + \sum_{k=1}^{i-1} p_{asc}(k) + 1$

        if $qtemp < \sigma$ **&&** $qtemp \leq \frac{p(i)}{2}$ **then**

            $w2 = \frac{p_{asc}(i)}{2}$

        else

            $w2 = 0$

        end if

        if $(qtemp > \sigma)$ **then**

            $qtemp = qtemp - \sigma$

        end if

        if $(\frac{p_{asc}(i)}{qtemp} > pbyq_{max})$ **then**

            $w1 = \mathrm{ceil}(\frac{p_{asc}(i) - qtemp*pbyq_{max}}{1 + pbyq_{max}})$

        else

            $w1 = 0$

        end if

        $w = \max(w1, w2)$

        $p_{asc}(i) = p_{asc}(i) - w$

        if $(remwt < p_{asc}(i))$ **then**

            $p_{asc}(i) = remwt$

            $distdone = 1$

        end if

        $i = i + 1$

    end while

    $M = i - 1$

    $P = \mathrm{sort}(p_{asc},' descend')$

TABLE II

$N = 10, R = 4, \sigma = 1, M = 14, pbyq_{max} = 2.87$

i	p(i)	q(i)	$\frac{p(i)}{q(i)}$
1	424	176	2.4
2	255	90	2.8
3	146	53	2.8
4	84	31	2.8
5	48	19	2.7
6	28	11	2.8
7	16	7	2.7
8	9	5	2.3
9	5	4	1.7
10	3	3	1.5
11	2	2	2
12	1	2	1
13	1	1	1
14	1	0	NA

TABLE III

$N = 10, R = 4, \sigma = 2, M = 15, pbyq_{max} = 2.87$

i	p(i)	q(i)	$\frac{p(i)}{q(i)}$
1	423	178	2.4
2	255	91	2.8
3	146	54	2.8
4	84	32	2.8
5	48	20	2.7
6	28	12	2.8
7	16	8	2.7
8	9	6	2.3
9	5	5	1.7
10	3	4	1.5
11	2	3	2
12	1	3	1
13	1	2	0.5
14	1	1	1
15	1	0	NA

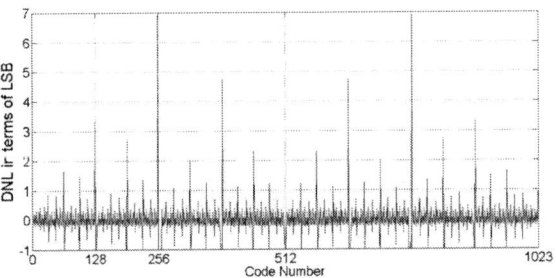

Fig. 5. DNL Plot of binary with bit-clk period of 800ps

VII. RESULTS

A. Improvement in DAC settling time

With t_{DACb} = 800ps, and R of 4, a non-binary distribution was obtained using the algorithm discussed above. Static characterization of both binary and non-binary redundant SAR ADCs was done by using a slowly varying full scale ramp. As shown in Fig.5, for a DAC settling time of 200ps, the binary SAR ADC gives a DNL of 7 LSB, whereas the DNL of the non-binary SAR ADC remains within 0.3LSB, as observed from Fig.6. INL of non-binary distribution remains within the allowed range of 0.5LSB. A sine wave of 1MHz was applied to both the binary and the non-binary redundant SAR ADC. The results are shown in Fig.7. The noise floor of the binary SAR ADC is $4dB$ above that of non-binary SAR ADC. These simulations were run using Cadence ADE setup.

It must be noted that in both binary and non-binary SAR ADCs, the reference DAC does not settle within $0.5 * LSB$ **of the final voltage.** However, non-binary SAR ADC recovers from the decision errors caused by incomplete the DAC settling.

B. Performance in noisy environment

To observe the performance of the binary and the non-binary redundant SAR ADC in noisy environment, normally distributed noise with standard deviation of $2LSB$ and mean of 0 was introduced at the input of the comparators used in

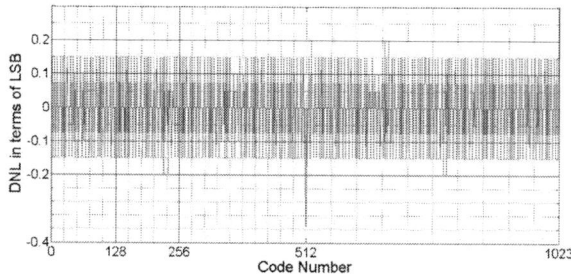

Fig. 6. DNL Plot of non-binary SAR ADC with bit-clk period of 800ps

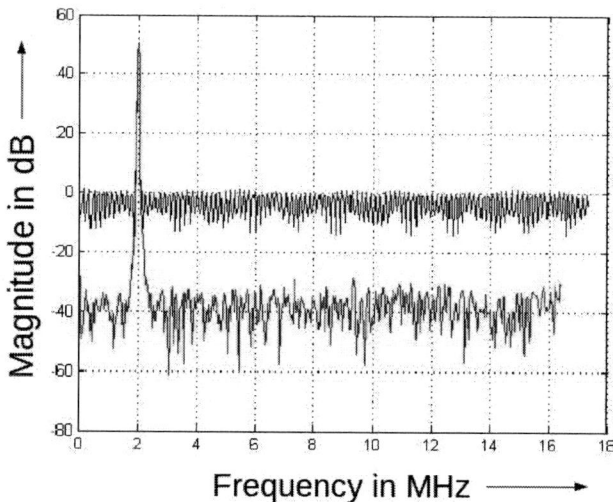

Fig. 7. FFT Plot of binary(in blue) and non-binary(in red) redundant SAR ADC, F_{in} 2MHz, $T_{bit-clk}$ = 800ps

both the ADCs. Two different distributions were simulated for non-binary redundant SAR ADC. The bit cycles with an overlap range of less than $2LSB$ were protected from noise as explained in Section V. As shown in Fig.8, the non-binary distribution shows an improvement of 5.5dB over the binary distribution.

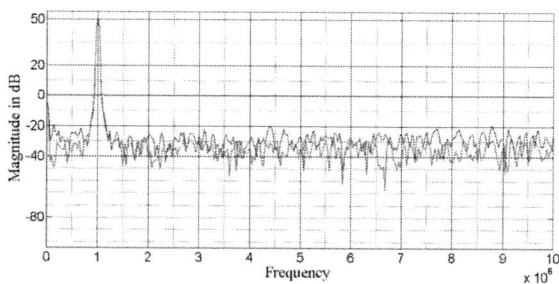

Fig. 8. FFT of the binary(blue) and the nonbinary(red, magenta) SAR ADC; f_{in} = 1MHz, f_s = 13MHz, using normalized distribution for simulating noise.

VIII. CONCLUSION

It is proven that non-binary redundant SAR ADCs offer SNR improvement, along-with speed improvement. The last few bit cycles of the conversion exhibit an overlap range of zero. These bit cycles should be re-sampled and comparator noise should be reduced during these bit cycles to ensure the improvement in SNR. The weights of the capacitors used in the DAC of the non-binary redundant SAR ADC could be determined using the algorithm discussed. These weights could be used irrespective of DAC switching scheme used and in both synchronous as well as asynchronous SAR logic.

REFERENCES

[1] F. Kuttner, "A 1.2-V 10-b 20-Msample/s nonbinary successive approximation ADC in 0.13-µm CMOS", in IEEE ISSCC Dig. Tech. Papers, pp.176-177, 2002.

[2] T. Ogawa, H. Tobayashi, M. Hotta, "SAR ADC Algorithm with Redundancy", in IEEE Circuits and Systems, APCCAS2008, pp.269-270, 2008.

[3] C C Liu, S J Chang, G Y Huang, Y Z Lin, "A 10-bit 50 MS/s SAR ADC with a Monotonic Capacitor Switching Procedure", in IEEE J. Solid-State Circuits, vol.50, No.11, pp.732-733, Nov. 2015.

[4] C C Liu, C H Kuo, Y Z Lin, "A 10 bit 320 MS/s Low-Cost SAR ADC for IEEE 802.11ac Applications in 20 nm CMOS", in IEEE J. Solid-State Circuits, IEEE J. Solid-State Circuits, vol.50, no.11, pp.2647-2648, Nov. 2015.

[5] S. W. M. Chen, R. W. Brodersen, "A 6-bit 600-MS/s 5.3-mW asynchronous ADC in 0.13-µm CMOS", in IEEE J. Solid-State Circuits, vol.41, no.12, pp.2669-2680, Dec. 2006.

[6] Martin Kramer, Kostas Doris, "A 14-Bit 30-MS/s 38-mW SAR ADC Using Noise Filter Gear Shifting", in IEEE Transactions on Circuits and Systems - II, vol.64, no.2, pp.116-120, Feb. 2017.

[7] Vito Giannini, Pierluigi Nuzzo, "An 820W 9b 40MS/s Noise-Tolerant Dynamic-SAR ADC in 90nm Digital CMOS", in ISSCC, Feb. 2008.

[8] Muhammad Ahmadi, Won Namgoong, "Comparator Power Minimization Analysis for SAR ADC Using Multiple Comparators", in IEEE Transactions on Circuits and Systems - I, vol.62, no.10, pp.2369-2379, Oct. 2015.

2018 31th International Conference on VLSI Design and 2018 17th International Conference on Embedded Systems

Mitigating Aperture Error in Pipelined ADCs Without a Front-end Sample-and-Hold Amplifier

Diego James[†], Abishek T. Kunnath[†], A. Purushothaman[†] and Bibhu Datta Sahoo[‡]

[†]Dept. of ECE, Amrita School of Engg., Amrita University, India

[‡]Dept. of ECE, University of Illinois at Urbana-Champaign, Urbana, IL 61801

e-mail: bsahoo@ieee.org

Abstract—Pipelined analog-to-digital converters (ADCs) without dedicated front-end sample-and-hold amplifier (SHA) suffer from *aperture-error*. This error is caused by the mismatch in the sampled signal between the sub-ADC and the multiplying digital-to-analog converter (MDAC) of the first stage. This paper describes a new architecture that eliminates this *aperture-error*. A 12-bit pipelined ADC was designed and simulated in UMC 65-nm SP process with the first stage being replaced by the proposed architecture. Simulations show that the performance of the proposed architecture is same as that of the conventional architecture provided the conventional architecture does not have any *aperture-error*.

Keywords—*Analog-to-digital converter (ADC), SHA-less, aperture-error, Clock skew, Path mismatch, capacitor sharing.*

I. INTRODUCTION

Pipelined analog-to-digital converters (ADCs) are widely used in applications where latency is not of concern and that require high speed and high resolution. Although, aggressive technology scaling has facilitated low-power high-speed pipelined ADCs, the power consumption of the pipelined ADCs is dominated by the front-end SHA which has to meet both the linearity and thermal noise requirements of the ADC. The nonlinearity of the front-end SHA cannot be overcome by digital calibration techniques [1]-[5] which are mainly used to overcome other nonidealities of the pipelined ADC, *viz.*, capacitor mismatch, finite op amp gain, and op amp non-linearity. In the absence of the front-end SHA, the input signal sampled in the multiplying digital-to-analog converter (MDAC) and sub-ADC of the first-stage of the pipeline could differ significantly as they go through different paths which could have mismatch. If the difference between the signals sampled on the MDAC and sub-ADC, also called *aperture-error*, exceeds the redundancy range of the first-stage of the pipeline then the subsequent stages of the pipelined ADC would saturate causing significant performance degradation. However, SHA-less architectures have been pursued [8]-[14] with great success at the expense of area penalty and significant design effort to match the two paths. There is therefore a need for a SHA-less architecture that does not suffer from *aperture-error*.

Fig. 1 shows the first stage of a conventional pipelined ADC without SHA circuit and Fig. 6 shows the timing diagram of the clocks being used. Without a SHA, the first stage sub-ADC and the MDAC sees a time varying input signal, which is getting sampled on to the capacitor C_S. Unless both these sampling networks sample the same value at exact time instant, it is going to cause an error referred

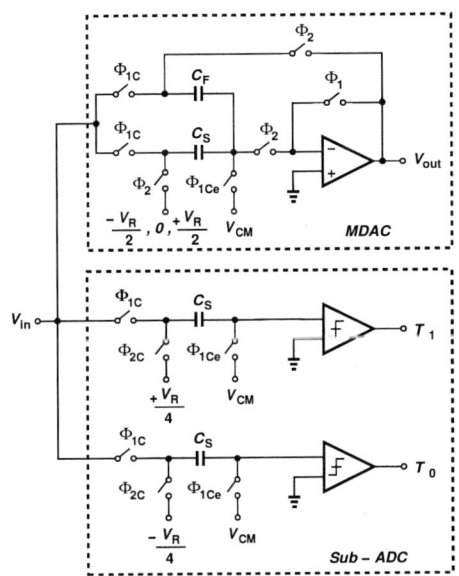

Fig. 1: First-stage of a conventional SHA-less pipelined ADC.

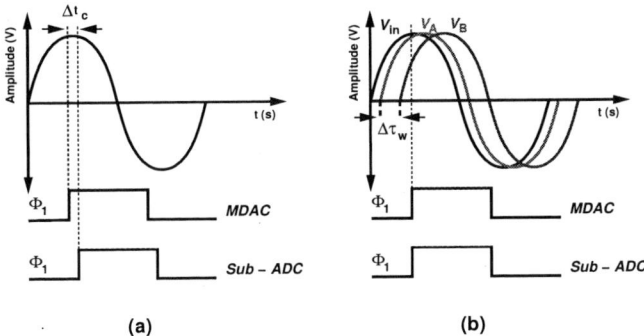

Fig. 2: Aperture-error due to a) Clock-skew b) RC-delay of sampling networks.

to as *aperture-error* [13]. Fig. 2 shows the two possible ways of *aperture-error*. The primary source of error is the clock skew (Δt_c) between the sampling clocks of sub-ADC and the MDAC as shown in Fig. 2(a). This error is brought down by symmetrically matching the path of the two clock signals by the addition of buffers. The other source of error is the bandwidth mismatch or RC-delay ($\Delta \tau_w$) between the sampling network of sub-ADC and the MDAC as shown in Fig. 2(b). In most designs, it is taken care by making the sub-ADC and the MDAC circuit symmetrical in terms

978-1-5386-3693-0/18 $31.00 © 2018 IEEE

of RC-delay by the addition of dummy switches wherever necessary. In the layout, we make sure that the input and clock lines to both the circuits are matched. But as the input frequency increases the error in the sampled value also increases since it is proportional to the slope of the input waveform [6]. As a result, for high-frequency input signals the above mentioned solutions would not be that effective.

This paper describes a prototype of a 12-bit pipelined ADC in which first stage sub-ADC and MDAC share a single capacitor for sampling the input. This way the *aperture-error* due to path mismatch is eliminated. The proposed architecture can sample high-frequency analog input signals.

Section II discusses the cause for *aperture-error* in detail and existing methods to correct it. Section III describes an existing architecture which shares the sampling capacitor between sub-ADC and MDAC for eliminating *aperture-error*. Section IV describes the proposed architecture and Section V validates the concept with simulation results.

II. APERTURE-ERROR AND CORRECTION METHODS

As mentioned in Section I, excluding SHA at the front end would result in input dependent dynamic offset as the first stage sees the time varying input signal [12]. A portion of the redundancy range is consumed by this dynamic offset leaving less margin for other offsets (e.g., comparator and amplifier offset). At high speed, this might exceed the digital redundancy range and result in out-of-range residues at the MDAC output. The source of *aperture-error* are the clock skew (Δt_c) and the bandwidth mismatch ($\Delta \tau_w$) as explained in Section I.

In [14] the authors have shown that for an input sinusoidal signal given by:

$$V_{in} = V_o sin(w_o t) \qquad (1)$$

where V_o is the peak amplitude and w_o is the angular frequency of the input signal, the worst case mismatch error, V_e occurs near the zero crossing of the signal and is given by:

$$V_e \cong V_o w_o (\Delta T) \qquad (2)$$

where $\Delta T = \Delta t_c + \Delta \tau_w$, is the total mismatch between sub-ADC and MDAC network.

For the correct functionality of the MDAC with built-in redundancy, the total error voltage, V_{et}, should be less than half LSB as expressed below:

$$V_{et} < \frac{V_{REF}}{2^{N+1}} \qquad (3)$$

where V_{REF} is the full-scale range of the ADC and N is the number of bits being resolved in the first-stage of the pipelined ADC. The total error voltage, V_{et}, is the combined error caused by comparator offset, amplifier offset, reference gain error, charge injection and path mismatch error.

From Eq. (2) and (3) it is evident that for an input frequency, f_{in}, the total mismatch error is given by:

$$\Delta T < \frac{1}{2^{N+2} \pi f_{in}} \qquad (4)$$

For an f_{in} of 500 MHz, from Eq. (4) the total mismatch should be less than 40 ps and 20 ps for 1.5-bit and 2.5-bit stages, respectively. Allocating one-third of the error budget for mismatch error makes the total mismatch to be

less than 14 ps and 7 ps in case of 1.5-bit and 2.5-bit stages, respectively. This requirement is demanding a perfect matching in layout which motivates our idea of using a single capacitor for sampling.

To overcome the above-mentioned problems there are different techniques. In [16] the authors match the sampling networks of sub-ADC and MDAC so that the time constants of both the networks are same with respect to the input. The disadvantage of the configuration is that it cannot be used for high frequency application as the non-linearity and mismatches in the input signal path increases with increase in input frequency. In [10] a split-sharing technique is being used where op amp does the sample-and-hold (S/H) function and the MDAC amplification. This technique is said to avoid the crosstalk and the *aperture-error*. Similar work is shown in [15]. The advantage of these techniques is that they reduce power consumption in the first stage as compared to a SHA incorporated stage. Another way of eliminating the path mismatch or *aperture-error* is by sharing a single sampling capacitor as shown in [11][17]. Section III describes an existing architecture based on this approach.

III. EXISTING ARCHITECTURE

In [11] a new configuration is used in which the sampling capacitor is being shared between the sub-ADC and the MDAC. The architecture is shown in Fig. 3 and the corresponding clocks being used is shown in Fig. 4. The working of the stage can be described in 4 phases: *sampling phase, op amp as Unity Gain Amplifier phase, comparator decision phase* and *amplification phase*.

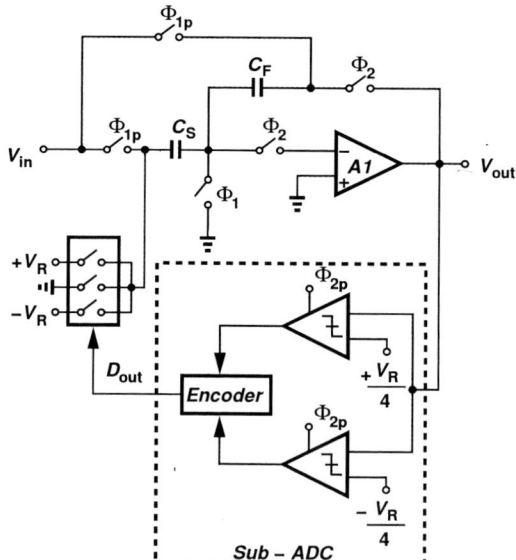

Fig. 3: Existing ADC architecture [11].

A. Sampling Phase

During Φ_1 phase, the input signal voltage is sampled on to the capacitors C_S and C_F.

B. Op Amp as Unity Gain Amplifier Phase

During Φ_2 phase, the capacitor C_F gets flipped around the op amp and the output, V_{out}, of the op amp approximately equals to V_{in} since the charge on the capacitor C_F must be conserved. Δt time is being used for settling by the op amp.

978-1-5386-3693-0/18 $31.00 © 2018 IEEE

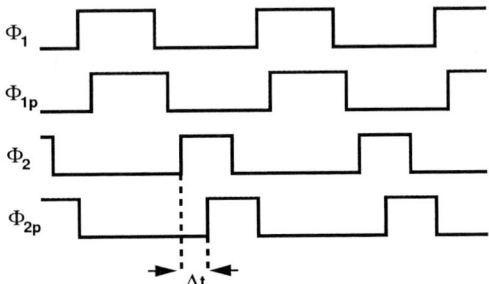

Fig. 4: Timing diagram for existing architecture [11].

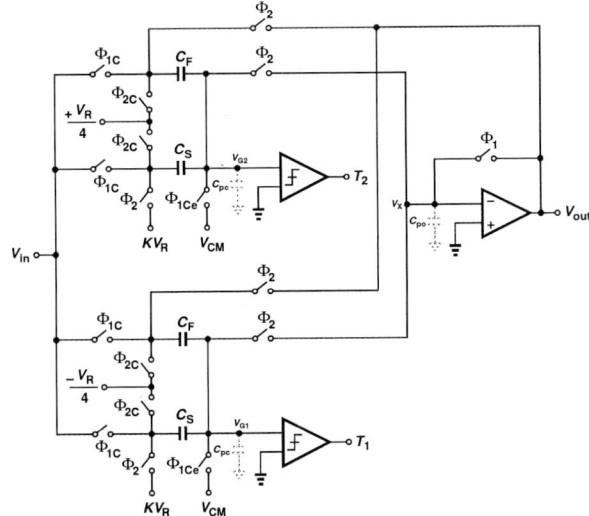

Fig. 5: Proposed 1.5-bit stage flip-around architecture.

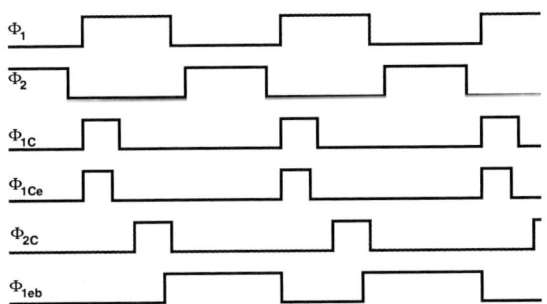

Fig. 6: Timing diagram for proposed architecture.

C. Comparator Decision Phase

After the op amp output gets settled the comparator clock Φ_{2p} is enabled. The comparator makes the decision according to the settled value and the corresponding D_{out} bits are generated by the encoder.

D. Amplification Phase

During Φ_2 phase and after Δt time, depending on the D_{out} bits, the left plate of C_S is switched to $+V_R$, ground or $-V_R$ to generate the residue signal.

The above mentioned architecture has several disadvantages:

  i. The output of the op amp has to settle within Δt time period with some amount of tolerable settling error, ε during *op amp as unity gain amplifier phase*. This means that the op amp needs to be faster with high bandwidth which makes it a power hungry circuit.

 ii. Needs extra time (Δt) for holding the input signal because of which the op amp gets less time to generate the residual signal during *amplification phase*.

iii. As the frequency of operation increases, the Δt becomes smaller and the op amp settling error increases which makes the comparison error well beyond the limit where it cannot be recovered by digital error correction.

In the following section, the proposed architecture is described which eliminates the shortcomings of the above architecture.

IV. PROPOSED ARCHITECTURE

The proposed architecture overcomes the speed limitation and relaxes the design requirement of the op amp. Bootstrap switches are used to remove the non-linearity due to the sampling switch and bottom plate sampling is done to eliminate the non-linearity due to charge injection.

A. 1.5-bit Stage

1) Flip around topology

Fig. 5 shows the proposed stage 1 with 1.5-bit flip-around architecture. Only one capacitor is being used to sample the input for both sub-ADC and the MDAC thus eliminating the path mismatch. The clocks used are non-overlapping clocks, as shown in Fig. 6, in which Φ_1 and Φ_2 are 50% duty cycle clocks and Φ_{1C}, Φ_{1Ce} and Φ_{2C} are the 25% duty cycle clocks. In this design, 25% of the clock period is used to sample the input and the rest 75% is used for sub-ADC conversion and MDAC settling in the first stage [3]. The full function of the stage can be divided into 4 phases : *sampling phase, threshold sampling phase, comparator decision phase* and *amplification phase*.

a) Input Sampling Phase

During this phase, Φ_{1C} and Φ_{1Ce} clocks are high making the input to get sampled on to the capacitors C_S and C_F. Φ_{1Ce} is being used for bottom plate sampling so that no input dependent charge gets stored on to the capacitor. The charge equation is given by :

$$Q_s = V_{in}C \tag{5}$$

where $C = C_S + C_F$.

At the same time during Φ_1 phase, the op amp is in reset mode, making the node voltage V_X and V_{out} to be at V_{CM} as shown in Fig. 5.

b) Threshold Sampling Phase

During Φ_{2C} phase, the threshold is applied to the left plate of the capacitors thus making the other plate of the capacitor to change accordingly. Due to the presence of parasitic capacitor C_{pc}, the RC delay increases and contributes to an extra term in the charge equation, as shown below :

$$Q_h = (V_{th} - V_{Gi})C - V_{Gi}C_{pc} \tag{6}$$

where V_{th} is the threshold voltage of each comparator and its value is $-V_R/4$ and $+V_R/4$, respectively, and V_{Gi}, for $i = 1, 2$ is the voltage at the input of the comparator.

c) Comparator Decision Phase

The comparator sees a value V_G given by :

$$V_{Gi} = (V_{th} - V_{in})\frac{C}{C + C_{pc}} \tag{7}$$

978-1-5386-3693-0/18 $31.00 © 2018 IEEE

and makes the decision during Φ_{1eb} phase. The parasitic capacitor has no effect on the comparator since it is implemented in differential manner and both inputs see the same fractional decrease in value.

d) Amplification Phase

Depending on the thermometer code, the left plate of the capacitor is connected to the corresponding reference value during Φ_2 phase. The op amp works in amplification phase with C_F being flipped and acting as the feedback capacitor. The charge sampled on to the capacitors after Φ_1 phase is given by :

$$Q_s = 2V_{in}C \qquad (8)$$

The charge equation during amplification phase Φ_2 is given by :

$$Q_{amp} = (V_{out} - V_X)2C_F + (KV_R - V_X)2C_S \\ + (0 - V_X)C_{pT} \qquad (9)$$

where $C_{pT} = 2C_{pc} + C_{po}$ denotes the sum of input parasitic capacitance of the op amp,C_{po}, and comparators,C_{pc}.
The conservation of charge during both the phases yields:

$$2V_{in}C = (V_{out} - V_X)2C_F + (KV_R - V_X)2C_S - V_X C_{pT} \qquad (10)$$

Substituting $V_X = -V_{out}/A$ to Eq. (10), where A is the open loop gain of the op amp, we get

$$V_{out} = \frac{C}{C_{eq}}V_{in} - \frac{C_S}{C_{eq}}KV_R \qquad (11)$$

where, $C_{eq} = C_F + (C_S + C_F + C_{pT}/2)/A$ and $K = -1/2, 0, +1/2$.

2) Non-Flip around topology

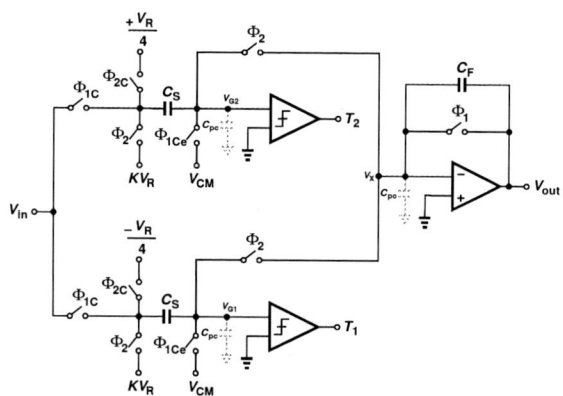

Fig. 7: Proposed 1.5-bit stage non-flip around architecture.

The non-flip around topology works in the same manner as the flip around. Fig. 7 shows the proposed 1.5-bit stage non-flip around architecture with feedback capacitor C_F, having a value half of that of the sampling capacitor C_S. The clocks used are also the same.
The charge sampled on to the capacitors after Φ_1 phase is given by :

$$Q_s = 2V_{in}C_S \qquad (12)$$

The charge equation during amplification phase Φ_2 is given by :

$$Q_{amp} = (V_{out} - V_X)2C_F + (KV_R - V_X)2C_S \\ + (0 - V_X)C_{pT} \qquad (13)$$

where, $C_{pT} = 2C_{pc} + C_{po}$ denotes the sum of input parasitic capacitance of the op amp, C_{po}, and comparators, C_{pc}.
The conservation of charge during both the phases yields:

$$V_{out} = \frac{C_S}{C_{eq}}V_{in} - \frac{C_S}{C_{eq}}KV_R \qquad (14)$$

where, $C_{eq} = C_F + (C_S + C_F + C_{pT}/2)/A$ and $K = -1, 0, +1$.

It is clear from Eq. (11) and Eq. (14) that any mismatch between C_S and C_F will result in unequal voltage gains for V_{in} and V_R in a flip-around topology whereas it remains same for both V_{in} and V_R in a non-flip-around topology making non-flip-around superior to flip-around topology. On the other hand, the non-flip-around topology has lower loop gain which leads to slower settling.

B. 2.5-bit Stage

1) Flip around topology

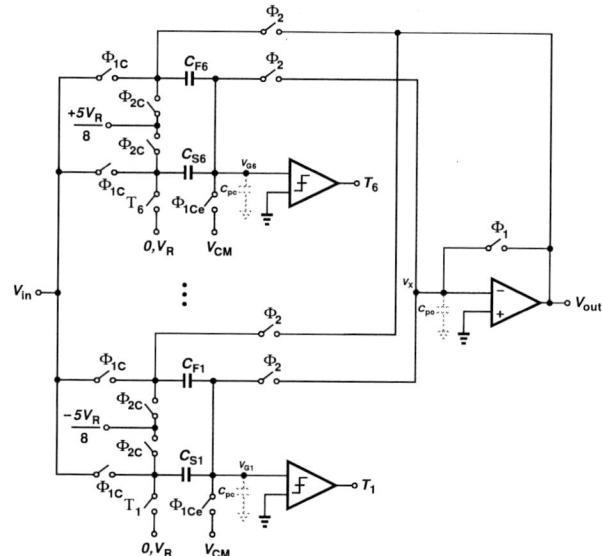

Fig. 8: Proposed 2.5-bit stage flip-around architecture

A 2.5-bit flip around architecture is shown in Fig. 8. By principle of charge conservation, the output voltage of a 2.5-bit/stage architecture is given by

$$V_{out} = \frac{\sum\limits_{i=1}^{6}[C_{Si} + C_{Fi}]V_{in} - \sum\limits_{i=1}^{6} T_i C_{Si} V_R}{\sum\limits_{i=1}^{6} C_{Fi} + \eta} \qquad (15)$$

where, $C_S = 3C_F$, $\eta = (\sum\limits_{i=1}^{6}[C_{Fi} + C_{Si}] + C_{pT})/A$, A is the finite op amp gain, T_i is the thermometer code obtained from the comparator and $C_{pT} = 6C_{pc} + C_{po}$ denotes the sum of input parasitic capacitance of the op amp, C_{po}, and comparators, C_{pc}.

2) Non-Flip around topology

The 2.5-bit non-flip around architecture, as shown in Fig. 9, has 8 sampling capacitors, C_{1-8} and a feedback capacitor, C_F. By principle of charge conservation, the output voltage

is given by

$$V_{out} = \frac{\sum_{i=1}^{8} C_i V_{in} - \sum_{i=1}^{6} T_i C_i V_R}{C_F + \eta} \qquad (16)$$

where, $C_{1-8} = C$, $C_F = 2C$, $\eta = (\sum_{i=1}^{8} C_i + C_{pT})/A$, A is the finite op amp gain, T_i is the thermometer code obtained from the comparator and $C_{pT} = 6C_{pc} + C_{po}$ denotes the sum of input parasitic capacitance of the op amp, C_{po}, and comparators, C_{pc}.

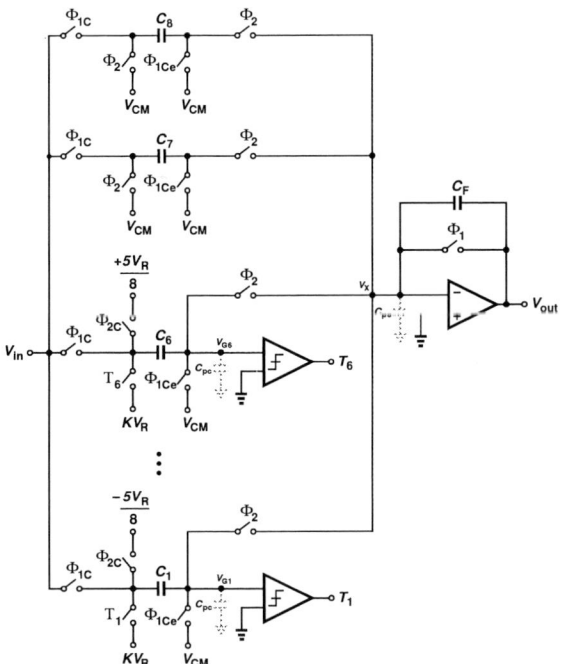

Fig. 9: Proposed 2.5-bit stage non-flip around architecture.

The proposed architectures has several advantages over existing architecture [11]:

i. The op amp gets more time to generate the residue signal during *amplification phase*.

ii. The power requirement of the opamp is relaxed as it gets more time to settle.

iii. Higher operational frequency.

V. SIMULATION RESULTS

In order to validate the concept, two 12-bit ADCs are designed in UMC $65nm$ SP technology. Each of the ADCs are designed to have 9, 1.5-bit stages with flip-around architecture followed by a 3-bit stage. The first stage of both the ADCs are implemented using conventional and proposed manner, respectively (ref. Fig. 1 and Fig. 5). Since the *aperture-error* increases as the input frequency increases, the SNR measurements are done with an input signal frequency, $f_{in} \approx f_S/2$, *i.e., nyquist-sampling*.

In a conventional ADC topology, the effect of *aperture-error* on SNDR is validated in simulation using delayed sampling clock and delayed input to the MDAC with respect to sub-ADC clock and input, respectively (ref. Fig. 2). Fig. 10 and Fig. 11 show the variation in SNDR as a function of

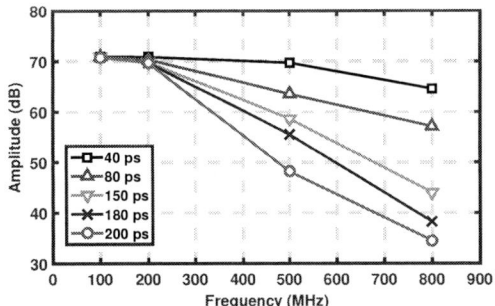

Fig. 10: SNDR versus sampling frequency (f_S) for *nyquist-sampling* for different Clock-skew.

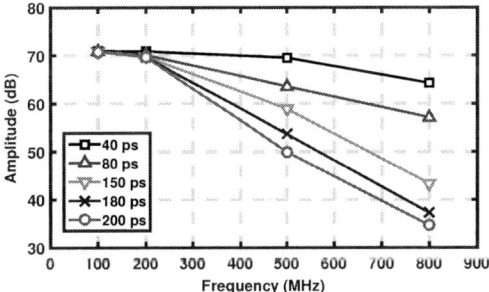

Fig. 11: SNDR versus sampling frequency (f_S) for *nyquist-sampling* for different RC-delay.

f_S for *nyquist-sampling*, when the clock-skew and RC-delay are varied. At a sample-rate of 500 MHz and delay of 80 ps (either in the form of clock-skew or RC-delay) there is a drop of 9 dB in SNDR, which leads to 1.5-bit loss. Since the proposed architecture uses a single capacitor for MDAC and sub-ADC the degradation in SNDR due to *aperture-error* is eliminated.

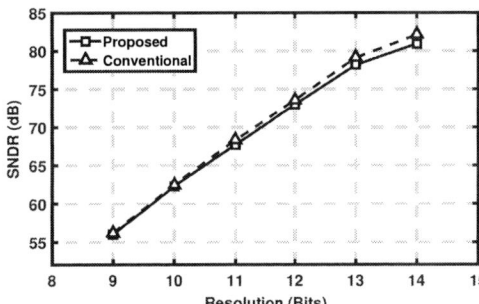

Fig. 12: SNDR versus Resolution of 1.5-bit architecture

To compare the performance characteristic of conventional and proposed topology a test set-up is made in which the first stage op amp of the ADCs are made ideal and an ideal ADC is kept as back-end. In the set-up, no delay in clock and input is given to the conventional ADC making it free from *aperture-error*. Fig. 12 and Fig. 13 show SNDR as a function of resolution for 1.5-bit and 2.5-bit flip-around architecture, respectively. We can clearly see that there is hardly any degradation in the SNDR of the proposed architecture. The minor degradation is due to the additional parasitic capacitance due to the additional switches used in the proposed architecture. Table I compares the number of

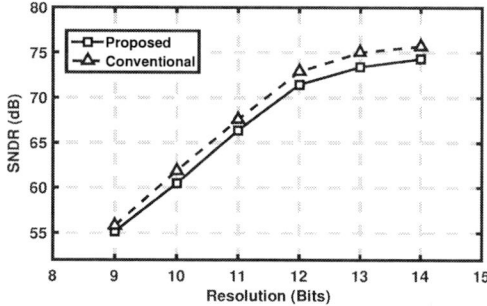

Fig. 13: SNDR versus Resolution of 2.5-bit architecture

switches used in the conventional and proposed architecture. Due to the scaling trend in CMOS, the switches are getting faster and tinier with every new technology. Although, the number of switches used to realize the *aperture-error* free proposed architecture is significantly large as compared to the conventional architecture, it still gives an area efficient solution as the size of the capacitor, which takes bulk of the area is halved and the switches are anyway tiny in scaled CMOS nodes.

TABLE I: ADC System Level Comparison

Stage-1 Architecture		*No. of Switches*	*Total Sampling Capacitor (pF)*
1.5-bit	Conventional	15	2
	Proposed	21	1
2.5-bit	Conventional	43	2.52
	Proposed	55	1.44
3.5-bit	Conventional	91	4.2
	Proposed	127	2.24

Table I shows the system requirements for different bit stages for proposed and conventional architectures. Even though the switches required increases, the total capacitor required gets reduced by more than 40% in the proposed architecture. This leads to reduction in area with almost similar performance compared to conventional as mentioned earlier.

VI.CONCLUSION

This paper proposes a 12-bit pipelined ADC based on capacitor sharing technique to overcome the *aperture-error*. The prototype ADC is implemented in $65nm$ CMOS process and uses the 1.5-bit/stage architecture in the first stage to validate the concept. The proposed architecture eliminates *aperture-error* and reduces the area requirement by more than 40% while giving almost similar performance compared to conventional.

REFERENCES

[1] H. S. Lee, "A 12-b 600 ks/s Digitally Self Calibrated Pipelined Algorithmic ADC," *IEEE J. Solid-State Circuits*, vol. 29, no. 4, pp. 509-515, April. 1994.

[2] C. R. Grace and P. J. Hurst and S. H. Lewis, "A 12-bit 80-Msample/s pipelined ADC with bootstrapped digital calibration," *IEEE J. Solid-State Circuits*, vol. 40, no. 5, pp. 1038-1046, May. 2005.

[3] B. Sahoo, and B. Razavi, "A 12-b 200-MHz CMOS ADC," *IEEE J. Solid-State Circuits*, vol. 44, no. 9, pp. 2366-2380, Sep. 2009.

[4] C. J. Tseng, H. W. Chen, W. T. Shen, W. C. Cheng and H. S. Chen, "A 10-b 320-MS/s Stage-Gain-Error Self-Calibration Pipeline ADC," *IEEE J. Solid-State Circuits*, vol. 47, no. 6, pp. 1334-1343, June 2012.

[5] B. Sahoo, "An overview of digital calibration techniques for pipelined ADCs," *IEEE 57th International Midwest Symposium on Circuits and Systems (MWSCAS), College Station, TX*, pp. 1061-1064, Aug. 2014.

[6] B. Razavi, "Problem of timing mismatch in interleaved ADCs," *Proceedings of the IEEE 2012 Custom Integrated Circuits Conference*, Sep. 2012, pp. 1-8.

[7] P. Bogner, "A 28mW 10b 80MS/s pipelined ADC in 0.13m CMOS," *Circuits and Systems, 2004. ISCAS '04. Proceedings of the 2004 International Symposium on*, 2004, pp. I-17-20, vol. 1.

[8] Dong-Young Chang, "Design techniques for a pipelined ADC without using a front-end sample-and-hold amplifier," *IEEE Trans. on Circuits and Systems I: Regular Papers*, vol. 51, no. 11, pp. 2123-2132, Nov. 2004.

[9] P. Huang et al., "SHA-Less Pipelined ADC With In Situ Background Clock-Skew Calibration," *IEEE J. Solid-State Circuits*, vol. 46, no. 8, pp. 1893-1903, Aug 2011.

[10] Z. Wang, M. Wang, W. Gu, C. Chen, F. Ye and J. Ren, "A High-Linearity Pipelined ADC With Op Amp Split-Sharing in a Combined Front-End of S/H and MDAC1," *IEEE Trans. on Circuits and Systems I: Regular Papers*, vol. 60, no. 11, pp. 2834-2844, Nov. 2013.

[11] J. Li, X. Zeng, L. Xie, J. Chen, J. Zhang and Y. Guo, "A 1.8-V 22-mW 10-bit 30-MS/s Pipelined CMOS ADC for Low-Power Subsampling Applications," *IEEE J. Solid-State Circuits*, vol. 43, no. 2, pp. 321-329, Feb. 2008.

[12] Ju. Hyungyu and Lee Minjae, "A 1.1 V 10-bit 62MS/s pipeline ADC with two-step non-overlapping clock generation for multi I-Q channel RF receivers," *Analog Integrated Circuits and Signal Processing*, vol. 87, no. 3, pp. 341-352, April. 2016.

[13] M. Brandolini et al., "A 5 GS/s 150 mW 10 b SHA-Less Pipelined/SAR Hybrid ADC for Direct-Sampling Systems in 28 nm CMOS," *IEEE J. Solid-State Circuits*, vol. 50, no. 12, pp. 2922-2934, Dec. 2015.

[14] A. M. A. Ali et al., "A 14-bit 125 MS/s IF/RF Sampling Pipelined ADC With 100 dB SFDR and 50 fs Jitter," *IEEE J. Solid-State Circuits*, vol. 41, no. 8, pp. 1846-1855, Aug. 2006.

[15] K. H. Lee, K. S. Kim and S. H. Lee, "A 12b 50 MS/s 21.6 mW 0.18 μ m CMOS ADC Maximally Sharing Capacitors and Op-Amps," *IEEE Transactions on Circuits and Systems I: Regular Papers*, vol. 58, no. 9, pp. 2127-2136, Sep. 2011.

[16] I. Mehr and L. Singer, "A 55-mW, 10-bit, 40-Msample/s Nyquist-rate CMOS ADC," *IEEE J. Solid-State Circuits*, vol. 35, no. 3, pp. 318-325, March. 2000.

[17] Ying Ju, Fule Li, XiuJu He, Chun Zhang and Zhihua Wang, "Aperture error reduction technique for sub-range SAR ADC," *IEEE 14th International New Circuits and Systems Conference (NEWCAS), Vancouver, BC*, pp. 1-4, June 2016.

A 12.5Gbps Transmitter for Multi-Standard SERDES in 40nm Low Leakage CMOS Process

Biman Chattopadhyay, Sharath N. Bhat, Gopalkrishna Nayak, Ravi Mehta
Silicon and Beyond Pvt. Ltd.
Bangalore, India

Abstract— **A source-series terminated (SST) transmitter that is compatible with the various swing, FFE and idle-mode requirements of SATA, PCIe, USB3.1, Ethernet 10G-Base-KR and JESD204B standards is presented. A novel technique for idle-mode operation is implemented. The transmitter is demonstrated in a 40nm Low Leakage CMOS process with an area of 0.09 sq. mm. The power dissipation is 36mW at 12.5Gbps.**

Keywords—SERDES; transmitter; SST; FFE; idle mode

I. Introduction

High speed SERDES design has benefitted from the raw speed improvement of silicon process in advanced low geometry technology nodes. However, the need for high bandwidth to transfer data is equally important for designs on slower and older technology nodes. The choice of process technology for a silicon chip is often driven by requirements such as cost and other functionality, and not necessarily by the ease of design of the SERDES. A low leakage variant of a silicon process poses further challenges due to slower speed.

A transmitter in a multi-standard SERDES is required to support a wide range of swing and feed-forward equalization levels [1-5] in a power-efficient manner. The termination resistor of such a transmitter needs to be programmable to take care of process variation as well as adjustment for impedance matching with board/cable.

Source-series-terminated (SST) driver [6-7, 9-10, 12] is a popular choice as a SERDES transmitter due to its various advantages including lower power, superior return loss performance and ease of portability when compared with an H-bridge or a CML type driver [8, 11]. However, achieving full flexibility of swing, equalization and termination resistor programmability in an SST driver requires non-trivial design considerations [12]. Several standards [1-3] define a transmitter idle mode where the output pads are required to drive a nominally zero differential voltage but maintain the same common mode voltage as that of the active mode. A voltage mode driver consumes the maximum current in order to generate a zero differential compared to non-zero differential voltage for a given common mode. This defeats the purpose of realizing lower power in the idle mode.

This work presents a transmitter having an SST driver topology with a wide range of programmability features. It is designed in low-leakage variant of a 40nm CMOS technology node for data rates ranging from 1Gbps to 12.5Gbps supporting multiple standards including SATA, PCIe, USB3.1, 10G-Base-KR and JESD204B. This work proposes a novel idle mode driver satisfying the requirements of output pad voltages while operating with low power.

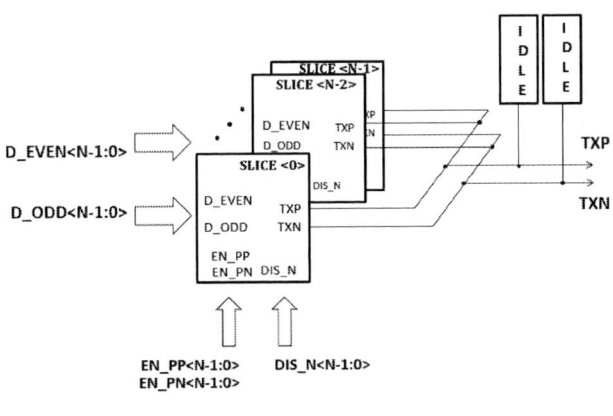

(a)

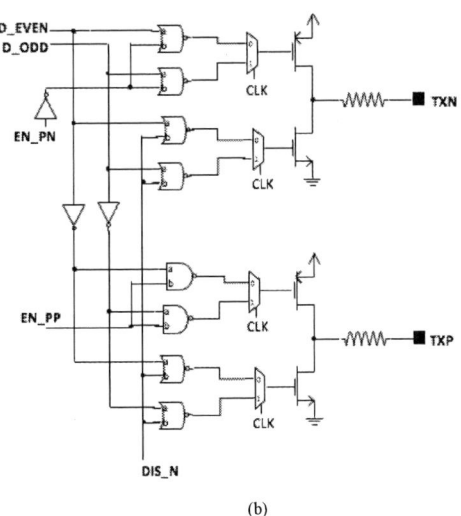

(b)

Fig. 1.   (a) Parallel SST Driver and (b) Sub-driver Implementation

978-1-5386-3693-0/18 $31.00 © 2018 IEEE

13

The paper is organized as follows: Section II gives an overview of the sub-driver (slice) based SST architecture and details through the design of a sub-driver. The process of the assignment of sub-driver weight and input data multiplexing scheme are explained in Section III. The idle mode implementation is presented in Section IV. Laboratory measurements on a low-leakage 40nm design are outlined in Section V. And finally conclusions are presented in Section VI.

II. Architecture

Fig. 1(a) shows the SST driver implemented as a set of parallel sub-drivers. Each sub-driver slice performs 2:1 serialization and drives the serialized data through an inverter with a series resistor.

The data buses D_EVEN<N-1:0> and D_ODD<N-1:0> are driven synchronously at half the data-rate by a digital parallel-to-serial converter. D_EVEN<N-1:0> is the even data stream while D_ODD<1:0> is the odd data stream. The even and the odd data streams are launched at opposite phase of the half-rate clock. Additionally, EN_PP<N-1:0>, EN_PN<N-1:0> and DIS_N<N-1:0> provide asynchronous control to every slice. EN_PP<N-1:0> and EN_PN<N-1:0> control PMOS of TXP and PMOS of TXN respectively. DIS_N<N-1:0> provides control for NMOS as explained in the next paragraph. Two idle-mode drivers are connected to the output pads.

The circuit level implementation of each slice is shown in Fig. 1(b). The assertion of DIS_N cuts off both the NMOS transistors independent of the data input while de-assertion of DIS_N allows the high speed data to be passed onto the NMOS gates. The EN_PP and EN_PN control can be used in conjunction with the data input to pass either the high speed data (EN_PP = 1, EN_PN=1) or drive both the PMOS gates high (EN_PP = 0, EN_PN = 1 and data driven high) or drive both the PMOS gates low (EN_PP = 1 and EN_PN = 0 and data driven low). The combinations of these signals enable the sub-driver slice to operate in high-speed data transfer mode or to pull up both resistors to voltage supply or to tri-state the slice. The data to each gate is retimed at the 2:1 multiplexer with a duty-cycle corrected half-rate clock. This cleans up any static skew in the various paths and dynamic supply-induced noise. Use of flip-flop for retiming is avoided since that would require a clock at the data rate. The final driver stage and the 2:1 multiplexers are run off a separate clean supply. The clock signal used for retiming runs off a regulated supply. The supply domain partition gives flexibility to get the required circuit speed in a low-leakage process.

The adjustment of differential impedance of the driver is done by tri-stating selected number of slices. The adjustment of swing is done by configuring selected slices in the pull-up resistor mode [7]. The driver common mode changes depending on the number of slices in pull-up mode. However, the common mode voltage is not permitted to change in a driver while in operation for data transfer. Hence, this method of swing reduction is not suitable for dynamically altering swing. It is however used as a static control for output swing

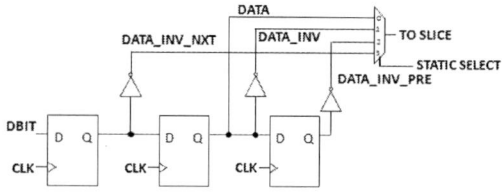

Fig. 2. Static Data Multiplexing for FFE

level. This mode is utilised by the transmitter when configured as a JESD204B-compliant PHY. The advantage of static swing control is lower power driver compared to conventional technique of swing reduction [6].

Data slices which get tri-stated or pulled-up for resistance calibration or swing control do not participate in driving high speed data. The remaining active slice elements participate in driving the high-speed data. The even/odd data source for each slice is independently selectable by means of a static multiplexer. In the most generic implementation shown in Fig. 2, a 4:1 multiplexer with selection for the data, the inverted data, the inverted one-clock-delayed data and the inverted one-clock-advanced data for each slice enables the most flexible programmability for cursor, post-cursor and pre-cursor swings. There are 2N multiplexers for even data D_EVEN<N-1:0> and D_ODD<N-1:0>. However, by grouping slices for specific functions, the 4:1 multiplexer can be reduced to 3:1 or even 2:1. The implementation example shown in Table II shows such a reduced requirement on the multiplexers and will be explained in the next section. A minimum number of available active slices are required to achieve a target granularity of swing and FFE programming.

The transmitter presented in this paper achieves a termination resistor variation of +/-7 ohms (post calibration) around a nominal of 100 ohms in presence of process and temperature induced variation of 70 to 130 ohms. The swing can be statically programmed by the pull-up resistor slices to one-half of the full-swing in steps of 32 mV differential.

Further, FFE programming allows getting more than 12dB de-emphasis and pre-shoot each, where the programming step is 16 mV differential. A direct implementation of slices would require more than 110 slices with each slice having two 4:1 selectable multiplexers for odd and even half-rate data. The partitioned VML driver proposed in [12] cannot be scaled to multiple tap FFE. The next section presents the principle of sub-driver assignment to optimize the number of slices to a quarter as well as simplifying the data multiplexing requirement.

III. Sub-driver Assignment

Binary weighted grouping of the sub-drivers is the most efficient way to reduce the number of controls. This reduces the number of controls from 'n' to $\log_2(n)$. But, the same set of 'n' sub-driver slices need to be controlled in a binary manner for various functions. For example, a certain number of sub-drivers are tri-stated by binary-weighted controls for adjusting the termination resistance. The tri-stated sub-drivers are no

978-1-5386-3693-0/18 $31.00 © 2018 IEEE

longer available in the pool. Hence, the remaining sub-drivers do not form a binary weighted array. Now if we wish to apply static swing control by enabling pull-ups, the non-availability of some of the slices does not give the full flexibility of programming. Further, the two-aforementioned programming leaves a highly-restricted set of available elements to do FFE programming. In general, it is not possible to do a binary weighted control for multiple overlapping functions if one programming precludes the usage of some of the elements thereby making the remaining programming non-binary.

We propose an assignment technique which allows concurrent granular control of sub-driver slices for programming (1) termination resistance, (2) static swing, (3) cursor co-efficient, (4) post-cursor co-efficient and (5) pre-cursor co-efficient. The basic principle is explained first followed by an implementation example.

Noting that only half of the total number of slices is sufficient to do resistor calibration and the other half is sufficient to do static swing control, we can do a non-overlapping mapping of slices to perform resistor calibration and static swing control and form two different sub-groups. Each of these sub-groups can be individually grouped in a binary weighted fashion and controlled by binary control words. However, this does not allow flexible programming of FFE coefficients. Hence, the each of the sub-groups is further sub-grouped into multiple slice groups such that several binary weighted control words can be used. In the next set of assignments, these slice groups are assigned data inputs. The data input to each slice group is from a multiplexer. The multiplexer can select a sub-set of the data, the inverted data, the inverted one-clock-delayed data and the inverted one-clock-advanced data. It is guaranteed in this assignment that each of the pre, post, current cursor programming can be done to its full dynamic range. At power-up, the transmitter termination resistor and static swing control are applied. These programming makes some of the slices not usable for FFE. However, by means of the assignment described above, the remaining slices are still available in binary weight for each of pre, post and current cursor programming. After completion of resistor calibration and static swing programming, a digital FSM scans through the available slices to make the optimum assignment.

TABLE I.   GROUPING FOR RESISTANCE AND STATIC SWING PROGRAMMING

Group #	No. of slices in Group	No. of Groups	Sub-total	Capability
1	2	2	4	Can be Tri-stated
2	4	4	16	,,
3	8	5	40	,,
4	2	4	8	Can be pulled up
5	4	3	12	,,
6	8	1	8	,,
7	Ungrouped		30	Not-used

We start with 118 sub-driver slices. First, the sub-driver slices are grouped according to Table I to perform resistance and swing programming such that no element used for resistance programming is used for swing programming. Non-overlap assignment guarantees no loss of granularity.

Second, the same set of sub-driver slices are grouped according to Table II to perform FFE co-efficient programming. Same slice can be programmed for multiple types of data inputs (current data, invert, post-invert, pre-invert).

TABLE II.   GROUPING FOR FEED FORWARD EQUALIZATION PROGRAMMING

Group #	No. of slices in Group	Static MUX selection for Input Data per slice
A	30	data/invert/post_inv
B	14	data/invert/pre_inv
C	15	data/invert
D	28	data/post_inv
E	31	data/post_inv

The third step is to do a cross mapping between Table I and Table II since they apply to the same set of driver slices. The mapping ensures that the resistor/static swing programming removes available elements from the pool such that FFE co-efficient programming elements invert, post-invert and pre-invert deplete as 2:1:1. The element assignment is made binary weighted wherever possible. Table III presents a cross-mapping scheme. Four types of slice groups are built with weight of 8, 4, 2 and 1. These slices are controlled by 26 controls.

TABLE III.   CROSS MAPPING BETWEEN GROUPS

	Group #1, #2, #3	Group # 4, #5, #6	Group #7
Group #A	(8+8)+8+4+2		
Group #B		8+4+2	
Grioup #C			8+4+2+1
Group #D	8+4+4+2	4	4+2
Group #E	8+4	4+4+2	4+2+(2+1)

While the preceding three steps define assignments which are implemented by circuit wiring, a final assignment is required for MUX selection for each of the 26 data inputs. A digital finite state-machine (FSM) performs this operation and is integrated within the transmitter. Based on the termination and swing programming, the FSM gives control for tri-state and pull-up using EN_P and DIS_N. Then the FSM proceeds to do an optimum MUX selection for sub-driver slices which are not tri-stated and not pulled-up. In addition to the above function, the FSM is used to find the optimum trim value for the termination resistor.

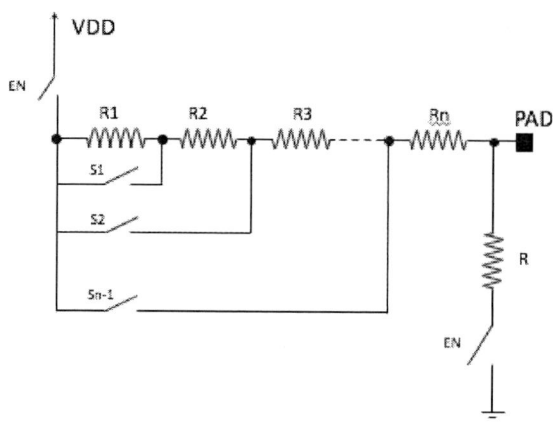

Fig. 3. Idle Mode Driver

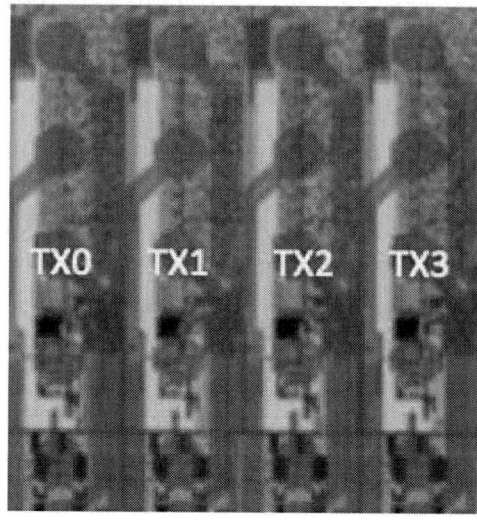

Fig. 4. Die Micrograph

IV. Idle Mode Driver

The common-mode output voltage of the SST driver is equal to mid-supply for full-swing. The static swing control by pull-up on both the differential pads causes the common mode to deviate from mid-supply towards the supply voltage. The common mode is decided by the resistor divider formed by parallel combination of the driver resistors connected to the supply and those connected to the ground [7]. The SERDES link is an AC-coupled system. Hence, the transmitter steady state common mode is not dependent on the receiver termination resistor.

The idle mode common-mode needs to match with the active mode common-mode voltage. The idle mode is implemented with a separate driver as shown in Fig. 3. The driver has a fixed pull-down resistor and a programmable pull-up resistor. The pull-up resistor is a DAC with series resistor elements. The DAC switches are controlled by a thermometric code. The thermometric code is generated by a digital FSM using a mapping table. The mapping table provides the thermometric code based on the static swing code and the resistor termination code. This DAC allows the idle mode to have an output common-mode voltage which is the same as that of the active mode. The high-speed driver slices are tri-stated by appropriate signal states of DIS_N and EN_P in the idle mode. The idle mode driver operates with a current of 100uA. Implementing the idle operation with the main driver by driving a zero-differential voltage would have consumed 10mA. The transitions into and out of idle mode are fast due to digital nature of implementation.

V. Measurement results

Four lanes of transmitters were fabricated as part of a multi-standard multi-lane SERDES in a low-leakage 40nm CMOS technology as shown in Fig. 4, where each transmitter occupies an area of $0.09mm^2$. The SST driver occupies 0.03 mm^2 while the remaining area is for clocking circuit, regulator and decoupling capacitor. The silicon die is housed in a flip-chip BGA package. The bumps for the transmitter output pads are placed symmetrically on top of the driver.

Fig. 5a shows the transmitter eye at the maximum data rate of 12.5Gbps at the end of a medium loss channel. Fig. 5b is the near-end eye-diagram at 10Gbps after de-embedding the channel as specified in JESD204B while the far-end eye-diagram with transmit FFE, 23dB lossy channel and USB3.1 reference CTLE is demonstrated in Fig. 5c. The reference CTLE is implemented in the oscilloscope. The jitter bath-tub for the transmitter in the PCIe Gen3 mode of operation is presented in Fig 6. The clock-recovery is done with software PLL whose cut-off frequency is set in accordance with the standard-specified value. The effect of board traces is de-embedded to produce the near-end eye while reference CTLE is embedded for generating the far-end eye.

The termination resistor is trimmed by a digital trim engine FSM. The measured post-trim termination resistor with process, temperature varies between 102 to 115 ohms as shown in Fig. 7. There is a systematic upward shift due to some top-level routing resistance. This can be corrected by modifying the trim target specified to the digital trim engine. The presented result is with the default implementation of the trim.

A comparison with previously published implementations of transmitters in 40nm/45nm is presented in Table IV, where [7-11] are arranged in reverse chronological order of publication. Compared to the SST implementations in [7, 9-10], this work compares favourably in pJ/bit efficiency. Compared to the CML driver-based transmitter implementations in [8, 11], this SST implementation gives better power due to lower load current, although a direct comparison could not be done as separate transmit power numbers are not reported in [8, 11]. While [7, 10] support higher speed, it is observed that [7] is not on a low-leakage process and [10] is on an SOI process.

978-1-5386-3693-0/18 $31.00 © 2018 IEEE   16

VI. Conclusions

An SST driver for a multi-standard SERDES in low-leakage 40nm CMOS was presented. The design technique for simultaneous trimming of termination resistance, power-efficient static swing control and dynamic equalization settings was elaborated. A novel low-power idle mode driver with same common mode as active mode of SST driver was realized. The silicon results showing eye-diagrams for various standards were demonstrated using a flip chip implementation.

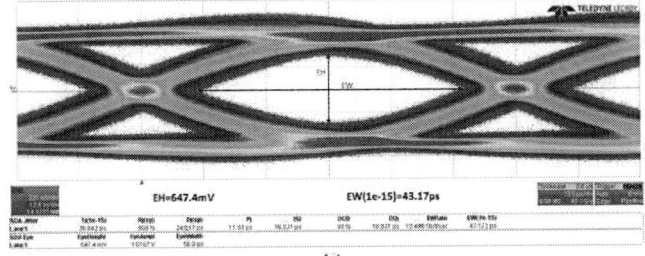

(a)

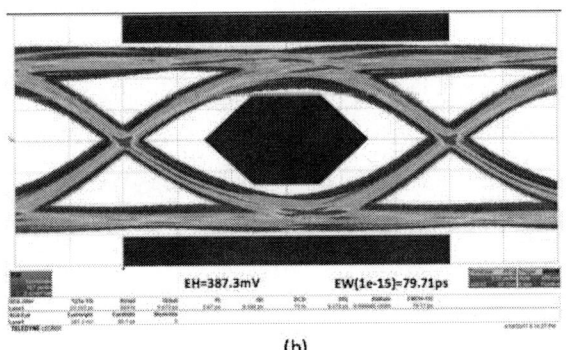

(b)

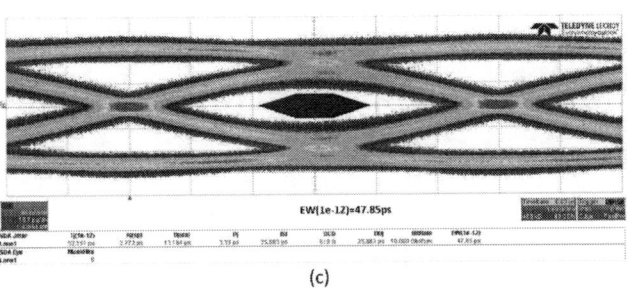

(c)

Fig. 5.   Measured Eye Diagram (a) 12.5Gbps (b) 10Gbps near end (JESD 204B) and (c) 10Gbps USB3.1 Far-end after 23dB loss and reference equalizer implemented in scope

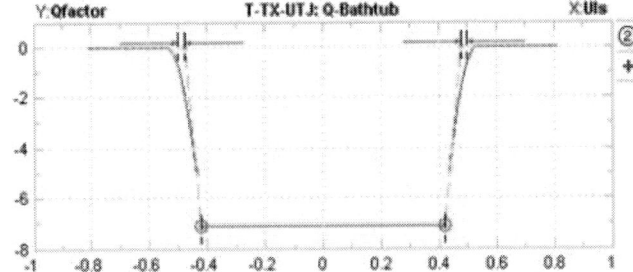

Fig. 6.   Measured Jitter Bath-tub for PCIe Gen3

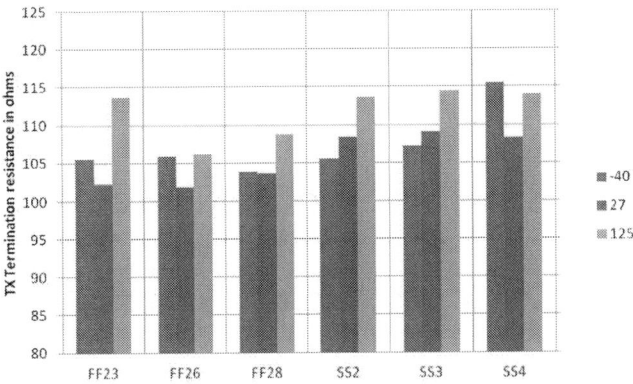

Fig. 7.   Measurement results for differential termination resistance across process and temperature

TABLE IV.   COMPARISON WITH OTHER SERDES TRANSMITTERS IN 40NM/45NM

	[7]	[8]	[9]	[10]	[11]	This Paper
Tech-nology	40nm CMOS	40nm LL CMOS	45nm SOI	45nm SOI	40nm CMOS	40nm LL CMOS
Speed (Gbps)	14	11.2	14	7.4	5	12.5
Power (mW)	39	336[1]	70.6	32	125[1]	36
Effici-ency (pJ/bit)	2.78	-	5.04	4.32	-	2.88
Driver	SST	CML	SST	SST	CML	SST
SST Driver Supply (V)	0.9	-	1	1.08	-	1.1
Static Swing Cntrl.	Yes	No	No	No	No	Yes
FFE Taps	3	3	4	2	2	3
Driver Area (mm²)	0.04	0.16[2]	0.03	0.03[2]	0.07[2]	0.03

1.   TX power is not separately reported. The reported total transceiver power includes RX and TX and may not be a fair comparison with transmit only power.

2.   TX area is not separately reported. Total transmitter area is estimated from the chip micrograph.

Acknowledgment

The authors would like to thank Vishal Nimbark and Manjunath Shet for their help in silicon characterization.

References

[1]   Universal Serial Bus 3.1 Specification, Rev. 1.0, July 2013

[2]   Serial ATA, Rev 3.2, August 2013

[3]   PCI Express base specification revision 3.0

[4]   IEEE Std 802.3-2012, Clause-72, 2012

[5]   JEDEC Standard JESD204B.01, January 2012

978-1-5386-3693-0/18 $31.00 © 2018 IEEE

[6] M. Kossel, et al. "A T-Coil-Enhanced 8.5Gb/s High-Swing SST Transmitter in 65 nm Bulk CMOS With <- 16dB Return Loss Over 10GHz Bandwidth", IEEE Journal of Solid-State Circuits, VOL. 43, NO.12, December 2008

[7] K. Frazan, M.Ramezani, A. McLaren, R. Pahuta, N. Amarasinghe, D. Cassan and S. Sadr, "A low jitter 2.7mW/Gbps 180Gb/s 12-Lane Transmitter in 40nm CMOS Technlolgy," IEEE European Solid State Conference, 2012, pp 225-228

[8] S. D. Vamvakos, et al, "A 2.488-11.2 Gb/s multi-protocol SerDes in 40nm low-leakage CMOS for FPGA", IEEE 55th International Midwest Symposium on Circuits and Systems, August 2012

[9] C. Menolfi, et al, "A 14 Gb/s High-Swing Thin-Oxide Device SST TX in 45nm SOI", IEEE International Solid-State Circuits Conference, February 2011, pp 156-158

[10] W. D. Dettloff, et al, "A 32mW 7.4Gb/s Protocol-Agile Source-SeriesTerminated Transmitter in 45nm CMOS SOI" , IEEE International Solid-State Circuits Conference, February 2010, pp 370-372

[11] Mu-shan Lin, et al, "A 5Gb/s Low-Power PCI Express/USB3.0 Ready PHY in 40nm CMOS technology with High-Jitter Immunity", IEEE Asian Solid-State Circuits Conference, Novemeber 2009, pp 177-180

[12] R. G. Raghavendra, B. S. Rathor, "A 1-tap 10.3125Gb/s Programmable Voltage Mode Line Driver in 28nm CMOS Technology", IEEE 29th International Conference on VLSI Design, January 2016, pp 174-178

978-1-5386-3693-0/18 $31.00 © 2018 IEEE

An Ultra Low Power, 10-bit Two-Step Flash ADC for Signal Processing Applications

Mahesh Kumar Adimulam, Amit Kapoor, Sreehari Veeramachaneni, M.B Srinivas

EE Department, Birla Institute of Technology and Science – Pilani, Hyderabad Campus, Hyderabad 500078, India

Abstract— An ultra low power, 10-bit two-step flash analog-to-digital converter (ADC) for communication and bio-potential signal processing applications is presented in this paper. In the proposed design the conventional open loop comparator is replaced with programmable bias inverter (PBI), the bias inverter (BI) consists of basic digital inverter with cascode PMOS and NMOS as bias transistors in the top and bottom. The switching threshold voltage of BI changes with different reference bias voltages, which compares the analog input voltage with BI switching threshold and provide respective digital outputs. The programmability of the BI makes the proposed ADC to operate from DC to 1 GS/s sampling frequency range and hence ADC power gets scaled accordingly. The major advantages of the PBI based two-step flash ADC is lower power consumption, smaller area and improved static/dynamic performance due to lower mismatch between the PBI comparators and smaller input capacitance. The proposed ADC is designed in 90nm standard CMOS process occupying a core area of 0.096 mm². The performance parameters of the proposed ADC are found to be, differential non-linearity (DNL) of ±0.38 LSB, integral non-linearity of ±0.54 LSB, signal-to-noise-and-distortion ratio (SNDR) of 57.38 dB, spurious free dynamic range (SFDR) of 69.3 dB, effective number of bits (ENOB) of 9.24 at 1.0 V supply voltage. The power consumption of this ADC at 100 kS/s sampling frequencies is 280 nW and at higher sampling frequencies upto 1 GS/s it is 5.6 mW.

Keywords- Programmable Bias Inverter (PBI), two-step flash ADC, programmable bandwidth, low power consumption and lower area.

I. INTRODUCTION

Analog-to-digital converters (ADCs) have been in demand for improving performance in applications such as high-speed Ethernets, high speed communication receivers, sensors networks, mobile communication, and bio-medical devices. While the lower technology nodes helped in improving operating speeds, the operating speed of the ADCs are still slow in comparison with digital designs. The technology scaling has improved the ADC speeds to some extent, but the performance parameters such as noise and offset voltages have degraded.

In recent years, high/low speed, low-to-medium resolution and low power ADCs demand has increased [1]-[3]. The CMOS flash/two-step flash ADCs is suitable for high/medium speed and low resolutions ($\leq$ 10-bits). The advantage of flash and two-step flash ADCs are fast operation (> 1GS/s) but the limitations are power consumption and area due to conventional open loop latch comparators [4]-[7].

The conventional flash and two-step flash ADC architectures [8] consists of resistor ladder, comparators and an encoder. The N-bit flash ADC requires 2^N resistors and 2^N-1 comparators, whereas the N-bit two-step flash ADC requires $2*2^{N/2}$ resistors and $2*2^{N/2}$-1 comparators. The power efficiency, area, and performance of the flash/two-step flash ADCs increases significantly at high resolutions (> 8-bits) [4]-[8]. At lower speed of operation and low-to-medium resolution ($\geq$ 10-bits) successive approximation register (SAR) ADCs are most commonly used, however the limitations of SAR ADCs are performance due to DAC capacitor mismatches and power consumption due to DAC switching logic [9]-[11].

In recent years, inverter based topologies [12]-[13] are used instead of open loop comparator [8] to reduce power consumption and area. However, the major limitation of the inverter based ADCs are dependency of input dynamic range on threshold voltage variation of the MOS devices over process and temperature. The inverter threshold variation can be controlled by using body bias techniques but the performance becomes worse than latched comparator based flash ADC due to implementation complexity.

In this paper, considering the limitations of comparator based flash/two-step flash ADC, SAR ADC and inverter based flash ADC at low-to-medium resolutions and high/low speed operations, a programmable bias inverter (PBI) based two-step flash ADC is designed. The proposed ADC uses a cascode biased PMOS device on the top (towards supply) and a biased NMOS device in the bottom (towards ground) in a digital inverter which controls the switching threshold of the BI. The BI dimensions are programmed through the select bit based on the low/medium sampling frequency band or medium/high sampling frequency band operation. The power, performance, and area of the proposed ADC is better than the comparator based two-step flash ADC since BI takes a smaller area and consumes lesser power. The power efficiency of the PBI circuits gets scaled based upon the sampling frequency band. The performance of the proposed ADC is better than the SAR ADC due to no DAC non-idealities & capacitor mismatches. However, the inverter area is smaller than the capacitor used in SAR ADC that results in a lower magnitude mismatch. The power consumption, area, and process variation of the proposed design are better than the inverter based two-step flash ADC since bias inverter dimensions are small and same for all inverters. The inverter based ADC requires different PMOS & NMOS dimensions [12] for defining different switching threshold voltages which result in more power and increased area, while the design complexity of inverter based ADC is high since each inverter design is unique.

The organization of the paper is done as follows: Section II describes the architecture of the proposed PBI two-step

flash ADC. Section III describes the Proposed PBI two-step flash ADC Circuit implementation. Section IV describes the simulation results and finally the conclusion is described in section V.

II. ARCHITECTURE OF THE PBI TWO-STEP FLASH ADC

The architecture of the proposed PBI two-step flash ADC design is shown in Fig.1. It consists of a 5-bit PBI flash ADC for coarse and fine stages, multiplying digital-to-analog converter (MDAC), a reference bias generator, and a programmable oscillator. In the proposed architecture, the fundamental idea is to replace the conventional open loop comparators with PBIs and use the resistor reference ladder for biasing different bias inverters. The analog input (V_{IN}) is applied to all the bias inverters of the coarse 5-bit flash ADC and MDAC, and the MDAC output is applied to bias inverters of fine 5-bit flash ADC. The bias inverters take the decision and provide the digital output based upon the input voltage level and the switching threshold set by bias transistors. The

output of the bias inverters is converted to a binary format by using an encoder. The proposed 10-bit two step flash ADC is programmed to operate for different sampling frequencies from DC to 1GS/s by using the static control bit (S). The power consumption of the ADC gets scaled based upon different sampling frequency bands.

The functional operation of the programmable bandwidth 10-bit PBI based two-step flash ADC along with calibration is described as follows. The programmable bias inverters are first calibrated after getting enabled and the supply voltage is used as reference during calibration. The process calibration is done for bias inverters PBI<0>, PBI<15>, and PBI<30> by incrementing or decrementing the reference voltage in bias circuit until the bias inverters changes the logic levels. This digital code is used during the data conversion process. After calibration, the PBI two-step flash ADC is programmed for low/medium sampling frequency (DC to 100 MS/s) or medium/high sampling frequency (100 MS/s to 1 GS/s) by using the Select bit (S) generated from the digital block.

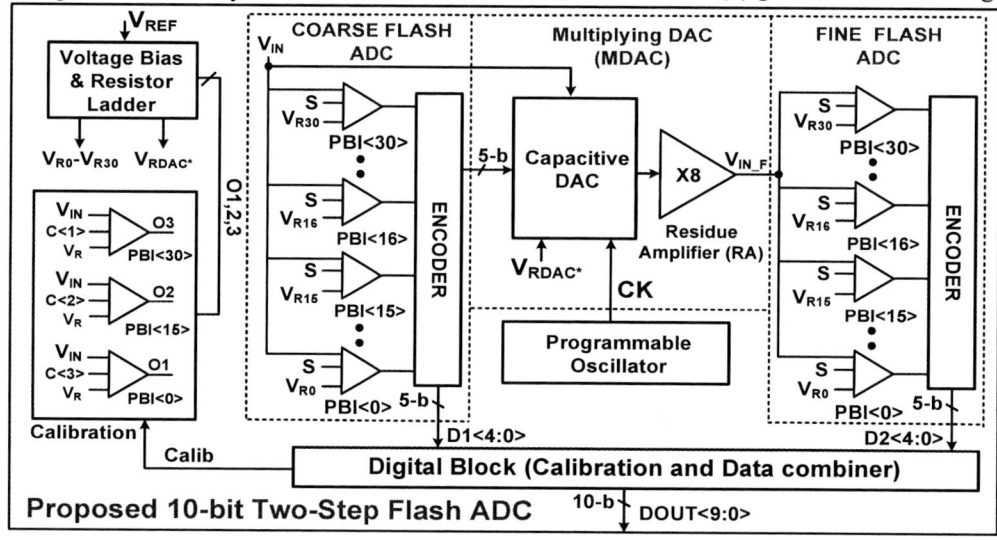

Figure 1. Block diagram of 10-bit PBI Two-Step Flash ADC

The analog input (V_{IN}) is given to the coarse 5-bit PBI flash ADC and MDAC. The coarse 5-bit PBI flash ADC convers analog data to thermometer code based upon the input voltage level, and the encoder provides a binary output for the same. The coarse 5-bit flash ADC digital outputs (D1<4:0>) are given to MDAC and digital block. The MDAC block consists of switched capacitor digital-to-analog converter (DAC) and residue amplifier (RA) with a gain of 8. The MDAC provides the scaled-up error voltage to fine 5-bit PBI flash ADC. The fine 5-bit PBI flash ADC convers analog data to thermometer code based upon the input voltage level and the encoder provides a binary output for the same. The fine 5-bit flash ADC digital outputs (D2<4:0>) are given to the digital block. The digital block provides 10-bit (DOUT<9:0>) output by combining data from the coarse and fine PBI flash ADCs.The remaining sub-blocks voltage bias & resistor ladder are used to generate reference voltages to the PBI flash ADCs and MDAC, while the programmable oscillator

(POSC) is used to generate the required sampling frequency based upon the input signal frequency.

III. CIRCUIT IMPLEMENTATION OF THE PROPOSED PBI TWO-STEP FLASH ADC

A. Programmable Bias Inverter (PBI) based Comparator

In two-step flash ADC architectures, a comparator [8] is a critical block which defines the ADC power consumption and area. The proposed PBI comparator and its voltage transfer characteristics (VTC) are shown in Fig. 2. The bias voltage (V_R) defines the VTC of PBI by keeping fixed dimensions for all the transistors which indicates that the input signal quantization is set by only bias voltage rather than changing the inverter (MP0, MP1, MN0 and MN1) dimensions as done by [12]. The bias voltage V_R is generated by using a programmable resistor ladder. The V_R voltage is set to V_{R0} which is the highest reference value in the resistor ladder to

978-1-5386-3693-0/18 $31.00 © 2018 IEEE

achieve $\beta_n/\beta_p \gg 1$ so that the PBI inverter quantizes lowest input voltage. Similarly, V_R voltage is set to V_{R30} which is the lowest reference value in the resistor ladder to achieve $\beta_p/\beta_n \gg 1$ so that the PBI inverter quantizes the highest input voltages. The other input voltages in the full-scale range (Vin $\sim 0.3\ V - 0.7\ V$) are quantized by the reference voltages V_{R1} - V_{R29} as illustrated in Fig. 2(b). The switch S is used to program the PBI to different sampling frequency bands (by increasing/decreasing the current flowing through the BI) by passing V_{RI} (=V_R) voltage to MPB1 and MNB1 transistors.

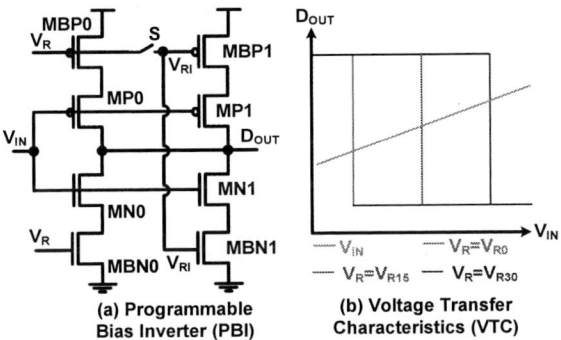

Figure 2. (a) Programmable Bias Inverter (PBI) comparator (b) Voltage Transfer Characterisctics (VTC)

The threshold voltage of the PBI can be defined by the following equations

The upper threshold voltage (Vth_H) point is given by

$$Vth_H = \frac{VDD}{2} + IR_{MBP0,1} \qquad (1)$$

The lower threshold voltage (Vth_L) point is given by

$$Vth_L = \frac{VDD}{2} - IR_{MBN0,1} \qquad (2)$$

The linear switching threshold voltage (Vsw) is given as

$$Vsw = \frac{VDD}{2} + I\frac{(R_{MBP0,1} - R_{MBN0,1})}{2} \qquad (3)$$

Where $R_{MBP0,1}$ and $R_{MBN0,1}$ are $MBP_{0,1}$ and $MBN_{0,1}$ transistors switch "ON" resistance respectively, and "I" is the current flowing through the PBI which is calculated by standard formula taken from [15].

The switching threshold voltage of the PBIs varies linearly with respect to (w.r.t) input voltage, the input dynamic voltage range is $0.3\ V - 0.7\ V$ considering the 280 mV MOSFET threshold voltage in 90nm standard CMOS technology. The process variation of the Vsw is controlled by adjusting the V_R voltage by trimming the resistor ladder during calibration and hence complicated body bias techniques are not needed. The $\pm 3\sigma$ monte carlo variation of the Vsw voltage is done for PBI<15>, the results are discussed in section IV.

B. Voltage Biasing and Resistor Ladder

Fig.3 shows the circuit diagram implementation of the voltage biasing and resistor ladder. The voltage-to-current converter technique adopted from [14] is used as reference for combining voltage biasing and resistor ladder for reference generation. The input reference voltage V_{REF} is taken from on-chip programmable bandgap reference generator circuit. The

operational transconductance amplifier (OTA) used is a single stage folded cascode [15] considering stability and power consumption.

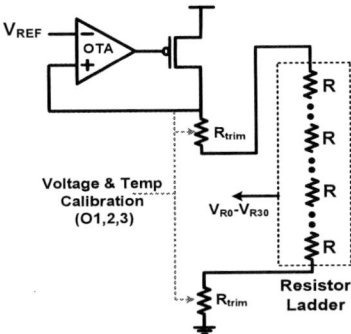

Figure 3. Voltage Biasing and Resistor Ladder Circuit

The resistor ladder consists of 32 resistors and trimming is done for reducing the process variation of PBIs. The reference voltages V_{R0} - V_{R30} are given to the 5-bit coarse/fine flash ADC and reference voltage (V_{RDAC}) is given to the MDAC. The performance parameters of the voltage biasing and resistor ladder are described as follows: the OTA is designed for 0.25 LSB (i.e., ~0.1 mV) offset voltage, and the remaining parameters are found to be DC gain of 82 dB, phase margin (PM) of 62°, and a settling time of 280 nS.

C. Encoder

The encoder circuits with different architectures are designed considering power efficiency, area, and latency [16]-[17]. The proposed PBI two-step flash ADC implements the encoder which is an extended version from [17] for 5-bit coarse/fine ADC operation.

The integration of the data conversion path, i.e., coarse flash ADC, fine flash ADC, and MDAC is done considering bias inverters matching with minimum parasitic routing capacitance and DAC capacitors common centroid matching. The voltage biasing with resistor ladder is integrated considering the resistor ladder matches with the uniform routing of reference voltages for every PBI cell. The fully differential amplifier architecture is used from [18] for RA implementation, and the programmable oscillator is made compact in layout to reduce parameters. This implementation technique minimizes the parasitics and enhances the performance at high frequencies. The simulation results of the proposed ADC are discussed in section IV.

IV. SIMULATION RESULTS

The simulation results of the proposed programmable 10-bit PBI two-step flash ADC is discussed in this section. The worst-case RC post layout simulations (includes PVT and $\pm 3\sigma$ monte carlo) are done so that the performance parameters can be compared with measured values. The circuit implementation of the proposed ADC is done in 90nm standard CMOS process occupying a core area of 0.096 mm². Fig.4 shows the layout view of the programmable 10-bit PBI two-step flash ADC.

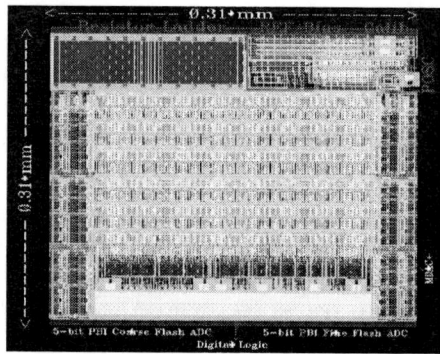

Figure 4. Layout view of the programmable 10-bit PBI two-step flash ADC

The functional operation of the proposed ADC design coarse/fine stages are verified by applying the analog ramp input range from 0.3 V- 0.7 V (i.e., ADC full scale voltage) with a ramp time of 500 ns. The digital outputs of the ADC coarse/fine stages are verified and no missing of data is observed. Fig. 5 shows the correct functionality of the coarse/fine stages with the digital code going from 0 to 31 at the output.

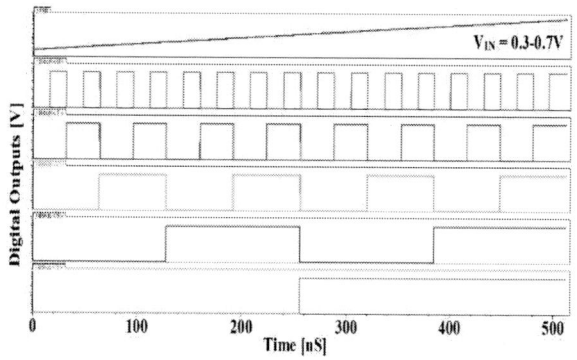

Figure 5. Functionality of the proposed PBI sub flash (coarse/fine) ADC

Fig.6 shows the worst case PBI transient monte carlo simulations to capture the threshold voltage variation considering only device mismatch (i.e., no process variation) and temperature range is from -40°C to 125°C, the threshold voltage variation of bias inverters due to process is minimized by reference voltage calibration. The ±3σ threshold voltage variation is ±130 μV, which is less than ~0.38 LSB (i.e., 1 LSB = $0.4/2^{10}$=0.34 mV) suitable for 10-bit operation.

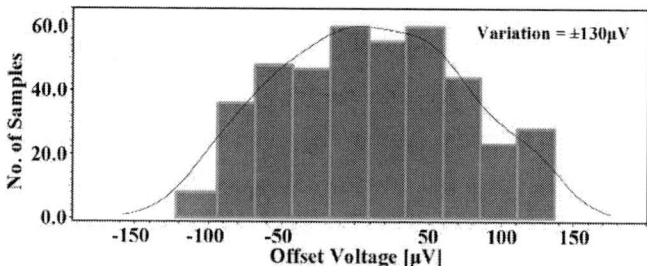

Figure 6. Worst PBI block ±3σ threshold variation

The static parameters (DNL & INL) are captured at ideal DAC by running transient simulations with slow varying ramp input applied to the PBI two-step flash ADC. Fig.7 shows the DNL and INL results. While the DNL of the proposed design is ± 0.38 LSB, the INL is ± 0.54 LSB for 10-bit operation.

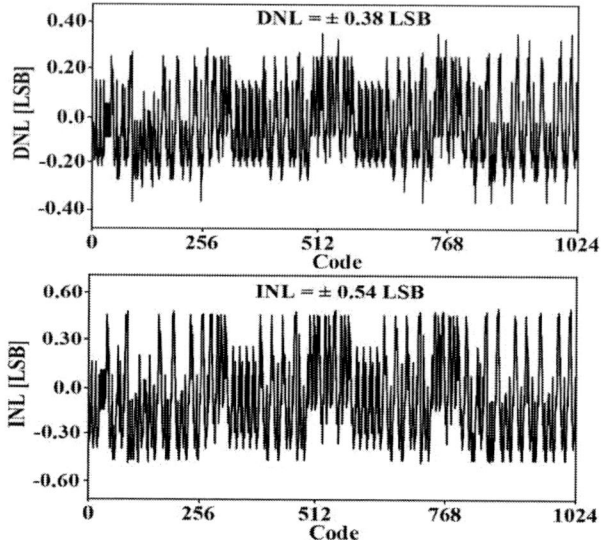

Figure 7. DNL and INL Results for 10-bit Operation

Transient simulations of the proposed ADC are performed for different input signal frequencies by applying 0.4 Vp-p sinusoidal input voltage to capture performance parameters such as SNDR, SFDR, and ENOB. Fig. 8 shows the power spectrum of the proposed ADC at 100 kS/s and 1 GS/s sampling frequencies. The SNDR of 57.38 dB, SFDR of 69.3 dB, and ENOB of 9.24 are captured at 1GS/s sampling frequency. Similarly, the SNDR of 59.31 dB, SFDR of 72.5 dB and ENOB of 9.56 are captured at 100 kS/s sampling frequency.

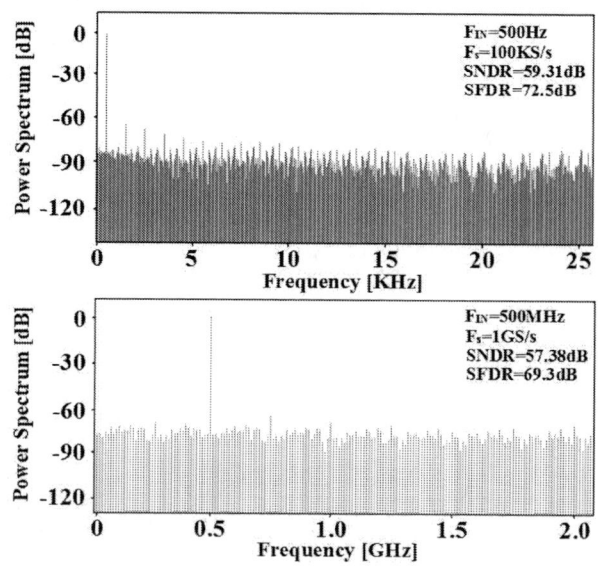

Figure 8. SFDR for 100 kS/s and 1 Gs/s sampling frequency

978-1-5386-3693-0/18 $31.00 © 2018 IEEE

The process variation of the proposed PBI two-step flash ADC is analyzed and compared with conventional comparator based [4]-[5] & inverter based [12] two-step flash ADCs. The proposed design is much better than the inverter based ADC design. Since the reference voltages generated from the resistor ladder are calibrated, the switching voltage variation of the bias inverters varies by 1.5 mV with respect to (w.r.t) temperature range from -40°C to 125°C. In the proposed design, fine calibration of the reference voltage is done at 27°C to reduce switching voltage (SW) variation to less than 0.7 mV which is 0.7% w.r.t input dynamic range.

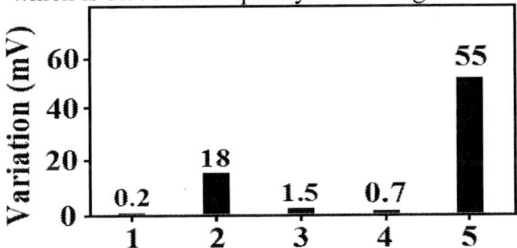

1 = Conventional Comp two-step flash ADC
2 = PBI two-step flash ADC PT Variation
3 = PBI two-step flash ADC Temp Variation
4 = PBI two-step flash ADC Calibrated
5 = Inverter two-step flash ADC PT variation

Figure 9. Switching Voltage variation of conventional comparator, inverter based and PBI Two-Step flash ADCs due to process and temperature

Fig. 9 shows the process and temperature (PT) variation of the conventional comparator based two-step flash ADC, inverter based two-step flash ADC, and PBI two-step flash

ADC. The PBI two-step flash ADC variation is less than 1 LSB for 10-bit operation.

Fig.10 shows the power consumption pie-chart in 10-bit operation mode for 100 kS/s and 1 GS/s sampling frequencies respectively. The power consumption is found to be 280 nW at 100 kS/s sampling frequency and 5.6 mW at 1 GS/s sampling frequency.

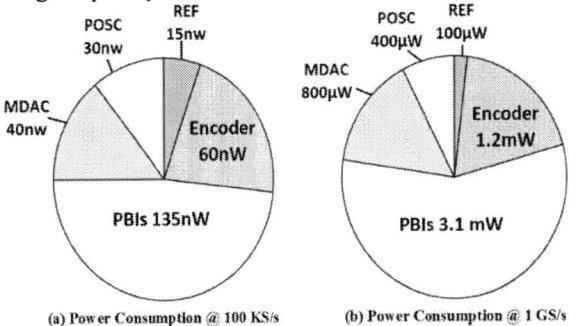

(a) Power Consumption @ 100 KS/s   (b) Power Consumption @ 1 GS/s

Figure 10. Power Consumption for 100 kS/s and 1 GS/s sampling frequencies

Table I shows the performance parameters comparison of PBI two-step flash ADC and comparison w.r.t the conventional comparator based two-step flash ADC and the inverter based two-step flash ADC. The optimized design for the conventional comparator based two-step flash ADC and inverter based two-step flash ADC are done in the same technology (i.e., 90nm) for precise comparison. The results show the performance parameters are better for PBI two-step flash ADC compared to the conventional comparator based and inverter based two-step flash ADCs.

TABLE I.   PERFORMANCE PARAMETERS OF THE PBI TWO-STEP FLASH ADC AND COMPARISON WITH COMPARATOR / INVERTER BASED ADCS

Parameters	Comparator Based Two-Step Flash ADC		Inverter Based Two-Step Flash ADC		PBI Two-Step Flash ADC	
Technology (nm)	90					
Supply voltage (V)	1.0					
Resolution (bit)	10					
Sampling Frequency	100 kS/s	1 GS/s	100 kS/s	1 GS/s	100 kS/s	1 GS/s
DNL (LSB)	±1.2	±1.2	±0.95	±0.95	±0.38	±0.38
INL (LSB)	±2.2	±2.2	±1.65	±1.65	±0.54	±0.54
SNDR (dB)	55.64	53.71	54.9	53.0	59.31	57.38
SFDR (dB)	66.51	64.85	65.4	64.1	72.5	69.3
ENOB (bits)	8.95	8.63	8.83	8.52	9.56	9.24
Power Consumption	2.8µW	24.6mW	1.9µW	16.73mW	280 nW	5.6 mW
Figure of merit (fJ/conv-step)	56.5	62.1	41.69	45.5	3.71	9.26
Area (mm²)	0.56		0.21		0.096	

Table II shows the comparison of the PBI two-step flash ADC performance parameters with recent state of art designs at 100 kS/s and 1 GS/s sampling frequencies. The proposed PBI two-step flash ADC at lower input frequency is compared with recent SAR ADC designs for better correlation and most of the performance parameters are better than those of recent designs. The proposed PBI two-step flash ADC at 1 GS/s sampling frequency is compared with recent flash/subrange ADC designs and most of the performance parameters of the proposed design are better than those of the recent designs.

V. CONCLUSION

In this paper, an ultra-low power programmable 10-bit PBI two-step flash ADC is presented. In the proposed design, PBIs are used as open loop comparators to reduce power consumption, area, and performance instead of conventional comparators and digital inverters. The input dynamic voltage variation of the PBI two-step flash ADC is reduced by using calibration which was a major limitation in the inverter based ADC design. The proposed design is implemented in 90nm standard CMOS process: worst-case RC post layout

978-1-5386-3693-0/18 $31.00 © 2018 IEEE

simulations are carried out to capture the performance parameters at 100 kS/s and 1 GS/s sampling frequencies. Table I & II shows the performance parameters comparison of the proposed design and comparison with recent state of art designs. The results show that the proposed ADC is highly suitable for bio-medical applications at low frequency. At higher sampling frequencies (upto 1 GS/s), the proposed design is highly suitable for communication applications.

TABLE II. PERFORMANCE PARAMETERS COMPARISON WITH RECENT STATE OF ART DESIGNS

Parameters	2013 ISSCC [10]*	2015 JSSC [11]*	2016 JSSC [19]*	2013 TCASI [20]*	2014 EDSSC [21]**	2012 A-SSCC [22] *	2014 VLSI [23] *	This Work**	
Technology (nm)	90	180	28	45	180	65	65	90	
Supply voltage (V)	0.4	0.6	1.1	1.15	1.8	1.2	1.2	1.0	
Architecture	SAR	SAR	SAR	Flash	Flash	Subrange	Subrange	PBI Two-Step Flash	
Resolution (bit)	10	10	12	7	6	8	7	10	
Calibration	NO	YES	YES	YES	NO	YES	YES	YES	
Sampling Frequency	500 kS/s	100 kS/s	104 MS/s	1.4 GS/s	1.4 GS/s	1.0 GS/s	1.0 GS/s	100 kS/s	1.0 GS/s
DNL (LSB)	0.34	-0.26/0.50	0.5	< 0.74	-0.67/0.25	0.6	0.54	±0.38	±0.38
INL (LSB)	0.62	-0.82/0.89	1.1	< 1	-0.48/0.36	0.7	1.07	±0.54	±0.54
SNDR (dB)	54.3	57.2	60.5	39.0	31.25	45.2	37.7	59.31	57.38
ENOB (bits)	8.72	9.2	9.76	6.18	4.9	7.2	5.97	9.56	9.24
Power Consumption (W)	500n	390n	3.06m	33.24m	98m	17.5m	5.91m	280n	5.6m
Figure of merit (fJ/conv-step)	2.37	6.7	34	330	760	125	94	3.71	9.26
Area (mm^2)	0.042	0.103	0.024	0.085	0.45	0.25	0.087	0.096	
*Measured Results and **Simulation Results									

REFERENCES

[1] Daly, Denis C., and Anantha P. Chandrakasan. "A 6-bit, 0.2 V to 0.9 V highly digital flash ADC with comparator redundancy." *IEEE Journal of Solid-State Circuits* 44.11 (2009): 3030-3038.

[2] Lin, Jin-Yi, and Chih-Cheng Hsieh. "A 0.3 V 10-bit SAR ADC With First 2-bit Guess in 90-nm CMOS." *IEEE Transactions on Circuits and Systems I: Regular Papers* 64.3 (2017): 562-572.

[3] Pernillo, Jorge, and Michael P. Flynn. "A 1.5-GS/s flash ADC With 57.7-dB SFDR and 6.4-bit ENOB in 90 nm digital CMOS." IEEE Transactions on Circuits and Systems II: Express Briefs 58.12 (2011): 837-841.

[4] Chahardori, Mohammad, Mohammad Sharifkhani, and Sirus Sadughi. "A 4-bit, 1.6 GS/s low power flash ADC, based on offset calibration and segmentation." *IEEE Transactions on Circuits and Systems I: Regular Papers* 60.9 (2013): 2285-2297.

[5] Proesel, Jonathan, et al. "An 8-bit 1.5 GS/s flash ADC using post-manufacturing statistical selection." *Custom Integrated Circuits Conference (CICC), 2010 IEEE.* IEEE, 2010.

[6] Razavi, Behzad, and Bruce A. Wooley. "A 12-b 5-Msample/s two-step CMOS A/D converter." *IEEE Journal of Solid-State Circuits* 27.12 (1992): 1667-1678.

[7] Wang, Guannan, et al. "14-bit 20 μW column-level two-step ADC for 640× 512 IRFPA." *Electronics Letters* 51.14 (2015): 1054-1056.

[8] Lin, James, et al. "Ultralow-voltage high-speed flash ADC design strategy based on FoM-delay product." *IEEE Transactions on Very Large Scale Integration (VLSI) Systems* 23.8 (2015): 1518-1527.

[9] Lin, Jin-Yi, and Chih-Cheng Hsieh. "A 0.3 V 10-bit 1.17f SAR ADC with merge and split switching in 90 nm CMOS." *IEEE Transactions on Circuits and Systems I: Regular Papers* 62.1 (2015): 70-79.

[10] Liou, Chang-Yuan, and Chih-Cheng Hsieh. "A 2.4-to-5.2 fJ/conversion-step 10b 0.5-to-4MS/s SAR ADC with charge-average switching DAC in 90nm CMOS." Solid-State Circuits Conference Digest of Technical Papers (ISSCC), 2013 IEEE international.

[11] Jin, Jiayi, Yang Gao, and Edgar Sanchez-Sinencio. "An energy-efficient time-domain asynchronous 2 b/step SAR ADC with a hybrid R-2R/C-3C DAC structure." *IEEE Journal of Solid-State Circuits* 49.6 (2014): 1383-1396.

[12] Tangel, Ali, and Kyusun Choi. "'The CMOS Inverter" as a comparator in ADC designs." *Analog Integrated Circuits and Signal Processing* 39.2 (2004): 147-155.

[13] Proesel, Jonathan E., and Lawrence T. Pileggi. "A 0.6-to-1V inverter-based 5-bit flash ADC in 90nm digital CMOS." *Custom Integrated Circuits Conference, 2008. CICC 2008. IEEE.* IEEE, 2008.

[14] Azcona, Cristina, et al. "Low-voltage low-power CMOS rail-to-rail voltage-to-current converters." *IEEE Transactions on Circuits and Systems I: Regular Papers* 60.9 (2013): 2333-2342.

[15] Behzad, Razavi. "Design of analog CMOS integrated circuits." *International Edition* (2001).

[16] Wang, Mingzhen, et al. "A 6-b 4-GSPS low-voltage flash ADC with a pipelined DCVSPG logic encoder." *Computational Problem-solving (ICCP), 2013 International Conference on.* IEEE, 2013.

[17] Chen, Yong, et al. "Comparator with built-in reference voltage generation and split-ROM encoder for a high-speed flash ADC." *Signals, Circuits and Systems (ISSCS), 2015 International Symposium on.* IEEE, 2015.

[18] Lim, Yong, and Michael P. Flynn. "A 1 mW 71.5 dB SNDR 50 MS/s 13 bit fully differential ring amplifier based SAR-assisted pipeline ADC." *IEEE Journal of Solid-State Circuits* 50.12 (2015): 2901-2911.

[19] Tseng, Wei-Hsin, et al. "A 12-bit 104 MS/s SAR ADC in 28 nm CMOS for digitally-assisted wireless transmitters." *IEEE Journal of Solid-State Circuits* 51.10 (2016): 2222-2231.

[20] Nakajima, Yuji, et al. "A 7-bit, 1.4 GS/s ADC with offset drift suppression techniques for one-time calibration." *IEEE Transactions on Circuits and Systems I: Regular Papers* 60.8 (2013): 1979-1990.

[21] Zhang, Sheng, et al. "A 6-bit low power flash ADC with a novel bubble error correction used in UWB communication systems." *Electron Devices and Solid-State Circuits (EDSSC), 2014 IEEE International Conference on.* IEEE, 2014.

[22] Ohhata, Kenichi. "1-GHz, 17.5-mW, 8-bit Subranging ADC Using Offset-Cancelling Charge-Steering Amplifier." IEICE Transactions on Electronics 97.4 (2014): 289-297.

[23] Yoshioka, Kentaro, et al. "7-bit 0.8–1.2 GS/s dynamic architecture and frequency scaling subrange ADC with binary-search/flash live configuring technique." *VLSI Circuits Digest of Technical Papers, 2014 Symposium on.* IEEE, 2014.

978-1-5386-3693-0/18 $31.00 © 2018 IEEE

Rescheduling of Power Gating instructions for reduction of In-rush current

Sumanta Pyne
Department of Computer Science and Engineering
National Institute of Technology, Rourkela
Sundergarh, Odisha - 769008, India
Email: pynes@nitrkl.ac.in

Abstract—**The present work introduces two compilation techniques for reduction of in-rush current in processors with power gating (*PG*) facility. These are done by rescheduling the *PG* instructions responsible in turning on multiple components from sleep to active mode at overlapped time intervals. The first method eliminates overlapped wake-up of the components resulting lesser in-rush current at the cost of performance degradation due to increase in program size. The second method allows overlapped wake-up as long as the resultant in-rush current is tolerable by the system with lesser increase in delay and program size. Algorithms are designed to automate these methods. The efficacy of the proposed methods are evaluated on MiBench and MediaBench benchmark programs. The original program with *PG* and their translated versions are executed on gem5 which simulates ARM Cortex-M4F processor enhanced with *PG*. McPAT is used to find the values of power consumed and in-rush current. The first and second methods reduce in-rush current by an average of 54% and 35%, respectively with corresponding average performance loss of 21% and 9%.**

I. INTRODUCTION

The emergence of deep submicron process technology with the decrease in dimensions of the transistor has increased the transistor count and speed of operation at the cost of greater device leakage currents. Power gating (*PG*) is a technique used to reduce power consumption of VLSI chips, by shutting off the blocks that are not in use, thus reducing stand-by or leakage power. Shutting down of the blocks can be initiated either by software or by hardware. Many modern processors are equipped with *PG* instructions to switch-off and switch-on different blocks to sleep and active modes, respectively. This allows the compiler and operating system to reduce runtime leakage power.

When a power gated block is switched on from sleep mode to active mode it draws a huge amount of in-rush current due to simultaneous charging of its internal capacitors. In-rush current is several times higher than the actual current required by the block to function in active mode. The flow of in-rush current may cause permanent damage to the circuit and also lead to higher power consumption. It can reduce the battery life for battery-operated systems due to rise in load current.

There exists several hardware schemes to reduce in-rush current. Some of them are usage of multiple power switches, daisy chaining, soft start with voltage regulators, discrete and integrated load switches. Most of them are not practical in

PG design industry since they require more space and higher design cost.

The present work proposes two software based approaches for reduction of in-rush current due to overlapped wake-up caused by *PG* instructions that simultaneously activates multiple components. The first method eliminates overlapped wake-up, while the second one allows overlapped wake-up with guaranteed tolerable in-rush current. These techniques can enable a compiler to generate target code that will reduce in-rush current to tolerable levels.

The existing works on management of in-rush current in *PG* systems are discussed in Sec. II. The proposed compilation techniques are explained in Sec. III. Section IV covers explaination of the experimental setup with analysis of the results. Finally, Sec. V concludes the present work with future directions.

II. RELATED WORKS

The existing research and development works on reduction of in-rush current in systems with *PG* are based on hardware techniques at circuit level. Some of the techniques are discussed in the following paragraph.

In [1] the authors proposed *PG* structures in which sleep transistors are turned on in a non-uniform stepwise manner to reduce the magnitude of peak current and voltage glitches in the power distribution network requiring minimum time to stabilize power and ground. In this case, the rush current gradually increases as the number of switches is turned on. However, the rush current can be large unless the daisy chain is very slow. On the other hand, such a long daisy chain can cause long propagation delay and the slowly rising voltage can introduce other problems such as hot electron effects [2]. Another approach is to separate the power switches into two passes: a weak transistor pass and a strong transistor pass [4]. At wake-up, the weak transistors are turned on first so as to slowly turn on the rush currents. When the design is discharged (charged) to a voltage close to zero (V_{dd}), the strong transistor pass is turned on ready for normal operation. In [2] the authors introduced a timing and voltage drop analysis tool named CoolTime. It can guide the designer in setting power switch structure and sequence for controlling wake-up (rush current and wake-up time). An in-rush current limiter circuit [3] invented by Ball can sense the increased load current and

produces sense current having a load current - sense current ratio of 1000:1, hence reducing the in-rush current. Kiong et al. introduced the in-rush current optimization power up flow analysis with PFET removal algorithm [5] to improve the in-rush current. In [6] the authors introduces two intermediate scheme to reduce wake-up time. During wake-up procedure their *PG* scheme implements a small transistor to control the sleep transistor (*ST*), has two stages: relaxation stage and completely turn-on stage. During the relaxation stage, the drain to source voltage (V_{ds}) of the *ST* reduces significantly while limiting the current exponentially as the V_{ds} of *ST* changes. During the complete turn-on stage, the small control transistor is turned off, and the *ST* acts like a current source. This two-stage transition method reduces the peak voltage fluctuations in the virtual ground and virtual power, and it also reduces the circuit wake-up time. They also introduced two circuit schemes with intermediate states to further reduce the ground bounce based on their proposed power gating circuit scheme. Meanwhile, the intermediate state saves more charges by charge recycling while allowing the virtual ground or virtual power floating between V_{dd} and ground. In [7] the authors proposed model memory access power gating (*MAPG*), a low-overhead technique to enable power gating of an active core when it stalls during a long memory access. They described a programmable two-stage power gating switch design that can vary a core's wake-up delay while maintaining voltage noise limits and leakage power savings with controlled in-rush current. A novel framework for generating a proper power-up sequence of the switches to control the in-rush current of a power-gated domain has been introduced in [8]. It also minimizes the power-up time and reduces the dynamic IR drop of the active domains. A detailed study of cause and effects of in-rush current with remedies to reduce it has been discussed in [9]. The authors explain in-rush current reduction techniques like soft-start with the help of voltage regulators to increase rise time. They explained that power switching with a controlled rise time can be accomplished by using discrete circuitry. Effective usage of integrated load switches in place of the discrete solution was also highlighted. In [10] Kim et al. discussed the reduction of in-rush current by turning on each switch cell at different times. They showed that in-rush current can be reduced even more if signal transition time to switch each cell is adjusted.

The existing software methodologies to reduce leakage current on systems with *PG* are mainly based on compilation techniques [11], [12], [13], [15], [14], [16] which deals in scheduling of *PG* instructions. Task scheduling with *PG* is proposed in [17]. Reduction of in-rush current has not been addressed in these works.

The current status of research carried in this field shows that there is a scope to perform investigation on software techniques for in-rush current management in systems with *PG*. Existing hardware techniques add extra circuitry for in-rush current management increasing design cost, design space and average power consumption. The software techniques can address these drawbacks.

III. Present Work

An arrangement for instruction controlled *PG* is shown in Fig. 1. It has n *PG* components $C_0, C_1, \cdots, C_{n-1}$. *PG* is done with the help of the header p-MOS transistors having higher threshold voltage (V_T). The header switches are controlled by an n-bit power gating control register (PGCR) placed in the power gating controller (PGC). The bits $0, 1, \cdots, n-1$ are the *PG* bits of $C_0, C_1, \cdots, C_{n-1}$, respectively. If any of these bits $\alpha \in \{0, 1 \cdots, n-1\}$ is '0', then the component C_α is in active mode, otherwise C_α is in sleep mode. Let there be two *PG* instructions *switch_off* and *switch_on* each consuming three clock cycles - one cycle in each of instruction fetch (*IF*), instruction decode (*ID*) and execution (*EX*) stages of the instruction pipeline. To put C_α in sleep (or power gated) mode the instruction *switch_off*(C_α) is used to set the value of αth bit of PGCR. C_α in sleep mode can be put to active mode with the help of the instruction *switch_on*(C_α) which resets the value of αth bit of PGCR. Hence, a program can use this *PG* facility.

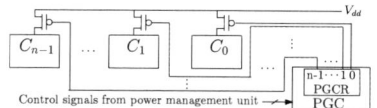

Fig. 1.   An arrangement for instruction controlled *PG* system

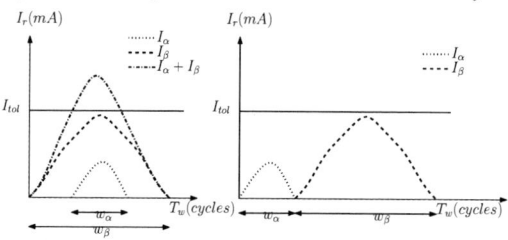

(a) Overlapped wake-up   (b) Non-overlapped wake-up

Fig. 2.   In-rush current for overlapped and non-overlapped wake-up

When C_α in sleep mode is switched on using *switch_on*(C_α) it draws in-rush current I_α for a period of w_α cycles where w_α is the wake-up time (T_w) of C_α. It is considered that $I_\alpha \leq I_{tol}$ for wake-up of any individual *PG* component C_α, where I_{tol} is the maximum tolerable in-rush current for a given system. The problem of intolerable in-rush current may arise during wake-up of multiple components during an overlapped time interval. Fig. 2(a) shows in-rush current (I_r) in milliampere (mA) for overlapped wake-up of two components C_α and C_β where $\beta \in \{0, 1, \cdots, n-1\}$ and $\beta \neq \alpha$. w_β and I_β are the wake-up time and in-rush current of C_β, respectively. The resultant in-rush current is $I_\alpha + I_\beta$. Simultaneous overlapped wake-up of several components can lead to higher flow of in-rush current resulting higher peak power dissipation and reduction of chip reliability. Hence, it is better to have non-overlapped wake-up as shown in Fig. 2(b).

A. PG with Non-overlapped Wake-up (PGNW)

A program with *switch_on* instructions may cause overlapped wake-up of *PG* components. Considering an assembly language program with M instructions where i^{th} instruction J_i is *switch_on*(C_α) and j^{th} instruction J_j is *switch_on*(C_β),

978-1-5386-3693-0/18 $31.00 © 2018 IEEE

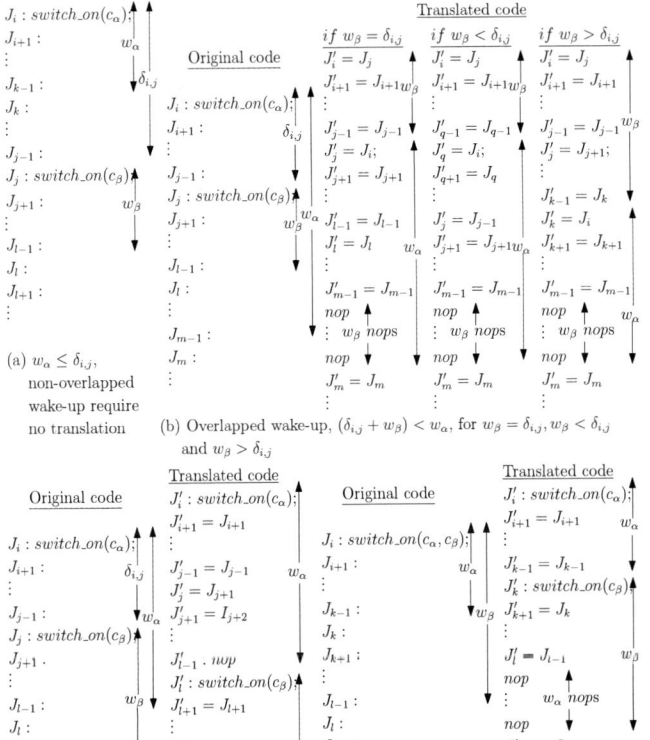

Fig. 3. Elimination of overlapped wake-up using *PGNW*

where $i, j \in \{0, 1, \cdots, M-1\}$ and $i \leq j$. Let it take $\delta_{i,j}$ cycles from J_i to reach J_j. If $i = j$, then $\delta_{i,j} = 0$. For $i < j$, $\delta_{i,j} = \sum_{p=i}^{j-1} cycles(I_p, I_{p+1})$, where $cycles(I_p, I_{p+1})$ is the time gap (in cycles) between the entry of J_p and J_{p+1} in the *EX* stage. J_i and J_j will result non-overlapped wake-up of C_α and C_β if $i < j$ and $w_\alpha \leq \delta_{i,j}$. Hence, no translation of the program is required as shown in Fig. 3(a). Overlapped wake-up occurs if $w_\alpha > \delta_{i,j}$. Rescheduling of *switch_on* instructions for elimination of overlapped wake-up are shown in figures 3(b), 3(c) and 3(d) where $0 \leq i \leq q \leq j \leq k \leq l \leq m < M$. No-operation (*nop*) instructions are inserted in the translated code to ensure the usage of a component after completion of its wake-up. An *nop* is assumed to consume one byte of memory space. It takes three clock cycles - one cycle in each of *IF*, *ID* and *EX* stages. An *nop* being an immediate successor of a *switch_on* acts as a delay of one clock cycle after the execution of *switch_on*. This adds time and space overheads of $O(w_\alpha)$ if $(\delta_{i,j} + w_\beta) < w_\alpha$. The time and space overheads are $O(w_\beta)$ for $(\delta_{i,j} + w_\beta) \geq w_\alpha$ as well as in case of $\delta_{i,j} = 0$ and $w_\beta \geq w_\alpha$. The space overhead can be reduced to $O(1)$ by replacing *nop*s with a loop having empty body or a single *nop* depending on the values of w_α and w_β.

The proposed Algorithm 1 (*PGNW* Algorithm) produces the code for *PG* with non-ovelapped wake-up. It takes an assembly language level code fragment of the source program

having $M'(M' \leq M)$ instructions with overlapped wake-up of $n'(n' \leq n)$ components. As output it produces an equivalent *PGNW* code fragment with rescheduled *switch_on* instructions leading to non-overlapped wake-up for n' components. The time taken by M' instructions of input code fragment is T clock cycles. The input and output code fragments always start with a *switch_on* instruction.

Algorithm 1 *PG* with non-overlapped wake-up (*PGNW*)

Input: A code fragment of M' instructions with overlapped wake-up of n' components, where $M' \leq M$ and $n' \leq n$.
Output: A code fragment with rescheduled *switch_on* instructions leading to non-overlapped wake-up of n components.
Initialization: (i) Store the *switch_on* instructions in S in non-decreasing order of earliest finishing time of the wake-up.
         (ii) Consider the input code fragment as *PGNW* code.

1: $\alpha \leftarrow S[1].component$
2: At cycle 1, replace the *switch_on* instruction with $switch\_on(C_\alpha)$.
3: $t \leftarrow w_\alpha + 1$
4: **for** $(u \leftarrow 1; u \leq n'; u \leftarrow u + 1)$ **do**
5:   $i \leftarrow S[u].instruction$
6:   $j \leftarrow S[u-1].instruction$
7:   $\alpha \leftarrow S[u].component$
8:   $\beta \leftarrow S[u-1].component$
9:   $\tau \leftarrow t - w_\beta + \delta_{i,j}$
10:   **if** $\delta_{i,j} + w_\beta < w_\alpha$ **then**
11:     At cycle τ, remove $switch\_on(C_\beta)$.
12:     Move all instructions starting from cycle $\tau + 1$ backward by one step.
13:     Move all instructions starting from cycle t forward by one step.
14:     At cycle t, insert $switch\_on(C_\alpha)$.
15:     Move all instructions starting from cycle $t + w_\alpha$ forward by w_β steps.
16:     Insert w_β *nops* from cycle $t + w_\alpha - w_\beta$ to cycle $t + w_\alpha - 1$.
17:   **else if** $\delta_{i,j} + w_\alpha \geq w_\beta$ **then**
18:     At cycle τ, remove $switch\_on(C_\beta)$.
19:     Move all instructions starting from cycle $\tau + 1$ backward by one step.
20:     **if** $\delta_{i,j} > 0$ **then**
21:       At cycle $t + w_\beta - 1$, insert an *nop*.
22:     **end if**
23:     Move all instructions starting from cycle t forward by one step.
24:     At cycle t, insert $switch\_on(C_\alpha)$.
25:     Move instructions starting from cycle $t + w_\alpha$ forward by $w_\beta - \delta_{i,j}$ steps.
26:     Insert $w_\beta - \delta_{i,j}$ *nops* from cycle $t + w_\alpha - (w_\beta - \delta_{i,j})$ to cycle $t + w_\alpha - 1$.
27:   **end if**
28:   $t \leftarrow t + w_\alpha$
29: **end for**

PGNW Algorithm considers an array data structure S having n' elements representing *switch_on* instructions for n' components. $S[u]$ is the uth *switch_on* instruction, where $u \in \{1, 2, \cdots, n'\}$. S stores the *switch_on* instructions belonging to the given code fragment in non-decreasing order of earliest finishing time of the wake-up of components involved with *switch_on* instructions. $S[u].instruction \in \{0, 1, \cdots, M'\}$ is the instruction number of *switch_on* instruction assigned to $S[u]$. The value representing the component turned on by $S[u]$ is denoted by $S[u].component \in \{0, 1, \cdots, n-1\}$. The algorithm begins by considering the given input code fragment *PGNW* code.

The input and output code fragments are considered to start at clock cycle $t = 1$. Here, t and τ denotes number of clock cycles starting from cycle 1. In step 2 the first instruction of the input code fragment which runs during cycle 1 is replaced by $switch\_on(C_\alpha)$. Steps 11 and 18 are involved in removal of *switch_on* instructions which run during cycle τ. This allow steps 12 and 18 to move all instructions starting from cycle $\tau + 1$ b bytes backward to lower memory addresses of *PGNW* code, where b is the size of the instructions $switch\_on(C_\alpha)$ and $switch\_on(C_\beta)$. Hence, in *PGNW* code these instructions will begin at cycle τ. Steps 13 and 23 moves all the instructions starting at cycle t forward to higher memory address by b bytes, delaying them by one cycle. This enable steps 14 and 24 to schedule $switch\_on(C_\alpha)$ at cycle t. Steps 15 and 25 move all instructions staring at cycle t forward by w_β and $w_\beta - \delta_{i,j}$ bytes, respectively. Thus delaying them by same number of cycles. This allows steps 16 and 26 to insert w_β and $w_\beta - \delta_{i,j}$ bytes, respectively.

Algorithm 1 considers all cases for overlapped wake-up as shown in Fig. 3. At the beginning of each iteration of the **for** loop of steps 4-29 indexed by u, the subarray consisting of elements $S[1], S[2], \cdots, S[u-1]$ constitutes the currently scheduled $switch\_on$ instructions that produces non-overlapped wake-up, the element $S[u]$ constitute the $switch\_on$ instruction to be scheduled in the current iteration u for non-overlapped wake-up with previous $u-1$ $switch\_ons$, and the remaining elements of subarray $S[u+1], S[u+2], \cdots, S[n']$ constitutes the $switch\_on$ instructions to be scheduled in next $n'-u$ iterations for non-overlapped wake-up. This property forms the loop invariant which is preserved during initialization (in steps 1-3), maintenance (in steps 4-29, for $u \in \{2, 3, \cdots, n'\}$) and termination (when $u = n'+1$) phases to ensure the correctness of the algorithm.

In each iteration the movement of $O(M')$ instructions in $O(M' \times w_\beta)$ time results insertion of w_β *nops* in the *PGNW* code. w_β is $O(w_{max})$, where $w_{max} = max(w_0, w_1, \cdots, w_{n-1})$. Hence, it takes $O(n' \times M' \times w_{max})$ time to generate a *PGNW* code with $O(n' \times w_{max})$ *nops*.

The proposed *PGNW* method is strict in elimination of overlapped wake-up. It is suitable for systems where reliability has higher priority than delay. It may not be suitable for safety-critical and real-time systems where apart from reliability lower delay is crucial. Sec. III-B introduces a method to deal with these issues.

B. PG with Tolerable In-rush current (PGTI)

PGNW guarantees tolerable in-rush current at the cost of increase in delay and program size. These overheads can be reduced by allowing overlapped wakepus within the limitation maximum tolerable in-rush current. For each *PG* component C_α an in-rush current table (IT_α) is maintained. The tuple $t \in \{1, 2, \cdots, w_\alpha\}$ of IT_α denoted by $IT_\alpha[t]$ stores the value of I_α during t^{th} cycle of wake-up of C_α. I_α is minimum during cycles 1 and w_α. I_α is maximum or at peak during cycle $\frac{w_\alpha}{2}$.

The proposed Algorithm 2 (*PGTI* Algorithm) produces the code for *PG* with tolerable in-rush current. It takes an assembly language level code fragment of the source program having $M'(M' \leq M)$ instructions with overlapped wake-up of $n'(n' \leq n)$ components, in-rush current tables (*ITs*) for each n' components and maximum tolerable in-rush current (I_{tol}) as inputs. As output it generates an equivalent *PGTI* code fragment with rescheduled $switch\_on$ instructions that guarantees atmost I_{tol} amount of in-rush current due to overlapped wake-up of n' components. The input and output code fragments always start with a $switch\_on$ instruction. *PGTI* Algorithm considers two array data structures I_{tot} and S. I_{tot} is an array

---

Algorithm 2 *PG* with tolerable in-rush current (*PGTI*)

Input: (i) A code fragment of M' instructions with overlapped wake-up of n' components, where $M' \leq M$ and $n' \leq n$.
    (ii) In-rush current tables (*ITs*) for each n' components.
    (iii) Maximum tolerable in-rush current (I_{tol}).
Output: A *PGTI* code with rescheduled $switch\_on$ instructions leading to overlapped wake-up of n components with tolerable in-rush current.
Initialization: (i) $I_{tot}[t] \leftarrow 0 \; \forall t | t \in \{1, 2, \cdots, T\}$.
    (ii) Store the $switch\_on$ instructions in S using following rules:
        (a) in order of their occurrences in the code fragment.
        (b) if a $switch\_on$ is having more than one component then store $switch\_on$ for each component in non-decreasing order of wake-up time.
    (iii) Consider the input code fragment as *PGTI* code.

1:   $\alpha \leftarrow S[1].component$
2:   **for** $(t \leftarrow S[1].start; t \leq S[1].end; t \leftarrow t+1)$ **do**
3:      $I_{tot}[t] \leftarrow IT_\alpha[t - S[1].start + 1]$
4:   **end for**
5:   **for** $(u \leftarrow 2; u \leq n'; u \leftarrow u+1)$ **do**
6:      $\alpha \leftarrow S[u].component$
7:      $\beta \leftarrow S[u-1].component$
8:      $\Delta t \leftarrow 0$
9:      $t \leftarrow S[u].start$
10:     **while** $(t \leq S[u].end)$ **do**
11:       **if** $(I_{tot}[t + \Delta t] + IT_\alpha[t - S[u].start + 1]) > I_{tol}$ **then**
12:         **if** $S[u].start = S[u-1].start$ **then**
13:           Move all successors of $switch\_on(C_\alpha)$ forward by one step.
14:           Insert $switch\_on(C_\alpha)$ next to $switch\_on(C_\beta)$.
15:         **else**
16:           Move $switch\_on(C_\alpha)$ and all its successors forward by one step.
17:           Insert an *nop* as immediate predecessor of $switch\_on(C_\alpha)$.
18:         **end if**
19:         $\Delta t \leftarrow \Delta t + 1$
20:         goto Step 9.
21:       **end if**
22:     **end while**
23:     **for** $(t \leftarrow S[u].start; t \leq S[u].end; t \leftarrow t+1)$ **do**
24:       $I_{tot}[t + \Delta t] \leftarrow I_{tot}[t + \Delta t] + IT_\alpha[t - S[u].start + 1]$
25:     **end for**
26: **end for**

---

of T elements, where T is the total number of clock cycles required by M' instructions belonging to the code fragment given as input. $I_{tot}[t] \in \mathbb{R}_{\geq 0}$ is the total in-rush current due to overlapped wake-up during cycle $t \in \{1, 2, \cdots, T\}$. Initially, all the elements of I_{tot} are assigned with zero. The array S has n' elements representing $switch\_on$ instructions for n' components. $S[u]$ is the uth $switch\_on$ instruction, where $u \in \{1, 2, \cdots, n'\}$. S stores the $switch\_on$ instructions in order of their occurrences in the given code fragment. In case of a $switch\_on$ instruction is having more than one component, the $switch\_on$ instruction for each component are stored in non-decreasing order of wake-up time. $S[u].start$ and $S[u].end$ are the respective staring and finishing times (or cycles) of wake-up due to $S[u]$, where $S[u].start, S[u].end \in \{1, 2, \cdots, T\}$ and $S[u].start \leq S[u].end$. The value representing the component turned on by $S[u]$ is denoted by $S[u].component \in \{0, 1, \cdots, n-1\}$. The algorithm begins by considering the given input code fragment as the *PGTI* code.

Steps 1-4 assigns the in-rush current values during the wake-up of $S[1]$ to the array I_{tot}. In each iteration of the **for** loop covering steps 5-26 reschedules $S[u]$ for tolerable in-rush current. The inner **while** loop comprising of steps 10-22 check the possiblity of intolerable in-rush current caused by wake-up of $S[u].component$ overlapped with the wake-up of $S[1].component, S[2].component, \cdots S[u-1].component$ considered in previous $u-1$ iterations. For a particular cycle

$t \in \{S[u].start, S[u].start + 1, \cdots, S[u].end\}$, if the total in-rush current is found to be greater than I_{tol}, then $S[u]$ is delayed by one cycle using an *nop* and the delay counter Δt is increased by 1. Another trial for checking of tolerable in-rush current for $S[u]$ is done. This goes on until $S[u]$ is rescheduled to guarantee tolerable in-rush current. On finding a tolerable in-rush current the steps 23-24 adds in-rush current produced by rescheduled $S[u]$ to the total in-rush in I_{tot}.

The steps 13 moves all successive instructions of *switch_on*(C_β) forward to higher addresses by b bytes in memory of *PGTI* code. Step 14 inserts a b byte instruction *switch_on*(C_α) as immediate succesor of *switch_on*(C_β). Similarly, step 16 moves *switch_on*(C_β) and all its successive instructions forward to higher addresses by 1 byte in memory of *PGTI* code. This creates room for an *nop* inserted as immediate predecessor of *switch_on*(C_α) in step 17.

At the beginning of each iteration of the **for** loop of steps 5-26 indexed by u, the subarray consisting of elements $S[1], S[2], \cdots, S[u-1]$ constitutes the currently scheduled *switch_on* instructions that produces tolerable in-rush current due to overlapped wake-up, the element $S[u]$ constitute the *switch_on* instruction to be scheduled in the current iteration u for tolerable in-rush current due to overlapping with previous $u - 1$ wake-up, and the remaining elements of subarray $S[u+1], S[u+2], \cdots, S[n']$ constitutes the *switch_on* instructions to be scheduled in next $n' - u$ iterations for tolerable in-rush current. This property forms the loop invariant which is preserved during initialization (in steps 1-4), maintenance (in steps 5-26, for $u \in \{2, 3, \cdots, n'\}$) and termination (when $u = n' + 1$) phases to ensure the correctness of the algorithm.

In each iteration the inner **while** loop of steps 10-22 take $O(M')$ time due to movement of $O(M')$ instructions to insert an *nop* in *PGTI* code. In each iteration of the outer **for** loop, the inner **while** takes $O(T \times M')$ time inserting $O(T)$ *nop*s and the inner **for** loop of steps 23-25 take $O(T)$ time. Hence, it takes $O(n' \times T \times M')$ time to generate a *PGTI* code with $O(n' \times T)$ *nop*s.

IV. EXPERIMENT AND RESULTS

To establish the efficacy of the proposed approach, simulations are carried out on **gem5** [18] architecture simulator. McPAT [20] is used for obtaining power values. **gem5** is configured with the instruction set and functional units (*FU*s) of the ARM Cortex-M4F processor [19]. The processor has seven *FU*s. Integer ALU (*ialu*) is not power gated because it is used in majority of the instructions. The bits 0, 1, 2, 3, 4 and 5 of PGCR are the *PG* bits of Floating Point Divider (*fpdiv*), Floating Point Multiplier (*fpmul*), Floating Point Adder (*fpadd*), Integer Divider (*idiv*), Integer Multiplier (*imul*), and Barrel Shifter (*bshf*), respectively as shown in Fig. 4. The size of the instruction cache is considered to be 32 KB.

McPAT is configured with the power model of ARM Cortex-M4F based on 32nm process technology, where the leakage power dissipation is almost 70% of the total power consumption. Here, the processor clock frequency $f_{clk} = 1.0$ GHz, and power supply voltage $V_{dd} = 0.9$ V. V_T of processor's n-MOS and p-MOS transistors are $V_{tn} = 0.18$ V and $V_{tp} = -0.18$ V, respectively. V_T of p-MOS transistors which act as header switches are -0.45 V. $I_{tol} = 200$ mA. Table I show the values of load capacitance ($c_l^{(\alpha)}$), maximum operating current ($I_{op}^{(\alpha)}$), w_α and peak I_α (I_α^{pk}) for each C_α belonging to ARM Cortex-M4F with *PG*.

TABLE I
VALUES OF $c_l^{(\alpha)}, I_{op}^{(\alpha)}, w_\alpha$ AND I_α^{pk}

C_α	fpdiv	fpmul	fpadd	idiv	imul	bshf
$c_l^{(\alpha)}$ (in nF)	6.58	5.9	3.89	2.24	1.81	0.8
$I_{op}^{(\alpha)}$ (in mA)	17.24	15.47	12.84	9.63	8.12	4.72
w_α (in *cycles*)	32	30	24	18	16	10
I_α^{pk} (in mA)	185	177	146	112	102	72

The high level *PG* instructions *switch_off* and *switch_on* are designed to support *PG* in high level languages. An assembly language level instruction **pg pgcr_bit_vector** has been added to the instruction set, where *pgcr_bit_vector* is a 32-bit vector representing the *PG* bits of PGCR. Its size is five bytes. It sets/resets corresponding *PG* bits of PGCR to switch OFF/ON the *FU*s with *PG* in three clock cycles - one cycle in each of *IF*, *ID* and *EX* stages. The GCC compiler for ARM Cortex-M4F [21] is extended to replace high level *switch_on* and *switch_off* instructions with equivalent **pg pgcr_bit_vector** instruction. The features leading to generation of basic *PG* (using [12], [14]), *PGNW* and *PGTI* codes are also added to the GCC compiler for ARM Cortex-M4F.

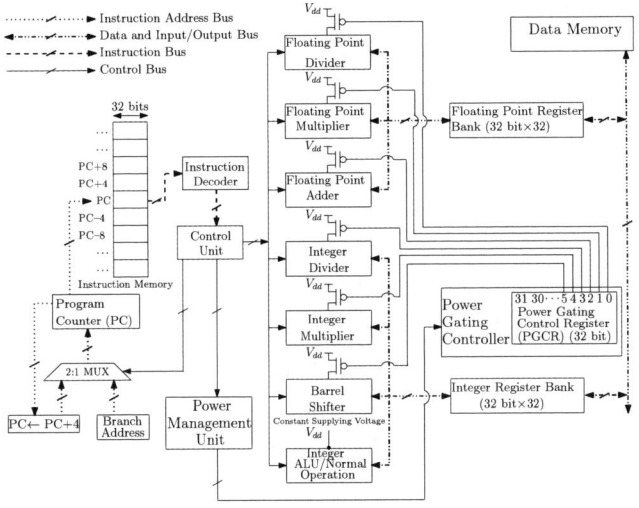

Fig. 4.   A machine architecture model with *PG* control

TABLE II
BENCHMARK DESCRIPTION

Program	fft	ffti	rsynth	mpeg2	jpeg	epic	gsm	pgp
Bench	MiBench	MiBench	MiBench	Media	Media	Media	Media	Media
Category	Telecomm	Telecomm	Office	Video	Image	Image	Speech	Crypto
#cfow	15	8	15	18	14	7	23	9

The proposed techniques are tested on MiBench [22] and MediaBench [23] benchmark programs as shown in Table II, where #cfow is the number of code fragments with overlapped wake-up. The benchmark programs are compiled using updated GCC compiler. The generated target code is executed on gem5 behaving as ARM Cortex-M4F processor. The performance values are generated by gem5. These performance values along with process technology and power related parameters of ARM Cortex-M4F act as input to McPAT in a prescribed XML file format. McPAT produces the power trace with the help of information in the XML file. The values of peak, average, dynamic and leakage power are produced by McPAT. The values of overlapped in-rush current are obtained from the peak power values.

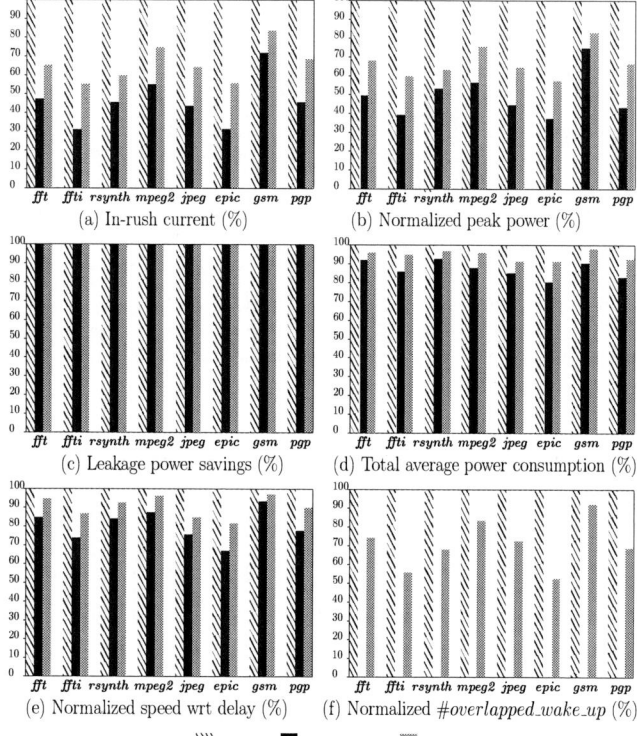

(a) In-rush current (%)   (b) Normalized peak power (%)

(c) Leakage power savings (%)   (d) Total average power consumption (%)

(e) Normalized speed wrt delay (%)   (f) Normalized #overlapped_wake_up (%)

▨ PG   ■ PGNW   ▨ PGTI

Fig. 5. Comparison of experimental results

The experimental results are shown in Fig. 5. *PGNW* and *PGTI* are compared with respect to the normalized values of power and performance related to basic *PG* program. Leakage power savings achieved by *PGNW* and *PGTI* are similar to that of *PG*. Peak power dissipation $\propto$ in-rush current $\propto$ number of overlapped wake-up (#overlapped_wake_up). For *PGNW* #overlapped_wake_up = 0 and for *PGTI* it is lesser than that of *PG*. Hence, peak power and in-rush current for *PGTI* are lesser than *PG* but higher than *PGNW*. Reduction of in-rush current and peak power dissipation experienced by (i) *PGNW* lies within 28-68% and 25-65%, respectively, and (ii) *PGTI* lies within 16-47% and 18-45%, respectively. This leads to reduction in total average power consumption for *PGNW* and *PGTI* which lies within 7-21% and 2-9%, respectively. The addition of *nop*s in *PGNW* and *PGTI* codes increase execution time. The loss in performance experienced by *PGNW* and *PGTI* lie within 5-35% and 2-18%, respectively.

V. CONCLUSION

The present work introduces two compilation techniques for reduction of in-rush current in *PG* sytems. The proposed method *PGNW* reduces in-rush current by eliminating overlapped wake-up at the cost increased delay and program size.

To address these issues *PGTI* has been introduced. *PGTI* allows ovelapped wake-up within the limitations of tolerable in-rush current. These methods are evaluated on standard benchmark programs. *PGNW* achieves higher reduction in in-rush current. *PGTI* is better in terms delay and increase in code size. The future work will investigate to reduce time and space overheads of *PGNW* and *PGTI* codes. Thus making them fit for real-time and safety-critical embedded systems.

REFERENCES

[1] S. Kim, S. V. Kosonocky and D. R. Knebel, "Understanding and minimizing ground bounce during mode transition of power gating structures," *Proc. of Int. Symp. on Low Power Electronics and Design (ISLPED '03)*, 27-27 Aug. 2003.

[2] K. Choi and J. Frenkil, "An Analysis Methodology for Dynamic Power Gating," Sequence Design Inc. https://pdfs.semanticscholar.org/09d3/db89fffc3dd250c6584b54a1925ba-fc1b27d.pdf

[3] A. Ball, "Integrated in-rush current limiter circuit and method," *US patent US20040090726 A1*, May 13, 2004.

[4] P. Royannez, H. Mair, F. Dahan and U. Ko, "90nm Low Leakage SoC Design Techniques for Wireless Applications", *Proc. of IEEE Int. Solid-State Circuits Conf.*, Feb. 2005, pp. 138 - 139.

[5] T. S. Kiong and U. C. Kong, "Power Gate Optimization Method for In-Rush Current and Power Up Time," *Intel Corporation*, https://dac.com/sites/default/files/App_Content/files/48/48_07U_1.pdf

[6] K. He, R. Luo and Y. Wang, "A power gating scheme for ground bounce reduction during mode transition," *Proc. of 25th Int. Conf. on Computer Design, ICCD 2007*, 7-10 Oct. 2007, Lake Tahoe, USA, pp. 388-394.

[7] K. Jeong, A. B. Kahng, S. Kang, T. S. Rosing and R. D. Strong, "MAPG: Memory access power gating," *Proc. of Design, Automation, Test & Exhibition in Europe Conf., DATE 2012*, Dresden, Germany, Mar. 12-16, 2012. pp. 1054-1059.

[8] S. H. Chen, Y. L. Lin and M. C. T. Chao, "Power-Up Sequence Control for MTCMOS Designs," *IEEE Transactions on Very Large Scale Integration (VLSI) Systems*, Vol. 21, No. 3, pp. 413-423 , Mar. 2013.

[9] A. Kaknevicius and A. Hoover, "Managing Inrush Current," *Application Report SLVA670A*, Texas Instruments, May 2015, www.ti.com/lit/an/slva670a/slva670a.pdf

[10] S. Kim, S. Paik, S. Kang and Y. Shin, "Wake-up scheduling and its buffered tree synthesis for power gating circuits," *Integration, the VLSI journal, Elsevier*, Vol. 53, pp. 157-170, 2016.

[11] Y. P. You, C. Lee and J. K. Lee, "Compiler Analysis and Supports for Leakage Power Reduction on Microprocessors," *Proc. of 15th Int. Conf. on Languages and Compilers for Parallel Computing (LCPC) 2002*, Springer LNCS, Vol. 2481, pp. 45-60, 2005.

[12] Y. P. You, C. Lee and J. K. Lee, "Compilers for Leakage Power Reduction," *ACM Transactions on Design automation of Electronic Systems (TODAES)*, Vol. 11, No. 1, pp. 147-164, 2006.

[13] S. Roy, S. Katkoori and N. Ranganathan, "A compiler-based leakage reduction technique by power-gating functional units in embedded microprocessors," *Proc. of 20th Int. Conf. on VLSI Design and 6th Int. Conf. on Embedded Systems*, pp. 215-220, 2007.

[14] S. Roy, S. Katkoori and N. Ranganathan, "A Framework for Power-Gating Functional Units in Embedded Microprocessors," *IEEE TVLSI*, Vol. 17, No. 11, pp. 1640-1649, 2009.

[15] Y. P. You, C. Lee and J. K. Lee, "Compilation for Compact Power-Gating Controls," *ACM TODAES*, Vol. 12, No. 4, Article No. 51, 2007.

[16] D. Park, J. Lee, N. S. Kim and T. Kim. "Optimal algorithm for profile-based power gating: a compiler technique for reducing leakage on execution units in microprocessors," *Proc. of Int. Conf. on Computer-Aided Design (ICCAD 2010)*, pp. 361-364, 2010.

[17] Y. Wang, J. Xu, Y. Xu, W. Liu and H. Yang, "Power Gating Aware Task Scheduling in MPSoC," *IEEE TVLSI*, Vol. 19, No. 10, pp. 1801-1812, October 2011.

[18] http://gem5.org/Main_Page

[19] http://infocenter.arm.com

[20] http://www.hpl.hp.com/research/mcpat/

[21] https://developer.arm.com/open-source/gnu-toolchain/gnu-rm

[22] http://vhosts.eecs.umich.edu/mibench//

[23] http://mathstat.slu.edu/˜fritts/mediabench/

Dynamic Thermal Management by using Task Migration in Conjunction with Frequency Scaling for Chip Multiprocessors

Alankar V. Umdekar, Arijit Nath
Dept. of CSE, IIT Guwahati
Assam, India, 781039
{maj.umedkar, arijitnath}@iitg.ernet.in

Shirshendu Das
Dept. of CSE, IIIT Guwahati
GNB Road, Ambari, Guwahati, 781001
shirshendu@iiitg.ac.in

Hemangee K. Kapoor
Dept of CSE, IIT Guwahati
Assam, India, 781039
hemangee@iitg.ernet.in

Abstract—**Reduction in transistor size has led to increase in transistor density allowing multiple cores to be placed on the same die; thus increasing the power dissipation of the chip. The heat dissipation by the cores may cause the chip temperatures to rise beyond a threshold limit causing it to malfunction. Dynamic Thermal Management (DTM) techniques address this issue by checking the rising temperature of the cores in the CMPs. Current techniques available for DTM use either Dynamic Frequency Scaling (DFS) or TM. The DFS based technique controls the chip temperature by reducing the cores operating frequency; however this may affect its performance. The technique of task rotation incurs the overheads of application migration.**

In this paper we propose a DTM technique that combines the benefits of both DFS and TM. The aim of the approach is to facilitate the efficient execution of high priority tasks. For this we allow other low priority tasks to sacrifice their performance to control the overall chip temperature. In particular, the high priority task is made to run at maximum allowable speed and, if needed, migrated to less heated cores while maintaining its performance. In this process, a low priority task runs at lower frequency to aid temperature reduction. Experimental analysis found that our approach reduces peak temperature of chip by 15.62°C with 0.24% performance (MIPS) degradation of high priority tasks compared to respective baseline.

I. INTRODUCTION

Chip Multiprocessors (CMP) are the main components for building today's multi-core systems [1]. In a CMP multiple processing and other components like cache memory and on-chip interconnects are placed on a single chip. As the technology advances the number cores and cache size in CMP increases. Also the transistor density in chip is increasing continuously [2]. Increasing transistor density and placing multiple cores in a single die increases the heat dissipation from the chip. Due to such heat dissipation the temperature of the chip could go beyond the maximum limit and may cause the chip to malfunction. Some components of the CMP may dissipate more heat and hence as a result may become the hot spots of the chip. The cost of cooling the hot spots is increasing in terms of both price and performance.

The cores are considered as the hottest components in the CMP, as all the computations are done in cores [3]. Reducing the temperature of cores is a major research area as it will also reduce the chip temperature. The amount of heat dissipated by each core is dependent on the factors like the frequency of the core and the task running on it [4]. A core running at high frequency dissipates more heat compared to a core running at a lower frequency. Additionally, if the executing task has more number of complex computations then it will keep all the functional units of the core busy, resulting in increased heat dissipation. The location of a core is also a factor of chip temperature. For example, in case of a 3D-integrated chip, placing a core in an intermediate layer may cause high temperature as the dissipated heat cannot ventilate properly. As the core temperature depends on frequency, task behavior and location, the cores have non-uniform thermal behavior and hence the temperature within the chip is not uniform.

To maintain the chip temperature under control without using the expensive cooling mechanism, various software and hardware based techniques can be used dynamically (i.e. at runtime). These techniques are termed as the Dynamic Thermal Management (DTM). DTM is a major research area for todays CMP [2]. Most of the work in DTM is done either in Dynamic Frequency Scaling (DFS) or Temperature Aware Task Migration (TM). In DFS, the temperature of the core (or other component) is controlled by scaling down the frequency. Lowering the frequency delays the execution of the task as the clock cycle time increases which results in less number of instructions getting executed in a particular time interval. Thus controlling chip temperature using DFS degrades the performance of CMP. Some DFS related techniques are [2], [3], [5], [6]. TM exploits the scenario when the cores have non-uniform thermal behavior, i.e. some core are hotter compared to others. For some critical tasks the core has to run in high frequency hence the core becomes hotter. TM migrates the task of such core to another cooler core. The frequency of the hotter core can now be scaled down to reduce its temperature. This re-mapping of tasks helps to balance the chip temperature. The main issue with TM is the swapping overhead; too many task swapping may degrade the performance. The swapping must be performed only when the difference between the temperatures of hot and cold core is reasonable [4].

In this work we propose a technique that combines the

advantages of both DFS and TM for better temperature reduction with lesser performance degradation. We call this technique as Dynamic <u>F</u>requency <u>S</u>caling with <u>T</u>ask <u>M</u>igration (FSTM). A better DTM technique should reduce chip temperature without degrading performance. In todays real time environment some tasks has to finish execution within the given deadline; such tasks are considered as high priority (HP) tasks by the operating system (OS). The tasks having less time restriction are called low priority task (LP). For a better DTM technique the priority of the task must be considered as an important factor. The HP task must be executed in high frequency as it has to complete within the given deadline. Thus it has a tendency to make the core hotter as scaling down frequency may not a be good choice for such tasks. The heating issue of cores running LP task can be managed by reducing their frequency. To manage the temperature of the core running an HP task, FSTM swaps the task with a core having an LP task. The hotter core gets an LP task to execute after swapping, hence the frequency of that core can be scaled down. This keeps the overall chip temperature under control. The swapping decision is done based on temperature of the core and is not done at the fixed intervals as in other existing methods [7]. This makes FSTM efficient in terms of number of task swapping. Experimental analysis found that FSTM reduces the peak temperature of chip by 15.62°C compared to baseline with just 0.24% performance degradation of HP task. The performance is calculated in terms of Million Instructions Per Second (MIPS). The proposed technique is also compared with DFS and TM [7]. The DFS based technique shows 28.8% degradation in performance to maintain similar peak temperature as FSTM. The number of tasks swapping compared to TM is reduced by 85% in FSTM.

The rest of the paper is organised as follows. Next section discusses the existing works in DTM. The proposed technique is discussed in section III. The experimental setup is explained in Section IV. Section V shows the result comparisons. Finally Section VI concludes the paper.

II. RELATED WORKS

A significant amount of work has already been performed in the area of dynamic thermal management and task scheduling for CMPs [8], [9] Since the cores runs different applications, the thermal behavior of the cores are not same which makes it difficult to manage the temperature of CMP. Dynamic allocation of task among the cores is also a major factor. The existing DTM techniques can be divided into two parts: DFS and TM. Some multi-core DTM techniques [10], [11] addresses the issue of power minimization but such schemes may not be optimal under thermal constraints. Many DFS based techniques are already proposed to reduced the chip temperature. Some of them reduces the energy consumption of a specific program during its execution [12], while other techniques consider the work load of cores to to dynamically scale the frequency [13]. The latter methods predict the impact of frequency transition, usually on performance, to decide on the frequency to apply. In a recent work [6] the author uses DFS technology to optimize performance, energy and temperature together. Other DTM based techniques proposed are [2], [3], [5]. One of the preliminary works to consider task migration is [7] which evaluates a simple technique of migrating hot tasks to cooler cores to improve performance. However in this paper we have combined the two approach to reduce their respective shortcomings.

III. PROPOSED TECHNIQUE FSTM

The main motive of this work is to optimize the chip temperature by combining the concept of DFS and TM. The proposed technique (FSTM) balances the performance loss due to DFS and TM. In this section we give a generalized description of our proposed FSTM. The technique can be applied to any CMP architecture having even number of cores. An on-chip network or bus can be used for communication among the components of the CMP. In the next section we experiment this technique for a particular CMP architecture.

A. *The Task Behavior:*

The tasks that run on CMP are either HP or LP tasks as mentioned in Section I. Categorizing the tasks as HP or LP is the responsibility of operating systems task scheduling. Note that, in this paper we assume that each core is running a separate task. An HP task needs more computation and high frequency so the functional units of the core consume more power and also dissipate more heat. Hence this type of task increases the core temperature very fast. In LP task the number of computation required is less or are not urgent and hence can be executed in low frequency. The main objective of this work is to reduce chip temperature without degrading the performance of HP tasks by allowing slight degradation (in MIPS) for LP tasks. FSTM is mainly proposed for the situations when the tasks executing on the CMP cores, are mixture of both HP and LP, i.e some cores are running HP tasks and some other running LP tasks. FSTM can easily handle the situation when all the tasks are LP. An extension of FSTM is also proposed when all the tasks are HP. At any time during execution the cores that run the HP tasks are called *HP-cores* and the cores executing LP tasks are called *LP-cores*. Note that a core may change from HP-core to LP-core or vice versa after task migration during execution.

B. FSTM

Consider a CMP having N number of homogeneous cores. Let $C = \{c_1, c_2, \ldots, c_N\}$ be the set of all cores in the CMP. The set of all tasks is defined as $T = \{t_1, t_2, \ldots, t_N\}$. We assume that the number of task is same as the number of cores such that each core can be assigned one task. Let $F = \{f_1, f_2, \ldots f_M\}$ be the set of all frequencies available in which the cores can operate. Here the total frequency available is M and $\forall i, f_i < f_{i+1}$, where $1 \leq i \leq M$. The task are assigned to the cores to execute. The function $\mathcal{H}:T \to C$ maps a task to the core where it is currently executing (*host-core*). The cores are grouped into set of pairs such that $P = \{p_1, p_2, \ldots, p_{N/2}\}$, where each element in P is a pair of cores. Each pair $p_i, 1 \leq$

$i \leq N/2$ has two unique cores $c_{2(i-1)}$ and $c_{2(i-1)+1}$. A pair in P is called a *core-pair*. The two cores in a core-pair are called the *peer-cores* of each other. The function $\mathscr{P}:C \to C$ maps a core to its peer-core. A task which is currently assigned to a core is allowed to migrate to its peer-core. That is, for any task t, the host-core $\mathscr{H}(t)$ can migrate the task to its peer-core $\mathscr{P}(\mathscr{H}(t))$. Function $\mathscr{F} : C \to F$ gives the current frequency of a core. Another function $\mathcal{T} : C \to T$ is used to get the currently executed task on a core.

For FSTM we assume that all the core-pairs are executing one HP task and one LP task. Other possibilities are discussed later. The technique of FSTM is implemented for each core-pair separately. The procedure of FSTM is performed on every core-pair after every fixed interval called the *epoch*. The selection of *epoch* is depend on some parameters which are discussed in Section III-E. An HP-core when run at the highest frequency gives the best performance but dissipate maximum heat. To reduce the temperature the core can be run at a lower frequency, however doing so may degrade the performance. As this is not acceptable for an HP task, instead of reducing the frequency of the core, migrating the task to its peer-core may allow it to execute at the same frequency.

Let the core-pair be $p(c_1, c_2)$ and the two assigned tasks are t_i and t_k; t_i is initially assigned to c_1 i.e. $\mathscr{H}(t_i) = c_1$ and t_k is assigned with c_2 i.e. $\mathscr{H}(t_j) = c_2$. Among these two tasks, t_i is considered as HP task and t_j as LP task. Hence initially c_1 must be HP-core and c_2 is LP-core. The core c_1 starts execution from an intermediate frequency (can be f_1) and keeps increasing the frequency after every *epoch* if the temperature is within threshold T_{max}. The main motive to start a HP-core from low frequency is to give high priority to temperature control. The LP-cores is expected to maintain lower temperature, hence even if it starts the task at higher frequency, it periodically reduces frequency to lower temperature. However this is limited by the performance degradation constraints (P_{max}) for the LP-task. In FSTM, when the temperature of HP-core reaches T_{max} the task can be swapped with LP-core. In a pure DFS based technique frequency must be decreased in such situation to control the temperature.

Algorithm 1 shows the detail process of FSTM. For each core c, T_{prv}^c and P_{prv}^c are assumed as global variables which stores the temperature and MIPS recorded for core c during the previous *epoch*. In real hardware, two counters will be required to implement them. After every *epoch* the function *FSTM(c)* is called for each core c. In this function line 9 to 15 are for HP-core and line 18 to 26 are for LP-core. In case of LP-core the frequency keeps decreasing (line 20) but if the MIPS degradation is going more than threshold P_{max} then the frequency of the core is increased to the next frequency available in F (line 24). The function *Swap-Tasks(c_1,c_2)* is used to swap the tasks of c_1 and c_2 which are peer-cores of each other. Note that after swapping (line 30) the frequency of the new HP-core (c_2) is set as the frequency of the old HP-core (c_1). But the frequency of the new-LP core (c_1) remains same as the frequency gradually decreases in LP-core.

Algorithm 1: FSTM Operations.

```
1  foreach epoch interval do
2        /* this executes after the end of every epoch.*/
3        foreach core c do
4              T_prv^c: temperature of c recorded in previous epoch.
5              P_prv^c: MIPS of c recorded in previous epoch.
6              Call FSTM(c)

7  Function FSTM(c)
8        /*The function executes for core c.*/
9        if c is HP-core then
10             T_new^c ← getCurrentTemparature(c)
11             if T_new^c > T_max then
12                   Call Swap-Tasks(c,𝒫(c));
13             else
14                   ℱ(c) = getNextFreq(c);
15             T_prv^c = T_new^c
16        else
17             /*When the core is LP-core.*/
18             P_new^c ← getCurrentMIPS(c)
19             if P_new^c − P_prv^c < P_max then
20                   ℱ(c) = getPrevFreq(c)
21                   flag=1
22             else
23                   if flag=1 then
24                         ℱ(c) = getNextFreq(c)
25                         flag=0;
26             P_prv^c = P_new^c

27 Function Swap-Tasks(c_1,c_2)
28       /*c_1 is HP-Core and c_2 is LP-core.*/
29       x = 𝒯(c_1), y = 𝒯(c_2)
30       ℋ(x) = c_2, ℋ(y) = c_1
31       /*host-core for both tasks are interchanged.*/
32       ℱ(c_2) = ℱ(c_1) /*c_2 now is the HP-core.*/

33 Function getNextFreq(c)
34       Returns next higher frequency in F after ℱ(c).
35       /* returns maximum frequency when current frequency is already
         maximum*/

36 Function getPrevFreq(c)
37       Returns the frequency in F just before ℱ(c).
38       /* returns minimum frequency when current frequency is already
         minimum*/
39
40 The function getCurrentTemparature(c) returns the current temperature
   of core c. getCurrentMIPS(c) returns the current MPIS of the core c.
   Sensors and counters are used to measure temperature and MIPS.
```

After the swapping c_2 becomes the new HP-core and c_1 becomes the new LP-core. But the current HP-core (c_2) will be less heated because of its pre-swap LP task execution. So it can be executed in a higher frequency. This keeps the temperature under control without any major performance degradation. The new LP-core (c_1) has high temperature because of its pre-swap HP task execution. But since LP task can be run at low frequency as per the algorithm the frequency of c_1 gradually goes down hence also brings down the temperature of that core. The task swapping are not done very often, it depends on the *epoch* time and the condition mention in line 11 of the algorithm. Hence the cores get enough time to settle after any task swapping. Too many swapping may not be beneficial as it will not allow to stabilize the cores. Section III-F explains how to select T_{max} and P_{max}.

C. Addition to FSTM

The FSTM discussed above assumes each core-pair running one HP task and one LP task. Situation may occur when both the tasks in a core-pair are either HP or LP tasks. When both tasks are LP there is very less chance of any temperature issue. But when both the tasks in a core-pair are HP tasks, swapping them may not be beneficial. To handle the issue we propose a technique which is slightly different than Algorithm 1. In this case the task swapping is performed only if the temperature difference between the two cores is more than a threshold T_{dif}; otherwise DFS is applied to reduce temperature. We call this technique as FSTM-hp.

D. Operations

To use FSTM the Algorithm 1 must be implemented in each core separately. The OS scheduler handles the entire mechanism. Some logic has to be implemented in each core to calculate the MIPS at a fixed interval of time. Also we need to maintain some counters for storing the previous temperature and MIPS. Since very less hardware is required for the work we assume the additional cost as negligible.

E. Task swapping and epoch time

The *epoch* time is considers as **10** milliseconds to support task migration. Task migration in FSTM is performed in terms of task swapping. In most of the recent operating systems, 10 ms is considered as the context switching time, hence the overheads associated with task swapping is overlapped with context switching of operating systems [7]. The cost of task swapping is reduced after combining with context switching and it results in less negative impact on the performance of the system. However to make the performance reduction even lesser, FSTM reduces the number of task swapping compared to other techniques like TM [7].

F. Calculating Threshold Values

Three threshold vales are used in FSTM: T_{max}, P_{max} and T_{dif} (cf. Algorithm 1). The purpose of these thresholds are already discussed. The value of these thresholds are hardware specific. The temperature thresholds are depend on the chip architecture. In the next section while using a particular CMP architecture (*Alpha 21264*) for experiment we use the values provided by the chip manufacturer. P_{max} can be calculated by some experimental analysis be set by the scheduler.

IV. EXPERIMENTAL SETUP AND PROCEDURE

Section III gives a generalized description of our proposed technique. In this section we experiment it with a specific CMP architecture. We choose **Alpha-21264** homogeneous quad core processor. The floor plan of the architecture we used is shown in Figure 1. According to the floor plan the L2 cache is in between the cores. For this particular processor $N = 4$, $C = \{c_1, c_2, c_3, c_4\}$, $F = \{660, 667, 700, 733\}$, all frequency values are in MHz. The two core-pairs are: $P = \{(c_1, c_2), (c_3, c_4)\}$. Four benchmarks from Parsec benchmark suite [14] are used for the experiment. We consider

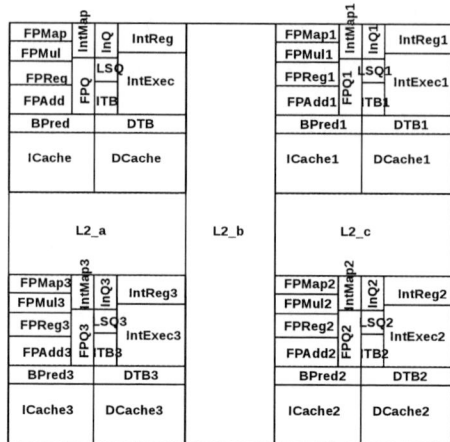

Fig. 1: Floor plan for **Alpha-21264** processor.

TABLE I: System Parameters.

Component	Parameters
No. of cores	4
Block size	64 bytes
L1 cache (Instruction + data)	64KB
Total LLC (L2) size	8MB
Memory bank	1GB, 4KB/page

body and *ferret* as HP task and *swap* and *freq* as LP task. Here $T = \{ferret, swap, body, freq\}$; $t_1 =$*ferret*, $t_2 =$*swap*, $t_3 =$*body* and $t_4 =$*freq*. The first core-pair is assigned with *ferret* and *swap* while the second core pair is assigned with *body* and *freq*. Hence $\mathcal{H}(t_1) = c_1$, $\mathcal{H}(t_2) = c_2$, $\mathcal{H}(t_3) = c_3$ and $\mathcal{H}(t_4) = c_4$. The maximum or the threshold temperature T_{max} of each core was specified from the data sheet of Alpha-21264 processor, it was known that the operating temperature of the processor at 600 MHz is 70.3°C. Accordingly, our experiments were carried out with the value of T_{max} as 60°C and 65°C. The acceptable performance degradation limit P_{max} was kept at different values for the experiments. The final value is chosen to be equal to 10%.

A core has many Functional Unit Blocks (FUB). We calculate the power trace of each FUB separately. We used full system simulator Gem5 [15] to simulate the CMP environment. Table I shows the system parameters detail. Gem5 gives the required number of accesses to each Functional Unit Block (FUB) in the cores after every *epoch*. These access traces are then fed into McPAT [16] to obtain the power trace file. The obtained power values were fed into HotSpot [17] along with the floor plan of the processor to obtain the temperature values of each core as well as the chip temperature. Figure 2 shows the setup of our simulation process.

The proposed technique is compared with three baseline architectures:

- Baseline (BS): In this architecture no DFS or task migration are applied on the cores. The cores always run at the highest frequency and hence gives the best performance but high chip temperature may be an issue.

978-1-5386-3693-0/18 $31.00 © 2018 IEEE

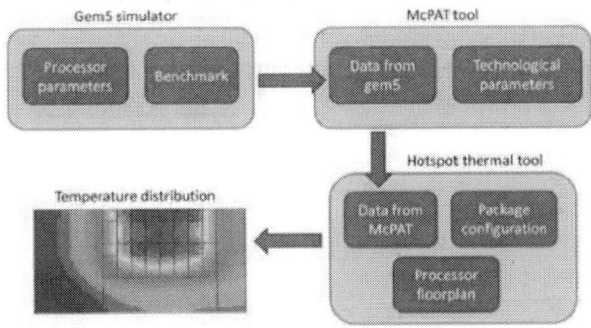

Fig. 2: Simulation setup.

Metric	$T_{max} = 60°C$			$T_{max} = 65°C$		
	BS	DS	FSTM	BS	DS	FSTM
Perf_Deg	0%	28.8%	0.24%	0%	23.77%	0.24%
Peak_Temp (°C)	88.34	74.21	72.72*	88.34	74.56	72.84*
Temp_Reduction over BS in °C		13.13	15.62		13.78	15.5

TABLE II: Comparison of FSTM with BS and DS. *Only for one epoch our algorithm brought down the temperature immediately.

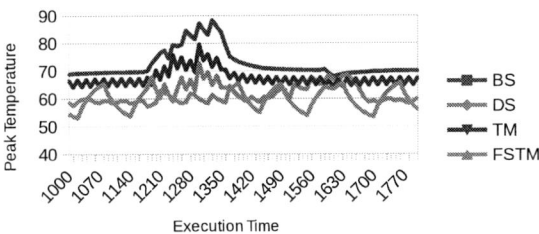

Fig. 3: Peak temperature at different execution time.

- DS: It is same as BS but DFS is applied on each core. Once the temperature of a core goes beyond threshold the frequency must be scaled down.
- Task Migration (TM) : In TM [7] the tasks are rotated after every *epoch*. No frequency scaling done here.

Our target is to reduce the chip temperature with minimum performance degradation of HP tasks.

V. EXPERIMENTAL ANALYSIS

Different experiments has been carried out for analysing the behavior of FSTM. The parameters consider for analysis are Peak Temperature, performance (MIPS) and number of task swapping. Peak temperature is considered as the highest temperature of the chip. We have done experiments for two T_{max} values: $T_{max} = 60°C$ and $T_{max} = 65°C$. The performance degradation limit is considered as P_{max} =10%. As mentioned above BS gives the best performance as it always runs at highest frequency. All the performance degradation mention in this paper are with respect to BS.

Table II compares FSTM with BS and DS for both $T_{max} = 60°C$ and $T_{max} = 65°C$. The peak temperature of DS is 74.21°C which is 13.13°C less than BS but it degrades the performance by 28.8%. This is because DS rigorously scales down the frequency to reduce temperature. FSTM has peak temperature of 72.72°C which is almost same as DS (15.62°C less than BS) but the performance degradation (of HP task) is just 0.24%. Note that peck temp is only for one epoch and it is immediately brought down by our algorithm. FSTM sacrifices the performance of LP task by 11% which does not create any issue as LP task can be delayed in execution. Both DS and FSTM has similar reduction in peak temperature but there is huge difference between performance degradation. For $T_{max} = 65°C$ the results are almost same for FSTM. In case of DS the performance degradation is less (23%) because the cores are allowed to run at high frequency as the T_{max} is increased. The 0.24% performance degradation in FSTM for both $T_{max} = 60°C$ and $T_{max} = 65°C$ is mainly because of task swapping.

FSTM is also compared with TM. The main advantage of FSTM over TM is that it restricts task swapping. Also TM executes each core at high frequency as no frequency scaling

mechanism is applied. The LP task may not required such high frequency but executing it in high frequency increases the temperature. Table III compares FSTM and TM over peak temperature and task swapping. It can be observed that for $T_{max} = 60°C$ the peak temperature of TM is 4.81°C more than FSTM. Compared to BS, both TM and FSTM reduces the peak temperature by 10.81°C and 15.62°C respectively. The controlled task swapping in FSTM reduces the number of swaps by 28% over TM. Though the task swapping is overlapped with the context switching of OS the cost of such swapping is not negligible. For $T_{max} = 65°C$ the number of task swapping required for FSTM is even reduced. Hence FSTM get 80% reduction in task swapping. FSTM does swapping only when it can reduce temperature without degrading performance. It can be observed that even after fixing T_{max} as 60°C or 65°C the peak temperature is going upto 72°C in FSTM. This may occur during only one *epoch* as the migration or frequency scaling can only possible at the end of every *epoch*.

Figure 3 shows the peak temperature of BS, DS,TM and FSTM during different execution time. The graph shows the values form 1000th to 1800th *epoch* for all the techniques. The peak temperature of every *epoch* is considered as the hottest component of the chip at that moment. It can be observed that BS and TM always have high peak temperature. DS and FSTM have almost similar patterns for peak temperature but DS shows high performance degradation as shown in table II.

A. When all cores are HP-Core

The proposed technique is also analysed for the situation when both the tasks in a core-pair are HP task. The technique is called FSTM-hp (cf. Section III-C). Table IV shows how the behavior of the proposed DTM technique changes according to the task behavior. The performance degradation is same for FSTM and FSTM-hp (0.24%) but FSTM-hp has higher peak temperature. This is because core frequency cannot be

978-1-5386-3693-0/18 $31.00 © 2018 IEEE

	$T_{max} = 60°C$			$T_{max} = 65°C$		
Metric	**TM**	**FSTM**	**Improvement in FSTM**	**TM**	**FSTM**	**Improvement in FSTM**
Peak_Temp (in °C)	77.53	72.72[*]	4.81	77.53	72.84[*]	4.69
No. of. swaps	520	374	28%	520	101	80%

TABLE III: Comparison of FSTM with TM. *Only for one epoch our algorithm brought down the temperature immediately.

	$T_{max} = 60°C$		$T_{max} = 65°C$	
Metric	**FSTM-hp**	**FSTM**	**FSTM-hp**	**FSTM**
Perf_Deg	0.24%	0.24%	0.24%	0.24%
Peak_Temp (in °C)	77.61	72.72	77.94	72.84
No. of swaps	110	374	80	101

TABLE IV: The effect of FSTM-hp.

scaled down more while running HP-task. Due to such reason FSTM-hp shows the same performance degradation as FSTM but increases the peak temperature. As both the cores in a core-pair are running HP tasks, the chances of getting a significant temperature difference (T_{dif}) between the two cores is less; which reduces the number of swaps in FSTM-hp. In case of $T_{max} = 60°C$ the FSTM-hp has only 110 task swapping compared to 374 swapping in FSTM.

VI. CONCLUSION

As the transistor size reduces the transistor density is increasing which led multiple cores to be placed in a single chip. Such processor architectures are called Chip-multiprocessor (CMP). The heat dissipation by the cores may cause the chip temperatures to rise beyond a threshold limit causing it to malfunction. The issue of temperature rise in chips can be controlled by a mechanism called Dynamic Thermal Management (DTM). Current DTM techniques applied to CMPs are either Dynamic Frequency Scaling (DFS) or Task Migration (TM) based. The DFS based technique controls the chip temperature by reducing the cores operating frequency; however this may affect its performance. The technique of task migration incurs the overheads of task migration.

In this paper we proposed a DTM technique called FSTM, to reduce the chip temperature of CMPs with minimum performance degradation. In FSTM, we allow some performance degradation of low priority tasks to reduce the overall chip temperature and maintain the performance of high priority task. In particular, the high priority task is made to run at maximum allowable frequency and, if needed, migrated to less heated cores while maintaining its performance. In this process, a low priority task runs at lower frequency to aid temperature reduction. Experimental analysis found that FSTM reduces the peak temperature of chip by 15.62°C compared to baseline with just 0.24% performance degradation of HP task. The proposed technique is also compared with DFS and TM [7]. The DFS based technique shows 28.8% degradation in performance to maintain similar peak temperature as FSTM.

REFERENCES

[1] R. Balasubramonian, N. P. Jouppi, and N. Muralimanohar, *Multi-Core Cache Hierarchies*. Morgan and Claypool Publishers, 2011.

[2] D. Brooks and M. Martonosi, "Dynamic Thermal Management for High-performance Microprocessors," in *High-Performance Computer Architecture, 2001. HPCA. The Seventh International Symposium on*. IEEE, 2001, pp. 171–182.

[3] V. Hanumaiah and S. Vrudhula, "Temperature-aware DVFS for Hard Real-time Applications on Multicore Processors," *IEEE Transactions on Computers*, vol. 61, no. 10, pp. 1484–1494, 2012.

[4] P. Zajc, M. Szermer, M. Janicki, C. Maj, P. Pietrzak, and A. Napieralski, "Analysis of the Effectiveness of Core Swapping in Modern Multicore Processors," in *19th International Workshop on Thermal Investigations of ICs and Systems (THERMINIC)*, Sept 2013, pp. 385–388.

[5] V. Hanumaiah and S. Vrudhula, "Energy-efficient Operation of Multicore Processors by DVFS, Task Migration, and Active Cooling," *IEEE Transactions on Computers*, vol. 63, no. 2, pp. 349–360, 2014.

[6] H. F. Sheikh, I. Ahmad, and D. Fan, "An Evolutionary Technique for Performance-Energy-Temperature Optimized Scheduling of Parallel Tasks on Multi-Core Processors," *IEEE Transactions on Parallel and Distributed Systems*, vol. 27, no. 3, pp. 668–681, March 2016.

[7] D. Zhao, H. Homayoun, and A. V. Veidenbaum, "Temperature Aware Thread Migration in 3D Architecture with Stacked DRAM," in *Quality Electronic Design (ISQED), 2013 14th International Symposium on*. IEEE, 2013, pp. 80–87.

[8] H. Sasaki, A. Buyuktosunoglu, A. Vega, and P. Bose, "Mitigating Power Contention: A Scheduling Based Approach," *IEEE Computer Architecture Letters*, vol. 16, no. 1, pp. 60–63, Jan 2017.

[9] M. Chen, X. Zhang, H. Gu, t. wei, and Q. Zhu, "Sustainability-Oriented Evaluation and Optimization for MPSoC Task Allocation and Scheduling Under Thermal and Energy Variations," *IEEE Transactions on Sustainable Computing*, vol. PP, no. 99, pp. 1–1, 2017.

[10] T. Chantem, X. S. Hu, and R. P. Dick, "Online Work Maximization Under a Peak Temperature Constraint," in *Proceedings of the 2009 ACM/IEEE International Symposium on Low Power Electronics and Design*, ser. ISLPED '09, 2009, pp. 105–110.

[11] J. Li and J. F. Martínez, "Power-performance Considerations of Parallel Computing on Chip Multiprocessors," *ACM Trans. Archit. Code Optim.*, vol. 2, no. 4, pp. 397–422, Dec. 2005.

[12] V. W. Freeh and D. K. Lowenthal, "Using Multiple Energy Gears in MPI Programs on a Power-scalable Cluster," in *Proceedings of the Tenth ACM SIGPLAN Symposium on Principles and Practice of Parallel Programming*, ser. PPoPP '05, 2005, pp. 164–173.

[13] J.-P. Halimi, B. Pradelle, A. Guermouche, N. Triquenaux, A. Laurent, J. C. Beyler, and W. Jalby, "Reactive DVFS Control for Multicore Processors," in *Green Computing and Communications (GreenCom), 2013 IEEE and Internet of Things (iThings/CPSCom), IEEE International Conference on and IEEE Cyber, Physical and Social Computing*. IEEE, 2013, pp. 102–109.

[14] C. Bienia, S. Kumar, J. P. Singh, and K. Li, "The PARSEC Benchmark Suite: Characterization and Architectural Implications," in *Proceedings of the 17th International Conference on Parallel Architectures and Compilation Techniques*, ser. PACT '08, 2008, pp. 72–81.

[15] N. Binkert, B. Beckmann, G. Black, S. K. Reinhardt, A. Saidi, A. Basu, J. Hestness, D. R. Hower, T. Krishna, S. Sardashti et al., "The Gem5 Simulator," *ACM SIGARCH Computer Architecture News*, vol. 39, no. 2, pp. 1–7, 2011.

[16] S. Li, J. H. Ahn, R. D. Strong, J. B. Brockman, D. M. Tullsen, and N. P. Jouppi, "McPAT: an Integrated Power, Area, and Timing Modeling Framework for Multicore and Manycore Architectures," in *Microarchitecture, 2009. MICRO-42. 42nd Annual IEEE/ACM International Symposium on*. IEEE, 2009, pp. 469–480.

[17] R. Zhang, M. R. Stan, and K. Skadron, "HotSpot 6.0: Validation, Acceleration and Extension," *University of Virginia, Tech. Report CS-2015-04*, 2015.

978-1-5386-3693-0/18 $31.00 © 2018 IEEE

Formal Methods for Coverage Analysis of Power Management Logic with Mixed-Signal Components

Sudipa Mandal*, Aritra Hazra*, Pallab Dasgupta*, Rama Mohan Chunduri.[†]

*Dept. of Computer Science and Engg., IIT Kharagpur, West Bengal, India-721302.
[†] Intel Technology India Pvt. Ltd., Bangalore, India-560066.
Email: *{sudipa.mandal, aritrah, pallab}@cse.iitkgp.ernet.in, [†]chunduri.r.mohan@intel.com.

Abstract—Due to the increasing complexity in the power management logic of low-power designs, formal validation of the architectural power intent, comprising of both digital and analog power management features, is becoming a crucial task. Consequently, the formal verification frontier has also been extended, in recent times, to ensure the correctness for analog as well as digital power intent. The quality of verification can be evaluated by formal coverage analysis which can be determined from the reachability of safe global power states by the power manager. This article proposes a novel formal method for computing the coverage of architectural power states for power management logic having analog components like LDOs and PLLs. The efficacy of the proposed method has been shown using an industry level case-study.

I. INTRODUCTION

The rapid growth in the number of functionalities in battery-operated low-power devices has made power consumption a primary concern for its designers. In order to cope up with the stringent power budget, the Power Management Logic (PML) of these designs are implemented with various sophisticated power reduction features, such as clock and power gating [1], [2], dynamic voltage and frequency scaling [3], [4], operand isolation [5] etc. Power-managed circuits are typically partitioned into several power domains, each of which may function in multiple power states having different voltage/frequency levels. The PML directs each of the power domains by providing necessary low-level signals for performing transitions between the power states. The PML comprises of a Global Power Manager (GPM) which coordinates with a collection of Local Power Managers (LPMs) to orchestrate the power states of individual power domains [6], [7].

Several research works [5], [8] have been carried out to make the power usage more efficient for low-power designs. Consequently, formal validation of the PML with respect to a specified high-level architectural power management strategy has also been carried out in the past to verify the proper sequencing among the power state transitions [6], [7], [9], [10], [11]. However, all these earlier works only consider the power management and power partitions of digital designs; whereas the analog components like LDOs, PLLs (Phase Locked Loops), Line Drivers etc. were external to the design in the past. Now-a-days, the power management fabric has become self-contained by incorporating the analog components which supply power and clock to the other domains. Power domains containing such analog components are called *Analog Mixed-Signal (AMS) power domains* and just as other power domains,

The authors thank Intel for partial support of this research.

components like the LDOs, PLLs, Line Drivers etc. in AMS power domains can be switched off when not in use thereby saving more power. In Fig. 1, the Power Management Unit (PMU) of D-PHY Tx Master implementing the PML consists of a GPM and 11 LPMs. LPMs directing AMS power domains are shown in blue and LPMs directing digital power domains are shown in pink. LPM-LDOM0 directs the AMS power domain LDOM0 containing an analog component LDO (Low DropOut regulator). The LDO supplies power (1.2V or 1.5V) to DLANE0 through the power rails of DLANE0. The state registers change the states of the analog component i.e. the LDO in LDOM0 resulting in variable voltage outputs by the LDO. LPM-DLANE0 manages a digital power domain DLANE0 which is a digital block.

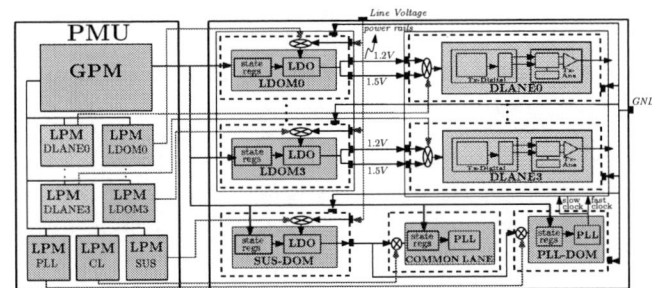

Fig. 1: Power Management Unit for *DP-PHY*

The onus of GPM is to confirm that the design always stays in safe combinations of power states. For example, in this design with several LDO power domains, all the LDOs should not be powered up together. Rather, they must be powered up in a staggered manner. This staggered rise of LDO domains are maintained by the GPM. In addition to this, the digital counterpart of these power-managed design may also have certain restrictions to follow, such as, non-occurrence of high/active state combinations for a particular set of power domains[1]. Therefore, the PMU, now-a-days, controls not only digital parts of a design but also the analog components of the AMS power domains.

Coverage analysis is an important aspect to ascertain the quality and completeness of validation. In the context of GPM, the notion of coverage analysis is as follows. Each power domain of a design consists of multiple local power states/modes controllable by their corresponding LPM. The

---
[1]For example, consider a cellphone activity where the audio-video play can not happen simultaneously with the phone-call activity; hence these two power domains can not remain in high voltage states together.

global power states are the possible combinations of all the local power states of the power domains. Among these, some combinations of the local power states are *Safe* (adhering to the given power requirements) and the rest are *Unsafe* (i.e., being in such states might cause unwanted scenarios like power glitch, power droop leading to erroneous output etc.). In the manual implementation of the global power management logic in GPM, some global power states will be *Reachable* during the state transitions, while the others will remain *Unreachable*. The purpose of formal verification is to establish that the GPM does not reach an *Unsafe* state. The purpose of coverage analysis is to determine which power states, specifically the *Safe* ones, are *Reachable*, so that all safe states supposed to be reachable are proven to be reachable.

Previous work [9] presents a coverage analysis of architectural power states of digital designs. With the introduction of AMS power domains in the design, revision in the existing methodology of formal verification and coverage analysis for the AMS power managers [12] is needed. The coverage analysis will be different from that presented in [9] because the local power states for an AMS power domain requires new annotations. The main contributions of this work are as follows:

- A methodology to define global power states for designs having AMS power domains.
- A format for expressing the behavioral aspects of analog power management components in terms of low level signals with the help of designer knowledge.
- A prototype tool-flow for formally ascertaining the coverage of power states.
- Proof-of-concept of our proposed methodology on an industrial case-study.

In our work, we are dealing with designs which contain both digital and AMS power domains and our objective is to perform the coverage analysis of the PML of such designs.

II. METHODOLOGY OVERVIEW

We first introduce the formal model of global power states for designs containing AMS power domains and then present the overview of the proposed methodology.

A. The Formal Model of Global Power States:

Auxiliary Power State Machines (APM) are used to represent the power states of a domain and the possible transitions among the states. Let us first define an APM, $\mathcal{M}$, as a tuple, $\mathcal{M} = \langle \mathcal{S}, \mathcal{O}, \mathcal{T}, s, \mathcal{F} \rangle$, where:

- $\mathcal{S}$ is a finite set of power states of $\mathcal{M}$.
- $\mathcal{O}$ is a finite set of outputs used in $\mathcal{M}$.
- $\mathcal{T} : \mathcal{S} \times \mathcal{P}(\mathcal{O}) \to \mathcal{S}$ is the transition function for $\mathcal{M}$, where $\mathcal{P}(\mathcal{O})$ is a bounded SystemVerilog Assertion (SVA) [13] sequence over $\mathcal{O}$ with Boolean and bounded delay operators.
- s is the start state of $\mathcal{M}$.
- $\mathcal{F} : \mathcal{S} \to 2^{\mathcal{O}}$ is a labeling of power states with control sequences true in that state.

Say, a design comprises of N power domains, $\mathcal{D}_1, \mathcal{D}_2, \ldots, \mathcal{D}_N$, where each $\mathcal{D}_i$ may be either a digital domain having digital components or an AMS power domain which contains analog components such as LDO, PLL etc. The APM $\mathcal{M}_i$, for domain $\mathcal{D}_i$, is extracted from UPF specifications and is represented as follows:

$$\mathcal{M}_{\mathcal{D}i} = \langle \mathcal{S}_{\mathcal{D}i}, \mathcal{O}_{\mathcal{D}i}, \mathcal{T}_{\mathcal{D}i}, s_{\mathcal{D}i}, \mathcal{F}_{\mathcal{D}i} \rangle$$

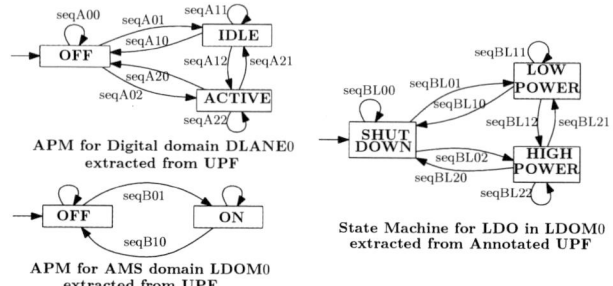

APM for Digital domain DLANE0 extracted from UPF

APM for AMS domain LDOM0 extracted from UPF

State Machine for LDO in LDOM0 extracted from Annotated UPF

Fig. 2: Auxiliary Power State Machines for Digital and Analog Power Domains

Typically, each AMS power domain contains one analog component in the design and this may supply voltage/power (for domains containing LDO) or clock with suitable frequency (for domain containing PLL) to other operational digital domains. An analog component in an AMS power domain may remain in a particular APM state, but operate in various behavioral states depending on its functional supply needs[2]. Moreover, general UPF specifications for a power domain is restrictive in capturing these behavioral states of AMS power domains and hence these are extracted from the extended annotated UPF. The behavioral state machine, $\mathcal{M}_{\mathcal{B}i}$, for the analog component present in $\mathcal{D}_i$, is extracted from the annotated UPF specifications and is represented as follows:

$$\mathcal{M}_{\mathcal{B}i} = \langle \mathcal{S}_{\mathcal{B}i}, \mathcal{O}_{\mathcal{B}i}, \mathcal{T}_{\mathcal{B}i}, s_{\mathcal{B}i}, \mathcal{F}_{\mathcal{B}i} \rangle$$

The APMs for domain LDOM0, DLANE0 and LDO in LDOM0 are shown in Fig. 2. For LDOM0, $\mathcal{S}_{\mathcal{D}} = \{ON, OFF\}$, $\mathcal{O}_{\mathcal{D}} = \{$pwr_good, in_preq, in_clk$\}$, $\mathcal{T}_{\mathcal{D}} = \{$seqB01, seqB10$\}$ (which are SVA sequences made with $\mathcal{O}_{\mathcal{D}}$) and $s_{\mathcal{D}} =$OFF.

The overall set of power states, $\mathcal{S}_i$, for any power domain, $\mathcal{D}_i$ ($i \in [1, N]$), is denoted as,

$$\mathcal{S}_i = \begin{cases} (\mathcal{S}_{\mathcal{D}i} \times \mathcal{S}_{\mathcal{B}i}), & \text{if } \mathcal{D}_i \text{ is an AMS power domain} \\ \mathcal{S}_{\mathcal{D}i}, & \text{if } \mathcal{D}_i \text{ is a digital domain} \end{cases}$$

Now, we denote the set of possible global power states as $\mathcal{S}_{\mathcal{G}}$, where: $\mathcal{S}_{\mathcal{G}} = (\mathcal{S}_1 \times \mathcal{S}_2 \times \cdots \times \mathcal{S}_N)$. As per the global power management logic, we can partition $\mathcal{S}_{\mathcal{G}}$ into *legal* ($\mathcal{S}_{\mathcal{G}}^L$) and *illegal* ($\mathcal{S}_{\mathcal{G}}^I$) set of global power states, i.e. $\mathcal{S}_{\mathcal{G}} = \mathcal{S}_{\mathcal{G}}^L \cup \mathcal{S}_{\mathcal{G}}^I$. In the next subsection, we present a summary of our proposed approach in analyzing the coverage of these legal and illegal global power states leveraging the state reachability procedure.

---

[2]For example, an AMS power domain having an LDO can be in ON state as per the APM, but it may supply V_1 voltage to domain A_1 and V_2 voltage to domain A_2 through the connected power rails.

B. Proposed Approach for AMS Coverage Analysis:

Let us outline our approach with an example. The design, *D-PHY*, as shown in Fig. 1, has a digital power domain DLANE0 and an AMS power domain LDOM0. The local power states of DLANE0 is determined by the input voltage (1.2V and 1.5V provided by the power rails which are the outputs of AMS power domain LDOM0) and clock frequency (f_1 and f_2) combinations and they can work in DVFS condition. The UPF specification of DLANE0 provides information about its local power states ($\mathcal{S}_{\mathcal{D}_A}$), namely OFF, ACTIVE (with frequency f_1 and voltage 1.5V) and IDLE (with frequency f_2 and voltage 1.2V). The construction of APM $\mathcal{M}_{\mathcal{D}_A}$ for DLANE0, as shown in Fig. 2, is as described in [9]. The extracted control sequence for driving DLANE0 from state ACTIVE to state OFF is as follows:

```
sequence seqA20;
    $rose(ret) ##[1:3] $rose(iso) ##[1:2] $fell(pwr);
endsequence
```

Here, the signal `ret` enables the retention, the signal `iso` enables the isolation and the signal `pwr` controls the power gate of the domain.

LDOM0 is an AMS power domain containing an LDO which supplies voltages 1.2V or 1.5V to DLANE0. Similar to the digital domains, the local power states of domain LDOM0 ($\mathcal{S}_{\mathcal{D}_B}$) are also dependent on its input voltage and frequency combination leading to two power states- ON and OFF i.e. it has single voltage (LineVoltage) and frequency as input as shown in Fig. 1. The behavioral states ($\mathcal{S}_{\mathcal{B}_B}$) of the LDO in LDOM0, maintained by the PML via state registers, are SHUT-DOWN ($0V$), LOW-POWER ($1.2V$) and HIGH-POWER ($1.5V$). For a design with only LDOM0 and DLANE0 domains,

$$\{< \texttt{ON, ACTIVE} >, < \texttt{ON, IDLE} >, < \texttt{ON, OFF} >\} \subset \mathcal{S}_{\mathcal{G}}^{\texttt{L}}$$

But an *Illegal* global state would occur if LDOM0 is in ON state, the LDO in LDOM0 is in HIGH-POWER state supplying high voltage $1.5V$ but DLANE0 is in IDLE state. Hence, $< (\texttt{ON, HIGHPOWER}), \texttt{IDLE} > \in \mathcal{S}_{\mathcal{G}}^{\texttt{I}}$.

Therefore, the legality of global states of designs with AMS power domains can not be captured only by the local power states of the domains but also by the behavioral states of the analog components. Our present work performs the coverage analysis for global power states of a power management logic. The automatic extraction of assertions to validate the PML is described in [14]. The coverage analysis of the same is not straight forward due to the fact that the power states and intermediate transition sequences of local power managers are to be extracted automatically from the UPF specifications. A preliminary methodology for extracting the mentioned sequences is described in [6], [7] which we have extended for AMS power domains.

Thus the global state space for such designs will be determined by the cross product of the APMs of digital domains, APMs of AMS power domains and Behavioral State Machines of analog components in AMS power domains. Capturing the behavioral states of analog components such as LDOs and PLLs along with the power states of AMS power domains can be done in the following way:

1) In [14], the authors introduced the notion of UPF annotations for analog domains. The behavioral states of analog components in AMS power domains can be specified in this annotated UPFs. Thus, these annotated UPFs can be used to extract the states of the behavioral state machine of analog components, whereas, the normal UPF provides the APM of the AMS power domain as shown in Fig. 2.

2) The control sequences can be gleaned out from the behavioral knowledge for an analog component provided by the designer. We shall explain this step with an example in the next section.

Now in our approach, for LDOM0, we get the annotated UPF as explained in [14] with the `contains_analog` construct which describes the power states of the analog component as described next. From this UPF construct, we can find out that the LDO in LDOM0 has three states, namely, SHUT-DOWN, LOW-POWER and HIGH-POWER. Hence, the state machine for the LDO in LDOM0 can be constructed with these three states as shown in Fig. 2.

```
contains_analog B_LDO
    -domain B
    -type LDO
    -input_supply_port { inhigh_B VDD_switch_B }
    -output_supply_port { out_B VDD_B_out }
    -mode_register_port { mode_BP mode_B }
    -control_port { Bcp B_pwr_off }
    -state_ana { HIGH-POWER 1.5 { mode_BP && Bcp }}
    -state_ana { LOW-POWER 1.2 { !mode_BP && Bcp }}
    -state_ana { SHUT-DOWN 0.0 { !Bcp } }
    ... ... ...
```

The transitions (for example, seqBL01, seqBL12 etc.) between the states of APMs happen by specific sequences of signals asserted by the PML and the value of the state registers. Though few of these low level signals can be gleaned out from the UPF specifications of the respective analog domains, these signals do not follow standard steps similar to the digital domains. But every analog component in the analog domain follow some generic template to power up or power down. For example, an LDO in an LDO-domain, while waking up always checks whether the input power and clock to the domain are up. Only after that, it sends enable signal to raise the voltage. These signals are necessary for waking up an LDO domain, no matter which type of LDO resides inside the domain. We plan to extract these low level signals and construct the corresponding control sequences from a pre-defined format containing low level analog signals provided by the designer. The control sequence to drive the LDO of LDOM0 from SHUT-DOWN state to LOW-POWER state is as follows:

```
sequence seqBL01;
    pwr_good ##[1:6] in_clk ##[1:2](ldo_en&& !state);
endsequence
```

The low-level signals such as `pwr_good` for input power, `in_clk` for input clock, `ldo_en` for enabling the LDO, `state` for mode selection will be extracted from our analog signal template.

These auxiliary state machines, derived from the normal and annotated UPF specifications as well as analog domain knowledge, only capture the local power states and the transitions among them. The APMs do not represent the knowledge of the global power states. For example, if DLANE0 is driven by LDOM0, DLANE0 can not be in OFF state while LDO in LDOM0 is in HIGH-POWER state. This requirement of not reaching this global power state, in this case, is specified in the following way:

```
property reach_A_OFF_B_HIGHPOWER;
@(posedge clk)  disable iff (rst)
   !((stateA == 'OFF) && (stateB == 'ON)
                     && (state_BL == 'HIGH-POWER));
endproperty
```

The pass of this property, in this case, means that this global power state is Unreachable and failure of this property means that the state is Reachable.

In this paper, we have constructed a tool which performs the coverage analysis of the PML of a design containing AMS power domains. The main idea of the tool flow is as follows:

- A top module is created for the design with the local power states (digital domain states are extracted from UPF and AMS power domain states are extracted from annotated UPF) as inputs to the module. For example, to represent the states of the LDO in LDOM0 (refer to Fig. 2), we may use a 2-bit state variable, stateB, with SHUT-DOWN, LOW-POWER and HIGH-POWER states represented by $2'b00$, $2'b01$ and $2'b10$, respectively. The PML i.e. the GPM and the LPMs are also instantiated inside the top module.
- We convert each state transition of the APMs into *assume properties* and add these properties in the top module. For example, the following assume property is included in the top module for the LDO in LDOM0 for the state transition from LOW-POWER to HIGH-POWER with control sequence seqBL12.

```
property assume_12;
   @(posedge clk)
   (stateBL == 2'b01) ##0 seqBL12 |->
                        ##1 (stateBL == 2'b10);
endproperty
prop_assume_12 : assume property(assume_12);
```

These assume properties help in labeling the global power states by monitoring the output power states of the PML and feeding them back to the module as inputs.

- Finally, with the help of a set of reachability assertion properties, the coverage analysis of the global power states is performed using a model checker [15].

III. THE WORK-FLOW OF THE TOOL

In this section, we present the work-flow of the tool for the coverage and reachability analysis of the architectural power intent of designs with digital and AMS power domains. The steps are described below:

(A) *Extracting APMs from Annotated UPF Specification:*
The steps for extracting APMs from normal and annotated UPFs are as follows:

(i) *Sequence Extraction for Digital domains –*
Let us consider the digital power domain DLANE0 and AMS power domain LDOM0 from Fig. 2. The UPF specification of DLANE0 contains all the operational local power state information (namely, OFF, ACTIVE and IDLE) along with the low level control signals (namely, ret, iso, pwr etc.) responsible for the state

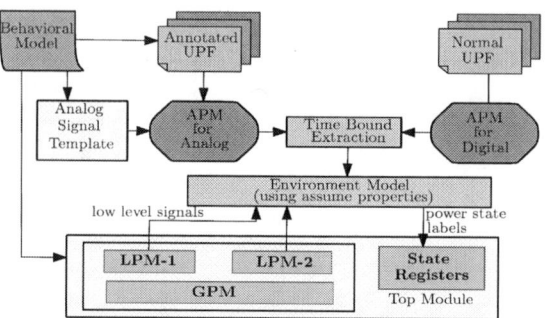

Fig. 3: Work-flow of the Tool

transitions. With the knowledge of the low power design of digital domains, the control sequences can easily be extracted. For example, the control sequence seqA20 for DLANE0 can be extracted directly from the UPF as follows:

```
sequence seqA20;
$rose(ret)##[1:3]$rose(iso)##[1:2]$fell(pwr);
endsequence
```

(ii) *Sequence Extraction for AMS power domains –*
The UPF of AMS power domains are same as the digital domains. Hence, the APM for LDOM0 will be extracted similarly like DLANE0. The sequence for LDOM0 for the transition from OFF to ON state is as follows:

```
sequence seqB01;
pwr_good##[1:2]in_preq##[1:5]in_clk;
endsequence
```

(iii) *Sequence Extraction for Analog components –*
In case of the AMS power domains, LDOM0, local states and control signals for analog components are not available in the UPF specification. UPF specification of every analog domain is annotated with the local states of its components in the contains_analog or similar kind of UPF construct as explained in the previous section. The behavioral model of the analog domain also helps in extracting out the low level signals responsible for the power state transitions in form of Analog Signal Template. It stores the generic signal template for each analog component and the signal name is provided by the designer. For example, the signal templates for the LDO of LDOM0 are – <domain_power>, <domain_clock>, <domain_enable>, <domain_mode> – which are filled by the designer, in this case, with pwr_good, in_clk, ldo_en and state. Auxiliary state machines are then built for the analog components with the annotated UPF and Analog Signal Template.
The control sequences responsible for the transitions of the APMs are temporal in nature. During the first level of extraction from the normal and annotated UPF specifications, the time bounds are in the form of [1:$] because UPF does not capture the timing details for the events occurring. For example, the control sequence for domain B extracted in this process is as follows:

978-1-5386-3693-0/18 $31.00 © 2018 IEEE

```
sequence seqBL01;
pwr_good##[1:$]in_clk##[1:$](ldo_en&& !state);
endsequence
```

(iv) *Determining Specific Timing Bounds* –

In the next level, these control sequences are processed for extracting accurate time bounds. The designer chooses a pessimistic upper bound $\mathcal{L}$ for the time bound. This upper bound is then refined by iterative successive approximation with bisection between 1 and $\mathcal{L}$ and verifying with a model checker tool on the LPM logic to validate its soundness. For the same example, after the time bound refinement, seqBL01 becomes:

```
sequence seqBL01;
pwr_good##[1:6]in_clk##[1:2](ldo_en &&!state);
endsequence
```

The evaluation of time bounds is an important step since model checkers suffer capacity limitations with increased sequential depth of properties. A significant amount of performance gain can be obtained by the reduction of the sequential depth of such properties.

(v) *APM Modeling Through Assume Properties* –

Fig. 3 illustrates the work flow of our automated tool for computing the reachable power states of the PML. The environment model consists of a set of assume properties extracted out from the UPF specifications of both the AMS and digital domains. In our framework, the PML is in Verilog-HDL [16] and the assume properties are developed as SystemVerilog Assertions (SVA) [13].

Three types of assume properties are developed for each power state of each power domain:

- *Assume Properties for Modeling State Transitions for Power Domains.* These assume properties capture the transitions of the APMs of the power domains. For example, transitions for LDOM0 from ON to OFF and from OFF to ON.

- *Assume Properties for Modeling State Transitions for Analog Components.* These assume properties capture the transitions from one state to other state of the analog component. For example, for state SHUT-DOWN of the LDO of LDOM0, there will be two assume properties as follows:

```
property BL_00_01;
  @(posedge clk)
  (stateB == 2'b00) ##0 BL_SD_LP |->
                    ##1 (stateB == 2'b01);
endproperty
assume_B1:assume property(BL_00_01);
property BL_00_10;
  @(posedge clk)
  (stateB == 2'b00) ##0 BL_SD_HP |->
                    ##1 (stateB == 2'b10);
endproperty
assume_B2:assume property(BL_00_10);
// Annotated UPF extracted seq
sequence BL_SD_LP;
  pwr_good ##[1:2] in_clk
          ##[1:5](ldo_en && !state);
endsequence
sequence BL_SD_HP;
  pwr_good ##[1:2] in_clk
          ##[1:5](ldo_en && state);
endsequence
```

- *Assume Properties for Modeling Self-loops.* These assume properties are written to make sure that the power domain stays in the current power state until any control sequence triggers it to another power state. Without these type of assume properties, the model checker may come up with false counter examples involving erroneous power state transitions. These assume properties are modeled by analyzing the control sequence from each power state. For instance, in case of LDOM0, the domain always enables the ldo_en signal while going to LOW-POWER or HIGH-POWER state. Hence, the negation of this signal ensures that LDOM0 is in SHUT-DOWN state. This is automatically extracted out and the assume property capturing this phenomenon is as follows:

```
property BL_00_00;
  @posedge(clk)
  (stateBL == 2'b00) && !ldo_en |->
                    ##1 (stateBL == 2'b00);
endproperty
assume_B3: assume property(BL_00_00);
```

(B) *Coverage and Reachability Analysis:*

The coverage analysis checks whether all the Safe and Reachable states are covered during the formal analysis (indicating completeness). Whereas, the reachability analysis checks whether all the reachable states are Safe (ensuring correctness). For example, a global state where DLANE0 is in OFF state, LDOM0 is in ON state and LDO in LDOM0 in in HIGH-POWER state should never be reached. This property is captured as follows:

```
property reach_A_OFF_B_HIGHPOWER;
  @(posedge clk)  disable iff (rst)
  !((stateA == 2'b00) && (stateB == 2'b01)
            && (stateBL == 2'b10));
endproperty
assert_P1:    assert
          property(reach_A_OFF_B_HIGHPOWER);
```

IV. COVERAGE ANALYSIS EXPERIMENT: D-PHY POWER MANAGEMENT FABRIC – A CASE STUDY

The D-PHY architecture [14] is designed for transmission and reception of video for camera and display interfaces. This test case is a good example because the LDOs and PLLs of its AMS power domains are regulated by its PML. Fig. 1 shows the power domains in the D-PHY TX Master, namely:

- Four *data lanes*, called DLANE0 through DLANE3 for transferring data to the display unit, having local power states - OFF, LOW-SPEED, and HIGH-SPEED.

- Four analog *LDO domains*, called LDOM0 through LDOM3, respectively, for supplying power to DLANE0 through DLANE3.

- A *PLL domain*, which contains a PLL for supplying clock to the four data lanes. This is also an analog domain having local power states – OFF, LOW-FREQUENCY, and HIGH-FREQUENCY.

- A *common lane*, which contains the PML and associated circuitry for powering up/down the LDO domains. The common lane is powered by an LDO which is part of a power domain called SUS_DOM. These two analog domains have two local power states - OFF and ON.

The D-PHY TX Master transmits data in two modes, namely a *High Speed Data Transmission Mode* (HS-mode) and a *Low Power Data Transmission Mode* (LP-mode). In HS-mode, the LDO of an LDOM domain operate in HIGH-POWER state,supplying high voltage to the respective DLANE domain and thus driving the DLANE domain into HIGH-SPEED state. The PLL domain must supply the high frequency clock in HS-mode and thus, stays at HIGH-FREQUENCY power state. Similarly, in LP-mode, the voltage, speed and frequency become LOW.

There are several architectural aspects of this design with analog components which must be covered and verified. Few of them are explained with example:

- *DLANE0 and DLANE1 should not be in LOW-SPEED (LS) state when LDOM0 and LDOM1 are in ON state and the LDOs are in HIGH-POWER (HP) state supplying high voltage through the power rails.*

```
property P3;
@(posedge clk)  disable iff (rst)
  !((stateDL0 == 'LS) && (stateDL1 == 'LS)
  &&(stateLD0 == 'ON) && (stateLD1 == 'ON)
  &&(stateL0 == 'HP) && (stateL1 == 'HP));
endproperty
```

- *The PLL domain should not be in LOW-FREQUENCY (LF) mode when DLANE0 is in HIGH-SPEED (HS) mode.*

```
property P2;
@(posedge clk)  disable iff (rst)
  !((state_PD == 'ON) && (state_PP == 'LF)
                     && (stateDL0 == 'HS));
endproperty
```

Similar properties can be written for other DLANE domains, i.e. for DLANE1, DLANE2 and DLANE3.

We now discuss the experimental results for the coverage and reachability analysis technique. The proposed methodology was applied on a small example, *AmsTest* and the D-PHY TX Master. All the experiments were performed on a 2.33 GHz Intel-Xeon server with 32GB RAM. Table I specifies the number of inputs, outputs, number of sequential and combinational elements for these test cases. One important point is to be noted that, Table I represents the size of the PML, not the whole design which is much larger in size.

Name of Design	Power Domains	# I/Ps	# O/Ps	# Seq Elmts.	#Comb. Elmts.	# Nets
AmsTest	2 (1 ams, 1 dig)	9	6	195	898	1102
D-PHY Tx Master	10 (5 ams, 5 dig)	22	18	11991	44290	56303

TABLE I: DESIGN STATISTICS

Table II shows the results for global state coverage by our approach. For each test case, Column 2 presents the size of the global power states which is basically the product of the local power states of each domain. Column 3 and Column 4 state the number of Reachable and Unreachable states. Column 5 presents the analysis time for the power state reachability test through Magellan using its coverage goal construct. It may be noted that the coverage analysis time for D-PHY design is significantly high because of the following two facts: (i) the state space of the power management design model is comparatively big; and (ii) the sequential depth of the assertions (constructed leveraging the extracted sequences) is also large. These two factors leads to higher analysis time during formal verification/analysis.

Table II also presents an interesting post-analysis of the reachable states. It shows the number of *Legal* and *Illegal* states reached by the PML. For example, an illegal state of D-PHY was found out where during the power-down, the LDOM0 is in OFF state but the DLANE0 is still in ACTIVE state.

Design Name	#Global States	#Rch. States	#Unrch. States	Time (secs)	#Legal States	#Ill. States
AmsTest	18	12	6	133	10	2
D-PHY	1296	520	776	15840	288	232

TABLE II: FORMAL COVERAGE ANALYSIS RESULTS

V. CONCLUSION

This paper presents a unique methodology for the formal coverage analysis of the global power management logic with AMS power domains. With the increasing incorporation of the analog components in SoCs and other devices, the coverage analysis of these has become a challenging problem. Our coverage analysis methodology has been verified on D-PHY TX-Master, an industry level mixed-signal design test-case. A possible extension of this work can be developing a more generic front-end for our tool which will raise the level of abstraction for the PML.

REFERENCES

[1] D. Flynn. et. al., *"Low power methodology manual: for system-on-chip design"*. Springer Science & Business Media, 2007.

[2] S. Jadcherla, *"Verification methodology manual for low power"*. Synopsys, 2009.

[3] A. P. Chandrakasan. et. al., "Low-power cmos digital design," *IEICE Transactions on Electronics*, pp. 371–382, 1992.

[4] A. Sinha. et. al., "Dynamic voltage scheduling using adaptive filtering of workload traces," in *VLSID*, 2001, pp. 221–226.

[5] A. Correale Jr, "Overview of the power minimization techniques employed in the ibm powerpc 4xx embedded controllers," in *Proceedings of the 1995 international symposium on Low power design.* ACM, 1995, pp. 75–80.

[6] A. Hazra. et. al., "Leveraging upf-extracted assertions for modeling and formal verification of architectural power intent," in *Design Automation Conference*, 2010, pp. 773–776.

[7] A. Hazra. et. al., "Formal verification of architectural power intent," *IEEE TVLSI*, vol. 21, no. 1, pp. 78–91, 2013.

[8] D. Macko. et. al., "Power-efficient power-management logic," in *PATMOS*. IEEE, 2014, pp. 1–7.

[9] A. Hazra. et. al., "Formal methods for coverage analysis of architectural power states in power-managed designs," in *(ASP-DAC), 2012 17th Asia and South Pacific*, 2012, pp. 585–590.

[10] A. Hazra. et. al., "Power-tructor: An integrated tool flow for formal verification and coverage of architectural power intent," *IEEE TCAD*, vol. 32, no. 11, pp. 1801–1813, 2013.

[11] P. Dasgupta. et. al., "Formal hardware/software co-verification of embedded power controllers," *IEEE TCAD*, vol. 33, no. 12, pp. 2025–2029, 2014.

[12] E. Barke et. al., "Embedded tutorial: Analog-/mixed-signal verification methods for ams coverage analysis," in *2016 DATE*, pp. 1102–1111.

[13] SystemVerilog LRM, "System Verilog LRM 3.1a by Accellera," 2004.

[14] S. Mandal. et. al., "Formal verification of power management logic with mixed-signal domains," in *VLSID*, 2017, pp. 239–244.

[15] Synopsys, "Magellan. An Industrial Formal Verification Tool from Synopsys." www.synopsys.com.

[16] IEEE VHDL Standard, "IEEE 1364-2005 Standard Verilog Hardware Description Language," 2006.

DPFair Scheduling with Slowdown and Suspension

Sanjay Moulik
Department of Computer Science
Indian Institute of Information Technology
Guwahati, India
Email: s.moulik@iitg.ac.in

Arnab Sarkar* and Hemangee K. Kapoor[†]
Department of Computer Science
Indian Institute of Technology
Guwahati, India
Email:*arnabsarkar@iitg.ac.in, [†]hemangee@iitg.ernet.in

Abstract—Energy has become a first class design criterion in many of today's real-time embedded systems which are often operated by limited energy sources like batteries. Processor slowdown (using DVFS) and processor suspension/sleep (using DPM) are the two primary techniques adopted to control energy dissipation at the operating system level. This work proposes a combined slowdown-sleep strategy to minimise energy dissipation in hard real-time multicore systems, scheduled using the DPFair technique. DPFair is an important optimal proportional fair scheduler which ensures lower design costs through full resource utilisation while guaranteeing the satisfaction of all deadlines, even for dynamically arriving task sets. While there exists a DVFS based state-of-the-art energy-aware DPFair scheduler, our proposed scheduling technique, *DPFair with slowdown and suspension (DPFair-SS)*, exhibits appreciable performance gains over the existing strategy in situations when the system experiences low workloads over significantly long durations.

I. INTRODUCTION

Proportional Fair Schedulers are often preferred in real-time embedded systems due to their inherent advantages such as ability to provide temporal isolation to a task in the face of possible anomalous behaviour of other tasks, high resource utilisation, seamless handling of dynamic task arrivals etc. The proportional fair algorithm Pfair [1] and it's work conserving version ERFair [2] are the first known optimal real-time multiprocessor schedulers. Given a set of n tasks to be executed on m processors, ERFair scheduler ensures that the minimum rate of progress for each task is proportional to a parameter called the task's weight wt_i, which is defined as a ratio of it's execution requirement e_i and period p_i. That is, in order to satisfy ERFairness, each task T_i should complete atleast $(\sum_{i=1}^{n}(e_i/p_i) \times t)$ part of it's total execution requirement e_i, at any time instant t, subsequent to the start of the task at time s_i. Following the above criterion, ERFair is able to deliver optimal/full resource utilisation; hence, any task set is guaranteed to be schedulable provided: $\sum_{i=1}^{n}(e_i/p_i) \leq m$. However, such a strict fairness criterion imposes heavy overheads including high task selection times and unrestricted preemptions/migrations.

More recently DPFair [3], an optimal but lower overhead approximate proportional fair scheduler has been proposed. DPFair divides time into frames demarcated by the periods/deadlines of all tasks in the system. At the beginning of each frame, tasks are partitioned based on their proportional shares among the available processors through a *Stacking and Slicing* mechanism, such that each task is mapped to either one

or two processors for execution within the frame. With such a semi-partitioned scheduling mechanism, DPFair is able to ensure atmost *m-1* task migrations within frames, where m is the number of processors available in the system. *This work assumes DPFair as the underlying scheduling mechanism.*

In addition to meeting temporal constraints, minimising energy dissipation has become a very important design objective in many modern real-time embedded systems. Even as the hardware becomes more and more energy efficient, the onus of extracting maximum efficiency out of the hardware very often falls on the software. Two widely used energy saving techniques employed at the software level are: Dynamic Voltage/Frequency Scaling (DVFS) [4] and Dynamic Power Management (DPM) [5].

The DVFS mechanism decreases dynamic energy consumption by reducing the voltage/frequency of a processor to the minimum value that is sufficient to handle a given workload. An offline DVFS technique has been proposed in [6] for systems in which voltages and frequencies may be controlled uniformly (in a continuous fashion) and independently among the available processors. This technique needs the scaling to be controlled very minutely which may not be always possible. An energy minimisation approach based on two-dimensional bin packing has been proposed in [7] for parallel independent task systems with discrete operation modes and under timing constraints. Both [6] and [7] being offline solutions cannot be applied for dynamic real-time task sets. Chen and Kuo [8] proposed an approximation algorithm based on the Kuhn-Tucker optimality conditions for homogeneous multiprocessors. A DVFS technique named Deterministic Stretch to Fit (DSF) based on the principle of inter-task slack shareability has been proposed in [9]. The authors have firstly proposed an online algorithm to reclaim energy by adapting to the variations in actual workload of target application tasks. Secondly, they have extended the algorithm with an adaptive and speculative speed adjustment mechanism. This mechanism anticipates early completion of future task instances based on the information of their average workload. In [10], the authors have proposed a DVFS based energy-aware scheduling strategy with DPFair as the base scheduling scheme. Similar to DPFair, this algorithm (which we refer to as *DVFS based DP-Wrap*) progresses frame by frame with the operating frequency in each frame being governed by the workload in the frame.

With exponential growth in chip transistor densities over

technology generations, static energy dissipation due to leak-age drain from transistors has steeply increased over the years [11]. Today, static energy dissipation has already become the major source of power wastage within a chip. Static energy wastage is mainly controlled using DPM, where a processor is put into inactive low power suspension/sleep states by procrastinating task executions while simultaneously guaranteeing their timely completion.

DPM techniques try to minimise static energy dissipation in the system by putting a processor in low-power suspen-sion/sleep mode for as long as possible while still guaran-teeing the tasks' timing constraints. Awan and Petters [12] proposed an energy efficient slack management approach to minimize leakage energy consumption for mixed-criticality uniprocessor systems. They presented this approach for dy-namic priority systems with multiple sleep states. Bhatti et al. [13] have proposed a DPM based energy-aware strategy for global multiprocessor systems called Assertive Dynamic Power Management (AsDPM). AsDPM first determines the minimum number of active processors needed to fulfill the execution requirement of released jobs at runtime. Then it attempts to cluster the distributed idleness existing on a subset of the active processors into longer continuous idle intervals, so that these obtained intervals may be employed to switch some of the processors to deeper low-power states for a longer duration of time. In [14], the authors have presented a two phased DPM based scheduling heuristic for sporadic tasks on heterogeneous multicores. The problem of online dynamic power management that provides hard real-time guarantees for multi-processor systems has been considered in [15]. Legout et al. [16] presented a DPM based online scheduling scheme by extending the existing approach *Fixed Priority until Zero Laxity (FPZL)* in order to handle both hard real-time and mixed criticality systems.

There are also existing works which combine both DVFS and DPM techniques together to achieve better energy savings in the system. These approaches provide higher energy savings because the slowdown caused by DVFS also increases static leakage power. Therefore, for any given technology, there exists an optimal frequency called *critical frequency* [17] at which the processor should be clocked to minimize overall energy consumption of the system. In [17], [18], combined slowdown-suspension techniques for real-time systems with fixed-priority tasks have been considered. Total system-wide energy minimisation mechanisms using dynamic slack recla-mation has been proposed by Jejurikar et al. in [19]. In [20], a technique has been proposed to directly model the idle intervals of individual cores such that both DVFS and DPM can be optimized at the same time. Based on this technique, the energy optimization problem has been formulated by means of mixed integrated linear programming. In [21], the authors have addressed the problem of task-to-core allocation onto heterogeneous multicore platforms such that the overall energy consumption of the system is minimised. To this end, they proposed a DVFS-cum-DPM based two-phased approach. In the first phase, tasks have been allocated to cores such that

Symbol	Description
n	Number of tasks
T_i	i^{th} task
s_i	Start time of ith task
e_i	Execution requirement of ith task
p_i	Period of ith task
wt_i	Weight of ith task
shr_i	Share of ith task
t	Current time
G_k	k^{th} frame
U	Utilisation Factor in the system
$fr_{critical}$	Normalised critical frequency
f_{max}	Maximum value of normalised operating frequency
f_i	i^{th} frequency level
fr_{T_i}	Minimum normalised frequency sufficient to complete T_i on a single core in a frame
TD	Set of tasks for which $fr_{T_i} > fr_1$
m	Number of cores in the system
V	Set of cores
V_k	k^{th} core
fr_1	Optimum operating frequency selected in a frame
fr_g	Global Operating frequency selected in a frame

TABLE I: Important Terminologies

dynamic energy consumption is reduced, while the second phase refines the prior task allocation in order to achieve better sleep states by trading off dynamic energy consumption with reduction in leakage energy consumption.

A major class of real-time systems typically experience short bursts of heavy CPU activity interleaved with long dura-tions of significantly lower workloads. The *DVFS based DP-Wrap* [10] algorithm discussed above provides an important scheduling solution for such resource constrained hard real-time systems due to it's ability to provide optimal resource utilisation, controlled migrations and minimised dynamic en-ergy dissipation. However being leakage drain ignorant, it allows significant static energy dissipation whenever system workloads are not high enough to demand operation above the critical frequency. This work therefore proposes an integrated DVFS-cum-DPM based DPFair based scheduling strategy called, *DPFair with slowdown and suspension*, in order to minimise overall system-wide energy consumption combining both static and dynamic energy dissipation. Our experimental analyses show promising results.

II. Specifications and Terminologies

A few important terminologies used in the later sections of our paper have been listed in Table I. Now, we describe the system and task specifications along with the power model used in our work.

A. System and Task Specifications

This work attempts to minimise system-wide energy con-sumption by designing an efficient scheduling strategy for a set of n periodic tasks $T = \{T_1, T_2, ..., T_n\}$ on m uniform cores $V = \{V_1, V_2, ..., V_m\}$. The cores can operate at *max* discrete frequency levels whose normalised values in the range [0,1] are represented as $F = \{f_1, f_2, ..., f_{max}\}$, with $f_{max} = 1$ being the maximum frequency. Also, the cores are assumed to be mutually independent in terms of their frequency of

978-1-5386-3693-0/18 $31.00 © 2018 IEEE

operation. Each task T_i has an execution requirement e_i (each e_i value is generated based on the tasks's execution time on a core operating at f_{max}) to be completed within a period p_i. The ratio $wt_i = e_i/p_i$ is denoted as the *weight* of T_i and utilisation factor $U = \sum_{i=1}^{n} wt_i/m$ represents the total workload the system should handle.

B. Power Specifications

We have adopted the analytical core energy model presented in [17]. Let total power consumption of a CMOS-based core be denoted by P during active operation and P_{sleep} when the core is suspended/sleeping. The active power consumption P further consists of dynamic power consumption (P_d), static power consumption (P_s) and an inherent power cost to keep the processor on (P_{on}).

$$P = P_d + P_s + P_{on} \qquad (1)$$

The dynamic power consumption (P_d) of CMOS circuits is given by:

$$P_d = C_{eff} \times V_{dd}^2 \times f \qquad (2)$$

where, V_{dd} is the supply voltage, C_{eff} is the average switched capacitance per cycle, and f is the clock frequency. Here, V_{dd} may be considered to be roughly proportional to f. The major components of static current in a standard inverter are due to reverse bias junction current [22] and subthreshold conduction [22]. Hence, the static power consumption, P_s, is given by:

$$P_s = (V_{dd} \times I_{subn} + |V_{bs}| \times I_j)L_g \qquad (3)$$

where, V_{bs} is the the body bias voltage, I_{subn} is the subthreshold leakage current, I_j reverse bias junction current in the NMOS device and L_g is the number of devices in the circuit. The technology constants for the 70nm technology have been taken from [23], [22]. According to this energy model, we can calculate the frequency f, dynamic power P_d and static power P_s for different voltage levels, as shown Table II. The value of V_{bs} in our experiments was taken to be -0.7. It may be observed from the voltage frequency relationships described in Table II that the minimum normalised frequency $f_1 = 0.19$ is obtained when $V_{dd} = 0.55V$ and $V_{dd} = 1.0V$ corresponding to the maximum normalised frequency $f_{max} = 1$. Following [22], values of P_{on} and P_{sleep} has been assumed to be $276mW$ and $80\mu W$ in our experiments.

Critical Frequency: As defined in Equation(2), the dynamic power consumption of a core (P_d) is a cubically increasing function of the operating frequency. However, since lowering the operating frequency of a core stretches the execution times of tasks scheduled on that core, frequency reduction results in higher static energy consumption. Due to this effect, beyond a certain degree of reduction in frequency, static power wastage overhauls the gain obtained through dynamic voltage/frequency scaling. Thus, there exists a critical frequency ($fr_{critical}$) beyond which further frequency reduction actually

V_{dd} (V)	f (GHz)	Normalised Freq.	P_d (mW)	P_s (mW)
0.55	0.58	0.19	75	171
0.60	0.79	0.26	122	204
0.65	1.01	0.33	185	245
0.70	1.26	0.41	267	289
0.75	1.53	0.49	370	339
0.80	1.81	0.58	499	396
0.85	2.10	0.68	655	461
0.90	2.42	0.78	843	535
0.95	2.74	0.88	1066	619
1.00	3.10	1.00	1327	714

TABLE II: Dynamic and Static power consumption for 70nm processor

leads to increased net energy (summation of dynamic and static energy) consumption. Following [17], which considers a 70nm technology based core design, we have assumed the critical frequency to be 1.26 GHz ($V_{dd} = 0.7\ V$). Hence, in our system which assumes a maximum frequency of 3.1 GHz (refer Table II), the normalised frequency becomes $fr_{critical} = 1.26/3.1 = 0.406$.

III. DPFair with Slowdown and Suspension (DPFair-SS)

The *DPFair-SS* scheduler is based on basic DPFair and hence, the schedule progresses in a frame by frame fashion, where frames are demarcated by the period/deadlines of all tasks at a given time. At the beginning of the next frame say G_k, the share shr_i (with respect to the maximum frequency $f_{max} = 1$) of each task T_i is determined as:

$$shr_i = wt_i \times |G_k| \qquad (4)$$

where, $|G_k|$ denotes the length of G_k. It may be noted, that by such an allocation and execution of shares within G_k, *DPFair-SS* maintains accurate proportional fairness at frame boundaries.

Next, *DPFair-SS* calls the function *Slowdown-Suspend-Schedule (SSS)* to determine appropriate core frequencies, suspension strategy and schedule of tasks in G_k. Algorithm 1 presents the pseudo-code of *DPFair-SS*.

Algorithm 1 DPFair-SS Algorithm
___

1: $k \leftarrow 0;\ time \leftarrow 0$
2: **while** true **do**
3:     Determine length $|G_k|$ of the next frame G_k based on the nearest deadline among all executing tasks
4:     **for** each task T_i **do**
5:         Determine share shr_i of T_i
6:     Call function *SSS* for G_k
7:     $k \leftarrow k + 1;\ time \leftarrow time + |G_k|$

The Slowdown-Suspend-Schedule Function (SSS): At the start of G_k, *SSS* determines the minimum frequency fr_1 required to execute the given workload in G_k in the following way:

Algorithm 2 Function Slowdown-Suspend-Schedule

1: Determine operating frequency fr_1
2: Allot a dedicated core to each task T_i having: $fr_{T_i} > \lceil fr_1 \rceil$. The core's operating frequency becomes: $\lceil fr_{T_i} \rceil$
3: Determine minimum operating frequency (fr_g) for the rest $(n - |TD|)$ tasks on the remaining $(m - |TD|)$ cores
4: **if** $fr_g > fr_{critical}$ **then**
5:     Modify shares of each task T_i: $shr_i = shr_i/fr_g$
6:     Assign task shares to the $(m - |TD|)$ cores in any order
7:     Determine the amount of time t_1 ($t1 \le |G_k|$) for which the $(m - |TD|)$ core should operate at frequency $\lfloor fr_g \rfloor$
8: **else**
9:     Calculate m', the exact system capacity that is necessary to execute the $(n - |TD|)$ tasks at frequency $\lceil fr_{critical} \rceil$.
10:     $\{\lfloor m' \rfloor$ cores must execute for entire frame duration$\}$
11:     $\{(m - |TD| - \lceil m' \rceil)$ cores may sleep for entire frame duration$\}$
12:     Determine t_2, the number of time slots for which the remaining core should execute
13:     Modify shares of each task T_i: $shr_i = shr_i/\lceil fr_{critical} \rceil$ and assign to the cores

$$fr_1 = \frac{\sum_{i=1}^{n} shr_i}{|G_k| \times m} \tag{5}$$

As the system under consideration contains a discrete set of alternative operating frequencies, the exact frequency fr_1 may not be available. In such a situation, we choose the next higher available frequency denoted by $\lceil fr_1 \rceil$ as the operating frequency.

Now, the frequency that is sufficient to complete execution of the share of any given task (say T_i), when executed for the entire duration of a frame on a dedicated core, is given by:

$$fr_{T_i} = shr_i/|G_k| \tag{6}$$

If $fr_{T_i} \ge \lceil fr_1 \rceil$, a dedicated core running at $\lceil fr_{T_i} \rceil$ must be allocated to T_i for the frame duration $|G_k|$. Let TD be the set of tasks for which $fr_{T_i} > \lceil fr_1 \rceil$ and let shr_{TD} be the summation of task shares in TD. When all the tasks in TD are allocated to dedicated cores and the rest of the tasks are allotted the remaining $(m - |TD|)$ cores, the minimum frequency at which these $(m - |TD|)$ cores must execute is given by:

$$fr_g = \frac{\sum_{i=1}^{|T|} shr_i - shr_{TD}}{(m - |TD|) \times |G_k|} \tag{7}$$

As $fr_{critical}$ denotes the frequency of execution which maximises net energy savings, all cores need to execute for the entire frame duration only if, $fr_{critical} \le fr_g$. In case $fr_{critical} \le fr_g$, the $(m - |TD|)$ cores execute at frequency $\lfloor fr_g \rfloor$ for a duration say, t_1 and at frequency $\lceil fr_g \rceil$

for the remaining duration of the frame ($|G_k| - t_1$), where $\lfloor fr_g \rfloor \times t_1 + (|G_k| - t_1) \times \lceil fr_g \rceil = fr_g \times |G_k|$. Hence,

$$t_1 = \frac{|G_k| \times (\lceil fr_g \rceil - fr_g)}{\lceil fr_g \rceil - \lfloor fr_g \rfloor} \tag{8}$$

In case $fr_{critical} > fr_g$, the exact system capacity that is sufficient to execute the task set $(T\text{-}TD)$ is given by:

$$m' = (m - |TD|) \times (fr_g/\lceil fr_{critical} \rceil) \tag{9}$$

Now, $\lfloor m' \rfloor$ cores must execute for the entire frame duration $|G_k|$ and one more core should execute for say, t_2 time slots at frequency $\lceil fr_{critical} \rceil$, where t_2 is given by:

$$t_2 = (m' - \lfloor m' \rfloor) \times |G_k| \tag{10}$$

This partially filled core can be suspended for the remaining time ($|G_k| - t_2$). The remaining $(m - |TD| - \lceil m' \rceil)$ cores can sleep for the whole frame duration $|G_k|$.

The pseudo-code of the function *Slowdown-Suspend-Schedule (SSS)* presented in Algorithm 2, depicts the overall energy-aware intra-frame scheduling strategy of *DPFair-SS*. We now present an example to illustrate the working principle of function *SSS*.

Example 1: Consider a list of 6 tasks T_1, T_2, T_3, T_4, T_5 and T_6 having execution requirements 90, 20, 20, 10, 10 and 10 respectively, to be scheduled within a frame of size 100 time slots, in a system of $m = 4$ cores.

Normal DP-Wrap: The task allocation using energy oblivious DP-Wrap has been shown in Fig. 1(a). We can observe that V_0 runs at f_{max} for 100 time slots, V_1 runs at f_{max} for 60 slots and idles for the remaining time in *power on* mode while V_2 and V_3 idles in *power on* mode for the entire frame. The energy consumption in a system using this strategy (from Equation 1) can be calculated as: $E_{sys} = (P_d + P_s + P_{on}) \times 160 + (P_{on} \times 240) = (1327 + 714 + 276) \times 160 + (276 \times 240) = 43.7W$.

DVFS based DP-Wrap: Using this strategy, the optimum frequency required to execute the workload is found out using Equation 5, $fr_1 = (90 + 20 + 20 + 10 + 10 + 10)/(4 \times 100) = 0.34$. But if the system executes at fr_1, then T_1 won't be able to complete it's required share in the frame. Hence, T_1 is allotted to V_0 with operating frequency of 0.9. For rest of tasks, required operating frequency is recalculated using Equation 7, which comes out to be 0.26. The modified shares for all the tasks are recalculated with respect to operating frequency of corresponding cores as: $shr_i = shr_i/fr_gS$. The tasks are allocated using DP-Wrap scheduling as shown in Fig. 1(b). Therefore, the energy consumption in the system will be: $E_{sys} = (1066 + 619 + 276) \times 100 \times 1 + (122 + 204 + 276) \times 100 \times 3 = 37.67W$. Although, the energy consumption is reduced significantly in comparison to the energy oblivious DP-Wrap scheduling strategy, but dynamic energy savings due to frequency reduction is overhauled by

978-1-5386-3693-0/18 $31.00 © 2018 IEEE

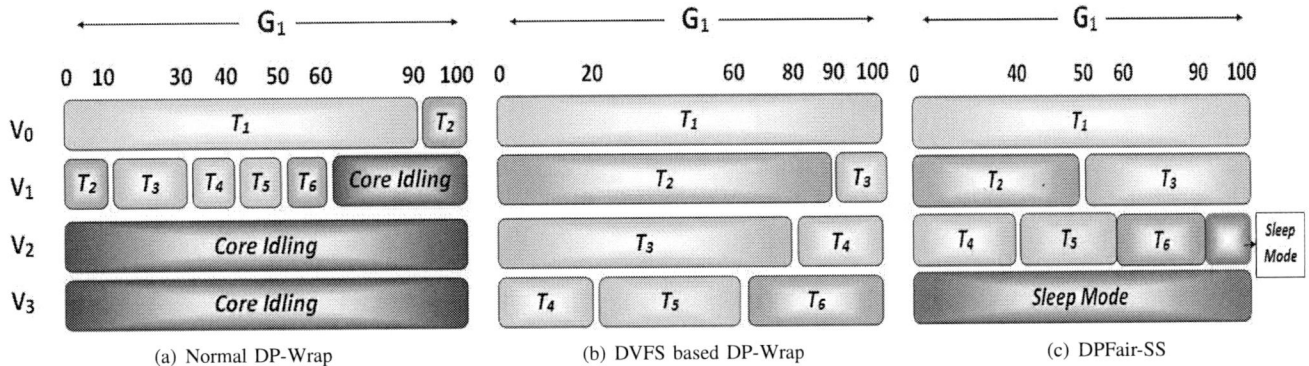

(a) Normal DP-Wrap   (b) DVFS based DP-Wrap   (c) DPFair-SS

Fig. 1: Task Allocation for Example 1

static energy wastage as we see next.

DPFair-SS: Similar to *DVFS based DP-Wrap*, T_1 is allocated to V_0 and frequency fr_g for the rest of the cores becomes 0.26 (using equation 7). For the system considered by us, $fr_{critical} = 0.406$ and so, $fr_g < fr_{critical}$. The value of m' becomes 1.9 using equation 9. Therefore, one core must execute for the entire frame at frequency $\lceil fr_{critical} \rceil$. One core may sleep for the entire frame duration. The remaining core will execute for $t_2 = 90$ time slots (refer equation 10) at frequency $\lceil fr_{critical} \rceil$. This core can sleep for the remaining 10 time slots in the frame. The tasks are then assigned to the cores using *DP-Wrap* as shown in Fig. 1(c). An estimate of the energy consumed may be found as: $E_{sys} = (1066 + 619 + 276) \times 100 \times 1 + (267 + 289 + 276) \times 190 + P_{sleep} \times 10 + (P_{sleep} \times 100) = 35.41W$. ∎

IV. EXPERIMENTAL SET UP AND RESULTS

In order to evaluate the performance of our algorithm, we have considered task sets consisting of 64 tasks to be scheduled on 8 cores. The execution times of individual tasks present in the set have been generated randomly from a normal distribution with standard deviation $\sigma_e = 20$ and mean $\mu_e = 100$. The task weights have also been generated from normal distributions with standard deviation $\sigma_{wt} = 0.1$ and mean $\mu_{wt} = 0.3$. The summation of the randomly generated task weights may not exactly be equal to the desired *Utilisation Factor* (U). Therefore, the individual weights are scaled so that the summation becomes equal to U. The execution time for each simulation was 100000 time slots. For each set of input parameters, we ran the simulation on 50 different test cases. The average of these 50 test cases has been considered as the final result. We compared performance of out proposed *DPFair-SS* algorithm against *DVFS based DP-Wrap* [10]. We now provide an analysis of the results obtained.

The system under consideration is subjected to two distinct average workload values over the simulation duration. The average *Utilisation Factor* of the system is U_{low} for t_{low}% of

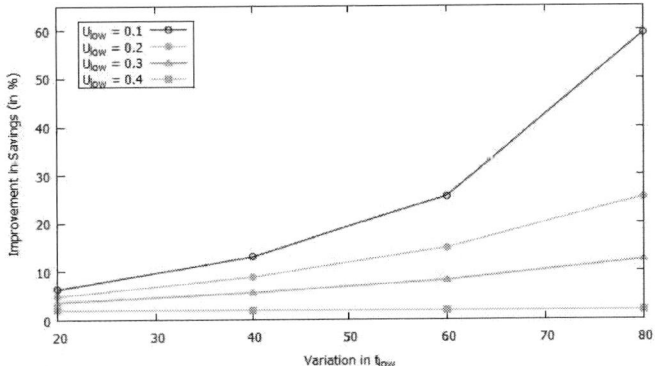

Fig. 2: % Improvement in Energy Savings for *DPFair-SS* over *DVFS based DP-Wrap* ($U = 0.7$, $m = 8$ and $n = 64$)

the total simulation duration, while for the other $(100 - t_{low})$% of the simulation duration, the average *Utilisation Factor* is fixed at 0.7. We have considered four distinct values (0.4, 0.3, 0.2 and 0.1) for U_{low} while t_{low} has been varied between 20% and 80%. Fig. 2 depicts results for the percentage improvement in energy savings achieved by *DPFair-SS* over *DVFS based DP-Wrap*. From the figure, it may be seen that for any given value of t_{low} energy savings increase as U_{low} decreases. Further, for a particular value of U_{low}, energy savings improve as the percentage of time under the lower workload, increases. Overall, the gains in energy savings obtained by *DPFair-SS* may be observed to be quite significant. This is especially because it is quite common for many systems to experience brief periods with heavily CPU intensive workloads along with significantly longer periods when computation demands are quite low. In particular, for $t_{low} = 80$%, improvements in energy savings may be observed to be 59.36%, 25.37%, 12.38% and 1.71% for U_{low} values 0.1, 0.2, 0.3 and 0.4, respectively. The normalised energy consumption values obtained by varying U_{low} and t_{low} has been presented in Table III. Here,

Time Spent under $fr_{critical}$: t_{low} (in %)	DVFS based DP-Wrap ($U_{low} = 0.1$)	DPFair-SS ($U_{low} = 0.1$)	DVFS based DP-Wrap ($U_{low} = 0.2$)	DPFair-SS ($U_{low} = 0.2$)	DVFS based DP-Wrap ($U_{low} = 0.3$)	DPFair-SS ($U_{low} = 0.3$)	DVFS based DP-Wrap ($U_{low} = 0.4$)	DPFair-SS ($U_{low} = 0.4$)
20	0.54	0.51	0.56	0.53	0.57	0.55	0.59	0.58
40	0.44	0.39	0.47	0.44	0.50	0.47	0.53	0.52
60	0.33	0.26	0.39	0.34	0.43	0.39	0.47	0.46
80	0.23	0.14	0.30	0.24	0.35	0.30	0.41	0.40

TABLE III: Normalised Energy Consumption for various values of t_{low} and U_{low} ($U = 0.7$, $m = 8$ and $n = 64$)

normalised energy consumption is determined by the ratio of the energy consumption value obtained for *DPFair-SS / DVFS based DP-Wrap*, with respect to the energy consumed when all cores run at the highest frequency f_{max} for the entire simulation duration.

V. CONCLUSION

In this paper, we have proposed a combined slowdown-suspension oriented energy-aware optimal hard real-time scheduler for multicore systems. The proposed algorithm uses *DPFair* as the underlying scheduling strategy. Combining the benefits of both DVFS and DPM, the proposed strategy named *DPFair-SS*, is able to perform significantly better than a purely DVFS based state-of-the-art *DPFair* scheduler in situations when the system experiences low workloads over significantly long durations. Our extensive simulation based experiments show promising results.

REFERENCES

[1] S. Baruah, N. Cohen, C. Plaxton, and D. Varvel, "Proportionate progress: A notion of fairness in resource allocation," *Algorithmica*, vol. 15, no. 6, pp. 600–625, 1996. [Online]. Available: http://dx.doi.org/10.1007/BF01940883

[2] J. Anderson and A. Srinivasan, "Early-release fair scheduling," in *Real-Time Systems, 2000. Euromicro RTS 2000. 12th Euromicro Conference on*, 2000, pp. 35–43.

[3] G. Levin, S. Funk, C. Sadowski, I. Pye, and S. Brandt, "DP-FAIR: A simple model for understanding optimal multiprocessor scheduling," in *Real-Time Systems (ECRTS), 2010 22nd Euromicro Conference on*, July 2010, pp. 3–13.

[4] Y. Shin, K. Choi, and T. Sakurai, "Power optimization of real-time embedded systems on variable speed processors," in *Computer Aided Design, 2000. ICCAD-2000. IEEE/ACM International Conference on*, Nov 2000, pp. 365–368.

[5] N. K. Jha, "Low power system scheduling and synthesis," in *Proceedings of the 2001 IEEE/ACM International Conference on Computer-aided Design*, ser. ICCAD '01. Piscataway, NJ, USA: IEEE Press, 2001, pp. 259–263. [Online]. Available: http://dl.acm.org/citation.cfm?id=603095.603147

[6] K. Funaoka, S. Kato, and N. Yamasaki, "Energy-efficient optimal real-time scheduling on multiprocessors," in *Object Oriented Real-Time Distributed Computing (ISORC), 2008 11th IEEE International Symposium on*, May 2008, pp. 23–30.

[7] J. E. G. Coffman, M. R. Garey, D. S. Johnson, and R. E. Tarjan, "Performance bounds for level-oriented two-dimensional packing algorithms," *SIAM Journal on Computing*, vol. 9, no. 4, pp. 808–826, 1980. [Online]. Available: https://doi.org/10.1137/0209062

[8] J.-J. Chen and T.-W. Kuo, "Multiprocessor energy-efficient scheduling for real-time tasks with different power characteristics," in *Parallel Processing, 2005. ICPP 2005. International Conference on*, June 2005, pp. 13–20.

[9] M. K. Bhatti, C. Belleudy, and M. Auguin, "An inter-task real time DVFS scheme for multiprocessor embedded systems," in *2010 Conference on Design and Architectures for Signal and Image Processing (DASIP)*, Oct 2010, pp. 136–143.

[10] S. Funk, V. Berten, C. Ho, and J. Goossens, "A global optimal scheduling algorithm for multiprocessor low-power platforms," in *Proceedings of the 20th International Conference on Real-Time and Network Systems*, ser. RTNS '12. New York, NY, USA: ACM, 2012, pp. 71–80. [Online]. Available: http://doi.acm.org/10.1145/2392987.2392996

[11] N. S. Kim, T. Austin, D. Baauw, T. Mudge, K. Flautner, J. S. Hu, M. J. Irwin, M. Kandemir, and V. Narayanan, "Leakage current: Moore's law meets static power," *Computer*, vol. 36, no. 12, pp. 68–75, Dec 2003.

[12] M. A. Awan and S. M. Petters, "Enhanced race-to-halt: A leakage-aware energy management approach for dynamic priority systems," in *2011 23rd Euromicro Conference on Real-Time Systems*, July 2011, pp. 92–101.

[13] M. Bhatti, M. Farooq, C. Belleudy, and M. Auguin, "Controlling energy profile of rt multiprocessor systems by anticipating workload at runtime," in *SYMPosium en Architectures nouvelles de machines*, 2009.

[14] M. A. Awan and S. M. Petters, "Energy-aware partitioning of tasks onto a heterogeneous multi-core platform," in *Real-Time and Embedded Technology and Applications Symposium (RTAS), 2013 IEEE 19th*, April 2013, pp. 205–214.

[15] J.-J. Chen, M.-J. Kao, D. Lee, I. Rutter, and D. Wagner, "Online dynamic power management with hard real-time guarantees," *Theor. Comput. Sci.*, vol. 595, no. C, pp. 46–64, Aug. 2015. [Online]. Available: http://dx.doi.org/10.1016/j.tcs.2015.06.014

[16] V. Legout, M. Jan, and L. Pautet, "Scheduling algorithms to reduce the static energy consumption of real-time systems," *Real-Time Systems*, vol. 51, no. 2, pp. 153–191, Mar 2015. [Online]. Available: https://doi.org/10.1007/s11241-014-9207-7

[17] R. Jejurikar, C. Pereira, and R. Gupta, "Leakage aware dynamic voltage scaling for real-time embedded systems," in *Design Automation Conference, 2004. Proceedings. 41st*, July 2004, pp. 275–280.

[18] J.-J. Chen and T.-W. Kuo, "Procrastination for leakage-aware rate-monotonic scheduling on a dynamic voltage scaling processor," in *Proceedings of the 2006 ACM SIGPLAN/SIGBED Conference on Language, Compilers, and Tool Support for Embedded Systems*, ser. LCTES '06. New York, NY, USA: ACM, 2006, pp. 153–162. [Online]. Available: http://doi.acm.org/10.1145/1134650.1134673

[19] R. Jejurikar and R. Gupta, "Dynamic slack reclamation with procrastination scheduling in real-time embedded systems," in *Proceedings of the 42Nd Annual Design Automation Conference*, ser. DAC '05. New York, NY, USA: ACM, 2005, pp. 111–116. [Online]. Available: http://doi.acm.org/10.1145/1065579.1065612

[20] G. Chen, K. Huang, and A. Knoll, "Energy optimization for real-time multiprocessor system-on-chip with optimal DVFS and DPM combination," *ACM Trans. Embed. Comput. Syst.*, vol. 13, no. 3s, pp. 111:1–111:21, Mar. 2014. [Online]. Available: http://doi.acm.org/10.1145/2567935

[21] M. A. Awan, P. M. Yomsi, G. Nelissen, and S. M. Petters, "Energy-aware task mapping onto heterogeneous platforms using dvfs and sleep states," *Real-Time Systems*, vol. 52, no. 4, pp. 450–485, Jul 2016. [Online]. Available: https://doi.org/10.1007/s11241-015-9236-x

[22] W. Wang and P. Mishra, "Leakage-aware energy minimization using dynamic voltage scaling and cache reconfiguration in real-time systems," in *2010 23rd International Conference on VLSI Design*, Jan 2010, pp. 357–362.

[23] S. M. Martin, K. Flautner, T. Mudge, and D. Blaauw, "Combined dynamic voltage scaling and adaptive body biasing for lower power microprocessors under dynamic workloads," in *Proceedings of the 2002 IEEE/ACM International Conference on Computer-aided Design*, ser. ICCAD '02. New York, NY, USA: ACM, 2002, pp. 721–725. [Online]. Available: http://doi.acm.org/10.1145/774572.774678

978-1-5386-3693-0/18 $31.00 © 2018 IEEE

Fault-tolerant Learning in Spiking Astrocyte-Neural Networks on FPGAs

Anju P. Johnson†, Junxiu Liu*, Alan G. Millard†, Shvan Karim*, Andy M. Tyrrell†,
Jim Harkin*, Jon Timmis†, Liam McDaid* and David M. Halliday†
†Department of Electronic Engineering, University of York, York YO10 5DD, UK
∗School of Computing and Intelligent Systems, Ulster University, Derry BT48 7JL, UK
Email: {anju.johnson, alan.millard, andy.tyrrell, jon.timmis, david.halliday}@york.ac.uk
Email: {j.liu1@, haji_karim-s@email., jg.harkin@, lj.mcdaid@}ulster.ac.uk

Abstract—**The paper presents a neuromorphic system implemented on a Field Programmable Gate Array (FPGA) device establishing fault tolerance using a learning method, which is a combination of the Spike-Timing-Dependent Plasticity (STDP) and Bienenstock, Cooper, and Munro (BCM) learning rules. The rule modulates the synaptic plasticity level by shifting the plasticity window, associated with STDP, up/down the vertical axis as a function of postsynaptic neural activity. Specifically when neurons are inactive, either early on in the normal learning phase or when a fault occurs, the window is shifted up the vertical axis (open), leading to an increase in firing rate of the postsynaptic neuron. As learning progresses, the plasticity window moves down the vertical axis until the desired postsynaptic neuron firing rate is established. Experimental results are presented to show the effectiveness of proposed approach in establishing fault tolerance. The system can maintain the network performance with at least one nonfaulty synapse. Finally, we discuss a robotic application utilizing the proposed architecture.**

Keywords—*Neuromorphic Computing, Fault Tolerance, Self-Repair, Spiking Neural Network, Astrocyte, Field Programmable Gate Array, Bio-inspired Engineering.*

I. INTRODUCTION

Neuromorphic computation is an emerging research domain, which derives inspiration from the architecture of nervous systems of living beings to solve complex tasks. Spiking neurons are core components of many computational models of the brain that aim to improve understanding of brain function. Although neuromorphic systems based on Very Large Scale Integration (VLSI) architecture were introduced in the late 1980s [1], use of FPGAs in the design of spiking neuromorphic architectures started to be used over the last ten years [2]. In Spiking Neural Networks (SNNs), communication and computation happen by an exchange of spatiotemporal patterns encoded as spikes as in biological neurons. Astrocytes have been shown to coexist with neurons [3] where these cells communicate with synapses and neurons, thereby regulating synaptic activity [4]. The astrocyte cells together with spiking neurons form a Spiking Astrocyte-Neural Network (SANN) with a distributed and fine-grained self-repair capability.

Neural activity levels in the nervous system rely on various plasticity mechanisms, environmental variations and developmental changes. Synaptic plasticity is a mechanism used by neurons to counteract excessive excitation or inhibition by adjusting synaptic strengths [5], [6]. In this work, we propose such a synaptic plasticity mechanism in hardware where the activity of a post synaptic neuron is used to modify synaptic weights to overcome faults in the system through learning mechanisms.

Hebbian learning is a synaptic plasticity rule where a synapse between two neurons is strengthened when its pre and post synaptic neurons have highly correlated outputs. Spike Timing Dependent Plasticity (STDP) is an asymmetric form of Hebbian learning induced by tight temporal correlations between the presynaptic and postsynaptic neuron spikes [7], [8]. The Bienenstock, Cooper, and Munro (BCM) synaptic modification rule modulates the postsynaptic activity if it deviates from the required response [9], [10]. In this paper, we use a self-repairing learning rule [11] that uses evidence [12] to explain how the STDP and BCM learning rules co-exist to give a learning function that is under the control of postsynaptic neuron activity.

When faults occur in the pathways between neurons the postsynaptic neurons can fall silent. Silent/near silent neuron presents a weak node in the system. This issue might be due to hardware failures in the system. The proposed work is a solution for faults leading to failures in the interconnections between the neurons. These may be sensor failures, SEUs, stuck-at fault, interconnect fracture, noise, etc. We define repair as the ability of the system to restore firing rates through synaptic weight modulations regulated by learning rules.

To summarize, this paper consists of the following contributions: Firstly, we propose a novel learning rule utilizing STDP and BCM rules, implemented on an FPGA. Secondly, an efficient fault tolerance is demonstrated on the proposed architecture, showing repair capability if at least one healthy synapse exists in the system. Finally, we use the strengths of biological SANN systems with their parallel processing and learning capabilities to solve a real world task.

The rest of the paper is organized as follows. Section II describes the biological background of the self-repairing learning rule. Section III presents a hardware implementation of the proposed self-repair learning methodology. In section IV, we analyze the repair capability of the proposed system implemented on an FPGA. Section V presents experimental results establishing the effectiveness of the proposed scheme. Section VI demonstrates an application of the proposed learning mechanism. Finally, conclusions and future work are discussed in Section VII.

II. SELF-REPAIRING LEARNING RULE

The network uses the Leaky Integrate and Fire (LIF) [13] neuron model (N_1 N_2, N_3 and N_4 in Figure 1), which is a simplified model of a biological neuron widely used in neuromorphic computing. It represents a good trade-off between computational complexity and biological realism. The representation of a LIF neuron is shown in Equation (1).

$$\tau_{mem}\frac{dv}{dt} = -v(t) + R_{mem}\sum_{i=1}^{m} I_{syn}^{i}(t) \qquad (1)$$

where τ_{mem}, R_{mem}, v and I_{syn} are the time constant, membrane resistance, membrane potential and current injected by a synapse respectively. m represents the total number of synapses connected to the neuron. On reaching the threshold voltage, the membrane potential is brought back and held at 0V following a nominal refractory period (2 clock cycles). The expression is evaluated using the Euler method.

A tripartite synapse is a junction between a presynaptic terminal, a postsynaptic terminal and an astrocyte cell. Let us consider a signaling pathway between these three terminals and a GABA interneuron terminal. We assume that the presynaptic spiking frequency (f_{pre}) is same as that of the GABA neuron. Interneuron-astrocyte signaling dynamically affects excitatory neurotransmission. Astrocytes, by the action of retrograde signaling [12], regulates transmission of spikes between two layers of the network (A^* in Figure 1). For example, under low f_{pre}, the interneuron-astrocyte interaction leads to an inhibitory effect representing a low transmission probability (PR) with the associated neuron. However, at high f_{pre}, the effect gets reversed leading to a high transmission probability. These outcomes are due to release of inositol 1, 4, 5-trisphosphate (IP_3), Calcium and Glutamate released within the astrocyte [14]. The above phenomenon controls the spikes arriving at the postsynaptic terminal by regulating the PR and can be mathematically modeled by Equation (2).

$$PR = exp\left(\frac{(f_{fre} - f_s)^2}{2\sigma^2}\right) \qquad (2)$$

In biological systems, dendrites of a presynaptic neuron make multiple connections with a postsynaptic neuron [15]. The propagation delays between these paths may differ. Hence, in the proposed SANN, each pre-post neural coupling has multiple synaptic pathways (8 in our experiments) where each pathway is assumed to be morphologically modeled by a different delay. This design of multiple pathways permits the post synaptic potential to build up in a realistic way. Also, more pathways gives greater fault tolerance, but the cost is in implementation overhead.

To implement a self-repairing mechanism for the SANN, a learning algorithm needs to be designed. In this approach, the STDP [7], [8], together with BCM learning rule [9], [10] are combined to develop the BCM-STDP rule. STDP uses the time difference between presynaptic and postsynaptic spikes to adjust the synaptic weights, where the equations in (3) cause long-term potentiation (LTP) for $\Delta t \leq 0$ and long-term depression (LTD) for $\Delta t > 0$. In this approach,

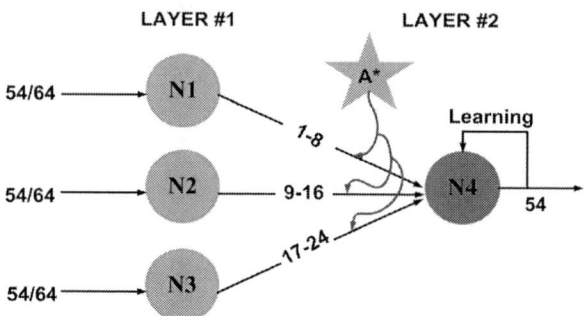

Fig. 1: **Basic unit for fault-tolerant learning mediated by an astrocyte**: Neurons N_1, N_2 and N_3 resides in layer 1 and N_4 resides in layer 2 of the SANN. Astrocyte A^* regulates the PR of inputs received by N_4 from N_1, N_2 and N_3. There are 8 parallel paths (synapses) between each pair of presynaptic and postsynaptic neuron. The input layer receives signals 54 or 64 spikes/window. A^* modulates PR so as to permit the selected pattern to the neuron N_4. Based on the input pattern, the system learns to achieve the required spike rate (54 spikes/window).

potentiation/depression is described by, Equation (3)

$$\delta w(\Delta t) = \begin{cases} +A_0.exp\left(\frac{\Delta t}{\tau_+}\right), \Delta t \leq 0 \\ -A_0.exp\left(\frac{-\Delta t}{\tau_-}\right), \Delta t > 0 \end{cases} \qquad (3)$$

where $\delta w(\Delta t)$ is the weight update, A_0 is the time varying height of STDP learning window controlling the maximum levels of weight potentiation and depression, τ_+ and τ_- control the decay rate of weight updating. $\Delta t = (t_{pre} - t_{post})$ is the time difference between presynaptic spike time (t_{pre}) and postsynaptic spike time (t_{post}). Symmetrical plasticity window is assumed and $\tau_+ = \tau_- = 5$ clock cycles. In addition, the BCM learning rule modulates the height of the STDP plasticity window as a function of the firing rate. This is modelled by Equation (4).

$$A_0 = \frac{A}{1 + exp\left(a\left(f - f_0\right)\right)} - A_- \qquad (4)$$

where f and f_0 are the actual and target firing rate of the postsynaptic neuron (N_4 in Figure 1), respectively. A is the maximum height of plasticity window and A_- is the maximum height of plasticity window for depression. The parameter a is constant which controls the opening/closing speed of plasticity window and is found experimentally to be 0.1. The BCM rule modulates the synaptic plasticity level by A_0, associated with STDP, as a function of postsynaptic neural activity. Specifically when neurons are inactive, either early on in the normal learning phase or when a fault occurs, the window is shifted up the vertical axis (A_0 increases) and as the postsynaptic neuron activity increases, as learning progresses, the plasticity window is moved down the vertical axis (A_0 decreases) until learning ceases.

III. HARDWARE IMPLEMENTATION OF SELF-LEARNING RULES

The basic sub-block implemented on the FPGA for detecting a three bit pattern associated with firing rates is shown in Figure 1. We use a window of size 2^{10} clock cycles to determine the spike rate of the post synaptic neuron. Moving average is used to determine the spike frequency. Input spike trains of frequency 54 spikes/window and 64 spikes/window are used to represent the input pattern. Neurons in layer 1 facilitate transmission of spike trains. There are 8 parallel variable delay paths between the presynaptic and postsynaptic neuron. Astrocyte A^* regulates the spikes received by the neuron in layer 2. For example, let us consider a case in which the sub-block presented in Figure 1 is used for detecting a pattern $(54, 54, 64)$. If pattern $(54, 54, 64)$ is selected, the neurons in layer 1 produces spikes of frequencies 54, 54 and 64 respectively. Astrocyte monitors the spikes and generates transmission PR depending on the pattern to permit the selected pattern in the neuron N_4. Considering a compact hardware implementation, the PR represented in equation 2 consumes more resources and should be approximated. We try to implement equation 2 using a set of piecewise linear equations (8 in our implementation). Due to the astrocyte PR regulation, If the pattern $(54, 54, 64)$ is received by the input neurons, the spike trains are delivered to the neuron N_4. Astrocyte restricts any other patterns of firing rates to be transited to N_4 by lowering PR as per equation (2).

The neurons in the system are LIF and have the following parameters. $R_{mem} = 1M\Omega$, V_{th} (threshold voltage) $= 15mV$, $\tau_{mem} = 10ms$ and resting membrane potential $V_r = 0v$. Euler method of integration evaluates the LIF expresion with a fixed time step of $\Delta t = 2^{-10}$s (an approximation for 1ms). The synapses used in this model are probabilistic in nature. We use a uniformly distributed pseudorandom number generator to produce a random number which is denoted by $rand$. We use a Linear Feedback Shift Register (LFSR) to generate $rand$. If $rand$ is less than or equal to the release PR, synapse injects a current to the neuron.

$$I_{syn} = \begin{cases} I_{inj}, rand \leq PR \\ 0, rand > PR \end{cases} \quad (5)$$

Based on the input pattern, neuron N_4 learns to achieve the required spike rate (54 spikes/window). Learning is achieved using STDP and BCM rules. As per equation 3 and equation 4, if output frequency deviates from the required output frequency (54 spikes/window for neuron N_4), the weights of synapses are updated by a certain amount. Equation 3 is approximated using the following relation.

$$\delta w^*(\Delta t) = \begin{cases} +A_0.2^{\left(\frac{\Delta t}{\tau_+}\right)}, \Delta t \leq 0 \\ -A_0.2^{\left(\frac{-\Delta t}{\tau_-}\right)}, \Delta t > 0 \end{cases} \quad (6)$$

Equation 6 is efficient for hardware implementation as this requires A_0 to be shifted by a certain amount controlled by Δt. Considering a compact hardware implementation, the BCM rule represented in equation 4 consumes more resources and should be approximated. We implement equation 4 using a set of 8 piecewise linear equations (Greater the number of piecewise linear equations higher is the accuracy of approximation). Based on the synaptic weights, a current is generated.

$$I_{inj} = (w + \delta w)^*.\varepsilon \quad (7)$$

where w is the weight of the synapse at the current time step and ε is a scaling factor determined empirically. In our implementation $\varepsilon = 2^{-6}$.

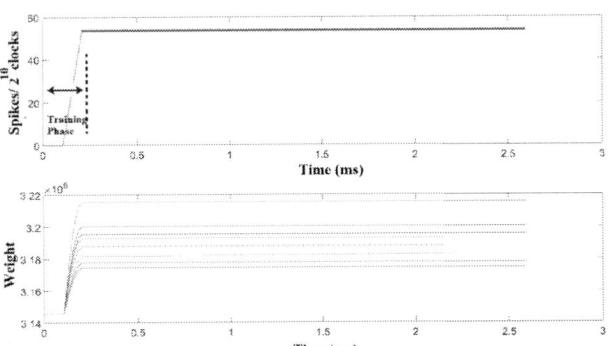

Fig. 2: **(A) Firing rate of N_4 under fault free condition**: All synapses between neurons in layer 1 and layer 2 are fault free enabling 8 parallel delayed paths between each connected neuron. A stable firing rate of 54 $spikes/window$ is established around $0.25ms$. **(B) Synaptic weights under fault free condition**: During the training phase, the synaptic weights (24 plots: 8 synapses $\times 3$) increases and achieves a stable value around $0.25ms$.

One aspect of our model is that it operates at an accelerated biological time scale similar to that in [16], proving to be an efficient realization of real-world tasks compared to [11].

IV. FAULT ANALYSIS

A. No Faults

Three input spike trains of frequency 54 spikes/window, 54 spikes/window, and 64 spikes/window are used for testing the basic functionality of architecture depicted in Figure 1. These input spike trains are compatible with the center frequency of the associated synapses and hence the astrocyte delivers a high PR for this pattern. The target frequency of neuron N_4 is set to be 54 spikes/window. A normal learning phase will occur and the spike rate gradually increases during the training phase and eventually stabilizes at the target frequency of 54 spikes/window in $0.25ms$ as shown in Figure 2-A. Figure 2-B shows the synaptic weights which, as expected, show a slow rise over the learning period and stabilize at $0.25ms$. Additionally, the system is tested with presynaptic spike train frequencies outside the filter window (patterns other than (54,54,64) spikes/window) (results not shown) showed that no learning occurred, which verified that the proposed network is selective to the input spike train patterns.

B. Partial Faults

To evaluate the self-repairing capability of the proposed SANN, the spike train frequency of both N_1, N_2, and N_3 was set to the center frequency of the Gaussian PR curve (54,54,64 spikes/window respectively) and the SANN was trained with a target frequency for N_4 of 54 spikes/window. Next, the system is subjected to faults gradually after every $100ms$ (Figure 3). For example, at $100ms$, one of the path between N_1 and N_4 is broken, we can see an activity drop here in Figure 3-A, which subsequently causes the learning window to re-open and the training process restarts. Likewise, we gradually increase the

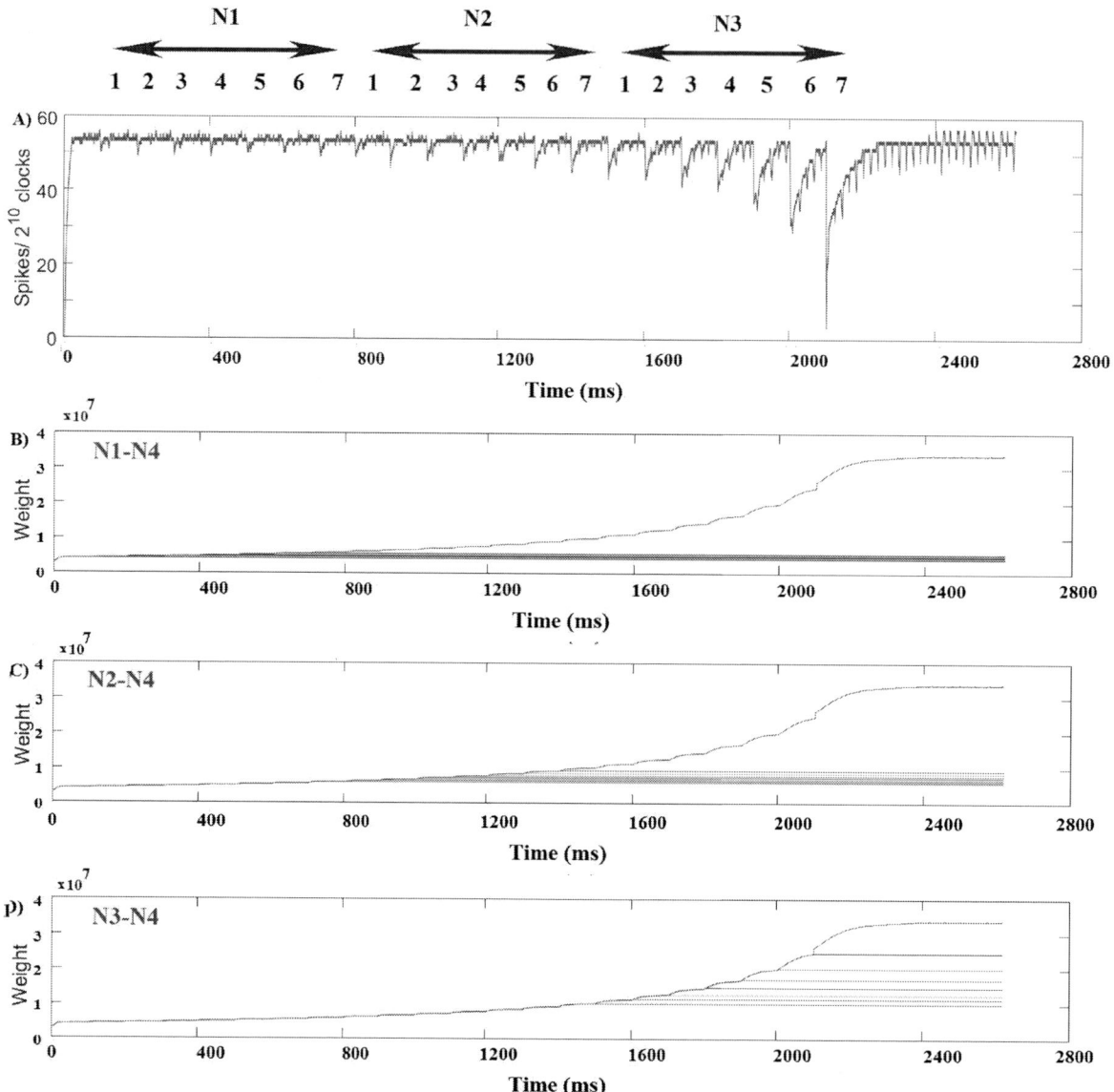

Fig. 3: **(A) Firing rate of N_4 under different faulty conditions**: Synapses between neurons in layer 1 and layer 2 are induced faults gradually (1-7 under neuron N_1 specifies the point in which each synapses between N_1 and N_4 is broken). A stable firing rate of $54 spikes/window$ is established in all faulty cases. **(B) Synaptic weights between N_1 and N_4 under faults of different percentage**: We induce gradual faults on the system after every $100ms$. Only one synapse of N_1 is left unbroken after $700ms$. **(C) Synaptic weights between N_2 and N_4 under faults of different percentage**: We induce gradual faults on the system after $700ms$. The only synapse of N_2 is left unbroken after $1400ms$. **(D) Synaptic weights between N_3 and N_4 under faults of different percentage**: We induce gradual faults on the system after $1400ms$. Only synapse of N_2 is left unbroken after $2100ms$. We can see that for each fault percentages, the systems learn and modulates the synaptic weights for establishing a constant firing rate similar to the fault-free results. The broken synapses do not increase the weight it just retains the weight.

faults in synapses between N_1 and N_4 leaving at least one (No repair happens if all paths between N_1 and N_4 break). At $700ms$, 7 synapses between N_1 and N_4 are broken. Then we induced faults in synapses between N_2 and N_4. At $1400ms$, 7 synapses between N_2 and N_4 are broken and at $2100ms$, 7 synapses between N_3 and N_4 are broken. Hence weights

of nonfaulty synapses increase to compensate for synaptic input to N_4. This process of re-training the network to recover the firing rate of N_4 defines the self-repairing process. The modulated weights of synapses associated with N_1, N_2, N_3 are shown in Figure 3-B, C and D respectively. It is evident from Figure 3-A that the fault repair happens at a rate of ms, proving

TABLE I: Hardware Overhead for the basic unit(Figure 1) implemented on the FPGA

Methodology/Components	Slice	Slice Reg	LUT	BRAM	DSP
8 delayed paths per input	3154	3407	9108	0	128
A_0 Generator	140	32	470	0	16
Moving Average	11	42	42	2	0
LIF Neuron	7	36	78	0	7

to be a very efficient scheme for real world applications. This is a very important capability for the fault-tolerant hardware systems due to the distributed, fine-grained repair capability which will yield a significantly enhanced performance over conventional approaches [17], [18].This enhancement is also due to ability of the system to run at an accelerated time scale.

V. HARDWARE RESULTS

The proposed architecture is implemented on a Xilinx Virtex-7 FPGA board. The firing rates of the output layer neurons are monitored using the *Xilinx ChipScope Pro* analyzer. Power estimation of the circuits was carried out using *Xilinx XPower Analyzer* and timing analysis using *Xilinx Timing Analyzer*. Table I reports the hardware resource footprint of the proposed models. Total on-chip power dissipation of the system is $0.535W$. As evident from these reports, the proposed architecture of neural self-learning can be implemented on the FPGA with minimal hardware overhead and power consumption. But, on considering large network architectures and applications, the proposed implementation should be improved in terms of hardware especially in the amount of DSP consumption. In our implementation, we have 8 equations for realizing PR/synapse and 8 equations for realizing A_0. Each equation consumes 2 DSPs, proving to be a constrain for larger SANNs. We are considering alternative approximation techniques to overcome this issue. The results presented in Figure 2 and Figure 3 corresponds to the system operating at a frequency of $10MHz$. The maximum operating frequency observed for the complete system is $60MHz$. Hence fault recovery can happen 6 times faster than the reported figures.

VI. APPLICATION

The main aim of the proposed work is to implement fault tolerant SANN in FPGA. Once this is established successfully, we apply the concept to a real-world task. There are some works which demonstrate the application of FPGA-based neural networks in solving real world tasks [19]–[21]. Compared to these works, the proposed system has enhanced fault-tolerance and learning capabilities. The emphasis of the selected application is on the fault recovery while establishing the task of robot navigation towards the direction of a colored target (say red). The task implemented by the proposed architecture can be explained in association with Table II. The image of the red target is sensed in forward, right and left direction. Based on the direction of the red target either in forward, right or left direction, the FPGA input pins reflect a value of logic 1 in F_c, R_c or L_c respectively. The system receives information on the presence of an obstacle and is presented a logic 1 in F_o, R_o or L_o on pins of the FPGA in the forward, right or left directions respectively. D represents the final direction of robot movement. In our experiment forward direction has the highest priority and the reverse direction has

TABLE II: Mapping of input sensor reading to output spikes

F_c	F_o	R_c	R_o	L_c	L_o	Direction (D)
0	0	x	x	x	x	Forward
1	0	x	x	x	x	Forward
1	1	x	x	x	x	Forward
0	1	0	0	x	x	Right
0	1	1	0	x	x	Right
0	1	1	1	x	x	Right
0	1	0	1	0	0	Left
0	1	0	1	1	0	Left
0	1	0	1	1	1	Left
0	1	0	1	0	1	Reverse

F_c=1 represents presence of the selected coloured target in the robot's forward direction. F_c=0 represents absence of selected coloured target in the robot's forward direction. F_o=1 represents an obstacle in robots forward direction. F_o=0 represents no obstacle in robots forward direction. F_c=1 and F_o=1 represents an obstacle in robots forward direction, the obstacle is the selected coloured target. Similar meaning holds of the directions Right (R) and Left (L)

the lowest priority (Forward > Right > Left > Reverse). To analyze this let us consider the first case reported in Table II. Here F_c and F_o have a logic 0, since there is no obstacle in the forward direction spikes are delivered in the forward neuron establishing movement in the forward direction. In the second case F_c, F_o=(1, 0) detects the presence of the colored target in the forward direction, hence robot navigates to forward direction. In the third case, we have F_c, F_o=(1, 1), detecting the red target and the obstacle. We assume that the red target is treated as the obstacle and hence the robot navigates towards the red target by having spikes in forward direction. Case F_c, F_o=(0, 1) depicts an obstacle in the forward direction which is not the red target, hence priority goes to the right direction. Similar interpretation holds for the right, left, and reverse directions.

Figure 4 represents the complete system required for navigation considering the specification in Table II. The SANN consists of three layers. The input layer, the hidden layer, and an output layer. The system consists of 6 neurons in the input layer, each detecting either the presence of a red target or an obstacle. These neurons are connected to the hidden layer neuron using 8 variable delay synapses. An astrocyte provides required PR tuning for each pattern and enables the encoded pattern to be passed to the hidden layer neuron. For example, there are 10 patterns to be distinguished by the hidden layer and hence there are 10 astrocytes (not shown in the figure) and 10 hidden layer neurons for detecting these patterns. If there is a spike in the forward direction, this disables spike generation in right, left and reverse direction. The output layer neuron combines the hidden layer neurons responsible for each direction. Since we consider 4 directions, we have 4 motor neurons in the output layer.

VII. CONCLUSION

In this paper, we discussed three contributions. Firstly, we built a fault tolerant neuromorphic architecture for SANNs. Based on this self-repairing mechanism, when faults occur and the synaptic connection is broken, the network still retains the capability to reorganize itself by re-training and consequently recover to the pre-fault mapping. Secondly, we discussed an approximate model of the system for FPGA-based implementations. The work presented represents an initial step towards a new form of fault tolerant designs, with low overhead and

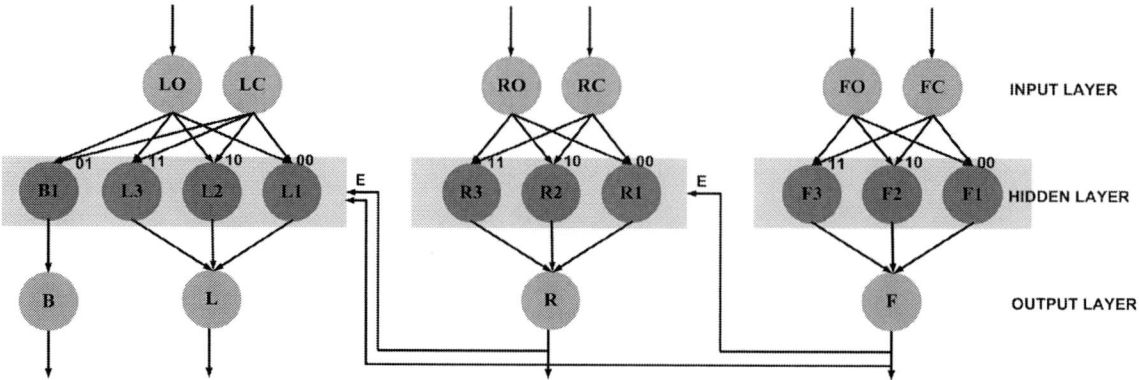

Fig. 4: **Application establishing robot navigation task** There are 6 input layer neurons, each receives information on the presence of a colored target or obstacle. Astrocytes enable specific patterns to be delivered to the hidden layer neuron (Astrocyte is not shown in the figure). There are 4 motor neurons in the output layer, each delivering directionality information for particular directions. The priority of movements is (Forward > Right > Left > Reverse), which is established using binary enable signals (E).

high performance. We are working further towards a more compact architecture based on the proposed methodology for large scale implementations of SANNs. Finally, the proposed idea is applied to a robotic application. The proposed architecture is appropriate for FPGA-based applications running in environments that induce faults in systems, where reliability is crucial.

VIII. ACKNOWLEDGEMENTS

The work is part of the SPANNER project and is funded by EPSRC grant(EP/N007050/1, EP/N00714X/1). Additionally, the authors would like to acknowledge the platform grant(EP/K040820/1) funded by EPSRC.

REFERENCES

[1] C. Mead and M. Ismail, *Analog VLSI implementation of neural systems*. Springer Science and Business Media, 2012, vol. 80.

[2] A. Cassidy, S. Denham, P. Kanold, and A. Andreou, "FPGA Based Silicon Spiking Neural Array," in *IEEE Biomedical Circuits and Systems Conference*, Nov. 2007, pp. 75–78.

[3] L. E. Clarke and B. A. Barres, "Emerging Roles of Astrocytes in Neural Circuit Development," *Nature Reviews Neuroscience*, vol. 14, no. 5, pp. 311–321, 2013.

[4] B. Stevens, "Neuron-astrocyte Signaling in the Development and Plasticity of Neural Circuits," *Neurosignals*, vol. 16, no. 4, pp. 278–288, 2008.

[5] G. G. Turrigiano, "The self-tuning neuron: Synaptic Scaling of Excitatory Synapses," *Cell*, vol. 135, no. 3, pp. 422–435, 2008.

[6] K. Pozo and Y. Goda, "Unraveling Mechanisms of Homeostatic Synaptic Plasticity," *Neuron*, vol. 66, no. 3, pp. 337–351, 2010.

[7] L. F. Abbott and S. B. Nelson, "Synaptic plasticity: taming the beast," *Nature neuroscience*, vol. 3, pp. 1178–1183, 2000.

[8] S. Song, K. D. Miller, and L. F. Abbott, "Competitive Hebbian Learning through Spike-Timing-Dependent Synaptic Plasticity," *Nature neuroscience*, vol. 3, no. 9, pp. 919–926, 2000.

[9] E. L. Bienenstock, L. N. Cooper, and P. W. Munro, "Theory for the development of neuron selectivity: orientation specificity and binocular interaction in visual cortex," *Journal of Neuroscience*, vol. 2, no. 1, pp. 32–48, 1982.

[10] M. Bear and F. Ebner, "A physiological basis for a theory of synapse modification," *WORLD SCIENTIFIC SERIES IN 20TH CENTURY PHYSICS*, vol. 10, pp. 121–130, 1995.

[11] J. Liu, L. McDaid, J. Harkin, J. Wade, S. Karim, A. P. Johnson, A. G. Millard, D. M. Halliday, A. M. Tyrrell, and J. Timmis, "Self-Repairing Learning Rule for Spiking Astrocyte-Neuron Networks (accepted)," in *Proceedings of the 9th International Conference on Neural Information Processing (ICONIP)*, 2017.

[12] J. Wade, L. McDaid, J. Harkin, V. Crunelli, and S. Kelso, "Self-repair in a Bidirectionally Coupled Astrocyte-Neuron (AN) System based on Retrograde Signaling," *Frontiers in computational neuroscience*, vol. 6, 2012.

[13] W. Gerstner and W. M. Kistler, *Spiking Neuron Models: Single Neurons, Populations, Plasticity*. Cambridge university press, 2002.

[14] G. Perea, R. Gómez, S. Mederos, A. Covelo, J. J. Ballesteros, L. Schlosser, A. Hernndez-Vivanco, M. Martín-Fernndez, R. Quintana, A. Rayan, A. Díez, M. Fuenzalida, A. Agarwal, D. E. Bergles, B. Bettler, D. Manahan-Vaughan, E. D. Martín, F. Kirchhoff, and A. Araque, "Activity-dependent Switch of GABAergic Inhibition into Glutamatergic Excitation in Astrocyte-neuron Networks," *Elife*, vol. 5, pp. 1–26, Dec. 2016.

[15] J. Smith and M. Martonosi, *Space-Time Computing with Temporal Neural Networks*, ser. Synthesis Lectures on Computer Architecture. Morgan & Claypool Publishers, 2017.

[16] N. Jing, J.-Y. Lee, Z. Feng, W. He, Z. Mao, and L. He, "SEU Fault Evaluation and Characteristics for SRAM-based FPGA Architectures and Synthesis Algorithms," *ACM Transactions on Design Automation of Electronic Systems (TODAES)*, vol. 18, no. 1, p. 13, 2013.

[17] A. P. Johnson, D. M. Halliday, A. G. Millard, A. M. Tyrrell, J. Timmis, J. Liu, J. Harkin, L. McDaid, and S. Karim, "An FPGA-based hardware-efficient fault-tolerant astrocyte-neuron network," in *IEEE Symposium Series on Computational Intelligence (SSCI)*. IEEE, 2016, pp. 1–8.

[18] J. Liu, J. Harkin, L. P. Maguire, L. J. McDaid, and J. J. Wade, "SPANNER: A Self-Repairing Spiking Neural Network Hardware Architecture," *IEEE Transactions on Neural Networks and Learning Systems*, 2017.

[19] A. P. Johnson, J. Liu, A. G. Millard, S. Karim, A. M. Tyrrell, J. Harkin, J. Timmis, L. J. McDaid, and D. M. Halliday, "Homeostatic Fault Tolerance in Spiking Neural Networks: A Dynamic Hardware Perspective," *IEEE Transactions on Circuits and Systems I: Regular Papers*, vol. PP, no. 99, pp. 1–13, Jul. 2017.

[20] A. P. Johnson, J. Liu, A. G. Millard, S. Karim, A. M. Tyrrell, J. Harkin, J. Timmis, L. McDaid, and D. M. Halliday, "Fault-tolerant Learning in Spiking Astrocyte-Neural Networks on FPGAs (accepted)," in *2017 International Conference on Field-Programmable Technology (FPT)*, Dec. 2017.

[21] J. B. George, G. M. Abraham, B. Amrutur, and S. K. Sikdar, "Robot Navigation Using Neuro-electronic Hybrid Systems," in *28th International Conference on VLSI Design*, Jan. 2015, pp. 93–98.

2018 31th International Conference on VLSI Design and 2018 17th International Conference on Embedded Systems

FPGA Implementation of an Improved Watchdog Timer for Safety-critical Applications

Ravi Krishnan Unni*, Vijayanand P.† and Y. Dilip‡

Aeronautical Development Establishment, Defence Research and Development Organization
Bangalore, Karnataka - 560075, India
Email: *rk_unni@ade.drdo.in, †vijayanand@ade.drdo.in, ‡ydilip@ade.drdo.in

Abstract—Embedded systems that are employed in safety-critical applications require highest reliability. External watchdog timers are used in such systems to automatically handle and recover from operation time related failures. Most of the available external watchdog timers use additional circuitry to adjust their timeout periods and provide only limited features in terms of their functionality. This paper describes the architecture and design of an improved configurable watchdog timer that can be employed in safety-critical applications. Several fault detection mechanisms are built into the watchdog, which adds to its robustness. The functionality and operations are rather general and it can be used to monitor the operations of any processor based real-time system. This paper also discusses the implementation of the proposed watchdog timer in a Field Programmable Gate Array (FPGA). This allows the design to be easily adaptable to different applications, while reducing the overall system cost. The effectiveness of the proposed watchdog timer to detect and respond to faults is first studied by analysing the simulation results. The design is validated in a real-time hardware by injecting faults through the software while the processor is executing, and conclusions are drawn.

I. INTRODUCTION

For applications where a system crash could lead to human injury, highest reliability is required. Such systems should have fault tolerance mechanisms that account for the unexpected to ensure proper safety of operation. These systems should also be able to recover from a crash without any human assistance. These fault tolerance mechanisms detect when a fault occurs in order to handle the fault and to limit the system downtime [1]. One way to achieve fault tolerance is by implementing system redundancy. By using multiple copies of the critical components of the system, the overall system reliability is enhanced [2]. However, this improved system reliability is achieved through increased hardware and software complexity, depending on the type of architecture used.

When developing a fault-tolerant system, one of the most cost effective ways of detecting and handling operation time related failures is the watchdog [3]. A watchdog timer (WDT) is a hardware subsystem that monitors the operations of the system and takes certain actions in the event of detecting a fault [4]. It typically consists of a timer circuit and the processor is required to periodically reset the timer. If the WDT expires, it is a secondary indication of some problem with the system under observation [5]. When the processor fails to reset the watchdog, a decision is made to restart the

system or put the system into a known state from which it can recover, thus preventing further damages.

A watchdog can be internal (on-chip) or external to the processor. Internal watchdog reduces the hardware complexity and cost, however, is not a robust solution. The software has control over it during runtime and a runaway code can disable the watchdog timer [3]. Moreover, since it is connected to the processor clock, a crystal failure will make the watchdog incapable of monitoring the hardware for faults [6]. When the reliability of an embedded system is crucial, external watchdogs become unavoidable. An external watchdog runs independent of the processor and does not share its clock with the processor. This overcomes the limitations of internal watchdogs and leads to much more robust fault-tolerant system architectures [7].

A class of standalone watchdog timer microchips offer only fixed timeout periods, which make them less generic. Other set of devices allow adjusting the timeout periods by using additional external circuitry. Though useful, this method adds to the complexity of the hardware and increases the overall system cost. The increased cost and complexity of external watchdogs can be managed to a certain extend by realizing the watchdog functionality within a Field Programmable Gate Array (FPGA). Many of the modern embedded systems incorporate one or more FPGA devices to accomplish the desired system functionality [8]. Accommodating the watchdog timer within a FPGA can yield an efficient and robust solution.

The work done by Giaconia et al. [9] considered the implementation of a custom concurrent watchdog processor in FPGA for real-time control systems. The design did not provide a timer for the processor; rather, it performed a reasonableness check on some variables and a basic program flow check. El-Attar et al. [10] proposed a sequenced watchdog timer that used time registers to determine whether or not a fault has occurred. However, it did not offer much configuration options and the fault detection features implemented were limited. In [11] the authors addressed the basic concepts of a multiple hardware watchdog timer system in FPGA, but kept the design of the watchdog simple.

In this paper, we propose the design of an improved windowed watchdog timer and its implementation in FPGA. Realizing the design in FPGA means that the same watchdog hardware can be interfaced to different processors and systems, with only minor modifications of the associated

978-1-5386-3693-0/18 $31.00 © 2018 IEEE

hardware description language (HDL) code [8]. It also allows for accommodating multiple watchdog timers for multicore architectures. The proposed watchdog timer is well suited for safety-critical embedded systems, where redundant channels are employed to enhance the system reliability. Designing the WDT as a reusable IP core also addresses the component obsolescence issues faced by many embedded systems, especially those in the aerospace and military applications [12]. The paper describes the architecture of the proposed watchdog timer, the fault detection features, and its implementation in FPGA.

The remainder of this paper is organized as follows. The following section introduces the architecture of the proposed watchdog timer. The fault detection mechanisms built into the watchdog is discussed in III. Section IV describes the implementation of the watchdog timer in FPGA. Simulation results and evaluation of the design in hardware are detailed in section V. Finally, section VI concludes this paper.

II. PROPOSED WATCHDOG TIMER ARCHITECTURE

An effective watchdog should be able to detect all abnormal software modes and bring the system back to a known state. It should have its own clock and should be capable of providing a hardware reset on timeout to all the peripherals [3]. The watchdog timer proposed in this paper operates independently of the processor and uses a dedicated clock for its functions. The architecture follows a windowed watchdog implementation, where the window periods can be configured by the software during initialization. A fail flag is raised when the watchdog timer expires and after a fixed amount of time from raising the flag, a reset is triggered. The time in-between can be used by the software to store valuable debugging information to a non-volatile medium.

A standard watchdog timer can catch problems in the system such as hanging because of endless loops in code execution. However, the main disadvantage of this watchdog is that if the system enters a fault state in which it continually resets the timer, the error state will never be detected. In other words, a standard watchdog timer can detect slow faults, but cannot detect fast faults which occur within the watchdog timer period [13]. However, a windowed architecture can handle this properly. Here the watchdog defines a small time window within which the watchdog must be reset in order to avoid a timeout. This provides protection against systems from running too fast and too slow [14], thus increasing the error recognition coverage.

A. I/O Interface and Configuration

Fig. 1 shows the input-output (I/O) interface of the proposed watchdog timer. The watchdog has two outputs, namely the watchdog fail output (WDFAIL) and the reset output (RSTOUT). When the SYSRESET input is low, the WDFAIL output remains asserted and the RSTOUT output stays deasserted. The design also consists of a configuration register with bit fields defined as in the figure. The register enables adjustments to the watchdog parameters and also provides

status information. The WDRST and WDSRVC fields are used respectively for resetting and servicing the watchdog. The state of the INIT input and the WDFAIL output are automatically updated in the configuration register. The SWSTAT field holds the state of the service window and the FLSTAT field logs the watchdog failure mode, if any. The control inputs to the watchdog timer, ENABLE and RD/WR, permit the read and write to the configuration register. The ABUS and DBUS signals in the figure indicate address bus and data bus, respectively.

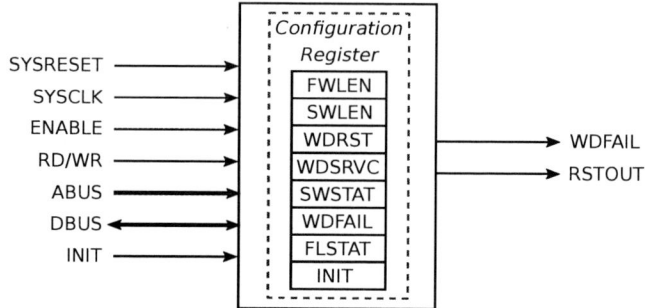

Fig. 1. Watchdog timer input-output interface and configuration register

The proposed windowed watchdog design constitutes a service window and a frame window. The service window duration will be much smaller compared to that of the frame window. The length of the two windows can be programmed by the software after power-up by writing to the bit fields, SWLEN and FWLEN, in the configuration register. Once the window periods are configured after power-up, modifying the values is disabled by design. If needed, the software will have to go through a stringent unlock procedure in order to be able to once again write to the configuration register. This prevents any accidental modification of the watchdog window parameters by a runaway code.

The INIT input to the watchdog timer initializes the service window. A high-to-low transition on this input will start the service window, provided the fail flag (WDFAIL) is not active. The processor is required to service the watchdog within the service window, in order to prevent a timeout. The watchdog timer is serviced using the watchdog service (WDSRVC) field in the configuration register. A rising edge on this bit inside the service window will immediately close the window and start the frame window. The frame window defines how periodically the watchdog should be serviced. Typically, the duration of this window is kept slightly more than the main loop of the embedded control system and the watchdog is serviced once in every cycle [15].

The INIT signal to the watchdog timer can be driven in different ways. One way is to trigger the INIT signal at the end of the main loop, after performing some sanity checks [16]. An external interval timer may be used to avoid any processor intervention in generating the INIT signal. The frame window in this case should be set for a length slightly more than the execution time of the main loop. This mode of configuration is particularly suitable for embedded systems that schedule their tasks in frames.

B. Watchdog Timer Initialization

On power-up or reset the watchdog wakes up in a failed state, *i.e.*, the WDFAIL output will be asserted high. It is the responsibility of the software to initialize the watchdog and keep it running. Fig. 2 illustrates the waveform for watchdog reset initialization and general operation. In order to bring the watchdog to a working state, first the watchdog reset (WDRST) field in the configuration register must be toggled from low-to-high. This, followed by servicing the watchdog inside the service window, will de-assert the WDFAIL flag and make it operational. Since the frame window is kept larger than the system frame time, another service window will start before the current frame window expires. When the watchdog is again properly serviced, the frame window will be reinitialized. As long as the frame window counters keep running, no failures will be flagged by the watchdog.

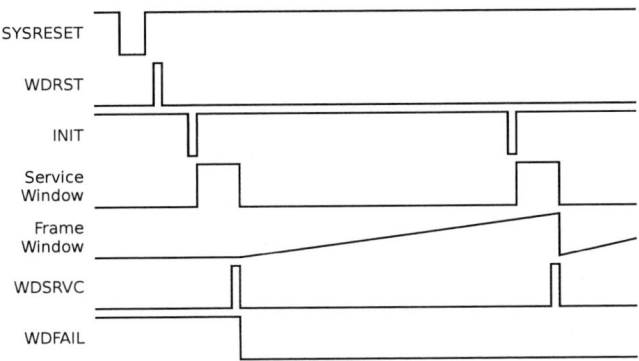

Fig. 2. Watchdog timer initialization and service operation

Critical real-time embedded systems make use of redundancy or diversity to achieve fault-tolerance [17]. Asserting the watchdog fail signal on power-up proves to be a useful feature for such systems. The fail state can be used to indicate that a particular channel is unavailable for computations. Once the watchdog is brought to a healthy state, the channel can be declared online. Moreover, during normal operations if a particular channel is found to be functioning abnormally, the redundancy management logic can activate the watchdog fail of that channel. This can effectively withdraw the faulty channel from taking part in any further computations.

III. FAULT DETECTION FEATURES

Several fault detection mechanisms are built into the proposed watchdog timer in order to improve its effectiveness in capturing erratic software modes. When the software fails to service the watchdog inside the service window, the window expires and sets a fail flag internally. In this case, the frame window does not reinitialize and expires upon reaching its terminal value. On the expiry of the frame window the watchdog asserts its WDFAIL signal, indicating a failure. This failure mode is depicted in Fig. 3.

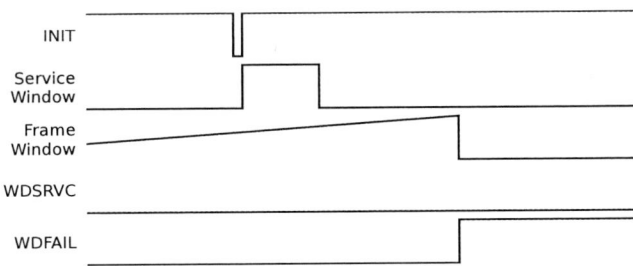

Fig. 3. Watchdog fail due to frame window expiry

A watchdog fail will occur when the software services the watchdog outside the service window, as shown in Fig. 4. It can be seen that the invalid service operation instantly terminates the frame window and asserts the WDFAIL signal. A favourable consequence of this feature is that two successive service operations will also lead to a watchdog fail. Here, the first service operation will immediately close the service window and the next one will invariably occur outside the window. This becomes equivalent to servicing the watchdog outside the service window and leads to a watchdog failure.

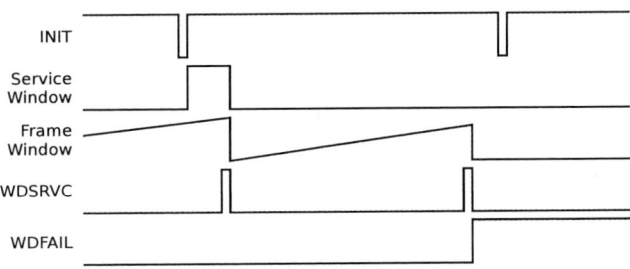

Fig. 4. Watchdog fail due to service outside the service window

Fig. 5 illustrates a scenario where the WDSRVC falling edge is occurring inside the service window. This is also considered as an illegal service operation and the watchdog fail signal is asserted. This implies that, after servicing the watchdog, the software is required to de-assert the WDSRVC signal before the start of the next service window. All of these fault detection mechanisms ensure that a software running haywire will not go undetected by the proposed watchdog timer.

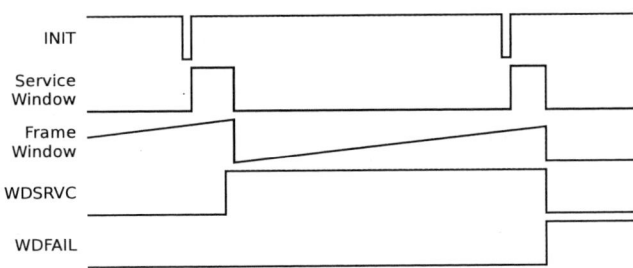

Fig. 5. Watchdog fail due to WDSRVC falling edge inside service window

The WDFAIL output from the watchdog timer can be used to activate a fail-safe state, or warn the processor of the fault by issuing a non-maskable interrupt (NMI) signal. After a predefined amount of time from asserting the WDFAIL output, the watchdog will assert its RSTOUT output. This signal can be tied to the reset pin of the processor, causing it to reset the embedded system automatically. The intervening time will give the software an opportunity to save information that can be valuable for debugging. In the event of a failure, the corresponding failure mode will be logged in the FLSTAT field in the watchdog configuration register. The software can attempt to save this information also to a non-volatile memory for debugging purposes.

IV. WATCHDOG TIMER IMPLEMENTATION IN FPGA

This section details the realization of the proposed watchdog timer in FPGA. The high-level diagram of the watchdog hardware is shown in Fig. 6. The design is clocked by its SYSCLK input, which is independent of the processor clock. The possible sets of window lengths are arrived based on the application and hard-coded in the design. These values can be selected by writing to the appropriate bits in the configuration register - SWLEN for the service window and FWLEN for the frame window - after power-on.

Once the values are selected, the window length configuration fields get locked automatically; *i.e.*, writes to these bits are disabled. For the cases where the window lengths have to be modified again, a 16-bit unlock register is provided in the design. In order to change the window lengths, the software will have to perform two successive writes to this register with data 0xAAAA and 0x5555. Subsequent to writing the first pattern the second one must be written within 10 μs, after which the software gets a 10 μs period to modify the length configuration fields. If these timings are not strictly met, writes to these bits will remain disabled.

The service window is started when a high-to-low transition is detected on the INIT signal. The service window uses a derived clock (SWCLK) that is much slower than the SYSCLK. The slower clock helps in reducing the number of comparators required, thus minimizing the resource utilization in FPGA. The service window has an offset up/down counter that are clocked by the SYSCLK, and a main counter that runs at SWCLK. The offset up counter finds the offset (T_{offset}) between the INIT input and the next rising edge of the SWCLK. This is necessary as the INIT signal may be asynchronously driven and can come at any time within the SWCLK period, T_{swclk}. The offset value is saved and the main counter is started, which then runs for (SWLEN - 1) times. Once the main counter expires, the offset down counter runs for a duration $T_{swclk} - T_{offset}$. This counting procedure allows for a precise control over the window length. The running status of the service window is also updated in the watchdog configuration register periodically.

When the watchdog is correctly serviced, the counters in the service window stop immediately and the frame window starts. The frame window also uses a derived slower clock (FWCLK) for its operations. It has an offset up/down counter and a main counter with functionalities similar to that of the service window. The offset up counter here finds the offset between the termination of the service window and the next rising edge of

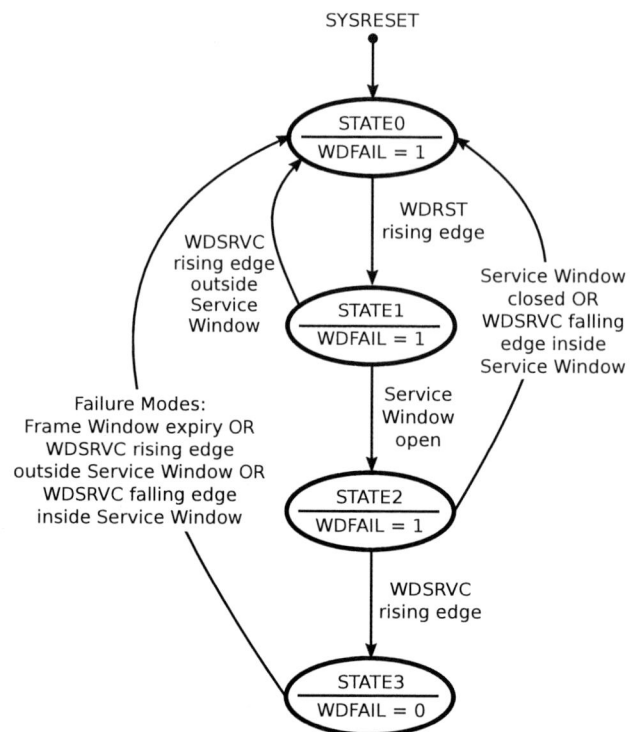

Fig. 7. Finite state machine design of watchdog fault detection logic

the FWCLK. The main counter counts for (FWLEN - 1) times which is then followed by the offset down counter. The frame window counters reset when a watchdog service operation occurs within the next service window duration, before the frame window expires.

A. Reset Initialization and Fault Detection

The diagram in Fig. 7 shows the finite state machine (FSM) implementation of watchdog reset initialization and fault detection logics. On power-up the WDFAIL output is asserted, indicating a watchdog failure. A rising edge on the WDRST bit prepares the watchdog timer for initialization. When the service window opens, a rising edge on the WDSRVC bit de-asserts the WDFAIL output and the window counters start running. However, if the watchdog is serviced incorrectly, the whole initialization process is discarded and the software will have to repeat the entire procedure. The WDFAIL signal gets de-asserted only when the watchdog is properly initialized.

While the watchdog is up and running if any of the failure modes described in section III occurs, the WDFAIL output is again asserted. The configuration register is updated with the failed status and the nature of the failure. Assertion of the watchdog fail also triggers a reset counter that runs for a predefined amount of time. The duration of the counter can be determined by considering the amount of debug information that needs to be stored. On the expiry of the counter, the WDT asserts its RSTOUT output high. The reset counter will be non-functional during power-up and the RSTOUT output will be set to low at this point. When the watchdog is initialized for the first time, the counter gets automatically enabled.

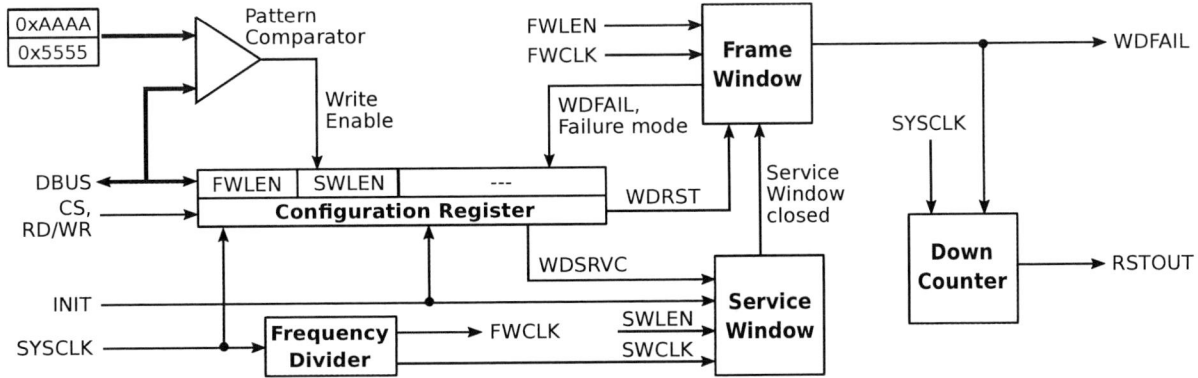

Fig. 6. Functional block diagram of the proposed watchdog timer

V. EXPERIMENTAL RESULTS

The proposed watchdog timer architecture has been implemented using VHDL and realized in a FPGA device. A dedicated 25 MHz clock signal was used for the SYSCLK input. Possible values for the window lengths were calculated based on the application and embedded in the design. In one particular implementation of WDT for an embedded control system, the service window duration could be $100\mu s$, $200\mu s$, $400\mu s$ or $800\mu s$. The frame window had eight selectable options - 1ms, 2ms, 5ms and then up to 30ms in steps of 5ms. The processor could select the desired window lengths by writing the appropriate value to the configuration register.

A programmable interval timer was implemented in the FPGA and the expiry of the timer was used to drive the INIT signal. The WDFAIL output from the watchdog was used as an interrupt request to the processor and the RSTOUT output was connected to the reset pin of the processor. The reset counter was designed to run for 3 milliseconds. This value was arrived after calculating the amount of fault log information that will have to be written to the NVRAM present in the system, in the case of a failure. The duration of the reset pulse from the watchdog timer was also set according to the reset input requirements of the processor.

The proposed watchdog timer design has been simulated using ModelSim software by creating adequate test benches and running the acceptance test procedures (ATP). A processor bus function model was used to access the configuration register and service the watchdog as per the ATP. Fig. 8 shows the simulated waveform for WDT reset initialization. On power-up, the WDFAIL output of the watchdog is asserted high to indicate a failure. It can be seen from the waveform that the service window opens (SWSTAT=1) when the INIT signal goes low. Inside the service window, the WDRST bit is set to high and then the WDSRVC bit is toggled from zero to one. This closes the service window immediately and de-asserts the WDFAIL output.

The functionality of the watchdog for all possible combinations of window lengths were simulated and verified. Using emulation based fault injection techniques, faults were introduced in the test bench models. All the three failure scenarios

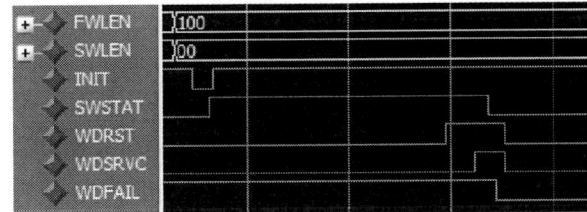

Fig. 8. Watchdog timer initialization from reset

mentioned in section III were created and the response of the design was analysed. In all the cases the watchdog detected the fault, asserted the WDFAIL signal, classified the failure mode and logged the fault in the configuration register, before initiating a system reset. The simulated waveform in Fig. 9 shows the response of the watchdog timer to an improper service operation. It can be seen that the watchdog fail signal, WDFAIL, is asserted within a short time of 81 ns.

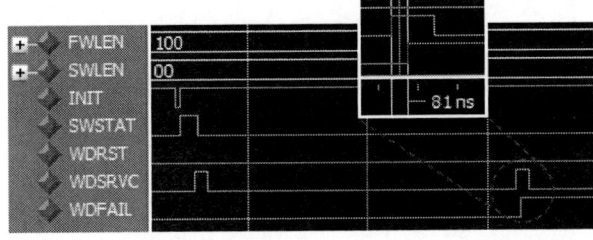

Fig. 9. Declaration of watchdog fail for an improper service operation

A. Design Verification in FPGA

The design has been synthesised and implemented on a Microsemi ProASIC3E series Flash based FPGA. A Flash based FPGA device was chosen for its greater immunity to single event upsets (SEUs), while allowing it to be in-system reprogrammable [18]. The implementation occupied 648 logic elements (3-input LUT equivalent), which amounts to only 1% of the selected device's capacity. The design complexity is slightly more than the work presented in [11]. Here, the authors kept the design of the watchdog very simple and used the expiry of a down counter to indicate watchdog fail. They also illustrated the implementation of such multiple WDTs in

978-1-5386-3693-0/18 $31.00 © 2018 IEEE

one FPGA. The design proposed in this paper, however, uses a windowed architecture and has several fault detection features that are absent in the existing systems. The design can also be extended to accommodate multiple watchdog timers within a single FPGA. The proposed design also has advantages over industry standard microprocessor supervisory circuits such as MAX693/MAX6323/TPS381X in terms of configuration options and fault detection features.

The design implementation has been proved in a real-time safety-critical embedded system equipped with a 32-bit NXP microcontroller. The window configurations for the watchdog were selected according to the system requirements. Software based fault injection method was used to validate the design. This method involves the modification of the software executing on the system in order to provide the capability to alter the system state [19]. A typical hardware failure is when the processor does not service the watchdog timer. This could happen because of faulty memory reads or a software bug. The processor may also mistime the servicing of watchdog due to interrupt overloading, intermittent failures or transient faults. Another case is when the processor services the watchdog too frequently. Based on these actual hardware failures, models were created for fault injection. Sufficient instructions were inserted in the software to enable fault injection. A hardware exception was raised to the processor during run-time to invoke them and to create various failure scenarios.

The software execution was modified to skip the watchdog service operation, allowing the frame window to expire and raise the watchdog fail flag. The software was also made to service the WDT outside the service window, which immediately caused the watchdog fail output to go high. Another scenario was created to service the watchdog two times successively. This was also promptly detected by the watchdog and asserted its fail flag. The setting depicted in Fig. 5 was created in the software and observed that the design behaved as expected. Finally, a hardware based fault injection was introduced to fail the INIT input to the watchdog. Consequently, the service window failed to start and caused the frame window to expire, resulting in a watchdog fail. In all the above cases, the proposed watchdog timer correctly classified the failure mode and logged it in its configuration register. The reset counter was also triggered and after 3ms the watchdog the asserted its RSTOUT output, causing the system to reset.

VI. Conclusion

This paper presented in detail the architecture and design of an improved windowed watchdog timer and its implementation in FPGA. The watchdog timer runs completely independent of the processor and permits adjusting the timer parameters according to the application. Several fault detection techniques are built into the watchdog for the early detection of erratic software modes. It has the capability to identify the failure type and log it, which can become valuable while debugging. Upon detecting a failure, the watchdog timer also allows the software sufficient time for saving the debug information, before initiating a reset.

Implementing the entire design in FPGA has the advantage of making it adaptable and reusable. HDL based designs are vendor-independent and can be used on different FPGA devices with low overhead. The same design can also be customised for different processors and applications with only minor HDL modifications. In addition, realizing the design in FPGA addresses the component obsolescence issues present in long life cycle embedded systems. The implementation has low complexity and takes up very less amount of hardware resources. The proposed design was tested in a real-time safety-critical embedded hardware using fault injection techniques and proved to be effective in handling various faults.

References

[1] S. N. Chau, L. Alkalai, A. T. Tai, and J. B. Burt, "Design of a fault-tolerant COTS-based bus architecture," *IEEE Transactions on Reliability*, vol. 48, no. 4, pp. 351–359, Dec. 1999.

[2] V. B. Prasad, "Fault tolerant digital systems," *IEEE Potentials*, vol. 8, no. 1, pp. 17–21, Feb. 1989.

[3] J. Beningo, "A review of watchdog architectures and their application to Cubesats," Apr. 2010.

[4] A. Mahmood and E. J. McCluskey, "Concurrent error detection using watchdog processors - a survey," *IEEE Transactions on Computers*, vol. 37, no. 2, pp. 160–174, Feb. 1988.

[5] B. Straka, "Implementing a microcontroller watchdog with a field-programmable gate array (FPGA)," Apr. 2013.

[6] J. Ganssle, "Great watchdogs," *V-1.2, The Ganssle Group, updated January 2004*, 2004.

[7] E. Schlaepfer, "Comparison of internal and external watchdog timers application note," *Maxim Integrated Products*, 2008.

[8] P. Garcia, K. Compton, M. Schulte, E. Blem, and W. Fu, "An overview of reconfigurable hardware in embedded systems," *EURASIP Journal on Embedded Systems*, vol. 2006, no. 1, pp. 13–13, Jan. 2006.

[9] G. C. Giaconia, A. Di Stefano, and G. Capponi, "FPGA-based concurrent watchdog for real-time control systems," *Electronics Letters*, vol. 39, no. 10, pp. 769–770, Jun. 2003.

[10] A. M. El-Attar and G. Fahmy, "An improved watchdog timer to enhance imaging system reliability in the presence of soft errors," in *Signal Processing and Information Technology, 2007 IEEE International Symposium on*. IEEE, Dec. 2007, pp. 1100–1104.

[11] M. Pohronská and T. Krajčovič, "FPGA implementation of multiple hardware watchdog timers for enhancing real-time systems security," in *EUROCON-International Conference on Computer as a Tool (EUROCON), 2011 IEEE*. IEEE, Apr. 2011, pp. 1–4.

[12] H. Guzman-Miranda, L. Sterpone, M. Violante, M. A. Aguirre, and M. Gutierrez-Rizo, "Coping with the obsolescence of safety- or mission-critical embedded systems using fpgas," *IEEE Transactions on Industrial Electronics*, vol. 58, no. 3, pp. 814–821, 2011.

[13] H. Amer and A. Sobeih, "Increasing the reliability of the motorola MC68HC11 in the presence of temporary failures," in *Electrotechnical Conference, 2002. MELECON 2002. 11th Mediterranean*. IEEE, May 2002, pp. 231–234.

[14] A. M. El-Attar and G. Fahmy, "A study of fault coverage of standard and windowed watchdog timers," in *Signal Processing and Communications, 2007. ICSPC 2007. IEEE International Conference on*. IEEE, Nov. 2007, pp. 325–328.

[15] M. Barr, "Introduction to watchdog timers," *Embedded Systems Design*, 2001.

[16] N. Murphy, "Watchdogs timers," *Embedded Systems Programming*, p. 112, 2000.

[17] F. Afonso, C. A. Silva, A. Tavares, and S. Montenegro, "Application-level fault tolerance in real-time embedded systems," in *Industrial Embedded Systems, 2008. SIES 2008. International Symposium on*. IEEE, Jun. 2008, pp. 126–133.

[18] M. Wirthlin, "High-reliability fpga-based systems: Space, high-energy physics, and beyond," *Proceedings of the IEEE*, vol. 103, no. 3, pp. 379–389, Mar. 2015.

[19] H. Ziade, R. A. Ayoubi, R. Velazco *et al.*, "A survey on fault injection techniques," *The International Arab Journal of Information Technology*, vol. 1, no. 2, pp. 171–186, Jul. 2004.

978-1-5386-3693-0/18 $31.00 © 2018 IEEE

Image Compression using 2D-Discrete Wavelet Transform on a Light-Weight Reconfigurable Hardware

Nupur Jain, Mansi Singh and Biswajit Mishra
VLSI and Embedded Systems Research Group
Dhirubhai Ambani Institute of Information and Communication Technology
Gandhinagar, 382007 Gujarat
Email: {nupur_jain,mansi_singh,biswajit_mishra}@daiict.ac.in

Abstract—This paper presents a hardware implementation of a 2D-DWT on a reconfigurable architecture targeting image processing applications. The architecture is capable of performing various other digital signal and image processing functions such as CORDIC, FIR filtering, 2D convolution and DCT to compute transforms, trigonometric functions *etc*. In this paper, we present the mapping of a configurable 2D-DWT algorithm using convolution method with separable filter approach having filter length upto 8 taps on the reconfigurable architecture hardware. The reconfigurable hardware architecture mapped with 2D-DWT is ported onto an FPGA that has a frequency of operation of 37.26 MHz. For a 1-level decomposition, the number of clock cycles are 496 per 8×8 block of the N×N image with a total clock cycles equals $31N^2/4$ with 75% compression and can be further improved by computing higher levels of decomposition of the 2D-DWT.

Keywords—2D-DWT, Image Processing, Reconfigurable Hardware, Wavelets.

I. INTRODUCTION

Continuous health monitoring systems require sensor nodes to send the useful information to an aggregator over a wireless network. These nodes, typically, consist of components namely: a controller, radio transmitter, sensor, power supply *etc*. The radio transmitter is responsible for transmission of the sensed data to a processing unit and is usually the most power consuming component of the system. To reduce this, onboard computing capabilities are enhanced; limiting the radio transmission to alarm and/or alerts as opposed to continuous raw data streaming [1]. The controller unit typically generates the data in the form of text, images or signals. For example, the medical images captured by the sensors require large amount of storage memory and huge bandwidth for transmission. Thus, the ability to perform on-node image compression or data manipulation to reduce the size of data for ease of storage and transfer is a desirable feature of the controller unit.

Various compression algorithms, including Discrete Cosine Transform (DCT) and Discrete Wavelet Transform (DWT), have been developed over the past years. The DWT is localized in both time and frequency thereby, decomposing the signal into various frequency sub-bands making it possible to achieve high compression ratio and perform scalable image transmission in terms of quality and resolution. Several architecture have been reported for computing 1-D DWT and 2D-DWT in [2–6] with two basic classification: first is convolution based,

second is lifting based. The lifting scheme breaks the high-pass and low-pass decomposition as well as reconstruction filters into upper and lower triangular matrices thereby converting filter implementation into banded matrix multiplications. The complete scheme is divided into three steps, namely: splitting, predicting and updating [2]. The main advantage of this scheme is that it decreases the number of computations as compared to convolution scheme and increases the hardware utilization. Even after such benefits, it has irregular structure, poor scalability and long critical paths [4].

Convolution based approach is based on filter banks. In contrast to lifting, it has regular structure, good scalability and can be pipelined but suffers from large hardware requirement. This has been overcome by folded structure proposed in [3] that has two types of architecture: word-level folded architecture and the bit-level folded or digit-serial architecture. In the word-level folded architecture, the computations of all wavelet levels are folded to one low-pass and high-pass filter. This gives low latency but it increases hardware area and results in complex routing, whereas in bit-level folded digit-serial architecture, a digit-serial processor processes more-than-one but not-all bits of a word or sample [3]. It has simpler routing and interconnections as well as higher hardware utilization with an increase in the system latency.

Furthermore, the 2D-DWT algorithm can be divided into separable filter approach and non-separable filter approach. In separable filter, the input gets divided into approximation and three detailed coefficients by low pass filtering and high pass filtering along both row and column separately followed by decimation by 2 as shown in Fig. 1. In contrast to this, the non-separable filter has input that gets divided into four set of coefficients in a single step using 2D filter. In [4], the authors have used a 2D non-separable, parallel, recursive architecture and distributive control that allows to change the filter coefficients and the number of levels. The architecture is dedicated to perform DWT and is not configurable for other functions or types. Also, it uses 32 dedicated multipliers increasing the overall gate count and is not suitable for low power applications. In [5], the authors propose a separable 2D-DWT architecture with serial input order. It uses parallel-pipelined computation with near 100% hardware utilization. In [6], the authors discuss a multi-level architecture designed for low power and area dedicated for 2D-Discrete Haar Wavelet Transform (2D-DHWT).

In this paper, we present 2D separable convolution on a reconfigurable multiplierless architecture that can be configured to perform several operations. These operations, such as, convolution, FIR filtering, addition, multiplication, moving average, CORDIC, and DCT, are frequently used in DSP applications [7]. It uses only shift-and-accumulation resulting in serial computation and significant saving in the gate-count making it feasible for ultra-low power applications [8] and is thus focussed towards light-weight processing applications.

The proposed design has novel communication interfaces, such as, a circular memory that provides the data in smaller blocks so as to eliminate the need of large storage memory. Additionally, a UART (Universal Asynchronous Receiver and Transmitter) interface is developed to facilitate analysis of computed data in PC for further post processing. The reconfigurable architecture mapped with 2D-DWT is ported on FPGA for functional verification and proof of concept.

This paper is organized as follows. Firstly, an overview of the 2D-DWT is given in Section II. Section III presents the reconfigurable architecture. Section IV describes the generic 2D-DWT algorithm. Results and comparisons of various DWT architectures are discussed in Section V followed by concluding remarks in Section VI.

II. DISCRETE WAVELET TRANSFORM

The wavelet transform gives time-frequency analysis of the signals while the discrete wavelet transform is defined as complex sequences of N-dimensional vectors over a Hilbert space and helps in multi-resolution analysis (MRA) and sub-band decomposition of signals [9]. The bases are formed by scaled functions of the mother wavelet.

A 1-D DWT decomposes any signal of one level into two signals, approximation and detail, of another higher level. For example, if $A_i(n)$ and $D_i(n)$ are approximation and detail coefficients at level i, respectively. Then the approximation and detail coefficients of next higher level $(i + 1)$ are given by eqn 1 and eqn 2, respectively, where $h(k)$ is the low pass filter and $g(k)$ is the high pass filter and L is the size (or length) of the filter.

$$A_{i+1}(n) = \sum_{k=0}^{L-1} h(k) A_i(2n - k) \qquad (1)$$

$$D_{i+1}(n) = \sum_{k=0}^{L-1} g(k) D_i(2n - k) \qquad (2)$$

In 2D-DWT, the same decomposition occurs in both row and column dimensions. The 2D-DWT decomposes a signal into four frequency sub-bands, namely LL (approximation (cA) matrix), LH (horizontal (cH) matrix), HL (vertical (cV) matrix) and HH (diagonal (cD) matrix). As an example, Fig. 1 depicts the decomposition of an $N \times N$ Lena image into four sub-bands and its reconstruction from those sub-bands by convolution and sampling. Here, h_d(n) and g_d(n) represent low pass and high pass decomposition filter coefficients, respectively, whereas h_r(n) and g_r(n) represent low pass and high pass reconstruction filter coefficients, respectively. The sampling done is dydadic i.e. selecting alternate samples.

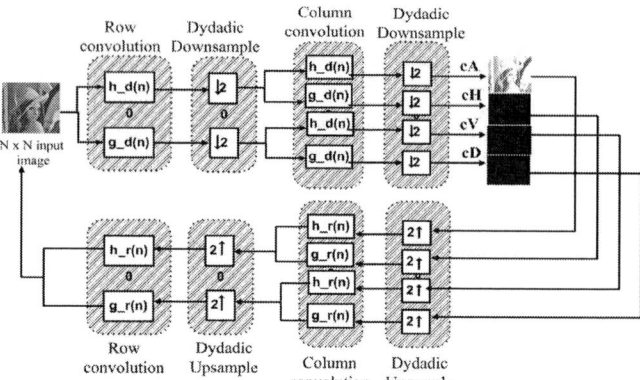

Fig. 1: Decomposition and Reconstruction of an Image using 2D-DWT

III. THE RECONFIGURABLE ARCHITECTURE

Authors in [8], have shown that biomedical signal processing algorithms are largely dominated by digital signal processing operations that can be decomposed into the fundamental function of multiply-accumulate. However, multipliers occupy large area and increases the overall gate count of the design. A multiplierless architecture similar to [10] is adopted here where N- tap filter replaces N multipliers with shift and accumulation (SA) blocks, with a configurable datapath. We propose a novel reconfigurable multiplierless architecture that perform operations such as convolution, FIR filtering, moving average, addition, multiplication, CORDIC and DCT. In this paper, the proposed design architecture serially calculates multiplication by means of generating one partial product per clock and has the latency of 8 clock cycles for a 8-b×8-b multiplication. The proposed design is switching most of the time due to serial execution and offers power advantage in low power, low voltage applications by leveraging the switching power over the dominant leakage power [8].

The proposed architecture contains an array of 36 functional units called Register Unit (RU). The Register Units (RUs) comprises of two 9-bit registers, namely, coefficient register and data register that gets the data from a circular memory. The coefficient register has the coefficients from the coefficient memory and sets the control word for the desired application in addition to the DWT operation. Other functions supported by the reconfigurable architecture are CORDIC, DCT, FIR, multiplication, addition, squaring *etc.* and will be discussed elsewhere. The RUs perform signed multiplication by means of AND-XOR operation between 9-bit (8-b magnitude + 1-b sign) coefficient and data. The serially generated partial products are forwarded to the computation unit. The Computation Unit (CU) consists of a 7-stage ripple carry (RCA) tree adder structure and an accumulator that shifts and accumulate the partial products according to their respective bit weights generating the final output in 8 clock cycles and storing it in an output register. For a detailed architectural design please refer [11]. The datapath and the memory interactions are discussed in the following section.

IV. A GENERIC 2D-DWT ALGORITHM

A. Targeting multiple wavelets

A block-wise convolution based generic 2D-DWT algorithm and its mapping methodology on the proposed reconfigurable architecture is described here. The algorithm targets 28 different wavelets with filter length less than or equal to eight with added benefit of having the same mapping methodology for each targeted wavelet. The block wise convoltion approach suffers from blocking effect which is generally removed by periodic padding that also ensures spatial correspondence between the filter coefficient and the image pixels during filtering operation. For 8×8 DWT computation, the 8×8 subimage is periodically padded on either sides where padding on left and right sides is given by eqn 3 and 4, respectively [9].

$$exthl = floor(lenh/2) \qquad (3)$$

$$exthr = lenh - exthl - 2 \qquad (4)$$

where, $lenh$ is length of the low pass decomposition filter. Since we are considering filter length to be fixed, $lenh$ is taken as 8. Thus, for an 8×8 block of an N×N image, the padded image dimension is 8×14. This is due to $lenh = 8$, $exthl = 4$ and $exthr = 2$ resulting in filter length 14 (8 + 4 + 2) for the padded design. For the wavelets with filter length < 8, zeros are padded on both sides at appropriate locations. For example, for Biorthogonal wavelet bior2.2 with $lenh = 5$, padded filter length increases to 8 by padding 1 zero on right and 2 zeros on left as can be seen in Fig. 2. The proposed architecture can

0	0	-0.1768	-0.3536	1.0607	-0.3536	-0.1768	0

Fig. 2: Coefficients of bior2.2 after zero padding

also perform wavelet transform for all the wavelets whose low pass decomposition filter length is less than or equal to eight. In this work the following 28 wavelets are targetted.

- Haar Wavelet
- Biorthogonal Wavelets : bior3.1, bior2.2, bior1.3, bior3.3
- Reverse Biorthogonal Wavelets: rbio1.1, rbio1.3, rbio1.5, rbio2.2, rbio2.4, rbio2.6, rbio2.8, rbio3.1, rbio3.3, rbio3.5, rbio3.7, rbio3.9, rbio4.4
- Fejer-Korovkin: fk4, fk6, fk8
- Coiflets : coif1
- Daubechies: db2, db3, db4
- Symlets: sym2, sym3, sym4

B. The 2D-DWT Algorithm interpretation for efficient implementation on RU-CU Reconfigurable Architecture

In order to implement image compression, we compute LL sub-band *i.e.* the approximation coefficients. Furthermore, we assume the other three set of coefficients namely horizontal, vertical and diagonal to be zero to achieve a compression of 75% after 1-level decomposition. The compression can be improved by increasing the number of levels of decomposition.

Following steps are performed while performing 2D-DWT on $N \times N$ image.

1) Step 1: Take 8×8 block of $N \times N$ image (I).
2) Step 2: Periodically pad the 8×8 block with four columns on left and two columns on right. Thus, the padded image P_i is of 8×14 dimension.
3) Step 3: To perform row convolution over P_i, take the first 8×8 block of P_i, termed as $P_{(8,1 \to 8)}$, *i.e* columns 1 to 8 of the 8×14 P_i. and multiply each element of the first row of $P_{(8,1 \to 8)}$ *i.e.* $I_{51}, I_{61}, \ldots, I_{21}$ with

the corresponding coefficient *i.e.* $C_8, C_7, \ldots, C_1$ and add all the products to form X_{11} as shown in Eq. 5.

$$X_{11} = I_{51} \cdot C_8 + I_{61} \cdot C_7 + \ldots + I_{21} \cdot C_1 \qquad (5)$$

Now, move the mask over second row of $P_{(8,1 \to 8)}$ and similarly multiply and add to form X_{21} as shown in Eq 6.

$$X_{21} = I_{52} \cdot C_8 + I_{62} \cdot C_7 + \ldots + I_{22} \cdot C_1 \qquad (6)$$

Repeating this for the remaining rows of $P_{(8,1 \to 8)}$, we get the first column of intermediate matrix, X.

4) Step 4: Repeat Step 3 for other three 8×8 blocks of $P_{(8,1 \to 14)}$, namely $P_{(8,3 \to 10)}$, $P_{(8,5 \to 12)}$ and $P_{(8,7 \to 14)}$. This generates the remaining three columns of intermediate matrix, X of order 8×4. We are taking alternate 8×8 block so to eliminate the need of downsampling thus reducing the number of computations.
5) Step 5: To perform column convolution over X, periodically pad four rows on top and two rows on bottom of X giving 14×4 P_x matrix .
6) Step 6: In matrix $P_{x(1 \to 8,1)}$, multiply each element *i.e.* $X_{51}, X_{61}, \ldots, X_{21}$ with the corresponding coefficient *i.e.* $C_8, C_7, \ldots, C_1$ and add all the products to form cA_{11} as shown in Eq. 7. .

$$cA_{11} = X_{51} \cdot C_8 + X_{61} \cdot C_7 + \ldots + X_{21} \cdot C_1 \qquad (7)$$

Move the mask over alternate 8×1 column vector of P_x, *i.e.* over $P_{x(3 \to 10,1)}$, $P_{x(5 \to 8,1)}$ and $P_{x(7 \to 14,1)}$, followed by multiply and add to form cA_{21} as shown in Eq. 8. .

$$cA_{21} = X_{52} \cdot C_8 + X_{62} \cdot C_7 + \ldots + X_{22} \cdot C_1 \qquad (8)$$

Repeat this for all the four 8×1 column vectors. Finally, we get the first column of DWT approximation matrix, cA.
7) Step 7: Repeat Step 6 for other three columns of P_x matrix. This provides the 4×4 approximation matrix cA.
8) Step 8: Repeat steps 1 - 7 for all other 8×8 blocks of $N \times N$ image. We get the approximation matrix of dimension $\frac{N}{2} \times \frac{N}{2}$.

C. Datapath

A 54-bit control word (shown in Fig. 3(a)) is used to configure the RU-CU architecture to perform the generic 2D-DWT. It consists of a code and a config field. The code field represents the operation under execution whereas the config field decides the datapath by means of configuration (CMx) and bypass multiplexers (BPx)(see Fig. 3(b)). In case of generic 2D-DWT the control word is set to "07C56B37EB8006H" where bits [53 − 52] are unused, bits [51 − 47] is the code field for 2D-DWT and is set to "11111". Bits [46 − 16] defines the datapath by forcing the select lines of the configuration and bypass multiplexers to 0 or 1. The bits [15 − 8] are used to define the select lines of the feedback multiplexers and bits [7 − 5] are unused whereas bits [4 − 0] are used to feedback the output value back into the RUs.

The partial products generated after row convolution are parallely fed back into RU# 19-26 to perform column convolution and then keep advancing every 8th clock cycle to subsequent RUs. Therefore, the mapping does not require

Control Word [53:0]

53-52	51-47	46	45	44	43	42	41	40	39	38	37	36	35	34	33	32	31	30	29	28	27	26	25	24	23	22	21	20	19	18	17	16	15	14	13	12	11	10	9	8	7-5	4	3	2	1	0
00	CODE	CM1	CM4A	CM4B	BP4	CM7	BP7	CM10A	CM10B	BP10	CM13	BP13	CM16A	CM16B	BP16	CM19A	CM19B	BP19	CM22A	CM22B	BP22	CM25	BP25	CM28A	CM28B	BP28	CM31	BP31	CM34A	CM34B	CM34C	BP34	FB1(1)	FB1(0)	FB4(1)	FB4(0)	FB19(1)	FB19(0)	FB22(1)	FB22(0)	Unused	fb1	fb4	fb19	fb22	fb34
00	11111	0	0	0	1	0	1	0	1	1	0	1	0	1	1	0	0	1	1	0	1	1	1	1	1	1	1	0	1	0	1	1	1	0	0	0	0	0	0	0	00	0	0	1	1	0

(a)

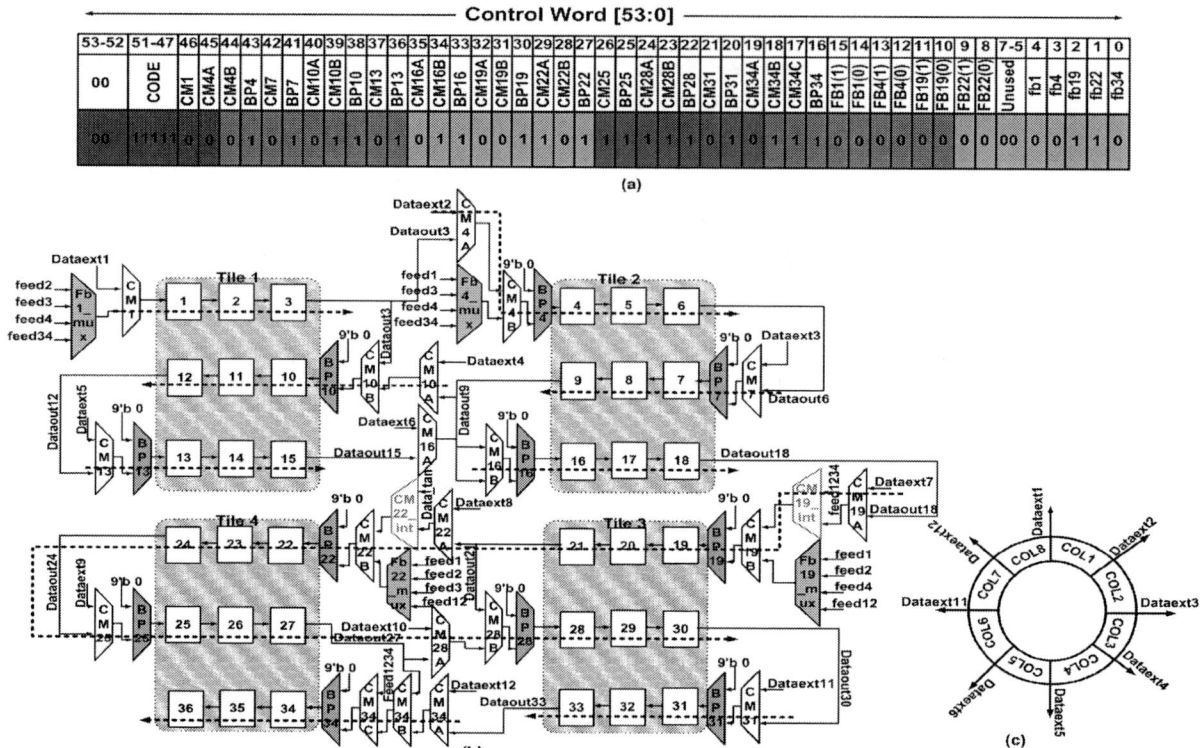

(b)

(c)

Fig. 3: (a) Control Word (b) Datapath for 2D-DWT (c) Circular Memory Interface

additional memory to store the 2D-DWT intermediate results. The control word and the complete datapath for 2D-DWT of the reconfigurable architecture is shown in Fig. 3(b). The RUs hatched are the active RUs and the arrow defines the flow of data.

D. Circular Memory for DWT

The circular data memory, 64 deep and 9-bit wide, configured for 2D-DWT operation has 8 read ports and 1 write port. It can be virtually partitioned into 8 sub-blocks, each having 8 locations. The image data is supplied column wise to the reconfigurable architecture by means of eight read ports as shown in Fig. 3(c).

For an 8×8 sub-image, we pad the sub-image according to the length of the filter. As the filter length is fixed to 8, the resultant padded image has a dimensions of 8×14. This padded image has to be loaded from the block ROM to the circular memory. To process 8×8 block of the 8×14 image, we first write the 8×8 block into column wise manner to 64 locations of the circular memory in 64 cycles. This data is read by the 8 read ports placed at every 8th memory location. The read operation happens once every eighth clock cycles to accommodate the 8 clock cycle latency of the architecture.

The 2D-DWT is computed by calculating alternate convolution using the decomposition filter. One of the four 8×8 sub-block of the padded 8×14 image block is fed to the RU-CU architecture from the circular memory and the first 8×1 column vector is generated as a result of row wise convolution. Due to alternate convolution computation, two subsections of the circular memory containing the first two columns of 8×14 padded image are replaced by 9th and 10th column as shown

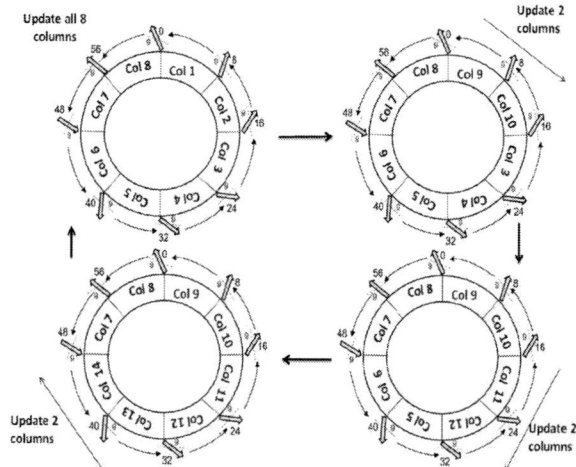

Fig. 4: Circular memory configuration for DWT

in Fig. 4. The architecture generates the second column of row wise convolution when this data *i.e.* column 3 to column 10 of 8×14 padded image block is fed to it. Similarly, the remaining two 8×1 column vectors of the row wise convolution are generated.

E. State Machine for Coefficient loading

The DWT wavelet decomposition filter coefficients are loaded in the coefficient memory of the RUs. This is done by a state machine with five states. The five states of the FSM describes which of the RUs are participating in the datapath and the flow of execution. The user defined wavelet coefficients

TABLE I: State Table.

State	Active RUs	Coefficients loaded	Clock cycles	Value generated
1	1, 4, 7, 10, 13, 16, 31, 34	Ext1, 4, 7, 10, 13, 16, 31, 34	64	Intermediate Matrix I (8×4)
2	19, 20, 21 ,22, 23, 24, 25, 26	Ext10, 7, 4, 1, 34, 31, 16, 13	8	1st Column (4×1) of cA
3	20, 21 ,22, 23, 24, 25, 26, 27	Ext4, 1, 34, 31, 16, 13, 10, 7	8	2nd Column (4×1) of cA
4	21 ,22, 23, 24, 25, 26, 27, 28	Ext34, 31, 16, 13, 10, 7, 4, 1	8	3rd Column (4×1) of cA
5	22, 23, 24, 25, 26, 27, 28, 29	Ext16, 13, 10, 7, 4, 1, 34, 31	8	4th Column (4×1) of cA

are supplied as external input of the coefficient multiplexer. The coefficient multiplexer selects between various pre-defined and user defined coefficients. User defined coefficients are applied to RU # 1, 4, 7, 13, 16, 31 and 34. Additionally, a 9-b 8:1 DWT Mux passes the necessary user defined coefficients to RU # 19-28. The five states and their functioning is described in Table I. State 1 computes the eight elements comprising first row of the intermediate matrix and takes 64 cycles whereas the other four states takes 8 cycles each. The active RUs are loaded with external coefficients listed in Table I. For example, in state 1 RU#1 is loaded with external coefficient 1. The intermediate matrix is computed in state 1 followed by sequential column calculation of matrix cA in the remaining states.

F. Cycle Profiling

The total number of clock cycles for writing of 8×14 block is $64 + 16 \times 3 = 112$ *i.e.* 64 cycles to write 8×8 block and three times 16 cycles for thrice updating two columns as explained in section IV-D. The number of clock cycles required for computation of 4×4 cA matrix for one 8×8 block of image is 384 since $64 \times 4 = 256$ cycles are required to compute the intermediate matrix X and $32 \times 4 = 128$ cycles are required to compute the cA matrix, refer section IV-B. Thus, a total of 496 (112 + 384) cycles are required to write and compute 2D-DWT coefficients per 8×8 block of an $N \times N$ image.

V. RESULTS AND DISCUSSION

The test setup with the circular memory, register unit, computation unit, UART interface, coefficient memory, block ROM and buffer for storing the computed values ported to FPGA is shown in Fig. 5(a) and the actual hardware setup is shown in Fig. 5(b).

A. Block ROM

The control word, wavelet filter coefficients and the image pixels are written to the FPGA ROM. The width is equal to 9 bits (8-bit data and 1-bit sign) and the depth is equal to $42 + 7N^2/4$ (6 bytes of control word, 36 bytes of filter coefficients and $7N^2/4$ bytes image pixels after padding the $N \times N$ image).The operating frequency of the architecture is 37.26 MHz. To match the UART frequency and avoid losing data due to overwriting the frequency of operation is reduced. The baud rate of the UART is set at 115200 bps or 14.4kHz. The architecture generates 128 bytes of data in 440 cycles for each 8×14 block of padded image. For simultaneous data writing to memory and reading from UART, the architecture takes 20.18 μs ((128 x 69.4) / 440) leading to a frequency of 50 kHz. An output buffer of size 256 bytes is used wherein UART starts reading data once 128 bytes of data is written in the buffer. Additionally, by the time UART is finished reading 128 bytes, the second batch of 128 bytes is written into the memory, thereby facilitating uninterrupted data writing and reading.

TABLE II: Comparison of the original and reconstructed Lena images

Image1	Image2	PSNR(dB)	MSE	L2-Norm
Original	MATLAB Reconstructed	74.8819	2e-3	9.9e-1
Original	H/W Reconstructed	74.6535	2.2e-3	9.5e-1
Matlab Reconstructed	H/W Reconstructed	87.7629	1.1e-4	9.6e-1

B. Reconstructed Image

We performed 2D-DWT on the architecture taking standard Lena image of size 128×128 as the input, with bior2.2 wavelet. The architecture computes the approximation coefficients (*i.e.* LL sub-band) for the input image. These coefficients are transferred from FPGA to PC through the RS-232 communication interface. Using these coefficients, we have reconstructed the image after 75% of compression. The original and the reconstructed images both from the MATLAB simulation and the hardware implemented are shown in Fig. 6. These images are compared with MATLAB compressed image in Table II for Lena image of size 128×128. From the given comparisons, one can conclude that the mean square error between the original and the hardware reconstructed image is 0.0022 and the PSNR is 74.6535 dB. The design takes 496 cycles to compute 2D-DWT per 8×8 block of an $N \times N$ image. Thus, for level-1 decomposition the total number of clock cycles is $31N^2/4$. For level-2 decomposition $31N^2/4 + 3N^2/2$ clock cycles are required. For level-3 decomposition $31N^2/4 + 3N^2/2 + 3N^2/8 = 77N^2/8$ clock cycles are required.

TABLE III: Comparison of 2D-DWT Architectures

Architecture	This work	[2]	[4]	[5]	[6]
Input Order	Serial	Serial	Serial-Parallel	Serial	Parallel
Computing Method	Separable	Lifting	Non-separable	Separable	Separable
Number of *	None	3	32	10	None
Number of +	3	4	32	10	4
Frequency(in MHz)	37.26	100	45	25	209
Clock cycles	10 N^2	N.A.	$N^2 + N$	$N^2 + N$	N.A.
Length of filter	upto 8	5 or 9	Any	Any	2

C. Comparison of Architectures

Table III provides the comparison of the proposed work with other similar hardware implementations. Architectures [2, 4, 5] including this work are serial or mixed implementations and makes use of multipliers and thus have high gate count as compared to shift-accumulate architecture implementation in this work. Architecture in [6] is parallel in nature and exhibit highest frequency of operation. The number of clock cycles is largely governed by the nature of execution and is of order N^2 in all the serial architecture. The proposed hardware architecture support DWT filter upto 8 taps, whereas [2] and [6] can support 5 and 2 taps, respectively. This work offers lower gate count and lower power savings due to multiplierless and serial execution.

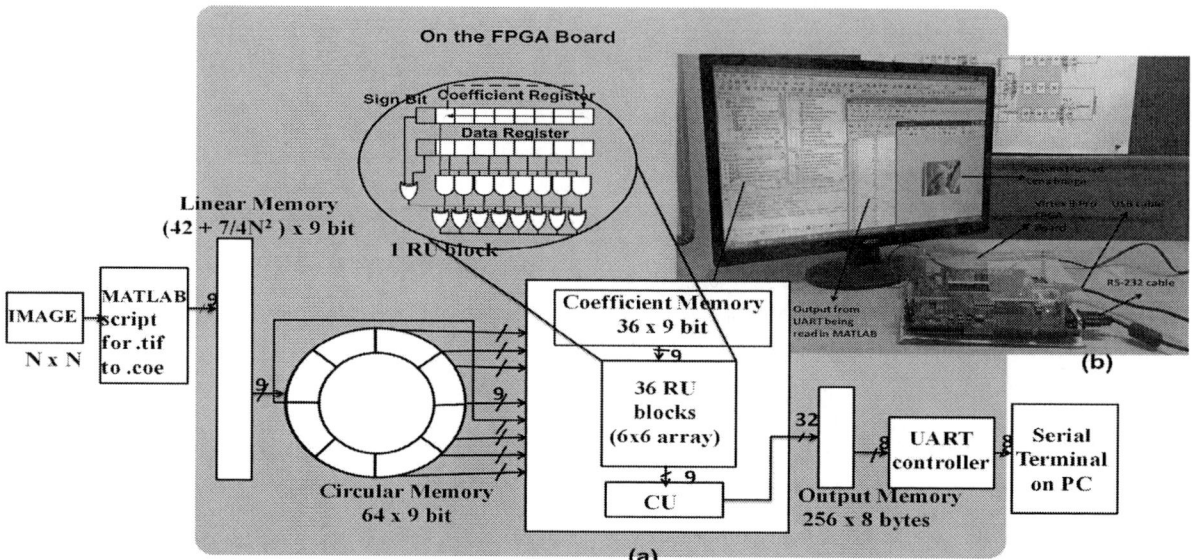

Fig. 5: (a) Test Setup (b) Actual setup.

Fig. 6: Result:(a) Original (b) MATLAB decomposed and reconstructed (c) Hardware decomposed MATLAB reconstructed

VI. CONCLUSION

This work presents 2D-Discrete Wavelet Transform (DWT) mapped on a reconfigurable architecture for image compression. Its functionality is verified by implementing it on the Virtex-II Pro FPGA board. The reconstructed images obtained are compared with the MATLAB counterpart to validate the accuracy of the algorithm. The PSNR obtained for original image and hardware image is $68.1271dB$ and $74.6535dB$ for checkerboard and Lena image respectively with 75% compression. The 2D-DWT computation of a 128×128 Lena image using Bior2.2 wavelet with 8×8 block size takes 496 clock cycles/block with a total clock cycles equal to $77N^2/8$ (= 157696 cycles) for 3 levels of decomposition. The operating frequency of the design is 37.26MHz.

In comparison with other architectures we conclude that the proposed architecture offers comparable performance for lightweight image and signal processing applications. Additionally, the mapping methodology does not require additional memory to store the intermediate results thereby maintaining light weight nature of the hardware.

REFERENCES

[1] V. Pop, "Human+ +: Wireless Autonomous Sensor Technology for Body Area Networks," in *Proc. 16th Asia and South Pacific Design Automation Conference (ASPDAC 11)*, pp. 561-566, January 2011.

[2] X. Chengyi, T. Jinwen, and L. Jian, "Low Complexity Reconfigurable Architecture for the 5/3 and 9/7 Discrete Wavelet Transform," in *Journal of Systems Engineering and Electronics*, vol. 17, pp. 303-308, June 2006.

[3] K. K. Parhi and T. Nishitani, "Folded VLSI Architectures for Discrete Wavelet Transforms," in *1993 IEEE International Symposium on Circuits and Systems*, vol. 3, pp. 1734-1737, May 1993.

[4] R. J. C. Palero, R. G. Gironés, and A. S. Cortes,"A Novel FPGA Architecture of 2D Wavelet Transform," in *Journal of VLSI signal processing systems for signal, image & video technology*, vol. 42, pp. 273-284, March 2006.

[5] M. H. Sheu, M. D. Shieh, and S. W. Liu, "A VLSI Architecture Design with Lower Hardware Cost and Less Memory for Separable 2D-Discrete Wavelet Transform," in *Circuits and Systems, 1998, ISCAS. Proceedings of the 1998 IEEE Intl Symposium on*, vol. 5, pp. 457-460, June 1998.

[6] S. Al-Azawi, "Low-Power, Low-Area Multi-Level 2D-Discrete Wavelet Transform Architecture," *Circuits, Systems, and Signal Processing*, pp. 1–15, April 2017.

[7] J. Kwong, and A. Chandrakasan,"An Energy-Efficient Biomedical Signal Processing Platform", *IEEE J. Solid-State Circuits*, vol. 46, no. 7, July 2011, pp. 1742-1753.

[8] B. Mishra, and B. M. Al-Hashimi, "Subthreshold FIR filter Architecture for Ultra Low Power Applications," in *Integrated Circuit & System Design. Power & Timing Modeling, Optimization and Simulation*, vol. 8, pp. 1-10, September 2008.

[9] A. Skodras, C. Christopoulos, and T. Ebrahimi, "The JPEG 2000 Still Image Compression Standard," *IEEE Signal processing magazine*, vol. 18, pp. 36-58, September 2001.

[10] S. Y. Eun and M. H. Sunwoo, "An Efficient 2D Convolver Chip for Real-time Image Processing," in *Proc. of IEEE Design Automation Conf.*, pp. 329-330, February 1998.

[11] N. Jain and B. Mishra,"DCT and CORDIC on a Novel Configurable Hardware," in *Asia Pacific Conference on Postgraduate Research in Microelectronics & Electronics (PRIME Asia)*, 2015, pp. 51-56.

978-1-5386-3693-0/18 $31.00 © 2018 IEEE

2018 31th International Conference on VLSI Design and 2018 17th International Conference on Embedded Systems

YaNoC: Yet another Network-on-Chip Simulation Acceleration Engine using FPGAs

Prabhu Prasad B M, Khyamling Parane and Basavaraj Talawar
SPARK Lab, National Institute of Technology Karnataka, India.
{ prabhu.cs15f10, khyamlingcs15fv05, basavaraj}@nitk.edu.in

Abstract—In this paper, we present an FPGA based NoC simulation framework, YaNoC, that supports the creation of standard and custom topologies, design of routing algorithms, generation of various synthetic traffic patterns, and exploration of a full set of microarchitectural parameters. The framework supports all standard minimal routing algorithms for conventional NoCs and implements table based routing to support the creation of new routing algorithm. A custom topology called Diagonal Mesh (DMesh) has been evaluated using table based and a modified version of the XY routing algorithm. Mesh and DMesh topologies saturate at the injection rates of 45 % and 55 %. We find that the Table based routing implementation consumes 0.98× fewer hardware resources than the conventional XY routing. We observed the speedup of 2548× compared to the Booksim software simulator. YaNoC achieves speedup of 2.54× and 25× with respect to CONNECT and DART FPGA based NoC simulators.

Keywords: Network-on-Chip, NoC, FPGA, Simulation acceleration, Custom topology

I. INTRODUCTION

As the number of communicating modules increase, bus based communication will not be efficient in terms of throughput, scalability, and performance. The communication time can influence the total turnaround time of the application significantly [1]. The Network-on-Chip (NoC) has become the tangible on-chip communication technique [2].

As thousands of cores play a vital role in the near future, there is a need to model and evaluate large-scale NoC designs quickly and accurately in order to explore the performance characteristics along with the effect on the overall system. This will help the architects to understand the impact of various design parameters before chip fabrication by reducing the total cost. NoC researchers have relied on cycle accurate power and performance simulators (viz. Orion[3], Garnet[4], SICOSYS[5], Booksim[6]) to explore the microarchitectural design space of on-chip networks. As the number of simulated cores is increased, software simulators will become slower [7].

There is a need to model and evaluate large NoC designs quickly and accurately as thousands of cores are targeted in the near future many-core architectures. This will help the architects to understand the impact of various design parameters during design time thus reducing the total cost.

Field Programmable Gate Array (FPGA) is made up of highly reconfigurable look up tables which can be programmed to realize arbitrary logic functions. Events can be executed in parallel as the FPGAs are hardware devices. These features enabled the researchers to propose FPGA based simulators to accelerate simulation by parallelizing various functionalities of a simulator.

In this paper, we present an FPGA based NoC simulation framework called YaNoC. YaNoC supports the creation of stan-dard and custom routing algorithms, generation of synthetic traffic patterns and exploration of a full set of microarchitectural parameters. Key features of the work:

- An FPGA framework for NoC design space exploration.
- Provision for designing custom topologies. The custom topology can be created by connecting the nodes in an arrangement. The framework provides an interface to create custom topologies by allowing the user to connect the router module as per the requirements.
- Provision for designing custom routing algorithms. The template provided for route computation can be modified as per the requirement. The modified module can be instantiated to route the incoming flits in router module.

The rest of the paper is organized as follows: Section II introduces state-of-the-art in the area of NoC software simulation and FPGA emulation. The experimental methodology is detailed in Section III. Design principles of the framework are briefed in IV. Section V explains about design of the custom topology and routing algorithms. Results and inferences are detailed in Section VI. Comparison of the proposed work with state-of-the-art FPGA based NoC simulators has been done in Section VII. The paper concludes in Section VIII.

II. RELATED WORK

In this section, we introduce state-of-the-art in the area of NoC software simulators and FPGA based emulators.

A. Software Simulators

To explore the microarchitectural design space of NoCs, researchers rely upon simulators to evaluate the power and performance. SICOSYS [5] is a general-purpose NoC simulator which allows modeling a wide variety of message routers in a precise way. The parameters such as applied load, traffic pattern, message length etc., can be provided as input for simulation. Noxim [8], is another NoC simulator which is implemented in SystemC. Booksim2.0 [6] is a cycle-accurate simulator. It is flexible in terms of modeling network components. A large set of network parameters which are configurable such as routing algorithm, topology, flow control and router microarchitecture are implemented. Booksim simulator has been validated against the RTL implementation of the NoC router in order to the verify the accuracy of the simulations. 3 × 3 Mesh network with single VC and 16-flit input buffers has been simulated in Booksim and RTL simulations. The maximum difference of 5% in network latency measurements was observed. Simulation speed of Booksim is determined by the complexity of the network configuration and the activity in the network.

Due to the phenomenon of *simulation wall*, the computer simulation performance is ever decreasing relative to the next generation computer being simulated [7]. The software simulation will be slower when there are more number of cores to be simulated. Hence, there is a need for the techniques which can accelerate the software simulators to get the simulation results quickly.

B. FPGA Based NoC Frameworks

FPGA based simulators have been proposed as an alternative to software simulators to enhance simulation speed.

In [9], a NoC emulation environment on FPGA called AcENoCs has been proposed. AcENoCs utilizes both software and hardware components of the FPGA. Traffic generators, clock generation and traffic sinks are implemented on the MicroBlaze softcore processor. The generation of clock on software is flexible but potentially slow. The hardware platform is the network-on-chip to be emulated. Also, it consists a register bank acting as an interface between the NoC under verification and software. This framework support for generic mesh topology, there is no provision for simulating the custom topologies.

An FPGA-based NoC emulator has been proposed in DART [10]. The synthetic and trace-driven workloads are supported. Global interconnect across all the nodes is provided. Any topology can be emulated by DART leaving out resynthesis of design utilizing these global interconnects and employing a software tool to configure the routing tables by configuring the routing tables properly. Most of the FPGA resources are consumed by the global interconnect.

A synthesizable NoC RTL generator called CONNECT has been proposed in [11]. The NoC design decision parameters such as topology, link width, router pipeline depth, network buffer sizing, and flow control have been parameterized. Table based routing algorithm is implemented to route the packets in all topologies.

Another FPGA emulation platform called Ultra-Fast [12] proposes two methods enabling swift emulations of larger NoC architectures on a single FPGA. Synthetic workloads are modeled accurately on FPGA by decoupling time of network being emulated from traffic generation units time. The TDM approach has been employed to emulate entire network utilizing more physical nodes. Ultra-Fast [12] supports design of Mesh topology only.

A fast and flexible FPGA based NoC simulator has been proposed in AdapNoC [13]. Various router microarchitectural parameters are configurable. AdapNoC supports transplantable Traffic Generators and Receptors running on the software side. Dual clock virtualization methodology has been employed to simulate larger topologies and reducing the simulation time drastically. AdapNoC supports Adaptive Toggle Dimension Order Routing (ATDOR) as a known adaptive routing algorithm. AdapNoC is capable of simulating Mesh and Torus topologies only.

The proposed YaNoC framework supports the design of custom topologies and routing algorithms. High-end FPGAs such as Xilinx Virtex 6 and Virtex 7 have been employed for exploring the NoC designs in previous works. YaNoC employs a Xilinx Artix 7 FPGA board efficiently. Mesh topology of size 10×10 was simulated successfully. As the custom Diagonal

TABLE I
CONFIGURABLE PARAMETERS IN YaNoC

Router Parameter	Range of values
Topology	Mesh based, Tree based, Ring based, Custom
Flit buffer depth	Variable
Flit width	Variable
Ports	2 to 16
Routing Algorithms	Standard minimal routing and Table based
Arbitration schemes	Round Robin and Priority based
Traffic patterns	Uniform random, Bit complement

Mesh (DMesh) topology consumes more hardware than Mesh topology, we were able to simulate 6×6 sized DMesh.

III. YaNoC - DESIGN AND IMPLEMENTATION

Fig. 1 shows the architecture of YaNoC which is highly modular and parameterizable. The parameters such as input and output ports, flit width, buffer depth, routing algorithms and arbitration scheme are all reconfigurable as per the designer's requirements. Hierarchy of the components is very well structured with a common interface design allowing plug-and-play of all the modules depending on the topology and routing algorithm. By integrating all the modules, any custom topology required by the designer can be simulated with minimum effort.

A. Router Architecture

Table I lists the architectural parameters that can be configured in YaNoC.

1) Flit Buffer: FIFO implementation has been used for the buffer to store the incoming flits. Depth of the buffer is parameterized to provide the flexibility for the user to explore various kinds of flit width.

2) Flit Structure: Flits with variable widths can be configured for the simulation purpose. Fig. 2(a) shows the structure of head, body and tail flits used in Section VI. Each flit is 32-bit wide. The header flit contains fields for flit type, destination address, timestamp and packet id. Body flit contains the fields for flit type and payload. The tail flit contains the timestamp similar to head flit in order to calculate the latency of the network. The packet format used in our experiments can be seen in Fig. 2(b). Variable flit widths as per the design decision can be configured.

3) Input Output Ports: Reconfigurable ports are advantageous in building various topologies. Mesh and DMesh topology have 5 and 9 ports for communication. Provision for variable ports has been provided to explore various custom topologies.

4) Routing Algorithms: Dimension order (XY) and custom routing algorithms have been employed for Mesh based topology. Table based routing is used in order to support the exploration of custom topologies.

There are look-up tables holding the output ports to all the destinations in the network. For large networks, the entries in routing look up tables will be large. The proposed architecture implements the routing look up tables in Distributed RAMs of FPGA. A single DRAM will be typically of single-bit wide memory with 16-64 elements constrained to a specific FPGA family. As the entries in routing tables are maximum of 3 bits wide, they are mapped very efficiently to DRAMs.

5) Arbitration Schemes: Round Robin and Priority based arbitration schemes have been implemented for fair selection of resources.

978-1-5386-3693-0/18 $31.00 © 2018 IEEE

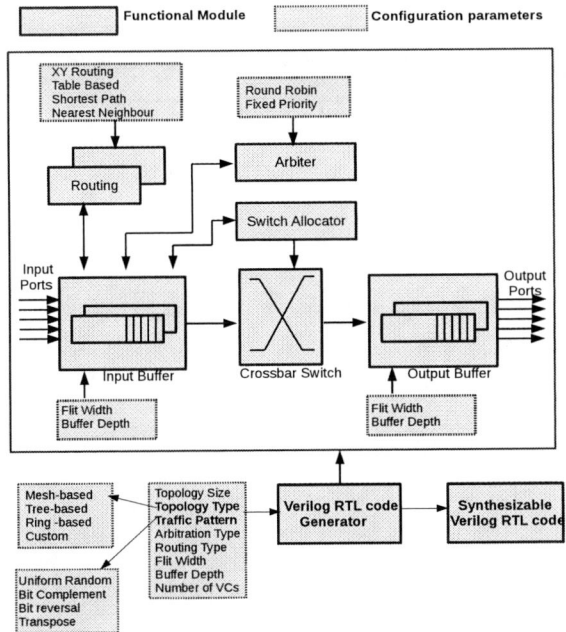

Fig. 1. Architecture of the proposed YaNoC FPGA based NoC simulation acceleration framework

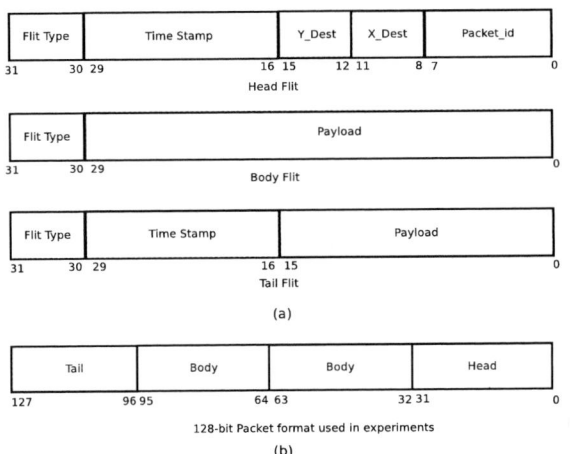

Fig. 2. (a)Flit types and (b)Packet structure used in experiments. (Time stamp field is useful in calculating the latency of a packet)

B. Traffic Generator

The traffic generation module of the simulation framework is capable of generating several types of synthetic traffic. We use Linear Feedback Shift Register (LFSR) mechanism to introduce randomness in the traffic being generated. This module is incorporated in each of the routers.

C. Verilog HDL Code Generator

To quickly get the synthesizable Verilog code, we have developed a Hardware Description Language (HDL) generator implemented in python to generate the synthesizable Verilog

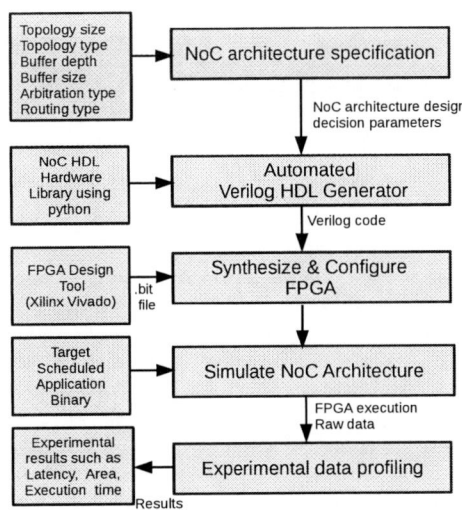

Fig. 3. Simulation Framework Flow

HDL-level code. The HDL generator based on the input parameters such as type and size of topology, link width, flit buffer depth, buffer width, routing algorithm and arbiter type generates the synthesizable Verilog HDL code.

IV. DESIGN PRINCIPLES AND OVERVIEW

Our design was split into two phases namely, correctness phase and implementation phase. This was done to ensure that a design being implemented on FPGA will be functionally correct. In this phase, the design to be simulated on FPGA was thoroughly analyzed considering clock by clock transitions. The clock transitions from each pipeline stage that is Buffer allocation, Route compute, Arbitration, Crossbar allocation, and Link traversal were analyzed. In the implementation phase, with the help of Verilog HDL generator, the HDL for required NoC design was obtained and it was burnt on the FPGA using Xilinx Vivado.

A. Design Overview

Proposed platform consists of a host PC, JTAG cable connecting the host PC and FPGA board and Xilinx Artix 7 FPGA (XC7A100T). The NoC simulation engine is hosted on Artix7 FPGA board.

In order to run the simulations, the required NoC design decision parameters are specified in a configuration file. Based upon these parameters, the HDL generator generates the Verilog code. This code is provided to Xilinx Vivado suite. The FPGA is configured using the bitstream (.bit) file generated from Vivado.

Complete flow of the proposed framework can be seen in Fig. 3. In the first step, microarchitectural parameters of NoC such as topology type and size, buffer depth, flit width, routing algorithm, arbitration type have to be specified in a configuration file. The Automated Verilog HDL generator implemented in Python extracts the information mentioned in configuration file to generate the synthesizable Verilog RTL code for the NoC architecture. The synthesizable Verilog code is then imported in Xilinx Vivado 2016.2. After the synthesis and implementation phases, the bitstream generated will be programmed on the FPGA to simulate the NoC architecture. The experimental

978-1-5386-3693-0/18 $31.00 © 2018 IEEE

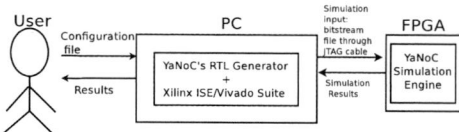

Fig. 4. High-level block diagram of YaNoC consisting of Host PC connected to a FPGA Board.

results of interest such as hardware resource consumption, latency and execution time are extracted. Synthesizable Verilog HDL code for the regular topologies such as Mesh based, Tree based and Ring based can be generated by specifying `topology` name along with other parameters in the configuration file.

For the design specific custom topology, below procedure has to be followed to specify the interconnection in between nodes and routing tables along with other parameters in the config.v file. This file is given as the input to the HDL generator to generate the Synthesizable Verilog HDL.

1) Network Topology: For constructing a topology, one has to interconnect the nodes in a particular arrangement. In YaNoC, the links from source and destination node along with the port numbers are enumerated to specify the network topology. Connection between two nodes 0 and 1 is specified in following lines:

```
#Router_Link_From       Router_Link_To
#(port_num:src_node)    (port_num:neighbor_node)
       1:R0                    2:R1
```

```
#Local_Conn_Port (Port number for connecting the
Processing Element)
#(port_num:node)
       0 : R0
       0 : R1
```

Above mentioned syntax can be followed to design any topology.

2) Routing Tables: Each router contains a routing table that stores mappings from source node to the destination output port. Below lines specify the routing table for the Router ID "0" in a 2×2 Mesh topology.

```
#Dest_Router_ID         Dest_Out_Port
       0                      0
       1                      1
       2                      4
```

B. Front-End Interface

YaNoC consists a JTAG connection in between host PC and the FPGA board. A portal has been developed for interaction with the Simulation engine located on the FPGA and the host PC. Simulation results from the FPGA can be accessed by using the portal as shown in Fig. 4.

V. CASE STUDY: MESH VS DIAGONAL MESH

Proposed framework is capable of simulating any custom topology. This is possible because of the table based routing approach. Apart from the supported Table based routing, if the user wants to design his own routing algorithm, template provided for the route computation (compute.v) can be modified according to the need.

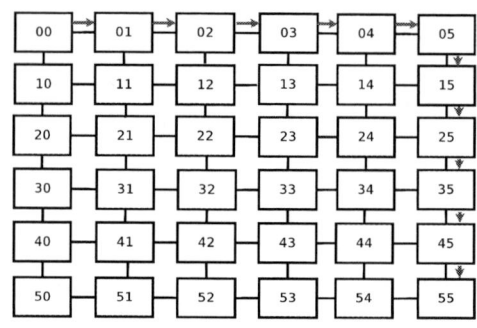

Fig. 5. 6x6 Mesh topology. (Red colored arrow indicates the route computed by XY routing algorithm.)

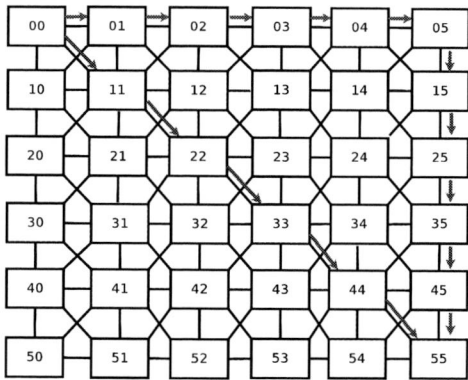

Fig. 6. 6x6 Diagonal Mesh topology.(Red and Green colored arrow indicate the routes computed by XY and modified XY routing algorithms)

For designing the shortest path Dimension order XY routing for DMesh, the logic for computing the shortest path can be implemented in "compute.v" template which can be modified according to the user logic. This compute.v has to be instantiated in the router module (router.v) for the route computation.

Fig. 5 and 6 show the 6×6 Mesh and DMesh topologies. Red line in Fig. 5 indicates the route followed by the conventional XY routing algorithm. In this case, it takes 10 hops to reach the destination "55" from the source "00". The proposed shortest path XY routing algorithm for DMesh topology chooses the shortest path from source to destination. In Fig. 6 it can be seen that it takes only 5 hops from "00" node to "55" node. The same is represented by green arrows.

Table based routing can be used to store routing information in the case of custom topologies whose route compute modules are complex to design.

VI. EXPERIMENTAL RESULTS

Synthesis results of the simulation are extracted from Xilinx Vivado 2016.2. Results include resource usage for Xilinx Artix-7 FPGA (XC7A100T, CSG324 package, speedgrade-3). The topologies were tested with injection rates of 0.005 to 0.8 using Uniform random pattern. Table II shows the experimental setup details used in this paper.

A. Area Utilization

Table III shows the area utilization of individual 5-port and 9-port routers. 9-port router consumes 2× resources than that

978-1-5386-3693-0/18 $31.00 © 2018 IEEE   70

TABLE II
EXPERIMENTAL SETUP DETAILS

Experimental Setup	
Topology	6 × 6 Mesh, Diagonal Mesh
Buffer type	FIFO buffer
Buffer depth	8
Arbiter type	Round-robin
Routing Algorithm	XY (Dimension-order), Modified XY routing and Table based
Router pipeline depth	5-stage
Flow control	Wormhole
Flit size	32-bit
Packet length	4-flits
Traffic pattern	Uniform Random

TABLE III
RESOURCE UTILIZATION OF A SINGLE ROUTER

Resource utilization of Router		
	5-port	9-port
LUT	775	2647
FF	550	1098
DRAM	120	216

of the 5-port router as a complex control logic is required to implement 9-port router.

Table IV shows the resource utilization breakdown of YaNoC router components on the Xilinx Artix7 XC7A100T device.

TABLE IV
RESOURCE UTILIZATION OF 5 AND 9 PORT ROUTER COMPONENTS

	5-port Router	9-port Router
Input buffer	240	522
Router logic	26	127
Arbiter	184	808
Crossbar	301	1093
Allocator	23	95

We were able to simulate 100 nodes of Mesh (10 × 10) on Xilinx Artix-7 FPGA board. For the fair comparison in between Mesh and DMesh topologies, results of simulating 36 nodes (6 × 6) have been presented. Table V shows the results considering XY and modified version of the XY routing algorithms for Mesh and DMesh topologies respectively. It can be seen that the resource consumption of DMesh topology is more compared to the normal Mesh as there are more number of ports (Mesh requires 5 and DMesh requires 9 ports) in all the components of the router. DMesh topology consumes 2.3× resources than the Mesh topology.

YaNoC is also capable of supporting table based routing algorithm for custom topologies. Table V shows the synthesis results of Mesh and DMesh topologies considering the table based routing. The table based routing consumes 12% and 20% fewer LUTs compared to XY and modified version of the XY routing algorithms for Mesh and DMesh topologies respectively. This is because of the replacement of compute logic in these algorithms by the routing tables. The routing tables store route to all the other nodes in the topology.

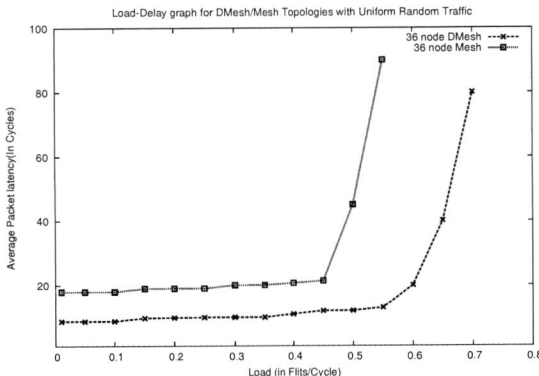

Fig. 7. Load-Delay graph for Mesh and DMesh topologies with various injection rates

B. Latency Analysis

Fig. 7 plots the behavior of average network packet latency vs. injection rate. Packets of length 128-bit comprising a head, two body and a tail flits each of length 32-bits are used in these experiments. It can be seen that the Mesh topology saturates at the injection rate of 45%. DMesh topology sustains the traffic load till injection rate of 55%. This is because of the higher Bisection bandwidth and connectivity of DMesh topology.

DMesh topology achieves the highest performance than Mesh topology with low latency considering various injection rates.

C. Speedup

The simulation time of Booksim simulator was measured on a computer with Core i7 4770 CPU and 8GB memory. The speedup is calculated as the ratio of simulation time in clock cycles of Booksim to the simulation time of YaNoC. The simulation for a 6 × 6 network was run on both Booksim and YaNoC. A speedup of 2548× is observed over Booksim simulator.

TABLE V
SYNTHESIS RESULTS OF YaNoC ON ARTIX-7 FPGA DEVICE
(XC7A100T, SPEED-3)

Flit width=32 -bits Flit buffer depth=8				
	XY Modified XY		Table Based	
	Mesh	DMesh	Mesh	DMesh
%LUTs	35.65	87.55	27.70	67.76
%DRAMs	11.37	20.46	11.12	20.02
%Flip Flops	15.14	20.62	13.08	19.89

TABLE VI
RESOURCE UTILIZATION OF CONNECT ON ARTIX-7 FPGA DEVICE
(XC7A100T, SPEED-3) FOR 6 × 6 MESH AND DMESH TOPOLOGIES

Flit width=32-bits Flit buffer depth=8					
		XY routing		Table based	
		Mesh	DMesh	Mesh	DMesh
CONNECT	%LUTs	44.94	Exceed	43.71	Exceed
	%DRAMs	27.54	Exceed	26.39	Exceed
	%FFs	6.71	Exceed	5.91	Exceed

TABLE VII
RESOURCE UTILIZATION OF DART AND YaNoC ON ARTIX-7 FPGA
DEVICE (XC7A100T, SPEED-3) FOR 3 × 3 MESH TOPOLOGY WITH
FLIT WIDTH=32-BITS AND FLIT BUFFER DEPTH=8

	DART	YaNoC
%LUTs	30	12.41
%DRAMs	21.74	5.68
%FFs	19.28	5.40

VII. YaNoC vs. THE STATE-OF-THE-ART

The CONNECT and DART frameworks were investigated to discover how their RTL code can be optimized for efficient FPGA mapping resulting in minimal area resource utilization.

A. YaNoC and CONNECT

The synthesizable Verilog HDL code of 6×6 Mesh and custom DMesh topologies are generated from CONNECT and YaNoC frameworks for comparing the hardware resource utilization employing Xilinx Vivado 2016.2.

Considering XY routing algorithm, it can be observed from the Tables V and VI that YaNoC's implementation of 6 × 6 Mesh topology consumes fewer resources (35.65% LUTs) than CONNECT's Mesh topology (44.94% LUTs). Similar behavior is observed for the Table based routing algorithm. YaNoC's Table based routing ensures that always the shortest path is chosen between the communicating routers.

Also, for custom DMesh topology, the synthesis will not succeed as there will be a resource crunch while employing CONNECT's implementation. The Input Output Blocks (IOBs) will be exceeding the limit of Artix7 FPGA board. Whereas, synthesis of YaNoC's RTL code succeeds and the corresponding resource utilization is shown in the table. The same behavior is observed for both the XY and table based routing algorithms.

Speedup of 500-1000× and 2548× has been observed in CONNECT and YaNoC respectively with respect to Booksim. YaNoC is 2.55× faster than CONNECT NoC simulation.

B. YaNoC and DART

3 × 3 network with XY routing algorithm of both DART [10] and YaNoC are compared in table VII. It can be observed that the % LUT consumption is 12.41 and 30 for YaNoC and DART implementations respectively. Large topologies (10 × 10 Mesh) can be analyzed by using YaNoC's implementation on a small FPGA board like Artix7. Whereas the DART implementation consumes more FPGA resources and hence it requires high end FPGA boards for the analysis.

DART simulation achieves over 100× speedup relative Booksim. YaNoC is 25× times faster than DART.

VIII. CONCLUSION

This paper presents YaNoC, a Network-on-Chip simulation acceleration framework using FPGAs. We were able to simulate 100 node Mesh topology (10 × 10) with conventional XY routing on a basic Xilinx Artix-7 FPGA board. Along with the conventional routing algorithms, custom routing algorithms can also be designed. YaNoC includes table based routing to support the evaluation of custom topologies. Using this feature, a custom topology called DMesh has been evaluated considering a modified XY routing and table based routing algorithms.

Mesh and DMesh topologies have been compared in terms of latency and hardware resource consumption. From our results, it is evident that the DMesh topology saturates at much higher injection rate (55%) than the Mesh topology (45%). Considering hardware resource utilization, DMesh topology consumes 2.3× resources than the Mesh topology. From these results, it can be concluded that DMesh topology suits well for latency performance sensitive applications. The Mesh topology can be used in the applications where performance can be compromised. The speedup of 2548× compared to the Booksim software simulator was observed using YaNoC. YaNoC is compared with the DART and CONNECT FPGA based NoC simulation frameworks. YaNoC consumes 9.29% fewer resources and is 2.5× faster than the CONNECT framework. Also, YaNoC consumes 17.59% fewer resources and 25× faster than the DART simulator. Employing YaNoC, large topologies of size 10 × 10 Mesh can be analyzed using a FPGA board like Xilinx Artix7.

REFERENCES

[1] P. P. Pande et al., "Performance evaluation and design trade-offs for network-on-chip interconnect architectures," *IEEE Transactions on Computers*, vol. 54, no. 8, pp. 1025–1040, 2005.

[2] W. J. Dally and B. Towles, "Route packets, not wires: on-chip interconnection networks," in *Proceedings of the 38th Design Automation Conference (IEEE Cat. No.01CH37232)*, 2001, pp. 684–689.

[3] A. B. Kahng et al., "ORION 2 . 0 : A Power-Area Simulator for Interconnection Networks," *Tvlsi*, vol. XX, no. 1, pp. 1–5, 2010.

[4] N. Agarwal et al., "Garnet: A detailed on-chip network model inside a full-system simulator," in *ISPASS 2009*, April 2009, pp. 33–42.

[5] V. Puente et al., "Sicosys: an integrated framework for studying interconnection network performance in multiprocessor systems," in *Parallel, Distributed and Network-based Processing, 2002*, 2002, pp. 15–22.

[6] N. Jiang et al., "A detailed and flexible cycle-accurate network-on-chip simulator," in *Performance Analysis of Systems and Software (ISPASS), 2013 IEEE International Symposium on*, April 2013, pp. 86–96.

[7] H. Angepat et al., *FPGA-Accelerated Simulation of Computer Systems*. Morgan & Claypool Publishers, 2014.

[8] Davide Patti. Noxim. [Online]. Available: https://github.com/davidepatti/noxim

[9] S. Lotlikar et al., "Acenocs: A configurable hw/sw platform for fpga accelerated noc emulation," in *VLSID 2011*, Jan 2011, pp. 147–152.

[10] D. Wang et al., "Dart: A programmable architecture for noc simulation on fpgas," in *NOCS 2011*, ser. NOCS '11. New York, NY, USA: ACM, 2011, pp. 145–152.

[11] M. K. Papamichael and J. C. Hoe, "The connect network-on-chip generator," *Computer*, vol. 48, no. 12, pp. 72–79, Dec 2015.

[12] T. V. Chu et al., "Ultra-fast noc emulation on a single fpga," in *FPL 2015*, Sept 2015, pp. 1–8.

[13] H. M. Kamali and S. Hessabi, "Adapnoc: A fast and flexible fpga-based noc simulator," in *2016 26th International Conference on Field Programmable Logic and Applications (FPL)*, Aug 2016, pp. 1–8.

2018 31th International Conference on VLSI Design and 2018 17th International Conference on Embedded Systems

An Energy-Efficient Trusted FSM Design Technique to Thwart Fault Injection and Trojan Attacks

Vijaypal Singh Rathor, Bharat Garg and G. K. Sharma

ABV-Indian Institute of Information Technology and Management, Gwalior 474015, India

Email: {vijaypal, bharat, gksharma}@iiitm.ac.in

Abstract—**Finite state machine (FSM) is a crucial component in most of the digital designs as it controls entire functionality of the system. In case an FSM is vulnerable from setup time violation based fault injection (STVFI) and Trojan attacks, the design becomes untrustable. The existing techniques are inefficient to control unauthorized access to all states from the above attacks and require significant overhead. Therefore, we propose a new energy-efficient trusted FSM design technique that effectively neutralizes the STVFI and Trojan attacks. We also introduce a new metric to identify all possible vulnerable transitions from the fault injection attacks. Further, a new light-weight vulnerabilities mitigation technique is proposed that protects the design from above attacks. The proposed mitigation technique comprises a trusted FSM design algorithm followed by an energy-efficient vulnerabilities mitigation architecture that only allows the access of reachable-set of present state to prevent unauthorized access. The simulation results on the FSMs of AES and RSA encryption modules show that the proposed metric identifies all possible vulnerable transitions, whereas proposed architecture reduces 38.7%, 47.4%, and 60.9% area, power and delay respectively over the existing AVFSM framework.**

Index Terms—**Fault Injection; Hardware Trojan; Trusted FSM Design; Low-Cost Architecture; Vulnerability;**

I. INTRODUCTION

The secure, trustworthy and reliable integrated circuits (ICs) play a major role in the mission-critical applications such as financial, health-care and national defense. The involvement of various untrusted third parties in IC design flow makes it vulnerable to hardware Trojan (HT) attacks [1], [2]. The inserted Trojan can leak sensitive information, reduce the reliability of the system and can also launch other attacks such as denial of service. Further, the presence of wide variety of emerging attacks also poses a major security challenge to the designer for developing a secure and trusted design. It is reported that the security of cryptosystems and microprocessor circuits can be compromised using timing analysis [4], power analysis [3] and fault injection [5] attacks. It is also reported that the fault injection in RSA controller circuit can leak a secret key which poses a severe threat to the security of RSA cryptosystem [6], [7]. Therefore, this paper mainly focuses on fault injection and Trojan attacks.

However, most of the current research primarily focuses on protecting the data path, very few research efforts concentrate on protecting the control path of the design from setup time violation based fault injection (STVFI), and Trojan attacks [8]. As the controller circuitry of the system is typically designed using finite state machines (FSMs), the attacker can compro-

mise the security of the entire system by injecting the fault in FSM [8]. Moreover, the synthesis tool inserts additional don't care states and/or transitions into the original FSM during optimization [9] due to deterministic nature. Therefore, the attacker can also use these don't care states and/or transitions to inject the fault and insert a Trojan to gain unauthorized access to some protected or secured states which consequently makes the design untrusted [9], [8]. However, a trusted FSM design using controlled flip-flops (FFs) is presented in [9]. But it is ineffective for STVFI attack and cannot control indirect unauthorized access of protected states. Further, this approach does not analyze the vulnerabilities arising due to don't-care states and transitions [8].

To quantitatively analyze the vulnerabilities in FSM from the STVFI and HT attacks, Nahiyan et al. presented a framework namely Analyzing Vulnerabilities in FSM (AVFSM) [8]. However, the AVFSM can only identify those vulnerable transitions which directly access some specific type of protected states. It cannot identify all types of vulnerable transitions to protected states from STVFI attacks. The AVFSM framework also fails to prevent unauthorized access to non-protected states and requires significant design overhead. Thus, the existing techniques are not suitable for developing a secure and trusted FSM against STVFI and Trojan Attacks. Therefore, a new energy-efficient secure and trusted FSM design technique is proposed that effectively identifies and mitigates the vulnerabilities against STVFI and hardware Trojan attacks. The major contributions of this paper are as follows:

1) We propose a new trusted FSM design algorithm to thwart STVFI and Trojan attacks.
2) A new metric namely vulnerable transition identification from fault injection (VTI_{FI}) is proposed that effectively identifies all direct and indirect vulnerable transitions to access the protected states in an FSM.
3) We propose a novel energy-efficient architecture to mitigate the vulnerabilities against fault injection and Trojan attacks.
4) Our results show 79.5% and 77.4% reduction in energy consumption over the AVFSM for AES and RSA based FSMs respectively.

The rest of the paper is organized as follows. Section II presents the background of FSM followed by attacks and related work. Section III presents the proposed metric to identify vulnerable transitions in an FSM. Section IV presents

978-1-5386-3693-0/18 $31.00 © 2018 IEEE

73

the proposed trusted FSM design algorithm followed by an energy-efficient vulnerabilities mitigation architecture for secure FSM design. The results and comparative analysis are presented in Section V. Section VI concludes the paper.

II. BACKGROUND, ATTACKS, AND RELATED WORK

This Section first presents the background and basic terminology related to FSM followed by the scenarios for STVFI and hardware Trojans attacks and related work.

A. Preliminaries and Definitions

An FSM is typically defined as a 6-tuple $(\Sigma, Q, \delta, q_0, O, \lambda)$ where Σ set of input symbols, Q is the set of states, $\delta : Q \times \Sigma \to Q$ is the transition function, $q_0 \in Q$ is the initial state, O is the set of output symbols, and $\lambda : Q \times \Sigma \to O$ is the output function [8], [9].

For convenience, an FSM can be represented as directed graph (also known as state transition graph (STG)) where the vertex represents a state $q \in Q$ and edge (labeled with a/b) represents the transition $\delta(x, a) = y$ from current state x to the next state y for the input symbol a while providing the output b. For simplicity, a transition from present state x to next state y can also be denoted as T(x,y) [8]. In an FSM, if both δ and λ are defined for all possible states and input pairs (x, a), the FSM is said to be completely specified; otherwise it is incompletely specified. The unspecified states/transitions are called as don't cares [9].

In the STG, if a state x is accessed by the set of states (y), then the *accessible set of state x* is defined as,

$$A(x) = \{y \in Q \mid x \text{ is accessible from } y\} \qquad (1)$$

Similarly, if a state x can reach to the set of states (y), then the *reachable set of state x* is defined as,

$$R(x) = \{y \in Q \mid x \text{ is reachable to } y\} \qquad (2)$$

Now, let us consider that M' is the implementation generated by the synthesis tool from an incomplete behavioral specification of the FSM M. Due to the deterministic nature and optimization, the synthesis tool introduces additional states (Q_D) and/or transitions (δ_D) for the don't care conditions which are given below [8].

$$Q_D = \{q' | (q' \in Q') \cap (q' \notin Q)\}; \delta_D = \{d' | (d' \in \delta') \cap (d' \notin \delta)\} \qquad (3)$$

where, the Q and Q' are the states whereas δ and δ' are the transitions in the M and M' respectively.

The circuit implementation of a original machine is trusted if and only if for each state $q \in Q$ and its corresponding state $q' \in Q'$, $A(q) = A(q')$ or $R(q) = R(q')$ [9]. If these conditions are violated during the transition from one state to another, then the FSM will become untrusted. Further, if L is a set of authorized states that are only allowed to access the protected state $(p \in Q)$, a trusted FSM can also be defined as $A(p) = \{L \mid L \in Q\}$ [8]. An FSM is untrusted if any p state is accessed by the state apart from the states in L. *However, the second definition [8] is not valid for unauthorized access of non-protected states.* Because, when a non-protected state is unauthorizedly accessed in an FSM, then the FSM is trusted according to second definition [8]. Whereas, this FSM is untrusted according to the first definition [9]. On the other hand, if an FSM is trusted by the first definition then it will also be trusted according to the second definition. Therefore, this paper considers the first definition $(R(q') = R(q))$ for trusted FSM design and also for the further analysis.

B. Fault and Trojan Attacks on FSM

To analyze the vulnerabilities into the FSM against the STVFI and Trojan attacks, we have considered the threat model similar to the one given in [8]. In this model, the attacker can violate the setup time constraints of flip-flops using overclocking, voltage starving, and/or heating the device to inject the faults [10]. These attacks can be performed easily with the low-cost equipment while posing most severe threats [8]. It is shown in [8] that the security of FSM of AES encryption module can be compromised through STVFI and Trojan attacks. Further, the attacker can also use the don't care states/transitions for STVFI and Trojan insertion attacks to gain unauthorized access to protected states [8]. For example, the transition T(10, 01) from present state "10" to next state "01" is vulnerable to fault injection attacks as shown in Figure 1(a). Where, the attacker can perform STVFI attack to go to state "11" by violating setup time constraint of MSB FF while maintaining the setup time constraint of LSB FF. It can be achieved by increasing the logic-path delay of MSB FF while maintaining the logic-path delay of LSB FF [8].

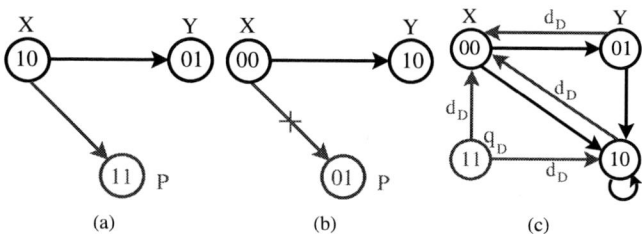

Figure 1. (a) Fault injection is possible for protected state (P) (b) Fault is not possible (c) Trojan insertion using don't care state (q_D) /transitions (d_D).

Similarly, if "00" is a protected state, the fault can be injected by increasing the path delay of LSB FF while maintaining the path delay of MSB FF. On the other hand, the fault cannot be injected for the transition T(00, 10) to go to the protected state "01" as the LSB bits of both present and next states are same. Apart from the fault attacks, the attacker can also exploit the don't care states/transitions to insert a hardware Trojan into the design which may provide unauthorized access to some secure states [9] to the attacker. For example, consider the STG as shown in Figure 1(c) where the state "00" is secure because it is originally inaccessible to any other states of the design. But the don't-care state "11" is directly or indirectly accessing this state in the synthesized STG. Since the attacker can utilize "11" as Trojan to gain unauthorized access of "00", this makes the FSM untrusted.

978-1-5386-3693-0/18 $31.00 © 2018 IEEE

Various techniques are reported in the literature to identify and mitigate the vulnerabilities to perform STVFI, and Trojan attacks in the design and same are discussed in the following subsection.

C. Related Work

The concurrent error detection technique protects the FSM from fault injection attack by considering the specific error model [6]. However, this technique does not consider the vulnerabilities created by synthesis tool and can not work for other adversarial model [11]. Dunbar et al. analyzed the security risk introduced by the synthesis tool and presented an approach for designing a trusted FSM against hardware Trojan attack [9]. However, this technique cannot prevent indirect access of protected states and fails to analyze the vulnerabilities arising due to don't care states/transitions [8].

A vulnerability analysis approach based on the statement hardness observability of circuit signals is presented that identifies the circuit's susceptibility against the Trojan insertion at the behavioral level [12]. Further, a Timing Violation Vulnerability Factor (TVVF) is reported that evaluates the vulnerability of data path of hardware cryptosystems against STVFI attacks [13]. It is difficult to model the fault injection attacks due to probabilistic nature of fault attack process [14]. Thus, a security metric named as System Security Factor (SSF) is proposed that evaluate the system vulnerability against fault attacks based on this nature [14]. Since the vulnerability analysis in FSM requires the consideration of don't care states/transitions, all these techniques cannot be used for identifying vulnerabilities in FSM as [8].

Thus, the AVFSM framework has been presented to identify and mitigate the vulnerabilities in the FSM [8] from STVFI and Trojan attacks. This framework extracts the STG from the synthesized netlist, identifies the don't care states/transitions and analyzes the vulnerabilities from STVFI using following metric [8]:

$$C = \Pi_{i=0}^{(n-1)}((b_{xi} \oplus b_{yi}) + (b_{xi}.b_{pi})) \tag{4}$$

where, the b_{xi}, b_{yi} and b_{pi} are the ith bit of the present states, next state and protected state respectively. If $C = 1$, fault attack is possible for the transition T(x,y) to go to protected state p else fault is not possible. For example, the value of C is 1 and 0 respectively for transitions T(10, 01) and T(00, 10) as shown in Figure 1(a) and (b). However, this metric cannot identify the STVFI attacks for all types of protected states and the states apart from protected, i.e., non–protected states (discussed in the next section). Further, the AVFSM frame–work also fails to mitigate the vulnerabilities for unauthorized access of non–protected states (indirect unauthorized access to protected states) and requires large design overhead. Whereas, a trusted FSM design requires the controlling of unauthorized access of protected as well as of non–protected states [9]. Therefore, a new vulnerability identification and a new low–overhead mitigation technique is required for a secure and trusted FSM design.

III. Proposed Vulnerability Identification Metric

The AVFSM framework cannot identify direct and indirect vulnerable transitions for all types of protected and non–protected states respectively. For example, consider the protected states "000" (P1), "111"(P2) and "011" (P3) as shown in Figure 2(a). The existing metric (Eq. (4)) reports C=0 (the fault is not possible) for the transition T(X, Y) to go to the protected states P1, P2, and P3. But actually, fault attack is possible for unauthorized access of protected states P1 and P2. The attacker can easily inject the faults to go to the protected state P1 ("000") from X ("010") by increasing the path delay of LSB FF while maintaining the path delay of remaining FFs. Similarly, the fault can be injected to go to the protected states P2 ("011") by increasing the path delay of the middle FF while maintaining the path delay of other FFs.

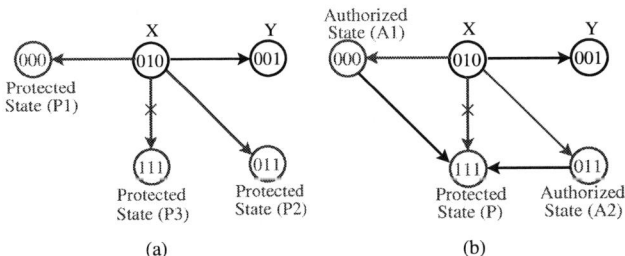

Figure 2. Fault is not detected when an unauthorized state X is (a) Directly accessing the protected states P1 and P2 and (b) Indirectly accessing the protected state P.

Further, consider another sample of STG as shown in Figure 2(b), where A1 ("000") and A2 ("011") are the authorized states to access a protected state P ("111"). However, the fault cannot be injected to directly access this protected state for the transition T(010, 001). The attacker can first inject a fault to go to either of an authorized state (A1/A2) and then he can easily reach to P from these authorized states. Thus, the existing metric is ineffective to identify all types of vulnerable transitions for unauthorized access to protected states. Therefore, we propose a new metric called vulnerable transition identification from fault injection (VTI_{FI}) to iden–tify all possible vulnerable transition from STVFI attacks. This new metric (VTI_{FI}) is derived from Eq. (4) as

$$VTI_{FI} = \prod_{i=0}^{(n-1)} ((b_{xi} \oplus b_{yi}) + (b_{xi} \odot b_{pi})) \tag{5}$$

where, the b_{xi}, b_{yi} and b_{pi} are the ith bits of the present state (x), next state (y) and protected state (p) respectively. The proposed VTI_{FI} metric checks whether a given transition $T(x,y)$ is vulnerable to unauthorized access of protected state p or not. Since fault attack is possible for transition T(x,y) either when b_x and b_y are different or when b_x and b_p are same. Therefore, when fault attack is possible, then $VTI_{FI} = 1$, otherwise, it is zero. However, the above metric can only check the vulnerabilities for a single protected state. Another form of VTI_{FI} for multiple protected states is expressed as

978-1-5386-3693-0/18 $31.00 © 2018 IEEE

$$VTI_{FI} = \sum_{s=0}^{m-1} \prod_{i=0}^{(n-1)} ((b_{xi} \oplus b_{yi}) + (b_{xi} \odot b_{psi})) \quad (6)$$

where m is the total number of protected states in the STG.

This metric identifies all types of vulnerable transitions for direct as well as indirect unauthorized access to protected states. For example, the $VTI_{FI} = 1$ (fault is possible) for Transition T(010, 001) to go to protected states "000", "011", whereas the $VTI_{FI} = 0$ (fault is not possible) for accessing the protected state "111" as shown in Figure 2(a). Further, the proposed metric can also identify the vulnerabilities for indirect unauthorized access of protected states by *considering all non-protected states (excluding present and next states) as protected states*. For example, if we consider the authorized states (A1 and A2 in Figure 2(b)) as protected states, the $VTI_{FI} = 1$ (fault is possible) for transition T(010, 001).

Next, we propose a new light–weight technique to effectively mitigate these vulnerabilities in FSM against STVFI and Trojan attacks.

IV. New Light–Weight Vulnerabilities Mitigation Technique

In this Section, we first presents a new trusted FSM design algorithm followed by an energy–efficient vulnerabilities mitigation architecture.

A. Trusted FSM Design Algorithm

The proposed algorithm as shown in Algorithm 1 provides a procedure to design a trusted FSM against STVFI and Trojan attacks. It accepts synthesized gate–level RTL design netlist (RTL_DN), reset state (RS) and the reachable sets of all the states as inputs. If no vulnerable transitions exist in the design, the algorithm provides original netlist as secure RTL design netlist ($sRTL\_DN$) to thwart STVFI and Trojan attacks.

The procedure of trusted FSM design starts from designing a control logic block (CLB) that can generate a control signal (CS) for each transition from present state (PS) to next state (NS). If NS does not belong to reachable set of PS ($R(PS)$), the transition T(PS, NS) is unauthorized and a high control signal ($CS = 1$) will be generated to reset the FFs. Otherwise CS will be set to low ($CS = 0$) for normal operation of FFs. Further, the 2x1 MUXes are inserted into the synthesized RTL design netlist (RTL_DN) using **Insert_MUX**(RTL_DN, RS, CS) procedure. The main use of MUXes is to pass NS (in case of normal operation) or reset state (for unauthorized transition) bits value to the D–input of FFs. The MUX insertion procedure first copies the RTL_DN in a temporary list ($tRTL\_DN$) and then inserts the MUXes at the input of each FF. The first input of each MUX is connected to the original input wire of respective FF whereas the second input is connected to the respective bit of RS. The selection lines of all the MUXes are connected to the CS. When an unauthorized transition takes place, then a high value of CS resets the D–input bit of each FF with the respective bit of reset state to prevent unauthorized access.

Algorithm 1 : Trusted FSM Design Algorithm

1: **Input:** RTL_DN: ▷ *Original synthesized RTL netlist*
2:     RS: ▷ *Reset state of the design*
3:     $R(Q)$:   ▷ *Reachable sets of all state*
4: **Output:** $sRTL\_DN$:   ▷ *Final secure RTL netlist*
5: **Begin**
6: Use proposed metric to check vulnerable transitions (VTs)
7: **if** ($VTI_{FI} = 0$ for all transitions) **then**   ▷ *No VT exists*
8:     $sRTL\_DN \leftarrow RTL\_DN$;   ▷ *Return original netlist*
9: **else**
10:     Design CLB to generate CS for a transition (7–11)
11:     **if** ($NS \notin R(PS)$) **then**
12:         $CS = 1$;   ▷ *For resetting the FFs*
13:     **else**
14:         $CS = 0$;   ▷ *For normal operation*
15:     **end if**
16:     $sRTL\_DN \leftarrow$ **Insert_MUX**(RTL_DN, RS, CS);
17: **end if**
18: **End**
19: **Return:** $sRTL\_DN$;   ▷ *Returns the secure RTL design*

1: **Procedure: Insert_MUX**(RTL_DN, RS, CS)
2: $tRTL\_DN \leftarrow RTL\_DN$; ▷ *Assign the original RTL netlist to temporary netlist*
3: Insert MUXes at the input of $tRTL\_DN$ Flip–flips
4: Connect first input of each MUX to original input wire of the flip–flop.
5: Connect the second input of the MUXes to the RS bits
6: Select line of each MUX is connected to the CS
7: **Return:** $tRTL\_DN$;

Finally, the algorithm assigns the modified $tRTL\_DN$ to secure synthesized design netlist ($sRTL\_DN$). The design obtained using the above procedure provides a low–cost and secure implementation of FSM against fault injection and Trojan attacks. This procedure of trusted FSM design is gener–alized and provides the energy–efficient design irrespective of FSM. An energy–efficient architecture for implementing secure and trusted FSM using above procedure is presented in next subsection.

B. Energy-Efficient Vulnerabilities Mitigation Architecture

The existing architecture only mitigates the vulnerability of direct unauthorized access to the protected states. It fails to prevent the indirect unauthorized access of protected states and also requires significant overhead. Whereas, the trusted FSM must control direct as well as indirect unauthorized accesses of protected states. In the AVFSM framework, a protected state can be indirectly accessed from an unauthorized state either using don't care or vulnerable transitions as shown in Figure 2(b). The attacker can utilize the following scenarios to indirectly access the protected state:

1) The attacker can inject the fault to redirect the vulnerable transitions from normal/don't care states to an authorized state which is directly accessing a protected state.

978-1-5386-3693-0/18 $31.00 © 2018 IEEE

2) The attacker can either manipulate the don't care tran–sitions of normal/don't–care states or inject the fault to redirect them at the authorized state which is directly connected to a protected state.

Therefore, to prevent the direct as well as indirect unauthorized access of protected states, we have proposed a new energy–efficient architecture that effectively mitigates the vulnerabilities of STVFI and Trojan attacks. The proposed generalized vulnerability mitigation architecture for designing a secure and trusted FSM is shown in Figure 3. In the proposed architecture, an additional 2x1 MUX is inserted at the input of each FF. Further, a control logic block is also added that generate a CS to controls the functionality of MUXes.

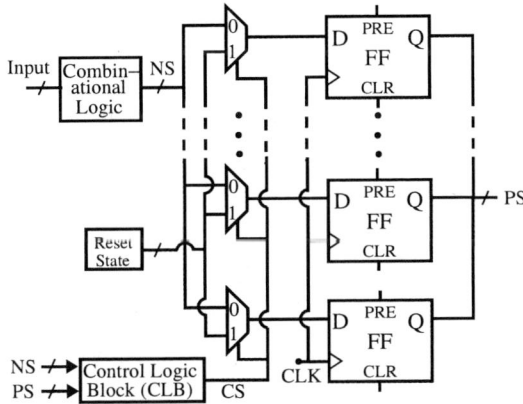

Figure 3. Proposed energy–efficient vulnerabilities mitigation architecture.

The behavioral model of CLB is defined in Algorithm 1 (line no. 7–11) where it checks each transition and identifies as unauthorized if $NS \notin R(PS)$. The CLB sets the value of CS to high when the next state does not belong to the reachable set of present state. The high value of CS redirects all such unauthorized transitions from the PS to reset state. For satisfying the definition of the trusted FSM, the proposed CLB only allows the access of those next states which belong to the reachable set of present state. Therefore, the proposed technique provides a highly secure and trusted FSM design against the STVFI and Trojan attacks while significantly reducing design overhead compares to the AVFSM framework.

V. EXPERIMENTAL RESULTS AND ANALYSIS

This section presents experimental setup followed by vulnerability and overhead analysis of proposed technique over existing.

A. Experimental Setup

The proposed vulnerability identification approach is implemented in C, and its efficacy is evaluated on FSMs of AES and RSA encryption modules over existing approach. Further, the efficacy of proposed mitigation architecture is evaluated by implementing the above benchmarks in Verilog. The design metrics are extracted for ASIC and FPGA implementation using Cadence with 180nm and Artix7 with "XC7A100T" device family respectively.

B. Vulnerability Analysis

To analyze the vulnerabilities in FSM, we also use STGs extracted from the synthesized netlist of AES and RSA encryption modules with two different encoding Scheme–1 (Schm–1) and Scheme–2 (Schm–2) [8]. The number of vulnerable transitions as shown in Table I are identified using proposed and existing AVFSM approaches for two encoding schemes while considering single and all state(s) as protected.

Table I
NUMBER OF VULNERABLE TRANSITIONS FOR DIFFERENT APPROACHES

Technique	Protected State	AES		RSA	
		Schm–1	Schm–2	Schm–1	Schm–2
AVFSM	Single	1	3	3	1
	All	15	27	15	27
Proposed	Single	5	10	4	5
	All	37	51	28	41

This table does not contain those reported vulnerable tran–sitions where a present state is authorized to directly access a protected state as the fault injection for these transitions will not have any impact. However, it includes those vulnerable transitions where a don't care state is directly accessing the protected states. It is observed from the Table that the proposed metric identifies more vulnerable transitions for all cases over the AVFSM [8]. Further, it is also verified that no other vulnerable transitions are left in extracted STG that are not identified with proposed metric while considering above threat model. Therefore, the proposed technique determines 100% vulnerable transitions whereas the AVFSM framework only able to identifies on an average 50% vulnerable transitions.

C. Overhead Analysis

It is observed from the Table I that the Scheme–2 is more vulnerable to fault injection attacks than the Scheme–1, as it has more vulnerable transitions. Therefore, we have implemented the synthesized FSM of AES and RSA with Scheme–2 in Verilog. The Verilog netlist of each implemented FSM is synthesized using Cadence RTL compiler and Xilinx ISE with Artix7 FPGA kit. The following subsections present the design metrics using ASIC and FPGA implementations.

1) ASIC Results: The area and power results for the ASIC implementation of AES and RSA encryption modules using different approaches are shown in Table II. The area is computed and represented in terms of # cells and total area (μm^2). Similarly, two different power dissipations namely leakage and dynamic are computed.

It is observed from the table that the implementation of AES and RSA controller circuit using the proposed technique requires minimal area and power over the AVFSM framework. The AVFSM technique requires significant area overhead due to the use of master–slave FFs (one additional FF), MUXes and a combinational logic with attached latches. Whereas, the proposed technique only requires a simple CLB and MUXes. The proposed CLB is a small light–weight combinational logic and requires minimal overhead. Therefore, the proposed technique provides a significant reduction in overhead as compared to

Table II
AREA, POWER RESULTS USING CADENCE RTL COMPILER

FSM Circuit	Techniques	Area		Power (nW)	
		# Cells	Total Area (μm^2)	Leakage	Dynamic
AES	Original	14	319	8.24	14596
	AVFSM	33	798	21.61	31016
	Proposed	26	489	14.60	16323
RSA	Original	16	356	10.47	12971
	AVFSM	35	832	23.18	29760
	Proposed	34	609	17.68	15593

the AVFSM irrespective to the FSM. The normalized area and power overhead required by the proposed technique is minimal compared to the AVFSM technique as shown in Figure 4. The proposed method provides the 38.7%, and 90% reduction in area and power for AES whereas it provide 26.7% and 90.7% reduction in area and power for the implementation of RSA controller circuit.

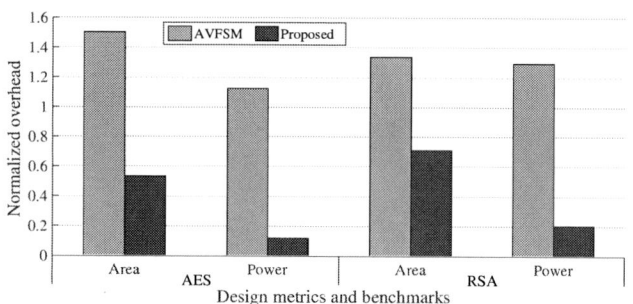

Figure 4. Overhead comparison of various techniques normalized from the original implementation.

2) FPGA Results: The FPGA results of the AES and RSA modules for the proposed and existing techniques using Artix7 with "XC7A100T" device family are shown in Table III. It is clear from the table that the proposed technique requires the same number of LUTs for both modules as the original implementation requires. The AVFSM requires 2× more number of LUTs in comparison to the proposed and original implementations. Similarly, the proposed implementation reduces the delay by 61.1% and 57% while providing 2.54× and 2.32× higher operating frequency for AES and RSA modules respectively.

Table III
AREA, PERFORMANCE RESULTS WITH USING XLINX

FSM Circuit	Techniques	# Slice LUTs	Delay (ns)	Performance (MHz)
AES	Original	3	0.98	1019
	AVFSM	13	2.54	394
	Proposed	3	0.99	1013
RSA	Original	3	0.98	1013
	AVFSM	8	2.30	434
	Proposed	3	0.99	1009

VI. CONCLUSION

In this paper, a new trusted FSM design technique is proposed to thwart STVFI and Trojan attacks. The existing techniques cannot identify and mitigate the indirect unauthorized access of protected states and require significant design overhead. Therefore, in the proposed technique, a new metric (VTI_{FI}) is introduced that identifies all vulnerable transitions for unauthorized access of protected as well as non–protected states. Further, a new light–weight vulnerabilities mitigation approach is proposed that prevents all direct as well as indirect unauthorized transitions to access the protected and non–protected states. The simulation results on AES and RSA based FSMs show that the proposed technique identifies and mitigates all the vulnerabilities against STVFI and Trojan attacks over the existing methods. Finally, ASIC and FPGA implementation of AES and RSA controller circuits show that on an average proposed technique reduces 32.7%, 90%, and 59% area, power, and delay respectively over the existing once.

REFERENCES

[1] S. Bhunia, M. S. Hsiao, M. Banga, and S. Narasimhan, "Hardware trojan attacks: threat analysis and countermeasures," *Proceedings of the IEEE*, vol. 102, no. 8, pp. 1229–1247, 2014.

[2] K. Xiao, D. Forte, Y. Jin, R. Karri, S. Bhunia, and M. Tehranipoor, "Hardware trojans: lessons learned after one decade of research," *ACM Transactions on Design Automation of Electronic Systems (TODAES)*, vol. 22, no. 1, p. 6, 2016.

[3] P. Kocher, J. Jaffe, and B. Jun, "Differential power analysis," in *Advances in cryptologyCRYPTO99*. Springer, 1999, pp. 789–789.

[4] P. Kocher, "Timing attacks on implementations of diffie–hellman, RSA, DSS, and other systems," in *Advances in CryptologyCRYPTO96*. Springer, 1996, pp. 104–113.

[5] E. Biham and A. Shamir, "Differential fault analysis of secret key cryptosystems," *Advances in CryptologyCRYPTO'97*, pp. 513–525, 1997.

[6] B. Sunar, G. Gaubatz, and E. Savas, "Sequential circuit design for embedded cryptographic applications resilient to adversarial faults," *IEEE Transactions on Computers*, vol. 57, no. 1, pp. 126–138, 2008.

[7] D. Karaklajić, J.–M. Schmidt, and I. Verbauwhede, "Hardware designer's guide to fault attacks," *IEEE Transactions on Very Large Scale Integration (VLSI) Systems*, vol. 21, no. 12, pp. 2295–2306, 2013.

[8] A. Nahiyan, K. Xiao, K. Yang, Y. Jin, D. Forte, and M. Tehranipoor, "AVFSM: a framework for identifying and mitigating vulnerabilities in FSMs," in *Design Automation Conference (DAC), 2016 53nd ACM/EDAC/IEEE*. IEEE, 2016, pp. 1–6.

[9] C. Dunbar and G. Qu, "Designing trusted embedded systems from finite state machines," *ACM Transactions on Embedded Computing Systems (TECS)*, vol. 13, no. 5s, pp. 153–172, 2014.

[10] J. Fan, X. Guo, E. De Mulder, P. Schaumont, B. Preneel, and I. Verbauwhede, "State–of–the–art of secure ECC implementations: a survey on known side–channel attacks and countermeasures," in *Hardware-Oriented Security and Trust (HOST), 2010 IEEE International Symposium on*. IEEE, 2010, pp. 76–87.

[11] Z. Wang and M. Karpovsky, "Robust FSMs for cryptographic devices resilient to strong fault injection attacks," in *On-Line Testing Symposium (IOLTS), 2010 IEEE 16th International*. IEEE, 2010, pp. 240–245.

[12] H. Salmani and M. Tehranipoor, "Analyzing circuit vulnerability to hardware trojan insertion at the behavioral level," in *Defect and Fault Tolerance in VLSI and Nanotechnology Systems (DFT), 2013 IEEE International Symposium on*. IEEE, 2013, pp. 190–195.

[13] B. Yuce, N. F. Ghalaty, and P. Schaumont, "TVVF: Estimating the vulnerability of hardware cryptosystems against timing violation attacks," in *Hardware Oriented Security and Trust (HOST), 2015 IEEE International Symposium on*. IEEE, 2015, pp. 72–77.

[14] M. Li, L. Lai, V. Chandra, and D. Z. Pan, "Cross–level monte carlo framework for system vulnerability evaluation against fault attack," in *Proceedings of the 54th Annual Design Automation Conference 2017*. ACM, 2017, pp. 17–22.

978-1-5386-3693-0/18 $31.00 © 2018 IEEE

2018 31th International Conference on VLSI Design and 2018 17th International Conference on Embedded Systems

Novel Variability Aware Path Selection for Self-Referencing Based Hardware Trojan Detection

Vaikuntapu Ramakrishna, Lava Bhargava, Vineet Sahula
Department of Electronics and Communication Engineering
Malaviya National Institute of Technology, Jaipur-302017, India
vramakrishna409@gmail.com, lavab@mnit.ac.in, sahula@ieee.org

Abstract—Hardware Trojans can be inserted by an adversary at untrusted third-party fabrication houses. Many Side-channel analysis based detection techniques have been proposed in past to detect such Trojans. However, their efficiency is highly affected by process variations. Hardware Trojan (HT) inserted in an IC affects the path delays within the IC. In this work, we exploit the fact that the path delays of topologically symmetric paths in an IC will be affected similarly by process variations. We tend to choose paths that are minimally affected by process variations. In this paper, we propose a path selection technique for delay based HT detection technique. We further use the concept of self-referencing to improve detection accuracy as well as to eliminate the requirement of golden ICs. Simulations performed using ISCAS-85 benchmarks establish that the proposed method is achieving a true positive rate of 100% with a false positive rate less than 5%. We have considered maximum of 10% intra-die and 15% inter-die variations in threshold voltage (V_{th}).

Index Terms—Hardware Trojan detection, Self-referencing, Process variation, Path delay.

I. INTRODUCTION

Hardware Trojan (HT) can be introduced by an adversary at an untrusted design or fabrication house. Depending on the interests of the adversary the HT can cause change in functionality, denial-of-service, information leakage or reliability reduction. The detailed taxonomy of hardware Trojans at various abstraction levels was presented in [1]. Different methods have been presented for detecting Trojans inserted at different abstraction levels [2]. Post-silicon detection methods aim to detect HTs in fabricated ICs inserted at untrusted fabrication foundries. Even though logic testing [3] based approaches for HT detection have been proposed their efficiency is limited by the stealthy nature of HT, test generation and execution time. Majority of existing post-silicon methods are devised based on side channel analysis (SCA) i.e analyzing the side channel parameters like power, path delay and electromagnetic measurements etc. Many existing SCA based HT detection techniques depend on golden ICs (i.e ICs without any HTs) used as a reference to compare the side-channel (SC) signatures of suspect IC to decide whether HTs exist in the IC under purview. Obtaining such golden ICs is expensive as it involves invasive techniques such as reverse engineering. Another challenge for SCA methods is the process variation (PV) which impacts the efficiency of detection even if golden ICs are available. In presence of PV direct comparison of the SC signatures of suspect chip with the golden chips may not accurately detect HTs. To mitigate the effects of PVs on SC

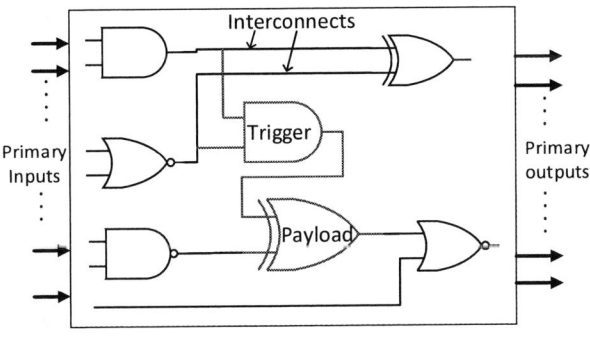

Fig. 1. Hardware Trojan structure

parameters researchers have proposed a technique known as Self-referencing [4]. In self-referencing based methods the SC parameter of an IC is compared with the SC parameter of the IC itself instead of comparing with golden signature. The major advantages of self-referencing methods are, requirement of golden IC is eliminated and the effect of inter-die process variation are mitigated as it affects the entire chip in a similar fashion. In this work, we chose path delay for Trojan detection because of its specific advantages such as, HT effect is local to the path and HT affects path delay irrespective of its activation status which eliminates the need to activate Trojans. Following are the contributions of this paper:

- It presents a methodology to detect hardware Trojans (HTs) inserted during fabrication. This method uses the concept of self-referencing and the delays of symmetric paths are analysed to detect Trojans.
- It proposes a path selection algorithm to select symmetric path pairs in such a way to mitigate the effect of inter-die and intra-die process variations on detection accuracy. It also presents a procedure to create a symmetric path pair if it does not exist in the design.

This paper is organized as follows, HT's effect on path delay, potential locations of HTs and brief overview of existing HT detection methods are presented in Section II. Section III explains the proposed methodology in detail and path selection procedure is explained in Section IV. Simulation results are reported in Section V and we conclude in section VI.

978-1-5386-3693-0/18 $31.00 © 2018 IEEE

79

II. BACKGROUND AND RELATED WORK

A. HT impact on path delay

HTs can impact the delay of paths in a circuit in two ways. A gate that acts as Trojan trigger connects to existing interconnects in the circuit and activates the HT when specific activation condition is met as shown in Fig. 1. These extra connections induce some additional capacitance at the interconnects which in turn impacts delay. The gate that acts as Trojan payload is inserted by restitching a net in the original circuit (see Fig. 1) and launches its attack when the HT is activated. An additional delay equal to the delay of inserted payload gate is added to the delay of paths passing through the restitched net in the original circuit. The additional delay introduced by payload gate is typically larger than that of trigger gate.

B. Potential locations of HT

An attacker may choose to insert HT in a way that it should be stealthy i.e undetectable by both logic testing and side-channel based detection techniques. To meet such requirements, it is normally assumed that attacker attempts to insert the HT at nodes of the circuit which exhibit very low switching activity. The reason behind this assumption is Trojans inserted at such nodes are expected to show minimal impact in logic-testing and side-channel based detection techniques [5]. We have considered low switching nodes/nets as possible and vulnerable locations for HT insertion.

C. Process Variations

In general HTs are small in size in comparison with the entire design, hence their effect on SC parameters may be masked by the effects of PVs. The inter-die variation affects all devices in an IC in a similar fashion, whereas, intra-die variation affects these devices differently depending on their locations on the IC [6]. In general inter-die variations are larger than intra-die variations [7]. We have considered variations in transistor threshold voltage (V_{th}). The variation model considered in this work is shown in eq. (1) and (2).

$$v_{th}(x,y) = v_{th,nom} + \Delta v_{inter} + \Delta v_{intra}(x,y) \qquad (1)$$

The intra-die component $\Delta v_{intra}(x,y)$ can further be divided as a correlated spatial component and a random component as shown in eq. (2)

$$\Delta v_{intra}(x,y) = \Delta v_{spatial}(x,y) + \Delta v_{random}(x,y) \qquad (2)$$

where $v_{th,nom}$ is the nominal threshold voltage, Δv_{inter} is the inter-die variation which which is same at all locations, $\Delta v_{spatial}(x,y)$ is a spatially correlated component which can be modeled as a correlated multivariate Gaussian random variable. $\Delta v_{random}(x,y)$ is the random component which can be modeled as an independent Gaussian random variable.

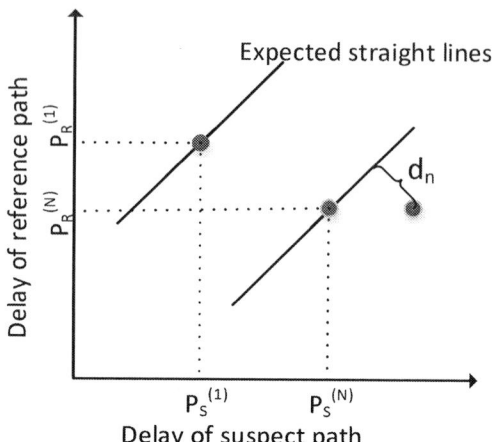

Fig. 2. Deviation from expected straight line

D. Related Work

The detailed taxonomy and classification of HTs is presented in [1]. To detect HTs, SCA based detection techniques have been considered to be more efficient than traditional functionality testing methods. For the first time, Agarwal et al. proposed to use side channel parameter (power) for Trojan detection [8] by constructing fingerprint using Golden-ICs. We present the work available in literature on detecting HT similar to the proposed work i.e. path delay analysis based HT detection. Path delays were used to generate the fingerprint of an IC family using principle component analysis in [9]. The limitations of these methods are they require a set of golden ICs and their detection accuracy is largely affected by process variations. In [10], the authors used shadow registers to measure path delays and to detect HTs and this method requires multiple clocks and requires extra registers inserted for each selected path. Efficiency of delay-based techniques to detect HTs in the presence of process variations was studied in [11]. In [12] authors proposed an embedded test structure by using existing scan structures to measure path delays so as to detect HTs. In [13], authors proposed to choose the shortest paths through each possible Trojan location to improve HT detection accuracy in the presence of process variations. A clock glitching method was proposed to measure path delays and statistical techniques are used to reduce the effects of process variations in [14]. In [15] authors proposed a clock sweeping based path delay measurement technique for detecting HTs. A self-referencing based golden-IC free HT detection method is proposed in [16]. It compares the delays of symmetric paths in a design. In [17] a pulse propagation technique is proposed to detect extra capacitance induced by HT on logical paths. A high resolution on-chip embedded test structure called a time-to-digital converter to measure path delays and a chip-averaging technique was proposed to detect delay anomalies introduced by HTs in [18]. Like any other delay based detection techniques, proposed method is also not able to detect Trojans that don't have any impact on path delays.

978-1-5386-3693-0/18 $31.00 © 2018 IEEE

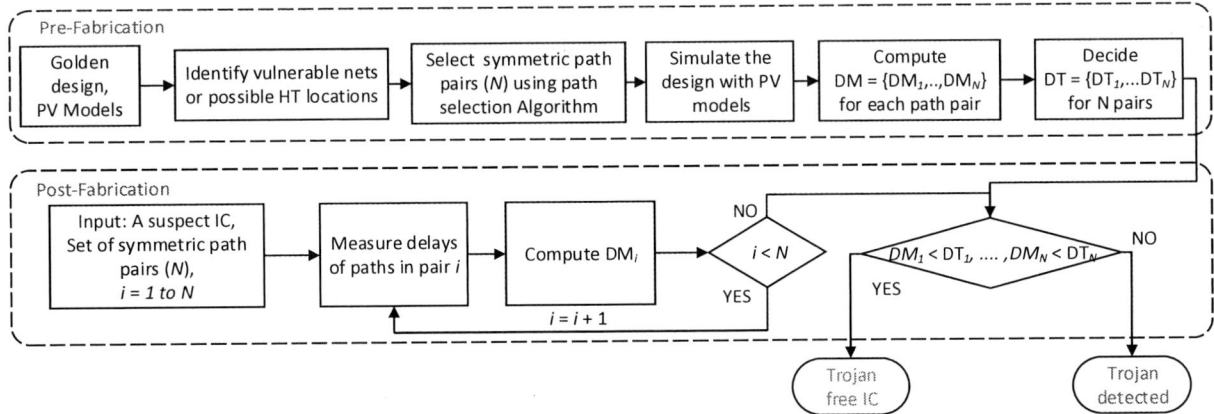

Fig. 3. Proposed Hardware Trojan Detection Methodology

III. PROPOSED DETECTION METHODOLOGY

In this section we present the proposed methodology in detail. Every IC contains many paths out of which some paths are topologically similar which we call as symmetric paths. Any two paths which have equal number of gates of same type can be defined as symmetric paths as shown in Fig.4. The design netlist has been analysed to find vulnerable nets which are considered to be potential locations of HT insertion. We chose a path through such vulnerable net which we call as suspect path and search for a symmetric path which is topologically similar which we call as reference path to form a suspect and reference path pair. The path delays of such paths may or may not be equal as it is dependent on the interconnection length and capacitive load of gates. As both the paths in a pair have similar gates they experience same inter-die process variation (global variation or die-to-die variation). We chose a symmetric path pair for each vulnerable net in the netlist. When the delays of such pairs are plotted on a two dimensional space they should follow strait lines corresponding to each pair as shown in Fig.2. If HT is inserted at a vulnerable net then the delay of suspect path will be increased by an amount of HT induced delay. Which in turn leads to some deviation from the expected straight line as shown in Fig. 2. This deviation has been used to calculate a detection metric (DM) for each selected pair in an IC under test. But, in the presence of intra-die process variations (local or within-die variations) the delays of pairs may deviate a little from the straight line. Therefore, this detection metric is compared with a pre-defined threshold (DT) to separate Trojan inserted ICs from Trojan free ICs. The proposed methodology consists two stages i.e (i) Pre-fabrication stage and (ii) Post-fabrication stage as shown in Fig. 3.

A. Pre-fabrication stage

We assume that a golden model of the design is available, as the HT is assumed to be inserted after design sign-off while fabrication. The golden design netlist is simulated using a large number of random test vectors to collect a set of vulnerable nets as $\{N_v\}$. The shortest path as reported in [13]

has been chosen through each vulnerable net which we call as suspect path P_s. For each such suspect path, a topologically symmetric path is identified which we call as reference path P_r. The path selection algorithm which is used to select suspect and reference paths is explained in section IV. A set of symmetric path pairs $\{N\}$ is formed by combining a suspect path and its corresponding reference path through each net n in $\{N_v\}$. $P^i_{s,nom}$ and $P^i_{r,nom}$ represent the nominal path delays of suspect path and reference path respectively, of ith symmetric path pair corresponding to i^{th} vulnerable net in $\{N_v\}$. The detection metric DM_i of ith pair is calculated as the normalised distance [19] from expected straight line through the point $(P^i_{s,nom}, P^i_{r,nom})$ as shown in eq. (3).

$$DM_i = \frac{d_i}{\sqrt{(P^i_s)^2 + (P^i_r)^2}} \tag{3}$$

Here, the distance d_i is given by eq. (4).

$$d_i = \frac{1}{\sqrt{2}}(|P^i_s - P^i_r + (P^i_{s,nom} - P^i_{r,nom})|) \tag{4}$$

Monte-Carlo simulations are performed using reliable process variation models provided by the foundry to generate a set of Trojan free ICs. The $DM = \{DM_1, DM_2, ..., DM_N\}$ of all Trojan free ICs is calculated and a detection threshold is decided for each pair as a set $DT = \{DT_1, DT_2, ..., DT_N\}$.

B. Post-fabrication stage

The path delays of all $2N$ paths in N pairs are measured for each suspected IC. Techniques presented in [10], [15] can be used to measure the delays of selected paths. The DM of each pair is calculated using eq. (3). Each DM_i is compared with pre-defined detection threshold DT_i to infer whether the IC under test is Trojan free as shown below.

$$if \begin{cases} DM_1 < DT_1 \ \ and \\ DM_2 < DT_2 \ \ and \\ ... \\ DM_N < DT_N. \end{cases} \text{then IC is Trojan free}$$

$$else \qquad \qquad \text{IC has Trojan inserted}$$

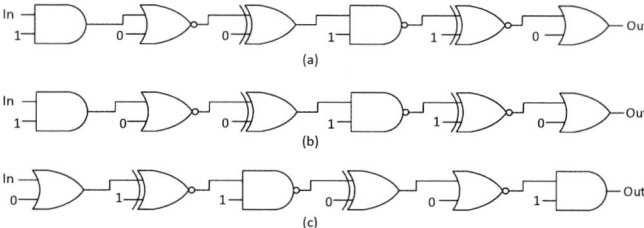

Fig. 4. Symmetric paths. (a) suspect path (b) type-1 reference path (c) type-2 reference path

IV. PATH SELECTION PROCEDURE

A. Symmetric paths

In this work, we have considered two paths as symmetric if they pass through similar types of logic gates as shown in Fig. 4. Assume Fig. 4(a) shows a suspect path through a vulnerable net. The reference path for the suspect path in Fig. 4(a) can be anyone of the two paths shown in Fig. 4(b) and 4(c). We call path shown in Fig. 4(b) as type-1 reference path and path in Fig. 4(c) as type-2 reference path. As all the three paths in Fig. 4 have same number of logic gates of same type, they all experience same inter-die process variations [20]. The delays of the paths shown in Fig. 4 may or may not be same as they also depend on interconnection delays and capacitive load experienced by gates in the respective paths of the design post layout.

B. Path selection algorithm

The proposed path selection technique is discussed in this section and is presented in Algorihm 1. This algorithm returns a suspect path if any, and its corresponding reference path for each vulnerable net in the design. We call a path with minimum delay as shortest path. All paths passing through a vulnerble net n_i are collected in set $S_{suspect}$. A shortest sensitizable path p_{short} is chosen from the set $S_{suspect}$. All paths that are symmetric to p_{short} are collected in set $S_{symmetric}$. If there are no symmetric paths available for p_{short} then next shortest path in $S_{suspect}$ is selected as p_{short}. If there are any nets for which there exist no symmetric path for at least one path in $S_{suspect}$ then a symmetric path is created as explained in Section IV-C. After finding out suspect path p_{susp} for each vulnerable net and its corresponding set of symmetric paths, final layout of the netlist is prepared. This layout is partitioned into 64 grids as 8 rows and 8 columns. The purpose of partitioning the layout is to exploit spatial correlation component in intra-die variation among the adjacent grids. We rank each path in $S_{symmetric}$ according to the distance between grids that consist same gates as shown in Fig.5. A path in $S_{symmetric}$ which is near (i.e with lowest rank) to p_{susp} is selected as reference path p_{ref}. A symmetric path pair $SymmetricPpair_i$ corresponding to net n_i is formed as (p_{susp}, p_{ref}). We used Synopsys Tetramax ATPG tool [21] to decide whether a path is sensitizable.

Algorithm 1: Path selection algorithm

Input : Netlist, N_v = set of vulnerable nets
Output: Symmetric path pairs

1 **foreach** *net n_i in N_v* **do**
2     $S_{suspect,i}$ = all paths passing through net n_i
3     $p_{short,i}$ = a sensitizable path with minimum delay in $S_{suspect,i}$
4     $S_{symmetric,i}$ = set of paths that are symmetric(type-1 or type-2) to $P_{short,i}$ which are not passing through net n_i
5   **if** $|S_{symmetric,i}| = 0$ **then**
     /* If there are no symmetric paths for $p_{short,i}$ then select next minimum delay path */
6       $S_{suspect,i} = S_{suspect,i}$ - $p_{short,i}$
7   go to: step 3
8   **end**
9   **if** $\sum_{j=1}^{|S_{suspect,i}|} |S_{symmetric,j}| = 0$ **then**
     /* If there are no symmetric paths for at least one path in $S_{suspect,i}$ then create a symmetric path */
10      $p_{short,i}$ = a sensitizable path with minimum delay in $S_{suspect,i}$
11   p_{sub} = a path that has a subset of total gates in the path $p_{short,i}$
12      $extragates = gatesin(p_{short,i}) - gatesin(p_{sub})$
13      $p_{ref,i}$ = path p_{sub} with $extragates$ inserted at its input
14   Connect off-path inputs of $extragates$ to non-controlling values
15   **end**
16   Generate the layout of (modified) netlist and partition it into 8×8 grids
17    $p_{susp,i} = p_{short,i}$
18    $p_{ref,i}$ = a path from $S_{symmetric,i}$ which is nearest to $p_{susp,i}$ i.e the path passing through grids near to that of $p_{susp,i}$ i.e path with lowest rank
19    $SymmetricPpair_i \leftarrow (p_{susp,i}, p_{ref,i})$
20 **end**

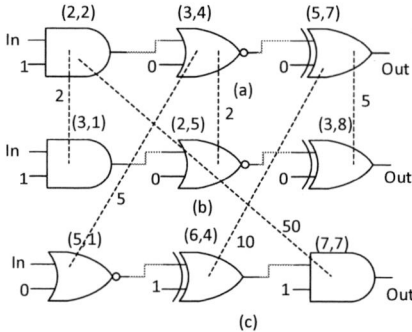

Fig. 5. Example for ranking symmetric paths and locations of gates in $8X8$ grids (a) suspect path (b) reference path with rank = 9 (2+2+5) (c) reference path with rank = 65 (50+5+10)

978-1-5386-3693-0/18 $31.00 © 2018 IEEE

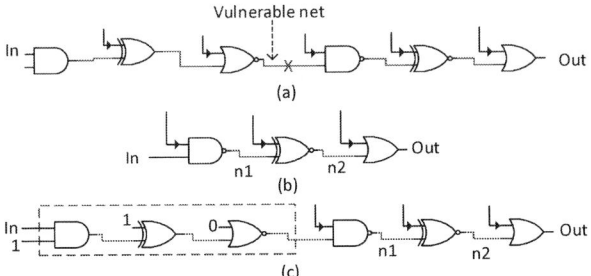

(a)

(b)

(c)

Fig. 6. Generating symmetric path (a) selected suspect path (b) a non-critical path (c) generated symmetric path to cover the vulnerable net

C. Reference path creation

There may be some vulnerable nets for which a symmetric pair does not exist, we call such nets as uncovered nets. In order cover such nets, we create a symmetric path to be used as reference path by adding few extra logic gates to the existing non-critical paths as shown in Fig. 6. We chose the shortest path through the uncovered net and search for a non-critical path which has gates that are subset of gates of selected path through the uncovered net. The remaining gates are inserted at the input of the non-critical path without changing the functionality to create a type-1 or type-2 symmetric path to be used as a reference path. The off-path inputs of extra inserted gates must be set to non-controlling values. The designer has to ensure that all paths through the input net of the created new path must be non-critical even after the addition of extra gates and functionality must remain same. The extra overhead because of adding extra gates is dependent on the number of uncovered nets and the size of selected suspect and reference paths.

V. RESULTS

Simulations are performed on ISCAS-85 benchmark circuits [22] to evaluate proposed methodology. These benchmarks are synthesized using Synopsys generic (SAED_EDK32nm) library. To perform SPICE level simulations 32nm CMOS Predictive Technology Models (PTM) [23] have been used. The netlists are simulated over a large number of random test vectors (i.e. $10e6$) and nets with switching activity less than predefined threshold ($10e-3$) are considered as vulnerable nets as explained in section II-B. Static timing analysis (STA) is performed using Synopsys primetime [24] to generate path data and Tetramax ATPG [21] is used to decide whether a path is sensitizable. Symmetric paths are created to cover the uncovered nets as explained in section IV-C. Physical layout of design is carried out using Synopsys IC Compiler and layout is partitioned into 8×8 grids. Symmetric path pairs are selected using path selection procedure explained in section IV-B. Selected paths are extracted in SPICE netlist form. Process variation on threshold voltage (V_{th}) is modeled as explained in section II-C. 3σ values for inter-die and intra-die variation are considered as 15% and 10% respectively. As the layout has been partitioned into 8×8 grids, 64 correlated multivariate random variables are generated to model the spatial variation.

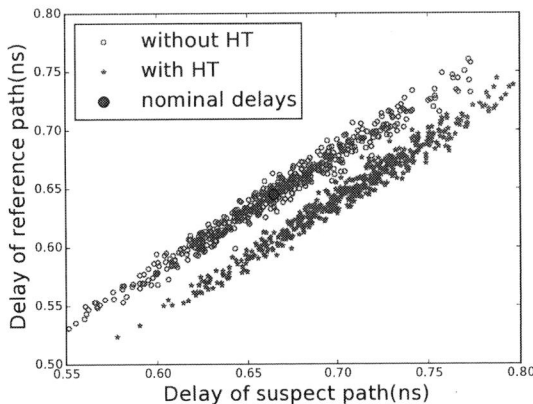

Fig. 7. Effect PV on a symmetric path pair and variation along expected straight line

Spatial correlation decreases with increase in distance between grids. 500 ICs without HT and 500 ICs with HT (shown in Fig. 1) are generated and SPICE level simulations are performed with the generated V_{th} profiles using HSPICE. The variation of delays of suspect and reference paths in a symmetric path pair of c880 circuit is shown in Fig. 7. The path delays vary along a straight line passing through the point of nominal delays due to inter-die variations, however, due to intra-die variations they deviate away from the straight line. It is observed that the delays in HT inserted ICs also follows a straight line but they are away from the expected straight line. Detection metric (DM) of all trojan free and with trojan ICs is calculated as per eq. (3) and shown in Fig. 8. True positive rate (TPR) is defined as number of ICs that are detected as HT inserted out of total HT inserted ICs. False positive rate (FPR) is defined as the number of ICs that are detected as HT inserted but which in fact are HT free. A TPR of 100% and a FPR of 3% is observed with a DT of 0.01. The detection threshold (DT) has been decided in a pessimistic way by allowing a false positive rate of 5% but, row 7 in table-I shows the minimum FPR that can be achieved for the TPR shown in row 6. Table-I shows the summary of simulation results. Row 2 shows total number of nets present in the circuit and row 3 presents the number of low-switching nets considered as vulnerable nets. Nets which do not have reference paths i.e uncovered nets are shown in row 4 and row 5 shows the hardware overhead (i.e number of gates) due to extra gates inserted to create a reference path for each uncovered net. Rows 6 and 7 show the TPR and FPR achieved by the proposed detection method. It is observed that maximum TPR possible is 95% even with 18% FPR for c432. It is due to the unavailability of shortest symmetric paths through the net n6 . For this net the delays of selected symmetric paths are higher than 75% of critical path delay. Even though paths are longer, the performance of the method would have been better if the symmetric paths are closer. But, in this case these paths pass through less spatially correlated grids. One way to improve the performance is to create a shorter symmetric path closer to any one of the paths through net n6.

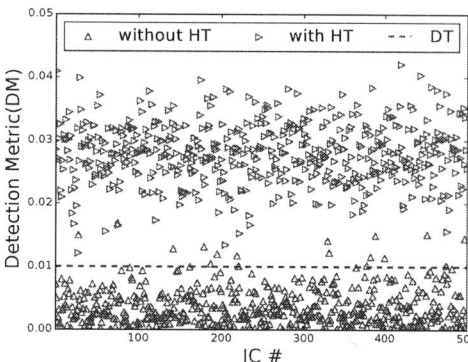

Fig. 8. Detection metric of 500 HT free and 500 HT inserted ICs for a symmetric path pair of c880

TABLE I
SIMULATION RESULTS ON ISCAS-85 BENCHMARKS

Circuit	c432	c499	c880	c1355	c1908	c2670	c5315
Total nets	120	200	238	229	398	152	734
Vulnerable nets	9	42	17	40	16	22	8
Uncovered nets	2	0	3	0	7	0	3
Overhead[†]	4	0	11	0	22	0	9
TPR(%)	95	99	100	100	100	98	100
FPR(%)	18	5	3	3	2	5	1

[†] In number of extra gates inserted.

VI. CONCLUSION

We have considered threat model of HT insertion while fabrication at untrusted foundries. In this work, we presented a self-referencing based HT detection method using path delays which eliminates the requirement of golden ICs. Further, we proposed a procedure to select paths that minimizes the effect of both inter-die and intra-die PV. We have used topologically symmetric paths to mitigate inter-die variations and selected closer paths to exploit the spatial correlation to reduce the impact of intra-die variations. This method uses simulation results to decide DT, on-chip PV monitors can be used to estimate the DT of each suspect IC which will be focused in our future work.

ACKNOWLEDGEMENT

This work is partly supported by SMDPC2SD project, sponsored by Ministry of Electronics & Information of Technology, Government of India.

REFERENCES

[1] M. Tehranipoor and F. Koushanfar, "A Survey of Hardware Trojan Taxonomy and Detection," *IEEE Design and Test of Computers*, vol. 27, no. 1, pp. 10–25, 2010.

[2] H. Li, Q. Liu, J. Zhang, and Y. Lyu, "A Survey of Hardware Trojan Detection, Diagnosis and Prevention," in *2015 14th International Conference on Computer-Aided Design and Computer Graphics (CAD/Graphics)*, 2015, pp. 173–180.

[3] A. G. Voyiatzis, K. G. Stefanidis, and P. Kitsos, "Efficient Triggering of Trojan Hardware Logic," in *2016 IEEE 19th International Symposium on Design and Diagnostics of Electronic Circuits Systems (DDECS)*, 2016, pp. 200–205.

[4] D. Du, S. Narasimhan, R. S. Chakraborty, and S. Bhunia, "Self-referencing: A Scalable Side-channel Approach for Hardware Trojan Detection," in *Proceedings of the 12th International Conference on Cryptographic Hardware and Embedded Systems*, vol. CHES'10, 2010, pp. 173–187.

[5] H. Salmani, M. M. Tehranipoor, and J. Plusquellic, "New Design Strategy for Improving Hardware Trojan Detection and Reducing Trojan Activation Time," in *Hardware-Oriented Security and Trust, 2009. HOST '09. IEEE International Workshop on*, 2009, pp. 66–73.

[6] C. Hongliang and S. S. Sapatnekar, "Statistical Timing Analysis Considering Spatial Correlations Using a Single PERT-like Traversal," in *ICCAD-2003. International Conference on Computer Aided Design (IEEE Cat. No.03CH37486)*, 2003, pp. 621–625.

[7] X. Chen, L. Wang, Y. Wang, Y. Liu, and H. Yang, "A General Framework for Hardware Trojan Detection in Digital Circuits by Statistical Learning Algorithms," *IEEE Transactions on Computer-Aided Design of Integrated Circuits and Systems*, vol. 0070, no. c, pp. 1–1, 2016.

[8] D. Agrawal, S. Baktir, D. Karakoyunlu, P. Rohatgi, and B. Sunar, "Trojan Detection using IC Fingerprinting," in *Proceedings - IEEE Symposium on Security and Privacy*, 2007, pp. 296–310.

[9] Y. Yier, Jin and Makris, "Hardware Trojan Detection Using Path Delay Fingerprint," in *Proceedings of the 2008 IEEE International Workshop on Hardware-Oriented Security and Trust*, 2008, pp. 51–57.

[10] J. Li and J. Lach, "At-Speed Delay Characterization for IC Authentication and Trojan Horse Detection," in *2008 IEEE International Workshop on Hardware-Oriented Security and Trust, HOST*, 2008, pp. 8–14.

[11] J. Rai, D. and Lach, "Performance of Delay-Based Trojan Detection Techniques under Parameter Variations," in *Hardware-Oriented Security and Trust, 2009. HOST '09. IEEE International Workshop on*, 2009, pp. 58–65.

[12] C. Lamech and J. Plusquellic, "Trojan Detection based on Delay Variations Measured using a High-Precision, Low-Overhead Embedded Test Structure," in *Hardware-Oriented Security and Trust (HOST), 2012 IEEE International Symposium on*, 2012, pp. 75–82.

[13] S. K. Cha, Byeongju and Gupta, "Trojan Detection via Delay Measurements: A New Approach to Select Paths and Vectors to Maximize Effectiveness and Minimize Cost," in *Design, Automation Test in Europe Conference Exhibition (DATE), 2013*, 2013, pp. 1265–1270.

[14] I. Exurville, L. Zussa, J.-b. Rigaud, and B. Robisson, "Resilient Hardware Trojans Detection based on Path Delay Measurements," in *Hardware Oriented Security and Trust (HOST), 2015 IEEE International Symposium on*, 2015, pp. 151–156.

[15] K. Xiao, X. Zhang, and M. Tehranipoor, "A Clock Sweeping Technique for Detecting Hardware Trojans Impacting Circuits Delay," *IEEE Design & Test*, vol. 30, no. 2, pp. 26–34, 2013.

[16] N. Yoshimizu, "Hardware Trojan Detection By Symmetry Breaking In Path Delays," in *IEEE International Symposium on Hardware-Oriented Security and Trust, HOST 2014*, 2014, pp. 107–111.

[17] S. Deyati, B. J. Muldrey, and A. Chatterjee, "Trojan Detection in Digital Systems Using Current Sensing of Pulse Propagation in Logic Gates," in *2016 17th International Symposium on Quality Electronic Design (ISQED)*, 2016, pp. 350–355.

[18] D. Ismari, J. Plusquellic, C. Lamech, S. Bhunia, and F. Saqib, "On Ddetecting Delay Anomalies Introduced by Hardware Trojans," in *Proceedings of the 35th International Conference on Computer-Aided Design - ICCAD '16*, 2016, pp. 1–7.

[19] V. Ramakrishna, L. Bhargava, and V. Sahula, "Golden IC free Methodology for Hardware Trojan Detection using Symmetric Path Delays," in *VLSI Design and Test (VDAT), 2016 20th International Symposium on, IEEE*, 2016, pp. 1–2.

[20] I. Sutherland, B. Sproull, and D. Harris, *Logical Effort: Designing Fast CMOS Circuits*. San Francisco, CA, USA: Morgan Kaufmann Publishers Inc., 1999.

[21] *TetraMAX ATPG User Guide, Version L-2016.03-SP2*, Synopsys, Inc.

[22] F. Brglez and H. Fujiwara, "A Neutral Netlist of 10 Combinational Benchmark Circuits and a Target Translator in Fortran," in *Proceedings of IEEE Int'l Symposium Circuits and Systems (ISCAS 85)*. IEEE Press, Piscataway, N.J., 1985, pp. 677–692.

[23] (2014) Predictive technology model. [Online]. Available: http://ptm.asu.edu/

[24] *PrimeTime User Guide, Version L-2016.06, June 2016*, Synopsys, Inc.

978-1-5386-3693-0/18 $31.00 © 2018 IEEE

2018 31th International Conference on VLSI Design and 2018 17th International Conference on Embedded Systems

A Secure Low-Cost Edge Device Authentication Scheme for the Internet of Things

Ujjwal Guin*, Adit Singh*, Mahabubul Alam*, Janice Cañedo[†], and Anthony Skjellum[†]

*Department of Electrical and Computer Engineering
[†]Department of Computer Science and Software Engineering
Auburn University, AL, USA
{ujjwal.guin, singhad, mahabubul.alam, canedo, skjellum}@auburn.edu

Abstract—Because of the enhanced capability of adversaries, edge devices of Internet of Things (IoT) infrastructure are now increasingly vulnerable to counterfeiting and piracy. Ensuring the authenticity of such devices is of great concern since an adversary can create a backdoor either to bypass the security, and/or to leak secret information over an unsecured communication channel. The reliability of such devices could also be called into question because they might be counterfeit, defective and/or of inferior quality. It is of prime importance to design and develop solutions for authenticating such edge devices. In this paper, we present a novel low-cost solution for authenticating edge devices. We use SRAM based PUF to generate unique "digital fingerprints" for every device, which can be used to generate a unique device ID. We propose a novel ID matching scheme to verify the identity of an edge device even though the PUF is extremely unreliable. We show that the probability of impersonating an ID by an adversary is extremely low. In addition, our proposed solution is resistant to various known attacks.

Index Terms—Internet of Things (IoT), Physically Unclonable Functions (PUF), Edge Device, Authentication.

I. INTRODUCTION

Being identified as the biggest business prospect for the next decade by International Technology Roadmap for Semiconductors (ITRS), the Internet of Things (IoT) offers huge potential for business and scientific development in the field of semiconductor design in the post Moore era [1]. IoT is an infrastructure in which billions of devices ("things") are connected to the Internet to enable direct interactions between the physical world and computer-based systems. We denote these things as *edge devices*, which are constituted from a wide variety of electronic and electromechanical devices such as smart thermostats, lights, watches, mobile phones, sensors, actuators, and many others. Recently, different reports indicate that the number of connected devices will well exceed several billions by 2020 [2], [3]. As ubiquitous sensing is the backbone of any IoT application, edge devices are distributed among large geographical areas and generate real time data for further analysis and decision making. The pervasive nature of these applications has resulted in severe resource constraints in these edge devices, such as, low power budget, low die area allocation, and low processing power.

Quite a few architectures have been proposed for IoT applications and almost all of them have similar communication scenario - edge devices connecting with server/applications through gateway devices [4]–[6]. Figure 1 shows a sample IoT system architecture. According to Cisco IoT framework, the

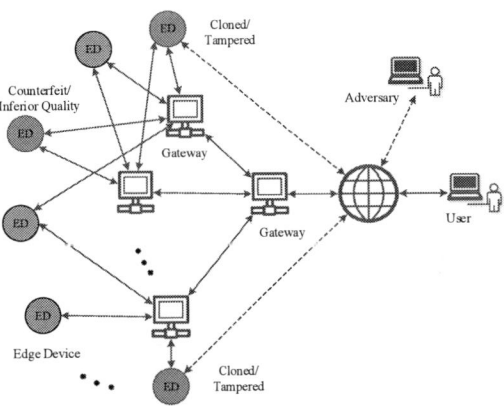

Figure 1: A standard IoT model with hardware vulnerabilities.

edge devices in IoT should have the capability of generating data, converting analog data to digital format and they should also have the ability of being queried/controlled over the network [4]. To accommodate these functions, the edge devices should be equipped with sensors, analog to digital converters, communication modules, memory and embedded processors. Importantly, edge devices in IoT applications are expected to be low-cost devices in order to facilitate widespread adoption [7]. Remote applications also limit the energy resource. Equipping edge devices with large batteries is not feasible for most applications because of the cost constraints [8]. So, minimizing power consumption in every possible ways has become a prime design concern. Energy constraint prohibits IoT edge devices to use standard cryptographic schemes [8] which has resulted in the development of several light-weight cryptographic schemes [9]. Still, a majority of the IoT devices do not use encryption due to power constraints. In a recent study, HP has found almost 70% of the tested IoT devices did not use encryption during communication with the server over the Internet [10].

Moreover, most of these devices are manufactured in environments of limited trust that in particular lack relevant government or other appropriate oversight, and then travel across the globe through intermediaries in the supply chain before being deployed. These factors make it virtually impossible to gauge the origin of these systems and their components, and to track their route in the supply chain. As a result, it is becoming increasingly difficult to ensure the security, integrity, and authenticity of these edge devices. Numerous

978-1-5386-3693-0/18 $31.00 © 2018 IEEE   85

incidents have highlighted the far-reaching penetration of such counterfeit devices into the electronics supply chain [11]–[13], including cloned systems into the United States defense supply chain [14]–[18]. Figure 1 shows an abstract view of an IoT System with various attacks originating from untrusted hardware. Hardware attacks are generally initiated at the physical location where the system is located. For instance, a rogue employee can hypothetically replace an authentic device with a counterfeit or cloned device in order to gain access to the system improperly. A user can also unknowingly add a cloned device to the system since it is nearly impossible to track the origin of a given device in today's convoluted and generally untrustworthy supply chains. In addition, various software attacks can be performed through networks such as Phishing, Denial of Service (DoS), and data spoofing [19], [20]. The scope of this paper is to address the hardware attacks by ensuring the authenticity of the edge devices.

In this paper, we have presented a lightweight communication protocol to verify the authenticity of a low-cost edge device by verifying an unclonable device ID. The communication protocol is designed to be cost effective as it can use on-board resources available in even a minimal IoT edge device, such as, a processor and SRAM memories. Furthermore, the unclonable device ID can be generated from an on-board SRAM memory (SRAM PUF [21]) to avoid the cost of programmable non-volatile memory in low cost edge devices. Recently, physically unclonable functions (PUFs) have received much attention in the hardware security community because they can generate unique and unclonable bits for the identification and authentication of ICs. PUFs use inherently uncontrollable and unpredictable variations from a manufacturing process to produce random and unclonable bits. Several PUF architectures have been proposed over the years that include arbiter PUF [22], ring oscillator PUF [23], SRAM PUF [21], among others. As the IoT edge devices have SRAM based memory and embedded processor, SRAM PUFs offer a better solution to produce device IDs with no additional cost. However, SRAM PUFs are often found to be unreliable which makes authentication through exact ID matching a difficult task [24]. So, in this paper, we have further proposed a novel repeated ID matching scheme which can be used to authenticate edge devices with unreliable IDs. The contributions of our paper are as follows:

- *Secure communication protocol:* We propose a new lightweight communication protocol that uses existing hardware resources of an IoT edge device to verify its identify. Our proposed solution uses a secure hash function [25], which can be implemented using the embedded processor and memory of the edge devices [26], [27]. Care needs to be taken such that the device ID never leaves the system without encryption. Our heuristic security evaluation shows that this protocol is resistant to various known attacks (see Section IV for details).
- *Novel device ID verification technique:* Due to the unreliable nature of SRAM PUF responses, we propose a repeated ID matching scheme (see details in Section III), which does not require expensive helper data and algorithms for error correction, currently in practice for SRAM PUFs. Our

proposed repeated ID matching scheme extracts the most reliable bits from the responses of an unreliable PUF. We show that it is extremely unlikely (see details in Table I) to impersonate an edge device by an adversary. With random trials, an adversary can pass simple ID matching scheme (see Section III-A) when a PUF is unreliable. However, this is unfeasible to impersonate an ID two times by an adversary using random guesses. Note that one can also implement an authentication scheme that verifies ID more than two times to further increase the difficulty of impersonating an authentic edge device.

The rest of the paper is organized as follows. Section II describes our proposed lightweight communication protocol for authenticating an edge device. We present our proposed ID matching scheme by repeated authentication in Section III. We perform a security evaluation in Section IV. We conclude the paper in Section V.

II. PROPOSED AUTHENTICATION SCHEME

Limited device resource has become one of the major constraints for implementing standard secure protocols for authenticating an edge device. Low die area, low power, and limited memory requirements severely limit the performance of the majority of edge devices. Trappe et al. showed that these constraints prohibit the low end IoT devices from using any standard secure protocols like TLS or IPsec [8]. It has resulted in the development of IoT devices with minimum security considering the generated information is of little value to the attackers. Recent studies conducted by HP, Symantec have found major portion of the IoT devices not to use any kind of encryption during communication over the Internet at all [10], [28]. But these seemingly harmless information can be used to get into any complex systems which has been showed in recent studies [28], [29]. Counterfeit or cloned devices in an IoT environment can cause significant damage to the security of the system which raises the need for edge device authentication. In this paper, we address the verification of an edge device by correctly identifying its origin. However, we agree that the data encryption, which is only practical for the gateways as they have higher resources, is required in order to prevent various run time attacks. Our paper does not address the identity verification of gateways, and treats it as authentic.

Figure 2 shows our proposed authentication approach. A true random number generator (TRNG) is necessary in the gateway device for generating a random nonce (n). We propose to use an area-efficient cryptographically secure pseudorandom number generator (CSPRNG) depending on the implementation choice [30] or [31]. A one-time-pad (OTP) [32], [33] is used for encrypting the key with n. OTP has extremely low area footprint since it only requires a simple XOR network. We require an on-chip memory to store secret key-ID pair, $\{K_i, ID_i\}$, corresponding to the edge device (ED_i) of an IoT system. In the edge device, an SRAM PUF can be used to generate this secret unclonable device ID, ID_i (see Section III for details).

The communication protocol for authenticating an edge device is described as follows:

978-1-5386-3693-0/18 $31.00 © 2018 IEEE

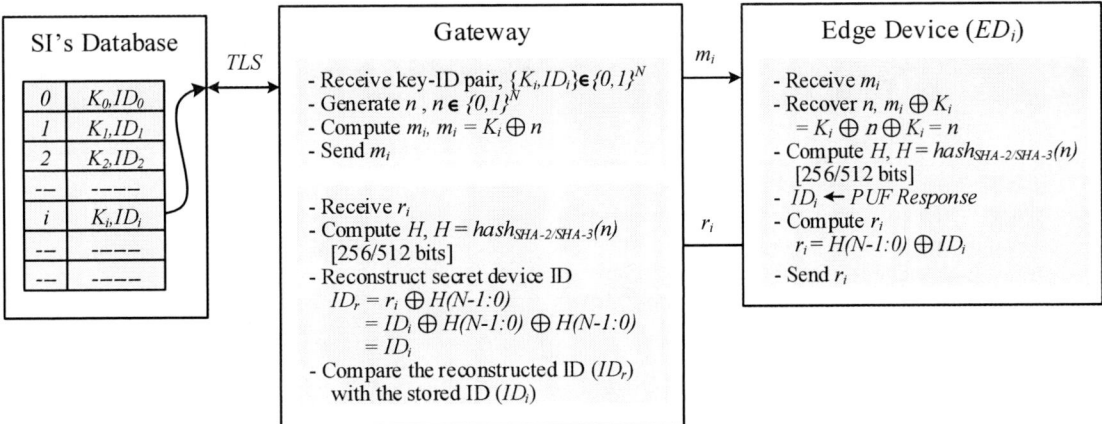

Figure 2: Proposed communication protocol for verifying the identity of an edge device.

1) The gateway uses an existing secure communication protocol (*e.g.*, TLS [34]) to receive the secret device key-ID pair $\{K_i, ID_i\}$, from the trusted system integrator (SI) who produced that edge device. During production, every device is registered in a secure database with a public ID, and a key-ID pair, $\{ID_i \in \{0,1\}^N,\ K_i \in \{0,1\}^N\}$. Here, N represents the size of the ID and the key. For example, it can be 128 depending on the level of security one requires. Now, the public ID is necessary to identify the edge device in the database. It is also possible to store the data in a tamper-proof NVM of the gateways. However, since the gateway is always connected to the Internet and the standard security measures in place as it does not have any resource constraints, we recommend receiving the key-ID pair from the trusted integrator rather than storing into NVM.

2) The gateway stores $\{K_i, ID_i\}$ in its on-chip (volatile or non-volatile) memory. We need to make sure that one cannot access this $\{K_i, ID_i\}$ through the input/output (I/O) pins of the gateway. This will prevent an adversary from getting access to the $\{K_i, ID_i\}$, which was generated during the registration phase of the edge device.

3) An on-chip CSPRNG generates a unique nonce (n), which is stored in the memory. This n will be used later for decrypting the secret device ID. A one-time pad (OTP) now encrypts the key (K_i) with this random nonce. The gateway then sends this encrypted key ($n \oplus K_i$) (depicted as (m_i) in the figure) to the ith edge device, ED_i, to request for its identification.

4) The unique nonce (n) is recovered at the edge device by XORing the m_i with the shared secret key (K_i). A cryographically secure hash (e.g., SHA-2 or SHA-3 [25]) is computed on this nonce (n) to produce a 256/512 bits hash output (H). We recommend to use the existing hardware resources (embedded processor and memory [26], [27]) of the edge devices to compute the hash.

$$H = hash_{SHA-2/SHA-3}(n)$$

5) The edge device encrypts the device ID, ID_i using N bits of computed H.

$$r_i = ID_i \oplus H(N-1:0)$$

The encrypted data $\{r_i\}$ is sent to the gateway.

6) After receiving r_i, the gateway computes the same hash (SHA-2 or SHA-3) using the stored random nonce, n. The gateway now reconstructs the secret device ID.

$$\begin{aligned} r_i \oplus H(N-1:0) \\ = ID_i \oplus H(N-1:0) \oplus H(N-1:0) \\ = ID_i \end{aligned} \tag{1}$$

7) This ID is then verified with the stored ID (see Section III for details). Steps 3-6 are followed for the second stage of the authentication to increase the confidence that the ID is originated from an authentic edge device.

III. DEVICE AUTHENTICATION BY ID MATCHING

In order to secure the semiconductor supply chain and prevent intrusion of counterfeit ICs in critical IoT applications, edge devices need to be equipped with unclonable IDs. In our proposed authentication scheme, we have considered an SRAM PUF for producing an unclonable device ID considering the IoT edge devices are expected to incorporate SRAM-based memories, which makes SRAM PUF a low cost solution for developing such IDs. However, reliability is a major concern for most PUF implementations today [21], [23], [35]. The response of a PUF must remain stable over a wide range of environmental variations such as temperature, noise generated from power supply, and electromagnetic interference, as well as unpredictable degradation that arises from aging. Therefore, it is necessary to perform error correction (e.g., fuzzy extraction [36]) to produce stable PUF responses. Helper data are often required to improve the accuracy and efficiency. However, the implementation of expensive error correction is challenging in the resource constrained edge devices.

As SRAM PUF outputs are unreliable (without error correction), authenticating an edge device by comparing IDs is challenging. SRAM PUF outputs may vary because of aging degradation, temperature variation and supply voltage fluctuations. So, if a PUF response is taken at two different environmental conditions, the responses can vary significantly. *How can we use a PUF to verify the identity of a device, if a PUF produces an unreliable ID?* There will be a mismatch among the stored response (obtained from PUFs during the registration) and the new response while authentication from the same PUF. The basic idea is to capture multiple responses from the PUF in a very short duration under almost identical environmental conditions to ensure the responses are similar. This led us to develop a repeated authentication scheme, where the gateway interrogates an edge device more than once in short duration (see details in Section III-B).

A. Simple ID Matching

The primary reason that we can overcome PUF uncertainty is that we do not need to match completely both stored and received IDs bit-by-bit to authenticate an edge device. We propose to use Hamming distance (HD) to measure the similarity between the stored ID (ID_S, PUF response during registration) and received ID (ID_R, PUF response during authentication). The Hamming distance between two IDs is defined as the number of bits at which the two IDs differ.

The authentication can be performed as follows:

$$HD(ID_S, ID_R) \rightarrow \left\{ \begin{array}{l} \leq HD_T, \text{ The device is authentic} \\ > HD_T, \text{ The device is counterfeit} \end{array} \right.$$

where, HD_T is the threshold that can be determined after characterizing the PUF. For example, one can set a threshold of 10 for a 128-bit device ID. This signifies that a device under authentication is certified as authentic when the mismatch of bits between ID_S and ID_R does not exceed 10.

Let us now determine the probability of authenticating an edge device as genuine by an adversary. We assume that the size of the stored ID (PUF response) is of N-bits and HD_T is of k-bits. Now, the probability of finding one vector with exactly $(N-k)$-bit match or k-bit mismatch is as follows:

$$p = \frac{{}^N C_{(N-k)}}{2^N} = \frac{\binom{N}{k}}{2^N}.$$

Thus, the probability of finding a vector with at most k-bit mismatch to pass the ID matching test becomes:

$$p = \frac{\sum_{i=0}^{k} \binom{N}{i}}{2^N}. \tag{2}$$

From Equation 2, it is clear (see Table I for details) that the attacker has a low probability of impersonating a device. However, this is not sufficient enough to prevent an adversary guessing an ID by random trials when there are many unreliable bits in an ID. This may occur when an edge device uses an SRAM PUF without any error correction. It is thus necessary to further improve simple ID matching scheme.

B. Repeated ID Matching

The reliability issue of the PUF primarily arises from the environmental variation, such as temperature variation and device aging. The environmental conditions and device age during registration can be significantly different when an edge device is requested for authentication. This results in bit flipping, where the values of a few bits are different between registration and subsequent authentication.

$$\{ID_R[i]\}_{T_R \neq T_S, t_R > t_S} \neq \{ID_S[i]\}_{T_S, t_S}$$

where, ID_R and ID_S are the received and stored PUF responses, respectively. T_R and T_S denote the temperature during the authentication and registration. Similarly, t_R and t_S are the device age during the authentication and registration, respectively.

The primary reason for obtaining different responses is that the PUF is challenged with different conditions. We should have received the same responses if we were to apply the same challenge with similar conditions. *The basic idea of extracting the stable bits from an unreliable PUF is to provide the same challenge repeatedly in short duration such that we can extract PUF responses at similar conditions.* The repeated authentication scheme utilizes this property of a PUF to verify identity of an edge device.

The repeated authentication scheme is described below:

1) The gateway device requests an edge device (ED_i) for its device ID by sending $n_1 \oplus K_i$. The edge device returns encrypted ID_R ($H_{n_1}(N-1:0) \oplus ID_R$). The gateway first decrypts the ID (see Equation 1), and then computes the mismatch locations of the received ID ($ID_R[i] \neq ID_S[i]$). It then contracts the robust ID (RID) by discarding the mismatch bits and keeps track of the mismatch locations.

$$RID[k] = ID_R[i], \text{ if } ID_R[i] = ID_S[i]; \quad 0 < k < i < N \tag{3}$$

2) The gateway again requests ED_i for its ID by sending $n_2 \oplus K_i$ ($n_1 \neq n_2$). ED_i then returns the encrypted ID_R ($H_{n_2}(N-1:0) \oplus ID_R$). The gateway decrypts the ID by using Equation 1, and then computes the new robust ID (RID^*) by using 3.

3) The two robust IDs are compared by using Hamming distance, which is described below:

$$HD(RID, RID^*) \rightarrow \left\{ \begin{array}{l} \leq HD_T^\dagger, \text{ The device is authentic} \\ > HD_T^\dagger, \text{ The device is counterfeit} \end{array} \right.$$

Note that $HD_T^\dagger$ is much less than HD_T because the PUF produces a similar response for similar conditions. It is important to keep in mind that one can implement an authentication scheme that uses more than two repeated IDs from the same device.

The probability of authenticating a device two times as genuine by an adversary can be determined. We assume that the size of the stored ID is of N-bits, HD_T is of k-bits and that $HD_T^\dagger$ is of r-bits ($r \ll k$). The probability of passing two repeated authentication becomes:

$$p = \left(\frac{\sum_{i=0}^{k} \binom{N}{i}}{2^N} \right) \times \left(\frac{\sum_{i=0}^{r} \binom{N-k}{i}}{2^{(N-k)}} \right) \tag{4}$$

Equation 4 provides an interesting observation that the probability of successful authentication is reduced significantly

978-1-5386-3693-0/18 $31.00 © 2018 IEEE

Table I: Probability of matching an ID.

	Simple ID Matching		Repeated ID Matching									
			$HD_T^\dagger=1$		$HD_T^\dagger=2$		$HD_T^\dagger=4$		$HD_T^\dagger=8$		$HD_T^\dagger=16$	
HD_T	ID =128	ID = 256	ID =128	ID = 256	ID =128	ID = 256	ID =128	ID = 256	ID =128	ID = 256	ID =128	ID = 256
1	3.8e-037	2.2e-075	2.9e-073	9.8e-150	1.8e-071	1.3e-147	2.4e-068	6.7e-144	3.2e-063	1.6e-137	2.1e-055	3.9e-127
2	2.4e-035	2.8e-073	3.6e-071	2.5e-147	2.3e-069	3.2e-145	3.0e-066	1.7e-141	3.8e-061	3.9e-135	2.4e-053	9.3e-125
4	3.2e-032	1.5e-069	1.9e-067	5.4e-143	1.2e-065	6.8e-141	1.5e-062	3.5e-137	1.8e-057	7.9e-131	9.7e-050	1.8e-120
8	4.5e-027	3.7e-063	4.1e-061	2.0e-135	2.5e-059	2.5e-133	2.9e-056	1.3e-129	3.1e-051	2.6e-123	1.2e-043	5.2e-113
16	3.2e-019	9.3e-053	6.9e-051	1.3e-122	3.9e-049	1.5e-120	4.0e-046	7.2e-117	3.2e-041	1.3e-110	7.0e-034	2.0e-100
32	6.4e-009	5.9e-037	7.9e-036	4.9e-102	3.8e-034	5.5e-100	2.8e-031	2.3e-096	1.2e-026	3.1e-090	6.7e-020	2.6e-080
64	5.4e-001	2.4e-016	1.9e-018	7.5e-072	6.0e-017	7.2e-070	2.0e-014	2.2e-066	1.5e-010	1.6e-060	2.1e-005	3.7e-051

if we perform repeated authentication. We can also prevent repeated failed authentication at the gateway by implementing a simple counter to keep track of that situation. If the count crosses some threshold, that can raise a flag.

IV. ANALYSIS

A. Replay Attack on Our Proposed Authentication Scheme

In this attack scenario, an attacker attempts to authenticate an edge device by impersonating the ID using prior communications. We assume that an attacker does not have access to the secret key (K_i), which is stored in the edge device. Let us assume that the attacker observes two prior communications. First, he/she observes $n_1 \oplus K_i$ from the gateway and $H_{n_1}(N-1:0) \oplus ID_i$ from the edge device. From this observation the attacker can compute $K_i \oplus ID_i \oplus n_1 \oplus H_{n_1}(N-1:0)$, which is shown below:

$$(n_1 \oplus K_i) \oplus (H_{n_1}(N-1:0) \oplus ID_i) \qquad (5)$$

From the second communication, the attacker observes $n_2(\neq n_1) \oplus K_i$ from the gateway and $H_{n_2}(N-1:0) \oplus ID_i$ from the edge device, and can compute $K_i \oplus ID_i \oplus n_2 \oplus H_{n_2}(N-1:0)$.

Now the attacker can perform the following operations:

$$(n_1 \oplus K_i) \oplus (n_2 \oplus K_i) = n_1 \oplus n_2 \qquad (6)$$

$$(H_{n_1}(N-1:0) \oplus ID_i) \oplus (H_{n_2}(N-1:0) \oplus ID_i)$$
$$= H_{n_1}(N-1:0) \oplus H_{n_2}(N-1:0) \qquad (7)$$

From equation 7, it is obvious that an adversary successfully replays a prior communication if it becomes zero, when $H_{n_1}(N-1:0) = H_{n_2}(N-1:0)$. This contradicts the collision property of a secure hash [25]. Thus, the communication protocol becomes resistant to replay attack, when a system designer implements a secure hash function (SHA-2 or SHA-3).

B. Probability Analysis for Proper ID Matching

Hamming distance plays an important role to reduce the chances of guessing an ID in the allowable ID spaces by an adversary. It becomes harder for guessing an ID by random trials, when the hamming distance is small. However, the choice of hamming distance largely depends on the reliability of the PUF. For a reliable PUF, simple ID matching is sufficient enough to provide enough protection against cloning. However, we can achieve significant improvement from our proposed repeated ID matching scheme even for a very unreliable PUF, which can be the case for a resource constrained IoT edge device. In this section, we will analyze two

different (reliable and unreliable) types of PUFs to evaluate the effectiveness of our ID matching schemes.

The Table I summarizes the probabilities of matching edge device IDs under simple ID matching scheme (see Section III-A) and proposed repeated ID matching scheme (see Section III-B). We perform our analysis considering 128-bit and 256-bit device IDs. The hamming distances (HD_Ts) chosen for the analysis are 1, 2, 4, 8, 16, 32, and 64. The PUFs are very reliable when HD_T are of 1, 2 and 4. On the other hand, we also consider very noisy PUFs where 32 or 64 bits may be flipped during authentication. As expected, the probabilities of ID matching increases with the increase of HD_T. This is intuitive as an adversary has lesser effort of matching an ID. For a reliable PUF, an attacker has an extremely low luck of matching an authentic ID. For example, the probability of finding a match becomes 4.5×10^{-27} and 3.7×10^{-63} considering HD_T of 8, when the IDs are 128 bit and 256 bit respectively. However, the probability increases significantly 5.4×10^{-1} and 2.5×10^{-16} considering HD_T of 64 for an ID of 128 and 256 bits, respectively.

For repeated ID matching scheme, the hamming distances ($HD_T^\dagger s$), chosen for the second stage of the authentication process, are of 1, 2, 4, 8, and 16 bits. As before, the probabilities of ID matching becomes higher with the increase of $HD_T^\dagger s$. However, it is much less compared to the simple ID matching scheme. We now have a significant improvement of not finding an ID as the probability has decreased significantly. For example, the probability of finding an ID is 1.9×10^{-18} for a PUF that is heavily impacted by aging (64 out of 128 bits are unstable) assuming the stable bits remains stable ($HD_T^\dagger$ is of 1). For a 128 bit ID, we also have a very low probability of 1.5×10^{-10} and 2.1×10^{-5} with $HD_T^\dagger$ of 8 and 16 bits, respectively. For a 256 bit ID, it is fairly impossible for an adversary to pass the repeated ID matching scheme even though PUF responses are unreliable.

C. Physical Attacks

The security of our proposed scheme largely depends on the shared secret key between the gateway and the edge devices which should be stored in tamper-proof memory and the SRAM PUF based device IDs for edge devices. These information can be stolen through sophisticated physical attacks or reverse engineering. Today's optical microscopes can produce 3D images of a microchip with superfine resolution. Scanning Electron Microscopes (SEM) and Transmission Electron Microscopes (TEM) can generate images of different inner layers of a microchip. Chipworks (now TechInsights) have successfully performed such experiments legitimately for the purpose of competitive analysis and patent research.

The physical layout of a chip can be reconstructed through destructive physical attacks as well. Data stored in a non-volatile memory (NVM) can be reconstructed through infrared backside imaging, which can be used to directly look at the memory contents. All these physical attacks can definitely be used to find the secret key or the device ID. However, an adversary can impersonate only one device through such physical attacks, which does not make any financial motivation for performing such attacks.

V. CONCLUSION

In this paper, we have presented a novel communication protocol to authenticate edge devices for an IoT system by using unclonable device IDs. As resource limitation prohibits standard cryptographic schemes to be followed in the edge devices, we have presented a light-weight encryption scheme that uses a secure hash function and can be implemented with existing resources of the IoT edge devices. The unclonable device ID can be created by using an on-board SRAM PUF, as it produces a unique device footprint based on the manufacturing variability. The unpredictable aging degradation and temperature variation make the output of the PUFs often unreliable. Thus, developing a verification scheme, which takes care of an ID generated from an unreliable PUF, is of prime importance. We address this reliability issues of PUFs by introducing repeated authentication. We have utilized the concept of hamming distance for our ID matching scheme. It is extremely difficult for an adversary to clone an ID to pass repeated ID matching scheme which is evident from our probability analysis. The larger device ID also provides higher protection against cloning since it poses increased difficulty for an attacker to pass ID matching test.

ACKNOWLEDGEMENT

This work was supported in part by an internal research grant from the Charles D. McCrary Institute to the first author. Additional support from the authors' respective departments and the Auburn Cyber Research Center is also acknowledged.

REFERENCES

[1] International Technology Roadmap for Semiconductors 2.0, 2015 Edition.
[2] G. Research, "Gartner says 6.4 billion connected "things" will be in use in 2016, up 30 percent from 2015," 2015, http://www.gartner.com/newsroom/id/3165317.
[3] D. Evans, "The Internet of Things: How the Next Evolution of the Internet Is Changing Everything," 2011.
[4] Cisco, "The internet of things reference model," 2014.
[5] L. Atzori, A. Iera, and G. Morabito, "The internet of things: A survey," *Computer networks*, vol. 54, no. 15, pp. 2787–2805, 2010.
[6] J. Gubbi, R. Buyya, S. Marusic, and M. Palaniswami, "Internet of things (iot): A vision, architectural elements, and future directions," *Future generation computer systems*, vol. 29, no. 7, pp. 1645–1660, 2013.
[7] J. Bryzek, "Roadmap for the trillion sensor universe," *Berkeley, CA, April*, vol. 2, 2013.
[8] W. Trappe, R. Howard, and R. S. Moore, "Low-energy security: Limits and opportunities in the internet of things," *IEEE Security & Privacy*, vol. 13, no. 1, pp. 14–21, 2015.
[9] M. Katagi and S. Moriai, "Lightweight cryptography for the internet of things," *Sony Corporation*, pp. 7–10, 2008.
[10] K. Rawlinson. Hp study reveals 70 percent of internet of things devices vulnerable to attack. [Online]. Available: http://www8.hp.com/us/en/hp-news/press-release.html?id=1744676.WUrrwWgrKM8

[11] M. M. Tehranipoor, U. Guin, and D. Forte, *Counterfeit Integrated Circuits: Detection and Avoidance.* Springer, 2015.
[12] U. Guin, K. Huang, D. DiMase, J. Carulli, M. Tehranipoor, and Y. Makris, "Counterfeit integrated circuits: A rising threat in the global semiconductor supply chain," *Proceedings of the IEEE*, vol. 102, no. 8, pp. 1207–1228, Aug 2014.
[13] U. Guin, D. DiMase, and M. Tehranipoor, "Counterfeit integrated circuits: Detection, avoidance, and the challenges ahead," *Journal of Electronic Testing*, vol. 30, no. 1, pp. 9–23, 2014.
[14] "Texas brothers plead guilty to selling counterfeit 'Cisco' computer products," News Releases, U.S. Immigration and Customs Enforcement, August 2009.
[15] The Federal Bureau of Investigation, "Departments of Justice and Homeland Security Announce 30 Convictions, More Than $143 Million in Seizures from Initiative Targeting Traffickers in Counterfeit Network Hardware ," May 2010.
[16] U. Guin, S. Bhunia, D. Forte, and M. Tehranipoor, "Sma: A system-level mutual authentication for protecting electronic hardware and firmware," *IEEE Transactions on Dependable and Secure Computing*, 2016.
[17] U. Guin, Q. Shi, D. Forte, and M. Tehranipoor, "FORTIS: A Comprehensive Solution for Establishing Forward Trust for Protecting IPs and ICs," *ACM Transactions on Design Automation of Electronic Systems (TODAES)*, 2016.
[18] M. M. Tehranipoor, U. Guin, and S. Bhunia, "Invasion of the hardware snatchers," *IEEE Spectrum*, vol. 54, no. 5, pp. 36–41, 2017.
[19] T. Borgohain, U. Kumar, and S. Sanyal, "Survey of security and privacy issues of internet of things," *CoRR*, vol. abs/1501.02211, 2015. [Online]. Available: http://arxiv.org/abs/1501.02211
[20] H. He, C. Maple, T. Watson, A. Tiwari, J. Mehnen, Y. Jin, and B. Gabrys, "The security challenges in the iot enabled cyber-physical systems and opportunities for evolutionary computing other computational intelligence," in *2016 IEEE Congress on Evolutionary Computation (CEC)*, July 2016, pp. 1015–1021.
[21] J. Guajardo, S. S. Kumar, G.-J. Schrijen, and P. Tuyls, "Fpga intrinsic pufs and their use for ip protection," in *International workshop on Cryptographic Hardware and Embedded Systems*. Springer, 2007, pp. 63–80.
[22] B. Gassend, D. Clarke, M. Van Dijk, and S. Devadas, "Silicon physical random functions," in *Proceedings of the ACM Conference on Computer and Communications Security (CCS)*. ACM, 2002, pp. 148–160.
[23] G. Suh and S. Devadas, "Physical Unclonable Functions for device authentication and secret key generation," in *Proc. of ACM/IEEE on Design Automation Conference*, 2007, pp. 9–14.
[24] K. Xiao, M. T. Rahman, D. Forte, Y. Huang, M. Su, and M. Tehranipoor, "Bit selection algorithm suitable for high-volume production of sram-puf," in *Hardware-Oriented Security and Trust (HOST), 2014 IEEE International Symposium on*. IEEE, 2014, pp. 101–106.
[25] NIST, "FIPS PUB 180-4: Secure Hash Standard (SHS)," August 2015.
[26] T. Eisenbarth, S. Heyse, I. von Maurich, T. Poeppelmann, J. Rave, C. Reuber, and A. Wild, "Evaluation of sha-3 candidates for 8-bit embedded processors," in *The Second SHA-3 Candidate Conference*, 2010.
[27] "Atmel AVR232: Authentication Using SHA-256," Tech. Rep. [Online]. Available: http://www.atmel.com/Images/doc8184.pdf
[28] M. B. Barcena and C. Wueest, "Insecurity in the internet of things."
[29] R. M. Ishtiaq Roufa, H. Mustafaa, S. O. Travis Taylora, W. Xua, M. Gruteserb, W. Trappeb, and I. Seskarb, "Security and privacy vulnerabilities of in-car wireless networks: A tire pressure monitoring system case study," in *19th USENIX Security Symposium, Washington DC*, 2010, pp. 11–13.
[30] D. E. Holcomb, W. P. Burleson, and K. Fu, "Initial sram state as a fingerprint and source of true random numbers for rfid tags," in *In Proceedings of the Conference on RFID Security*, 2007.
[31] B. Sunar, W. Martin, and D. Stinson, "A provably secure true random number generator with built-in tolerance to active attacks," *Computers, IEEE Transactions on*, vol. 56, no. 1, pp. 109–119, Jan 2007.
[32] G. S. Vernam, "Secret signaling system," 1919, US Patent 1,310,719.
[33] J. Katz and Y. Lindell, *Introduction to modern cryptography*. CRC press, 2014.
[34] T. Dierks, "The transport layer security (tls) protocol version 1.2," 2008.
[35] A. Maiti and P. Schaumont, "The impact of aging on a physical unclonable function," *IEEE Transactions on Very Large Scale Integration (VLSI) Systems*, vol. 22, no. 9, pp. 1854–1864, 2014.
[36] Y. Dodis, R. Ostrovsky, L. Reyzin, and A. Smith, "Fuzzy extractors: How to generate strong keys from biometrics and other noisy data," *SIAM journal on computing*, vol. 38, no. 1, pp. 97–139, 2008.

978-1-5386-3693-0/18 $31.00 © 2018 IEEE

2018 31th International Conference on VLSI Design and 2018 17th International Conference on Embedded Systems

Hardware Trojan Detection using ATPG and Model Checking

Jonathan Cruz[1], Farimah Farahmandi[2], Alif Ahmed[2], and Prabhat Mishra[2]

[1]Department of Electrical and Computer Engineering
[2]Department of Computer and Information Science and Engineering
University of Florida, Gainesville FL, USA

Abstract—The threat of hardware Trojans' existence in integrated circuits has become a major concern in System-on-Chip (SoC) design industry as well as in military/defense organizations. There is an increased emphasis on finding effective ways to detect and activate hardware Trojans in current research efforts. However, state-of-the-art approaches suffer from the lack of completeness and scalability. Moreover, most of the existing methods cannot generate efficient tests to activate the potential hidden Trojan. In this paper, we propose an effective test generation approach which is capable of activating malicious functionality hidden in large sequential designs. Automatic test pattern generation (ATPG) works well on full-scan designs, whereas model checking is suitable for logic blocks without scan chain. Due to overhead considerations, partial-scan chain insertion is the standard practice today. Unfortunately, neither ATPG nor model checking is suitable for partial-scan designs. Our proposed hardware Trojan detection technique utilizes the combination of ATPG and model checking approaches. We use model checking on a subset of non-scan elements and ATPG on scan elements to avoid common pitfalls of running the original design using any one of these techniques. Experimental results demonstrate the effectiveness of tests generated by our proposed approach to detect Trojans on Trust-hub benchmarks.

I. INTRODUCTION

Designing today's system-on-chips (SoC) is a highly complex process that is subject to stringent time-to-market constraints. It is a common practice to integrate third-party Intellectual Property (IP) blocks in the production of SoCs in order to remain competitive in today's global market. However, interfacing with untrusted third-party IP affects a design's trustworthiness and security. An adversary can tamper with the design or insert malicious components, known as hardware Trojans, within third-party facilities. Hardware Trojans are a small modification in the design that can be triggered in an extremely rare input event. As a result, hardware Trojans are dormant during most of the run-time and can escape detection during conventional functional validation techniques (such as simulation-based validation using random or constraint random tests). The effects of hardware Trojan insertion attacks range from information leakage to complete chip malfunction [1]. Therefore, it is extremely important to find efficient validation approaches to detect and activate hardware Trojans if they exist.

Existing logic testing methodologies for Trojan detection target rare node activation while monitoring the designs ob-

This work was partially supported by grants from National Science Foundation (CNS-1441667), SRC (2014-TS-2554), and Cisco.

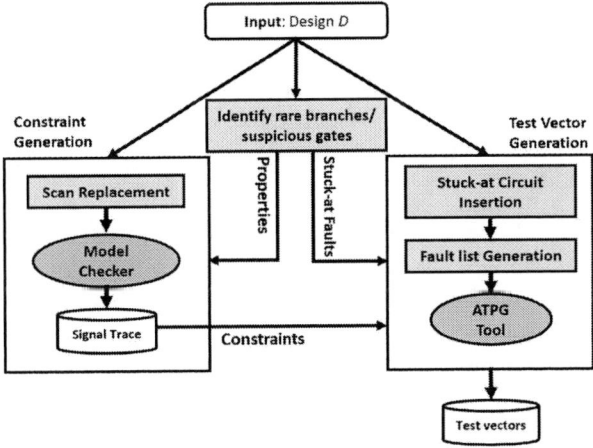

Fig. 1: The overview of our proposed approach.

servable outputs [25]. Some approaches use N-detect testing with an ATPG tool for generating test vectors leading to an increased likelihood for Trojan activation [2], [24]. Most of these techniques assume a full scan chain when dealing with IP that contain sequential elements [3], [2]. However, partial-scan chains offer an attractive alternative to designs with area or testing constraints. Therefore, with the introduction of partial-scan inserted IP, it becomes even harder to detect Trojans that rely on these verification tools alone since the depth and testable complexity of sequential parts will be increased [4], [5]. Formal methods have also been used in Trojan detection and verifying design security. With model checking, security properties can be specified by a designer to verify the security of the design [6]. However, model checking is known to suffer from state explosion which limits its effectiveness on verifying security in large designs.

Motivated by the limitations of these tools, we propose a test generation framework that combines the strengths of these two common design verification techniques, already a part of the normal verification design flow. Our approach increases the efficiency for Trojan detection in partial-scan designs by reducing the potential state space for model checker and removing the complexity of non-scan sequential ATPG in ATPG tools. Figure 1 shows the overview of our approach. As shown in Figure 1, our method identifies suspicious branches/gates which may be used as triggering conditions for hardware Trojans. In order to generate tests to activate rare nodes, scan replacement is done in the next step. We generate security

properties that targets activation of equivalent signals/gates of rare nodes in the gate-level netlist. The scan replaced netlist as well as the security properties are used by the model checker. We generate a set of constraints using model checker to facilitate directed test generation using ATPG tool. To the best of our knowledge, there are no previous attempts to address Trojan detection particularly in partial-scan designs using a combination of model checking and ATPG tools. We have applied our methodology on Trust-Hub and custom benchmarks and have demonstrated that our approach not only detects the hidden Trojans but also activates them in an effective manner.

The rest of the paper is organized as follows. Section II introduces the background in ATPG and model checking. Section III describes the related work using these techniques. Section IV describes our hardware Trojan detection approach of using ATPG and model checking. Section V describes the experimental setup, analysis, and results. Finally, Section VI concludes the paper.

II. BACKGROUND

A. Design for Test

Design for Testability (DFT) is a technique employed in designing ICs for the purposes of reducing test costs and associated time [1]. A very common technique for DFT is scan-chain insertion. The idea here is to replace regular flip-flops (FFs) in the design with scan flip-flops. Scan FFs can be chained together to form a scan-chain. The purpose of these scan-chains is to reduce the loading time when testing sequential elements of a design, increasing a circuit's overall gate controllability and observability. While including a full scan-chain is ideal, it may not be feasible due to the incurred area overhead or delays that invalidate various design constraints.

Partial-scan chain insertion is used as a cost-effective alternative [7] to achieve acceptable test coverage, while maintaining design constraints. In partial-scan designs, not all FFs are included in the scan-chain. However, these non-scan FFs introduce branches in the circuit with low controllability and observability, which can be exploited as a rare trigger condition by a malicious attacker. Controllability measures the difficulty associated with setting a particular signal to a logic value, and observability measures the difficulty of observing a signal at an observable point in the circuit.

B. Automatic Test Pattern Generation

ATPG is a test methodology used to identify faulty behavior(s) in circuits due to design defects. The goal of ATPG is to create a set of test patterns that achieve a desired test coverage, TC, and fault coverage, FC, through fault simulation. Test coverage is the ratio of detected faults over testable faults, while fault coverage is the ratio of detected faults over all faults. With the use of DFT and scan-chains, ATPG can efficiently generate test vectors by treating any design as combinational logic, consequently reducing the complexity. This is no longer the case in partial-scan designs. ATPG tools must now consider a sequential set of test vectors to activate a target fault. The sequential ATPG complexity is linearly proportional to the sequential depth and exponentially porportional to the number of sequential cycles (more complex than feedback loops) [8]. The worst-case complexity (cyclic sequential designs) of sequential ATPG becomes 9^N, in a 9-valued logic system where N is the number of flip-flops [9].

C. Model Checking

Model checking is a formal method used to verify a design against functional properties (expected design behavior). To verify a design using model checkers, users must first either manually, or through the use of a program, translate their design into a model specification language understood by the tool. Design properties are then described in a temporal language using either computational tree logic (CTL) or linear temporal logic (LTL). Once the tool has both the design and properties to be verified, it begins to unroll the state space. A Boolean satisfiability assignment is extracted from the unrolled states and checked using an internal SAT solver. If unsatisfiable, the model checker produces a counterexample computation path [10].

Functional design verification with model checking can be extended for use in verifying design security [15]. Properties can be written and verified with security features in mind, such as monitoring the primary output for leaking secret keys [6]. However, as previously mentioned, a common problem of the model checking approach is state explosion caused by the exponential nature of exploring a designs state space. This fact limits the practicality of model checking on larger designs.

III. RELATED WORK

ATPG and formal methods have been used for Trojan detection and security verification in recent years [15]. As with any ATPG tool, sequential test pattern generation is a complex process [9], [8]. While scan-chains are implemented to mitigate the complexity, designs with a significant amount of non-scan cells can greatly reduce ATPG performance, resulting in ATPG-untestable faults [12] that an attacker can exploit.

One common approach to Trojan detection involves logic testing to trigger a set of rare nodes. These methods generate test patterns with high probability of activating Trojans. Common to most of these approaches is the use of an ATPG tool, SAT solver, or some combination for test generation. Both Wolff et al. and Zhang et al. propose Trojan detection approaches that incorporate ATPG tools for generating test patterns [13], [3]. Yet, with the introduction of partial-scan designs, the effectiveness of full-sequential ATPG for generating test patterns is greatly reduced due to the complexity of full-sequential ATPG on non-scan FFs. The method proposed in [14] utilizes N-detect full scan ATPG and SAT solver for Trojans detection. However, this approach also fails to effectively consider designs that have a significant non-scan regions which will limit the effectiveness of ATPG.

Functional design verification with model checking can be extended for use in verifying design security [15]. Properties

---

978-1-5386-3693-0/18 $31.00 © 2018 IEEE

can be written and verified with security features in mind, such as monitoring the primary output for leaking secret keys [6]. However, as previously mentioned, a common problem of the model checking approach is state explosion caused by the exponential nature of exploring a designs state space. This fact limits the practicality of model checking on larger designs. Although approaches based on symbolic algebra detect the existence of hardware Trojans, they cannot generate tests to activate it [16].

IV. TROJAN DETECTION USING ATPG AND MODEL CHECKING

Many Trojan detection techniques utilize ATPG, model checking, or both in generating test patterns for detection; however, they do not consider partial-scan instances of a third-party IP (3PIP). ATPG is expected to encounter notable execution overhead in most partial-scan designs with significant sequential depth or cyclic sequential structures due to their complexity [8]. Additionally, previously detectable faults can be unintentionally rendered undetectable with the removal of scan FFs from the scan-chain. This makes it much harder to generate tests for suspicious regions as it is no longer within the realm of combinational complexity. With model checking alone, medium to large designs are expected to suffer from state explosion, preventing efficient test generation.

We propose a framework to improve the test generation efficiency by combining the benefits of the two approaches. To the best of our knowledge, our approach is the first attempt to use both ATPG and model checking for efficient test pattern generation in partial-scan designs for Trojan detection. Algorithm 1 shows the major steps in our proposed framework shown in Figure 1. Algorithm 1 takes a design D, and outputs a set of test vectors T. In line 5, the set of rare nodes R are identified in the design which are used by the *constraintGeneration* and *testVectorGeneration* procedures. The *constraintGeneration* procedure (Algorithm 2) uses model checking to produce a set of signal traces S. Finally the *testVectorGeneration* method (Algorithm 3) uses ATPG with the design, rare nodes and signal traces to produce a set of test vectors for activating each rare node.

To maximize the benefits of each tool, our framework also includes parallel execution of model checking and ATPG. The final test pattern is taken from whichever execution generates results first. The remainder of this section describes the steps involved in the proposed hybrid approach: rare branch identification, constraint generation, and test vector generation.

A. Rare Branch Identification

For each IP, initial analysis is performed at the RTL level to determine suspicious gates in the design. In a design, rare branches are branches that are not covered after running random tests up to millions of cycles. Mapping the RTL branches to gate-level netlist after synthesis is done in two phases. The first phase identifies any suspicious boundary and register nets and uses the synthesis tool commands to attempt

---

Algorithm 1 Trojan Detection Algorithm

1: **Input:** Design D
2: **Output:** Set of testvectors T
3: **procedure** TROJANDETECTION(D)
4:     $R, S, T = \{\}$
5:     $R = \text{identifyRareBranches}(D)$
6:     $S = \text{constraintGeneration}(D,R)$
7:     $T = \text{testVectorGeneration}(D,R,S)$
8:     **return** T

---

to preserve suspicious signal nets. In these cases, identifiable naming will be preserved after synthesis. If any rare branch is not accounted for, then, the second phase constructs a structural dependency graph of the two representations and attempts to match these graphs using approximate graph matching heuristics as suggested in [17].

Other statistical or functional methods for determining rare nodes at RTL or gate-level such as FANCI [18] and MERS [19] are equally applicable. These circuit branches identified as rare will be used in model checking property generation and ATPG stuck-at faults/ node justification. The rationale here is that a Trojan can be activated as a result of a rare sequence of inputs and/or state transitions; otherwise, the malicious insertion becomes a triviality that can be detected during traditional design testing and verification [13]. By focusing on activating hard-to-trigger or rare nodes, we are increasing the likelihood of Trojan detection. With partial-scan designs, the non-scan FFs can be ideal candidates for embedding Trojans due to the low controllability and observability values. These Trojan instances are generally much harder to detect and will be the threat model we use for the remainder of the paper.

Algorithm 2 Constraint Generation Algorithm

1: **Input:** Design D, set of rare nodes R
2: **Output:** set of signal traces S
3: **procedure** CONSTRAINTGENERATION(D, R)
4:     $S, P = \{\}$
5:     *replace scan FFs with pseudo-primary inputs*
6:     **for each** $r \in R$ **do**
7:         $P_i = assert\ G\ !(r\_cond)$
8:         $S_i = \text{modelChecker}(P_i,D)$
9:     **return** S

B. Constraint Generation

Because ATPG performance generally suffers in the presence of non-scan sequential elements, we use model checking to generate traces transformed into constraint structures for the ATPG tool and facilitate test generation for rare nodes with non-scan FFs in their fan-in. Suppose a design D, has m scan elements and n non-scan elements. The potential state space is 2^{m+n}. To reduce the state space for use in model checking, we create that has all scan FFs replaced with pseudo-primary inputs. This is equivalent to breaking up a scan-chain of m elements into m separate scan-chains. As a result of this

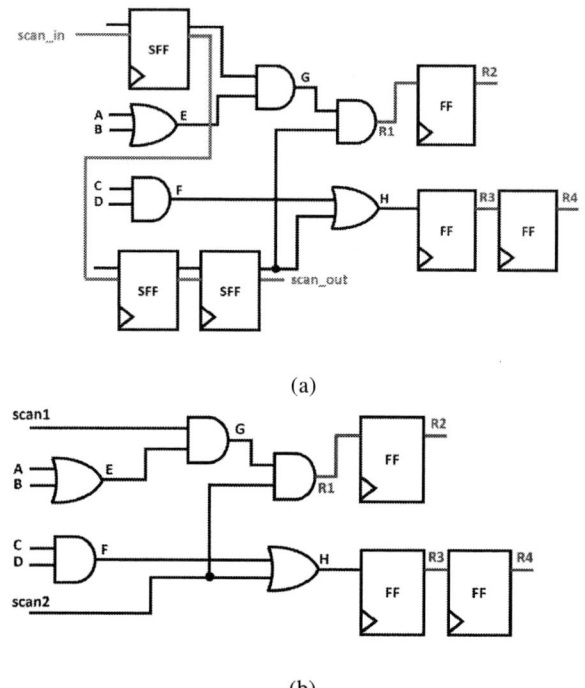

(a)

(b)

Fig. 2: (a) Design before scan replacement. (b) Design after scan replacement with pseudo-primary inputs scan1 and scan2

replacement, the model checking tool now has an effective state space of 2^n mitigating the state explosion issue. It is important to note that design can still have a significant amount of state space after the scan replacement. In such cases, the design can be fed to the model checking tool at the component level, helping reduce the state space even further.

Example 1: Let us consider a sample design A given in Figure 2 (a). After simulating the design for millions of cycles with random input, the rare branches are identified and the corresponding nodes are marked in red R1-R4. All scan FFs are then replaced with pseudo-primary inputs as shown in Figure 2 (b) for use in constraint generation.

The synthesized design is automatically converted to the tool's intermediate representation. For each rare node in the design identified from the first step, we generate properties (P_i) expressed in LTL as:

$$P_i : \; assert \; G \; !(ActivationCond)$$

The negation of the rare node activation is specified as properties in order to output the trigger condition as a counter-example trace from the model checker tool.

Algorithm 2 generates a set of signal traces to be used in ATPG. The algorithm takes the design, D, and replaces the scan FFs with pseudo-primary inputs. A property specified as the negation of the activation is generated for each rare node $r \in R$. Model checker then outputs a signal trace for each property, which is used in Algorithm 3. Note, if the traces from this algorithm are invalid, ATPG will not generate a valid test vector due to justification conflicts.

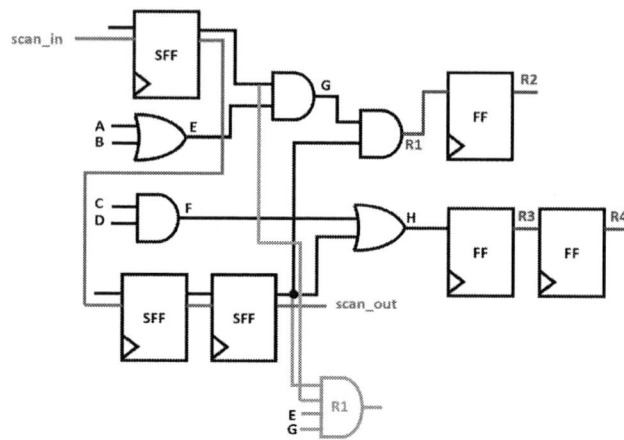

Fig. 3: Rare node R1 Counter-example trace from model checker captured in stuck-at circuit

Example 2: Suppose we want to generate a trace for rare node R1 (low 1 signal probability). From Algorithm 2, the property $assert \; G \; !(R1)$ ($G ==$ always temporal operator) along with the scan replaced design from Figure 2 (b) are run through the model checker. One possible signal trace for the activation levels would be {1,1,1,X,1,1} ("X" = don't care) for signals {scan1, scan2, A, B, E, G}.

Algorithm 3 Test Vector Generation Algorithm
___
1: **Input:** Design D, Rare nodes R, signal traces S
2: **Output:** set of test vectors T
3: **procedure** TESTVECTORGENERATION(D, R, S)
4:      $T, F = \{\}$
5:      *add ATPG AND primitive from S*
6:      **for each** $r \in R$ **do**
7:          $F_i = $ addStuckAtFault(r_i)
8:          $T_i = $ runFullSequentialATPG(F_i)
9:          **if** $F_i == $ AU **then**
10:             $T_i = $ runJustification(r_i)
11:             $T = T \cup T_i$
12:      **return** T
___

C. Test Vector Generation

We cannot expect random test patterns to reliably activate rare or suspicious nodes in a circuit. Therefore, we use ATPG with N-detect testing to generate directed tests patterns for the remaining circuit. The activation levels of all relevant internal signals from the suspicious node's fan-in cone and scan replacements are extracted from the trace and combined together with an ATPG primitive AND gate referred to as stuck-at circuit. The addition of these primitives are for test generation purposes only and have no effect on the design functionality. A stuck-at 0 fault is added to the tool's fault list for each stuck-at circuit. The ATPG tool is then run using full-sequential ATPG to generate test vectors that trigger each fault. If the stuck-at faults from the modified design are

978-1-5386-3693-0/18 $31.00 © 2018 IEEE

undetectable by the ATPG tool, we then attempt to justify the rare node trigger condition to generate a pattern. With our framework, the ATPG tool experiences a much faster execution time because the complexity of non-scan sequential ATPG is removed by utilizing a model checker for that portion shown in the results. In the event that no test pattern can be generated for a rare node due to justification conflicts, we cannot say anything about the existence of a Trojan in the design.

In Algorithm 3, we use the design D, rare nodes R, and signal trace S from model checking as input and output a set of test vectors T. The test vectors and fault list, F, are initially empty sets. We build the design in the ATPG tool then create stuck-at circuits using ATPG primitives and the traces generated in the previous step. Faults for each rare node r stuck-at circuit are added to the ATPG's fault list F. Full-sequential ATPG is run with the current fault list to generate patterns. If the ATPG tool returns AU (ATPG untestable) no test pattern is generated. We then attempt to run justification to generate a test pattern.

Example 3: Let us consider the design after the constraints are generated from model checking. We provide the design from Figure 1, the list of rare nodes, and signal traces to the ATPG tool. The relevant traces are transformed into a stuck-at circuit with ATPG AND primitives as shown in Figure 3. A stuck-at 0 fault is added for each stuck-at circuit and the ATPG tool is run. An example test vector from ATPG for a stuck-at 0 fault at stuck-at circuit R1 would be:
scan-in: $\{1, 1, 1\}$ *primary inputs* (A,B,C,D) : $\{1,1,1,1\}$

V. EXPERIMENTS

A. Experimental Setup

To evaluate our approach, we implemented the framework described above and applied it on two AES-128 and RS232 benchmarks from Trust-Hub benchmark suite. More information on the Trojan circuits and their implementation can be found on Trust-Hub [20], [21]. Additionally, two modified AES-128 benchmarks (cb_aes) are used to showcase the limitations of model checking and ATPG. Both custom AES benchmarks are a subset of the AES module and only include 15 and 20 key rounds, respectively. The Trojan (a comparator and FF) checks the final output of the last round against a predetermined output. If they match, the secret key is leaked through the primary output.

A machine with Intel Core i5-3470 CPU @ 3.20GHz and 8 GB of RAM is used for testing. For each benchmark, the design is simulated for millions of cycles to identify the rare branches. The benchmarks are then synthesized using design compiler with DFT scan insertion with no area or power optimization. In mapping rare branches to rare nodes for our experiments, identifying boundary signal naming from primary inouts, and registers was sufficient.

A subset of the FFs are randomly selected for scan in an iterative process to maintain a high test coverage for partial-scan insertion. To make it harder to detect and showcase the efficiency of our approach, Trojan activation and payload

FFs are excluded from the scan-chain effectively simulating a scenario in which an adversary would insert a Trojan in hard to detect areas after scan-chain insertion. With the rare nodes identified and scan chain inserted, the scan FFs are then replaced with primary inputs and given to the SMV tool [22] for model checking. Properties for each rare node are written in LTL and given to the model checker along with the design, which is converted from Verilog to .smv format. The resulting counter-example traces are then given to Synopsys TetraMAX [23] for ATPG N-detect with N = 10. Stuck-at circuits are constructed from the traces using ATPG AND primitives and corresponding stuck-at 0 faults are added to the tool's fault list. For a Trojan to be detected, its effect must be propagated to an observable output. Therefore, the final phase in our approach is to translate the test vectors into testbenches. By targeting rare nodes, the test vectors generated from N-detect ATPG have an increased likelihood of activating a Trojan. The suspicious design is simulated using ModelSim with the resulting testbench and its output is compared (XOR) with either a golden model or functional specification from the IP vendor to detect the presence of a hard-to-detect functional Trojan. Some constraints were imposed on the design due to tool limitations. For example, in SMV results are not accurate in designs with multiple clock domains and in TetraMAX sequential elements with multiplexed clocking are not allowed. In order to maximize the benefits of using ATPG and model checking approaches, our framework also includes parallel execution of them. The test pattern is taken from whichever execution generates results first.

B. Results

The experimental results from designs with partial-scan insertion are described in Table 1. The benchmarks are listed in the first column. Columns 2 and 3 describe the percent of FFs that are included in the scan-chain and the corresponding test coverage. Column 4 shows the number of rare branches identified from our initial analysis simulation. Columns 6, 8, 10, and 12 report the CPU time in generating test vectors for each approach. In columns 5, 7, 9, and 11 we show whether the test vectors detected the Trojan. Finally the last three columns show the improvement of our framework over ATPG, model checking, and MERO.

For AES and RS232 circuits, sequential ATPG alone outperforms model checking and the combined approach. Note that the combined approach still generates a test vector in comparable time, yet we take the test pattern resulting from ATPG as it finishes first. The execution time difference between AES and RS232 can be attributed to the location of the Trojans. Both AES benchmarks have Trojans that compare the state (primary input) to a predetermined value. On the other hand, in RS232, the Trojans are activated as a result of internal sequential signal combination.

The custom benchmarks are used to illustrate the weaknesses in both sequential ATPG and model checking. cb_aes_15 has a total of 5889 FFs, 883 non-scan FFs, and a max non-scan sequential depth of 3. cb_aes_20 has a total

TABLE I: Comparison of our approach against ATPG, Model Checking, and MERO for Trojan Detection with Partial-Scan. MO indicates memory overflow with 8GB of RAM. TO indicates timeout after 57600s

Benchmarks	Scan FFs (Scan/Total)	Test Cov.	#Rare Bran.	ATPG		Model Chk. (MC)		MERO		Our Approach		Improvement over		
				Detect	Time	Detect	Time	Detect	Time	Detect	Time	ATPG	MC	MERO
AES-T1000	6448/6933	99%	2	✓	0.02s	✓	85.6s	X	TO	✓	8.80s	1x	4280x	–
AES-T2000	6468/7108	99%	5	✓	0.90s	✓	216.5s	X	TO	✓	22.0s	1x	241x	–
RS232-T400	30/59	97%	2	✓	0.24s	✓	3600s	✓	2810s	✓	0.52s	1x	15000x	11708x
RS232-T800	26/58	97%	1	✓	0.06s	✓	7.23s	✓	3157s	✓	0.12s	1x	120x	52617x
cb_aes_15	5006/5889	99%	1	✓	28800s	X	MO	X	15720s	✓	7.85s	3669x	–	2003x
cb_aes_20	7262/7809	99%	1	✓	28800s	X	MO	X	16740s	✓	38.3s	752x	–	437x

"✓" indicates Trojan detected, "X" indicates Trojan not detected, "–" indicates not applicable

of 7809 FFs, 547 non-scan FFs, and a max non-scan sequential depth of 3. We can see the ATPG tool took a significant amount of time in generating test patterns to trigger the rare branch. Similarly model checking fails to generate a pattern due to state explosion and the tool experiencing a memory overflow (MO). Our proposed hybrid technique is able to generate test vectors even when the circuit structure has sufficient non-scan FF depth and structure.

We also compared our approach to MERO [2]. From Table 1, it is evident that our framework outperforms this approach in partial-scan designs. Because MERO uses ATPG to justify trigger conditions, it experiences significant execution overhead. This fact is illustrated in the AES benchmarks which experience a timeout (TO). In the case of the custom AES benchmarks, MERO did not detect the Trojans because the justification portion could not activate the trigger condition given the amount of non-scan sequential elements in the design.

Our experimental results highlight two important aspects of our test vector generation framework. First, we achieve acceptable CPU times for all the test benches covered. Second, our approach is able to generate test vectors that ATPG and model checking were unable to finish in a reasonable time. Our results show that we can achieve up to four orders of magnitude faster execution times than state-of-the-art methods. This speed up is achieved by leveraging the strengths of each tool. Specifically, reducing the state space of model checking and removing the non-scan sequential complexity encountered using ATPG.

VI. CONCLUSION

Trust in SoC design is an ever increasing concern as more companies are including third party IPs gathered from untrusted vendors. Current Trojan detection methods that utilize ATPG and model checking tools cannot effectively handle partial-scan designs. This limitation can cause a significant execution time penalty due to the complexity of non-scan sequential ATPG or complete failure in model checking from state explosion. We proposed a framework that combines ATPG and model checking for hardware Trojan detection in partial-scan designs. Experimental results demonstrated the merits and weaknesses in both approaches and the effectiveness of combining them in case of partial-scan designs for generating test vectors targeted at hardware Trojan detection. We plan to extend our approach using Groebner basis based

formal verification to help with larger circuits that model checking cannot handle [16]. Future work will also include investigating more partial-scan benchmarks with differing sequential structure and depth to further explore the effectiveness of the proposed framework.

REFERENCES

[1] M. Tehranipoor and C. Wang, *Introduction to Hardware Security and Trust*, Springer, 2011.
[2] R. Chakraborty et al., "MERO: A statistical approach for hardware trojan detection", CHES, 2009.
[3] X. Zhang and M. Tehranipoor, "Case Study: Detecting Hardware Trojans in Third-Party Digital IP Cores", HOST, 2011.
[4] S. Bhunia et al., "Hardware Trojan Attacks: Threat Analysis and Countermeasures", IEEE Special Issue on Trustworthy Hardware, 2014.
[5] M. Tehranipoor and F. Koushanfar, "A Survey of Hardware Trojan Taxonomy and Detection", IEEE Design & Test, 2010.
[6] J. Rajendran et al., "Formal Security Verification of Third Party Intellectual Property Cores for Information Leakage" , VLSI Design, 2016.
[7] V. Chickermane and J. H. Patel, "An Optimization Based Approach to the Partial Scan Design Problem", Test Conference, 1990.
[8] T. E. Marchok et al., "Complexity of Sequential ATPG" , European Design and Test Conference, 1995.
[9] M. Bushnell and V. Agrawal, *Essentials of Electronic Testing for Digital, Memory and Mixed-signal VLSI Circuits*, Springer, 2004.
[10] E. M. Clarke et al., *Model Checking*, MIT press, 1999.
[11] H. M. Koo and P. Mishra, "Test Generation using SAT-based Bounded Model Checking for Validation of Pipelined Processor", GLSVLSI, 2006.
[12] I. Pomeranz and S. M. Reddy, "On Undetectable Faults in Partial Scan Circuits" , ICCAD, 2002.
[13] F. Wolff et al., "Towards Trojan-free Trusted ICs: Problem Analysis and Detection Scheme", DATE, 2008.
[14] M. Banga and M. S. Hsiao, "Trusted RTL: Trojan Detection Methodology in Pre-silicon Designs", HOST, 2010.
[15] P. Mishra et al., *Hardware IP Security and Trust*, Springer, 2016.
[16] F. Farahmandi et al., "Trojan Localization using Symbolic Algebra", ASP-DAC, 2017.
[17] Y. C. Hsu et al., "Visibility Enhancement for Silicon Debug", DAC, 2006.
[18] A. Waksman et al., "FANCI: Identification of Stealthy Malicious Logic using Boolean Functional Analysis,", CCS, 2013.
[19] Y. Huang et al., "MERS: Statistical Test Generation for Side-channel Analysis based Trojan Detection", CCS, 2016.
[20] B. Shakya et al., "Benchmarking of Hardware Trojans and Maliciously Affected Circuits", HaSS, 2017.
[21] H. Salmani et al., "On design vulnerability analysis and trust benchmarks development", ICCD, 2013.
[22] K. McMillan, "Symbolic Model Checking", Kluwer, 1993.
[23] Synopsys, "TetraMAX User Guide," *Version H-2013.03-SP4*, 2013.
[24] S. Saha et al., "Improved Test Pattern Generation for Hardware Trojan Detection using Genetic Algorithm and Boolean Satisfiability", CHES, 2015.
[25] F. Farahmandi et al., "Cost-Effective Analysis of Post-Silicon Functional Coverage Events", DATE, 2017.

978-1-5386-3693-0/18 $31.00 © 2018 IEEE

Towards Single Pin Scan for Extremely Low Pin Count Test

Mudasir S. Kawoosa, Rajesh K. Mittal, Maheedhar Jalasuthram and Rubin A. Parekhji

Texas Instruments (India) Pvt. Ltd., Bangalore, India.

[mudasir, rajmittal, maheedhar, parekhji]@ti.com

Abstract—Design-for-Testability (DFT) techniques for test cost reduction of digital circuits rely on efficient deployment of scan test through smart DUT (device under test) partitions, matching of scan test I/O needs to the DUT needs for increased scan data throughput, and optimal test pattern generation methods. The test cost is further reduced by testing multiple devices in parallel using the same set of ATE resources. As a result, the number of scan pins available in a DUT for test application is reduced. The need for managing different DUT partitions and ensuring that the scan I/O interface matches with the DUT requirements further aggravates the need for scan test pins. In this paper, a generic methodology for reduced pin count scan test is presented. Specific solutions are described for internal generation of all scan mode control signals, e.g. scan enable, clock control, X-mask control, switching between slow speed shift clock and high speed capture clock, etc., across one or more DUT partitions. It is shown how this method can ultimately lead to a single pin scan test solution. These methods have been implemented in cost sensitive SOC designs and, as a result, the scan test time has been reduced by 2x – 3x over what has been achieved using other aggressive test cost reduction methods for scan compression. These methods are also independent of the specific scan compression solution used.

Keywords—*Low pin count scan test, scan compression, multi-site test.*

I. INTRODUCTION

As technology node shrinks and design complexity scales up, the requirement for test quality scales up as well. A continued improvement in testing of highly complex SOCs is needed alongside so as to contain the test cost of the chip within affordable limits. With ATPG (Automatic Test Pattern Generation) tool generated test patterns, testing of highly complex SOCs (system on chips) using scan based logic test methods has been highly successful. To reduce the test data volume, scan compression techniques are an efficient and optimal way [1],[2]. Several compression techniques have already been proposed in the literature and are being widely used in the industry [3],[4],[5],[6]. In scan compression, the assembly of a decompressor and a compactor built-in around the DUT (device under test) is commonly termed as CoDec. These compaction schemes rely either on compressing the test stimuli, the test response or both and therefore can be lossy, resulting in either the inability to encode all the DUT care bits for a given test pattern, or the inability to uniquely capture the test response under a fault condition (e.g. aliasing effect). Hence the efficiency of such compression methods directly impacts the pattern volume incurred and the fault coverage attained. Besides, the circuit entropy sets bounds on the attainable compression [7].

Implementing scan compression in a large SOC designs has always been challenging. Two methods in use are: (i) apportioning dedicated scan test I/Os to individual DUT partitions and their associated scan CoDecs, and (ii) using hierarchical scan compression, where the same set of scan I/Os are shared across multiple partitions, each with their dedicated scan CoDecs [8],[9],[10]. However, with reduced package sizes and fewer pins being made available for test, DFT solutions for reduced pin count scan test are critically important. High end testers cost significantly higher and hence are often used with the incumbent requirement of supporting much higher multi-site factor, that is the number of devices which are tested in parallel during one test touch-down either at wafer probe or for packaged part testing or both [11],[12],[13]. Moreover these requirements for higher multi-site factor are also driven by the need to enable high test throughput in high volume production. For example, with a multi-site factor of 128 and ATE digital resource count of 1024, each DUT must be tested using only 8 pins for all of data, control and clocks. Requirements with lower test cost configurations are even more stringent where fewer pins, (e.g. 2, 4, 6) are available for test.

These requirements have motivated the development of low pin count test (LPCT) solutions which are presented in this paper. The main contributions of this paper are: (i) It explains the requirement of different scan control signals across various scan compression architectures. (ii) It then explains how on-chip generation of these control signals significantly reduces overall test cost of the SOC. (iii) It also presents multiple techniques that increase the scan channel bandwidth, or conversely, reduce the ATE contacted scan pins. These methods have been implemented in extremely cost sensitive SOC designs, and have helped significantly reduce the number of test pins as well as have enabled migration to higher multi-site configurations. They also provide a path to a single pin scan test solution.

The paper is organized into five sections. Section II provides an overview of existing scan optimizations techniques for LPCT enabled through DFT inside DUT. It also lists the various requirements for scan I/Os with different scan compression solutions. Section III describes novel solutions for reducing the number of scan I/Os for data, control and clock functions as well as increasing the scan bandwidth within the available pin budget. Section IV provides insight into diagnostics. Section V concludes the paper.

II. OVERVIEW OF EXISTING REDUCED PIN SCAN COMPRESSION TECHNIQUES

The effectiveness of scan compression is evaluated on the basis of scan compression ratio. This compression ratio, in a good implementation, tends to track the ratio of the number of decompressor outputs to the number of decompressor inputs.

However there are some bounds for this ratio: (i) Larger the number of decompressor outputs, higher will be the (targeted) compression ratio, but also higher can be the correlation in the outputs of the decompressor (in case of combinational decompressors). This can result in a significant pattern volume increase offsetting the gains of compression. (ii) In an extreme case, with very few inputs, independent of the targeted compression ratio, the correlation at the decompressor outputs may prevent in achieving the desired coverage. Different tradeoffs have been presented in [14] to show how compression results vary with scan I/O bandwidth (between ATE and DUT).

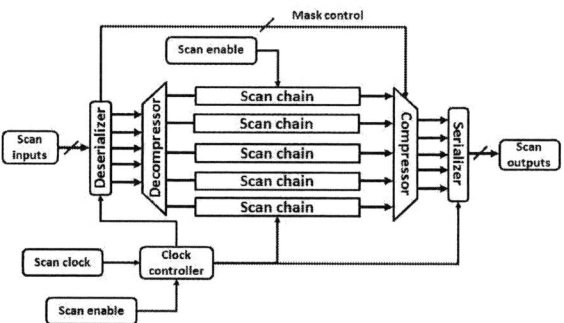

Figure 1 A generic serializer based scan compression solution.

Different LPCT based scan compression solutions have been proposed in the literature where the main goal is to reduce test cost without compromising the quality of test with different implementations that focus on a variable number of pins. These are reviewed in the following sub-sections.

A. Serializer-Deserializer based scan compression

The serializer structure in scan compression enables LPCT by using the principle of time division multiplexing of compacted scan test vectors [15],[16],[17].

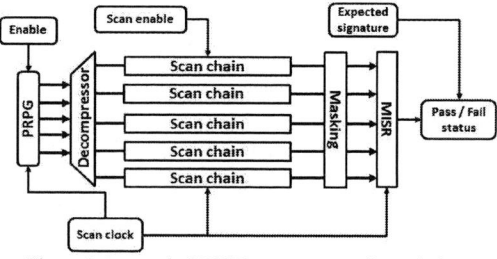

Figure 2 A generic LBIST scan compression solution.

The implementation involves packing and unpacking of scan data from the device I/Os into the CoDec I/Os using a set of deserializer and serializer registers on a per cycle basis. As a result, the test application time for a given volume of scan test data is increased. Nonetheless, data on different designs indicates that the pattern volume increase and coverage drop with fewer input pins is completely offset by using the serializer structure [16]. Figure 1 shows a generic serializer based scan compression solution. In its simplest form, the solution will at least require a one top-level scan input and one top-level scan output, in addition to regular clock and control signals. The test application time inflation is also compensated

due to the feasibility of higher multi-site test configuration now enabled with fewer pins.

B. LBIST (Logic Built-in Self Test) based scan compression

The test time increase incurred with serializer-deserializer CoDec architecture can alternately be offset using a logic BIST (LBIST) architecture as shown in Figure 2. LBIST is commonly implemented using a sequential decompressor (PRPG) and a sequential compactor (MISR). This combines the benefits of scan compression with a very low pin count test interface, positively impacting both the test data volume compression as well as the test application time [18]. In traditional LBIST, patterns are continually generated with a given PRPG initial state. However, with standard implementations of scan compression targeting pattern generation for high coverage for production test, it is necessary to re-seed the PRPG, (i.e. provide it with different initial states), and then generate a new set of patterns for each such initial state. There exist well-known schemes for re-seeding based LBIST [4] compression architectures.

C. OPMISR (On-Product Multiple Input Signature Register) based scan compression

Conventional single PRPG initial state based LBIST patterns suffer from low fault coverage. Hence re-seeding of the PRPG has been considered. Another method is to replace the PRPG with a combinational decompressor while retaining the MISR to capture the STUMPS outputs. The scan outputs thus saved are now re-used as scan input pins, thereby providing additional bandwidth into the decompressor inputs. Such a method, first proposed in [3] and termed OPMISR (On-Product Multiple Input Signature Register), offers the advantages of coverage improvement as well as test throughput improvement (lesser test application for given test data volume) over classical LBIST.

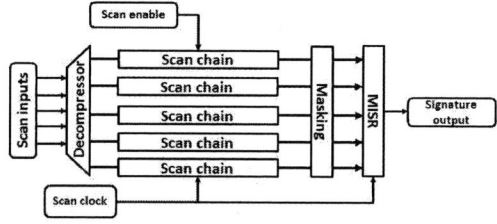

Figure 3 A generic OPMISR scan compression solution.

Figure 3 shows a generic OPMISR based scan compression implementation. MISR based compactors have inherent limitations with debugging ability and handling Xs in the values captured into the scan flip-flops. EDA tools address these limitations either by bypassing the MISR and observe the STUMPS directly on the DUT output pins or through a combinational compactor, or control is provided to mask Xs from entering the MISR on a per-STUMPS per-cycle basis. The granularity of such control determines the additional data transfer and control operations required between the ATE and DUT.

The above methods indicate how different scan compression architectures influence the choice of and need for scan input and output pins in the DUT. Test pins required for scan compression implementation are broadly classified into three categories viz. clock, control and data pins. These pin

978-1-5386-3693-0/18 $31.00 © 2018 IEEE

requirements and pin definitions vary for different scan compression architectures available across multiple EDA tools. In general, the number of test control pins required for a sequential CoDec architecture / implementation is higher than for the combinational one. Table I lists the pin description for OPMISR based scan compression (as available in one EDA tool). These signal requirements are an impediment when implementing an LPCT solution. The simplest implementation with one scan input and one scan output still requires a minimum of 6 pins for clock and control purposes.

TABLE I. SIGNALS FOR OPMISR BASED SCAN COMPRESSION.

Pin Name	Pin Type	Pin Description
Scan Clock	Clock	Shift and capture clock
At-speed Clock	Clock	Free runinng clock
Scan Inputs	Data	Scan input pins
Scan Outputs	Data	Scan output pins
CME	Data	X-masking control pin
Scan Enable	Control	Scan shift enable
MRE	Control	MISR registers reset control
MISR_OBS	Control	MISR signature read control
CMLE	Control	Mask register load control

III. NOVEL LPCT IMPLEMENTATIONS FOR SCAN COMPRESSION

The main focus of this paper is to present methods to reduce the number of tester contacted pins during scan test, thereby completely eliminating the need for dedicated ATE resources to drive them. This is accomplished through novel on-chip generation of such signals for which a dedicated pin is no longer assigned. The targeted pins are mostly pseudo-static i.e. they don't toggle on a per cycle basis unlike test data pins. These includes Scan enable (*SE*), Mask register control (*CMLE*), Scan reset (*SR*), MISR reset control (*MRE*), Signature observe control (*MISR OBS*) and LFSR or PRPG reset. We now explain how all of the above control signals can be generated on-chip without using any resources from the ATE to drive them. The ATPG tool continues to use these pins as are defined in the scan compression architecture and drives them suitably during pattern generation. However, since these pins are no longer available in the circuit in silicon, on-chip controls are provided and the pattern set are suitably modified to include their functionality. The proposed on-chip implementations have been devised keeping in mind a few considerations.

a) While the illustrations may consider a representative scan compression architecture from a particular tool vendor, these methods are independent of the scan compression architecture, and can be generically applied for generating on-chip controls for scan test.

b) Controls for clocks required for scan test are generated using on-chip logic. The clock pin requirements are simplified to need just one pin, which can be used for both shift as well as capture clock, including for at-speed capture pulses from the ATE free funning clock for application of delay fault ATPG patterns.

c) These optimizations allow the application of both launch-off-shift (LOS) as well as launch-off-capture (LOC) at-speed ATPG patterns.

d) The combined methods provide a path to a single pin scan solution. This pin is required for scan data input alone.

We propose five different solutions to handle all static scan control and clock signals. The first two solutions rely upon on-chip generation of scan control signals thereby eliminating them. The next two solutions increase the scan channel bandwidth through smarter pin allocation during scan test. Finally the last solution targets re-use of the existing clock pin for slow speed shift as well as at-speed capture.

A. Pattern match detection based scan control signal generation

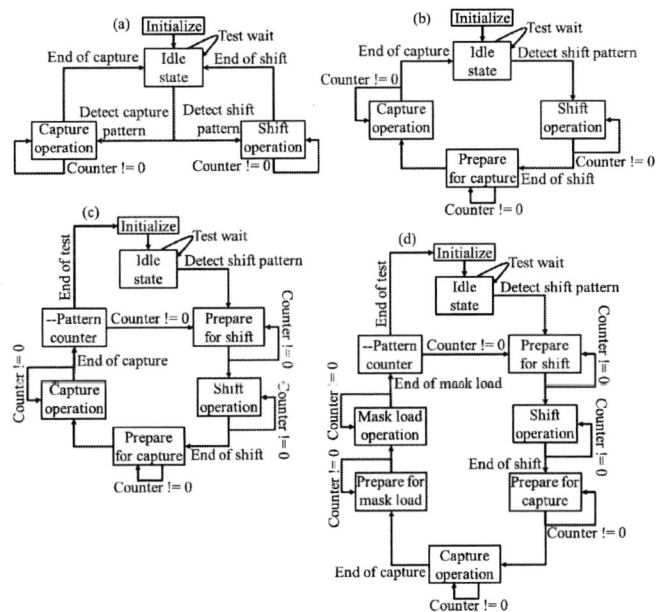

Figure 4 Pattern match detection based scan control signal generation.

Pattern match detection on scan input pins to generate the scan enable signal internally was introduced in [19]. However, the implementation is not scalable when several such control signals have to be generated. That implementation uses a single scan input pin to identify the beginning of the scan shift operation. Additionally, there are two sets of programmable counters which are initially loaded with the pre-defined shift and capture cycle counts. The shift and capture phases are identified using a matching pattern on the available scan data pin. For every shift or capture cycle, these counters decrement and an all zero value indicates the corresponding end of shift or end of capture phase. This scheme in its simplest form is shown in Figure 4(a). Upon device initialization, the state machine waits in the idle state till shift operation gets detected. Once the state machine enters into the shift operation, the shift counter starts decrementing till the end of shift. The state machine then transitions to the capture preparation state and another counter starts decrementing. At the end of capture preparation phase, the state machine transitions to actual capture operation and the third counter triggers. During this phase, single or multiple scan capture pulses can be provided. Finally the state machine returns back to the idle state. With an additional counter, it is possible to handle the entire operation with single pattern match detection only. Such an implementation is shown in Figure 4(b). With the addition of a fourth counter which is pre-loaded with the total test pattern count, the intermediate transition to the idle state after each test vector is eliminated. Such an implementation is shown in Figure 4(c).

A similar mechanism is used to support internal generation of other pseudo-static control signals for which an additional counter controls the operation as shown in Figure 4(d). In addition to eliminating the need of dedicated control pins, this solution also reduces the test application time by 10-15% on account of the removal of the extra clock cycles consumed by the ATE when switching from streaming mode to event mode, and vice versa.

B. 1149.1 JTAG based scan control signal generation

In the case of advanced scan compression architectures like OPMISR, the number of scan control signals are larger and therefore are difficult to implement on-chip using the counter and pattern match detection mechanism. Also in case of sequential decompressor or streaming compression based CoDec architectures wherein the length of scan shift cycles during the test can be varying, the pattern match detection with fixed counter values is not suitable.

TABLE II. TAP CONTROLLER STATE DECODING FOR INTERNAL SCAN CONTROL SIGNAL GENERATION.

Scan control pin	TAP controller state/s
Scan Enable	Shift_DR state and TMS=0
Capture Mode	Run_Test_Idle state and TMS=0
CMLE	Pause_DR state and TMS=0
Scan Reset	Select_DR > Capture_DR > Exit-1_DR > Update_DR
MRE	Shift_DR > Exit-1_DR after first scan load only or subsequently after every assertion and de-assertion of MISR_OBS signal
MISR_OBS	Select_DR > Capture_DR > Exit-1_DR > Pause_DR and TMS=0

TABLE III. IMPACT OF INTERNAL SCAN CONTROL SIGNAL GENERATION ON SCAN DATA BANDWIDTH. (NOTE THAT FIVE SCAN DATA PINS ARE RECOVERED).

IOs	External Control	Internal Control
GPIO-1	Scan Enable	Scan Input 1
GPIO-2	CMLE	Scan Input 2
GPIO-3	Scan Reset	Scan Input 3
GPIO-4	MRE	Scan Input 4
GPIO-5	MISR_OBS	Scan Input 5
GPIO-6	Smart Scan Input 1	Scan Input 6
GPIO-7	TCK / Scan clock	TCK / Scan clock
GPIO-8	TMS	TMS
GPIO-9	At-speed Clock	At-speed clock
GPIO-10	POR (Power on reset)	POR (Power on reset)

A novel technique for internal scan control signal generation is described in this section. The proposed solution makes use of on-chip 1149.1 TAP (Test Access Port) controller state machine to generate all the scan control signals internally in a highly optimal way. The TAP state machine has 16 states and few of them are suitably combined to generate these signals. To enable the internal scan control generation logic, a dedicated IR (instruction register) is decoded. After that the TAP FSM (finite state machine) stays in the *Run_Test_Idle* state. The *Run_Test_Idle* state is considered as a scan capture phase. Upon transitioning to *Shift_DR* state, scan enable (*SE*) gets asserted and scan shift operation begins. It is a common practice to share JTAG clock (*TCK*) directly as a scan clock.

However in the proposed solution, *TCK* is internally gated using a combination of TAP states and JTAG *TMS* pin to avoid any leakage of clock pulses during transition of JTAG states. Scan shift operation terminates upon exiting the *Shift_DR* state. In one form of scan compression, to enable loading of the mask chain (whose values are used to control X propagation into the compactor), the corresponding *CMLE* control signal is decoded by configuring the TAP controller in the *Pause_DR* state. The *Run_Test_Idle*, *Shift_DR* and *Pause_DR* states are carefully chosen for scan capture, scan shift and mask chain load operations, respectively, so as to allow any number of scan capture, scan shift and mask load cycles, as long as *TMS* is held at logic 0.The chosen three states can be interchangeable and the implementation becomes fully independent of scan shift length or mask chain length. In addition to this, additional control signals like *MRE, MISR OBS* and *SR* are also generated using combinations of other TAP states when executed in a specified order. Table II lists the sequence for internal generation of all scan control signals. With the specifically chosen JTAG state or a combination of states, the solution is optimal as not only does it generate all the control signals required across various commercially available scan compression schemes, but it also consumes the least number of clock cycles to generate them while transiting from one state to another. Moreover compared to the previous pattern match detection based solution, this implementation does not need any internal counters or pattern detection on scan inputs to decode any state. The transition sequences are provided as input to the ATPG tool during pattern generation and therefore no post processing of the ATPG tool generated patterns is required.

With the possibility of eliminating the need of dedicated scan control pins for scan test operation, it is indeed possible to map the savings to increased scan channel bandwidth or conversely enable the LPCT. In one of the designs that has OPMISR based scan compression and a deserializer for achieving higher compression ratio, the above solution increases the scan data bandwidth by 6X making it possible to eliminate the deserializer completely, as is shown in Table III.

C. Increasing scan channel bandwidth by sharing all JTAG pins

The on-chip IEEE 1149.1 JTAG interface and TAP offer an option to share some of its pins during scan test. These includes *TCK, TDI* and *TDO* for clock, scan input and scan output respectively. An additional pin viz. *TMS* is not shared for scan test as it is used to sequence the TAP controller. (The fifth pin *TRSTN* is optional). In many cases, these pins are used to configure the device into different test modes, thereby making it difficult to share them for scan test. A simple solution is to put the TAP controller in *Run_Test_Idle* state upon entering into the scan test mode and forcing the internal *TMS* and *TRSTN* signals of TAP controller to an appropriate value, while releasing the DUT pins. However, the control of the TAP FSM is now lost. Re-powering up is required to reset it and get control of the pins again. Such a power-up has many implications in an SOC, including ATE and DUT internal test time due to initialization of several on-chip functions, including embedded power management. (However, our focus is not as

978-1-5386-3693-0/18 $31.00 © 2018 IEEE

much on reduction of the test mode control pins (namely JTAG), as is on elimination of scan control pins). A novel mechanism has been developed through which these two pins, *TMS* and *TRSTN* pins are also shared dynamically for scan test. At any point during or after scan test, the functional control on these two pins can be regained back without need of any additional power-up. In our proposed solution, the TAP controller is kept in *Run_Test_Idle* state and when the scan enable pin is asserted, internal TMS and TRSTN signals at the TAP controller levels are asserted to suitable logic levels, thereby allowing these two pins to be shared for scan test. During the capture phase when scan enable is de-asserted, functional control on these two pins is regained back and they are kept at the desired functional values to improve coverage. With such an implementation, it is possible to share these pins for scan test thereby reducing the number of tester contacted pins or being able to increase the scan channel bandwidth. Additionally, it is possible to combine the pattern detection based generation of SE signal internally with the proposed method of sharing of all JTAG pins dynamically.

D. Non-overlapped scan implementation for specific design

Normally in a combinational only scan compression architecture, scan load and unload operations are overlapping i.e. both happen simultaneously. The frequency at which scan test operates is either determined by speed of data transfer through ATE or the speed at which the DUT I/Os can operate or else the DUT internal shift speed. However low power SOCs use ultra-low leakage cells and the DUT I/Os associated with such devices are therefore much slower. Implementing scan test with such I/Os limits the frequency at which scan data can be shifted. In one of such low power designs, the output buffer could shift out at a much lower frequency (20MHz) than the input buffer (100MHz). Since scan input and output data shift happens simultaneously and synchronously, the frequency at which scan test operates in such devices is determined by the output buffer speed.

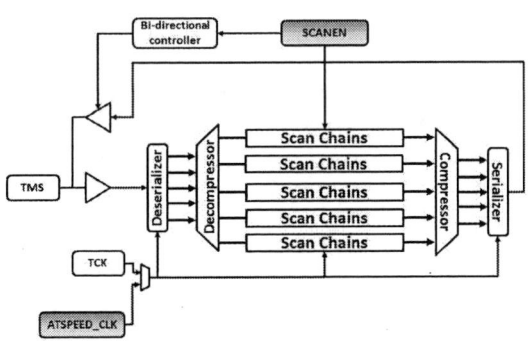

Figure 5 Non-overlapped scan implementation.

A novel mechanism has been implemented wherein the scan operation is made non-overlapping but still achieving the lower test time. This is done by shifting in the scan data at a much higher frequency (as compared to the minimum possible frequency when both shift in and shift out happen simultaneously). Consider the ATE shift speed and input buffer shift speed of 100 MHz and the output buffer shift speed of 20

MHz, in case of regular overlapped scan operation, the maximum speed for scan data transfer gets limited to 20 MHz. With the proposed solution, during scan load time, all scan data pins will be used to load scan data at 100 MHz and same will be used to unload the scan data at 20 MHz, thereby significantly increasing the scan data bandwidth on both sides.

TABLE IV. EXPLOITING NON-OVERLAPPED SCAN OPERATION FOR TEST TIME REDUCTION.

	Existing Solution	Proposed Solution	k=1	k=2	k=5
Ext. load frequency	F	kF	F	2F	5F
Ext. unload frequency	F	F	F	F	F
Int. load frequency	F/8	kF/4	F/4	F/2	5F/4
Int. unload frequency	F/8	F/4	F/4	F/4	F/4
Load time	2T	T/k	T	T/2	T/5
Unload time	2T	T	T	T	T
Total time	2T	T/k+T	2T	1.5T	6T/5
Test time savings(%)	NA	(k-1)/2k	0	25	40

Such a design implementation combined with previous solution is shown in Figure 5. In one of the designs where the serializer ratio was eight with one scan input and one scan output, with this implementation it was reduced to four and with the increase in the input frequency, significant test time reduction was achieved. Assuming maximum input buffer frequency is *k* times the maximum output buffer frequency, the overall test time savings compared to existing scan shift frequency is listed in Table IV. For *k*=5, around 40% test time reduction is achieved.

E. Single pin scan solution

The functional clock (as distinguished from the test clock applied to the *TCK* pin in the JTAG interface) must be used for at-speed ATPG patterns, since the capture pulse (after the launch pulse) must be issued at functional timing. This is accomplished by using the functional clock (also termed as the system clock) as a free running clock to the PLL or as an at-speed clock directly to the on-chip clock shaper network for the issue of two (or more) at-speed pulses across one or more clock domains.

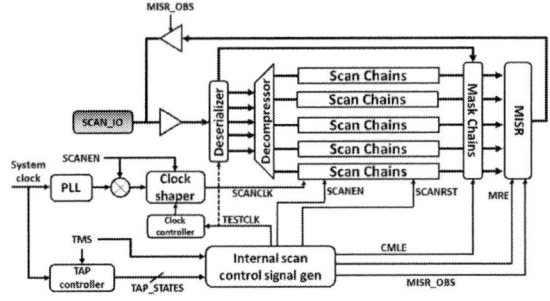

Figure 6 Single pin scan solution.

We now show how *TCK* clock pin of the JTAG interface can be used for the application of three types of clocks, namely as the TAP FSM clock, as the clock for the logic module generating the internal scan control signals and as the high frequency clock which is input to the on-chip PLL. The implementation begins with the device powering up in a regular functional or test mode, then configuring the device into scan test mode, loading the required PLL divider and multiplier register values and finally enabling the internal scan control logic by loading the specified IR. All this is accomplished with a single clock and the same is shared as a reference clock to the PLL. Once the PLL is enabled, its input clock cannot be halted at any time during scan test. Hence, unlike in earlier implementations where the ATE clock was controlled, any form of clock controls needed to meet timing requirements must be managed internally to the design.

As shown in Figure 6, with above approach, it is now possible to execute the entire scan operation with a single scan pin only, considering the fact that TMS is to be tester contacted irrespective of its usage for scan test. Clock IO feeds the on-chip PLL, TAP controller and internal scan control signal generation module. A deserializer is used on the input side and an OPMISR on the output side. The final signature is read back on same scan data pin only. Clock to the deserializer is controlled from internal scan control generation module while the clock to the design is controlled through clock controller for stuck-at as well as delay fault test patterns. In one design, the at-speed clock requirement was 200 MHz. This was met with a 100 MHz input clock and with the PLL divider and multiplier programmed to values 4 and 8 respectively. (Since a single clock pin is used to serve multiple purposes, it is important to close the timing of the associated logic at the rated frequency).

IV. DIAGNOSTICS

ATPG pattern driven diagnostics is an important consideration for scan design and scan pattern generation. Since all the scan control signals are generated internally, it is necessary to expose them to the DUT pins during debug mode. In such case, each generated signal is multiplexed to one or more scan I/Os to observe them during different scan test phases. Hence additional pins can be used, since not all sites must be active during debug. The proposed solutions have been implemented on different devices and production test has been successfully performed. They have demonstrated how reducing the number of tester contacted pins has enabled higher multi-site testing and reduced test cost. Across multiple devices, a 2x – 3x test cost reduction has been realized compared to standard implementations with all scan clock, control and data pins being contacted from the ATE.

V. CONCLUSION

This paper presents many novel scan test solutions suitable for low pin count test. It is shown that by embedding the generation of all scan control signals internal to the circuit, as well as having a common functional and scan shift and capture clock pin, the number of test pins required is significantly reduced, thereby creating a path to a single pin scan test solution. These techniques are independent of different scan compression architectures that are supported in EDA tools, thereby making it possible to integrate them as part of the CoDec insertion flow as well as to leverage them during test pattern generation.

REFERENCES

[1] N.A.Touba, "Survey of Test Vector Compression Techniques", IEEE Design and Test of Computers, pp. 294-303, 2006.

[2] R.Kapur, S.Mitra and T.Williams, "Historical Perspective on Scan Compression", IEEE Design & Test of Computers, pp. 114-120, 2008.

[3] C.Barnhart, V.Brunkhorst, F.Distler, O.Farnsworth, B.Keller and B.Koenemann, "OPMISR: The Foundation for Compressed ATPG Vectors", Intl. Test Conf., 2001.

[4] J.Rajski, J.Tyszer, M.Kassab, N.Mukherjee, R.Thompson, K.H.Tsai, A.Hertwig, N.Tamarapalli, G.Mrugalski, G.Eide, J.Qian, "Embedded deterministic test for low cost manufacturing test", Intl. Test Conf., 2002.

[5] P.Wohl, J.Waicukauski, S.Patel and M.Amin, "Efficient compression and application of deterministic patterns in a logic BIST architecture", Design Automation Conf., 2003.

[6] P.Wohl, J.A.Waicukauski, F.Neuveux and E.Gizdarski, "Fully X-tolerant Very High Scan Compression", Design Automation Conf., 2010.

[7] K.J.Balakrishnan and N.A.Touba, "Relationship between entropy and test data compression", IEEE Trans. Comput.-Aided Design Integr. Circuits Syst., pp. 386-395, Feb. 2007.

[8] A.Jain and S.Subramanian, "Multi-CoDec Configurations for Low Power and High Quality Scan Test", VLSI Int. Des. Conf., 2011.

[9] P.Wohl, J.A.Waicukauski, J.E.Colburn and M.Sonawane, "Achieving extreme scan compression for SoC Designs", Intl. Test Conf., 2014.

[10] B.Keller, K.Chakravadhanula, B.Foutz, V.Chickermane, A.Garg, R.Schoonover, J.Sage, D.Pearl and T.Snethen, "Efficient testing of hierarchical core-based SOCs", Intl. Test Conf., 2014.

[11] J.Bedsole, R.Raina, A.Crouch and M.S.Abadir, "Very Low Cost Testers: Opportunities and Challenges", IEEE Design & Test of Computers, pp. 60-69, Sept.-Oct. 2001.

[12] J.Rivoir, "Lowering cost of test: parallel test or low-cost ATE", Asian Test Symp., 2003.

[13] H.Hashempour, F.J.Meyer and F.Lombardi, "Analysis and evaluation of multisite testing for VLSI", IEEE Trans. Instrum. Meas., Vol. 54, No. 5, pp. 1770-1778, Oct. 2005.

[14] S.Alampally, J.Abraham, R.A.Parekhji, R.Kapur and T.W.Williams, "Evaluation of Entropy Driven Compression Bounds on Industrial Designs", Asian Test Symp., 2008.

[15] J. Moreau, T. Droniou, P. Lebourg, P. Amagnat, "Running scan test on three pins: yes we can!", Intl. Test Conf., 2009.

[16] A.Sanghani, B.Yang, K.Natarajan and C.Liu, "Design and implementation of a time division multiplexing scan architecture using serializer and deserializer in GPU chips", VLSI Test Symp., 2011.

[17] K.Chakravadhanula, V.Chickermane, D.Pearl, A.Garg, R.Khurana, S.Mukherjee and P.Nagaraj, "SmartScan - Hierarchical Test Compression for Pin-Limited Low Power Designs", Intl. Test Conf., 2013.

[18] C.E.Stroud, A Designer's Guide to Built-In Self-Test, Springer, 2002.

[19] R.Mittal, L.Balasubramanian, A.Sontakke, H.Parthasarthy, P.Narayanan, P.Sabbarwal and R.A.Parekhji, "DFT for Extremely Low Cost Test of Mixed Signal SoC with Integrated RF and Power Management", Intl. Test Conf., 2011.

978-1-5386-3693-0/18 $31.00 © 2018 IEEE

Test-Time Reduction for Power-Aware 3D-SoC

Sabyasachee Banerjee, Subhashis Majumder
Department of Computer Science and Engineering
Heritage Institute of Technology
West Bengal, Kolkata 700107
Email: sabyasachi.banerjee@heritageit.edu
subhashis.majumder@heritageit.edu

Bhargab B. Bhattacharya
Nanotechnology Research Triangle
Advanced Computing and Microelectronics Unit
Indian Statistical Institute
West Bengal, Kolkata 700108
Email: bhargab@isical.ac.in

Abstract—**Minimization of overall test time is one of the primary concerns in the design of 3D-SoCs, whereas satisfying the thermal constraints and bounding the number of inter-layer TSVs are also of critical importance. This paper presents a scheduling-based technique to reduce test-time for core-based 3D-SoCs, under certain constraints on TAM-width and the number of TSVs. A partitioning technique is also suggested to assign the layers to cores under TSV and power constraints. The proposed methods have been tested on several SoC benchmarks. Experimental results reveal an improvement in test time for most of the circuits, while satisfying the above-mentioned constraints.**

Keywords: SoC, TSV, TAM, Wrapper.

I. INTRODUCTION

Continuous advancements in process technologies as well as design tools over the last few decades have led to successful and cost-effective implementation of many-core system-on-chip (SoC) integrated circuits (ICs). A number of pre-designed cores like CPUs, GPUs, DSP cores, application-specific IP-blocks, memories, network controllers and peripherals are embedded into an SoC [2]. Customized Test Access Mechanism (TAM) and core-wrappers have been introduced as many of these cores are usually inaccessible via chip inputs and outputs. Test stimuli from the source to the core-under-test, and test responses from the core-under-test to the test-response sink are transported via TAMs, and the core wrapper acts as an interface between the core and the TAM. With the introduction of 3D-SoC, testing of various cores that are placed on various layers has become a major concern. The 3D-TAMs consist of a number of Through-Silicon-Vias (TSV), which are generally used as vertical connectors for transmitting signals across the layers or dies. The TSVs are also used for signal-routing, power/ground routing, thermal sinking, and transmission of clock signals. Since all the pins and sinks are placed at the bottom of a 3D-IC, TSVs are required to transmit the test stimulus from pins to cores-under-test that are placed across layers. These TSVs are quite costly and occupy considerable area on a chip. Therefore, the test-access architecture is required to be designed in such a way that the number of TSVs that are used to test all the cores of the SoC is minimized, and at the same time, testing should be completed as early as possible.

In this work, we ignore the overhead for pre-bond testing. We propose two algorithms: the first one attempts to identify multiple cores that can be tested in parallel satisfying the given constraints on TAM-width under the same time-stamp so as to minimize the overall test time. The second algorithm provides a placement-solution for assigning layers to different cores on the basis of two optimization criteria, namely power and TSV-count. Comparative results with a few earlier work [4] [12] demonstrate the superiority of the proposed method.

The rest of this paper is organized as follows: Section II presents related prior work on 3D SoCs. Section III illustrates the TAM-architecture with a simple example. Section IV describes the problem formulation. Section V elaborates the proposed algorithms and Section VI presents experimental results on ITC-02 SoC benchmarks [1]. Finally, Section VII provides concluding remarks and future work.

II. PRIOR WORK

In the past, Integer Linear Program (ILP) and mixed-ILP (MILP) [3] [7] [12] solvers have been used for test time optimization. However, ILP techniques are computationally quite expensive for large SoCs since they require lot of CPU-time, which might be unrealistic for solving the problem for larger SoCs in real time. Based on greedy approaches, some heuristics have been suggested to solve this problem, which perform much better in terms of CPU-time as compared to the earlier methods such as ILP. Pradhan et al. [9] had suggested such a heuristic; however, they dealt with only two layers and their solutions did not address the power issue. The Bin Packing method [6] [5] has also been used to solve this problem. Another work done by Chandran et al. [4] suggests a thermally efficient scheduling technique to reduce the total test time of the 3D SoC. Similarly, Roy et al. [10] had suggested a heuristic to schedule the tests for each core in such a way that the overall test time for the SoC is reduced. However, none of these papers address the impact of the area constraints. Note that the advantage of designing a 3D-IC will be lost if the placement of cores on different layers is made just satisfying the other constraints and disregarding the area factor.

In this paper, we propose a power-aware layer assignment policy and a test scheduling technique considering the constraints on TAM-width and TSVs. The proposed algorithms run faster compared to previous approaches [4] [9] [12] and offer improved test time for most of the benchmark SoCs.

III. AN EXAMPLE OF MULTI-LAYER SoC

Consider Fig. 1. Let us assume that the two cores M_1 and M_4 can be tested in minimum time when connected with a TAM of width W_1, i.e., increasing the TAM width does not reduce the test time any further. Similarly, core M_2 requires a TAM of width W_2 to be tested in optimal time, whereas core M_3 requires a width of $W_1 + W_2$. Note that if we want to test each of the cores in minimum time, we have to have the provision for a TAM of width at least $W_1 + W_2$. Then M_1 and M_2 can be tested in parallel in the same timestamp say t_1, core M_4 may be tested after test of M_1 is finished, say in timestamp t_2 and finally core M_3 in timestamp t_3. Due to area constraint, it is naturally expected that we may not be able to place all the cores in one layer. Now, if a core consumes more

978-1-5386-3693-0/18 $31.00 © 2018 IEEE

103

power, it will generate more heat and should be placed closer to the heat sinks, which are normally available at Layer 1 (bottommost layer). Following this argument, suppose cores M_1 and M_2 are placed at Layer 1, whereas M_3 and M_4 are placed at Layer 2. From Fig. 1, it is apparent that we require a total $2(W_1 + W_2)$ number of TSVs. The number of TSVs that participate in the test process is always going to be a multiple of 2, since if a certain number of wires are used to transfer the stimulus to different layers, the same number of wires are also needed to bring down the responses to the bottommost layer where the pins reside [11]. In this example, the total number of wires required is $W_1 + W_2$ and the total number of test pins required is $2(W_1 + W_2)$. Note that, unlike this example, in some cases all the TAM wires may not need to go beyond Layer 1, and in such cases number of TSVs required will be less.

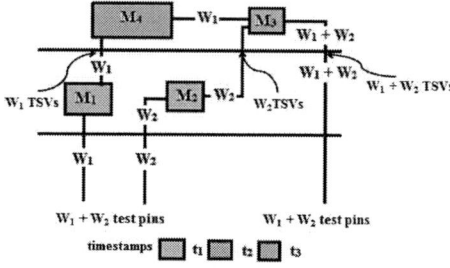

Fig. 1: Example of TAM usage with TSV

IV. PROBLEM FORMULATION

We model the problem of partitioning the cores of an SoC into its different layers as a graph partitioning problem. An SoC is typically modeled as a graph $G = (V, E)$, where $V = \{V_1, V_2, ..., V_n\}$ denotes the set of cores and E denotes the set of TAM-wires.

(i) We have to partition the set V in such a way that

$$\cup_{j=1}^{l} Partition_j = V,$$

$$\forall_{1 \le i, j \le l,\ i \ne j} Partition_i \cap Partition_j = \emptyset,$$

where l is the number of layers and $Partition_j$ represents the set of cores present in the layer j. Let $Area_{Partition_j}$ be the total area required to place the set of cores in layer j. We need to make sure that

$$Area(Partition_j) \le (1 + x) \frac{Total Area}{l}$$

where $Total Area = \sum_{j=1}^{l} Area(Partition_j) = \sum_{i=1}^{n} Area(V_i)$ and x is the fractional tolerance beyond which the area of a particular layer is not allowed to grow over and above the average. Finally, $Area(SoC) = max_j Area(Partition_j)$. Ideally $x = 0$ may ensure minimum-area usage for the SoC, but this condition is not practically viable.

(ii) Let $Pwr_{Partition_j}$ be the total amount of power consumed by the cores placed at layer j. Our objective is to place the cores in the layers in such a way that $Pwr_{Partition_j}$ will decrease with increase of j, where $j = 1$ represents the bottommost layer and $j = l$ represents the topmost layer.

(iii) Let $Time_{V_i}$ be the time required to test the core V_i in minimum time and $Time_{t_j}$ be the total test time required to test the cores that are being tested under the timestamp t_j. Then $Time_{t_j} = max\{Time_{V_i} :$ core V_i is tested within the timestamp $t_j\}$. Our final objective is to minimize $\sum_j Time_{t_j}$ so that we can test the SoC in minimum time.

(iv) We finally define a normalized cost function for mixed optimization with a parameter $\alpha \in [0, 1]$ for convex combination, as follows:

$$Cost_{V_i} = \frac{\alpha * Width_{V_i}}{max_j(Width_{V_j})} + \frac{(1 - \alpha) * Pwr_{V_i}}{max_j(Pwr_{V_j})}$$

Here $Width_{V_i}$ and Pwr_{V_i} respectively represents the number of TAM wires required to test core V_i and the amount of power consumed while it is being tested. When $\alpha = 0$, we emphasize on the power consumed by the core and when it is 1, we emphasize on the TSV count. It is easy to observe that the maximum value that $Cost_{V_i}$ can attain is 1. In the next section, we illustrate the implications of the above cost definition.

V. PROPOSED ALGORITHM

In order to optimize the overall test time, we schedule the cores in a way so that we can optimally use the available TAM width for testing more than one core in the same timestamp. Generally, the test time for a core reduces when the available TAM width increases, but only up to a certain limit [7]. In fact, the test time decreases akin to a falling-staircase pattern with the increasing TAM width, i.e., holding a constant value for a certain interval and then suddenly falling by a step for just an increase in the TAM width by unity. Since the test time of a core cannot be reduced indefinitely, we attempt to utilize available TAM wires for testing yet untested cores whenever extra TAM wires are available after scheduling tests for a core. The proposed approach is shown in Fig. 2 on the ITC'02 benchmark SoC d695 for a total TAM width of 24.

At first, we sort the cores in non-increasing order of their test times under the available TAM width W of the SoC, ties being resolved randomly. We then start to schedule the cores one by one in that order. In each timestamp, while scheduling the first core V_f, we assign it as much TAM width as possible say w ($\le W$), so that its test time is minimized to the extent possible say tt_f. However, we consider the pareto-optimal points in the TAM width vs. test time plot of V_f also, so that w corresponds to the minimum TAM width that is required to guarantee a test time of tt_f. If V_f does not require the full available TAM width, i.e., $w < W$, we try to schedule the next core in the sorted order say V_n, in the same timestamp, if it can be tested within the remaining TAM width available

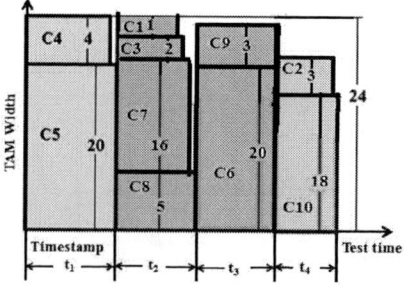

Fig. 2: Scheduling of cores for d695 with TAM width 24

978-1-5386-3693-0/18 $31.00 © 2018 IEEE

```
input  : Total number of cores present in the SOC (n),
         Total TAM width of the SOC (W),
         Test time of cores under various TAM width
         (Time_w V_i: test time for core i for width w)
output : Total time needed to test the SOC (TotTime),
         TAM width used for each core (Width_{V_i}),
         Timestamp in which each core is tested (t_{V_i})
1  for i ← 1 To n do
2  |   isTest_{V_i} ← 0;              /* 1 ⇒ core i is tested */
3  end
4  Sort(V[ ]);     /* Cores sorted in descending order of
   test times for TAM width W */
5  CurrentTstamp ← 0;
6  TotTime ← 0;
7  while (All cores are not yet tested) do
8  |   w = 0;
9  |   W_available ← W;
10 |   CurrentTstamp ← CurrentTstamp + 1;
11 |   for i ← 1 To n do
12 |   |   /* consider cores in order            */
13 |   |   if (isTest_{V_i} = 0) then
14 |   |   |   w ← W_available;
15 |   |   |   if (w = W) then
16 |   |   |   |   /* 1st core in timestamp      */
17 |   |   |   |   TstampSpan ← Time_w V_i
   |   |   |   |   TotTime ← TotTime + TstampSpan;
18 |   |   |   end
19 |   |   |   else
20 |   |   |   |   if (Time_w V_i > TstampSpan) then
21 |   |   |   |   |   continue;     /* width left not
   |   |   |   |   |   enough, try next core */
22 |   |   |   |   end
23 |   |   |   |   while (Time_{w-1} V_i ≤ TstampSpan) do
24 |   |   |   |   |   w ← w - 1;    /* save width */
25 |   |   |   |   end
26 |   |   |   end
27 |   |   |   while (Time_w V_i = Time_{w-1} V_i) do
28 |   |   |   |   w ← w - 1;        /* take advantage of
   |   |   |   |   pareto optimal points */
   |   |   |   |   if (w = 1) then
29 |   |   |   |   |   break;        /* from while */
30 |   |   |   |   end
31 |   |   |   end
32 |   |   |   Time_{V_i} ← Time_w V_i;
33 |   |   |   isTest_{V_i} ← 1;
34 |   |   |   Width_{V_i} ← w;
35 |   |   |   t_{V_i} ← CurrentTstamp;
36 |   |   |   W_available ← W_available - w;
37 |   |   |   if (W_available = 0) then
38 |   |   |   |   break;            /* out of for loop */
39 |   |   |   end
40 |   |   end
41 |   end
42 end
43 end
```

Algorithm 1: Scheduling cores in diff. timestamps

```
input  : Area of each core (Area_{V_i}),
         Power consumption of each core(Pwr_{V_i}),
         TAM width of each core(Width_{V_i}),
         timestamp of each core(t_{V_i}),
         Number of layers (l),
         parameter for convex optimization α
output : for each core, layer where placed(Layer_{V_i})
1  for i ← 1 To n do
2  |   TotalArea ← TotalArea + Area_{V_i};
3  |   Cost_{V_i} ← α * (Width_{V_i}/max_i(Width_{V_i})) + (1 -
   |   α) * ((Pwr_{V_i}/max_i(Pwr_{V_i}));
4  end
5  Sort(Core[ ]);                /* Cores are sorted in
   non-increasing order of their Cost */
6  Layer_{V_1} ← 1;  /* Core with highest cost goes to
   layer 1 */
7  Area(Part_1) ← Area(Part_1) + Area_{V_1};
8  for i ← 2 To n do
9  |   isPlaced ← False;
10 |   for j ← 1 To i - 1 do
11 |   |   if (t_{V_i} = t_{V_j}) then
12 |   |   |   /* if possible, place cores with
   |   |   |   same timestamp in same layer   */
13 |   |   |   if (Area(Layer_{V_j}) + Area_{V_i} <
   |   |   |   ((1 + x) * TotalArea/l)) then
14 |   |   |   |   Layer_{V_i} ← Layer_{V_j};
15 |   |   |   |   Area(Layer_{V_j}) ←
   |   |   |   |   Area(Layer_{V_j}) + Area_{V_i};
16 |   |   |   |   isPlaced ← True;
17 |   |   |   |   break;           /* out of for loop */
18 |   |   |   end
19 |   |   end
20 |   end
21 |   if (isPlaced = True) then
22 |   |   continue;        /* to next unplaced core */
23 |   end
24 |   /* not yet: place as low as possible     */
25 |   for c ← 1 To l do
26 |   |   if ((Area(Part_c) + Area_{V_i}) <
   |   |   ((1 + x) * TotalArea/l)) then
27 |   |   |   Layer_{V_i} ← c;
28 |   |   |   Area(Part_c) ← Area(Part_c) + Area_{V_i};
29 |   |   |   break;
30 |   |   end
31 |   end
32 end
```

Algorithm 2: Placing cores at different layers

$W - w$, and in time $\leq tt_f$. Furthermore, considering the pareto-optimal points in the plot of V_n, we try to reduce the TAM width assigned to that core as much as possible. This in turn might increase its test time tt_n, but we ensure that it remains $\leq tt_f$. For example, in Fig. 2, as many as 4 cores got scheduled in timestamp t_2, all of which will be tested in parallel. The test times of the cores $C1$, $C3$ and $C7$ have been stretched to fit within the test time of $C8$, which also happens to be the length of timestamp t_2, $C8$ being the first scheduled core of that particular timestamp. The details of how the entire scheduling is done is given in Algorithm 1. In fact, the scheduling of the cores described above is totally independent of the 3D architecture. So, after the scheduling is over, we try to put the cores in the different layers of the 3D-SoC. Since all the heat sinks are placed at the bottom of the 3D IC, we try to place the more power hungry cores either

in the bottommost layer or in the layers close to it. However, it begets another problem - the cores with the requirement of higher TAM width may get placed in the upper layers, which might increase the number of TSVs. To keep a balance, we have introduced a cost function with a tuning factor α (as described in Section IV), which basically balances between the two important optimization criteria - power dissipation and TSV count. Apart from optimizing test time, power and TSV count, we have also intended to maintain an area balance by not allowing the area of any of the layers to go beyond a certain bound, which was mostly ignored in almost all the earlier related works. On the basis of the cost values of the cores, we first sort all the cores in non-increasing fashion. We then place the cores in the sorted order starting from layer 1. We continue placing the cores in a layer till the sum total area of the cores placed in that layer crosses a particular threshold, related to the average area per layer. We then shift to the next upper layer. The above procedure is described in Algorithm 2.

Finally, after the placement is finished, we run a procedure to calculate the total number of TSVs required in the SoC. An example architecture of SoC d695 with TAM width 24 is shown in Fig. 3. Note that the number of TSVs required to bring up the test stimuli from Layer 1 to Layer 2 is 24. The same number of TSVs will be required again to bring down the test stimuli from Layer 2 to Layer 1. It makes the total $24 * 2 = 48$ and hence the total number of TSVs is always going to be even. Similarly between Layer 2 and layer 3 we require $3 * 2 = 6$ TSVs in total. Finally, a total of 6 TSVs is necessary between Layer 3 and Layer 4. Hence a total of $(48 + 6 + 6 = 60)$ TSVs will be required to test d695 under TAM width of 24.

A. Complexity Analysis:

Let n be the number of cores and l be the number of layers present in the SoC. In Algorithm 1, Step 1 requires $O(n)$ time and Step 4 runs in $O(n \log_2 n)$ time for sorting. The while loop in Step 6 may execute at most n times and the for loop in Step 10 (nested inside the while loop), may also execute n times in the worst-case. Each of the two while loops in Steps 25 and 29 may execute at most W (TAM width) times. The rest of the steps can run in $O(1)$ time. Hence, the worst-case time complexity of Algorithm 1 will be $O(n \log_2 n + n^2 W) = O(n^2 W)$, though in practice, it runs much faster. Note that using a linked-list for storing the sorted cores might improve the runtime slightly but will not improve the asymptotic complexity.

Likewise in Algorithm 2, the running time of Steps 1 to 5 will be bounded by $O(n \log_2 n)$. The two nested for loops in Steps 8 and 10 will together require $O(n^2)$ time. Note that since the number of layers will be mostly less than the number of cores, the for loop in Step 25 will not affect the final complexity, which will be clearly $O(n^2)$. We state here without proof that the counting of TSVs can be performed in $O(n l)$ time, where l is the number of layers. So, the proposed algorithm will have an overall time complexity of $O(n^2 W)$.

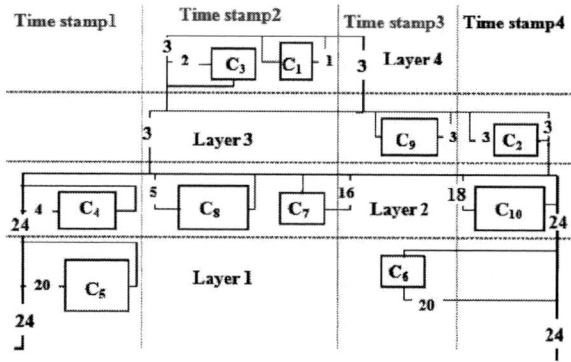

Fig. 3: Architecture of d695 with TAM width 24

VI. EXPERIMENTAL RESULTS

The proposed algorithms have been implemented in C (Linux environment) and tested on an Intel Corei3 machine with a primary memory of 2GB. Extensive experiments were conducted for 7 different ITC02 [1] benchmarks, by varying the layers from 2 to 4 for each SoC, using 8 different TAM widths and by varying α from 0 to 1 with an interval of 0.1. However, for brevity, we present results only for some values of α; like for all the bar diagrams given below, $\alpha = 0.8$. The CPU times noted were negligible (always in the range of 0.0 ms) even for the larger benchmarks. In our experiments, when

the cores are placed in different layers, we noted that on an average, the maximum area of any layer lied within 20% of the average area of layers. As done by Wu et al. [12], we have calculated the test times for the individual cores using a concept given by Marinisen et al. [8] and calculated the individual power of each core from the weblink [1].

TABLE I: Comparison of Test times with [4]

| TAM Width (W) | 3D Test Time | | | | | |
| | d695 | | p22810 | | p93791 | |
	Ours	[4]	Ours	[4]	Ours	[4]
16	42833	44027	485728	469487	1856606	1864872
24	31569	32542	371217	376819	1256380	1257312
32	23297	24133	252887	270592	1108271	1185928
40	18862	19718	191580	240542	1103105	1107394
48	18333	18837	195269	179755	634441	987669
56	15555	15237	165352	157402	604257	701307
64	13732	13816	142428	146630	514652	659045

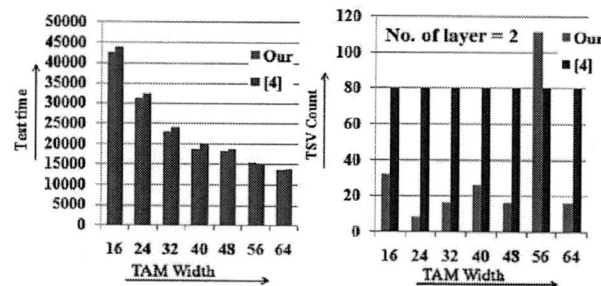

Fig. 4: Comparison with Chandran et al. [4] for d695

We first compare our results with a previous work by Chandran et al. [4], where equal number of cores were placed in each layer. Also, they considered only three benchmarks using only one set of layers in each case. Table I presents comparative test-times for seven different TAM widths. We also plot them as bar-charts for the ease of comparison, and Fig. 4, 5 and 6 show that test-time and TSV-count reported by our algorithms are consistently better with a few exceptions.

Next, we compare our results with that of Wu et al. [12]

TABLE II: Experimental Results of $p22810$

Number of cores: 28							
TSV-Limit for Wu et al. [12] throughout is 80 CPU time for our experiment is 0.0secs throughout							
Lyrs	TAM Width	TSV-Limit			3D Test Time		CPU time
		Ours $\alpha = 0.6$	Ours $\alpha = 0.7$	Ours $\alpha = 0.8$	Ours	[12]	[12]
2	16	14	14	14	485728	324902	0.2
	24	14	14	14	371217	308674	1.1
	32	14	14	14	252887	270379	0.3
	40	28	28	28	191580	265452	0.7
	48	34	26	26	195269	265452	0.5
	56	12	12	12	165352	245234	0.3
	64	28	28	28	142428	245234	0.7
3	16	38	38	38	485728	321172	1.6
	24	50	50	50	371217	282849	0.5
	32	50	50	50	252887	258778	78.9
	40	66	66	50	191580	230462	47.8
	48	68	68	72	195269	226844	37.4
	56	52	50	50	165352	226813	17.6
	64	52	52	40	142428	223079	6.5
4	16	48	48	48	485728	309905	0.4
	24	68	60	60	371217	271146	4.2
	32	60	60	60	252887	240120	45.5
	40	106	80	74	191580	220194	73.5
	48	130	120	88	195269	214379	297.5
	56	48	52	52	165352	189146	315
	64	78	78	78	142428	181698	407

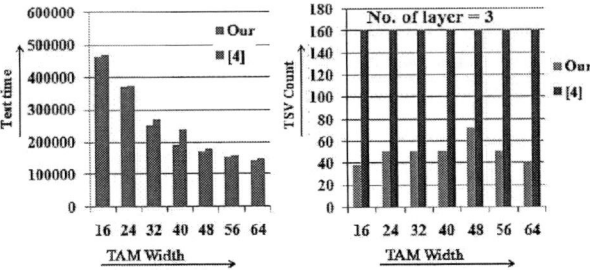

Fig. 5: Comparison with Chandran et al. [4] for p22810

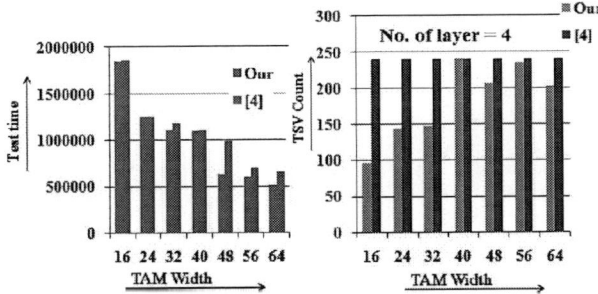

Fig. 6: Comparison with Chandran et al. [4] for p93791

for all four benchmark SoCs that they have used in their experiments. These benchmarks were also studied by some other authors [9] [10]. In our case, the observed test-time is comparable to theirs and the CPU-time is much improved. We have made a detailed comparison with the work of Chandran et al. [4] and Wu et al. [12] from the perspective of power-awareness. Also, Pradhan et al. [9] reported experimental results for only two layers.

TABLE III: Number of TSVs of $p22810$ under various α

| TAM Width | TSV-Limit | | | | 3D Test Time |
	$\alpha = 0\%$	$\alpha = 30\%$	$\alpha = 80\%$	$\alpha = 100\%$	
Layer = 2					
16	32	14	14	14	485728
24	48	24	14	12	371217
32	64	36	14	12	252887
40	72	26	28	28	191580
48	94	48	26	20	195269
56	104	22	12	12	165352
64	120	28	28	28	142428
Layer = 3					
16	64	38	38	38	485728
24	96	54	50	38	371217
32	126	50	50	50	252887
40	144	94	50	50	191580
48	188	112	72	68	195269
56	216	106	50	40	165352
64	246	52	40	34	142428
Layer = 4					
16	82	48	48	48	485728
24	140	78	60	62	871217
32	164	104	60	58	252887
40	224	112	74	70	191580
48	272	134	88	74	195269
56	216	106	50	40	165352
64	330	78	78	50	142428

In Table II, we present an extensive comparison of test-times, TSV counts (for 3 different values of α) and CPU times for p22810, one of the large benchmarks. Note that our test-times are better in almost all the cases except for very low TAM widths, whereas our TSV counts are less in more than 90% of the cases. Especially for higher values of α (≥ 0.75), TSV counts are far less. For all cases, our CPU times are

TABLE IV: Power Values at various layers of $p22810$

| α | Power Consumption | |
	Layer 1	Layer 2
Layer = 2, TAM Width = 16		
0	8834.08	472.34
0.3	8790.01	516.40
0.8	8790.01	516.40
1	8720.00	586.40

α	Layer 1	Layer 2	Layer 3
Layer = 3, Tam Width = 16			
0	8524.20	628.26	153.94
0.3	8457.32	585.42	263.66
0.8	8083.51	959.24	263.66
1	5463.37	3560.73	282.3

α	Layer 1	Layer 2	Layer 3	Layer 4
Layer = 4, Tam Width = 16				
0	7866.04	1026.11	384.75	29.51
0.3	6921.48	1868.53	503.23	13.17
0.8	7941.33	848.68	503.23	13.17
1	5405.29	3404.74	442.06	54.32

less by several orders throughout. Note that our test-time is obtained using Algorithm 1 and is hence independent of α. In Table III, we present some more results that clearly show the variation of TSV counts with various values of α for the same benchmark and it shows that except for $\alpha = 0$, i.e., where the only optimization objective is power, the TSV counts are always within reasonable limits for practical usage. Next, we present a typical power profile with the variation of α for just one TAM width in Table IV. Note that the layer-wise power values fall monotonically as we go up through the layers, as we intended. We could not compare our power profile with the earlier works as none of them reported any.

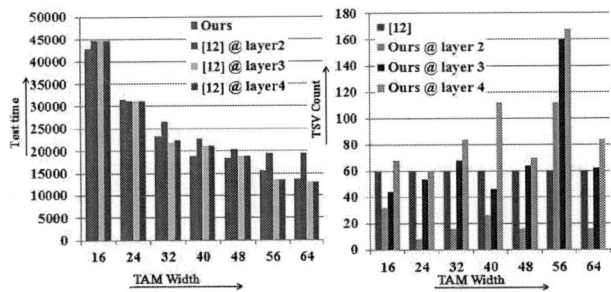

Fig. 7: Comparison with Wu et al. [12] for d695

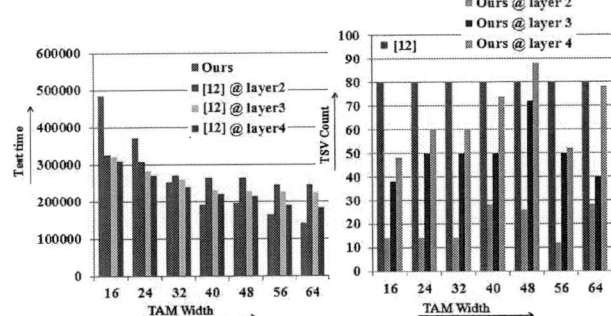

Fig. 8: Comparison with Wu et al. [12] for p22810

For brevity, we present the results for other three benchmarks d695, p34392, and p93791 in a concise manner. In Figs. 7-10, we show the plots after comparing the test times

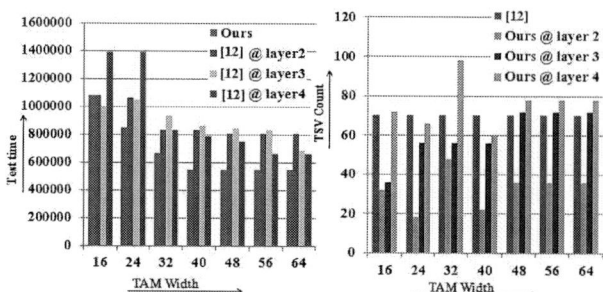

Fig. 9: Comparison with Wu et al. [12] for p34392

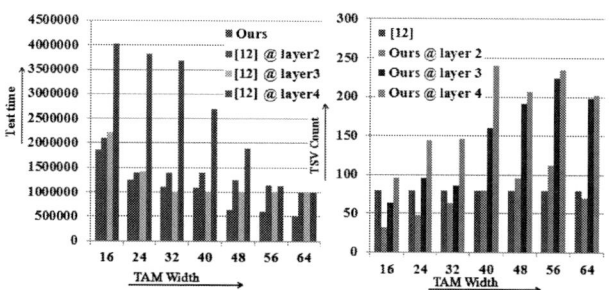

Fig. 10: Comparison with Wu et al. [12] for p93791

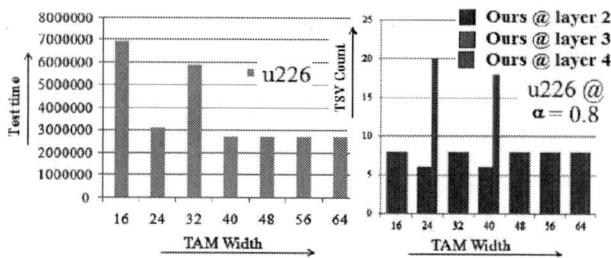

Fig. 12: Experimental results for u226 (9 cores)

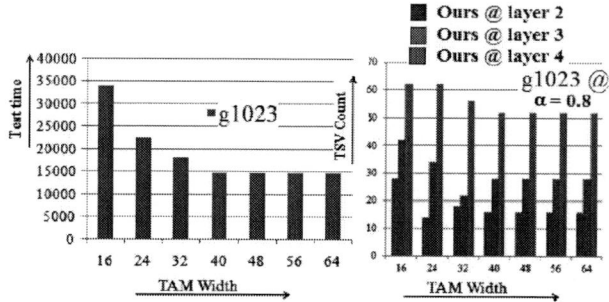

Fig. 13: Experimental results for g1023 (14 cores)

and the TSV counts with Wu et al., where we have chosen $\alpha = 0.8$, as we noted that this particular value gives the most balanced optimization between the two conflicting objectives. Note that for all the 4 benchmarks, our test times are less in most cases, whereas the run-times are always distinctly less. Our TSV counts compete well when we use 2 or 3 layers.

In Figs. 11-13, we present the results for three other benchmarks t512505, u226, and g1023.

VII. CONCLUSION AND FUTURE WORK

We have proposed an efficient framework for TAM architecture optimization of 3D-SoCs, where minimization of the test time is achieved while maintaining the area and thermal constraints. It also reduces the number of thermal TSVs. The experimental results presented show the efficacy of our method. We look forward to deal with some additional constraints like delay issues in conjunction with power and area.

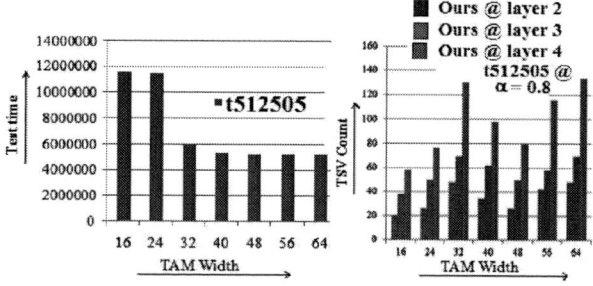

Fig. 11: Experimental results for t512505 (31 cores)

REFERENCES

[1] 3d SoC test benchmarks http://3dsocbench.ece.wisc.edu/.

[2] Itc 2002 SoC benchmarking initiative.

[3] K. Chakrabarty. Test scheduling for core-based systems using mixed-integer linear programming. *IEEE Trans. Computer-Aided Design of Integrated Circuits and Systems*, 19(10):1163–1174, 2000.

[4] U. Chandran and D. Zhao. Thermal driven test access routing in hyper-interconnected three-dimensional system-on-chip. In *Proc. Defect and Fault Tolerance in VLSI Systems*, pages 410–418, 2009.

[5] Y. Huang, S. M. Reddy, W. T. Cheng, P. Reuter, N. Mukherjee, C. C. Tsai, O. Samman, and Y. Zaidan. Optimal core wrapper width selection and SoC test scheduling based on 3-D bin packing algorithm. In *Proc. International Test Conf.*, pages 74–82, 2002.

[6] V. Iyengar, K. Chakrabarty, and E. J. Marinissen. On using rectangle packing for SoC Wrapper/TAM co-optimization. In *Proc. 20th IEEE VLSI Test Symp.*, pages 253–258.

[7] V. Iyengar, K. Chakrabarty, and E. J. Marinissen. Test wrapper and test access mechanism co-optimization for system-on-chip. *J. Electronic Testing: Theory and Applications*, 18:211–228, 2002.

[8] E. J. Marinissen, S. K. Goyel, and M. Loucbergi. Wrapper design for embeded core test. In *Proc. International Test Conference 2000*, pages 911–920, 2000.

[9] M. Pradhan, C. Giri, H. Rahaman, and D. K. Das. Optimizing test time for core-based 3-D integrated circuits by a technique of bi-partitioning. In *Proc. Design and Test Symposium (EWDTS), 2014 East-West*, pages 1–4, 2014.

[10] S. Roy, P. Ghosh, H. Rahaman, and C. Giri. Session based core test scheduling for minimizing the testing time of 3D SoC. In *Proc. Electronics and Communication Systems (ICECS)*, pages 1–5, 2014.

[11] R. Weerasekera, L. Zheng, D. Pamunuwa, and H. Tenhunen. Extending systems-on-chip to the third dimension: Performance, cost and technological tradeoffs. In *Proc. Int.Conference on Computer-Aided Design*, pages 212–219, 2007.

[12] X. Wu, Y. Chen, K. Chakrabarty, and Y. Xie. Test-access mechanism optimization for core-based three-dimensional SoCs. *Microelectronics Journal*, 41:601–615, 2010.

2018 31th International Conference on VLSI Design and 2018 17th International Conference on Embedded Systems

Identification of Faulty TSVs in 3D IC during Pre-bond Testing

Dilip Kumar Maity[1], Surajit Kumar Roy[2] and Chandan Giri[3]

[1]Dept. of Computer Science and Engineering, Academy of Technology, Hooghly 712121, India
[2,3]Dept. of Information Technology, Indian Institute of Engineering Science & Technology, Shibpur, Howrah - 711103, India
[1]dilip.maity@aot.edu.in, [2]suraroy@gmail.com, [3]chandangiri@gmail.com

Abstract— Manufacturing of three-dimensional (3D) integrated circuit (IC) using through-silicon vias (TSVs) passes through a complex process and testing of TSVs is a critical issue to the researchers. Pre-bond testing eliminates bad dies before bonding. In this paper, we propose a heuristic approach for pre-bond TSV testing which reduces the test time considerably. The proposed method uses recursive bi-partitioning and padding of test sessions and runs in linear time.

Keywords-3D IC; Pre-bond testing; TSV test

I.   INTRODUCTION

Three dimensional integrated circuit (3D IC) is evolving as an area of great interest to overcome the challenges of current 2D IC. A 3D IC is formed by stacking multiple dies vertically and interconnecting the dies using through-silicon vias (TSVs). 3D IC provides several benefits than conventional 2D IC higher performance, reduced interconnection length, reduced power consumption and smaller footprint [1]. Despite these advantages, there are several challenges related to testing of 3D IC [2].

TSV provides communication link for the upper layer dies of 3D IC. Fabrication of TSVs is a complex process and defects may be introduced during the manufacturing of 3D ICs. Different types of TSV faults are presented in [3]. Faults like pinhole or microvoid may arise during manufacturing of 3D IC and these faults are identified by pre-bond testing. The resistance of a TSV increases due to microvoid whereas the capacitance between TSV and the substrate increases due to pinhole. Hence, capacitance measurement is required for pinhole fault, but resistance measurement is used for microvoid fault [4]. Few defects can be identified by post-bond testing evolved due to alignment, bonding or stress. A single faulty TSV can paralyze the whole chip. So, testing of TSVs is necessary for proper functionality of the chip. This paper considers pre-bond TSV testing.

A pre-bond TSV probing method is presented in [4] where multiple TSVs are shorted together with a probe needle to form a TSV network. The number of TSVs within a network is typically less than *20* and depends on the relative diameter of the probe needle and the pitch of TSVs [7, 8]. The probing method is modeled by adding an active driver in the probe needle and forming a charge sharing circuit between single (multiple) TSV(s) and the probe

needle. The capacitor of the charge-sharing circuit is charged through the connected TSV and the charging time is recorded. It is then compared with a non-faulty TSV. The charging time of the capacitor is considered as the test time of TSV for resistance measurement of TSV. Charging time varies with the number of TSVs tested in parallel [7].

Recently several research works are being undertaken on the pre-bond TSV testing. In [5], an on-chip sense amplifier is proposed for detecting capacitive TSV faults. A solution for identifying resistive open faults and leakage faults in TSVs using ring oscillators is addressed in [6]. A pre-bond TSV test method is presented in [7] where multiple TSVs can be tested simultaneously. But only small number of defective TSVs are identified in reduced test time. The authors present heuristic methods to identify defective TSVs in reduced test time in [9, 10]. These two methods improve the results. An integer linear programming (ILP) based method to detect faulty TSVs is proposed in [8].   In this method some sessions are unnecessarily tested during the identification process and time complexity is also higher than [9, 10]. The literatures in [11, 12] also propose other techniques for pre-bond TSV probing but these techniques require prior information about maximum number of faulty TSVs to be identified. Thus as seen above, the various solutions proposed in the research papers still have scope for improvement and pre-bond TSV testing remains a major challenge. In this paper, we have proposed a fast heuristic method that identifies the defective TSVs efficiently in terms of test session generation and significantly reduces the overall test time. The major contributions of this work are:

a) Using our proposed method the test time is reduced by 35% compared to prior works.

b) Our proposed method runs in linear time which is better than the prior works.

c) There is no need of any constraint regarding maximum number of faulty TSV to be identified before testing. Also, proposed algorithm is efficient for known number of redundant TSVs.

The rest of this paper is organized as follows: Section II describes the problem formulation of the proposed methodology. Proposed algorithm for identifying the defective TSVs uniquely is presented in Section III. Proposed methodology is explained using an illustrative example in Section IV. Analysis of the proposed methodology is described in Section V. Section VI presents

978-1-5386-3693-0/18 $31.00 © 2018 IEEE   109

the experimental results and finally Section VII concludes the paper.

II. PROBLEM FORMULATION

In general, most of the pre-bond TSV faults are resistive in nature [3]. So, proper resistance measurement is required to detect faulty TSVs. In this paper we have considered resistive TSV fault as in [7, 8, 9]. Pre-bond TSV testing helps to identify defective dies early before bonding. Also, pre-bond TSV testing ensures that dies to be stacked include only good TSV interconnects which is required for known good die (KGD) information.

Objective of TSV probing is to identify the defective TSVs within a TSV network. A TSV network is formed by all the TSVs that are shorted together to the single probe needle. Identification of defective TSVs can be done by testing one TSV at a time and that requires longer test time. The test time can be reduced if the TSVs are tested in parallel. The proposed pre-bond TSV test method generates sequence of test sessions to detect faulty TSVs in reduced test time. A test session is formed by TSVs that are tested simultaneously and number of TSVs within a session indicates the session size *(q)*. The difference in capacitor charging time between faulty and non-faulty TSVs decreases when size of the session increases which affect the resolution [4]. So the size of the session cannot be increased beyond a certain limit. Resolution constraint p indicates the upper bound of session size.

Now, formally the problem can be stated as follows: *Given n untested TSVs within a n-TSV network, resolution constraint p and the test time t(q), 1≤q≤p, for the test session containing q number of TSVs, determine the sequence of test sessions to identify faulty TSVs of the n-TSV network such that the test time and test session are minimized as much as possible.* We have proposed a heuristic based algorithm to solve the above mentioned problem.

III. PROPOSED METHODOLOGY

Parallel TSV testing reduces the test time compared to the sequential TSV testing. Sequential testing is advantageous when there is a large number of defective TSVs within a TSV network. But it is impractical because each TSV network has limited number of redundant TSVs (*r*) to repair the network. The TSV network is not repairable if the number of defective TSVs is greater than *r*. Otherwise, the redundant TSVs are sufficient to replace the defective TSVs. Finding *r+1* number of defective TSVs implies that the network is not repairable; so further test is useless. The proposed algorithm can identify *m* number of defective TSVs, where *0≤m≤ r+1*. Also the test process can be done without knowing any prior information about the number of faults or number of redundant TSVs.

The proposed algorithm tries to generate a sequence of test sessions for identifying faulty TSVs uniquely in a TSV network. The test session generation technique is described in Algorithm 1. This algorithm generates a session by

picking *p* number of TSVs from the set of n untested TSVs. The last session may have less than *p* number of TSVs. All the TSVs of a session can be tested simultaneously. Each generated session is tested and the corresponding test time is recorded. Here, we are considering only the charging time as the test time.

Algorithm 1:Test_session_generation(n, p)

Input: number of TSVs (n), number of test pins (p).
Output: Set of test sessions, and total test time.
if count=r+1**or** n=0 **then return**;
if n>0 **then**
    Create a session s by taking all n (if n<p) or p(if n≥p) number of TSVs;
    Modify the session s;   //padding non-faulty tested TSVs
    //test time accumulation
    total_test_time:=total_test_time+t(length(s));
    if session s is tested as being faulty **then**
        j:= Defective_TSV_identification(length(s), p);
        //defective TSV is identified at location j
        n:=n-length(s)+(j+1); count:=count+1;
        Test_session_generation (n, p);
    else   // the session s is fault free
        n= n-length(s); Test_session_generation(n, p);
end.

Algorithm 2: Defective_TSV_identification(s, p)

Input: The faulty session (s), number of probe pins (p).
Output: Position of defective TSV.
if length(s)=1 **then**   //defective TSV is identified
    return the position of the defective TSV;
else
    Bi-partition the session s into two new sessions s1 and s2;
    Modify the session s1 and s2;//padding non-faulty tested TSVs
    //test time accumulation
    total_test_time:=total_test_time+t(length(s$_1$));
    if session s1 is tested as being faulty **then**
        Defective_TSV_identification(s$_1$,p);
    else   // session s$_1$ is fault free so session s2 must be faulty
        Defective_TSV_identification(s$_2$,p);
end.

If a session is identified as fault-free then all the TSVs of the session are considered as good TSV. But if the session is faulty, then Algorithm 2 is invoked to find the exact position of the first occurrence of the defective TSV within the current session. Algorithm 2 divides a faulty session until a single faulty TSV is identified. The faulty session of length p (number of TSVs of the session) is bi-partitioned into two smaller sessions (as left half and right half). The length of each half will be *p/2* if p is even or $\lceil p/2 \rceil$ for left half and $\lfloor p/2 \rfloor$ for right half if p is odd. Now, the left half is tested. If the left half is found to be faulty, subsequent testing is done on this half itself. On the other hand, if the left half is found as fault free, the right half is bi-partitioned and tested. In this fashion a faulty session is bi-partitioned until the length of session is one. Finally the position of the first defective TSV is identified by the algorithm. When the left half is identified as faulty, we have no concrete information about the right half. In a session p TSVs are tested in parallel and more than one faults can be present within the p TSVs. So after bi-partitioning each partition may contain the fault. Therefore, this right half is considered as untested TSVs. During the test process, non-faulty tested TSVs are padded with the smaller sessions to utilize the maximum resolution constraint for reducing the test time.

IV. ILLUSTRATIVE EXAMPLE

In this section the proposed method is elaborated with an example. Let us consider a problem instance where number of TSVs (n) is16 and resolution constraint (p) is 4. TSVs are represented with the numbers 1, 2...16 and initially all are untested. Assume that the defective TSVs are at position 6 and 14. To further reduce the test time, if the length of any session is less than the value of the resolution constraint, non-faulty tested TSVs are randomly appended to such sessions.

Step 1: Pick the TSVs at position 1, 2, 3, 4 for the first session {1, 2, 3, 4}. Now the session is tested to check whether there is any faulty TSV or not. Since the defective TSVs are at position 6 and 14, the session is detected as fault free, so the next session will be formed from the remaining untested TSVs.

Step 2: Now the next session {5, 6, 7, 8} is created from the remaining untested TSVs (5 to 16) and is identified as faulty. This faulty session is bi-partitioned and two new sessions {5, 6}, and {7, 8} are generated. The session {5, 6} is tested and detected as faulty. So it is decomposed as {5} and {6}. After testing {5}, it is detected as fault-free. Obviously the TSV at position 6 must be faulty. We cannot make any decision about whether TSVs at position 7 and 8 are faulty or not i.e. they are considered as untested.

Step 3: Similarly next session {7, 8, 9, 10} is generated and detected as fault free.

Step 4: Again, session {11, 12, 13, 14} is identified as faulty and split as {11, 12} and {13, 14}. The session {11, 12} is detected as non-faulty and TSVs at position 11 and 12 are non-faulty. So the fault is within the session {13, 14} which is to be bi-partitioned next. Similarly the session {13} is tested and identified as fault-free. Hence, TSV at position 14 is defective as other TSVs of this session are non-faulty.

Step 5: The last session {15, 16} is generated and detected as fault-free.

V. ANALYSIS OF PROPOSED METHODOLOGY

A. Analysis of Test Session in Worst Case

This section presents the number of sessions in worst case. The number of sessions $T_S(n)$ depends on the number of TSVs n, number of faults m, and resolution constraint p. *Algorithm 2* recursively finds out the position of the defective TSVs from a session of TSVs of length p. This algorithm divides the session of size p into two smaller

sessions of size approximately $p/2$ and detect the faulty TSV from one of the sessions depending on the test result. The number of test sessions $S_S(p)$ required to identify the position of a defective TSV within a particular session by *Algorithm 2* is represented as,

$$S_S(p) = \begin{cases} S_S(p/2)+1 & if \quad p > 1 \\ 0 & if \quad p = 1 \end{cases} \tag{1}$$

Solving the equation, we get, $S_S(p) = \lceil \log_2 p \rceil$.

Let the first fault is identified at the ith position. So, total number of test sessions generated by the proposed method is calculated as,

$$T_S(n) = \begin{cases} \lceil i/p \rceil + \lceil \log_2 p \rceil + T_S(n-i) & if \quad m > 0 \\ 0 & if \quad n = 0 \\ \lceil n/p \rceil & if \quad m = 0 \end{cases} \tag{2}$$

The first term $\lceil i/p \rceil$ indicates the number of sessions tested upto the session containing the ith TSV. The second term $\lceil \log_2 p \rceil$ indicates the number of tests within a session to identify the ith TSV. The last term $T_S(n-i)$ indicates the number of sessions tested for $(n-i)$ TSVs in the network.

For our proposed algorithm, the worst case scenario occur for two reasons. Firstly, if the defective TSVs are located consecutively and a number of faulty TSVs are present in test session. In such a scenario, consecutive test sessions have a large number of TSVs from previously tested sessions and same TSVs are tested in different sessions repeatedly. Therefore, the number of test sessions increases. As a result test time increases. Secondly, our proposed algorithm recursively bi-partitions the faulty session to identify the defective TSVs. There is a possibility that singleton TSV is tested due to insufficient known good TSVs. This situation occurs if faulty TSVs are identified at the starting of the process. Therefore, without loss of generality we assume that the first defective TSV be located at position 1 *(i.e. i =1)*. We get,

$$T_S(n) = 1 + \lceil \log_2 p \rceil + T_S(n-1) \tag{3}$$

Now we consider that the TSV network has m number of defective TSVs. These m defective TSVs are concentrated at the initial test sessions. So the number of test sessions required to identify m defective TSVs uniquely,

$$T_S(n) = m(1 + \lceil \log_2 p \rceil) + (n-m)/p \tag{4}$$

B. Time Complexity of the Algorithm

The major task of the proposed algorithm is parallel testing which depends on the number of test sessions required to identify m number of defective TSVs. From the analysis we have seen that the number of test sessions required is $T(n) = T_s(n)$. If m is very large $(m \approx n)$ then the complexity of the algorithm is $O(m \log_2 p)$. If m is small then the complexity of the algorithm is $O(n)$. In real application, for an n-TSV network, the number of defective TSVs is very small compared to n; so running time for our proposed algorithm is $O(n)$.

VI. EXPERIMENTAL RESULTS

In this section, the experimental results of the proposed algorithm have been analysed and compared with [8], [10], [11] and [12]. The proposed algorithm is coded and compiled in *gcc* compiler and executed on a Intel Core i5 processor with 3GB RAM. HSPICE simulations are considered to find the test time for different test sessions. The resistance and capacitance of each TSV are considered as 1Ω and 20 fF respectively as in [4].

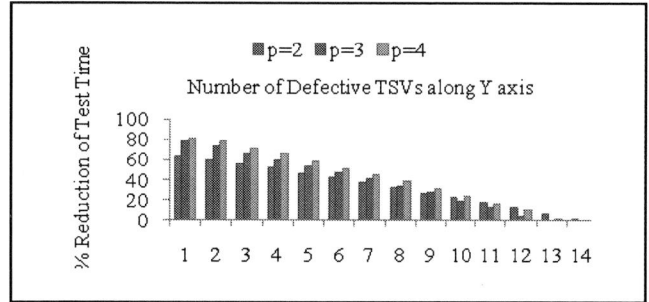

Fig. 1. Percentage Reduction of Test Time for 20- TSV

Figure 1 shows the variation of reduction in test time with the number of faulty TSVs identified for 20-TSV network. It is seen from the figure that for a given value of resolution constraint the reduction in test time decreases with increasing value of number of faulty TSVs. 13 faults can be identified with reduced test time for 20-TSV network. The reduction of test time will be large for greater resolution constraint. Because the number of test sessions will be less for large value of p. There is 15% reduction of test time than serial testing for 20-TSV network such that 12 faulty TSVs can be detected uniquely. From the Figure 1 it is also observed that, for small m, the differences of reduction in test time with $p=2$, $p=3$ and $p=4$ are significant, but for large value of m, the test time reduction is quite small. In Figure 2, for 20-TSV network and resolution constraint 3, our experimental result is compared with [8] and [10]. It is seen that the proposed methodology can identify four faults with reduced test time of 60% and above. It can also identify eight faults with reduced test time of 35% and above. Interestingly to detect six faults (as seen in Figure 2)

our proposed heuristic approach takes slightly more time than [10]. However, this single anomaly can be overlooked when compared with the reduction of time achieved in all other cases. So it can be concluded that the proposed heuristic approach provides an efficient solution for identifying defective TSVs for 3D ICs for larger TSV network.

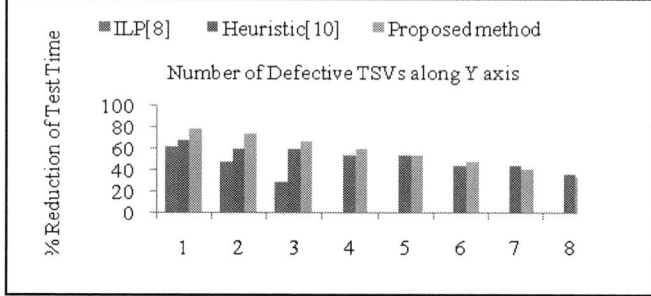

Fig. 2. Comparison in Test Time for 20-TSV Network for p = 3

Table 1 shows the comparisons between the dynamically optimized test in [11] and the proposed method for various TSV networks. Column 1 indicates the parameters n (network size), r (the redundant TSVs in the network) and p (resolution constraint). Here, we have considered all possible faults for a given r. So, average number of test sessions and average test time are considered for a given number of faulty TSVs within a TSV network. The parameter m in the Column 2 indicates the exact number of defective TSVs that has been identified within the network. Column 3 and Column 4 indicate the total number of sessions and total test time (in μs) used by the method in [11] for average and worst cases respectively. Column 5 and Column 6 indicate the number of tested sessions and test time to identify m defective TSVs for average and worst cases respectively. Column 7 indicates the relative reduction in column 5 over column 3. Column 8 indicates the relative reduction in column 6 over column 4. From Table 1, the following observations can be made. First, it is seen from the table that obtained results are better than [11] in almost all cases. Interestingly for input parameter (8, 3, 3) (as seen in Table 1) our proposed heuristic approach takes slightly more time compared to the dynamically optimized test [11]. In this case, it is possible that due to the small number of tested TSVs, a single faulty TSV is tested several times in different sessions. So, the number of sessions may be increased. However, this single anomaly can be overlooked when compared the reduction of time achieved in all other cases. Second, as expected, number tested session increases as number of fault increases. Third, as m increases the average percentage reduction decreases. This is expected as large number of sessions is tested to identify more defective TSVs within a TSV network.

In general, the faulty TSVs are distributed in cluster. The proposed method can be applied for clustered defect distribution. If $\gamma(m)$ be the test time to identify m number of

Table 1. Comparative Study of Test Sessions and Test Times Constructed by [11] and Proposed Heuristic Algorithm

(Number of TSVs, Number of redundant TSVs, Resolution constraint)	Number of defective TSVs identified (m)	Dynamically optimized test in [11]		Proposed method			
		Average case (Number of test sessions, time in μs)	Worst case (Number of test sessions, time in μs)	Average case (Number of test sessions, time in μs)	Worst case (Number of test sessions, time in μs)	Average case reduction (sessions, time)	Worst case reduction (sessions, time)
(8, 3, 3)	0	(5.0, 2.10)	(5, 2.10)	(3.0, 1.26)	(3, 1.26)	(40.00%, 40.00%)	(40.00%, 40.00%)
	1	(5.3, 2.25)	(6, 2.52)	(4.0, 1.72)	(6, 2.77)	(24.53%, 23.78%)	(0.00%, -9.92%)
	2	(6.4, 2.71)	(8, 3.36)	(5.4, 2.55)	(8, 3.77)	(15.63%, 5.90%)	(0.00%, -12.20%)
	3	(7.5, 3.17)	(8, 8.36)	(6.71, 3.09)	(10, 4.20)	(10.53%, 2.52%)	(-25.00%, -25.00%)
(12, 4, 3)	0	(7.0, 2.94)	(7, 2.94)	(4.0, 1.68)	(4, 1.68)	(42.86%, 42.86%)	(42.86%, 42.86%)
	1	(7.5, 3.14)	(9, 3.78)	(5.0, 2.12)	(7, 3.05)	(33.33%, 32.52%)	(22.22%, 19.31%)
	2	(8.7, 3.65)	(12, 5.04)	(7.7, 3.38)	(10, 4.48)	(11.49%, 7.51%)	(16.67%, 11.11%)
	3	(10.3, 4.32)	(14, 5.88)	(9.3, 4.19)	(12, 5.44)	(9.71%, 3.01%)	(14.29%, 7.48%)
	4	(11.8, 4.97)	(16, 6.72)	(10.45, 4.75)	(14, 6.41)	(11.44%, 4.43%)	(12.50%, 4.61%)
(15, 5, 3)	0	(8.0, 3.36)	(8, 3.36)	(5.0, 2.10)	(5, 2.10)	(37.50%, 37.50%)	(37.50%, 37.50%)
	1	(9.6, 4.03)	(14, 5.88)	(6.0, 2.55)	(8, 3.46)	(37.50%, 36.77%)	(42.86%, 41.16%)
	2	(11.1, 4.68)	(17, 7.14)	(8.7, 3.67)	(11, 4.92)	(21.62%, 21.52%)	(35.29%, 31.09%)
	3	(12.6, 5.33)	(20, 8.40)	(10.2, 4.53)	(13, 5.79)	(19.05%, 15.08%)	(35.00%, 31.07%)
	4	(14.3, 6.03)	(23, 9.66)	(12.5, 5.53)	(14, 6.4)	(12.59%, 8.29%)	(39.13%, 34.16%)
	5	(15.8, 6.66)	(25, 10.50)	(14.1, 6.32)	(16, 7.21)	(10.70%, 5.11%)	(36.00%, 31.33%)
(20, 5, 4)	0	(9.0, 3.42)	(9, 3.42)	(5.0, 1.90)	(5, 1.90)	(44.44%, 44.44%)	(44.44%, 44.44%)
	1	(10.8, 4.10)	(15, 5.69)	(7.8, 3.02)	(8, 3.13)	(27.78%, 26.34%)	(46.67%, 44.99%)
	2	(12.3, 4.68)	(18, 6.83)	(8.6, 3.37)	(11, 4.44)	(29.92%, 27.93%)	(38.89%, 34.99%)
	3	(13.9, 5.31)	(21, 7.97)	(11.6, 4.64)	(14, 5.8)	(16.55%, 12.67%)	(33.33%, 27.23%)
	4	(15.1, 5.76)	(24, 9.11)	(13.5, 5.52)	(16, 6.52)	(10.60%, 4.17%)	(33.33%, 28.43%)
	5	(18.0, 6.85)	(25, 9.49)	(14.95, 6.11)	(18, 7.31)	(16.94%, 10.80%)	(28.00%, 22.97%)

defective TSVs, $\eta(m)$ be the number of sessions for identifying m defective TSVs and $P(m)$ be the probability of m defective TSVs within a n-TSV network, then the expectation of test time $(E(T))$ and number of test sessions $(E(S))$ are obtained using the equations of literature in [12] as follows.

$$E(T) = \begin{cases} \sum_{m<2} \gamma(m)P(m) + T_t \sum_{m \geq 2} P(m) & \text{if } r = 1 \\ \sum_{m \leq 2} \gamma(m)P(m) + T_t \sum_{m \geq 3} P(m) & \text{if } r \geq 2 \end{cases} \quad (5)$$

$$E(S) = \begin{cases} \sum_{m<2} \eta(m)P(m) + T_t \sum_{m \geq 2} P(m) & \text{if } r = 1 \\ \sum_{m \leq 2} \eta(m)P(m) + T_t \sum_{m \geq 3} P(m) & \text{if } r \geq 2 \end{cases} \quad (6)$$

Where T_t and T_s represents the total test time and tested sessions respectively to find $m=r+1$ defective TSVs.

Table 2 shows the expectation of number of tested sessions and tested time for both SOS3 in [12] and the proposed method based for various TSV networks. The result is shown by varying TSV yield from 98.0% to a practically expected value 99.5% with defect clustering coefficient =1 [12]. The first column of Table 2 indicates the parameter network size, redundant TSVs in the network and resolution constraint. The second column indicates the parameters like total number of tested sessions and percentage reduction of tested session with TSV yield 99.5% and 98.0% respectively. The third column is used to represent the same parameters with respect to expectation of test time (10^{-7}s). From the Table 2 the following observation can be made: a) the expectation of tested sessions and expectation of tested time decrease for higher TSV yield. This is expected because the probability of having larger number of defective TSVs decreases as TSV yield increases.

Table 2. Comparative Study of Expectation of Test Sessions and Test Times $(10^{-7} s)$ Constructed by [12] and Heuristic Algorithm

(Number of TSVs, Number of redundant TSVs, Resolution constraint)	Expected number of tested sessions Defect clustering coefficient = 1		Expectation of test time $(10^{-7}sec)$ Defect clustering coefficient = 1	
	TSV yield= 99.5% (Sessions for SOS3[12], Sessions for proposed method, reduction over SOS3)	TSV yield= 98.0% (Sessions for SOS3[12], Sessions for proposed method , reduction over SOS3)	TSV yield= 99.5% (Test time for SOS3[12], Test time for proposed method , reduction over SOS3)	TSV yield= 98.0% (Test time for SOS3[12], Test time for proposed method , reduction over SOS3)
(8, 1, 2)	(4.0, 3.9, 2.5%)	(4.3, 3.6, 16.3%)	(21.5, 21, 2.3%)	(22.8, 19.4, 14.9%)
(8, 2, 3)	(4.0, 3.1, 22.5%)	(4.2, 3.4, 19.0%)	(16.9, 13.2, 21.9%)	(17.7, 14.7, 16.9%)
(11, 1, 2)	(6.0, 5.8, 3.3%)	(6.4, 5.3, 17.2%)	(32.1, 31, 3.4%)	(34.1, 27.9, 18.2%)
(11, 2, 3)	(4.1, 4.1, 0.0%)	(4.6, 4.4, 4.3%)	17.3, 17.3, 0.0%)	(19.3, 18.7, 3.1%)
(15, 2, 2)	(8.1, 8.0, 1.2%)	(8.7, 8.0, 8.0%)	(43.0, 42.7, 0.7%)	(46.5, 42.4, 8.8%)
(15, 3, 3)	(6.1, 5.1, 16.4%)	(6.9, 5.8, 15.9%)	(25.8, 21.8, 15.5%)	(29.0, 24.4, 15.9%)
(16, 3, 4)	(5.2, 4.2, 19.2%)	(6.1, 4.9, 19.7%)	(19.8, 16.1, 18.7%)	(23.2, 19.0, 18.1%)
(16, 4, 4)	(4.3, 4.2, 2.3%)	(5.4, 5.2, 3.7%)	(16.3, 16.2, 0.6%)	(20.8, 20.3, 2.4%)
(20, 3, 4)	(6.3, 5.2, 17.5%)	(7.5, 5.9, 21.3%)	(23.9, 19.9, 16.7%)	(28.7, 22.6, 21.3%)
(20, 4, 4)	(6.3, 5.2, 17.5%)	(7.8, 6.3, 19.2%)	(24.0, 20.0, 16.7%)	(29.7, 24.1, 18.9%)

b) The obtained results in the proposed method are better than SOS3 [12] in all cases. Our proposed method has an average reduction in expectation of test session and expectation of test time by 10% compared to [12].

VII. CONCLUSION

In this paper, we have proposed a faster heuristic model to uniquely identify the defective TSVs. The proposed heuristics algorithm has some advantages. First, as we have seen, the proposed algorithm showed a better result compared to [8, 10, 11, 12] in terms of the percentage of reduction of test time. Besides this, the proposed model was able to identify more than 50% defective TSVs in the reduced test time. Second, the test time reduction remains consistent for various TSV networks. Third, time saving can occur even in worst case scenarios compared to [11]. Fourth, the running time complexity is linear which is better than the prior works.

REFERENCES

[1] W. R. Davis et al., "Demystifying 3DICs: the Pros and Cons of Going Vertical", IEEE Design and Test of Computers, vol. 22, no. 6, pp. 498-510, 2005.

[2] H. Lee and K. Chakrabarty, "Test Challenges for 3D Integrated Circuits", IEEE Design and Test of Computers, vol. 26, no. 6, pp. 26 - 35, 2009.

[3] H. Chen, J. Shih, S. W. Li, H.C. Lin, M. Wang, and C. Peng.,"Electrical tests for Three-Dimensional ICs (3DICs) with

TSVs", in International Test Conference 3D-Test Workshop, pp. 1-6, 2010.

[4] B. Noia and K. Chakrabarty, "Pre-bond Probing of TSVs in 3D Stacked ICs", in Proc. IEEE International Test Conference, pp. 1 - 10, 2011.

[5] P. Chen, C. Wu. and D. Kwai, "On-chip Testing of Blind and Open-sleeve TSVs for 3D IC Before Bonding", in IEEE 28th VLSI Test Symposium (VTS), pp. 263–268, 2010.

[6] L.-R. Huang, S.-Y. Huang and K.-H. Tsai, "Oscillation-Based Pre-Bond TSV Test", in IEEE Transactions on Computer-Aided Design, vol. 32, no. 9, pp.140–144, 2013.

[7] B. Noia and K. Chakrabarty, "Identification of Defective TSVs in Pre-Bond Testing of 3D ICs", in Proc. IEEE Asian Test Symposium (ATS), pp. 187 - 194, 2011.

[8] B. Zhang and V. D. Agrawal, "Diagnostic Tests for Pre-Bond TSV Defects", in Proc. of IEEE International Conference on VLSI Design, pp. 387–392,2014.

[9] S. K. Roy, S. Chatterjee and C. Giri, "Identifying Faulty TSVs in 3D Stacked IC During Pre-bond Testing", in Proc. IEEE International Symposium on Electronic System Design (ISED), pp. 162 - 166, 2012.

[10] S. K. Roy, S. Chatterjee, C. Giri, and H. Rahaman, "Faulty TSVs Identification and Recovery in 3D Stacked ICs During Pre-bond Testing", in Proc. of International 3D Systems Integration Conference (3DIC), pp. 1-6, 2013.

[11] B. Zhang and V. D. Agrawal, "An Optimal Probing Method of Pre-Bond TSV Fault Identification for 3D Stacked ICs", In IEEE SOI-3D-Subthreshold Microelectronics Technology Unified Conference (S3S), 2014.

[12] B. Zhang and V. D. Agrawal, "An Optimized Diagnostic Procedure for Pre-bond TSV Defects", in Proc. of 32nd IEEE International Conference on Computer Design (ICCD), 2014.

978-1-5386-3693-0/18 $31.00 © 2018 IEEE

2018 31th International Conference on VLSI Design and 2018 17th International Conference on Embedded Systems

Modeling & Analysis of Redundancy based Fault Tolerance for Permanent Faults in Chip Multiprocessor Cache

Avishek Choudhury
Department of Computer Science
New Alipore College
Kolkata, India
avishek.nac.cs@gmail.com

Biplab K Sikdar
Department of Computer Science & Technology
Indian Institute of Engineering, Science & Technology
Shibpur, India
Email: biplab@cs.iiests.ac.in

Abstract—With increased number of cores in Multicore Chips, power consumption raises. Voltage scaling is applied largely but it causes cell failure in cache. For that, various techniques for fault tolerance in cache have been introduced. Among these techniques, the redundancy based technique is the most effective one but no analysis has yet been done to assess its performance. This work attempts to propose and evaluate the parameters like Expected Miss Ratio for Multicore (EMR_{MC}) and Expected Latency in Multicore (EL_{MC}), as the function of Probability of Cell Failure (P_{fail}) for redundancy based technique. It is found that both (EMR_{MC}) and (EL_{MC}) decrease with increased number of cores. The lower bound of P_{fail} i.e. 1e-4, up to which both (EMR_{MC}) and (EL_{MC}) in all architectures remain same. Above that both (EMR_{MC}) and (EL_{MC}) increases except for octacore. The (EMR_{MC}) for octacore starts increasing above P_{fail} of 5E-4 and converges with others for high P_{fail} like 0.005. On the other hand, the (EL_{MC}) for octacore is hardly affected up to high P_{fail} like 0.001 and above that though it increases but remains lower than others. Performance gain of up to 34.63% for (EMR_{MC}) and 45.67% for (EL_{MC}) is observed with increased number of cores.

Keywords-Chip multiprocessors; Fault tolerance; Redundancy; Permanent fault; Multicore;

I. INTRODUCTION

Processing demand influenced introduction of Chip Multiprocessors (CMPs). However, the power consumption increased exponentially [1]. The only remedy was Dynamic Voltage and Frequency Scaling (DVFS). But unfortunately, aggressive voltage scaling causes process variation induced failures in the SRAM cells [2] as shown in Fig.1.

To achieve error free processing despite of faults, fault tolerance design became necessity. The cache being most susceptible to failure [3], various techniques for fault tolerance in cache had been introduced. But, these techniques came with significant overhead and the analysis of their impacts became necessary for fitness comparison.

Among these techniques, redundancy based technique is the most effective one that ensures lowest miss rates in single core. But its impact in multicore is never been analysed. This influenced to analyse the performance of redundancy based

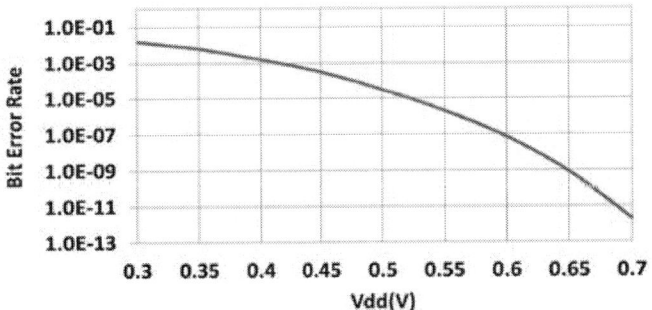

Figure 1.   Probability of SRAM cell bit failure rate in 90nm [2]

fault tolerance in Multicore chips with distributed Last Level Cache (LLC).

In this work, the parameters like Expected Miss Ratio for Multicore (EMR_{MC}) and Expected Latency in Multicore (EL_{MC}), the function of Probability of Cell Failure (P_{fail}), are proposed and evaluated for redundancy based techniques. It is found that both the EMR and EL decrease with increased number of cores. The lower bound of P_{fail} i.e. 1e-4, upto which both the EMR and EL in all architectures remain same. Above that both EMR and EL start increasing except for Octacore. EMR for Octacore starts increasing above P_{fail} of 5E-4 and converges to other EMRs for high P_{fail} like 0.005. EL for Octacore is hardly affected upto high P_{fail} like 0.001 and above that though it increases but remains lower than others. Performance gain of upto 34.63% for EMR and 45.67% for EL is observed with increased number of cores. All the simulations are carried out in the Multi2Sim simulation framework with benchmark programs included in the SPLASH-2 and PARSEC 3.0 benchmarks suites.

II. RELATED WORK

Fault tolerance techniques have been introduced in different level. At circuit level, variations on the traditional 6T (6 transistors) SRAM cell design or designs including 8T, 10T, 11T[4] cells as well as a Schmidt trigger based 10T

978-1-5386-3693-0/18 $31.00 © 2018 IEEE

115

SRAM cell have been proposed. However, these techniques cause significant area overheads. At the system level, Error Correcting Codes (ECC) [5] is used. However, multiple bit ECC needed for large number of faults. But they suffer from storage overheads, long decoding times and complex decoders. ECC are best suitable for handling soft errors. The disabling of faulty cache parts (ways, sets, blocks or sub-blocks) has been studied in several works[6, 7, 8]. But they suffer from increased number of additional misses as the number of the faulty cache parts increases (graceful degradation). Redundancy based techniques is another class of techniques where reduction of the misses caused by the disabled faulty parts of the cache is achieved by using re-dundant word-lines in the cache [9], a small fully associative spare cache [10] or the use of the victim cache [11].

III. BASELINE ARCHITECTURE

This work deals with the Tiled-CMP architecture that gives a scalable solution for Chip Multiprocessor designs. A tile-based CMP architecture is comprised of a number of identical tiles connected via a switched direct network as shown in Fig. 2. Each tile contains a processing core with L1 caches (both Instruction and Data caches), a shared L2 cache, distributed on-chip directory and a connection to the on-chip network. The L2 cache is shared among the different processing cores by physical distribution among them.

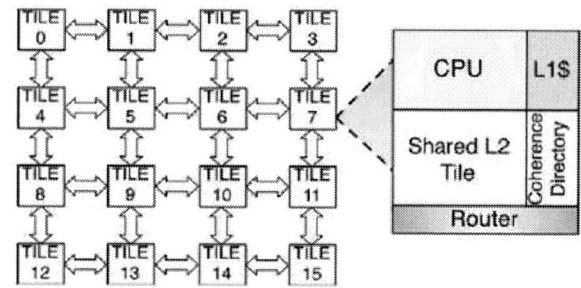

Figure 2.   Tiled CMP architecture

IV. THE ANALYTICAL MODEL

This model for uncorrelated faults in Chip Multiproces-sors (CMPs) determines different parameters that describe the behavior of the cache in presence of uncorrelated faults with permanent cell failure probability P_{fail}. The mathe-matical formulations of the parameters are introduced in the following subsections.

A. Probability of block failure and fault remapping

If there are C cores, N banks, S sets / bank, B blocks/ set, k bits / block and the Probability of cell failure is P_{fail}, then the probability of block failure is:

$$P_{bf} = [1 - (1 - P_{fail})^k] \qquad (1)$$

The Probability Distribution of i block failures consider-ing all possible case is:

$$P_{bf}(i) = \binom{B}{i} P_{bf}^i (1 - P_{bf})^{(B-i)} \qquad (2)$$

Similarly, probability distribution of j blocks remapped to non-conflicting faulty blocks among i faulty block is:

$$P_{fr}(j) = \binom{i}{j} P_{fr}^j (1 - P_{fr})^{(i-j)} \qquad (3)$$

B. Expected miss ratio in multicore (EMR_{MC})

For assessing the impact of redundancy in multicore, we have incrementally disabled blocks in all cores simul-taneously and tried to map them in non-conflicting faulty block and then analysed the impact. The Expected Miss in Multicore can be defined, using Eq. 2 and 3 as

$$
\begin{aligned}
EM_{MC} &= \sum_{i=0}^{B} \sum_{j=0}^{i} E_{Miss}^{ij} \\
&= \sum_{i=0}^{B} \sum_{j=0}^{i} M_{ij} \cdot P_{bf}(i) \cdot P_{fr}(j)
\end{aligned}
\qquad (4)
$$

The EM_{MC} is the expected miss of the whole cache and E_{Miss}^{ij} is the number of miss for i disabled and j remapped blocks in all cores. The M_i is the number of miss for j remapped blocks among i faulty blocks in all cores. M_{ij} can further be defined as

$$
M_{ij} = \begin{cases} \sum_{x=0}^{C-1} m_x & \text{if } i = 0 \\ \sum_{x=0}^{C-1} \sum_{k=j}^{i-1} A_{xk} + \sum_{x=0}^{C-1} m_x & \text{if } i \neq 0 \end{cases}
\qquad (5)
$$

where m_x is the miss for all healthy way in j^{th} core, A_{xk} is the number of access to k^{th} way of xth core. Finally, the EMR is calculated as

$$EMR_{MC} = EM_{MC}/A \qquad (6)$$

where A is the number of access to all cores.

C. Expected Latency in Multicore (EL_{MC})

The expected latency in multicore architecture can be expressed as:

$$
\begin{aligned}
EL_{MC} &= \sum_{i=0}^{B} \sum_{j=0}^{i} E_{Lat}^{ij} \\
&= \sum_{i=0}^{B} \sum_{j=0}^{i} [P_{bf}(i)[P_{fr}(j) \cdot 2L_{L2} + (1 - P_{fr}(j)) \\
&\quad \cdot (L_{L2} + L_{MM})] + (1 - P_{bf}(i)) \cdot 2L_{L2}]
\end{aligned}
\qquad (7)
$$

where EL_{MC} is the expected latency of the whole cache, E_{Lat}^{ij} is the latency for i disabled and j remapped blocks in all cores, L_{L2} is the L2 cache access latency and L_{MM}

978-1-5386-3693-0/18 $31.00 © 2018 IEEE

is the main memory access latency. Auxiliary L2 access latency considered for both hit and miss due to the fault map and victim map accesses.

V. METHODOLOGY

For finding the parameters in Multicore, we have extended the methodology used in [12] for analysing disabling technique in single cache bank to the Multicore distributed cache bank environment.

A. EMR calculation

The expected miss in multicore is calculated as shown in Table I. The first two columns show the number of cores and sets respectively. Third column shows, the number of disabled ways (cumulatively). The fourth column gives the remapped ways (cumulatively) and fifth column describes how the miss is calculated.

Table I
CUMULATIVE MISS MATRIX

Cores	Sets	Disabled ways	Remapped ways	Miss
		way 0	-	$\sum_{i=0}^{c-1}\sum_{j=0}^{s-1}\sum_{k=0}^{0} A_{ijk} + \sum_{i=0}^{c-1} M_i$
			way 0	$\sum_{i=0}^{c-1} M_i$
		way 0,1	-	$\sum_{i=0}^{c-1}\sum_{j=0}^{s-1}\sum_{k=0}^{1} A_{ijk} + \sum_{i=0}^{c-1} M_i$
			way 0	$\sum_{i=0}^{c-1}\sum_{j=0}^{s-1}\sum_{k=0}^{0} A_{ijk} + \sum_{i=0}^{c-1} M_i$
			way 0,1	$\sum_{i=0}^{c-1} M_i$
Core 0 to c-1	Set 0 to s-1	..	..	..
			-	$\sum_{i=0}^{c-1}\sum_{j=0}^{s-1}\sum_{k=0}^{w-1} A_{ijk} + \sum_{i=0}^{c-1} M_i$
		way 0		$\sum_{i=0}^{c-1}\sum_{j=0}^{s-1}\sum_{k=0}^{w-2} A_{ijk} + \sum_{i=0}^{c-1} M_i$
		..	..	
			way 0,1..w	$\sum_{i=0}^{c-1} M_i$

The misses are multiplied with corresponding probability of bock failure and failure remapping respectively from Eq. 2 and 3 and added to get the overall expected miss using Eq. 4. It is divided by the number of accesses for whole system (A) to get the EMR for whole system (Eq. 6).

B. EL calculation

The expected latency is calculated following Eq. 7. The block accesses are available as in Table I. If a block is remapped, then twice the L2 latency is multiplied with number of accesses and added to the whole latency, else the main memory access latency plus L2 access latency is considered.

VI. EVALUATION

The evaluation is done using the Multi2Sim 5.0 simulation framework with benchmark programs included in the SPLASH-2 and PARSEC 3.0 benchmark suites. Simulation results for fft program is given here for lookup.

A. System configuration

The Multi2sim has been configured with 1,2,4 and 8 cores of x86 model of frequency 1k. Each of these cores is associated with a dedicated L1 cache and a shared L2 cache. Both L1 and L2 cache have 8 sets and 16 ways having block size 512B. Data latency is kept as 2 and 20 units for L1 and L2 repectively. The L2 cache is shared among the cores and connected with L1 caches. Both L1 and L2 uses LRU replacement technique and cache writeback policy with cache coherence protocol used is NMOESI. The main memory is also having blocks of same size having data access latency is of 200 units. There are two levels of networks. The high level network connects the L1 modules to lower level L2 modules and the low level network connects the L2 modules to main memory.

B. Results

The evaluation is done separately by varying the number of cores and also varying the P_{fail} for a set of cores. As we assume 16 ways in L2, we need to consider 17 possible cases starting from 0 faulty ways among 16 to 16 faulty ways among 16 ways.

With a certain P_{fail}, say 0.000001, where P_{bf} becomes 0.0005118692, the probability distribution resulted in is shown in Table II.

Table II
PROBABILITY DISTRIBUTION OF BLOCK FAILURE

i	$\binom{B}{i}$	P_{bf}^{i}	$1 - P_{bf}^{(B-i)}$	$PD_{bf}(i)$
0	1	1	0.9918	0.9918
1	16	0.0005	0.9923	0.0081
:	:	:	:	:
16	1	2E-53	1	2E-53

The first column of Table II represents the number of blocks failed due to permanent fault. The second column represents the number of ways the 16 blocks can become faulty. The third column shows the probability of i Blocks getting faulty. The fourth column gives the probability of other blocks remain healthy. The last column shows the probability distribution of i block failure using Binomial distribution.

The probability of each number of faulty ways is again distributed to number of remapped ways. Table III shows the probability distribution of fault remapping of the probability of four ways becoming faulty i.e. 1E-10.

978-1-5386-3693-0/18 $31.00 © 2018 IEEE

Table III
PROBABILITY DISTRIBUTION OF FAULT REMAPPING

i	$\binom{i}{j}$	P_{fr}^j	$1-P_{fr}^{(i-j)}$	$PD_{fr}(j)$ (A)	$PD_{bf}(i)$ (B)	A X B
0	1	1	0.998	0.998	1.24E-10	1.24E-10
1	4	0.0005	0.9985	0.002	1.24E-10	2.54E-13
2	6	3E-07	0.999	2E-06	1.24E-10	2E-16
3	4	1E-50	0.9995	5E-10	1.24E-10	6.66E-20
4	1	7E-14	1	7E-14	1.24E-10	8.52E-24

In Table III, 1st to 5th column contains data similar to Table 2 except remapping distribution instead of fault distribution. The sixth column contains the probability of 4 ways becoming faulty. This probability is distributed in the last column according to the distribution of fault remapping in the 5th column.

1) Calculation of Expected Miss and EMR: -

Given the probability distribution of faulty blocks in Table II, fault remapping distribution in Table III and number of misses using Eq. 4, the expected miss is calculated for all cores. The calculation of expected miss is shown for octacore in Table IV. Same process is used for other cores also.

Table IV
CALCULATION OF EXPECTED MISS IN OCTACORE

Faulty ways	Remapped ways	P_{bf}	P_{fr}	Access + Miss	E_{miss}^i
0	0	0.9918	1	544	539.5618
1	0	0.0081	0.9949	544	4.398645
1	1	0.0081	0.0051	544	0.022579
:	:	:	:	:	:
16	0	2E-53	0.9918	1383	3E-50
:	:	:	:	:	:
16	16	2E-53	2E-53	544	3E-103

The first column of Table IV contains the number of faulty blocks say i. The second column contains how many of the faulty ways have been remapped. The third and fourth column contains the probability of i block failures and remapped respectively. The fifth and sixth column contains the actual miss and expected miss respectively. P_{fail} is taken as 0.000001.

- Using Eq. 3 and Eq. 4, we get the overall expected miss i.e. 544 in this case and the expected miss rate for all cores as 0.3936

2) Calculation of expected latency: -

Given the probability distribution of faulty blocks in Table II, fault remapping distribution in Table III and number of misses using Eq. 4, the expected latency is calculated for all cores using Eq. 7. The calculation of expected latency is shown for octacore in Table V. Same process is used for other cores also.

Table V
CALCULATION OF EXPECTED LATENCY IN OCTACORE

Faulty ways	Remapped ways	P_{bf}	P_{fr}	Access + Miss	E_{lat}^{ij}
0	0	0.9918	1	544	151950.11
1	0	0.0081	0.9949	544	1238.74
1	1	0.0081	0.0051	544	6.36
:	:	:	:	:	:
16	0	2E-53	0.9918	1383	6.70E-48
:	:	:	:	:	:
16	16	2E-53	2E-53	544	7.56E-101

The first column of Table V contains the number of faulty blocks say i. The second column contains how many of the faulty ways have been remapped. The third and fourth column contains the probability of i block failures and remapped respectively. The fifth and sixth column contains the actual miss and expected latency respectively. P_{fail} is taken as 0.000001.

- Using Eq. 3 and Eq. 4, we get the overall expected latency i.e. 153200 in this case.

C. Analysis

Analysis of Expected Miss Ratio and Expected Latency is done on the results obtained by simulation in the following:

1) Analysis of EMR: Expected Miss Ratios are obtained by varying P_{fail} in the following Table VI.

Table VI
COMPARISON OF EMR

P_{fail}	$EMR_{c=1}$	$EMR_{c=2}$	$EMR_{c=4}$	$EMR_{c=8}$	% gain for c=1 over c=8
0.0000001	0.4565	0.4570	0.4092	0.3936	13.778
0.0000005	0.4566	0.4571	0.4093	0.3936	13.797
:	:	:	:	:	:
0.001	0.7204	0.7189	0.6776	0.4709	34.633
0.005	1	1	1	0.9870	01.297

Table VI shows the comparison of Expected Miss Ratio in various cores. The first column contains the Probability of Call failure. The second to fifth column contains Expected Miss Ratio for Singlecore, Dual core, Quadcore and Octacore respectively. The last column contains the gain in Expected Miss Ratio in octacore over singlecore.

From the data in Table VI, we get the graph of EMR for probability of failures in Fig. 4.

978-1-5386-3693-0/18 $31.00 © 2018 IEEE

EXPECTED MISS RATIO IN VARIOUS CORES FOR DIFFERENT Pfail

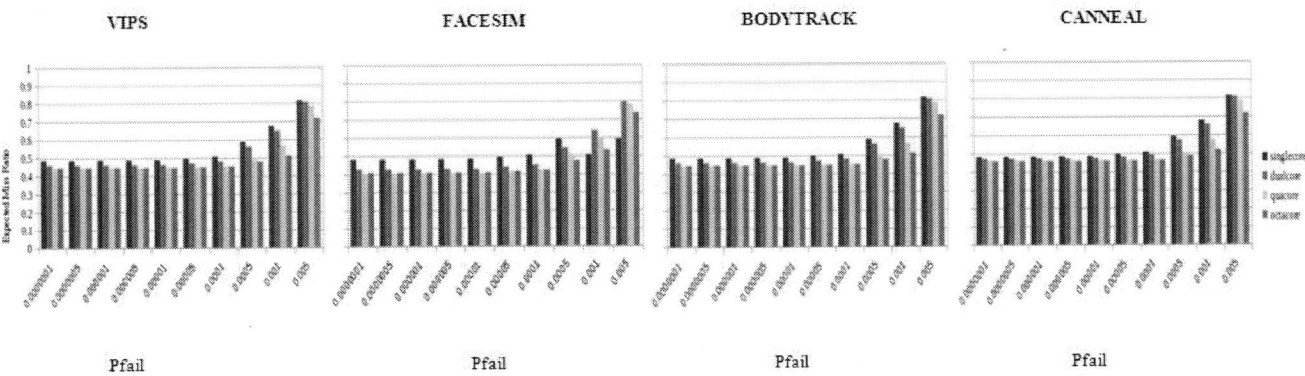

Figure 3.   Comparison of Expected Miss Ratio in different cores for different benchmarks

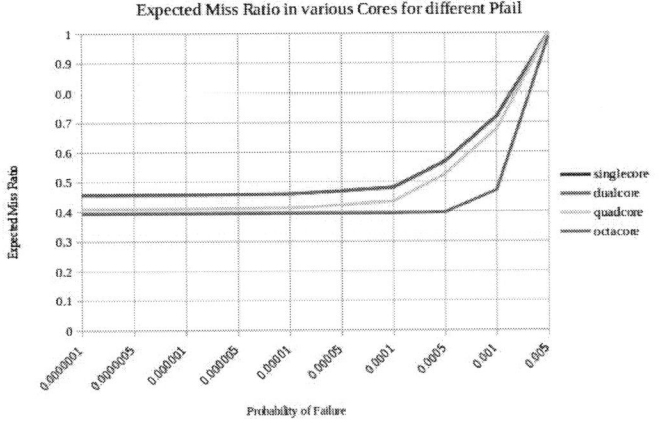

Figure 4.   EMR in different Cores

Fig. 4 shows the comparison between Expected Miss Ratios in various cores. The x-axis contains Probability of cell failure and the y-axis contains the Expected Miss Ratio.

• From Fig. 4, it is observed that the EMR decreases with increased number of cores. Significant benefit comes if number of cores is at least four.

• The lower bound of P_{fail} i.e. 1e-4, upto which the EMR in all cases remain same. Above that bound the EMR starts increasing for all except octacore.

• EMR for octacore starts increasing above P_{fail} of 0.0005 and converges with other EMRs for high P_{fail} as 0.005.

Fig. 3 shows the Expected Miss Ratio for different cores for different benchmark programs of PARSEC 3.0 benchmark suite. The x-axis contains different probability of cell failures and the y-axis contains the Expected Miss Ratio. It can be stated that the EMR in multicore redundancy stays

lower than the EMR in singlecore redundancy.

Table VII
COMPARISON OF EXPECTED LATENCY

P_{fail}	$EL_{c=1}$	$EL_{c=2}$	$EL_{c=4}$	$EL_{c=8}$	% gain for c=8 over c=1
0.0000001	226148.36	218587.77	162766.45	153200	32.26
0.0000005	226181.80	218618.87	162792.27	153200	32.27
:	:	:	:	:	:
0.001	292198.32	281788.17	216731.48	158728.10	45.68
0.005	340244.52	329587.79	261258.68	230284.90	32.32

2) Analysis of Expected Latency: Table VII shows the comparison of Expected Latency in various cores. The first column contains the Probability of cell failure. The second, third, fourth and fifth column contains Expected Latency for different cores. The last column contains the gain in Expected Latency in octacore over singlecore.

From the data in Table VII, we get the graph of Expected Latency for Probability of cell failures in Fig. 6.

• From Fig. 6, it is observed that the EL decreases with increased number of cores. Significant benefit comes if number of cores is at least four.

• The lower bound of P_{fail} i.e. 1e-4, upto which the EL in all cases remain same. Above that bound the EL starts increasing for all except octacore.

• EL for octacore is hardly affected upto high P_{fail} like 0.001 and though it increases but remains lower than others.

• In Fig. 6, it is also observed that in octacore upto 45.67% gain in Latency is achieved over singlecore.

Fig. 5 shows the Expected Latency for different cores for different benchmark programs of PARSEC 3.0 benchmark suite. The x-axis contains different probability of cell failures and the y-axis contains the Expected Latency. It can be stated that the EL decreases with increased number of cores.

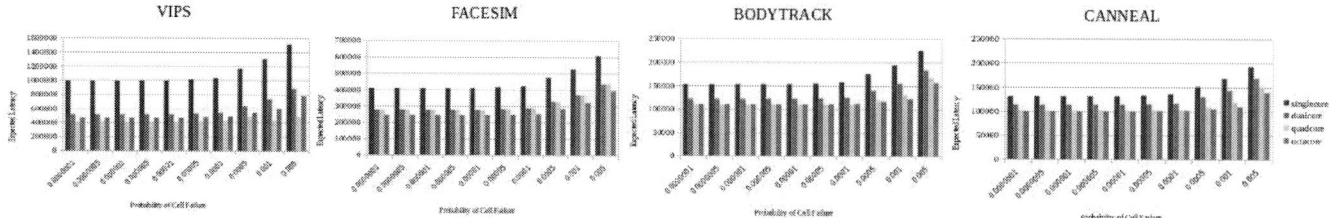

Figure 5. Comparison of Expected Latency in different cores for different benchmarks

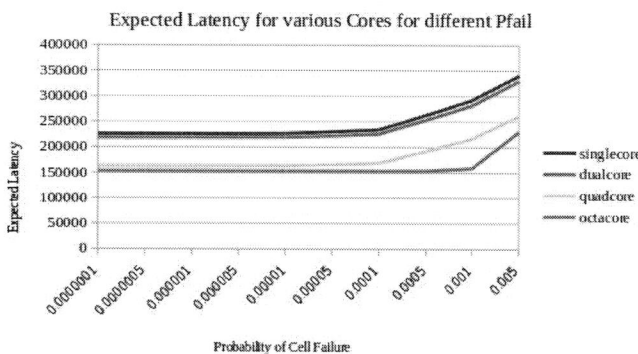

Figure 6. Expected Latency in different Cores

VII. CONCLUSION

This work evaluates the performance of redundancy based fault tolerance in multicore chips with distributed cache banks and compares the performances by varying the number of cores and probability of cell failure. For the performance evaluation, parameters like EMR_{MC} and Expected Latency in multicore (EL_{MC}) which are the functions of probability of cell failure P_{fail}, are proposed and evaluated for different values of P_{fail}.

It pairs to the fact that both EMR and EL decreases with increased number of cores for at least four. The lower bound of P_{fail} i.e. 1e-4, upto which both EMR and EL in all architectures remain same. Above that both EMR and EL starts increasing except for octacore. EMR for Octacore starts increasing above P_{fail} of 5E-4 and converges with other EMRs for high P_{fail} like 0.005. EL for Octacore is hardly affected upto high P_{fail} like 0.001 and above that though it increases but remains lower than others. Performance gain of upto 34.63% for EMR and 45.67% for EL is observed.

REFERENCES

[1] T. Skotnicki, J. Hutchby, T.-J. King, H.-S. Wong, and F. Boeuf, *The end of cmos scaling: toward the introduction of new materials and structural changes to improve mosfet performance*, Circuits and Devices Magazine, IEEE, vol. 21, no. 1, pp. 16 26, jan.-feb. 2005.

[2] A. Banaiyanmofrad, H. Homayoun, N Dutt, *Using a Flexible Fault-Tolerant Cache to Improve Reliability for Ultra Low Voltage Operation*, ACM Transactions on Embedded Computing Systems, Vol. 14, No. 2, Article 32, Publication date: February 2015

[3] H. R. Ghasemi, et al., *Low-voltage on-chip cache architecture using heterogeneous cell sizes for multi-core processors*, IEEE International Symposium on High-Performance Computer Architecture, pages 3849, February 2011

[4] F. Moradi, D. Wisland, S. Aunet, H. Mahmoodi, and T. Cao, *65nmsub-threshold 11t-sram for ultra low voltage applications*, Intl. Symposium on System-on-a-Chip, pages 113118, Sept. 2008.

[5] S. Lin and D. J. Costello, *Error Control Coding (2nd Edition)*, Prentice Hall, 2004.

[6] S. Ozdemir et al., *Yield-Aware Cache Architectures*, Proc. Of Intl. Symposium on Microarchitecture, 2006.

[7] F. Pour and M. D. Hill, *Performance implications of tolerating cache faults*, Trans. on Computers, 1993.

[8] G. Sohi, *Cache Memory Organization to Enhance the Yield of High-Performance VLSI Processors*, Trans. on Computers, 1989.

[9] A. Ansari et al., *Enabling ultra low voltage system operation by tolerating on-chip cache failures*, Proc. of Intl. Symposium on Low power electronics and design, 2009.

[10] H.T. Vergos and D. Nikolos., *Performance Recovery in Direct- Mapped Faulty Caches via the Use of a Very Small Fully Associative Spare Cache*, Proc. of Intl. Computer Performance and Dependability Symposium, 1995.

[11] N. Ladas, Y. Sazeides, and V. Desmet, *Performance-Effective Operation below Vcc-min*, Proc of Intl Symposium on Performance Analysis of Systems & Software, 2010.

[12] D. Sanchez, Y. Sazeides, J. Cebrian, J. M. Garcia, J. L. Arago N, *Modeling the Impact of Permanent Faults in Caches*, ACM Transactions on Architecture and Code Optimization, Vol. 10, No. 4, Article 29, Publication date: December 2013.

978-1-5386-3693-0/18 $31.00 © 2018 IEEE

Exact Synthesis of Biomolecular Protocols for Multiple Sample Pathways on Digital Microfluidic Biochips

Oliver Keszocze*[†], Mohamed Ibrahim[‡], Robert Wille[†§], Krishnendu Chakrabarty[‡], Rolf Drechsler*[†]

*Institute of Computer Science, University of Bremen, Bremen, Germany
[†]Cyber-Physical Systems, DFKI GmbH, Bremen, Germany,
[‡]Department of Electrical and Computer Engineering, Duke University, Durham, NC
[§]Institute for Integrated Circuits, Johannes Kepler University, Linz, Austria

{keszocze,drechsler}@informatik.uni-bremen.de {mohamed.s.ibrahim, krish}@duke.edu robert.wille@jku.at

Abstract—*Digital-microfluidic biochips* (DMFBs) allow laboratory steps associated with genomic bioassays to be carried out in an automated and highly efficient fashion. However, to support complex (multi-sample) biomolecular protocols for quantitative analysis, conventional design-automation methods are ineffective, thereby motivating the need for more suitable solutions. While first methods to address this problem have recently been introduced, they neglect architectural details and technology-related constraints, and use heuristics to oversimplify the search for a design solution. These simplifications result in designs that may either not be realizable on a DMFB or it is not possible to evaluate how far they are from being optimal. In this work, we propose an exact synthesis method that does not rely on simplistic models and heuristics, but solves the underlying design problem in an exact fashion. The proposed method allows for the determination of exact and realistic results for the realization of quantitative-analysis biomolecular protocols on DMFBs.

Fig. 1: Illumina's ring-based NeoPrep solution [12].

I. INTRODUCTION

Digital-microfluidic biochips (DMFBs) have emerged as an enabling lab-on-a-chip technology for contemporary biomolecular research [1]. In contrast to conventional benchtop procedures, DMFBs can precisely control a large number of droplets and biomolecular assay parameters, e.g., protein or messenger RNA expression levels, in a high-throughput manner [2]. Using digital-microfluidics technology, cells are encapsulated in picoliter droplets, electroporated, mixed with reagents such as reverse-transcription components, and heated for biological-signal amplification using a grid of electrodes that are driven by a sequence of actuation voltages from a computer [3]. As a result, genomic bioassays such as nucleic-acid isolation, DNA purification, and DNA amplification can be conducted on a DMFB [4], [5].

In practice, many DMFBs are partitioned into dedicated reaction chambers that are capable of performing different fluid-handling operations. This architecture allows for a modular configuration that facilitates the design of plug-and-play¡ devices, improves fault tolerance, and increases system throughput by scaling up the number of reactions while avoiding reaction interference among different samples [6], [7]. The chambers themselves are then connected by a corresponding topology that is tailored for the target application, whereas the ring-bus architecture is commonly used for infection disease tests or sample preparation applications. For example, both GenMark's recent ePlex system for diagnostics as well as Illumina's NeoPrep solution employ this topology. Fig. 1 shows a commercialized chip by Illumina, where the reaction chambers and the ring are highlighted by red rectangles and a line, respectively.

Today's microbiology applications (e.g., quantitative gene-expression analysis) may include tens of independent sample pathways running through a sequence of procedural decision points (see e.g., [8]). Moreover, multiple reaction pathways, which enable high-resolution reaction-time control, are necessary for producing synthesized organic compounds. Such a laboratory framework, known as flash chemistry [9], [10], requires accelerated reactions and it is extremely valuable for the pharmaceutical industry. Even a small reduction in reaction time obtained using an exact method is significant enabler for flash chemistry Therefore, design automation for integrated biomolecular analysis and support for multiple sample pathways have emerged as a challenging problem. To address this problem, Ibrahim et al. introduced the first design-automation ("synthesis") methodology to support non-trivial biomolecular analysis [11] based on a ring-bus architecture. The proposed synthesis method conducts greedy resource allocation and dynamic mapping of multiple sample pathways to biochip resources.

However, despite the novelty of [11], there are several drawbacks and shortcomings that limit its applicability:

- The method relies on an abstract view of the underlying DMFB, which ignores important architectural details. Because of this, the obtained synthesis results are often infeasible for the given DMFB architecture.
- The method ignores realistic technology-related constraints, such as the biochip's lifetime or the degradation level of the electrodes. As a result. the synthesized designs are likely to lead to faults on the DMFB and, consequently, incorrect outcomes.
- The method relies on heuristics that oversimplify the search for a design solution. Accordingly, it is not possible to evaluate how far the results are from being optimal. Therefore, the heuristic method does not support fine-grained control of pathway reaction time in flash chemistry.

In this work, we propose a synthesis method that overcomes these problems. Instead of relying on greedy heuristics and simplistic abstractions, an *exact* methodology is provided that is capable of determining optimal results while, at the same time, satisfying architectural constraints. Moreover, the pro-

978-1-5386-3693-0/18 $31.00 © 2018 IEEE

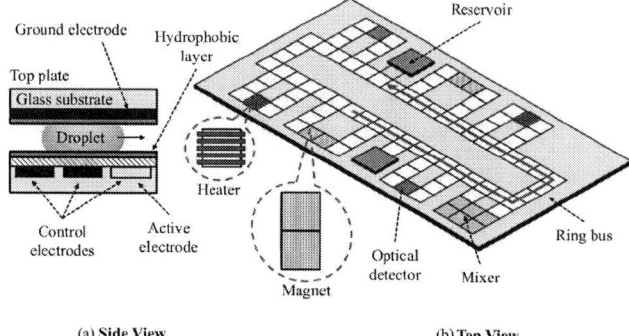

(a) Side View   (b) Top View

Fig. 2: Schematic view of a DMFB: (a) a side view; (b) a ring-based DMFB with 2D array and on-chip resources.

posed solution can easily be enriched by additional constraints that incorporate realistic technology-related constraints.

Experimental evaluations demonstrate the improvements accomplished with this methodology. For the first time, exact results for the realization of quantitative-analysis biomolecular protocols (with multiple sample pathways) on DMFBs are obtained, which allow to evaluate the impact of abstracting the architectural characteristics as well as the greedy schemes from the previous solution. Our results show that the gap between exact and heuristic solutions is significant and it is magnified for a larger number of pathways in realistic applications. In addition, the proposed approach also allows us to evaluate, for the first time, how ignoring technology-related constraints affects the resulting design.

II. BACKGROUND AND MOTIVATION

This section describes the related background on DMFBs as well as synthesis for biomolecular protocols. It also discusses the shortcomings of the state-of-the-art and motivates the contribution of this work.

DMFBs execute biomolecular protocols by manipulating nanoliter droplets of samples and reagents through a set of elementary fluid-handling operations, such as mixing, incubation, dilution, and heating. As shown in Fig. 2(a), a typical design of an DMFBs consists of two parallel plates coated with a hydrophobic layer. The top plate represents a single conducting layer that is always grounded, whereas the bottom plate comprises an array of control electrodes that can be programmed to move sandwiched droplets using electrowetting [1]. Similar to the field-programmable gate arrays, the array-based design of DMFBs embraces their flexibility, since the majority of fluid-handling operations can be performed anywhere on the array.

Early synthesis methods aimed at leveraging the full flexibility of DMFBs by introducing solutions for scheduling of fluid-handling operations, resource binding, and droplet routing [13], [14], [15], [16], [17], [18]. Such solutions attempt to efficiently synthesize any given protocol in a sample-in-result-out manner, while dynamic re-synthesis is only limited to counteract error occurrences observed in a single sample pathway [19]. Although these methods allowed for the exploitation of the inherent flexibility of DMFBs, they are restricted to simple droplet manipulation only. That is, they are not adequate for realistic biomolecular analysis.

In order to address this problem, a new synthesis methodology has recently been introduced in [11]. It tackles the complexity of microbiology protocols (e.g., gene-expression analysis) by cyber-physical adaptation [20], [21] and abstraction. A framework has been proposed that concurrently controls multiple sample pathways and promptly responds to decisions about the protocol flow at each pathway. To provide architectural support, the concept of virtual topologies from [22] has been borrowed and the DMFB has been assumed to have a unidirectional ring-based topology as shown in Fig. 2(b). Here, a circular array of electrodes forms a ring that is connected to dedicated reaction chambers to perform fluid-handling operations[1]. By this, the actual synthesis problem has been simplified to a resource-allocation problem. However, while the methodology of [11] was able to narrow the gap between DMFBs and biomolecular analysis, it suffers from drawbacks that limit its applicability. More precisely:

A. Abstraction from Architectural Details The synthesis method from [11] relies on an abstract view of the DMFB that ignores architectural details. In particular, the routing between the reaction chambers has been abstracted away. As a result, neither the time needed to move droplets from one reaction chamber to the next one nor collisions of droplets in the ring are considered.

B. Realistic Technology Constraints As DMFBs are getting more involved in point-of-care biomolecular research [24], recent studies have been directed to investigate technology constraints (e.g., lifetime or degradation level of array electrodes [25], [26]) that have to be satisfied to guarantee reliable biochip operations. As a result, design automation researchers are burdened with an extra level of complexity to handle these constraints. In [11], the authors tackled this challenge by presenting a degradation-aware resource allocation, wherein the allocation scheme only restricts the usage of an on-chip resource when its degradation level exceeds the safety margin. That is, the degradation constraint is considered as a soft constraint. This, however, does not guarantee that resources with completely degraded (i.e., broken) electrodes are avoided— leading to possible failures in protocol execution.

C. Greedy Results The solution proposed in [11] relies on heuristics and greedy methods in order to efficiently generate a design. Therefore, the obtained results are often far from being optimal. As an example, consider the protocol to be realized as shown in Fig. 3(a). A ring-based DMFB is available which integrates on-chip resources as shown in Fig. 3(b). The greedy method introduced in [11] scans all pathways and allocates resources to bioassays in a sequential manner such that resources with a lower degradation level are given a higher priority during allocation. Note that, due to the greedy nature of the solution, the algorithm enforces the degradation-aware policy by ensuring that the resources previously selected to execute Bioassay I, such as Detector 1, cannot be reserved to execute an operation in Bioassay II. This yields a solution as shown on the top of Fig. 3(c) with a total completion time of 15 time units[2]. In contrast, an optimal solution only requires 14 time units (as shown on the bottom of Fig. 3(c)). Furthermore, the greedy resource allocation causes the magnet to exceed its safety threshold, potentially leading to a resource breakdown during the execution. The optimal solution, on the other hand, reliably allocates resources so that a safety margin is never exceeded.

Because of these shortcomings, the method proposed in [11] often leads to inefficient solutions that violate technology-related constraints and are far from being optimal.

III. PROBLEM FORMULATION AND GENERAL IDEA

The above discussion motivates the following design problem:

Given: A multi-sample biomolecular protocol and an DMFB architecture.

---

[1]This topology is accordingly applied in commercial solutions e.g. provided by Illumina or GenMark [12], [23].

[2]For the sake of clarity, scheduling time and degradation levels are computed in terms of time units (e.g., seconds).

978-1-5386-3693-0/18 $31.00 © 2018 IEEE

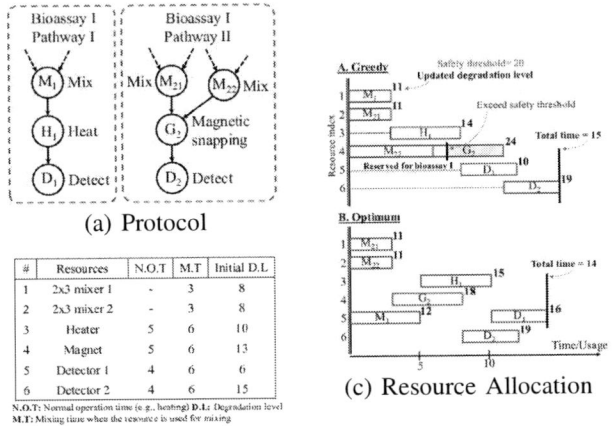

(a) Protocol

#	Resources	N.O.T	M.T	Initial D.L
1	2x3 mixer 1	-	3	8
2	2x3 mixer 2	-	3	8
3	Heater	5	6	10
4	Magnet	5	6	13
5	Detector 1	4	6	6
6	Detector 2	4	6	15

N.O.T: Normal operation time (e.g., heating) D.L: Degradation level
M.T: Mixing time when the resource is used for mixing

(b) Chip Resources

(c) Resource Allocation

Fig. 3: An example showing a comparison between the performance of the greedy method in [11] and the optimal behaviour.

Design Goal: A DMFB design realizing the protocol (i.e., a resource allocation and scheduling).
Constraint: Technology constraints (e.g., electrode degradation).
Objective: Minimize the protocol completion time.

The protocol is specified by fluid-handling operations that are arranged in multiple sequencing graphs (associated with different bioassays), as shown in Fig. 3(a). Note that, although two independent sequencing graphs are shown here, they may interact with each other when they both have to be implemented on a resource-limited DMFB, i.e., they compete for on-chip resources. In addition, each sample pathway comprises a tree of sequencing graphs—a sequencing graph is selected for execution at a certain stage based on an intermediate decision point within the protocol flow.

The DMFB architecture is described in terms of a ring-based system, as shown in Fig. 2(b). Using this topology, distances between the resources are known and have to be incorporated when determining a solution. Moreover, droplets can not move past each other on the bus. They can only overtake another droplet that has left the bus to be used in a resource. Finally, technology-related constraints, like e.g., electrode degradation, have to be taken into account. Resources can not be used infinitely but will break after a certain amount of usages. Resource allocation methods need to respect these bounds.

Based on these inputs and objectives, the general idea of the proposed methodology is as follows: First, it is checked whether the desired protocol can be realized on the ring-based DMFB and with the applied constraints at a fix maximal completion time of $T = 1$ (i.e. within a single time step). To this end, a corresponding SMT formulation is created and passed to an SMT solver [27]. If the SMT solver proves that no such realization is possible, T is increased by 1 and the process is repeated. This continues until the SMT solver shows that the desired protocol can be realized under these conditions at a certain time step T. Then, the precise realization is obtained from the solution of the SMT solver. By increasing the value of T in each iteration and by starting with $T = 1$, minimality with respect to the number of time steps is guaranteed. Similar approaches have been successfully employed for, e.g., routing and synthesis [15], [17], [18].

IV. SMT FORMULATION OF THE DESIGN PROBLEM

Based on the general idea sketched above, we describe how the design problem is formulated in SMT. More precisely, an SMT formulation is required to represent the following question: Is it possible to realize the given protocol on a ring-based

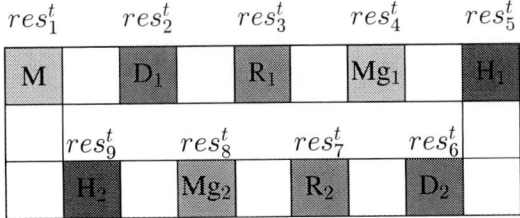

Fig. 4: Variables for the ring-based DMFB from Fig. 2(b).

DMFB while, at the same time, satisfying all architectural and technology-related constraints. In this section, details of the corresponding SMT formulation are provided. First, all variables utilized for this purpose are introduced. Next, constraints restricting the set of all possible solutions to only valid ones are described. Passing the resulting formulation to an SMT solver either yields an assignment satisfying all constraints or proves that such an assignment is not possible. In the former case, a realization of the protocol on the DMFB can be obtained. In the latter case, it has been proven that no such realization exists for the currently considered number T of time-steps.

A. Utilized Variables

In order to formulate the considered design problem as an SMT instance, we introduce an integer variable res_i^t for each resource $i \in R$ (with R representing all possible resources) and for each time step t (with $0 \leq t < T$). Hence, for a protocol composed of n different operations which is to be realized on a ring-based DMFB with k different resources and which should be completed after at most T time steps, we have $0 \leq res_i^t \leq n$ for $1 \leq i \leq n, 0 \leq t < T$. The value of res_i^t describes what operation is executed in resource i at time step t. The value 0 is a special value indicating that no operation is executed in this resource at that time step.

As an example, consider again the ring-based DMFB shown in Fig. 2(b). As can be seen, nine different resources are available. For each of them, a res_i^t variable is introduced as sketched in Fig. 4. Using these variables, the assignment $res_9^2 = 5$ $res_9^3 = 5$ $res_9^4 = 5$ $res_9^5 = 5$ states that operation 5 is executed in time steps 2 to 5 by the heating resource with the identifier 9. Similarly, the variable assignment $res_9^6 = 0$ states that this heating resource is not used in time step 6 (i.e. the previous operation has been completed).

Using these variables, all possible realizations of operations on the DMFB are represented. However, without any further constraints, this formulation permits invalid realizations that, e.g., allow an operation to be executed multiple times (even at the same time step). To prevent this, the assignments of the res_i^t variables have to be restricted.

B. Helper Formulations

In order to properly describe the constraints enforced on the res_i^t variables, we first define a set of *helper formulations*. They describe certain aspects of the overall SMT instance, e.g., whether a certain operation has already been completed. These helpers serve as building blocks for the SMT constraints modeling the overall problem. Besides that, we utilize the notation as summarized in Table I. At the same time, we always provide an intuitive description to ease the readability. Note that, also for sake of readability, corner cases are not explicitly discussed. For example, we are always referring to a predecessor time step by $t - 1$ and do not explicitly discuss the corner case when $t = 0$.

The following helper formulations are used:

- *Executing:* The term $exec(o, t) := \bigvee_{i \in Res(o)} res_i^t = o$, indicates whether the operation o is being executed in time step t.

Table I: Notation used in the SMT formulation.

Symbol	Meaning
i, R	Resources identifier & set of all resources
o, O	Operation identifier and set of all Operations
res_i^t	Variable for usage of resource i at time step t
$Ops(i)$	Operations that can be executed in resource i
$Res(o)$	Resources that can execute operation o
$Prev(o)$	Operations whose result is used by operation o
$Next(o)$	Operations using the results of operation o
$dist(i, j)$	Distances between the resources i and j

- *Starting:* The term $starting(o, t) := \neg exec(o, t - 1) \wedge exec(o, t)$ indicates that the operation o is being started in time step t.
- *Completing:* The term

$$completing(o, t) := \neg exec(o, t) \wedge \bigvee_{t'=0}^{t-1} exec(o, t')$$

indicates whether the operation o has completed its operation in time step t.[3]

- *Schedulable:* The term

$$schedulable(o, t) := \bigwedge_{o' \in Prev(o)} completing(o', t),$$

indicates whether the operation o can be scheduled in time step t.

- *Allocation:* We define the term representing an allocation of an operation o to a resource i starting in step t as

$$alloc(o, i, t) := usage(o, i, t) \wedge block(o, i, t),$$

where $usage(o, i, t)$ describes the resource being used and $block(o, i, t)$ ensures that the succeeding nodes of o will not start too early (i.e., respect the time needed to overcome the distances between the resources). Using the abbreviation $t_{end} := t + duration(i) - 1$, $usage$ and $block$ are defined as

$$usage(o, i, t) := \bigwedge_{t'=t}^{t_{end}} res_i^{t'} = o$$

and

$$block(o, i, t) := \bigwedge_{\substack{o' \in Next(o) \\ i' \in Res(o')}} \bigwedge_{t=end}^{end+dist(i,i')} res_{i'}^t \neq o',$$

respectively.

C. Formulation of the Design Problem

Using the variables as well as the helper formulations introduced above, the design problem now can be formulated by means of the following five constraints:

1) Correct Assignment: To ensure that the resources are correctly assigned, e.g., detecting operations are executed in a detecting device, the following constraints are added to the SMT instance

$$\bigwedge_{t=0}^{T} \bigwedge_{i \in R} \left(res_i^t = 0 \vee \bigvee_{o \in Ops(i)} res_i^t = o \right).$$

2) Correct Ordering: To ensure that the operations are scheduled as defined by the sequencing graphs, the following constraints are added

$$\bigwedge_{t=0}^{T-1} \bigwedge_{o \in O} starting(o, t) \Rightarrow schedulable(o, t).$$

This enforces that all predecessors of o have successfully completed before o can start.

---

[3] Note that this formulation currently ignores the duration of the operation. The duration will be enforced by another formulation introduced later.

3) Correct Allocation: When an operation is starting, a choice has to be made what resource shall be utilized for it. To make sure that exactly one allocation is made (and, at the same time, the chosen resource is blocked for the respective amount of time steps), the constraints

$$\bigwedge_{t=0}^{T-1} \bigwedge_{o \in O} starting(o, t) \Rightarrow \sum_{i \in Res(o)} alloc(o, i, t) = 1$$

are added. The sum ensures that the operation o is allocated in exactly one of the resources i in $Res(o)$.

4) Respect Degradation: To respect the hard technology constraints (here, with respect to degradation), resources can only be used up to a threshold. Each resource i has an associated usage threshold $thres_i$. To enforce this threshold, the constraints

$$\bigwedge_{i \in R} \left(\sum_{t=0}^{T-1} res_i^t \neq 0 \right) \leq thres_i$$

are added. This enforces that the number of time steps in which the resource is occupied by an actual operation (recall that 0 is the special nil operation) stays below the threshold.

5) Enforce Starting: The previous formulations ensure that all solutions adhere to the constraints of the assay. However, up to this point, assigning 0 to all resources for every time step would constitute a valid assignment. Hence, to ensure that operations are actually executed, the constraints

$$\bigwedge_{o \in O} \sum_{t=0}^{T-1} started(o, t) = 1$$

are added. This requires that every operation indeed has to be started exactly once and, by this, enforces the actual execution of the protocol.

V. APPLICATIONS AND EVALUATIONS

The methodology described above significantly improves the previously proposed solution with respect to accuracy as well as by guaranteeing the satisfaction of hard technology constraints. In this section, we demonstrate this by three case studies. We first show how the results obtained by the proposed method (which are optimal and respect architectural characteristics) compare to the results obtained by [11] (which relies on a greedy scheme and ignores architectural characteristics). The next experiment explicitly considers the effect of employing hard technology constraints. Afterwards, the applicability of the proposed method in the context of flash chemistry is evaluated.

Using the proposed methodology, these evaluations can be performed for the first time. To actually conduct the experiments, we make use of the Z3 SMT solver [27].

A. Comparison to Previous Work

In a first series of evaluations, we compared the results obtained by the proposed methodology to previous work [11]. As discussed in Section II, [11] relies on an abstract view of the DMFB that ignores the architectural details and works with greedy schemes. This significantly oversimplifies the search for a design solution. The proposed methodology addresses these shortcomings and, hence, yields more realistic designs.

This has been confirmed on a wide variety of different examples. However, due to page limitations, we summarize explicit results only for a representative scenario, namely, for *Gene Expression Analysis* (GEA), which provides quantitative measures of the transcriptional behavior of a reporter gene under specific epigenetic conditions [8]. This scenario has been considered in [11] as well. We vary the GEA by performing

---

978-1-5386-3693-0/18 $31.00 © 2018 IEEE

Table II: Resources used in the experiment

(a) General resources

Resource	Amount
Heater	3
Camera	1
Detector	2
Waste Reservoir	5
Magnet	1

(b) Dispensers

Dispenser	Amount
Wash	6
Sample	2
$Buffer_1$	2
$Buffer_2$	2
Beads	1
Elution	1
Primers	1
dNTP	1
CNT	1
NTC	1

Table III: Results for the Gene Expression Analysis.

# Pathways	Assay size	Distance	Method from [11]		Proposed method	
			RRS	URS	Steps	Time (s)
1	1	1	102	102	90	61
1	2	1	128	121	99	101
2	1	1	145	145	108	330
1	3	1	151	137	115	844
3	1	1	175	170	127	1432
2	2	1	183	169	124	1920
1	4	1	177	158	130	6129
1	1	2		n/a	115	208
1	2	2		n/a	125	320
2	1	2		n/a	131	1303
1	3	2		n/a	134	551
3	1	2		n/a	152	5550
2	2	2		n/a	150	6916
1	4	2		n/a	152	4892
1	1	3		n/a	140	548
1	2	3		n/a	151	728
2	1	3		n/a	156	3287
1	3	3		n/a	161	1233
3	1	3		n/a	174	16549
2	2	3		n/a	173	8449
1	4	3		n/a	170	2966
1	1	4		n/a	165	1181
1	2	4		n/a	177	1568
2	1	4		n/a	181	7331
1	3	4		n/a	189	3075

multiple experiments in parallel (# Pathways) and/or running the experiment with a different level of precision (Assay size). As architecture, a ring-based DMFB with resources as listed in Table II has been considered.

Table III summarizes the results obtained by the solution presented in [11] and the methodology proposed in this work. The columns *# Pathways* and *Assay size* denote the number of parallel experiments and the level of precision, respectively. The column *Distance* denotes the distance between the resources on the ring by counting electrodes. The most important values are provided in the columns denoted *Steps* which provide the number of time steps needed by the respectively obtained design in order to complete the assay. These numbers are provided for the designs obtained by the solution proposed in [11] – using two different settings, namely with *Restricted Resource Sharing* (RRS) and with *Unrestricted Resource Sharing* (URS) – as well as for the designs obtained by the proposed solution. In the latter case, we additionally provide the run-time required to generate the design in CPU seconds (see column *Time*)[4].

As the method from [11] does not consider distances between resources, we compare both methods assuming a distance of 1 cell for both. This comes closest to the inherent restriction employed in [11]. Besides that, we additionally provide results obtained with larger distances in the bottom

---

[4]These experiments have been conducted on an Linux machine with 3.5 GHz of processor speed and 32 GB of main memory.

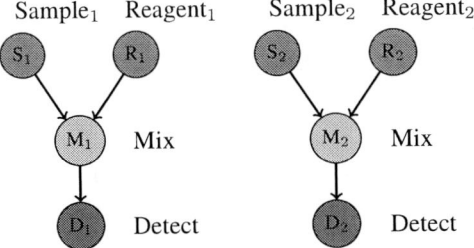

Fig. 5: Protocol used to demonstrate the influence of the degradation threshold.

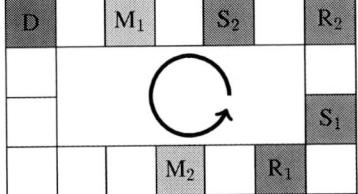

Fig. 6: Architecture used to demonstrate the influence of the degradation threshold.

rows of Table III[5]. The results clearly show that the proposed method determines solutions with completion times that are significantly smaller. On average, just 76% and 80% of the steps required by the designs obtained with the previously proposed method in configuration RRS and URS, respectively, are needed. Moreover, the results obtained by the proposed method are guaranteed to be optimal, i.e., no faster realization will be possible under the applied constraints. The applicability of the proposed method shows that there is no need to resort to using heuristic approaches.

B. Effect of Technology Constraints

Using the proposed methodology, technology constraints such as degradation can be enforced during the design of DMFBs for the first time. The advantages of that have been evaluated in a second series of evaluations. Again, a representative is discussed in the following.

More precisely, consider the protocol shown in Fig. 5 to be executed on a ring-based DMFB as sketched in Fig. 6. Furthermore, consider the case where mixers can be applied an arbitrary number of times (no consideration of technology constraints) and where they can be applied only once (i.e. considering a technology constraint). Table IV sketches the obtained results.

When no technology constraint is considered (cf. Table IV(a)), $Mixer_1$ is simply used twice. This avoids the long run to the second mixer and, hence, yields a significantly faster completion time (namely 16 time steps only). In contrast, when the technology constraint that mixers are to be used only once is enforced (cf. Table IV(b)) droplets may have to take the longer route to $Mixer_2$. Overall, this leads to a significantly larger completion time of 23 time steps[6].

Overall, studies like that can be conducted using the proposed approach for the first time. In contrast, the state-of-the-art solution from [11] does not consider technology constraints at all, i.e., realizations generated using this approach are more prone to faults or defects. This is avoided with the solution proposed in this work.

---

[5]Note that distances have not been considered in [11], and, therefore, these rows are marked "n/a". Considering distances yields much more realistic designs as resources are usually not right next to each other in an DMFB.

[6]Recall that both results are optimal under the given constraints.

Table IV: (a) Solution for the protocol using 16 time steps. $Mixer_1$ is used twice. (b) Solution for the protocol with harsh degradation constraints using 23 time steps. $Mixer_1$ can be used only once.

(a) No threshold		(b) Use mixer once	
Resource	Usage (steps)	Resource	Usage (steps)
$Mixer_1$	5–8; 9–12	$Mixer_1$	11–14
Detector	11–13; 14–16	Detector	18–20; 21–23
$Mixer_2$	–	$Mixer_2$	11–14
$Sample_1$	2–3	$Sample_1$	3–4
$Reagent_1$	3–4	$Reagent_1$	6–7
$Sample_2$	1–2	$Sample_2$	1–2
$Reagent_2$	1–2	$Reagent_2$	1–2

Table V: Results for the parallel GEA experiments.

Benchmark	Time Steps
short_2_3	208
short_1_3	223
short_1_2	198
medium_2_3	273
medium_1_3	239
medium_1_2	273
long_2_3	343
long_1_3	484
long_1_2	347

C. Completion-Time Profile for Two-Pathway Reactions

Next, we aim to show the significance of the exact method in the context of flash chemistry. Recall that biochemical pathways within a flash chemistry environment require fine-grained control and minimization of protocol completion time, even with increasing the pathway length. For this purpose, there is a need to investigate the completion-time profile of reaction pathways when the reaction complexity is increased.

To perform this study, we use the GEA benchmark to formulate a two-pathway reaction. We gradually increase the length of pathways to investigate the growth of protocol time obtained by the exact method. We consider three different cases in terms of pathway length (number of bioassays): (1) a pathway with 6 bioassays; (2) a pathway with 8 bioassays; (3) a pathway with 10 bioassays. Table V lists some of the obtained results.

These results are invaluable to accurately characterize the steps in flash chemistry (while heuristic methods are not useful in this context since they ignore realistic constraints and are they not guaranteed to minimize the reaction time). Our results show that the proposed exact method is feasible for these realistic test cases. The CPU time to generate each of these solutions was in the range of 35 hours; this run time is not a concern because it is a one-time cost and carried out well in advance of the launch of the on-chip experiment. Heuristic methods, which offer reduced CPU time but are associated with longer protocol-completion time, are not necessary because the exact method can be used in practice for these protocols.

VI. CONCLUSION

In this work, we proposed a methodology supporting complex biomolecular protocols on DMFBs that respects technology constraints while, at the same time, does not make use of an oversimplified model of the DMFB. Experimental results confirm the applicability of the approach. Using the proposed exact methodology, we were able to produce protocol completion times that are at approximately 70% of those of previous work, showing that there still is room for improvement for heuristic approaches or even no need to resort to using heuristics at all. This is noteworthy as in previous work no distances were considered, i.e., one would expected even shorter solutions. We further were able to prove that considering technology constraints is necessary as it can have a significant influence on the protocol completion time.

REFERENCES

[1] R. B. Fair, "Digital microfluidics: is a true lab-on-a-chip possible?" *Microfluidics and Nanofluidics*, vol. 3, no. 3, pp. 245–281, 2007.

[2] T. Xu, K. Chakrabarty, and V. K. Pamula, "Defect-tolerant design and optimization of a digital microfluidic biochip for protein crystallization," *Trans. on CAD*, vol. 29, no. 4, pp. 552–565, 2010.

[3] M. Pollack, A. Shenderov, and R. Fair, "Electrowetting-based actuation of droplets for integrated microfluidics," *Lab on a Chip*, vol. 2, no. 2, pp. 96–101, 2002.

[4] A. Rival *et al.*, "An EWOD-based microfluidic chip for single-cell isolation, mRNA purification and subsequent multiplex qPCR," *Lab on a Chip*, vol. 14, no. 19, pp. 3739–3749, 2014.

[5] Z. Hua *et al.*, "Multiplexed real-time polymerase chain reaction on a digital microfluidic platform," *Analytical chemistry*, vol. 82, no. 6, pp. 2310–2316, 2010.

[6] L. R. Volpatti and A. K. Yetisen, "Commercialization of microfluidic devices," *Trends in biotechnology*, vol. 32, no. 7, pp. 347–350, 2014.

[7] S. Miller, K. Henthorn, D. Osato, J. Lips, A. Schroeder, R. Humphries, and L. Freeman-cook, "Multiplex detection of respiratory pathogens with genmark's eplex sample-to-answer system," *Clinical Chemistry and Laboratory Medicine*, vol. 54, no. 5, pp. eA5–eA6, 2016.

[8] B. S. Wheeler, B. T. Ruderman, H. F. Willard, and K. C. Scott, "Uncoupling of genomic and epigenetic signals in the maintenance and inheritance of heterochromatin domains in fission yeast," *Genetics*, vol. 190, no. 2, pp. 549–557, 2012. [Online]. Available: http://www.genetics.org/content/190/2/549

[9] J.-i. Yoshida, A. Nagaki, and T. Yamada, "Flash chemistry: fast chemical synthesis by using microreactors," *Chemistry–A European Journal*, vol. 14, no. 25, pp. 7450–7459, 2008.

[10] J.-i. Yoshida, Y. Takahashi, and A. Nagaki, "Flash chemistry: flow chemistry that cannot be done in batch," *Chemical Communications*, vol. 49, no. 85, pp. 9896–9904, 2013.

[11] M. Ibrahim, K. Chakrabarty, and K. Scott, "Integrated and real-time quantitative analysis using cyberphysical digital-microfluidic biochips," in *DATE*, 2016, pp. 630–635.

[12] "Illumina," https://www.illumina.com/systems/neoprep-library-system.html.

[13] F. Su and K. Chakrabarty, "High-level synthesis of digital microfluidic biochips," *JETC*, vol. 3, no. 4, p. 1, 2008.

[14] E. Maftei, P. Pop, and J. Madsen, "Module-based synthesis of digital microfluidic biochips with droplet-aware operation execution," *JETC*, vol. 9, no. 1, p. 2, 2013.

[15] O. Keszocze, R. Wille, and R. Drechsler, "Exact routing for digital microfluidic biochips with temporary blockages," in *ICCAD*, 2014, pp. 405–410.

[16] J.-W. Chang, S.-H. Yeh, T.-W. Huang, and T.-Y. Ho, "Integrated fluidic-chip co-design methodology for digital microfluidic biochips," *Trans. on CAD*, vol. 32, no. 2, pp. 216–227, 2013.

[17] O. Keszocze, R. Wille, T.-Y. Ho, and R. Drechsler, "Exact One-pass Synthesis of Digital Microfluidic Biochips," in *DAC*, ser. DAC, 2014, pp. 142:1–142:6.

[18] O. Keszocze, R. Wille, K. Chakrabarty, and R. Drechsler, "A General and Exact Routing Methodology for Digital Microfluidic Biochips," ser. ICCAD, 2015, pp. 874–881.

[19] Y. Luo, K. Chakrabarty, and T.-Y. Ho, "Error recovery in cyberphysical digital microfluidic biochips," *Trans. on CAD*, vol. 32, no. 1, pp. 59–72, 2013.

[20] K. Hu, M. Ibrahim, L. Chen, Z. Li, K. Chakrabarty, and R. Fair, "Experimental demonstration of error recovery in an integrated cyberphysical digital-microfluidic platform," in *BIOCAS*, 2015, pp. 1–4.

[21] K. Hu, B.-N. Hsu, A. Madison, K. Chakrabarty, and R. Fair, "Fault detection, real-time error recovery, and experimental demonstration for digital microfluidic biochips," in *DATE*, 2013, pp. 559–564.

[22] D. T. Grissom and P. Brisk, "Fast online synthesis of digital microfluidic biochips," *Trans. on CAD*, vol. 33, no. 3, pp. 356–369, 2014.

[23] "Genmark," https://www.genmarkdx.com/solutions/systems/eplex-system/.

[24] E. Samiei, M. Tabrizian, and M. Hoorfar, "A review of digital microfluidics as portable platforms for lab-on a-chip applications," *Lab on a Chip*, vol. 16, no. 13, pp. 2376–2396, 2016.

[25] H. Norian *et al.*, "An integrated CMOS quantitative-polymerase-chain-reaction lab-on-chip for point-of-care diagnostics," *Lab on a Chip*, vol. 14, no. 20, pp. 4076–4084, 2014.

[26] M. Mibus, X. Hu, C. Knospe, M. L. Reed, and G. Zangari, "Failure modes during low voltage electrowetting," *ACS Applied Materials & Interfaces*, vol. 8, no. 24, pp. 15 767–15 777, 2016.

[27] L. De Moura and N. Bjørner, "Z3: An efficient SMT solver." Springer, 2008, pp. 337–340, Z3 is available at http://z3.codeplex.com/.

978-1-5386-3693-0/18 $31.00 © 2018 IEEE

2018 31th International Conference on VLSI Design and 2018 17th International Conference on Embedded Systems

Design Optimization at the Fluid-level Synthesis for Safe and Low-Cost Droplet-based Microfluidic Biochips

Arpan Chakraborty[*], Piyali Datta[†], and Rajat Kumar Pal[*]

[*]Department of Computer Science and Engineering, University of Calcutta, JD– 2, Sector – III, Saltlake, Kolkata – 700 098, West Bengal, India

[†]Department of Computer Science and Engineering, Heritage Institute of Technology, Anadapur, Kolkata- 700 107, West Bengal, India

Email: arpanc250506@gmail.com[*], piyalidatta150888@gmail.com[†], pal.rajatk@gmail.com[*]

Abstract—Droplet-based Digital Microfluidic Biochips (or DMFBs) are now being the prime platform for a number of point of care diagnostics, clinical studies, and sample preparations. Several design optimization methods exist at the fluid-level that automate the tasks of a DMFB. However, in high frequency applications, its performance heavily deteriorates in the order of degrading electrodes producing incorrect outcomes. Vague results may also be generated if cross contamination during droplet routing is not avoided. We define a DMFB to be *safe* if it is capable of handling reliability and also free from cross contamination. Alongside, as the fluid-level of DMFBs comprises several tasks that altogether introduces design cycles and leads to higher cost, a low-cost platform is urgently required. This paper proposes a fluid-level design for DMFBs that considers the above facts together. A graph model has been used to tackle them. An exact algorithm is presented. The obtained results are validated with several benchmarks.

Keywords—automation, design optimization, microfluidics, NP-completeness, reliability, safety

I. Introduction

As Digital Microfluidic Biochips (DMFBs) revolutionized the biomedical analysis procedures, it has become a prime choice for several market leaders in biochip industries. This typical lab-on-chip system integrates various biochemical protocols with high precision control, lesser fluid consumption, and less human intervention [1]. A 2D array of electrodes together with some peripheral devices (i.e. detector, dispenser) constructs the basic architecture of a DMFB. The endeavour to automate the fundamental fluidic tasks (e.g. dispensing, mixing, dilution, and detection) needs an external voltage driver under a clock control. The basic droplet movement is based on the EWOD principle [2] and is carried out by actuation and de-actuation of the key electrodes. Hence, any operation can be executed anywhere in the array by consuming a group of cells in reconfigurable manner [3].

A number of algorithms in designing the fluid-level (in Fig. 1) of the DMFBs exists that prepares a synthesis decision, i.e. *binding, scheduling, placement,* and *routing* tasks from the input *bioassay graph* together with the chip architecture and design specification [4], [5]. The needful minimization of the bioassay-completion time is performed at this stage. Besides, determination of proper assay outcome is an urgent requirement in medical diagnosis [6]. For ensuring safety and

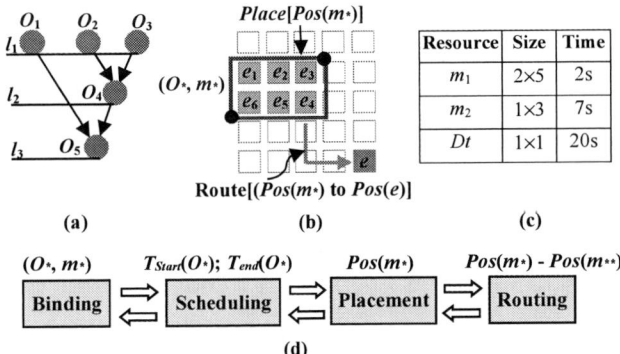

Figure 1. (a) A bioassay graph, (b) A DMFB executing an operation O_* in a mixer m_*, (c) Resource library, (d) Fluid-level synthesis with design cycles.

criticality, the bioassay operations must be tactfully examined so that the result remains at some satisfactory level. However, a bioassay run may produce an erroneous outcome and become *unsafe* for a number of reasons. The lifetime of an electrode varies inversely with the number times it is being actuated consecutively. Now, due to inherent reconfigurable nature of the DMFBs, a degraded cell may have to be reused while affecting a correct mixing or splitting task [7]. However, at any time during execution, an exceedingly actuated cell may encounter a permanent failure disrupting the bioassay run. This interruption leads to iteration of the fluid-level synthesis for the hampered operations to avoid the faulty cell. Indeed, the completion time becomes longer violating the deadline which is not desirable.

Another imperative issue for which a DMFB results in erroneous outcome is the cross contamination problem during droplet routing [8]. Some residue of a sample or reagent droplet (i.e., functional-droplet) may be stuck at the electrode wall during transportation. Now, while moving two dissimilar type droplets, the path of one may be intersected with that of the other causing contamination. For a high throughput biochip, multiple droplet routes may possess multiple contaminated locations (CL). Hence, for a safe assay outcome, the synthesis phase must ensure that it avoids all such locations. It is worth mentioning that each CL must be washed before any functional droplet passes through it. For a multiple CL, sometimes route deadlock may occur for not finding a suitable route (for functional droplet and wash droplet) to bound the assay deadline within a moderate level. Hence, the

978-1-5386-3693-0/18 $31.00 © 2018 IEEE 127

fluid-level synthesis must avoid all *CL* while minimizing bioassay completion time.

Though electrode reliability and cross contamination avoidance measure the *safeness*, the associated cost in DMFB realization must not be neglected. Most of the previous works in fluid-level design pass through unified synthesis schemes that generally solve the sub-tasks one at a time [9]. Evidently, outcome of a task may become infeasible for a later task for not satisfying the required constraints (or some design criteria) at this phase and hence the ancestor tasks may again have to be iterated until a desirable solution is achieved. This procedure indeed adds an indefinite cost which is not acceptable in low cost design of DMFBs.

Till now two works [7] and [10], have raised the electrode degradation problem and accordingly proposed a placement synthesis. A categorization is done on the electrodes based on their quality and in the synthesis phase only the better ones are deployed. The assay completion time is also minimized while maximizing the lifetime of a cell. A considerable amount of researches have been made to avoid cross contamination at the routing phase [8], [11]. Here, the bioassay is formulated as a number of routing sub-problems. Each sub-problem is solved in order to get a contamination free routing by reduction from disjoint-path (either vertex or edge disjoint) problem [8]. Moreover, a washing scheme is also integrated with functional droplet routing. All the works use a unified synthesis method to generate a final outcome [9]. For an example, to find a feasible route for functional droplets previously generated placements may have to be refined or readjusted implying a design cycle between the placement and routing tasks.

Reliability and cross contamination avoidance are important parameters for a safe biochip platform. Current design optimization methods at the fluid-level synthesis separately consider these issues. Existing error checking protocol with integrated cyber-physical technology serves these purposes through re-execution of the erroneous operations [12]. This leads to rise in the assay completion time. On the other hand, it fails to meet the aim for a low-cost design platform. Thus, this paper offers a complete fluid-level synthesis scheme which proposes a low-cost design that remains safe as well. A graph theoretic model is proposed. The considered problem is proved as NP-complete and an exact solution is presented.

II. PROBLEM FORMULATION

A. Illustrative Example

Fig. 2 depicts how an example bioassay (as in Fig. 1(a)) is synthesized onto a given sized chip. A bioassay is essentially a directed acyclic polar graph where nodes and edges respectively represent mixing/dilution and dependency relations. At the beginning, with the help of a characterized module library, a binding solution, i.e. (O_i, m_j), between an operation (O_i) and a resource (m_j) is generated for the input bioassay. Next, each O_i is given a start time-stamp. Considering the ordering relations, a schedule, i.e. decision of start and end time stamps of the O_is is performed.

Accordingly, a placement solution for each m_j is acquired. Finally, the droplet transportation path considering the nets, i.e. (m_j, m_k) is found out.

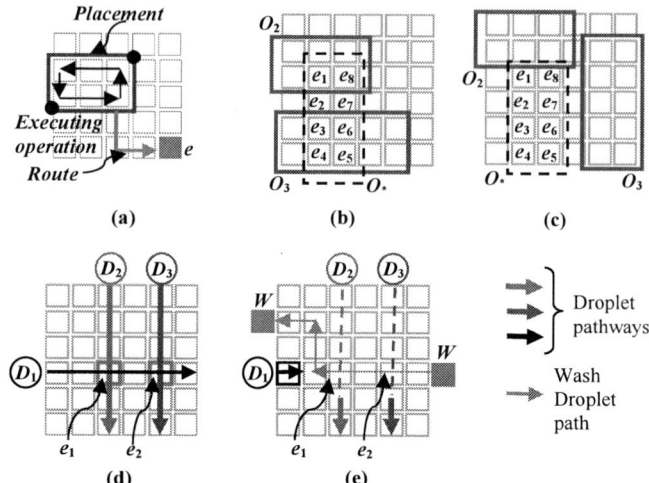

Figure 2. (a) Placement and route between the modules, (b) Usage of electrodes between the previous operation O_*, and next operations, (c) An alternative placement option, (d) Two contaminated cells as droplet D_1 travels after D_2 and D_3, (e) Incorporation of washing between the route pathways.

Observation 1: Due to size constraint, only a certain number of concurrent O_i can be executed simultaneously, implying a halt state for a group of operations. Also, each sub-task has its own set of constraints; e.g., during a placement task a group of modules output from the previous binding and scheduling are to be placed on the chip satisfying the safe separation constraint. If this is not maintained, a new binding and/or scheduling solution have to be acquired. This inherent design cycle lengthens time to achieve the final outcome.

Fig. 2(b)-(c) show how reliability of electrodes should be taken care of. Let O_* is an executing operation after whose completion modules for O_2 and O_3 could be placed. If the synthesis solution is as in Fig. 2(b), then certainly the electrodes that have just passed through actuation and deactuation for O_* now will be required in execution for the later ones (i.e., e_1 and e_8 for O_2 and e_3, e_4, e_5, and e_6 for O_3).

Observation 2: If $F_{rel}(e_i)$ measures the reliability of e_i and $N(e_i)$ is the total number of actuations on it, then $F_{rel}(e_i) \propto 1/N(e_i)$. It implies that the placement solution in Fig. 2(d) for O_2 and O_3 consumes more degraded cells and hence the synthesis outcome is not reliable. For a more reliable solution, Fig. 2(c) can be well accepted.

Fig. 2(d)-(e) explain the contamination problem during droplet transportation of three droplets D_1, D_2, and D_3. Let, as per the time stamp, D_1 be aimed to route after routing of D_2 and D_3. Accordingly, two contaminated locations (*CL*) occur at e_1 and e_2. Hence, the contaminated droplet D_1 must not be used. Hence, a washing droplet washes the *CL* before D_1 uses the cells. At this point in time, synchronization of wash droplet together with the functional droplet must be incorporated, i.e. D_1 is being halt for some time while the fluidic constraints are maintained among the droplets.

978-1-5386-3693-0/18 $31.00 © 2018 IEEE

Observation 3: Washing time increases with the increasing number of CL. As a result, the needful assay completion time also lengthens. Hence, the functional-droplets should be routed in such a way so that crossings in between are minimized and a lesser number of CL is produced.

Observation 4: From Observations 1, 2, and 3, it can be readily understood that the existing design cycles among the synthesis sub-tasks increases the time to DMFB realization. For an example, if no feasible routes exist at the time of synchronization of wash and functional droplets, then a new placement solution (or at least its alteration) is needed. Also, for each infeasible placement new binding and/or schedule are generated. These facts imply how a DMFB process may become costly in terms of generating final outcome.

From the previous study, the considered design optimization problem can be briefed as follows.

Input: A bioassay graph (G), a module library (L) holding resource types and requisite times, design specification, i.e. maximum allowable deadline (T_{finish}) of G, the size of the chip (C), and the target reliability factor $F_{rel}(C)$ of the biochip.

Constraints:

a) Area constraint: Area used by the resources must be less than or equal to the total chip area.

b) Safe separation constraint: At any time, any two resources must be separated by a band of electrodes.

c) Fixed resource constraint: At any time, the total number of required non-reconfigurable resources in execution of G must remain within the bound of their availability.

d) Operational precedence constraint: Any successor operation O_i cannot start executing unless its predecessor O_j completes its task.

e) Fluidic constraint: During any static or dynamic provision of more than one droplet a safe separation of electrodes is to be maintained to avoid undesirable mixing.

f) Reliability constraint: To avoid being unreliable, no electrode can be actuated for more than the number of limited consecutive times.

Objective: To prepare a fluid-level synthesis solution that maximizes the *safeness*, i.e. highly reliable and free from cross contamination, while minimizing the bioassay completion time and reducing the design cycles in process.

III. PROPOSED STRATEGY

Due to the size restriction of working chip, only a limited number of *disjunctive* operations (i.e. having no dependencies) can be executed simultaneously. Factually, the fluid-level synthesis can be observed as a series of sub-problems where each sub-problem solves the binding, scheduling, placement, and routing for an intermediate set of disjunctive operations. A 'good' synthesis outcome results from an efficient solution of all these sub-problems. Moreover, this necessitates that any two sub-problems must be well linked to each other constituting a 'good' solution as a whole.

Let, two sets S_1 and S_2 hold the disjunctive operations respectively for level l_1 and l_2 of the bioassay G. An operation $O_{ij} \in S_j$, must be executing in a suitable resource (mixer or storage cells) for their accomplishment. However, the

available resource type determines how fast each set S_j finishes its jobs; e.g., a 2×5 mixer is superior to a 2×3 mixer. Again, after completion of S_j the next set S_{j+1} begins. At this time, if any operation $O_{k(j+1)}$ is dependent on O_{ij}, the corresponding droplet transportation between the placed modules must be incorporated. Hence, the completion time $T_{comp}(S_j)$ of an S_j can be determined from the mixing and droplet routing times as follows.

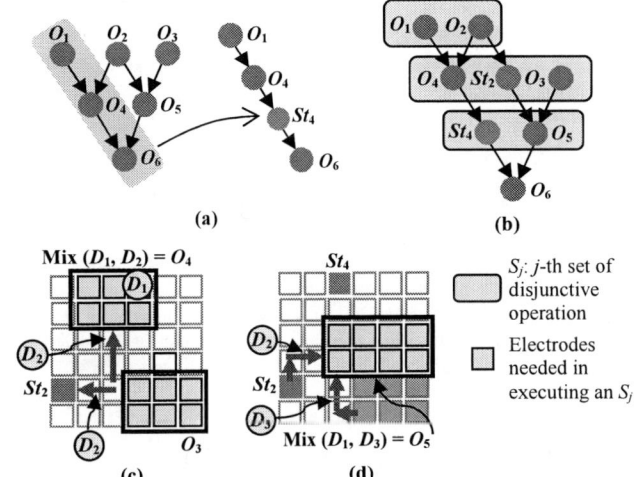

Figure 3. (a) Mixing and store operations in a branch, (b) Disjunctive operations (mixing and store) at each level, (c) Performing j-th set of disjunctive operations, (d) Performing a (j+1)-th set of disjunctive operations.

$$T_{comp}(S_j) = \max\{T_{mix}(m_{kj} \in S_j)\} + \max\{T_{route}(S_{j-1}, S_j)\},$$

where m_{kj} is a mixer associated for O_{kj}. Here, we assume that the time required in holding a droplet in a storage cell is implicitly defined by the equation $\{T_{mix}(m_{kj} \in S_j)\}$. Also, an S_{j+1} starts in execution after obtaining all the required droplets. Based on this fact, we may explicitly define a j-th sub-problem of the fluid-level as the combination of j-th placement (i.e. *place$_j$*) and j-th routing (i.e. *route$_j$*) problems. Evidently, a *place$_j$* solves the module placement required for all $O_{ij} \in S_j$ whereas a *route$_j$* performs the required droplet routing among the modules of S_j and S_{j+1}. Fig. 3 explains the above facts.

A. Fluid-level Search Space

To determine the decisive modules that solves a *place$_j$* and *route$_j$* in order to obtain a minimized $T_{comp}(S_j)$, we follow a greedy approach. Firstly, to maintain the availability constraint, a depth first traversal (DFS) is made on the input bioassay, for which, each S_j holds the requisite number of mixing and store operations that in turn determine the number of modules required in an S_j. The position of the modules on the chip can be defined by satisfying the safe separation constraint. For an example, let there be two mixings (O_1 and O_2) and one store (St) in an S_j. The modules, i.e. m_1, m_2, and St can be physically placed on the chip. Now, we have two binding options for the mixings, i.e. either $O_1 \rightarrow m_1$ and $O_2 \rightarrow m_2$ or $O_1 \rightarrow m_2$ and $O_2 \rightarrow m_1$. Similarly, for a later S_{j+1} several such options can be readily generated. It is worth mentioning that the binding and scheduling sub-problems are now reduced to a single placement problem. Also, out of

978-1-5386-3693-0/18 $31.00 © 2018 IEEE   129

several options for an S_j, only a single one is considered to solve the $place_j$. The most suitable combination that optimizes $T_{comp}(S_{j+1})$ can now be obtained by searching the entire search space which comprises of several options for S_j and S_{j+1}.

For n sub-problems, the fluid-level search space can be reduced into an n level staged graph (G_{fluid}) (in Fig. 4). In G_{fluid} a j-th level corresponds to the S_j. Each vertex denotes an option op_{kj} solving a $place_j$. A directed edge $ed_{krj(j+1)}$ from op_{kj} and $op_{r(j+1)}$ denotes the corresponding $route_j$. Evidently, there is no edge between any two options belonging to the same $place_j$. Also, any two consecutive levels in the G_{fluid} form a bipartition. Each combination, i.e. (op_{kj}, $ed_{krj(j+1)}$) evidently represents a sub-problem ($place_j$, $route_j$). In summary, n combinations, each solving a distinct sub-problem, are to be extracted from G_{fluid} for solving the fluid-level synthesis.

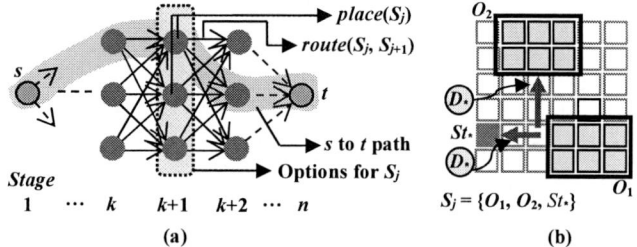

Figure 4. (a) Fluid-level staged graph (G_{fluid}) and corresponding options in a stage forming a sub-problem, (b) Placement and routing in a sub-problem.

B. Guiding the Fluid-level Solution

To have a reliable and contamination free synthesis outcome, two weights are classified for each edge. The reliability constraint allows us to place the modules for an op_{kj} $\in S_j$ using the 'good' electrodes, i.e. whose consecutive number of actuation is within some threshold. However, for size restriction always only the 'good' ones cannot be deployed. Hence, we should have some relaxation over the threshold value unless the bioassay completion time will be indefinitely longer. If E and E_j denote respectively the total electrode set and that of used in an op_{kj}, the reliability measure during an $op_{r(j+1)}$ is defined by $R(op_{kj}, op_{r(j+1)}) = (E_j \cap E_{j+1}) / E$. Let, $W1_{krj(j+1)} = R(op_{kj}, op_{r(j+1)})$. Since, lesser $R(op_{kj}, op_{r(j+1)})$ implies more reliability of the chip during S_{j+1}, a lesser $W1_{krj(j+1)}$ guides in selecting an $ed_{krj(j+1)}$ from the G_{fluid}.

As stated, each $route_j$ is to perform the required droplet routing between the modules of S_j and S_{j+1}. Following [8], an edge disjoint path connecting the modules is found out. However, an edge disjoint path in a $route_j$ may consist of one or more CL (i.e. contaminated cells). If so, the needful washing is also necessary to begin the next $op_{r(j+1)}$. It implies that for higher number of CL, the beginning time of $op_{r(j+1)}$ is more delayed. Hence, $T_{comp}(S_j)$ gets also longer. A modified Lee's algorithm [8] can be effectively applied to find out the edge disjoint routes in $route_j$. If a $route_j$ consists of p number of module nets (i.e. 2-pin net implies a module at each end), then a net priority can be incorporated based on the bounding rectangle connecting the end points of each net [8]. The box with the lowest perimeter has the highest priority and is routed

first. Subsequent routing can be determined by taking the previous routes as obstacles. Hence, an edge disjoint path can be found out for each $route_j$. Within each $route_j$, let $N_{CL}(op_{kj}, op_{r(j+1)})$ be the number of CL generated from the Lee's algorithm. We define another edge weight $W2_{krj(j+1)} = N_{CL}(op_{kj}, op_{r(j+1)})$. It implies that lower the value of $W2_{krj(j+1)}$ lesser is the contamination and lesser the completion time. Hence, a next $op_{r(j+1)}$ can be acquired based on the minimum value of $W2_{krj(j+1)}$.

Lemma 1: *A path in the fluid-level staged graph (G_{fluid}, n) of stage n, is a feasible synthesis realization, if the followings hold true. (i) The path is simple and (ii) There are exactly n nodes in the path.*

Proof: Let the bioassay *start* and *end* are denoted respectively as s and t vertices. By construction of (G_{fluid}, n), each sub-problem ($place_j$, $route_j$) is denoted as a pair (op_{kj}, $ed_{krj(j+1)}$), where op_{kj} is the j-th vertex at stage k. By definition, solving a bioassay essentially involves n sub-problems, i.e. n combinations of (op_{kj}, $ed_{krj(j+1)}$), where no two sub-problems overlap. Then, exactly a single vertex op_{kj} is chosen from each stage between s and t implying a simple path avoiding non-repeating vertices. Now, there are exactly n nodes in an s to t path which solves all the n sub-problems satisfying the needful constraints. Hence, an s to t simple path is a feasible synthesis realization.

Lemma 2: *Let Δ_{DE} and Δ_{CL} are two non-negative real numbers. Then, for a feasible synthesis path in G_{fluid} the followings hold true. (i) The chip is dependable up to Δ_{DE}, if $\Sigma W1_{krj(j+1)} \leq \Delta_{DE}$ and (ii) The maximum number of contaminated locations is at most Δ_{CL}, if $\Sigma W2_{krj(j+1)} \leq \Delta_{CL}$.*

Proof: From Observation 2, the reliability of an electrode is inversely proportional to the total number of actuations during the bioassay run. From Lemma 1, it is obvious that a path from s to t in (G_{fluid}, n) consists of n vertices. Each vertex op_{kj} is solving a $place_j$. From the earlier discussion we have that the reliability measure $W1_{krj(j+1)}$ between consecutive $place_j$ and $place_{j+1}$ must avoid recently used cells. Hence, the total number of activated cells in a bioassay run equals to $\Sigma W1_{krj(j+1)}$ which should be desirably less than the target reliability Δ_{DE} of the chip. Similarly, it holds true for $\Sigma W2_{krj(j+1)} \leq \Delta_{CL}$.

Theorem 1: *A feasible synthesis path in G_{fluid} reduces the design cycles and hence asserts a low cost fluid-level design.*

Proof: From Lemmas 1 and 2, it is evident that if we have an s to t path in (G_{fluid}, n) meeting $\Sigma W1_{krj(j+1)} \leq \Delta_{DE}$ and $\Sigma W2_{krj(j+1)} \leq \Delta_{CL}$, then the iterations needed in the existing methods for refinement of one or more sub-tasks (i.e. binding, scheduling, placements, and routing) can be avoided. Hence, design cycles can be reduced asserting a low cost design.

Problem 1: Minimum Cost Fluid-level Synthesis (*MCFS*).

Instance: A fluid-level n-staged graph (G_{fluid}, n); designated start (s) and end (t) vertices respectively at level l_1 and l_n; a k-th vertex at l_j is denoted as op_{kj}; each directed edge $ed_{krj(j+1)}$ between op_{kj} and $op_{r(j+1)}$ is allied with two weights $W1_{krj(j+1)}$ and $W2_{krj(j+1)}$; two non-negative real numbers Δ_{DE} and Δ_{CL}.

978-1-5386-3693-0/18 $31.00 © 2018 IEEE

Question: Is there a path $P(s, t)$ with non-repeating vertices in G_{fluid} such that $\Sigma W1_{krj(j+1)} \leq \Delta_{DE}$ and $\Sigma W2_{krj(j+1)} \leq \Delta_{CL}$?

Theorem 2: *MCFS is NP-complete.*

Proof: For a G_{fluid} the certificate for *MCFS* is a simple path consisting of n vertices (comprising $(n-2)$ number of op_{kj}, s and t) taking exactly a single vertex from each level. On this path for all $W1_{krj(j+1)}$ and $W2_{krj(j+1)}$, the summation $\Sigma W1_{krj(j+1)}$ and $\Sigma W2_{krj(j+1)}$ can be performed and checked in polynomial time, whether $\Sigma W1_{krj(j+1)} \leq \Delta_{DE}$ and $\Sigma W2_{krj(j+1)} \leq \Delta_{CL}$, respectively. Hence, *MCFS* is in *NP*.

To prove NP-hardness, an instance of PARTITION [13] is reduced to an instance of *MCFS* in polynomial time. For this, consider an instance of PARTITION: a set A, *size* of each element $x \in A$ is $s(x)$, a subset $A1$ such that $\Sigma_{x \in A1} s(x) = \Sigma_{x \in A-A1} s(x)$. Here, $|A| = 2 \times |A1| = 2 \times |A-A1|$.

Let $A = \{x_1, x_2, ..., x_n\}$, where n is an even number. Also, let G be a directed acyclic graph having $(n+3)$ levels, $2(n+1)+2$ nodes, and two nodes at each level except at the first and the last level. If V is the vertex set of G then $V = \{s, v_{11}, v_{21}, v_{12}, v_{22}, ..., v_{1(n+1)}, v_{2(n+1)}, t\}$. If E is the edge set then $E = \{(s, v_{p1}), (v_{pi}, v_{q(i+1)}), (v_{p(n+1)}, t) \mid 1 \leq p, q \leq 2$ and $1 \leq i \leq (n+1)\}$. For $i = 1$ to $(n+1)$, let, $W1(s, v_{p1}) = W2(s, v_{p1}) = 0$; $W1(v_{1i}, v_{1(i+1)}) = W1(v_{2i}, v_{1(i+1)}) = x_i$; $W2(v_{1i}, v_{1(i+1)}) = W2(v_{2i}, v_{1(i+1)}) = 1$; $W2(v_{2i}, v_{2(i+1)}) = W2(v_{1i}, v_{2(i+1)}) = x_i$; $W1(v_{2i}, v_{2(i+1)}) = W1(v_{1i}, v_{2(i+1)}) = 1$; ... ...; $W1(v_{p(n+1)}, t) = W2(v_{p(n+1)}, t) = 0$.

Let $\Delta_{DE} = \Delta_{CL} = [\Sigma_{i=1 \text{ to } n} s(x_i)+n]/2$. It implies that there is a path from s to t if and only if the PARTTION problem has a solution with $2 \times |A1| = n$.

C. Achieving an Exact Solution for the Fluid-level

Evidently, a viable solution of *MCFS* comprises a path $P(s, t)$ in G_{fluid}, where both the classified weights (as defined in Section B) maintain their respective constraints. Hence, Problem 1 is equivalent to the well visited Shortest Weight-Constrained-Path Problem (SWCPP) [13]. Several approaches effectively solve SWCPP. Among them the Label Setting Algorithm (LSA) based on dynamic programming approach that achieves a pseudo-polynomial time solution is more convincing [14].

For the two cost metrics, degree of reliability ($\Sigma W1_{krj(j+1)} \leq \Delta_{DE}$) and contamination ($\Sigma W2_{krj(j+1)} \leq \Delta_{CL}$), LSA outputs a path with minimized contaminated spots while satisfying Δ_{DE}. In doing so, LSA maintains a set of labels L_p on each intermediate node p. Each label q can be identified as a pair of two weights ($W1_{pq}$, $W2_{pq}$) corresponding to a particular path from the source vertex (s) to p. Hence, a q-th label actually represents the degree of reliability (i.e. $W1_{pq}$) and contamination (i.e. $W2_{pq}$) for node p following the path $P(s, p)$. Moreover, LSA only preserves the non-dominated labels in forming an L_p. Let there are two s to p paths $P1(s, p)$ and $P2(s, p)$. Also, let ($W1_{pq}$, $W2_{pq}$) $\in P1(s, p)$ and ($W1_{px}$, $W2_{px}$) $\in P2(s, p)$. Then, the non-domination is defined if and only if either $W1_{pq} \leq W1_{px}$ and $W2_{pq} \geq W2_{px}$ or $W1_{pq} \geq W1_{px}$ and $W2_{pq} \leq W2_{px}$. As only non-dominated labels are collected, any path that would violate the weight constraint is avoided. It is worth mentioning that LSA follows the principle of Dijkstra's

algorithm, i.e. starting from s it spreads out searching while examining the already searched neighbour nodes. The run time complexity of the algorithm is O($OPT \times |E|$), where $|E|$ is the arc set of G_{fluid} and OPT is cost of $\Sigma W1_{krj(j+1)}$ while satisfying $\Sigma W1_{krj(j+1)} \leq \Delta_{DE}$ for which total number of contaminated locations is minimum.

D. Dealing with Wash Droplet Routing

After solving *MCFS*, we have a path $P(s, t)$ in G_{fluid} that represents a fluid-level synthesis implementation with minimized contamination and maximized reliability. Now, all the *CL* must be effectively washed to have a contaminated free solution. Washing indeed increases the overall time. Hence, as mentioned in [8], the wash droplets must be effectively synchronized with the functional droplets such that it minimizes the overall completion time.

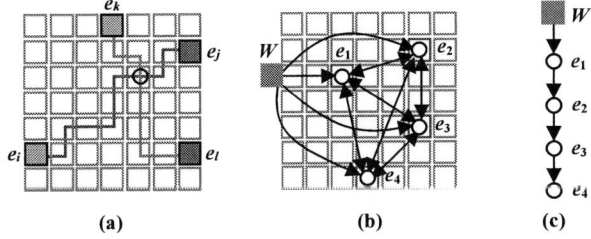

Figure 5. (a) Routing nets and single contaminated location problem, (b) Washing port, multiple contaminated locations and reachability graph, (c) Topological sorting based on path-length.

Fig. 5(a)-(c) explains the washing scheme. Within each sub-problem ($place_j$, $route_j$), a number of *CL* remains in the droplet routes. A wash droplet (D_w) from the outside port is dispensed that washes each *CL* in the path of any functional droplet (D_f) before D_f reaches there. Also, in solving $route_j$ for each D_f we have its required route-path or routing cells and the time-stamps between the source and destination. Firstly, a *reachability graph* connecting the washing-port and *CL* can be formed. A topological ordering on this graph results in a path between washing-port and a *CL* on the basis of their path-lengths. Hence, the *CL* with the shortest path length will be selected first for required washing. At this time, for all the D_f those are not satisfying fluidic constraint after insertion of a D_w, must be stalled in a cell ensuring minimum stalling. Hence, within each ($place_j$, $route_j$) all the *CL* can be effectively washed by observing the movement of D_f and D_w, incorporating necessary stalls in D_f, and deriving a shortest route between each *CL* and washing-port. After solving n number of ($place_j$, $route_j$) we have a *CL* free solution.

IV. EXPERIMENTAL RESULTS

The proposed fluid-level synthesis methodology concerning reliability with minimized cross contamination have been implemented in C language on a 1.99 GHz Core i3 machine with 4 GB RAM. In-Vitro Diagnosis (IVD), Polymerase Chain Reaction (PCR), Colorimetric Protein Assay with Exponentiation (CPA_Exp), and Interpolated Colorimetric Protein Assay (CPA_Int) have been chosen as benchmarks for performance assessment [12]. Two individual chip-dimensions have been considered for evaluating the effectiveness of the proposed method in terms of reliability, contamination reducibility, and ability to minimize bioassay completion time. The number of permissible electrodes for consecutive actuation has been considered as a constraint. To solve the NP-complete problem, we use greedy heuristics for placement of the modules after which Lee's algorithm finds the paths for each source-destination pair. After forming a staged graph, LSA is applied to find a set of non-dominated solutions that satisfies the reliability constraints.

TABLE I. Obtained results through the proposed synthesis method.

Bioassay	Chip Size	(# Allowable electrodes in consecutive actuation) in percentage	#contaminated cells throughout the assay	Completion time (s)	Execution time (s)
IVD	8×8	(13) 20%	2	13.50	0.435
		(9) 15%	3	19.58	0.419
	10×10	(20) 20%	2	13.50	0.372
		(15) 15%	4	13.92	0.369
PCR	8×8	(16) 25%	5	18.83	0.329
		(13) 20%	7	21.23	0.324
	10×10	(25) 25%	4	17.00	0.345
		(20) 20%	7	17.05	0.335
CPA_Exp	8×8	(16) 25%	23	34.68	0.679
		(13) 20%	24	36.77	0.699
	10×10	(25) 25%	22	35.22	0.897
		(20) 20%	24	39.64	0.950
CPA_Int	8×8	(16) 25%	25	39.16	0.668
		(13) 20%	27	43.16	0.732
	10×10	(25) 25%	22	37.83	0.712
		(20) 20%	25	42.86	0.825

Figure 6. Performance of the benchmarks with respect to bioassay completion time, allowable electrodes, and contaminated cells on varying chip dimensions.

Table I depicts the obtained results. In column three, the number of allowable electrodes in successive executions is considered. With the increase of this number, the completion time for the bioassay is reduced as more disjunctive operations can then be simultaneously executed. Also, the number of CL is getting smaller as more droplet routes are now available during Lee's algorithm. For an example, on an 8×8 biochip, the required time for CPA_Exp is 34.68s when the number of allowable electrodes for consecutive actuation is 25%. Also, the number of produced CL is 23. On the other hand, the completion time is larger and becomes 36.77s when we allow 20% electrodes for consecutive actuation. Indeed, we get one more extra CL. Now, as previously discussed, the ratio of

allowable sharing of electrodes to that of the total electrodes on the chip, measures the reliability of the chip during an execution. Fig. 6 shows the pictorial view of all these variations. Observably, if the designer wants a more reliable platform, then the completion time has to be compromised while the chances of cross contamination will be higher. On the contrary, a less degree of reliability reduces the completion time and results lesser contamination.

V. CONCLUSION

This paper proposes a fluid-level synthesis algorithm considering the safeness property of a biochip. For the need of safety-critical operation, a biochip is safer if it is more reliable and free from contamination during execution. However, bioassay completion time is also an important system attribute. By considering the involved reliability and contamination problem during execution, the bioassay completion time is minimized. This study investigates the variation of electrode reliability with the contaminated cells and bioassay completion time. Several benchmarks are used to evaluate the efficacy of the proposed fluid-level synthesis method.

REFERENCES

[1] Chakrabarty K. Design automation and test solutions for digital microfluidic biochips. IEEE Transactions on Circuits and Systems, 2010, 57(1), (4-17).
[2] Huang TW, Lin YY, Chang JW, Ho TY. Chip-level design and optimization for digital microfluidic biochips. IEEE Circuits and Systems, 54th International Midwest Symposium, 2011, (1-4).
[3] Ho TY. Design automation for digital microfluidic biochips: From fluidic-level toward chip-level. IEEE Solid-State and Integrated Circuit Technology (ICSICT), 11th International Conference, 2012, (1-4).
[4] Ho TY. Design automation for digital microfluidic biochips. IPSJ Transactions on System LSI Design Methodology, 2014, (16-26).
[5] Chakrabarty K. Design automation and test solutions for digital microfluidic biochips. IEEE Transactions on Circuits and Systems I, 2010, 57(1), (4-17).
[6] Chakrabarty K. Toward fault-tolerant digital microfluidic lab-on-chip: Defects, fault modeling, testing, and reconfiguration. Proceedings *IEEE ICBCS*, 2008, (329–332).
[7] Chen YH, Hsu CL, Tsai LC, Huang TW, Ho TY. A reliability-oriented placement algorithm for reconfigurable digital microfluidic biochips using 3-D deferred decision making technique. IEEE Transactions on Computer-Aided Design of Integrated Circuits and Systems, 2013, 32(8), (1151-1162).
[8] Zhao Y, Chakrabarty K. Cross-contamination avoidance for droplet routing in digital microfluidic biochips. IEEE Transactions on Computer-Aided Design of Integrated Circuits and Systems. 2012, 31(6), (817-830).
[9] Su F, Chakrabarty K. High-level synthesis of digital microfluidic biochips. ACM Journal on Emerging Technologies in Computing Systems (JETC). 2008 Jan 1; 3(4): 1.
[10] Yu ST, Yeh SH, Ho TY. Reliability-driven chip-level design for high-frequency digital microfluidic biochips. IEEE Transactions on Computer-Aided Design of Integrated Circuits and Systems, 2015, 34(4), (529-539).
[11] Lin CC, Chang YW. Cross-contamination aware design methodology for pin-constrained digital microfluidic biochips. IEEE Transactions on Computer-Aided Design of Integrated Circuits and Systems. 2011, 30(6): 817-828.
[12] Luo Y, Chakrabarty K, Ho TY. Hardware/Software co-design and optimization for cyberphysical integration in digital microfluidic biochips. Springer, 2015.
[13] Garey MR, Johnson DS. Computers and intractability, vol. 174.
[14] Dumitrescu I, Boland N. Algorithms for the weight constrained shortest path problem. International Transactions in Operational Research, 2001, 8(1):15-29.

Hysteresis Free sub-60 mV/dec Subthreshold Swing in Junctionless MOSFETs

Manish Gupta and Abhinav Kranti

Low Power Nanoelectronics Research Group, Discipline of Electrical Engineering,
Indian Institute of Technology Indore, Simrol 453552, India
akranti@iiti.ac.in

Abstract—**In this work, we report on a methodology to suppress hysteresis in current-voltage characteristics while retaining steep sub-60 mV/decade switching in *n*-type Double Gate (DG) Junctionless (JL) transistors. Hysteresis, which occurs due to impact ionization results in two different threshold voltages for forward and reverse gate voltage sweeps, can be effectively suppressed by using independent gate operation. It is shown that hysteresis free drain current with Subthreshold swing (*S*-swing) ~18 mV/decade can be achieved with a negative back gate (V_{bg}) of -0.9 V. The sub-kT/q *S*-swing implies negative values of total gate capacitance. The limit on back gate bias is imposed by the extent of Band-to-Band Tunneling (BTBT) which can potentially increase off-current. An optimization methodology is highlighted to suppress off-state BTBT while preserving the effectiveness of impact ionization to achieve sharp hysteresis free drain current transition from off-to-on state.**

Keywords—**Junctionless, MOSFET, Impact ionization, Band-to-Band tunneling.**

I. INTRODUCTION

In a Metal Oxide Semiconductor (MOS) Field Effect Transistor (FET), the transition from off-state to on-state is governed by a parameter known as Subthreshold swing (*S*-swing), which is defined as the minimum gate voltage required to change the drain current by one decade, and is given by ~$(kT/q)\ln(10)$, where k is Boltzmann constant, T is temperature, and q is electron charge [1-2]. The minimum theoretical minimum value of *S*-swing at room temperature (60 mV/decade) [3] can potentially limit the downscaling of MOSFETs. Several device architectures based on different conduction mechanism like Band-to-Band Tunneling (BTBT) [2] and Impact Ionization (II) have been proposed in the literature [4-6] to yield *S*-swing below 60 mV/decade. Such devices, offering *S*-swing < 60 mV/decade at low supply voltages, are ideally suited for the design of logic circuits with minimum power consumption [2]. Although tunneling based FETs have been considered as promising candidates for achieving *S*-swing < 60 mV/decade, their performance is limited by lower current drive and presence of traps [7]. In general, impact ionization requires higher drain bias to trigger strong floating body effects necessary for steep current transition [8]. The higher value of supply voltages makes impact ionization based devices non conducive for low power logic circuits.

Recently, a heavily doped gated resistor, eliminating the requirement of the fabrication of idealized *pn* junction, known as Junctionless (JL) transistor has been proposed [9]. JL

devices are relatively immune towards short channel effects due to the longer effective channel length in the subthreshold region [10]. Moreover, experimental results on JL device demonstrate that these devices can trigger impact ionization at relatively lower values of drain bias in comparison to inversion mode devices, and achieve sharp rise in drain current with nearly ideal *S*-swing ~1 mV/decade [11]. The reason for the enhanced degree of impact ionization is the higher current density and wider area of impact ionization which results in an enhanced Impact Generation Rate (*IGR*) [11].

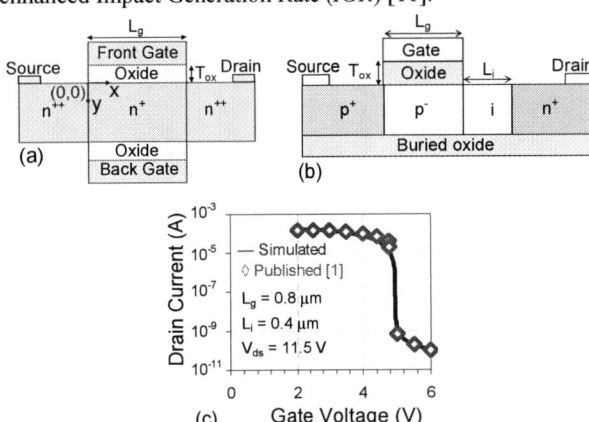

Fig. 1 Schematic diagram of (a) Double Gate Junctionless (JL) transistor, (b) Depleted Impact Ionization MOS transistor (DIMOS) [1] and (c) Comparison of our simulation result with the published experimental data for impact ionization induced steep switching Si based DIMOS transistor [1].

The previously reported work [12-14] on steep switching JL devices has focused on the understanding of impact ionization induced steep drain current transition and floating body effects. The problem associated with impact ionization induced ideal *S*-swing ~1 mV/decade in JL transistor is the occurrence of hysteresis, i.e. drain current traverses different paths during forward and reverse sweeps of gate voltage [13]. This effect has been previously observed in inversion mode [15], Punch through Impact Ionization MOS (PIMOS) transistor [16] and JL devices [13-14]. The occurrence of hysteresis with steep (*S*-swing ~1 mV/decade) drain current transition is not desirable for logic applications as it can result in two threshold voltages (V_{th}) corresponding to forward (V_{fth}) and reverse (V_{rth}) gate voltage sweeps [14]. While the hysteresis effect in the transfer characteristic can be certainly

978-1-5386-3693-0/18 $31.00 © 2018 IEEE

used for designing of dynamic memories [14], it leads to undesirable conditions such as racing, variations in drain current and propagation delays depending on device switching history which can cause instabilities like bit reversal for logic applications [17]. As hysteresis results in undesirable effects in circuit applications, its suppression while preserving S-swing < 60 mV/decade is necessary for steep switching logic devices and circuits.

In this work, we investigate the significance of independent gate operation in suppressing hysteresis in transfer characteristics of JL transistor while preserving sub-60 mV/decade S-swing at room temperature. Biasing the back gate with a negative value reduces the current density which critically limits hysteresis and increases the usable range of the device. Physical insights to the negative value of total gate capacitance in a JL transistor are presented and physical explanation for the occurrence of the same is discussed. In the later part of the paper, effect of back gate bias on the off-state BTBT in JL transistor is evaluated, and an optimization technique presented to reduce the off-state tunneling while preserving hysteresis free S-swing < (15-20) mV/decade.

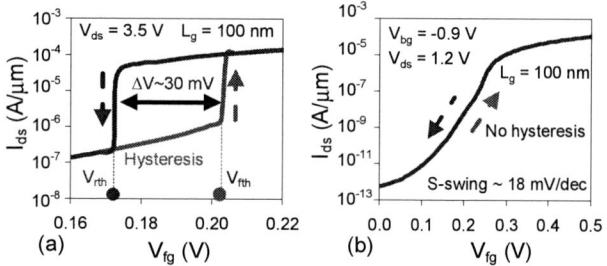

Fig. 2 Drain Current (I_{ds}) – front gate voltage (V_{fg}) characteristics with (a) hysteresis, and (b) without hysteresis in symmetric and asymmetric gate JL transistor, respectively for forward and reverse gate voltage sweep.

II. SIMULATION

In order to understand the occurrence of hysteresis, sub-60 mV/decade S-swing and total negative gate capacitance, Double Gate (DG) JL transistors were analyzed with ATLAS simulation tool [18]. The schematic diagram of Si JL transistor used in this work is shown in fig. 1a which consisted of a gate length (L_g) of 100 nm, film thickness (T_{si}) of 8 nm and gate oxide thickness (T_{ox}) of 1 nm. The channel doping (N_d) of JL MOSFETs was fixed at 10^{19} cm^{-3} whereas source/drain regions were heavily doped 10^{20} cm^{-3} to reduce the series resistance effect. The drain bias (V_{ds}) of 1.2 V was used for the analysis for asymmetric mode operation whereas a higher V_{ds} of 3.5 V was used for symmetric mode operation. The models used to capture the occurrence of impact ionization are similar to that suggested by Mayer *et al.* [19] and were calibrated with the published experimental results for Si based Depleted Impact Ionization MOS (DIMOS) transistor (fig. 1b) [1]. Our simulations results shown in fig. 1c are able to predict impact ionization induced steep drain current transition from off-to-on state at 300 K [1]. All simulations have been performed with field and concentration dependent mobility models [18] along with the modules of bandgap narrowing, impact ionization and bipolar effects. Scattering mechanisms affecting carrier mobility [20] are considered through Lombardy mobility model. The main focus of this work is to limit the impact

ionization generated power per unit volume in the semiconductor film to eliminate hysteresis from the transfer characteristics while reducing the off-state tunneling, and preserving S-swing < 60 mV/decade with negative values of total gate capacitance.

III. RESULTS AND DISCUSSION

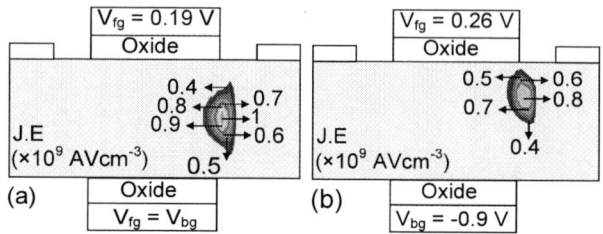

Fig. 3 2D contour plot of product of current density (J) and electric field (E) i.e. $J.E$ shown for reverse sweep of gate voltage at $V_{fg} = V_{rth} + 20$ mV for (a) symmetric gate ($V_{ds} = 3.5$ V) and (b) asymmetric gate operation ($V_{ds} = 1.2$ V) of JL transistor. Reverse sweep threshold voltage (V_{rth}) for symmetric and asymmetric gate JL devices are 0.17 V and 0.24 V, respectively.

Fig. 2a shows the drain current (I_{ds}) - gate voltage (V_{gs}) characteristics for symmetric gate operation of JL transistor at $V_{ds} = 3.5$ V. When a sufficiently higher drain bias (~ 3.5 V) is applied, electrons gain sufficient energy and collide with the crystal lattice to generate electron-hole pairs. The generated electrons contribute to drain current while holes accumulate at the lower potential region in the film and trigger strong floating body effect which induces positive feedback loop in parallel with normal MOS operation. This positive feedback increases the current density in the device which further enhances the impact ionization and raises the drain current sharply from off-state to on-state along with near ideal S-swing ~1 mV/decade for forward sweep and reverse sweep of gate voltage. The two different sweeps of the gate voltage result in two different threshold voltage for forward sweep (V_{fth} ~ 0.17 V) and reverse sweep (V_{rth} ~ 0.20 V) and consequently, hysteresis window ($\Delta V = V_{fth} - V_{rth}$) of ~30 mV. The hysteresis in I_{ds}-V_{fg} characteristics (fig. 2a) occurs due to a more negative gate voltage being required to deplete the excess impact generated carriers to overcome the positive feedback loop during reverse gate voltage sweep. Although the sharp rise in drain current is beneficial, the hysteresis associated with steep transition is not suitable for logic circuits. Hence, it is necessary to suppress hysteresis while preserving the sub-60 mV/decade at lower applied voltages.

The impact ionization generated power per unit volume in the semiconductor film can be limited by operating the JL device in the asymmetric mode. Fig. 2b shows the I_{ds}-V_{fg} curve for asymmetric mode JL transistor for forward and reverse gate voltage sweep. The transfer characteristics are obtained at back gate bias (V_{bg}) of -0.9 V. It can be observed that the no hysteresis is observed in the transfer characteristics which exhibit S-swing of ~18 mV/decade. While negative voltage at the back gate significantly depletes the electrons from the film underneath the gate and reduces the current density, it increases the potential difference between the channel and drain to a level sufficient to trigger impact ionization. Since impact generation rate is governed by the product of current density

978-1-5386-3693-0/18 $31.00 © 2018 IEEE   134

(J) and electric field (E) i.e. $J.E$ [12], a back bias of -0.9 V results in depletion of electrons which reduces current density while enhancing the field at the gate edge near to drain. The sufficiently high electric field ensures the occurrence of impact ionization so as to obtain hysteresis free S-swing of sub-60 mV/decade. The 2D contour plot of $J.E$ is shown in fig. 3 at V_{fg} = V_{rth} + 20 mV for symmetric (fig. 3a) and asymmetric (fig. 3b) gate operation. The V_{rth} obtained for reverse sweep of gate voltage is 0.17 V and 0.24 V for symmetric and asymmetric gate operation, respectively. The peak magnitude of $J.E$ in symmetrically biased JL device is observed at the center of the semiconductor film at the gate edge near to drain side whereas, for asymmetrically biased JL device, the maximum value of $J.E$ is observed above the centre of the film towards the drain. The reduced current density due to the application of negative back bias reduces the $J.E$ values from 1×10^9 AVcm^{-3} (symmetric JL, V_{ds} = 3.5 V) to 0.8×10^9 AVcm^{-3} (asymmetric JL, V_{ds} = 1.2 V, V_{bg} = -0.9 V). The lower $J.E$ value limits the degree of impact ionization in the device to achieve hysteresis free transfer characteristics. It should be noted that operating the JL device in symmetric mode at V_{ds} = 1.2 V will suppress impact ionization and will result in the classical S-swing of 60 mV/decade, and hence, the same is not beneficial for steep switching applications. In the remaining part of the paper, we shall focus on the asymmetric operation of JL transistor.

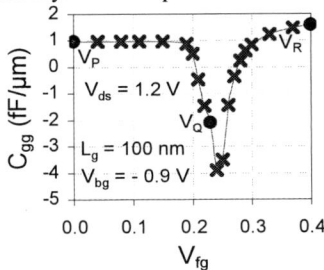

Fig. 4 Variation of total gate capacitance (C_{gg}) with the varying front gate voltage at V_{bg} = -0.9 V and V_{ds} = 1.2 V.

The unique attribute of the asymmetric mode operation of JL transistor is the negative value of total gate capacitance (C_{gg}) shown in fig. 3 which corresponds to sub-60 mV/decade current transition. The peak negative value of C_{gg} occurs at the gate voltage corresponding to the threshold voltage of the device. The physical affect leading to the negative total gate capacitance can be explained by analyzing the 2D contour plots for electron and hole concentration as shown in fig. 5a-f. These contour plots for asymmetrically JL device have been shown at three different front gate biases i.e. at V_{fg} = 0 V (represented by V_P), at V_{fg} = 0.24 V (represented by V_Q) and at V_{fg} = 0.4 V (represented by V_R).

When the front gate voltage is less than the reverse sweep threshold voltage i.e. V_{fg} = V_P, impact ionization in the film is not significant as electrons are depleted and lower values of C_{gg} ~1 fF/μm at V_{fg} = 0 V correspond to the off-state of the device. This can also be confirmed by the contour plot of electron concentration shown in fig. 5a where the electron concentration ($\leq 10^{13}$ cm^{-3}) is well below the channel doping (10^{19} cm^{-3}). Since the semiconductor film exhibits lower current density due to the depletion of electrons, the higher hole concentration (> 10^{18} cm^{-3}) shown in fig. 5b at V_{bg} = -0.9 V at the back

surface will not be able to initiate the impact ionization in the film.

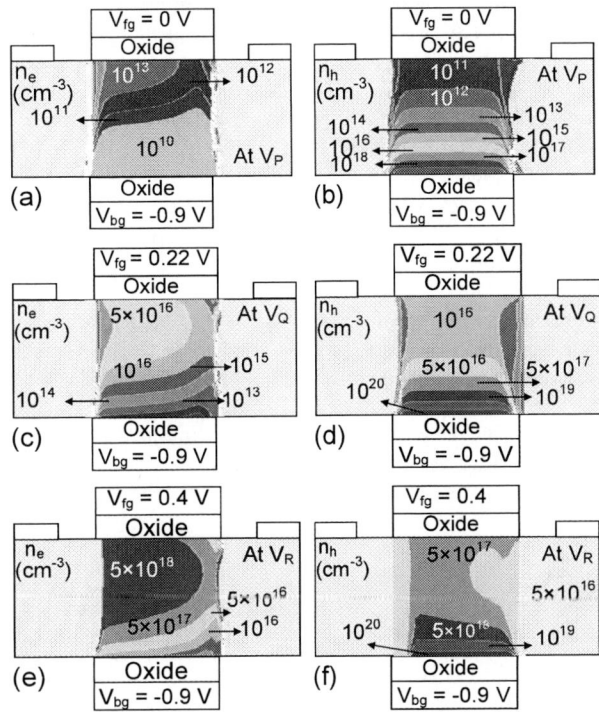

Fig. 5 2D contour plot of JL transistor showing (a) electron (n_e) and (b) hole (n_h) concentration at V_{fg} = 0 V (V_{fg} = V_P), (c) electron (n_e) and (d) hole (n_h) concentration at V_{fg} = 0.22 V (V_{fg} = V_Q), (e) electron (n_e) and (f) hole (n_h) concentration at V_{fg} = 0.4 V (V_{fg} = V_R).

As V_{fg} approaches to V_Q, electron concentration (fig. 5c) in semiconductor film increases (~ 10^{16} cm^{-3}) which further increases the current density and initiates impact ionization to generate more electron-hole pairs. Since lower potential region is maintained at the back surface, maximum hole concentration in the film at the back surface reaches to ~10^{20} cm^{-3} (fig. 5d). The successive accumulation of holes with increasing V_{fg} before the onset of steep transition changes a portion of the n-type film into a pseudo 'p-type' region as (n_h) > $10^4(n_e)$ at the back surface. The enhanced value of hole concentration forces the capacitance to a negative value, and the observed C_{gg} (fig. 4) corresponding to steep transition is -4 fF/μm. While JL transistor operating at relatively lower V_{ds} (~50 mV) do not exhibit impact ionization and electron concentration increases with an increase in front gate voltage, JL devices operated at higher V_{ds} (~1.2 V) show a significant change in hole concentration due to impact ionization (fig. 5d) which gives rise to negative value of total gate capacitance. The negative C_{gg} values in asymmetric JL transistor can also be understood with the classical definition of intrinsic capacitance between two terminals u and v, which is given by $C_{uv} = (\sigma)(\partial Q_u / \partial V_v)$ where u and v corresponds to source, gate and drain terminals, σ = -1 if $u \neq v$ and 1 if $u = v$, Q_u is the terminal electrode charge and V_v is the terminal voltage [21]. The gate-to-gate capacitance is defined as $C_{gg}= (\sigma)(\partial Q_g / \partial V_{fg})$. In asymmetrically biased JL device operating at V_{bg} = -0.9 V, the back surface of semiconductor film is accumulated by the hole for V_{fg} < 0.24 V

978-1-5386-3693-0/18 $31.00 © 2018 IEEE

(fig. 5d), and with an increase in V_{fg}, the hole concentration further rises in the film (~ 10^{20} cm^{-3}) whereas, electron concentration remains lower than the doping concentration (10^{19} cm^{-3}). This significant change in the magnitude of hole concentration from 10^{18} cm^{-3} (V_{fg} = 0 V) to 10^{20} cm^{-3} (V_{fg} = 0.22 V) with increasing front gate voltage forces ∂Q_g at the front gate electrode to be equally negative and, hence, C_{gg} switches its polarity from positive to negative around V_{th} = 0.24 V. Similar physical phenomena for negative values of total gate capacitance has been reported in the literature for power transistor [22]. For gate voltage higher than reverse sweep threshold voltage (V_{fg} = V_{rth}), the conduction channel in the asymmetric mode spreads across the film and the electron concentration (n_e) becomes high ~5×10^{18} cm^{-3}. As gate voltage increases beyond V_{fg} = 0.4 V, the conduction channel spreads even more and electron concentration in the film approaches to the channel doping (10^{19} cm^{-3}). Thus, electrons become the dominant (majority) carriers inside the film and C_{gg} switches back to positive values.

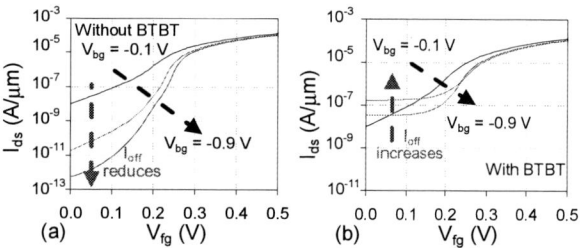

Fig. 6 I_{ds}-V_{gs} characteristics for asymmetrically biased JL transistor (a) without BTBT and (b) with BTBT for varying V_{bg} from -0.1 V to -0.9 V at V_{ds} = 1.2 V.

Previous studies [12-14, 23] have primarily focused on the understanding [13] and optimization [23] of the impact ionization induced steep switching JL devices and have not considered the effect of off-state tunneling which can be considerable in JL devices. Various research groups [24-29] have highlighted that BTBT can degrade the device performance at lower gate voltages. In fig. 6, I_{ds}-V_{fg} characteristics for asymmetric gate operation of JL transistor are shown without (fig. 6a) and with (fig. 6b) BTBT. The asymmetric mode operation of JL transistor without considering BTBT shows a reduction in off-current with a simultaneous reduction in S-swing with back gate bias. However, considering BTBT at more negative V_{bg}, off-current strongly deteriorates the performance of the JL devices. As shown in fig. 6b, BTBT increases the off-current and limits on-to-off current ratio, a crucial parameter for steep switching devices.

In fig. 7, I_{off} (fig. 7a) and S-swing (fig. 7b) are plotted with and without BTBT for varying back gate bias from -0.1 V to -0.9 V. The off-current (without BTBT) reduces from ~10 nA (V_{bg} = 0 V) to 10^{-3} nA (V_{bg} = -0.9 V) along with a desirable reduction in S-swing from 55 mV/decade to 18 mV/decade, respectively (fig. 7b). The significant reduction in off-current is due to the greater depletion of the electrons at more negative values of V_{bg}. This shows that asymmetric mode operation is indeed beneficial to achieve S-swing ~18 mV/decade. However, considering BTBT, S-swing and I_{off} both degrade with the increasing back gate bias. At V_{bg} = -0.1 V, S-swing

and I_{off} values obtained in JL device with BTBT and without BTBT are nearly equal (figs. 7a-b). This is due to the lower depletion of electrons from the back surface which result in reduced tunneling. While increasing V_{bg} to -0.3 V will reduce the S-swing value to ~36 mV/decade, it results in higher I_{off} i.e. I_{off} increases from 10^{-2} nA/μm to 50 nA/μm. Further increment in the back bias will considerably strengthen BTBT tunneling and deteriorates S-swing which degrades from 36 mV/decade (V_{bg} = -0.3 V) to 51 mV/decade (V_{bg} = -0.9 V) along with an increase in I_{off} from 10^{-3} nA/μm to 180 nA/μm, respectively.

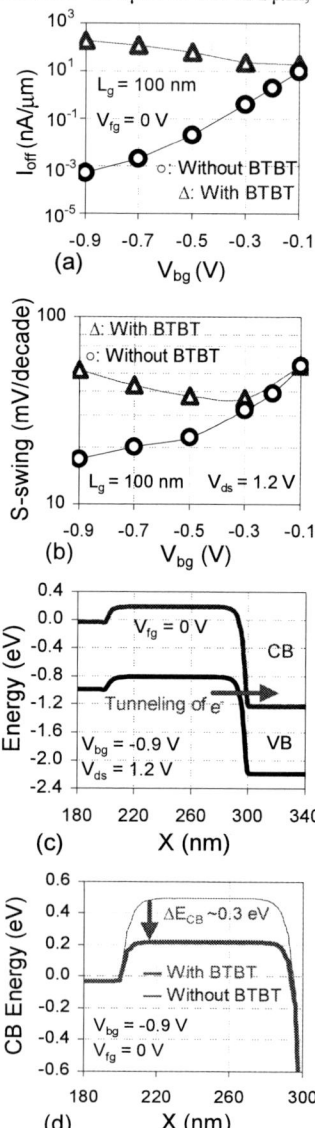

Fig. 7 Variation of (a) S-swing, (b) Off-state (I_{off}) current with varying absolute of V_{bg} from 0.1 V to 0.9 V, (c) Conduction Band (CB) and Valence Band (VB) showing the tunneling of electron at V_{bg} = -0.9 V and (d) CB energy profile at V_{bg} = -0.9 V with and without BTBT. Energy band diagrams are extracted along the x-direction at the channel location at V_{fg} = 0 V and V_{ds} = 1.2 V.

In order to understand the reason behind the increased off-current due to BTBT, Conduction band (CB) and Valence Band (VB) energy profile shown in fig. 7c is extracted along the x-direction at the channel location in the film at V_{ds} = 1.2 V and $V_{fg} = 0$ V. The negative voltages at the back gate deplete the electrons and cause bands to come in close proximity (channel-drain edge). This reduces the tunneling width (fig. 7c) along with an increased overlap between the filled states in VB and empty states in CB [29] facilitates tunneling of electrons from VB to CB and increases the generation of holes underneath the gate. The successive accumulation of holes at $V_{fg} = 0$ V with the more negative V_{bg} contribute to positive potential and lowers the conduction band energy (fig. 7d) from 0.5 eV (without BTBT) to 0.2 eV (with BTBT). This lowering in the barrier ($\Delta E_{CB} \sim 0.3$ eV) between the source and channel region underneath the gate significantly increase off-current and degrades the advantage achieved through sub-60 mV/decade S-swing.

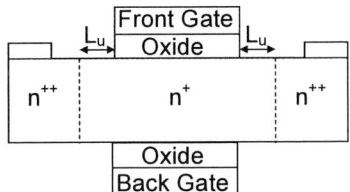

Fig. 8 Schematic diagram of DG JL transistor with an underlap (L_u) design.

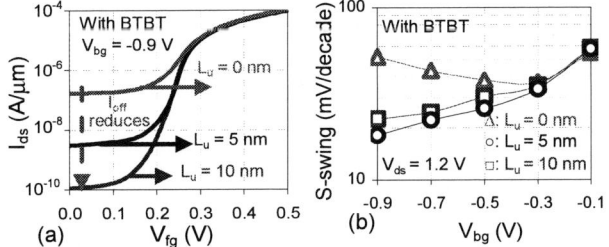

Fig. 9 (a) I_{ds}-V_{gs} characteristics for asymmetrically biased JL transistor with an underlap of 5 nm and 10 nm and (b) Comparison of S-swing values for an L_u = 0 nm and 10 nm. All the results are plotted with BTBT model at V_{bg} = -0.9 V and V_{ds} = 1.2 V.

In order to restore proper device functionality and to reduce off-state BTBT, a possible option proposed in the literature [29] is to reduce channel doping (N_d). Although, designing JL device with lower doping limits BTBT, it considerably lowers the current density in the device which is required to trigger impact ionization. Therefore, reducing channel doping is not a viable option. Alternatively, BTBT can be suppressed in JL transistor by incorporating gate-source/drain underlap region. The JL topology with underlap is shown in fig. 8. Underlap length (L_u) of 5 nm and 10 nm was considered at the source/drain side. JL device with an underlap region between gate and drain can reduce the degree of impact ionization [23] by increasing the effective channel length in the subthreshold region [10]. Hence, the selection of optimal underlap length is very crucial. I_{ds}-V_{gs} characteristics for L_u= 0 nm, L_u = 5 nm and L_u = 10 nm are shown in fig. 9a. Results depict that a 5 nm underlap reduces the tunneling current by nearly an order i.e. from 10^{-7} A to 10^{-8} A along with an improved S-swing i.e. S-swing reduces from 51 mV/decade to 18 mV/decade (fig. 9b). Further increase in the underlap length to 10 nm, reduces the off-current by ~2 orders i.e. 10^{-8} A to 10^{-10} A (fig. 9a) with a marginal increase in S-swing value from 18 mV/decade to 22 mV/decade (fig. 9b) due to a reduction in the strength of impact ionization. The asymmetrical gate operation of JL

transistor with an optimal underlap can be useful to significantly reduce the off-state BTBT and improve off-current with hysteresis free sub-60 mV/decade off-to-on transition.

IV. CONCLUSION

A systematic study for the suppression hysteresis in steep switching JL transistor is proposed through insightful investigation of asymmetric gate operation of JL device. The main reason for hysteresis free transition is the relatively lower value of $J.E$ which potentially limits the impact ionization generated power in the film to a level sufficient to preserve sub-60 mV/decade S-swing along with negative value of total gate capacitance. The potential advantage of operating JL device in the asymmetric mode is low S-swing values ~ (15-20) mV/decade achieved at relatively lower V_{ds} as compared to symmetric gate operation. It is show that off-state BTBT can limit the advantages gained through asymmetric operation. An optimal underlap of 10 nm can reduce BTBT without significantly degradation S-swing and hysteresis free transfer characteristics. The work provides insights into the functionality, design and optimization of steep switching JL MOSFETs.

ACKNOWLEDGMENT

This work is supported by the Council of Scientific and Industrial Research (CSIR), Government of India, under Grant 22(0688)/15/EMR-II, and by the University Grants Commission (UCG), Government of India, through the Junior Research Fellowship (JRF) award to Manish Gupta (Ref.: 4015/NET-June 2013).

REFERENCES

[1] C. Onal, R. Woo, S. H-Y Koh, P. B. Griffin and J.D. Plummer, A novel depletion-IMOS (DIMOS) device with improved reliability and reduced operating voltage," *IEEE Electron Device Lett.*, vol. 30, no. 1, pp. 64-67, Jan. 2009.

[2] A.M. Ionescu and H. Riel, "Tunnel field-effect transistors as energy-efficient electronic switches," *Nature*, vol. 479 pp. 329-37, 2011.

[3] I. Ferain, C.A. Colinge, and J. Colinge, "Multigate transistors as the future of classical metal–oxide–semiconductor field-effect transistors," *Nature*, vol. 479 pp. 310-16, 2011.

[4] C.-Y. Chen, J.-T. Lin and M.-H. Chiang "Subthreshold kink effect revisited and optimized for Si nanowire MOSFETs," *IEEE Trans. Electron Devices* vol. 63 pp. 903-909, 2016.

[5] *J.R. Davis, A.E. Glaccum, K. Reeson and P.L.F. Hemment, "Improved subthreshold characteristics of n-channel SOI transistors," IEEE Electron Device Letters, vol. 7, pp. 570-572, 1986.*

[6] B.Y. Mao, R. Sundaresan, C.E.D. Chen, M. Matloubain and G. Pollack, "The characteristics of CMOS devices in oxygen-implanted silicon-on-insulator structures," *IEEE Trans. on Electron Devices*, vol. 35, no. 5, pp. 629-633, 1988.

[7] Y. Qiu, R. Wang, Q. Huang and R. Huang "A comparative study on the impacts of interface traps on tunneling FET and MOSFET," *IEEE Trans. Electron Devices* vol. 61 pp. 1284-1291, 2014.

[8] J.-P. Colinge, "Hot-electron effects in Silicon-On-Insulator n-channel MOSFET's," *IEEE Trans. Electron Devices* vol. 34 pp. 2173-2177, 1987.

[9] J.-P. Colinge, C.-W. Lee, A. Afzalian, N. Dehdashti Akhavan, R. Yan, I. Ferain, P. Razavi, B. O'Neill, A. Blake, M. White, A.-M. Kelleher, B. McCarthy and R. Murphy, "Nanowire transistors without junctions," *Nature Nanotechnology*, vol. 5 no. 3, pp. 225-229, 2010.

[10] M.S. Parihar, D. Ghosh and A. Kranti, 2013 "Ultra-low power junctionless MOSFETs for subthreshold logic applications," *IEEE Trans. Electron Devices*, vol. 60, pp. 1540-46, 2013.

[11] C. W. Lee, A. N. Nazarov, I. Ferain, N. D. Akhavan, R. Yan, P. Razavi, R. Yu, R. T. Doria, and J. P. Colinge, "Low subthreshold slope in junctionless multigate transistors," *Appl. Phys. Lett.*, vol. 96, pp. 102106, 2010.

[12] M.S. Parihar, D. Ghosh, and A. Kranti, "Single transistor latch phenomena in junctionless transistors", *J. Appl. Phys.*, vol. 113, pp. 184503, 2013.

[13] M.S. Parihar, D. Ghosh, G. A. Armstrong, R. Yu, P. Razavi, and A. Kranti, "Bipolar effects in unipolar junctionless transistors," *Appl. Phys. Lett.*, vol. 101, pp. 093507, 2012.

[14] M.S. Parihar, D. Ghosh, G. Alastair Armstrong, and A. Kranti, "Bipolar snapback in junctionless transistors for capacitorless dynamic random access memory," *Appl. Phys. Lett.*, vol. 101, pp. 263503, 2012.

[15] A.Boudou and B.S. Doyle, "Hysteresis *I-V* effects in short channel silicon MOSFET's," *IEEE Electron Device Lett.*, vol. 8, pp. 300-2, 1987.

[16] K.E. Moselund, D. Bouvet, V. Pott, C. Meinen, M. Kayal, and A. M.Ionescu, "Punch-through imapct ionization MOSFET (PIMOS): From devcie principle to applications," *Solid-State Electron.*, vol. 52, 1336-44, 2008.

[17] S. Krishnan and J.G. Fossum, "Grasping SOI floating-body effects" *IEEE Circuits and Devices Magazine* vol. 14, pp. 32-37, 1998.

[18] ATLAS Users Manual, Silvaco.

[19] F. Mayer, C. Le Royer, G. Le Carval, L. Clavelier and S. Deleonibus, "Static and dynamic TCAD analysis of IMOS performance: from the single device to the circuit," *IEEE Trans. Electron Devices*, vol. 53 pp. 1852-57, 2006.

[20] C. Lombardi, S. Manzini, A. Saporito, and M. Vanzi, "A physically based mobility model for numerical simulation of nonplanar devices," *IEEE Trans. Computer Aided Design of Integrated Circuits*, vol. 7, no. 11, pp. 1164-1171, 1988.

[21] D. Flandre "Analysis of floating substrate effects on the intrinsic gate capacitance of SO1 MOSFET's using two-dimensional device simulation," *IEEE Trans. Electron Devices* vol. 40 pp. 1789-1796, 1993.

[22] I. Omura, H. Ohashi and W. Fichtner "IGBT negative gate capacitance and related instability effects," *IEEE Electron Device Lett.* vol. 18 pp. 622-624, 1997.

[23] M. Gupta and A. Kranti, "Sidewall spacer optimization for steep switching junctionless transistors," *Semicond Sci. Technol.* vol. 31, no. 6, pp. 065017, 2016.

[24] J. Hur, B.-H. Lee, M.-Ho Kang, D.-C. Ahn, T. Bang, S-B. Jeon, and Y.-K. Choi, "Comprehensive analysis of gate-induced drain leakage in vertically stacked nanowire FETs: Inversion-mode versus junctionless mode," *IEEE Electron Device Lett.*, vol. 37, no. 5, pp. 541-544, May 2016.

[25] S. Sahay and M.J. Kumar, "Insight into lateral band-to-band-tunneling in nanowire junctionless FETs," *IEEE Trans. on Electron Devices*, vol. 63, no. 10, pp. 4138-42, Oct. 2016.

[26] T.A. Oproglidis, T.A. Karatsori, S. Barraud, G. Ghibaudo, C.A. Dimitriadis, "Leakage current conduction in metal gate junctionless nanowire transistors," *Solid. State. Electron.*, vol. 131, pp. 20–23, May 2017.

[27] C. Sun, R. Liang, L. Liu, J. Wang, and J. Xu, "Leakage current of germanium-on-insulator-based junctionless nanowire transistors," *Appl. Phys. Lett.*, vol. 107, no. 26, pp. 132105, Sep. 2015.

[28] H. Lou, D. Li, Y. Dong, X. Lin, J. He, S. Yang and M. Chan, "Suppression of tunneling leakage current in junctionless nanowire transistors," *Semicond Sci. Technol.* vol. 28, no. 6, pp. 125016, Nov. 2013.

[29] S. Gundapaneni, M. Bajaj, R.K. Pandey, Kota V. R. M. Murali, S. Ganguly and A. Kottantharayil, "Effect of Band-to-Band Tunneling on Junctionless Transistors," *IEEE Trans. Electron Devices*, vol. 59, no. 4, pp. 1023-29, Apr. 2012.

978-1-5386-3693-0/18 $31.00 © 2018 IEEE

2018 31th International Conference on VLSI Design and 2018 17th International Conference on Embedded Systems

New Asymmetric Atomistic Model for the Analysis of Phase-engineered MoS$_2$-Gold Top Contact

Richa Chakravarty[1,2], Dipankar Saha[1], *Member, IEEE* and Santanu Mahapatra[1], *Senior Member, IEEE*
[1]*Nano-Scale Device Research Laboratory, Department of Electronic Systems Engineering,*
Indian Institute of Science (IISc) Bangalore, Bangalore-560012, India
[2]*Indian Space Research Organisation (ISRO) Headquarters, Antariksh Bhavan, Bangalore- 560094, India*
Email: richa.chak@gmail.com, dipsah_etc@yahoo.co.in, and santanu@iisc.ac.in

Abstract—Realizing low contact resistance at source/drain is one of the key challenges for obtaining high ON-current in 2D (two dimensional)-material channel based Metal-Oxide-Semiconductor transistors. In order to form ultra-low resistance contacts, the experimental techniques which involve inducing local metallic phases in the usual semiconducting MoS$_2$ crystal, have attracted much attention. Howbeit, the density functional theory (DFT) based atomistic modeling of metal-2D material interface is important to get insights on the charge transfer through such systems, which is difficult to access by experiments. Though numerous studies have been reported for various semiconducting MoS$_2$-metal interfaces, no such work exists for phase-engineered MoS$_2$. In this work, first we compare the Schottky barrier height of both the systems, by analysing their electronic structures obtained via the DFT calculations.

Next we propose a novel asymmetric Au-MoS$_2$-Au atomistic model to assess the carrier transport and estimate the resistances offered by such hetero interfaces using non equilibrium Green's function (NEGF) formalism. Our study reveals almost a three-fold decrease in the resistance of the Au-1T$'$-Au device compared to that of the Au-2H-Au one.

Keywords-Density functional theory; phase-engineered MoS$_2$; atomistic model; non equilibrium Green's function; local density of states.

I. INTRODUCTION

Research on two dimensional (2D) transition metal dichalcogenides (TMDs) propelled to a new height with the successful fabrication of monolayer MoS$_2$ transistors by Radisavljevic et al. [1]. According to the ITRS (International Technology Roadmap for Semiconductors) report [2] on emerging research materials, MoS$_2$ could be the potential candidate to replace silicon-based channels in sub-decananometer next-generation transistors. However, the biggest challenge in designing ultra-thin transistors with single layer MoS$_2$ lies in forming the low resistance source/drain contacts [2], [3]. A significantly high ON-current (I$_{ON}$) and a higher value of mobility are the two major criteria which need to be satisfied in order to achieve the goal of designing the ultra-thin sub-10 nm solid state digital switches [1], [3], [4]. Thus far, the reported I$_{ON}$ values of the MoS$_2$ transistors are nowhere near the expectation [1], [5]. This is mainly because of the suppression of carrier

injection at the metal-semiconductor interfaces and the scattering events which occur due to the surface-roughness of the substrate beneath. Several techniques, e.g., chemical doping, insertion of graphene and TiO$_2$ in between the MoS$_2$ and the metal contact have been deployed to overcome the problem of suppressed carrier injection. R. Kappera et al. recently proposed a strategy, where the locally patterned 1T metallic phases (on a monolayer 2H MoS$_2$ flake) have been treated as the electrodes [3]. They have demonstrated a significant improvement in the device performance (including I$_{ON}$ / I$_{OFF}$ ratio, subthreshold swing, ON-current, etc.), when the metals are contacted via the 1T electrodes.

However, for a better understanding of the charge transfer trough the metal- phase-engineered MoS$_2$ contacts, as well as to know the electronic properties which get locally modulated, we need to opt for the first-principles based DFT study.

In this work, we consider two dynamically stable ploytypes of MoS$_2$ crystal which are 2H and 1T$'$. Nonetheless, we take Gold (Au) as the metal electrode which forms physisorbed contact with MoS$_2$ [6]. The 2H phase of MoS$_2$ ("Mo" surrounded by the "S" atoms in triangular prismatic form) is energetically a stable structure [7], [8]. From a 2H sample, the phase transformed metallic 1T MoS$_2$ can be obtained using the chemical method called "Li$^+$" intercalation [9], [10]. However, the energetic stability of the 1T phase of MoS$_2$ is quite poor [8], [10], [11]. Thermodynamically, a more stable state is the distorted counterpart of 1T polytype which is known as 1T$'$ MoS$_2$ [10], [11]. Further, as reported in [12], [13], the dynamical stability of the 1T$'$ phase is significantly better than that of the 1T MoS$_2$. Hence, in this paper, we opt for the phase transformed 1T$'$ MoS$_2$ (instead of 1T) and compare the electronic properties of the 1T$'$-Au interface with that of the 2H-Au one. We propose the asymmetric Au-2H-Au and Au-1T$'$-Au devices and probe into the carrier transport using DFT-NEGF combination. The modeling methodology described in this work could be useful for understanding the charge carrier transfer in other vertically stacked 2D material interfaces too.

978-1-5386-3693-0/18 $31.00 © 2018 IEEE   139

II. METHODOLOGY

In this study, the first-principles based density functional theory calculations are carried out using Atomistix Tool Kit (ATK) [14]. Electronic-structure calculations and geometry optimizations are done using the generalized gradient approximation (GGA) exchange correlation with Perdew-Burke-Ernzerhof (PBE) functional [15] in conjunction with the OPENMX (Open source package for Material eXplorer) code [16], [17]. The basis sets for "Mo" and "S" are taken as s3p2d1 and s2p2d1 respectively. The k-points in Monkhorst-Pack grid is set to $9 \times 9 \times 1$. The optimized in-plane lattice constants for the hexagonal 2H unit cell of MoS$_2$ have been computed as a = b = 3.197 Å. Apart from that, for the 1T′ unit cell, the optimized values of lattice constants are obtained as a = 3.18 Å, b = 5.75 Å and c = 18 Å. For the 2H unit cell, we find a direct band gap of 1.7 eV at K-point of the Brillouin zone. Nonetheless, considering the optimized unit cell of 1T′ MoS$_2$, we observe a negligible band gap opening near the Γ point (which is consistent with the previously reported results [12], [13]). We have opted for a value of 90 Hartree as the density mesh cut off in numerical accuracy settings. The Grimme's dispersion correction (DFT-D2) has been included in order to consider the van der Waals (vdW) interactions [18].

In order to carry out the bulk as well as the device analyses, we take hexagonal 2H and 1T′ MoS$_2$ supercells with the in-plane lattice constant values ~ 12.7 Å. Each of those supercells contains 32 "S" atoms and 16 "Mo" atoms. For designing the gold electrodes, we take (111)-surface cleaved FCC-Au (of lattice constant 4.07 Å).

It is important to mention here that, for both the 2H and the 1T′ supercells, the strain mismatch with Au(111) is restricted to a very low value (mean absolute strain <1.4 %). Besides, we pick a basis set of s2p2d1 for the Au atoms of the metal contacts. Fig. 1(a-b) illustrate the Au(111)-MoS$_2$ hetero structures, with 6 layers of gold (sufficient for screening the effect of any surface related phenomenon). Howbeit, the calculated interlayer distances are slightly different for those two structures (Fig. 1(a-b)). We find that the Au(111) binds with the monolayer 2H and 1T′ MoS$_2$ surfaces at the distances of 2.85 Å and 2.57Å respectively.

In order to conduct the electrical transport calculations, we take help of the DFT-NEGF combination and set the Monkhorst-Pack grid to $3 \times 3 \times 99$ (in X, Y and Z directions respectively). Dirichlet boundary condition along the transport direction (i.e., the Z direction), and periodic boundary conditions in the other two directions have been used for solving the Poisson's equation. Other than that, the lower bound of the of contour integral is taken as 3 Hartree.

III. RESULTS AND DISCUSSIONS

First, we investigate the electronic-structures of the Au(111)- MoS$_2$ hetero interfaces of Fig. 1(a-b). Fig. 1(c) shows the projected density of states (PDOS) diagram of 2H MoS$_2$, whereas Fig. 1(d) delineates the same for the 1T′ MoS$_2$ sample. The Fermi energy is denoted by E$_F$. Following the conventional way, from the PDOS of 2H monolayer MoS$_2$, we identify the Schottky barrier (SB) ~ 0.51 eV. On the other hand, for the case of 1T′ MoS$_2$, there exist a significant amount of states near the Fermi energy reflecting the conductive nature of the material.

However, the goal of this study is to accurately determine

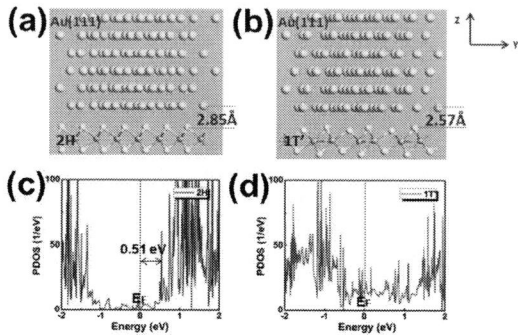

Figure 1.   (a) Interface structure of Au(111) and 2H MoS$_2$ flake. (b) The same for Au(111) and 1T′ MoS$_2$. We illustrate the cross-plane views of the structures. Structures are periodic in the X-Y directions. PDOS diagrams for the MoS$_2$ layers, from Au(111)-2H and the Au(111)-1T′ hetero structures are shown in (c) and (d) respectively.

the SB with the help of transport calculation. In order to find the barrier height at the vdW interface (between Au and phase-engineered monolayer MoS$_2$), we propose the device architectures as shown in Fig. 2. Here, we purposefully

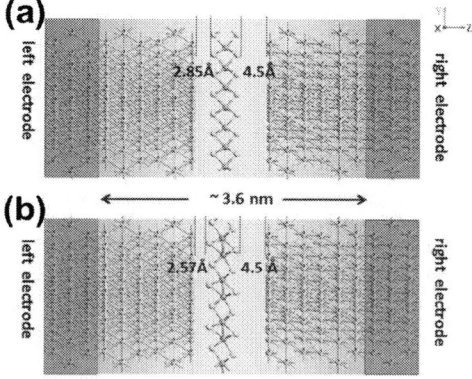

Figure 2.   Proposed architectures of the devices, with the single layer MoS$_2$ flakes of (a) 2H-phase, and (b) 1T′-phase. The left and the right electrodes are of same length (7.06 Å).

shift the right electrodes by an amount of 4.5 Å from the MoS$_2$ surfaces, such that the coupling effects get minimized. However, the left-Au(111)-electrodes are still binding with the monolayer 2H and 1T′ MoS$_2$ surfaces at the optimized equilibrium distances of 2.85 Å and 2.57 Å respectively.

978-1-5386-3693-0/18 $31.00 © 2018 IEEE

So that the metal-MoS$_2$ interaction at the vdW interfaces mainly occur due to the Au(111) layers of the left electrodes. The right-Au(111)-electrodes artifact as pseudo terminals which help to provide the desired electric field at the overlapping regions of 2H-Au(111) and 1T'-Au(111).

Next, we obtain the energy-position resolved local density of states (LDOS) diagrams for both the structures of Fig. 2(a) and Fig. 2(b), at zero bias. We observe a precise energy barrier of 0.52 eV at the interface region of the left-Au(111)-electrode and the 2H MoS$_2$ (shown in Fig. 3(a)). This is consistent with the PDOS result as shown earlier in Fig. 1(c).

Howbeit, for the other structure we find that the 1T' flake is quite transparent to the charge carriers, and the carriers find almost no barrier at the 1T'-Au(111) interface. As shown in Fig. 3(b), the LDOS values corresponding to 1T' region describe the ease of tunnelling of carriers from the left electrode to the right one.

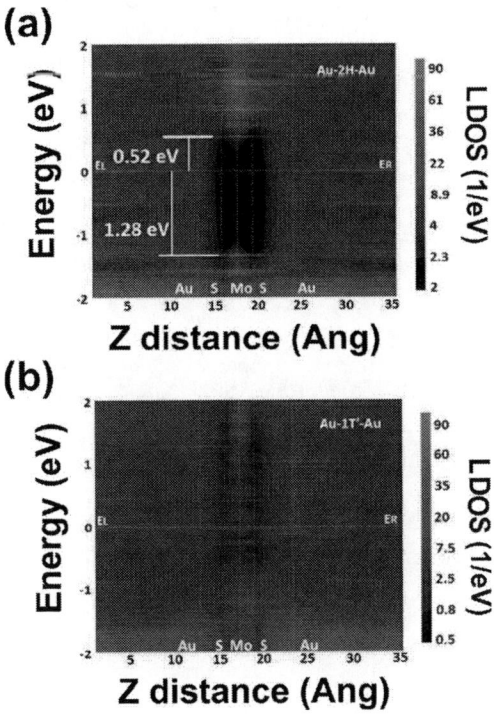

(a)

(b)

Figure 3. Energy-position resolved local density of states diagrams of the (a) Au-2H-Au, and (b) Au-1T'-Au device architectures, for the energy range of -2 eV to +2 eV. E$_L$ and E$_R$ are the Fermi levels of the left and the right electrodes respectively.

To further justify this qualitative behaviour, we find out the transmission spectra of both the structures (i.e., Au-2H-Au and Au-1T'-Au). Considering the energy window of -2 eV to +2 eV (around the Fermi level), we find the transmission states of Au-1T'-Au structure is significantly larger than that of the Au-2H-Au device. Fig. 4 shows the transmission spectra (up-spin components) for the Au-2H-

Au and Au-1T'-Au devices for the bias voltages (V$_{bias}$) of 0 Volt and 0.5 Volt.

Finally, we compute the current conduction of the two architectures (Fig. 2(a) and Fig. 2(b)) at the finite bias of 0.5 Volt. As per the Landauer's picture, current is computed using the expression [19]–[21],

$$I = \frac{2q}{h} \int_{-\infty}^{+\infty} T_e(E) \left[f\left(\frac{E - \mu_R}{k_B T}\right) - f\left(\frac{E - \mu_L}{k_B T}\right) \right] dE,$$ with

the transmission function ($T_e(E)$). '$f(E, \mu)$' denotes the Fermi-Dirac distribution function. The chemical potentials of the right and the left electrodes are represented as μ_R and μ_L respectively. At 0.5 Volt, the current through the Au-2H-Au is obtained as 1634 nA, whereas the same for the Au-1T'-Au is found as 4305 nA. However, at near-equilibrium (utilizing the linear response approximation [20], for temperature $\sim$ 300 K) the resistance values of the Au-2H-Au and the Au-1T'-Au structures are computed as $\sim$ 261.4 K-Ω and $\sim$ 106.2 K-Ω respectively.

Thus, almost a three-fold decrease in the resistance of the Au-1T'-Au device is evinced, irrespective of the V$_{bias}$ (for small bias voltages). Moreover, the results derived form transport studies are consistent with the experimental findings [3].

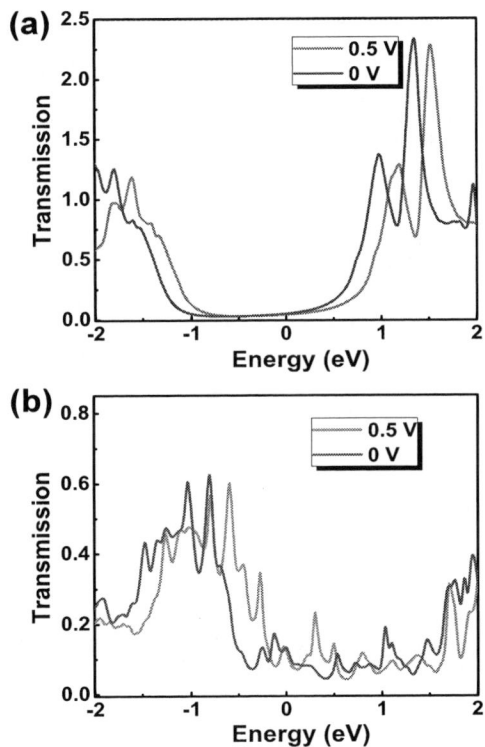

(a)

(b)

Figure 4. Zero bias and the finite bias transmission spectra obtained for the (a) Au-2H-Au, and (b) Au-1T'-Au devices. The energy axes are denoting E-E$_F$.

IV. CONCLUSION

In this paper, we propose a new atomistic model to obtain the SB at the interface of the Au(111)- MoS_2 vdW structure. Our results show that the phase-engineered $1T'$ MoS_2 can provide significantly less resistance compared to the 2H polytype, when contacted with Au electrode. The near ohmic nature of the Au-$1T'$ contact could be a viable solution to form the source/drain electrodes for the next-generation nano-scaled transistors.

ACKNOWLEDGMENT

This work was supported by the Science and Engineering Research Board (SERB), Department of Science and Technology (DST), Government of India, under Grant No: SB/S3/EECE/0209/2015. R.C. acknowledges Guoping Gao of Queensland University of Technology, Brisbane, Australia, for the useful discussions. Authors would also like to thank Dr. Umberto Martinez Pozzoni, Dr. Jess Wellendorff and Dr. Daniele Stradi of QuantumWise A/S Copenhagen, Denmark, for the fruitful technical discussions.

REFERENCES

[1] B. Radisavljevic et al., "Single-layer MoS_2 transistors," *Nature Nanotechnology*, vol. 6, pp. 147-150, 2011.

[2] International Technology Roadmap for Semiconductors (ITRS), 2013, *Available: http://www.itrs2.net/2013-itrs.html*

[3] R. Kappera et al., "Phase-engineered low-resistance contacts for ultrathin MoS_2 transistors," *Nature materials*, vol. 13, pp. 1128-1134, 2014.

[4] H. Schmidt et al., "Transport Properties of Monolayer MoS_2 Grown by Chemical Vapor Deposition," *Nano letters*, vol. 14, pp. 1909-1913, 2014.

[5] Han Wang et al., "Integrated Circuits Based on Bilayer MoS_2 Transistors," *Nano Lett.*, 12, 4674-4680, 2012.

[6] C. Gong, L. Colombo, R. M. Wallace, and K. Cho, "The Unusual Mechanism of Partial Fermi Level Pinning at Metal-MoS_2 Interfaces," *nano Lett.*, 14, 1714-1720, 2014.

[7] Goki Eda et al., "Coherent Atomic and Electronic Heterostructures of Single-Layer MoS_2," *ACS Nano*, vol. 6, no. 8, 7311-7317, 2012.

[8] M. Calandra, "Chemically exfoliated single-layer MoS_2: Stability, lattice dynamics, and catalytic adsorption from first principles," *Phys. Rev. B*, 88, 245428, 2013.

[9] D. Voiry et al., "Covalent functionalization of monolayered transition metal dichalcogenides by phase engineering," *Nature Chem.*, vol. 7, 45-49, 2015.

[10] G. Gao et al., "Charge Mediated Semiconducting-to-Metallic Phase Transition in Molybdenum Disulphide Monolayer and Hydrogen Evolution Reaction in New $1T'$ Phase," *J. Phys. Chem. C*, 119, 13124-13128, 2015.

[11] D. B. Putungan, S-H. Lin, and J-L. Kuo, "A first-principles examination of conducting monolayer $1T'$-MX2 (M = Mo, W; X = S, Se, Te): promising catalysts for hydrogen evolution reaction and its enhancement by strain," *Phys.Chem.Chem.Phys*, 17, 21702, 2015.

[12] X. Qian, J. Liu, L. Fu, and J. Li, "Quantum spin Hall effect in two-dimensional transition metal dichalcogenides," *Science*, vol. 346, 6215, pp. 1344-1347, 2014.

[13] D. Saha and S. Mahapatra, "Atomistic modeling of the metallic-to-semiconducting phase boundaries in monolayer MoS_2," *Appl. Phys. Lett.*, 108, 253106, 2016.

[14] QuantumWise Atomistix ToolKit (ATK) with Virtual NanoLab, Available at http://quantumwise.com/.

[15] J. P. Perdew, K. Burke, and M. Ernzerhof, "Generalized gradient approximation made simple," *Phys. Rev. Letters*, 77, 18, 3865, 1996.

[16] T. Ozaki, "Variationally optimized atomic orbitals for large-scale electronic structures," *Phys. Rev. B*, 67, 155108, 2003.

[17] T. Ozaki and H. Kino, "Numerical atomic basis orbitals from H to Kr," *Phys. Rev. B*, 69, 195113, 2004.

[18] S. Grimme, "Semiempirical GGA-type density functional constructed with a long-range dispersion correction," *J. Comput. Chem.*, vol. 27, no. 15, pp. 1787-1799, 2006.

[19] M. Büttiker, Y. Imry, R. Landauer, and S. Pinhas, "Generalized many-channel conductance formula with application to small rings," *Phys. Rev. B*, 31, 10, 6207-6215, 1985.

[20] D. Saha and S. Mahapatra, "Theoretical insights on the electro-thermal transport properties of monolayer MoS_2 with line defects," *J. Appl. Phys.*, 119, 134304, 2016.

[21] M. Brandbyge et al., "Density-functional method for nonequilibrium electron transport," *Phys. Rev. B*, 65, 165401, 2002.

Power Side Channel Resistance of RNS Secure Logic

Ravikumar Selvam[*], and Akhilesh Tyagi[†]

Dept. of Electrical & Computer Engineering
Iowa State University
Ames, Iowa - 50011, USA
Email: {[*] rkselvam, [†] tyagi}@iastate.edu

Abstract—Over the last decade, significant research effort has gone into secret sharing schemes to secure cryptographic implementations to thwart power side-channel attacks. Higher-order side-channel attacks can correlate the behavior of multiple shares of a bit that leads to learning the bit state. This violates the power side-channel privacy of cryptographic logic families such as t-private logic. The only recourse is to increase the number of secret shares t, which results in excessive hardware (quadratic in t) needs in area, energy and time for providing the desired level of security. In this paper, we present a new secure logic family based on secret sharing concepts using a residue number system. This technique maps the input from binary space into multiple un-correlated shares in the residue domain. These shares are processed independently in independent hardware lanes. The results are decoded back to binary space using the Chinese Remainder theorem. This technique increases the computational complexity for a side channel adversary through proper selection of random mask and residual moduli - which increase both side-channel privacy and cryptographic privacy. Further, we implemented the secure RNS logic and computed the SCA metrics. Finally, we evaluated the power SCA resistance using ML-classifiers. The results show that our RNS secure logic provides better resistance against power side-channel attacks both in terms of power distribution uniformity and success rates of power side channel attack root kits.

I. INTRODUCTION

Side-channel attacks (SCA) are hardware cryptanalytic techniques to reveal the secret data value such as a key embedded into an algorithm by exploiting the implementation vulnerabilities. If two different values for a key or a subkey result in different measurements of a physical attribute such as power, timing, electromagnetic radiation or even acoustics, the side channel privacy is lost through this physical leakage. In [1], Paul Kocher reported the first side-channel attack and showed that the power consumption of the device is highly dependent on intermediate values of the cryptographic algorithm. To prevent such attacks, it is important to randomize or mask the intermediate values to decouple them from the device power consumption. Several countermeasure techniques have been proposed to counteract side-channel attacks in [2], [3]. Many of these approaches use random values to mask intermediate values to make the power consumption independent of the original binary data. In the mean time, secret sharing schemes [4] were developed in cryptography for multiparty computation and for sharing secrets. An adaptation of secret sharing schema for SCA splits each original bit into multiple uncorrelated shares in order to prevent the device side channel leakage. The main idea behind secret sharing schemes is to split the input data into multiple shares. All the data shares are processed independently in parallel. The result of computation at the primary output end contains multiple result shares for each expected primary result. These result shares are combined at the output end to reconstruct the primary output. These techniques improve the resistance against power

analysis attacks by providing uniformity & data independence in power consumption of individual shares.

Perhaps the best known secret sharing schema, Ishai *et al.* [5] developed a bit level secret sharing technique by splitting each input bit into $t + 1$ shares. For each input value x, t shares are derived from t random values r_{x1}, r_{x2}, r_{x3}, ..., r_{xt}. $(t + 1)$st share is computed as $x \oplus r_{x1} \oplus r_{x2} \oplus r_{x3} \oplus ... \oplus r_{xt}$. Their adversary model is a t-probing adversary which is a stronger adversary than a power side channel adversary. A t-probing adversary can probe up to any t circuit nodes per cycle. A t-private circuit does not reveal any information about any bit x, even with a t-probing adversary. This provides both power side channel privacy as well as limited (to t nodes probing) cryptographic privacy. Park *et al.* [6], [7] showed several practical constructions of t-private gates that optimize its area, energy, and number of random bits. Mangard [8] discussed a security flaw in private circuits. They stated that glitches contribute significant power consumption and showed how such glitches weaken the security of private circuits. Later, in [9], Zachary *et al.* showed a practical power analysis attack using correlation enhanced collision attack. A secret sharing scheme similar to t-private circuits is proposed in Svetla *et al.* [10]. This secret sharing technique is called Threshold implementation. It is based on multiparty computation and provably secure against DPA with fewer assumptions over hardware leakage. However, the threshold implementation techniques are still vulnerable to higher-order power analysis attacks described in [11].

Higher-Order Side-channel analysis is physical cryptanalysis that exploits the combined leakage through power consumption of multiple individual shares. This analysis uses higher-order statistical moments to recover the secret value of a cryptographic algorithm [12]. Most of the existing countermeasures are still vulnerable to such higher order power analysis attacks for two reasons. First, the leakage of intermediate values is distributed over shares, which is the primary SCA mitigation technique rather than masking the share values. Further, these shares utilize a linear function to reconstruct the original data. Hence it is relatively easy for an adversary to model the leakage of the shared secret implementation. Second, if the shares are processed together with common Vdd and Ground pins, the combined power consumption leads to leakage from such a susceptible implementation on actual intermediate values. Further, if the secure implementation were still in Boolean space, then the adversary is able to model the leakage with a hypothetical secret value along with some additional mask bits to correlate with the target implementation leakage.

Logic design styles to make power consumption independent of data values with dual rail logic include SABL [13], [14] and WDDL [15]. These design styles offer power side channel privacy, but not cryptographic privacy. The data is in open,

non-encrypted form. The stronger countermeasure techniques like t-private scheme provide both power side-channel privacy and limited cryptographic privacy. A cryptographic adversary needs to observe $t+1$ shares in order to decrypt original values. Out of practical considerations, the value of t cannot be very large. This opens up space for a secure design style that is both power side channel private and cryptographically private within the design space for secure system implementations. Our proposed RNS secure logic fills this need.

In this paper, we propose a new secure design style based on [16] and [17]. Our approach is to transform a bit in the Boolean domain into multiple encrypted shares derived from residues in a residue number system (RNS). These residue shares exhibit homomorphism with respect to the bit-wise operations like AND and XOR. Our proposed scheme is well suited for a multi-core platform, where an application can exploit parallelism in security related applications. Each encrypted share can be processed in a separate core independently. There are many variations to the base schema for residue generation depending on the adversary model and the desired resource overhead. We explore this schema space to come up with three possible secure design styles with varying characteristics. We evaluate the resistance of secure RNS circuits against various side-channel adversary models. Further we implement the secure RNS logic and report its power side channel resistance through power uniformity based metrics and success rates of power side channel root kits. The root kits use power analysis attack with machine learning classifiers such as LDA, QDA and Naive Bayes. RNS secure logic exhibits lowest success rates for machine learning based attacks compared to t-private logic.

The switching uniformity can be evaluated either analytically or through a distance metric like Kullback-Leibler divergence [18]. A natural conclusion seems to be that as switching gets more uniform or KL-divergence of power reduces, the success rate for power side channel attack root kits should go down. We have however observed that even with an increase in KL-divergence for power, the power side-channel root kit success rate has gone down. We conjecture that cryptographic privacy, even if not directly addressing power uniformity, thwarts power side-channel attacks. An interesting trade-off between power side-channel privacy and cryptographic privacy to minimize the success rate of a power side-channel adversary exists, which we are exploring.

II. BASIC PRINCIPLES

In this section, some basic principles for our approach are discussed. Our proposed scheme maps from the message space to the residue code space. Message space consists of binary values ('0' or '1') and corresponding bit level operations/gates. Residue code space consists of residue values represented with l-bits. These residues use modulo operations such as modular addition and modular multiplication.

In message space, we use $\oplus$ and & to denote the logical addition (XOR) and multiplication (AND) operations over $\mathbb{Z}_2$. Similarly, we denote $+$ for addition and $\cdot$ for multiplication in residue space over $\mathbb{Z}_n$. A q bit vector $m = (x_1, x_2, x_3, \dots, x_q)$ denoted by $\bar{x}$ represents data in message space and its equivalent residue code is represented by $(X_{1,m}, X_{2,m}, X_{3,m}, \dots, X_{q,m})$ denoted as $\overline{X}$.

Secure RNS logic is based on a combination of homomorphic encryption and residue number system. We use homomorphic encryption to create the encrypted shares. The binary input values are transformed from message space to residue code space. Additionally, the homomorphism preserves the mathematical integrity of binary message space in the residual value space. An input encoding stage, which need not be on the chip implemented with the secure RNS logic shares, performs the binary message space to residue space conversion. Any computing host can perform this conversion and transmit the residue shares over any link including a network. The binary gates have equivalent modulo operations which are applied over the encrypted shares. Once the results in residue space are computed, they are decoded into the binary space. Once again, decoding need not occur at the secure chip. The residue shares can be transmitted back to a client over a link, where the decoding can be performed. We start by describing the construction of RNS secret sharing scheme. Our approach comprises of three stages as *Input encoder*, *RNS circuit*, and *Output decoder*.

Input encoder: The homomorphic secret sharing scheme encodes the input message using a function called Input encoder (`Enc`). The encoder `Enc` maps each binary input x to an l-bit residue code denoted by X_{m_i}, where m_i is the chosen moduli. Moduli choice has an important role in recovering the output back in binary value from residue code space which will be described in the output decoder function. The variable l defines the size of residue space. We first choose an l-bit random value r_x and moduli m_i from the relatively prime moduli set $\mathbb{M} = \{m_1, m_2, m_3, \dots, m_n\}$. The encoding function is modulo addition of random value r_x with binary input x over m_i.

$$X_{m_i} = x + r_x (mod\ m_i) \qquad (1)$$

The security of the RNS secret shares fully depends on the random value r_x and moduli m_i. Note that without the random value r_x, the input binary bit x is exposed in the residue domain. The moduli m_i is typically chosen per chip implementation, whereas the random values r_x are assumed to be refreshed for every instantiation. They are generated by a statistically tested random number generator.

A. Switching Uniformity

Note that the two main goals of a secure logic family are (1) uniform switching or power distribution so that it is not data dependent, and (2) remove any correlation between intermediate values. Note that the t-private logic achieves both these goals. Through an induction based proof, the inductive hypothesis establishes that input encoder output has these properties. 1-prob(x) denotes the probability that node x state is 1. 1-prob(x) is a fairly good indicator of its switching probability: $2 * 1\text{-prob}(x) * (1 - 1\text{-prob}(x))$. Note that 1-prob($r_i$) of a random bit is 0.5, a random bit holds state 1 with probability 1/2. Also note that when an input bit x with arbitrary $0 \leq 1\text{-prob}(x) \leq 1$ is XORed with a random bit r_i, $1\text{-prob}(x \oplus r_i) = 0.5$. These two facts establish that all of the $(t + 1)$ shares output by an encoder have 1-prob equal to 0.5. The entropy of any two random bits r_i and r_j is 2 bits since they are not correlated (distribution of states 00, 01, 10, 11 is uniform). By this token, the entropy of t random bits is t, establishing the other property. For inductive hypothesis, consider the t-private gate for AND ($\wedge$). The two incoming vectors $X = (x_0, x_1, \dots, x_t)$ and $Y = (y_0, y_1, \dots, y_t)$ have these two properties by inductive hypothesis. If each row of shift and add multiplication of X and Y forms a share, the $1\text{-prob}(x_i y_j \oplus x_i y_k)$ is 0.5 given that each share has 1-prob equal to 0.5. However, there is correlation between rows reducing their entropy. In fact, all the shares of X are revealed within a row along with one share of Y - y_i thereby losing cryptographic privacy. By using an additional random bit per row, the entropy is restored to $t + 1$. We will like to show similar analytical uniform switching for secure RNS logic style.

Theorem 1. *The output of the input encoder (Enc) is uniformly distributed over moduli m_i or the set $\{0, 1, \dots, m_i - 1\}$.*

Proof. Let $\mathbb{P}$ denote the plaintext in the binary space, $\mathbb{X}$ denote the encrypted share in the residue space. The residue space $M_X = \{0, 1, \dots, m_i - 1\}$. $X = Enc_{r_x}(x)$ where, the random value r_x is uniformly distributed over M_X.

$$P(\mathbb{R} = r_x) = P(\mathbb{X} = X) = \frac{1}{\alpha},$$

$$\forall \; r_x \; and \; X \; \epsilon \; M_X$$

where, $\alpha = |M_X|$. To prove this statement,

$$P(\mathbb{X} = X | \mathbb{P} = x) = \frac{P(\mathbb{X} = X) \cdot P(\mathbb{P} = x)}{P(\mathbb{P} = x)}$$

$$= P(\mathbb{X} = X) \; = \; P(\mathbb{R} = r_x)$$

$\square$

Thus the input encoder function maps the binary input without any bias on the residue code space. For a given message, the output of input encoder is equi-probable for the chosen moduli m_i. The same encoder function can be used to generate different shares by choosing different moduli m_i with the same random value r_x.

RNS circuit: Our goal is to transform the binary operators such as AND and XOR into an equivalent residue operator using the composition of modulo multiplication and modulo addition in order to perform the operation securely. We construct an RNS circuit that computes the residue space equivalent of a Boolean AND as shown in Figure 1. The size of this circuit is independent of number of shares. It depends only on the modulus size (l).

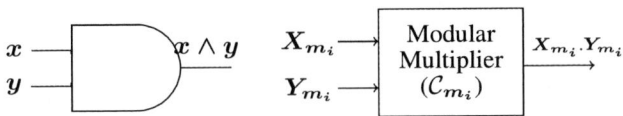

Fig. 1: RNS circuit AND

Consider an AND ($\wedge$) gate in the binary space with inputs x, y and output z. i.e., $z = x \wedge y$. In our model, the Boolean AND operation is performed with modular multiplication of X_{m_i} and Y_{m_i} over moduli m_i.

$$Z_{m_i} = X_{m_i} \cdot Y_{m_i} \; (mod \; m_i)$$

The perfect privacy of our proposed scheme requires that the intermediate values or the output values be uniformly distributed with respect to the moduli m_i. This leads to uniform switching distribution as well with 1-prob of each of the output bits of a residue output equal to **0.5**.

Theorem 2. *Let f be any modulo function over m_i with inputs X_{m_i}, Y_{m_i} and output Z_{m_i}. Then, the output Z_{m_i} is uniformly distributed over residue code space, given that inputs are generated by an Input encoder (Enc).*

Output Decoder: Each output share is computed independently for the given input vector for each modulus m_i. The output residue code Z is defined as linear congruence to output of binary value z with respect to moduli m_i. To compute the resultant binary output bit, we apply Chinese Remainder Theorem (CRT) on the output shares obtained from the RNS circuit.

Theorem 3. *(Chinese Remainder Theorem) Suppose that $\mho \subset \mathbb{M}$, where all the elements are pairwise co-prime. let $Z_{m_1}, Z_{m_2}, \dots, Z_{m_k}$ be integers $\epsilon \; \mho$. Then the system of congruences, $z \equiv Z_{m_i} \; (mod \; m_i)$ for $1 \leq i \leq k$, has a unique solution modulo $M = m_1 \times m_2 \times \dots \times m_k$, which is given by:*

$$z \equiv Z_{m_1} \cdot M_1 \cdot M_1^{\star} + Z_{m_2} \cdot M_2 \cdot M_2^{\star} + \dots + Z_{m_k} \cdot M_k \cdot M_k^{\star},$$

where $M_i = M/m_i$ and $M_i^{\star} \equiv (M_i)^{-1} \; (mod \; m_i)$ for $1 \leq i \leq k$.

Proof. Notice that $\gcd(M_i, m_i) = 1$ for $1 \leq i \leq k$. Therefore, the y_i's all exist. Now, notice that since $M_i \cdot M_i^{\star} \equiv 1 \; (mod \; m_i)$, we have $Z_{m_i} \cdot M_i \cdot M_i^{\star} \equiv Z_{m_i} \; (mod \; m_i)$ for $1 \leq i \leq k$. On the other hand, $Z_{m_i} \cdot M_i \cdot M_i^{\star} \equiv 0 \; (mod \; m_j)$ if $j \neq i$. Thus, we see that $z \equiv Z_{m_i} \; (mod \; m_i)$ for $1 \leq i \leq k$. $\square$

To apply Chinese Remainder Theorem, it is important that the shares chosen to reconstruct the value use moduli m_i relatively prime to each other. Further, in order to remove the mask, the value e has to be subtracted from the output of CRT followed by mod 2 operation. For this example, the value e is calculated as $r_x y + x r_y + r_x r_y$.

Definition 1. *(2,k,n) threshold secret sharing scheme let n be an integer, $n \geq 3$, and $3 \leq k \leq n$. A $(2, k, n)$-threshold secret sharing scheme is a method for generating shares for x as $P = \{X_{m_1}, X_{m_2}, \dots, X_{m_n}\}$ such that*

- *for any $A \subset P$ such that $|A| < 2$, learning the element x should be difficult.*
- *for any $A \subset P$ such that $|A| = 2$, reconstruction of element x is possible, given that $\gcd(m_i, m_j) = 1$.*
- *for any $A \subset P$ such that $|A| \geq k$, reconstruction of the element x becomes easier, given the set $\{X_{m_i} | i \epsilon A\}$ are relatively prime.*

The RNS secret sharing scheme follows a variant of (k, t, n)-threshold scheme [19]. Our threshold scheme is defined in Definition 1. The RNS secret sharing scheme requires a minimum of 2 shares to decode the result residue shares to binary output. Also, the shares chosen for decoding must be computed with moduli that are co-prime.

III. RNS CIRCUIT RESILIENCE CHARACTERISTICS

In this section, we discuss the resilience characteristics of our proposed scheme. We first review the more general definition of masking technique and then we will show how our proposed approach is resilient to side-channel attacks.

Definition 2. *(Masking) An intermediate value v masked with r results in a masked value $v_r = f(v, r)$ which is independent of v. The intermediate value is said to be masked, if the power consumption of v_r is independent of v.*

In RNS secret sharing scheme, the shares are created with a selection of moduli m_i set. The masking function, along the lines of Definition 2, is the encoding function Enc, which takes two inputs x and r_x. The input value x is masked using random value r_x. We first define the side channel leakage $\mathcal{L}$ from RNS circuit $\mathcal{C}$. The leakage of shares is captured

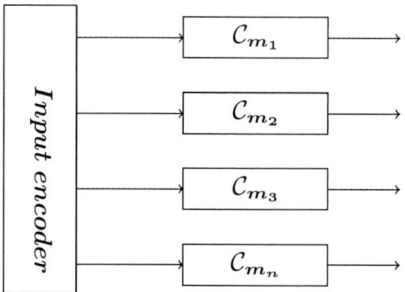

Fig. 2: Independent computation over encrypted share

by $\mathcal{L}_{m_1}$, $\mathcal{L}_{m_2}$, $\mathcal{L}_{m_3}$, ..., $\mathcal{L}_{m_n}$ from the circuits $\mathcal{C}_{m_1}$, $\mathcal{C}_{m_2}$, $\mathcal{C}_{m_3}$, ..., $\mathcal{C}_{m_n}$ respectively. All the circuits are functionally equivalent and homomorphic to a binary function in the binary message space. Each share is processed individually, which means that there is zero dependence between shares during computation as shown in Figure 2. In a multi-core device, each encrypted share can be processed independently on separate cores with staggered unpredictable schedule. Moreover, power pins for different cores are isolated. This leads to close to zero probability of leaking any information about the input binary value x.

In addition to this power independence characteristic of our scheme, there is another interesting property called data indistinguishability. Let us assume $l = 2$ implying 2-bit shares. Then a valid moduli set is $M = \{2, 3\}$ with $gcd(2, 3) = 1$. The input encoding for $l = 2$ is shown in Table I. The columns X_{m_1}, X_{m_2} are the shares created for m_1 and m_2 as 2 and 3 respectively. Based on the input binary value, the residue codes are classified into two for x =0 and x =1. The $X_{m_1} \cup X_{m_2}$ contains $\{0,1,2\}$ for x=0. Similarly for x=1, the $X_{m_1} \cup X_{m_2}$ contains $\{0,1,2\}$. The residue sets for $x = 0$ and $x = 1$ contain the same data. Hence, the leakage $\mathcal{L}_{m_i}$ obtained from the $\mathcal{C}_{m_i}$ for a given X_{m_i} makes it hard to distinguish the input binary value x. By considering the worst case scenario, even if the adversary observes the residue shares at circuit $\mathcal{C}_{m_i}$, it is hard for an adversary to identify the actual binary input value x without any knowledge of r_x and m_i. Even the distributions of the residue shares for each row for $x = 0$ and $x = 1$ shows very similar patterns: 00, 02, 11, 10 for $x = 0$ and 11, 02, 10, 01 for $x = 1$ making it difficult to correlate the value of x with the observed shares.

TABLE I: Sample Residue computation for $l = 2$

x	r_x	X_{m_1}	X_{m_2}
0	0	0	0
0	1	1	1
0	2	0	2
0	3	1	0
1	0	1	1
1	1	0	2
1	2	1	0
1	3	0	1

IV. POWER SIDE CHANNEL ADVERSARY

In this section, we discuss the strength of RNS secret sharing scheme and introduce random renewal techniques to

achieve better resistance against power side-channel attack. We first define our basic assumptions for SCA target circuit. We assume that the adversary can probe or guess only the residue input values. The binary inputs leading up to the residue shares are not exposed to adversary. This model is more suitable for RNS logic since an adversary is likely to guess the intermediate binary values from the power leakage.

A power analysis attack is a type of side-channel attack that exploits the leakage obtained in the form of power consumption from the target circuit. Masking techniques are used to randomize the power consumption to make sure that the measured leakage are independent of processed data. RNS secret sharing scheme is also a type of masking scheme, which uses homomorphic encryption to mask the intermediate values. Secure RNS circuits are highly resistant to power analysis because of their resilience characteristics defined in section III. More formally, we describe strength of RNS secret sharing scheme as follows.

Definition 3. *[1] Let $\mathcal{C}$ be an circuit under investigation with secret value $\dot{y}$. Then the differential power analysis is defined by*
$$\triangle_{y,\dot{y}}(N, j) =$$
$$\frac{\sum_{i=1}^{N} \mathcal{D}(X_i, y_j)\mathcal{C}_{\dot{y}}(X_i)}{\sum_{i=1}^{N} \mathcal{D}(X_i, y_j)} - \frac{\sum_{i=1}^{N}(1 - \mathcal{D}(X_i, y_j))\mathcal{C}_{\dot{y}}(X_i)}{\sum_{i=1}^{N}(1 - \mathcal{D}(X_i, y_j))}$$

where, $\mathcal{D}$ is a function for key hypothesis.

The definition 3 says that the adversary can successfully model the leakage to distinguish the intermediate values between 0 and 1. This can be achieved only if the input value strongly determines the intermediate value. Our RNS secret sharing scheme, applies homomorphic encryption on input values using random value r_x and moduli m_i. Our encoding scheme completely weakens the control of input binary values over intermediate values. Hence, the power analysis adversary is unable to model the leakage $\mathcal{D}(X_i, y_j)$ for a successful attack. Additionally, The data indistinguishability characteristics of secure RNS circuit more or less equalize the power consumption values $\mathcal{C}_{\dot{y}}(X_i)$ between all transitions. Hence, the adversary is not able to distinguish the leakages with respect to output binary level transitions. We believe that the cryptographic privacy of our proposed scheme also makes it difficult to distinguish on the basis of power leakage.

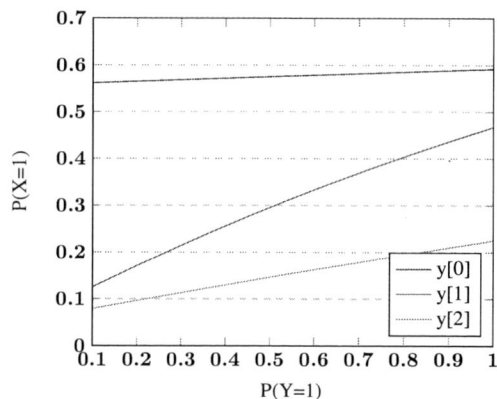

Fig. 3: Transition Probability of RNS encoding scheme

In order to study the power leakage characteristics of our RNS circuit, we computed the transition probability for each output bit of encoding scheme as defined in Equation 1 with

978-1-5386-3693-0/18 $31.00 © 2018 IEEE

$l = 3$. The input signal probabilities were propagated in a gate level description of encoding scheme and the results are plotted in Figure 3. The results show that the output transition probability of our encoding scheme is skewed with input signal probability. We found that the moduli reduction reduces the effect of random value r_x and makes the transition probability biased. Hence, it is more likely to be vulnerable against power analysis attack with larger circuits.

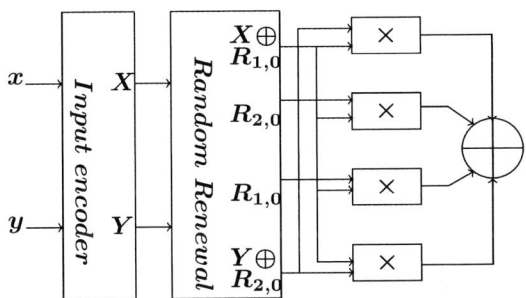

Fig. 4: Secure RNS Circuit with random renewal scheme

To make the transition probability of the RNS circuit unbiased we introduce a random renewal scheme as in an AND gate of t-private logic. In random renewal scheme, we perform bitwise exclusive-OR function between a random value $R_{i,j}$ and output of encoder. The random variable $R_{i,j}$ is l-bits wide with each bit distributed independently and uniformly. This makes the output transition probability of an RNS encoder to be 0.5 and unbiased. The variables i and j refer to the input and the circuit stage respectively. The modified secure RNS circuit is shown in Figure 4. The random renewal exclusive-OR operation maintains homomorphism over the residue values only with true multiplication. Therefore no moduli reduction is performed. Once the recovery exclusive-OR operation has been done, the residue values are obtained by modulo reduction with appropriate moduli value m_i. To maintain the unbiased transition probabilities, the random renewal techniques should be applied at each stage with independent random values.

A. Probing and Fault Attack Adversary

1) Resistance against Probing attack: A probing adversary can probe specific wires/nodes of a circuit each clock cycle. In [5], Ishai et al proposed t-private circuits that resist probing attacks. t-private circuit can withstand upto t-probes per cycle. In RNS secret sharing scheme, each share is processed by circuit C_{m_i}. All the shares in residue domain and the output of C_{m_i} are l-bits wide. If adversary tries to learn values of a single share, the adversary has to probe l nodes. To probe all the n shares, the adversary requires $l \times n$ probes per cycle which is infeasible due to high cost.

The main difference between t-private logic and secure RNS logic is that the t-private logic carries its own decoding forward - or it is a self-decoding scheme. All the $t + 1$ shares are XORed for decoding. This makes it vulnerable to $t + 1$ probes. Secure RNS logic, on the other hand, relies on secrets based on moduli set M and r_x for decoding. These secrets are not built into the circuits C_{m_i} making it difficult for a probing adversary.

2) Resistance against fault attack: Fault attack is a subclass of active side-channel attacks. In this attack the adversary injects fault and causes either temporary or permanent errors in the computation. The fault injection techniques such as clock glitch, power glitch or light beam are chosen accordingly based on the precision and cost requirements. In a typical fault based side-channel attack, the adversary tries to exploit the difference in computation caused by the fault and the correct computation. We model an adversary that can inject $n - k$ faults at a time. In [20], the author described the characteristics of fault resilience implementation as follows:

1) Inspite of the effect of fault, the user is able to retrieve the correct result.
2) User should be able to track the fault occurrence and discontinue/interrupt the faulty computation.
3) To retrieve the secret, the implementation should impose difficulty for an adversary, even returning a faulty result.

With respect to these characteristics, our RNS secure circuit scheme provides a better fault resilience implementation. Our scheme follows the $(2, k, n)$-threshold implementation, which means that any 2 shares are sufficient to retrieve the correct result of the function. A fault can be easily tracked by comparing pairwise results of shares. The adversary only knows the faulty binary values of the corresponding input injected into the system. With this information, the adversary is not able to back propagate the binary value without knowing the random value r_x. Hence, our scheme delivers stronger resistance against fault analysis attack.

V. EXPERIMENTAL RESULTS

We implemented RNS circuit of Boolean AND gate with $l = 3$ using 45nm FreePDK Standard Cell library and Cadence analogue simulator (Spectre). We conducted exhaustive simulations over residue space using the ocean script. We measured the peak current and average power consumption of all possible input transitions (2^{12}). We performed two styles of analysis over the simulated data, one with common random value for both the inputs and the other one with separate random values for the inputs. We first computed the average values for each class of output transition. We obtained the probability density function of each class and plotted it in Figure 5. Then, we calculated the normalized standard deviation for each analysis which is given in Table II.

TABLE II: SCA metrics computation

Transition	encoding with single r		encoding with separate r	
	Power Consumption (mW)	Peak Current (mA)	Power Consumption (mW)	Peak Current (mA)
$0 \rightarrow 0$	103.06	1.04	102.78	0.92
$0 \rightarrow 1$	101.49	0.88	102.00	1.05
$1 \rightarrow 0$	102.25	0.65	102.07	1.03
$1 \rightarrow 1$	103.65	0.65	104.90	0.77
Average (μ)	102.618	0.862	102.93	0.947
Standard Deviation (σ)	0.945	0.158	1.351	0.128
Normalized SD ($\frac{\sigma}{\mu}$)	0.0092	0.1835	0.0131	0.1348
Kullback Divergence (DL_{max})	-	4.539	-	1.212

The normalized standard deviation is a well known SCA metric to quantify the effectiveness of the countermeasures. Lower the value, better resistance against power analysis attacks. In our scheme the normalized standard deviation is likely to converge towards lower values for larger circuits. We also computed SCA metric using Kullback-Leibler divergence for our analysis which defines failure probability of the attack. The analysis with separate random values for inputs shows better results compared to a single common random value.We expect similar results for the random renewal scheme. The

978-1-5386-3693-0/18 $31.00 © 2018 IEEE

secure RNS circuits with random renewal are expected to achieve lower standard deviation and KL divergence due to more uniform switching distribution.

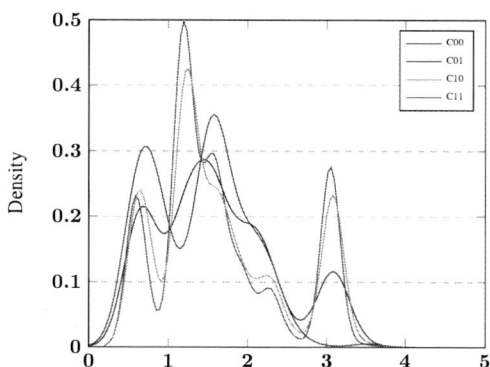

Fig. 5: Probability Density of Maximum Current

We also applied machine learning based classification techniques such as Linear Discriminant Analysis (LDA), Quadratic Discriminant Analysis (QDA), and Naive Bayes to evaluate the security of our circuit. For each classifier, we created a training set with 4000 measurements followed by validation test data set that includes 1000 measurements. The success rate was computed for each classifier and its results are given in Table III.

TABLE III: Success Rate of Classifier

Classifier	t-private	Secure RNS Circuit
LDA	36.9%	25.28%
QDA	31.4%	25.55%
Naives Bayes	40.3%	27.64%

Form the success rate results table, it is clear that the RNS circuits provide good resistance against power side-channel adversary. RNS circuits have lower success rate for these classifiers than t-private logic.

VI. CONCLUSIONS

We have presented a new secure logic based on secret sharing and residue number system. We showed how the secure RNS circuit transforms the boolean function into residue operations like modular multiplication and modular addition. We also investigated the resistance of secure RNS circuits using analytical transition probability distribution, normalized variance and Kullback-Leibler divergence based metrics, and the success rates of common machine learning classifiers such as QDA, LDA and Naive Bayes. Secure RNS logic has lower success rates than t-private logic.

We enhanced secure RNS logic with random renewal technique to further mitigate power side-channel adversary. The analytical switching distribution is improved for the random renewal RNS logic. We expect our experimental data to validate these analytical results.

ACKNOWLEDGMENT

We are grateful to National Science Foundation support under Grant CNS 1441640 and Semiconductor Research Corporation for their support under contract 2014-TS-2556.

REFERENCES

[1] P. C. Kocher, J. Jaffe, and B. Jun, "Differential power analysis," in *Advances in Cryptology - CRYPTO '99, 19th Annual International Cryptology Conference, Santa Barbara, California, USA, August 15-19, 1999, Proceedings*, 1999, pp. 388–397.

[2] T. S. Messerges, "Securing the AES finalists against power analysis attacks," in *Fast Software Encryption, 7th International Workshop, FSE 2000, New York, NY, USA, April 10-12, 2000, Proceedings*, 2000, pp. 150–164.

[3] E. Trichina and T. Korkishko, "Secure AES hardware module for resource constrained devices," in *Security in Ad-hoc and Sensor Networks, First European Workshop, ESAS 2004, Heidelberg, Germany, August 6, 2004, Revised Selected Papers*, 2004, pp. 215–230.

[4] A. Shamir, "How to share a secret," *Commun. ACM*, vol. 22, no. 11, pp. 612–613, 1979.

[5] Y. Ishai, A. Sahai, and D. A. Wagner, "Private circuits: Securing hardware against probing attacks," in *Advances in Cryptology - CRYPTO 2003, 23rd Annual International Cryptology Conference, Santa Barbara, California, USA, August 17-21, 2003, Proceedings*, 2003, pp. 463–481.

[6] J. Park and A. Tyagi, "t-private logic synthesis on fpgas," in *2012 IEEE International Symposium on Hardware-Oriented Security and Trust, HOST 2012, San Francisco, CA, USA, June 3-4, 2012*, 2012, pp. 63–68.

[7] ——, "Towards making private circuits practical: DPA resistant private circuits," in *IEEE Computer Society Annual Symposium on VLSI, ISVLSI 2014, Tampa, FL, USA, July 9-11, 2014*, 2014, pp. 528–533.

[8] S. Mangard, T. Popp, and B. M. Gammel, "Side-channel leakage of masked CMOS gates," in *Topics in Cryptology - CT-RSA 2005, The Cryptographers' Track at the RSA Conference 2005, San Francisco, CA, USA, February 14-18, 2005, Proceedings*, 2005, pp. 351–365.

[9] Z. N. Goddard, N. LaJeunesse, and T. Eisenbarth, "Power analysis of the t-private logic style for fpgas," in *IEEE International Symposium on Hardware Oriented Security and Trust, HOST 2015, Washington, DC, USA, 5-7 May, 2015*, 2015, pp. 68–71.

[10] S. Nikova, C. Rechberger, and V. Rijmen, "Threshold implementations against side-channel attacks and glitches," in *Information and Communications Security, 8th International Conference, ICICS 2006, Raleigh, NC, USA, December 4-7, 2006, Proceedings*, 2006, pp. 529–545.

[11] A. Moradi, "Statistical tools flavor side-channel collision attacks," in *Advances in Cryptology - EUROCRYPT 2012 - 31st Annual International Conference on the Theory and Applications of Cryptographic Techniques, Cambridge, UK, April 15-19, 2012. Proceedings*, 2012, pp. 428–445.

[12] T. S. Messerges, "Using second-order power analysis to attack DPA resistant software," in *Cryptographic Hardware and Embedded Systems - CHES 2000, Second International Workshop, Worcester, MA, USA, August 17-18, 2000, Proceedings*, 2000, pp. 238–251.

[13] M. Bucci, L. Giancane, R. Luzzi, and A. Trifiletti, "Three-phase dual-rail pre-charge logic," in *Proceedings of the 8th International Conference on Cryptographic Hardware and Embedded Systems*, ser. CHES'06. Berlin, Heidelberg: Springer-Verlag, 2006, pp. 232–241. [Online]. Available: http://dx.doi.org/10.1007/11894063_19

[14] K. Tiri and I. Verbauwhede, "Design method for constant power consumption of differential logic circuits," *CoRR*, vol. abs/0710.4756, 2007. [Online]. Available: http://arxiv.org/abs/0710.4756

[15] S. Guilley, S. Chaudhuri, L. Sauvage, T. Graba, J.-L. Danger, P. Hoogvorst, V.-N. Vong, M. Nassar, and F. Flament, *Shall we trust WDDL?* Wiesbaden: Vieweg+Teubner, 2009, pp. 208–215. [Online]. Available: https://doi.org/10.1007/978-3-8348-9324-6_22

[16] M. Gomathisankaran and A. Tyagi, "A novel design of secure and private circuits," in *IEEE Computer Society Annual Symposium on VLSI, ISVLSI 2012, Amherst, MA, USA, August 19-21, 2012*, 2012, pp. 362–367.

[17] M. Gomathisankaran, K. Namuduri, and A. Tyagi, "Horns: A semi-perfectly secret homomorphic encryption system," in *Computing Communication Networking Technologies (ICCCNT), 2012 Third International Conference on*, 2012, pp. 1–5.

[18] J. Park and A. Tyagi, "Security metrics for power based sca resistant hardware implementation," in *2016 29th International Conference on VLSI Design and 2016 15th International Conference on Embedded Systems (VLSID)*, Jan 2016, pp. 541–546.

[19] G. R. Blakley and C. A. Meadows, "Security of ramp schemes," in *Advances in Cryptology, Proceedings of CRYPTO '84, Santa Barbara, California, USA, August 19-22, 1984, Proceedings*, 1984, pp. 242–268. [Online]. Available: https://doi.org/10.1007/3-540-39568-7_20

[20] A. Barenghi, L. Breveglieri, I. Koren, and D. Naccache, "Fault injection attacks on cryptographic devices: Theory, practice, and countermeasures," *Proceedings of the IEEE*, vol. 100, no. 11, pp. 3056–3076, 2012.

Positive Feedback Symmetric Adiabatic Logic against Differential Power Attack

Bhuvana B P, Kanchana Bhaaskaran V S*, Senior Member, IEEE*

School of Electronics Engineering,
VIT University Chennai, India.
bhuvana13490@gmail.com
vskanchana@ieee.org*

Abstract—In recent years, side channel attacks such as Differential Power Analysis (DPA) and Simple Power Analysis (SPA) are the most common in cryptographic hardware systems. Smart Cards, RFID tags, military communications and electronic commerce are some of the applications where cryptography plays a significant role. Adiabatic logic design is one among various low power VLSI design techniques used in the design of cryptographic hardware systems. In this paper, we propose the Positive Feedback Symmetric Adiabatic Logic (PFSAL) which reduces the energy dissipated by the S-Box circuit used in the cryptographic designs. Proposed PFSAL is validated by the design of the PFSAL based Rijndael S-Box. The efficiency of the proposed design is validated by comparing it with Rijndael S-Box designed using other existing security based adiabatic logic circuits. Results are simulated in Cadence Virtuoso. Proposed PFSAL is competent in terms of energy consumption, uniform power and current traces when compared with the existing counterparts.

Keywords—Adiabatic computing; Positive Feedback Symmetric Adiabatic Logic; Rijndael S-Box; DPA Resistant Cryptographic Hardware

I. INTRODUCTION

Internet of Things (IoT) is widespread in the domains of manufacturing, medical, communication, automotive, etc. IoT is based on wireless sensor networks and Radio Frequency Tagged IDs (RFID). These hardware modules store undisclosable information, which is vulnerable to side channel attacks. These attacks can be executed if data can be gained in the form of electromagnetic radiation, power consumption, etc. These side channel attacks are classified into invasive and non-invasive attacks. Jamming, tag tracking and cloning categorize into invasive attacks, while the electromagnetic attack, power attack and timing attack fit into non-invasive attacks. One of the most prominent non-invasive attacks is the Differential power analysis (DPA) attack. DPA attack relates the data and cipher key to the instantaneous power consumed by the particular device. Commonly used cryptographic algorithms are Advanced Encryption Standard (AES) and Data Encryption Standard (DES). S-Box used in the hardware design of the cryptographic algorithm consumes very high power and this in turn leads to high power consumption of the cryptographic hardware.

Several literatures are found portraying many countermeasures for the DPA attack at various levels, such as cell level, gate level and architectural level. Some of such widespread countermeasures for the DPA attack include Sense Amplifier Based Logic [1], Wave dynamic differential logic [2], Random Switch Logic [3], Dual-rail Transition Logic [4] and Energy Recovery or the so called Adiabatic Logic [5]. Among all these logic circuit design methodologies, the recent eras find the adiabatic logic to be a promising approach for the design of DPA resistant hardware even while incurring lower power consumption.

Adiabatic logic design is the most frequently used methods for the design of ultra-low power hardware of optimal performance applications. These circuits recover a major part of the energy stored in the output nodal capacitances which can be reused for the successive computations. Appropriately, they are also called as the energy recovery or charge recovery logic. These adiabatic logic families can operate efficiently below 1GHz frequency. Hence, they can be utilized to implement IoT based devices which operates at comparatively low frequency levels. Some of the efficient adiabatic logic designs found in the literatures are 2N2P, 2N-2N2P [6], Differential Cascode Pre-resolve Adiabatic Logic (DCPAL) [7] and Pre-resolve and Sense Adiabatic Logic (PSAL) [8]. Various DPA resistant logic families found in the literature are Secure Adiabatic Logic [9], Symmetric Adiabatic Logic [10], Charge Sharing Symmetric Adiabatic Logic [11], Secured Quasi Adiabatic Logic [12] and Symmetric Pass Gate Adiabatic Logic [13]. Symmetric Pass Gate Adiabatic Logic is found to be more competent in terms of power consumption. However, it suffers from non-adiabatic power loss. This paper proposes a novel Positive Feedback Symmetric Adiabatic Logic (PFSAL) which predominantly focuses on reduction in non-adiabatic loss incurred by the circuit. The use of power clock sources powering the cascaded stages of the adiabatic circuit ensures uniform current consumption by the adiabatic circuit modules. The current waveforms are ensured to be uniform for all input combinations. To demonstrate and validate the design of proposed PFSAL, the XOR gate, NAND gate and Buffer gates are described in this work. Rijndael S-Box has been implemented using various DPA resistant adiabatic logic families and the proposed PFSAL is found to be comparatively more efficient in terms of energy consumption than the counterpart circuit designs found in the literature.

The paper is systematized as follows. Section II describes the adiabatic logic and various losses associated with it. Section III elaborates the Positive Feedback Symmetric Adiabatic Logic (PFSAL). Section IV presents the design of Rijndael S-Box

using PFSAL. Section V discusses the results and section VI concludes.

II. ADIABATIC LOGIC

Adiabatic logic methodology is a low power VLSI design technique, which recovers a part of the energy stored in the output nodal capacitances and these can be used in the successive stages of circuit operation. Recovering the stored charges reduces the power consumption to a greater extent. The basic structure of an adiabatic logic circuit is shown in Fig. 1. These circuits utilize AC or time varying power clock signals as sources instead of a conventional DC power supply. This enables the recovery of the energy. Theoretically, an adiabatic logic circuit consumes zero power. However, in practice, they show an inherent loss in energy due to the nonzero resistance in the switching devices, leakage of stored charge from the circuit nodes through leakage paths and incomplete energy recovery from logic nodes [11]. However, the maximum switching frequency of operation of an adiabatic circuit is lower than the conventional CMOS circuits [14] primarily due to the nature of operation of adiabatic circuits.

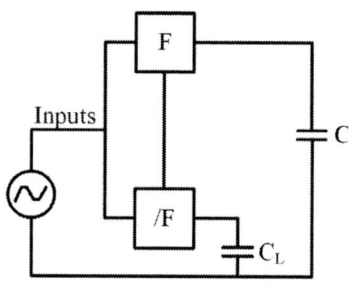

Fig. 1. Adiabatic Logic Circuit

Energy stored on the output capacitance in the conventional CMOS design can be given as

$$E_C = \frac{1}{2}CV_{DD}^2 \tag{1}$$

The energy dissipated in an adiabatic logic circuit when a constant current source is supplied is given by

$$E_{diss} = \frac{RC}{T}CV_{DD}^2 \tag{2}$$

Here, T is the charging/discharging time of the capacitor, C is the load capacitor, R is the channel resistance of the transistors, V_{DD} is the full swing voltage value of power clock. When $T> 2RC$ (time constant), the energy dissipated by the adiabatic circuit is less than that of the conventional CMOS circuit. Various losses in adiabatic logic circuit include 1) adiabatic loss, 2) Non-adiabatic loss and 3) Leakage loss.

A. Adiabatic Loss

Energy dissipated during the charging event is given by

$$E = \frac{RC}{T}CV_{DD}^2 \tag{3}$$

An adiabatic cycle consists of charging and recovery processes. Both the charging and recovery processes dissipate certain amount of energy determined as determined by the switching paths. Hence, the total energy dissipated in adiabatic logic can be approximated as given by

$$E_{AL} = 2\frac{RC}{T}CV_{DD}^2 \tag{4}$$

where R is the resistance of the charging/discharging paths, C is the load capacitance and T is the total time period. From the above equation, it is clear that slower the circuit gets charged, lower is the energy consumed. Fig. 2 shows the equivalent circuit to determine adiabatic loss [14].

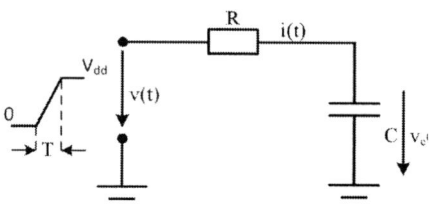

Fig. 2. Equivalent Circuit to Determine Adiabatic Loss

B. Non-Adiabatic Loss

Non-adiabatic loss is given by

$$E_{non-adiabatic} = \frac{1}{2}CV_{th,p}^2 \tag{5}$$

Non-adiabatic loss is independent of operating frequency.

C. Leakage Loss

During Evaluation, Hold and Recovery phases, leakage current flows from V_{DD} to GND. This leads to dissipation of charge that cannot be recovered. Leakage loss can be expressed as

$$E_{leak} = V_{DD}\overline{I_{leak}}\frac{1}{f} \tag{6}$$

III. PROPOSED POSITIVE FEEDBACK SYMMETRIC ADIABATIC LOGIC (PFSAL)

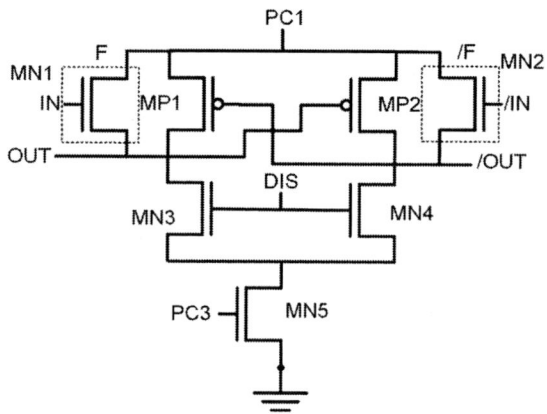

Fig. 3. Proposed Positive Feedback Secure Adiabatic Logic

Fig. 3 shows the proposed Positive Feedback Symmetric Adiabatic Logic (PFSAL) inverter/buffer circuit. The logic utilizes 4-phase power clock for powering the adiabatic circuit

978-1-5386-3693-0/18 $31.00 © 2018 IEEE   150

and for recovery of charge stored in the output nodal capacitances. It consists of two PMOS transistors MP1-MP2 and two NMOS transistors MN1-MN2 in the pull-up network. MN3 and MN4 are the discharge transistors and MN5 is the stacking transistor. Stacking transistor present in PFSAL reduces the leakage currents to a greater extent. F and /F are the functional blocks, which can be replaced by the desired function and its complement. The length of both NMOS and PMOS are kept constant as 180nm for all circuits. However, the width of the transistor is fixed as 600nm. Each stage of PFSAL utilizes two power clock supplies. PC3 lags PC1 by 90°. To discuss the operation of the circuit through its various phases of operation, consider the output nodes are at GND. Fig. 4 shows the input, output and power clock transients of the proposed PFSAL.

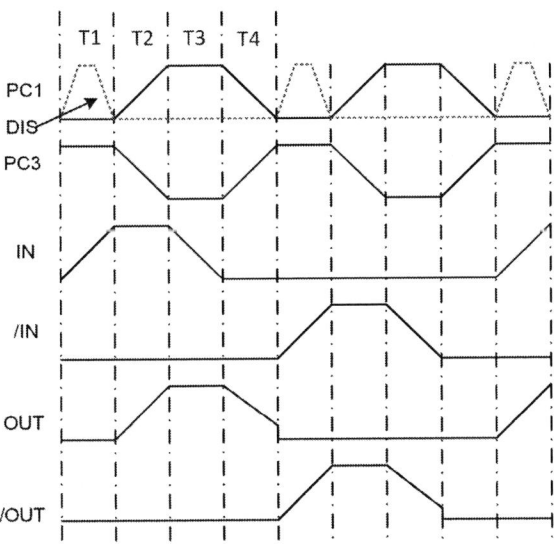

Fig. 4. Input-Output Transients of PFSAL

Wait Phase:

At T1, power clock PC1 is at GND, the PC3 is at V_{DD}. This causes MN5 to be switched ON. Discharge signal is also at V_{DD}. This makes the transistor MN3 and MN4 to be switched ON. They are used to discharge remnant charges stored in the output nodal capacitances. When input IN ramps up from 0 to V_{DD} as shown and when the gate to source voltage V_{GS} is larger than the NMOS device threshold voltage V_{tn}, i.e., $V_{GS} > V_{tn}$, then MN1 is turned ON. Current flow through MN1 is NIL due to the fact that the source and the drain of MN1 are at GND. Fig. 5 describes the operation of PFSAL wait phase.

Evaluate Phase:

At T2, input IN is at V_{DD} and /IN is at GND. Discharge signal is also at GND. Power clock PC1 slowly rises from 0 to V_{DD} and PC3 ramps down to GND. OUT node follows PC1 as the potential of PC1 is higher than the potential at OUT node. For PMOS device MP1 to be switched ON, gate to source voltage V_{sg} should be higher than the threshold voltage V_{tp} of PMOS. When the PC1 rises from 0 to V_{tp}, transistor MN1 conducts. For MN1, PC1 acts as the drain terminal and OUT node acts as the

source terminal. Meanwhile MP1 switches ON when PC1 rises above V_{tp}. Both MN1 and MP1 conduct when PC1 rises above V_{tp} and reaches till $V_{DD}-V_{tn}$. Further, when PC1 is increased, MN1 is switched OFF as $V_{gs} < V_{tn}$ and OUT follows PC1 through MP1 alone. Fig. 6 describes the operation of PFSAL evaluation phase.

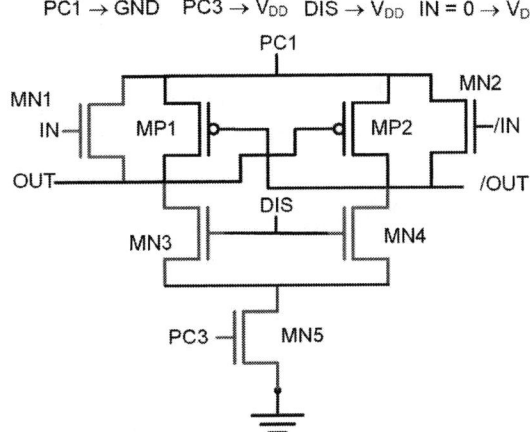

Fig. 5. PFSAL Operation During T1 (IN=1, /IN=0)

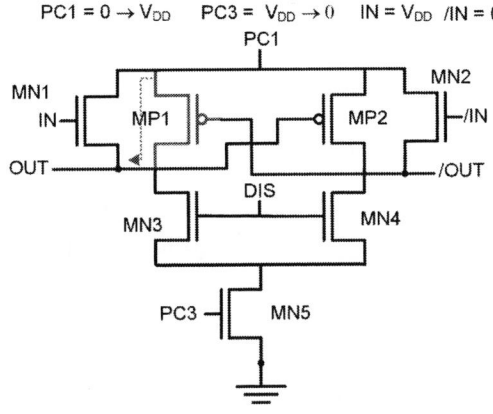

Fig. 6. PFSAL Operation During T2 (PC1 increases to VDD, PC3 falls to 0, IN=1, /IN=0)

Hold Phase:

During T3, PC1 is HIGH. IN decreases to GND and OUT node follows PC1 through MP1 transistor. Hold phase operation is depicted in Fig. 7.

Recovery Phase:

During T4, PC1 decreases from V_{DD} to GND. Charges stored in the output capacitor are recovered back to the power clock PC1 through MP1 transistor. MP1 conducts till the voltage of OUT node falls below V_{tp}. Charges that are stored in the output capacitor find its path to GND in T5 phase, where discharge

978-1-5386-3693-0/18 $31.00 © 2018 IEEE

transistors MN3, MN4 are switched ON along with the stacking transistor MN5. PFSAL recovery phase is shown in Fig. 8.

$$PC1 = V_{DD} \quad PC3 = 0 \quad IN = V_{DD} \rightarrow 0 \quad /IN = 0$$

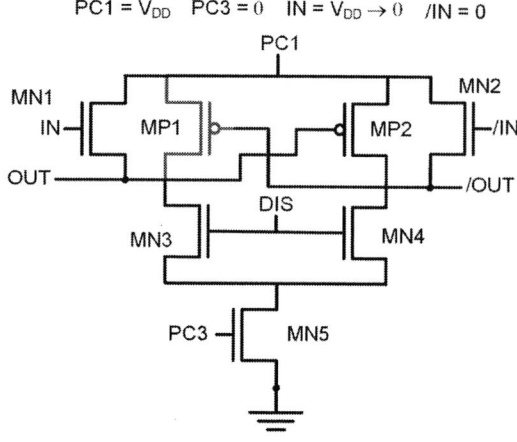

Fig. 7. PFSAL Operation During T3 (PC1 = V_{DD}, PC3 = 0, IN =1, /IN=0)

$$PC1 = V_{DD} \rightarrow 0 \quad PC3 = 0 \rightarrow V_{DD} \quad IN = 0 \quad /IN = 0$$

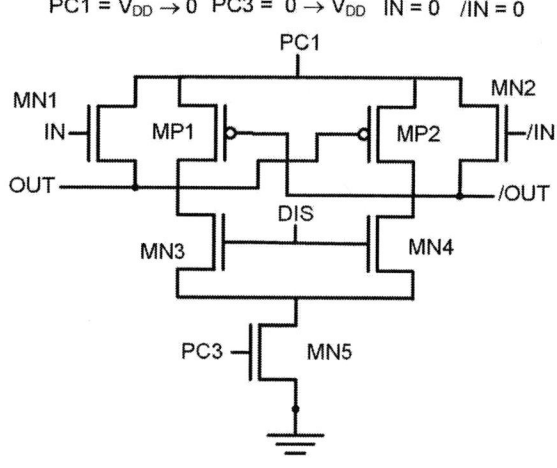

Fig. 8. PFSAL Operation During T4 (PC1 = 0, PC3 rises to V_{DD}, IN=1, /IN=0)

IV. DESIGN OF RIJNDAEL S-BOX

In the AES algorithm, S-Box plays a major role in encryption operations. The AES algorithm involves four steps, namely, Sub-Byte, Shift Rows, Mix columns and Add Round Key. Among them, Sub-byte transformation is a non-linear byte substitution, where each and every byte in the state array is replaced with S-Box. AES has a block size of 128 bits and it uses 128, 192 and 256 bits of key. AES functions using an array of bytes called as a state. Rijndael S-Box has been implemented for sub byte transformation using PFSAL. Sub byte transformation is designed by using multiplicative feature of Galois Field arithmetic (GF) 2^8 and it is followed by an affine

transformation. It also involves a multiplicative inversion operation. The individual bits in a byte representing a GF (2^8) element can be viewed as coefficients to each power term in the GF (2^8) polynomial. Rijndael S-Box is shown in Fig. 9(a). Legends of the blocks are shown in Fig. 9(b).

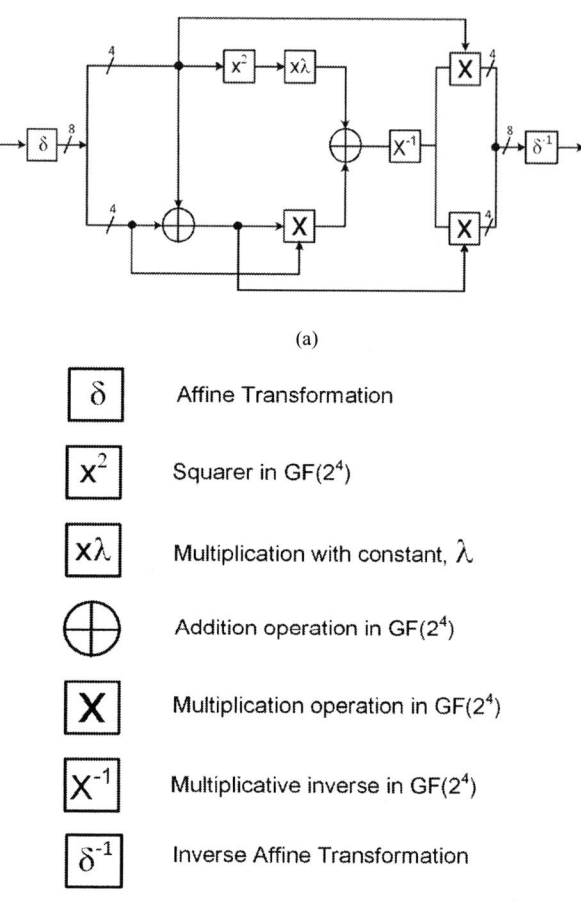

(a)

δ	Affine Transformation
X^2	Squarer in GF(2^4)
$X\lambda$	Multiplication with constant, λ
$\oplus$	Addition operation in GF(2^4)
X	Multiplication operation in GF(2^4)
X^{-1}	Multiplicative inverse in GF(2^4)
δ^{-1}	Inverse Affine Transformation

Fig. 9(a). Rijndael S-Box and (b). Building Blocks of Rijndael S-Box

V. RESULTS AND DISCUSSION

This section describes the features of PFSAL and validates through comparison with other existing counterpart designs. All circuits discussed in the work are simulated in Cadence® Virtuoso using GPDK 180nm technology. For justifiable comparisons, the transistors have been sized similarly.

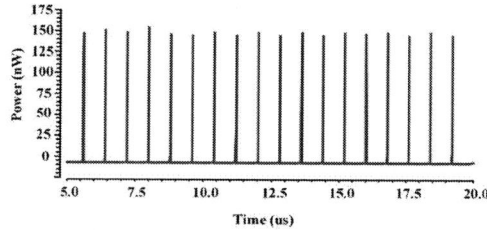

Fig. 10(a). Uniform Power Traces of PFSAL Buffer

978-1-5386-3693-0/18 $31.00 © 2018 IEEE   152

The average energy consumption for various adiabatic circuits have been calculated for the benchmark circuits of XOR and NAND designs and DPA-resistant adiabatic logic operating at 1.25MHz, 12.5MHz and 125MHz.Fig. 10(a) and (b) show the uniform power and current traces of the proposed PFSAL Buffer circuit. Uniform power current waveforms for the PFSAL based XOR gates are illustrated in Fig. 10 (c). Uniform power traces realized depict the fact that the circuits designed using PFSAL XOR gates can neutralize the DPA attack in the circuit level.

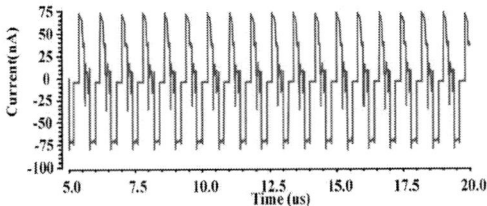

Fig. 10(b). Uniform Current Trace of PFSAL Buffer

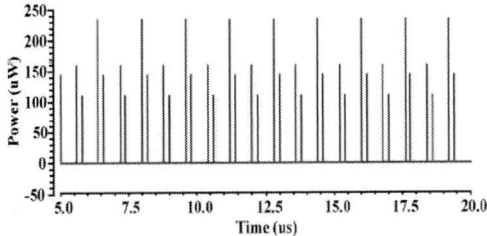

Fig. 10(c). Uniform Power Traces of PFSAL XOR Circuit for All Combinations of Input.

TABLE I. COMPARISON OF TRANSISTOR COUNT OF DPA RESISTANT ADIABATIC LOGIC FAMILIES

Adiabatic Logic	Logic Functions	Total number of transistors per gate
PFSAL	BUFFER	7
	XOR/XNOR	11
	AND/NAND	13
SPGAL	BUFFER	6
	XOR/XNOR	10
	AND/NAND	12
SQAL	BUFFER	5
	XOR/XNOR	9
	AND/NAND	13
CSSAL	BUFFER	11
	XOR/XNOR	21
	AND/NAND	21
SAL	BUFFER	14
	XOR/XNOR	18
	AND/NAND	18

Table I depicts the comparison of the transistor count in various adiabatic logic families. SQAL employs less number of transistors for the buffer circuit. However, the non-adiabatic loss incurred by SQAL causes more energy dissipation. The proposed PFSAL provides a tradeoff between the total number of transistors used and energy consumed by the circuit. The PFSAL buffer circuit has an improvement in energy performance by 50% and 36% when compared with SAL and CSSAL, respectively. However, it has the area overhead of about 40% and 14% when compared with SQAL and SPGAL.

TABLE II. AVERAGE ENERGY CONSUMPTION OF XOR GATES OF DPA RESISTANT ADIABATIC LOGIC FAMILIES

Adiabatic Logic	1.25 MHZ	12.5MHz	125MHz
PFSAL (fJ)	9.108	12.89	19.54
SPGAL (fJ)	10.64	13.78	21.08
SQAL (fJ)	28.66	36.25	44.98
CSSAL (fJ)	41.91	56.23	98.18
SAL (fJ)	19.39	19.48	33.32

TABLE III. AVERAGE ENERGY CONSUMPTION OF AND GATES OF DPA RESISTANT ADIABATIC LOGIC FAMILIES

Adiabatic Logic	1.25 MHZ	12.5MHz	125MHz
PFSAL (fJ)	11.00	15.96	27.27
SPGAL (fJ)	12.82	16.47	28.35
SQAL (fJ)	21.72	24.93	30.17
CSSAL (fJ)	35.58	48.04	74.16
SAL (fJ)	14.23	15.28	31.32

Table II shows the average energy consumption of XOR based adiabatic design. At 1.25MHz, PFSAL has an improvement of 53%, 78%, 68% and 14% when compared with SAL, CSSAL, SQAL and SPGAL, respectively. Table III depicts the energy consumption for NANA/AND logic, where PFSAL consumes 11.0 fJ, which is too less when compared to other existing counterparts.

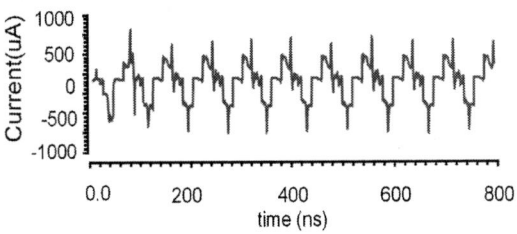

Fig. 11. Uniform Current Traces of PFSAL Based Rijndael S-Box

Fig. 11 shows the uniform current traces of the PFSAL based Rijndael circuit. Table IV presents the average energy dissipation of Rijndael S-Box. PFSAL consumes 1.780 pJ whereas SPGAL, SQAL, CSSAL and SAL consumes 5.876 pJ, 39.66 pJ, 45.67 pJ and 9.324 pJ. Hence proposed design consumes very less energy when compared with existing counterparts. Graphical representation of Table IV is shown in Fig.12.

TABLE IV. AVERAGE ENERGY CONSUMPTION OF RIJNDAEL S-BOX OF DPA RESISTANT ADIABATIC LOGIC FAMILIES

Adiabatic Logic	1.25 MHZ	12.5MHz	125MHz
PFSAL (pJ)	1.780	7.648	24.30
SPGAL (pJ)	5.876	10.35	26.99
SQAL (pJ)	39.66	540.3	577.0
CSSAL (pJ)	45.67	631.9	694.1
SAL (pJ)	9.324	78.93	193.2

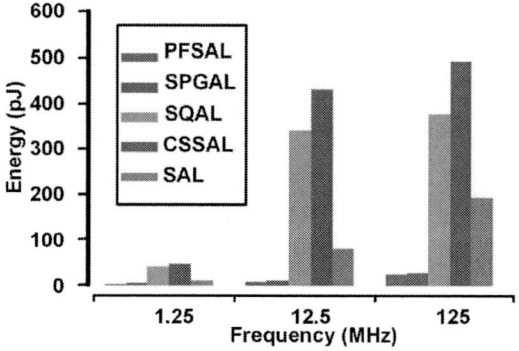

Fig. 12. Rijndael S-Box - Average Energy Consumption of DPA-Resistant Adiabatic Families

CONCLUSION

DPA is the most commonly encountered side channel attack in cryptography. Adiabatic logic provides a solution to this issue by generating uniform current traces. Attacker performs a DPA attack by analyzing the power and current waveforms of the cryptographic hardware. In this paper, a novel Positive feedback symmetric adiabatic logic (PFSAL) is proposed, which consumes very less energy when compared with the existing security based adiabatic logic circuits. PFSAL has an improvement of about 80.9%, 96%, 95% and 65% when compared with the SAL, CSSAL, SQAL and SAL Circuits. Also, PFSAL reduces the transistor count when compared with SAL and CSSAL. All these features lead PFSAL to be appropriate for the design of cryptographic hardware devices.

ACKNOWLEDGEMENT

I would like to thank my supervisor Dr. V. S. Kanchana Bhaaskaran for her constant support for my research activities. Also, I would like to thank Prof. A. Prathiba for her valuable suggestions and contribution in the design of S-Box.

FUTURE WORK

Proposed PFSAL will be used to design a complete cipher block and DPA attack will be performed to prove the efficiency of the circuit.

REFERENCES

[1] K. Tiri, M. Akmal, and I. Verbauwhede, "A dynamic and differential cmos logic with signal independent power consumption to withstand differential power analysis on smart cards," in Solid-State Circuits Conference, 2002. ESSCIRC 2002. Proceedings of the 28th European. IEEE, 2002, pp. 403–406.

[2] K. Tiri and I. Verbauwhede, "A logic level design methodology for a secure dpa resistant asic or fpga implementation," in Proceedings of the conference on Design, automation and test in Europe-Volume 1. IEEE Computer Society, 2004, p. 10246.

[3] D. Suzuki and M. Saeki, "Security evaluation of dpa countermeasures using dual-rail pre-charge logic style," in Cryptographic Hardware and Embedded Systems-CHES 2006. Springer, 2006, pp. 255–269.

[4] A. Moradi, M. T. M. Shalmani, and M. Salmasizadeh, "Dual-rail transition logic: A logic style for counteracting power analysis attacks," Computers & Electrical Engineering, vol. 35, no. 2, pp. 359–369, 2009.

[5] A. Moradi, M. Khatir, M. Salmasizadeh, and M. T. M. Shalmani, "Charge recovery logic as a side channel attack countermeasure," in Quality of Electronic Design, 2009. ISQED 2009. Quality Electronic Design. IEEE, 2009, pp. 686–691.

[6] A. Kramer, J. S. Denker, B. Flower and J. Moroney, "2nd order adiabatic computation with 2N-2P and 2N-2N2P logic circuits." Proceedings of the 1995 international symposium on Low power design., pp.191-196,1995.

[7] V. S. Kanchana Bhaaskaran and J. P. Raina, "Differential cascode adiabatic logic structure for low power," Journal of low powerelectronics. Vol. 4, No.2, 2008.

[8] V. S. Kanchana Bhaaskaran and J. P. Raina, "Pre-Resolve and Sense Adiabatic logic for 100 Khz to 500 Mhz frequency classes," Journal of Circuits, Systems, and Computers, Vol. 21, No. 5, 2012.

[9] M. Khatir and A. Moradi, "Secure adiabatic logic: a low-energy dpa resistant logic style." IACR Cryptology ePrint Archive, vol. 2008, p. 123,2008.

[10] B.-D. Choi, K. E. Kim, K.-S. Chung, and D. K. Kim, "Symmetric adiabatic logic circuits against differential power analysis," ETRI journal, vol. 32, no. 1, pp. 166–168, 2010.

[11] C. Monteiro, Y. Takahashi, and T. Sekine, "Charge-sharing symmetric adiabatic logic in countermeasure against power analysis attacks at cell level," Microelectronics Journal, vol. 44, no. 6, pp. 496–503, 2013.

[12] M. Avital, H. Dagan, I. Levi, O. Keren, and A. Fish, "DPA-secured quasi-adiabatic logic (SQAL) for low-power passive RFID tags employing s-boxes," Circuits and Systems I: Regular Papers, IEEE Transactions on, vol. 62, no. 1, pp. 149–156, 2015.

[13] Kumar, S. Dinesh, Himanshu Thapliyal, Azhar Mohammad, and Kalyan S. Perumalla. "Design exploration of a symmetric pass gate adiabatic logic for energy-efficient and secure hardware." Integration, the VLSI Journal 58 (2017): 369-377.

[14] P. Teichmann, Adiabatic logic: future trend and system level perspective. Springer Science & Business Media, 2011, Vol.34.

2018 31th International Conference on VLSI Design and 2018 17th International Conference on Embedded Systems

Online Detection and Reactive Countermeasure for leakage from BPU using TVLA

Sarani Bhattacharya
Department of Computer Science and Engg
IIT Kharagpur
sarani.bhattacharya@cse.iitkgp.ernet.in

Shivam Bhasin
Temasek Laboratories
NTU Singapore
sbhasin@ntu.edu.sg

Debdeep Mukhopadhyay
Department of Computer Science and Engg
IIT Kharagpur
debdeep@cse.iitkgp.ernet.in

Abstract—**Branch Prediction Units (BPUs) of computing systems have been targeted by several side channel analysis of public key encryptions. In recent years, performance counters have been used as a side channel source for the branch mispredictions which can be used to attack ciphers with user privileges. In this paper we propose an online leakage detection tool Branch-Monitor for branch misprediction traces which does an online detection of the leakage and raises an alarm if there exists a significant difference in the distribution of branch misses for selected inputs. The Monitor triggers a randomization module on detecting such leakage which effectively runs a software module to confuse the branch predictor unit such that it inherently prevents the information leakage. We have practically validated our detection module on Intel systems and is easily scalable to other platforms and processors.**

I. INTRODUCTION

Micro-architectural side-channel threats have gained importance manifold in the last decade since the cloud service providers allow several users to share the same hardware. These attacks target information leakages with respect to micro-architectural events of the system such as cache misses and branch misses. These leakages are considered to be benign for the normal applications, but, if monitored precisely, they result in revealing sensitive information of the cryptographic algorithm. Cryptographic algorithms, in spite of being mathematically strong, can leak secret keys through such micro-architectural events since the implementations of such cryptographic algorithms leave their execution footprints on the shared system resources.

Hardware Performance Counters are a set of special-purpose registers storing the counts of hardware-related activities within the microprocessor. These counters are affected by the internal activities of the processor and hence can be utilized as a source of information leakage. In [1], the authors exploited these HPCs as side-channels for time based cache attacks. HPC L1 and L2 D-cache miss counters have been exploited as side-channels in [1] for performing cache attacks on symmetric-key algorithms, like AES as in [2].

Asymmetric-key cipher implementations typically have key-dependent conditional branching statements which when implemented on systems with branch predictors, are subjected to side-channel attacks exploiting the deterministic branch predictor behavior due to their key-dependent input sequences. In this paper, we propose Branch-Monitor which is an online

detection tool for attacks and threats which exploit the information leakage from the Branch Predictor Unit (BPU).

The first attack targeting the BPU appears in [3], where the penalty for mispredicted branches in number of clock cycles is observed as side-channel to identify the data dependent operations of the public-key cryptosystem. On a standard RSA implementation, four different types of attacks were performed exploiting the Branch Prediction Unit (BPU) by using both synchronous and asynchronous techniques. A further improved version of this attack [4], [5] has also been carried out with proper knowledge of underlying hierarchical Branch Target Buffer (BTB) architecture of the target system. In [6], the authors introduced a new covert channel to perform secret communication between the Trojan and the spy processes which exploits the residual state of the dynamic branch predictor behavior of the system. While in [7], authors describe an implementation of a Contention-based Covert channel. An attack has been developed in [8] to derive kernel and user-level ASLR(Address Space Layout Randomization) offset which exploits the BTB collisions between the branch instructions. In [9], [10], it was first established that branch misses from HPCs can reveal the secret key in RSA. In [11], techniques for implementing binary exponentiation algorithms without requiring branch instructions have been proposed. However, the study of using Hardware Performance Counters (HPCs) to exploit the cipher codes implemented with branch statements is vital because there still exist several standard implementations using branches (as in OpenSSL [12], [13]).

Our approach needs no knowledge of the implementation of the algorithm under consideration. We perform hypothesis testing as proposed in the Test Vector Leakage Analysis(TVLA) [14] methodology on the output misprediction traces to check for data dependent leakage. In [15] TVLA is introduced as a reliable, quick and easy test to detect leakage between two specific distributions. This paper clearly shows the formulation of TVLA and validates their analysis on various modes of AES. A better analysis and leakage detection on any higher orders was formalized in [16]. The authors in [17] improved the original formulation of TVLA to make it more robust to environmental noise and observed that paired-t test works significantly well under such circumstances. While in a more recent paper [18], the authors have a fast leakage assessment methodology which can detect leakage orders of

978-1-5386-3693-0/18 $31.00 © 2018 IEEE

155

magnitude faster than the original TVLA formulation. This approach computes histograms for separate classes and separate time samples and then performs the statistical moments calculation on these histograms. In all of these works, the leakage analysis has been applied over power measurement traces, but in [19] this statistical t-test is applied to the timing side channel. The implementation design described in the paper determines whether a particular software module run in constant time or not.

In this paper, we propose an online branch misprediction leakage detection tool as Branch-monitor which can detect leakages by observing the branch misprediction traces and starts a defence mechanism if the distributions are significantly different. The contributions of our paper are:

- In this paper, we present a thorough analysis of the branch misprediction traces, leakage from those traces and detecting these leakage points using the Branch-monitor implementation.
- We evaluate our analysis on two setups of Intel processors and show the effectiveness of Branch-Monitor to detect leakage using a significantly lower number of traces.
- Following this, we propose a randomization module which is triggered as and when the Branch Monitor detects the leakage and, the experiments show that the countermeasure thus proposed to confuse the BPU hardware is efficient to prevent the leakage.
- Lastly, we analyze the timing overhead of the normal process computation in the presence of the randomization module.

II. PRELIMINARIES

In this section, we provide a background on some key-concepts, which include some implementation algorithms for public-key ciphers and some well-known branch predictors which have been subjected to attack.

A. Understanding Branch Mispredictions

Commonly, the implementations of public-key exponentiation algorithms and the scalar multiplications algorithms in ECC are such that the sequence of operations executed in every run of the algorithm is dependent on the secret bits. Both the exponentiation and scalar multiplication algorithms are commonly implemented with a set of statements in `if-else` block and the execution of the if-else statements are conditionally dependent on the secret key bits. The relation between these conditional sequences and branch misses is the following. Let the n-bit secret scalar in ECC be denoted as $(k_0, k_1, \cdots, k_i, \cdots, k_{n-1})$. The double and add operations of the double-and-add algorithm or the SPA resistant Montgomery Ladder algorithm being conditioned on the secret scalar bits, the trace of taken or not-taken branches as conditioned on scalar bits and expressed as $(b_0, b_1, \cdots, b_{n-1})$.

- If a particular key bit k_j is 1 then the conditional addition statement in the double and add algorithm gets executed. Thus, the condition is checked first, and if the particular key bit is set then its immediate next statement, i.e.,

addition gets performed. Since this is a normal flow of execution, the branch is considered as not-taken i.e., $b_j = 0$ in this case.
- While when $k_j = 0$, the addition operation is skipped and the execution continues with the next squaring statement. Thus, in this case, the branch is taken i.e., $b_j = 1$.

Thus for any `if-else` block, we consider the respective branch statement to be `not-taken` if the `if` conditional statement satisfies. On the other hand, if the `else` block is executed then we consider the respective branch to be `taken`.

The history of `taken` and `not-taken` branches are available to the branch predictor, and the predictor predicts next branches based on the history of branches that have already been encountered. Whenever, the predictor encounters a conditional statement, it predicts based on the history and predicted instructions gets fetched in the instruction pipeline. It is only during the execute stage that the condition gets evaluated and if there is a mismatch in the predicted and the evaluated branch then the corresponding instruction is flushed from the instruction pipeline resulting in pipeline stall, which is commonly referred to as branch misses.

B. Test Vector Leakage analysis (TVLA)

Test Vector Leakage Analysis(TVLA) was first proposed in [14] to identify if a distribution of power trace from a cipher implementation is statistically different from the another set of traces using the Welch's t-test in which the test statistic follows a Student's t distribution. The t statistic calculation takes power traces in two groups and performing the calculation independently on both the groups. Both the groups take power traces in two subsets A and B.

- N_A and N_B can be assumed as the size of the subsets A and B respectively.
- Compute X_A the average of all the traces in group A, X_B the average of all traces in group B.
- S_A be the standard deviation of the traces in group A and S_B is the sample standard deviation of the traces in group B.
- Computes the t-statistics for the particular trace as:

$$\frac{X_A - X_B}{\sqrt{\frac{S_A^2}{N_A} + \frac{S_B^2}{N_B}}} \tag{1}$$

Each trace is composed of an array of measurements sampled across different timestamps. The average and sample standard deviations of the traces are calculated for each sample point vertically across the set of traces. Thus after calculating mean and standard deviation they are also vectors over the same points in time.

The same analysis is repeated for the other group for their subsets. If at any point in time the t-test statistic exceeds $+/-4.5$ for both of the groups, the device fails the security test. This effectively claims that the device under test leaks information such that the two separate subsets are significantly distinct from each other and the null hypothesis is rejected.

In the next section, we understand the leakage from the branch prediction unit and in the following sections we propose the new online leakage detector to thwart information leakage through branch misprediction HPC event counts.

III. LEAKAGE FROM BRANCH PREDICTION UNIT

The values of the event counters which are available to user level processes through Linux perf utility actually leak a significant amount of information about the concurrently running processes in the system. The event counters get affected by the concurrent processes running in the system, and surprisingly they have a huge impact on the event counts as well. We emphasize on this with a simple experimental scenario, where two users share the same hardware and have two different processes running.

- Process 1: An unprivileged user running multiplication operation.
- Process 2: A privileged user performing exponentiation which is dominated mostly with conditional if-else statements.
- The unprivileged user observes Perf statistics for the user level multiplication process concurrent to the privileged exponentiation process.

As illustrated in Figure 1, it has been observed on an Ubuntu 16.04 system that there is an unexpected increase in the number of branch misses in the perf stat of the user process while the execution of exponentiation is performed. If the unprivileged user runs the multiplication process and observes the number of branch misses for the multiplication process then it encounters around $200 - 220$ branch misses. There has been a sharp increase in branch misses on and from 30 ms in Figure 1 which is only getting affected due to the exponentiation process being run by the privileged process. As in Figure 1, we observe that the time from which there is an increase in the number of branch misses (as observed by the user process), coincides with the time when the exponentiation algorithm begins. This experiment shows:

- An unprivileged user process residing on the same system as the privileged process can gain access to sensitive information of the privileged, or more generally any other user's process execution.
- Thus the increase or decrease of branch misses of a privileged process can be monitored by some spy process having only user level privileges.

IV. BRANCH-MONITOR: IMPLEMENTATION AND ANALYSIS

We propose Branch Monitor as a watchdog to the information leakage from the hardware performance counters. In this paper, we demonstrate that the information leakage for a shared user platform from the branch misprediction hardware is significant and our proposed detection tool can detect the leakage by applying a statistical test on the captured traces of the branch mispredictions using the perf 'ioctl' calls. We tabulate the steps of the detection as follows:

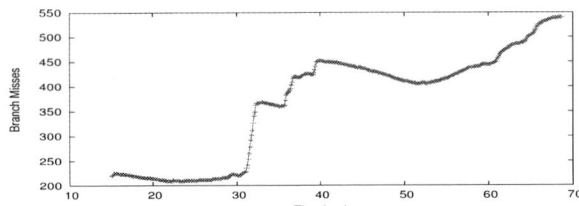

Fig. 1. Abrupt increase in branch miss observed by unprivileged user due to exponentiations from privileged process

- Branch-Monitor tool observes the branch misprediction traces for an executable and performs an online t-statistic test over time.
- For each set of points in the trace, the statistic is modified accordingly, and an alarm is raised if the value exceeds the threshold $+or - 4.5$.

In [9], the authors showed that the information leakage from the branch misprediction traces is much powerful to leak the secret key bits in as low as 100 traces. Our Branch-monitor detection tool can detect data-dependant leakage in only 40 traces.

In the next subsection, we explain the trace collection procedure for branch misprediction.

A. Obtaining branch misprediction traces using perf ioctl calls

Branch-Monitor is an online branch-misprediction analysis tool which observes the branch mispredictions from Hardware Performance counters (HPCs). The measurement procedure with Branch-Monitor uses a dummy code and observes the misprediction traces over the dummy code for the concurrently running encryption or decryption operation. Granularly observing the branch misses over each iteration of the if-else block of the code under consideration is only possible when the monitor is very closely observing the underlying algorithm, or the tool has the control of the underlying code. Since our tool Branch-Monitor is assumed to perform an online analysis, we present a much practical solution based on sampling perf event counter values using ioctl calls. The idea is such that the perf object is instantiated with an event which is used as a sampler. For example, if instruction count event is used as the sampling event, branch miss event is measured from HPCs for the particular sampling period.

B. Implementing the Branch-Monitor

We have implemented our prototype for Branch-Monitor as illustrated in Figure 2 using a c-code which runs on all variety of systems. We have targeted the Edward-1174 curve implementation written with long integer multiplications in c. The code computes scalar multiplication using the secret scalar by performing doubling for each bit and conditional addition for the bits which are set. For the statistical t-test to detect leakage, we need two sets of inputs to be fed to the algorithm under test. The test setup is described as follows:

978-1-5386-3693-0/18 $31.00 © 2018 IEEE

- Branch-Monitor collects branch misprediction event counts over a periodic interval of instruction count which is entirely handled by the perf ioctl system calls.
- The measurements of branch misses are observed on a dummy code snippet, while the secret scalar multiplication is getting performed concurrently in background.
- We obtained two sets of misprediction traces where the concurrently running Edward curve scalar multiplication was once provided a fixed basepoint as input and in the second case the input basepoint is varied randomly.
- Online mean and variance of the two sets are calculated separately and the t statistic value is measured.

Figure 3(a) and 3(b) represents two scenarios where the leakage is detected in as low as 30 traces on an Intel-i5-5200U processor running Ubuntu 16.04 LTS, and a particular part of the trace is observed to leak more as the number of traces are increased. The t statistic is calculated over each point on the trace, and the figures clearly show that both of the traces leaks information since the t statistic is having a value above 4.5 and below -4.5. The x-axis in the figures represent the individual sampling point of a branch misprediction trace and the y-axis values represent their appropriate t-statistic value. We have replicated our experiments on a comparatively older processor Intel-i5-3470 running Ubuntu-12.04 LTS, and the leakage detection scenario is illustrated in the Figures 4(a),(b). The figures clearly show that there are multiple points of leakage where the t statistic crossed the range(4.5,-4.5).

In the following section, we propose an interesting countermeasure to prevent this form of leakage.

V. BPU RANDOMIZATION BY BRANCH-MONITOR TOOL

We present a practical countermeasure to thwart these forms of leakages through branch misprediction information by randomizing the internal states of the branch prediction hardware. The BPU being shared by all processes of the same processor core, the effect of one process is predominant on the concurrently running execution on the same processor core. This property has been used by the attackers in [9] to attack secure implementation. In this paper, we use the same philosophy to prevent the leakage from these branch misprediction event counters. Branch-Monitor, as shown in Figure 2 being an online detection tool, the BPU randomization countermeasure is not triggered unless an alarm is raised.

- The monitor raises an alarm when the online detection mechanism encounters the branch misprediction leakage.
- On an alarm, the Branch-Monitor starts its defence mechanism of randomizing the state of the BPU such that the branch misprediction traces observed with such randomized design looses its correlation with the key.

The randomization module design is very easy to implement and highly effective to randomize the internal state of the BPU such that the information monitored over the BPU cannot be used to reconstruct the states of the secret computation. The Branch-monitor on detecting a leakage triggers on a particular randomization module within the code, which is typically written using a conditional if-else structure of code. The conditional if-else structure on execution results in conditional branching instructions, which are made to be dependent over randomly generated $0/1$ binary sequences. Since these random sequences bear no correlation with the secret scalar and the randomization module executes concurrent to the cipher module execution, the internal BPU states are modified by both the cipher module and the randomization module. This prevents the leakage through the event counters, which normally leaks a huge amount of information through these tools which are available in the user space.

The experiment as performed on Ubuntu 16.04 LTS is illustrated in Figure 5(a),(b) which plots the leakage in the presence of the BPU randomization. The leakage detection is performed using HPC event counts with the randomization being performed along with the cipher execution. Figure 5(a) shows that the randomization works really well such that the t-statistic is well within the range $+/-4.5$ for 2000 traces. We just made the test run overnight and figure 5(b) plots the leakage from 20000 traces, which also shows that the t statistic value is not crossing the region of -4.5 to 4.5. Again, we replicate the same experiment for the Intel-i5-3470 running Ubuntu-12.04 LTS and the Figures 6(a),(b) shows that for a significantly higher number of traces the leakage is not exceeding the safe limit thus preventing the leakage.

A. Impact of this randomization on timing side-channel

Fixing a side-channel leakage sometimes leads to a leakage through some other channels. In our design, the Branch-Monitor incorporates randomization to the Branch Predictor intermediate state by running a piece of code concurrent to a sensitive cipher execution. This randomization being due to concurrent executions, result in a number of context switches of the processes and thus can affect the execution time of the sensitive application. We plotted the frequency distribution of the execution time of the cipher implementation in two different scenario: when only the cipher implementation is running and second case when both the Branch Monitor and the randomization modules are running. The distributions as shown in Figure 7 absolutely overlap on each other and thus there is indeed minimal or no overhead in execution time on the running when Branch-monitor is allowed to run on

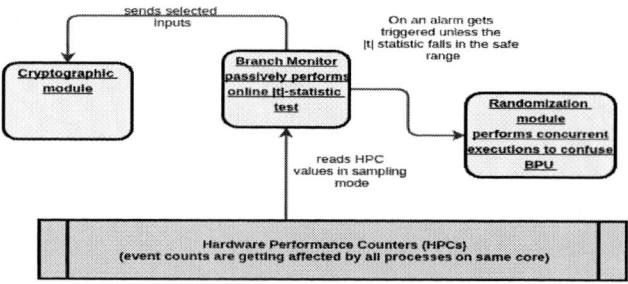

Fig. 2. Branch-Monitor design proposal

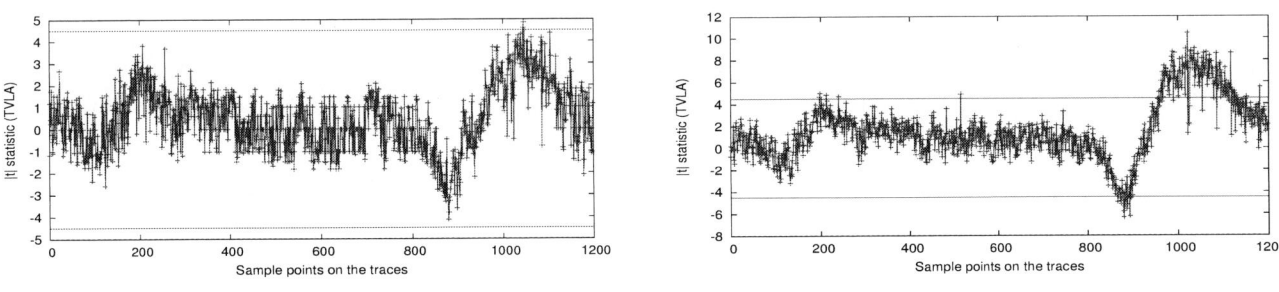

(a) **TVLA on 30 traces**　　　　　　　　　(b) **TVLA on 200 traces**

Fig. 3. Test Vector Leakage Analysis on Edward-1174 implementation on Ubuntu 16.04 in Intel Core i5-5200

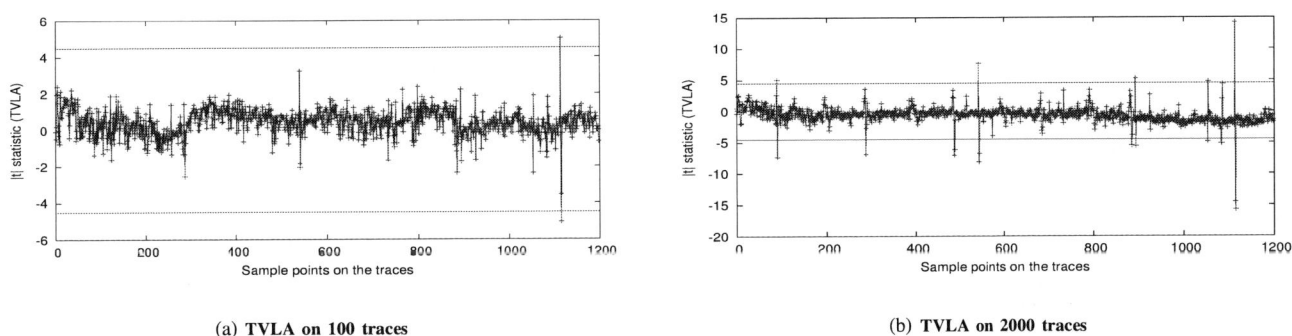

(a) **TVLA on 100 traces**　　　　　　　　　(b) **TVLA on 2000 traces**

Fig. 4. Test Vector Leakage Analysis on Edward-1174 implementation on Ubuntu 12.04 in Intel Core i5-3470

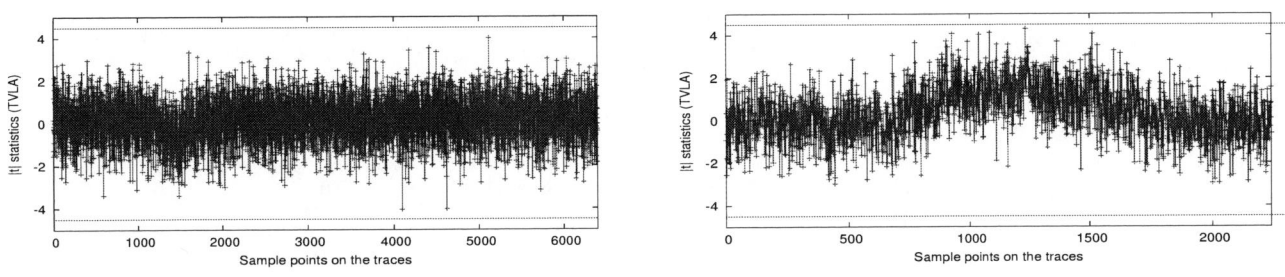

(a) **TVLA on 2000 traces in presence of randomization**　　　　　　(b) **TVLA on 20000 traces in presence of randomization**

Fig. 5. Test Vector Leakage Analysis on Edward-1174 implementation on Ubuntu 16.04 Intel Core-i5 5200U with concurrently running randomization module

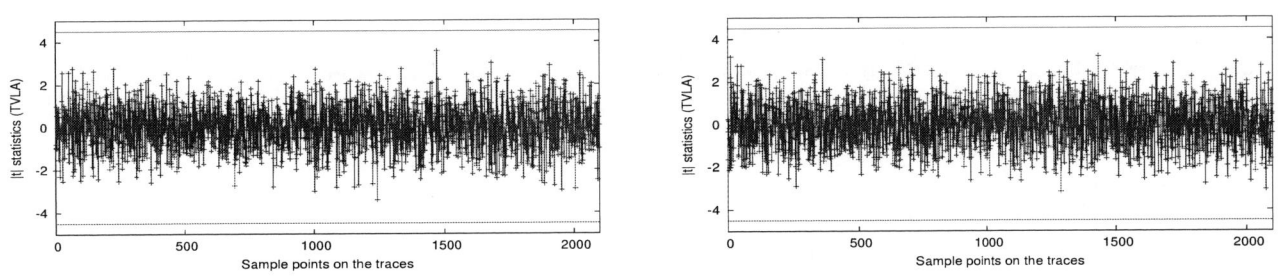

(a) **TVLA on 2000 traces in presence of randomization**　　　　　　(b) **TVLA on 20000 traces in presence of randomization**

Fig. 6. Test Vector Leakage Analysis on Edward-1174 implementation on Ubuntu 12.04 Intel Core-i5 3470 with concurrently running randomization module

978-1-5386-3693-0/18 $31.00 © 2018 IEEE　　　　159

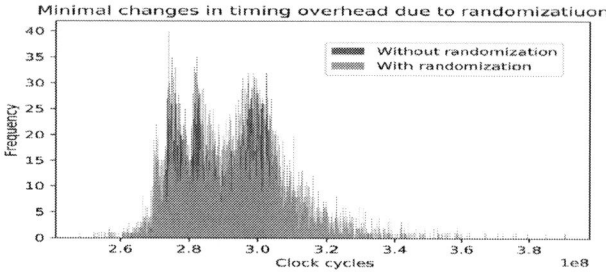

Fig. 7. Histograms of execution time for Edward-1174 curve while detection (without randomization) and with randomization shows minimal overhead on execution time of these two scenarios

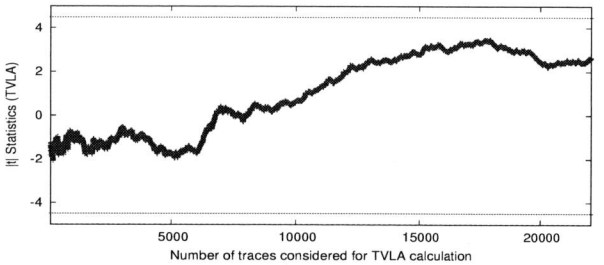

Fig. 8. Variation of timing-TVLA with increasing number of traces

the system. Next, we extended our experiments to test that the countermeasure proposed here leaks information through timing channel or not. We performed the t-statistic test on the execution time of the Edward-1174 curve implementation when the randomization module is also getting executed in parallel to the cipher implementation. Figure 8 illustrates how the t-statistic is changing as the number of traces are increased, and even after observing for 25000 traces the TVLA value indicates that there is no leakage through timing channels. This makes our claim even stronger, that the BPU randomization is an effective countermeasure in preventing the leakage and the timing-TVLA result adds up to its confidence.

VI. CONCLUSION

In this paper, we present an effective detection mechanism for the cipher implementations with respect to branch misprediction leakage. The branch misprediction traces are less noisy and much powerful to known conventional side channels and thus requires fewer traces to reveal secret. Thus we needed a detection mechanism which could detect the leakage much faster and start the defense mechanism so as to prevent further leaking of sensitive information. We validated all our experiments for the proposed, very efficient Branch-Monitor on Intel platforms. In this paper, we have only addressed the leakage assessment for branch mispredictions, but there exists a group of performance counter events which are equally vulnerable. Extending this analysis for the leakage assessment test of the whole bunch of performance counters would the future objective of this research.

REFERENCES

[1] L. Uhsadel, A. Georges, and I. Verbauwhede, "Exploiting hardware performance counters," in *FDTC*, L. Breveglieri, S. Gueron, I. Koren, D. Naccache, and J.-P. Seifert, Eds. IEEE Computer Society, 2008, pp. 59–67.

[2] J. Bonneau and I. Mironov, "Cache-Collision Timing Attacks Against AES," in *CHES*, ser. Lecture Notes in Computer Science, L. Goubin and M. Matsui, Eds., vol. 4249. Springer, 2006, pp. 201–215.

[3] O. Aciiçmez, Çetin Kaya Koç, and J.-P. Seifert, "Predicting Secret Keys Via Branch Prediction," in *CT-RSA*, ser. Lecture Notes in Computer Science, M. Abe, Ed., vol. 4377. Springer, 2007, pp. 225–242.

[4] O. Aciiçmez, S. Gueron, and J.-P. Seifert, "New Branch Prediction Vulnerabilities in OpenSSL and Necessary Software Countermeasures," in *IMA Int. Conf.*, ser. Lecture Notes in Computer Science, S. D. Galbraith, Ed., vol. 4887. Springer, 2007, pp. 185–203.

[5] O. Aciiçmez, J.-P. Seifert, and Çetin Kaya Koç, "Micro-Architectural Cryptanalysis," *IEEE Security & Privacy*, vol. 5, no. 4, pp. 62–64, 2007.

[6] D. Evtyushkin, D. V. Ponomarev, and N. B. Abu-Ghazaleh, "Covert channels through branch predictors: a feasibility study," in *Proceedings of the Fourth Workshop on Hardware and Architectural Support for Security and Privacy, HASP@ISCA 2015, Portland, OR, USA, June 14, 2015*, 2015, pp. 5:1–5:8.

[7] D. Evtyushkin, D. Ponomarev, and N. B. Abu-Ghazaleh, "Understanding and mitigating covert channels through branch predictors," *TACO*, vol. 13, no. 1, pp. 10:1–10:23, 2016.

[8] D. Evtyushkin, D. V. Ponomarev, and N. B. Abu-Ghazaleh, "Jump over ASLR: attacking branch predictors to bypass ASLR," in *49th Annual IEEE/ACM International Symposium on Microarchitecture, MICRO 2016, Taipei, Taiwan, October 15-19, 2016*, 2016, pp. 1–13.

[9] S. Bhattacharya and D. Mukhopadhyay, "Who watches the watchmen?: Utilizing performance monitors for compromising keys of RSA on intel platforms," in *Cryptographic Hardware and Embedded Systems - CHES 2015 - 17th International Workshop, Saint-Malo, France, September 13-16, 2015, Proceedings*, 2015, pp. 248–266.

[10] ——, "Formal fault analysis of branch predictors: attacking countermeasures of asymmetric key ciphers," *Journal of Cryptographic Engineering*, vol. 2016, pp. 1–12.

[11] D. Molnar, M. Piotrowski, D. Schultz, and D. Wagner, "The program counter security model: Automatic detection and removal of control-flow side channel attacks," in *Information Security and Cryptology - ICISC 2005, 8th International Conference, Seoul, Korea, December 1-2, 2005, Revised Selected Papers*, 2005, pp. 156–168.

[12] Y. Yarom and N. Benger, "Recovering openssl ECDSA nonces using the FLUSH+RELOAD cache side-channel attack," *IACR Cryptology ePrint Archive*, vol. 2014, p. 140, 2014.

[13] B. B. Brumley and R. M. Hakala, "Cache-timing template attacks," in *Advances in Cryptology - ASIACRYPT 2009, 15th International Conference on the Theory and Application of Cryptology and Information Security, Tokyo, Japan, December 6-10, 2009. Proceedings*, 2009, pp. 667–684.

[14] G. Goodwill, B. Jun, J. Jaffe, and P. Rohatgi, "A testing methodology for sidechannel resistance validation," *Non-invasive Attack Testing Workshop*, vol. 2011.

[15] G. Becker, J. Cooper, E. DeMulder, G. Goodwill, J. Jaffe, G. Kenworthy, T. Kouzminov, A. Leiserson, M. Marson, P. Rohatgi, and S. Saab, "Test vector leakage assessment (tvla) methodology in practice," in *International Cryptographic Module Conference 1001,2013*, 2013.

[16] T. Schneider and A. Moradi, "Leakage assessment methodology - A clear roadmap for side-channel evaluations," in *Cryptographic Hardware and Embedded Systems - CHES 2015 - 17th International Workshop, Saint-Malo, France, September 13-16, 2015, Proceedings*, 2015, pp. 495–513.

[17] A. A. Ding, C. Chen, and T. Eisenbarth, "Simpler, faster, and more robust t-test based leakage detection," in *Constructive Side-Channel Analysis and Secure Design - 7th International Workshop, COSADE 2016, Graz, Austria, April 14-15, 2016, Revised Selected Papers*, 2016, pp. 163–183.

[18] O. Reparaz, B. Gierlichs, and I. Verbauwhede, "Fast leakage assessment," *IACR Cryptology ePrint Archive*, vol. 2017, p. 624, 2017.

[19] O. Reparaz, J. Balasch, and I. Verbauwhede, "Dude, is my code constant time?" in *Design, Automation & Test in Europe Conference & Exhibition, DATE 2017, Lausanne, Switzerland, March 27-31, 2017*, 2017, pp. 1697–1702.

978-1-5386-3693-0/18 $31.00 © 2018 IEEE

Secure Neural Circuits to Mitigate Correlation Power Analysis on SHA-3 Hash Function

James Thesing, Dhireesha Kudithipudi

NanoComputing Research Lab, Rochester Institute of Technology

Abstract—Correlation Power Analysis, is recently demonstrated as a viable attack model for Keccak, which is the new hash function selected by NIST for SHA-3. Early studies show that CPA attacks can be launched on this algorithm with conventional CMOS implementations. To mitigate such power attacks, in this research we propose secure neural primitives using memristor neural logic blocks. Five different mitigation techniques are proposed, including baseline dualcore design, θ Plane Masking, neural logic block based θ Plane Masking, Analog neural logic block θ Plane Masking, and Analog dual-neural logic block θ Plane Masking. A framework for the CPA attack was designed and the mitigation techniques were assessed based on the number of power traces used, correlation coefficients, confidence ratios, and transistor count. Success rate of guessing a key during SIIA-3 operations, while configured as a MAC, is used as a system benchmark. Secure neural primitives are shown to be robust to the CPA attacks.

Index Terms—Side Channel Attacks, Correlation Power Analysis, Neuromorphic Computing, Neural Logic Block, SHA-3, Keccak.

I. INTRODUCTION

IN 2012, Keccak was chosen as the winner of the SHA-3 competition. Key Keccak features include a different construction, sponge, and efficient hardware implementations [1].Keccak won the SHA-3 competition due to its well defined mathematical model that allows in-depth analysis on its strengths against common attack strategies; however, Side Channel Attacks(SCA) do not attack an algorithm's construction, rather its physical implementation. One of the most potent attacks within SCA is the power attacks. As defined by Kocher [2], there are three commonly used methodologies of power analysis attacks, simple power analysis (SPA), differential power analysis (DPA), and correlation power analysis (CPA). SHA-3 is inherently resistant to SPA and DPA. However, early analysis has shown that it is vulnerable to CPA attacks [3]. This early work [3] also presents the CPA attack framework models for the standard CMOS technologies.

SHA-3 authors proposed various architectures for SCA mitigation. These architectures focus on the non-linear permutation in SHA-3, similar to the S-Boxes in AES. In fact, this non-linear permutation is a weak point for SHA-3, and the architectures presented add significant hardware overhead and overall performance degradation. The focus of this article is to increase SCA resistance of SHA-3 by focusing on a different section of the algorithm, Theta operation, that will not affect speed, power consumption, and area as much as commonly proposed mitigation techniques [4]. Popular hardware based SCA mitigation techniques that have been proposed to work

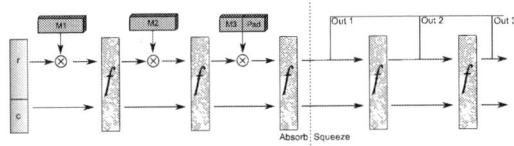

Fig. 1. Sponge Construction as can be seen in SHA-3. The f symbol denotes a full SHA-3 permuation while r and c denote the SHA-3 algorithmic parameters. The left side is the 'absorbing' phase while the right side is the 'squeezing' phase.

for general cryptographic algorithms are [5] and dual-rail logic: Wave dynamic differential logic [6], Masked Dual-Rail pre-charge Logic [7], Secure Triple Track Logic [8], and balanced cell-based differential Logic [9]. The downside of these methods is larger circuit area, higher power consumption and larger delays than the baseline circuits when realized in conventional CMOS. Emerging devices like memristors offer inherent power and area advantages that can alleviate these constraints. There were several early demonstrations of using memristors to address the power attacks and also for realizing PUFs [10], [11]. In this research, we propose secure neural circuits designed using memristor neural logic blocks (NLB) to mitigate the CPA attacks. Specifically we propose five different strategies, baseline dualcore design, θ Plane Masking, Multithreshold NLB based θ Plane Masking, Analog NLB based θ Plane Masking, and Analog dual-NLB θ Plane Masking. An attack framework was also developed in HDL to validate the CPA attacks on all the proposed techniques. Confidence ratio is used as a metric to measure the efficacy of the proposed techniques.

II. SHA-3

Fig. 1 shows the construction of SHA-3. SHA-3 uses a sponge construction where every hash operation is denoted by an f in the figure. SHA-3 is defined by specifying the values of r and c. The value r refers to the rate of the algorithm and is the rate that messages can be consumed and digests are generated. The value c refers to the internal capacity, and is never propagated as an output. The internal capacity is closely related to the overall security of the algorithm as this information is never propagated to a user. The authors of SHA-3 prove that the strength of SHA-3 is equal to $2^{c/2}$ [1]. One SHA-3 hash permutation is broken into $12 + 2\ell$ rounds where ℓ is found from $w = 2^{\ell}$, with w being the size of a word in the implementation.

The suggested SHA-3 implementation in FIPS 202 is used, which has $r = 1024$, $c = 576$, and $w = 64$. The SHA-3 permutation will thus have 24 rounds for any block of input that needs to be absorbed.

SHA-3 focuses on manipulation of an internal state at the bit level. A SHA-3 state is a 5x5xw, or 5x5x64 in this case. The state is filled starting from $(0, 0, 0)$ along the z-axis until full, then along the x-axis, and finally the y-axis. Once r bits have been filled, the remaining state is initialized to 0 to represent c. Similar to chaining algorithms, there must be a padding function that makes sure the input data is a multiple of r. The padding function used in SHA-3 is a 10x1 structure where the bits 1, 1 to $r - 1$ zeros, and a final 1 bit are appended to the message to ensure a multiple of r.

SHA-3 can be further broken down into five distinct operations: θ, π, ρ, χ and ι. The implementations are all assumed to be fully parallel where one SHA-3 permutation can be completed each clock cycle in the high-speed core.

A. Theta operation (θ)

θ is an operation on 11-bits of the SHA-3 state that provides column diffusion by calculating each column's parity and using an XOR function to combine two neighboring columns:

$$S[x][y][z] = s[x][y][z] \oplus (\oplus_{i=0}^{4}(s[x-1][i][z]) \\ \oplus (\oplus_{i=0}^{4}(s[x+1][i][z-1]) \quad (1)$$

For all equations in this section: $x, y \in \{0, 1, ..., 4\}$, $z \in \{0, 1, ..., 63\}$. To reduce hardware complexity of 11-bit XOR operations for every state bit, θ is typically implemented in two phases: the θ plane creation and the θ output phase. The θ plane creation is as follows:

$$\theta_{plane}(x, z) = \oplus_{y=0}^{4}(s[x][y][z]) \quad (2)$$

The θ plane is created with 5-bit XOR operations. θ_{out} can be implemented with 3-bit XOR operations as follows:

$$\theta_{out}(x, y, z) = s[x][y][z] \oplus \theta_{plane}(s[x-1][i][z]) \\ \oplus \theta_{plane}(s[x+1][i][z-1]) \quad (3)$$

There is a mixing of known and unknown bits in θ, making this permutation a prime target for SCAs. Specifically, this attack will be focused on the θ_{plane} creation step which mixes four known bits to one unknown bit.

ρ provides inter-slice diffusion inside the state. ρ is useful to bit diffusion of a slice, but does not involve mixing known and unknown bits, which limits its usefulness for SCA. Since π operation is just a remapping of bits, this is a bad SCA target for the same reason as mentioned for ρ. χ can be thought of as $5w$ parallel S-Boxes operating on the rows of the SHA-3 state. [12] shows that when keys get larger than one plane, χ is a required attack point for finding the full key.In this work, χ will not be looked at as an SCA attack point as it is not needed for keys that reside only in the first SHA-3 plane. If mitigations work for the simple 1-plane key case, they are inherently valid for the more complicated cases. ι is the last permutation of the SHA-3 where a constant value is XOR'd with the first lane $(0, 0)$. -1 x=y=0

B. SHA-3 MAC

The use case that will be explored in this work will be hash functions used in MAC implementations. MACs are used to determine if a message has originated from a known party. The key is used as the secret information that will provide a unique signature. This can be further applied to authenticated encryption, for which SHA-3 is used as a base [13]. In this work, the bottom plane of the state will contain a 320-bit key which in hex format is: $\{00, 01, 02, 03, ..., 39\}$. The planes at $y = 1$ and $y = 2$ will be filled with message bits along with the lane at $x = 0, y = 3$. The final bits are initialized to the initial SHA-3 state of '0'.

III. CORRELATION POWER ANALYSIS (CPA)

CPA as described in [14] is based off the Pearson correlation coefficient that determines the relationship between how two random variables change. Commonly used Pearson correlation coefficient, shown in 4, uses covariance of the data, power traces and hamming distances/weights, divided by the variance of the two distributions.

$$\rho = \frac{cov(x, y)}{\sigma_x \sigma_y} \quad (4)$$

For linear functions, CPA takes into account the state of all bits, for every key guess section, as the hamming weight or distance is compared to power data, instead of the state of individual bits. In the above equation, x is the power data measured from simulations and y is the hamming weights of the θ plane creation using one key byte at a time (shown in 5), where p is the input vector and i is the key byte, and V is the max number of power traces.

$$HW_{p,i} = \left(\sum_{z=0}^{7} \theta_{plane,p} \left(\left\lfloor \frac{i}{8} \right\rfloor, ((8 * i) \mod 64) + z \right) \right) \\ p \in \{1, 2, ..., V\}, i \in \{1, 2, ..., 39\}$$

$$\quad (5)$$

IV. PROPOSED SECURE NEURAL CIRCUITS FOR CPA MITIGATION

A Multithreshold Neural Logic Block (MTNLB) similar to the one proposed in [15] is used as a baseline. The n-input NLB is a single layer component that is capable of learning any basic logic function and any n input functions. It is important to note that this is capable of learning non-linearly separable operations. Since the NLB does not use simple monotonic activation functions, it can implement any n input function as long as the activation function has $n + 1$ boundaries. This component can be looked at as a trainable, analog look up table.

The circuits in this work are split into two groups: digital and analog. For the digital secure circuits, full SHA-3 algorithm is implemented for validation. For the analog circuits, only the θ plane was instantiated due to the length of the simulation time. To provide accurate comparisons between the analog circuits, that only have θ implemented, to the full digital implementations, all circuits discussed will have another circuit that only has θ implemented. This digital θ only circuit will not be presented as it is a sub-circuit of the presented digital circuits.

978-1-5386-3693-0/18 $31.00 © 2018 IEEE

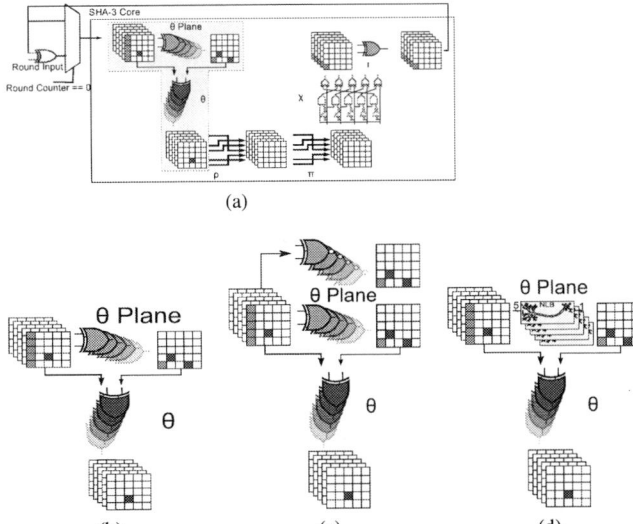

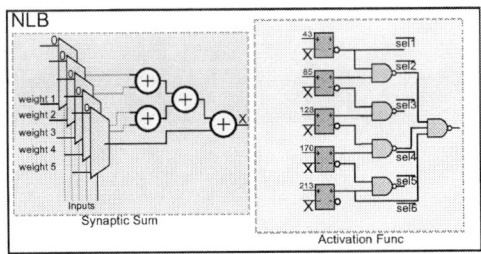

Fig. 3. RTL Diagram of Multi-Threshold Neuron Logic Block. The left shows how synaptic summation is implemented while the right shows how the MTNLB activation function is implemented.

(a)

(b)   (c)   (d)

Fig. 2. Block diagram of the basic SHA-3 high-speed core implementation, where the state is represented by a 5x5x5 cube while the actual implementations use 5x5x64. Mitigations are applied in the dotted box of (a). (b) Shows an unchanged θ for the dual core implementation since the dual-core replicates all of (a) with inverted data. (c) Shows the XNOR mitigation which adds an inverse θ creation step using XNOR gates. (d) shows the MTNLB mitigation which replaces the XOR gates in the θ plane creation with MTNLBs.

A. Digital Techniques

1) High-Speed Core - Base Circuit: For analysis of the SHA-3 algorithm, the SHA-3 high-speed core implementation is used [16]. To provide a good comparison, mitigations were applied to this base circuit's HDL model. Fig. 2a shows the block level diagram of the high-speed core. Starting from the left, the input into a round is either the previous state, in the case of squeezing, or the previous state XOR'd with a new block of input, in the case of absorbing. The columns of the 1600-bit state are XOR'd to generate a 5x64 plane that is called the θ plane. To generate the resulting θ output, the bit at (x, y, z) is XOR'd with two-bits in the θ plane to obtain $\theta(x, y, z)$. The next phases of ρ and π are remappings that were discussed in earlier sections. The χ operation takes each input bit and applies a non-linear function comprised of 3 logic gates, AND, XOR and NOT, per bit. The final operation is a 64-bit XOR on the first lane of the state using the current round constants.

2) Mitigation One — Dual-Core Design: The first circuit for SCA mitigation testing was to use two simultaneously instantiated SHA-3 high-speed cores. The first core would preform the SHA-3 hash algorithm while the second core would compute the SHA-3 hash with inverted input data. This design is inspired from the dual-rail logic family where for every '0' to '1' bit transition, there also exists a '1' to '0' transition. Since the θ plane creation uses a linear function, XOR, dual-rail logic can be implemented by using the inverse input data and preforming the hash permutation. This is not full dual-rail logic as it does not hold for the χ permutation; however, since the attack is happening on the θ plane, the methodology is valid, sown in Fig. 2b. Two SHA-3 cores are instantiated and one is fed the normal input data while the

second gets the inverse data by using 1600 NOT gates, one for each bit.

3) Mitigation Two — θ Plane Masking: For a more compact digital implementation, mitigation two uses the inverse operation only on the creation of the θ plane. This is done by modifying the base SHA-3 core and adding a set of XNOR gates to calculate the inverse θ plane. This saves $1600 * 3$ gates that are used in the χ permutation and the one 64-bit gate used during ι. Fig. 2c shows the block diagram of this circuit.

4) Mitigation Three — MTNLB θ Plane Masking : The implementation of NLBs presented in previous work has only been implemented in analog design. This work will implement both an analog and a digital MTNLB circuit to compare mitigation in different design domains. Implementing the MTNLB in digital logic can show decorrelating power consumption from input data while still using full digital CMOS logic.

Fig. 2d shows the high level block diagram of this mitigation. The 5-bit XOR gates in the θ plane creation step are replaced with 5-bit MTNLBs. Fig. 3 shows the RTL description of one MTNLB. Weight values represent the synaptic connections between the input and the MTNLB, which are trained outside of this RTL model using Matlab. The weights are represented using fixed point notation (Q_m, Q_n) where Q_m is the number of integer bits and Q_n is the number of fractional bits. The (Q_m, Q_n) used in this implementation is $(3, 4)$ for a total of 8-bits.

Normally the synaptic response between the input and the neuron is found by multiplying the synaptic weight with the corresponding input. In hardware this would require a (Q_m, Q_n) x (Q_m, Q_n) multiplier for every input. In this design, there is a need for 320 5-bit MTNLBs which would result in $320 * 5$ multiplier units. To simplify this, a 8-bit 2-1 MUX is used for every synaptic input. The select lines are the normal XOR inputs, which are either '0' or '1'. If the input is low, it will pass on the value of '0'. If the input is high, it will pass on the value of the synaptic weight. The outputs of these 8-bit 2-1 MUXs will then be summed using an adder tree of (Q_m, Q_n) bit adders. For the neuron activation function, the non-linear function will be implemented much like the analog version which used comparators to determine what range the synaptic sum falls into. In this case, there are 256 possible values as the fixed-point number is 8 bits. The 256 values are split into 6 sections fulfilling the $n + 1$ decision boundaries requirement to ensure a possible solution.

978-1-5386-3693-0/18 $31.00 © 2018 IEEE   163

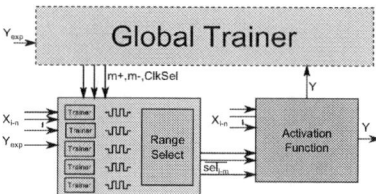

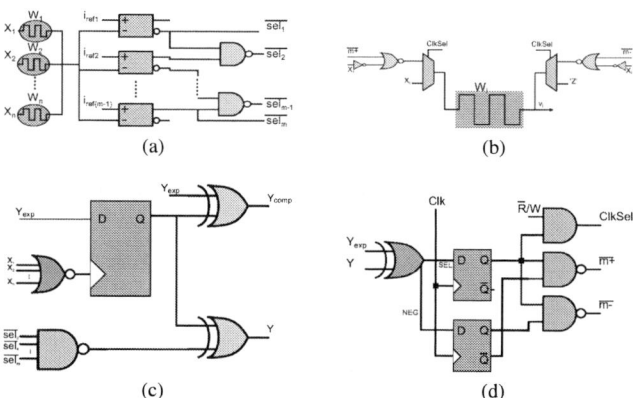

Fig. 4. Block level diagram of the analog MTNLB. For 1 MTNLB there is a single global trainer, n local trainings, n memristors, one range select component and one activation function.

Fig. 5. (a) the range select component. (b) the local trainer circuit. (c) the activation circuit. (d) the global trainer.

The final output of the MTNLB will then be either '0' or '1' dependent on an even boundary being active (active low in this implementation). This is equivalent to the activation function in the MTNLB.

B. Analog Secure Neural Circuits

1) Mitigation Four — NLB θ Plane Masking: Digital circuits, as discussed before, are not as efficient for implementing some of the basic features found in neuromorphic designs, such as addition, weight storage, and synaptic response calculation. The analog MTNLB solves these issues with memristive components that efficiently calculate synaptic responses, and shared memristor nodes to allow for summing of those responses. The MTNLB can be broken down into three main components, the global trainer, the range-select/local trainer and the activation function as can be seen in Fig. 4.

The range-select component can be seen in Fig. 5a. It takes a voltage input and performs the input and synaptic weight multiplication by the current that flows through the memristor. The currents are summed by a shared node instead of adders, as in the digital implementation. This summed current is sent to n comparators where the number of comparators is equal to the number of inputs of the MTNLB. The comparators find which range, m, that the summed current falls into by having one of the active-low select lines being active.

Inside the range-select block are n local trainers, whose responsibility is adjusting individual synapse weights, memristors, based on control signals from the global trainer. Fig. 5b shows the circuit diagram for the local trainer. The signals $ClkSel$, $\overline{m+}$ and $\overline{m-}$ all come from the global trainer. When

$ClkSel$ is set high, the values of $\overline{m+}$ and $\overline{m-}$ are valid for training. If the input, X_i, is low, the value of '0' is passed through the MUX and the weight of the memristor is unchanged. If the value of X_i is high, the values of $\overline{m+}$ and $\overline{m-}$ are applied to the memristor to change its internal state. $\overline{m+}$ and $\overline{m-}$ are set by the global trainer to correctly increment or decrement the weight. When $ClkSel$ is set low, no training is done, and the voltage on the input of the memristor is X_i, while the output is summed together with the rest of the memristors.

The activation function receives the active-low select lines from the range select block, $\overline{sel_i}$ and the inputs to the MTNLB, X_i. Fig. 5c shows the circuit diagram for the activation function. The activation function is a generic function where the desired activation can be configured by changing the reference currents in the range-select block. To get the non-monotonic function, the even select lines are fed into a NAND gate. If all the inputs are low, the D-flip-flop will have the value of Y_{exp} clocked to the output. If Y_{exp} is '0' while all inputs are low the values of Y and Y_{comp} will be the normal expected outputs. If Y_{exp} is '1' while all the inputs are low, the *bias* signal will be set to '1'. This has the effect of training the circuit on the inverse of the desired function, and then outputting that inverse whenever a future input propagates through the circuit. The final block is the global trainer, shown in Fig. 5d, which generates training signals when the MTNLB output does not match the expected output. A special XOR is used for the Y_{exp} and Y signals that outputs the XOR of the two signals (SEL) and the complement of Y (NEG). If NEG is high, Y is less then Y_{exp} which sets $\overline{m+}$ and $\overline{m-}$ to cause a negative drop across the memristor (increased resistance). If NEG is low, Y is greater than Y_{exp} setting $\overline{m+}$ and $\overline{m-}$ to cause a positive drop across the memristor. When SEL and the $\bar{R}/W$ signal are high, training is on, so $ClkSel$ goes high. If either SEL is low, correct output, or $\bar{R}/W$ is low, currently reading a value, the value of $ClkSel$ is set low [1].

C. Mitigation Five — Analog Dual-MTNLB θ Plane Masking

Mitigation five follows the same idea as the SHA-3 digital dual-core circuit implementation. For every MTNLB trained to compute an XOR function, there will be a complementary MTNLB that will learn the XOR function with inverted inputs. There will be a total of ten memristor weights trained to the XOR function with five (as ten total inputs with complements) inputs always active for any given input combination. This can be considered as the dual-rail logic version of the MTNLB.

V. RESULTS AND ANALYSIS

For comparison between mitigation techniques and a base circuit, the full SHA-3 algorithm was simulated using the HDL model at 45 nm. Then, just θ was simulated, the shaded box in Fig. 2, to provide a circuit that has less noise from power consumption than the other sections in SHA-3. Since mitigations are targeted at masking the data being manipulated

---

[1]The amount of time the $\bar{R}/W$ signal is high affects how much the memristor weights change, which in turn affects the training rule used.

TABLE I

RESULTS FOR ALL THE PRIMITIVES AND MITIGATION TECHNIQUES. AVERAGE CORRELATION IS THE AVERAGE OF THE TOP CORRELATION FROM EACH TOP KEY BYTE GUESS.

Circuit	#Trans	Pow. (mW)	Avg. Conf	Avg. Corr	Success Rates			
					Top1	Top2	Top4	Top8
Initial	77312	17.874	1.202	0.0317	97.5%	100%	100%	100%
Initial θ	24320	0.237	1.258	0.0314	100%	100%	100%	100%
Dual-Core θ	48640	1.989	1.038	0.0106	5%	7.5%	20%	22.5%
XNOR θ	32640	1.011	1.265	0.0404	100%	100%	100%	100%
Dig NLB θ	550400	6.977	1.013	0.0607	7.5%	7.5%	7.5%	15%
Analog NLB	101760	2.36e-05	1.014	0.0837	32.5%	40%	60%	70%
Analog NLB 16 nm	101760	1.35e-05	1.223	0.1076	100%	100%	100%	100%
Dual NLB	203520	5.42e-05	1.047	0.0879	32.5%	50%	65%	82.5%
Dual NLB 16 nm	203520	3.10e-05	1.062	0.0427	5%	10%	15%	22.5%

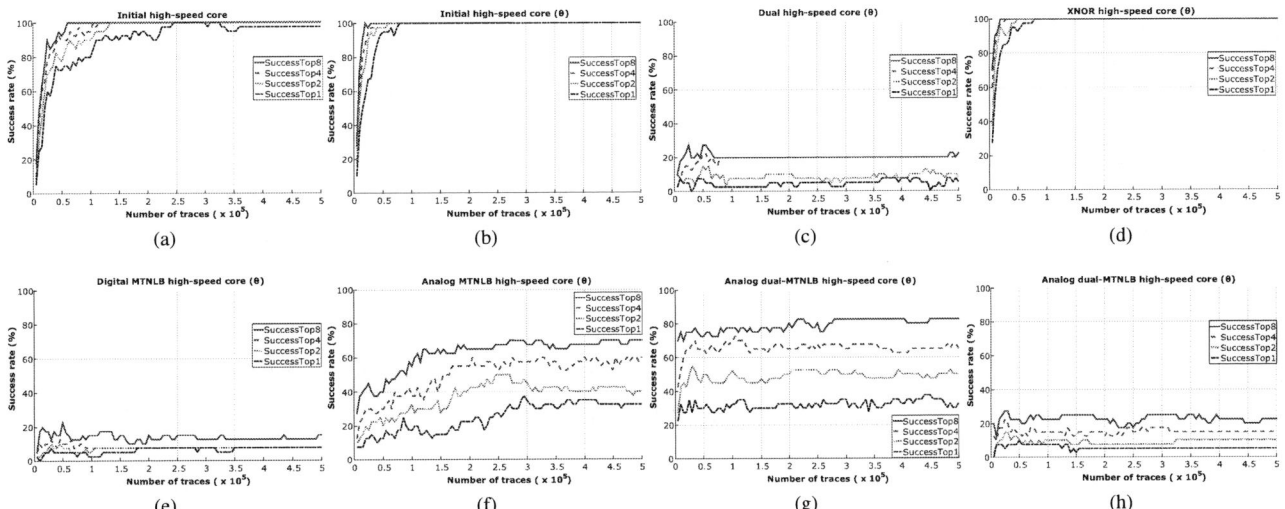

Fig. 6. Benchmark results for SHA-3 implementations. (a) initial SHA-3 circuit, (b) initial SHA-3 circuit with just θ, (c) dual-core mitigation , (d) XNOR mitigation, (e) Digital MTNLB mitigation, (f) analog MTNLB mitigation, (g) dual-MTNB mitigation, (h) dual-MTNLB mitigation at 16 nm.

in θ, the result of θ is a better comparison tool. Fig. 6 shows the success rates for guessing the correct SHA-3 key over multiple power traces of the circuits.

The plots are broken into groups based on the correct key being in the top number of guesses for all 40 bytes in that group. The *top1* group at 100% accuracy means the key was correctly guessed for all 40 key bytes. The *top8* group at 100% would mean the correct key was in the *top8* 100% of the time. From *top1* to *top8*, the key search space is reduced to: 2^0, 2^{1*40}, 2^{2*40}, and 2^{3*40}. AES-128 is seen as computationally secure with its 128-bit key size from brute force attacks. Because of this, it is safe to assume that the *top1* or *top2* group reaching 100% accuracy represents a successful attack while *top3* and *top4* gives an idea about what is happening during the attack, but does not indicate a computationally feasible attack.

Fig. 6a and 6b show the attack becoming easier when just θ is implemented. This comparison shows that implementing only the θ portion of SHA-3 allows an easier attack, thus the results are a good indicator of the highest accuracy achievable. Fig. 6d shows that preforming logic and its inverse can be done in such a way as to not mask power consumption. An XNOR gate is the inverse of an XOR gate, but it can be seen that it

does not mask power as well as Fig. 6c, only XOR gates with complimentary data. Fig. 6c, 6e, 6f, and 6g show dual-core, digital MTNLB, analog MTNLB, and analog dual MTNLB mitigations all effectively mask key data.

Fig. 7 shows the confidence ratio, defined by the ratio between the magnitude of the highest correlation value over the magnitude of the next highest correlation value. This gives a sense on how effective an attack is, as a high confidence value means the hypothesized guess is more probable the correct guess. When confidence values approach one, the correlation coefficients of the data converges on a single value, thus masking the underlying key data.

Table I shows the results of transistor count, power consumption, average confidence, average max correlation values and success rates for all the circuits. The best preforming circuit was the digital NLB implementation with the top8 group accuracy of 15%. The benefit of this implementation is that it uses common cell libraries, with overhead of 20x greater than the base circuit. The smallest form factor mitigation is the dual-core circuit which adds 2x the overhead but with a 7x increase in power. For the lowest power circuit, the MTNLB in both configurations provides very small power consumption with top8 group accuracy at 70% for the single MTNLB case

978-1-5386-3693-0/18 $31.00 © 2018 IEEE

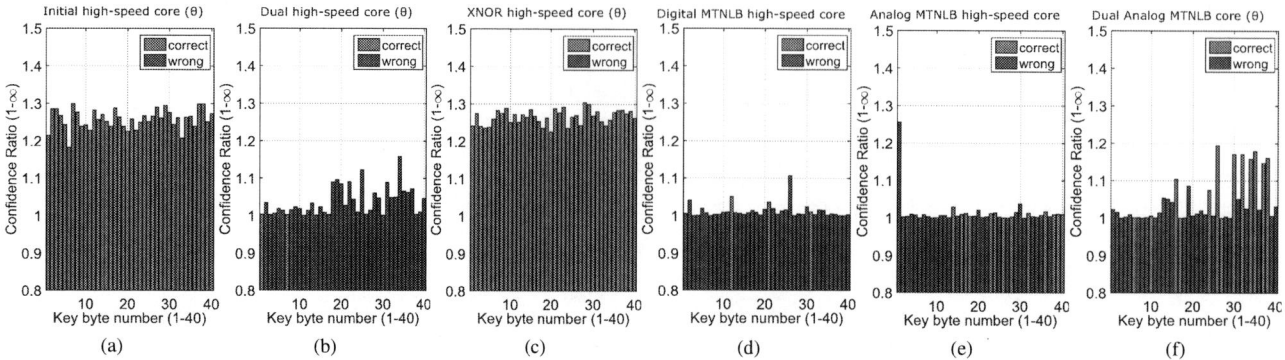

Fig. 7. Confidence ratio results for SHA-3 implementations. (a) initial SHA-3 circuit, (b) dual-core mitigation , (c) XNOR mitigation, (d) Digital MTNLB mitigation, (e) analog MTNLB mitigation, (f) dual-MTNB mitigation. Green bars represent when the key was guessed correctly for that byte while red represents incorrect guesses.

and 82.5% for the dual-NLB case. At 16 nm node size, the success rates goes to 100% for the single MTNLB case but drops to 22.5% for the dual-MTNLB case. The area overhead is $\approx 4x$ and $\approx 8x$ respectively ($\approx 2x$ and $\approx 3x$ when the full circuit is considered).

A. Reconfigurable circuits

Table I shows a custom design benefits the digital dual-core mitigation technique. Since, this uses both the harder to attack linear function, XORs, and the symmetry in correlation coefficients from the inverse data. The MTNLB, however performs better when deployed on FPGAs.

A minimum sized LUT is defined as: $2^n * T_m + 4 * (2^n - 1)$ where n is the number of inputs and T_m is the number of transistors in the memory element implementation. The size of one MTNLB is $2nm + 20n + 12m + 30 + 2ceil(m/2)$ where n is the number of inputs and m is the number of decision boundaries needed, $n + 1$ in this case. This places the size of one 5-input MTNLB at 268 transistors and one 5-input LUT at 316. An MTNLB would be $\approx 84\%$ the size of a LUT. The dual-MTNLB implementation would add an overhead of approximately 1.6x versus the much higher overhead found in current dual-rail logic (2x-4x).

VI. Conclusions

Emerging technologies offer unique opportunities in enhancing the robustness of the CMOS circuits. In this case, the NLB blocks were conducive to seamlessly integrate in the SHA-3 algorithm and offer power masking capabilities to mitigate CPA attacks. It was shown that MultiThreshold NLBs can be used effectively to mitigate CPA on linear functions, θ in this case. Single MTNLBs demonstrate 100% success rate at 16nm node sizes, and demonstrate high energy and area efficiency in reconfigurable platforms. Dual MTNLB mitigation techniques are very power efficient but increase significant area overhead. Depending on the resource availability and the sensitivity to CPA attacks, one of the five proposed mitigation techniques can be applied.

References

[1] G. Bertoni, J. Daemen, M. Peeters, and G. Van Assche, "The keccak reference," 2011.

[2] "Introduction to differential power analysis," *Journal of Cryptographic Engineering*, vol. 1, no. 1, 2011.

[3] M. Taha and P. Schaumont, "Differential power analysis of mac-keccak at any key-length," in *Advances in Information and Computer Security*. Springer, 2013, pp. 68–82.

[4] G. Bertoni, J. Daemen, M. Peeters, and G. Van Assche, "Building power analysis resistant implementations of keccak," in *Second SHA-3 candidate conference*, vol. 142. Citeseer, 2010.

[5] J. Ambrose, R. Ragel, and S. Parameswaran, "Rijid: Random code injection to mask power analysis based side channel attacks," in *Design Automation Conference, 2007. DAC '07. 44th ACM/IEEE*, June 2007, pp. 489–492.

[6] S. Guilley, L. Sauvage, J.-L. Danger, T. Graba, and Y. Mathieu, "Evaluation of power-constant dual-rail logic as a protection of cryptographic applications in fpgas," in *Secure System Integration and Reliability Improvement, 2008. SSIRI '08. Second International Conference on*, July 2008, pp. 16–23.

[7] "Masked dual-rail pre-charge logic: Dpa-resistance without routing constraints," in *Cryptographic Hardware and Embedded Systems CHES 2005*, ser. Lecture Notes in Computer Science, J. Rao and B. Sunar, Eds., 2005, vol. 3659.

[8] R. Soares, N. Calazans, V. Lomné, P. Maurine, L. Torres, and M. Robert, "Evaluating the robustness of secure triple track logic through prototyping," in *Proceedings of the 21st Annual Symposium on Integrated Circuits and System Design*, ser. SBCCI '08, 2008, pp. 193–198.

[9] M. Nassar, S. Bhasin, J.-L. Danger, G. Duc, and S. Guilley, "Bcdl: A high speed balanced dpl for fpga with global precharge and no early evaluation," in *Proceedings of the Conference on Design, Automation and Test in Europe*, ser. DATE '10, 2010, pp. 849–854.

[10] G. S. Rose, J. Rajendran, N. McDonald, R. Karri, M. Potkonjak, and B. Wysocki, "Hardware security strategies exploiting nanoelectronic circuits," in *Design Automation Conference (ASP-DAC), 2013 18th Asia and South Pacific*. IEEE, 2013, pp. 368–372.

[11] P. Koeberl, Ü. Kocabaş, and A.-R. Sadeghi, "Memristor pufs: a new generation of memory-based physically unclonable functions," in *Proceedings of the Conference on Design, Automation and Test in Europe*. EDA Consortium, 2013, pp. 428–431.

[12] M. M. Taha and P. Schaumont, "Differential power analysis of mac-keccak at any key-length." in *IWSEC*. Springer, 2013, pp. 68–82.

[13] G. Bertoni, J. Daemen, M. Peeters, G. Van Assche, and R. Van Keer, "Submission to the caesar competition." *CAESAR submission KEYAK*, vol. v1, 2014.

[14] "Correlation power analysis with a leakage model," in *Cryptographic Hardware and Embedded Systems - CHES 2004*, ser. Lecture Notes in Computer Science, M. Joye and J.-J. Quisquater, Eds., 2004, vol. 3156.

[15] M. Soltiz, D. Kudithipudi, C. Merkel, G. Rose, and R. Pino, "Memristor-based neural logic blocks for nonlinearly separable functions," *Computers, IEEE Transactions on*, vol. 62, no. 8, pp. 1597–1606, Aug 2013.

[16] G. Bertoni, J. Daemen, M. Peeters, and G. Van Assche, "Hardware implementation in vhdl," 2011, http://keccak.noekeon.org/files.html.

978-1-5386-3693-0/18 $31.00 © 2018 IEEE

A Novel Zero Blind Zone Phase Frequency Detector for Fast Acquisition in Phase Locked Loops

Sucheth S Kuncham, Manasa Gadiyar, Sushmitha Din K, Kiran Kumar Lad, Tonse Laxminidhi

Department of Electronics and Communication Engineering

National Institute of Technology Karnataka, Surathkal

Email: kunchamsucheth@gmail.com, laxminidhi_t@yahoo.com

Abstract—**The inability to sense the transitions in the input by conventional phase frequency detector (PFD) during the reset operation leads to blind zone, which reduces the acquisition speed and the detection range. The pull down network in proposed design is modified so as to eliminate the reset pulse for phase difference beyond the dead zone in order to have a full detection range and less cycle slippage. As the design gives the right polarity for phase differences close to $\pm 2\pi$, the acquisition time is reduced substantially. The Transfer characteristic of the PFD manifests an identical response. The PFD design is implemented in 180nm CMOS technology and consumes 1.36mW at an operating frequency of 1GHz.**

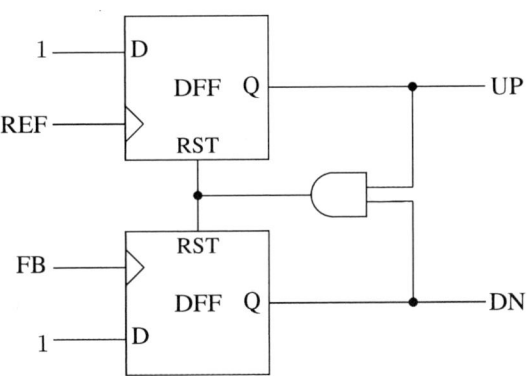

Fig. 2. Conventional Phase Frequency Detector

I. Introduction

Charge Pump Phase Locked Loops are widely used in high speed frequency transceivers - clock and data recovery, frequency synthesizers and high speed clock generators of microprocessors. The PFD serves as an essential integral block as it carries out the phase and frequency comparison, thereby generating suitable control signals for the loop to acquire and maintain its lock. The sensitivity of the PFD along with the nature of the control signals play an essential role in the performance of the overall PLL system. The development in high speed communication applications have motivated the need for fast acquisition PLLs with very low steady state phase error.

Phase Frequency Detector is a simple tri-state machine consisting of basic memory elements, the state diagram of which is shown in Fig 1. Extensive research has been carried out to increase the effective input range and linearity of PFD operation. The simplest and conventional phase frequency detector is shown in Fig 2, has been presented in [1]. Further, many architectures have been developed which include pre-charge type PFD [2], ncPFD [3], latch based PFD [4]- [7], and a couple of designs with non-linear transfer characteristics [8] [9].

The transfer characteristics of an ideal PFD is shown in Fig. 3(a). However, in reality, the effective detection range of PFD is less than 2π due to practical limitations. The PFD operation faces limitations at two major phase intervals called dead zone and blind zone. The two non idealities are explained in detail in section II.The proposed design modifies the PFD architecture so as to generate the reset pulse only when the two inputs are close to each other. The proposed PFD removes the dependency of blind zone on the reset pulse which also removes the lower bound defining the blind zone. This design topology aims at low power, highly sensitive (REF and FB slew) and a linear design with zero blind zone for high frequency operations.

The paper is organized as follows, section II discusses the non-idealities of the conventional PFD. Section III discusses the proposed design and working mechanism. Section IV discusses the results obtained during simulations followed by conclusion.

II. Non idealities of Conventional PFD

The *dead zone* is seen in the vicinity of zero phase difference. Here, the pulse width of PFD output signals is not enough for settling of charge pump currents. This creates static phase error and also contributes to jitter. In order to avoid this, both the outputs of PFD are kept ON with the help of a reset pulse. The lower bound of the reset pulse width is a function of the settling time of the charge pump current. The minimal pulse width of reset signal has not decreased significantly with the scaling of CMOS technology. As the

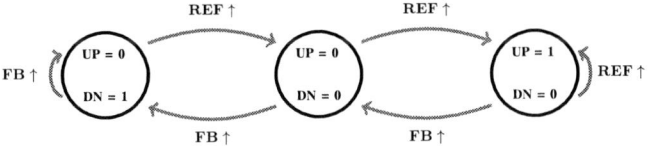

Fig. 1. Tri-state FSM Logic for PFD

978-1-5386-3693-0/18 $31.00 © 2018 IEEE

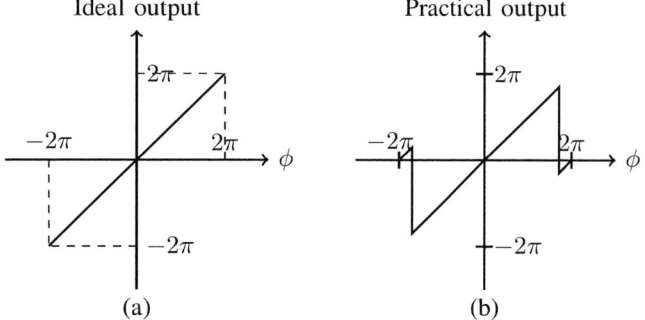

Fig. 3. Normalized characteristic curves of (a) ideal and (b) practical PFDs

frequency of operation increases, the reset pulse occupies a major portion of time period and thereby limits the frequency of operation. The maximum frequency of operation of PFD is given by the width of reset pulse (T_{res}) [11].

$$F_{max} = \frac{1}{2T_{res}} \quad (1)$$

The *blind zone* becomes significant when the phase difference between the two input signals is in the vicinity of 2π which is $[2\pi - \Delta, 2\pi]$ and $[-2\pi, -2\pi + \Delta]$. When the phase difference between the inputs falls into the blind zone and one of the outputs is high, the PFD detects the lagging signal edge first and fails to detect the rising edge of leading phase signal which results in reversed phase information. A phase difference of $2\pi - \Delta$ will be reversed and detected as $-\Delta$ which slows down the acquisition process. Therefore, it is important to alleviate the blind zone which increases the effective operation range of PFD. The blind zone (T_{BZ}) is defined by two components in the conventional design of PFD, the width of the reset pulse (T_{res}) and the precharge time (T_{pch}) of the pull up network [10].

$$T_{BZ} = T_{res} + T_{pch} \quad (2)$$

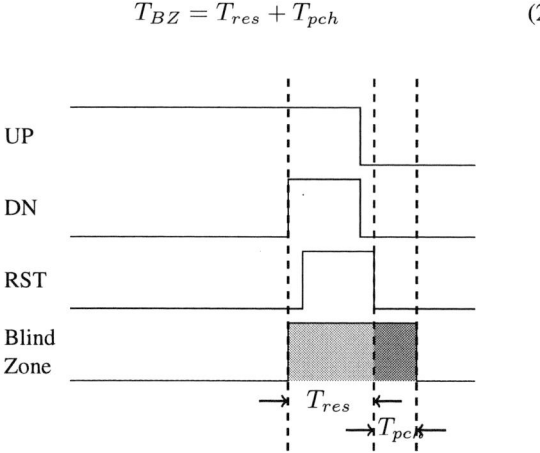

Fig. 4. Blind Zone

The conventional PFD fails to detect transitions in the input signals when the phase difference is large, near 2π. The reset

pulse masks the transition of either of the two signals REF and FB. This interval of phase difference is the *blind zone*. At high frequencies, the blind zone occupies major portion of the reference time period which deteriorates the performance of the PLL. This introduces cycle slippage which slows down the frequency acquisition and also defines an upper limit to the maximum operational frequency, F_{max}. The lower bound of reset pulse width mainly depends on the settling time of the charge pump which defines the lower bound of blind zone. Most architectures attempt to push the rising edges out of the reset pulse thus pushing the transition out of the blind zone. But these attempts to alleviate blind zone effects are limited by PVT variations [10].

III. PROPOSED DESIGN

A number of designs have been developed in the past to suppress the effects of the reset pulse and precharge time to alleviate blind zone range, the common techniques include adding delay elements to the inputs of D Flip Flop [5] which are equal to reset pulse width, thereby reducing the blind zone. The precharge effect can be reduced by adding suitable delayed input signals to the pull up branch [10].

A close look at the PFD operation reveals that reset pulse is essential when the phase difference is extremely small and is not required throughout the input phase difference $[-2\pi, 2\pi]$. Therefore, if the PFD can be designed to incorporate this fact, one can avoid complicated circuits that are used to minimize the blind zone effect. The proposed PFD is designed to provide reset pulse only at small phase differences, but still maintaining its simplicity in architecture.

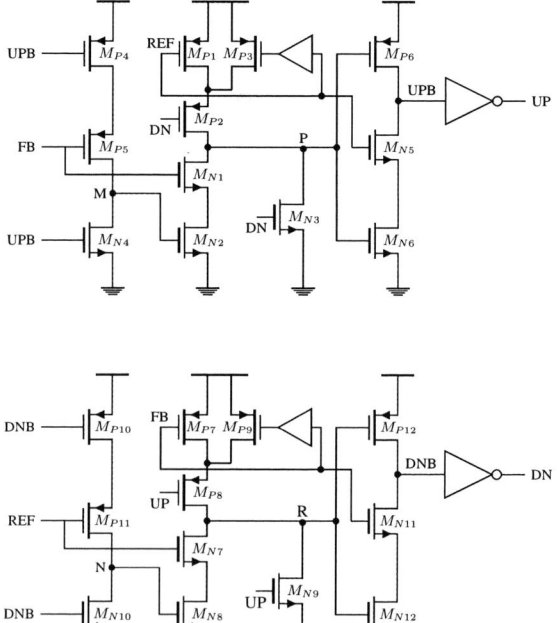

Fig. 5. Proposed Phase Frequency Detector

978-1-5386-3693-0/18 $31.00 © 2018 IEEE

The proposed PFD is shown in Fig 5 and the corresponding timing diagram in Fig. 6. The pull down branch and the pull up networks of the conventional dynamic PFD have been modified suitably to get the desired operation. The nodes P and R are the decisive nodes in the schematic. The transistors M_{P1}, M_{P2} and M_{P7}, M_{P8} form the pull up networks. The transistors M_{N1}, M_{N2} and M_{N7}, M_{N8} forms one set of pull down network whereas M_{N3} and M_{N9} acts as an additional pull down network.

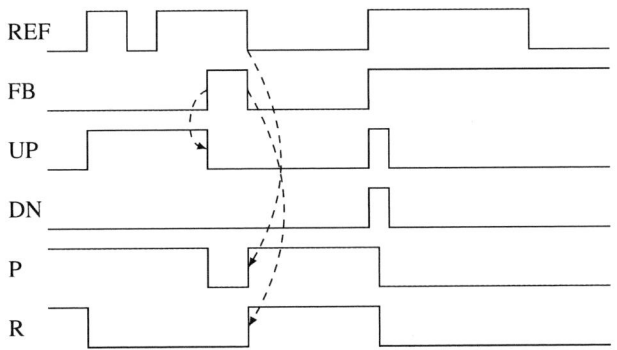

Fig. 6.   Timing diagram for the proposed PFD

Initially the inputs (REF and FB) are at logic low and PFD is in the state where UP and DN are at logic low. The nodes P and R are precharged to V_{DD} through their corresponding pull up networks.

A transition in REF signal from low to high will pull down the node UPB through M_{N5}, M_{N6} and UP signal goes to logic high. Thus, the state machine has made a transition to state UP=1 and DN=0. Following this, the node R gets discharged through M_{N9} as soon as UP is driven to logic high and the pull up network M_{P7}, M_{P8} is disabled as gate of M_{P8} is driven high. Further, the node M is charged to V_{DD} through M_{P4}, M_{P5}. Currently, the PFD is in the state UP=1 and DN=0. Any transition in the REF signal will not change the state of the PFD.

In this state, a transition in FB signal from low to high discharges node P through M_{N1}, M_{N2} and UP is pulled down to logic low. As node R had been discharged when UP=1, there is no transition in DN signal. Thus, the state machine makes a direct transition from state UP=1, DN=0 to UP=0, DN=0 unlike conventional designs. The signals UP and DN were pulled down through different sets of pull down branches which eliminated the need for explicit reset pulse at phase difference greater than dead zone range. A similar working explanation can be provided transition to state UP=0, DN=1 by transition in signal FB followed by signal REF. This happens when the delay between the two input signals is greater than the dead zone range.

In the above case, if the delay between the two signals is very small that the phase difference is within the dead zone, P and R being precharged to high will pull up both UP and DN signals. As soon as UP and DN make a transition to logic high, the nodes P and R are pulled down through M_{N3}

and M_{N9} transistors correspondingly. Thus, we get a minimal width Reset pulse without the need for an explicit reset pulse generation circuit. With suitable scaling of M_{N3} and M_{N9}, the width of reset pulse can be modified. The strength of pull down path formed by M_{P1}, M_{P2} and M_{P7}, M_{P8} should be more than pull up network to ensure proper operation for phase difference in the range $[\pi, 2\pi]$ and $[-2\pi, -\pi]$ ensuring no reversal in phase information.

The blind zone of the network is still present because of large precharge time of internal nodes P and R. A general technique to alleviate this effect is to add precharge transistors. We have added two transistors M_{P3} and M_{P9} driven by delayed input signals to turn on the pull up paths at the rising transition of the inputs. The delay added should be equal to the precharge time (T_{pch}), this completely eliminates the blind zone. Further the signals UP and DN fed to the pull down branches M_{N3} and M_{N9} can also be delayed to increase the stability of the system and modify the width of reset pulse.

IV. SIMULATION RESULTS

The design is implemented in 180 nm CMOS technology. The PFD consumes power of 1.36 mW when operated at an input frequency of 1 GHz with a supply of 1.8 V. The design was subjected to multiple tests to verify its functionality and to benchmark its improvement.

To examine the performance of the proposed design a Phase Locked Loop test bench was set up using Verilog-AMS to provide an identical environment to the blocks under test as shown in Fig 7. To test the functionality, the conventional D-Flip Flop based PFD and the proposed PFD were subjected to 500 MHz reference within the loop. The proposed PFD outperformed the conventional PFD by faster acquisition and lesser cycle slippage.

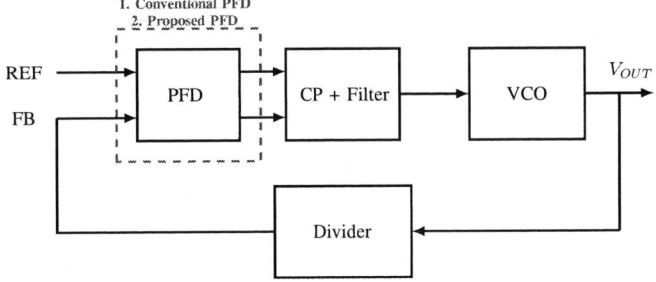

Fig. 7.   Test bench for PFD functionality in the loop

A similar test bench was set up for a frequency synthesiser with the PFD input at a frequency of 1 GHz, from another frequency synthesiser. It is observed that the cycle slippage has significantly reduced in the proposed PFD due to the eliminated blind zone. Acquisition of PLL with proposed and conventional PFDs is compared in Fig. 8. Clearly, PLL with proposed PFD offers faster frequency acquisition. The decrease in frequency acquisition time is found to be around 18% for 1GHz which is in good coherence with the literature.

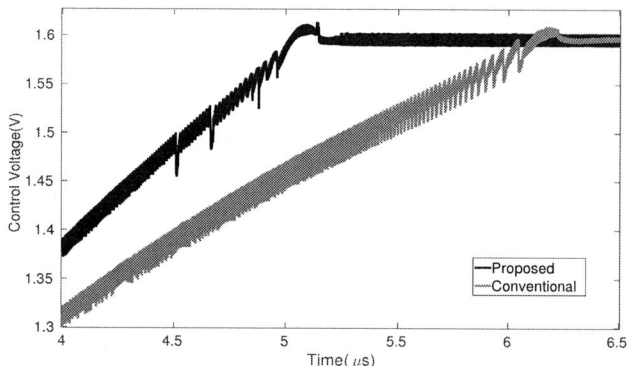

Fig. 8. Around 18% faster frequency acquisition is observed in the proposed PFD as compared to the proposed PFD

The proposed PFD was also simulated for all corners, with 1 GHz input. The normalized transfer characteristics of the proposed PFD for typical (TT) corner and the two extremes slow-slow (SS) and fast-fast (FF) corners are shown in Fig 9. The design maintained a good performance to yield faster acquisition.

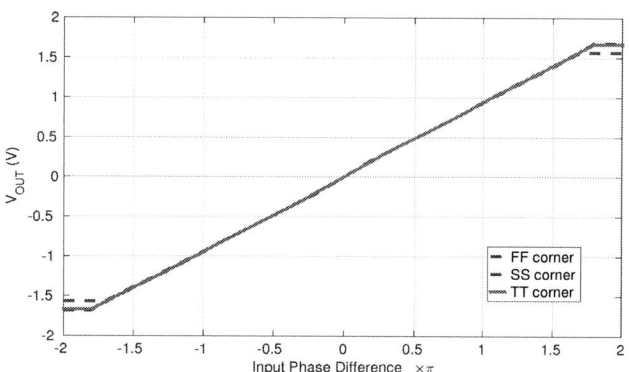

Fig. 9. Transfer curves for corners for an operating frequency of 500MHz

The proposed PFD is compared with some of the designs found in literature w.r.t. power in Table I.

TABLE I
SIMULATION RESULTS AND COMPARISON.

Ref	Supply	Technology(μm)	Power(mW)
[5]	-	0.25	1.4 @500MHz
[6]	1.8	0.18	1.56 @500MHz
[7]	1.2	0.13	1.44 @1GHz
This work	1.8	0.18	1.36 @1GHz

V. CONCLUSION

The explicit reset mechanism is the major contributor to the blind zone. The input detection range and linearity is deteriorated by the blind zone. As the mechanisms to alleviate effect of reset pulse on blind zone is limited by PVT variations, a new circuit topology has been proposed which removes the need for an explicit reset pulse. The effect of reset pulse on blind zone has been completely removed. The effect due to precharge time is eliminated through application of delayed inputs which nullifies the blind zone completely. With improved linearity, this topology reduces the cycle slippage and improves the acquisition rate of locking. The proposed design involves a simple design modification in the pull down network and operates at very high frequency with relatively low power consumption.

ACKNOWLEDGMENT

The authors would like to thank The Ministry of Electronics and Information Technology(MeitY), Government of India for the necessary tool support provided to carry out this work under SMDP-C2SD project.

REFERENCES

[1] C. A. Sharpe, "A 3-state phase detector can improve your next PLL design", *EDN Magazine*, pp. 55-59, Sep. 1976.
[2] H. Kondoh, H. Notani, T. Yoshimura, H. Shibata, and Y. Matsuda, "A 1.5-V 250-MHz to 3.0-V 622-MHz operation CMOS phase-locked loop with precharge type phase-frequency detector", *IEICE Trans. Electron.*, vol. E78-C, no. 4, pp. 381388, Apr. 1995.
[3] H. Johansson, "A simple precharged CMOS phase frequency detector", IEEE J. Solid-State Circuits, vol. 33, no. 2, pp. 295299, Feb. 1998.
[4] G.-Y. Tak, S.-B. Hyun, T. Y. Kang, B. G. Choi, and S. S. Park, "A 6.3-9-GHz CMOS fast settling PLL for MB-OFDM UWB applications", *IEEE J. Solid-State Circuits*, vol. 40, no. 8, pp. 16711679, Aug. 2005.
[5] M. Mansuri, D. Liu, and C.-K. Yang, "Fast frequency acquisition phase frequency detectors for Gsamples/s phase-locked loops", *IEEE J. Solid-State Circuits*, vol. 37, no. 10, pp. 13311334, Oct. 2002.
[6] R.Y. Chen, Wen-Yan Chen "A High-speed Fast-acquisition CMOS Phase/Frequency Detector for MB-OFDM UWB", *IEEE Trans. on Consumer Electronics*, vol. 53, no. 1, pp 2326, Feb 2007.
[7] Zaira Zahir and Gaurab Banerjee, "A Fast Acquisition Phase Frequency Detector for High Frequency PLLs ", *IEEE International WIE Conf. on Electrical and Computer Engg.*, pp. 366-369, Dec. 2015.
[8] Yi He, Xiaole Cui, Chung Len Lee, and Dongmei Xue, "An Improved Fast Acquisition PFD with Zero Blind Zone for the PLL Application", *IEEE International Conf. on Electron Devices and Solid-State Circuits*, pp. 1-2, 2014.
[9] W. Hu, L. Chunglen, and X. Wang, "Fast frequency acquisition phase-frequency detector with zero blind zone in PLL", *Electronic Letters, IET Journals & Magazines*, vol. 43, issue 19, pp. 1018-1020, 2007.
[10] Wu-Hsin Chen, Maciej E. Inerowicz, and Byunghoo Jung, "Phase Frequency Detector With Minimal Blind Zone for Fast Frequency Acquisition", *IEEE Trans. Circuits Syst. II, Exp. Briefs*, vol.57, no.12, pp. 936940, Dec. 2010.
[11] M.Soyuer and R.G.Meyer, "Frequency limitations of a conventional phase frequency detector", *IEEE J. Solid-State Circuits*, vol.25, n.4, pp. 1019-1022, Aug. 1990.

2018 31th International Conference on VLSI Design and 2018 17th International Conference on Embedded Systems

A 1.2 pJ/cycle KHz Timer Circuit for Heavily Duty-Cycled Systems

Manikandan R R, Vipul Singhal, Rajat Chauhan, Vinod Menezes, and Mahesh Mehendale

Kilby Labs, Texas Instruments, Bangalore, India

Abstract—An ultra-low energy KHz timer/oscillator circuit suitable for heavily duty cycled systems is presented. The proposed oscillator design utilizes a 3-stage ring topology made of CMOS Schmitt trigger delay cells biased with sub-nA currents. To extract a maximum delay per stage, high and low threshold voltages of the delay cells are set at supply and ground potentials, respectively. An ultra low power non-overlap buffer circuit that consumes less than 500 pA current and provides faster transition times of <10ns is designed to minimize the short circuit current in the digital load circuits.

The proposed oscillator circuit is fabricated in a 130 nm CMOS process and occupies an area of 0.027 mm^2. The oscillator circuit consume a total current of 2.18nA from 1V supply, for a 1.87 KHz frequency of oscillation (1.16pJ/cycle). The proposed design operates across a wide supply voltage range of 1.0 V to 3.3 V, and a temperature range of -40°C to 125°C. Measured oscillator performance in this wide operating range is presented in detail.

Index Terms—Ultra low power, relaxation, subthreshold, duty-cycle, ring oscillator, non-overlap, short circuit current, timing.

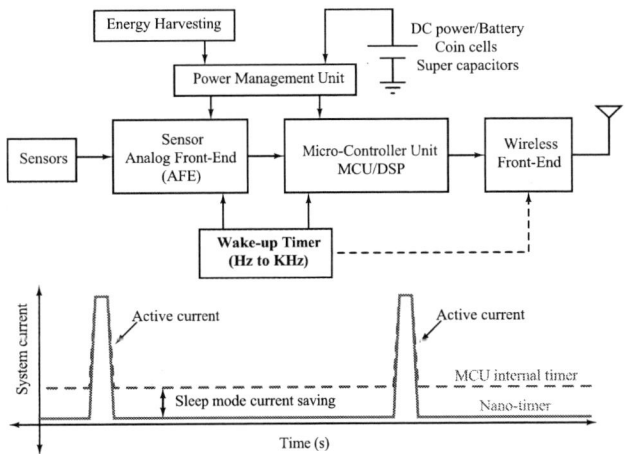

Fig. 1. An ultra-low power sensor platform with wake-up timer.

I. INTRODUCTION

Recent advances in the ultra low power sensor platforms and Internet-of-Things (IoT) led to a development of many new applications of environmental (temperature/gas/humidity sensors, thermostats, intruder detection) and health monitoring systems (blood glucose and eye pressure monitors) [1]. These systems typically operate from limited energy sources such as a coin cell or an energy harvested source. Heavily duty-cycled operation with extreme power gating techniques are often used in these systems to minimize the average power consumption (in the order of nW's) and to extend their lifetime.

Fig. 1 shows the block level description of a sensor system. Wake-up timer, in this system, performs the duty cycling of sensors/AFE/MCU and wireless transmission circuits. Data sensing and transmission circuits can operate with different duty cycles and accuracies, depending on the end target application. With the duty cycled active power in the order of low nW's, the total system power is dominated by the timer and standby power consumption. MCUs have an integrated low power timer circuit to perform system timing during its standby state. However, a best-in-class integrated system timing in MCUs today consume over 300 nA [2]. A nano timer IC built targeting these applications [3] provides timing signals ranging from 1s to 64s with 10% accuracy, consuming a 30 nA of current from 2.5V supply voltage (75nJ/cycle).

Nano power crystal oscillator circuits with stable output frequencies against PVT variations are reported in [4] - [7] (45

to 170 pJ/cycle). However, crystals due to their off-chip and bulky nature, lead to excessive system cost and form factor.

RC oscillators are typically used to generate frequencies in the order of 10's of KHz and have relatively good accuracies. A few nano-power circuit implementations of these relaxation oscillators are reported in [8] - [12] (2.8 to 530 pJ/cycle). Power consumption of these oscillator circuits are typically dominated by the current & voltage reference circuits and comparators. In [10], a 3.3 pJ/Cycle RC oscillator circuit using current-mode comparator biased with 750 pA currents is demonstrated. [12] used transistor OFF currents to generate current & voltage references and demonstrated a 530 pJ/cycle, nano-Watt timer circuit.

Relaxed accuracy timer circuits having ultra low energy performance are useful in charge-pumps, event monitoring, coarse wake-up timers in sensor platforms, and slow clock sources [13] [14] [15]. Targeting these applications, [16] (14 pJ/cycle) & [17] presented pico-Watt timer circuits with bias currents derived from transistor gate leakages. A 150 pW timer circuit (9.1 pJ/cycle) using a program and hold technique for the bias voltages is presented in [18]. In this design, the bias generation circuits are turned ON periodically to refresh the held bias voltages on capacitors and thereby minimizes the power consumption in bias generation operation.

Generation of large time constants (ms $\rightarrow$ s) on-chip at sub-10 pJ/cycle of energy in a fast CMOS process is extremely challenging. In these designs, techniques need to be developed to limit or avoid short circuit currents and to minimize the

978-1-5386-3693-0/18 $31.00 © 2018 IEEE

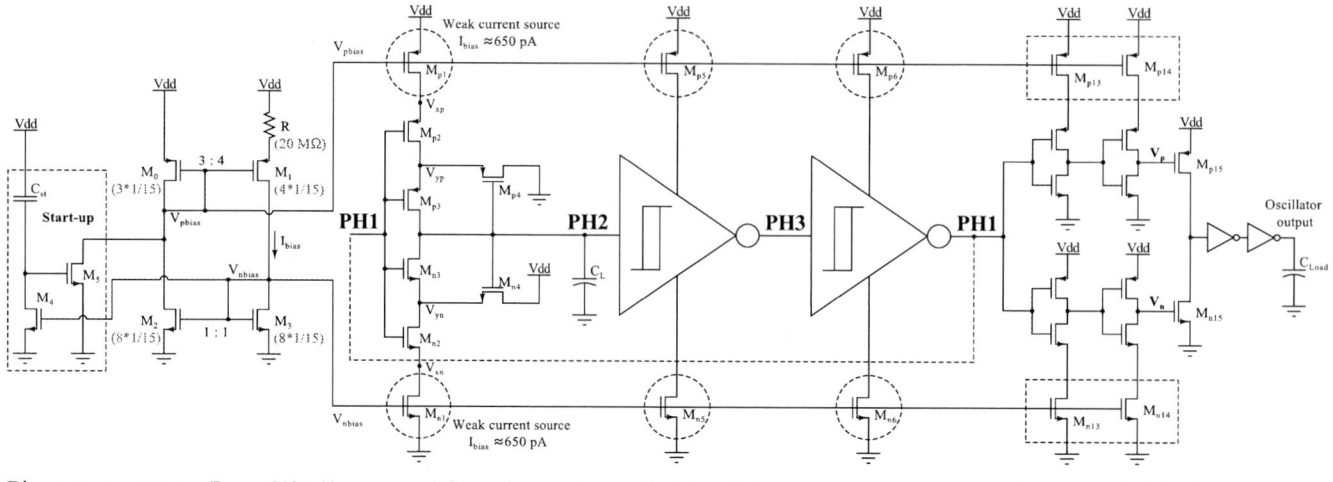

Bias current generator ($I_{bias} \approx 650$ pA)   3-Stage ring topology with Schmitt trigger delay cells   Non-overlap buffer circuit

Fig. 2. Proposed ultra-low energy oscillator circuit using weakly biased Schmitt trigger delay cells and non-overlap buffer.

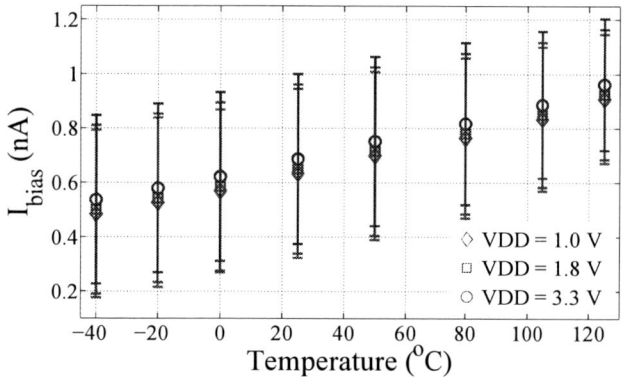

Fig. 3. I_{bias} (mean and 3-sigma) Vs temperature and supply voltage.

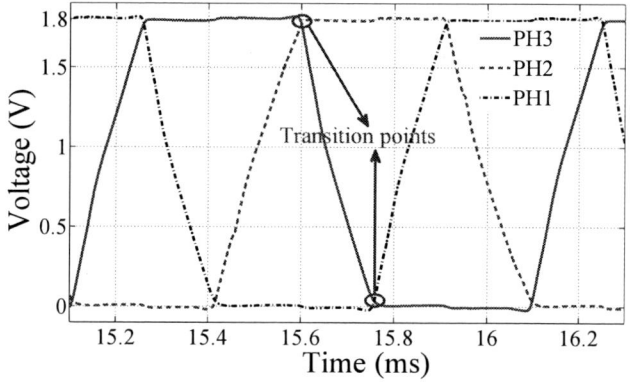

Fig. 4. Oscillator core : transient waveforms (PH1, PH2, & PH3).

power overhead in bias generation circuits. Additional requirements such as wide operation range across supply voltage & temperature, with ease of portability across process nodes add further complexity to these ultra-low power designs.

In this work, we present an ultra-low energy KHz oscillator circuit addressing the above mentioned design challenges. Circuit techniques to minimize the oscillator energy consumption and to realize larger time-periods using a ring oscillator core are discussed. A non-overlap buffer circuit that provide faster transition times at ultra-low currents is also proposed. Implemented in a 130 nm CMOS process, the proposed design consumes a total current of 2.18 nA from 1 V supply to generate a 1.87 KHz timing signal (1.16 pJ/cycle).

II. PROPOSED LOW POWER OSCILLATOR TOPOLOGY

A current starved ring oscillator topology is best suited for ultra low power operations [13] [19]. The proposed design utilizes a three stage ring oscillator core made of Schmitt trigger delay cells biased using ultra low magnitude currents. Fig. 2 shows the proposed oscillator circuit. The bias current

generator provide a 650 pA reference current (I_{bias}) to the oscillator stages and buffer circuit. The 3-stage oscillator core generates timing signals ranging from 500 Hz to 4 KHz and consume a 975 pA (for I_{bias} = 650 pA) of current. High and low threshold levels of the Schmitt trigger delay cells are set at supply and ground potentials, respectively, to extract a maximum delay per stage for a given I_{bias} & C_L. A non-overlap buffer circuit generate an output signal with faster transition times of less than 10 ns consuming a 500 pA of current. Description of building blocks are as follows,

A. Bias Current Generator

Bias current generator circuit is a critical component in this design, as it determines the power consumption of the oscillator stages and adds to the power overhead as well. A very low value for I_{bias} in the order of sub-nA's is preferred to achieve an ultra low power operation and to generate larger delay values. Transistor gate leakage and OFF current based pico-Ampere bias current generation circuits are reported in [12] [16] [17]. However, these circuit techniques work only

978-1-5386-3693-0/18 $31.00 © 2018 IEEE   172

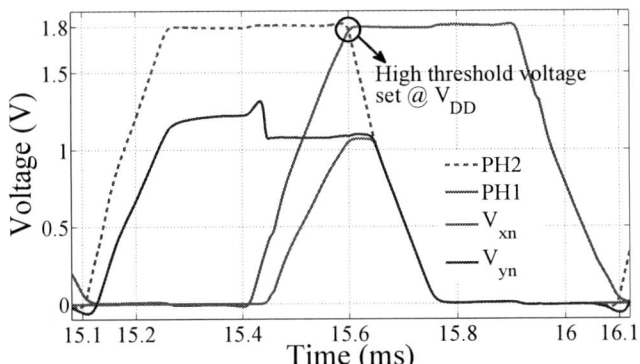

Fig. 5. Schmitt trigger delay cell : High threshold point set @ V_{DD}.

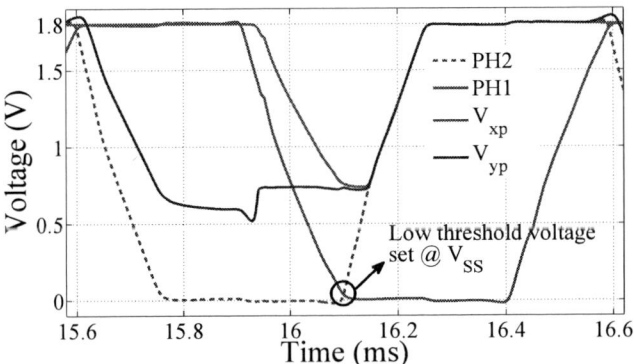

Fig. 6. Schmitt trigger delay cell : Low threshold point set @ V_{SS}.

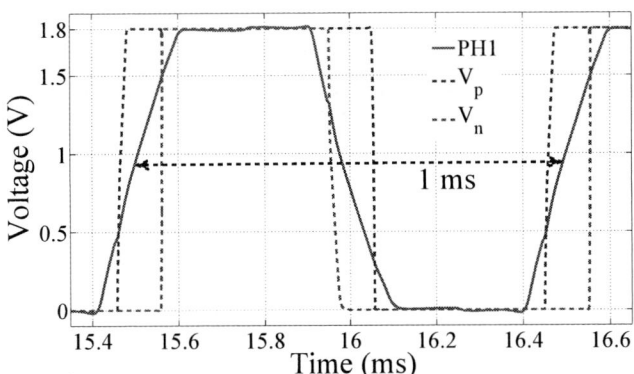

Fig. 7. Non-overlap buffer internal node waveforms.

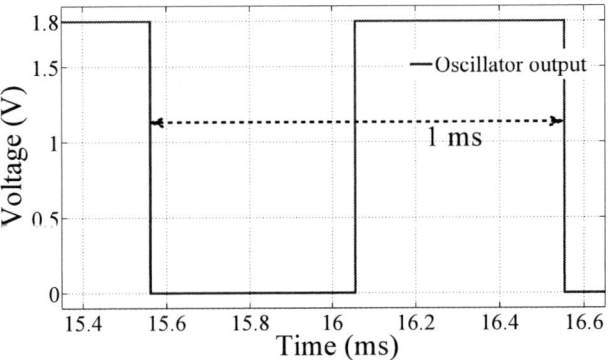

Fig. 8. Non-overlap buffer output signal with transition times <10 ns.

across a narrow temperature range due to the negligible OFF current magnitudes at lower temperature ranges and are limited by its process specific nature with modeling inaccuracies.

A β-multiplier reference current generator circuit shown in Fig. 2 is used in this design to generate a bias current (I_{bias}) of 650 pA . Transistors M_0 - M_3 operate in the subthreshold region and I_{bias} generated can be expressed as,

$$I_{bias} = \frac{nV_T}{R} \cdot ln(K) \qquad (1)$$

where, n is the subthreshold slope factor, V_T is the thermal voltage, and K is ratio of aspect ratios of the transistors M_0 & M_1. For K = 4/3, n $\simeq$ 1.721, V_T = 25.85 mV at 27°C, and R = 20 MΩ, I_{bias} can be calculated as 640 pA from the Eqn. 1. Resistor, R in Fig. 2 is realized in a binary weighted form to trim for process ($\pm$ 90 pA) and mismatch ($\pm$ 300 pA, 3-sigma) variations in I_{bias}. Fig. 3 shows the simulated I_{bias} characteristics with mismatch across the operating range.

B. Oscillator Core

A 3-stage current starved oscillator core and the detailed circuit of a delay cell are shown in Fig. 2. Inverting Schmitt trigger delay cells are preferred in this design due to their superior immunity to noise and disturbances [20]. Generation of Hz to KHz timing signals at an ultra low power consumption is a key requirement for this design. A N-stage oscillator core,

on biased with a reference current of I_{bias} (650 pA), consume a total average current of N ∗ I_{bias}/2 (975 pA).

To generate a very low frequency timing signal using a ring oscillator topology the following design options are traditionally used, 1. low magnitude bias currents [13] [19], 2. larger number of delay stages (increased current consumption), and 3. a large C_L (pF's, at the cost of area). In this design, we generate larger delay per stage by using an ultra low magnitude I_{bias} of 650 pA accompanied with the CMOS Schmitt trigger high and low threshold levels set at supply and ground voltages, respectively.

Fig. 4 shows the transient waveforms of 3 phases (PH1, PH2, & PH3) generated from the oscillator. PH2 begins its charging and discharging phases only after PH1 reaches the V_{SS} and V_{DD} voltage levels, respectively. Fig. 5 & 6 explains the high and low threshold levels set in the delay cell.

1) High threshold point: A high to low transition in PH2 happens through the set of transistors M_{n1} to M_{n4}. To start with, while PH1 is at ground potential, PH2 is held at V_{DD} and the internal nodes V_{xn} & V_{yn} are respectively at V_{SS} and ($V_{DD} - V_{th,n}$) as shown in Fig. 5. As PH1 rises slowly from a V_{SS} level, node V_{xn} charges to ($V_{DD} - V_{th,n}$) through the subthreshold conduction of M_{n2}. In this time duration, transistor M_{n3} is in its super cut-off state with a negative V_{gs} and the PH2 signal is held at V_{DD}.

As PH1 voltage rises above ($V_{DD} - V_{th,n}$) level, M_{n3} begins

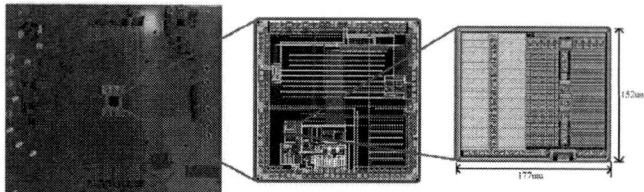

Fig. 9. Test-chip fabricated in 130 nm CMOS process and the PCB used in measurements.

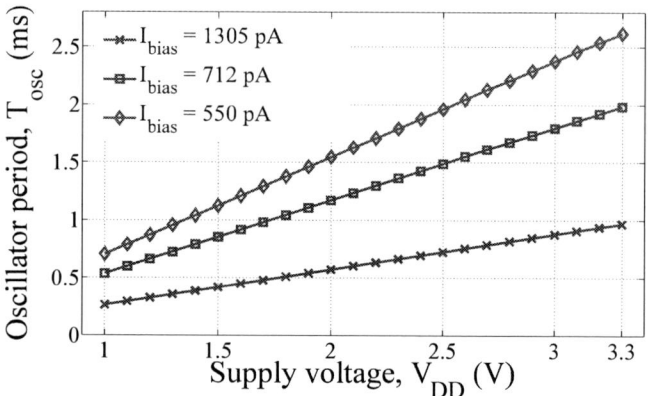

Fig. 10. Measured time period of oscillation versus supply voltage.

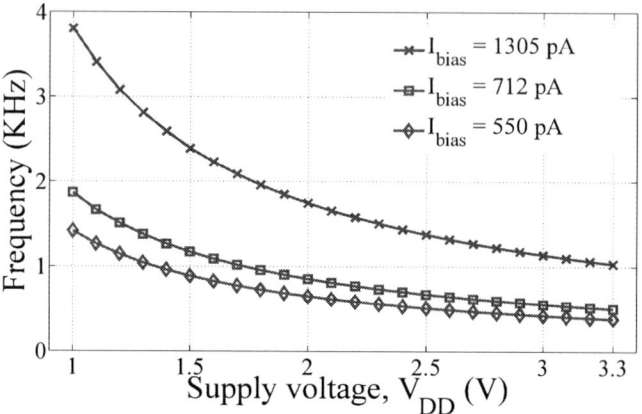

Fig. 11. Measured oscillator frequency versus supply voltage.

to enter into the subthreshold region of operation. However, nodes V_{yn} and PH2 remain undisturbed at $(V_{DD} - V_{th,n})$ and V_{DD}, respectively, as the ultra low bias current (650 pA) required by M_{n1} is now provided by the transistors M_{n2} & M_{n4}. When PH1 reach a V_{DD} level ($V_{th,n}$ above V_{yn} voltage), M_{n3} begins to conduct and initiates the discharge phase of PH2. This defines the high threshold point of the Schmitt trigger delay cell used in this design.

2) Low threshold point: Similarly, a low trip point of the Schmitt trigger delay cell set at V_{SS} can be explained through the set of transistors M_{p1} to M_{p4}. While PH1 is at V_{DD} voltage, PH2 is held at V_{SS} and the internal nodes V_{xp} & V_{yp} are respectively at V_{DD} and $V_{th,p}$, as shown in Fig. 6. When, PH1 reaches a V_{SS} voltage, M_{p3} starts to conduct and initiates the charging phase of PH2.

The oscillation frequency of the proposed design with the CMOS Schmitt trigger delay cells is given by,

$$F_{osc} = \frac{I_{bias}}{2 * N * C_L * V_{DD}} \quad (2)$$

For I_{bias} = 650 pA, C_L = 60 fF, V_{DD} = 1.8 V, and N = 3, frequency of oscillation (F_{osc}) can be calculated from Eqn. 2 as 1 KHz. From Eqn. 2, frequency of the proposed oscillator design is sensitive to process and temperature only through the I_{bias}, and is inversely proportional to V_{DD}. In this design, we used a PTAT current source for I_{bias} and the supply voltage is allowed to vary from 1.0 V to 3.3 V. For high performance applications, a constant bias current source or an I_{bias} tracking V_{DD} variations, and coarse supply regulation techniques [8] [9] can be used with an additional power overhead.

C. Non-Overlap Buffer Circuit

Ultra-low current designs suffer from poor rise and fall times, due to a very low magnitude of charging and discharging currents. The oscillator output signals in these designs need to be buffered strongly to provide faster transition times in the order of $<$10ns before driven to the digital load circuits, to avoid causing large short circuit currents in them. PH1, PH2, & PH3 signals (in this design), on buffered using conventional digital circuits cause large short circuit currents in them or impose a current limited operation on these buffer circuits using a larger I_{bias}. In this work, we solve this problem through a non-overlap buffer configuration shown in Fig. 2.

Current starved (I_{bias} = 650 pA) pre-drivers having skewed switching thresholds generate non-overlapping signals V_p & V_n from PH1. The PMOS (M_{p15}) and NMOS (M_{n15}) transistors in the buffer circuit on driven by these signals, turn ON at non-overlapping times and avoid short circuit current from V_{DD} to V_{SS}. This operation also provide faster signal transitions to the final inverter stages in this design. Fig. 7 shows the internal node waveforms of the buffer circuit. The non-overlap buffer circuit including the final inverter stages (in Fig. 2), consume a total average current of $\simeq$500 pA and provide an output signal with faster transition times ($<$10ns, Fig. 8) at the load capacitor, C_{Load} ($\simeq$50 fF).

III. MEASUREMENT RESULTS

The proposed ultra low power oscillator circuit is fabricated in a 130 nm CMOS process and Fig. 9 shows the test chip with PCB used in the measurements. The oscillator occupies an active area of 0.027 mm² (177 μm X 152 μm).

At room temperture, the proposed oscillator design including bias generation, oscillator core, and non-overlap buffer consume a total current of 2.8 nA from 1.8 V supply voltage to generate a 1 KHz timing signal (5 pJ/cycle). The bias current generator consume a 1.3 nA (2*I_{bias}) current, oscillator core consume 975 pA (3 * I_{bias}/2) of current, and the non-overlap buffer circuit consume a current of 525 pA. The proposed oscillator circuit achieves an ultra low energy performance of 1.16 pJ/cycle at 1 V supply voltage consuming 2.18 nA current

978-1-5386-3693-0/18 $31.00 © 2018 IEEE

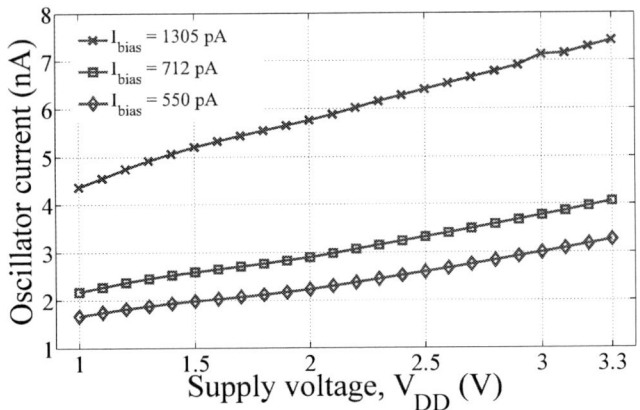

Fig. 12.   Measured oscillator current consumption Vs supply voltage.

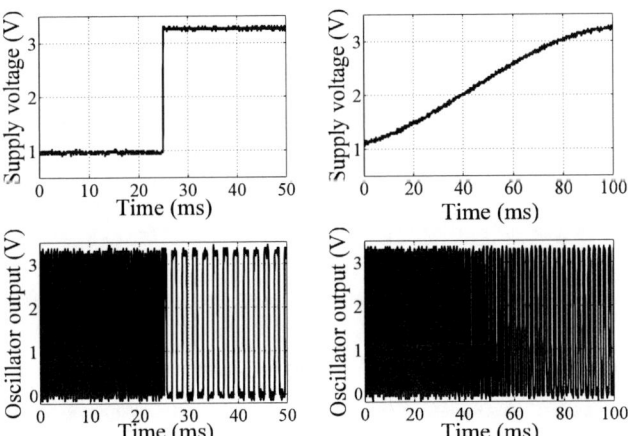

Fig. 13.   Measured oscillator output waveforms under supply transients (varied from 1.0 V to 3.3 V).

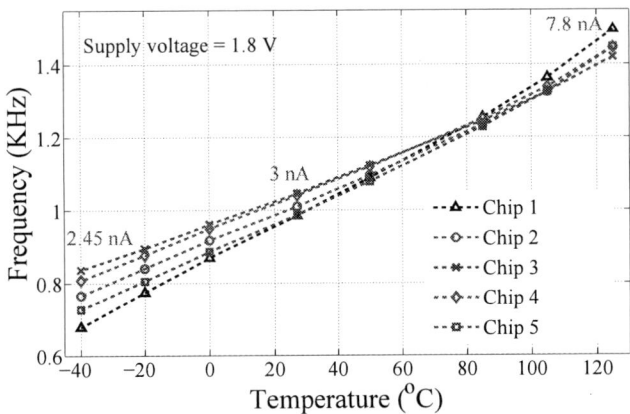

Fig. 14.   Measured frequency of oscillation versus temperature.

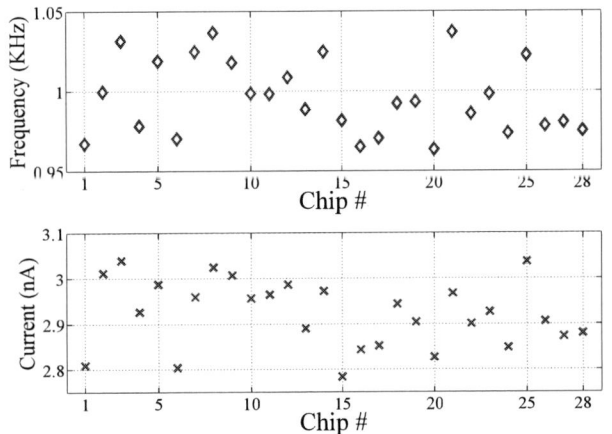

Fig. 15.   Oscillator frequency and current consumption measured from 28 chips (F_{osc} = 1 KHz, V_{DD} = 1.8 V).

to generate a 1.87 KHz frequency of oscillation. The proposed oscillator design operates across a wide supply voltage range of 1.0 V to 3.3 V, and a temperature range of -40°C to 125°C.

Time period of oscillation of the proposed oscillator circuit can be expressed as, $T_{osc} = \left(\frac{2 * N * C_L * V_{DD}}{I_{bias}} \right)$. Fig. 10 shows the measured time period of oscillation versus the supply voltage. The measured time period versus supply voltage characteristics is in perfect agreement with the T_{osc} expression. Effect of I_{bias} on the time period/frequency and current consumtption of the oscillator circuit are measured under three different I_{bias} settings of 550 pA, 712 pA, & 1305 pA.

Fig. 11 & 12 shows the measured frequency and current versus the supply voltage. Frequency of oscillation is inversely proportional to V_{DD} and the measured frequency characteristics shown in Fig. 11 are in agreement with Eqn. 2. The proposed oscillator circuit has a measured supply voltage dependence of 0.0613%/mV over the supply voltage range of 1.0 V to 3.3 V, which is acceptable in the relaxed performance applications [14] [15]. For stringent application requirements, a pseudo regulation [8] [9] can be added to the proposed design at the cost of additional power consumption. The osillator

current consumption shown in Fig. 12 increases with the supply voltage due to a reduction in the non-overlap time window designed in the ouput buffer circuit. The robustness of this ultra low current design against supply transients are verified with step and sinusoidal supply voltage characteristics (from 1.0 V to 3.3 V), and the results are shown in Fig. 13.

Fig. 14 shows the frequency versus temperature characteristics of the proposed circuit measured from five chips. The measured frequency rises almost linearly with the temperature due to a PTAT characteristic of the bias current used in this design and shows a temperature sensitivity of 0.417%/°C. The oscillator consume a total current of 2.45 nA, 3 nA, and 7.8 nA at -40°C, 27°C, and 125°C, respectively, from a 1.8V supply.

Fig. 15 shows the measured frequencies and currents from 28 chips. Table I & Fig. 16 presents the summary of measured performance of the proposed circuit and its comparison with recently published low power timer circuits.

IV. CONCLUSION

An ultra low energy, KHz oscillator circuit suitable for heavily duty cycled applications is presented. The proposed

978-1-5386-3693-0/18 $31.00 © 2018 IEEE

TABLE I
PERFORMANCE SUMMARY AND COMPARISON AGAINST RECENTLY PUBLISHED LOW POWER TIMER CIRCUITS.

	[13]	[16]	[18]	[21]	[12]	[9]	[8]	[10]	[11]	[19]	This Work
Process, (nm)	180	130	130	130	180	65	65	350	90	130	130
Supply, (V)	1.8	0.3	0.6	1.2	1.2	1.0	1.2	1.0	0.8	1.1	1.0
Frequency, (Hz)	18	0.07	11	6	11	18500	33000	3300	100000	100000	2000
Current consumption, (nA)	0.021	0.0033	0.167	0.55	4.83	120	158	11	350	137	2.18
Power, (nW)	0.038	0.001	0.1	0.66	5.8	120	190	11	280	150	2.18
Energy consumption, (pJ/cycle)	2.11	14	9.1	110	530	6.5	5.8	3.3	2.8	1.5	1.16
Temperature stability, (%/$^{\circ}$C)	0.85	0.6	0.067	0.0031	0.0045	0.0038	0.0038	0.05	0.0105	0.0005	0.417
Temperature range, ($^{\circ}$C)	-30 to 80	0 to 80	0 to 90	-20 to 60	-20 to 90	-40 to 90	-20 to 90	-20 to 80	-40 to 90	20 to 70	-40 to 125
Supply sensitivity, (%/mV)	0.0091	0.15	0.06	0.42	0.001	0.0010	0.0001	0.0023	0.0094	0.1	0.0613
Supply voltage range, (V)	0.6-1.8	0.3-0.65	0.55-0.65	0.65-1.25	1.2-2.2	1.5-3.3	1.15-1.45	1-3.6	0.725-0.9	-	1.0-3.3
Area, (mm^2)	0.18	0.0005	-	0.0153	0.24	0.032	0.015	0.1	0.12	0.25	0.0271

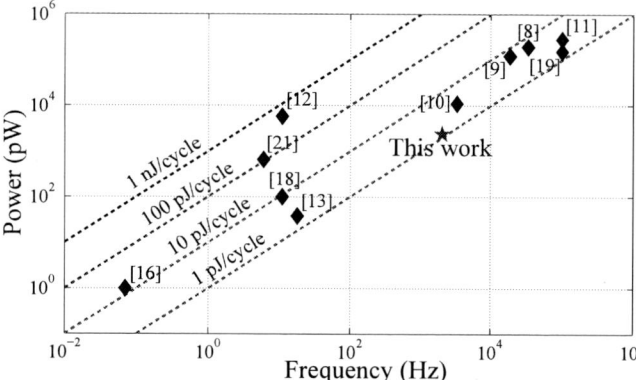

Fig. 16. Design space for recently published low-power oscillators (Table I)

design employs an ultra low current biasing scheme, larger delay generation techniques and non overlap buffer circuits to minimize the oscillator total power consumption and to generate KHz timing signals. A chip prototype of the proposed circuit is fabricated in a 130 nm CMOS process. The oscillator achieves an ultra low energy performance of 1.16 pJ/cycle consuming a 2.18 nA current from 1 V supply to generate 1.87 KHz frequency of oscillation. The performance of the proposed ultra low power design is verified across a wide supply voltage range of 1.0 V to 3.3 V, and a temperature range of -40°C to 125°C, and the measured performance from the test-chips are presented.

V. ACKNOWLEDGEMENTS

The authors thank Dinesh Hegde, Rashmi Bhushi, and Vinoth Durai Manickam for the layout design and device characterization support.

REFERENCES

[1] X. Zou, X. Xu, L. Yao, and Y. Lian, "A 1-v 450-nw fully integrated programmable biomedical sensor interface chip," *IEEE Journal of Solid-State Circuits*, vol. 44, no. 4, pp. 1067–1077, April 2009.

[2] Texas Instruments, MSP 430/432 product series, 2017.

[3] Texas Instruments, Nano-Power Programmable Timer with Watchdog Functionality, TPL 5000.

[4] T. M. Wang, M. D. Ker, and H. T. Liao, "Design of mixed-voltage-tolerant crystal oscillator circuit in low-voltage cmos technology," *IEEE Transactions on Circuits and Systems I: Regular Papers*, vol. 56, no. 5, pp. 966–974, May 2009.

[5] D. Yoon, T. Jang, D. Sylvester, and D. Blaauw, "A 5.58 nw crystal oscillator using pulsed driver for real-time clocks," *IEEE Journal of Solid-State Circuits*, vol. 51, no. 2, pp. 509–522, Feb 2016.

[6] K. J. Hsiao, "17.7 a 1.89nw/0.15v self-charged xo for real-time clock generation," in *2014 IEEE International Solid-State Circuits Conference Digest of Technical Papers (ISSCC)*, Feb 2014, pp. 298–299.

[7] A. Shrivastava, D. A. Kamakshi, and B. H. Calhoun, "A 1.5 nw, 32.768 khz xtal oscillator operational from a 0.3 v supply," *IEEE Journal of Solid-State Circuits*, vol. 51, no. 3, pp. 686–696, March 2016.

[8] D. Griffith, P. T. Rine, J. Murdock, and R. Smith, "17.8 a 190nw 33khz rc oscillator with ±0.21% temperature stability and 4ppm long-term stability," in *2014 IEEE International Solid-State Circuits Conference Digest of Technical Papers (ISSCC)*, Feb 2014, pp. 300–301.

[9] A. Paidimarri, D. Griffith, A. Wang, G. Burra, and A. P. Chandrakasan, "An rc oscillator with comparator offset cancellation," *IEEE Journal of Solid-State Circuits*, vol. 51, no. 8, pp. 1866–1877, Aug 2016.

[10] U. Denier, "Analysis and design of an ultralow-power cmos relaxation oscillator," *IEEE Transactions on Circuits and Systems I: Regular Papers*, vol. 57, no. 8, pp. 1973–1982, Aug 2010.

[11] T. Tokairin *et al.*, "A 280nw, 100khz, 1-cycle start-up time, on-chip cmos relaxation oscillator employing a feedforward period control scheme," in *2012 Symposium on VLSI Circuits (VLSIC)*, June 2012, pp. 16–17.

[12] S. Jeong, I. Lee, D. Blaauw, and D. Sylvester, "A 5.8 nw cmos wake-up timer for ultra-low-power wireless applications," *IEEE Journal of Solid-State Circuits*, vol. 50, no. 8, pp. 1754–1763, Aug 2015.

[13] P. M. Nadeau, A. Paidimarri, and A. P. Chandrakasan, "Ultra low-energy relaxation oscillator with 230 fj/cycle efficiency," *IEEE Journal of Solid-State Circuits*, vol. 51, no. 4, pp. 789–799, April 2016.

[14] V. Ivanov, R. Brederlow, and J. Gerber, "An ultra low power bandgap operational at supply from 0.75 v," *IEEE Journal of Solid-State Circuits*, vol. 47, no. 7, pp. 1515–1523, July 2012.

[15] Y. P. Chen, M. Fojtik, D. Blaauw, and D. Sylvester, "A 2.98nw bandgap voltage reference using a self-tuning low leakage sample and hold," in *2012 Symposium on VLSI Circuits (VLSIC)*, June 2012, pp. 200–201.

[16] Y. S. Lin, D. Sylvester, and D. Blaauw, "A sub-pw timer using gate leakage for ultra low-power sub-hz monitoring systems," in *2007 IEEE Custom Integrated Circuits Conference*, Sept 2007, pp. 397–400.

[17] Y. Lee, B. Giridhar, Z. Foo, D. Sylvester, and D. Blaauw, "A 660pw multi-stage temperature-compensated timer for ultra-low-power wireless sensor node synchronization," in *2011 IEEE International Solid-State Circuits Conference*, Feb 2011, pp. 46–48.

[18] Y. S. Lin, D. M. Sylvester, and D. T. Blaauw, "A 150pw program-and-hold timer for ultra-low-power sensor platforms," in *2009 IEEE International Solid-State Circuits Conference - Digest of Technical Papers*, Feb 2009, pp. 326–327,327a.

[19] A. Shrivastava and B. H. Calhoun, "A 150nw, 5ppm/o c, 100khz on-chip clock source for ultra low power socs," in *Proceedings of the IEEE 2012 Custom Integrated Circuits Conference*, Sept 2012, pp. 1–4.

[20] L. A. P. Melek, A. L. da Silva, M. C. Schneider, and C. Galup-Montoro, "Analysis and design of the classical cmos schmitt trigger in subthreshold operation," *IEEE Transactions on Circuits and Systems I: Regular Papers*, vol. 64, no. 4, pp. 869–878, April 2017.

[21] Y. Lee, B. Giridhar, Z. Foo, D. Sylvester, and D. B. Blaauw, "A sub-nw multi-stage temperature compensated timer for ultra-low-power sensor nodes," *IEEE Journal of Solid-State Circuits*, vol. 48, no. 10, pp. 2511–2521, Oct 2013.

2018 31th International Conference on VLSI Design and 2018 17th International Conference on Embedded Systems

A Quadrature-Phase Voltage Controlled Oscillator for Offset Phase and Frequency Compensation

Pragya Maheshwari[1], Pavan Kumar Sadhu[1], Mukesh Kumar Deharia[1],
Nandakumar Nambath[2], and Shalabh Gupta[1]
[1]Department of Electrical Engineering, IIT Bombay, Mumbai – 400076, India
[2]Department of Electrical Engineering, IIT Goa, Ponda – 403401, India
shalabh@ee.iitb.ac.in

Abstract—**This paper presents a quadrature-phase voltage controlled oscillator (QVCO) for offset phase and frequency compensation in various applications. A method of adjusting the QVCO output amplitude using tunable negative-g_m cells is proposed. Such a QVCO can be very useful as a controller for phase interpolator in clock and data recovery circuits to reduce per lane power consumption and overall jitter. Designed in 65 nm CMOS technology, the QVCO has a gain of $\sim$34 MHz/V, while drawing a current of 3.9 mA from a 1 V supply.**

Index Terms—**Quadrature-phase voltage controlled oscillator, analog quadrature phase interpolator, clock and data recovery, carrier synchronization, analog signal processing.**

I. INTRODUCTION

Applications such as clock and data recovery (CDR) in high-speed serial links, and carrier phase recovery and compensation (CPRC) in communication links require a QVCO whose frequency can be swept from positive to negative through zero. A QVCO can be used to generate weights for the phase interpolator in a phase interpolator based CDR circuit such as the one reported in [1], based on a phase detector output voltage. In a digitally controlled phase interpolator the recovered clock will be having more jitter due to finite phase steps. Whereas, QVCO based architecture helps in reducing the jitter since the phase is continuous. Moreover this architecture helps in reducing per lane power consumption compared to PLL-based CDR circuit, since the high frequency VCO is shared by all lanes. In CPRC, a QVCO can be used to eliminate the phase and frequency offsets between the transmitter and receiver local oscillators [2]. This paper presents the architecture of a QVCO which is capable of generating positive and negative offset frequencies, including zero frequency which can efficiently be implemented using analog integrated circuits. A method of adjusting the amplitude of the QVCO output using tunable negative-g_m cells is also presented.

II. TWO INTEGRATOR QUADRATURE-PHASE VOLTAGE CONTROLLED OSCILLATOR

Two integrator loop oscillator, shown in Fig. 1(a), is one of the simpler models of a quadrature-phase oscillator [3], wherein each integrator provides a phase shift of 90° and negative feedback provides a phase shift of 180°, making the total phase shift around the loop to 360°. Since the integrators are having a DC gain of ω_0 each, the unity gain frequency

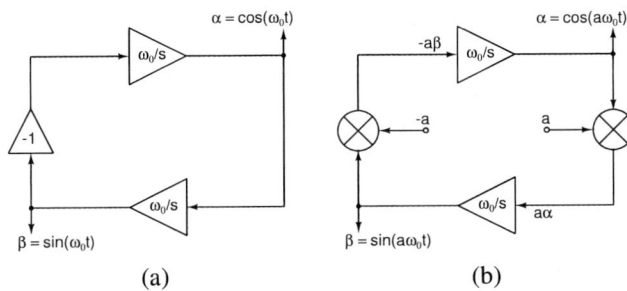

Fig. 1: Architecture of the two integrator loop oscillator: (a) With frequency of oscillation as ω_0; (b) Voltage controlled oscillator with frequency of oscillation $a\omega_0$. Here, α and β are the quadrature outputs of the oscillator.

and hence the natural frequency of the loop will be ω_0. This loop is governed by the differential equation

$$\frac{d^2 x}{dt^2} + \omega_0^2 x = 0. \tag{1}$$

The solution of which can be represented as

$$x(t) = \exp(\pm j\omega_0 t). \tag{2}$$

The two integrator loop oscillator can be modified into a QVCO as shown in Fig. 1(b), which is governed by the second order differential equation [4]

$$\frac{d^2 x}{dt^2} + a^2 \omega_0^2 x = 0. \tag{3}$$

The solution of this differential equation is given by

$$x(t) = \exp(\pm j a\omega_0 t). \tag{4}$$

This oscillator generates in-phase and quadrature-phase signals with a frequency of $a\omega_0$. Based on the polarity of the control signal, a this oscillator generates both positive and negative frequencies. Architecture of the QVCO, proposed in this work, is shown in Fig. 2. High gain amplifiers are used in the place of integrators. These amplifiers can be single or cascaded differential amplifier stages. When the circuit starts oscillating, phase difference between α and β will be exactly 90°. Negative-g_m cells, shown in Fig. 2, make the output impedance of the amplifiers infinite and ensure the pole remains very close to the origin. Non-ideal integrators suffer

978-1-5386-3693-0/18 $31.00 © 2018 IEEE   177

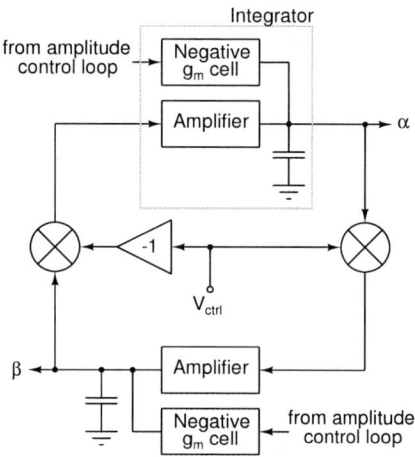

Fig. 2: Architecture of a QVCO with negative frequency sweep using amplifiers and mixers. Frequency of oscillation depends on V_{ctrl} and capacitor values. Negative-g_m cells make the real part of impedance at the amplifier output infinite.

from leakage issues due to non-zero pole location which may lead to damped oscillations, especially at very low frequencies. This necessitates the use of an amplitude control loop in the oscillator, the operation of which is discussed in the following section.

III. AMPLITUDE CONTROL

In the circuit shown in Fig. 2, if the control voltage is zero the oscillator loop is broken. In the absence of an amplitude control loop, the values of α and β slowly decay to zero because of the finite output impedance of the differential amplifiers. In CDR application, the differential amplifiers need to hold the values of α and β when the control voltage becomes zero, otherwise the phase of recovered clock changes continuously. To avoid this decay an amplitude control is needed in this oscillator to maintain the output amplitude. The amplitude control loop serves two purposes–holds the values of α and β when the control voltage becomes zero and keeps the amplitude constant for all other values of the control voltage.

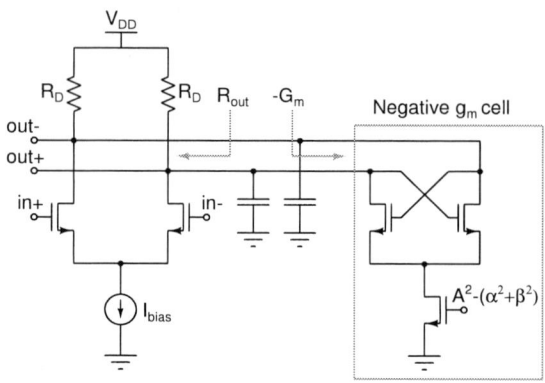

Fig. 4: Schematic of the integrator used in the oscillator. Parallel combination of R_{out} and $-G_m$ makes the output impedance infinite.

Different methods of amplitude control have been reported in [5], but for quadrature signals the most appropriate method is to make use of Pythagoras' law method reported in [6]. Here an error signal is generated by subtracting the sum of α^2 and β^2 from A^2 where A is the expected amplitude of oscillation. Block diagram of the amplitude control loop is shown in Fig. 3. For adjusting the amplitude of α and β, we propose to use negative-g_m cells with tunable g_m. As shown in Fig. 4, the negative g_m value is adjusted to make the real part of impedance at the amplifier output infinite in the steady state. This adjustment is done by changing current in the negative g_m cell using the amplitude control loop. When $(\alpha^2 + \beta^2)$ becomes less than A^2, the error voltage increases, which in-turn increases the value of g_m in the negative-g_m cell. This increases the output impedance of the differential amplifier and increases the amplitude of both α and β.

IV. SIMULATION RESULTS

In the design of QVCO, a cascade of two differential amplifier stages are used for implementing the high gain amplifier. The output impedance and gain of the amplifier in the second stage are higher than that of the first stage so that

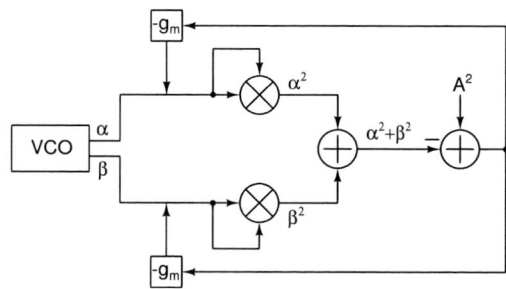

Fig. 3: Block diagram of the amplitude control loop. α and β are the VCO's quadrature outputs and A is the expected amplitude of oscillations.

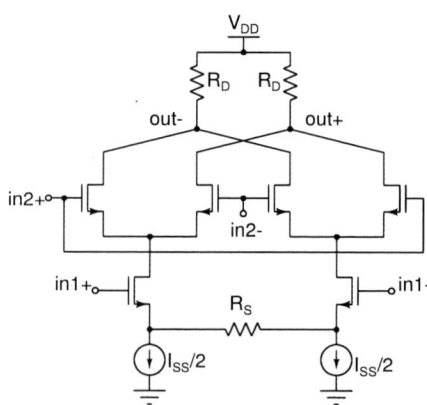

Fig. 5: Schematic of the Gilbert cell mixer that is used as the multiplier in the QVCO. A degeneration resistor R_s is used to improve the linearity.

978-1-5386-3693-0/18 $31.00 © 2018 IEEE   178

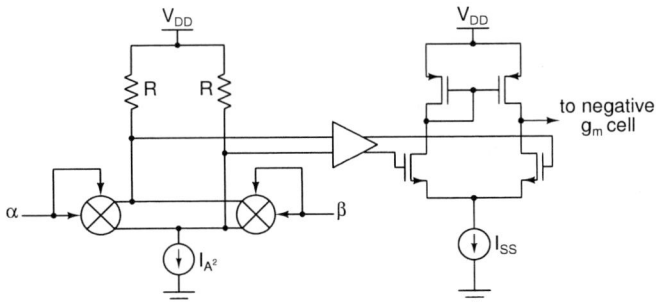

Fig. 6: Schematic of the circuit that generates error voltage in the amplitude control loop. An amplifier is used to increase the loop gain and the current mirror load converts the differential control voltage to single ended.

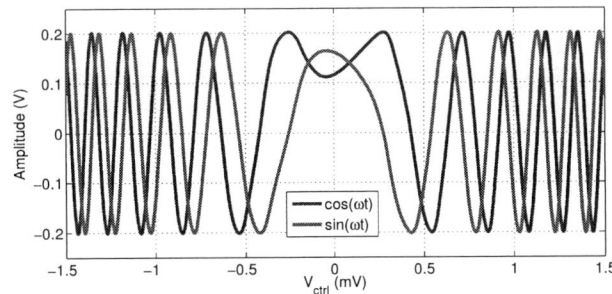

Fig. 8: $\cos(\omega t)$ and $\sin(\omega t)$ signals for a control voltage sweep from -1.5 mV to 1.5 mV. Frequency changes its polarity when control voltage changes from negative to positive. For positive frequencies $\cos(\omega t)$ leads $\sin(\omega t)$ and for negative frequencies $\sin(\omega t)$ leads $\cos(\omega t)$. The amplitude of oscillation remains constant because of amplitude control loop.

the integrator has only one dominant pole. The schematic of the differential amplifier with negative-g_m cell is shown in Fig. 4. Fig. 5 shows the schematic of the multiplier used in the design. A source degeneration resistor is used in the mixer to improve linearity.

Schematic of the circuit which generates error voltage for the amplitude control loop is shown in Fig. 6. In this circuit, the first stage subtracts $\alpha^2 + \beta^2$ from A^2. Second stage amplifies the output of the first stage so that loop gain of the amplitude control loop is high. The output of the second stage is given to negative-g_m cell for amplitude correction. Loop gain of the amplitude control loop should be very high in order to minimize the steady state error. The QVCO was implemented in 65 nm CMOS technology. The QVCO, along with the amplitude control loop, consumes a power of $\sim$3.9 mW.

Variation of the QVCO frequency with respect to the control voltage is shown in Fig. 7. The QVCO exhibits linear characteristics in -20 mV to 20 mV control voltage range and has a gain of approximately 34 MHz/V. The amplitude control loop adjusts the amplitude of α and β such that the amplitudes remain constant with varying the control voltage. Hence, one can say that by varying the control voltage input of the QVCO that has the amplitude control loop only the frequency of

operation can be controlled but not the amplitude. Fig. 8 shows the response of the QVCO with a control voltage sweep from -1.5 mV to 1.5 mV. It can be observed that the polarity of frequency changes when the control voltage changes its polarity. The amplitudes remain constant for all values of the control voltage. To lower the steady state error of the amplitude control loop the loop gain of the amplitude control loop is made high.

Fig. 9 shows settling of the phase interpolator based CDR loop for a phase step, where phase interpolator changes the phase of the input clock till the recovered clock is aligned to the middle of the input data bit. The phase interpolator is controlled by α and β signals generated by the QVCO, such that the recovered clock is given by: $\alpha Clk_I + \beta Clk_Q$, where Clk_I and Clk_Q are the quadrature clocks in the CDR. For a phase step, CDR loop takes approximately 1 μs to reach the steady state and α, β signals settle to their DC values.

V. Conclusion

A low frequency QVCO whose frequency can be swept from positive to negative through zero has been shown. A novel method of controlling the amplitude of this QVCO, using tunable negative-g_m cells has also been presented. Use

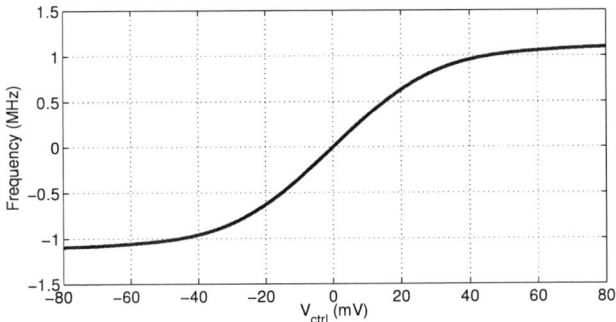

Fig. 7: Variation of QVCO frequency of oscillation with respect to V_{ctrl}. In the linear region of the transfer characteristics the QVCO has a gain of $\sim$34 MHz/V.

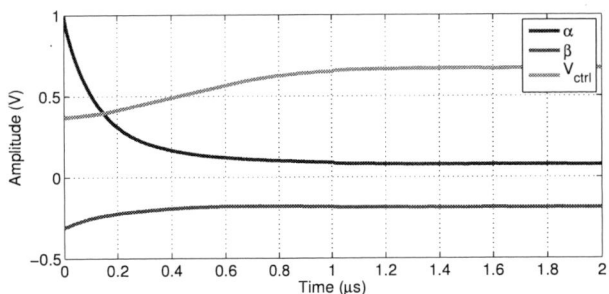

Fig. 9: Phase loop and amplitude loop settling in a QVCO controlled phase interpolator based CDR for a phase step. α and β signals, and amplitude loop error voltage of QVCO are shown.

of this QVCO as a controller for the phase interpolator in a CDR circuit has two advantages–(i) per lane power consumption is lower compared to PLL-based CDR circuit, since the high frequency VCO is shared by all lanes and (ii) unlike digitally controlled phase interpolator, which has finite phase steps, this QVCO controlled phase interpolator has less jitter. The QVCO implemented using 65 nm CMOS technology consumes 3.9 mW of power.

ACKNOWLEDGEMENT

The authors would like to thank Ministry of Electronics and Information Technology (MeitY), Government of India for funding the project.

REFERENCES

[1] R. Kreienkamp, U. Langmann, C. Zimmermann, T. Aoyama, and H. Siedhoff, "A 10-Gb/s CMOS Clock and Data Recovery Circuit with an Analog Phase Interpolator," *IEEE Journal of Solid-State Circuits*, vol. 40, no. 3, pp. 736–743, March 2005.

[2] N. Nambath, R. Raveendranath, D. Banerjee, A. Sharma, A. Sankar, and S. Gupta, "Analog Domain Signal Processing-Based Low-Power 100 Gb/s DP-QPSK Receiver," *Journal of Lightwave Technology*, vol. 33, no. 15, pp. 3189–3197, August 2015.

[3] A. M. Soliman, "Two Integrator Loop Quadrature Oscillators: A Review," *Journal of Advanced Research*, vol. 4, no. 1, pp. 1–11, October 2013.

[4] M. S. Williamsen, "RC Oscillator Admits Negative Frequency," *International Journal of Circuit Theory and Applications*, vol. 39, no. 6, pp. 687–695, 2011.

[5] W. Mikhael and S. Tu, "Continuous and Switched-Capacitor Multiphase Oscillators," *IEEE Transactions on Circuits and Systems*, vol. 31, no. 3, pp. 280–293, March 1984.

[6] F. Doorenbosch and Y. Goinga, "Integrable, Wideband, Automatic Volume Control (a.v.c.) using Pythagoras's Law for Amplitude Detection," *Electronics Letters*, vol. 12, no. 16, pp. 418–420, 1976.

2018 31th International Conference on VLSI Design and 2018 17th International Conference on Embedded Systems

CMOS Oscillator having Stable Frequency with Process, Temperature and Voltage Variation

Vikas Rana
STMicroelectronics, Greater Noida, India
Vikas18feb@gmail.com

Abstract—**Clock generation is an extremely important part of any electronic system. At present Crystal oscillator is the most stable and reliable clock generation technique used as clock generation for any type of SoC. But this solution is very expensive and inconvenient for chip integration. Second option can be PLL based clock generation technique. Again PLL based techniques uses huge silicon area and high on-chip power. Further PLL based clocks need a reference clock for locking system. So we need a solution which can provide a stable clock against process, temperature and supply variations. CMOS Ring Oscillator provides a ready solution, but the biggest challenge with this kind of circuit is to achieve stable clock with temperature, process and supply voltage variations. This paper describes a symmetric oscillator structure, which provides an on-chip compensated clock against process, temperature and supply variations. This architecture is not one to one replacement of crystal oscillator or PLL but is very useful for many applications like on-chip charge-pump or DC-DC converters, clock required for modify pulse in Phase Change Memory etc. The oscillator is designed in BCD9S (110nm) process, to produce a stable frequency of 20 MHz, within a temperature range of -40 to 160°C and supply varying from 3V to 5.5V. The variation of frequency is within ±4.5% range and the circuit overall consumes an average current of 68μA.**

Keywords—*Oscillator, Temperature insensitive*

I. INTRODUCTION

Clock generators are one of the most crucial block of any SoC, because all the internal operations and sequence of event are timed using clock. Hence, the accuracy of clock frequency is of prime importance. The typical approach is to use crystal oscillators, which have extremely stable frequency. Further Phase lock loop (PLL) circuit uses this clock (from crystal oscillator) to generate high frequency clock, which is reconfigurable depending on the requirement [1]. As the crystal and its interface to the SoC is not the part of CMOS process integration, so it requires extra cost to use it. Although PLL is a completely CMOS based design, but it requires a lot of area and consumes lot of power. In application requiring not so much accurate clock, the entire cost of integrating crystal oscillator and the PLL, might not be justified. Hence, there is a growing trend in building CMOS oscillator, which have low variation in output frequency and require less silicon area and on-chip power. [2-6].

Most CMOS Ring oscillators consist of series of odd number of inverters which are back coupled (as shown in Figure 1) to provide an Astable state and lead to oscillations. The delay of components used in feedback loop determines the frequency of such system. The transition traveling around the

loop has to pass through each inverter twice to arrive at initial stage.

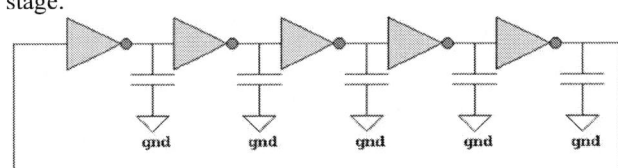

Figure 1. Conventional Ring Oscillator

Therefore, frequency of oscillation is determined as:

$$F = 1/2N\tau \qquad (1)$$

Here N is number of stages used in oscillator and τ is delay of each single inverter stage.

This delay varies a lot with process, supply voltage and temperature hence making the output frequency of the oscillator vary significantly. Equation 2 indicates the fall time delay of inverter.

$$\zeta = \frac{C_{load}}{K_n(Vdd-Vtn)}\left[\frac{2\,Vtn}{Vdd-Vtn} + \ln\left(\frac{4(Vdd-Vtn)}{Vdd} - 1\right)\right] \quad (2)$$

Here C_{load} is the output capacitance seen at the inverter, K_n is the conduction parameter of the n-channel MOSFET and V_{dd} is power supply [7]. In a chain of inverters, one-inverter drives another inverter, so C_{load} also becomes supply depended thus total loop delay varies a lot with different parameters.

There have been several important researches to reduce frequency variation in CMOS oscillators with process, temperature and supply variation. Few relevant researches are discussed below:

a) The work in reference [12] describe a methodology where a supply and temperature insensitive current is generated and this reference current is used to make current controlled ring oscillator and make the ring oscillator less sensitive to supply and temperature variations. In his research work, author reports -15% to +20% frequency variation with 10% variation in supply and insensitivity to temperature is 404 ppm/°C over the temperature range of -40°C to 100°C for a 4MHz oscillator designed in 0.35um technology.

b) The work in reference [13] describes a frequency locked loop technique where oscillation frequency of a ring oscillator is locked to {$I_{REF}/(C*V_{BG})$}. In this architecture

978-1-5386-3693-0/18 $31.00 © 2018 IEEE   181

both I_{REF} and V_{BG} are independently temperature compensated. In this research work, author claims to achieve ±0.05% clock frequency variation with respect to supply variation of 1.2V to 3V and ±0.4% variation against temperature variation of -20°C to 120°C. Design is made for 10MHz clock frequency in 0.18um CMOS technology.

c) The work in reference [14] describes a RC oscillator targeting 33 KHz of clock frequency. In this work author has used a regulator that provides the supply to oscillator and an external reference current is used to bias the oscillator. Author claims to achieve 0.21% temperature stability for a range of -40 to 90°C with 4-ppm long-term stability.

This paper is organize as follows. Section II explain the architecture, operating principle of the design, circuit implementation at block level and layout implementation. Section III, discuss the post-layout spice simulation results of the design and compare them with other report implementations. Finally, section IV draws the conclusion.

II. ARCHITECTURE

A. Overview

The proposed design is related to a new CMOS oscillator architecture that provides a stable output frequency with supply, process and temperature variations. Figure 2 shows the block diagram of the entire circuit. The circuit works on a regulated supply voltage *VOSC*, which is generated by a voltage regulator. The voltage regulator uses one OPAMP with one input as Vbg (bandgap reference voltage [8]) other input is connected in feedback from output to input through resistive network.

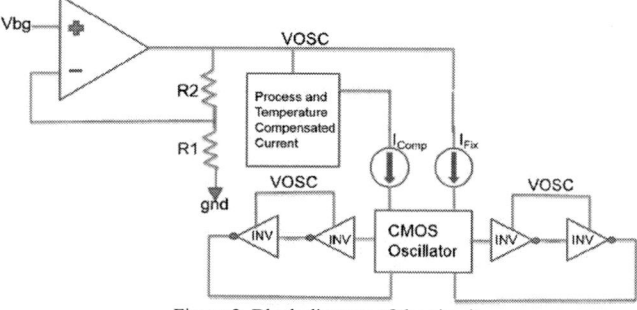

Figure 2. Block diagram of the circuit

VOSC voltage can be written by equation 3:

$$VOSC = Vbg\left(1 + \frac{R2}{R1}\right) \quad (3)$$

Here Vbg is the bandgap voltage generated on-chip having +/-1% variation (after trimming) thus causing +/-1% variation in *VOSC* voltage.

This scheme of oscillator, operating at regulated voltage makes the oscillator immune to supply variations, it is because this close loop in regulator keeps the output voltage *VOSC*

constant even if supply voltage is varying (3V to 5.5V in our case). The loop is designed such that it has low output impedance and there is low drop in the output (*VOSC*) voltage whenever there is sharp demand in current. This regulator also powers a process and temperature compensation network, which generates a current (*Icorr*) based on process and temperature variations. There is also a fixed current, *Ifix* to be provided by a bandgap current reference that can be configured to create a fixed output frequency.

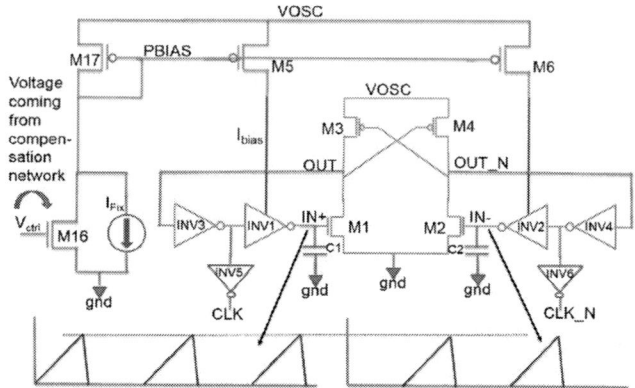

Figure 3. Current starved level-shifter based ring oscillator

B. Operating Principle

Figure 3 shows the circuit diagram of level-shifter based ring oscillator. This circuit works at regulated supply (*VOSC*) thus insensitive to supply variation. Design consists of one level shifter (half latch based differential amplifier made using four transistors M1, M2, M3 and M4), two current starved inverters (INV1 and INV2), two delay inverters (INV3 and INV4) and 2 inverters (INV5 and INV6) to strengthen CLK and CLK_N signals. In this architecture, the level shifter's branches (either using M1 and M3 or using M2 and M4) acts like an inverter, thus making the loop Astable like the odd number of stages in the ring oscillator.

As shown in Figure 3, the capacitors C1 and C2 are sequentially charged by a constant current sourced by PMOS transistors M5 and M6 respectively, depending upon the active pull-up path. Among the nodes IN+ or IN-, one is static at ground while the other charges in ramp fashion till the switching threshold of NMOS transistor (either M1 or M2) is reached. As soon as the voltage at IN+/IN- reaches the threshold voltage of NMOS, OUT/OUT_N node reaches to "GND" level via NMOS (either M1 or M2) and it helps half latch (made of M3 and M4) to change its state quickly from GND to *VOSC* level. As the level shifter is designed such that the NMOS is stronger than the cross coupled PMOS, the switching point becomes a strong function of the threshold voltage of the NMOS M1/M2. Thus the ramp time can be shown as equation 4:

$$Tramp = \frac{C1,2(Vth-0)}{Ibias} \quad (4)$$

978-1-5386-3693-0/18 $31.00 © 2018 IEEE

Where the *Vth* is the threshold voltage of M1 or M2. As this phenomenon of oscillation is periodic so (*Tramp + Tdelay*) become half period of the oscillation's time period. Where *Tdelay* is the delay of the other elements (two Inverters and one branch of level-shifter) in the signal chain. Capacitors C1 and C2 are implemented using accumulation mode NMOS, so that the capacitance is fixed during the periodic activity on nodes IN+/IN-. INV5 buffers the CLK signal out of the circuit and INV6 acts as a dummy to provide balanced load in the signal chain and hence symmetry is maintained.

Now, as per equation 4, the time period of the oscillation is proportional to the ratio of *Ibias* and *Vth,* hence the frequency of this circuit can be kept constant, if *Ibias* can be generated with a strong correlation with *Vth* during process and temperature variation. The equation governing threshold voltage, *Vth* is

$$V_{th} = V_{fb} - 2\varphi_f + \frac{\sqrt{4\varepsilon_s q N_a \varphi_f}}{C_{ox}} \qquad (5)$$

Here *Vfb* is the flat band voltage, φ_f is the surface potential, N_a is the doping concentration of the substrate, ε_s is the permittivity of silicon, q is the charge of electron and C_{ox} is the Capacitance density of the oxide. As

$$\varphi_{f} = \frac{KT}{q} \ln \frac{N_a}{N_i} \qquad (6)$$

Where K is the Boltzmann's Constant, Ni is the intrinsic concentration of the carriers [7]. From equation 5, it is clear that *Vth* decreases linearly with temperature because it has a strong dependence on the derivative of the $-2\varphi_f$ term. It proves that the threshold voltage (*Vth*) can be used as a signature of process and temperature; thus, the proposed oscillator architecture opens the opportunity to compensate the output frequency by generating a compensating current.

The desired output frequency in proposed architecture is given by equation 7 as below:

$$F_{out} = \frac{((I_{corr}[typ]) * M + I_{fix})}{N \, V_{TH[typ]} \, C_{1,2}} \qquad (7)$$

Here $I_{corr}[typ]$ is the DC current value of the correction circuit at typical process and room temperature. *Ifix* is the DC current value from an external reference; *M* is the mirror ratio of the current mirror at the compensation circuit (M16 in **Figure 3** and M15 in **Figure 4**). N is the mirror ratio of the current at the oscillator (M17 and M5 in **Figure 3**), $C_{1,2}$ is the value of the capacitor in the oscillator and $V_{TH[typ]}$ is the threshold ratio of the NMOS at typical process and room temperature.

Here conclusion is that *Ibias* (used in equation 4) is a combination of two parameters: *Icorr* and *Ifix*. *Ifix* is an external fixed current and *Icorr* is a compensated current used to make frequency stable across conditions.

C. Compensation Network

Figure 4 shows the temperature and process compensation current bias generator used in the proposed architecture. In the circuit, drains of NMOS M10 and M11 are kept at *VOSC* voltage and low current is drawn from M10 and M11. Size of M10 and M11 are chosen in such a way that these transistors operate in saturation region. In this way source of M10 settles at "*VOSC - Vth*" and source of M11 at "*VOSC - 2Vth*" hence these nodes are *VOSC* referred version of *Vth* and *2Vth* respectively. Finally this "*VOSC - 2Vth*" voltage is converted to a current using a conventional common source with source degeneration stage, such that the trans-conductance is linear over the voltage range, hence making the current proportional to *Vth.*

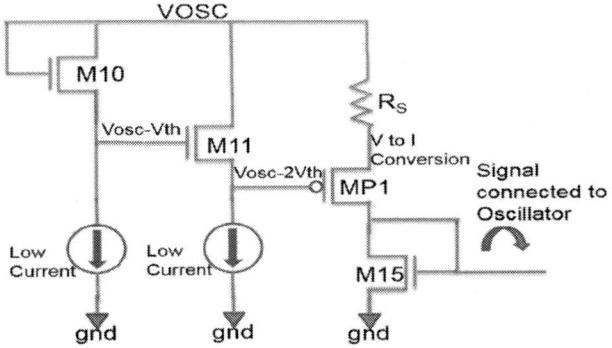

Figure 4. Temperature & process compensating bias generator

This correction current is added to the fixed current and mirrored appropriately, such that the *Ibias* tracks the variation in V_{th}. This third stage linearly converts voltage into current thus making the correction current a function of the *Vth.*

$$\frac{\Delta I_{corr}}{\Delta V_{in}} = -\frac{g_m}{1 + g_m R_s} \approx -\frac{1}{R_s} \text{ When } g_m R_s \gg 1 \quad (8)$$

Hence equation 8 proves that *Icorr* can be generated linearly from (*Vosc – 2Vth*) voltage across different conditions. Trimming options needs to be added to keep Rs constant so that *Icorr* tracks Vth variation across different conditions.

Figure 5 shows the schematic implementation of entire system, comprising of the oscillator and the compensation network. The biasing current (*Ibias*) is used to charge/discharge the nodes IN+/IN- which is consists of two components: *Ifix* (which is an external current generated from a reference block) and *Icorr* (which is generated from compensation network depending upon the process and temperature variation). The compensation network track the variation in process and temperature, accordingly modify the "*VOSC - 2Vth*" voltage. This voltage changes the *Icorr* current thus changing the *Ibias* current such that frequency of oscillator remains at stable value.

The proposed clock oscillator is designed and layout in 110nm BCD9S technology. C1, 2 are implemented with NMOS based gate capacitors (biased in accumulation region). Inside the compensation network P-Poly resistor is used which has low temperature variations as compare to other resistors available in technology.

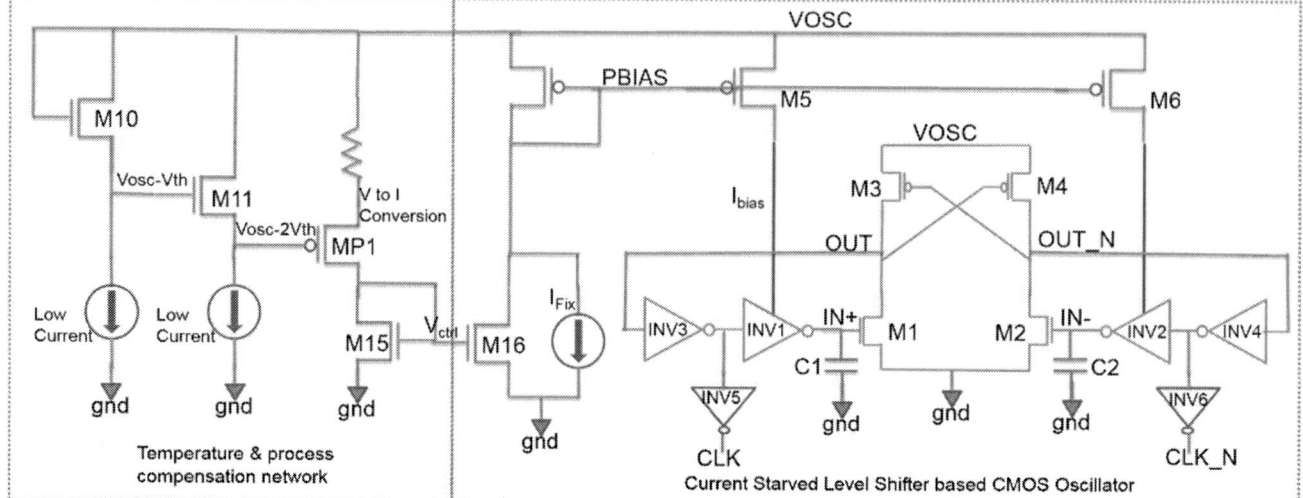

Figure 5. Complete Oscillator with compensation network

III. SIMULATION RESULTS

Following section discuss the simulation results of proposed circuit in different conditions.

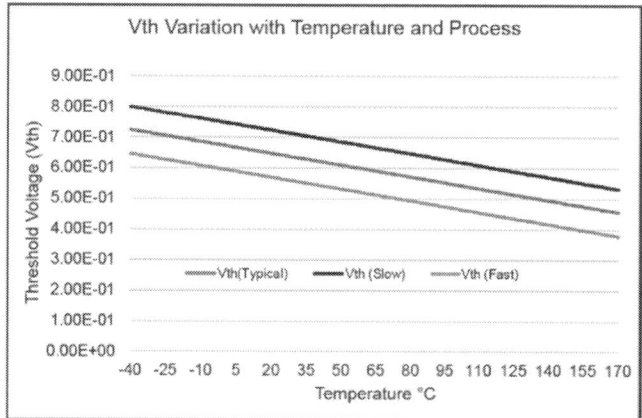

Figure 6. Threshold voltage variation with Temperature and Process

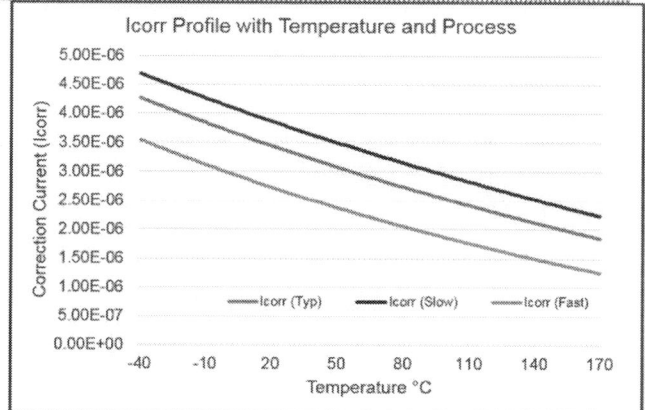

Figure 7. Compensation current (*Icorr*) with Temperature and Process

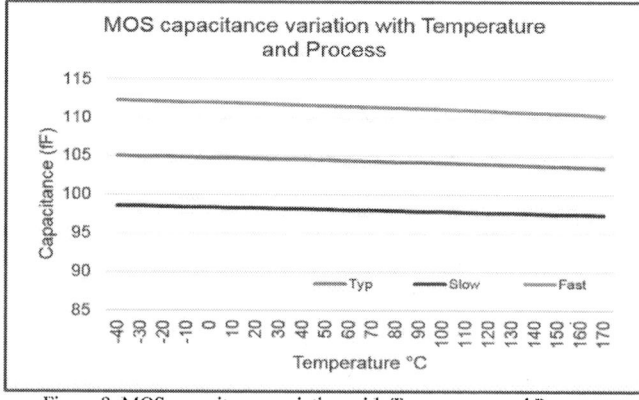

Figure 8. MOS capacitance variation with Temperature and Process

Figure 6, shows the simulation results of threshold voltage (*Vth*) variation with temperature. As discussed in equation 5, threshold voltage decreases linearly with temperature and results are confirmed in Figure 6. Results are shown for NMOS threshold voltage variation in Slow, typical and fast process corners for a temperature range of -40 to 160 °C.

Figure 7, shows the simulation results for compensation network used in proposed architecture. As discussed that *Vth* decreases linearly with temperature across the different process, accordingly compensation current (*Icorr*) also decreases almost linearly with temperature across the different process.

Figure 8, shows the simulation result for variation in MOS capacitance with temperature and process. NMOS is biased in accumulation region to obtain this capacitance. It is evident from the simulation results that variation in capacitance value across the temperature is very less hence frequency remains stable across temperature range. With different process corner,

978-1-5386-3693-0/18 $31.00 © 2018 IEEE

there is a small offset in capacitance value. But this variation in capacitance across process is correlated because same kind of NMOS is used in oscillator circuit and compensation network. Thus the compensation current (*Icorr*) is modified in correlation with variation in capacitance.

The oscillator is designed to be centred at 20.4MHz (although this is not the limit of architecture). Figure 9, shows the frequency variation across all process corners and temperature range from -40 to 160 °C.

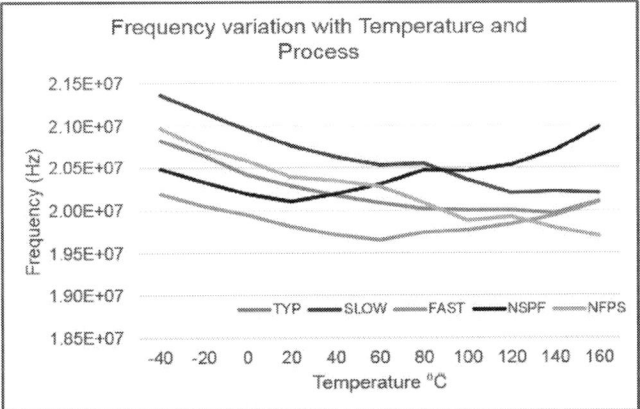

Figure 9. Frequency variation with Temperature and Process

The oscillator achieves a frequency variation of $\pm 0.1\%$ against supply variation of 3V to 5.5V (Although even if we further increase supply voltage it will not have any impact on oscillator frequency because circuit is working at regulated supply) and $\pm 2\%$ with temperature variation from -40°C to 160°C. Due to Vbg variation of +/-1%, *VOSC* varies by +/-1% and frequency varies by +/-0.2%.

The circuit consumes around 68uA (including voltage regulator) and around 20uA without regulator. The performance of the oscillator presented and compared with some recently reported reference clock generators in Table1.

TABLE 1. COMPARISON WITH EXISTING REFERENCES

	[9]	[10]	[11]	[12]	This work
Process	*0.5μm CMOS*	*0.18μm CMOS*	*0.35μm CMOS*	*0.35μm CMOS*	*0.11μm BCD*
Output Frequency (MHz)	11.6 – 21.4	10	30	4	20
Temperature range (°C)	-40 to 125	-20 to 120	-20 to 100	-40 to 100	-40 to 160
Supply Range (V)	3 - 5	1.2 - 3	1.8 - 3	3.3	3 -5.5
Power (μW)	400	80	180	221	60
Line regulation	$\pm 1.6\%$	$\pm 0.05\%$	4%/V	Not mentioned	$\pm 0.1\%$
Temperature coefficient (ppm/°C)	150	67	90	Not mentioned	100
Chip Area (mm²)	0.2	0.22	0.08	0.02	0.001754

IV. CONCLUSION

A 20MHz clock oscillator is design in 110nm-BCD technology targeting stable frequency across PVT conditions. Total spread in frequency is 100 ppm/°C. Due to current starved architecture, total power consumed is 60μW that is much lesser than many references. Area required for this circuit is 0.001754mm² that is much lesser than other references.

REFERENCES

[1] K. Kurita, T. Hotta, T. Nakano and N. Kitamura, "PLL-BiCMOS on-chip clock generator for very high speed microprocessor," IEEE J. Solid-Stage Circuits, vol. 26, no. 4, pp. 585-589, Apr. 1991.

[2] Michael P. Flynn and Sverre U. Lidholm. "A 1.2-um CMOS Current-Controlled Oscillator" IEEE J. Solid-State Circuits, Vol. 27. No. 7 1992.

[3] Abhirup Lahiri and Anurag Tiwari, "A 140μA 34ppm/°C 30 MHz clock oscillator in 28nm CMOS bulk process". 26th Int. Conference on VLSI Design.Jan 2013.

[4] F. Sebastiano and B. Nauta et al: "A 65nm CMOS Temperature-Compensated Mobility-Based Frequency Reference for Wireless Sensor Network," in Proc. ESSCIRCs Sept. 2010. Pp. 102-105.

[5] Keng-Jan Hsio, "A 32.4 ppm/°C 3.2-1.6 V Self-Chopped Relaxation Oscillator with Adaptive Supply Generation" Proc. of 2012 Symposium on VLSI Circuits Digest of Technical Papers.

[6] K.K. Huang, J. Brown, et al. " An Ultra-Low-Power 9.8 Ghz Crystal-Less UWB Transceiver With Digital Baseband integrated in 0.18um BiCMOS" IEEE J. of Solid State Cicruits vol 48 no. 12 Dec 2013

[7] Sung-Mo Kang and Yusuf Leblebici., "CMOS Digital Integrated Circuits," McGraw-Hill, Second Edition.

[8] K. Sundaresan, P. Allen and F. Ayazi,. "Process and Temperature compensation in a 7-MHz CMOS clock oscillator" IEEE J. Solid State Circuits, vol. 41, no. 2, pp. 433-442, Feb 2006.

[9] A. Vilas Boas and A. Olmos, "A Temperature compensated digitally trimmable on-chip IC Oscillator with low voltage inhibit capability," Proc. of the 2004 International Symposium on Circuits and Systems, 2004.

[10] Junghyup Lee and SeongHwan, "A 10MHz 80μm 67 ppm/°C CMOS Reference Clock Oscillator with a Temperature Compensated Feedback Loop in 0.18μm CMOS" Proc. on 2009 Symposium on VLSO Circuits Digest of Technical Papers.

[11] K. Ueno, Tetsuya Asai, and Y.Amemiya, "A 30MHz, 90-ppm/°C Fully-Integrated Clock Reference Generator with Frequency-locked Loop," Proc. of ESSCIRC 2009.

[12] Chi-Hsiung wang and Cheng-Feng Lin "Supply Voltage and Temperature insensitive Current Reference for 4MHz Oscillator" Published in Integrated Circuits (ISIC), 2011 13th International Symposium on

[13] J. Lee and S. Cho, "A 10MHz 80μW 67 ppm/°C CMOS reference clock Oscillator with a temperature compensatd feedback loop in 0.18 μm CMOS, "in Peoc. VLSI, June 2009, pp. 226-227.

[14] D. Griffith, P.T.Roine, J. Murdock, R. Smith "A 190nW 33kHz RC oscillator with ±0.21% temperature stability and 4ppm long-term stability" Published in ISSCC, 2014, doi 10.1109/ISSCC.2014.6757443.

2018 31th International Conference on VLSI Design and 2018 17th International Conference on Embedded Systems

High Speed FPGA Fabric Aware CSD Recoding with Run-time Support for Fault Localization

Ayan Palchaudhuri and Anindya Sundar Dhar

Department of Electronics & Electrical Communication Engineering

Indian Institute of Technology Kharagpur, Kharagpur, West Bengal, India, PIN: 721302

E-mail: ayanpalchaudhuri@gmail.com, asd@ece.iitkgp.ernet.in

Abstract—**Canonic signed digit (CSD) representation involves very few non-zero bits, making it an ideal candidate for aiding speed and area optimization in VLSI based DSP architectures. In this paper, we present high speed, FPGA fabric aware pipelined architectures for CSD recoding circuits starting from a two's complement number or a redundant signed digit input. Using the configured, yet under-utilized logic elements of the FPGA that realized the CSD recoding circuit, we additionally introduce fault localization circuitry to trace the source of any hard or soft errors that often creep in owing to various reliability hazards. The entire exercise has been carried out through target FPGA specific primitive instantiation coupled with placement constraints that guarantees high speed design, as well as eases the exercise to locate faulty FPGA slice coordinates, if any.**

Index Terms—**Canonic signed digit; redundant signed digit; carry look-ahead; FPGA; carry chain; primitive instantiation; C-testability; self dual; scan chain**

I. INTRODUCTION

Canonic signed digit (CSD) representation is characterized by sparse and non-adjacent occurrence of non-zero bits, thereby leading to high speed, area efficient VLSI DSP architectures [1]. This property has attracted significant research efforts in designing efficient VLSI architectures to facilitate conversion of two's complement or redundant signed digit (SD) numbers to their equivalent CSD form [2], [3]. However, much of the work has been carried out for ASIC designs, with no significant discussion for efficient FPGA implementations.

Primitive instantiation based approach for a given target FPGA platform is often suitable for design optimization as it facilitates the designer to exercise a superior control over dictating the appropriate configuration of the target technology nodes [4], [5]. It also eases the exercise of ensuring physical adjacency of the logically related elements, often dictated by the proximity of bit indices or the cascading order of circuits in a multi-logic level order, through issuing of appropriate relative or absolute placement coordinates. However, often due to the fixed topological geometry and fixed positional arrangement of the FPGA logic elements, it may be observed that the configured logic elements lie fairly underutilized. This leaves a scope for appendage of relevant functionalities into the underutilized logic elements without logic overhead.

Modern day FPGAs are susceptible to manufacturing defects and transistor aging owing to bias temperature instability and hot carrier injection, primarily due to substantial downsizing of logic nodes for increased logic integration within a single FPGA chip without appreciable augmentation in the chip size [6]. Thus, implementation of error detection and fault localization techniques on FPGAs have gained momentum [7]–[9]. In this paper, we have targeted the configured, yet under-utilized technology nodes to append fault localization circuitry without any logic overhead for the CSD recoding circuits. The following are the main contributions of the paper:

- We have proposed high speed ripple carry (RC) and carry look-ahead (CLA) schemes for CSD recoding for two's complement or redundant signed digit (SD) numbers.
- Fault localization circuitry is realized for the CSD recoders using four techniques, namely, alternating logic, scan insertion, signal monitoring and C-testability.
- Our proposed architectures outperform previously proposed CSD recoding circuits on FPGAs.

The rest of the paper is organized as follows. In Sec. II, we elaborate on the proposed CSD recoding architectures amenable to FPGA implementation. The results and discussions are presented in Sec. III. We conclude in Sec. IV.

II. PROPOSED ARCHITECTURES

Xilinx Virtex-7 FPGA, the chosen target platform, is composed of several Configurable Logic Blocks (CLBs), with each CLB comprising of two slices. Each slice is composed of four 6-input single output/ 5-input dual output Look-Up Tables (LUTs), three wide function multiplexers, a carry chain and eight flip-flops (FFs) [10]. The carry chain fabric serves as a dedicated routing fabric which has been thoroughly exploited in our designs for high speed, compact implementation.

A. Original Two's Complement to CSD Recoding Architecture

The CSD encoding $\{Y_i^d, Y_i^s\}$ = "00" for '0', "01" for '1' and "10" for '-1' is followed [11]. The conversion algorithm dictates replacement of a string of 1's ranging from index position i to $i + j - 1$ with a "1" at position $i + j$ and a $\bar{1}$ at position i. If X is a two's complement number, the Boolean logic governing the above functionality necessitates generation of a carry equivalent signal $c_{i+1} = X_{i+1}X_i + (X_{i+1}+X_i)c_i = \overline{(X_{i+1} \oplus X_i)}X_{i+1} + (X_{i+1} \oplus X_i)c_i$. Such logic forms are amenable to multiplexer based mapping of the carry chain, as shown in Fig. 1(a). The second logic level computes the CSD recoded bits as $Y_i^d = \overline{X_{i+1}}((X_i \oplus X_{i+1} \oplus c_i) \oplus X_{i+1}) = \overline{X_{i+1}}(X_i \oplus c_i)$ and $Y_i^s = X_{i+1}((X_i \oplus X_{i+1} \oplus c_i) \oplus X_{i+1}) = X_{i+1}(X_i \oplus c_i)$ using dual output LUTs whose outputs are

978-1-5386-3693-0/18 $31.00 © 2018 IEEE

186

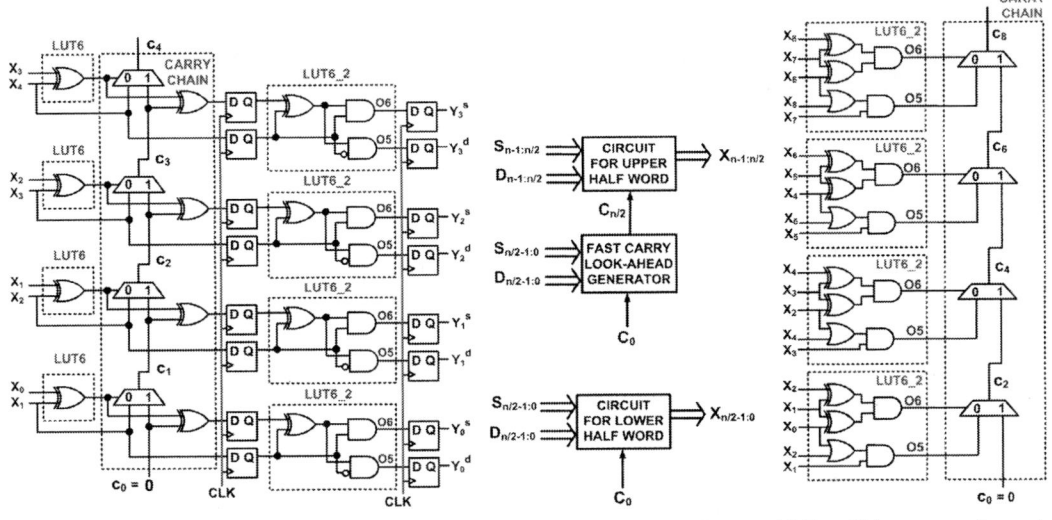

(a) Pipelined two's complement to CSD recoding circuit using RC scheme

(b) Carry look-ahead scheme

(c) Fast CLA generator for two's complement to CSD recoding

Fig. 1. Proposed pipelined architectures for two's complement to CSD recoding using RC and CLA schemes.

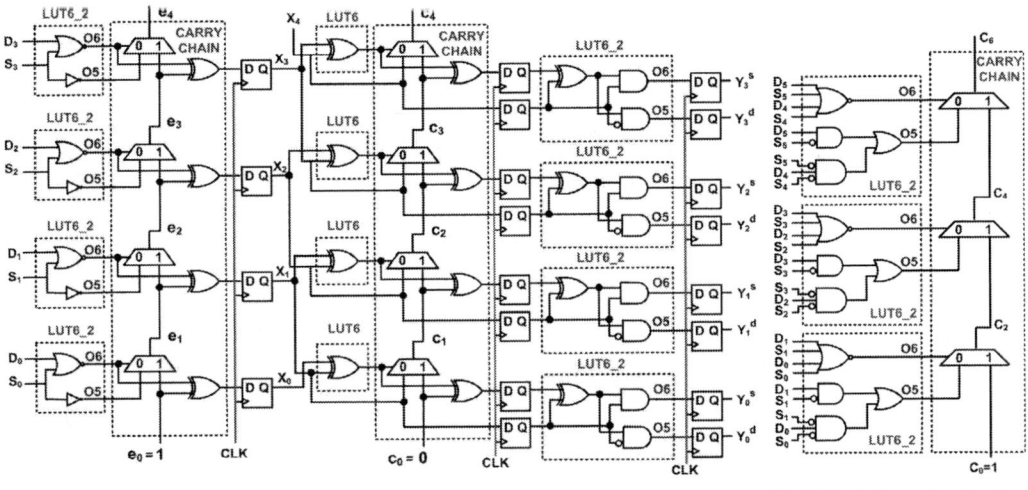

(a) Pipelined architecture for redundant binary to CSD recoding using RC scheme

(b) CLA design of redundant binary to two's complement converter circuit

Fig. 2. Proposed pipelined architectures for redundant binary to CSD recoding using RC and CLA schemes.

registered using FFs available in the same slice as that of the LUTs, as also shown in Fig. 1(a). As evident, the circuit is also amenable to two stages of forward path pipelining.

CLA based implementation for FPGA adders was proposed in [12]. We adopt a similar philosophy to realize a CLA based two's complement to CSD converter. In this CSD converter, the RC based carry chain sub-circuit is partitioned into two identical halves: one accepting the lower half word as its inputs and the other accepting the upper half word as its inputs. The upper half of the circuit receives its carry input from the output carry of the fast CLA generator as shown in Fig. 1(b), twice as fast compared to the RC scheme. The fast CLA generator

computes the carry signal c_{i+1} from c_{i-1} governed by the Boolean logic $c_{i+1} = \overline{(X_{i+1} \oplus X_i)(X_i \oplus X_{i-1})}(X_{i+1} + X_{i-1})X_i + (X_{i+1} \oplus X_i)(X_i \oplus X_{i-1})c_{i-1}$, making it amenable to carry chain implementation. The corresponding fast CLA generator requiring three inputs per LUT is shown in Fig. 1(c).

B. Original Redundant SD to CSD Recoding Architecture

Redundant SD to CSD recoding necessitates a two step conversion: redundant binary SD to two's complement number followed by CSD recoding. We assume the input bit representation (S_i, D_i) for the RB digits $X_i \in \{0, 1, -1\}$ as $\{(0, 0), (0, 1), (1, 0)\}$ respectively [13]. The circuit may be realized as an adder with initial carry $e_0 = 1$, and

978-1-5386-3693-0/18 $31.00 © 2018 IEEE

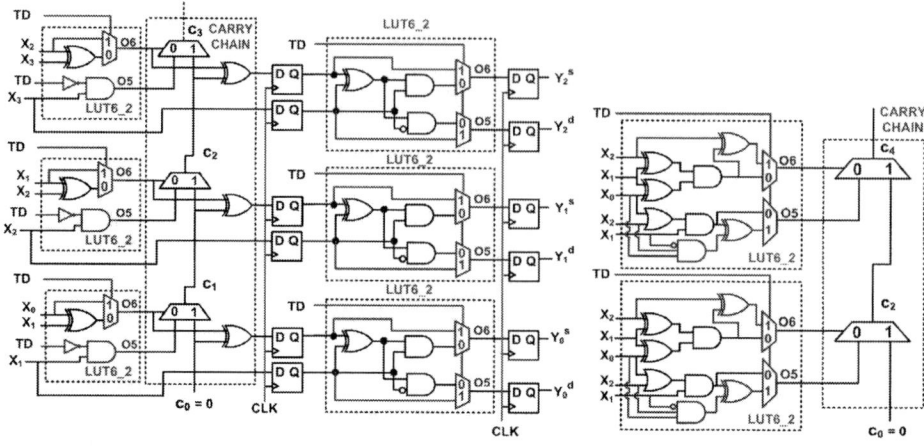

(a) Pipelined architecture for RC based two's complement to CSD recoding with fault localization circuitry using alternating logic

(b) Fast CLA design of two's complement to CSD converter circuit with alternating logic for fault localization

Fig. 3. Proposed pipelined architectures for two's complement to CSD recoding with fault localization circuitry using alternating logic following RC and CLA schemes. (The added functionality of the LUTs is shown in red.)

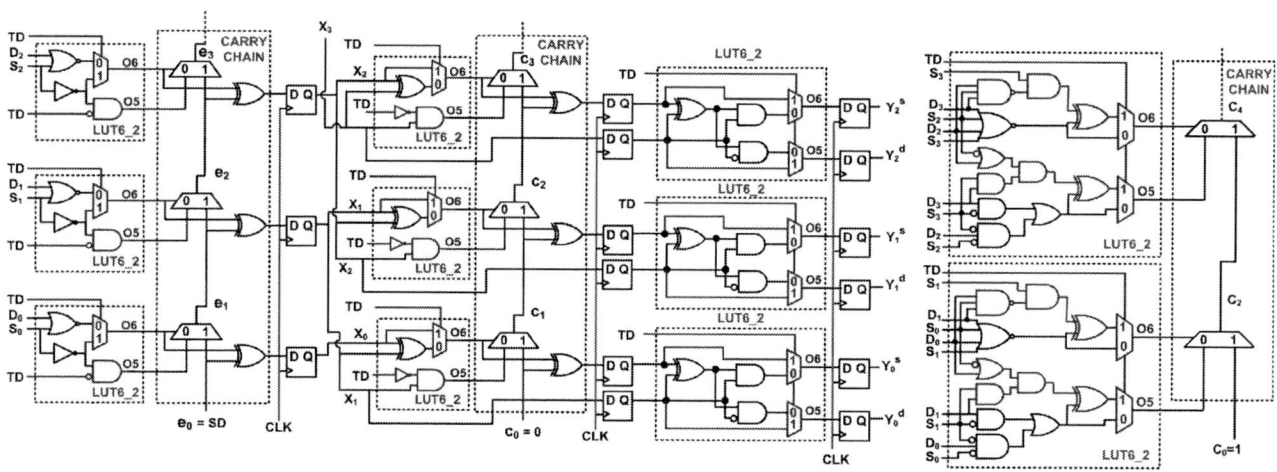

(a) Pipelined architecture for RC based redundant binary to CSD recoding with fault localization circuitry using alternating logic

(b) Fast CLA design of redundant binary to two's complement converter with alternating logic for fault localization

Fig. 4. Proposed pipelined architectures for redundant binary to CSD recoding with fault localization circuitry through alternating logic using RC and CLA schemes. (The added functionality of the LUTs is shown in red.)

input operands $(\overline{S_i}D_i, \overline{S_i})$. The output carry computation $e_{i+1} = \overline{S_i}D_i(S_i + D_i) + e_i(\overline{S_i + D_i})$ is amenable to carry chain implementation. The sum bits are computed as $X_i = (\overline{S_i + D_i}) \oplus e_i$, and are registered prior to sending them as inputs to the two's complement to CSD recoder circuit, which is implemented identically as described in Sec. II-A. The complete three stage pipelined architecture is shown in Fig. 2(a). CLA based redundant SD to two's complement conversion follows the same scheme as shown in Fig. 1(b). For the fast CLA generator, $c_{i+1} = \overline{S_i}D_i + (\overline{S_i + D_i})c_i =$
$\overline{S_i}D_i + (\overline{S_i + D_i})[\overline{S_{i-1}}D_{i-1} + (\overline{S_{i-1} + D_{i-1}})c_{i-1}] =$
$\overline{S_i}(D_i + \overline{D_i} \ \overline{S_{i-1}}D_{i-1}) + \overline{S_i} \ \overline{D_i} \ \overline{S_{i-1}} \ \overline{D_{i-1}}c_{i-1} =$
$\overline{S_i}(D_i + \overline{S_{i-1}}D_{i-1}) + (\overline{S_i + D_i + S_{i-1} + D_{i-1}})c_{i-1} =$
$\overline{S_i}(D_i + \overline{S_{i-1}}D_{i-1})(S_i + D_i + S_{i-1} + D_{i-1}) +$

$(\overline{S_i + D_i + S_{i-1} + D_{i-1}})c_{i-1}$, which can be mapped onto the carry chain, as shown in Fig. 2(b). The remaining recoding circuitry is identical to that described in Sec. II-A.

C. Fault Localization Circuitry for Two's Complement to CSD Recoding Architecture using Alternating Logic

A Boolean function $f(X)$ is self dual if $f(X) = \overline{f}(\overline{X})$ [14]. Another function $g(X)$ may be defined as $g(X) = f(X) \oplus \delta(X)$. $\delta(X)$ is regarded as a self dual complement (SDC) of $f(X)$ if $g(X)$ is self-dual, where $\delta(X) = x_i(f(X) \oplus \overline{f}(\overline{X}))$; $(1 \leq i \leq m)$ and $X = \{x_1, \ldots, x_m\}$. Thus $g(X)$ shall alternate with alternating inputs. For RC based two's complement to CSD converter, the LUTs in the first logic level realizes $f(X) = X_i \oplus X_{i+1}$. Hence, $\delta(X) = X_i$ and $g(X) = X_i$.

978-1-5386-3693-0/18 $31.00 © 2018 IEEE

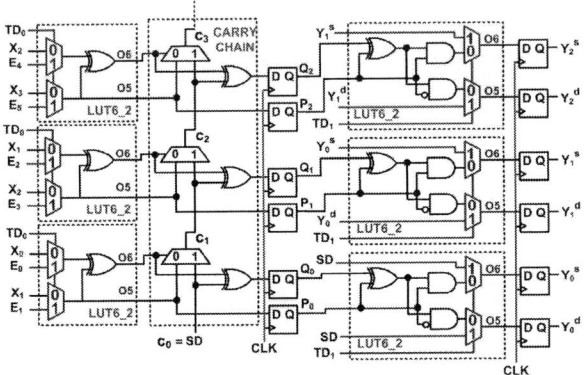

Fig. 5. Proposed architecture for C-testable and scan FF based fault localization architecture for two's complement to CSD recoding. (The added functionality of the LUTs is shown in red.)

For propagating the alternating outputs to the second logic level, the carry chain multiplexer outputs are ensured to be $c_i = 0$. Similarly for the LUTs in the second logic level, $g(X)$ may be any one of the input variables. The circuit has two operating modes: *normal* ($TD = 0$) and *test* mode ($TD = 1$). In the *normal* mode, the circuit functions as the original recoder. In the *test* mode, the circuit output for both the logic levels alternates with alternating primary inputs. To implement this circuit without logic overhead, the LUTs must have atleast 20% under-utilized inputs, which in our case is 60%. Apart from the LUTs, carry chain and FFs through which the alternating signal propagates also get tested for any possible faults. The overall circuitry is shown in Fig. 3(a). For LUTs realizing the CLA generator having 40% under-utilized inputs, $\delta(X)$ for the $O6$ output is computed as $\delta(X) = X_{i+2}$. For the $O5$ output, $\delta(X) = X_{i+2}\overline{X_{i+1}}X_i$. The corresponding circuitry is shown in Fig. 3(b).

D. Fault Localization Circuitry for Redundant SD to CSD Recoding Architecture using Alternating Logic

In the redundant SD to two's complement converter for the RC scheme constituting the first level of logic, $\delta(X)$ for the $O6$ LUT output may be computed as $\delta(X) = D_i\overline{S_i}$, for which $g(X) = \overline{S_i}$. The $O5$ output of the LUTs are set to zero for reliable transmission of alternating outputs to successive logic levels. Here, $e_0 = SD$, where $SD = 1$ in the *normal* mode and $SD = 0$ in the *test* mode. The remaining circuitry is identical to two's complement to CSD recoder. The overall RC based implementation is shown in Fig. 4(a). For the CLA based circuitry which has 20% under-utilized inputs in each configured LUT, $\delta(X)$ for the $O6$ output may be computed as $\delta(X) = S_{i+1}\overline{D_{i+1}S_iD_i}$, whereas for the $O5$ output, $\delta(X) = S_{i+1}D_{i+1}(\overline{S_i} + D_i)$. The corresponding architecture for the fast CLA generator is shown in Fig. 4(b).

E. Fault Localization Circuitry for Two's Complement to CSD Recoding using Signal Monitoring and Scan Logic

Consider a 1-bit iterative logic arrays (ILA) of the first level of logic comprising of LUTs with carry chains,

with X_i, X_{i+1} and c_{in} as inputs and c_{out} as output. For $\{X_i, X_{i+1}, c_{in}\}=\{0, 0, 0\}$ or $\{0, 1, X\}$ or $\{1, 0, X\}$ or $\{1, 1, 1\}$, $c_{out} = c_{in}$; and for $\{X_i, X_{i+1}, c_{in}\}=\{1, 1, 0\}$ or $\{0, 0, 1\}$, $c_{out} \neq c_{in}$. In this way, the signal c_{out} may be monitored through appropriate application of test vectors. Following this process, the intermediate multiplexer outputs of the carry chain logic may be tapped out to test for faults. A constant number of test vectors can check any arbitrary sized ILA. If the ILA inputs are not directly controllable, we introduce a multiplexing arrangement to apply external test vector inputs, for which 60% vacant LUT inputs are required, which is satisfied for this example. The remaining dual output LUTs and FFs in the successive logic level may be monitored by initiating a scan path of registers through proper multiplexing arrangement using the vacant LUT inputs. The desired percentage of under-utilized LUT inputs for establishing the scan path without logic overhead is 60% which is satisfied in the present scenario. The corresponding architecture is shown in Fig. 5. Multi-scan path is also feasible to optimize test time.

For an ILA cell in the CLA generator (refer Fig. 1(c)), with $\{X_{i+2}, X_{i+1}, X_i, c_{in}\}$ as inputs and c_{out} as output, $c_{out} = c_{in}$ if $\{X_{i+2}, X_{i+1}, X_i, c_{in}\}=\{0, 0, X, 0\}$ or $\{0, 1, 0, X\}$ or $\{0, 1, 1, 1\}$ or $\{1, 0, 0, 0\}$ or $\{1, 0, 1, X\}$ or $\{1, 1, X, 1\}$; and $c_{out} \neq c_{in}$ otherwise. These 16 test vectors may be applied to the circuit accordingly. It must be noted that the LUTs do not have sufficient under-utilized inputs to introduce external inputs, so we assume that either the X_i inputs are directly controllable, or some scan register arrangement feeding the input vectors to the CLA is available. The remaining circuitry, apart from CLA generator, remains unchanged.

F. Fault Localization Circuitry for Redundant SD to CSD Recoding Architecture using C-Testability and Scan Logic

Certain ILAs are said to be C-testable if they can be pseudo-exhaustively tested with only a certain number of test vectors independent of the number of cells in the ILA. The carry chain sub-circuits of the CSD recoders qualify for C-testability. The redundant SD to two's complement converter sub-circuit is composed of several 1-bit ILAs accepting $\{S_i, D_i, c_{in}\}$ as inputs and $\{X_i, c_{out}\}$ as outputs. For $\{D_i, S_i, c_{in}\} = \{0, 0, X\}$ or $\{0, 1, 0\}$ or $\{1, 0, 1\}$ or $\{1, 1, 0\}$, $c_{out} = c_{in}$; and these test vectors may be applied concurrently to consecutive ILA cells. For $\{D_i, S_i, c_{in}\} = \{0, 1, 1\}$ or $\{1, 0, 0\}$ or $\{1, 1, 1\}$, $c_{out} \neq c_{in}$; which must be alternately applied to alternate ILA cells. In order to test the ILA of the two's complement to CSD converter, the requisite test vectors may be initialized through the scan register chain of the redundant SD to two's complement converter. The remaining circuitry is similar to what is discussed in Sec. II-E. The entire architecture following the RC scheme is shown in Fig. 6(a).

The CLA design for the redundant SD to two's complement converter with scan path integration is realizable as shown in Fig. 6(b) with minor modification in the LUT configuration realizing the sub-circuit accepting the upper half word as its inputs to perpetuate the scan chain. For $\{D_{i+1}, S_{i+1}, D_i, S_i, c_{in}\}=\{0, 0, 0, 0, X\}$ or $\{0, 0, 0, 1, 0\}$

978-1-5386-3693-0/18 $31.00 © 2018 IEEE

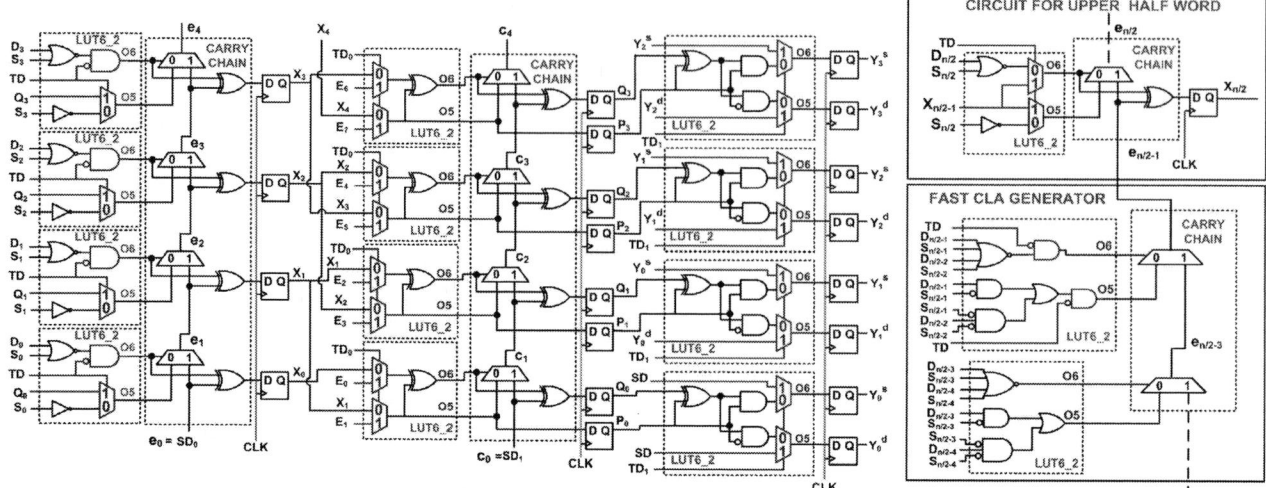

(a) Pipelined architecture for RC based redundant binary to CSD recoding with fault localization circuitry using C-testability and scan FF based design

(b) Fast CLA design of redundant binary to two's complement converter circuit using C-testability and scan FF based design

Fig. 6. Proposed pipelined architectures for redundant binary to CSD recoding with fault localization circuitry using using C-testability and scan FF based design for RC and CLA schemes. (The added functionality of the LUTs is shown in red.)

TABLE I
IMPLEMENTATION RESULTS FOR TWO'S COMPLEMENT TO CSD CONVERTER

Op. Width	Design Style	#FF	#LUT	#Slice	# Pipeline stages	Freq. (MHz)
32	Behav. design of [2]	–	46	29	Not pipelined	§ 401.77
	Prop. RC	127	64	16	2	852.51
	Prop. CLA	127	72	18	2	937.21
48	Behav. design of [2]	–	70	44	Not pipelined	§ 316.25
	Prop. RC	191	96	24	2	720.46
	Prop. CLA	191	108	27	2	813.67
64	Behav. design of [2]	–	94	61	Not pipelined	§ 292.48
	Prop. RC	255	128	32	2	624.22
	Prop. CLA	255	144	36	2	720.46
96	Behav. design of [2]	–	142	85	Not pipelined	§ 278.78
	Prop. RC	383	192	48	2	491.64
	Prop. CLA	383	216	54	2	583.09
128	Behav. design of [2]	–	190	112	Not pipelined	§ 223.26
	Prop. RC	511	256	64	2	402.41
	Prop. CLA	511	288	72	2	485.67

§ The behavioral designs are essentially combinational circuits proposed in [2] whose frequency of operation was obtained by inserting FFs at the input and outputs.

TABLE II
IMPLEMENTATION RESULTS FOR REDUNDANT BINARY TO CSD CONVERTER

Op. Width	Design Style	#FF	#LUT	#Slice	# Pipeline stages	Freq. (MHz)
32	Behav. design of [3]	–	131	90	Not pipelined	§ 245.46
	Prop. RC	158	96	24	3	735.29
	Prop. CLA	158	112	28	3	773.99
48	Behav. design of [3]	–	195	126	Not pipelined	§ 199.52
	Prop. RC	238	144	36	3	633.31
	Prop. CLA	238	168	42	3	688.71
64	Behav. design of [3]	–	259	171	Not pipelined	§ 181.40
	Prop. RC	318	192	48	3	556.48
	Prop. CLA	318	224	56	3	606.43
96	Behav. design of [3]	–	387	245	Not pipelined	§ 175.38
	Prop. RC	478	288	72	3	447.03
	Prop. CLA	478	336	84	3	496.28
128	Behav. design of [3]	–	515	350	Not pipelined	§ 172.03
	Prop. RC	638	384	96	3	370.92
	Prop. CLA	638	448	112	3	413.91

§ The behavioral designs are essentially combinational circuits proposed in [3] whose frequency of operation was obtained by inserting FFs at the input and outputs.

or $\{0, 0, 1, 1, 0\}$ or $\{X, 1, X, X, 0\}$ or $\{0, 0, 1, 0, 1\}$ or $\{1, 0, X, X, 1\}$, $c_{out} = c_{in}$; else $c_{out} \neq c_{in}$. A maximum of 32 test vectors are thus necessary for pseudo-exhaustively testing the entire ILA topology of the CLA of any arbitrary size by monitoring carry responses. The remaining sub-circuits are identical to what is discussed in Sec. II-E.

III. RESULTS AND DISCUSSIONS

All the above circuits were implemented on Xilinx Virtex-7 FPGA, device family XC7VX330T, package FFG1157, speed grade -2 using the Xilinx ISE 14.7 design environment. For the two's complement to CSD recoding circuit, our proposed architecture has been compared with a binary to CSD encoder [2] (Table I), where the authors claim that each of the processing element map efficiently into one slice comprising of two LUTs for Virtex-4 FPGAs, a platform which is several

generations out of date. Also, the Boolean logic governing the generation and propagation of the "bypass" (carry) signal [2] is not amenable for mapping on to the carry chain of the modern day FPGAs. In the process, the circuit relies upon the general purpose routing fabric instead of the fast, dedicated, hardwired carry chain to route the bypass signal, thereby leading to a slow speed realization. Moreover, the circuit is also not amenable to forward path pipelining on FPGAs, as was the case for our proposed architectures. Our proposed redundant binary to CSD converter architecture has been compared with another CSD recoder circuit proposed in [3] (Table II) in which the proposed circuit was shown to be amenable for ASIC platform. However, its performance seriously deteriorates when mapped onto the FPGA, owing to its inability to utilize the carry chain fabric on account of the Boolean equations governing the generation and propagation of the carry signal. The same

978-1-5386-3693-0/18 $31.00 © 2018 IEEE

TABLE III
IMPLEMENTATION RESULTS FOR TWO'S COMPLEMENT TO CSD
CONVERTER WITH FAULT LOCALIZATION CIRCUITRY

Op. Width	Mode	Fault Loc. Technique	#FF	#LUT	#Slice	# Scan Paths	Freq. (MHz)
32	RC	Alt. Logic	127	64	16	–	852.51
		Sig. monitor+Scan	127	64	16	1	852.51
	CLA	Alt. Logic	127	72	18	–	937.21
		Sig. monitor+Scan	127	72	18	1	937.21
48	RC	Alt. Logic	191	96	24	–	720.46
		Sig. monitor+Scan	191	96	24	1	720.46
	CLA	Alt. Logic	191	108	27	–	813.67
		Sig. monitor+Scan	191	108	27	1	813.67
64	RC	Alt. Logic	255	128	32	–	624.22
		Sig. monitor+Scan	255	128	32	1	624.22
			255	128	32	2	624.22
	CLA	Alt. Logic	255	144	36	–	720.46
		Sig. monitor+Scan	255	144	36	1	720.46
			255	144	36	2	720.46
96	RC	Alt. Logic	383	192	48	–	491.64
			383	192	48	1	491.64
		Sig. monitor+Scan	383	192	48	2	491.64
			383	192	48	3	491.64
	CLA	Alt. Logic	383	216	54	–	583.09
			383	216	54	1	583.09
		Sig. monitor+Scan	383	216	54	2	583.09
			383	216	54	3	583.09
128	RC	Alt. Logic	511	256	64	–	402.41
			511	256	64	1	402.41
		Sig. monitor+Scan	511	256	64	2	402.41
			511	256	64	4	402.41
	CLA	Alt. Logic	511	288	72	–	485.67
			511	288	72	1	485.67
		Sig. monitor+Scan	511	288	72	2	485.67
			511	288	72	4	485.67

TABLE IV
IMPLEMENTATION RESULTS FOR REDUNDANT SD TO CSD CONVERTER
WITH FAULT LOCALIZATION CIRCUITRY

Op. Width	Mode	Fault Loc. Technique	#FF	#LUT	#Slice	# Scan Paths	Freq. (MHz)
32	RC	Alt. Logic	158	96	24	–	731.53
		C-Test+Scan	158	96	24	–	735.29
	CLA	Alt. Logic	158	112	28	–	773.99
		C-Test+Scan	158	112	28	1	768.64
48	RC	Alt. Logic	238	144	36	–	630.52
		C-Test+Scan	238	144	36	1	633.31
	CLA	Alt. Logic	238	168	42	–	688.71
		C-Test+Scan	238	168	42	1	668.90
64	RC	Alt. Logic	318	192	48	–	554.32
		C-Test+Scan	318	192	48	1	556.48
			318	192	48	2	556.48
	CLA	Alt. Logic	318	224	56	–	603.86
		C-Test+Scan	318	224	56	1	602.77
			318	224	56	2	596.66
96	RC	Alt. Logic	478	288	72	–	445.63
			478	288	72	1	447.03
		C-Test+Scan	478	288	72	2	447.03
			478	288	72	3	447.03
	CLA	Alt. Logic	478	336	84	–	496.28
			478	336	84	1	484.50
		C-Test+Scan	478	336	84	2	495.79
			478	336	84	3	490.44
128	RC	Alt. Logic	638	384	96	–	369.96
			638	384	96	1	370.92
		C-Test+Scan	638	384	96	2	370.92
			638	384	96	4	370.92
	CLA	Alt. Logic	638	448	112	–	413.91
			638	448	112	1	413.56
		C-Test+Scan	638	448	112	2	413.56
			638	448	112	4	413.56

circuit [3] was also not very amenable to forward path pipelining in comparison to our proposed pipelined architecture of Fig. 2(a). Our proposed CLA schemes for both the recoders outperform their RC counterparts in speed at the cost of 12.5% and 16.67% hardware overhead for the two's complement and redundant SD to CSD recoder respectively.

The two CSD recoders with built-in fault localization architectures consume the same logic hardware and operate at almost the same speed as the original CSD recoders (refer Table III and Table IV). Such circuits can be re-configured at run time from *normal* to *test* mode with activation of the test signal TD. As the circuits are conceived through primitive instantiation and constrained placement based approach, any faulty output generated may be traced easily from the output vector index, as the placement locations of the configured primitives are known apriori. The design descriptions of such circuits can be automated by writing simple C programs.

IV. CONCLUSIONS

The proposed CSD recoding circuits are simple and attractive from the point of view of their architectural description. They can be realized following the bit-sliced design paradigm in the form of ILAs and are amenable to forward path pipelining with no requirement of staging delays to synchronize input and output arrival times. They extensively use the carry chain circuitry leading to high speed realization. The percentage of under-utilization of the inputs to the configured LUTs was optimum to insert fault localization circuitry, which is of emerging importance owing to the susceptibility of modern day ICs to various reliability hazards. Choice of the appropriate fault localization circuitry without incurring logic overhead is strongly dependent upon how the original circuit is configured in the FPGA fabric logic.

REFERENCES

[1] K. K. Parhi, *"VLSI Digital Signal Processing Systems: Design and Implementation"*. Wiley India Pvt. Limited, 2007.

[2] M. Faust, O. Gustafsson, and C.-H. Chang, "Fast and VLSI efficient-to-CSD encoder using bypass signal," *Electronics Letters*, vol. 47, no. 1, pp. 18–20, Jan. 2011.

[3] Y. Tanaka, "Efficient signed-digit-to-canonical-signed-digit recoding circuits," *Microelectronics Journal*, vol. 57, pp. 21–25, Nov. 2016.

[4] A. Ehliar, "Optimizing Xilinx designs through primitive instantiation," in *Proceedings of the 7th FPGAworld Conference*, 2010, pp. 20–27.

[5] A. Palchaudhuri and R. S. Chakraborty, *High Performance Integer Arithmetic Circuit Design on FPGA: Architecture, Implementation and Design Automation*. Springer India, 2016.

[6] A. Gupte, S. Vyas, and P. H. Jones, "A Fault-Aware Toolchain Approach for FPGA Fault Tolerance," *ACM Transactions on Design Automation of Electronic Systems (TODAES)*, vol. 20, no. 2, pp. 32:1–32:22, 2015.

[7] A. Palchaudhuri and A. S. Dhar, "Efficient Implementation of Scan Register Insertion on Integer Arithmetic Cores for FPGAs," in *29th International Conference on VLSI Design*, Jan. 2016, pp. 433–438.

[8] B. C. Basha, S. Pillement, and S. J. Piestrak, "Fault-aware Configurable Logic Block for Reliable Reconfigurable FPGAs," in *IEEE International Symposium on Circuits and Systems (ISCAS)*, May 2015, pp. 2732 – 2735.

[9] A. Palchaudhuri and A. S. Dhar, "Built-In Fault Localization Circuitry for High Performance FPGA Based Implementations," *Journal of Electronic Testing*, vol. 33, no. 4, pp. 529–537, Aug. 2017.

[10] Xilinx Inc., *"7 Series FPGAs Configurable Logic Block, User Guide, UG474 (v1.7)."* [Online]., Sep. 27 2016. [Online]. Available: http://www.xilinx.com/support/documentation/user_guides/ug474_7Series_CLB.pdf

[11] G. A. Ruiz and M. Granda, "Efficient canonic signed digit recoding," *Microelectronics Journal*, vol. 42, no. 9, pp. 1090–1097, Sep. 2011.

[12] P. Zicari and S. Perri, "A Fast Carry Chain Adder for Virtex-5 FPGAs," in *15th IEEE Mediterranean Electrotechnical Conference (MELECON)*, Apr. 2010, pp. 304–308.

[13] K. Hwang, *Computer Arithmetic: Principles, Architecture and Design*. John Wiley & Sons, 1979.

[14] V. V. Saposhnikov, V. V. Saposhnikov, A. Dmitriev, and M. Goessel, "Self-dual duplication for Error Detection," in *Seventh ATS*, Dec. 1998, pp. 296–300.

A Low-Power Circuit for Adaptive Dynamic Programming

Nan Zheng, Pinaki Mazumder
Electrical Engineering and Computer Science Department
University of Michigan
Ann Arbor, Michigan, United States
Email: {zhengn, mazum}@umich.com

Abstract—**This paper presents a low-power CMOS design for accelerating an adaptive dynamic programming algorithm, called action-dependent heuristic dynamic programming, which is widely employed in many real-life control problems. The objective of this work is to solve Bellman equation approximately using efficient hardware in order to generate near-optimal real-time control policies for many control applications. The hardware exploits the data-level parallelism exists in both inference and learning of neural networks in order to improve throughput as well as to provide good scalability. The circuit is realized in a 65-nm technology. It is shown with simulations that the design is two orders of magnitude faster than a software running on a general-purpose processor thanks to the parallelization of the algorithm and the reduction in unnecessary control overheads. Performance of the CMOS circuit is benchmarked with two popular control tasks. Successful learning can be achieved with a power consumption of 28 mW.**

Keywords-**Reinforcement learning, adaptive dynamic programming, neural networks, low-power CMOS circuit.**

I. INTRODUCTION

Recently, the development of microrobots has drawn many researcher's attention [1]–[3]. The microrobots can be deployed in many promising applications, such as remote monitoring and mapping of environment [4], [5], navigating in unknown dynamic environment [6], [7], and inspecting buildings for structural damages [8]. Despite various potential applications these robots can be employed for, controlling these robots can be quite challenging. Furthermore, for the tiny robots that chiefly rely on energy scavenging, power consumption is one of the most important design considerations. Many real-life control problems can be tackled through solving Bellman equation, the optimal equality. Solving the Bellman equation directly is in general intractable, attributed to the well-known curse-of-the-dimensionality. Fortunately, the adaptive dynamic programming (ADP) algorithm [9]–[16] provides an alternative to approach the solution adaptively and approximately with a reasonable computational complexity. The ADP algorithm is one type of reinforcement-learning algorithm, which has been used in many applications such as aircraft control [11], [13] and power system compensation [12], [14]. Even though the ADP algorithm has different forms and derivatives, the action-dependent heuristic dynamic programming

(ADHDP) is one of the most powerful ADP algorithms [9], [10], [15], [16]. The ADHDP algorithm is a model-free algorithm. In other words, no knowledge about the model of the system is required. The algorithm learns this information in the process of interacting with the system. Traditionally, ADP algorithms are programmed and run on general-purpose processors to take advantage of the flexibility of the processors. This software approach, however, suffers from poor energy efficiency, as most general-purpose processors are designed for programmability and flexibility instead of power efficiency. A big portion of power is wasted by the unnecessary control flows in the processors. Therefore, to satisfy the stringent requirement posed by the tight energy budget of the microrobots, specialized hardware are required.

In this paper, we demonstrate an accelerator design for the ADHDP algorithm. a single-instruction-multiple-data (SIMD) architecture is leveraged to increase the throughput and energy efficiency of the hardware. The design is implemented in a 65-nm technology. Two commonly used benchmark tasks are employed to examine the designed circuit. Simulation results are demonstrated to show the performance of the customized hardware.

II. CHIP ARCHITECTURE

The configuration of the actor-critic based ADHDP algorithm is illustrated in Fig. 1. $\mathbf{x}(t)$ is the state vector of the plant that needs to be controlled. $\mathbf{a}(t)$ is the action vector applied to the plant. The objective of the ADP algorithm is to maximize the reward-to-go J, expressed as follows

$$J\left[\mathbf{x}(t)\right] = \sum_{k=1}^{\infty} \gamma^{k-1} r\left[\mathbf{x}(t+k)\right] \tag{1}$$

where γ is the discount factor used to promote the short-term reward over long-term reward and $r\left[\mathbf{x}(t)\right]$ is the reward received at state $\mathbf{x}(t)$.

One can maximize $J\left[\mathbf{x}(t)\right]$ through solving the optimal equality

$$J^*\left[\mathbf{x}(t)\right] = \max_{\mathbf{a}(t)}\left(r\left[\mathbf{x}(t+1)\right] + \gamma J^*\left[\mathbf{x}(t+1)\right]\right) \tag{2}$$

where $J^*\left[\mathbf{x}(t)\right]$ denotes the optimal state-value function achieved with the optimal policy. The optimal policy $\mathbf{a}^*(t)$ can be obtained by solving (2).

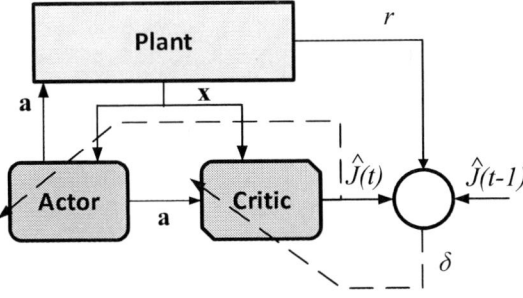

Figure 1. Configuration of the actor-critic based ADHDP algorithm.

By utilizing neural networks as function approximators, the ADP algorithm is able to adaptively approach the final solution of the problem. In Fig. 1, a critic network is used to estimate $J^*[\mathbf{x}(t)]$ with its output $\hat{J}[\mathbf{x}(t)]$. An actor network then attempts to select the action that can result in the maximum $\hat{J}[\mathbf{x}(t)]$. In order to approximate the solution to (2), both of these two networks need to adapt their weights iteratively, with the aim to minimize/maximize certain cost/utility functions. For the critic network, the cost function is the temporal difference error

$$\delta(t) = \hat{J}[\mathbf{x}(t-1)] - \gamma \hat{J}[\mathbf{x}(t)] - r[\mathbf{x}(t)] \qquad (3)$$

Clearly, the necessary condition for the critic network to learn the true reward-to-go, $J^*[\mathbf{x}(t)]$, is that the temporal difference shown in (3) becomes zero. It is this temporal coherence in estimating the future reward that trains the critic network.

For the actor network, the utility function is the estimated reward-to-go because the objective of the actor is to choose the action that maximizes the future rewards. Note that the output from the actor also serves as the input to the critic, because the estimated future reward also depends on the action taken at the moment. This is the reason that this algorithm is action-dependent. Training of both critic and actor networks can be accomplished through backpropagation, which is the standard way to train neural networks. Detailed learning algorithms for ADHDP can be found in [9], [10], [15], [16], and they are not presented in this paper for brevity. Interested readers are also referred to the excellent review papers on ADP [17]–[19] for more information.

The top-level block diagram of the ADP circuit is shown in Fig. 2. There are three main parts in the figure, namely controller, datapath, memory, and controller. The main role of the controller is to determine the instruction flow. A SIMD architecture is used to leverage the inherent parallelism in neural network-based computing. To provide flexibility and reconfigurability, the controller is implemented with a finite-state machine (FSM) that transitions based on instructions fetched from the instruction memory. To reduce the large

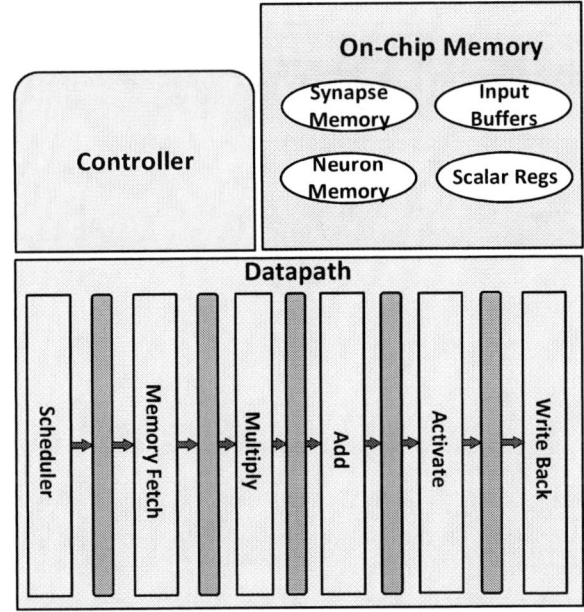

Figure 2. Block diagram of the proposed accelerator

control overhead incurred in general-purpose processors, most operations are matrix-based. For example, in order to conduct one matrix-vector multiplication, only one instruction is needed. The datapath maintains the data flow and parallel computation until the whole matrix-vector operation is completed.

The datapath is the core for conducting computations in the ADP algorithm. There are several pipeline stages in the datapath. They are schedule, fetch, multiply, add, activate, and write back. Instructions are fetched from the instruction memory in the schedule stage. The instruction decoder decodes the information needed for scheduling operations, such as addresses for the source data, addresses for write-back, and the type of operations. All data hazards need to be spotted out at this stage. After decoding, data are latched in pipeline registers according to the address obtained in the schedule stage. In this stage, data might come from the memory, the write-back buffer, or the input buffer if it is previously loaded.

Core arithmetic operations of the system are conducted in the multiply stage, add stage, and the activate stage. The hyperbolic tangent function is used in this design as the activation function, as shown in Fig. 3(a). We implement the activation function as a look-up table with linear interpolation, similar to those employed in [20], [21]. As shown in Fig. 3(b), coefficients, k_i and b_i that are used for interpolation are stored in the look-up table. The corresponding coefficients are first read out, and a multiply-and-add operation is conducted to perform the linear interpolation. To reduce the critical path, the readout of coefficients are

978-1-5386-3693-0/18 $31.00 © 2018 IEEE

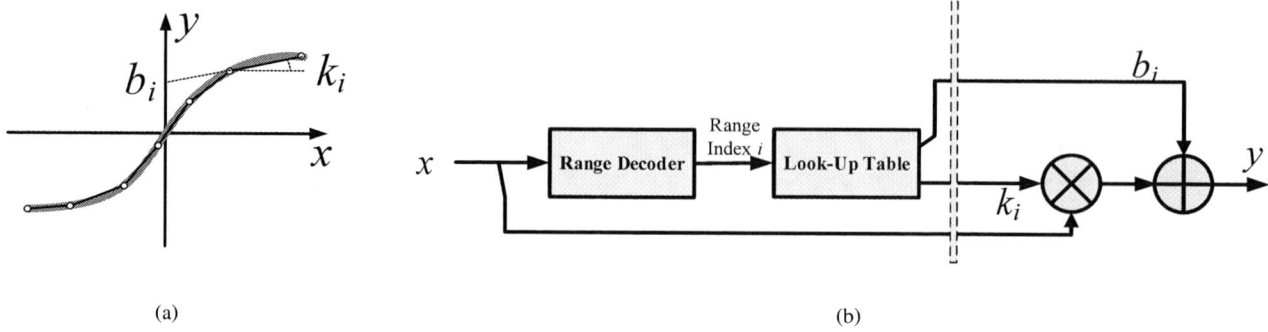

Figure 3. (a) Piecewise linear approximation of the hyperbolic tangent activation function. (b) Circuit diagram implementing the approximate activation function.

actually implemented in the preceding add stage. In the write-back stage, computed data are stored back to storage units based on the destination address provided in the instruction.

Most operations needed in the ADHDP algorithm are matrix-vector multiplication. A tile-based strategy [20], [22], [23] is adopted for better scalability and data locality. The size of the tile is four in this design, which is determined by the number of lanes available in the accelerator. Synapse weights are the most memory-demanding data in a neural network. Therefore, they are stored in a dense SRAM array. In order to effectively utilize the limited number of read ports of the SRAM array, the weights are arranged in a proper form to achieve a sequential and uniform access pattern during the inference and learning phases.

III. RESULTS

In this section, two most popular benchmark tasks that are widely used in the literature, as shown in Fig. 4, are employed to examine the performance of the designed ADP circuit. In the cart-pole balancing problem [9], [10], [15], [24], the target is to balance the pole on the cart while not moving the cart out of a certain range. The beam balancing task [25], [26] deals with a problem of balancing a ball that sits on a beam which can be rotated by a motor. In both of these two tasks, we want to control the system such that states of the plant stay in some pre-defined ranges. For the cart-pole problem, states need to be controlled are the angle θ between the pole and the vertical direction and the location of the cart, x. In the beam-balancing problem, the states to be regulated are the angle θ between the beam and the horizontal direction and the displacement of the ball, x.

In our design, we adopt a fixed-point number representation. Fixed-point numbers are often used in neural network accelerators for good energy efficiency and ease of implementation [20], [22], [23]. To find out what is a reasonable bitwidth that is required to represent the data in our system, parametric simulations are conducted. The

Table I
SUMMARY OF SPECIFICATIONS OF THE ADHDP ACCELERATOR

Technology	65nm
Area	550 μm × 550 μm
Number of lanes	4
Arithmetic precision	24-bit fixed-point
Supply voltage	1.2 V
Clock frequency	175MHz
Power consumption	28 mW

Table II
COMPARISON OF THE LEARNING PERFORMANCES

Benchmark task	Normalized performance
Cart-pole balancing	0.995 ± 0.1581
Beam balancing	1.044 ± 0.2382

accumulated time steps that the agent is able to regulate the states of the plants are used as the metric for evaluating performances. The results obtained with different bitwidths for the fractional part of the fixed-point numbers are shown in Fig. 5. In this figure, the obtained data are normalized with respect to the results obtained with a double-precision floating-point computation, and the 95% confidence interval are marked with the error bars. Based on the results in Fig. 5, we use 18 bits to represent the fractional part of the numbers in our system.

The circuit is implemented in a 65-nm technology. It is found through simulation that the maximum clock frequency the circuit can operate on is 175 MHz. To save simulation time and facilitate the evaluation of the designed circuit, simulators are developed and employed to model the input-output relationship and the numbers of clock cycles taken to accomplish the benchmark tasks. In the simulation, compared to the software approach running on an Intel Xeon processor, the customized hardware can achieve two orders of magnitude speedup on average for both of the two benchmark tasks, thanks to the parallelism in the computation and simplifications in the control flow. Specifications of the designed accelerator are summarized in Table I.

978-1-5386-3693-0/18 $31.00 © 2018 IEEE

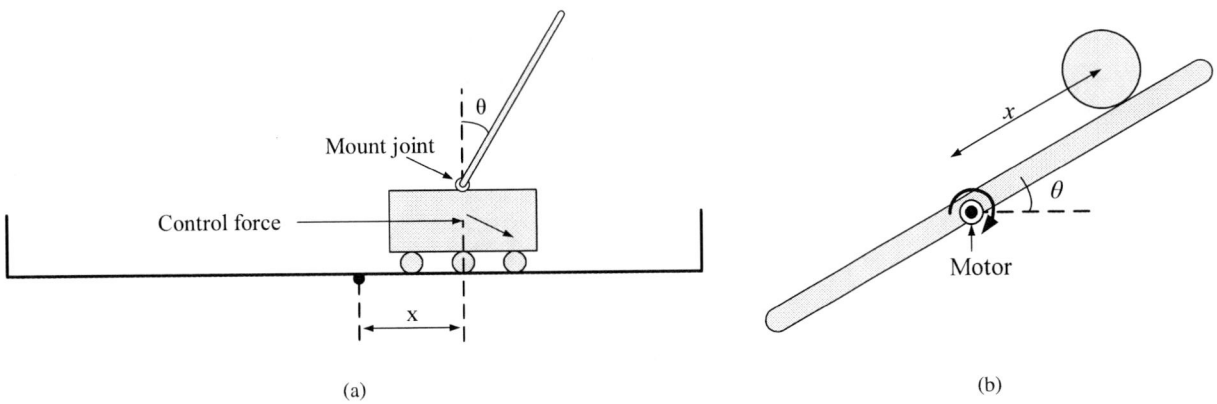

(a)   (b)

Figure 4.   Configuration of the (a) cart-pole balancing task and (b) beam balancing task

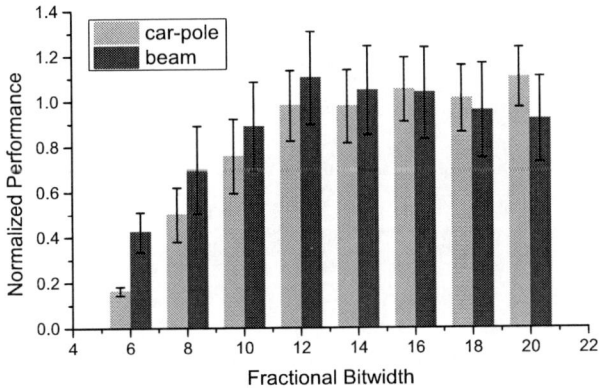

Figure 5.   The learning performances achieved with different levels of data quantization for two classic ADP benchmarks. The obtained performances are normalized with those obtained from double-precision floating point computations.

Table II compares the performances achieved by the accelerator and that obtained through the software approach running on an Intel Xeon Processor. The accelerator performances are normalized with respect to the software performances in the figure. The accumulated time duration in which states are successfully regulated are used for measuring performance. It can be observed from the table that the customized hardware can achieve a similar performance compared to the software approach.

To provide a closer look on how well the proposed accelerator performs on complicated control problems, the transient waveforms of the states in the two benchmark tasks shown in Fig. 4, are illustrated. Four both of these two tasks, four state variables are used as input to the actor network: the offset x, the angle θ, and the corresponding velocities x' and θ'. The action in the cart-pole task is a binary quantity, which determines the polarity of the force. The action in the beam-task, on the other hand, is a continuous quantity. The target of the cart-pole balancing task is to maintain the states

x and θ within the rage of $[-2.4\ \mathrm{m}, 2.4\ \mathrm{m}]$ and $[-12°, 12°]$, respectively. Similarly, the target of the beam-balancing task is to regulate the states such that x and θ can stay in the range of $[-0.48\ \mathrm{m}, 0.48\ \mathrm{m}]$ and $[-13.75°, 13.75°]$. Detailed models for these benchmark tasks are complex and lengthy, therefor they are not presented in this paper for brevity. Interested readers are referred to [9], [10], [15], [24]–[26] for more details. Typical waveforms obtained from the two benchmark tasks for successful learning are demonstrated in Fig. 6(a) and Fig. 6(b), respectively. It is shown that the accelerator successfully learns the control policy and maintains the states of the system well within the target range. In addition, the histograms that record the states of the plants under control are illustrated in Fig. 7. Results obtained from ten typical successful trials are demonstrated in the figure. These results further demonstrate the proposed ADP circuit can successfuly regulate the states of the plant as desired.

IV. CONCLUSION

A low-power CMOS circuit for the ADHDP algorithm is presented in this paper. A SIMD architecture is used to exploit the inherent data-level parallelism exists in the algorithm in order to improve the throughput as well as to enhance the energy efficiency. The effect of quantizing data on the learning performance is studied. The bitwidth of the system is then chosen based on results obtained from the parametric simulations. The proposed circuit is implemented in 65-nm technology. Two most common benchmark tasks are employed to examine the customized hardware design. Successful learning are observed in the simulation results.

ACKNOWLEDGMENT

This work was supported by the National Science Foundation under the Grant CCF 1421467.

978-1-5386-3693-0/18 $31.00 © 2018 IEEE

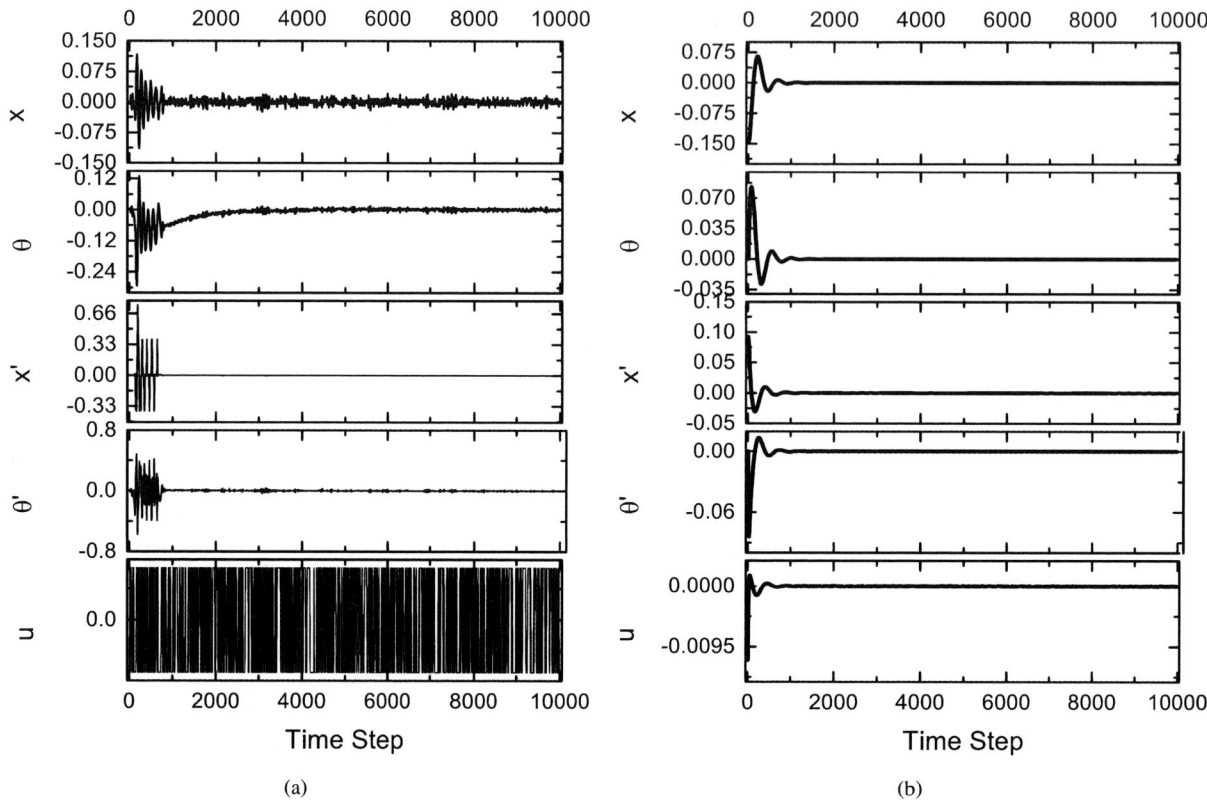

Figure 6.   Typical trajectories of the states that are under control for (a) the cart-pole balancing task, (b) the beam-balancing problem.

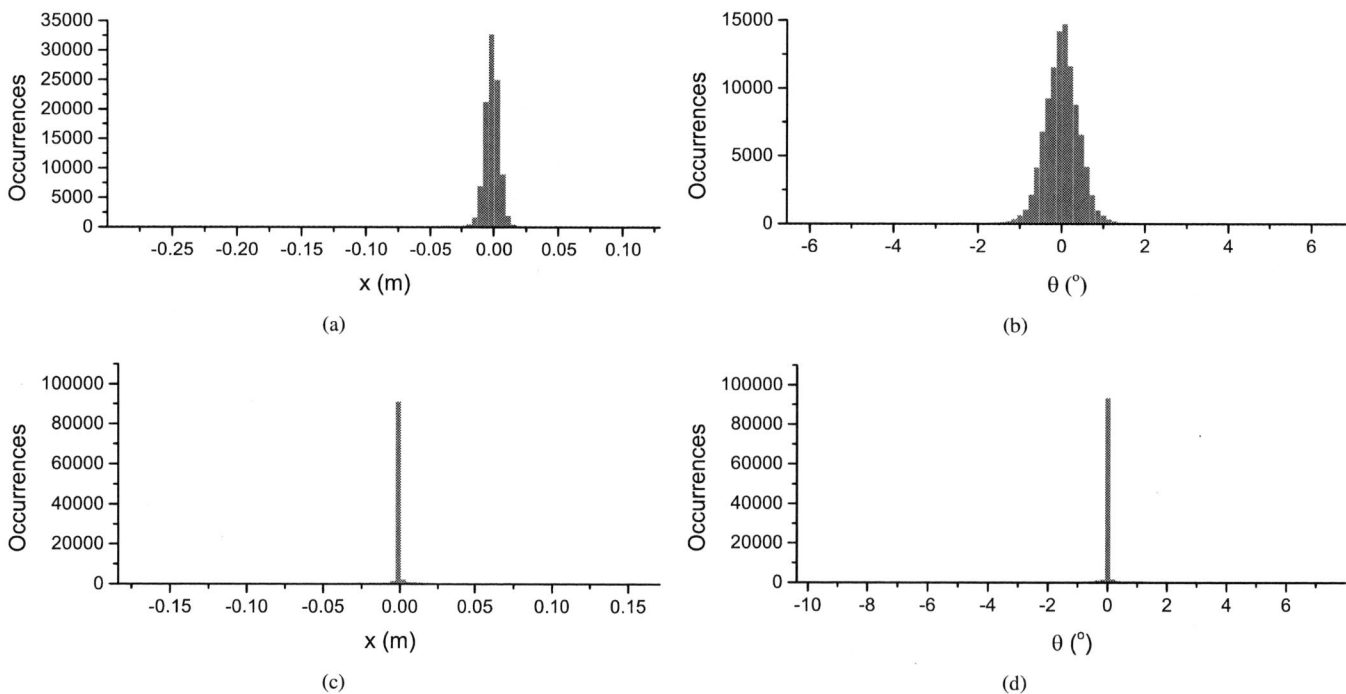

Figure 7.   Typical distributions of the controlled states in successful trials. (a) The histogram of the position x of the cart in the cart-pole balancing task. (b) The histogram of the angle θ in the cart-pole balancing task. (c) The histogram of the position x of the ball in the beam-balancing task. (d) The histogram of the angle θ in the beam-balancing task.

REFERENCES

[1] R. J. Wood, "The first takeoff of a biologically inspired at-scale robotic insect," *IEEE transactions on robotics*, vol. 24, no. 2, pp. 341–347, 2008.

[2] N. O. Pérez-Arancibia, K. Y. Ma, K. C. Galloway, J. D. Greenberg, and R. J. Wood, "First controlled vertical flight of a biologically inspired microrobot," *Bioinspiration & Biomimetics*, vol. 6, no. 3, p. 036009, 2011.

[3] P. Mazumder, D. Hu, I. Ebong, X. Zhang, Z. Xu, and S. Ferrari, "Digital implementation of a virtual insect trained by spike-timing dependent plasticity," *Integration, the VLSI Journal*, vol. 54, pp. 109–117, 2016.

[4] H. S. Ahn, I.-K. Sa, and J. Y. Choi, "Pda-based mobile robot system with remote monitoring for home environment," *IEEE Transactions on Consumer Electronics*, vol. 55, no. 3, 2009.

[5] S. W. Moon, Y. J. Kim, H. J. Myeong, C. S. Kim, N. J. Cha, and D. H. Kim, "Implementation of smartphone environment remote control and monitoring system for android operating system-based robot platform," in *Ubiquitous Robots and Ambient Intelligence (URAI), 2011 8th International Conference on*. IEEE, 2011, pp. 211–214.

[6] C. Ye, N. H. Yung, and D. Wang, "A fuzzy controller with supervised learning assisted reinforcement learning algorithm for obstacle avoidance," *IEEE Transactions on Systems, Man, and Cybernetics, Part B (Cybernetics)*, vol. 33, no. 1, pp. 17–27, 2003.

[7] A. Jayasiri, G. K. Mann, and R. G. Gosine, "Behavior coordination of mobile robotics using supervisory control of fuzzy discrete event systems," *IEEE Transactions on Systems, Man, and Cybernetics, Part B (Cybernetics)*, vol. 41, no. 5, pp. 1224–1238, 2011.

[8] D. Longo and G. Muscato, "The alicia/sup 3/climbing robot: a three-module robot for automatic wall inspection," *IEEE Robotics & Automation Magazine*, vol. 13, no. 1, pp. 42–50, 2006.

[9] J. Si and Y.-T. Wang, "Online learning control by association and reinforcement," *IEEE Transactions on Neural Networks*, vol. 12, no. 2, pp. 264–276, 2001.

[10] D. Liu, X. Xiong, and Y. Zhang, "Action-dependent adaptive critic designs," in *Neural Networks, 2001. Proceedings. IJCNN'01. International Joint Conference on*, vol. 2. IEEE, 2001, pp. 990–995.

[11] S. Ferrari and R. F. Stengel, "Online adaptive critic flight control," *Journal of Guidance, Control, and Dynamics*, vol. 27, no. 5, pp. 777–786, 2004.

[12] S. Mohagheghi, Y. del Valle, G. K. Venayagamoorthy, and R. G. Harley, "A proportional-integrator type adaptive critic design-based neurocontroller for a static compensator in a multimachine power system," *IEEE Transactions on Industrial Electronics*, vol. 54, no. 1, pp. 86–96, 2007.

[13] S. Ferrari, J. E. Steck, and R. Chandramohan, "Adaptive feedback control by constrained approximate dynamic programming," *IEEE Transactions on Systems, Man, and Cybernetics, Part B (Cybernetics)*, vol. 38, no. 4, pp. 982–987, 2008.

[14] W. Qiao, R. G. Harley, and G. K. Venayagamoorthy, "Co-ordinated reactive power control of a large wind farm and a statcom using heuristic dynamic programming," *IEEE Transactions on Energy Conversion*, vol. 24, no. 2, pp. 493–503, 2009.

[15] H. He, Z. Ni, and J. Fu, "A three-network architecture for on-line learning and optimization based on adaptive dynamic programming," *Neurocomputing*, vol. 78, no. 1, pp. 3–13, 2012.

[16] C. Mu, Z. Ni, C. Sun, and H. He, "Air-breathing hypersonic vehicle tracking control based on adaptive dynamic programming," *IEEE transactions on neural networks and learning systems*, vol. 28, no. 3, pp. 584–598, 2017.

[17] D. V. Prokhorov and D. C. Wunsch, "Adaptive critic designs," *IEEE transactions on Neural Networks*, vol. 8, no. 5, pp. 997–1007, 1997.

[18] F. L. Lewis and D. Vrabie, "Reinforcement learning and adaptive dynamic programming for feedback control," *IEEE circuits and systems magazine*, vol. 9, no. 3, 2009.

[19] F.-Y. Wang, H. Zhang, and D. Liu, "Adaptive dynamic programming: An introduction," *IEEE computational intelligence magazine*, vol. 4, no. 2, 2009.

[20] T. Chen, Z. Du, N. Sun, J. Wang, C. Wu, Y. Chen, and O. Temam, "Diannao: A small-footprint high-throughput accelerator for ubiquitous machine-learning," in *ACM Sigplan Notices*, vol. 49, no. 4. ACM, 2014, pp. 269–284.

[21] D. Larkin, A. Kinane, V. Muresan, and N. OConnor, "An efficient hardware architecture for a neural network activation function generator," in *International Symposium on Neural Networks*. Springer, 2006, pp. 1319–1327.

[22] Y. Chen, T. Luo, S. Liu, S. Zhang, L. He, J. Wang, L. Li, T. Chen, Z. Xu, N. Sun *et al.*, "Dadiannao: A machine-learning supercomputer," in *Proceedings of the 47th Annual IEEE/ACM International Symposium on Microarchitecture*. IEEE Computer Society, 2014, pp. 609–622.

[23] Y.-H. Chen, T. Krishna, J. S. Emer, and V. Sze, "Eyeriss: An energy-efficient reconfigurable accelerator for deep convolutional neural networks," *IEEE Journal of Solid-State Circuits*, vol. 52, no. 1, pp. 127–138, 2017.

[24] Y. Sokolov, R. Kozma, L. D. Werbos, and P. J. Werbos, "Complete stability analysis of a heuristic approximate dynamic programming control design," *Automatica*, vol. 59, pp. 9–18, 2015.

[25] Z. Ni, H. He, and J. Wen, "Adaptive learning in tracking control based on the dual critic network design," *IEEE transactions on neural networks and learning systems*, vol. 24, no. 6, pp. 913–928, 2013.

[26] Z. Ni, H. He, X. Zhong, and D. V. Prokhorov, "Model-free dual heuristic dynamic programming," *IEEE transactions on neural networks and learning systems*, vol. 26, no. 8, pp. 1834–1839, 2015.

Low Power Configurable Readout Integrated Circuit for Infrared Detector

Hari Shanker Gupta, A S Kiran Kumar
Space Applications Centre, ISRO
Ahmedabad, India
hari@sac.isro.gov.in,kiran@sac.isro.gov.in

Pranoy Datta, Maryam Shojaei Baghini, Dinesh K. Sharma
Dept. of Electrical Engineering
Indian Institute of Technology, Bombay
Mumbai, India
mshojaei@ee.iitb.ac.in,dinesh@ee.iitb.ac.in

Abstract—In general, Readout Integrated circuits (ROICs) are made of using standard silicon technology to collect signals from the infrared photo detectors arrays. Design of the ROIC is challenging because we have to ensure that the performance of the whole system is limited by the detector and not by the read out circuitry. Read while integration (snapshot) type ROIC's are required for the critical applications like space, astronomy and defense. For snapshot type ROIC's integration time needs to be at least as much as that required for reading the full array. This may be too long and may saturate the input circuits. In order to avoid saturation of ROIC output, Windowing, i.e. reading of a selected window of programmable size from the focal plan array is useful. The implementation of the windowing feature requires programmable control signals for the pixel selection. In this paper we have proposed low power reconfigurable ROIC circuit for dynamic windowing. The selected window can be as small as 32 × 48 to the full array of 320 × 256.

The design and post layout simulation results of ROIC test chip of 320×256 array of 30 μm × 30 μm pixel size in 180 nm, 3.3V compatible CMOS SCL process is discussed in this paper. Critical specifications are charge handling capacity of greater than 10 M$\bar{e}$ and low readout noise of less than 523 electrons, have been able to achieve at room temperature.

Keywords-ROIC, detectors, direct injection (DI), snapshot, Array, Semiconductor Laboratory (SCL), million electron (M$\bar{e}$), focal plane array (FPA).

I. INTRODUCTION

Infrared detectors are mainly of two types namely monolithic devices and hybrid devices [1]. Monolithic means pixel and associated circuitry are on the same chip. In hybrid, detectors are formed in a seperated material like GaAs and connected electrically through indium bump to the CMOS readout integrated circuit (ROIC)[1], [2]. Hybridization allows optimizing the detector and the read out circuit separately which is its advantage over the monolithic devices[3]. For hybrid sensor array detectors, read out circuit design is one of the critical module in determining the overall detector performance[4], [5]. Fig. 1 shows the basic block diagram of a ROIC. The ROIC circuit consists of unit cells for charge integration and charge to voltage conversion, timing and control circuits and pixel voltage multiplexers and amplification stages [6], [7], [8]. Fig. 1 shows an array of 320×256 pixels with pixel size of 30 μm×30 μm.

As the pixel pitch is very small, hence multiple control lines running through each pixels reduces the available

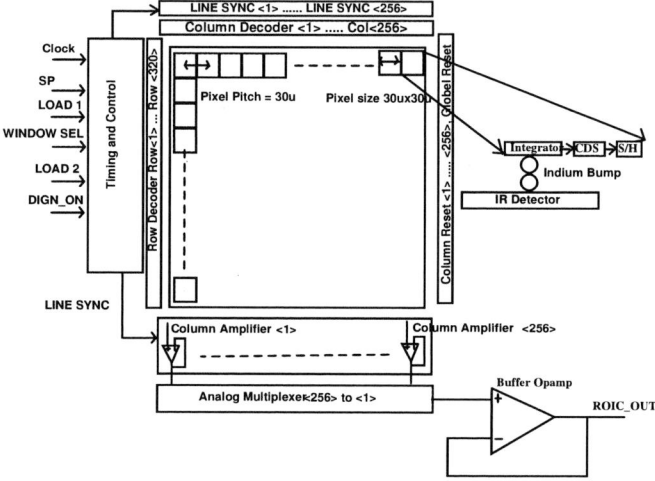

Figure 1.   Basic Block Diagram of a Readout Integrated Circuit

area for actual pixel circuitry. The importance of control circuit is to generate minimal control signals and provides maximum features to control ROIC performance. There are number of sequential events going on in each of the pixels and we need to synchronize all the events in each frame for all the pixels. High-performance infrared imaging systems are also required to operate at low temperatures (around 70K) to minimize noise[1].

Here in this paper, we mainly focus on designing the control circuit for the ROIC. This paper proposed a novel method for the power efficient timing and control circuit with an objective of optimized area and power. Major focus is to provide more programmable features for dynamic windowing, i.e. reading of a selected window of programmable size from the focal plan array. This is a feature which is useful for avoiding saturation of ROIC output, when integration and readout proceed simultaneously. The integration time has to be at least as much as that required for reading the full array. This may be too long and may saturate the input circuits. The optimization of the timing and the control circuit for reconfigurable timing sequence with minimum power and silicon area has been attempted.

978-1-5386-3693-0/18 $31.00 © 2018 IEEE

II. TYPICAL ROIC OPERATION

Typical ROIC operation requires a master clock at readout rate and a start pulse for the frame synchronization. Other control signals need to be generated internally for reading the ROIC array. Other signals like a frame synchronization (FSYNC) clock controls the start and stop of the integration process. The rising edge of the frame sync clock initiates a series of events on the ROIC. It includes:

1) Global sample and hold transfer clock for simultaneous readout (SH_TR),
2) Global reset for reset to all unit cell integration capacitors simultaneously(RST),
3) The unit cell integration capacitors are held in the reset state until the falling edge of the FSYNC clock pulse
4) The integration of photo-current starts with the falling edge of reset (RST) pulse.

The Integration continues until the end of the next global unit cell sample and hold clock pulse generated by the next rising edge of the FSYNC clock. The timing diagram provides the timing requirements for the read while integration mode of operation. An integration and a read cycles are shown in Fig. 2

In summary, the timing and control circuit plays a crucial role in ROIC design and performance optimization. For example, to control integration followed by the pixel by pixel reading of the ROIC array, we require the following control signals:

- Pixel Signals:
  - FSYNC : Frame synchronization : output pulse (falling edge indicate start of pixel array reading)
  - RST : Resetting the pixels
  - SH : Sample and hold transfer
  - Read_start : Internal synchronization
- Column controlling signals : COL$< 1 : 256 >$
- Row controlling signals : ROW$< 1 : 320 >$
- Line synchronizing signals: LSYNC$< 1 : 256 >$
- Column resetting signals : COL_RST$< 1 : 256 >$

III. PROPOSED ARCHITECTURE

ROIC operation requires a frame synchronization (FSYNC) clock to controls the start and stop of the integration process. As shown in Fig. 3 the rising edge of the frame sync clock initiates a series of events. It includes the RESET control switch resets the integration capacitor after the result of previous integration has been transferred to the hold capacitor. The sampled information is then serially transfered to column amplifiers and then to the output buffer stage.

Integration for the next frame is initiated concurrently just after the reset control switch goes OFF. Timing of the control signals has to be adjusted carefully, so that the operations controlled by these signals are completed properly. For example the sample control signal (SH_TR) should be high

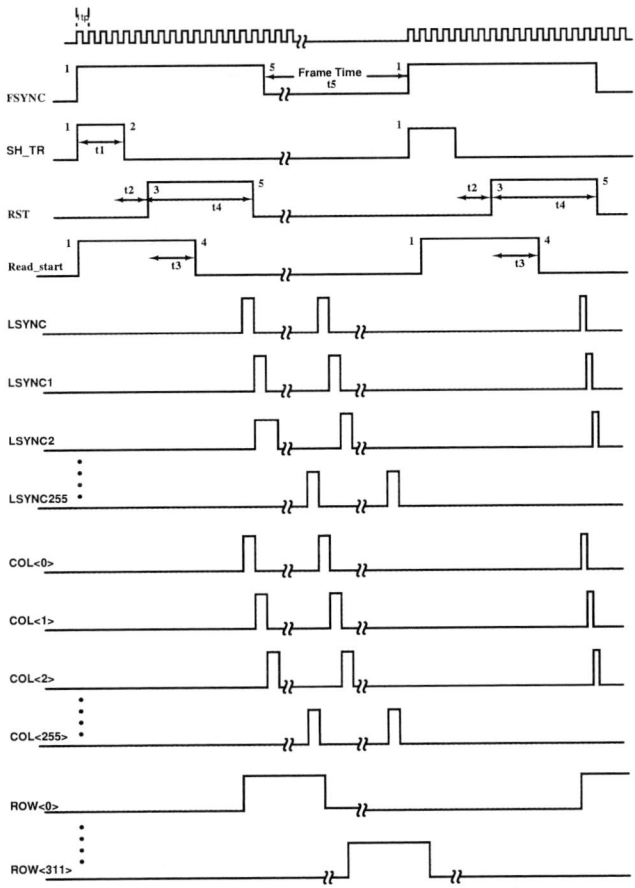

Figure 2. Timing Diagram

long enough to transfer the integrated pixel information to the hold capacitor. The reset control signal (RST) should become True only after the transfer is complete and SH_TR has been turned off. The sequence of control signals is shown in Fig. 4. The time taken to transfer charge or to reset a capacitor depends on the switch resistance and the time given for performing the action. Due to pixel area constraints, we can not make the analog switches very wide. Therefore, we have to make sure that the time given for each operation is adequate for maintaining the desired accuracy.

In order to reduce frame to frame interference, proper reset of each pixel's integrating capacitor is necessary. For a given frame rate, the reset time subtracts from the total time available for integrating the signal. This may not matter too much when the frame rate is low. However, for higher frame rates, and consequently shorter total frame durations, we may want to adjust the reset time such that it is long enough to give adequate resetting, but no longer. This makes it desirable to make the reset time programmable.

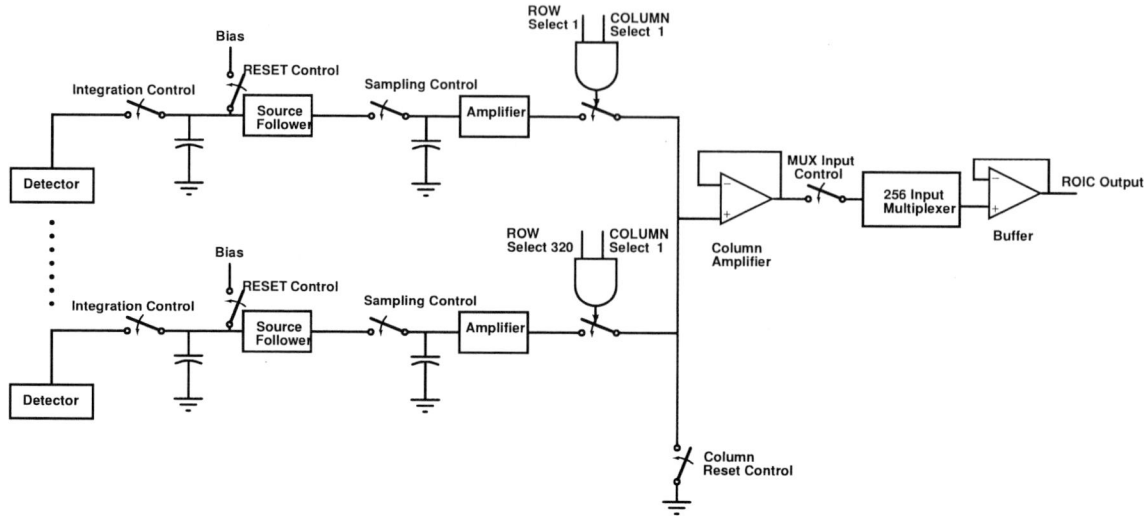

Figure 3. Control Signals in the ROIC

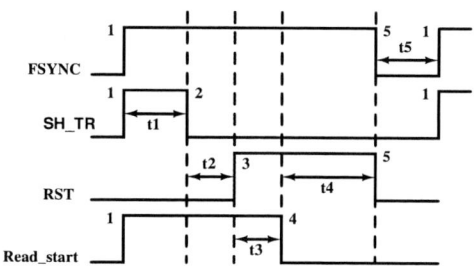

Figure 4. Pixel Control Signals

Similarly sampling control switch determines the transfer accuracy of integrated signal to the hold capacitor. This switch is controlled by SH_TR signal. The sampled control switch is also implemented with minimum size due to the pixel area constraints. It is desirable to make the duration of the sampling time programmable to ensure that the sample and hold capacitor is charged to the desired accuracy.

Process variations may also vary the ON resistance of analog switches. In view of these considerations, it is desirable that the duration of various control signals be made programmable.

We have implemented programmable durations for these control signals to achieve better settling accuracy. programmable parameters are loaded serially into a shift register and then transferred to the relevant counter/timer when needed. These parameters can also be changed between frames and will be effective from the next frame onwards. Default parameters are assumed if no parameters are loaded into the chip.

For an array of 320×256, the control signals are periodic

with a period of 320×256 = 81920 cycles. One way of implementing programmability would be to use a 17 bit counter to count these 81920 cycles and decode specific counts to turn on or off various control signals. However, we have adopted a different scheme to implement programmability with simpler hardware. Since we need to control the duration of different actions sequentially, we have used a common 8 bit down counter to control the interval between successive events. The simplest architecture to implement variable interval timing is through a down counter in which we load the count value and count down to zero. As shown in Fig. 5, 5 interval values are loaded in an 8 bit wide 5 stage shift register. An 8 bit counter effectively acts as the sixth stage of this shift register. Interval values are successively loaded into the 8 bit counter and counted down to generate the desired programmable intervals.

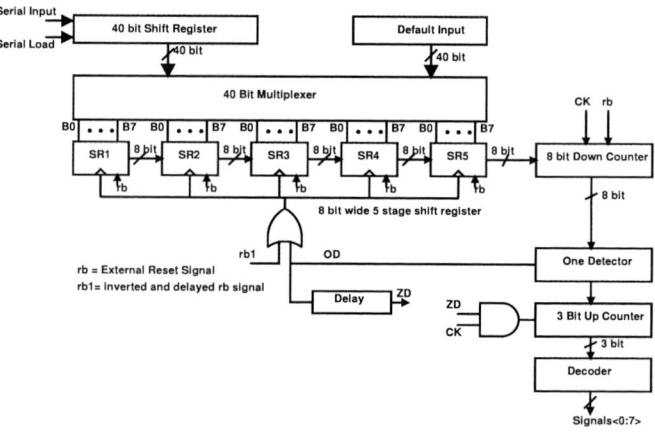

Figure 5. Proposed Architecture for Programmable intervals

Stages of the 8 bit wide shift register contain the interval durations t_5 to t_1 as shown in Fig. 4. To load this information from an external source, we have used a 40 bit shift register, where the user can serially load the input through Serial_input1 and Serial_Load1. The eight bit wide five stage shift register is loaded in parallel through the 40 bit multiplexer, which selects either the user provided values from the 40 bit shift register or the default values, depending on the status of the external Serial Load signal as follows.

- If external programming data is to be used then it is shifted in using serial input as data input and serial load as a serial load enable. The data is shifted in the 40 bit shift register using the master clock.
- If serial load is high, the 40 bit multiplexer presents the shifted in data at its output. Otherwise its output contains the default values.
- The Reset signal rb(active low) clears the counters. A signal rb1 is generated, which is the inverted reset signal with a delay of 1 clock period. The output of the 40 bit multiplexer parallel loads the 8×5 SR stages at the falling edge of the rb1 signal
- The down counter counts with the master clock. We detect the occurrence of count = 1 instead of 0 in the down counter, to compensate for the clock cycle need to shift the next interval value through the 8bit wide 5 stage shift register. When the count reaches one, the 'One Detect (OD)' signal become TRUE. It advances the 8×5 SR at the next clock, thus loading the next interval value into the counter when it reaches zero.
- The 'One Detect (OD)' signal delayed by a clock cycle is called The 'Zero Detect (ZD)' signal. It advances the 3 bit counter which is then decoded to generate various control signals.

A. Generation of Column and Row Control Signals

The master clock is used to read out ROIC pixels successively at the clock rate (fs). The time required to read a single pixel is 1tp $\equiv$ 1/fs and the total time need to read the entire frame is $320\times256\times$tp. There are 256 column control signals, COL< 0 : 255 > and 320 row control signals ROW< 0 : 255 >. A pixel is selected when its column select as well as row select signals become TRUE. The timing diagram for these signals is shown in Fig. 2. The basic design for 320×256 ROIC as shown in Fig. 6.

Instead of using a counter/decoder, the column and row control lines can be generated using shift registers. This is more efficient for the larger array, since large decoders consume power and area, and are slow. We use a 256 bit shift register with a single moving '1' to generate the column select signals. Similarly, a 320 bit shift register with a single moving '1' is used for generating row select signals. When the '1' bit arrives at the end of 256 bit columns select shift

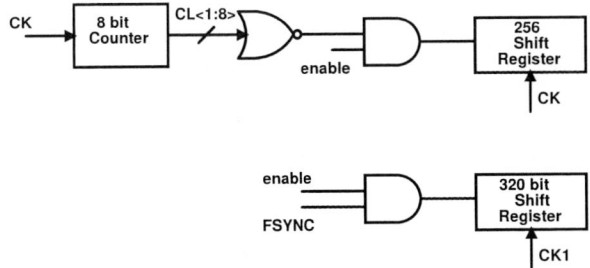

Figure 6. Power optimized architecture for column and row control signals

register, the 320 bit row select shift register with a moving '1' is advanced. Simultaneously, a fresh '1' is inserted in the column select shift register to start scanning the next row. Thus the column select signals move at the master clock frequency whereas the row select signals are clocked at a rate which is the master clock frequency divided by 256.

The insertion of a fresh '1' in the column select shift register is managed through an 8 bit down counter clocked by the master clock. When it reaches zero and if the enable signal is high, a fresh '1' is inserted into the 256 bit column select shift register. (Enable is LOW only during the reset period, after the last row has been reached and during inactive times, otherwise it is HIGH).

The enable line is derived from the master clock and is set high at the negative edge of the Read_start signal and remains high for 320 × 256 clock periods. The design for the generation of column and row select signals is shown in Fig. 6. For windowing design requires that the start and end column or row numbers can be programmed by the user. To keep the design simple, we allow the user to select the start and end row and column number which are multiples of 16. This minimizes the overhead for windowing.

The row control lines are also generated using shift registers, with inputs and outputs accessible at every sixteenth position. Thus there are 20 blocks of 16 rows and we need a 1:20 demultiplexer and a 20:1 multiplexer. The col_done line is used as a clock to advance the '1' in the row select shift register. Since column end signal is not available to start the very first row, we use the negative edge of the Read_start line to insert a '1' for the first row. In order to select a window size, a 5 bit wide staring address is required to select one out of 20 blocks of 16 rows each. Hence ten bit are required to address the start and end positions of rows in a window. Thus, a total of 18 bits is required to specify the window out of which 8 bits specify the start and end position of columns 10 bits for the start and end addresses of rows.

IV. OTHER CONTROL SIGNALS GENERATION

A control signal is required for selecting the output of a particular column amplifier 16 clock cycles after the pixel was selected. This reduces the settling time requirement of column amplifiers as described earlier. These delayed column select lines are called LSYNC< 0 : 255 >. Also, for resetting the column amplifier inputs after the data has been read, 256 control lines called col_rst< 0 : 255 > are required. Both these sets of signals will repeat like the corresponding column control lines.

The simplistic solution of putting a 16 stage delay line after every column is obviously not practical as it will require too much hardware (16 bit delay stages for each of the 256 columns) and will consume too much area and power. We can consider the 8 bit column number to be composed of two 4 bit numbers. The more significant nibble represents the group number, while the second one identifies the specific column within this group. We use two sixteen stage ring counters: One of these is clocked by the master clock divided by 16 while the other is clocked by the master clock itself. We number the stages of these ring counters from 0 to 15. The first counter is initialized with a '1' in the position corresponding to the group number of the start column of the window. (This is the 4 bit value fed as the start column address for programming the window). Thus, if the start address is 64, the group address is 4 and so the stage numbered 4 of the first 16 stage ring counter will be set to '1'. This '1' will advance every 16 cycles to the next position. When this reaches the group corresponding to the end column of the window, a '1' is re-introduced at the position corresponding to the starting group. These delayed signals are called LR< 0 : 15 >.

The position within the group is counted by the second ring counter, which is clocked by the master clock. It is always initialized with a '1' at the start position (numbered 0) and cycles through 16 positions (0 to 15). The location of '1' in this ring counter represents the column number within a given group. These sixteen stages are labeled as LC< 0 : 15 >.

Area and power optimization concept of LSYNC signal is generated through the AND gates whose inputs is LR lines and LC lines.

V. TIMING AND CONTROL CIRCUIT SIMULATION RESULTS

The serial load lines are deactivated by connecting it to '0'. Following post layout simulation has been performed in Cadence spectre tools. The 18 bit fixed input for the Window selection is '010001110111001111'used for simulation. Hence the window is from COL< 65 > to COL< 128 > (corresponding to inputs 0100 and 0111 i.e. 4 and 7) and ROW< 225 > to ROW< 256 > (corresponding to inputs

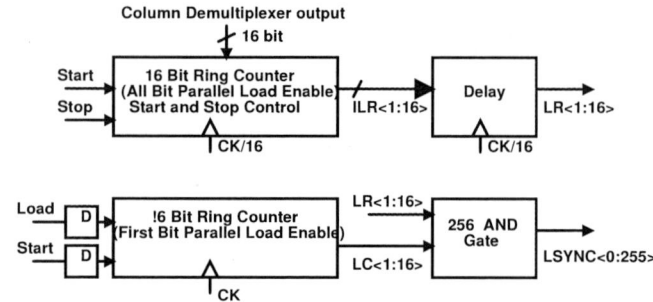

Figure 7.   Optimized Implementation for LSYNC Signal Generation

01110 and 01111 i.e. 14 and 15). The Spectre simulation results is shown in Fig. 8.

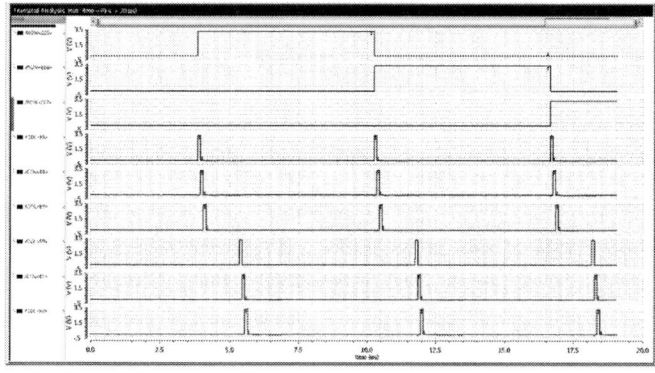

Figure 8.   Repeating Column Signals with Row Signals

VI. 320×256 READOUT INTEGRATED CIRCUIT

A Direct Injection (DI) circuit topology with CDS unit cell is implemented for our specifications. Selected topology [9] has been re-designed with SCL process as shown in Fig. 9.

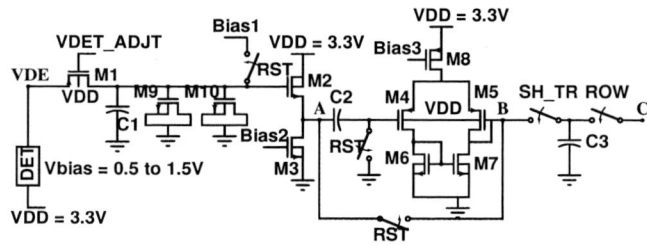

Figure 9.   Implemented Direct Injection (DI) architecture [9]

Integrated full chip view of 320×256 ROIC is shown in Fig. 10. The ROIC pixel is highlighted in large chip area of 9.5mm×10.5mm. Post layout simulation shows that this large area array provides a charge handling capacity of $10M\overline{e}$, readout noise of < 523 electrons and low power of <

150 mW at room temperature. The ROIC buffer can drive off-chip load of 25 pF. The buffer power is also included in the ROIC power. The expected performance is better at liquid Nitrogen temperature with charge handling capacity of 10M$\bar{e}$, readout noise of < 397 electrons and low power of < 128 mW.

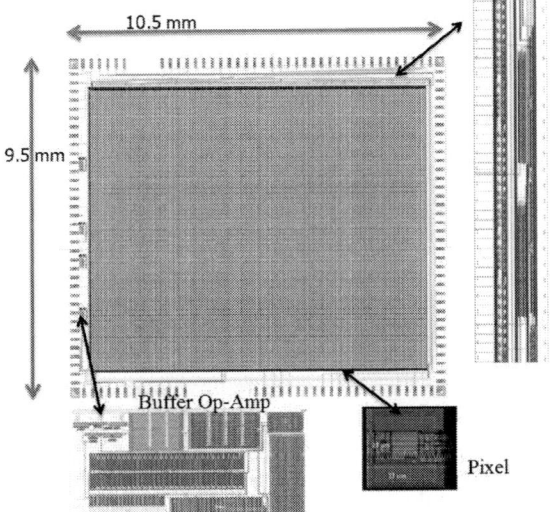

Figure 10. Integrated ROIC Chip for 320×256 array size

Table IV
ROIC Performance Comparison

ROIC Parameters	This Work	[10]	[11]	[7]	[12]
Array size	320×256	8×10	64×64	1024×1024	640×512
Pixel size (μm^2)	30×30	30×30	50×50	30×30	15×15
Input cell	DI	CTIA	CMI	CTIA	CTIA
Input Current (Nominal Range)	1 pA to 10nA	10 pA to 10nA	1 nA to 50nA	1 nA	NA
Charge handling capacity (M$\bar{e}$)	10	2	280	NA	0.5
Frame rate	0.1 to 5 KHz	110 Hz	110 Hz	NA	500Hz
Pixel rate	5 MHz	3 MHz	2 MHz	10 MHz	33 MHz
Integration type	snapshot	snapshot	NA	NA	NA
Power/pixel	900 nW	2.2 μW	NA	NA	NA
Readout noise @300K ($\bar{e}$)	523	6373	3mW	NA	NA
Temperature	77 to 310K	77 to 300K	300K	300K	300K
Linearity (%)	99.9	99	NA	99	99
Windowing	Smallest window size 32×48	-	-	NA	NA

VII. Conclusion

The novel re-configurable architecture of ROIC circuit is proposed and post-layout simulation results have been presented in this paper. The minimum window of size of 32 × 48 to full array selection was achieved through serial interface. This design is currently under fabrication.

Acknowledgment

The authors acknowledge with thanks the support extended by Shri Tapan Mishra, Director, Space Applications Centre, Ahmedabad, India. Authors would like to thanks Shri S S Sarkar DD SEDA, Shri Arup Roy Chowdhury, GD SEG and Shri Sanjeev Mehta Head SFED for valuable suggestions. We would like to thank VLSI Design Group, IIT-Bombay for insightful discussions.

References

[1] J.L. Vampola, "Readout electronics for infrared sensors," in *Infrared and Electro-Optical Systems Handbook, edited by W.D. Rogatto, Billingham, Washington*, 1993, p. Ch. 3.

[2] Francisco Serra-Graells et al, "Low-power and compact cmos aps circuits for hybrid cryogenic infrared fast imaging," *IEEE Transaction on Circuits and SystemsII Express Briefs*, vol. 54, no. 12, DEC. 2007.

[3] Antonio Rogalski, *Infrared Detectors*, Actuators A: Phys, second edition, 2010.

[4] A Rogalski P Martyniuk, "Quantum-dot infrared photodetectors: Status and outlook," *Progress in Quantum Electronics*, vol. 32, pp. 89120, 2008.

[5] D Das et al, "Optimization of the number of stacks in the submonolayer quantum dot heterostructure for infrared photodetectors," *IEEE Transactions on Nanotechnology*, vol. 99, pp. 1–1, 2015.

[6] James et al Beletic, "Teledyne imaging sensors: infrared imaging technologies for astronomy and civil space," .

[7] L. Xie et al., "Readout circuit with novel background suppression for long wavelength infrared focal plane arrays," *Int. J. Electron*, vol. 98(2), pp. 207–222, 2011.

[8] H. S. Gupta et al, "Design of large dynamic range, low-power, high-precision roic for quantum dot infrared photo-detector," *Electronics Letters*, vol. 49, no. 16, pp. 1018–1020, 2013.

[9] H. S. Gupta et al, "Design of high-precision roic for quantum dot infrared photodetector," *IEEE Photonics Technology Letters*, vol. 28, no. 15, pp. 1673–1676, Aug 2016.

[10] Yi-Chuan; Shieh Hsiu-Li Sun, Tai-Ping; Lu, "A novel readout integrated circuit with a dual-mode design for single and dual-band infrared focal plane array," *Infrared Physics Technology*, vol. 60, September. 2013.

[11] HalukKulah and Tayfun Akin, "A current mirroring integration based readout circuit for high performance infrared fpa applications," *IEEE Transaction on Circuits and SystemsII: Analog and digital Signal Processing*, vol. 50, no. 4, April. 2003.

[12] Genki Mizuno et al, "High performance digital read out integrated circuit (DROIC) for infrared imaging," 2016.

2018 31th International Conference on VLSI Design and 2018 17th International Conference on Embedded Systems

Feedback Biasing based Adjustable Gain Ultrasound Preamplifier for CMUTs in 45nm CMOS

Linga Reddy Cenkeramaddi

Intelligent Signal Processing and Wireless Networks (WISENET) Lab.
Department of Information and Communication Technology
University of Agder, 4879, Grimstad, Norway

Abstract— **As CMOS technology is scaled down, supply voltages are decreasing and intrinsic gain of the nanoscale CMOS transistors is dropping while the threshold voltages of transistors are remaining relatively constant. In such scaled down nanoscale CMOS technologies, conventional vertical stacking architectures (for example. cascode architectures) for high-gain becomes no more attractive. In this paper we present the analysis and design of a feedback biasing based adjustable gain ultrasound preamplifier which is capable of amplifying signals from 15 MHz to 45 MHz from Capacitive Micromachined Ultrasound Transducers (CMUTs) in 45nm CMOS technology for medical ultrasound imaging applications. From the simulations, the proposed preamplifier achieves a voltage gain 27.47 dB at 30 MHz from the CMUT signal source to the output and this transfer gain can easily be varied from 27.47 dB to 3.21 dB by varying the control voltage without altering the power consumption. It exhibits an output noise power spectral density of 4.8 μV/SQRT(Hz) at frequency of 30 MHz at maximum transfer gain. It consumes only 107 μA from a 900 mV power supply excluding bias currents.**

Keywords—Ultrasound preamplifier; CMUT; Adjustable gain amplifier; CMUT-CMOS.

I. Introduction

In medical ultrasound imaging, Capacitive Micromachined Ultrasonic Transducers (CMUTs) have recently emerged as an alternative to conventional piezoelectric transducers [1–4]. There are several advantages of using CMUTs instead of conventional piezoelectric transducers, such as increased bandwidth and suitability for high frequency applications. CMUTs also offer a great advantage of integration of CMOS front-end electronics [5-7] with them compared to the conventional piezoelectric transducers. Such integration of CMUTs with CMOS front-end electronics will greatly enhance the overall performance of the ultrasound systems at reduced cost.

For the intended 1mm diameter catheter based medical ultrasound imaging system [8], CMOS front-end amplifiers requirements are small area and ultra-low power consumption as these front-end amplifiers will be directly integrated with CMUTs to enhance the system performance. Gain adjustability in the preamplifiers will improve system performance as the gain of the preamplifiers can be adjusted according to the transducers signal levels. To achieve the

ultra-low power and small area requirements for the preamplifiers, we have chosen the 45nm CMOS technology with a 900mV power supply. Required gain from the CMUT signal source to output of the front-end amplifier is also very high (attenuation of signal current with respect to signal source voltage is on the order of -60 dB at 30 MHz center frequency). Conventional cascoding architectures could not be utilized because of low power supply voltage and with relatively high threshold voltage of the transistors in 45nm CMOS. A simple feedback biasing based adjustable gain preamplifier is proposed and is designed based on a simple common-source amplifier with resistive feedback. Feedback is utilized for biasing without affecting the frequency response so that the need for the additional biasing circuits is eliminated.

This paper is organized into five sections. In section II, analysis and design of the feedback biasing based adjustable gain ultrasound preamplifier is presented. In section III, Capacitive Micromachined Ultrasound Transducer, CMUT is described in brief. In section IV, performance of the preamplifier is discussed in detail when a single CMUT element is connected to it and compared the results with the published [9, 14-15] results in 90nm and 0.35μm CMOS technology for similar applications at similar frequency ranges. Finally, conclusion in section V.

II. Proposed Feedback Biasing Based Adjustable Gain Preamplifier

The proposed feedback biasing based adjustable gain ultrasound preamplifier is shown in Fig. 1. There are three stages in this preamplifier. First stage is formed by m1, m2 and m7. This stage is a simple common source amplifier with active source load with feedback resistor (m7). By varying the voltage on the gate of the transistor m7, VGC, gain can be varied. Also, this feedback resistor biases the transistor m1.

Feedback biasing is chosen here as it has been proven [9] that deep submicron technologies favor it in the designing of compact amplifiers. Second stage is formed by m3 and m4. This stage is DC coupled to the first stage. Transistors m5, m6 and m8 form the third stage. Third stage is same as that of the first stage. To minimize the area, minimum channel length and aspect ratios are used for all transistors by utilizing the small feature sizes of the 45nm CMOS technology.

978-1-5386-3693-0/18 $31.00 © 2018 IEEE

204

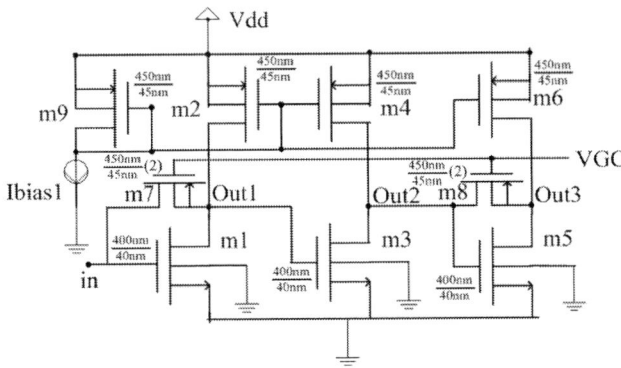

Fig. 1. Proposed feedback biasing based adjustable gain preamplifier.

Feedback resistors simply biases the transistors (m1 and m5 here), influence gain and will not have any influence on frequency response (until unity gain frequency) [9]. This means amplifiers operate in open loop [9]. The overall gain of this amplifier is the product of gains of the three stages. It is the gain which is the product of three simple common source amplifiers. This gain is given by:

$$(g_{m1}*z_{out1}) \times (g_{m2}*z_{out2}) \times (g_{m3}*z_{out3})$$

Where z_{out1}, z_{out2} and z_{out3} are impedances at the nodes Out1, Out2 and Out3 respectively.

III. CMUT

In this section, the Capacitive Micromachined Ultrasound Transducer (CMUT) is briefly described. A CMUT is an ultrasound transducer like the conventional piezoelectric transducer and it converts ultrasound signals into electrical signals and vice versa.

Fig. 2 shows the small signal electrical equivalent circuit of a single CMUT element.

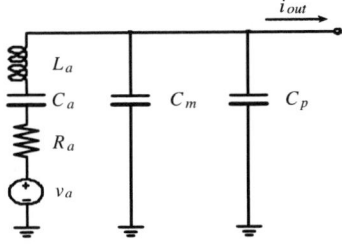

. Fig. 2. CMUT equivalent circuit.

The detailed description of the parameters shown in the above model can be found in the references [10-13]. Va is the voltage that represents an incoming acoustic wave, in other words it is the voltage generated by CMUT when it is hit by an ultrasound signal. For the noise analysis, Va is interpreted as the noise source of Ra. First round of fabricated CMUTs is shown in Fig. 3 [10]. It shows the SEM of a close view of one CMUT element with metal contacts and interconnections.

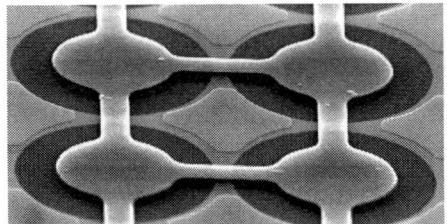

Fig. 3. SEM of a close view of one CMUT element.

In each CMUT element, there are four CMUTs. Two sets of CMUTs were fabricated. One with a membrane radius of 5.7 μm with a pitch of 12.5 μm, the second with a membrane radius of 11.4 μm and a pitch of 25 μm.

IV. DISCUSSION OF SIMULATION RESULTS

In this section, performance of the feedback biasing based adjustable gain ultrasound preamplifier is discussed in detail when it is connected to a single CMUT element. Performance parameters of the proposed amplifier are also compared with the amplifiers [9, 14-15] designed for similar applications at similar frequency range.

CMUT elements for the intended application [8] are designed with a center frequency of 30MHZ and bandwidth of 15MHz to 45MHz. Typical values for the electrical model parameters (see Fig. 2) of a single CMUT element are tabulated in table 1.

TABLE 1. CMUT MODEL PARAMETERS

Parameter	Set #1	Set #2	Units
L_a	0.29	0.937	mH
C_a	140	30	fF
R_a	106	159	kΩ
C_m	60	30	fF
C_p	30	25	fF

For the evaluation of the preamplifier, CMUT electrical equivalent circuit shown in Fig. 2 is used as the transducer with the model parameters listed in Table. 1 and connected it to the preamplifier as shown in Fig. 4.

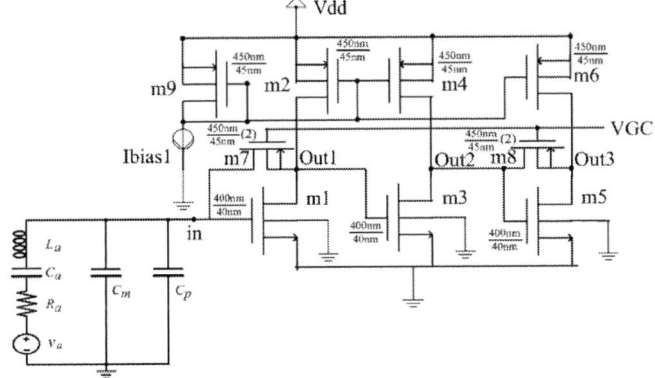

Fig. 4. CMUT connected to the proposed preamplifier.

978-1-5386-3693-0/18 $31.00 © 2018 IEEE

From the above figure, it can be observed that there is a dominant pole located at the node "in". If the control voltage (VGC), on the gate of the transistor, m7, is varied then the resistance of it changes. This causes gain and bandwidth to vary for the first stage formed by the transistors, m1, m2 and m7. The same principle applies to the third stage (formed by the transistors, m5, m6 and m8) but the first stage has the dominant pole.

Fig. 5 shows the transfer gain from the CMUT signal source, Va, shown in Fig. 4 to the output (Out3) of the preamplifier for various gain control voltages. It can be observed from this figure that gain varies from 27.47 dB to 3.21 dB at 30 MHz as the control voltage, VGC, has been varied from 0.5 V to 0.675 V. The flexibility of gain adjustability in the preamplifier stage will give more freedom to choose gain according to the impedance of the transducers.

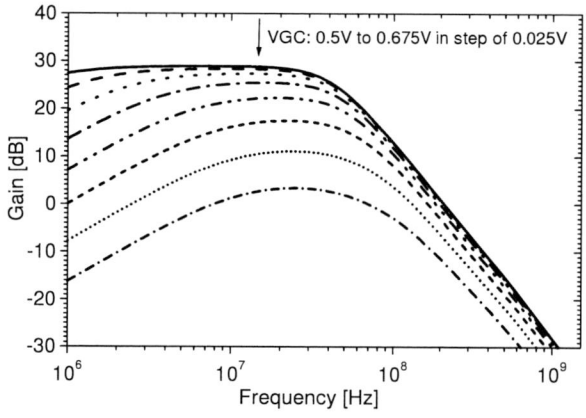

Fig. 5. Gain versus input signal frequency.

Regarding noise, noise power spectral density at the output (Out3) at 30 MHz is plotted in Fig. 6 by varying the gain control voltage, VGC. As shown in Fig. 6, noise power spectral density decreases as the gain control voltage, VGC, is increased. This is quite intuitive as the gain control voltage increases the feedback resistance of m7 and m8 decreases. It can also be observed from Fig. 6 that the proposed preamplifier exhibits an output noise power spectral density of 4.8 µV/SQRT(Hz) at maximum gain of 27.47 dB at 30 MHz and 1.53 µV/SQRT(Hz) at a gain of 3.21 dB at 30 MHz. It consumes only 107 µA from a 900mV power supply.

To compare the proposed preamplifier with the reported similar amplifiers [9, 14-15] for similar applications, main performance parameters are summarized in Table 2.

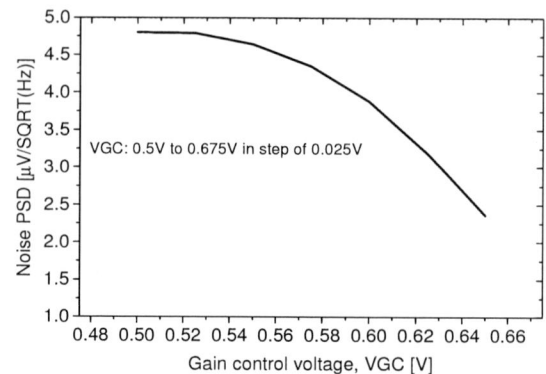

Fig. 6. Output noise PSD versus gain control voltage, VGC.

Table 2. COMPARISON OF THE MAIN PERFORMANCE PARAMETERS

Parameter	Proposed preamplifier	Ref. [9]	Ref. [14-15]
Technology	45nm CMOS	90nm CMOS	0.35µm CMOS
Supply voltage	900 mV	1 V	not mentioned clearly but this 0.35 µm CMOS has nominal supply voltage of 3.3 V
Gain at 30 MHz input signal frequency	27.47 dB	27.50 dB	10 dB
Output noise PSD at 30 MHz	4.8 (µV)/(Hz)$^{1/2}$	0.33 (µV)/(Hz)$^{1/2}$	-
Power consumption	96.3 µW	125 µW	35 mW
Gain adjustability	27.47 to 3.21 (dB)	-	-

From table2, it can be observed that the proposed ultrasound preamplifier has the gain adjustability from 27.47 dB to 3.21 dB and it draws only 107 μA from a 900-mV power supply. Compared to the designs presented in [14-15], the proposed preamplifier has better performance in terms of gain and power consumption. When it is compared to the reported design [9], the proposed preamplifier has the same gain but at reduced power consumption. Also the proposed preamplifier occupies small area as it is designed by utilizing the smaller feature sizes of the 45nm CMOS technology and feedback biasing compared to the designs reported in [9, 14-15].

V. CONCLUSION

From the designed proposed feedback biasing based ultrasound preamplifier, we conclude that it achieves a voltage gain of 27.47 dB at 30 MHz input signal frequency. Gain of the proposed preamplifier shown to be varied from 27.47 dB to 3.21 dB by varying the control voltage, VGC, from 0.5 V to 0.675 V. It draws only 107 µA from a 900 mV power supply. The proposed preamplifier shows an output noise power spectral density of 4.8 (μV)/SQRT(Hz) at center-frequency of 30 MHz. By using feedback biasing, necessity of additional biasing circuits is eliminated which will subsequently reduce the area. It has also been shown that the proposed adjustable gain ultrasound preamplifier exhibits better performance in terms of gain and power consumption compared to the reported designs for similar applications at similar frequency ranges. Such a high performance low power, low area adjustable gain ultrasound preamplifier will be suited best as an analog front-end amplifier to the IVUS medical imaging systems.

ACKNOWLEDGMENT

The authors would like to thank the Norwegian Research Council through the project Smart Microsystems for Diagnostic Imaging in Medicine (159559/130) for the financial help.

REFERENCES

[1] M.I. Haller, B.T. Khuri-Yakub, "A surface micromachined electrostatic ultrasonic air transducer," *Proc. IEEE Ultrason. Symp.* pp. 1241–1244, 1994.

[2] O. Oralkan, A.S. Ergun, J.A. Johnson, M. Karaman, U. Demirci, K. Kaviani, T.H. Lee, "Capacitive micromachined ultrasonic transducers: next-generation arrays for acoustic imaging?," *IEEE Trans. Ultrason Ferroelectr. Freq. Control* 49 (11), pp. 1596–1610, 2002.

[3] D.M. Mills, L.S. Smith, "Real-time in-vivo imaging with capacitive micromachined ultrasound transducer (cMUT) linear arrays," *Proceedings IEEE Ultrasonics Symposium,* pp. 568–571, 2003.

[4] S. Panda, C. Daft, P. Wagner, I. Ladabaum, P. Pellegretti, F. Bertora, "Capacitive microfabricated ultrasonic transducer (CMUT) probes: imaging advantages over PZT probes," *3rd International Workshop on Micromachined Ultrasonic Transducers*, Lausanne, Switzerland, June 26–27, 2003.

[5] Wygant, I. O., et al., "Integrated Ultrasonic Imaging Systems Based on CMUT Arrays: Recent Progress," *IEEE Ultasonics Symposium*, pp. 391-394, 2004.

[6] Noble, R. A., et al., "A Cost-effective and Manufacturable Route to the Fabrication of High-Density 2D Micromachined Ultrasonic Transducer Arrays and (CMOS) Signal Conditioning Electronics on the same Silicon Substrate," *IEEE Ultasonics Symposium*, pp. 941-945, 2001.

[7] Noble, R. A., et al., "Low-Temperature Micromachined cMUTs with Fully-Intrgrated Analogue Front-End Electronics," *IEEE Ultasonics Symposium*, pp. 1045-1050, 2002.

[8] http://www.iet.ntnu.no/projects/smida/

[9] Tageshwar Singh, Trond Sæther, Trond Ytterdal, "Feedback Biasing in Nanoscale CMOS Technologies," *IEEE Transactions on Circuits and Systems-II*: Express Briefs, Vol. 56, No. 5, pp. 349-353, May 2009.

[10] Kjersti Midtbø and Arne Rønnekleiv, "Fabrication and Characterization of CMUTs realized by wafer bonding," *IEEE Ultrasonics Symposium*, pp. 934-937, 2006.

[11] Linga Reddy Cenkeramaddi and Trond Ytterdal, "*Analysis and Design of a 1V Charge* Sampling Readout Amplifier in 90nm CMOS for Medical Imaging," *VLSI-DAT*, pp. 1-4, 2007.

[12] Linga Reddy Cenkeramaddi, Tajeshwar Singh and Trond Ytterdal, "Self-biased charge sampling amplifier in 90nm CMOS for medical imaging," *GLS-VLSI*, pp. 168-171, 2007.

[13] Linga Reddy Cenkeramaddi and Trond Ytterdal, "Clock Jitter impact on the Performance of General Charge Sampling Amplifiers," *Analog Integrated Circuits and Signal Processing*, DOI: 10.1007/s10470-009-9367-x, 2009.

[14] J. Fan, J. Talman, A. Fleischman, and S.L. Garverick, "Integrated Amplifier with Active Limiter for Intravascular Ultrasound Imaging," *IASTED International Conference on Circuits, Signals, and Systems*, Clearwater, FL, USA, 2004.

[15] Aaron Fleischman, Chaitanya Chandrana, Jin Fan, Jim Talman, Steve Garverick, Geoff Lockwood, and Shuvo Roy, "Components for Focused Integrated pMUTs for High-Resolution Medical Imaging", *IEEE Ultrasonics Symposium*, pp. 787-791, 2005.

2018 31th International Conference on VLSI Design and 2018 17th International Conference on Embedded Systems

A High efficiency body injected differential power amplifier at 2.4GHz for low power applications

M.Naga Sasikanth[1], Tarun Kanti Bhattacharyya[2]

Department of Electronics and Electrical Communication engineering

IIT Kharagpur

Kharagpur, West Bengal-721302, India

[1]kanthsasi66@gmail.com, [2]tkb@ece.iitkgp.ernet.in

Abstract—**In this paper, we propose a novel topology for an injection locked differential power amplifier. This power amplifier is targeted for low power wireless applications such as ZigBee at 2.4GHz band. This topology involving body injection locking for higher efficiency is designed in TSMC 65nm RF-CMOS process and it achieves a Power added efficiency of 61.8% and a Drain efficiency of 63.6% with an output power level of 12.3dBm at a supply voltage of 0.5V. General theory of injection locking has been discussed before presenting our topology. Reasons have also been presented to justify the adoption of body injection locking to improve the efficiency.**

I. INTRODUCTION

Two of the most significant issues in modern day low power communication systems such as wireless sensor nodes are efficiency and linearity. Power amplifier being the most power hungry block in a wireless transmitter system is responsible for undesirable power loss in the system, hence there is a need to design high efficiency power amplifier. The efficiency of a power amplifier is generally expressed in terms of PAE (power added efficiency) or DE (drain efficiency) as shown below.

$$PAE = \frac{P_{out} - P_{in}}{P_{dc}}$$
$$DE = \frac{P_{out}}{P_{dc}}$$

Power amplifiers can be divided into many classes such as A,B,AB,D,E,F etc. While linear power amplifiers such as class A produce very less distortion(High linearity), they suffer in Efficiency. Hence switching kind of non-linear power amplifiers such as D,E,F etc have been used to achieve high efficiency. But traditional switching power amplifiers suffer in linearity owing to their inherent non-linearity. Hence differential topologies have been employed to reduce the even order harmonic hence resulting lower THD(Total Harmonic Distortion). As the required output power of the power amplifier reduces, the driver power loss becomes significant. Consequently it becomes even more challenging to achieve higher efficiencies for low output power. In quest for achieving higher efficiencies few new topologies have been proposed involving class E operation[7],[11],[13-14]. But as mentioned priorly, single ended switching topologies suffer in linearity. The topologies that have been presented involving class E operation mainly focus on high power whose efficiency would degrade at low output power levels. Other works include Injection

locking technique[15],[6] which demands much lesser driving requirement hence lowering driver loss, thereby obtaining a better efficiency. Injection locking technique has been employed in both single ended[6] and differential topologies[4-5],[12],[15]. In [12] Drain injection locking technique with CMOS cross coupled pair has been employed. Regular differential injection locking techniques[4-5],[12] involve drain injection techniques, which need switching of transistors and hence require a higher driving voltage requirement. Bulk drive technique has been used previously in [11] to achieve better efficiency by lowering the effective on resistance of transistor. In this paper we present a new topology named body injected power amplifier that achieves a higher PAE when compared to existing power amplifiers for low power applications. This is done by exploiting the body drive technique instead of drain driving in a conventional drain injection locked power amplifier to reduce the driver loss. As body driving requires lower driving voltage, it helps in improving the efficiency. The rest of the paper is organized as follows, In section-II class E power amplifier's operation is discussed, In section-III injection locking technique has been explained, section-IV describes the design and implementations of Body injection power amplifier, section-V presents the results obtained together with a comparative study with existing topologies, sections-VI provides conclusion.

II. CLASS E POWER AMPLIFIER

A conventional class E power amplifier is as shown in fig.1. Here the transistor acts as a switch. When switch is off, the inductor pumps current into the capacitor, load. And while the switch is on, current from inductor and load flows through the transistor. Now if the switch and inductors are ideal only possible power loss can happen at the time of switching. The capacitor and inductor are to be chosen such that they satisfy ZVS (zero voltage switching) i.e. capacitor's voltage should be zero at the time switch is turned on and ZSVS (zero slope voltage switching) the current through the capacitor should be zero at the time of switching[1]. Hence the current and voltage across the switch will never be non-zero simultaneously. Consequently, there would not be any loss in the switch(ideal case). Based on these two conditions the components can be found from [2] as

$$L = \frac{0.732 * R_L}{\omega}, C = \frac{1.45}{\omega R_L}.$$

978-1-5386-3693-0/18 $31.00 © 2018 IEEE

208

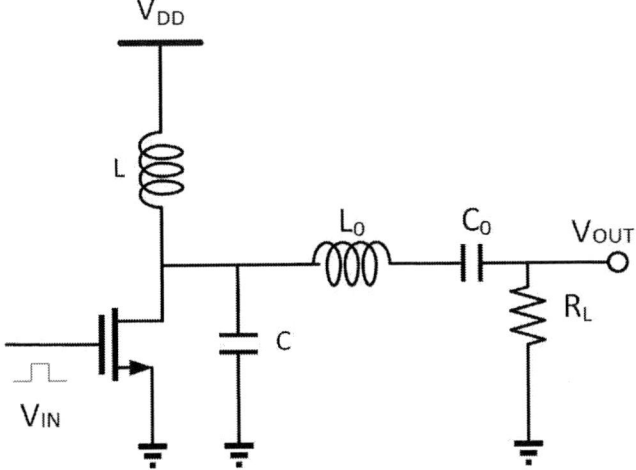

Fig. 1. Class E power amplifier

capacitance C can be written as

$$C = C_{cross} + C_{inj} + C_{sub} + C_{shunt}$$
$$C_{cross} = C_{gs3,4} + 2 * C_{gd3,4} + C_{db3,4} + C_{db1,2}$$
$$C_{inj} = 2 * C_{gd1,2} + C_{db1,2}$$

Here due to miller effect with gain close to be -1, C_{gd} components will have effective capacitance doubled. If the input injection signal is within the locking range of the system, the LC tank provides a phase change such that the sum of cross coupled transistor current, LC tank would be at the same frequency as input injection current[16]. In-order to obtain a better efficiency when compared to drain injection locked power amplifier, body injection locking technique is proposed in the next section.

L_0 and C_0 corresponding to output filter have to be tuned at input frequency.

The disadvantage of a class E power amplifier is that it has high degree of non-linearity due to the presence of even harmonic terms which can be only reduced if we increase the quality factor of output filter, which can be accomplished either by reducing effective load resistance(use of impedance transformation network) or increasing L_0 of output filter, either of the cases would lead to an increase of area. A differential configuration of a class E power amplifier can be employed to lower THD by eliminating even harmonics. This also reduces substrate coupling between power amplifier and other blocks of the system due to perfectly differential nature of currents. Injection locking technique involving differential configuration is described in the next section, it can be employed to enhance the efficiency of the system further as it requires lesser driving voltage.

III. INJECTION LOCKING TECHNIQUE

One of the primary reasons to employ injection locking technique is because of it's lesser driving voltage requirement when compared to a differential class E power amplifier at the same output power. Lower driving voltage implies a lower driver power loss hence achieving a higher efficiency when compared to a differential class-E power amplifier. The conventional injection locked power amplifier is as shown in fig.2. Here injected current is injected through drain of additional transistors (M1,M2) hence it is drain injection locked power amplifier. In the figure the pair of cross coupled transistors together with LC filter forms an oscillating circuit. The injection locking system in fig.2 can be effectively replaced as system in fig.3 with transistors acting as current sources. Here

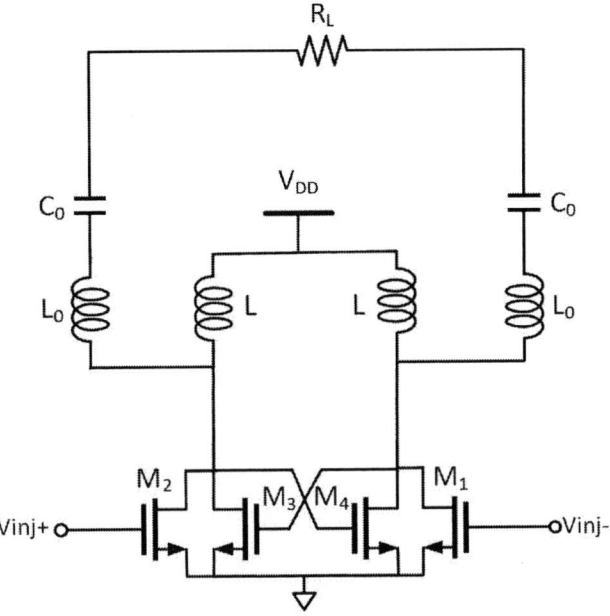

Fig. 2. Drain injection locked power amplifier

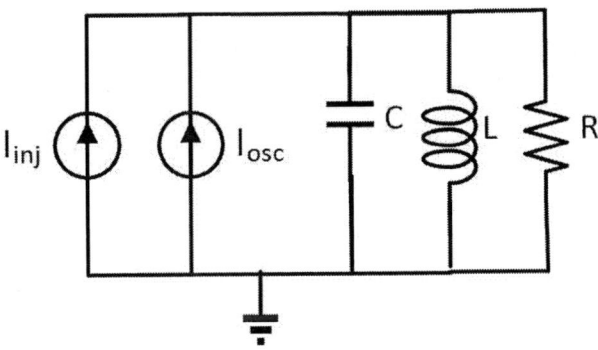

Fig. 3. Equivalent circuit of injection locked power amplifier

978-1-5386-3693-0/18 $31.00 © 2018 IEEE

IV. PROPOSED BODY INJECTED PA

The Body injection power amplifier is as shown in Fig.4. In this topology instead of using two additional power transistors for drain switching we exploit the body terminal of cross coupled transistor pair for input signal injection. The body injection PA achieves a higher efficiency when compared to conventional injection locking PA because of the reasons stated below

i) Driver power can be further reduced in body injection technique because the transistor in drain injected power amplifier needs to be switched thus requiring a higher driver voltage when compared to a body injected power amplifier which doesn't require such a large input voltage.

ii) As bodies of cross coupled transistor pair are kept at positive voltage this results in lower switch on resistance of cross coupled pair, hence would result in lower loss.This has been elucidated in appendix section.

The Body injected power amplifier proposed has the same cross coupled pair together with LC to give an LC oscillating circuit while the injection current is included with in the cross coupled transistor's current. The total capacitance at the drain of each cross coupled transistor for our design is $\approx$ 9.5pF, with $C_{gs} = 2.43pF$, $C_{gd} = 2.68pF$ and remaining capacitance contributed by parasitic capacitances due to metal lines(post-layout), inductors. The free running frequency of oscillator is 2.5GHz with a drain inductance of 414.4pH. The part of the current corresponding to the variation in V_{Th} (threshold voltage) of cross coupled pair is equivalent to drain injected current in a drain injection locked power amplifier. The variation of V_{Th} of a mosfet with body bias can be expressed as

$$V_{Th} = V_{Th0} + \gamma(\sqrt{-2\phi_f - V_{bs}} - \sqrt{-2\phi_f})$$

The driver stage inverter produces approximately a square wave at the body of cross coupled transistors.Injection current for body injection corresponds to the difference in drain currents between the cases when body input is high and when body input is low. The cross coupled transistors are sized so that process corner variations would still keep the input signal with in the locking range of oscillator. Triple well nmos transistors have been chosen for better isolation between bodies of two cross coupled transistors.

Driver stage: A driver stage is needed for the proposed power amplifier because of the difficulty to match the body of the power amplifier to 50Ω in the absence of driver, because of parasitic drain to body capacitances of power amplifier. A simple inverter based driver is used for the power amplifier taking care that the body source junctions would not be forward biased. The circuit diagram of the driver is as shown in Fig.6. Here nmos transistor is biased at VDD where as pmos is biased at 0V resulting in same V_{gs} for both the devices. An inductor is used to tune out the net capacitance at the output of driver to lower the driver loss. Blocking capacitor is used to prevent DC power loss in non-ideal inductor.

Input matching: In the input matching network shown, the

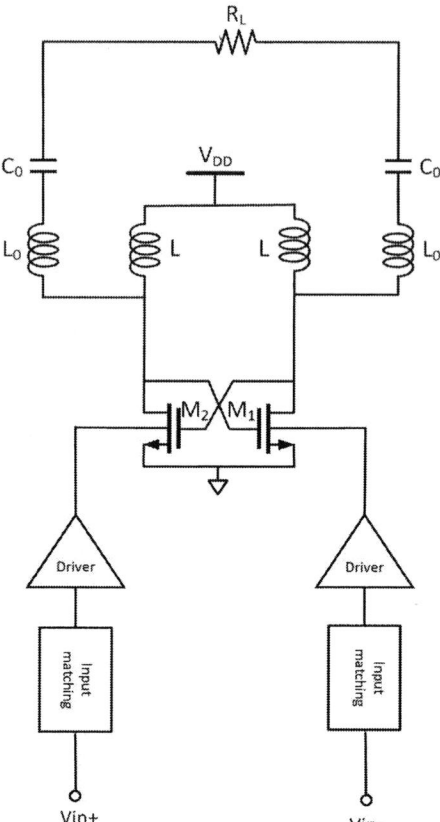

Fig. 4. Body injected power amplifier

R_{bias} are chosen so as to effectively produce 50Ω input impedance at the input of driver. A 20pF capacitance is used for DC blocking. Input filter is a simple series LC filter.

Layout: The layout of the core of power amplifier designed is as shown in Fig.5. For the layout only two of the inductors in the design are chosen to be on chip because of area constraints. The capacitive cancellation inductors from driver stage and input, output filter inductors are realized through bond wires. The transistors are placed following inter digitization layout to reduce effects due to process variations. The core area of the chip is $0.5mm^2$. The bond wire model used is shown in fig.7.

V. SIMULATION RESULTS

The proposed power amplifier topology is implemented in 65nm RFCMOS process. The power amplifier has achieved a PAE of 61.8% and DE of 63.6% at a VDD of 0.5V and output power of 12.3dBm for an input power of -1dBm. For all the simulations output load has been kept constant at 50 ohm, the quality factor of on chip spiral inductances are 17.88 with inductance of 414.4pH. At 0.5V VDD the power loss in each of the components are mentioned in Table-I. In Table-III power amplifier designed has been compared to some other low power topologies proposed earlier. Fig.8 shows the PAE

978-1-5386-3693-0/18 $31.00 © 2018 IEEE   210

TABLE I
HARMONIC DISTORTION

Harmonic	3	5	7
THD in dBC	-37.5	-47	-58.5

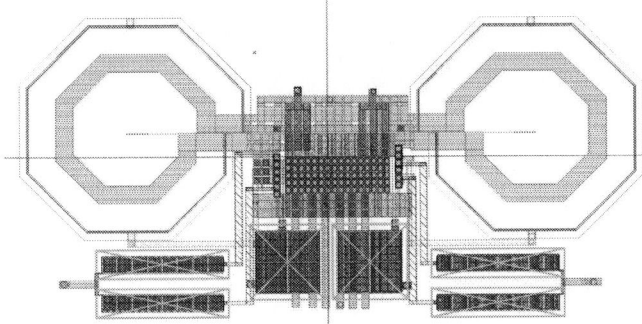

Fig. 5. Layout of body injected power amplifier

TABLE II
POWER LOSS DIVISION IN VARIOUS COMPONENTS

Component	Inductor	switch	Driver	Bond wires
Power loss in mW	2.3	1.9	0.9	0.76

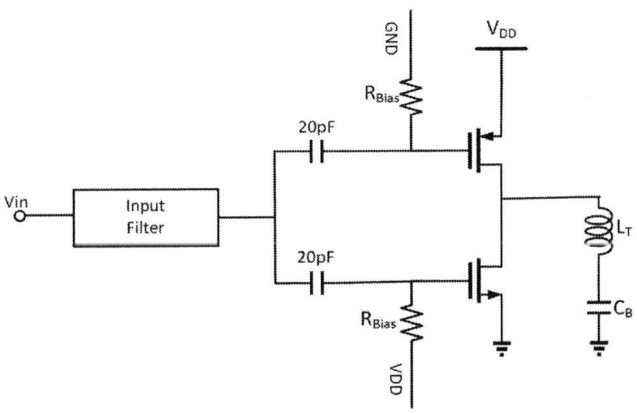

Fig. 6. Driver for body injected power amplifier

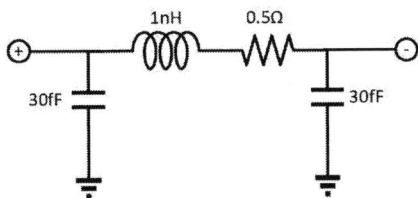

Fig. 7. Bond wire model used

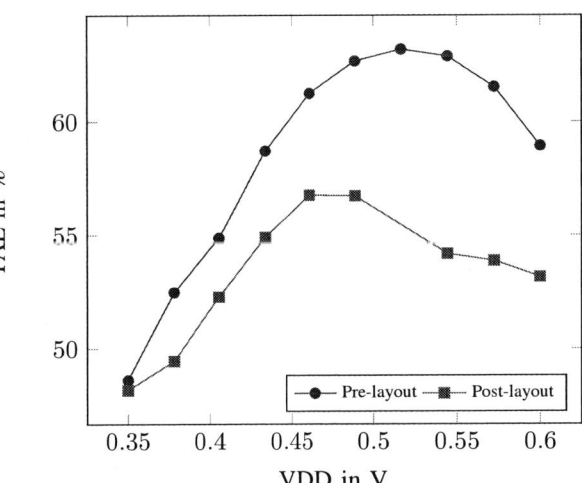

Fig. 8. PAE VS VDD

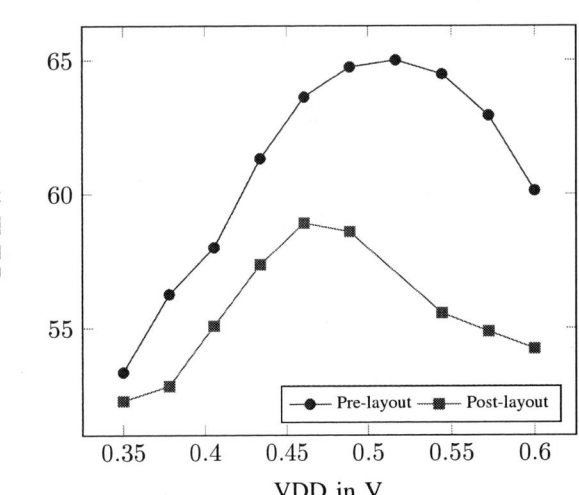

Fig. 9. DE VS VDD

vs VDD, Fig.9 shows DE vs VDD for the power amplifier in schematic level (pre-layout) and post-layout level simulations. As it can be seen the maximum power added efficiency of the power amplifier is 63.1% and maximum drain efficiency of the power amplifier is 65%. The power amplifier achieves a THD of 37dBc or 1.4%. The distribution of harmonic distortion is shown in Table.2.

Post layout simulation results resulted in a small degradation in efficiency due to the presence of parasitic resistances. The power amplifier has achieved a PAE of 56.4% and DE of 58.3% at a VDD of 500mV from post layout. The power amplifier has been tested to work across process corner, voltage and temperature variations.

In table-III the power amplifier designed has been compared with some of the existing topologies.

978-1-5386-3693-0/18 $31.00 © 2018 IEEE

TABLE III

COMPARISON OF PROPOSED PA WITH EXISTING TOPOLOGIES

Work	Type	Tech	Supply(V)	Pout(dBm)	PAE(%)	DE(%)	freq(Hz)
[15]	ILO	0.18u	1.5	7.6	36	-	2.4G
[6]	E-ILO	0.18u	1.2	11	44.5	49.3	2.35G
[7]^s	E	0.13u	0.35-0.8	0-5.7	53.5-55	-	2.45G
[11]^s	E	0.18u	0.5-0.85	2.7-7.2	-	57.3-60.7	2.4G
[13]	E	0.18u	1	11	30.5	-	2.5G
[14]	E	0.18u	0.6-1.4	6-13	50-53	43-51	2.4G
[10]^s	B	0.13u	1.2	12.5	47	-	2.45G
Thiswork^s	Body ILO	65n	0.5	12.3	61.8	63.4	2.4G

^{s} simulated results

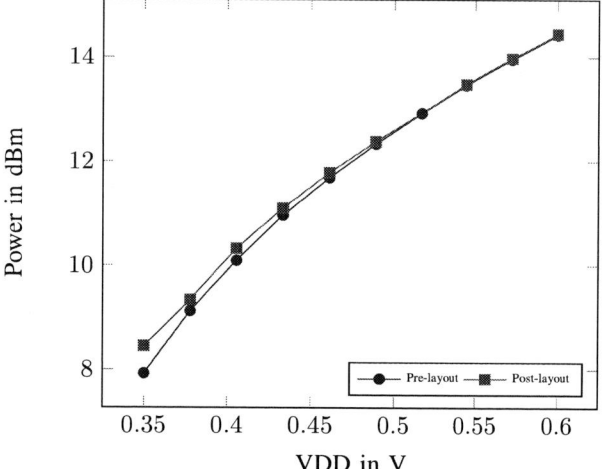

Fig. 10. Output power VS VDD

VI. CONCLUSION

A differential body injected power amplifier has been proposed with a PAE of 61.8% and DE of 63.4% at an output power level of 12.3dBm with 0.5V supply voltage. Justifications have also been presented as to why this topology helps in improving efficiency.

Future scope of work: Although differential configuration helps to improve linearity, it increases the output power level. If linearity is not a concern then one may use a single ended injection locking technique with body driving for achieving high efficiency at lesser output power levels.

REFERENCES

[1] N. O. Sokal and A. D. Sokal, "Class E-A new class of high-efficiency tuned single-ended switching power amplifiers", IEEE J. Solid-State Circuits, vol. 10, no. 3, pp. 168176, Jun. 1975. 2

[2] M. Acar, A. J. Annema, and B. Nauta, "Analytical design equations for class-E power amplifiers", IEEE Trans. Circuits Syst. I, Reg. Papers, vol. 54, no. 12, pp. 27062717, Dec. 2007. 3

[3] Steve.C Cripps, "RF Power Amplifiers for Wireless Communications". 4

[4] R. Brama et al., "A 1.7-GHz 31dBm differential CMOS Class-E Power Amplifier with 58% PAE", IEEE Custom Integrated Circuits Conf., pp. 551-554, September 2007. 5

[5] K. L. R. Mertens and M. S. J. Steyaert, "A 700-MHz 1-W Fully Differential CMOS Class-E Power Amplifier", IEEE Journal of Solid State Circuits, vol. 37, pp. 137-141, 2002. 6

[6] H.-S. Oh, T. Song, E. Yoon, and C.-K. Kim, "A power-efficient injection-locked class-E power amplifier for wireless sensor network", IEEE Microw. Wireless Compon. Lett., vol. 16, no. 4, pp. 173175, Apr. 2006. 7

[7] J. Tan, C.-H. Heng, and Y. Lian, "Design of Efficient Class-E Power Amplifiers for Short-Distance Communications", IEEE Transactions on Circuits and Systems I: Regular Papers, vol. 59, no. 10, pp. 22102220, 2012. 8

[8] K. C. Tsai and P. R. Gray, "A 1.9-GHz 1-W CMOS class-E power amplifier for wireless communications", IEEE J. Solid-State Circuits, vol.34, pp. 962970, July 1999. 9

[9] C.-H. Lin and H.-Y. Chang, "A Broadband Injection-Locking Class-E Power Amplifier", IEEE Transactions on Microwave Theory and Techniques, vol. 60, no. 10, pp. 32323242, 2012. 10

[10] P. Saffari, M. Taherzadeh-Sani, A. Basaligheh, F. Nabki, and M. Sawan, "Low-energy CMOS common-drain power amplifier for short-range applications ". IEEE 13th international conference on new circuits and systems (NEWCAS) (pp. 14), 2015. 11

[11] M. Modava, A. Sahafi, J. Sobhi, and Z. D. Koozehkanani, "Design of efficient power amplifier for low power transmitters", Analog Integrated Circuits and Signal Processing, vol. 90, no. 3, pp. 563571, 2016. 12

[12] P. Heydari and Y. Zhang, "A novel high frequency, high-efficiency, differential class-E power amplifier in 0.18 m CMOS", Proceedings of the 2003 International Symposium on Low Power Electronics and Design, 2003. ISLPED 03. 13

[13] X. Xu, Z. Sun, K. Xu, X. Yang, T. Kurniawan, and T. Yoshimasu, "A 2.5-GHz band low-voltage class-E power amplifier IC for short-range wireless communications in 180-nm CMOS", 2014 IEEE International Symposium on Radio-Frequency Integration Technology, 2014. 14

[14] M. J. Deen, M. M. El-Desouki, H. M. Jafari, and S. Asgaran, "Low-power integrated CMOS RF transceiver circuits for short-range applications", 2007 50th Midwest Symposium on Circuits and Systems, 2007. 15

[15] M. M. El-Desouki, M. J. Deen, Y. M. Haddara, and O. Marinov, "A Fully Integrated CMOS Power Amplifier Using Superharmonic Injection-Locking for Short-Range Applications", IEEE Sensors Journal, vol. 11, no. 9, pp. 21492158, 2011. 16

[16] B. Razavi, "A study of injection locking and pulling in oscillators", IEEE Journal of Solid-State Circuits, vol. 39, no. 9, pp. 1415-1424, 2004. 17

[17] C. Yoo and Q. Huang, "A Common-Gate Switched, 0.9W Class-E Power Amplifier with 41% PAE in 0.25m CMOS", VLSI Symp., pp. 56-7, June 2000. 18

[18] M. M. El-Desouki, M. J. Deen, Y. M. Haddara, "A low-power CMOS class-E amplifier for biotelemetry applications", Proc. 35th European Microwave Conf., pp. 441-444, Oct. 2005.

APPENDIX

The reasons for improvement of efficiency in a body injected power amplifier with respect to a drain injected power amplifier is presented below.

1) Switch loss due to finite switch on resistance: Due to finite switch on resistance power loss occurs in the switch and is one of the significant sources of power loss in a power amplifier. If we consider that when V_{gs} and I_d are same for both

978-1-5386-3693-0/18 $31.00 © 2018 IEEE

the cases, then the resistance offered by the switch can be calculated from current equation given earlier. i.e. $I_{injd} + I_{oscd} = I_{oscb}$ where I_{injd} indicates drain injected current in drain injection locked power amplifier, where I_{oscd}, I_{oscb} indicates cross coupled transistor's on current in drain injected and body injected power amplifiers respectively.

$$I_d = \frac{\beta V_{ds}(V_{gs} - V_{TH} - V_{ds}/2)}{(1 + \theta(V_{gs} - V_{TH}))(1 + \frac{V_{ds}}{E_C L})}$$

$$r_{on}^{-1} = \frac{\partial I_{ds}}{\partial V_{ds}}|_{V_{ds} \approx 0} = \frac{\beta(V_{gs} - V_{TH})}{1 + \theta(V_{gs} - V_{TH})} \approx \frac{I_{ds}}{V_{ds}}$$

$$r_{on} \approx \frac{V_{ds}}{I_{ds}}$$

Power loss P_{SL} in switch is proportional to r_{on}. Now the improvement in power loss for a body injected with respect to a drain injected power amplifier is

$$P_{SL\Delta} \propto \left(\frac{V_{dsd}}{I_{oscd}} \middle\| \frac{V_{dsd}}{I_{injd}} - \frac{V_{dsb}}{I_{oscb}}\right)$$

$$\propto \left(\frac{V_{dsd}}{I_{oscd} + I_{injd}} - \frac{V_{dsb}}{I_{oscb}}\right)$$

$$\propto \frac{V_{dsd} - V_{dsb}}{I_{oscb}}$$

$$\propto \frac{1 + \theta(V_{gs} - V_{TH0})}{V_{gs} - V_{TH0}} - \frac{1 + \theta(V_{gs} - V_{TH})}{V_{gs} - V_{TH}}$$

$$\propto (V_{TH0} - V_{TH})$$

Hence the power loss improvement due to switching loss is proportional to lowering of threshold voltage. From our simulations switch resistance is close to 0.12Ω with no body bias and 0.11Ω with body bias. Although it might not account for much improvement in efficiency, the second reason presented below explains improvement in efficiency due to body injection.

2) The driver loss can also be substantially reduced in the case of body injection power amplifier. This is because of lowering of power supply voltage, and power loss in inverter is $\propto VDD^2$. Now the power loss is low for the case of body injection because, it doesn't require switching as opposed to conventional drain injected power amplifier thus reducing the needed supply voltage for driver.

ACKNOWLEDGMENT

Authors would like to thank Ministry of electronics and information technology (meitY), government of India for providing necessary support under c2s program.

2018 31th International Conference on VLSI Design and 2018 17th International Conference on Embedded Systems

A Novel Low Power G_m-C Continuous-Time Analog Filter With Wide Tuning Range

P.S.Veerendranath*, Vasantha M.H.*, Nithin Kumar Y.B.*, Edoardo Bonizzoni[†]

* National Institute of Technology Goa, India
[†] University of Pavia, Pavia, Italy

Email:*veerendra.ece50@gmail.com, *(vasanthmh, nithin.shastri)@nitgoa.ac.in, [†]edoardo.bonizzoni@unipv.it

Abstract— This paper proposes a CMOS Operational Transconductance Amplifier (OTA) for wide tuning range filter applications with reduced power dissipation. The tuning range can be extended by operating the OTA in weak inversion region as well as in strong inversion region. This can be attained by tuning the bias current, which results in the linear variation of transconductance (g_m). A fifth order elliptical low pass filter is simulated with cutoff frequency tuning from 1.5 kHz to 1.6 MHz. Wide tuning range filters can be employed for multimode applications, which in-turns help in saving silicon area. A novel tunable resistor is simulated using twin well technology and its value ranges from 40 kΩ to 30 MΩ. The proposed circuit is simulated in SCL 180 nm standard CMOS technology with 1.8 V as supply voltage and achieves a dynamic range (DR) of 47 dB. Total power consumption of the circuit is 670 μW at 1.6 MHz.

Index Terms—G_m - C, OTA, Weak inversion (subthreshold), Strong inversion, Tuning, Low power

I. INTRODUCTION

Now-a-days the biggest challenge for engineers is to design circuits, which are functional and operable at Ultra Low Power (ULP) as, the demand for hand-held devices with multi-functionality like mobile phones, tablets, laptops and sensors is exponentially increasing. In designing circuits for ULP applications the core agenda is to minimize power consumption and cost effectiveness. However that in digital circuits the power consumption scales down appropriately as the supply voltage reduces unlike in analog circuits, where the threshold voltage of the devices dictates the lower limit for the supply voltage. There are two ways to counter balance the lowering of the supply voltage by biasing the transistors either in the subthreshold region or applying an appropriate biasing potential between source and bulk (V_{SB}), as the effective threshold voltage ($V_{th_{eff}}$) scales down accordingly ($V_{th_{eff}} \propto \sqrt{V_{SB}}$). It is simple to operate the circuit in sub-threshold region as compared to its operation in bulk driven mode as latter has the probability of forward biasing the source to bulk and drain to bulk junction (leading to huge leakage current through the bulk of MOSFET). Hence, the sub-threshold region provides an optimum solution against the scaling of the supply voltage, while maintaining the performance of the system.

In signal processing, low pass filters play a vital role in transceiver architecture. MOS varactors [3] were demonstrated in tuning active RC structures, which uses an operational amplifier (op-amp) as the basic building block and large discrete components (RC) for the implementation of low pass filters. It finds limited operation in low-frequency applications because of large parasitic capacitances. Therefore active RC circuits are not recommended for frequencies above 5 or 10 times that of the unity gain frequency of the op-amp. Another approach for implementing filters is through switched capacitor circuits, but finds limitation in high frequency applications due to their large time constant (τ) as $\tau \propto$ RC. Alternatively, an integrator in continuous time analog filters can be realized using transconductors and capacitors, and also the cut-off frequency of the filter does not depends on their RC product. The transconductance cells consumes less power as it just drives the capacitive loads.

Due to the absence of negative feedback, G_m - C cells are fast and have stable frequency response [10]. The advantage of operating the transconductance cell in the subthreshold region is that its g_m is directly proportional to its bias current and this aspect is explored in this paper for low power applications. In literature, various transconductance cells are designed to enhance speed, linearity and dynamic range as reported by [6] - [7]. Analog filters are used for band-limiting the signal in order to achieve better signal to noise ratio (SNR). At the transmitter side low pass filter is used as an anti-aliasing and on the receiver side it removes high frequency noise from the message band. Moreover, continuous time filters act as a front-end interface between the real world and digital systems.

This paper shows the CMOS implementation of a fifth order low power elliptic low-pass G_m-C filter that can be tuned for a wide range of frequency applications (i.e. as a selection filter for speech, audio, video, bio-medical and wireless applications). The transconductor design operates in different regions to maximize the transconductance range by tuning the bias current. The paper is organized as follows: Section II discusses the design of the modified transconductance cell for wide tuning range. Section III develops the equivalent resistor used in the OTA design. The proposed G_m-C filter is discussed in Section IV. Simulation results of OTA, resistor, and filter are shown in Section V followed by Conclusion in Section VI.

II. TRANSCONDUCTANCE CELL

In order to increase the tuning range of the filter, the transconductance range has to be adaptable. The conventional transconductance cell is shown in Fig. 1. The range extension can be achieved by operating the MOSFETS M_1 to M_8 of

978-1-5386-3693-0/18 $31.00 © 2018 IEEE

the modified transconductor cell [2], from subthreshold to saturation region with the help of tunable dc bias circuits. The main requirement is to provide linearity to the output response while tuning the transconductance.

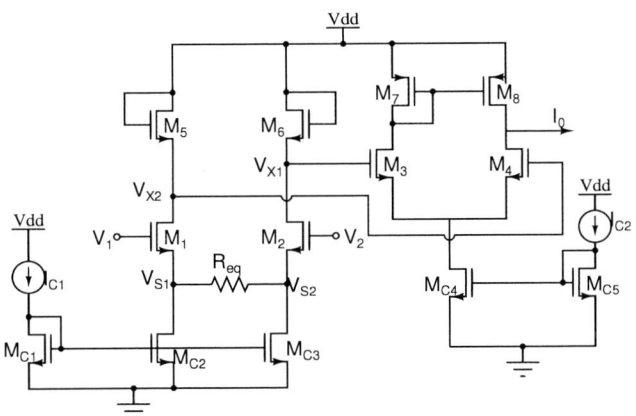

Fig. 1. Modified Transconductance cell with transistors M_1 to M_8 switching from subthreshold region to saturation region.

A. Transconductor cell operating in the subthreshold region:

To operate the MOSFETs in the subthreshold region, the gate to source voltage should be less than the threshold voltage and the drain to source voltage (V_{DS}) should be four times larger than the thermal voltage. Two features are attracting to operate the transistor in the subthreshold region is low power and transconductance is linear to the bias current and independent of device parameters. In sub-threshold region, the relationship between the gate to source voltage V_{GS} and the drain current I_D of a transistor is given by [1]

$$I_D = I_{D0} \frac{W}{L} exp\left(\frac{V_{GS}}{\eta U_T}\right) \qquad (1)$$

where W and L represent the width and length of the transistor respectively, I_{D0} is the reverse saturation current, η is the subthreshold slope factor whose value is approximately equal to 1.4 and U_T represents the thermal voltage. From equation (1) it can be seen that the output current varies exponentially with respect to the input voltage. It is evident from equation (1), potential V_{x2} is logarithmically related to the potential V_1. I_0 varies exponentially with respect to potential V_{x2}, hence I_0 depends linearly on the potential V_1. The sizing of transistors for M_1 to M_6 are all same in weak inversion region for both the input and output stages of the transconductor cell. M_5 and M_6 are diode-connected transistors, which carry the currents I_{D5} and I_{D6} respectively. The bulk terminals of NMOS and PMOS are connected to ground and supply voltage respectively. Consider the gate voltages of transistors M_3 and M_4 given by V_{X1} and V_{X2} respectively. V_{X2} - V_{X1} can be obtained by the relationship between output current and input voltage as given by [2]

$$I_0 = \frac{I_{C2}}{I_{C1}}\left(\frac{V_1 - V_2}{R_{eq}}\right). \qquad (2)$$

The output current is varied by tuning the two bias currents I_{C1} and I_{C2} for various transconductance values. In the subthreshold region, the output current is very small due to a high equivalent resistor (R_{eq}) and output dc voltage, which strongly affects the output current.

B. Transconductor cell operating in the saturation region:

The output current and input voltage relationship is given in [2], when all the transistors from M_1 to M_6 having same sizes are operating in saturation region as

$$I_0 \approx \frac{k}{2}(\Delta V_X)\sqrt{\frac{4I_{C2}}{k}} = \frac{\sqrt{\frac{2I_{C2}}{I_{C1}}}}{\sqrt{\frac{2}{kI_{C1}}} + R_{eq}}(V_1 - V_2). \qquad (3)$$

here $k = \mu_n C_{ox}\frac{W}{L}$, μ_n represents the mobility of electrons, C_{ox} is the oxide capacitance per unit area, and W, L indicate the width and length of the corresponding device respectively.

To increase the output current, the voltage difference ΔV_X should be large enough and to tune the transconductance, properly choose the value of I_{C1} and increase the value of I_{C2}. The equivalent resistor (R_{eq}) should be carefully chosen in both weak and strong inversion regions. The value of R_{eq} should be larger in weak inversion as compared to its value in strong inversion region to increase the transconductance range. If the R_{eq} value is very small, the linearity of the transconductor cell is affected. Hence R_{eq} is an important parameter in achieving linearity and also in tuning the transconductance.

The output current and the input voltage are related by nonlinear terms, therefore the transconductance is tuned by selecting the suitable I_{C1}, and then tuning I_{C2}. The non-linearity is a big concern in strong inversion and also in transition region hence input voltage should be maintained properly.

III. PROPOSED RESISTOR

The requirement of two resistor values are necessary for linearity consideration. Therefore designing the resistor of different values with the same architecture is possible either by tuning [15] or by switching. There are different approaches for implementing the resistor operating the transistors either in subthreshold or in saturation [11] region. Implementing the resistor in saturation mode leads to high power dissipation and it adds to an additional non-linearity in to the system, which makes it less favourable. Large resistance are realized using MOSFETs in sub-threshold region [13], but these resistors have limited region (range of applied potential) of operation and has prominent PVT variation, which makes the circuit prone to dysfunctioning. However, realizing large resistances is a bottleneck Therefore proper care must be taken while implementing resistor in the subthreshold mode for large resistance. Implementing small resistor is achieved by operating transistors in linear region which makes suitable for source generation circuits. Hence an appropriate combination of sub-threshold and linear region implementation of resistor

978-1-5386-3693-0/18 $31.00 © 2018 IEEE   215

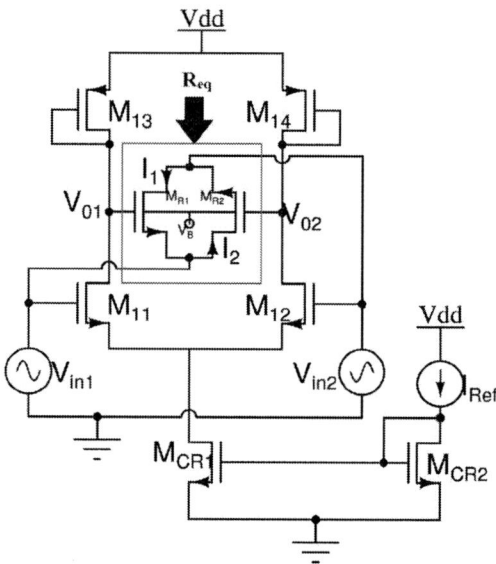

Fig. 2. Implementation of linearly variable equivalent resistor realized with active elements.

$$I_2 = \mu_n C_{ox}\frac{W}{L}[((V_g - \Delta V) - (V_s + \Delta V) - V_{th})((V_d - \Delta V)$$
$$-(V_s + \Delta V)) - \frac{1}{2}[(V_d - \Delta V) - (V_s + \Delta V)]^2, \quad (7)$$

$$I_1 = \mu_n C_{ox}\frac{W}{L}[(V_{gs} + 2\Delta V - V_{th})(V_{ds} + 2\Delta V) - \frac{1}{2}[(V_{ds} + 2\Delta V)^2],$$
$$(8)$$
$$I_2 = \mu_n C_{ox}\frac{W}{L}[(V_{gs} - 2\Delta V - V_{th})(V_{ds} - 2\Delta V) - \frac{1}{2}[(V_{ds} - 2\Delta V)^2],$$
$$(9)$$

Under dc operating conditions, no current flows through R_{eq} then the voltage drop across V_{ds} is zero. Hence the above equation becomes

$$I_1 - I_2 = \mu_n C_{ox}\frac{W}{L}[4\Delta V(V_{gs} - V_{th})], \quad (10)$$

$$\frac{I_1 - I_2}{2\Delta V} = \mu_n C_{ox}\frac{W}{L}[2(V_{gs} - V_{th})], \quad (11)$$

$$\frac{1}{R_{eq}} = 2\mu_n C_{ox}\frac{W}{L}(V_{gs} - V_{th}), \quad (12)$$

$$R_{eq} = \frac{1}{2\mu_n C_{ox}\frac{W}{L}(V_{gs} - V_{th})}. \quad (13)$$

ΔV represents the change in corresponding input voltage across terminals, V_{gs} indicates gate to source voltage, V_{ds} indicates drain to source voltage under dc biasing condition and V_{th} represents the threshold voltage. In subthreshold region the resistance value is proportional to $\frac{\eta U_T}{I_D}$, in order to have a large resistance, the drain current should be low. To attain large resistance, operate the transistors M_{R1} and M_{R2} in weak inversion region by increasing back gate voltage. The back gate voltage is maintained around –0.5 V, which increases the depletion region width, as well as the threshold voltage. Similarly to achieve low resistance in strong inversion region the substrate voltage should be equal to 0V. Additionally, in this architecture, even harmonic distortions get canceled and it also gives limited output swing to ensure linearity.

IV. PROPOSED FILTER ARCHITECTURE

A fifth order low power elliptic low-pass filter is proposed, which consists of five identical OTA's with seven capacitors, including two floating capacitors. The fifth order elliptic low power low pass filter [9] design starts from a standard fifth order LC-ladder prototype as shown in Fig. 3. The proposed design is obtained by minimizing the number of transconductance elements using signal flow graph [8]. A 0 dB dc gain is achieved for low frequency, when resistor R_L equals R_S. The transconductance of every cell is equal to $(1/R_S)$ or $1/R_L$. The final implementation of G_m-C filter is shown in Fig. 4. The challenge in designing low pass filters is a large time constant, which is a function of RC. The implementation of a large value of resistors and capacitors is impractical. In case of an OTA the cutoff frequency of low pass filter is proportional to g_m/C, where g_m represents the transconductance and C is the output capacitance. Hence OTAs are optimum in an integration of analog filters because no resistors are used and are independent of RC products. To

is employed in the proposed topology to tune the g_m of the transconductance cell.

The equivalent circuit of a resistor is shown in Fig. 2. which is connected across the source terminals of M_1 and M_2. The gate terminals of M_{R1} and M_{R2} are connected to the output terminals of V_{01} and V_{02} respectively, where all the transistors of biasing circuit (M_{11}, M_{12}, M_{13}, M_{14}, M_{CR1} and M_{CR2}) are working in saturation region. A level shifter is required in order to provide the sufficient voltage across the source terminals of M_1 and M_2.

The resistor is designed in twin well technology by the parallel connection of two transistors. Here resistor is tuned by varying the substrate voltage (V_B) of M_{R1} and M_{R2}. Increasing the substrate voltage results in an increase in the source to bulk voltage thus making it more reverse biased. Therefore, the threshold voltage increases with increase in depletion width, hence resistor can be tuned as shown in (13), but this equation is not valid for large biasing of the substrate voltage because transistors M_{R1} and M_{R2} enters into the subthreshold region.

A. Derivation for R_{eq}

The drain currents when the transistors M_{R1} and M_{R2} are operating in the linear region are as follows:

$$I_1 = \mu_n C_{ox}\frac{W}{L}[(V_{gs1} - V_{th})(V_{ds1}) - \frac{1}{2}V_{ds1}^2], \quad (4)$$

$$I_2 = \mu_n C_{ox}\frac{W}{L}[(V_{gs2} - V_{th})(V_{ds2}) - \frac{1}{2}V_{ds2}^2], \quad (5)$$

$$I_1 = \mu_n C_{ox}\frac{W}{L}[((V_g + \Delta V) - (V_s - \Delta V) - V_{th})((V_d + \Delta V)$$
$$-(V_s - \Delta V)) - \frac{1}{2}[(V_d + \Delta V) - (V_s - \Delta V)]^2], \quad (6)$$

978-1-5386-3693-0/18 $31.00 © 2018 IEEE

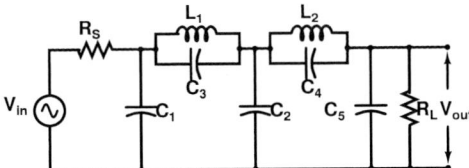

Fig. 3. Fifth order elliptic RLC ladder network protype.

obtain low cutoff frequency, the transconductor cell should operate in the weak inversion region and to achieve higher cutoff frequency, maximize the transconductance by operating the OTA in strong inversion region.

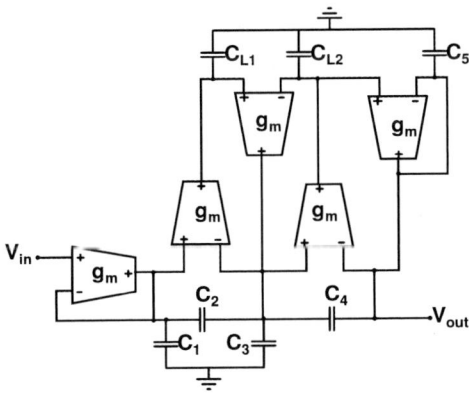

Fig. 4. Fifth order elliptic low pass G_m-C filter using only five g_m cells and seven capacitors.

V. NOISE ANALYSIS OF THE PROPOSED RESISTOR

At low frequency, the dominant input referred noise source in noise spectrum is flicker (1/f) noise. This noise exhibits an inverse frequency power density curve. The total input referred noise and its equivalent noise power spectral density per unit bandwidth is calculated by flicker noise model as [1]

$$\frac{V_{f-in}^2}{\Delta f} = \frac{2k_N}{C_{ox}W_nL_n}\frac{1}{f} + \frac{2k_P}{C_{ox}W_pL_p}\frac{g_{M13}^2}{g_{M11}^2}\frac{1}{f} \quad (14)$$

where g_{M11} and g_{M13} are the small signal transconductance of NMOS and PMOS transistors respectively, k_N and k_P are process-dependent constant for NMOS and PMOS respectively, f is frequency, W and L are the width and length of corresponding transistors. Thus low-frequency flicker noise can be minimized by properly setting the transistor dimensions.

Flicker noise is often characterized by the corner frequency f_c at which both thermal and flicker noise is equal. At frequency higher than the noise corner frequency, which estimated around 9 kHz, then the total noise is dominated by the thermal noise. The thermal noise is modeled using [1], the total input referred thermal noise is given by

$$\frac{V_{th-in}^2}{\Delta f} = 8kT\gamma\left[\frac{1}{g_{M11}} + \frac{g_{M13}}{g_{M11}^2}\right] \quad (15)$$

where k is the Boltzmann constant, T is the temperature and γ is the noise parameter depending on device bias condition and from [13] its value is equal to $1/2$ and $2/3$ at weak and strong inversion regions respectively. From the above equation, it affirms that the large value of g_{M11} reduces the input referred noise, which is accomplished by the proper design of transistor dimensions.

The contribution of noise due to transistors M_{R1} and M_{R2} is neglected because of the low transconductance. Since flicker noise is dominated at low frequency and by keeping the small value of transconductance, the effect of flicker noise in the total noise can be reduced. Thermal noise is the dominant noise factor at high frequencies, due to the parallel combination of transistor design in the proposed resistor, the effect of thermal noise is nullified. The total input referred noise per unit bandwidth is $1.13 * 10^{-5} \left[\frac{V}{\sqrt{Hz}}\right]$.

Noise analysis of the transconductance cell [2] is given by input referred noise per unit bandwidth is $9.88 * 10^{-5} \left[\frac{V}{\sqrt{Hz}}\right]$ and $1.773 * 10^{-4} \left[\frac{V}{\sqrt{Hz}}\right]$, when OTA is working in weak inversion and in saturation regions respectively.

VI. SIMULATION RESULTS

In this paper, the G_m - C filter is implemented in standard CMOS 180 nm technology process. The simulated results are shown in Fig. 5 - Fig. 10. The transconductance varies from 60 nS to 420 nS by changing the bias current I_{C2} from 10 nA to 70 nA with I_{C1} equal to 10 nA and varies from 35 μS to 87 μS by changing the bias current I_{C2} from 10 μA to 40 μA with I_{C1} equal to 10 μA, when both input and output stages are working in weak inversion and strong inversion region respectively, which is as shown in Fig. 5 - 6. The filter dissipates 0.67 mW for a higher cutoff frequency at 1.8 V supply voltage. The proposed resistor value ranges from 40 kΩ to 30 MΩ by varying the back gate voltage from 0 V to –0.5 V respectively as shown in Fig. 7. Table 1 shows the simulated results of the proposed filter.

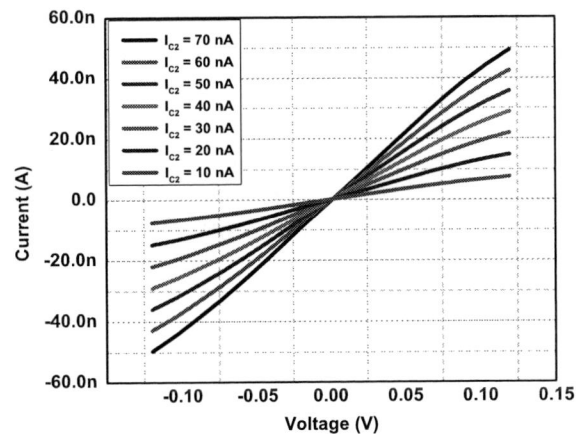

Fig. 5. Transconductance in weak inversion region.

TABLE I
PERFORMANCE SUMMARY

	[2]	[6]	[8]	Proposed
Technology	TSMC 0.18μm CMOS	1.2μm CMOS	1.2μm CMOS	Cadence 0.18μm CMOS
Supply Voltage	1.8V	3V	3V	1.8V
Order	5	5	5	5
Filter type	Elliptic low-pass	Elliptic low-pass	Elliptic low-pass	Elliptic low-pass
Tuning Range	250 Hz - 1 MHz	140 Hz - 3.5 kHz	280 - 405 kHz	1.5 kHz - 1.6 MHz
Dc gain	-6 dB	0 dB	0 dB	0 dB
Dynamic Range	48 dB	-	55 dB	47 dB
Power Consumption	0.8 mW	13 μW	2.48 mW	0.67 mW
FOM	152 dB	-	147 dB	151 dB

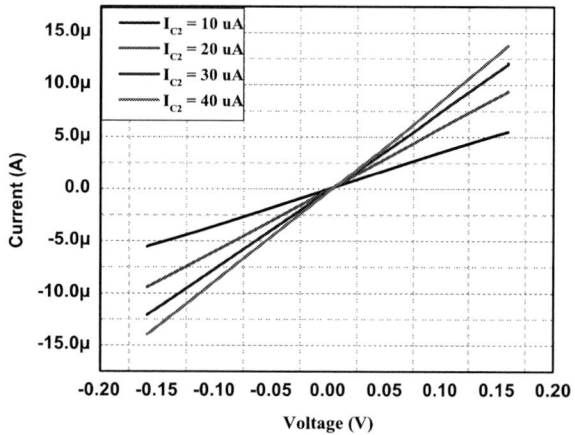

Fig. 6. Transconductance in strong inversion region.

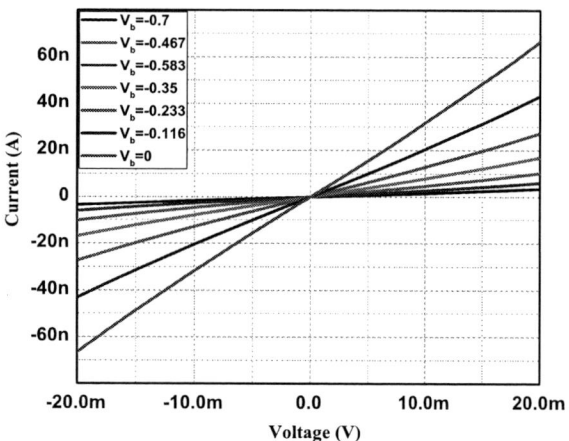

Fig. 7. Characterstics of equivalent resistor.

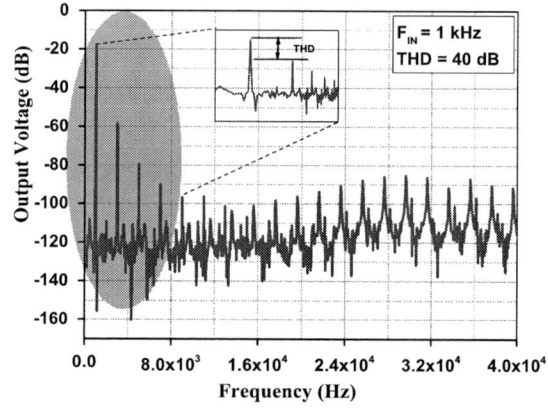

Fig. 8. Discrete Fourier Transform analysis of filter response output of Input signal at 1 kHz.

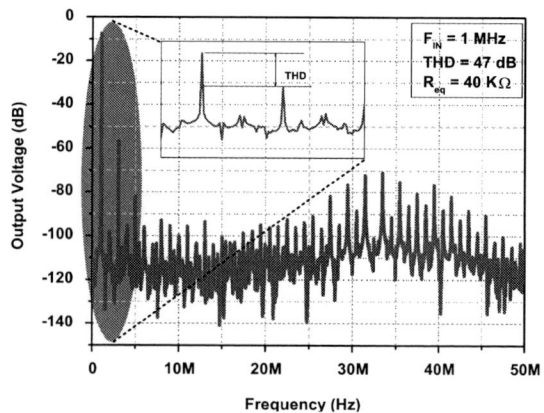

Fig. 9. Discrete Fourier Transform analysis of filter response output of input signal at 1 MHz.

Fig. 8 and Fig. 9 shows the DFT analysis of the proposed filter is shown in at 1 kHz and 1 MHz respectively. The third harmonic distortion is 45 dB for 90 mV input signal at 1kHz and 47 dB for 100 mV input signal at 1 MHz. Fig.10 shows the frequency response of the filter output tuning from 1.5 kHz to 1.6 MHz with I_{C1} regulating from 10 nA to 10 μA. When

the cutoff frequency of the filter is 1.5 kHz, a dynamic range of 48.6 dB is achieved with the 100 mV input signal with substrate voltage (V_B) set to –0.5V. On the other hand, a dynamic range of 47 dB is achieved with the 100 mV input signal with substrate voltage (V_B) set to 0V, when the cutoff frequency is 1.6 MHz.

978-1-5386-3693-0/18 $31.00 © 2018 IEEE

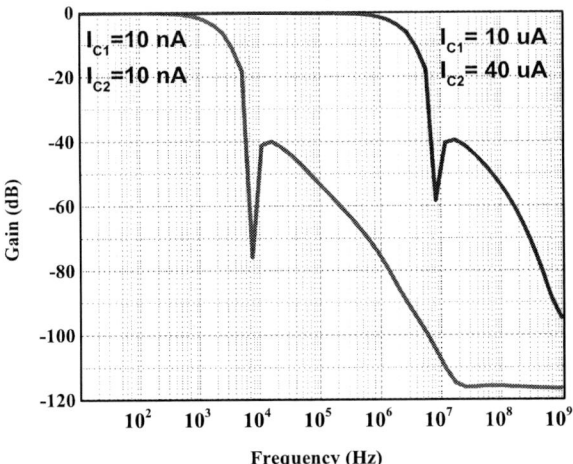

Fig. 10. Frequency response of filter output tuning from 1.5 kHz to 1.6 MHz with I_{C1} equal to 10 nA to I_{C1} equal to 10 μA.

A. Filter characteristics

The zeros of this filter lie on the imaginary axis and depend on the impedance of inductors and floating capacitors. By increasing the inductor value there is a decrease in cutoff frequency, quality factor, an increase in the ripple in the stop band. Floating capacitors decide the cutoff frequency, stopband ripple but has no effect on the quality factor. C_1, C_2 decide the quality factor and C_5 changes the passband frequency of a filter. The cutoff frequency can be tuned from 1.5 kHz to 1.6 MHz with the proposed architecture. Quality factor and gain depend on the capacitor ratios whereas, pole and zero locations depend on the absolute component values, so designer should carefully consider the parasitic capacitances which degrade the circuit performance. A digital circuit can be used on chip to tune the transconductance by regulating the current sources.

VII. Conclusion

A fifth order low pass elliptic filter with improved linearity and low power consumption is proposed in this paper. The idea has been validated by simulating the design in standard CMOS in 180nm technology. For low power, OTA operates in weak inversion region and then operates in strong inversion region to extend the tunability of filter. It can be used for multimode applications such as audio, speech, bio-medical and wireless applications as the filter can tune from 1.5 kHz to 1.6 MHz by setting the dc bias currents I_{C1} and I_{C2} respectively. In the proposed design the number of active elements are reduced to five as compared to seven, employed in [2] as this helps in saving silicon area as well as power dissipation.

VIII. Acknowledgment

This publication is an outcome of the Research work undertaken in the project under SMDP - C2SD, Department of Electronics and Information Technology, Ministry of Communication IT, Government of India. The author would also like to thank Vivek Sharma for assisting in writing this paper.

References

[1] B. Razavi, Design of Analog CMOS Integrated Circuit. New York:McGraw-Hill, 2001.

[2] T. Y. Lo and C. C. Hung, "A Wide Tuning Range G_m-C Continuous-Time Analog Filter," in *IEEE Trans. Circuits Syst.* I: Regular Papers, vol. 54, no. 4, pp. 713-722, Apr. 2007.

[3] S. Chatterjee, Y. Tsividis and P. Kinget, "0.5-V analog circuit techniques and their application in OTA and filter design," *IEEE J. Solid-State Circuits*, vol. 40, no. 12, pp. 2373-2387, Dec. 2005.

[4] J. Silva-Martnez, J. Adut, J. M. Rocha-Perez, M. Robinson, and S.Rokhsaz, "A 60-mW 200-MHz continuous-time seventh-order linearphase filter with on-chip automatic tuning system," *IEEE J. Solid-State Circuits*, vol. 38, no. 2, pp. 216-225, Feb. 2003.

[5] A. N. Mohieldin, E. Snchez-Sinencio, and J. Silva-Martinez, "A fully balanced pseudodifferential OTA with common-mode feed forward and inherent common-mode feedback detector," *IEEE J. Solid-State Circuits*, vol. 38, no. 4, pp. 499-531, Apr. 2003.

[6] C. C. Hung, M. Ismail, K. A. Halonen, and V. Porra, "Low-Voltage, micropower weakinversion CMOS G_M-C filter," *IEEE Trans. Circuits Syst.* II, Analog Digit. Signal Process., vol. 46, no. 6, pp. 816-820, Jun. 1999.

[7] A. J. Lopez-Martin, S. Baswa, J. Ramirez-Angulo, and R. G. Carvajal, "Low-voltage super class AB CMOS OTA cells with very high slew rate and power efficiency," *IEEE J. Solid-State Circuits*, vol. 40, no. 3,pp. 10681077, Mar. 2005.

[8] C. C. Hung, K. A. Halonen, M. Ismail, V. Porra, and A. Hyogo, "A low-voltage, low-power CMOS fifth-order elliptic G_m-C filter for baseband mobile, wireless communication," *IEEE Trans. Circuits Syst. for Video Technology*, vol. 7, pp. 584-593, Aug. 1997.

[9] L.P. Huelsman, Active and Passive Analog Filter Design. New York: McGraw-Hill, 1993.

[10] K. Kuo and A. Leuciuc, "A linear MOS transconductor using source degeneration and adaptive biasing," in *IEEE Trans. Circuits Syst.* II: Analog and Digital Signal Processing, vol. 48, no. 10, pp. 937-943, Oct 2001.

[11] M. Steyaert, J. Silva-Martinez, and W. Sansen, "High-frequency saturated CMOS floating resistor for fully differential analogue signal processors," *Electron. Lett.*, pp. 1609-1611, Aug. 1991.

[12] S. Pavan, Y. P. Tsividis and K. Nagaraj, "Widely programmable high-frequency continuous-time filters in digital CMOS technology," in *IEEE J. Solid-State Circuits*, vol. 35, no. 4, pp. 503-511, Apr. 2000.

[13] A. Tajalli, E. Vittoz, Y. Leblebici and E. J. Brauer, "Ultra-low power subthreshold current-mode logic utilising PMOS load device," in *Electron. Lett.*, vol. 43, no. 17, pp. 911-913, Aug. 2007.

[14] S. Hori, T. Maeda, H. Yano, et al.,"A widely tunable CMOS G_m-C filter with a negative source degeneration resistor transconductor," *ESSCIRC* 2004 - 29th European *Solid-State Circuits Conference IEEE* Cat. No.03EX705, Estoril, Portugal, 2003, pp. 449-452.

[15] D. Ma, B. M. Wilamowski and F. F. Dai, "A Tunable CMOS Resistor with Wide Tuning Range for Low Pass Filter Application," 2009 *IEEE* Topical Meeting on *Silicon Monolithic Integrated Circuits* in RF Systems, San Diego, CA, 2009, pp. 1-4.

0.36 nJ/bit MedRadio Band OOK Transmitter for Wearable Healthcare Applications

Abhishek Srivastava*, *Member, IEEE* and Maryam Shojaei Baghini, *Senior Member, IEEE*

Department of Electrical Engineering, Indian Institute of Technology Bombay, Mumbai, India.

*Email: abhisheks@ee.iitb.ac.in

Abstract—In this paper, an energy efficient low power on-off keying (OOK) transmitter (TX) is proposed for wearable healthcare applications. TX is suitable for biosignal communication from wearable healthcare devices (WHD) in Federal Communication Commission recommended 400 MHz Medical Device Radiocommunication band. The proposed TX uses the principle of edge combination for low power RF carrier generation. The proposed TX has been implemented in 180 nm mixed-mode CMOS technology. Measurements were done with a custom designed test chip. Measurement results show that in OOK mode the proposed TX achieves an energy efficiency of 0.36 nJ/bit. TX consumes an average power of 71 μW and radiates about -24 dBm power at 200 kb/s. Successful wireless communication of pseudo random bit sequence is also shown with the TX and a commercial off-the-shelf receiver at a data rate of 200 kb/s for a maximum communication distance of 2 m.

Index Terms—Transmitter, MedRadio, 400 MHz, OOK, wearable, remote health monitoring.

I. INTRODUCTION

With the advancements in wireless communication, wearable sensing and radio frequency (RF) technologies, concept of remote health monitoring is gaining a lot of attention for providing healthcare services to large populations. Fig. 1 depicts a scenario for remote health monitoring system, where a body sensor node (BSN) communicates to a near by (upto 3 m) local base station device (BSD). The BSD acts as a gateway to send the received information from the BSN to a remote central monitoring station (CMS) for clinical analysis. BSN is essentially a wearable healthcare device (WHD) which contains a sensor to capture biosignals, an analog conditioning circuit followed by an analog to digital converter followed by a wireless transmitter (TX). Communication range for the first wireless link between BSN and BSD is small because the radiated power from the BSN must be low enough in order to avoid any harm to human body. The other link from BSD to CMS can be established for longer communication ranges with Internet or Wi-Fi technologies. For short range biosignal communication, the key desirable features are: 1. there should be a dedicated band for communication in order to avoid interference from other wireless devices, 2. it should have low power transmitters (TX) and receivers (RX) to minimize the battery replacements at the BSN and BSD.

In year 1999, Federal Communication Commission (FCC) announced Medical Implant Communication Services (MICS) band in 402-405 MHz frequency range, dedicated for implantable healthcare applications [1]. In 2009, modifications

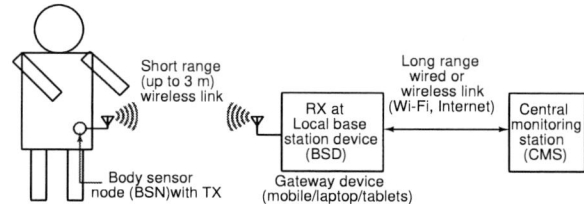

Fig. 1: Depiction of communication links for remote health monitoring applications

were done to MICS and Medical Device Radiocommunication (MedRadio) band was announced between 401-406 MHz frequency range for wearable and implantable medical applications [2]. The Wireless Planning Commission (WPC) in India has also allowed to use the unlicensed 402-405 MHz band for biomedical applications [3]. In MedRadio band the maximum radiated effective isotropic power ($EIRP_{max}$) is set to -16 dBm (25 μW), maximum allowable channel bandwidth is 300 kHz and the maximum communication range is specified as 3 m [4]. MedRadio band has become very popular for short range biosignal communication [5] and many works have been reported recently for low power TX suitable for BSN [6]- [11].

Architectural decisions help significantly in reducing the power consumption in MedRadio band TX. Due to the extremely low $EIRP_{max}$, linearity of power amplifier (PA), which is usually the output stage of a TX, is not an issue and therefore low power non-linear PA can be utilized in these TX [12]. However, RF carrier generation causes the major power consumption in MedRadio TX. Moreover, TX in bio-medical radios are often operated at very low duty cycles to conserve power. For example, MedRadio band uses the listen before talk (LBT) and low-power low-duty cycle (LPLDC) protocols which allow duty cycles of <1% for signal transmission [4]. Therefore, architectures which ensures low power RF carrier generation are the most suitable for MedRadio band TX. Another aspect for reducing power is to simplify the TX architecture with appropriate choice of modulation scheme. Due to the implementation simplicity, frequency shift keying (FSK) and on-off keying (OOK) have become the most popular choice for low power MedRadio band TX [6]- [11].

For RF carrier generation, typically phase locked loop (PLL) is used in transmitters but it is power hungry and not suitable for low power bio-medical applications. Another issue with

978-1-5386-3693-0/18 $31.00 © 2018 IEEE

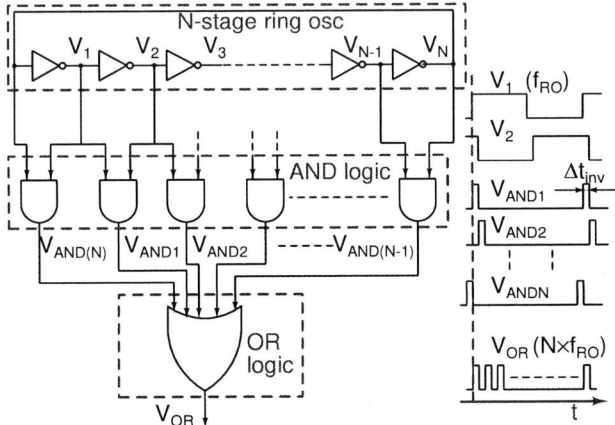

Fig. 2: Principle of edge combination to generate RF carrier

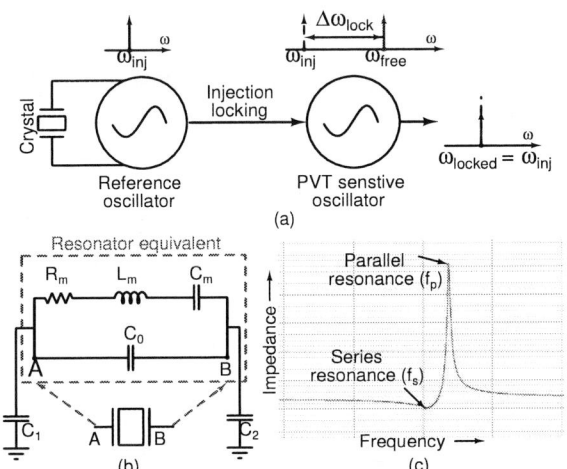

Fig. 3: (a) Depiction of injection locking, (b) electrical equivalent of a crystal resonator and (b) impedance characteristics of resonator showing series and parallel resonance modes

PLL is its high settling time which sometime poses difficulties to the low duty cycled MedRadio band TX for healthcare applications. In the recent past, injection locking principle has been used as the low power alternative for PLL [8], [9], [13], [14]. Injection locked systems have very less settling time and their noise performance is equivalent to type-1 PLL [15], which makes them suitable for highly duty cycled MedRadio band TX.

In this work, an energy efficient OOK TX is presented for a WHD in MedRadio band. The proposed TX is PLL-less and uses low power frequency multiplication technique for RF carrier generation. The proposed design is implemented in 180 nm mixed-mode CMOS technology. For proof of concept, utility of proposed design is demonstrated with the communication of pseudo random bit sequence (PRBS) by using a custom designed test chip [16] and a commercial off-the-shelf (COTS) receiver by Texas Instrument [17].

The paper is organized as follows. In section II, principles of edge combining and injection locking are described and TX architecture is presented. Section III presents implementation details of the proposed TX. In section IV, simulation and measurement results are shown. Section V concludes the paper.

II. Theory and Architecture

The proposed TX utilizes edge combination as the frequency multiplication technique for low power RF carrier generation. Moreover, injection locking technique has been also used in the TX to stabilize the RF carrier. In this section, principle of edge combination and review of injection locking technique are presented. A brief discussion on the choice of modulation scheme is also given followed by the architecture of the proposed TX.

A. Principle of edge combination

The principle of edge combination for RF carrier generation is shown in Fig. 2. As shown in the figure, every two consecutive phases of an N-stage ring oscillator (RO) are ANDed to generate N waveforms with the same frequency (f_{RO}) as that of RO. The pulse duration of an ANDed waveform is equal

to the delay of an inverter (Δt_{inv}) of the RO. Inverters are designed such that $\Delta t_{inv} = \frac{1}{2 \times N \times f_{RO}}$. All these pulses are combined with an OR logic which results in N short pulses in the duration of one cycle of RO signal and frequency multiplication is achieved ($f_{RF} = N \times f_{RO}$). Since RO is very susceptible to the PVT variations, it is highly desirable to stabilize its frequency. Typically, for edge combiners RO is used in a PLL or delay locked loop (DLL) which stabilizes the RO frequency and improves the phase noise of the edge combined signal [18]- [21]. However, PLL/DLL based approach is expensive in terms of power and not suitable for low duty cycled bio-medical TX. Injection locking principle can be used as a low power alternative to PLL/DLL, which is described in the following subsection.

B. Injection locking

As depicted in Fig. 3(a), injection locking refers to the phenomenon in which a self oscillating circuit (ω_{free}) is force to oscillate at the frequency (ω_{inj}) of a reference injection signal if the difference between the two frequencies is within a specified locking range ($\Delta\omega_{lock}$) [15]. Recently, many works have utilized the principle of injection locking for low power BSN transmitters [8], [9], [14].

Usually, crystal oscillators (CO) are used as the stable reference for injection locking because crystal resonators have inherently high quality factors ($\approx 10^5$) and very high frequency stability. The electrical equivalent of a crystal resonator is shown in Fig. 3(b), where L_m, C_m and R_m are the motional inductance, capacitance and resistance, respectively. C_0 is the parasitic capacitance of the crystal package, C_1 and C_2 are the load capacitances due to the CO circuit [22]. The crystal has two modes of resonance - series and parallel. The impedance characteristics of the crystal are plotted in Fig. 3(c), where $f_s \left(= \frac{1}{2\pi\sqrt{L_m C_m}} \right)$ and $f_p \left(= f_s \left\{ 1 + \frac{C_m}{2(C_0 + C_{load})} \right\} \right)$ are the series-resonance or resonant and parallel-resonance or anti-

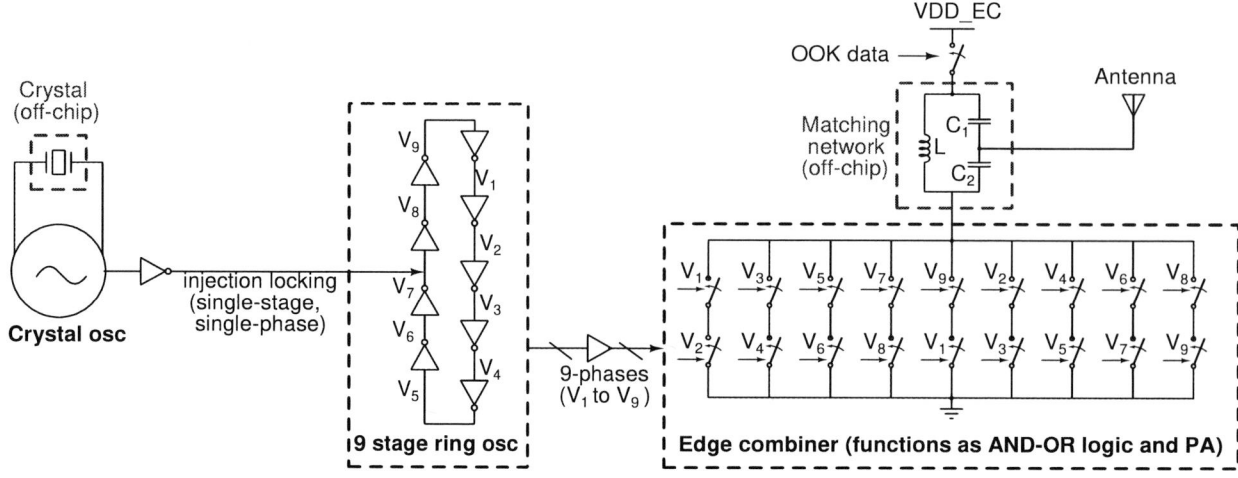

Fig. 4: Block diagram of the proposed OOK TX architecture

resonant frequencies, respectively, where $C_{load} = \frac{C_1 C_2}{C_1 + C_2}$ [22].

C. Modulation

Due to the implementation simplicity, FSK and OOK modulation are often used in low power bio-telemetry systems [6] - [11]. In edge combiner based FSK TX, modulation is done before the frequency multiplication process [8], [21]. For this, frequency of the low frequency RO is deviated by Δf_{REF} which gets multiplied by the factor N by the edge combiner and overall frequency deviation for FSK becomes $N \times \Delta f_{REF}$ at the TX output. In PLL-less injection locked systems, frequency of RO is changed by changing the frequency of reference CO. Deviation of Δf_{REF} in CO is achieved by pulling its anti-resonant frequency (f_p) with the help of an extra capacitor (ΔC_{FSK}), which is kept parallel to the crystal resonator. Using expressions of f_s and f_p, Δf_{REF} can be given by the following equation:

$$\Delta f_{REF} = \frac{f_s C_m \Delta C_{FSK}}{2(C_0 + C_{load})(C_0 + C_{load} + \Delta C_{FSK})} \quad (1)$$

Minimum transconductance ($g_{m_{critical}}$) required for sustained oscillations of CO is given by Eq. (2) [22].

$$g_{m_{critical}} = \frac{2\pi f_s C_m}{Q.P^2} \quad (2)$$

where, Q is the resonator's quality factor and P is the frequency pulling parameter defined by Eq. (3) [22].

$$P = \frac{C_m}{2(C_0 + C_{load} + \Delta C_{FSK})} \quad (3)$$

P reduces in presence of ΔC_{FSK}, which increases $g_{m_{critical}}$ significantly resulting in many time increase in power consumption. For example, with f_s = 44.5 MHz (C_0 = 4 pF, R_m = 12.34 Ω, C_m = 10.4 fF and L_m =1.23 mH) [23] and C_{load} = 2.5 pF (considering IC pin and package parasitics [24]), the required $\Delta C_{FSK} \approx$ 8 pF for $\Delta f_{REF} \approx$ 20 kHz. P reduces by about 2.2 times in the presence of ΔC_{FSK} (Eq.

(3)) and $g_{m_{critical}}$ increases by about 5 times (Eq. (2)), which results in significant increase in power consumption in CO.

For OOK operation, frequency pulling is not required as OOK can be done simply by turning on and off the TX output stage. Moreover, due to the transmission of two different RF carriers (for data '1' and '0') the output spectrum spread is more in FSK as compared to OOK where only single RF carrier is transmitted for data '1' and no carrier is transmitted for data '0'. Therefore, tighter injection requirements, such as multi-stage multi-phase locking, are needed for improved phase noise performance in FSK systems [8], [25]. Above facts indicate that the power consumption in OOK is lower as compared to FSK operation for MedRadio band TX and therefore OOK is used in this work.

D. TX architecture

Block diagram of the proposed OOK TX is shown Fig. 4. TX has a 9 stage RO which runs near 45 MHz. 9 phases of the RO are fed to an edge combiner, which results into the frequency multiplication by a factor of 9 and generates RF carrier in 400 MHz MedRadio band. However, due to the PVT variations, the RO frequency drifts from the desirable value. Therefore, in order to stabilize the RO frequency it is injection locked to an on-chip crystal oscillator ($f_{REF} \approx 45 MHz$). As shown in [25], multiple injection stages can improve the phase noise performance of the injection locked systems. However, the phase noise requirements are very relaxed ($\approx$ -90 dBc/Hz at an offset of 300 kHz) for MedRadio band TX [2] and therefore single-stage, single-phase injection locking is used for reduced power consumption in this work. Edge combination is achieved by combination of switches as shown in the Fig. 4 [26]. The switches in series of each branch accomplishes the AND operation and direct wire connection of all the branches add the switched currents and completes the OR operation of the edge combiner as discussed in section II-A. The ORed current pulses are passed through an off-chip capacitor tapped LC matching network which transfers

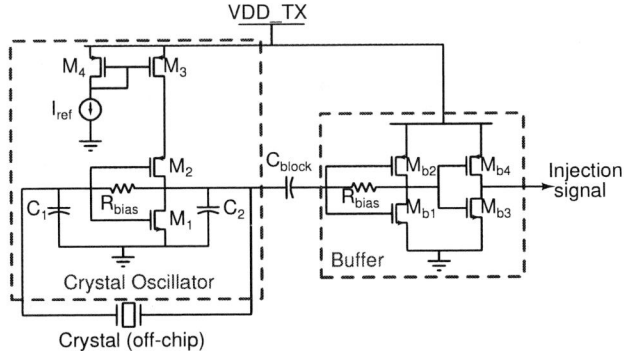

Fig. 5: Schematic of crystal oscillator followed by a buffer

the output power to 50 Ω antenna while providing adequate filtering [12]. OOK operation is achieved by modulating the supply of edge combiner with the base band data as shown in Fig. 4.

III. DESIGN IMPLEMENTATION

As shown in Fig. 5, complementary NMOS-PMOS Pierce oscillator topology is chosen for crystal oscillator implementation. Crystal oscillator is followed by a buffer which makes its output rail-to-rail. The buffer output goes to the RO as the injection signal through an injection device. The injection device has been implemented as a CMOS inverter. As shown in Fig. 6(a), RO is implemented by 9 CMOS inverters. For characterization of RO, an on-chip buffer is designed which can drive 20 pF of capacitive loads of the measuring instruments such as digital oscilloscopes. As shown in Fig. 6(b), edge combiner is implemented with NMOS switches. OOK operation is achieved by modulating the edge combiner supply with the input data. For OOK operation with square wave at a data rate of R_b, along with the main component spurs are also produced at $f_{RF} \pm \frac{k \times R_b}{2}$, where k is an integer. This happens due to the mixing action between edge combiner output and OOK data. Capacitor tapped LC network topology is used for off-chip matching network design, which provides sufficient rejection to unwanted out-of-band spurs and at the same time transfers the maximum power to the 50 Ω antenna [12].

IV. SIMULATION AND MEASUREMENT RESULTS

The proposed TX has been designed in 180 nm mixed-mode CMOS technology. Simulations were performed on the extracted design after the placement and routing of the TX layout on the IC padframe. Transient response of the TX output and the ring oscillator are shown in Fig. 7(a). It shows that the f_{RO} under injection locked condition is 44.54 MHz, which is exactly equal to the crystal oscillator frequency and TX output is at 401 MHz which is about 9 times of 44.54 MHz. From simulations, the phase noise of the TX output under injection locked condition was found to be about -120 dBc/Hz at an offset of 300 kHz which is much better than the desired -90 dBc/Hz at 300 kHz offset for MedRadio band

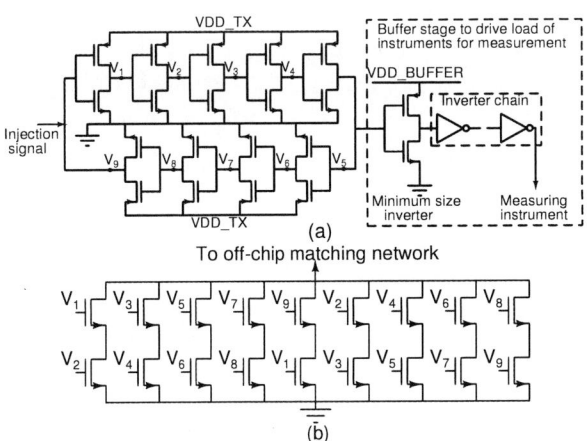

Fig. 6: Schematics of (a) 9 stage ring oscillator and (b) edge combiner

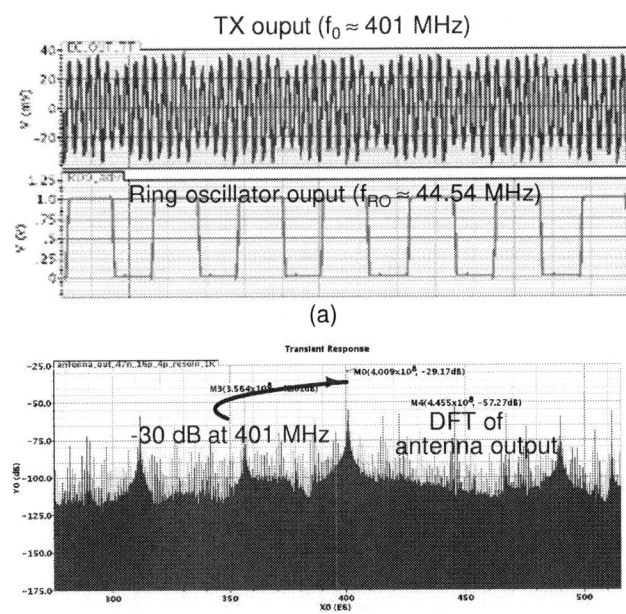

Fig. 7: Post layout transient simulation results showing (a) edge combiner and ring oscillator outputs under injection locked condition and (b) DFT of signal at antenna output

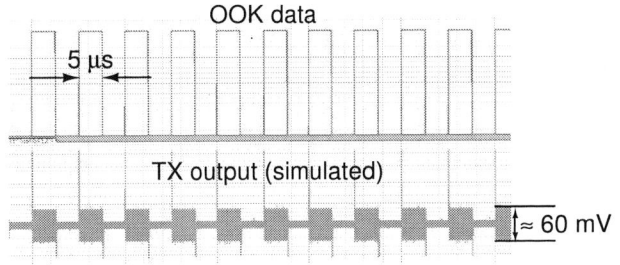

Fig. 8: Post layout simulation results showing transient output of TX for OOK operation at 200 kb/s

978-1-5386-3693-0/18 $31.00 © 2018 IEEE   223

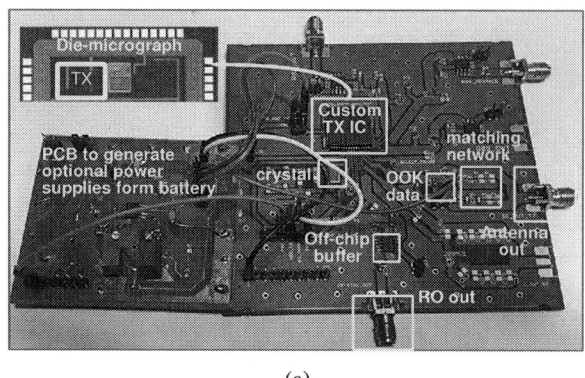 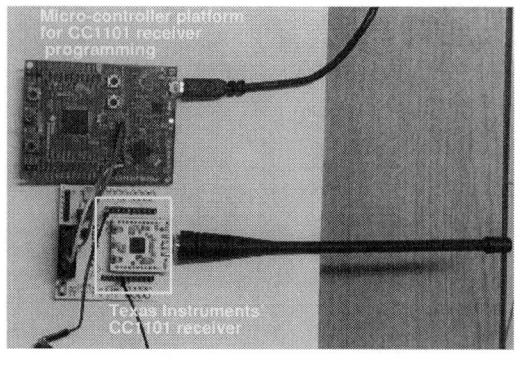

(a)   (b)

Fig. 9: (a) PCB and photo of the part of the die micrograph for OOK TX, (b) Texas Instruments' CC1101 receiver

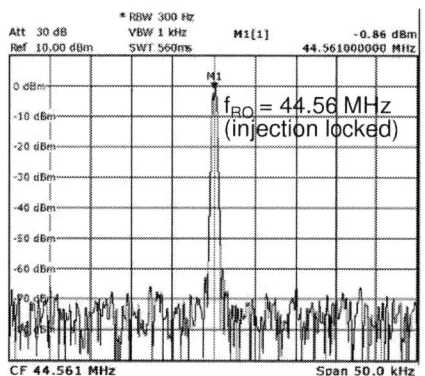

Fig. 10: Measured output spectrum of RO under locked condition

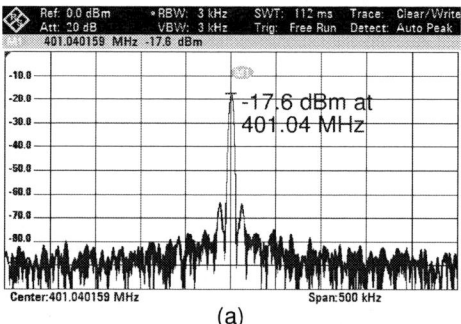

(a)

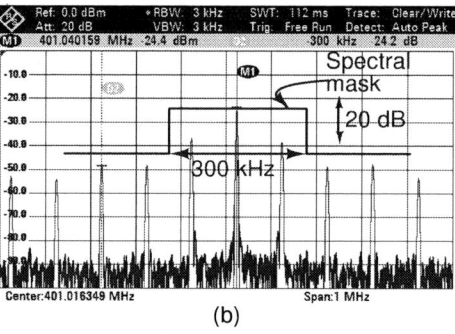

(b)

Fig. 11: Measured output spectrum of the TX (a) without OOK data (b) with OOK data at 200 kb/s

applications [2]. Fig. 8 shows the transient response of the TX output for OOK operation at data rate of 200 kb/s which is sufficiently high for biosignal communication.

For proof of concept, operation of the proposed TX has been shown with a custom designed IC, fabricated in 180 nm mixed mode CMOS technology [16]. Fig. 9 (a) shows the PCB used for the measurements of the proposed TX along with the part of the die micrograph. PCB has been designed carefully in order to reduce various parasitics as described in [27]. All the measurements were performed on ceramic quad flat packaged (CQFP) IC. For measuring the spectrum of RO, an off-chip buffer was used which can drive the 50 Ω load of a spectrum analyzer [28]. A crystal resonator with 44.56 MHz frequency was used for the measurements [23].

Measurement results show that the proposed TX consumes an average power of 71 μW in OOK operation (50% duty cycle) at a data rate of 200 kb/s with an energy efficiency of 0.36 nJ/bit. Power supply of EC was modulated with the baseband data for OOK operation. Fig. 10 shows the measured RO output which is locked to the crystal oscillator frequency of 44.56 MHz. Fig. 11 shows the measured output spectrum of the TX at 401.04 MHz, which confirms the desired frequency multiplication by a factor of 9. As shown in Fig. 11(a), without OOK data, TX output is at -17.6 dBm. For OOK operation,

TX output easily meets the spectral mask requirement at a data rate of 200 kb/s, as shown in Fig. 11(b). OOK operation of TX was also tested with commercial-off-the-shelf receiver (COTS-RX) by TI as shown in Fig. 9(b). For this, a wireless communication link was established between TX and COTS-RX for a maximum distance of 2 m. A pseudo random bit sequence (PRBS) was given to the TX as OOK data and communication was done. Fig. 12 shows the measured transmitted and received PRBS data at 200 kb/s. The measured bit error rate for the experiment was less than 10^{-3} which is good enough for the communication of biosignals.

A comparison of the proposed OOK TX with other OOK TX near 400 MHz frequency are presented in table I. As

978-1-5386-3693-0/18 $31.00 © 2018 IEEE   224

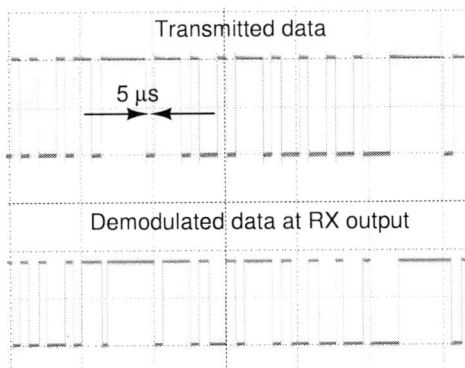

Fig. 12: Measured transmitted and received PRBS data for OOK operation at 200 kb/s for communication range of 2 m

shown in the table, TX of [9] has very low energy/bit, however power consumption of off-chip injection source is not reported. TX presented in [11] operates at 54 Mb/s but does not meet spectral mask requirement. As compared to other works, the proposed TX in this work consumes extremely less power and operates at sufficiently higher data rate of 200 kb/s while meeting the spectral mask requirement of the MedRadio band.

TABLE I: Comparison of proposed work with other OOK TX near 400 MHz

Parameter	[6]	[7]	[9]	[10]	[11]	This work
Tech (nm)	180	180	90	130	180	180
Freq. (MHz)	400	400	400	403	434	401
Output power	−2.17 dBm	−17.2 dBm	−17 dBm	−20.9 dBm	−10 dBm	−24 dBm
Supply	1.8 V	1.8 V	0.6 V	Battery less	1.8 V	1 V
Power	3.1 mW	840 μW	160 μW†	3.32 mW	5 mW	71 μW
Data rate	2 Mb/s‡	2 Mb/s‡	1 Mb/s	100 kb/s	54.2 Mb/s‡	200 kb/s
Energy/bit	1.53 nJ/bit	0.42 nJ/bit	0.16 nJ/bit	33.2 nJ/bit	0.092 nJ/bit	0.36 nJ/bit

†Injection locking is done by using an off-chip instrument and power consumption of injection locking reference is not included. ‡Does not meet spectral mask requirement.

V. CONCLUSION

In this work, an ultra low power sub-100 μW OOK TX is presented for wearable healthcare devices in 400 MHz MedRadio band. The proposed TX is based upon the principle of edge combination for RF carrier generation. To reduce the power, single phase injection locking is proposed for 9 stage ring oscillator. TX is implemented in 180 nm mixed-mode CMOS technology. Measurement results show that the proposed TX consumes an average power of 71 μW for OOK operation and meets the spectral mask requirement at 200 kb/s while radiating -24 dBm power. PRBS data communication is also demonstrated successfully with the proposed TX and a COTS receiver for maximum communication range of 2 m with a BER $< 10^{-3}$.

REFERENCES

[1] MICS Band Plan Federal Communication Commission, Part 95, FCC Rules and Regulations, Jan. 2003.

[2] Medical Device Radiocommunications Service. FCC [Online].

[3] "National Frequency Allocation Plan 2011," WPC, Govt. of India.

[4] MedRadio transmitters in the 401-406 MHz band. Federal Communication Commission, Part 95.627, FCC Rules and Regulations,June 2014.

[5] A. Srivastava et al., "Bio-Telemetry and Bio-Instrumentation Technologies for Healthcare Monitoring Systems," in IEEE Region 10 Humanitarian Technology Conference, Dec 2016, pp. 1–6.

[6] J. Liu et al., "An Ultra-Low Power 400MHz OOK Transceiver For Medical Implanted Applications," in Proc. of the ESSCIRC, Sept 2011.

[7] L.C. Liu et al., "A Medradio-Band Low-Energy-Per-Bit CMOS OOK Transceiver for Implantable Medical Devices," in Proc. IEEE Biomed. Circuits Syst,, Nov 2011, pp. 153–156.

[8] J. Pandey et al., "A Sub-100 μW MICS/ISM Band Transmitter Based on Injection-Locking and Frequency Multiplication," IEEE JSSC, vol. 46, no. 5, pp. 1049–1058, May 2011.

[9] C. Ma et al., "A Near-Threshold, 0.16 nJ/b OOK-Transmitter With 0.18 nJ/b Noise-Cancelling Super-Regenerative Receiver for the Medical Implant Communications Service," IEEE Trans. Biomed. Circuits Syst., vol. 7, no. 6, pp. 841–850, Dec 2013.

[10] H.C. Chen et al., "Batteryless Transceiver Prototype for Medical Implant in 0.18 μm CMOS Technology," IEEE Trans. Microw. Theory Techn., vol. 62, no. 1, pp. 137–147, Jan 2014.

[11] C.K. Ho et al., "A High Datarate Wideband OOK Transmitter For Wireless Neural Signal Recording," in IEEE EDSSC, June 2015.

[12] T. Lee, "The Design of CMOS Radio-Frequency Integrated Circuits,," Second edition. Cambridge publications.

[13] Y.H. Chee et al., "An Ultra-Low-Power Injection Locked Transmitter for Wireless Sensor Networks," IEEE JSSC, Aug 2006.

[14] X. Liu et al., "A 13 pJ/bit 900 MHz QPSK/16-QAM Band Shaped Transmitter Based on Injection Locking and Digital PA for Biomedical Applications," IEEE JSSC, vol. 49, no. 11, pp. 2408–2421, Nov 2014.

[15] R. Adler, "A Study Of Locking Phenomenon In Oscillators," Proc. IEEE, vol. 61, pp. 1380-1385, Oct. 1973.

[16] A. Srivastava et al., "Bio-WiTel: A Low Power Integrated Wireless Telemetry System for Healthcare Applications in 401-406 MHz Band of MedRadio Spectrum," IEEE J. of Biomed. and Health Informat., vol. PP, no. 99, pp. 1–1, 2016.

[17] "CC1101 Low-Power Sub-GHz RF Transceiver." Texas Instruments.

[18] G. Chien et al., "A 900 MHz local oscillator using a DLL-based frequency multiplier technique for PCS applications," in IEEE ISSCC, Feb 2000, pp. 202–203.

[19] O. Casha et al., "Analysis of the Spur Characteristics of Edge-Combining DLL-Based Frequency Multipliers," IEEE TCAS II: Express Briefs, vol. 56, no. 2, pp. 132–136, Feb 2009.

[20] F.R. Liao et al., "A Waveform-Dependent Phase-Noise Analysis for Edge-Combining DLL Frequency Multipliers," IEEE Trans. Microw. Theory Tech, vol. 60, no. 4, pp. 1086–1096, April 2012.

[21] R.R. Manikandan et al., "A Digital Frequency Multiplication Technique for Energy Efficient Transmitters," IEEE Trans. VLSI) System, vol. 23, no. 4, pp. 781–785, April 2015.

[22] E. Vittoz et al., "High-Performance Crystal Oscillator Circuits: Theory and Application," IEEE JSSC, 1988.

[23] "Model 406, Surface Mount Quartz Crystal." CTS Electronic Components. Available from http://www.ctscorp.com/components/Datasheets/008-0260-0.pdf.

[24] "Performance Characteristics of IC Packages." Packaging Databook, Intel. Available from http://www.intel.in/content/dam/www/public/us/en/documents/packaging-databooks/packaging-chapter-04-databook.pdf.

[25] P. Kinget et al., "An injection-locking scheme for precision quadrature generation," IEEE Journal of Solid-State Circuits, vol. 37, no. 7, pp. 845–851, Jul 2002.

[26] J.N. Pandey et al., "A Low Power Frequency Multiplication Technique for ZigBee Transciever," in Proc. of VLSID, Jan 2007, pp. 150–155.

[27] A. Srivastava et al., "LNA-LO Co-design Considerations for Low Intermediate Frequency Receivers in 401-406 MHz MedRadio Spectrum for Healthcare Applications," in Proc. of VLSID, Jan 2017, pp. 175–180.

[28] "BUF-602, High-Speed, Closed-Loop Buffer." Texas Instruments. Available from http://www.ti.com/lit/ds/symlink/buf602.pdf.

2018 31th International Conference on VLSI Design and 2018 17th International Conference on Embedded Systems

Time-Based PWM Controller for Fully Integrated High Speed Switching DC-DC Converters – An Alternative to Conventional Analog and Digital Controllers

Qadeer A. Khan
Dept. of Electrical Engineering
Indian Institute of Technology Madras
Chennai, INDIA

Seong-Joong Kim, Pavan Kumar Hanumolu
Dept. of Electrical and Computer Engineering
University of Illinois Urbana-Champaign
Illinois, USA

Abstract—**This tutorial discusses design of a highly integrated, low quiescent current continuous time PWM controller using time-based signal processing. By virtue of the continuous-time digital nature of the time-based PWM controller, it is capable of achieving very high resolution and speed without using any error amplifier and large compensation capacitor or any high resolution A/D converter and digital PWM while preserving all the benefits of both analog and digital PWM controllers. Using time as the processing variable, the controller operates with CMOS-level digital-like signals but without adding any quantization error. A voltage or current controlled ring oscillator is used as an integrator in place of conventional opamp-RC or Gm-C integrator while a voltage or current controlled delay line is used to perform voltage-to-time conversion. Starting with trade-offs with high speed design of conventional analog and digital PWM controllers, the concept of time-based proportional-integral-derivative (PID) controller with complete architecture of a time-based buck converter is presented. The technique was successfully demonstrated on silicon with implementation of 10-25MHz single phase and 30-70MHz 4-phase buck converters in 180nm and 65nm CMOS processes, respectively.**

Keywords-**power management, voltage regulator, dc-dc converter, pulse width modulation (PWM), buck converter, digital pulse width modulation (DPWM), PID controller, voltage controlled oscillator (VCO), voltage controlled delay line (VCDL), VCO-based integrator, VCO-based control, time-based control, high switching frequency, ring oscillator.**

I. INTRODUCTION

As technology is advancing with an effort of integrating loads of features in portable and hand-held devices such as smartphones and tablet PCs, the power density requirement is growing exponentially due to limited board size available on these gadgets. The increasing power demand while maintaining the small form factor has made power management ICs as one of the most challenging and complex modules in a portable/handheld device. For instance battery capacity of 100mAh-200mAh used in a feature phone now increased to 3000mA-4000mAh in a smartphone. Similarly, number of power supplies used in a smartphone are much larger compared to power supplies used in a feature phone. The power requirement of such devices is usually met by voltage regulators or dc-dc converters which could be either linear or switching. A voltage regulator ensures constant and stable voltage supply across varying operating conditions

such as load current, line voltage, temperature, process, etc. Figure 1 shows the power management system of a smartphone which consists of various switching and linear regulators.

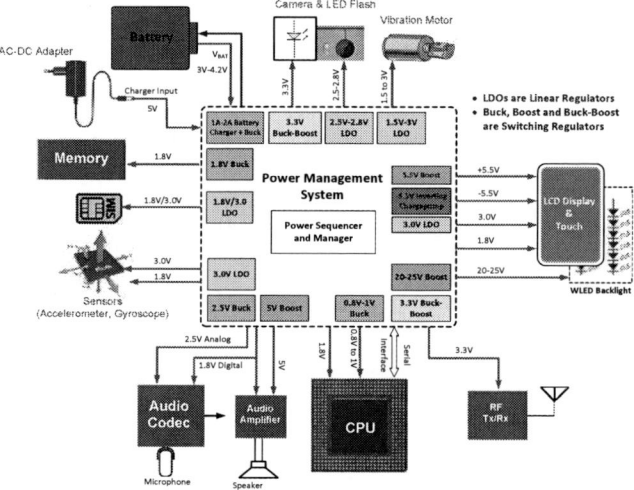

Figure 1. Power Management System in a Smartphone

Simpler design and least external components count make linear regulators highly cost effective but their application is limited to either low power demand (such as sensors, camera modules, etc.) or low drop-out, mainly due to poor efficiency at higher drop-out voltage [1]. On the other hand, high efficiency and capability of operating across wide load and input/output voltage range make switching dc-dc converters superior for higher power demand such as processors, memories, displays, audio amplifiers, RF power amplifiers, etc. Therefore switching dc-dc converters become ideal choice over linear regulators for a battery powered portable device where area and efficiency are the most important aspects. A switching converter can be implemented with a fixed frequency based pulse width modulation (PWM) or a variable frequency hysteretic controller. In applications where wide variation in switching frequency is tolerable, hysteretic controller is preferred because of its simpler design and faster dynamic response. However, in noise sensitive applications such as mobile phones and other wireless applications, fixed frequency PWM controllers are proffered over hysteretic. These PWM controllers often require large capacitors that are either

978-1-5386-3693-0/18 $31.00 © 2018 IEEE
226

impossible to integrate on-die or incur prohibitively large area penalty. Using external components takes away premium board space and increases system cost. With the ease of integration provided by state-of-the-art semiconductor device technologies, power MOSFETs can easily be integrated with the controller on a single chip. Therefore, size of a power module is mainly limited by its on-board passive components such as LC filter (inductor and capacitor). Figure 2 shows the power management module of Samsung's latest flagship smartphone Galaxy S8 where about 2/3rd of the module area is occupied by off-chip inductors and capacitors.

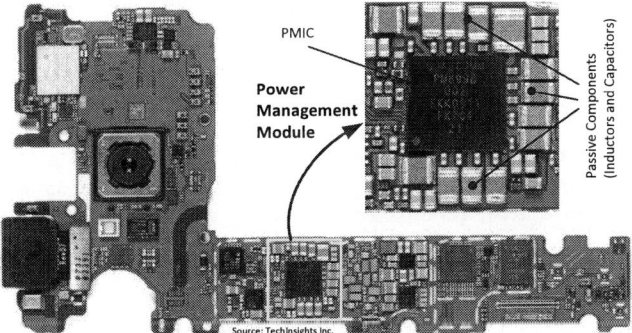

Figure 2.   Power Management in Samsung Galaxy S8 Smartphone
(*Source: Techinsights Inc.*)

II.   SPEED TREND IN DC-DC CONVERTERS

Size of off-chip passive components scale inversely proportional to F_{SW}. In particular, inductor whose value is typically in the range of few µH occupies large area on the board. Figure 3 shows trend in size reduction of a dc-dc converter module with increasing F_{SW}. For the same output ripple, value of inductor can be reduced by four times if F_{SW} is doubled.

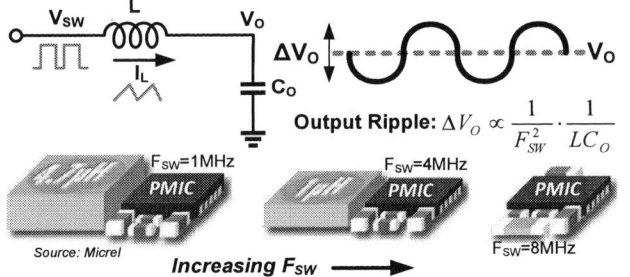

Figure 3.   Speed Trend in Switching DC-DC Converters
(*Source: Micrel Inc.*)

This size benefit is usually compromised with efficiency due to increased switching losses when switched at higher speed. However, with parasitic reduction in state-of-the-art semiconductor process, packaging, and passive component technologies, switching losses are minimized significantly and remain less dominant when converter is operated at full load [2]. Light load efficiency can be improved by using burst mode control and segmented output control can also be used to further improve the efficiency [3], [4]. Hence most

viable approach to achieve both small form factor and fast transient response is increasing the switching frequency. Simulations indicate that around 90% peak efficiency is achievable at F_{SW} up to 25MHz at a load current of few hundred mA [7] at input voltage of 1.8V. Therefore most of the challenges in designing switching converter beyond 10MHz is pushed towards the design of the PWM controller.

III.   TRADE-OFFS IN CONVENTIONAL PWM CONTROLLER

Figure 4 shows a buck converter with Type-III or proportional-integral-derivative (PID) compensator which occupies most of the controller's area due to large size of capacitors. The need of high gain-bandwidth error amplifier (EA) also puts a constraint on quiescent power when converter is designed for higher bandwidth which is usually 1/10th of F_{SW}. For instance, a converter switching at 10MHz with 1MHz bandwidth would require error amplifier of more than 10MHz gain-bandwidth (GBW) to ensure stable operation and high output accuracy. As transistors used in analog circuits are usually kept large, to operate in saturation region with high output resistance, we do not benefit much from process scaling. Therefore, designing an error amplifier with such a high GBW becomes quite challenging even with state-of-the art process technologies.

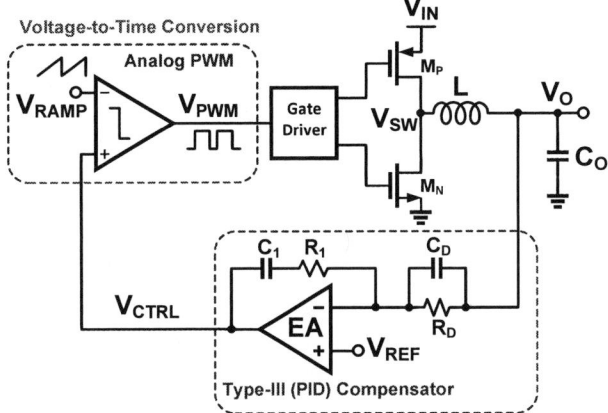

Figure 4.   An Analog PWM Controlled Buck Converter

Another limiting factor in high speed dc-dc converter is the finite delay associated with the PWM comparator and power MOSFETs. Assuming t_p is the propagation delay in PWM comparator, t_{rst} is the reset time of PWM ramp and t_{d_sw} is the delay of power MOSFETs, M_P and M_N (including gate driver delay), the total loop delay can be expressed as:

$$t_d = t_p + t_{rst} + t_{d\_sw} \qquad (1)$$

In order to operate at 10MHz with the duty cycle range of 10%-90%, the required delay should be less than 5ns which is quite difficult to achieve with low power consumption. In such a case, the converter's duty cycle range is limited. Limiting the duty cycles to a narrow range not only affects the transient response but also limits the input/output operating voltage range.

978-1-5386-3693-0/18 $31.00 © 2018 IEEE   227

In a digitally controlled dc-dc converter on the other hand, as shown in Figure 5, error voltage, V_e is digitized and processed in a digital PID compensator. The digital-to-time conversion takes place in a digital PWM (DPWM) which generates the desired duty cycle. It requires high precision analog-to-digital converter (ADC) and DPWM to avoid limit cycling [5] and maintain good accuracy in the regulated output voltage. Since power consumption of both ADC and DPWM grows exponentially with speed and resolution, it becomes hard to meet the low power requirement when converter is switching at high speed.

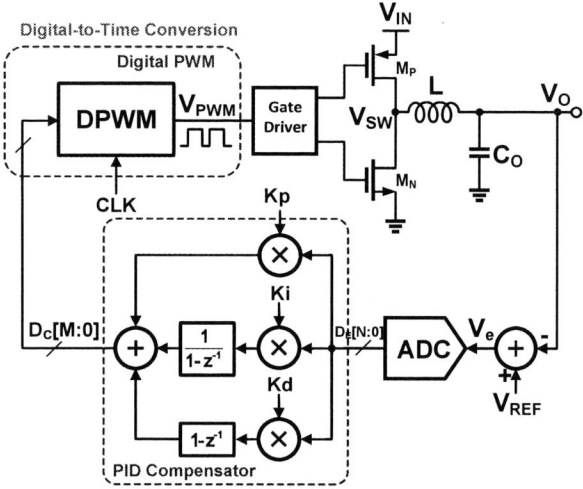

Figure 5. A Digital PWM Controlled Buck Converter

In an effort to overcome the above issues with conventional analog and digital PWM controllers, a highly integrated, continuous time PWM controller based on time-based signal processing has been proposed [6], [7]. By using time as the processing variable, it eliminates the need for wide bandwidth amplifiers, large on-chip compensation capacitor, PWM modulator, high resolution ADC and digital pulse width modulator (DPWM), while still operating with CMOS-level digital-like signals. In other words, time-based approach combines the advantages of both analog (high accuracy, low quiescent current) and digital (low voltage operation, smaller area and process scalability) controllers.

IV. TIME-BASED SIGNAL PROCESSING

The concept of time-based signal processing (TSP) was originally derived from phase-locked loop (PLL) where voltage-controlled oscillator (VCO) acts as a phase integrator. So far, the application of TSP has been mostly in ADCs where op-amp based integrators are replaced with VCOs [8]. Let's try to first understand the basic concept of how a VCO is used as an integrator. We know that the relationship between phase and frequency of an oscillator can be expressed as:

$$w = d\phi / dt \quad \text{or} \quad \phi = \int w\,dt \quad (2)$$

If oscillator is a VCO with gain K_{VCO} rad/sec and controlled by voltage V_{IN} then frequency of VCO can be defined as:

$$w = K_{VCO}V_{IN}$$

Substituting w in (2), we can write:

$$\phi(t) = K_{VCO}\int V_{IN}(t)\,dt \quad (3)$$

Which can be expressed in frequency domain with transfer function:

$$\frac{\phi(s)}{v_{IN}(s)} = \frac{K_{VCO}}{s} \quad (4)$$

We can easily conclude from (3) and (4) that a VCO can be used as a phase or time-based integrator with gain of K_{VCO} where output phase represents integral of the input voltage. In order to replace a conventional opamp-RC or Gm-C integrator with time-based integrator, we need differential input and phase output needs to be converted to voltage. This can be achieved by using two matched VCOs representing inverting and non-inverting inputs followed by a phase detector (PD) and low-pass filter (LPF) as shown in Figure 6. If frequency of VCOs is kept much higher than signal bandwidth then V_{OUT} will simply be integral of differential input, V_{IN}-V_{REF} and behaves exactly same as conventional opamp-RC or Gm-C integrator.

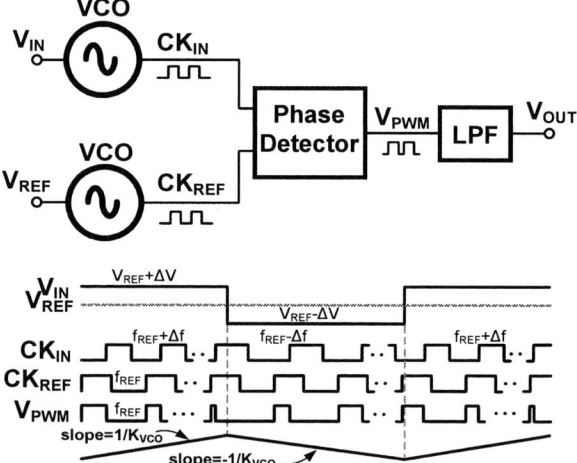

Figure 6. A time-based integrator using differiantial VCOs.

Time-based integrator can be used to transform a conventional voltage-based type-I buck converter into a time-based type-I buck converter as shown in Figure 7. It is interesting to note that time-based buck converter completely eliminates the use of PWM modulator due to the fact that output of phase detector (PD), which can be implemented using a simple SR-latch, is nothing but a PWM signal with duty cycle proportional to phase difference between the outputs of two VCOs. The duty cycle of V_{PWM} is set at every positive edge of reference

978-1-5386-3693-0/18 $31.00 © 2018 IEEE   228

clock (CK$_{REF}$) and reset at positive edge of control clock (CK$_{CTRL}$). Time-based buck converter shown in Figure 7 can also be viewed as a type-I phase-locked loop (PLL). Assuming the reference clock (CK$_{REF}$) frequency is within pull-in range, the feedback loop forces feedback VCO (FVCO) and reference VCO (RVCO) to lock in frequency, which results in regulated output voltage, V$_O$=V$_{REF}$. Under this condition, switching frequency is equal to the free running frequency of RVCO and V$_{PWM}$ duty cycle (D=T$_{ON}$/T$_{SW}$) is equal to (Φ_{CTRL}-Φ_{REF})/2π.

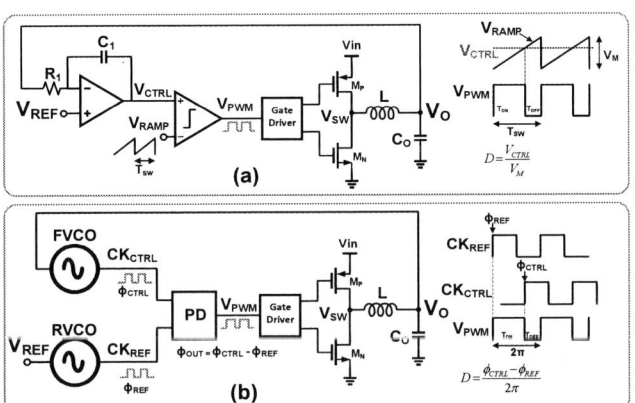

Figure 7. (a) Conventional voltage-based type-I buck converter (b) time-based type-I buck converter.

A time-based integrator performs two main functions, (1) voltage-to-time conversion, and (2) time-based integration. Unlike conventional controller shown in Figure 7 (a), which performs integration in voltage domain and therefore requires a PWM modulator for voltage-to-time conversion, time-based controller does not require a PWM modulator. This is mainly due to the inherent voltage-to-time conversion that happens in time-based integrator.

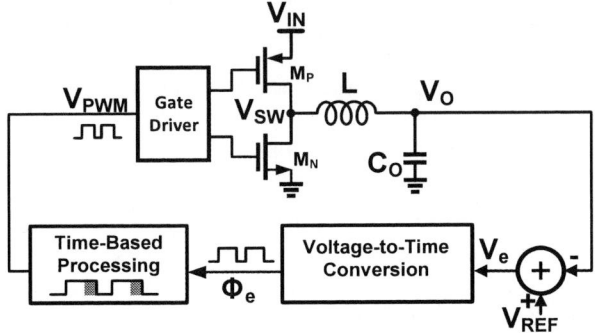

Figure 8. Conceptual block diagram of buck converter using time-based controller

Figure 8 shows the conceptual block diagram of a buck converter with time-based controller which initially performs voltage-to-time conversion and then signal is processed in time domain resulting in a PWM signal required to regulate the output voltage.

V. TIME-BASED PID CONTROLLER

The concept of time-based type-I controller (Figure 7b) can be extended to type-III or PID controller by adding proportional and derivative controls implemented using voltage controlled delay lines (VCDL) as shown in Figure 9.

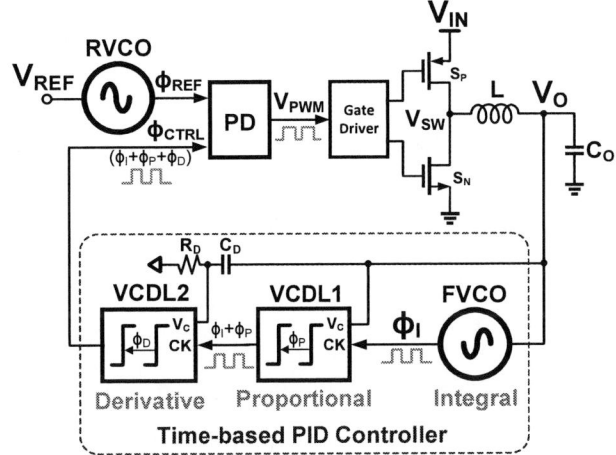

Figure 9. Buck converter using time-based PID controller.

Phase-to-voltage transfer function of time-based PID (TPID) controller can be expressed as:

$$H_{TPID}(s) = \frac{\phi_{CTRL}(s)}{v_O(s)} \qquad (5)$$

$$\phi_{CTRL}(s) = \phi_P(s) + \phi_I(s) + \phi_D(s) \qquad (6)$$

Where, $\phi_P(s)$, $\phi_I(s)$, $\phi_D(s)$ are phase transfer functions of proportional, integral and derivative controls, respectively and can be derived as:

$$\phi_P(s) = K_{VCDL1} \cdot v_O(s) \ ; \ \phi_I(s) = \frac{K_{VCO}}{s} \cdot v_O(s)$$

$$\phi_D(s) = K_{VCDL2} \cdot \frac{R_D C_D s}{1 + R_D C_D s} \cdot v_O(s) \text{ , which can be simplified}$$

as: $\phi_D(s) = K_{VCDL2} \cdot R_D C_D s \cdot v_O(s)$

Assuming pole $1/R_D C_D$ is placed at much higher frequency and outside the UGB.

Substituting $\phi_P(s)$, $\phi_I(s)$, $\phi_D(s)$ in (6), we can derive the final transfer function from (5) as:

$$H_{TPID}(s) = K_{VCDL1} + \frac{K_{VCO}}{s} + K_{VCDL2} \cdot R_D C_D s \qquad (7)$$

With PID coefficients:

$$K_P = K_{VCDL1} \ ; \ K_I = K_{VCO} \ ; \ K_D = K_{VCDL2} \cdot R_D C_D \qquad (8)$$

Transfer function of a conventional voltage mode PID controller of Figure 4 can be expressed as:

$$H_{PID}(s) = K_I \frac{(1 + s/w_{z1})(1 + s/w_{z2})}{s} \ ; \ \text{Which can be}$$

simplified in terms of PID coefficients as:

978-1-5386-3693-0/18 $31.00 © 2018 IEEE   229

$$H_{PID}(s) = K_I \left(\frac{1}{w_{z1}} + \frac{1}{w_{z2}} \right) + \frac{K_I}{s} + \frac{K_I}{w_{z1} \cdot w_{z2}} s \quad (9)$$

Comparing (8) and (9), we get:

$$K_{VCO} = K_I ; \quad K_{VCDL1} = K_I \left(\frac{1}{w_{z1}} + \frac{1}{w_{z2}} \right) ; K_{VCDL2} = \frac{1}{R_D C_D} \cdot \frac{1}{w_{z1} \cdot w_{z2}}$$

Once K_I, w_{z1} and w_{z2} are calculated based on converter design specifications [9], corresponding time based parameters, K_{VCO}, K_{VCDL1} and K_{VCDL2} can be determined.

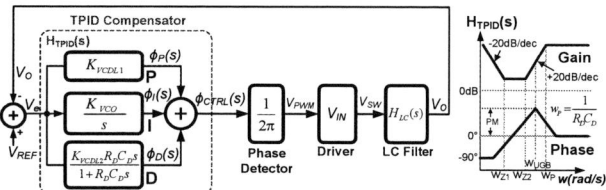

Figure 10. A time-based integrator using differiantial VCOs.

The complete buck converter with TPID controller can be modelled in frequency domain as shown in Figure 10. Gain of phase detector is modelled as *1/2π* and $H_{LC}(s)$ is the transfer function of output stage LC filter.

Performance of TPID based buck converter was demonstrated successfully in 180nm CMOS process [6], [7] with 10MHz to 25MHz switching frequency while consuming quiescent current of only 2µA/MHz. Power density of 3.75Watt/mm² with peak efficiency of 95% was achieved.

VI. MULTI-PHASE TIME-BASED CONTROLLER

Another major advantage of time-based controller is in multi-phase dc-dc converter. Unlike conventional voltage mode controller, which requires multiple PWM modulators and corresponding ramp signals, ring based VCO comes with built-in multiple phases thus does not require exclusive multi-phase generator. A conceptual block diagram of a 5-phase buck converter using 5-phase ring VCO is shown in Figure 11.

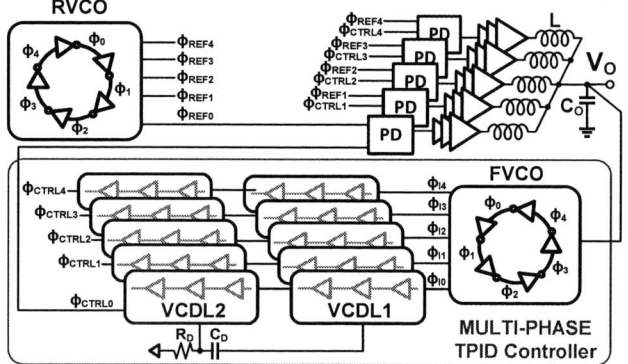

Figure 11. A time-based integrator using differiantial VCOs.

Matching between VCDLs of each phase must be ensured to achieve good phase balancing. Alternatively, for N-Phase controller, VCOs can be run at N-time higher frequency and multiple phases can simply be generated by using a 1/N frequency divider. This guarantees phase matching between phases and does not require active phase balancing used in conventional voltage based multi-phase dc-dc converters. A 4-phase buck converter using time-based PID controller was implemented in 90nm CMOS process [10] with 30MHz to 70MHz switching frequency while consuming quiescent current of only 3µA/MHz. Power density of 2.5Watt/mm² with peak efficiency of 87% was achieved.

VII. EFFECT OF VCO MISMATCH AND CYCLE SLIP IN TIME-BASED CONTROLLER

Even though the time-based controller has many advantages over conventional analog and digital controllers, it has some inherent issues which need to be addressed. These problems are discussed below with their possible solutions:

A. VCO Mismatch

Unlike in voltage mode integrators where the systematic offset due to transistors mismatch is suppressed by the transconductance, the offset in time-based integrator depends upon mismatch and K_{VCO}. Lower K_{VCO} makes the offset due to mismatch appear even larger at the output, V_O. This can be easily understood by considering the requirement of K_{VCO} to design the buck converter shown in Figure 9 at F_{SW}=10MHz and UGB=1MHz [6], [7]. The required K_{VCO} of RVCO and FVCO is ~1MHz/V with free running frequency of 10MHz. Mismatch of 1% between RVCO and FVCO causes a frequency difference of 100kHz. Therefore, according to K_{VCO}=1MHz/V, this 100kHz frequency difference will appear at the output as *100KHz/K_{VCO}=100mV* offset. A generalized equation for offset voltage can be written as:

$$V_{OS} = \frac{\Delta f_{VCO}}{K_{VCO}}$$

Where, Δf_{VCO} is the frequency error between RVCO and FVCO due to mismatch when $V_O = V_{REF}$.

Since K_{VCO} is set by the design specifications and usually much lower compared to switching frequency, even a smaller mismatch of 1% can cause significant error in the output voltage. This offset must be corrected to regulate the output voltage within desired accuracy. Thanks to the phase integration property of VCO, a very small frequency difference could be detected and corrected by running the two VCOs in frequency locked loop (FLL) mode. As shown in Figure 12, the FLL mode is enabled with EN_FLL bit during startup before enabling the dc-dc converter. During FLL mode, the V_O control to FVCO is disconnected and connected to V_{REF}. This forces error input to integrator equal to 0, any mismatch between frequencies of RVCO and FVCO will account to mismatch between the VCOs. This frequency error is accumulated in an accumulator (ACC) to generate a digital control code F_{TUNE}[N:0] to tune one of the VCOs (or both) and lock them in frequency. Once VCOs are

978-1-5386-3693-0/18 $31.00 © 2018 IEEE

locked, tuning code $F_{TUNE}[N:0]$ is latched and FLL mode is disabled by making EN_FLL = 0 which re-connects control input of FVCO back to V_O and closes the feedback loop.

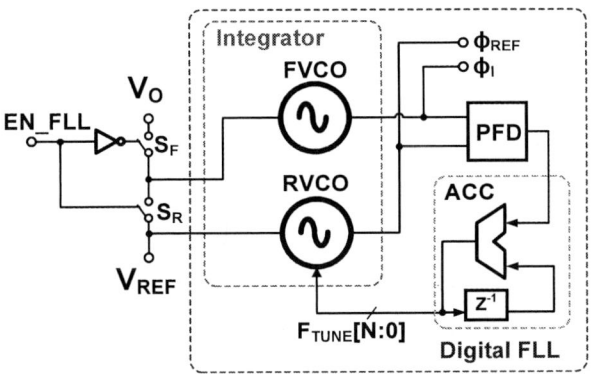

Figure 12. A time-based integrator using differiantial VCOs.

Experimental results indicate that 170mV offset was brought down to under 5mV using this offset correction technique [10]. FLL mode can also be used to synchronize the switching frequency of the converter to an external clock.

B. Cycle Slip

Cycle slip is a well-known behavior in phase-locked loop and by the virtue of time based signaling employed in time-based PWM controller, it also suffers from the same. The main reason for cycle slipping is limited range of the phase detector. Any linear phase detector has a maximum phase detector range of 2π (i.e. one clock cycle) due to phase folding nature of a clock ($2n\pi = 2\pi$). Therefore phase output gets reset if input phase difference becomes multiple of 2π. This causes duty cycle of PWM to go to 0 if the phase difference becomes 2π, as shown in Figure 13a, and phase integrator loses the steady state operating point. In a voltage mode integrator, the output of the integrator is stored on a capacitor so it will cause the duty cycle to saturate at 100%. Since phase cannot be stored like a voltage, the resetting of duty cycle causes the dc-dc converter to lose the output regulation. This will cause an undesirable transient behavior at the output as depicted in Figure 13b.

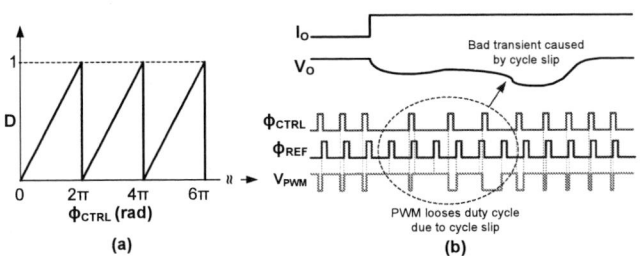

Figure 13. (a) Duty-cycle vs. control phase and (b) duty-cycle slipping in time-based controller.

The problem of cycle slipping can be mitigated by using cycle slip detector and forcing the duty cycle to 100% (or 0%) when duty cycle slip is detected [10].

VIII. CONCLUSION AND FUTURE RESEARCH

Time-based control techniques for the design of high frequency switching buck converters were presented. Using time as the processing variable, the controller combines the advantages of conventional analog and digital controllers. It operates with CMOS-level digital-like signals but without adding any quantization error. Using simple circuits such as ring oscillators, delay lines, and flip-flops, time-based controller eliminates the need for wide bandwidth error amplifier, PWM block used in analog controllers or high resolution ADC and digital PWM block used in digital controllers. As a result, it can be implemented in small area with minimal power consumption.

With the capability of running at much higher speed and very low power consumption, time-based PWM controller provides new research opportunities in the area of dc-dc converter. Time-based concept could also be explored in a current mode control where VCO based integrator can be used as current sensor. The concept could also be explored with other converter topologies such as boost and buck-boost. Implementation of PFM mode to improve the light load efficiency could be another interesting topic for future research.

REFERENCES

[1] Henry J. Zhang, "Basic Concepts of Linear Regulator and Switching Mode Power Supplies," *Linear Technology Application Note*, AN-140, Oct. 2013.

[2] C. Ó. Mathúna et al, "Review of Integrated Magnetics for Power Supply on Chip (PwrSoC)," in *IEEE Transactions on Power Electronics*, vol. 27, no. 11, pp. 4799-4816, Nov. 2012.

[3] H. Krishnamurthy, et al, A 500MHz, 68% efficient, fully on-die digitally controlled buck voltage regulator on 22 nm tri-gate CMOS," in *Proc. Symposium on VLSI Circuits (VLSIC)*, June 2014, pp. 210–211.

[4] J. Xiao, A. Peterchev, J. Zhang, and S. Sanders, "A 4- A quiescent current dual-mode digitally controlled buck converter IC for cellular phone applications," *IEEE Journal of Solid-State Circuits*, vol. 39, no. 12, pp. 2342–2348, Dec. 2004.

[5] A. Peterchev and S. Sanders, "Quantization resolution and limit cycling in digitally controlled PWM converters", *IEEE Trans. on Power Electronics*, pp. 301-308, Jan. 2003.

[6] Q. Khan *et al*, "A 10–25MHz, 600mA buck converter using time-based PID compensator with 2µA/MHz quiescent current, 94% peak efficiency, and 1 MHz BW," in *Proc. Symposium on VLSI Circuits (VLSIC)*, June 2014, pp. 212–213.

[7] S. J. Kim *et al.*, "High Frequency Buck Converter Design Using Time-Based Control Techniques," in *IEEE Journal of Solid-State Circuits*, vol. 50, no. 4, pp. 990-1001, April 2015.

[8] M. Park and M.H. Perrott, "A 78 dB SNDR 87 mW 20 MHz Bandwidth Continuous-Time Delta-Sigma ADC With VCO-Based Integrator and Quantizer Implemented in 0.13 um CMOS," *IEEE Journal of Solid-State Circuits*, vol. 44, pp. 3344-3358, Dec 2009.

[9] T. Hegarty, "Voltage-mode control and compensation: Intricacies for buck regulators" *Article, EDN Network*, June 20, 2008.

[10] S. J. Kim, R. K. Nandwana, Q. Khan, R. C. N. Pilawa-Podgurski and P. K. Hanumolu, "A 4-Phase 30–70 MHz Switching Frequency Buck Converter Using a Time-Based Compensator," in *IEEE Journal of Solid-State Circuits*, vol. 50, no. 12, pp. 2814-2824, Dec. 2015.

Flipped Voltage Follower Low Dropout (LDO) Voltage Regulators: A Tutorial Overview
(Invited Tutorial)

Punith R. Surkanti, Annajirao Garimella and Paul M. Furth
VLSI Laboratory, Klipsch School of Electrical and Computer Engineering
New Mexico State University, Las Cruces, NM, USA
punith@nmsu.edu

Abstract—This tutorial introduces flipped voltage follower (FVF) based low dropout voltage regulators (LDOs) and their recent advances in the literature. Conventional stand-alone LDO ICs often deploy an external output capacitor in the range of hundreds of nanofarads to tens of microfarads with an inherent equivalent series resistance (ESR) in order to attain stability and low overshoot and undershoot during load transients. Because of the large value of the output capacitor, the bandwidth of the LDO is often limited, thus limiting the power supply rejection ratio (PSRR). Higher SoC integration and the necessity to reduce bill of materials cost has resulted in the proliferation of low-value on-chip capacitor LDOs. One of the major requirements of on-chip capacitor LDOs is to achieve high bandwidth and, thereby, good PSRR.

An FVF is an analog voltage buffer circuit with fast local feedback, resulting in high bandwidth. It is capable of sourcing heavy current loads, which makes it an ideal candidate for LDOs. The architecture and design details, such as pole-zero equations, stability, and output impedance, of FVF LDOs are dealt with in detail in this tutorial.

Index Terms—Low dropout voltage regulators (LDOs), Flipped voltage follower (FVF), common-drain transistor amplifier, pole-zero analysis, stability, output impedance, PSRR.

I. INTRODUCTION

The objective of this paper is to elucidate the basic architecture and some of the recent advances in flipped voltage follower (FVF) based low dropout voltage regulators (LDOs). FVF LDOs are also compared to conventional multi-stage LDO architectures.

A. Conventional Output Dominant Pole LDO

Conventional stand-alone LDO ICs require a huge off-chip output capacitor C_{OUT} in the range of hundreds of nanofarads to tens of microfarads in order to attain stability and low overshoot and undershoot during load transients [1]–[14]. C_{OUT} creates a load tracking low-frequency dominant pole, whereas a non-dominant pole is created at the gate of the pass transistor. The equivalent series resistance (ESR) of the off-chip output capacitor R_{ESR} creates a left-half-plane (LHP) zero, which helps in cancelling the non-dominant pole and improve the stability of the LDO. Because of the high output capacitor value, the bandwidth of the LDO is lower and the transient response is slow.

In a closed-loop system, power supply rejection ratio (PSRR) is proportional to the open-loop gain of the system,

to a first order. In conventional LDOs, the DC open-loop gain is high; therefore, the low frequency PSRR is high. Because of the LDO's low bandwidth, the PSRR begins to drop at a low frequency [13]. Beyond the unity gain frequency (UGF), where the open-loop gain is less than unity, the PSRR depends on the output capacitor C_{OUT} alone.

In modern systems the usage of high switching frequency DC-DC converters are increasing in order to reduce the size of the passive elements. These switching converters convert the main variable supply to a regulated voltage and thereafter multiple LDOs are used to generate a quieter low-ripple supply voltage to several points-of-load (POLs). Therefore, LDOs are often required to have good PSRR at high frequencies in order to attenuate the high-frequency switching noise introduced by the DC-DC converters.

Most circuits in portable applications operate with low input supply voltage V_{LINE}. Therefore, POL supply generators operating with low-voltage headroom are gaining importance. Due to bill of materials (BOM) cost, size and integration requirements, SoC designers are restricted to architectures which require fewer off-chip components and input-output (IO) pins. Integrated on-chip capacitor LDOs satisfy these demands [15]–[21].

B. Integrated On-Chip Capacitor LDO with Internal Dominant Pole

The architecture of an integrated on-chip output capacitor LDO is shown in Fig. 1. It consists of a bandgap reference, an

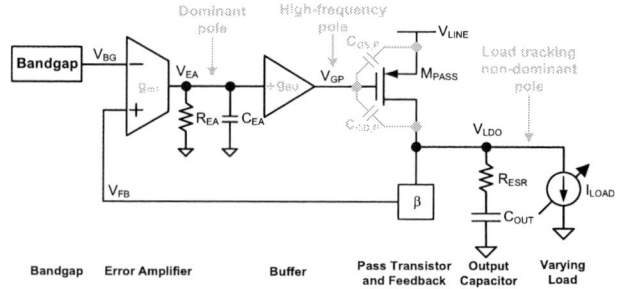

Fig. 1: On-chip capacitor LDO architecture with internal dominant pole situated at the output of error amplifier g_{m1}.

operational transconductance amplifier (OTA) with transconductance g_{m1}, cascaded with a buffer g_{BU} driving the gate of a pass transistor M_{PASS}. The scaled output voltage V_{FB} is fed back to the OTA in a negative feedback loop that regulates the output voltage V_{LDO} based on the bandgap voltage V_{BG} and scale factor β. The regulation of the LDO output voltage depends on the open-loop gain. Since the LDO drives heavy output loads, the gain of the output-stage common-source amplifier created by M_{PASS} is low. Therefore, the input-stage OTA needs to have high gain in order to achieve a high loop gain, even at heavy loads [5], [22]. In addition, high loop bandwidth helps reduce the load transient overshoot and undershoot and allows the LDO output to settle faster.

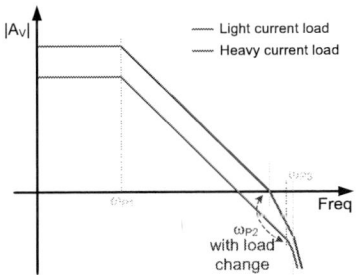

Fig. 2: Frequency magnitude response illustration of an integrated on-chip capacitor LDO of Fig. 1 with pole-zero locations for light and heavy load current conditions. Not drawn to scale.

The pole associated with the output node is non-dominant and tracks with the load. The gate capacitance associated with M_{PASS} is often high. Depending on the resistance at node V_{GP}, the pole associated with that node could become a low frequency dominant pole, limiting the bandwidth. Inserting a buffer g_{BU} with a low output impedance moves the pole associated with the gate of the M_{PASS} to high frequencies [5], [9]. Due to the low input capacitance of g_{BU}, the dominant pole associated with the output node of the OTA, V_{EA} moves to a higher frequency, resulting in an overall bandwidth increase. The location of poles and zeros for heavy and light load conditions are illustrated in Fig. 2. Since the bandwidth of the on-chip capacitor LDO is increased, the PSRR is high at low and moderate frequencies, but still low at higher frequencies, limiting this LDO's usage for wide-band applications.

II. Flipped Voltage Follower LDO Architectures

A flipped voltage follower (FVF) is a voltage buffer, with lower output impedance compared to its common-drain transistor amplifier counterpart [1], [23], [24]. A PMOS FVF has large current sourcing capability with low-voltage dropout from V_{LINE} to V_{LDO} and, hence, is an ideal candidate for LDO implementation [1], [22], [25]–[32].

The schematic of an FVF LDO is shown in Fig. 3. The FVF LDO architecture comprises a bandgap circuit (not shown) which generates V_{BG}, control voltage V_{CTRL} generation circuitry, PMOS pass transistor M_{PASS}, FVF control transistor M_C and bias current I_{B_LDO}. V_{CTRL} drives the gate of M_C. The output of the FVF LDO V_{LDO} is obtained at the source

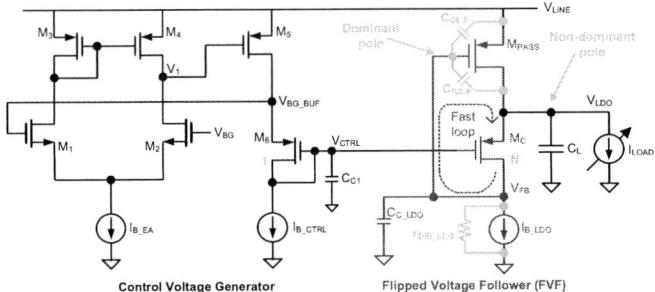

Fig. 3: Flipped voltage follower (FVF) LDO with on-chip output capacitor. The control voltage generator circuit is also shown.

of M_C, which is also the drain of M_{PASS}. Since the current through control transistor is fixed at I_{B_LDO}, the V_{SG} of M_C is ideally constant. Thus, the output V_{LDO} will be one V_{SG} above V_{CTRL}.

The fast local analog feedback loop created by M_C and M_{PASS} of the FVF regulates V_{LDO}. M_{PASS} is configured as a source follower and transistor M_C is a common-gate amplifier, resulting in a moderate open-loop gain.

Because of the local feedback loop, the output range of the FVF is limited on the low side, both by the supply voltage V_{LINE}, V_{CTRL}, and $V_{SG,PASS}$. The output voltage V_{LDO} satisfies

$$
\begin{aligned}
V_{LDO} &\leq V_{LINE} - V_{SDsat,PASS} \\
V_{LDO} &\geq V_{LINE} - V_{SG,PASS} + V_{SDsat,C} \quad (1) \\
V_{LDO} &\geq V_{CTRL} + |V_{THP,C}|
\end{aligned}
$$

where $|V_{THP,C}|$ is the PMOS threshold voltage. This output range is lower than the conventional LDO. This limited range impacts the pass transistor at light load currents, when $V_{SG,PASS}$ is small. Indeed, it is not possible to turn off the pass transistor completely.

The minimum bias current through M_{PASS} is I_{B_LDO} and occurs when the load current is zero. During a load transient from light to heavy load current, V_{LDO} will drop. When V_{LDO} drops, the V_{SG} of control transistor M_C decreases, which pulls down the gate of M_{PASS} to source higher current to the load. As such, V_{LDO} is restored to one V_{SG} above V_{CTRL}. Similarly, during a load transient from heavy to light output current, V_{LDO} will increase because of the high M_{PASS} current. This increases the V_{SG} of M_C and pulls up the gate of M_{PASS} to decrease the current through it. Consequently, V_{LDO} pulls back to one V_{SG} above V_{CTRL}. The slew rate and bandwidth of the FVF loop determines the settling time of the output voltage.

The V_{CTRL} generator circuit is a CMOS two-stage amplifier in buffer configuration with a level shifter in the output stage. The bandgap V_{BG} reference voltage is buffered at the output of the two-stage amplifier as $V_{BG,BUF}$. A PMOS diode-connected transistor M_6 is used as a level shifter to move $V_{BG,BUF}$ down by one V_{SG} to generate V_{CTRL}. V_{CTRL} is applied to the gate of the FVF control transistor M_C, which level shifts up to V_{LDO} by one V_{SG}. Thus, the goal is to generate an LDO output voltage V_{LDO} that is equal to V_{BG}. Grounded capacitor C_{C1} is added at node V_{CTRL} to achieve stability for the control voltage generator circuit.

The open-loop gain of the FVF LDO is given by

$$A_{DC} \approx -g_{mP}\left[r_{OIB\_LDO}||\left(g_{mC}\,r_{OC}\,R_{LDO}\right)\right] \quad (2)$$

where g_{mP}, r_{OP}, g_{mC}, and r_{OC} are the transconductance and output resistance of transistors M_{PASS} and M_C, respectively. R_{LDO} is the load resistor R_L (not shown) in parallel with r_{OP}. Parameter r_{OIB_LDO} is the output resistance of current source I_{B_LDO}. The open-loop gain of the FVF has a maximum value of $-g_{mP}r_{OIB\_LDO}$ at light load currents, where $g_{mC}r_{OC}R_{LDO} >> r_{OIB\_LDO}$. At heavy load currents, the open-loop gain reduces to $-g_{mP}\left(g_{mC}\,r_{OC}\,R_{LDO}\right)$.

Since the gate capacitance of M_{PASS} is large and the resistance at feedback node V_{FB} is also high, the pole associated with that node will be the dominant pole. The impedance at the output node is low and thus the pole associated with that node will be non-dominant. The expression for the poles are given by

$$\omega_{P1} \approx -\frac{1}{\left[r_{OIB\_LDO}||\left(g_{mC}\,r_{OC}\,R_{LDO}\right)\right]C_{GS,P}}$$
$$\omega_{P2} \approx -\frac{1}{\left(\frac{1}{g_{mC}}||R_{LDO}\right)C_{LDO}} \quad (3)$$

where, $C_{GS,P}$ and $C_{GD,P}$ are the gate-to-source and gate-to-drain parasitic capacitance of pass transistor M_{PASS}. C_{LDO} is the total equivalent grounded capacitance at node V_{LDO}, mainly dominated by the load capacitance C_L. From (3), we conclude that both poles move with load current. At light load currents, the poles are located at

$$\omega_{P1} \approx -\frac{1}{r_{OIB\_LDO}\,C_{GS,P}}$$
$$\omega_{P2} \approx -\frac{g_{mC}}{C_{LDO}} \quad (4)$$

Since $r_{OIB\_LDO} >> 1/g_{mC}$, the poles are located far from each other, resulting in good stability. The bandwidth of the FVF loop is set by ω_{P1}.

At heavy load currents, the poles move to

$$\omega_{P1} \approx -\frac{1}{\left(g_{mC}\,r_{OC}\,R_{LDO}\right)C_{GS,P}}$$
$$\omega_{P2} \approx -\frac{1}{R_{LDO}\,C_{LDO}} \quad (5)$$

In this case, the separation between the two poles is defined by the ratio $g_{mC}r_{OC}C_{GS,P}/C_{LDO}$. Careful design ensures these two poles are separated far enough to meet the phase margin requirements. Adding a suitable grounded compensation capacitor C_{C_LDO} at node V_{FB} will make sure the two poles are far from each other to ensure stability. The magnitude response of the FVF LDO with pole-zero locations at heavy and light load currents is illustrated in Fig. 4.

One of the issues with the FVF LDO is relatively poor load regulation. The lower the output resistance, the better the output regulation. Output resistance is inversely proportional

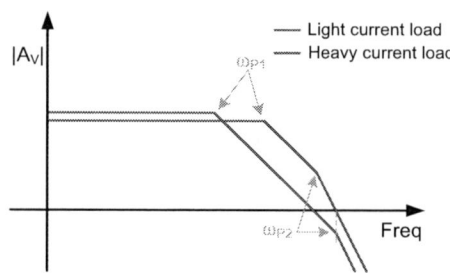

Fig. 4: Frequency response of FVF LDO with pole-zero locations for light and heavy load current conditions.

to the open-loop gain. The output resistance of the FVF LDO is expressed as [31]

$$R_{OUT} = \frac{\left(\frac{1}{g_{mC}}||r_{OP}\right)}{g_{mP}\left[r_{OIB\_LDO}||\left(g_{mC}\,r_{OC}\,R_{LDO}\right)\right]} \quad (6)$$

The output resistance of the FVF LDO varies with load current. This variation impacts load regulation.

III. FOLDED FVF LDO

To overcome FVF LDO's loop gain limitation and to improve the output regulation, the FVF can be replaced by a folded FVF, as illustrated in Fig. 5, with the addition of cascode transistor M_{CAS} in the feedback loop. Bias voltage V_{CN} is selected such that node V_{FB} is the minimum voltage required across current source I_{B_LDO}.

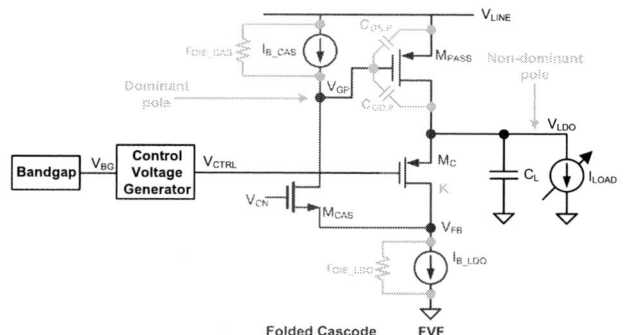

Fig. 5: Schematic of folded flipped voltage follower based LDO.

The major effect of M_{CAS}, high loop gain, is possible only if current source I_{B_CAS} is similarly cascoded. Nevertheless, the equivalent resistance at the gate of M_{PASS} is large and thus the pole associated with node V_{GP} is dominant and limits the bandwidth of the FVF loop.

The output voltage range is high in the folded FVF architecture. The inequality containing $V_{SG,PASS}$ in (1) is no longer valid. As such, it is possible to turn off transistor M_{PASS} almost completely.

In [25], the authors introduce a common-drain buffer [1], [2] M_{BUF} in the FVF loop, to drive the gate of the pass transistor, as shown in Fig. 6(a). Because of the low output resistance of the buffer, the pole associated with node V_{GP} moves to

978-1-5386-3693-0/18 $31.00 © 2018 IEEE

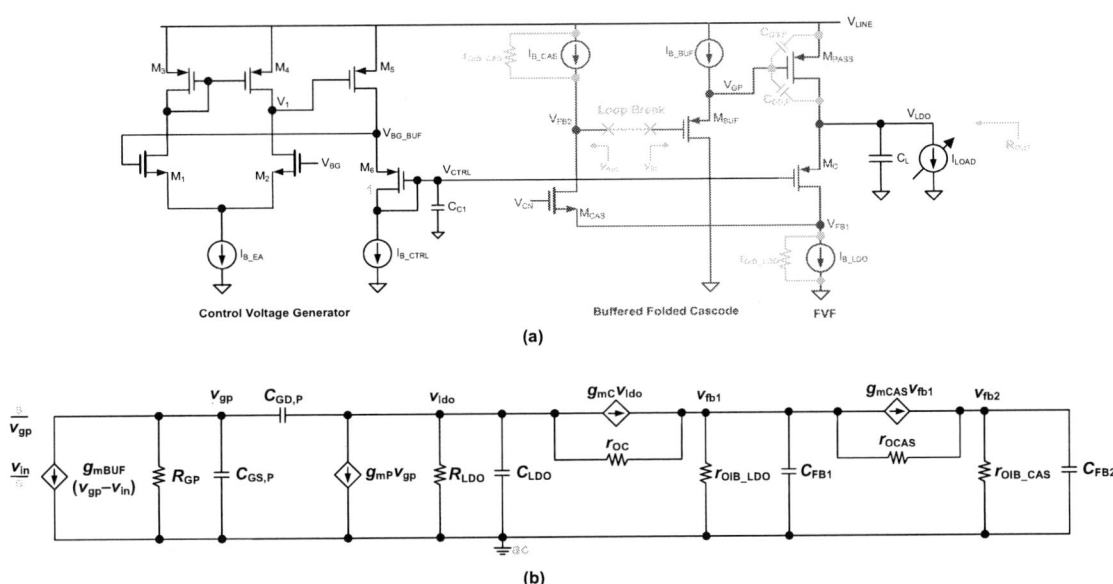

Fig. 6: (a) Transistor level schematic of buffered folded FVF based LDO from [25] and (b) its small-signal model.

higher frequencies. Now the dominant pole is determined by the high output impedance node V_{FB2}, which is at the drain of cascode transistor M_{CAS}. Since the input capacitance of buffer transistor M_{BUF} is low, the dominant pole moves to higher frequencies by nearly two decades. This increase in the bandwidth of the FVF loop has no impact on the low-frequency loop gain, since the buffer has a nominal gain of one.

The small-signal model of the buffered folded FVF LDO is shown in Fig. 6(b). Assuming current I_{B_CAS} attached to node V_{FB2} is a cascoded current source with high output impedance r_{OIB_CAS}, whereas current source I_{B_LDO} is simple, the small-signal gain and pole locations of the LDO of Fig. 6 are approximately

$$A_{DC} \approx -g_{mP}\left[r_{OIB\_CAS}||(g_{mCAS}r_{OCAS})r_{OIB\_LDO}\right]$$

$$\omega_{P1} \approx -\frac{1}{\left[r_{OIB\_CAS}||(g_{mCAS}r_{OCAS})r_{OIB\_LDO}\right]C_{FB2}}$$

$$\omega_{P2} \approx -\frac{1}{\left(\frac{1}{g_{mC}}||R_{LDO}\right)C_{LDO}}$$

$$\omega_{P3} \approx -\frac{1}{\left(\frac{1}{g_{mBUF}}\right)C_{GS,P}} \qquad (7)$$

$$\omega_{Z1} \approx +\frac{1}{\left(\frac{1}{g_{mP}}\right)C_{GD,P}}$$

where C_{FB2} is the equivalent grounded capacitance at node V_{FB2} and other parameters follow naming conventions.

From the DC gain expression, we note that the gain varies with g_{mP}. At light load currents, g_{mP} is lower and the gain is reduced. As the load current increases, g_{mP} increases with the square-root of load current and, thus, at higher load currents, the gain is higher. This higher loop gain reduces the output

resistance of the FVF to a much lower value and improves output regulation.

The dominant pole is created by the resistance and capacitance at node V_{FB2}. As the capacitance C_{FB2} is small, the location of the pole will be at higher frequencies but dominant by the high cascoded resistance of I_{B_CAS}. The second pole is defined by C_{LDO}, R_{LDO} and $1/g_{mC}$, Since $1/g_{mC}$ is small, this pole is also located at higher frequencies. These higher frequency poles compete for dominance as they might be closely located. Adding a compensation capacitor at node V_{FB2} to ground will help in achieving stability with little compromise in bandwidth. The third pole and the RHP zero are defined by $C_{GS,P}$ & $1/g_{mBUF}$ and $C_{GD,P}$ & $1/g_{mP}$, respectively; they are located at high frequencies.

One drawback of this FVF LDO configuration is that the additional buffer M_{BUF} increases the minimum voltage at node V_{GP} by $V_{SG,BUF}$. It is no longer possible to turn on M_{PASS} completely, which increases the dropout voltage of the LDO.

IV. MULTI-LOOP BUFFERED FVF LDO

The bandwidth of the buffered folded FVF LDO is limited by the dominant pole of the FVF loop that is created by the high impedance node V_{FB2} shown in Fig. 6. This bandwidth may be limited to tens of megahertz. In order to increase the bandwidth even higher, all the nodes in the FVF loop need to have high-frequency poles without compromising output voltage regulation. The authors in [28] introduce a multi-loop buffered FVF LDO that achieves high bandwidth; however, the regulation improvement is nominal. An advancement of the multi-loop LDO of [28] is presented in [32], in which the voltage regulation is improved significantly with no compromise in bandwidth. The schematic of the multi-loop buffered FVF LDO is shown in Fig. 7.

978-1-5386-3693-0/18 $31.00 © 2018 IEEE

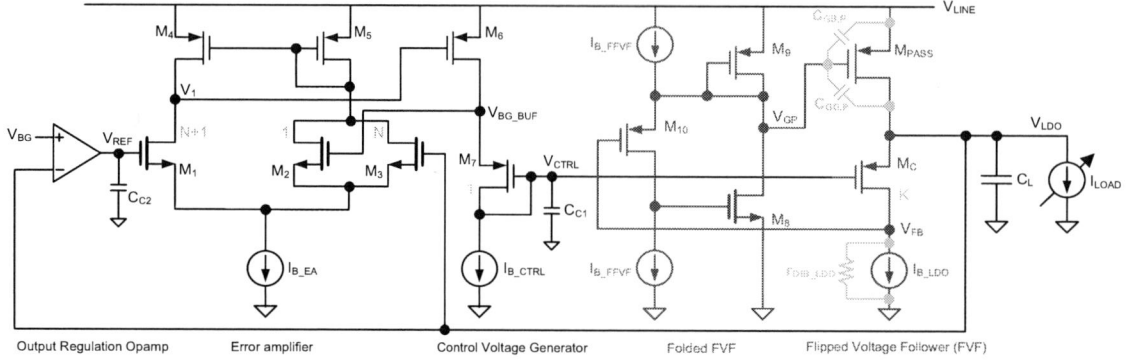

Fig. 7: Schematic of buffered cascoded FVF based LDO for accurate output regulation and high bandwidth, adapted from [6], [28] and [32].

The buffered FVF loop consists of an FVF output stage $M_{PASS} - M_C - I_{B\_LDO}$ buffered with a folded FVF formed by transistors $M_8 - M_{10}$ with two bias currents labeled I_{B_FFVF}. The advantage of a folded FVF over a common-drain voltage follower is lower output resistance. Thus, the pole at the gate of M_{PASS} is moved to higher frequencies. Transistor M_9 is added as a diode-connected transistor load of the folded FVF, which further reduces the impedance at node V_{GP}, particularly at high frequencies when the folded FVF output impedance begins to rise [31]. Note also that transistor M_9 forms the input side of a current mirror with transistor M_{PASS}. Thus, the ground current in the output stage increases with load current.

The loop gain of the FVF loop is designed to be relatively low, such that its bandwidth is high and, without the presence of the outer loops, would result in poor phase margin and poor load regulation. The dominant pole is moved to the output node, where a minimum load capacitance, C_L, is required for stability. The value of output capacitance is in the hundreds of picofarads and thus can easily be integrated on-chip.

The control voltage generation circuit is similar to the circuit shown in Fig. 6, with the exception that one side of the input differential pair is split into two transistors, M_2 and M_3, described below.

In order to improve load regulation and phase margin, one outer loop from the LDO output, V_{LDO}, to the control voltage generation circuit is added. The outer loop feedback is made stronger than the inner loop feedback from $V_{BG,BUF}$. This is implemented by splitting the input differential pair transistor into two transistors with ratio $N : 1$. The outer loop is connected to the higher-weighted transistor and the inner-loop to the smaller one. If the output deviates far from the reference voltage, then the control voltage V_{CTRL} is adjusted in order to pull the output voltage back to the reference value.

A second outer loop is added in [32] to improve the output regulation significantly by integrating the error between the bandgap voltage V_{BG} and the output voltage V_{LDO} using a single-stage opamp and capacitor C_{C2}. Opamp output signal V_{REF} is now the input to the control voltage generation block. In steady-state, V_{REF} equals V_{BG}. However, during a transient, if the output drops below V_{BG}, V_{REF} increases so as to restore the output voltage to V_{BG}.

The outer loop bandwidths are slower than the main FVF loop, such that excellent DC output regulation is achieved by the slow loops and excellent transient response is made possible by the high-frequency FVF loop.

V. DISCUSSION AND CONCLUSION

This tutorial began with a discussion of the conventional stand-alone LDO IC, which used an external output capacitor in the range of hundreds of nanofarads to tens of microfarads. It then described the recent trend in SoCs toward point-of-load integrated LDOs, requiring a low-value on-chip output capacitor. These LDOs achieve high open-loop gain for excellent output voltage regulation, but their bandwidth is limited by an internal dominant pole. This limited bandwidth resulted in reduced PSRR at high frequencies.

The tutorial then introduced the basic FVF circuit as a promising candidate for implementing high bandwidth LDOs due to its fast local feedback loop. Several iterations, including folding and buffering the FVF loop led up to a multi-loop design that simultaneously achieved high open-loop gain for good load regulation and high bandwidth. A peculiar characteristic of this integrated multi-loop FVF LDO architecture is the dominance of the output pole, which is set at a high frequency by an output capacitor in the hundreds of picofarads. This high bandwidth resulted in good PSRR over a wide frequency range.

A present challenge in FVF LDOs is the maximum achievable load current. Indeed, the multi-loop folded FVF LDOs of [28], [32] are limited to a load current of 10 mA. This limit appears to be caused, at least in part, by the limited voltage range at the gate of the pass transistor. Whereas the pass transistor can be fully turned off, it cannot be fully turned on, as the minimum gate voltage is at least one V_{SG} above ground.

REFERENCES

[1] A. Garimella and P. M. Furth, Eds., *High Performance Analog and Power Management Circuit Design Techniques for Modern SoCs*. Springer International Publishing, manuscript in preparation.

[2] G. Rincon-Mora, *Analog IC Design with Low-Dropout Regulators*. McGraw-Hill, Inc., 2006.

978-1-5386-3693-0/18 $31.00 © 2018 IEEE

[3] J. Falin, "ESR, Stability, and the LDO Regulator," *Texas Instruments Application Report SLVA115*, May 2002.

[4] "TLV743P 300-mA, Low-Dropout Regulator," *Texas Instruments TLV743P 300-mA, Low-Dropout Regulator datasheet*, Jul. 2017.

[5] A. Garimella, M. W. Rashid, and P. M. Furth, "Reverse Nested Miller Compensation Using Current Buffers in a Three-Stage LDO," *IEEE Transactions on Circuits and Systems II: Express Briefs*, vol. 57, no. 4, pp. 250–254, Apr. 2010.

[6] Y. Lu, Y. Wang, Q. Pan, W. H. Ki, and C. P. Yue, "A Fully-Integrated Low-Dropout Regulator With Full-Spectrum Power Supply Rejection," *IEEE Transactions on Circuits and Systems I: Regular Papers*, vol. 62, no. 3, pp. 707–716, Mar. 2015.

[7] K. Wong and D. Evans, "A 150mA Low Noise, High PSRR Low-Dropout Linear Regulator in 0.13μm Technology for RF SoC Applications," in *2006 Proceedings of the 32nd European Solid-State Circuits Conference*, Sep. 2006, pp. 532–535.

[8] G. A. Rincon-Mora and P. E. Allen, "A Low-Voltage, Low Quiescent Current, Low Drop-out Regulator," *IEEE Journal of Solid-State Circuits*, vol. 33, no. 1, pp. 36–44, Jan. 1998.

[9] M. Al-Shyoukh, H. Lee, and R. Perez, "A Transient-Enhanced Low-Quiescent Current Low-Dropout Regulator With Buffer Impedance Attenuation," *IEEE Journal of Solid-State Circuits*, vol. 42, no. 8, pp. 1732–1742, Aug. 2007.

[10] G. A. Rincon-Mora and P. E. Allen, "Optimized Frequency-Shaping Circuit Topologies for LDO's," *IEEE Transactions on Circuits and Systems II: Analog and Digital Signal Processing*, vol. 45, no. 6, pp. 703–708, Jun. 1998.

[11] K. N. Leung and P. K. T. Mok, "A Capacitor-free CMOS Low-Dropout Regulator with Damping-Factor-Control Frequency Compensation," *IEEE Journal of Solid-State Circuits*, vol. 38, no. 10, pp. 1691–1702, Oct. 2003.

[12] S. K. Lau, P. K. T. Mok, and K. N. Leung, "A Low-Dropout Regulator for SoC with Q-Reduction," *IEEE Journal of Solid-State Circuits*, vol. 42, no. 3, pp. 658–664, Mar. 2007.

[13] A. P. Patel and G. A. Rincon-Mora, "High Power-Supply-Rejection (PSR) Current-Mode Low-Dropout (LDO) Regulator," *IEEE Transactions on Circuits and Systems II: Express Briefs*, vol. 57, no. 11, pp. 868–873, Nov. 2010.

[14] G. A. Rincon-Mora, "Active Capacitor Multiplier in Miller-Compensated Circuits," *IEEE Journal of Solid-State Circuits*, vol. 35, no. 1, pp. 26–32, Jan. 2000.

[15] P. Hazucha, T. Karnik, B. A. Bloechel, C. Parsons, D. Finan, and S. Borkar, "Area-Efficient Linear Regulator with Ultra-Fast Load Regulation," *IEEE Journal of Solid-State Circuits*, vol. 40, no. 4, pp. 933–940, Apr. 2005.

[16] X. Qu, Z. K. Zhou, B. Zhang, and Z. J. Li, "An Ultralow-Power Fast-Transient Capacitor-Free Low-Dropout Regulator With Assistant Push-Pull Output Stage," *IEEE Transactions on Circuits and Systems II: Express Briefs*, vol. 60, no. 2, pp. 96–100, Feb. 2013.

[17] J. Torres, M. El-Nozahi, A. Amer, S. Gopalraju, R. Abdullah, K. Entesari, and E. Sanchez-Sinencio, "Low Drop-Out Voltage Regulators: Capacitor-less Architecture Comparison," *IEEE Circuits and Systems Magazine*, vol. 14, no. 2, pp. 6–26, Secondquarter 2014.

[18] Y. i. Kim and S. s. Lee, "A Capacitorless LDO Regulator With Fast Feedback Technique and Low-Quiescent Current Error Amplifier," *IEEE Transactions on Circuits and Systems II: Express Briefs*, vol. 60, no. 6, pp. 326–330, Jun. 2013.

[19] V. Gupta and G. A. Rincon-Mora, "A 5mA 0.6 CMOS Miller-Compensated LDO Regulator with -27dB Worst-Case Power-Supply Rejection Using 60pF of On-Chip Capacitance," in *2007 IEEE International Solid-State Circuits Conference. Digest of Technical Papers*, Feb. 2007, pp. 520–521.

[20] T. Y. Man, P. K. T. Mok, and M. Chan, "A High Slew-Rate Push-Pull Output Amplifier for Low-Quiescent Current Low-Dropout Regulators with Transient-Response Improvement," *IEEE Transactions on Circuits and Systems II: Express Briefs*, vol. 54, no. 9, pp. 755–759, Sep. 2007.

[21] R. J. Milliken, J. Silva-Martinez, and E. Sanchez-Sinencio, "Full On-Chip CMOS Low-Dropout Voltage Regulator," *IEEE Transactions on Circuits and Systems I: Regular Papers*, vol. 54, no. 9, pp. 1879–1890, Sep. 2007.

[22] T. Y. Man, K. N. Leung, C. Y. Leung, P. K. T. Mok, and M. Chan, "Development of Single-Transistor-Control LDO Based on Flipped Voltage Follower for SoC," *IEEE Transactions on Circuits and Systems I: Regular Papers*, vol. 55, no. 5, pp. 1392–1401, Jun. 2008.

[23] R. G. Carvajal, J. R.-. Angulo, A. J. Lopez-Martin, A. Torralba, J. A. G. Galan, A. Carlosena, and e. a. F. M. Chavero "The Flipped Voltage Follower: A Useful Cell for Low-Voltage Low-Power Circuit Design," *IEEE Transactions on Circuits and Systems I: Regular Papers*, vol. 52, no. 7, pp. 1276–1291, Jul. 2005.

[24] V. Peluso, P. Vancorenland, A. M. Marques, M. S. J. Steyaert, and W. Sansen, "A 900-mV Low-Power $\Delta\Sigma$ A/D Converter with 77-dB Dynamic Range," *IEEE Journal of Solid-State Circuits*, vol. 33, no. 12, pp. 1887–1897, Dec. 1998.

[25] H. Chen and K. N. Leung, "A Fast-Transient LDO Based on Buffered Flipped Voltage Follower," in *2010 IEEE International Conference of Electron Devices and Solid-State Circuits (EDSSC)*, Dec. 2010, pp. 1–4.

[26] P. Y. Or and K. N. Leung, "An Output-Capacitorless Low-Dropout Regulator With Direct Voltage-Spike Detection," *IEEE Journal of Solid-State Circuits*, vol. 45, no. 2, pp. 458–466, Feb. 2010.

[27] W. M. Chen, H. Chiueh, T. J. Chen, C. L. Ho, C. Jeng, M. D. Ker, C. Y. Lin, Y. C. Huang, C. W. Chou, T. Y. Fan, M. S. Cheng, Y. L. Hsin, S. F. Liang, Y. L. Wang, F. Z. Shaw, Y. H. Huang, C. H. Yang, and C. Y. Wu, "A Fully Integrated 8-Channel Closed-Loop Neural-Prosthetic CMOS SoC for Real-Time Epileptic Seizure Control," *IEEE Journal of Solid-State Circuits*, vol. 49, no. 1, pp. 232–247, Jan. 2014.

[28] Y. Lu, W. H. Ki, and C. P. Yue, "A 0.65ns-Response-Time 3.01ps FOM Fully-Integrated Low-Dropout Regulator with Full-Spectrum Power-Supply-Rejection for Wideband Communication Systems," in *2014 IEEE International Solid-State Circuits Conference Digest of Technical Papers (ISSCC)*, Feb. 2014, pp. 306–307.

[29] X. L. Tan, K. C. Koay, S. S. Chong, and P. K. Chan, "A FVF LDO Regulator With Dual-Summed Miller Frequency Compensation for Wide Load Capacitance Range Applications," *IEEE Transactions on Circuits and Systems I: Regular Papers*, vol. 61, no. 5, pp. 1304–1312, May 2014.

[30] S. H. Pakala, M. Manda, P. R. Surkanti, A. Garimella, and P. M. Furth, "Voltage Buffer Compensation Using Flipped Voltage Follower in a Two-Stage CMOS Op-amp," in *2015 IEEE 58th International Midwest Symposium on Circuits and Systems*, Aug. 2015, pp. 1–4.

[31] P. R. Surkanti, A. Garimella, M. Manda, and P. M. Furth, "On the Analysis of Low Output Impedance Characteristic of Flipped Voltage Follower (FVF) and FVF LDOs," in *2017 IEEE 60th International Midwest Symposium on Circuits and Systems (MWSCAS)*, Aug. 2017, pp. 17–20.

[32] M. Manda, S. H. Pakala, and P. M. Furth, "A Multi-Loop Low-Dropout FVF Voltage Regulator with Enhanced Load Regulation," in *2017 IEEE 60th International Midwest Symposium on Circuits and Systems (MWSCAS)*, Aug. 2017, pp. 9–12.

Overcoming Energy and Reliability Challenges for IoT and Mobile Devices with Data Analytics

Sudeep Pasricha

Department of Electrical and Computer Engineering
Colorado State University, Fort Collins, Colorado, 80523
sudeep@colostate.edu

Abstract – A very large amount of data is produced by mobile and Internet-of-Thing (IoT) devices today. Increasing computational abilities and more sophisticated operating systems (OS) on these devices have allowed us to create applications that are able to leverage this data to deliver better services. But today's mobile and IoT solutions are heavily limited by low battery capacity and limited cooling capabilities. This motivates a search for new ways to optimize for energy-efficiency. Advanced data analytics and machine-learning techniques are becoming increasingly popular to analyze and extract meaning from Big Data. In this paper, we make the case for designing and deploying data analytics and learning mechanisms to improve energy-efficiency in IoT and mobile devices with minimal overheads. We focus on middleware for inserting energy-efficient data analytics-driven solutions and optimizations in a robust manner, without altering the OS or application code. We discuss several case studies of powerful and promising developments in deploying data analytics middleware for energy-efficient and robust execution of a variety of applications on commodity mobile devices.

Keywords: data analytics, middleware, mobile computing, Internet-of-Things (IoT), energy-efficiency, robustness

I. INTRODUCTION

We are well into the era of 'Big Data', with several zetta-bytes (ZB) of data being handled by datacenters annually, and the volume of this data expected to double every two years [1]. The increase in data over the past decade has been fueled by the proliferation of embedded Internet of Things (IoT) devices and smart mobile computing. For instance, a study from CISCO in 2016 [2] suggests that on an average 10.7 exabytes of mobile data traffic is offloaded each month from mobile phones to cloud datacenters. The study further forecasts that this value is set to increase 8-fold by 2021.

The generated Big Data can be structured (e.g., financial records), semi-structured (e.g., tweets), unstructured (e.g., audio, video), or real-time (e.g., monitoring logs). All of these types of data have the potential to provide invaluable insights, if organized and analyzed appropriately. The analyses of such large and diverse datasets is fast emerging as an indispensable tool for innovations in various domains such as healthcare, business process optimization, and social-network-based recommendations.

Unfortunately, the sheer volume of Big Data and its projected growth in the coming years will largely outpace foreseeable improvements in the cost, reliability, energy-efficiency, and performance of computing infrastructure such as mobile platforms and IoT devices. Thus, the design of these computing platforms requires significant innovations in the near future, to overcome increased energy costs due to the high computational power when processing and analyzing large datasets. As an example of the severity of the problem, consider the battery life of mobile and IoT devices. Although lithium-ion battery technology and capacity has improved over the years, it still cannot keep pace with the energy demands of today's mobile devices. Figure 1 shows the battery limits, thermal power limits, and desired power dissipation of mobile chipsets with respect to developments in mobile communication capabilities and chip technology. The gap between desired power (to meet growing performance needs) and battery limits is only getting larger, e.g., over a period of seven years, the processing capability of the Samsung Galaxy S series mobile device has grown by ~20×, whereas, the battery capacity has grown only by ~2× [3]. Battery life limitations today inevitably constrain performance for most applications on mobile devices.

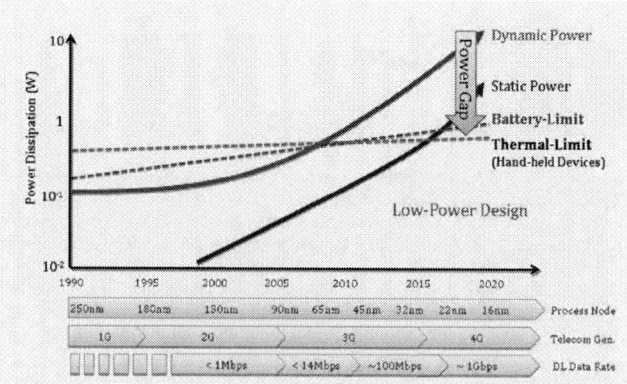

Figure 1: Trends for power dissipation as compared to battery and thermal limits in mobile devices since 1990 [4]

Fortunately, system designers that are today grappling with the challenge of efficiently supporting large datasets and data analytics workloads can themselves benefit from data analytics techniques to architect better computing platforms. For instance, in mobile computing, massive amount of data is typically collected about user-device interactions and usage. Can this data be analyzed to improve energy efficiency and robustness of applications executing on the mobile device? Even if it is possible to intelligently analyze and perform optimizations with the knowledge of such data, where exactly in the software stack should the data analytics techniques be implemented: in the apps, OS, or elsewhere?

The answer to the latter question may lie with middleware, which is becoming an increasingly important component in modern mobile and IoT device software stacks [5]. Middleware is intended to provide services or functionalities to the application developer that are not already a part of the OS. A salient feature of middleware is the abstraction level that it can create for the application layer. An application developer does not need to be aware of how different modules come together in the middleware for a new service to work. Thus middleware has been growing in popularity, and has especially found widespread application in distributed and cloud based applications. An example is Microsoft Azure [6], that is a growing collection of integrated services (applications) and platforms (operating system) combined using middleware to deliver numerous services-on-demand in manner that is personalized per user.

In this paper, we explore the design and implementation of intelligent data-analytics based middleware for mobile and IoT devices that can assist with improving the energy-efficiency and robustness of these devices. The abstraction advantage and modularity of middleware allows it to be a very good solution to swiftly create and deploy services that enhance the existing features of an OS. Several case studies of powerful and promising developments in prototyping data analytics middleware for energy-efficient and robust execution of a variety of applications on commodity mobile computing devices are discussed in this paper.

The rest of this paper is organized as follows. Section 2 discusses middleware to enable mobile-to-cloud offloading. Section 3 discusses middleware for user-interaction aware execution of applications on mobile devices. Section 4 presents middleware that captures spatio-temporal context for various optimizations. Section 5 explores middleware for mobile indoor localization.

II. MOBILE-TO-CLOUD OFFLOADING

The collection and processing of data on smartphones can significantly hamper the battery lifetime of the device. A promising solution that is being considered to support high end mobile data processing applications is to offload mobile computations to the cloud [7]-[9]. Offloading is an opportunistic process that relies on cloud servers to execute the functionality of an application that typically runs on a mobile device. In many cases, such opportunistic offloading can not only improve energy-efficiency but also makes computation performance more robust.

Kumar et al. [8] presented a mathematical analysis of offloading. Broadly, the energy saved by computation offloading depends on the amount of computation to be performed (C), the amount of data to be transmitted (D), and the wireless network bandwidth (B). If (D/C) is low, then it was claimed that offloading can save energy. Our experiments have shown that this is a simplistic view of the problem, e.g., energy-efficiency is also highly effected by the

type of wireless interface being used for the transmission. Cuervo et.al [7] proposed a framework called MAUI, based on code annotations to specify which methods from a software class can be offloaded to the cloud. Annotations are introduced in the source code by the developer during the development phase. At runtime, methods are identified by the MAUI profiler, which performs the offloading of the methods, if the bandwidth of the network and data transfer conditions are ideal. However, this annotation method puts an extra burden on the already complex mobile application development process. A better approach would be to have a middleware that is able to automatically make intelligent offloading decisions on the fly, without manual annotations.

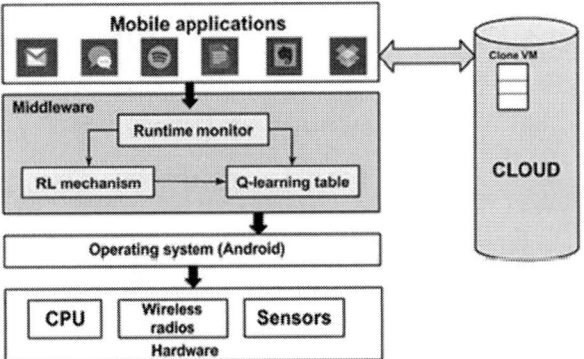

Figure 2. Reinforcement Learning (RL) based middleware framework for efficient application offloading from mobile device to the cloud [10]

In [10], we proposed a novel middleware framework for mobile devices that utilized various sources of data such as the app communication/computation intensiveness, type and status of available wireless networks, and capabilities of cloud servers, to dynamically make decisions on when and how to offload an application from the mobile device to the cloud, with the goal of improving energy-efficiency and performance robustness. Figure 2 shows an overview of our data analytics decision engine on the mobile device that works together with a clone virtual machine (VM) of the mobile software environment to execute apps on cloud servers. The middleware was deployed on the Android OS, and ran in the background as an Android service. The framework utilized an unsupervised Q-learning machine learning technique that analyzed the data for the app type, network type, network conditions, and cloud capabilities to select the optimal network type and decide when and what to offload.

Figure 3 shows an example of how the Android based torrent app Flud [11] can benefit from our middleware-based offloading framework. In this experiment, a cloud based service (Amazon Web Services EC2 instance) first downloads and aggregates the file to be received through torrent, and then the smartphone downloads the file in a single process. The experiment was conducted on an LG G3 Android smartphone. From the bars in the figure, it can be observed that different network types, the state of the network, and data transfer sizes result in varying improvements in energy and response time. For instance, 4G performs slightly better

978-1-5386-3693-0/18 $31.00 © 2018 IEEE

than 3G in terms of energy consumption for higher data sizes (45-85 MBs), but for smaller data sizes 3G is more efficient. The colored lines in Figure 3 indicate the performance of our middleware framework (green line) and a framework based on fuzzy logic for making offloading decisions (red line) [9]. In all cases, it can be observed that our framework is able to provide better offloading performance and greater energy efficiency. This is because our framework employs a more sophisticated and powerful data analytics-based learning algorithm and considers many more variables related to device and network context when making decisions, than prior work. Our experiments with real smartphones showed savings of up to 30% in battery life with up to 25% better response time when using our middleware framework compared to a state-of-the-art fuzzy logic based offloading approach [9]. For certain applications, e.g., voice recognition, offloading also improved recognition robustness (accuracy) by approx. 10%.

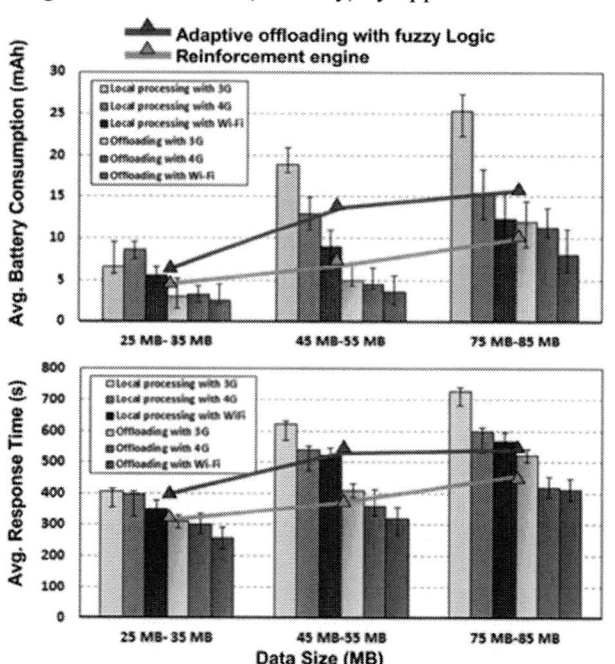

Figure 3: Average battery consumption and response time on a mobile device for a torrent file download application [10]

III. USER-INTERACTION AWARE OPTIMIZATIONS

Mobile devices today are seen as a personal tool and their typical usage can vary across different users. OS- and hardware-driven energy-optimization techniques, such as CPU dynamic voltage and frequency scaling (DVFS), are not smart enough to make decisions based on the user's behavior. To enable more aggressive energy optimizations in mobile devices, we developed a novel application- and user-interaction aware energy management middleware framework (AURA) for mobile devices [12], [13]. AURA takes advantage of user idle time between interaction events of the foreground application to optimize CPU and backlight energy consumption. Most interestingly, AURA is able to

adapt to changing behavior and learn from the individual user over time, to achieve longer battery lifetime without any user intervention.

To create an energy-efficient middleware solution, it is important to first identify the components of the mobile device that are major contributors towards battery life. Our preliminary analysis revealed that the display, the processing (CPU/GPU) subsystem, and the various wireless radios (e.g., Wi-Fi, GPS, 4G/LTE) have a significant impact on battery life. Any framework that aims to optimize energy-efficiency must address the energy inefficiencies in these components. Our AURA framework [12], [13] was one of the first to reduce energy costs for both the display and the processing subsystems in an integrated manner. The framework consists of an app-aware and user-aware energy optimization middleware that uses powerful machine learning techniques on user-device interaction data. More specifically, AURA includes a runtime monitor that captures data related to user-specific and app-specific interaction slack to reduce energy costs.

Interaction slack refers to the sum of the unused times between when a user first perceives a change on the display (perceptual slack) due to a previous interaction (e.g., a button on the screen changes color), then comprehends what the response "actually" represents (cognitive slack), and finally interacts with application again by touching the screen using his/her fingers (motor slack). By predicting this interaction slack interval on a per-app and per-user basis, AURA opportunistically reduce CPU voltage/frequency at the start of the interval and then increase the voltage/frequency just before the interval ends, to save energy without impacting user quality of service (QoS).

AURA's middleware was prototyped as a service that constantly runs in the background and creates an automated control system for CPU voltage/frequency scaling and display backlight modifications. The middleware consist of three main components: a runtime monitor, a Bayesian app classifier, and a power manager. The runtime monitor checks if the current foreground app has an entry in an 'interaction database', and if so then the interaction data (standard deviation and mean of a user's observed slack values from previous interactions) in it can be used for slack prediction. If a database entry does not exist, the middleware creates a new entry and starts collecting interaction statistics. A Bayesian classifier is then used to classify the interaction profile for the app using the collected data. Bayesian learning is a form of supervised machine learning that involves using evidence or observations along with prior outcome probabilities to calculate the probability of an outcome. The power manager runs a MDP (Markov Decision Process) to classify the apps. All apps were classified into seven categories ranging from very-low-interaction to very-high-interaction. The class of an app decides how to opportunistically decrease CPU voltage/frequency in between slack intervals. MDPs are discrete time stochastic control processes that are widely used as decision-making

978-1-5386-3693-0/18 $31.00 © 2018 IEEE   240

models for systems in which outcomes are partly random and partly controlled.

In [12], we explored two derivatives of the (normal) MDP to dynamically adapt to real-time user-interaction during each invocation of an application. The E-ADAPT variant is event-driven and uses the most recent window of events to predict future interaction events whereas T-ADAPT makes use of a moving average window of a predetermined size, to dynamically track and predict user interaction events in a temporal context. In [13], we prototyped a new Q-learning based power manager. Q-Learning does not require a model of the environment and has the advantage that the next state probability distributions that are used in MDPs are not required. Using the Android services based middleware approach allowed us to the rapidly develop, update, deploy, and test new versions of AURA.

The AURA middleware also exploits the idea of change blindness [14] as identified by research into human psychology and perception. Change blindness refers to the inability of humans to notice large changes in their environments, especially if changes occur in small increments. Multiple studies have shown this as a limitation of human perception, e.g., a majority of observers in one study failed to observe when a building in a photograph gradually disappeared over the course of 12 seconds. We used a similar approach by gradually reducing screen brightness over time using user-device and app-specific interaction data. In doing so, the power manager in AURA is able to achieve higher energy-efficiency without any noticeable loss in QoS.

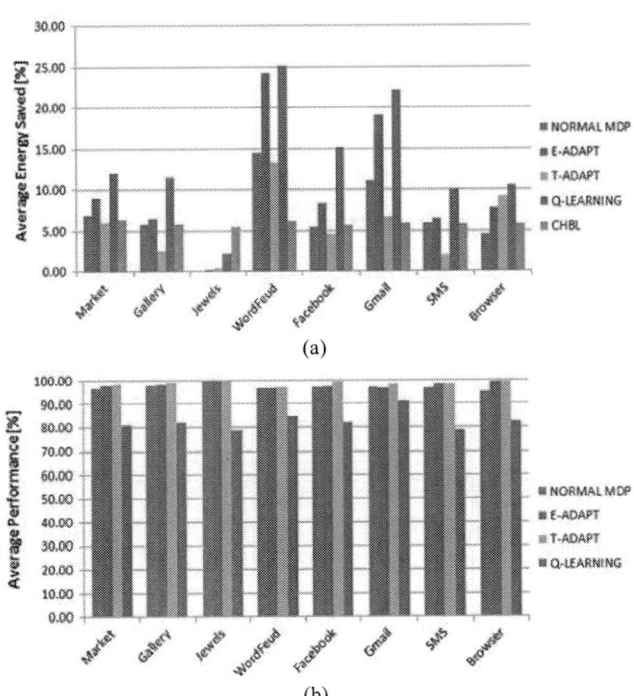

(a)

(b)

Figure 4: Real user study results on HTC Dream mobile device [13]

We deployed our AURA middleware framework on several smartphones such as the HTC Dream and Google Nexus One. Figure 4 shows the results for energy savings and interaction slack prediction on the HTC Dream smartphone across various common mobile applications, averaged for several real users. In addition to the four variants of AURA (with power managers based on NORMAL-MDP, E-ADAPT, T-ADAPT, and Q-LEARNING) we compare against Shye et al.'s algorithm, CHBL [15], which was the best known algorithm for energy savings on mobile devices. It can be seen from Figure 4(a) that our user-aware and application-aware algorithms (particularly Q-LEARNING) offer higher energy savings than CHBL because unlike CHBL they can dynamically adapt to the user-interaction patterns and take full advantage of user idle time. Figure 4(b) shows the average successful prediction rates for the real user interaction patterns. CHBL was not included in the results because it does not contain defined states or prediction mechanisms, making determining mispredictions impossible. The figures show high successful prediction rates with AURA that result in high QoS for users. Our extensive experiments indicated 17% energy savings on average for AURA compared to the baseline Android device manager and approximately 2.5× more energy savings over the best known prior work (CHBL [15]) on mobile CPU and display energy optimization.

IV. SPATIOTEMPORAL CONTEXT-AWARE OPTIMIZATION

Today mobile devices come with a variety of wireless interfaces such as GPS, Wi-Fi and 3G/4G. Our experimental analysis on various smartphones [16] indicated that even when 3G/4G, Wi-Fi, and GPS interfaces are all enabled and idle, they account for more than 25% of total system power dissipation. Furthermore, when only one of the interfaces is active, the other idle interfaces still consume a non-negligible amount of power. This is the motivation behind efforts to enable intelligent management of such wireless interfaces. It is important to note that activation of wireless interfaces, for location or data, is directly correlated with the context of the device itself, e.g., the type of application running, device motion, Wi-Fi availability, time of day, location, etc. This observation opens up an opportunity to realize a context-aware solution that is able to more efficiently manage wireless interfaces without human intervention.

Our middleware framework in [17], [18] represents one of the first efforts towards seamless wireless interface energy management in mobile devices. The first step in our approach is to collect and learn from the contextual data of the device, the user, as well as the state of wireless interfaces. Our framework is able to transparently capture contextual data attributes such as temporal use data (e.g., day of week and time), spatial environment data (e.g., ambient light, Wi-Fi RSSI, 3G/4G signal strength), and device state (e.g., battery status, CPU utilization). To prune the massive amount of data captured, we employed Principal Component Analysis (PCA), a form of dimensionality reduction, by projecting the captured data from various sources onto a

978-1-5386-3693-0/18 $31.00 © 2018 IEEE

fewer number of optimally selected eigenvectors, effectively reducing the attribute space to the (limited) number of eigenvectors, to enable efficient prediction on resource-constrained mobile devices.

We then explored the use of five different classes of machine learning algorithms to learn from this contextual data. The algorithms we considered included LDA (Linear Discriminant Analysis), LLR (Linear Logistic Regression), NN (Shallow and Deep Neural Networks), KNN (K-Nearest Neighbor), and SVM (Support Vector Machines). These algorithms allowed us to predict user data/location usage requirements (e.g., is data transfer needed? is coarse-grained location needed? is fine-grained location needed?) based on the spatiotemporal user and device context data collected.

We found that SVM and deeper NNs (with more hidden layers) resulted in the highest context prediction accuracy (85 – 99%). KNN, LLR, and LDA also performed fairly well, with prediction accuracies in the range of 60 – 90 %. The highest energy savings were achieved with LDA and LLR. However, these higher savings come at the cost of degraded user satisfaction (highlighted by their lower context prediction accuracy). SVM and KNN overall performed fairly well in terms of both prediction accuracy and energy savings potential, as did the NN approach. But KNN's run time on a mobile device is several orders of magnitude larger than any of the other algorithms, because all computations are deferred until its classification phase. Therefore, although KNN is as good as or better than the SVM and NN based approaches in terms of energy savings and prediction accuracy, it is not the most practical for deployment on mobile devices. Our SVM based middleware prototype provides good accuracy, good energy savings, and demonstrates the best adaptation to various unique user usage patterns, while maintaining a low implementation overhead on mobile devices. We showed approximately 85% energy savings with SVM for minimally active users

V. MOBILE INDOOR LOCALIZATION

Location tracking has found various applications outdoors. One can not only use GPS based services for navigation purposes, but companies such as Google have been providing users location based services such as locating the best places to eat in their vicinity, local news, local weather, reminders to get an item when near a grocery store [19], etc. The outdoor location based services available today are extremely helpful, yet, in most cases they are limited once a user moves indoors. For instance, in the previously suggested example of reminders when near grocery stores, the application can only remind a user to buy the item from the store but is unable to provide any guidance on how to locate that item within the store. There are several other use cases that remain unrealized, such as being reminded to go to a certain store within a large indoor mall, notifications to the user when they are close to specific items/aisles in a store, or navigation help to reach specific rooms in a large building. These and many other examples make a strong case for

the creation of indoor localization solutions on devices that most people carry with them everywhere: smartphones.

Indoor localization is a challenge that cannot be resolved through a conventional outdoor solution such as GPS. This is because GPS signals are weak and ineffective in indoor environments, and the wireless signal-based infrastructure for indoor localization is diverse, prone to interference, and often entirely non-existent [20]. A possible approach to overcome this challenge is fingerprinting, where the goal is to use data captured through smartphone radio interfaces and sensors to estimate the location of the user indoors (inside of buildings, caves, etc.) in real time. However, continuous monitoring of radio and sensor data drains battery life, thus indoor localization solutions must be energy-aware.

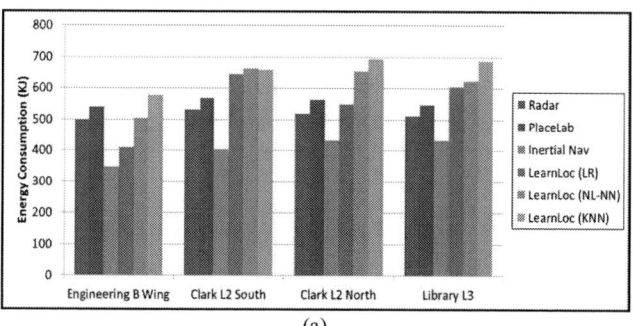

(a)

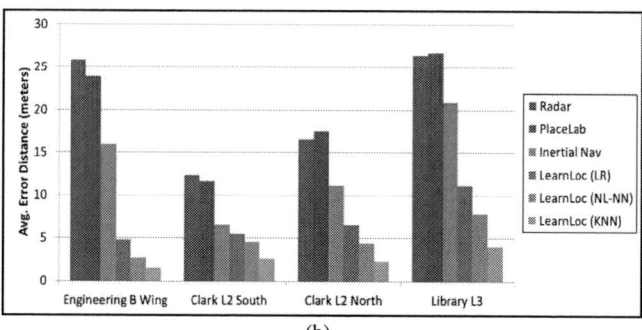

(b)

Figure 5: Comparison of indoor localization techniques [21].

We devised the LearnLoc middleware framework for mobile devices in [21] to improve indoor localization reliability (accuracy) and energy efficiency in a variety of indoor environments. We capture Wi-Fi "fingerprint" data (signal strength, signal type, access point ID) in a continually updated database for indoor locales together with inertial sensing readings (accelerometer, gyroscope, and magnetometer sensor data) on mobile devices. This data is then analyzed and parsed with machine learning techniques to estimate location in real time. The framework is implemented as a middleware service on mobile devices to provide indoor location based updates and suggestions in a non-intrusive and energy-efficient manner. The most important feature of our approach is that it does not require any additional hardware setup indoors, as Wi-Fi access points have become increasingly common in indoor public spaces which brings down the cost of realizing indoor localization.

978-1-5386-3693-0/18 $31.00 © 2018 IEEE

In [21], we prototyped and compared three variants of the LearnLoc framework that use Linear Regression (LR), non-linear neural networks (NL-NN) and K-nearest neighbor (KNN) techniques. The use of smart machine learning techniques helps LearnLoc significantly improve prediction accuracy and overcome noisy data readings compared to prior work. Figure 5 shows a comparison of energy consumption and localization accuracy of the three variants of the LearnLoc framework with well-known techniques from prior work (Radar [22], PlaceLab [23], Inertial_Nav [24]) along four indoor paths of 110m-140m lengths in various buildings at Colorado State University. We observed that PlaceLab and Radar consumed much less energy but that comes at a price of accuracy. The experimental results suggest that KNN delivers the most accurate location estimates, but also consumes the most energy. The LR and NL-NN variants perform considerably well overall. It is also important to note that Wi-Fi scan interval also plays a significant role in accuracy and energy-efficiency. The lowest Wi-Fi scan interval (1 second) delivers the best result, but a balance between energy consumption and accuracy can be achieved through the selection of a balanced scan interval. By prototyping a middleware framework that enables trade-offs between energy and accuracy, our work has brought viable indoor localization solutions that can be implemented on commodity mobile devices closer to reality.

VI. CONCLUSIONS

The rise of the data-driven science paradigm, in which massive amounts of data are produced and processed by mobile and IoT devices on a very strict power budget, requires new solutions to sustainably use these devices. Energy-efficient and robust data analytics approaches can help make the most out of available hardware resources. In this paper, we have shown, through various real-world case studies, that the flexibility and capabilities provided by such smart data analytics and machine learning middleware solutions can make a significant impact towards enhancing energy-efficiency, reliability, and performance robustness in a variety of mobile and IoT devices.

ACKNOWLEDGEMENTS

This work was supported by grants from the National Science Foundation (ECCS- 1646562, CCF-1252500).

REFERENCES

[1] 'Mobile data will skyrocket 700% by 2021', 2017 [online] Available: http://www.businessinsider.com/mobile-data-will-skyrocket-700-by-2021-2017-2

[2] Cisco Global Cloud Index: Forecast and Methodology, 2015–2020, White paper, [Accessed: Oct 2017]

[3] 'Samsung Galaxy S specifications', [Online], Available: https://www.phonearena.com/phones/Samsung-Galaxy-S_id4522

[4] 'System level Power Budgeting', 2017 [Online] Available: http://chipdesignmag.com/sld/blog/2014/03/12/system-level-power-budgeting/ [Accessed: Oct 2017]

[5] S. Tiku, S. Pasricha, "Energy-Efficient and Robust Middleware Prototyping for Smart Mobile Computing," IEEE International Symposium on Rapid System Prototyping (RSP), Oct 2017.

[6] 'Microsoft Azure, [online], Available: https://azure.microsoft.com/en-us/overview/what-is-middleware/

[7] E. Cuervo, A. Balasubramanian, D. K. Cho, A.Wolman, S. Saroiu, R. Chandra, and P. Bahl, "Maui: making smartphones last longer with code offload," in Proc. ACM Mobisys, 2010.

[8] K. Kumar and Y.-H. Lu, "Cloud computing for mobile users: Can offloading computation save energy?" Computer, vol. 43, 2010.

[9] H. R. Flores and S. Srirama, "Adaptive code offloading for mobile cloud applications: Exploiting fuzzy sets and evidence-based learning," Proc. ACM Mobisys, 2013.

[10] A. Khune, S. Pasricha, "Mobile Network-Aware Middleware Framework for Energy Efficient Cloud Offloading of Smartphone Applications", to appear, IEEE Consumer Electronics, 2017.

[11] 'Fuld – Torrent Downloader app', 2016, [Online]. Available: https://play.google.com/store/apps/details?id=com.delphicoder.flud&hl=en, [Accessed: Oct 2017].

[12] B. Donohoo, C. Ohlsen, S. Pasricha, "AURA: An Application and User Interaction Aware Middleware Framework for Energy Optimization in Mobile Devices", IEEE International Conference on Computer Design (ICCD 2011), Oct. 2011.

[13] B. Donohoo, C. Ohlsen, S. Pasricha, "A Middleware Framework for Application-aware and User-specific Energy Optimization in Smart Mobile Devices", Journal of Pervasive and Mobile Computing, vol. 20, pp. 47-63, Jul 2015.

[14] D.J. Simons, R.A. Rensink, "Change blindness: Past, present, and future", Trends Cogn. Sci. 9 (1) (2005).

[15] A. Shye, B. Scholbrock and G. Memik, "Into the wild: Studying real user activity patterns to guide power optimizations for mobile architectures," 42nd Annual IEEE/ACM International Symposium on Microarchitecture (MICRO), pp. 168-178, 2009.

[16] "Google Nexus One Tech Specs" [Online]. Available: http://www.htc.com/us/support/nexus-one-google/tech-specs

[17] B. Donohoo, C. Ohlsen, S. Pasricha, C. Anderson, "Exploiting Spatiotemporal and Device Contexts for Energy-Efficient Mobile Embedded Systems", IEEE/ACM Design Automation Conference (DAC 2012), Jul. 2012.

[18] B. Donohoo, C. Ohlsen, S. Pasricha, C. Anderson, Y. Xiang, "Context-Aware Energy Enhancements for Smart Mobile Devices", IEEE Trans. on Mobile Computing (TMC), 13(8):1720-1732, 2014.

[19] 'Guide to Google Cards, 2017 [Online] Available: https://www.androidcentral.com/ultimate-guide-google-now-cards

[20] C. Langlois, S. Tiku, S. Pasricha, "Indoor localization with smartphones", IEEE Consumer Electronics, Oct 2017.

[21] S. Pasricha, V. Ugave, Q. Han and C. Anderson, "LearnLoc: A Framework for Smart Indoor Localization with Embedded Mobile Devices," ACM/IEEE CODES+ISSS, Oct 2015.

[22] P. Bahl, and V. Padmanabhan, "RADAR: An in-building RF-based user location and tracking system," IEEE INFOCOM. 2000.

[23] A. LaMarca, et al., "Place Lab: Device Positioning Using Radio Beacons in the Wild," Proc. PERCOM, 2005, pp. 116-133.

[24] J. Á. B. Link, P. Smith, N. Viol and K. Wehrle, "FootPath: Accurate map-based indoor navigation using smartphones," IPIN 2011.

Data-Driven Resiliency Solutions for Boards and Systems

Shi Jin and Krishnendu Chakrabarty
Department of Electrical and Computer Engineering
Duke University, Durham, NC, USA

Abstract – Data analytics and real-time monitoring can be used to ensure that boards and systems operate as intended. This paper first describes how machine learning, statistical techniques, and information-theoretic analysis can be used to close the gap between working silicon and a working system. Next, it describes how time-series analysis can be used to analyze health status and detect anomalies in complex core router systems. Traditional techniques fail to identify abnormal or suspect patterns when the monitored data involves temporal measurements and exhibits significantly different statistical characteristics for its constituent features. This paper thus not only describes a feature-categorization-based hybrid method and a changepoint-based method to detect anomalies in time-varying features with different statistical characteristics, but also proposes a symbol-based health analyzer to obtain a full picture of the health status of monitored core routers. A comprehensive set of experimental results is presented for data collected during 30 days of field operation from over 20 core routers deployed by customers of a major telecom company.

1. INTRODUCTION

We are in the midst of a Big-Data revolution with a rapidly expanding ecosystem of diverse sources of massive datasets. Recent conservative studies estimate that global datacenter traffic was 4.7 zettabytes (ZB) in 2015, and it is expected to double every two years [1]. The generated data can be structured (e.g., financial, electronic medical records), semi-structured (e.g., tweets, emails), unstructured (e.g., audio, video), or real-time (e.g., network traces, monitoring logs). All of these types of data have the potential to provide invaluable insights, if organized and analyzed appropriately. The analyses of such large and diverse datasets is fast emerging as an indispensable tool for innovations in various domains such as healthcare, business process optimization, and social-network-based recommendations.

In this paper, we argue that data analytics and real-time monitoring can be used to ensure that boards and systems operate as intended. For instance, in the domain of board-level fault diagnosis, massive data of functional tests and measurements is typically collected from advanced chips. How can this data be analyzed to help solve resilience problems of these boards? Similarly, in the system testing domain, a large amount of data is monitored in real time. How can we utilize this data to assess health status of complex system and identify abnormal or suspect operations? Therefore, in the rest of the paper, we discuss and explore the data-driven resiliency solutions for both boards and systems.

2. DATA-DRIVEN BOARD-LEVEL FAULT DIAGNOSIS

Fault diagnosis isolates the root cause of a malfunction system by collecting and analyzing information on system status using measurements, tests, and other information sources [2]. It is important at all stages of the product life cycle, but particularly crucial during manufacture and field maintenance. The diagnosis process can be hierarchically carried out as a top-down process (system → boards → chips) or a bottom-up process (chips → boards → system). Bboard-level diagnosis is much more challenging than chip-level diagnosis because there is no effective flow to locate the root cause of a failure for functional tests. Typically, technicians run additional functional tests and measurements based on their personal experience. This process is time-consuming, and there is no guarantee of the success of repair. In addition, a board has to be scrapped after a few unsuccessful repair attempts.

In reasoning-based diagnosis system [2-3], the fault isolation process is driven by an inference engine based on failure-symptoms correlation rules, expressed in the form of labels. Reasoning-based methods have been advocated for diagnosis via data-driven techniques. Artificial neural networks (ANNs) have been used for fault diagnosis in digital systems. For instance, in [3], the authors developed a diagnosing system using multi-layer ANNs for combinational circuits. The training of this diagnosis system is based on a database of identified potential faults, which are generated by inserting one fault in the circuit followed by simulation. In [5], Zhang et al. proposed a single-layer ANN for board-level diagnosis to achieve a significant improvement over previous ANN-based diagnosis methods. However, the training time becomes prohibitively long as the system complexity increases.

Butcher et al. in [4] presented a diagnosis system based on Bayesian networks. This diagnosis system can be either derived from domain knowledge or learned from actual tests, which are extracted from debugging and maintenance data. A statistically significant (large) dataset is required to reflect an appropriate relationship between faults and test outcomes. However, since different components (with nonuniform failure rates) are used in different production lines, and equipment used for testing can vary, training a diagnosis system using these aggregated data may lead to incorrect learning. Therefore, correction techniques based on conditional probabilities are used to alleviate data corruption [4].

Ye et al. presented a fine-grained and adaptive diagnosis system for high-volume products [6]. This diagnosis system is based on multi-kernel support vector machines (MK-SVMs) and incremental learning. The MK-SVM method leverages a linear combination of single kernels to achieve accurate faulty component classification based on the errors observed. The MK-SVMs thus generated can also be updated based on incremental learning, which allows the diagnosis system to quickly adapt to new error observations and provide even more accurate fault diagnosis. A data fusion technique is also presented, namely majority-weight voting (MWV), to leverage multiple classifiers in our diagnosis system. The proposed MWV takes advantage of both ANNs and SVMs to provide an optimal repair-suggestion set.

A decision tree-based diagnosis system has been proposed to accelerate the diagnosis process [7]. The diagnosis procedure is constructed as a binary tree, with the most discriminative syndrome as root and final repair suggestions available as the leaf nodes of the tree. The number of syndromes required for diagnosis

978-1-5386-3693-0/18 $31.00 © 2018 IEEE   244

can also be significantly reduced compared to the number of syndromes used for system training. Furthermore, an incremental version of DTs is used to facilitate online learning, so as to bridge the knowledge obtained at test-design stage with the knowledge gained during volume production.

Shi et al. describe the design of a preprocessing step in a diagnosis system to handle missing syndromes [8]. Traditional diagnosis systems fail to provide appropriate repair suggestions when the diagnostic logs are fragmented and some error outcomes, or syndromes, are not available during diagnosis. Missing syndromes can be handled by using imputation methods. Several imputation methods have been evaluated in terms of their efficiency in handling missing syndromes. Moreover, syndrome analysis based on subset selection has been used to select a representative set of syndromes with minimum redundancy and maximum relevance. Root-cause analysis has been used to measure the discriminative ability of differentiating a given root cause from others.

During the initial product ramp-up phase, reasoning-based diagnosis is not feasible for yield learning, since the required database is not available due to lack of volume. A knowledge-discovery and a knowledge-transfer method has thus been developed to facilitate board-level functional fault diagnosis [9]. First, an analysis technique based on machine learning is used to discover knowledge from syndromes, which can be used for training a diagnosis engine. Second, knowledge from diagnosis engines used for earlier-generation products can be automatically transferred through root-cause mapping and syndrome mapping based on keywords and board-structure similarities.

3. DATA ANALYTICS IN CORE ROUTER SYSTEM

A three-layer hierarchical design is widely used in modern telecommunication systems to achieve high performance and reliability [10]. The three layers, namely *core*, *distribution*, and *access*, perform different roles for service fulfillment [10]. The core layer is also referred to as the network backbone, and it is responsible for the transfer of a large amount of traffic in a reliable and timely manner. The network devices (such as routers) used in the core layer are complex systems that contain both software and hardware (Figure 1), making them more vulnerable to hard-to-detect/hard-to-recover errors [11].

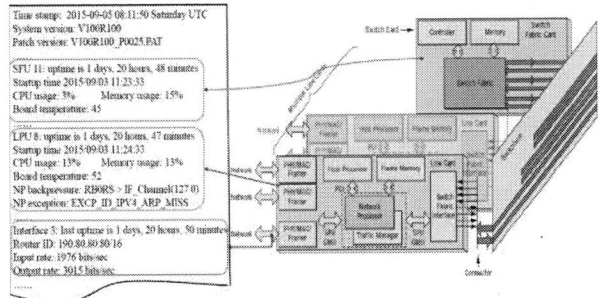

Figure 1. A multi-card chassis core router system and a snapshot of extracted (monitored) features [16-18].

Proactive fault tolerance is promising because it takes preventive action before a failure occurs [12]. The state of the system is monitored in a real-time manner. When system degradation is determined via health assessment, proactive repair actions such as job migration are executed to avoid errors, thereby ensuring the non-stop utilization of the entire system [13]. The effectiveness of proactive fault-tolerance solutions depends on whether abnormal patterns of core routers can be accurately pinpointed in a timely

manner [15]. In this section, we show how data analytics can be used to analyze health status and detect anomalies in a high-performance and complex communication system [16-18].

3.1. Time-Series-Based Anomaly Detection

A common way to identify a system's health status is to feed its features to an anomaly detector to see whether any data points are statistical outliers. The difficulty of developing an efficient anomaly detector for a complex communication system can be attributed to two reasons. The first reason is that features extracted from communication systems are far more complex than those from a general computing system. For example, a multi-card chassis core router system uses monitors to log a large number of features from different functional units. These features include performance metrics (e.g., events, bandwidth, throughput, latency, jitter, error rate), resource usage (e.g., CPU, memory, pool, thread, queue length), low-level hardware information (e.g., voltage, temperature, interrupts), configuration status of different network devices, and so on. Each of these features can have significantly different statistical characteristics, making it difficult for a single type of anomaly-detection technique to be effective. The second reason is that the monitored data in communication systems involves temporal measurements. Most existing anomaly-detection methods are not designed to address time-series data [14], hence they may not be able to detect time-series-specific anomalies such as the trend anomaly.

In complex communication systems such as a core router, data is collected in the form of time-series. A key challenge here is to detect anomalies in time-series data to determine whether the system is entering a degraded state or is likely to fail. Therefore, we have studied a range of techniques that may be used to detect anomalies in time-series data [15].

The first one is unsupervised distance-based anomaly detection, which utilizes a distance measure between a pair of time-series instances to represent the similarity between these two time-series [23]. Instances far away from others will be identified as being abnormal. The second one is window-based anomaly detection [15]. This method divides time-series instances into overlapping windows. Anomaly scores are first calculated per window, and then aggregated to be compared with a predefined threshold. Only when the overall anomaly score of a single time-series instance significantly exceeds a predefined threshold, will this instance be identified as being abnormal. The third method is supervised prediction-based anomaly detection [15]. First, a machine-learning-based predictive model is learned from historical logs. Next, predicted values are obtained by feeding test data to this predictive model. These predicted values are then compared with the actual measured data points. The accumulated difference between these predicted and the actual observations is defined as the anomaly score for each test time-series instance.

However, a single class of anomaly detection methods is effective for only a limited number of time-series types. Therefore, we have also developed a feature-categorization-based hybrid method (Figure 2) [16]. First, features that exhibit similar statistical characteristics are placed in the same category. Next, each group of features is fed to the anomaly detector that is most suitable for these types of features. Finally, the results provided by different anomaly detectors are aggregated so that an anomaly can be detected in terms of the entire feature space.

The number of feature dimensions will increase from hundreds to tens of thousands when more new features are identified and extracted from raw log data, making it more difficult and time-consuming to detect anomalies [15]. Therefore, a correlation analyzer has been developed to remove irrelevant and redundant features. As shown in Figure 3, the correlation analyzer consists of three components: the linear correlation component, the feature selection component, and the non-linear correlation component. The linear correlation component is used to find linear-dependent feature pairs. The feature selection component is developed to select an effective, but reduced, set of non-linear-dependent feature groups [16]. The non-linear correlation component is implemented for fine-grained non-linear dependence analysis [23]. Finally, the correlation analyzer outputs a number of correlated feature groups. An effective feature subset can be generated by selecting the most representative features from these correlated groups.

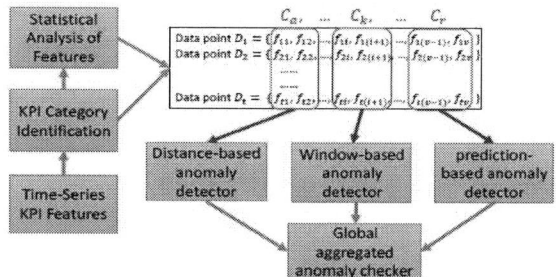

Figure 2. A depiction of feature-categorization-based hybrid anomaly detection.

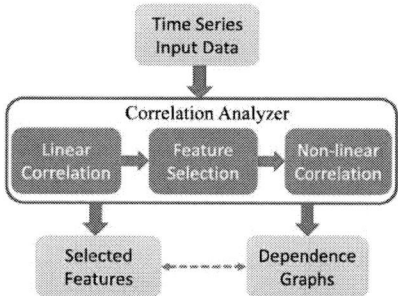

Figure 3. Architecture of the proposed correlation analyzer. Features within each group can be represented by a dependence graph $G = (V, E)$, where the set of vertices V represent feature candidates and the set of edges E represent dependent relationships between features. A dependence graph is generated for each group of features [16].

3.2. Changepoint-Based Anomaly Detection

A drawback of the above methods is that they make assumptions that are not valid in realistic scenarios. For example, the supervised prediction-based methods assume that we know labels (normal or abnormal) in advance for the training data. However, raw data collected from commercial core routers is unlabeled. Labeling such raw data requires manual checking by expert technicians, which needs a considerable amount of time and labor. In contrast, the unsupervised distance-based approaches do not require any labels. However, they assume that abnormal instances are rare events that are significantly different from all other historical instances. Such an assumption is not valid for a core router system because some normal instances are also rare. The window-based methods are infeasible for long-term time series instances because their time cost increases significantly with the size of the temporal dimension [14-15]. The feature-categorization-based hybrid method makes the assumption that the statistical properties of normal features remain constant across the entire temporal domain.

However, new normal patterns can appear as time proceeds and the statistical characteristics of features can change significantly even if no anomalies occur. Therefore, in this section, we address the problem of detecting anomalies in complex core router systems without relying on the unrealistic simplifying assumptions that were made in prior work.

We have therefore developed a two-step changepoint-based anomaly detection scheme [18], as shown in Figure 4. The key idea is that instead of directly detecting anomalies from a large volume of time-series data, we first detect all the changepoints, which indicate significant scenario changes. The changepoint (CP) windows are built from useful data around each changepoint. Machine-learning algorithms are then applied to these CP windows to determine normal window patterns and anomalous window patterns. When new data arrives, it goes through the changepoint detection procedure, and then the following actions are taken:

1) If no changepoint is detected, the new data will be identified as being normal and no further actions will be taken.

2) As long as any changepoints are detected, a set of new CP windows will be determined and fed to the previously learned normal/abnormal window pattern library. Then for each new CP window, the following actions will be taken: (i) If the new CP window is classified as an abnormal window pattern, an anomaly will be reported; (ii) If the new CP window belongs to the category of normal window patterns, no alert is generated; (iii) If the new CP window lies outside any existing window patterns, it is identified as a suspect window, and used to update the historical window-pattern library.

The advantages of the proposed anomaly-detection method are:

1) Only time points around changepoints are considered during model training and testing, significantly reducing the dimensionality of temporal domain.

2) Both abnormal and new/rare normal patterns can be identified, thereby reducing the number of false alarms caused by new/rare normal patterns.

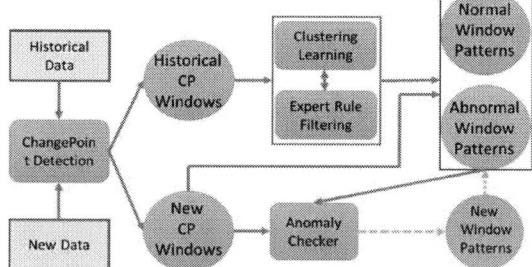

Figure 4.   Illustration of changepoint-based anomaly detection

1) Changepoint Detection: Changepoints are locations where abrupt changes occur in time-series data. A changepoint τ splits a time series into two disjoint segments d_1, d_2 with probability density functions $f(\mu_1, \sigma_1^2)$, $f(\mu_2, \sigma_2^2)$. The segment d_1 before τ and the segment d_2 after τ have significantly different statistical properties such as mean and variance. Therefore, the problem of changepoint detection can be described as identifying the time points where the statistical properties of the time sequence show heterogeneity or discontinuity.

The maximum likelihood estimation is a widely-used technique that can identify these statistical properties given actual observations [19]. In this case, the objective of detecting changepoints is to find an optimal partition that maximizes the product of likelihood functions of all disjoint segments. The objective function can thus be formulated as:

$$CP^* = \underset{\tau_1=1,...,n}{\text{argmax}}(\alpha L(CP) - \beta P(m))$$

$$= \underset{\tau_1=1,...,n}{\text{argmax}}(\alpha \sum_{i=1}^{m+1} \log p(s_i | \hat{q}_i, \hat{\theta}_i, \hat{\varepsilon}_i) - \beta P(m)) \quad (1)$$

where CP^* represents an optimal (i.e., maximum likelihood estimate) combination of changepoints, including both the number of changepoints m and the locations of changepoints $\{\tau_1, \tau_2, ..., \tau_m\}$. The parameter $\alpha L(CP)$ is the weighted sum of the log-likelihood of $m+1$ segments split by m changepoints, and $\beta P(m)$ is the weighted penalty function associated with the number of changepoints m to avoid overfitting.

Several heuristic algorithms can be utilized to reduce the search space for the multiple changepoint detection problem. Among these techniques, we consider binary segmentation and the Pruned Exact Linear Time (PELT) algorithm due to their relatively low computational complexity [19-20]. The basic idea of binary segmentation is divide-and-conquer. It first applies the single changepoint detection procedure on the entire time series. If a single changepoint is identified, the entire time series is divided into two segments. This procedure is repeated on all newly-generated segments until no new changepoints can be found. Binary segmentation reduces the computational complexity from $O(2^n)$ to $O(n \log n)$. The PELT algorithm applies dynamic programming to reduce the search space. Starting from the last changepoint τ_m, the entire time series D is divided into a subset D' before τ_m and a segment s_{m+1} after τ_m. This procedure is executed recursively until we reach the first changepoint τ_1; the optimum CP^* is recorded during this process. A pruning process, which removes solutions that are far from optimality, is also executed at each iteration to further speed up computation, yielding $O(n)$ computational complexity.

2) Changepoint Window Learning: After all changepoints have been identified, the next step is to identify changepoint windows so that all useful information before and after scenario changes can be incorporated. After all changepoint windows have been constructed from historical data, we can now apply machine-learning techniques to learn normal/abnormal patterns. Since labels of the CPs are unknown, unsupervised clustering methods are needed to automatically group similar changepoint window patterns. Among different kinds of clustering techniques, density-based spatial clustering of applications with noise (DBSCAN) is promising because it does not require the number of clusters as its prior and it can find arbitrarily shaped clusters [21].

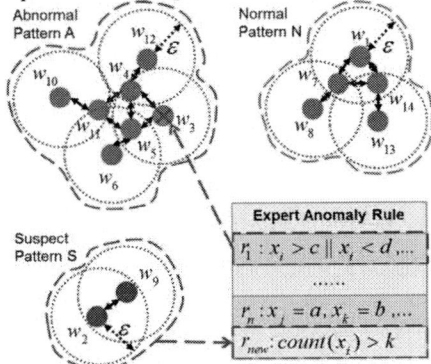

Figure 5. Illustration of DBSCAN-based changepoint window learning

Figure 5 shows the improved DBSCAN-based changepoint-window learning method. First, for each cluster identified by DBSCAN, its core point set will be fed to the expert anomaly rule table. If any core point satisfies any of the expert anomaly rules,

the entire cluster will be identified as being an abnormal window pattern. For example, the core point w_3 in the top-left cluster satisfies expert anomaly rule r_1. Therefore, the entire cluster is identified as "Abnormal Pattern A". In contrast, none of the core points in the top-right cluster are found in the expert anomaly rule table. Thus, this cluster is labeled as "Normal Pattern N". Moreover, neighboring noise points together are considered as a suspect window pattern. If any matched expert anomaly rules are found, this suspect pattern will be labeled as an abnormal window pattern. Otherwise, a new expert rule will be generated from this suspect pattern and then be inserted into the expert anomaly rule table. For example, the two neighboring noise points in the bottom-left corner form a "Suspect Pattern S". Since no related rules are found in the table, a new rule r_{new} is generated and inserted into the table.

3) Anomaly Detection: After the library of changepoint window patterns is generated, the next step is to detect anomalies. Each changepoint window pattern p_i in the library can be classified into three general categories: normal, abnormal, and suspect. When new time-series data $D_{new} = \{d_1, ..., d_l\}$ is extracted, it is first fed to the changepoint detection component. If no changepoints are detected, D_{new} is labeled as normal and no further checking is executed. Otherwise, for each changepoint window w_i, both 1NN and SVM are used to determine its normal/abnormal conditions.

3.3. Symbol-Based Health Analysis

An efficient time-series-based anomaly detector is not adequate to obtain a full picture of the health status of monitored core routers. First, an anomaly detector can only provide information about the anomalous points; patterns before or after anomalies are not revealed, which may also be necessary for predicting failures. Second, an anomaly detector can provide little useful information if no anomalies are identified. However, learning different normal patterns are also important because it can reveal how healthy a core router system is and how different task scenarios can affect the system. Therefore, in this section, we address the problem of analyzing the health status of complex core router systems. We use multiple symbolization techniques to discretize and compress long-term time series. We describe several symbol-based clustering and classification methods to analyze the health status of core routers in a more comprehensive way.

We propose a time-series-based health-status analysis scheme [17], as shown in Figure 6. The key idea is that instead of directly analyzing health status from a large volume of raw time series data, we first transform and segment the high-dimensional time series into a meaningful low-dimensional representation. The transformed time series is then fed to our pattern-learning component and health-analysis component for further fine-grained analysis. Pattern-learning component: This component consists of two parts: clustering and rule discovery. The objective of clustering is to group time series with similar shapes. Therefore, a number of different global patterns are learned after clustering. In contrast, the objective of rule discovery is to find repeating local subsequences within long-term time series. Therefore, a number of local patterns are learned after rule discovery.

1) Pattern-learning component: This component consists of two parts: clustering and rule discovery. The objective of clustering is to group time series with similar shapes. Therefore, a number of different global patterns are learned after clustering. In contrast, the objective of rule discovery is to find repeating local subsequences within long-term time series. Therefore, a number of local patterns are learned after rule discovery.

2) Health-analysis component: This component consists of two parts: classification and prediction. The classification part is used to identify categorical health-status level, and the prediction part is utilized to predict some specific numerical health-status metrics.

3) *Expert rule table*: This component is maintained and updated by an expert team. The expert rules can be used to either label patterns learned from the pattern-learning component or validate the health status identified by the health-analysis component.

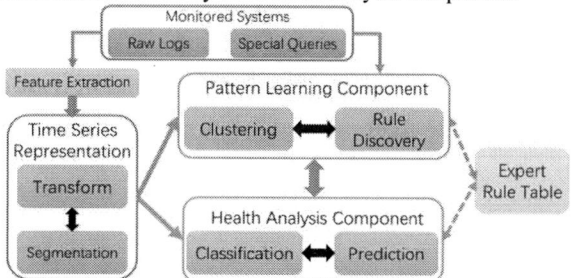

Figure 6.   The proposed health-status analyzer in core router systems.

3.3.1. Time Series Symbolization

1) Symbolic Aggregate Approximation: a PAA-based symbolic aggregation approximation (SAX) has been proposed in [22] to discretize time-series data into a sequence of symbols, where each symbol represents a predefined data region. The procedure of SAX symbolization consists of three main steps: normalization, piecewise aggregation approximation, and symbol mapping. Assume that we have a time-series data $D = \{d_1, ..., d_n\}$, where n is the length of D. First, a normalization process is executed to transform the original time series into a time series $DN = \{dn_1, ..., dn_n\}$ with a mean of zero and a standard deviation of one. Next, a piecewise aggregation approximation is used to convert the n-length normalized time series DN to a w-length discretized time series $DP = \{dp_1, ..., dp_w\}$. It divides the original time series into a set of equal-size segments, and then represents each segment by the mean value of all elements within that segment.

The final step is to map the w-length discretized time series DP into a w-length symbol sequence $DS = (ds_1, ..., ds_w)$. The key idea here is to map "close" discretized data points into the same data region, and each data region is represented by a unique symbol. Since the original time series D has been normalized in the first step, its data region can be divided into α equal-sized are based on the statistical probability property of the Gaussian distribution. A list of $\alpha + 1$ "breakpoints" $B = \{b_0, ..., b_\alpha\}$ can be precomputed to help partition the N(0,1) Gaussian curve into α equal-sized areas. The resulting equal-sized areas are then encoded by unique characters in a sorted English alphabet L = $\{l_1 = a, l_2 = b, ..., l_{26} = z\}$.

2) Moving-Average-based Trend Approximation: An important advantage of the SAX representation is that it can transform a long-term complex time series into a much shorter and concise symbol sequence without losing the global shape information of the original data. However, important local information such as shapes within segments and trends spanning neighboring segments are lost in SAX because each symbol in the SAX representation only contains the mean value information of its corresponding segment. Therefore, we propose a moving-average-based trend approximation method to extract, transform, and preserve critical local trend information within a long-term time series.

The moving average (MA) is a commonly used technique to smooth the original real-valued time series [23]. The basic idea of MA is to represent the original time series by mean values calculated from a set of overlapped data sub-windows. After moving-average-based smoothing, the next step is to identify local trend information of a time series with symbol sequences. Here, we want to represent a time series with a sequence of digits via the 3σ rule. Data points lying within the region between the moving max and moving min are represented by "0", data points lying above the moving max are marked by "1", and data points lying below the moving min are represented by "-1".

3.3.2. Symbol-based Pattern Learning and Health Analysis

After the original high-dimensional time-series data has been transformed to a low-dimensional symbol sequence, clustering and rule discovery algorithms can be applied to learn different patterns from these symbol sequences.

1) Hierarchical Clustering: Flat clustering is unsuitable for analyzing long-term complex time-series data because it treats all clusters equally, and thus does not reveal useful structure information within and among clusters. In contrast, hierarchical clustering can not only divide the dataset into a set of clusters, but also output a dendrogram, which is a tree diagram to illustrate the arrangement of the clusters [23].

2) Rule Discovery: Hierarchical clustering is efficient in building a multi-level structure of clusters, which enables us to analyze and label global patterns from different granularities. However, patterns generated from clustering are sometimes overly complex, which masks some simple but truly meaningful local patterns. Therefore, the rule-discovery component is necessary to learn patterns or rules that are hidden or not obvious in time-series data.

We utilize the sequitur algorithm [24] to discover hidden patterns because it can learn a hierarchical structure of rules from any discretized symbol sequence in linear time. The basic idea of sequitur is to recursively substitute repeating subsequence in the input symbol sequence with new rules and thus reconstruct the symbol sequence with a hierarchical-rule-based representation. For example, a new rule R1 → dc is discovered from the symbol sequence "bdeaaaddcabedcc" and the new hierarchical-rule-based symbol sequence is (bdeaaad"R1"abe"R1"c).

3) Symbol-based Classification and Prediction: After informative patterns and rules have been learned from historical time-series data, efficient classification and prediction techniques are needed to identify or predict the health status of new time-series data with these learned patterns.

Distance-based techniques are widely used to classify time series, where the overall "distance" between a pair of time-series instances is used to measure the similarities between them [23]. Therefore, time-series instances are classified into the same category if the overall "distance" between them, $Dist(i, j)$, is less than a threshold t. In addition to the distance-based methods, the decision tree (DT) model is also used here for two reasons: (i) it can handle both real-valued and categorical features, which makes it suitable for both raw data and symbolized data; (ii) its output results are easy to interpret and explain, which facilitates interactions with the expert team [25].

3.4. Experiments and Results

The commercial core router system used in our experiments consists of a number of different functional units such as the main processing unit, line processing unit, and switch fabric unit. A total of 602 features were monitored and sampled every 30 minutes for 15 days of operation of the core router system, generating a set of multivariate time-series data consisting of 720 time points.

Table 1 shows the features corresponding to various components in the core router. The success ratio (SR) is the ratio of the number of correctly detected anomalies to the total number of

978-1-5386-3693-0/18 $31.00 © 2018 IEEE

anomalies in the testing set while the non-false-alarm ratio (NFAR) is defined as the ratio of the number of correctly detected anomalies to the total number of alarms flagged by the anomaly detector. We found that that for six anomaly detection methods, i.e., rule-based approach, k-nearest neighbors (KNN), window-based KNN, support-vector regression, the feature-categorization-based hybrid method, and the changepoint-based SVM, the success ratios are 68.4%, 66.5%, 70.4%, 75.2%, 78.6% and 81.1%, respectively, and the non-false-alarm ratios are 65.5%, 56.3%, 64.7%, 72.1%, 71.6% and 82.1%. Although the feature-categorization-based hybrid method achieves higher success ratio and non-false alarm ratio than the KNN and rule-based methods, it is outperformed by the proposed changepoint-based method.

Table 1: Features corresponding to various components.

Component	Number of Components	Number of features	Representative features
MPU	2 (1+1 backup)	112	CPU/Mem Usage Board Temperature
SFU	4 (3+1 backup)	196	Route Age Board Uptime
LPU	16	832	Lost Packet Count
Interface	93	1004	Input/Output Rate Utilization Ratio
Others	124	748	Exception/Assertion

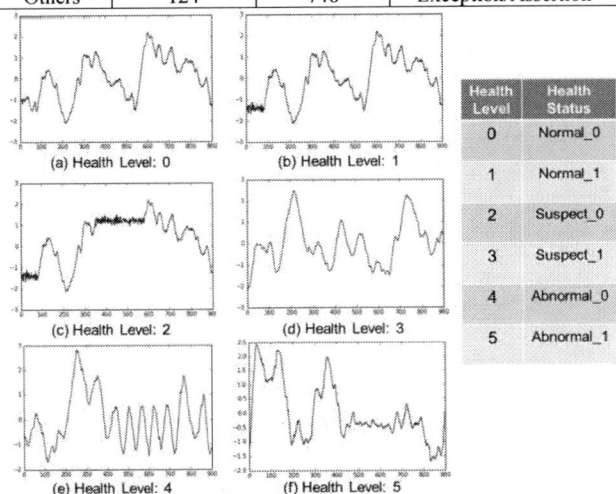

Health Level	Health Status
0	Normal_0
1	Normal_1
2	Suspect_0
3	Suspect_1
4	Abnormal_0
5	Abnormal_1

Figure 8. Illustration of health levels used in our experiments.

Six health levels were defined to represent the overall health status of core routers, as shown in Figure 8. Two information-theoretic metrics, called precision and recall, have been used to comprehensively evaluate the health-analysis system. For the long-term time-series instances extracted in our experiment, the two raw-data-based baseline algorithms perform well for the two normal health classes "Health Level 0" and "Health Level 1". However, they do not achieve satisfactory precision and recall levels for "Health Level 2", "Health Level 3", "Health Level 4", and "Health Level 5". In contrast, the proposed symbol-based methods perform much better for classifying the suspect and abnormal health classes.

4. CONCLUSIONS

We are witnessing the rise of the data science paradigm, in which massive amounts of data can be analyzed to make useful predictions. In this paper, we have argued that data analytics and real-time monitoring can be used to ensure that boards and systems operate as intended.

ACKNOWLEDGMENTS

This work was supported by grants from Huawei. The authors thank F. Ye, Z. Zhang, and X. Gu for their contributions.

REFERENCES

[1] Cisco Global Cloud Index: Forecast and Methodology, 2015–2020, White paper, [Accessed July 2017].

[2] W. R. Simpson and J. W. Sheppard, System Test and Diagnosis. Springer, 1994.

[3] A. A. Al-Jumah and T. Arslan, "Artificial neural network based multiple fault diagnosis in digital circuits," in Proc. ISCAS, vol. 2, pp. 304 - 307, 1998.

[4] S. Butcher and J. W. Sheppard, "Distributional smoothing in Bayesian fault diagnosis," in IEEE Transactions on Instrumentation and Measurement, vol. 58, no. 2, pp. 342 - 349, 2009.

[5] Z. Zhang et al., "Smart diagnosis: Efficient board-level diagnosis and repair using artificial neural networks," in Proc. ITC, 2011.

[6] F. Ye et al., "Adaptive board-level functional fault diagnosis using decision trees," in Proc. ATS, pp. 202 - 207, 2012.

[7] F. Ye et al., "Board-level functional fault diagnosis using artificial neural networks, support-vector machines, and weighted-majority voting," in IEEE TCAD, vol. 32, pp. 723–736, 2013.

[8] S. Jin et al., "Efficient board-level functional-fault diagnosis with missing syndromes," in IEEE TCAD, vol. 35, pp. 985-998, 2016.

[9] F. Ye et al., "Information-theoretic syndrome and root-cause analysis for guiding board-level fault diagnosis," in Proc. ETS, 2014.

[10] V. Antonenko and R. Smelyanskiy, "Global network modelling based on mininet approach," in Proc. ACM SIGCOMM Workshop on Hot Topics in Software Defined Networking, pp. 145–146, 2013.

[11] M. Me'dard and S. S. Lumetta, "Network reliability and fault tolerance," Encyclopedia of Telecommunications, 2003.

[12] P. K. Patra, H. Singh, and G. Singh, "Fault tolerance techniques and comparative implementation in cloud computing," in International Journal of Computer Applications, vol. 64, pp. 1–6, 2013.

[13] A. Gainaru et al., "Fault prediction under the microscope: A closer look into hpc systems," in Proc. Int. Conf. High Performance Computing, Networking, Storage and Analysis, 2012.

[14] A. Patcha and J.-M. Park, "An overview of anomaly detection techniques: Existing solutions and latest technological trends," in Computer Networks, vol. 51, pp. 3448–3470, 2007.

[15] V. Chandola, A. Banerjee, and V. Kumar, "Anomaly detection: A survey," ACM Computing Surveys, vol. 41, pp. 15:1–15:58, 2008.

[16] S. Jin et al., "Accurate anomaly detection using correlation-based time-series analysis in a core router system," in Proc. ITC, 2016.

[17] S. Jin et al., "Symbol-based Health-Status Analysis in a Core Router System," accepted for publication in Proc. ITC, 2017.

[18] S. Jin et al., "Changepoint-based Anomaly Detection in a Core Router System," accepted for publication in Proc. ITC, 2017.

[19] A. J. Scott and M. Knott, "A cluster analysis method for grouping means in the analysis of variance," Biometrics, 1974, pp. 507–512.

[20] R. Killick, P. Fearnhead, and I. Eckley, "Optimal detection of changepoints with a linear computational cost," Journal of the American Statistical Association, vol. 107, pp. 1590–1598, 2012.

[21] M. Ester et al., "A density-based algorithm for discovering clusters in large spatial databases with noise," In Kdd. vol. 96, 1996.

[22] J. Lin, et al., "Experiencing sax: a novel symbolic representation of time series," in Data Mining and knowledge discovery, vol. 15, pp. 107–144, 2007.

[23] O. Maimon and L. Rokach, Data Mining and Knowledge Discovery Handbook. Springer-Verlag New York, Inc., 2005.

[24] C. G. Nevill-Manning and I. H. Witten, "Identifying hierarchical structure in sequences: A linear-time algorithm," in J. Artif. Int. Res., vol. 7, pp. 67–82, 1997.

[25] J. Quinlan, "Induction of decision trees," Machine Learning, 1986.

Single-Error Hardened and Multiple-Error Tolerant Guarded Dual Modular Redundancy Technique

Sai Aparna Aketi, Joycee Mekie and Hemal Shah
Department of Electrical Engineering
Indian Institute of Technology Gandhinagar, India
Email: {aketi.sai, joycee, hemal.shah} @iitgn.ac.in

Abstract—Circuits designed for space applications need special consideration to tolerate radiations. Guarded dual modular redundancy (GDMR), a radiation hardened by design (RHBD) technique for single event transients (SETs) is presented in this paper. We present a neat mathematical procedure that captures the effects of multiple event transients (METs) in any given radiation-hard by design (RHBD) technique. We analyze the effectiveness of GDMR multiple event transients (METs) against the well-known triple-modular redundancy (TMR) technique using this procedure. Our results show that GDMR logic gates exhibit far better tolerance to METs as compared to TMR gates, except for some logic gates. We have implemented several logic gates and a benchmark circuit (C17) using unhardened, GDMR and TMR techniques in UMC 65nm technology and compared them. Our simulations of various logic gates show that GDMR gates consume about 50% less power, 3× less area and about 50% less delay compared to their TMR counterparts, and yet, GDMR outperforms TMR in terms of error-tolerance to METs by about 3×, except for some gates. For C17 ISCAS-85 Benchmark circuit implemented in UMC 65nm, we find that GDMR implementation consumes about 58% less area, has 31% less delay, 19% less power and 32% less probability of error due to METs than TMR implementation.

Keywords-Dual rail logic, guard gate, single event transients, multiple event transients, probability of error

I. INTRODUCTION

Designing electronic components that will be able to sustain harsh environmental conditions in space and military applications has been an important field of research since many decades. Prolonged bombardment of heavy-ions, protons, neutrons and other particles and irradiation on electronic equipment and circuits can result into permanent damage of the circuits or soft errors causing temporary failure. When high-energy neutrons (present in terrestrial cosmic radiations) or alpha particles (that are originated from impurities in the packaging materials) strike a sensitive node in a CMOS circuit, they generate a dense local track of additional electron-hole pairs in the substrate. In particular, the sensitive node in a CMOS semiconductor circuit is the drain of OFF-transistors. This additional charge is collected by the drain of an OFF transistor and a current pulse is observed that results in a transient voltage pulse. To circumvent the effects of these radiations, several radiation hardened by design (RHBD) techniques have been proposed in the literature mainly to deal with soft errors such as single event transients (SETs), single event upsets (SEUs), single event multiple transients, multiple

event transients (METs) and upsets. SETs and SEUs are more commonly observed errors, while METs are more likely to occur in circuits fabricated at deep sub-micron and ultra-deep sub-micron technology nodes.

In this paper we present a Guarded Dual Modular Redundancy (GDMR) circuit design technique where the duplicated outputs connect to a guard gate instead of voter in conventional designs. Conventional dual modular redundancy technique [1] derives the third input to the voter either as a time-delayed signal from one of the circuits or previous output. Thus, in trying to save area, the designer needs to pay in terms of additional delay. The property of the guard gate [7] is that its output changes only when both its inputs have the same value. The guard gate is also referred to as a glitch filter. If radiation strike affects only one of the gates at a given time instant, the guard gate output will not have a soft error due to the dual modular redundancy (DMR). The guard gate, therefore, blocks the error from being propagated further in the circuit. The GDMR technique is technology independent and radiation hard by design for SETs. One of the other important considerations of RHBD techniques is their tolerance toward multiple event transients (METs). In this paper, we have carried out a detailed analysis for METs for GDMR, TMR and unhardened gates. The analysis of TMR can be easily extended to conventional DMR. We have carried out systematic analysis to capture all possible scenarios of multiple event transients and assume equal likelihood of each to occur. Our analyses show that the probability of error in the final output for GDMR gates due to multiple event transients is independent of the logic of the gate. But in case of TMR, the probability of the error depends on the logic of the gate. Further, the area requirements of GDMR is small compared to TMR logic. We have done this analysis over more than fifteen different logic gates, and it is clearly evident that GDMR logic is more tolerant to multiple event transients as compared to TMR in most of the cases, and is also area efficient. The main contributions of this work are:

- Guarded dual modular redundancy is proposed as a new radiation hard by design technique to mitigate the effects of SETs.
- A generic probability model to analyse the effects of multiple-event transients in circuits. This model is used to compare effect of METs on unhardened, proposed

978-1-5386-3693-0/18 $31.00 © 2018 IEEE

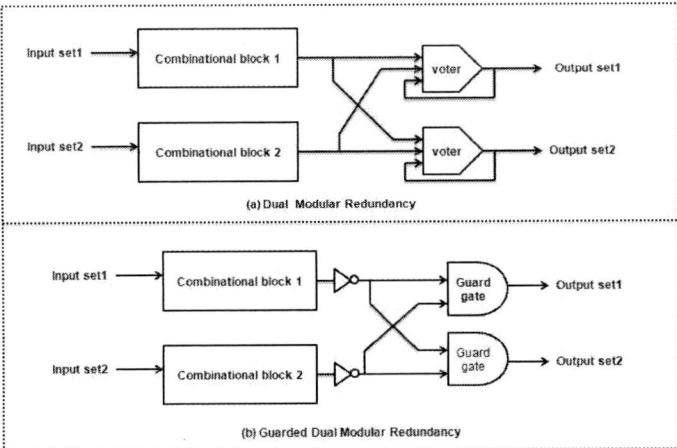

Fig. 1. DMR and GDMR techniques

GDMR and TMR designs.

- Power, area and delay comparison of different logic gates designed as unhardened, GDMR and TMR gates in UMC 65 nm technology node.

II. RELATED WORK

Radiation hard by designs (RHBD) such as dual modular redundancy [1], triple modular redundancy (TMR) [9], Double-Rail Redundant Approach (DR2A) [10], cascade voltage switch logic [11], gate sizing [12] etc. have been proposed in the earlier works mainly to deal with single event transients (SETs) occurring in combinational circuits. Single events in combinational circuits can also result in single event upset (SEU) if this values is latched at a wrong time. SEU-tolerant latches to mitigate the effects of single event upsets include the Dual Interlocked storage cell (DICE) [13], TAG4 latch [22], TMR latch [14], Quatro latch and similar designs. Built-in soft error resilience (BISER) technique for correcting radiation-induced soft errors in latches and flip-flops is mentioned in [6]. Technology scaling has increased the problem of charge sharing between neighboring transistors resulting in a variety of transient effects such as single events causing multiple transients [15], single events causing multiple upsets (SEMU) or multiple event transients [20]. A variety of techniques to mitigate METs have been suggested including the use of guard bands between transistors [17], [18] and automated layout arrangement using CAD support [21].

The Guard gates have previously appeared in radiation effects literature, where they are used to filter SETs [7], [8]. The authors of [7], [8] use temporal redundancy to mitigate SETs and this technique has less area overhead as compared to TMR but the performance degrades. Similar temporal redundant approaches were suggested in [22], [23] using Transition-And-Gates (TAG). Dual Modular Redundancy technique (DMR) using self-voter has been discussed in [1]. The authors also discuss the replacement of self-voter with C-element or TAG gate. Dual modular redundancy (DMR) has been used to design SET-tolerant circuits [1-5] that have less area overhead than

TMR logic and higher performance than temporal redundant circuits. Guarded Dual Rail Logic was proposed in [24] which uses dual modular redundancy and guard gates. In contrast, this work concentrates on Double Modular Redundancy in combinational logic using guard gate i.e Muller C Element without keeper instead of self-voter [1]. The proposed GDMR technique is resilient to SETs. We analyze the area, power and delay overheads of GDMR technique and the effects of METs on GDMR as compared to TMR using a detailed mathematical approach.

III. GUARDED DUAL MODULAR REDUNDANCY (GDMR) TECHNIQUE

Guarded Dual Modular Redundancy uses dual spatial redundancy and the dual outputs terminate in the guard gate. The guard gate filters all SETs occurring at the output of any gate. As shown in Fig. 2, if there is any SET at A or B, the output of the guard-gate holds the previous value as the SET is for a short duration, about 150 ps in 65 nm technology. When $A = B$, the output Y is a proper 1 or a 0 depending on the value of A and B as shown in the table. The GDMR NAND gate is shown in Fig. 3. In a similar manner, several other RHBD GDMR logic gates have been realized. It is clear from the arrangement of Fig. 3, that any SET at A1, B1, A2, B2 or any other internal node, will be filtered by the guard gate. Y1 or Y2 in the hold state for the duration of SET, which is quite small.

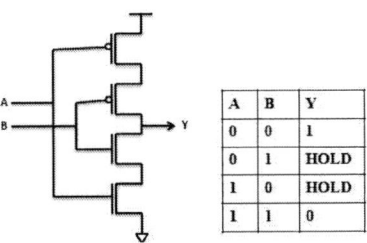

Fig. 2. Guard Gate as SET Filter

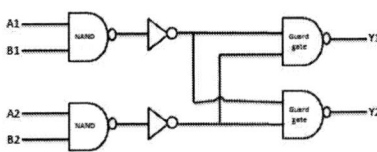

Fig. 3. GDMR NAND logic gate

A. Analysis of single event transients in GDMR

Guarded Dual Modular Redundancy is resilient to SETs. The two rails are independent of each other and this introduces spatial redundancy. Whenever there is an SET, only one rail (say rail1) is affected, and may eventually propagate up to the guard gate, if there was no masking. But the other input to the guard gate which is comes from rail2 will not have an error. So, both the inputs to the guard gate will not match and the guard gate will *hold* the previous value. Fig. 4

978-1-5386-3693-0/18 $31.00 © 2018 IEEE   251

shows the GDMR implementation of C17 from ISCAS85 benchmark. The GDMR C17 is implemented using $1\times$ drive strength gates in UMC 65nm technology. The two sets of

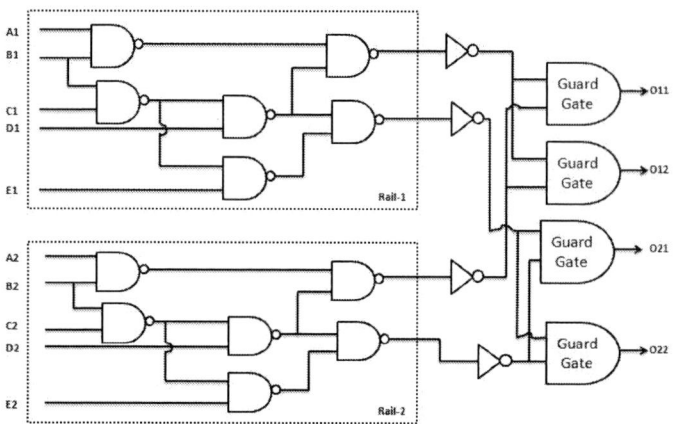

Fig. 4.   GDMR implementation of C17

input vector for GDMR C17 are $[A1, B1, C1, D1, E1]$ and $[A2, B2, C2, D2, E2]$ which are independent of each other. There are two sets of output vectors i.e. $[O11, O21]$ and $[O12, O22]$, which would eventually feed to the next stages of GDMR or flip-flops. Fig. 5 shows the simulation results when an SET occurs at $B1$ node of one of the dual structure in Fig. 4. It is clear that although the SET will affect two NAND gates simultaneously, resulting into a multiple transient, the final outputs remain intact.

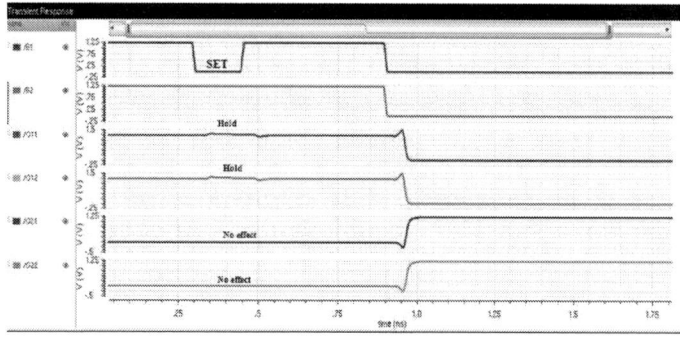

Fig. 5.   Cadence simulation of C17 benchmark implemented in GDMR showing mitigation of SETs

IV. MULTIPLE EVENT TRANSIENTS ANALYSIS

In this section, we consider the effect of multiple event transients (METs). METs may occur either due to packing density or single event causing multiple transients. While several SET mitigation techniques have been reported in the earlier works, we have not come across any work where multiple event transients have been analyzed in a formal setting. In this section, we show a generic approach to effectively compute the probability of failure or error in any circuit with single or multiple outputs in the presence of METs when RHBD technique is applied. Needless to say, the method can

seamlessly be applied to unhardened circuits as well. The notations used for the analysis of effects of METs are listed below.

- n is the total number of non-redundant inputs, m is the total number of non-redundant outputs and q is the total number of internal nodes and inputs.
- I: Set of all possible correct input vectors ($| I |= 2^n$)
- O: Set of all possible correct output vectors ($| O |\le 2^m$).
- Q: Set of all possible patterns for all internal and input nodes ($| Q |= 2^q$).
- A_i: Event that the output vector is expected to be i where $i \in O$.
- $P(A_i)$: The expected probability of having i as final output pattern where $i \in O$.
- M_i: Event that the output vector is equal to i where $i \in O$ in all possible conditions i.e. in the presence or absence of radiation strike. Note that radiation strike can result either in SET or METs or no error (in case of masking).
- $P(M_i)$: Probability of having i as final output where $i \in O$ in all possible conditions.
- $P(ERR_{RHBD})$: Probability of having an error for a given RHBD technique when there is a radiation strike. Note that the error may be because of METs or SET.
- $P(CORR_{RHBD})$: Probability of getting the correct output for a given RHBD technique in all possible conditions.

As an example consider the GDMR NAND gate shown in Fig. 3, where $n = 2$, $m = 1$, $q = 8$, $| I |= 2^2 = 4$, $| O |= 2^1 = 2$, $| Q |= 2^8 = 256$. The set $I = \{(0,0,0,0),(0,1,0,1),(1,0,1,0),(1,1,1,1)\}$ where each vector in I represents the inputs of GDMR NAND gate i.e. $(A1, B1, A2, B2)$ in Fig. 3. The set $O = \{(0,0),(1,1)\}$ where each vector in O represents the outputs of GDMR NAND gate i.e. $(O1, O2)$ in Fig 3. For a NAND gate the expected probability of 1 as output is $\frac{3}{4}$, i.e. $P(A_{11}) = \frac{3}{4}$. Similarly, $P(A_{00}) = \frac{1}{4}$. For a GDMR NAND gate, since we have redundant outputs the corresponding probabilities are represented as $P(A_{00})$ and $P(A_{11})$. Probabilities $P(A_i)$'s are dependent on the *logic* of the circuit where as $P(M_i)$'s are dependent on the specific RHBD technique used to mitigate errors. In general. M_i is defined as

$$P(M_i) = \frac{\text{no. of patterns which result in i as output}}{2^q}$$

To calculate $P(M_i)$'s, we consider all the sensitive nodes (i.e. q) and all possible 2^q different patterns. This number generally varies with the RHBD technique. We assume equal probability of occurrence for all 2^q patterns. Out of this 2^q patterns, only 2^n are logically correct. Thus for one logically correct pattern $j \in Q$, the rest of the patterns (say belongs to set R_j, $| R_j |= (2^q - 1)$) will capture the effects of SET and METs. For each pattern in R_j. We compute the number of patterns belonging to R_j which resulted in the correct output (i.e. output expected for i) despite of SET and METs. Therefore $P(M_i)$ is equal to the ratio of number of patterns in set Q which have i as the output pattern and the cardinality of set Q. Based on this

978-1-5386-3693-0/18 $31.00 © 2018 IEEE

approach, for a GDMR NAND gate, $P(M_{00}) = 0.75$ and $P(M_{11}) = 0.75$. These calculations are mentioned in detail in section IV-C. Now we will derive the probability of obtaining a correct output for a given RHBD technique, $P(CORR_{RHBD})$, using the rule of total probability or rule of marginalization.

$$P(CORR_{RHBD}) = \sum_{\forall\, i \in O} P(CORR_{RHBD}|A_i) * P(A_i)$$
$$= \sum_{\forall\, i \in O} P(M_i) * P(A_i)$$

For a GDMR NAND gate,

$$P(CORR_{GDMR\_NAND})$$
$$= P(CORR_{GDMR\_NAND}|A_{00}) * P(A_{00})$$
$$+ P(CORR_{GDMR\_NAND}|A_{11}) * P(A_{11})$$
$$= P(M_{00}) * P(A_{00}) + P(M_{11}) * P(A_{11})$$
$$= 0.75$$

Therefore the error probability for a given RHBD technique is

$$P(ERR_{RHBD}) = 1 - \sum_{\forall\, i \in O} P(M_i)P(A_i) \qquad (1)$$

A. Analyzing MET effects for Unhardened designs

For unhardened logic gates $P(M_i) = P(A_i)$. To simplify the MET analysis, we will consider basic logic gates with $m = 1$. However, this can be extended to any number of internal nodes and outputs using Eqn. 1. The equation for $P(ERR_{unhard-gate})$ for m=1 is given as,

$$P(ERR_{unhard-gate})$$
$$= 1 - [(P(A_1) * P(M_1)) + (P(A_0) * P(M_0))]$$
$$= 1 - [(P(A_1))^2 + (P(A_0))^2]$$

For example, $P(ERR_{unhard-NAND}) = 1-(0.25^2+0.75^2) = 0.375$

B. Analyzing MET effects in Triple Modular Redundancy Technique

In TMR technique, there are 3 sets of independent inputs and outputs. We will consider simple 1-output logic circuits with TMR technique for the MET analysis. A 2-input TMR NAND gate is shown in Fig. 6 as an example. In the TMR NAND gate, $q = 12$ (i.e. A1, A2, A3, B1, B2, B3, A, B, C, X, Y, Z in fig 6), $n = 2$ and $m = 1$. Let us consider a

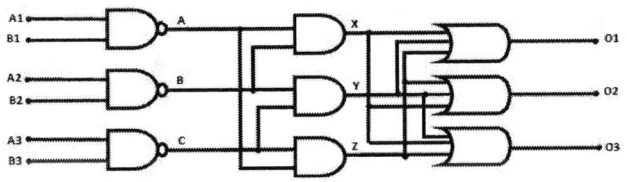

Fig. 6.   TMR NAND logic gate

logic gate with $m = 1$ and q internal nodes. As discussed in section III-A for each logically correct pattern, there are $2^q - 1$ alternate patterns possible as a result of METs. The patterns which result in $O1 \neq O2 \neq O3$ are considered as error. Now we are left with the two sets of patterns which either have $O1 = O2 = O3 = 1$ or $O1 = O2 = O3 = 0$. If the expected output is logic 1, then all the patterns with $O1 = O2 = O3 = 1$ are considered as error free even though they have SET or METs. Similar is the case if expected output is logic 0. For the TMR NAND gate, out of 2^{12} patterns, 512 patterns have $O1 = O2 = O3 = 0$, 3584 patterns have $O1 = O2 = O3 = 1$ and none of the patterns will result in other possible combinations of $O1,O2$ and $O3$. Therefore for TMR NAND $P(M_{111}) = 0.875$ and $P(M_{000}) = 0.125$. These probabilities remain same for any value of n and q given $m = 1$. Therefore the equation for the probability of the error for any logic circuit (with $m = 1$) implemented using TMR is as follows

$$P(ERR_{TMR})$$
$$= 1 - [(P(A_{111}) * P(M_{111})) + (P(A_{000}) * P(M_{000}))]$$
$$= 1 - [(P(A_{111}) * 0.875) + (P(A_{000}) * 0.125)]$$

When there are METs at the internal nodes of TMR gates, the probability for the output to be one is more compared to the probability for the output to be zero. The probabilities for different basic TMR logic gates are shown in Table I.

TABLE I
PROBABILITY OF ERROR IN CASE OF METs FOR TMR BASIC LOGIC GATES

TMR logic gates	P(1)	P(0)	P(error)
2-input NAND	0.75	0.25	0.3125
2-input OR	0.75	0.25	0.3125
Inverter	0.5	0.5	0.5
2-input XOR	0.5	0.5	0.5
2-input XNOR	0.5	0.5	0.5
2-input AND	0.25	0.75	0.6875
2-input NOR	0.25	0.75	0.6875
3-input NAND	0.875	0.125	0.21875
3-input OR	0.875	0.125	0.21875
3-input XOR	0.5	0.5	0.5
3-input XNOR	0.5	0.5	0.5
3-input AND	0.125	0.875	0.78125
3-input NOR	0.125	0.875	0.78125

C. Analyzing MET effects in GDMR technique

The analysis is similar to the TMR's MET analysis. A 2-input GDMR NAND gate is shown in Fig. 3, for which $m = 1$, $n = 2$ and $q = 8$. For GDMR NAND gate, we calculate $[Y1, Y2]$ for each $i \in Q$. When the expected output is logical 1, then the cases for which $[Y1, Y2] = [1, 1], [hold, 1], [1, hold]$ and $[hold, hold]$ are considered as correct cases i.e.considered towards $P(M_1)$. Similarly, When the expected output is 0 then the cases for which $[Y1, Y2] = [0, 0], [hold, 0], [0, hold]$ and $[hold, hold]$ are considered as error free i.e. considered towards $P(M_0)$. In case of GDMR NAND gate, out of 2^8 patterns belonging to Q, 64 patterns have $[Y1, Y2] = [1, 1]$, 64 patterns have $[Y1, Y2] = [0, 0]$ and 128 patterns have $[Y1, Y2] = [hold, hold]$. Therefore for GDMR NAND $P(M_{11}) = 0.75$ and $P(M_{00}) = 0.75$. These probabilities remain same for any value of n and q, given

$m = 1$. The probability for different possible output patterns in the presence of METs are given in Table II. Therefore the expression for probability of the error for any logic circuit (with $m = 1$) implemented using GDMR is as follows

$$P(CORR_{GDMR}) = [P(A_{11}) * (0.75)] + [P(A_{00}) * (0.75)]$$
$$= 0.75 \quad (\because P(A_{11}) + P(A_{00}) = 1)$$

$$P(ERR_{GDMR}) = 0.25$$

When there are METs at the internal nodes of GDMR gates, the probability for the output to be one is same as the probability for the output to be zero. Therefore probability of error in the presence of METs for a given m is independent of the logic of gates. This is because of the symmetric nature of guard gate. It was also observed that by changing the number of internal nodes the probabilities does not change. For multiple outputs we need to use Eqn. 1. For example, in case of C17, we have $n = 5$, $m = 2$ and $q = 26$. In the set Q there are 32 logically correct patterns. The output vector $[O11, O12, O21, O22]$ can take 4 different patterns. If the expected output is $[0, 0, 0, 0]$ then all the patterns which result in outputs to be either 0 or hold are considered as error free cases. In case of C17 we found that for each logically correct pattern in Q, METs or SET can result in $(2^{26} - 1)$ different patterns out of which $9*(2^{22})$ result in correct output. Therefore the probability of error for GDMR implementation of C17 is 0.4375.

TABLE II
PROBABILITY OF OCCURRENCE OF DIFFERENT OUTPUT PATTERNS FOR GDMR GATES IN CASE OF METs

Y1	Y2	Probability of occurrence
0	0	0.25
1	1	0.25
1	0	0.0
0	1	0.0
1	hold	0.0
0	hold	0.0
hold	1	0.0
hold	0	0.0
hold	hold	0.5

V. RESULTS

We have analyzed different logic gates to study the effect of METs on GDMR, TMR and unhardened gates. We have included the following gates: 2-input NAND, 2-input XOR, 3-input NAND, 3-input XOR, AOI21 ($Y = \overline{ab + c}$), 4-input NAND, AOI211 ($Y = \overline{ab + c + d}$) and 4-input XOR gate for the analysis. Figs [7-10] show the comparisons of error probability, area, power and delay in a variety of logic gates implemented using GDMR, TMR and unhardened techniques. Our simulations of various logic gates show that GDMR gates consume about 50% less power, 3× less area and have about 50% less delay compared to their TMR counterpart, and yet, GDMR outperform TMR in terms of error-tolerance to METs by about 3×, except for some gates which have very high

probability for logic '1' such as 3-input OR, 4-input OR, 4-input NAND etc. Table III shows the comparison between unhardened, GDMR and TMR implementation of C17 ISCAS'85 benchmark. The advantages of GDMR implementation over TMR are clearly visible from the table. Table III shows that GDMR consumes less area, power and delay compared to TMR, and still has better error tolerance compared with TMR. We expect that in bigger designs, these margins will only improve as more complex designs are implemented in GDMR and compared with TMR. For C17 ISCAS Benchmark circuits implemented in UMC 65nm (Fig. 4), we find that GDMR consumes about 58% less area, has 31% less delay, 19% less power and 32% less probability of error due to METs than TMR implementation. One interesting observation to be made from Fig. 7 is that TMR does marginally better in terms of error tolerance for METs when the final output has more cases of logic '1'. In other words, TMR might do better for OR-like gates.

TABLE III
ERROR, AREA, DELAY AND POWER COMPARISONS FOR C17 BENCHMARK DESIGN

RHBD technique	Area (μm^2)	Delay (ps)	Power (μW)	$P(ERR_{RHBD})$
unhardened	0.54	40	24.35	0.70
GDMR	**1.46**	**70**	**71.08**	**0.44**
TMR	3.46	102	87.95	0.65

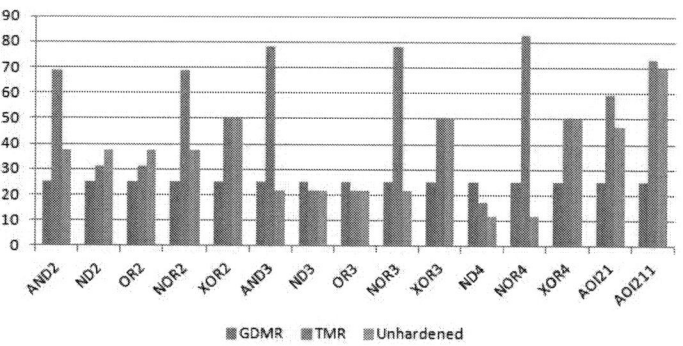

Fig. 7.   Percentage error in the presence of METs

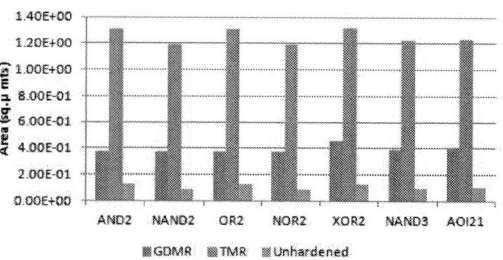

Fig. 8.   Comparison of area for different logic gates

VI. CONCLUSIONS AND FUTURE WORK

In this paper, we have proposed a new RHBD technique called Guarded Dual Modular Redundancy technique which

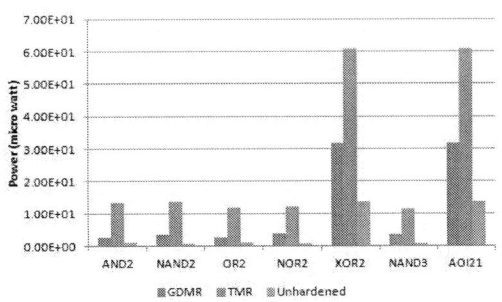

Fig. 9.   Comparison of power for different logic gates

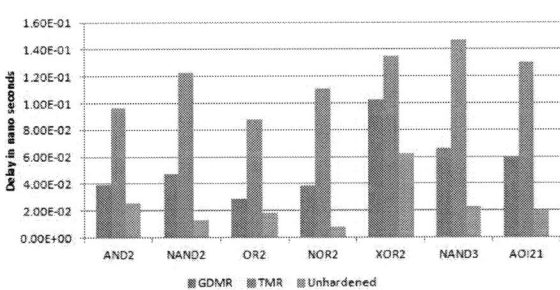

Fig. 10.   Comparison of delay for different logic gates

is resilient to SETs. We have also analyzed the effects of multiple-event transients on Guarded Dual Modular Redundancy (GDMR) and Triple Modular Redundancy (TMR) using a formal mathematical procedure developed for MET analysis. We have carried out exhaustive circuit level simulations for MET analysis on about 15 different basic logic gates of which some have been reported here. The probability of error for GDMR is independent on the logic of the gate. We have reported that GDMR technique has less area, delay and power overheads as compared to TMR. In fact, GDMR gates require nearly half area, one-thirds power and one-half delay compared to TMR gates, with much higher tolerance to METs. In future, we would like to fully automate the MET analysis and use complex benchmarks to evaluate the overall benefit of GDMR at a system level, as compared to TMR technique.

ACKNOWLEDGMENT

This work is supported by grant received from Ministry of Electronics and Information Technology (MEITY), Government of India for Special Manpower Development Project for Chips to System Design (SMDP- C2SD) and grant received from Department of Science and Technology (SR/FTP/ETA-95/2010).

REFERENCES

[1] John Teifel, "Self-Voting Dual-Modular-Redundancy Circuits for Single-Event-Transient Mitigation," IEEE Transactions on Nuclear Science, Vol. 55, NO. 6, December 2008.

[2] J. Gambles, "An ultra-low-power, radiation-tolerant reed solomon encoder for space applications, in Proc. IEEE Custom Integr. Circuits Conf., 2003, pp. 631634.

[3] W. Jang and A. J. Martin, "SEU-tolerant QDI circuits, in Proc. IEEE Int. Symp. Asynchronous Syst. Circuits, 2005, pp. 156165.

[4] M. P. Baze, "Propagating SET characterization technique for digital CMOS libraries, IEEE Trans. Nucl. Sci., vol. 53, no. 6, pp. 34723478, Dec. 2006.

[5] S. Rezgui, "New methodologies for SET characterization and mitigation in flash-based FPGAs, IEEE Trans. Nucl. Sci., vol. 54, no. 6, pp.25122524, Dec. 2007.

[6] M. Zhang,"Sequential element design with built-in soft error resilience, IEEE Trans. Very Large Scale Integr. (VLSI) Syst., vol. 14, no. 12, pp. 13681378, Dec. 2006.

[7] A. Balasubramanian, B. L. Bhuva, J. D. Black and L. W. Massengill, "RHBD techniques for mitigating effects of single-event hits using guard-gates," in IEEE Transactions on Nuclear Science, vol. 52, no. 6, pp. 2531-2535, Dec. 2005.

[8] R. L. Shuler, A. Balasubramanian, B. Narasimham, B. L. Bhuva, P. M. ONeill, and C. Kouba,"The effectiveness of TAG or guard-gates in SET suppression using delay and dual-rail configurations at 0.35 μm, IEEE Trans. Nucl. Sci., vol. 53, no. 6, pp. 34283431, Dec. 2006.

[9] D. G. Mavis and P. H. Eaton, "Soft error rate mitigation techniques for modern microcircuits," Reliability Physics Symposium Proceedings, 2002. 40th Annual, 2002, pp. 216-225.

[10] V. Ciriani, L. Frontini, V. Liberali, S. Shojaii, A. Stabile and G. Trucco, "Radiation-tolerant standard cell synthesis using double-rail redundant approach," Electronics, Circuits and Systems (ICECS), 2014 21st IEEE International Conference on, Marseille, 2014, pp. 626-629.

[11] M. C. Casey, B. L. Bhuva, J. D. Black and L. W. Massengill, "RHBD using cascode-Voltage switch logic gates for SET tolerant digital designs," in IEEE Transactions on Nuclear Science, vol. 52, no. 6, Dec. 2005.

[12] Quming Zhou and K. Mohanram, "Gate sizing to radiation harden combinational logic," in IEEE Transactions on Computer-Aided Design of Integrated Circuits and Systems, vol. 25, no. 1, pp. 155-166, Jan. 2006.

[13] T. Calin, M. Nicolaidis and R. Velazco, "Upset hardened memory design for submicron CMOS technology," in IEEE Transactions on Nuclear Science, vol. 43, no. 6, pp. 2874-2878, Dec 1996.

[14] D. G. Mavis, P. H. Eaton, M. D. Sibley, R. C. Lacoe, E. J. Smith, and K. A. Avery, "Multiple Bit Upsets and Error Mitigation in Ultra-Deep Submicron SRAMS, in IEEE Transactions on Nuclear Science, vol. 55, no. 6, pp. 2531-2535, Dec. 2008.

[15] O. A. Amusan et al., "Single Event Upsets in Deep-Submicrometer Technologies Due to Charge Sharing," in IEEE Transactions on Device and Materials Reliability, vol. 8, no. 3, pp. 582-589, Sept. 2008.

[16] Mahdi Fazeli, Seyed Nematollah Ahmadian, Seyed Ghassem Miremadi, Hossein Asadi and Mehdi B. Tahoori, "Soft error rate estimation of digital circuits in the presence of Multiple Event Transients (METs)," in Design, Automation and Test in Europe Conference and Exhibition (DATE), May 2011.

[17] J. D. Black et al., "RHBD layout isolation techniques for multiple node charge collection mitigation," in IEEE Transactions on Nuclear Science, vol. 52, no. 6, pp. 2536-2541, Dec. 2005.

[18] Hsiao-HengKelin Lee, Circuit and Layout Techniques for soft error resilient circuits, PhD Thesis, Stanford University, 2011.

[19] Reed et al, "Heavy Ion and Proton-Induced Single Event Multiple Upset, in IEEE Transactions on Nuclear Science, vol. 44, no. 6, pp. 2224-2229, Dec. 1997.

[20] C. Rusu, A. Bougerol, L. Anghel, C. Weulerse, N. Buard, S. Benhammadi, N. Renaud,G. Hubert, F. Wrobel, T. Carriere and R. Gaillard, "Multiple Event Transient Induced by Nuclear Reactions in CMOS Logic Cells," in On-Line Testing Symposium, IOLTS 07. 13th IEEE International, Jul. 2007.

[21] Mojtaba Ebrahimi,Hossein Asadi and Mehdi B. Tahoori,"A layout-based approach for multiple event transient analysis," in Design Automation Conference, May 2013.

[22] R. L. Shuler, C. Kouba and P. M. O'Neill," SEU performance of TAG based flip-flops," in IEEE Transactions on Nuclear Science, vol. 52, no. 6, pp. 2550-2553, Dec. 2005.

[23] R. L. Shuler,"Method and apparatus for reducing the vulnerability of latches to SEU, U.S. Patent 6 492 857, Dec. 10, 2002.

[24] R. Kaur, N. Surana and J. Mekie, Guarded dual rail logic for soft error tolerant standard cell library, IEEE RADECS 2016.

[25] H.B. Wang, J.S. Bi, M.L. Li, L. Chen, R. Liu, Y.Q. Li, A.L. He, and G. Guo,"An Area Efficient SEU-Tolerant Latch Design",IEEE Transactions On Nuclear Science, Vol. 61, No. 6, December 2014

Impact of Device Aging on Early Mode Failures in Pulsed Latches

Ankur Shukla, Rahul M Rao, James D Warnock

IBM Systems, Bangalore, India

Abstract— **Pulsed latches that are widely used in high performance circuits are susceptible to early mode failures. High performance circuits also experience higher voltage and temperature conditions that exacerbates device degradation due to mechanisms such as BTI and HCI. In this paper, the impact of aging on hold slack in pulsed latches is presented. Analysis of clock gating that leads to AC or DC stress in the pulse generation circuits is illustrated, and the need to provide additional hold margin for these effects is demonstrated.**

Keywords— *NBTI, PBTI, pulse based latch, hold time, setup time, hold slack, burn-in test, End of Life (EoL)*

I. INTRODUCTION

Improved device density due to a move to FinFETs along with the non-scaling of supply voltage as per Dennard's law [1] with device scaling have resulted in an increased power and thermal density. This is especially true for high performance designs that operate at high voltage and frequency conditions. One of the side-effects of increased thermal and power stress, is the degrading device and circuit reliability. Aging due to Bias Temperature Instability (BTI) and Hot Carrier Injection (HCI) have become significant source of concern in modern circuits. Both Negative Bias Temperature Instability (NBTI), which affects PFET devise, and Positive Bias Temperature Instability (PBTI) which affects NFET devices have a significant dependence on the voltage and temperature profile experienced by the devices, and results in device aging even under static (non-switching) conditions.

HCI on the other hand degrades devices due to charge tunneling into the oxide during device switching and is dependent on the frequency of operation, in addition to the voltage and temperature. Both these effects are typically modelled as degradation in threshold voltage (V_t), the amount of V_t degradation depends upon supply voltage, bias conditions, temperature, switching factor (HCI only) and the time of operation.

Additionally, device aging due to BTI exhibits a partial recovery, when the stress is relaxed. As a result, device aging can be categorized into two types, as described in [2]:

(a) Alternating Stress: This represents devices that are stressed by input signals that continue to toggle. As a result, the bias conditions on the devices alternate between stress and recovery, and the extent of device aging depends on the ratio of stress time to recovery time

(b) DC Stress: This represents devices that are stressed by input signals that are held at a constant value. As a result, the devices see no recovery, and the extent of degradation is larger than during alternating stress.

Reliability analysis of digital circuits has been a topic of significant interest in recent years, specifically in memory designs as presented in [2-4] and in digital circuits [5-6]. Analysis of device aging (symmetric and asymmetric) in clock networks and its impact on timing of master-slave sequential elements, for late as well as early mode arcs were presented in [7-10]. However, none of these publications address the impact of device degradation on early mode on latches with overlapping clocks or pulsed latches [11], which are widely used in high performance designs.

In this article, we present the impact of aging on hold violation in pulsed based latches. To the best of our knowledge, this is the first work to address this topic. The remainder of the paper is organized as follows: Section II explains the basics of pulse based latches. Section III has experimental setup. The simulation results are discussed in section IV followed by conclusions in V.

II. PULSE BASED LATCHES

Pulse based latches are widely used in high performance circuits due their various advantages as described in [11]. In addition to the power advantage, the pulse based latch also offer improved latency and hence reduce critical path delays. An additional pulse generation circuit is required in pulse based latches that generates a narrow pulse at every sampling edge of clock as shown in Fig.1. Pulse generating circuit shown in Fig.2(a) has a built-in capabilities of clock buffering, clock gating, providing scan clocks (*d1_clk, d2_clk*).

The setup slack is measured with the rising edge of the pulse clock. However, the pulsed latches provide a transparent window for the latch, during which the latch is open, allowing input data to transition in the middle of the window and thereby steal 'time' from the next pipeline stage.

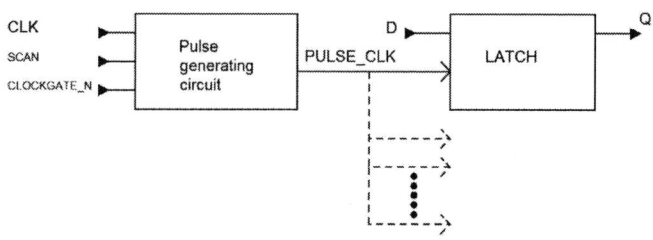

Fig.1 Pulse based latch

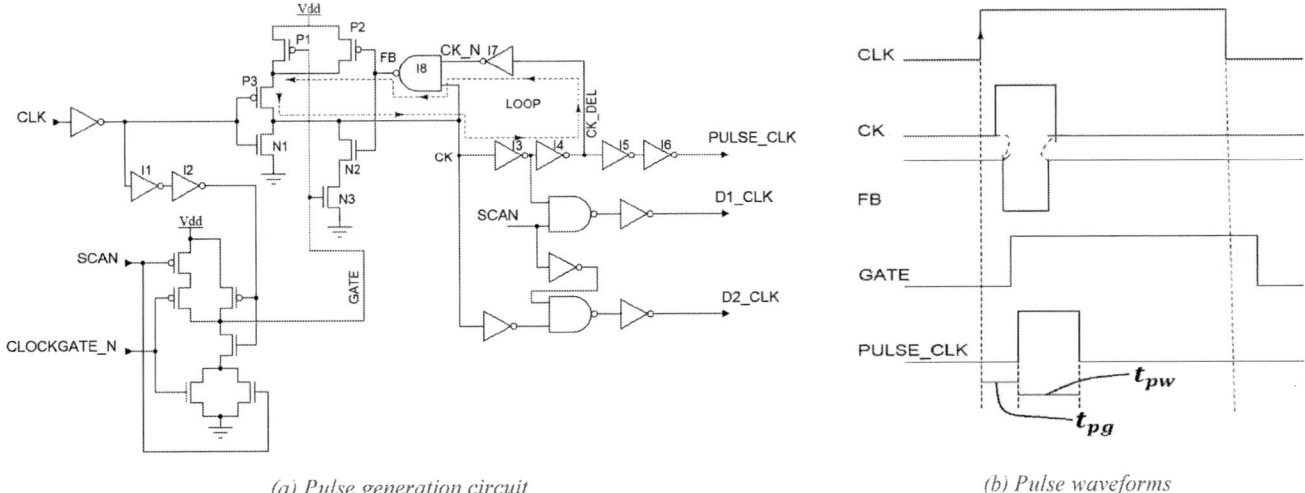

(a) Pulse generation circuit

(b) Pulse waveforms

Fig.2 Pulse generation circuit and waveform

The hold slack is measured with reference to the falling edge of the pulse clock. The pulse width of pulse clock ((t_{pw} or sampling window), as shown in Fig. 2(b), should be wide enough for the input data to propagate through the feedback loop of the latch. Large pulse width enables increased cycle stealing from the next cycle, but also exacerbates chances of race condition (hold violation). The pulse width of *pulse_clk* (t_{pw}) depends on the LOOP delay, and delay of pulse generation circuit (t_{pg}) depends on feed-forward path of the LOOP. t_{pg} plays an important role in hold slack calculation mentioned in section IV.

Fig.2(b) explains that *ck* goes high when a rising *clk* comes, which then brings *fb* down to low. This rising *ck* makes *ck_n* low after delay from I3, I4 and I7 inverters that forces *fb* to go high. The rising *fb* is the feedback node that makes *ck* to fall again (gate node already becomes high after 3 inverter delay that ensures minimum required pulse width), and results a pulsed clock with pulse width of LOOP delay. When the pulse generation circuitry is active, and the *pulse_clk* is being generated every cycle, the corresponding devices experience a AC stress. The duty cycle of the pulse clock, which is fairly small, determines the extent of aging of the devices in the path (i.e. devices in the gates labelled I3 to I8).

The *clockgate_n* signal in Fig. 2(a) also provides the ability to gate the pulse generation circuit for power savings, when appropriate. When *clockgate_n* is set to a logic low value, the intermediate gate signal is high, forcing the *pulse_clk* signal to a logic low value. In this scenario, a majority of the devices in the pulse generation circuitry, experience DC stress, resulting in increased device aging.

III. EXPERIMENTAL SETUP

In order to assess the impact of device aging on hold failures at pulsed latches, a representative hold path is constructed as shown in Fig.3. The data path consists of a finite number of books inserted for early mode padding (inverters in this case) followed by an XOR gate. The source and the sink latches are driven by structurally similar, yet distinct pulse generation circuits.

The hold slack at design time is assessed by introducing a phase difference between the source and the sink clocks and creating a hold fail. This will require multiple iterations to get the slack. In this work, a multi-frequency approach is used, where the sink clock is at a slower frequency than the source clock. The phase of sink clock with respect to source clock increases thereby reducing the hold slack each cycle. This setup will result in a hold fail when phase difference between the clocks become greater than hold slack. This setup doesn't require multiple iterations, the accuracy of hold slack depends on the frequency difference between source and sink clock (as opposed to adjusting the reducing the data path delay to create a hold fail, which ends up having the granularity of an inverter delay).

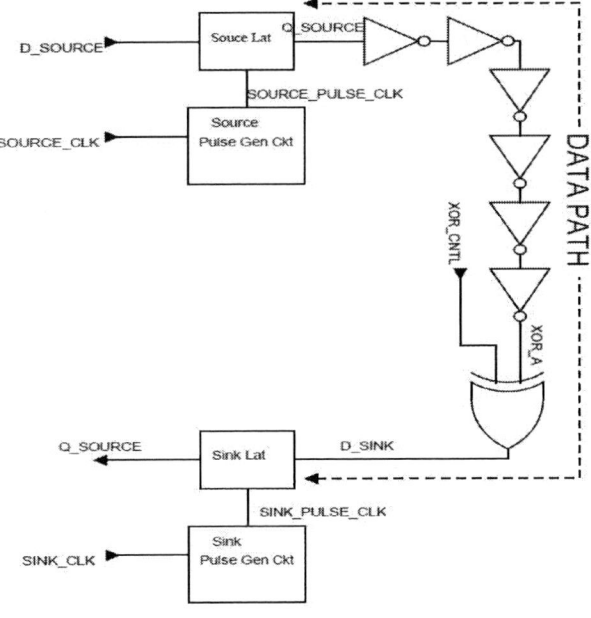

Fig.3 Schematic setup

A) Worst hold violation case and significance of XOR gate in data path

A 50% duty cycle data input (*d_source*) is applied which changes its state on alternate cycles. *Xor_a* in Fig.4 shows the delayed captured data (source latch delay + data path delay). Hold margin indicated in Fig.4 is in consideration of *sink_clk* being same as SOURCE_CLK.

The worst case of hold is when the latch is holding some state 'S' and an early latch input comes which switches from $\bar{S}$ to S, the latch doesn't need to change all its internal states for hold fail. In this schematic XOR serves this purpose where in every alternate cycle the output of XOR is driven by *xor_cntl* (which is maintained to be reliably sampled) and in other alternate cycles the output of XOR gate is driven by the data path (*xor_a*).

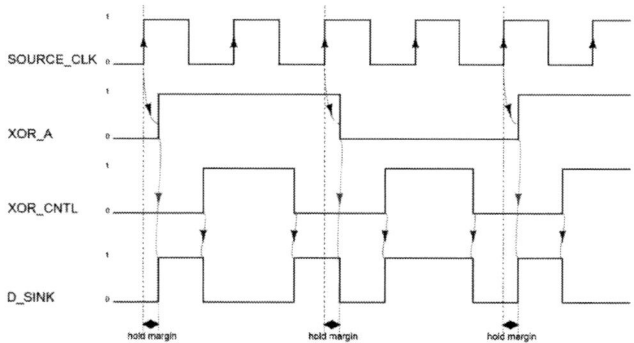

Fig.4 waveform explaining significance of XOR gate

B) Hold slack

The hold slack for the above setup depends on the clock to Q delay of the source latch, the delay of the data path (which includes time to write the latch) in comparison to the clock to the falling edge of the pulse clock at the sink latch, plus the clock skew between the source and the sink latch.

$$hold\_slack = t_{source_{clkQ}} + t_{datapath}$$
$$- \left\{ t_{sink_{pg}} + t_{sink_{pw}} \right\}$$
$$- t_{sink_{hold}} - t_{skew} \qquad Eq.(1)$$

Where,

$t_{source_{clkQ}}$: clock to Q delay of source latch, also called as launch delay.

$t_{datapath}$: data path delay i.e. delays of 6 inverters and XOR gate in our experimental setup

$t_{sink_{pg}}$: delay of pulse generation circuit at sink as shown in Fig.2(b)

$t_{sink_{pw}}$: pulse width of *sink_pulse_clk*,

$t_{hold_{sink}}$: hold time of sink latch

t_{skew} : clock skew and jitter

With the continued operation of the circuit, various devices in the circuit degrade at different rates, depending on their activity factors, due to the combined effect of BTI and HCI. When the circuit undergoes aging, the amount of hold slack degradation ($\Delta hold\_slack$) can be represented as:

$$\Delta hold_{slack} = \Delta t_{source_{clkQ}} + \Delta t_{datapath}$$
$$- \left\{ \Delta t_{sink_{pg}} + \Delta t_{sink_{pw}} \right\}$$
$$- \Delta t_{sink_{hold}} - \Delta t_{skew} \qquad Eq.(2)$$

The degradation in skew depends on the level of clock gating in a clock tree and common clock path. If the clock gating is applied at source of clock generator then maximum degradation in clock duty cycle and skew can be seen after stress. Alternately, $\Delta t_{skew} = 0$ if source and sink pulse generating circuit are driven from same clock source.

For this analysis, it is assumed that the same clock is feeding to source and sink sequential elements, i.e. the source and sink pulse generation circuits tap of adjacent node of a clock mesh. Further, we assume that the data in the source latch does not

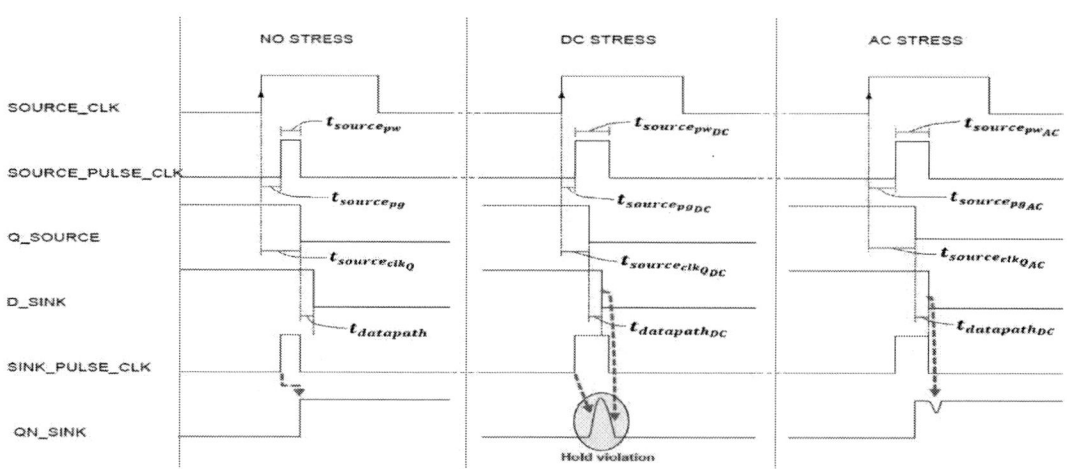

Fig.5 Impact of DC and AC stress

change during the duration of stress, and changes subsequently, potentially triggering a hold violation.

Fig. 5 explains how various delays in the experimental setup change following a period of DC or AC stress when a falling data transition is launched from source latch (similarly, rising data transition can also be taken).

(C) Clock Gated Pulse Generation Circuit (DC Stress)
When the pulse generation circuits are clock gated (i.e. clockgate_n = 0), the 'gate' signal is at logic high, which drives the ck node to logic 0. As a result, the PFET devices in *I3, I5, I7*, and the NFET devices in *I4, I6*, and one of the NFET devices in *I8* experience DC stress. This results in a maximum asymmetry in rise and fall delay causing an increase in the pulse width. When the clock gate is removed, the first rising edge of the *ck* signal causes a faster fall of the signal at the output of *I3* due to the reduced contention from the *I3* PFET (which has degraded). A similar behavior occurs through the other devices, resulting in an earlier transition time for the *pulse_clk*, than before. On the other hand, the delays for the transition of the falling edge of *ck* through to the output increases, since all the devices causing this transition have undergone DC stress. This results in a *pulse_clk* edge that has an earlier launch time and a wider pulse width. This coupled with the faster data launch from the source latch and lesser data path delay results in a worst-case hold slack degradation. DC stress section of Fig.5 is graphical representation of above analysis where $\Delta t_{source_{clkQ}}$, $\Delta t_{datapath}$ are negative and $\Delta t_{sink_{pg}}$, $\Delta t_{sink_{pw}}$ are positive that results maximum negative $\Delta hold\_slack$.

(D) Clock Enabled Pulse Generation Circuit (AC Stress)
In case of free running pulse clock (AC stress), the devices on the data path are still DC stressed, since the source latch is assumed to be at constant value during stress. However, the devices in the pulse generating circuit and the clocking circuitry inside a latch are in AC stress. Contrary to the DC stress condition, both the devices in the set of gates {I3 to I8} degrade. As a result, both the rising and falling edge of the *pulse_clk* degrade in their transition times with respect to the clk. Hence from AC stress section of Fig. 5, $\Delta t_{source_{clkQ}}$, $\Delta t_{sink_{pg}}$,

s$\Delta t_{sink_{pw}}$ are positive. The degradation of the data path degradation is similar to that in the case of DC stress.

The overall impact on hold criticality can be seen in Fig.5 where DC stress mode seems worse resulting hold violation as *d_sink* is transitioning during transparent (active) window of *sink_pulse_clk*. The magnitude of these degradations are presented in results section.

IV. RESULTS

The experimental setup was simulated in an advanced FinFET technology. As explained in section III, a slower clock frequency at sink is applied in our experimental setup that will result a hold violation when phase difference of sink clock becomes greater than hold slack. Therefore, the hold_slack is measured by measuring the phase difference between *source_clk* and *sink_clk* when a hold violation occurs at *qn_sink* (An example of such hold violation was shown in Fig. 5).

Seven sets of simulations were performed :
a) Initial condition (i.e. without any stress).
b) After a large duration (equal to expected product life span) of DC stress (all source, sink and data path).
c) After a large duration of all source and sink at AC stress, and data path at DC stress.
d) After a large duration of source at DC stress, sink at AC stress and data path at DC stress.
e) After a large duration of source and sink at AC stress, and data path being at AC stress (50% duty cycle) too.
f) After a large duration of source and sink at 50% clock gated (i.e. 50% AC and DC stress), and data path at DC stress.
g) After a large duration of source at AC stress, sink at DC stress, and data path with 50% duty cycle.

The worst hold slack along with corresponding dependent factors as per equation (1), pulse generation circuit's delay ($t_{pg_{source}}$, $t_{pg_{sink}}$), clock-Q-delay of source latch ($t_{source_{clkQ}}$), pulse width of PULSE_CLK ($t_{pw_{source}}$, $t_{pw_{sink}}$) and data path delay ($t_{datapath}$) are mentioned in Table 1. Wherever source and sink pulse generation circuits are stressed at identical conditions, t_{pg} and t_{pw} of source, sink go through similar degradation.

		Stress conditions			Hold slack	$t_{pg_{source}}$	$t_{pg_{sink}}$	$t_{pw_{sink}}$	$t_{source_{clkQ}}$	$t_{datapath}$
		Data path	Source	Sink						
a)	**Initial**	NA	NA	NA	1.17	5.65	5.65	6.40	8.30	3.57
b)	**EoL**	DC	DC	DC	-0.03	5.55	5.55	7.03	8.00	3.53
c)	**EoL**	DC	AC	AC	0.1	5.67	5.67	7.03	8.17	3.53
d)	**EoL**	DC	DC	AC	-0.17	5.55	5.67	7.03	8.00	3.53
e)	**EoL**	AC (50% duty cycle)	AC	AC	0.5	5.67	5.67	7.03	8.27	3.78
f)	**EoL**	DC	50% gated	50% gated	0.1	5.65	5.65	7.02	8.13	3.53
g)	**EOL**	AC	AC	DC	0.63	3.87	3.75	5.23	6.47	3.78

Table 1 Results of various stress conditions in FO4 delay

Row d) in Table 1 is found to be worst, where the hold slack of +1.17 FO4 degrades to -0.17 FO4 after EoL in our experimental setup, i.e. a total degradation of 1.34 FO4. This degradation is mainly coming from $\Delta t_{source_{clk_Q}}$, Δt_{pw}, Δt_{pg} and $\Delta t_{sink_{hold}}$, whereas the data path degradation is negligible (0.04 FO4). Similarly, the second worst case is observed when all source, sink and data path undergo DC stress (b) where EoL results out the total degradation of 1.2 FO4. On the contrary, the least degradation in hold slack, occurs in a scenario when the data path is switching every cycle, with the source clk being in free running (AC stress) mode, while the sink clk is DC stressed (clock-gated most of the time). However, even in this case, the hold slack degrades by nearly 0.54 FO4 as compared to the initial condition.

Degradation in pulse width (Δt_{pw}) of *pulse_clk* is nearly the same under both DC and AC stress and is a big contributor to the degradation in hold slack. t_{pg} degradation doesn't change hold slack in cases where pulse generating circuit at source and sink go through identical conditions of stress (b, c, e and f). The interesting observation here is the fact that in DC stress (b) t_{pg} is improving by 0.1FO4, and AC stress (c) degrades it by 0.08 FO4. Hence, the worst hold scenario is observed when source end is DC stressed and sink end is AC stressed (d).

If the circuit is timed initially with worst hold slack of less than 1.34 FO4, then the hold violation may occur prior to desired end of life. A sufficient guard time in hold slack should be added in order to maintain the reliable operation of any circuit for desired end of life.

V. CONCLUSIONS

In this paper, we have discussed the impact of DC and AC stress impact on pulse based latches.

- Significant hold margin is required to maintain reliable operation of circuit for given life of chip.

- Fully clock gated launch and free running capture clock with stacked devices in the path is the worst-case scenario. Whereas the free running launch clock and fully clock gated capture represents the best-case scenario.

- The majority of hold slack degradation is due to the pulse width modulation as opposed to data path delay modification.

REFERENCES

[1] R.H. Dennard et al., "Design of Ion-Implanted MOS-FETs with Very Small Physical Dimensions," Journal of Solid State Circuits, vol. SC-9, no. 5, pp. 256–268, 1974.

[2] A. Bansal et. al., "Impacts of NBTI and PBTI on SRAM static/dynamic noise margins and cell failure probability," Journal on Microelectronics. Reliability, pp. 642-649, 2009.

[3] Kunhyuk Kang et al., "Impact of negative-bias temperature instability in nanoscale SRAM array: modeling and analysis," Transactions on Computer Aided Design. p. 1770–81, 2007.

[4] J.C. Lin et. al., "Time Dependent Vccmin Degradation of SRAM Fabricated with High-k Gate Dielectrics," International Reliability Physics Symposium, pp. 439-444, 2007.

[5] J. Fang and S.S. Sapatnekar. "The impact of BTI variations on timing in digital logic circuits." Transactions on Device & Material Reliability, vol. 13, no.1, pp. 277-286, 2013.

[6] Sanjay V. Kumar, Chris H. Kim and Sachin S. Sapatnekar, "NBTI-Aware Synthesis of Digital Circuits," Design Automation Conference, pp. 370-375, 2007.

[7] Senthil et al., "Asymmetric aging of clock networks in power efficient designs", International Conference on Quality Electronic Design, pp. 484 – 489, 2014

[8] Velamala, J.B., et al., "Failure diagnosis of asymmetric aging under NBTI," International Conference on Computer Aided Design, pp. 428-433, 2011.

[9] Vikram G Rao, "Analysis of Reliability of Flip-Flops under Transistor Aging Effects in Nano-scale CMOS Technology", International Conference on Circuit Design, pp. 439-440, 2011.

[10] K. Ramakrishnan, X. Wu, N. Vijaykrishnan, Y. Xie, "Comparative Analysis of NBTI Effects on Low Power and High-Performance Flip-Flops", International Conf. on Computer Design, pp. 200-205, 2008.

[11] J. Warnock et al., "Circuit design techniques for a first-generation cell broadband engine processor," Journal of Solid-State Circuits, vol. 41, no. 8, pp. 1692–1706, Aug. 2006.

978-1-5386-3693-0/18 $31.00 © 2018 IEEE

A 0.6V Retention V_{MIN} Ultra-Low Leakage High Density 6T SRAM in 40nm CMOS Technology using Adaptive Source Bias

Ashish Kumar[1], G.S.Visweswaran[2]

[1]STMicroelectronics Pvt. Ltd., India, [2]PES University, Bengaluru, India

ashish.kumar@st.com, gvisweswaran@pes.edu

Abstract— **A low retention Vmin 6T-SRAM is realized in 40nm CMOS technology under source bias condition. A high density (HD), ultra-low leakage (ULL) 6T SRAM cell with an area of 0.242um2 is used. We could reduce the retention Vmin of SRAM array to 0.6V using adaptive source bias scheme. Source bias is applied to achieve ultra-low leakage during standby mode of operation. Source bias results in loss of stability due to reduced effective rail-to-rail voltage of memory cell and hence becomes difficult to reduce the retention Vmin of SRAM. An adaptive source bias scheme is used to apply reduced source bias selectively for low retention noise margin (RNM) conditions. We could reduce the retention Vmin by 100mV to 0.6V compared to 0.7V realized with the conventional scheme. We could reduce the leakage at FF/125°C by 30 percent due to reduction of Vmin from 0.7V to 0.6V. At TT/0.6V/25°C, we could achieve the target leakage of 0.5pA/Cell during standby mode of operation.**

Keywords— *Low leakage, IoT, SRAM, low Vmin, low voltage, source bias, RNM.*

I. INTRODUCTION

As the technology has scaled to nanometer regime, device density of System on Chip (SoC) has also increased. Increasing number of transistors on a SoC results in higher power density. This motivates for the reduction in supply voltage to reduce power consumption. Need for low voltage operation comes also for the IoT (Internet of things) applications where low power is a key consideration. Along with dynamic power consumption, standby power consumption is also a key issue. Leakage power constitutes a major portion of overall power consumption of SoC due to increased device count. Reduction of leakage both during active and standby mode of operation is important. However, for battery operated and devices kept in standby mode for larger duration of their operational life, leakage power becomes a key consideration. Hence, this is desired to keep SoC at minimum supply voltage (Vmin) during standby mode of operation to minimize leakage power. This Vmin is limited mostly by the retention Vmin of SRAM. Retention Vmin of SRAM is the minimum supply voltage that ensures data in SRAM to be retained.

Low supply voltage results in near sub-threshold device operation and hence causes large statistical variation in device characteristics. The 6T-SRAM cell using smallest geometry and packed with the highest density, is affected most by the statistical variation. Ensuring data retention of SRAM at low voltage becomes a challenge, putting a limitation on minimum retention voltage (Retention Vmin) for SoC. This is due to reduced Retention Noise Margin (RNM) of the SRAM cell at low supply voltage. Reduction in RNM with supply voltage for a SRAM cell without source bias (SB) is illustrated in Fig.1. To ensure data retention for a large SRAM array, we target a six-sigma qualification for RNM. As evident from Fig.1, RNM fails to qualify six-sigma below 420mV. We have used a 0.242um2 high density 6T-SRAM cell in 40nm Ultra Low Power (ULP) CMOS technology. This cell uses high threshold voltage (V_{th}) devices to achieve ultra-low leakage. To achieve further reduction in leakage, source bias is applied to the memory array [1-2]. Details of the SB scheme are described in section II. Application of SB reduces the rail-to-rail voltage of SRAM array and thus reduces the RNM. Source bias is applied only in the standby mode of operation and helps to reduce the leakage by a factor of three to five depending on the Process, Voltage and Temperature (PVT). Applying SB becomes essential for the SRAM array as leakage of SRAM is a major portion of the total SoC leakage in standby mode. Ensuring low retention Vmin becomes difficult under SB. This work presents a method to reduce retention Vmin while operating under SB condition for the SRAM array by selectively reducing or removing the source bias for low RNM and low leakage PVT conditions.

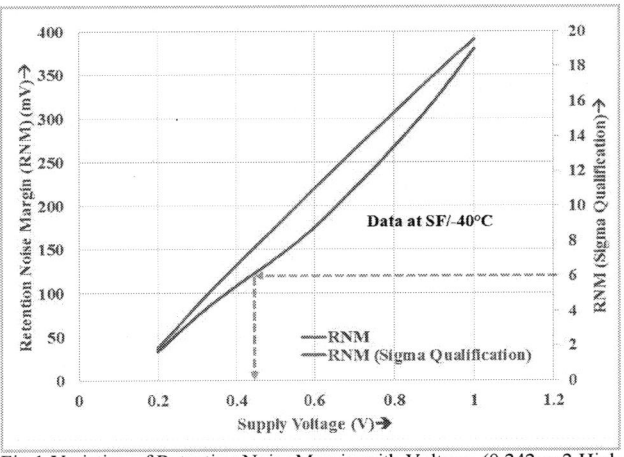

Fig.1 Variation of Retention Noise Margin with Voltage. (0.242um2 High-Density Ultra-Low Leakage 6T-SRAM cell in 40nm ULP CMOS)

II. BACKGROUND AND PROBLEM DESCRIPTION

A six-transistor SRAM (6T-SRAM) cell consists of pairs of PU, PD and PG devices as shown in Fig.2. Major components of Leakage at moderate and high temperature for a 6T-SRAM cell are illustrated in Fig.2. I_1, I_2 and I_3 representing the sub-threshold leakage while I_3 and I_4 represents the major gate leakage components. Apart from the displayed components, there are other components of leakage also present in SRAM cell but their contribution is negligible at moderate and high temperature conditions [3-5].

In order to reduce the leakage in standby condition, source bias (SB) is applied. Fig.3 illustrates the method to put memory array under SB. Memory array ground is gated through NMOS transistor and is controlled by an enable signal En. En is kept at logic '0' level in standby mode of operation. Leakage current of the array passes through the diodes M1 and M2. Due to leakage current, virtual ground level raises to the threshold voltage of diode structure and memory array goes into source bias condition. Stability of the cell is proportional to the rail-to-rail voltage, where $V_{rail\text{-}to\text{-}rail}$ is defined as,

$$V_{rail\text{-}to\text{-}rail} = VDD - \text{Virtual Ground} \quad (1)$$

Hence, lower the virtual ground level, more stable the memory cell is. However, lower virtual ground level reduces the source bias and less reduction in leakage is achieved. A combination of PMOS and NMOS is used to form diode structure in order to provide better cross-corner stability [6]. In cross corner situation, SF or FS, the faster device largely decides virtual ground level or by the device having a lower threshold voltage. RNM of memory cell is lowest at SF corner where NMOS is slow and PMOS is fast. Thus, by using a PN combination diode, virtual ground level is lowered by the faster PMOS diode in SF corner, increasing the stability. Bit-lines BL and BLB are kept floating to reduce bit-line leakage current.

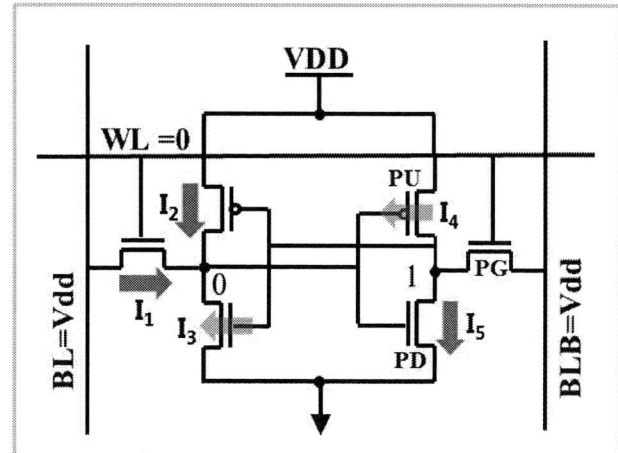

Fig.2. Leakage Components of a 6T-SRAM cell.

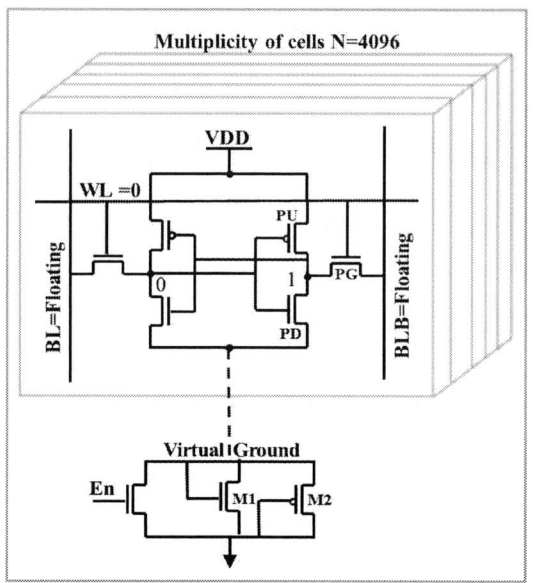

Fig.3. Source Bias (SB) scheme for SRAM array using Pmos-Nmos diodes.

Use of a combination of PMOS and NMOS as source biasing diodes helps to gain on stability for the memory array. However, SF condition still remains critical for stability. Fig.4 illustrates the effect of source bias on memory array leakage and RNM sigma qualification. RNM sigma qualification is the ratio of RNM and its standard deviation and is a measure of stability of SRAM cell in retention. A stability above six-sigma is required to ensure stability of the array. Leakage is reduced almost by a factor of three by SB at a supply voltage of 0.7V. We also observe that SF/-40°C and SF/125°C remains two most unstable conditions for the memory array, whereas leakage is highest for FF/125°C. SF/-40°C is the worst RNM condition although having negligible leakage. This is due to the fact that threshold voltage of the devices increases at a lower temperature. The increase in threshold voltage causes a higher virtual ground level and hence lowering the rail-to-rail voltage of the SRAM array despite of negligible array leakage. This results in lower RNM for the SRAM array.

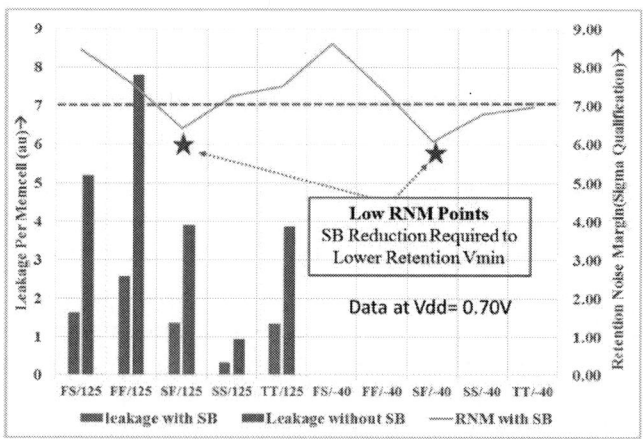

Fig.4 Effect of Source Bias on Leakage and Stability at Supply = 0.7V

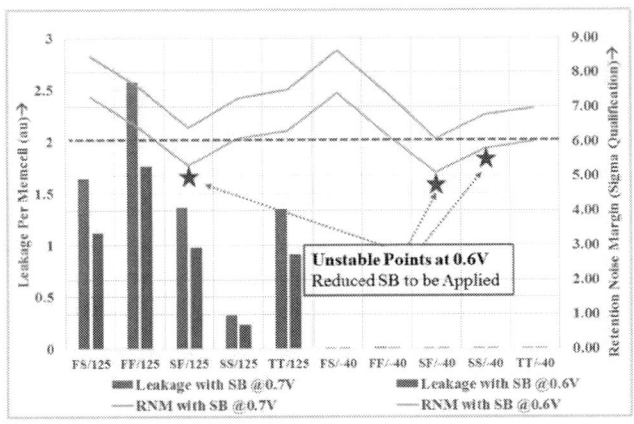

Fig.5 Leakage reduction and loss of Stability with SB at 0.6V

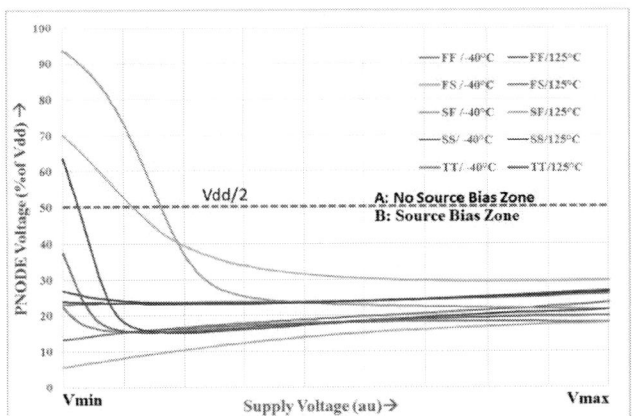

Fig.7. PNODE Voltage Variation with PVT. PNODE is compared with Vdd/2 and SB is selectively reduced for SF/-40°C,SF/125°C and SS/-40°C.

An illustration of leakage gain by lowering the retention Vmin is provided in Fig.5. We observe that approximately 30 percent reduction in leakage is possible at FF/125°C if we can lower the retention Vmin. However, this reduction in Vmin is not possible due to the fact that RNM sigma qualification goes below six-sigma for SF/125°C, SF/-40°C and SS/-40°C. All these limiting conditions are non-critical in terms of leakage current. Hence, we need to find a PVT adaptive solution to enhance the stability of SRAM cell for these conditions selectively.

III. DESCRIPTION OF THE PROPOSED SCHEME

To reduce or optimize the retention Vmin for the SRAM array under source bias, we intend to selectively apply different source bias for the RNM critical conditions. To achieve this a PVT detector is used as shown in Fig.6 [7]. A modified SRAM cell is used to detect conditions where slow PG device is coupled with low voltage of operation. Another PG device is added in series to make the access path weak. Control signal EN is activated to flow current through PU and PG devices transiently for a short duration. Due to this, PNODE rises with extent of rise depending on the relative strengths of PU and series connected PG devices.

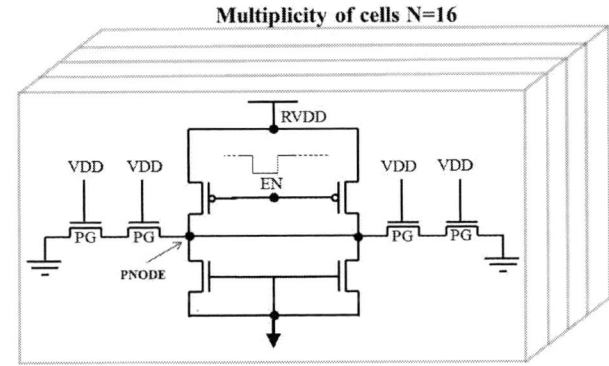

Fig.6. PVT Detector Cell Array for Selective SB

Variation of PNODE voltage with PVT condition is illustrated in Fig.7. For SF/-40°C, SF/125°C and SS/-40°C conditions, PNODE tends to be at higher voltage level than Vdd/2 at lower supply voltage. Zone 'A' refers to the PVT conditions where a reduced SB is required. Zone 'B' consists of PVT conditions where normal SB can be applied. We observe that at higher supply voltages, normal SB can be applied uniformly across all PVTs. This is the need of few conditions only at lower supply voltage where we need to apply differential SB to ensure stability at reduced retention Vmin.

PNODE voltage is compared with a reference voltage of Vdd/2 using a half-latch sense amplifier (Fig.8). SAT and SAF nodes are pre-charged at Vdd before evaluation. Evaluation of PNODE is carried out by SaEn signal and used to correct the SB level if required. Evaluation needs to be done at large time intervals due to the fact that temperature conditions do not change abruptly. For any change in operating voltage a re-evaluation of PNODE is carried out and source bias is ensured to be at correct level.

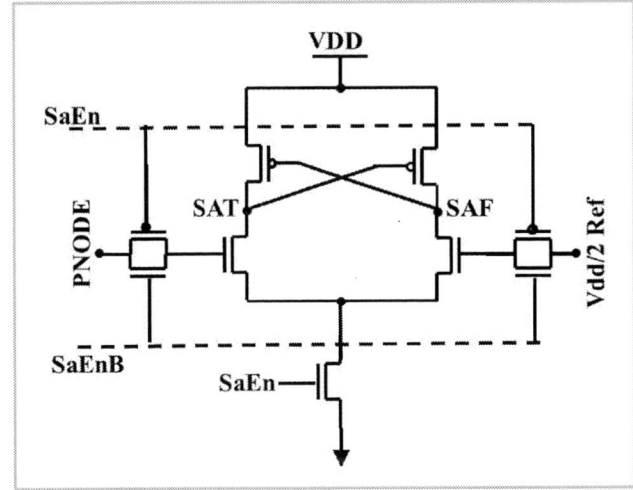

Fig.8. A half-latch Sense Amplifier is used to Compare PNODE Voltage with a reference Voltage of Vdd/2.

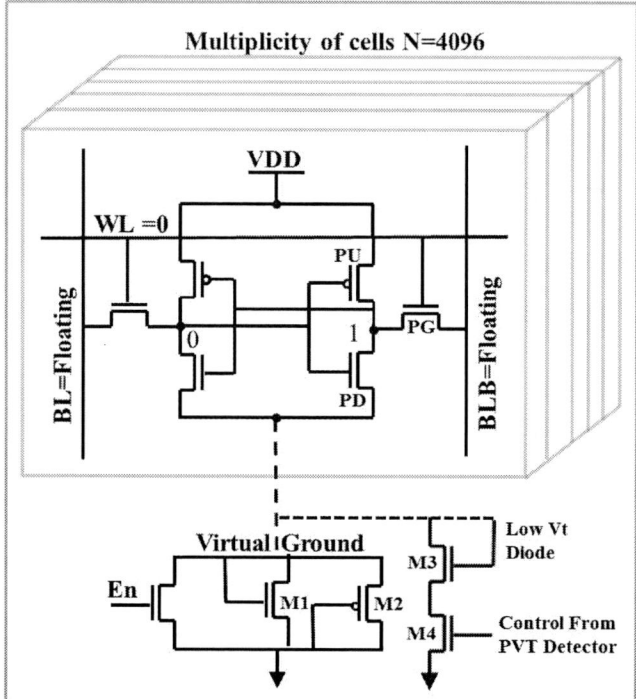

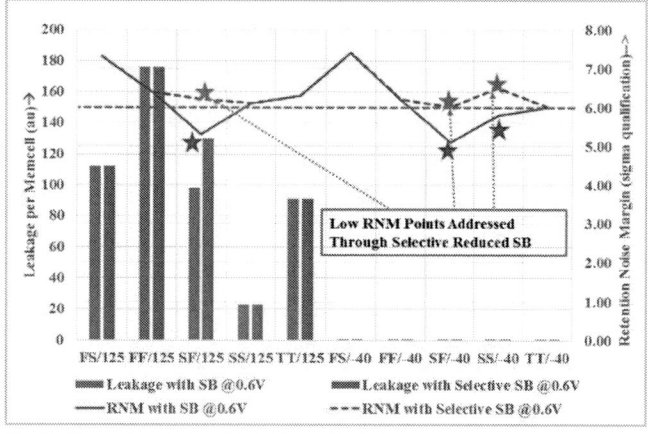

Fig.9 Proposed Source Bias Scheme with Additional Low-Vt Diode Controlled by PVT Detector.

Output of PVT detector is used to selectively reduce the extent of source bias for RNM critical conditions. The modified source bias scheme is illustrated in Fig.9. An additional low threshold voltage (Low-Vt) diode M3 is added in parallel to the standard-Vt diodes M1 and M2. M3 is enabled through M4 controlled by the PVT detector. The PVT detector detects Low RNM condition and M4 is enabled. Virtual Ground level is lowered significantly by the Low-Vt diode, raising the rail-to-rail voltage for the memory array and thus increasing the RNM.

Fig.10 illustrated the results of the adaptive source bias scheme. Reduced source bias is applied selectively at low RNM conditions and six-sigma stability is ensured for the SRAM array. Increased leakage current at these conditions are well below the acceptable maximum value at FF/125°C.

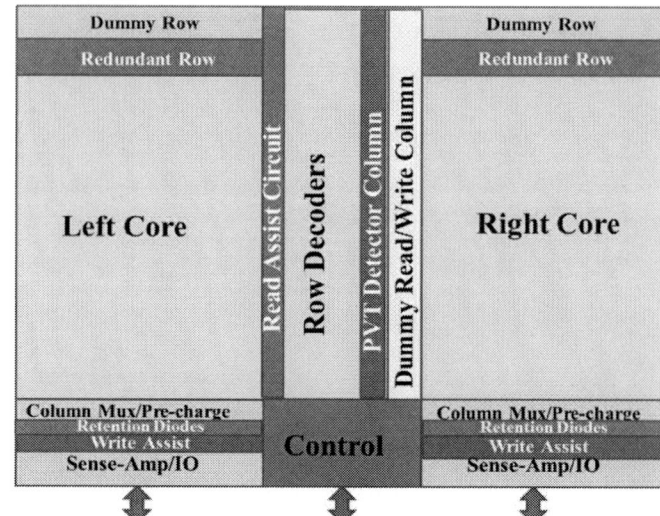

Fig.11. Floor-plan of SRAM using Adaptive Source Bias.

Floor plan of the memory is presented in Fig.11. A split core SRAM architecture is used with read and writes assist circuits to enable low voltage functionality. A column of detector SRAM cells with PNODEs connected together is placed besides Row Decoder column. Sense amplifier for the evaluation of PNODE is placed in control unit. Reference of Vdd/2 is realized through a Poly-silicon resistor divider for stability across PVTs.

IV. RESULTS AND DISCUSSION

A PVT adaptive source bias scheme is presented to reduce and optimize the retention Vmin of SRAM array. We could reduce the retention Vmin of SRAM array from 0.7V to 0.6V ensuring six-sigma robustness. A memory instance of 4K words and 32 bits is designed and validated on CAD. Validation is performed on post-layout RC extracted net-list of the memory instance. Stability of memory cells under source-biased condition is analyzed through statistical simulations using Eldo simulator. Silicon extracted spice models of 40nm ULP CMOS is used. Lowering of Vmin helped to reduce leakage by 30% at FF/125°C. At typical conditions, TT/0.6V/25°C, leakage of 0.5pA/cell was achieved. Leakage figures are with bit-lines in floating condition, minimizing bit-line leakage. All leakage figures used for comparison are with bit-lines in floating condition. The proposed scheme helps to reduce leakage further by enabling a lower Vmin. Reduction in retention Vmin is extremely important for overall reduction in SoC leakage during standby mode of operation. Area overhead of the scheme is less than one percent. PVT detector circuit requires an area of approximately 20um x 15um. Retention Vmin could have been increased by increasing the sizes if source biasing diodes M1 and M2 in Fig.9 without the need of having another diode M3. However, this would have resulted in reduced source bias across all PVTs and hence higher leakage. Selective activation of diode M3 helps to

Fig.10. Retention Vmin of 0.6V is enabled through Adaptive Source Bias.

978-1-5386-3693-0/18 $31.00 © 2018 IEEE

optimize leakage at critical PVTs, where leakage is higher and to selectively recover the loss of stability at low RNM conditions.

Kim et al [8] have proposed a PVT aware leakage reduction scheme. However, this scheme focuses mainly on the frequency of sleep mode activation. Dray et al [9] have also proposed a similar approach. Dynamic supply lowering to reduce leakage is discussed in [10-11]. However, the extent of lowering will depend on the retention Vmin of the cell. Importance of retention Vmin lowering has been realized only in deep sub-micron regime, where very high integration density and increased leakage has forced to look into this possibility of lowering retention Vmin by even a few tens of millivolts. Hsu et al [12] has proposed the use of a current limiter for optimizing leakage across PVTs. However, design of a current limiter puts complexity on design and difficult to optimize for deviation between SRAM and logic process. Further, the order of leakage reduction we want to achieve may not allow the usage of analog blocks due to required continuous bias currents. Yokoyama et al [13] has used device size optimization for leakage reduction. This is an effective method to reduce the leakage of SRAM array while also reducing the retention Vmin. An enlarged cell can support lower retention Vmin. However, due to area constraints in a SoC using large amount of memory, SRAM cell is required to be kept at smallest possible size. The proposed scheme is demonstrated in a 40nm CMOS technology. However, scheme can easily be extended to FinFET and Fully Depleted Silicon-On-Insulator (FD-SOI) technologies as well.

V. ACKNOWLEDGEMENTS

Authors are indebted to the design team for providing layout, extraction and simulation data for the evaluation of scheme. Authors are also indebted to STMicroelectronics for supporting this research work.

REFERENCES

[1] Y Takeyama et al, "A low leakage SRAM macro with replica cell biasing scheme", IEEE Journal of Solid-State Circuits (Volume:41 , Issue: 4 2006, p.p. 815-822.

[2] Fabio Frustaci et al, "Techniques for Leakage Energy Reduction in Deep Submicrometer Cache Memories", IEEE Transactions on Very Large Scale Integration Systems, Volume 14 , Issue: 11 ,2006, pp. 1238-1249.

[3] Shengqi Yang et al, "Low-leakage robust SRAM cell design for sub-100nm technologies", Proceedings of the Asia and South Pacific Design Automation Conference, 2005, Vol-1, p.p. 539-544.

[4] Lijun Zhang et al, "Integrated SRAM compiler with clamping diode to reduce leakage and dynamic power in nano-CMOS process", Institution of Engineering and Technology Micro & Nano Letters, Volume 7, Issue 2 , 2012, p.p. 171-173.

[5] Jeren Samandari-Rad et al, "Power/Energy Minimization Techniques for Variability-Aware High-Performance 16-nm 6T-SRAM", IEEE Access, Vol 4, 2016, pp. 594-613

[6] A. Kumar et al, "Sleep Circuit for SRAM Core with Improved Noise-Margin", IEEE International Conference on Integrated Circuit Design and Technology, 2008, pp. 139-142.

[7] A. Kumar et al, "A 6T-SRAM in 28nm FDSOI technology with Vmin of 0.52V using assisted read and write operation", ICICDT 2015

[8] Chris Hyung-il Kim et al, "PVT-Aware Leakage Reduction for On-Die Caches With Improved Read Stability", IEEE Journal of Solid-State Circuits ,Volume:41 , Issue 1, 2006, pp. 170-178.

[9] Cyrille Dray et al, "A 40nm low power SRAM retention circuit with PVT-aware self-refreshing virtual VDD regulation", IEEE International Memory Workshop, 2010, p.p. 1-4.

[10] Nobuaki Kobayashi et al, "A high stability, low supply voltage and low standby power six-transistor CMOS SRAM", The 20th Asia and South Pacific Design Automation Conference, 2015, pp. 10-11.

[11] Fatih Hamzaoglu et al, "A 3.8 GHz 153 Mb SRAM Design With Dynamic Stability Enhancement and Leakage Reduction in 45 nm High-k Metal Gate CMOS Technology" IEEE Journal of Solid-State Circuits, Volume 44, Issue 1, 2009, p.p. 148-154.

[12] Peter Kuoyuan Hsu et al., "A SRAM cell array with adaptive leakage reduction scheme for data retention in 28nm high-k metal-gate CMOS" ,Symposium on VLSI Circuits (VLSIC) 2012, pp 62-63.

[13] Yoshisato Yokoyama, et al., "40nm Ultra-low leakage SRAM at 170 deg.C operation for embedded flash MCU", Fifteenth International Symposium on Quality Electronic Design, 2014, pp. 24-31.

2018 31th International Conference on VLSI Design and 2018 17th International Conference on Embedded Systems

A 7-nm Dual Port 8T SRAM with Duplicated Inter-Port Write Data to Mitigate Write Disturbance

M. Sultan M. Siddiqui, Sumit Srivastav, Dattatray Ramrao Wanjul,
Manankumar Suthar and Sudhir Kumar.
Solutions Group, SYNOPSYS, Noida, India.
Email: msultan@synopsys.com; msultanms@gmail.com

Abstract— Two read-write 8T dual port static random access memories (SRAMs) suffer write disturb issue when both of its ports are accessed simultaneously. Write disturb is detrimental at low voltages in deep submicron technologies due to increased variations. This paper proposes a duplicated inter-port write data to mitigate write disturb in dual port SRAM design targeted on 7-nm FinFET TSMC technology. We have implemented duplicated inter-port write by employing two NMOS switches per bit line. We choose moderate sizes for these switches to have a balanced trade-off between area and duplication of inter-port write data. With NMOS switch size of nfin=4, we significantly suppress the write disturb and improve the write time by 2.16x at low voltage of 0.63V. Moreover, at further low voltages of 0.585V and 0.54V, our design achieves write time improvement over single port access.

Keywords— 7-nm; FinFET; Embedded SRAM; Dual port SRAM; Write time improvement; Psuedo read; Write disturb; Half-selected SRAM cell; Single port access; Dual port access; Access conflict; Write time pulse width; Monte Carlo; Cache memories; Dynamic voltage and frequency scaling (DVFS); Low voltage operation.

I. INTRODUCTION

Embedded single port and dual port static random access memories (SRAMs) are the key IPs in contemporary memory-rich SoCs. In deep submicron technologies, read and write margins of these embedded SRAMs are degraded due to increased variations at low voltage conditions. Embedded dual port SRAMs are widely used as buffer memories in multimedia and graphical processing chips [1-5]; as a data cache in a multi-core processor [6-7]. Dual port SRAMs suffer write time push out when it's both ports are accessed by same row addresses for its respective two ports. In literature, this is referred as write disturb issue in dual port SRAMs [8-10].

State-of-art designs address this issue and try either to suppress it or combat against it through the following circuit designs and/or schemes: (a) by active bit line equalizing circuitry [8]; (b) by a write-assist 8T cell [9]; (c) by employing a priority row decoder and a shifted bit line access scheme [10]. Circuit designs and scheme in (a) and (c) almost requires double the hardware for the write driver circuitry and two separate "same row" address decoders for Port A and Port B. Circuit design with write-assist 8T cell in (b) only assist single-sided write '0' in inter-port writing, limiting its influence to suppress the write disturb. It also requires two separate "same row" address decoders for Port A and Port B. Both circuit designs, (a) and (b) supports synchronous and asynchronous

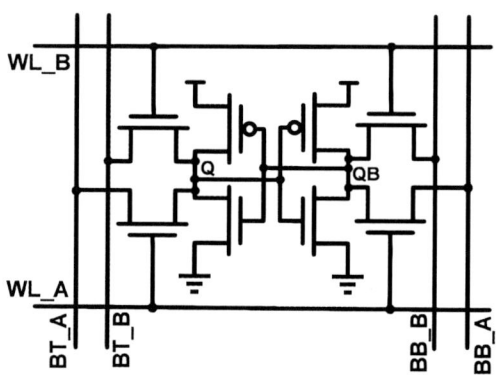

Fig. 1: 2RW 8T Dual port SRAM cell.

dual port SRAM operations. Whereas, only synchronous dual port SRAM operation is possible with circuit design (c).

In this paper, we propose a circuit designed to mitigate write issue in dual port SRAMs by moderately duplicating inter-port write data during dual port access. In our proposed design, we simply employ two NMOS switches to duplicate inter-port write data per bit line. Proposed design support double-ended write '0' in inter-port writing in contrast to single-ended write '0' in [9]. One NMOS switch is controlled by write disturb enable (WDEN) signal and the other NMOS switch is controlled by write enable signal. For our design, we require only one "same row" address decoder to generate single WDEN signal for both ports compared to two "same row" address decoders in [8-10]. Our design supports both synchronous and asynchronous dual port SRAM operations.

Proposed design employs moderate sizes for the NMOS switches to have a balanced trade-off between area and duplication of inter-port write data per bit line. Our design incurs little area penalty in the write driver circuitry compared to [8, 10]. With NMOS switch sizes of nfin=4, our circuit significantly mitigate the write disturb and improve the write time by 2.16x at 0.63V versus conventional dual port SRAM design.

This paper is organized as follows. In Section II, overview of dual port SRAM cell and write disturb issue in dual port access is presented. Section III describes the proposed design concept and its implementation to mitigate the write disturb issue. Next, we present the simulation results supporting our claim on 7-nm FinFET TSMC technology in Section IV. Finally, in Section V, we summarized the results and highlight the conclusions.

978-1-5386-3693-0/18 $31.00 © 2018 IEEE

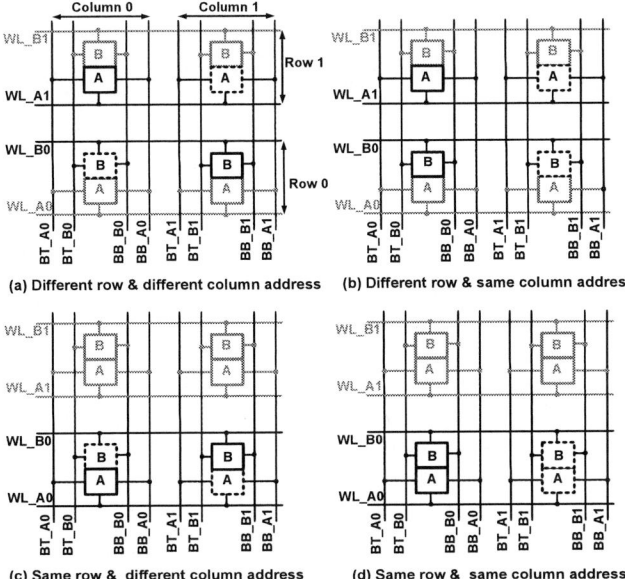

Fig. 2: Row and Column access in dual port SRAM with column multiplexed architecture.

II. OVERVIEW OF DUAL PORT SRAM CELL AND WRITE DISTURB ISSUE IN DUAL PORT ACCESS

Fig. 1 depicts a two read-write 8T dual port SRAM cell. Port A pass gates are controlled by word line A (WL_A) and connects the bit lines (BT_A, BB_A) to the internal storage nodes (Q, QB) of SRAM bit cell formed by the two cross-coupled inverters. Similarly, Port B pass gates are driven by word line B (WL_B) and connects bit lines (BT_B, BB_B) to the internal storage nodes (Q, QB). During read operation, precharged bit lines are discharged through internal storage nodes to replicate the SRAM cell data. In write operation, data to be written to the SRAM cell are loaded on the bit lines and storage nodes are forced to get updated with the data. Both Port A and B are independently available for reading and writing and hence called as two read-write (2RW) 8T dual port SRAM cell.

Contemporary SRAM architecture employs column multiplexed or bit-interleaving to avoid multi-bit errors. For instance, a column multiplexed 2 (CMUX=2) architecture allows 2 different columns (Column 0 and 1) to share the same row with common word line for SRAM array (Fig. 2). By virtue of this architecture, a selection of a particular address accesses the SRAM array with Selection, Half-selection and unselection of SRAM cells. A Selected SRAM cell in the SRAM array is the actual addressed cell (row and column address) with selected row and selected column. A Half-selected SRAM cell in the SRAM array is the one which is selected by virtue of sharing the same row i.e. selected row with common word line and unselected column. Unselected SRAM cells in the SRAM array are the ones which are not addressed for access.

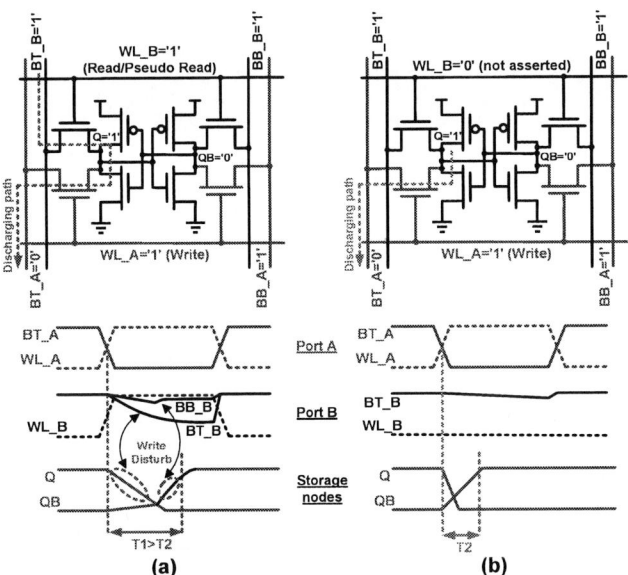

Fig. 3: Write operation. (a) Write disturb in case of dual port access (DPA). (b) No write disturb in single port access (SPA).

Fig. 2 portrays different sort of access in 2RW dual port SRAM cell in a column multiplexed or bit-interleaving SRAM architecture (CMUX=2). For simplicity and illustrative purpose, we have chosen 2 X 2 dual port SRAM array with 2 Rows: Row 0 and 1; 2 Columns: Column 0 and 1. Each dual port SRAM cell is shown as a rectangle box divided into two halves with Port A and B. Row 0 (Row 1) word line for: Port A is WL_A0 (WL_A1); Port B is WL_B0 (WL_B1). Column 0 (Column 1) bit lines for: Port A are BT_A0 (BT_A1) and BB_A0 (BB_A1); Port B are BT_B0 (BT_B1) and BB_B0 (BB_B1). Selected SRAM cell with selected Port, either A or B are shown in solid black color box. Half-selected SRAM cell with half-selected Port, either A or B are shown in dotted black color box. Unselected SRAM cells are shown in gray color. Activated signals are marked in black color and deactivated signals are marked in gray color.

Fig. 2a depicts a sort of access in which we access different row and different column addresses for Port A and B. Fig. 2b delineates a sort of access in which we access different row and same column addresses for Port A and B. In both of these sort of accesses, only single port of the dual port SRAM cell is accessed either as selected port or half-selected port. Therefore, we refer these sort of accesses (Fig. 2a, 2b) as Single Port Access (SPA) in dual port SRAM. Simultaneous writing or reading or reading (writing) and writing (reading) for both ports are possible. SPA does not possess any access conflict for both ports in selected dual port SRAM cell as only single port is accessed. Fig. 2c represents a sort of access in which we access same row and different column addresses for Port A and B. Fig. 2d shows a sort of access in which we access same row and same column addresses for Port A and B. In this case, an access conflict would arise if complementary write data is to be written to the dual port SRAM cell through Port A and B simultaneously. However, same write data to be written to the

978-1-5386-3693-0/18 $31.00 © 2018 IEEE   267

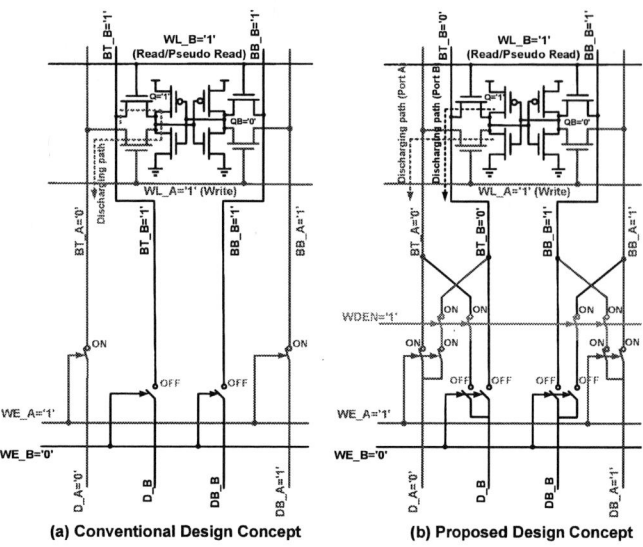

Fig. 4: Write operation in DPA: Conventional vs. Proposed design concept.

dual port SRAM cell through Port A and B simultaneously does not possess any access conflict. Still, the simultaneous reading or reading (writing) and writing (reading) from both ports are possible. Thus, we refer these sort of dual accesses (Fig. 2c, 2d) as Double Port Access (DPA) in dual port SRAM. Essentially DPA is accessing same row address with same/different column address for both ports. Therefore, DPA can be understood as a "same row" address access for both ports in a selected dual port SRAM cell.

DPA in dual port SRAM cell calls for a worst case write, pseudo read and worst case read conditions. Worst read and pseudo read affects the cell's stability in dual port SRAM cell and worst case write demands increase in word line pulse width for a successful write operation. Fig. 3a depicts a DPA in dual port SRAM cell with write operation in Port A and Read/Pseudo read operation in Port B. We consider a worst case overlapping operation in Port A and B by considering synchronous firing of word lines for both ports. Asynchronous firing of word lines will not be the worst scenario. In Fig. 3a with DPA (WL_A=WL_B= '1'), write operation through Port A cannot immediately discharge Q storage node to ground level because of the disturbance from Port B bit line BT_B. Also QB node rising to VDD level is also hampered by the disturbance from Port B bit line BB_B. These collective disturbances in write operation penalizing the write time (T1) is called the write disturb in dual port SRAM cell. Fig. 3b shows a SPA in dual port SRAM cell with write operation through Port A and no operation in Port B. Here, the write operation through Port A immediately discharged Q storage node to ground level because of no write disturb from Port B. The write operation finishes in write time, T2 which is very much improved compared to T1 (T1>T2). In order to mitigate, the write disturb issue in DPA dual port SRAM cell, we proposed to duplicate inter-port write data during write and read operation in both ports.

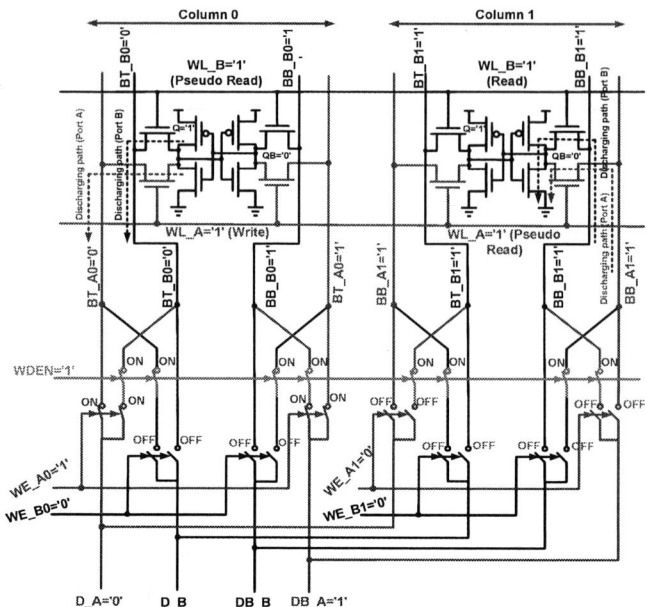

Fig. 5: Proposed design concept for dual port SRAM architecture with CMUX=2 in DPA.

III. PROPOSED DESIGN CONCEPT AND ITS IMPLEMENTATION TO MITIGATE THE WRITE DISTURB ISSUE

Fig. 4a depicts a write operation for Port A in conventional design with dual port SRAM cell in DPA (WL_A=WL_B= '1'). Port B may be either in read or pseudo read operation. WE_A and WE_B signals are write enable signal for Port A and B. A write operation through Port A (WE_A='1') cannot immediately discharged Q storage node to ground level because of the disturbance from Port B (WE_B='0') due to write disturb. In Fig. 4b, we proposed a design concept to duplicate the data from Port A to Port B by two additional switches in cascade which are ON when one or both ports undergo a write operation in DPA. One switch is controlled by write enable signal (WE_A/WE_B) and the other switch is controlled by write disturb enable (WDEN) signal which is asserted when DPA for dual port SRAM cell is selected.

Due to duplication of inter-port write data, Q storage node start discharging through BT_B also (Fig. 4b: two discharging paths through BT_A and BT_B) instead of write disturb from BT_B in contrast to conventional design (Fig. 4a: Discharging path only through BT_A and write disturb from BT_B). Thus it is very clear that by duplicating inter-port write data, we would mitigate the write disturb issue and consequently improve write time in DPA. Using the proposed design concept in DPA, we have the following for the selected dual port SRAM cell: (i) Non-corrupted read operations on both ports and; (ii) Corrupted read operation when write (read) and read (write) operations are performed on Port A and B. Fig. 5 depicts the proposed design concept for dual port SRAM architecture with CMUX=2 in DPA. Same row address is accessed for Port A and B. Next, Port A is undergoing write operation with selected Column 0 address and Port B is undergoing read operation with selected Column 1 address. Read enable switches are not

978-1-5386-3693-0/18 $31.00 © 2018 IEEE   268

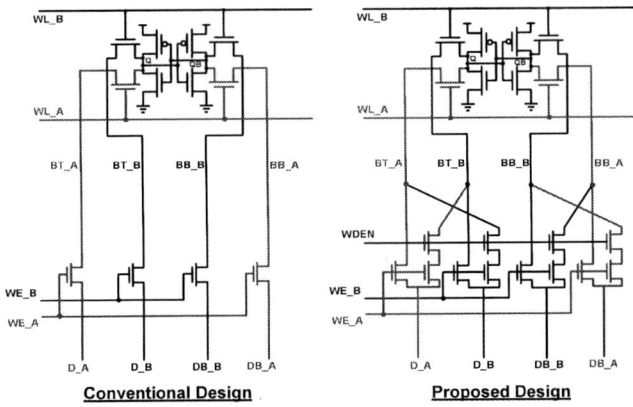

Fig. 6: Proposed design implementation and Conventional design.

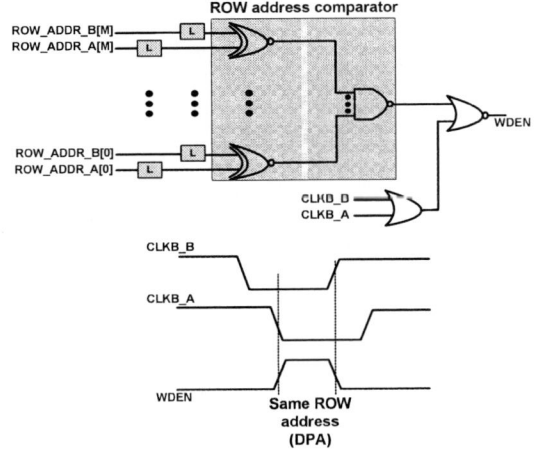

Fig. 7: Row address comparator

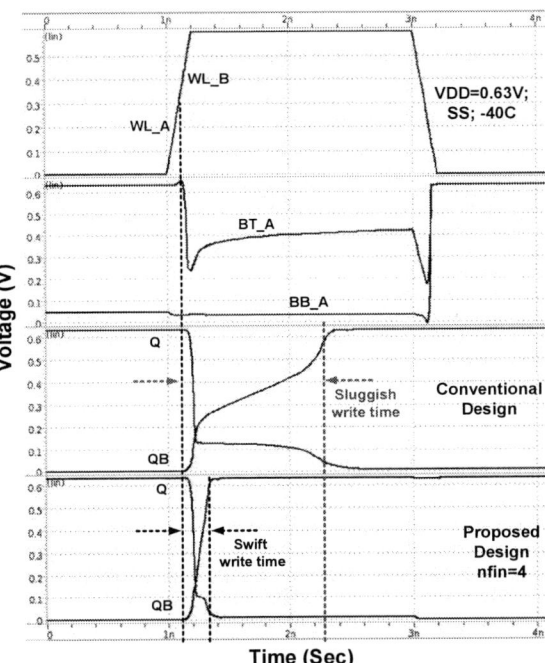

Fig. 8: Simulated waveform of write operation in 6σ dual port SRAM cell: Conventional vs. Proposed design.

shown which connect the bit lines to sense amplifier for read operation. Fig. 6 portrays the proposed design implementation using NMOS pass switches. The sizing of these NMOS switches play a pivotal role in duplicating the inter-port write data. Choosing very small size NMOS switches would duplicate the inter-port write data weakly. Whereas choosing very big sizes will incur area penalty and would duplicate the inter-port write data strongly. Thus, a moderate NMOS switch size is required to have a balanced trade-off between area and duplication of inter-port write data. In this paper, we employ NMOS switches with moderate sizes of nfin={4, 6}. We also generate common write disturb enable (WDEN) signal for both ports whereas in [8-10] individual write disturb enable signals are generated and controlled. Fig. 7 shows how we generate common WDEN signal for both ports. Here, we compare the row address of Port A (ROW_ADDR_A[0:M]) and Port B (ROW_ADDR_B[0:M]) and gate this signal with the overlap of the write/read operation for Port A (CLKB_A) and Port B (CLKB_B) to generate the WDEN. WDEN is asserted only for same Row address access (DPA) cases and remains inactive for different Row address access (SPA) cases.

The overall area penalty with our proposed design would be around 1% for 32Kb (256 x 128; 4 bit-interleaving; single bank) dual port SRAM compared to the overhead reported in

[8]. At SRAM array level in our design, we are adding 8 transistors in one column accounting to area overhead of 8/(256*8) equals 1/256. At Peripheral level, we are generating common write disturb enable (WDEN) signal for both ports whereas in [8-10] individual write disturb enable signals are generated and controlled. Here, area overhead is due to Row address comparator (Fig. 7). Therefore, total overall area comes around 1% of 256 x 128 dual port SRAM. Leakage overhead would also translate to be around 1% approximately. Moreover, as the memory size grows our proposed design area and leakage penalty would diminish.

Dynamic power overhead is essentially activity dependent which rely on the probability of accessing the same Row address in dual port SRAM memory. Our dual port SRAM 256 x 128 possess 256 rows and the probability that Port A and Port B accesses the same row boils down to 1/256. Therefore, dynamic power overhead would be around 0.8 (Actual dynamic power overhead) * 1/256 (Activity factor) =0.3125%. The dynamic power overhead would diminish further as the number of rows per column increases.

IV. SIMULATION RESULTS

Assume that a SRAM bit cell is holding a value Q='1' and QB='0'. In order to a write Q='0', we need to flip the data in the SRAM cell through a write operation. We define write time as the maximum time required from word line rise 50%VDD to storage node Q falling to 10%VDD and storage node QB rising to 90%VDD. Fig. 8 depicts simulation waveform of a write operation in Port A and Pseudo read operation in Port B for worst 6σ dual port SRAM cell with DPA (Conventional vs. Proposed design). We obtain worst 6σ

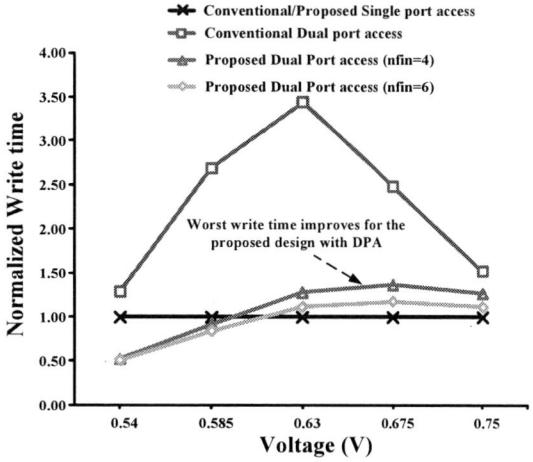

Fig. 9: Worst 6σ dual port SRAM cell's normalized write time with SPA and DPA.

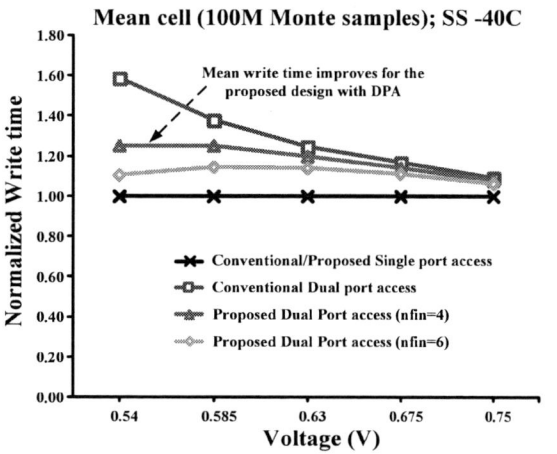

Fig. 10: Dual port SRAM cell's normalized mean write time with SPA and DPA

dual port SRAM cell operating at 0.63V, -40°C in worst case process corner SS (Slow NMOS Slow PMOS) by running 100M Monte Carlo simulations using Synopsys HSPICE Simulator. A write operation in Port A is writing Q='0' in the dual port SRAM cell. For conventional design, write time is very sluggish due to write disturb issue. Whereas for the proposed design with NMOS switch sizes nfin=4, swift write time is possible due to mitigation of write disturb issue.

Fig. 9 presents worst 6σ dual port SRAM cell's normalized write time with SPA and DPA as a function of different operating voltages {0.54, 0.585, 0.63, 0.675, 0.75} in volts (V) for conventional and proposed (nfin={4, 6}) designs. Also worst process corner SS and temperature of -40°C are chosen. Our proposed design with nfin=4 (nfin=6) achieves improvement of 2.16x (2.3x) at 0.63V compared to conventional design in DPA with write issue. It should be noted that at low voltages of 0.585V and 0.54V the proposed design show significant improvement in write time with DPA compared to SPA write time. Thus at low voltage of 0.585V and 0.54V the write operation timing closure can be done by considering the worst SPA timing rather than DPA. Fig. 10

portrays the mean dual port SRAM cell's normalised write time (100M Monte samples) with SPA and DPA as a function of different operating voltages {0.54, 0.585, 0.63, 0.675, 0.75} in volts (V) for conventional and proposed (nfin={4, 6}) designs. Also worst process corner SS and temperature of -40°C are chosen. Our proposed design achieves improvement in the mean write time over 100M monte samples compared to conventional design in DPA with write issue for complete voltage range. Different supply voltages are considered for performance evaluation of dual port SRAM to support dynamic voltage and frequency scaling (DVFS).

V.   CONCLUSION

We have presented a dual port SRAM design in 7-nm FinFET TSMC technology with duplicated inter-port write data to mitigate write disturb issue. We simply employ two NMOS switches to duplicate inter-port write data per bit line. One switch is controlled by write disturb enable (WDEN) signal which is asserted whenever write disturb occurs and the other is controlled by write enable signal. In our design, we choose moderate sizes for these switches to have a balanced trade-off between area and duplication of inter-port write data. With NMOS switch size of nfin=4, we significantly suppress the write disturb and improve the write time by 2.16x at low voltage of 0.63V. Moreover, at further low voltages of 0.585V and 0.54V, the proposed design write time with dual port access (DPA) is improved compared to single port access (SPA).

REFERENCES

[1]   J. Kim, Y. Choi, J. Jeong, S. Lee, and S. Kim, "The v2.0 + EDR Bluetooth SoC architecture for multimedia," *IEEE Trans. Consum. Electron.*, vol. 52, no. 2, pp. 436–444, May 2006.

[2]   C. C. Cheng, T. S. Chang, and K. B. Lee, "An in-place architecture for the deblocking filter in H.264/AVC," *IEEE Trans. Circuits Syst. II, Exp. Briefs*, vol. 53, no. 7, pp. 530–534, Jul. 2006.

[3]   H. B. Yin, H. G. Qi, H. Jia, D. Xie, and W. Gao, "Efficient macroblock pipeline structure in high definition AVS video encoder VLSI architecture," in *Proc. IEEE ISCAS*, 2010, pp. 669–672.

[4]   M. Inamori, J. Naganuma, and M. Endo, "A memory-based architecture for MPEG2 system protocol LSIs," *IEEE Trans. Very Large Scale Integr. (VLSI) Syst.*, vol. 7, no. 3, pp. 339–344, Sep. 1999.

[5]   M. Miyama, J. Miyakoshi, Y. Kuroda, K. Imamura, H. Hashimoto, and M. Yoshimoto, "A sub-mW MPEG-4 motion estimation processor core for mobile video application," *IEEE J. Solid-State Circuits*, vol. 39, no. 9, pp. 1562–1570, Sep. 2004.

[6]   Shiota et al., "A 51.2 GOPS 1.0 GB/s-DMA single-chip multi-processor integrating quadruple 8-way VLIW processors," in *IEEE ISSCC Dig. Tech. Papers*, Feb. 2005, pp. 194–593.

[7]   M. Nakajima, T. Yamamoto, M. Yamasaki, K. Kaneko, and T. Hosoki, "Homogeneous dual-processor core with shared L1 cache for mobile multimedia SoC," in *Symp. VLSI Circuits 2007 Dig. Tech. Papers*, Jun. 2007, pp. 216–217.

[8]   Ishii Y., Fujiwara H., Nii K., Chigasaki H., Kuromiya O., Saiki T., Miyanishi A. and Kihara Y., "A 28-nm dual port SRAM macro with active bitline equalizing circuitry against write disturb issue," in *Symp. VLSI Circuits 2010 Dig. Tech. Papers*, Jun. 2010, pp. 99–100.

[9]   Wu J. J., Chang M. F., Lu S. W., Lo R., and Li Q., "A 45-nm Dual-Port SRAM Utilizing Write-Assist cells against simultaneous access disturbances," TCAS II, Nov. 2012, pp. 790–994.

[10]   Nii et al., "Synchrounous ultra-high-density 2RW dual-port-8T-SRAM with circumvention of simultaneous common-ro-access," *IEEE J. Solid-State Circuits*, vol. 44, no. 3, pp. 977–986, March 2009.

978-1-5386-3693-0/18 $31.00 © 2018 IEEE

2018 31th International Conference on VLSI Design and 2018 17th International Conference on Embedded Systems

An Efficient VLSI Architecture for Convolution Based DWT Using MAC

Mohamed Asan Basiri M,
Department of CSE, IIT Kanpur,
Email: asanbasiri@gmail.com

Noor Mahammad Sk,
Department of CSE, IIITD&M Kancheepuram,
Email: noorse@gmail.com

Abstract—The modern real time applications related to image processing and etc., demand high performance discrete wavelet transform (DWT). This paper proposes the floating point multiply accumulate circuit (MAC) based 1D/2D-DWT, where the MAC is used to find the outputs of high/low pass FIR filters. The proposed technique is implemented with $45\ nm$ CMOS technology and the results are compared with various existing techniques. The proposed 8×8-point floating point 2-*levels* 2D-DWT achieves 27.6% and 83.7% of reduction in total area and net power respectively as compared with existing DWT [9].

Index Terms—Convolution, DSP Processor, FIR filter, and Wavelet transform.

I. INTRODUCTION

In image processing, DWT can be used in image compression [1], image reconstruction [2], image coding [3], and image fusion [4]. In general, VLSI architecture for DWT is classified into two categories, they are 1) convolution based [5]; and 2) lifting based [6]. Fig. 1 shows the architecture of convolution based DWT [7] with 3 stages, where low pass and high pass filters are represented as H and G respectively. Each filter output samples are decomposed down by the factor of 2. So, at each stage, the number of samples is equal to the half of the previous stage. Here, the input samples are a_0, $a_1, ... a_7$ and the number of input samples is 8. The coefficients of filter G are named as g_0, g_1, g_2, and g_3. The coefficients of filter H are h_0, h_1, h_2, and h_3. So, the transfer functions of G and H can be written as $G(z)=g_0+g_1z^{-1}+g_2z^{-2}+g_3z^{-3}$ and $H(z)=h_0+h_1z^{-1}+h_2z^{-2}+h_3z^{-3}$ respectively. The equations (1) and (2) show the high pass and low pass filter outputs in N-point convolution based DWT respectively, where P is the length of the filter and x is input sample. The high pass and low pass filter co-efficients are represented as g and h respectively.

$$y_h[k] = \sum_{n=0}^{P-1} x[-n+2k].g[n]; 0 \leq k \leq \frac{N}{2}-1 \quad (1)$$

$$y_l[k] = \sum_{n=0}^{P-1} x[-n+2k].h[n]; 0 \leq k \leq \frac{N}{2}-1 \quad (2)$$

In Fig. 1, the first stage outputs b and c have 4 samples. The second stage outputs d and e have 2 samples. The last stage outputs f and f' have one sample. The high pass outputs are,

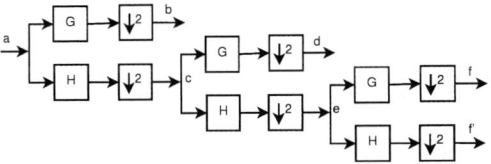

Fig. 1. Convolution based 1D DWT with 3 stages

$$\begin{bmatrix} b_0 \\ b_1 \\ b_2 \\ b_3 \end{bmatrix} = \begin{bmatrix} a_0 & a_{-1} & a_{-2} & a_{-3} \\ a_2 & a_1 & a_0 & a_{-1} \\ a_4 & a_3 & a_2 & a_1 \\ a_6 & a_5 & a_4 & a_3 \end{bmatrix} \begin{bmatrix} g_0 \\ g_1 \\ g_2 \\ g_3 \end{bmatrix} \quad (3)$$

$$\begin{bmatrix} d_0 \\ d_1 \end{bmatrix} = \begin{bmatrix} c_0 & c_{-1} & c_{-2} & c_{-3} \\ c_2 & c_1 & c_0 & c_{-1} \end{bmatrix} \begin{bmatrix} g_0 \\ g_1 \\ g_2 \\ g_3 \end{bmatrix} \quad (4)$$

$$f_0 = g_0e_0 + g_1e_{-1} + g_2e_{-2} + g_3e_{-3} \quad (5)$$

The 2D discrete wavelet transform can be found in 2 steps, they are row process and column process. Here, the input signal sample values are represented as a $N \times N$ matrix. During the row process, each row of the input signal matrix is 1D transformed and the results are stored in $N \times \frac{N}{2}$ buffer. After completing all the N rows of input signal matrix, transpose matrix of the buffer is taken for column process. In column process, each row of transposed buffer matrix is 1D transformed and results are the required 2D-transformed values. Fig. 2(a) shows the example for 2D-DWT using $8X8$ image signal with 1 *level* decomposition. Fig. 2(b) shows the example for 2D-DWT with 3 *levels* of decomposition. Fig. 3(a) and 3(b) show the 1D and 2D folded convolution based DWTs respectively.

A. Related Works

The following works are found in the VLSI architectures for 1D/2D DWT. The paper [16] shows the VLSI architecture of 2D-DWT. The non-separable convolution based DWTs are shown in [13] [14], where the transpose buffer is not used because the column process is combined with row process.

978-1-5386-3693-0/18 $31.00 © 2018 IEEE   271

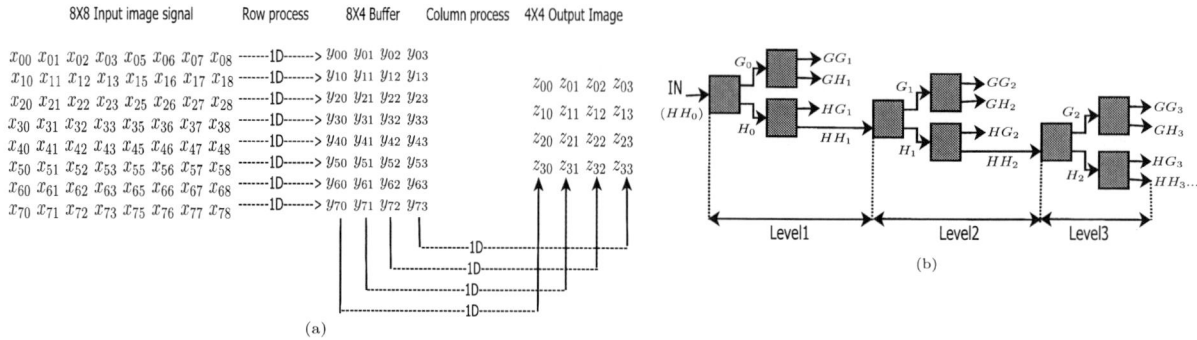

Fig. 2. Convolution based 2D-DWT (a) Example: $8X8$ image signal with 1 $level$ decomposition and (b) Example: decomposition with 3 $levels$

Therefore, multiplication with each filter co-efficient requires two multiplications. The papers [9][10][11] show the convolution based DWTs, where the critical path path involves one multiplier followed by log_2b levels of CLA (carry look ahead adder) tree. The multiplier involves log_2p levels of CSA (carry save adder) tree and one CLA. Here, b and p are the number of filter co-efficients and number of bits to represent each co-efficient respectively. In [12], separable convolution based 2D-DWT using odd/even decomposition is explained. In [14], non separable parallel convolution based 2D-DWT using the odd/even decomposition is explained. The lifting based parallel architectures are shown in [18][20][22][23], where the transpose buffer is not used and the critical path delay equal to two adders and one multiplier. The multiply accumulate circuit (MAC) based DWT is shown in [19], where the critical path contains two add-shift based multipliers and four adders. In the folded recursive [17] lifting based DWT, the half of the direct form (9, 7) is used. So, the whole operation takes more cycles to complete as compared with direct form (9, 7) DWT [16]. In the flipping based [25], the co-efficients used in direct form are inverted. In all these lifting based DWT, the drawback is the critical path delay, which increases the energy per operation and decreases the operating frequency.

B. Contribution of this paper

The major objective of this work is to improve the performance of the DWT for DSP applications. This paper proposes the floating point MAC based 1D/2D-DWT, where the high/low pass filter outputs are found by MAC. The experimental results show the proposed design requires less delay, area, and power dissipation than existing systems. In this paper, IEEE-754 based floating point single precision (32 bits) is considered because high dynamic range (HDR) image processing applications require IEEE-754 single/half precision arithmetic [27].

The rest of the paper is organized as, Section II states the proposed convolution based folded DWT. Design modeling, implementation, and results are discussed in Section III, followed by a Section IV conclusion.

II. THE PROPOSED CONVOLUTION BASED FLOATING POINT 2D-DWT ARCHITECTURE

In this section, convolution based floating point 2D-discrete wavelet transform architecture is proposed, which is designed with floating point multiply accumulate circuit (MAC) [21]. The MAC operation can be defined as multiplication and repeated addition. This means that the present multiplication result is added with previous MAC result ($z[j] = z[j-1] + (A[j].B[j])$), where $A[j]$ and $B[j]$ are present input values, $z[j]$ and $z[j-1]$ are present and previous MAC results respectively.

Fig. 4(b) shows proposed 8-point floating point high pass filter (G_1) for convolution based 2D-DWT. Here, one stage pipelined floating point MAC is used. If $in = 0$ then, MAC operations will be performed otherwise multiplication will be performed. The select lines s_0 and s_1 are used to select the proper inputs based on equations (3) to (5). During the row process of floating point 8×8-input in $level$ 1, the equation (3), requires 12 clock cycles. So, during 1st, 2nd, 5th, and 9th clock cycles of row process in $level$ 1, $in = 1$ and $in = 0$ during other clock cycles. Therefore, for each of 12 clock cycles of ith row for 8×8-input matrix of row process in $level$ 1, the corresponding $en_i = 1$ and others are 0 in 8×4 buffer as shown in Fig. 4(a). So, totally 96 clock cycles ($12 \times 8=96$) are required to finish the row process of 8×8-point input matrix. During the row process of $level$ 2, the equation (4) required 4 clock cycles. Here, $in = 1$ during 1st and 2nd clock cycles and $in = 0$ during other clock cycles. So, totally 16 clock cycles ($4 \times 4=16$) are required to finish the row process. Here, fa_0, fa_1, and fa_2 are the outputs of 4th column of HH_1 4×4-buffer. If two floating point MACs are used in each filter of Fig. 3(b), then half of the above mentioned clock cycles will be reduced. This way of implementing convolution based floating point 2D-DWT requires less area/power than conventional design, where 4 floating point multipliers and 3 floating point adders are used for each low/high pass filter in row/column processes.

The number of floating point MAC operations in the first $level$ of 1D-DWT is $\frac{3b^2}{4}$, where b is the length of low/high pass filters. From the second level onward the number of MAC operations would be $\frac{b^2}{4^{i-1}}$, where $i \geq 2$. The total number of

978-1-5386-3693-0/18 $31.00 © 2018 IEEE

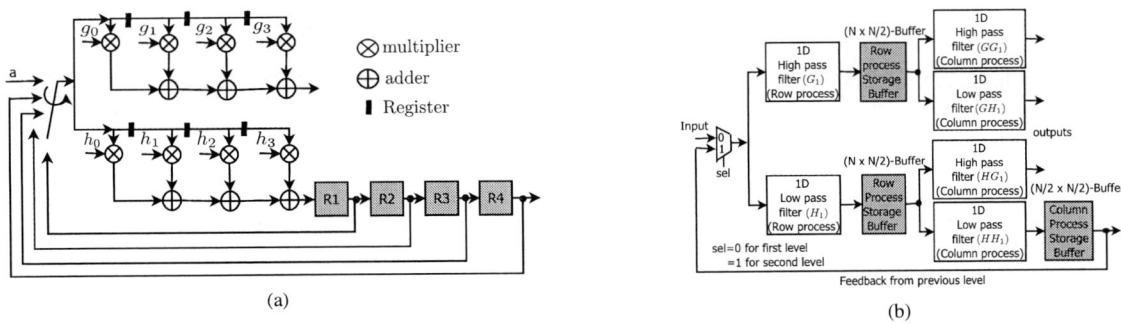

(a)   (b)

Fig. 3. (a) Convolution based folded 1D-DWT and (b) The $N \times N$-point convolution based folded 2D-DWT architecture

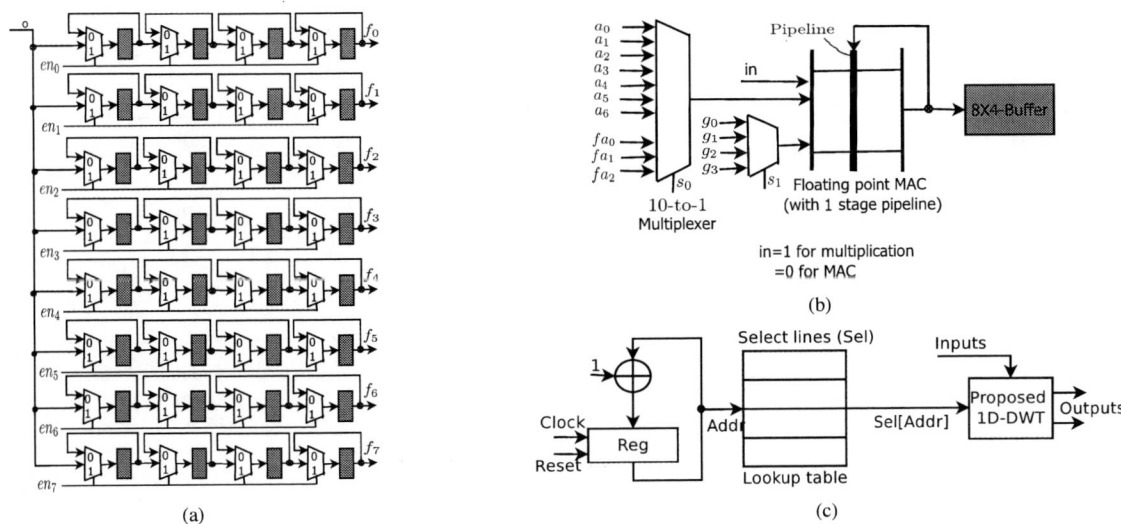

(a)   (c)

Fig. 4. (a) 8×4-Buffer used for the row process in proposed floating point convolution based 8×8 2D-DWT with 2 *levels*, (b) Proposed 8-point floating point high pass filter (G_1) for convolution based 2D-DWT, and (c) Overall architecture of proposed 1D-DWT for 2D implementation.

TABLE I
COMPARISON OF BETWEEN EXISTING AND PROPOSED N-POINT 1D-DWT ARCHITECTURES

	Number of multipliers	Number of cycles	Number of adders	critical path depth
[8]	$2b$-numbers of conventional multipliers	$1 + \frac{3b}{4} + \sum_{i=2}^{L}(\frac{b}{4^{i-1}})$	$2b$	$T_{MUL} + (log_2 b)T_{ADD}$
Parallel [10]	$2bL$ conventional multipliers	$\approx N^2$	$2bL$	$T_{MUL} + (log_2 b)T_{ADD}$
Floating point 1D-DWT proposed	Two proposed floating point MAC	$\frac{3b^2}{4} + \sum_{i=2}^{L} \frac{b^2}{4^{i-1}}$	0	T_{FMAC}
Floating point 1D-DWT proposed 2-parallel	Four proposed floating point MACs	$\frac{1}{2}(\frac{3b^2}{4} + \sum_{i=2}^{L} \frac{b^2}{4^{i-1}})$	0	T_{FMAC}
(5, 3) direct lifting [15]	2	$\sum_{i=0}^{L-1} \frac{\frac{N}{2^i}+1}{2}$	4	$T_{MUL}+2T_{ADD}$
(9, 7) direct lifting [16]	6	$\sum_{i=0}^{L-1} \frac{3+\frac{N}{2^i}}{2}$	8	$T_{MUL}+2T_{ADD}$
(9, 7) recurse lifting [17]	3	$\sum_{i=0}^{L-1} \frac{1+2\frac{N}{2^i}}{2}$	4	$T_{MUL}+2T_{ADD}$

Here, the length of the filter is b, where $N = 2b$, the number of *levels* of 1D-DWT is L. T_{MUL} is the critical path depth of floating point multiplier. T_{ADD} is the critical path depth of the floating point adder. T_{FMAC} is the critical path depth of proposed floating point MAC [21].

floating point MAC operations (number of cycles) (N_{MAC}^{1D}) required for the N-point 1D-DWT with L *levels* is shown in (6), where $N = 2b$. Similarly, the total number of floating

point MAC operations (number of cycles) (N_{MAC}^{2D}) required for the $N \times N$-point 2D-DWT with L *levels* is shown in (8), where $N = 2b$. Here, the number of N-point sequences

TABLE II

COMPARISON OF BETWEEN EXISTING AND PROPOSED $N \times N$-POINT 2D-DWT ARCHITECTURES

	Number of multipliers	Number of cycles	Number of adders	critical path depth	Transpose Buffer	Temporal Buffer
[9]	6b conventional multipliers	$\frac{N.3b}{4} + \frac{N.3b}{2.4} + \sum_{i=2}^{L}(\frac{N}{2^{i-1}}\frac{b}{4^{i-1}} + \frac{N}{2.2^{i-1}}\frac{b}{4^{i-1}}) = \frac{9Nb}{8} + \sum_{i=2}^{L}\frac{3Nb}{2^{3i-2}}$	$6(\frac{b}{2} + \frac{b}{4} + \frac{b}{8} + ...1) = 6(2^b - 1)$	$T_{MUL} + (log_2 b)T_{ADD}$	$2(N \times \frac{N}{2})$	$\frac{N}{2} \times \frac{N}{2}$
non separable parallel [10]	8bL conventional multipliers	$\approx N^2$	$8L(2^b - 1)$	$T_{MUL} + (log_2 b)T_{ADD}$	-	-
[11]	4b	$\frac{9Nb}{8} + \sum_{i=2}^{L}\frac{3Nb}{2^{3i-2}} + 2 + b + log_2 b$	$4(b-1)$	T_{MUL}	$2(N \times \frac{N}{2})$	$\frac{N}{2} \times \frac{N}{2}$
Odd-even decomposed [12]	4b	$\frac{2}{3}(1 - 4^{-L})N^2$	$4b$	$T_{MUL} + \frac{b}{2}T_{ADD}$	$2(N \times \frac{N}{2})$	$\frac{N}{2} \times \frac{N}{2}$
non separable folded [13]	8b conventional multipliers	$\approx N^2$	$4(2^b - 1) + 4$	$T_{MUL} + (log_2 b)T_{ADD}$	-	$\frac{N}{2} \times \frac{N}{2}$
[14]	8bL	$\approx N^2$	$8L(2^b - 1)$	$T_{MUL} + T_{ADD}$	-	-
Floating point proposed	Six proposed floating point MACs	$\frac{9Nb^2}{8} + \sum_{i=2}^{L}\frac{3Nb^2}{2^{3i-2}}$	0	T_{FMAC}	$2(N \times \frac{N}{2})$	$\frac{N}{2} \times \frac{N}{2}$
(5, 3) direct [15]	6	$\sum_{i=1}^{L}(\frac{3}{4}\frac{N}{2^{i-1}}(\frac{N}{2^{i-1}} + 1))$	12	$T_{MUL} + 2T_{ADD}$	$N \times \frac{N}{2}$	$\frac{N}{2} \times \frac{N}{2}$
(9, 7) direct [16]	18	$\sum_{i=1}^{L}(\frac{3}{4}\frac{N}{2^{i-1}}(3 + \frac{N}{2^{i-1}}))$	24	$T_{MUL} + 2T_{ADD}$	$N \times \frac{N}{2}$	$\frac{N}{2} \times \frac{N}{2}$
(9, 7) recurse [17]	9	$\sum_{i=1}^{L}(\frac{3}{4}\frac{N}{2^{i-1}}(1 + 2\frac{N}{2^{i-1}}))$	12	$T_{MUL} + 2T_{ADD}$	$N \times \frac{N}{2}$	$\frac{N}{2} \times \frac{N}{2}$
(9, 7) High speed [18]	18	$\sum_{i=1}^{L}(\frac{(\frac{N}{2^{i-1}})^2}{4})$	32	$T_{MUL} + 2T_{ADD}$	-	$\frac{N}{2} \times \frac{N}{2}$
(9, 7) Fast [18]	10	$\sum_{i=1}^{L}(\frac{(\frac{N}{2^{i-1}})^2}{2})$	16	$T_{MUL} + 2T_{ADD}$	-	$\frac{N}{2} \times \frac{N}{2}$
(9, 7) MAC based [19]	0	$\sum_{i=1}^{L}(\frac{3}{2}\frac{N}{2^{i-1}} + 3)$	12	$4T_{ADD} + 2T_{MUL}$	-	$\frac{N}{2} \times \frac{N}{2}$
(9, 7) M-parallel [20]	6M	$\sum_{i=1}^{L}(\frac{(\frac{N}{2^{i-1}})^2}{M})$	12M	$T_{MUL} + 2T_{ADD}$	-	$\frac{N}{2} \times \frac{N}{2}$
(9, 7) 2-parallel [22]	10	$\sum_{i=1}^{L}(\frac{(\frac{N}{2^{i-1}})^2}{2})$	16	$T_{MUL} + 2T_{ADD}$	-	$\frac{N}{2} \times \frac{N}{2}$
(9, 7) P-block parallel [23]	4P	$\sum_{i=1}^{L}(\frac{(\frac{N}{2^{i-1}})^2}{P})$	8P	$2T_{ADD}$	-	$\frac{N}{2} \times \frac{N}{2}$
(9, 7) [24]	4	$\sum_{i=1}^{L}(\frac{3(\frac{N}{2^{i-1}})^2}{4})$	8	$T_{MUL} + 2T_{ADD}$	4N	$\frac{N}{2} \times \frac{N}{2}$
(9, 7) flipping based [25]	10	$\sum_{i=1}^{L}(\frac{(\frac{N}{2^{i-1}})^2}{2})$	16	T_{MUL}	4N	$\frac{N}{2} \times \frac{N}{2}$

Here, the length of the filter is b, where $N = 2b$, the number of *levels* of 2D-DWT is L. T_{MUL} is the critical path depth of floating point multiplier. T_{ADD} is the critical path depth of the floating point adder. T_{FMAC} is the critical path depth of the proposed floating point MAC [21]. Temporal buffer is used to send the outputs from previous *level* to present *level* inputs. In case of proposed floating point 2-parallel 2D-DWT, two floating point MACs will be used parallel in each filter of row and column processes (totally 12 floating point MACs). Therefore, the number of cycles is $\frac{1}{2}(\frac{9Nb^2}{8} + \sum_{i=2}^{L}\frac{3Nb^2}{2^{3i-2}})$.

involved in row and column processes of the first level are N and $\frac{N}{2}$ respectively. The number of $\frac{N}{2^{i-1}}$-point sequences involved in the row and column processes of the ith level are $\frac{N}{2^{i-1}}$ and $\frac{N}{2.2^{i-1}}$ respectively, where $i \geq 2$.

$$N_{MAC}^{1D} = \frac{3b^2}{4} + \sum_{i=2}^{L}\frac{b^2}{4^{i-1}} \quad (6)$$

$$N_{MAC}^{2D} = N\frac{3b^2}{4} + \frac{N}{2}\frac{3b^2}{4} + \sum_{i=2}^{L}(\frac{N}{2^{i-1}} + \frac{N}{2.2^{i-1}})(\frac{b^2}{4^{i-1}}) \quad (7)$$

$$N_{MAC}^{2D} = \frac{9Nb^2}{8} + \sum_{i=2}^{L}\frac{3Nb^2}{2^{3i-2}} \quad (8)$$

III. DESIGN MODELING, IMPLEMENTATION, AND RESULTS

The proposed and existing designs are modeled in Verilog HDL. These Verilog HDL models are simulated/verified using the Xilinx ISE simulator and synthesized Cadence 6.1 ASIC design tool. All the designs are implemented for 45 nm technology, where the operating voltage is 0.88 v. Tables I and II show the theoretical analysis of various DWT architectures for 1D and 2D respectively. In general, the appropriate select

lines of multiplexers are used to perform the required 1D-DWT using proposed architecture. Fig. 4(c) shows the overall architecture of proposed 1D-DWT, where the select lines of multiplexers are stored in a lookup table with corresponding address. The *Addr* will be increased by one in every clock cycle and initially it is 0. The appropriate select lines $Sel[Addr]$ are obtained from memory during each clock cycle to perform 1D-DWT. Table III shows the comparison of critical path delay, total area, net power, and energy per operation between various 1D/2D-DWTs, where the proposed designs are represented with bold letters. Since 4-*taps* low/high pass filters are used in all the DWTs, 4 multiplications can be done in existing 1D-DWTs at a time. The total number of multiplications in a 8-point 1D-DWT with 4-*taps* low/high pass filters for 3-*levels* will be 17, which is shown in the equations from (3) to (5). Therefore, the total number of cycles required for existing 1D-DWT is $(17/4) + 1 = 4.25 + 1 = 6$.

The 8×8-point existing 2D-DWT with 4-*taps* low/high pass filters requires eight 8-point 1D-DWT operations in row and column process and hence it requires $(8 \times 3) + (4 \times 3) = 36$ cycles in first *level* because first *level* contains 12 multiplications per 1D-DWT operation (which requires $12/4 = 3$ cycles). The second *level* 1D-DWT needs 4 multiplications,

978-1-5386-3693-0/18 $31.00 © 2018 IEEE

TABLE III
PERFORMANCE ANALYSIS OF DIFFERENT ARCHITECTURES FOR 1D/2D-DWT WITH $45nm$ CMOS TECHNOLOGY USING CADENCE

	Critical path delay (ps)	Frequency (MHz)	Total area (μm²)	Net power (nw)	Switching power (nw)	Leakage power (nw)	EOP (fJ)	Number of cycles	Total cycle delay (ps)
Floating point 1D-DWT [8]	5219	191.6	28633	1807654.5	5980565.5	2821451.2	45937.7	6	31314
Floating point 1D-DWT [10]	5623.9	177.8	49516.2	1430660.6	2743570.6	4533061.6	40923.1	7	39367
Floating point 1D-DWT proposed	**1249**	**800.6**	**17341**	**74459.9**	**241076.1**	**966078.4**	**1507.7**	**17**	**21233**
Floating point 2D-DWT [9]	5271.4	189.7	58342	1193726.7	989010.0	3104497.3	21578.5	42	221398
Floating point 2D-DWT [10]	5623.9	177.8	53816.9	1455607.5	2924024.4	4634948.5	42510.9	42	236203
Floating point 2D-DWT [11]	3061.0	326.7	59138.9	1080896.3	3085137.6	3704450.1	20782.9	51	156111
Floating point 2D-DWT [12]	5289.0	189.1	49732.4	1183300.1	2596899.5	3822666.6	33953.0	42	222138
Floating point 2D-DWT [13]	5653.9	176.9	27679.6	1003397.4	1508324.8	3181276.6	26514.5	42	237463
Floating point 2D-DWT [14]	5226.5	191.3	53107.9	1417515.7	2890047.1	4533458.8	38798.9	42	219513
Floating point 2D-DWT proposed	**1249.0**	**800.6**	**42549**	**194794.1**	**704788.9**	**2202262.3**	**3630.9**	**168**	**209832**
(5, 3) direct lifting 1D-DWT [15]	4597.8	217.5	9782.3	79890	221072	433129	3007.8	5	436791
(9, 7) direct lifting 1D-DWT [16]	4561.7	219.2	19202	180350	491729	863615	6182.6	9	41055.3
(9, 7) recurse lifting 1D-DWT [17]	4792.8	208.6	10921.4	93948.6	269910.7	488251.9	3633.7	12	57513.6
(5, 3) direct lifting 2D-DWT [15]	4508.3	221.8	51254.1	262620.7	1022748.6	2206115.5	14556.6	208+70=278	1253307.4
(9, 7) direct lifting 2D-DWT [16]	4633.6	215.8	54808.5	327960.1	1060725.1	2422821.2	161413.6	162+70=232	1074995.2
(9, 7) recurse lifting 2D-DWT [17]	4667.7	214.2	37836.7	254023.5	888948.2	1599349.6	11614.6	216+40=256	1194931.2
(9, 7) High speed 1D-DWT [18]	4637.4	215.6	83965.3	501684.2	1458554.4	3736259.1	24090.4	21+7=28	129847.2
(9, 7) Fast 2D-DWT [18]	4679.8	213.7	38613.4	299117.5	907245.8	1718699.4	12288.8	42+14=56	262068.8
(9, 7) adder based MAC [19]	4909.9	203.7	34882.9	233462.8	785764.2	1368409.1	10576.7	18+12=30	147297.0
(9, 7) M-parallel 2D-DWT [20]	4715.6	212.1	182170.9	1196879.2	3353252.0	8240150.6	54669.8	13+6=19	89596.4
(9, 7) 2-parallel [22]	4777.2	209.3	38537.1	265122.5	895213.3	1697369.3	12385.2	42+27=69	329281.8
(9, 7) 4-block parallel [23]	4615.3	216.7	88442.1	272592.7	783395.4	4018359.7	22161.5	22+7=29	133843.7
(9, 7) [24]	4500.1	225.4	53454.3	271621.5	1123241.1	2110011.2	14549.9	162+70=232	1044023.2
(9, 7) flipping based [25]	4660.2	214.6	38543.2	309137.5	898235.3	1807519.1	12609.35	42+14=56	260971.2

Number of *levels* for all the 8-point convolution based 1D-DWTs and 8×8-point convolution based 2D-DWTs are 3 and 2 respectively. Filter length in all the cases is 4. Floating point inputs signal sample values and filter co-efficients are represented as IEEE-754 single precision (32-bits) format. $N = 13$ for (5, 3) 2D-DWT, $N = 9$ for (9, 7) 2D-DWT, and $M = 7$ for (9, 7) 2D-DWT. Total cycle delay=critical path delay × number of cycles. In proposed floating point 1D/2D-DWT, one proposed floating point MAC is used in each filter of row/column processes. Since these implementations are based on IEEE-754 single precision (32-bits) format, the PSNR for all the designs is <1. Energy per operation (EOP)=(switching power+leakage power)×critical path delay.

which requires 1 cycle. The second *level* of 8×8-point 2D-DWT with 4-*taps* low/high pass filters requires four 4-point 1D-DWT operations in row and column process and hence it requires $(4 \times 1) + (2 \times 1) = 6$ cycles. Therefore 8×8-point (fixed/floating point existing) 2D-DWT with 4-*taps* low/high pass filters for 2-*levels* requires $36 + 6 = 42$ cycles.

In proposed floating point MAC based DWT, the number of cycles depends on total number of multiplications. The proposed 8-point 3-*level* floating point 1D-DWT requires 12, 4, and 1 cycles in the first, second, and third levels (total 17 cycles). In the same way, the proposed floating point 8×8-point 2-*level* 2D-DWT requires 168 cycles, because first *level* of row and column processes requires $(8 \times 12) + (4 \times 12) = 144$ cycles. The second *level* of row and column processes requires $(4 \times 4) + (2 \times 4) = 24$ cycles.

The proposed 8-point floating point 3-*levels* 1D-DWT achieves 32.12%, 39.4%, and 95.8% of reduction in total cycle delay, total area, and net power respectively, compared with existing DWT [8]. The proposed 8×8-point floating point 2-*levels* 2D-DWT achieves 27.6% and 83.7% of reduction in total area and net power respectively, compared with existing DWT [9]. Fig. 5 shows the layout chip diagram for proposed 8×8-point floating point convolution based 2-*levels* 2D-DWT using $45\ nm$ technology, where the length of the low/high pass filters in row/column processes is 4, the floating point input signal sample values and filter co-efficients are represented as single precision IEEE-754 format. In proposed floating point

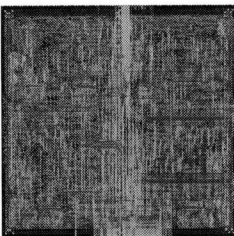

Fig. 5. Layout chip diagram for proposed floating point 8×8-point convolution based 2-*levels* 2D-DWT (input signal sample values and 4-*taps* filter co-efficients are represented as single precision IEEE-754 format) with a core area as $51058.8 \mu m^2$, die space around core as $5 \mu m$ and total chip area as $55678.03 \mu m^2$ using 45-nm CMOS technology.

1D/2D-DWT implementation, one proposed floating point MAC is used in each filter of the row/column process. Here, the critical path delay of [11] is less than others, because the multipliers used in [11] are inner pipelined. Similarly, the critical path delay of all the lifting based DWTs are almost around 4600ps, because the critical path in these existing designs include a floating point multiplier followed by an adder.

According to [26], the mean square error (MSE) is defined as,

$$\sigma^2 = \frac{1}{W} \sum_{n=1}^{W} ((x_n - y_n)^2) \qquad (9)$$

where x_n is the input sequence, y_n is the output sequence, and W is the length of a data sequence. Peak-Signal-to-Noise ratio (PSNR) measures the size of error relative to peak value x_{peak} (for 8 bit pixel x_{peak}^2 equals 255) of the signal and it is given by:

$$PSNR = 10log_{10}\frac{x_{peak}^2}{\sigma^2} \qquad (10)$$

In case of floating point MAC based proposed convolution DWT, IEEE-754 rounding technique is used. Here, the following three bits are used beyond the least significant bit of the mantissa for rounding. They are (1) Guard bit (G); (2) Round bit (R); and (3) sticky bit (S). These are used to perform round up, down, and even. The maximum of weight of the bits beyond the mantissa during IEEE-754 based rounding [21] is $W = 2^{-1}.G + 2^{-2}.R + 2^{-3}.S$. Therefore, the value of $W<1$ (it is always true). Therefore, $MSE<1$ for floating point rounding. Therefore, the PSNR value is always high and nearly equal to the peak signal value. In other words, the output image using IEEE-754 based rounding technique will give very less rounding noise (<1).

IV. CONCLUSION

In this paper, efficient VLSI architectures for convolution based folded 1D/2D-DWTs are proposed. This paper proposes the floating point MAC based 1D/2D-DWT, where the low/high pass FIR filter outputs are found using a MAC. The proposed techniques are implemented with $45\ nm$ CMOS technology. The proposed 8×8-point floating point $2\text{-}levels$ 2D-DWT achieves 27.6% and 83.7% of reduction in total area and net power respectively compared with existing DWT [9].

REFERENCES

[1] Ronald A. DeVore, Bjorn Jawerth and Bradely J. Lucier, "*Image Compression Through Wavelet Transform Coding*", IEEE Transactions on Information Theory, 1992, 38 (2), pp. 719-746.

[2] Yong Choi, Ja-Yong Koo, and Nam-Yong Lee, "*Image Reconstruction Using the Wavelet Transform for Positron Emission Tomography*", IEEE Transactions on Medical Imaging, 2001, 20 (11), pp. 1188-1193.

[3] Marc Antonini, Michael Barlaud, Pierre Mathieu and Ingrid Daubechies, "*Image Coding Using Wavelet Transform*", IEEE Transactions on Image Processing, 1992, 1 (2), pp. 205-220.

[4] Jorge Nunez, Xavier Otazu, Octavi Fors, Albert Prades, Vicenc Pala, and Roman Arbiol, "*Multiresolution-Based Image Fusion with Additive Wavelet Decomposition*", IEEE Transactions on Geoscience and Remote Sensing, vol. 37, no. 3, pp. 1204-1211, May 1994.

[5] Nikos D. Zervas, Giorgos P. Anagnostopoulos, Vassilis Spiliotopoulos, Yiannis Andreopoulos, and Costas E. Goutis, "*Evaluation of Design Alternatives for the 2-D-Discrete Wavelet Transform*", IEEE Transactions on Circuits and Systems for Video Technology, 2001, 11 (12), pp. 1246-1262.

[6] Chao-Tsung Huang, Po-Chih Tseng, and Liang-Gee Chen, "*Flipping Structure: An Efficient VLSI Architecture for Lifting-Based Discrete Wavelet Transform*", IEEE Transactions on Signal Processing, 2004, 52 (4), pp. 1080-1089.

[7] Si Jung Chang, Moon Ho Lee and Ju Yong Park, "*A High Speed VLSI Architecture of Discrete Wavelet Transform for MPEG-4*", IEEE Transactions on Consumer Electronics, 1997, 43 (3), pp. 623-627.

[8] Keshab K. Parhi and Takao Nishitani, "*Folded VLSI architectures for discrete wavelet transforms*", IEEE International Symposium on Circuits and Systems, May 1993, pp. 1734-1737.

[9] Basant Kumar Mohanty, and Pramod Kumar Meher,"*Memory-Efficient High-Speed Convolution-based Generic Structure for Multilevel 2-D DWT*", IEEE Transactions on Circuits and Systems for Video Technology, 2013, 23 (2), pp. 353-363.

[10] Chaitali Chakrabarti, and Mohan Vishwanath, "*Efficient Realizations of the Discrete and Continuous Wavelet Transforms: From Single Chip Implementations to Mappings on SIMD Array Computers*", IEEE Transactions on Signal Processing, 1995, 43 (3), pp. 759-771.

[11] Pramod Kumar Meher, Basant Kumar Mohanty, and Jagdish Chandra Patra, "*Hardware-Efficient Systolic-Like Modular Design for Two-Dimensional Discrete Wavelet Transform*", IEEE Transactions on Circuits and Systems - II, 2008, 55 (2), pp. 151-155.

[12] Po-Cheng Wu and Liang-Gee Chen, "*An Efficient Architecture for Two-Dimensional Discrete Wavelet Transform*", IEEE Transactions on Circuits and Systems for Video Technology, 2001, 11 (4), 536-545.

[13] Chu Yu and Sao-Jie Chen, "*VLSI Implementation of 2-D Discrete Wavelet Transform for Real-time Video Signal Processing*", IEEE Transactions on Consumer Electronics, 1997, 43 (4), pp. 1270-1279.

[14] Francescomaria Marino, "*Efficient High-Speed/Low-Power Pipelined Architecture for the Direct 2-D Discrete Wavelet Transform*", IEEE Transactions on Circuits and Systems - II: Analog and Digital Signal Processing, 2000, 47 (12), pp. 1476-1491.

[15] Tinku Acharya and Chaitali Chakrabarti, "*A Survey on Lifting-based Discrete Wavelet Transform Architectures*", Journal of VLSI Signal Processing, 2005, 42, pp. 321-339.

[16] Liu Hong-jin, Shao Yang, He Xing, Zhang Tie-jun, Wang Dong-hui and Hou Chao-huan, "*A Novel VLSI Architecture for 2-D Discrete Wavelet Transform*", IEEE International Conference on ASIC, Oct. 2007, pp. 40-43.

[17] Guangming Shi, Weifeng Liu, Li Zhang, and Fu Lii, "*An Efficient Folded Architecture for Lifting-Based Discrete Wavelet Transform*", IEEE Transactions on Circuits and Systems-II: Express Briefs, 2009, 56 (4), pp. 290-294.

[18] Chengyi Xiong, Jinwen Tian, and Jian Liu, "*Efficient Architectures for Two-Dimensional Discrete Wavelet Transform Using Lifting Scheme*", IEEE Transactions on Image Processing, 2007, 16 (3), pp. 607-614.

[19] Chih-Hsien Hsia, Jen-Shiun Chiang, Member, and Jing-Ming Guo, "*Memory-Efficient Hardware Architecture of 2-D Dual-Mode Lifting-Based Discrete Wavelet Transform*", IEEE Transactions on Circuits and Systems for Video Technology, 2013, 23 (4), pp. 671-683.

[20] Xin Tian, Lin Wu, Yi-Hua Tan, and Jin-Wen Tian, "*Efficient Multi-Input/Multi-Output VLSI Architecture for Two-Dimensional Lifting-Based Discrete Wavelet Transform*", IEEE Transactions on Computers, 2011, 60 (8), pp. 1207-1211.

[21] Mohamed Asan Basiri M and Noor Mahammad Sk, "*An efficient Hardware Based Higher Radix Floating Point MAC Design*", ACM Transactions on Design Automation of Electronic Systems (TODAES), 2014, 20 (1), pp. 15:1-15:25.

[22] Wei Zhang, Zhe Jiang, Zhiyu Gao, and Yanyan Liu, "*An Efficient VLSI Architecture for Lifting-Based Discrete Wavelet Transform*", IEEE Transactions on Circuits and Systems-II: Express Briefs, 2012, 59 (3), pp. 158-162, Mar. 2012.

[23] Basant K. Mohanty, Anurag Mahajan, and Pramod K. Meher, "*Area-and Power-Efficient Architecture for High-Throughput Implementation of Lifting 2-D DWT*", IEEE Transactions on Circuits and Systems-II: Express Briefs, 2012, 59 (7), pp. 434-438.

[24] Boon Hui Ang, Usman Ullah Sheikh, and Muhammad Nadzir Marsono, "*2-D DWT System Architecture for Image Compression*", Journal of Signal Processing Systems, Springer, 2015, (78), pp. 131-137.

[25] Anand Darji, Member, Shubham Agrawal, Ankit Oza, Vipul Sinha, Aditya Verma, S. N. Merchant, and A. N. Chandorkar, "*Dual-Scan Parallel Flipping Architecture for a Lifting-Based 2-D Discrete Wavelet Transform*", IEEE Transactions on Circuits and Systems-II: Express Briefs, 2014, 61(6), pp. 433-437.

[26] Hanaa ZainEldin, Mostafa A. Elhosseini, Hesham A. Ali, "*Image compression algorithms in wireless multimedia sensor networks: A survey*", Ain Shams Engineering Journal, Elsevier, 2015 (6), pp. 481-490.

[27] Tatsuya Murofushi, Toshiyuki Dobashi, Masahiro Iwahashi, and Hitoshi Kiya, "*An Integer Tone Mapping Operation for HDR Images in Openexr With Denormalized Numbers*", IEEE International Conference on Image Processing (ICIP), (2014) 4497-4501.

2018 31th International Conference on VLSI Design and 2018 17th International Conference on Embedded Systems

Hardware-Efficient and Wide-Band Frequency-Domain Energy Detector for Cognitive-Radio Wireless Network

Mahesh S. Murty
Center for VLSI & Embedded System Technologies
IIIT-Hyderabad
Hyderabad, India
Email: mahesh.murty@research.iiit.ac.in

Rahul Shrestha, *Member, IEEE*
School of Computing & Electrical Engineering
IIT-Mandi
Mandi, Himachal Pradesh, India
Email: rahul_shrestha@iitmandi.ac.in

Abstract—Cognitive radio is an innovative technology that possesses profound potential to mitigate spectrum scarcity in the next-generation wireless networks. In this technology, unlicensed users are capable of accessing unused portions of the spectrum, popularly known as *white spaces*, for communication that enhances that spectrum usage. Each of these unlicensed user nodes must incorporate spectrum sensor to detect white space for transmitting and receiving the information. Such spectrum sensors must be portable and it needs to support wide bandwidth suitable for contemporary wireless networks. This paper presents a new frequency-domain energy-detector for spectrum sensing that is hardware efficient. We propose VLSI architecture for this detector where the area and power consuming FFT unit has been replaced by an iterative FFT-computation unit with simple architecture. This area-efficient design also improves the performance of detector in real-world noisy environment. In addition, we extend the conventional single-band detection-approach to a wide-band detection approach and present its VLSI architecture. Performance analysis of such detection in AWGN channel environment showed that the detection probability of 0.16 is achieved at a SNR of -20 dB. Hardware implementation of the proposed energy detector is carried out using FPGA and its real world testing plus verification is performed using USRP software-defined radio-platform. It could operate with the bandwidth as high as 150 MHz with a sensing time of 7 ms. Compared to the state-of-the-art implementation, our prototype consumes 46% lesser hardware resources and does not require any memory in its architecture.

Keywords-Cognitive radio; spectrum sensing; energy detector; wireless network; OFDM; VLSI; FPGA; USRP;

I. INTRODUCTION

With a surge in the demand for higher data rate and rapid increase in the number of wireless devices, fixed spectrum allocation is a major limitation for the evolution of wireless technologies. Cognitive radio (CR) is a technology which coined the idea of opportunistic spectrum access. It proposes the utilization of temporally and spatially unused radio spectrum by the secondary users (SUs). In such CR network, primary users (PUs) and SUs must co-exist where the SUs make use of the radio spectrum opportunistically. In order to opportunistically access the spectrum, SU should be capable of determining whether the communication channel is free

to access or busy. Therefore, spectrum sensor aids such SU device to detect the presence of PU that keeps radio spectrum busy and unaccessible. Several spectrum sensing approaches have been reported in the literature [7], [11], [12], [8], [9] and among them the energy detection approach is widely employed in CR devices due to its lower implementation-complexity and shorter sensing-time. However, apart from these desirable features, energy detection based spectrum sensors deliver degraded performance under noisy channel environment [7], [6]. Furthermore, in order to reliably detect the PU signal, SU requires exact information of the noise variance to determine the detection threshold. Therefore, computation of noise variance in real time requires on-the-fly noise calibration [10].

There are several reported works on the hardware implementation of energy detector for spectrum sensing and their detection algorithms are broadly classified into time-domain and frequency-domain approaches. The time-domain based energy detection algorithms [11], [12] are implementation friendly and consume lesser hardware resources. However, it is highly prone to noise and its usage is deferred in practical applications. On the other side, energy detection algorithm for spectrum sensing in frequency domain results complex implementation consuming relatively more hardware resources. This is due to the fast Fourier transform (FFT) computation involved in this algorithm that estimates frequency content of the signal. It is robust towards noisy environments and delivers adequate performance in real world applications. However, frequency-domain approach is only applicable for narrow-band spectrum sensing and to sense wide band it requires more time. Albeit, wide range of frequency bands can be sensed using the time-domain approach with degraded noise performance. Thus, implementing reliable and efficient spectrum sensor imposes profound hardware-design challenge.

In this work, we intend to bridge this tradeoff between implementation and performance of energy detector for spectrum sensing. Firstly, we use Goertzel's algorithm [5] to recursively determine frequency-domain samples from the time domain signal that mitigates requirement of FFT

978-1-5386-3693-0/18 $31.00 © 2018 IEEE

277

computation in frequency-domain energy detector (FDED). Thus, this approach alleviates the hardware consumption of FDED drastically. Secondly, this work presents a new method of using adaptive-threshold computation approach to reliably estimate the threshold of proposed FDED that enhances its robustness. Thirdly, overall very-large scale-integration (VLSI) architecture of FDED has been presented that can be employed for various wide-band spectrum-sensing applications. Subsequently, we have analyzed the performance of suggested wide-band FDED using extensive simulation and its architecture has been hardware implemented using field-programmable gate-array (FPGA) device. Eventually, universal-software radio-peripheral (USRP) has been used in software-defined-radio (SDR) platform to verify the functionality of our FDED by processing real-world signals.

The remaining parts of this paper have been organized as follow: section II presents the brief mathematical background of spectrum sensing based on energy detection. Further, the proposed VLSI architectures based on iterative FFT-computation technique for FDED is presented in section III. Subsequently, section IV includes performance analysis, hardware implementation as well as verification in real-world scenario of our FDED design, followed by its comparison with the reported implementations. Finally, our paper is concluded in section V.

II. PRELIMINARIES

A digital signal $x(n)$ can be transformed from time to frequency domain by computing its discrete Fourier transform (DFT) and is given by

$$X(k) = \sum_{n=0}^{N-1} x(n) \cdot W^{nk} \qquad (1)$$

where W is twiddle factor which is expressed as

$$W = e^{-j(2\pi/N)}. \qquad (2)$$

The test statistic, used for sensing the presence of PU signal by FDED, is the power of frequency domain signal that is alternatively referred as power spectral density (PSD) of the signal. Thereby, such test statistic can be computed as

$$T(k) = \sum_{m=0}^{M-1} |\hat{X}_m(k)|^2 \qquad (3)$$

where this $T(k)$ represents the signal power in one FFT bin. Hence to sense the presence of PU signal on the entire bandwidth, it is necessary to run the detection test one-by-one on all frequency bins. The detection problem for each frequency bin can be modeled as a binary hypothesis test where hypothesis $H_0(k)$ stands for the presence of noise only and another hypothesis $H_1(k)$ indicates the presence of both signal and noise. Finally, the aforementioned test statistics $T(k)$ value is compared with the detection threshold

λ_k in order to decide between $H_0(k)$ and $H_1(k)$. Thus, the overall sensing time τ can be estimated as

$$\tau = (N \times M \times 1/F_S) \qquad (4)$$

where F_S is the sampling rate.

Energy detector performance is limited by the signal-to-noise ratio (SNR)-wall phenomenon which states that *in the presence of uncertainty, it is impossible to robustly distinguish signal from noise at SNR values lower than the SNR wall even if the sensing time tends to infinity* [6]. A major challenge involved in the realization of energy detection is to determine an appropriate threshold value that enables spectrum sensor to classify the test statistic result $T(k)$ in either $H_0(k)$ or $H_1(k)$ hypothesis. At the same time, determination of proper threshold value is also crucial. Otherwise, the energy detector may raise false alarms frequently which indeed degrades the detector performance. Thereby, in order to compute such threshold value reliably, we use an adaptive threshold calculation process with minimal requirement of hardware resources in our FDED design.

III. PROPOSED VLSI ARCHITECTURES

As discussed earlier, conventional FDED architecture incorporates FFT computation unit to transform the received time-domain data-samples into its equivalent frequency-domain samples. Such FFT unit must support very large size of N, which typically ranges from 1024 to 4096 points, in order to achieve adequate frequency resolution. Conventional VLSI architectures for FFT computation either consume huge hardware resources (butterfly units) or requires very high memory [3], [1], [2]. Additionally, butterfly units require complex multiplier which further increases area and power consumptions of the design. Therefore, several implementations have preferred time-domain energy-detection approach, though the frequency-domain energy detection being more robust. It is to be noted that running frequency-domain detection test over a complete bandwidth is not often preferred in practice, as it increases overall sensing time drastically. Instead, the detection tests are run over frequencies of interest which are small subset of the overall bandwidth. Hence in such application, high frequency resolution is desired that implies a higher point FFT is computed but a large number of frequency domain samples are under utilized.

A. Iterative FFT Computation

In this work, we propose an iterative method of computing FFT for the design of FDED using Goertzel's algorithm [5]. It is traditionally applied in the field of dual-tone multi-frequency (DTMF) signal detection [4] where the number of frequencies to be detected are very less. Moreover, this algorithm can iteratively compute the FFT output corresponding to any particular frequency component of any time-domain complex-input signal in N iterations

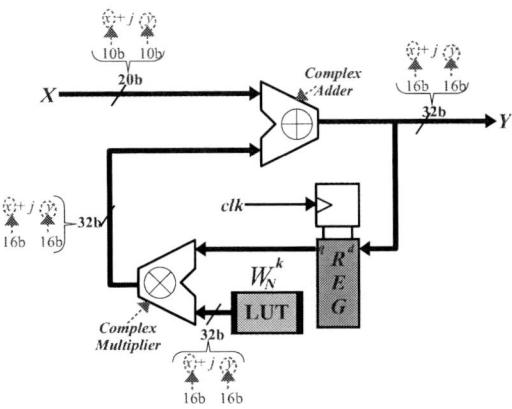

Figure 1. Proposed VLSI architecture for the iterative FFT computation unit.

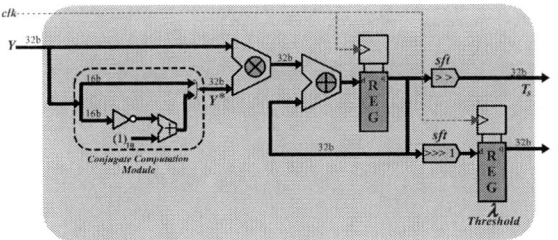

Figure 2. Proposed VLSI architecture for the test statistic & adaptive-threshold computation unit.

where N is the number of FFT points. From architectural perspective, it basically needs a multiply-and-accumulate (MAC) hardware unit comprising of a complex adder and a complex multiplier. This algorithm requires one memory element, as its present output depends on both present as well as past inputs. Subsequently, it requires only one twiddle factor, as an input and thereby, the size of read-only-memory (ROM) or look-up-table (LUT) to be used in our FDED design is also significantly reduced. Fig. 1 shows the proposed architecture for realizing the iterative FFT computation module. Furthermore, Algorithm 1 illustrates

---

Algorithm 1 : Iterative FFT computation algorithm

1: **procedure** FFT$_I$(X, w, N)
2:     $Y \leftarrow 0$
3:     $prevY \leftarrow 0$
4:     **for** $i \leftarrow 0$ to N **do**
5:         $Y \leftarrow prevY * w + X[i]$
6:         $prevY \leftarrow Y$
7:     **end for**
8:     **return** Y
9: **end procedure**

---

this iterative-FFT computation algorithm in detail where the variable X is an array of input signal, w is the constant twiddle factor and Y represents FFT output. The number of iterations N required to compute the output is dependent on the order of FFT computation required. This value of N can be decided at run time by the user. The FFT computation module incorporates a control unit internally which takes this N value as input and generates the required FFT output. Additionally, iterative FFT computation is beneficial for the VLSI implementation of FDED, as it does not impact the detection time (τ) of the spectrum sensor, given in (4), as the number of cycles required to compute FFT output equals to N which is the order of desired FFT. Thus,

this simple optimization reduces the amount of hardware required to implement FDED architecture, by significant amount, without degrading its performance.

B. Test Statistic & Adaptive-Threshold Computations

The FFT signals generated by iterative computation hardware-unit, as prior discussed, are fed to conjugate complex multiplier and accumulator. The output of accumulator after M samples is the required test-statistic value, as described in (3). This computed value of test statistic is normalized by dividing the accumulator by M which is the number of averages. It has been found that the value of M is always expressed in powers of two and hence we exclude the complex-division hardware unit and simply replaced it by a right bit-shifter module to perform division. The threshold value λ_k is computed by using the same hardware that is used for determining the test statistic value. Such λ_k value is to be computed by transmitting a known pilot signal through radio communication channel. Once the receiver is tuned and the detector is in noise calibration mode, the test statistic module will compute average test-statistic value and latches it using the output register. After M iterations, the value of threshold will be considered as $1/2$ the test statistic value and will be latched in the adaptive threshold register. Fig. 2 shows the detailed architecture of this module. Noise calibration mode can be selected by applying logic high to the *calib* control pin of this overall FDED architecture that will be discussed in next subsection.

C. Overall VLSI Architecture of FDED for Wideband Spectrum Sensing

Rather than scanning the whole frequency band for wideband spectrum sensing, we have chosen five widely-spaced different frequency bins. Here, the frequencies to be sensed can be selected by the end user by programming an appropriate value of the twiddle factor in the 16-bit twiddle-factor register. In our design, the FFT order to be computed is configured by programming a suitable value for the number of iterations (N) in the register. Thereby, the iterative FFT computation and test-statistic calculation hardware units those were discussed earlier are replicated 5× in order to compute the test statistic value $T(k)$. Fig. 3 shows the proposed

978-1-5386-3693-0/18 $31.00 © 2018 IEEE

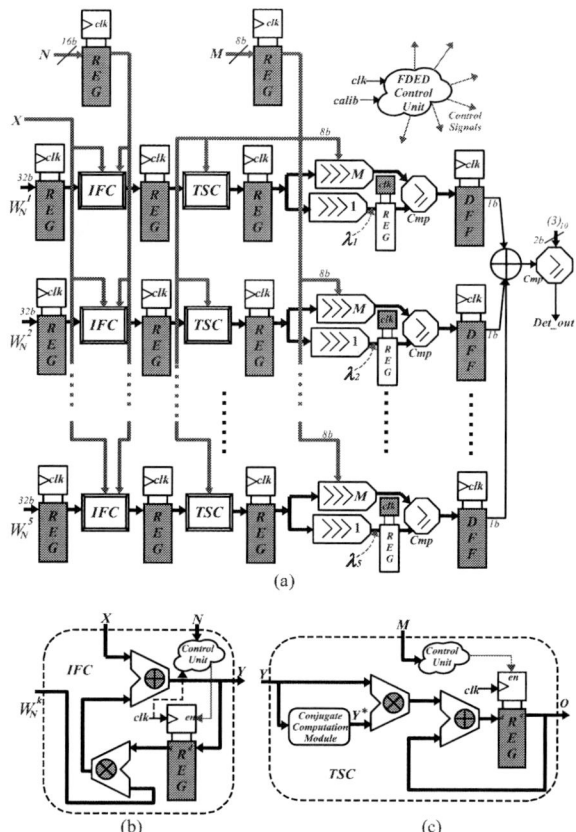

(a)

(b)   (c)

Figure 3. (a) Proposed overall VLSI-architecture of the hardware-efficient & wideband FDED for spectrum sensing in cognitive radio network. (b) Internal architecture of iterative FFT computation (IFC) unit. (c) Internal architecture of test statistics computation (TSC) unit.

and integrated VLSI-architecture of wide-band FDED for spectrum sensing in cognitive radio network. It includes five replicated chains of interconnected units like iterative FFT computation (IFC) unit, test statistics computation (TSC) unit, shifter, register (REG) and comparator (cmp). The configurable twiddle factors, N and M values are stored in registers and fed to IFC units, TSC units and shifters, as shown in Fig. 3. Additionally, *calib* control pin has been fed to the control unit of our FDED to generate required control signal to operate in noise calibration mode, as discussed earlier. Here, the test statistic output is compared with the adaptive threshold value (λ) that is computed using the noise calibration process. Therefore, the presence of primary user signal is considered only if at-least three out of those five-replicated hardware units report the presence of primary user, as shown in Fig. 3. Sensing a wideband spectrum in the conventional way would either require scanning the the complete FFT output $N\times$ where N is the FFT order which increases the wide-band sensing time by an order of N. In the proposed approach, the wide-band sensing time remains unaffected and the complete bandwidth can be sensed in τ

amount of time. However, the choice of frequency bins plays a significant role. If the frequency bins chosen are nearly spaced and the received signal is subjected to frequency selective fading, then the detector will fail to sense the spectrum. On the other side, if the frequencies selected are widely spaced then there are very less chances that at-least three frequencies will be subjected to frequency selective fading and hence the FDED can perform the detection process successfully. Also the number of frequency bins to be sensed is a critical topic of discussion. The number of bins sensed should be an odd value in order to make an distinct decision. If less number of frequency bins (for ex three) are selected, the detector will be prone to sensing errors and the wide-band sensing aim could not be accomplished, whereas of the number of sensing bins are increased, the hardware resources required will also increase by the same amount as the sensing chain is replicated those many times. Hence in this implementation we choose five frequency bins for wide-band spectrum sensing as it seems to be a suitable mid value both from performance and hardware resource consumption perspective.

IV. EXPERIMENTAL RESULTS AND COMPARISON

1) Performance Analysis: Extensive simulation has been carried out at different noise levels to analyze the performance of proposed wide-band FDED for spectrum sensing. The transmission side of this simulation model comprises of an orthogonal-frequency division-multiplexing (OFDM) transmitter that generates quadrature-phase shift-keying (QPSK) modulated signals using 512 subcarriers. Subsequently, it appends the cyclic prefix and transmits this OFDM signal. In order to simulate the effect of noise, the transmitted signal is passed through additive-white Gaussian-noise (AWGN) channel and then the channel output has been fed to the FDED algorithm. Here, the receiver processes received-complex digital I & Q samples as input and runs the proposed wideband frequency-domain energy detection algorithm. Such simulation process has been performed for different values of SNR to observe the effect of noise on detection performance. During this simulation, the FDED makes use of an N = 1024-point FFT and the averaging module calculates the average over 1024 (M=1024) samples. For every SNR value, Monte-Carlo simulations for 1000 iterations has been performed to obtain the reliable estimate of the probability of detection (P_d). Fig. 4 shows the P_d versus SNR plot generated from this simulation for N = 1024 and M = 1024. It can be observed that the proposed FDED algorithm delivers P_d of 0.16 and 1 at -20 dB and -11 dB, respectively, which are adequate for the cognitive-radio wireless network.

2) Hardware Implementation, Verification & Comparison: The hardware implementation of proposed wideband-FDED architecture has been carried out using an ALTERA

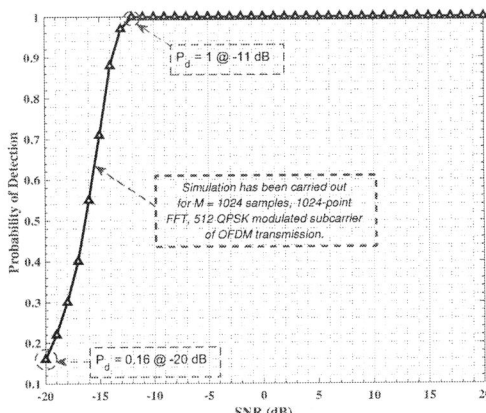

Figure 4. P_d versus SNR plot showing the performance analysis of proposed FDED algorithm in AWGN channel environment at various noise levels.

DE-1 FPGA board (with Cyclone-II device). The implementation results are reported in Table I. Additionally, it includes the comparison of our design with the reported FDED implementations in the literature. It can be observed that the suggested detector consumes least hardware resources among all. Specifically, our design requires 46% lesser LUT slices, 38.5% less number of multipliers and requires no memory. In addition, this prototype delivers a sensing time of 7 ms and bandwidth of 150 MHz at $N = M = 1024$. The analog radio-frequency (RF) sections of transmitter and receiver are implemented using two USRP SDR devices, as shown in Fig. 5. It also shows the system architecture of USRP transmitter that continuously transmits OFDM signal constructed using 512 subcarriers and cyclic prefix of length 64. The center frequency of this transmitter is 868 MHz and its sampling rate is 200 KHz. Additionally, Fig. 5 shows the system architecture of OFDM receiver implemented on another USRP device. This receiver is also tuned at the same center frequency of 868 MHz. Here, the received signal has been sampled at the rate of 1 MHz and decimated by a ratio of 5 to produce 200 KHz sampling rate. Further, the decimated signal is fed to a 128-tap finite-impulse-response (FIR) notch-filter to remove the DC component added to the received signal by amplifiers and other RF circuitry. Subsequently, this filtered IQ samples are stored in a file on disk. In order to verify the functionality of the detector samples of real world received signal captured using a USRP device are transferred from the file to on-chip read only memory (ROM) of the FPGA. The FPGA implemented detector fetches these values from ROM to process and generate the detection result. The hardware test-setup for this prototype has been shown in Fig. 6. It includes the above mentioned FPGA platform coupled with a Universal Serial Bus (USB) logic analyzer whose output can be seen on the

Table I
COMPARISON OF THE PROPOSED FDED IMPLEMENTATION WITH THE REPORTED WORKS

	[8]	[9]	**This Work**
Sensing Time (ms)	≤ 50	≤ 50	**7**
Sensing Bandwidth (MHz)	200	200	**150**
Gate Count	1.46M	-NA-	**-NA-**
Slices	-NA-	17.6K	**9.5K**
Multipliers	-NA-	39	**24**
Memory (bits)	106K	205K	**0**

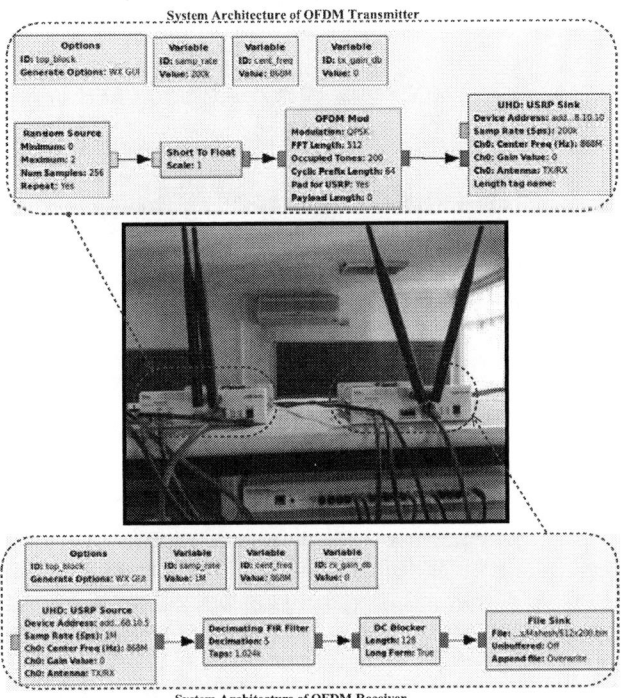

Figure 5. Test setup to capture real-world signals using USRPs as OFDM transmitter and receiver of cognitive-radio wireless network.

Personal Computer (PC) connected to it. A closer view of Altera DE-I FPGA shows that the on board switch is used as enable pin to initiate the detection process, as shown in Fig. 6(a). The detector-out (DetOut) and status (BusyPin) are mapped to pins on the General Purpose Input/Output (GPIO) header of the of FPGA and a USB logic-analyzer probe has been connected to generate output waveform, as shown in Fig. 6. Here, three channels represent clock, status (BusyPin) and detector-output (DetOut) signals. Once the detection begins, BusyPin outputs logic high and resets only after the detection is complete. After this period, the logic state of DetOut switches high or low indicating the presence

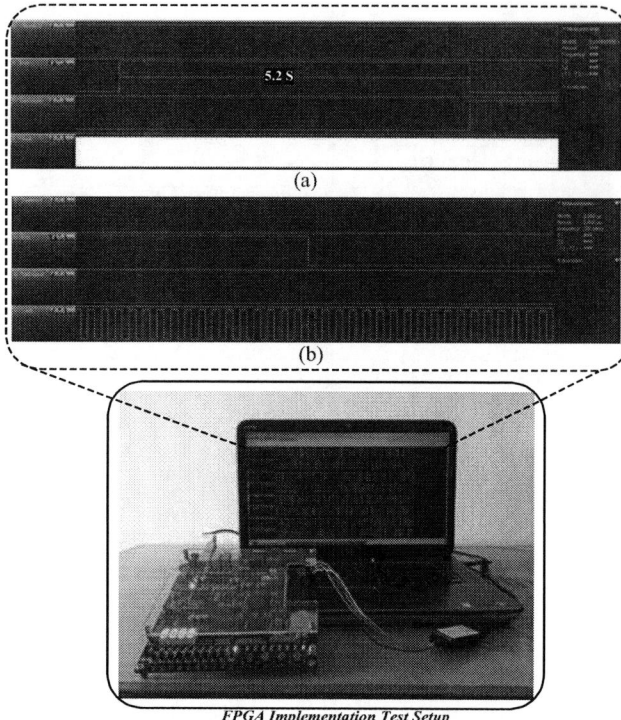

Figure 6. FPGA implementation test setup and functional verification of the proposed FDED for spectrum sensing.

or absence of PU respectively. In this work, the values of N and M are chosen to be 1024 and hence the total detection time required is around 5.2 S (denoted by logic high of BusyPin signal) at 200 KHz clock frequency, as shown in logic analyzer output-waveform from Fig. 6 (a). Logic low of DetOut signal after the detection period indicates the absence of PU, as shown in Fig. 6 (b). The detection time seems to be very large because the sample rate selected is very low (i.e. 200 KHz) due to the limitations of the measurement instruments. However the proposed detector architecture is capable of handling sample rates as high as 150 MHz and at such high sample rates the detection time would fall around 7 ms for the designed values of N and M.

V. CONCLUSION

This work proposed a wide-band spectrum sensor architecture based on the frequency-domain energy-detection algorithm for cognitive-radio wireless networks. Several architectural optimizations of the conventional energy-detection architectures were proposed that enabled to lower the implementation complexity and overall hardware consumption of the detector. This work primarily suggested the use of iterative FFT computation algorithm which aided to enhance the hardware efficiency drastically. Further, we proposed an adaptive threshold calculation technique whose implementation had minimal impact on the hardware and boosted the detector performance. Eventually, our architecture was

implemented on FPGA platform and the real world testing was carried out by utilizing USRPs as the RF front ends. The proposed architecture performs better than the conventional implementations, as it requires 46% lesser resources (logic plus registers) and excludes the usage of memory.

REFERENCES

[1] Y. N. Chang, "An Efficient VLSI Architecture for Normal I/O Order Pipeline FFT Design," *IEEE Transactions on Circuits and Systems II: Express Briefs*, vol. 55, no. 12, pp. 1234-1238, 2008.

[2] J. Garcia, J. A. Michell, and A. M. Buron, "VLSI Configurable Delay Commutator for a Pipeline Split Radix FFT Architecture," *IEEE Transactions on Signal Processing*, vol. 47, no. 11, pp. 3098-3107, 1999.

[3] D. Coelho, R. Cintra, N. Rajapaksha, G. Mendis, A. Madanayake, and V. Dimitrov, "DFT Computation using Gauss-Eisenstein Basis: FFT Algorithms and VLSI Architectures," *IEEE Transactions on Computers*, vol. 66, no. 8, pp. 1442-1448, 2017.

[4] M. J. Park, S. J. Lee, and D. H. Yoon, "Signal Detection and Analysis of DTMF Receiver with Quick Fourier Transform," *Industrial Electronics Society, 2004. IECON 2004. 30th Annual Conference of IEEE*, vol. 3, pp. 2058-2064, 2004.

[5] Sanjit K Mitra, *Digital Signal Processing. A Computer Based Approach*, 3rd ed., McGraw-Hill Companies, 2008.

[6] R. Tandra and A. Sahai, "SNR Walls for Signal Detection," *IEEE Journal of selected topics in Signal Processing*, vol. 2, no. 1, pp. 4-17, 2008.

[7] P. Sepidband and K. Entesari, "A CMOS Spectrum Sensor Based on Quasi-Cyclostationary Feature Detection for Cognitive Radios," *IEEE Transactions on Microwave Theory and Techniques*, vol. 63, no. 12, pp. 4098-4109, 2015.

[8] T. H. Yu, C. H. Yang, D. Cabric, and D. Markovic, "A 7.4-mW 200-MS/s wideband spectrum sensing digital baseband processor for cognitive radios," *IEEE Journal of Solid-State Circuits*, vol. 47, no. 9, pp. 2235-2245, 2012.

[9] T. H. Yu, O. Sekkat, S. Rodriguez-Parera, D. Markovic, and D. Cabric, "A Wideband Spectrum Sensing Processor with Adaptive Detection Threshold and Sensing Time," *IEEE Transactions on Circuits and Systems I: Regular Papers*, vol. 58, no. 11, pp. 2765-2775, 2011.

[10] R. Tandra and A. Sahai, "Noise Calibration, Delay Coherence and SNR Walls for Signal Detection," *New Frontiers in Dynamic Spectrum Access Networks, 2008. DySPAN 2008. 3rd IEEE Symposium on*, pp. 1-11, 2008.

[11] V. Khatri and G. Banerjee, "A 0.25-3.25GHz Wideband CMOS-RF Spectrum Sensor for Narrowband Energy Detection," *IEEE Transactions on Very Large Scale Integration (VLSI) Systems*, vol. 24, no. 9, pp. 2887-2898, 2016.

[12] P. D. Patangrao and P. P. Tasgaonkar, "VLSI Implementation of Energy Detection Algorithm for Cognitive Radio," *International Conference on Communication and Signal Processing*, pp. 1947-1949, 2016.

978-1-5386-3693-0/18 $31.00 © 2018 IEEE

Area and Power Efficient VLSI Architecture of Distributed Arithmetic Based LMS Adaptive Filter

M. Tasleem Khan
Student Member, IEEE
EEE Department
Indian Institute of Technology
Guwahati, India 781039
Email: tasleem@iitg.ernet.in

Dr. Shaik Rafi Ahamed
Associate Professor
EEE Department
Indian Institute of Technology
Guwahati, India 781039
Email: rafiahamed@iitg.ernet.in

Abstract—This paper presents a new area and power efficient VLSI architecture for least-mean-square (LMS) adaptive filter using distributed arithmetic (DA). Conventionally, DA based LMS adaptive filter requires look-up tables (LUTs) for filtering and weight updating operations. The size of LUTs grows exponentially with filter order. The proposed scheme has reduced the LUT size to half by storing the offset-binary-coding (OBC) combinations of filter weights and input samples. To make the adaptive filter more area and power efficient, it is not necessary to decompose LUT into two smaller LUTs. Hence, by using the non-decomposed LUT the proposed design achieves significant savings in area and power over the best existing scheme. In addition, the proposed architecture involves comparatively lesser hardware complexity for the same LUT-size. From synthesis results, it is found that the proposed design with 32nd order filter offers 19.83 % less area and consumes 20.54 % less power; utilizes 16.67 % and 19.04 % less number of LUT and FF respectively over the best existing scheme.

Index Terms—Distributed Arithmetic (DA), finite impulse response (FIR), look up table (LUT), offset binary coding (OBC).

I. INTRODUCTION

Adaptive filter is extensively used in noise and echo cancellation, system identification, channel estimation and equalization [1]. It comprises of a linear finite-impulse-response (FIR) filter whose transfer function is adjusted by changing the filter weights according to an optimization algorithm. Usually, least-mean-square (LMS) algorithm is preferred to update the filter weights due to its simplicity and ease of implementation. The combined FIR filter and the weight update unit consists of several multiply-and-accumulate (MAC) units depending upon the filter order. The computational efficiency of MAC based LMS adaptive filter is lower due to the large size of multipliers in MAC units. In most practical applications, the computational efficiency of any system can be improved by reducing the hardware complexity. One of the fundamental approach is *sequential processing* which reduces implementation complexity by using single computational unit over several number of clock cycles. But, it involves more latency, for example, if the filter order is N, then it would require at most N clock cycles.

Recently, distributed arithmetic (DA) is becoming popular due to its higher computational efficiency for the realization

of LMS adaptive filter. It was introduced by Croisier *et al.* [2]. Unlike MAC based FIR filter, DA has a look-up-table (LUT) and a shift-accumulate (SA) unit. The filter partial products are stored in LUT at various address locations. The filtering operation is performed by successive reading of LUT contents followed by shift-accumulation for some clock cycles depending on the precision of input sample. It is observed that the size of LUT grows exponentially with filter order. The LUT complexity is further increased when filter is made adaptive since each address location is need to be updated time-to-time. Hence, the complexity of DA based adaptive filter is mainly determined by the LUT size. Recently, several works [3]–[8] have been proposed for efficient implementation of DA based adaptive filter. In [3], an auxiliary LUT is employed to update the stored partial products of main LUT. This scheme has double LUT complexity for the implementation of DA based adaptive filter. To overcome this issue, authors of [4], [5] have proposed single LUT architecture for DA based adaptive filter. Since LUT complexity has been reduced, but it requires higher hardware complexity over [3]. To achieve the performance benefits of the works [3], [4], a new design has been suggested in [6] which is based on storing the offset binary coding (OBC) combinations of input samples and filter weights. But, it additionally decomposed the large sized LUT into two small multiplexed LUT for achieving the higher throughput over the designs [3], [4]. Since the multiplexed LUTs have been operated concurrently due to which power consumption of such design goes up. Moreover, the extra hardware for the LUT decomposition caused further increase in area and power. Recently, a pipelined implementation of DA based adaptive filter has been proposed [7], [8], which is based on the concept of parallel LUT. But, it has relatively higher hardware complexity of parallel LUT especially for large filter order. Recently, many new architectures have been investigated for DA based block LMS algorithm (BLMS) to improve their area and power efficiencies [9], [10]. To the best of our knowledge, no one has discussed the problems of area and power consumptions of DA based LMS adaptive filter when LUT decomposition is applied. Motivated by the works [3]–[8], a non-decomposed LUT architecture is proposed based on storing the OBC combinations of input samples and

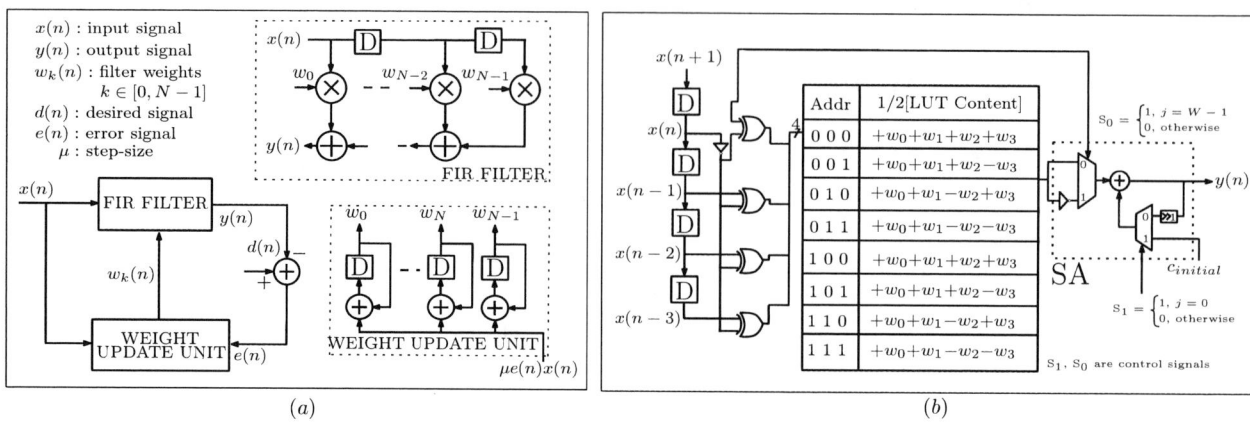

Fig. 1. (a) Block schematic of Nth order multiply-accumulate (MAC) based LMS adaptive filter (b) An 4th order implementation of LMS adaptive filter using offset-binary-coding (OBC) based distributed arithmetic (DA).

filter weights in two separate LUTs. In addition, we have also suggested a new scheme adaptation strategy to update the filter coefficients. The rest of the paper is organized as follows: In Section II, we present mathematical formulation of proposed design for DA based adaptive filter. In next Section, the architectural description of proposed design is presented. Section IV compares the performance of proposed and existing designs in terms of throughput, area and power. Conclusions are provided in Section V.

II. MATHEMATICAL FORMULATION

Consider a MAC based LMS adaptive filter in which the input samples $x(n-k)$ with $k \in [0, N-1]$ are processed by FIR filter, as shown in Fig. 1(a). And, the corresponding output $y(n)$ can be obtained, according to

$$y(n) = \sum_{k=0}^{N-1} w_k(n)x(n-k) \qquad (1)$$

where $w_k(n)$ are filter weights at time instant n. From (1), it can be noted that $y(n)$ is nothing but the multiplication of $w_k(n)$ and $x(n-k)$ followed by N successive accumulation. This has to be subtracted from the desired signal $d(n)$ to produce the error signal $e(n)$ as follows

$$e(n) = d(n) - y(n) \qquad (2)$$

By using (2) and input samples $x(n-k)$, the filter weights for the next iteration can be obtained by LMS criterion, as per

$$w_k(n+1) = w_k(n) + \mu e(n)x(n-k) \qquad (3)$$

where μ is step-size which adjusts convergence and mean-steady-state-error of adaptive filter. Usually, μ is selected in negative powers of two, so that the multiplication of μ and $e(n)$ in (3) can be performed by right-shift operation.

In order to implement the adaptive filter using DA, the input samples or the filter weights are to be represented in twos complement or offset binary coding (OBC). The proposed approach exploits the OBC combination of input samples. By

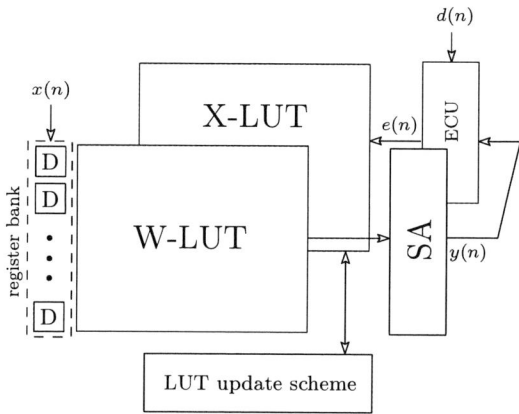

Fig. 2. System level diagram of proposed DA based LMS adaptive filter.

doing so, area and power efficient implementation of adaptive filter can be realized. Let us first consider twos complement representation of input samples which is given as

$$x_k = x(n-k) = -x_{k,W-1} + \sum_{j=1}^{W-1} x_{k,W-1-j} 2^{-j} \qquad (4)$$

From (4), it can be written as $x_k = 1/2[x_k - (-x_k)] = 1/2[x_k - \overline{x}_k]$ where $\overline{x}_k$ is twos complement of x_k. Hence, the expression for x_k becomes

$$x_k = \frac{1}{2}\Big[-(x_{k,W-1} - \overline{x}_{k,W-1}) + \\ \sum_{j=1}^{W-1} (x_{k,W-1-j} - \overline{x}_{k,W-1-j})2^{-j} - 2^{-(W-1)} \Big] \qquad (5)$$

the above expression commonly known as OBC scheme. Now, choose

$$d_{k,j} = \begin{cases} -(x_{k,j} - \overline{x_{k,j}}), & j \neq W-1 \\ -(x_{k,W-1} - \overline{x_{k,W-1}}), & j = W-1. \end{cases} \qquad (6)$$

978-1-5386-3693-0/18 $31.00 © 2018 IEEE

W-LUT	
Address	$1/2$[LUT Contents(n)]
0 0 0	$+w_0(n)+w_1(n)+w_2(n)+w_3(n)$
0 0 1	$+w_0(n)+w_1(n)+w_2(n)-w_3(n)$
0 1 0	$+w_0(n)+w_1(n)-w_2(n)+w_3(n)$
0 1 1	$+w_0(n)+w_1(n)-w_2(n)-w_3(n)$
1 0 0	$+w_0(n)-w_1(n)+w_2(n)+w_3(n)$
1 0 1	$+w_0(n)-w_1(n)+w_2(n)-w_3(n)$
1 1 0	$+w_0(n)-w_1(n)-w_2(n)+w_3(n)$
1 1 1	$+w_0(n)-w_1(n)-w_2(n)-w_3(n)$

X-LUT	
Address	$1/2$[LUT Contents(n)]
0 0 0	$x(n)+x(n-1)+x(n-2)+x(n-3)$
0 0 1	$x(n)+x(n-1)+x(n-2)-x(n-3)$
0 1 0	$x(n)+x(n-1)-x(n-2)+x(n-3)$
0 1 1	$x(n)+x(n-1)-x(n-2)-x(n-3)$
1 0 0	$x(n)-x(n-1)+x(n-2)+x(n-3)$
1 0 1	$x(n)-x(n-1)+x(n-2)-x(n-3)$
1 1 0	$x(n)-x(n-1)-x(n-2)+x(n-3)$
1 1 1	$x(n)-x(n-1)-x(n-2)-x(n-3)$

Fig. 3. Contents of W-LUT and X-LUT for 4th order FIR filter at time instant n.

By substituting (5) and (6) in (1) and re-arranging, we get

$$y(n) = \sum_{j=0}^{W-1} \left(\frac{1}{2} \sum_{k=0}^{N-1} w_k d_{k,j} \right) 2^{-j} - \frac{1}{2} \left(\sum_{k=0}^{N-1} w_k \right) 2^{-(W-1)}$$

(7)

Define

$$c_{W-1-j} = \frac{1}{2} \sum_{k=0}^{N-1} w_k d_{k,j}, \qquad 0 \le j \le W-1 \quad (8)$$

$$c_{initial} = -\frac{1}{2} \sum_{k=0}^{N-1} w_k \quad (9)$$

By substituting (8) and (9) in (7), we have

$$y(n) = \sum_{j=0}^{W-1} c_{W-1-j} 2^{-j} + c_{initial} 2^{-(W-1)} \quad (10)$$

From (8), it is clear that the term c_{W-1-j} represents the possible combinations of filter weights which are stored in LUT. For instance, if filter order is N, then c_{W-1-j} could take 2^N binary possible combinations of filter weights. However, due to the symmetry in OBC combinations, only 2^{N-1} half terms are required to store in LUT. The remaining half OBC combination can be obtained with the help of XOR gates, as shown in Fig. 1(b). This technique has an advantage in terms of low area, less power and less LUT access time. Unlike [3], the OBC combinations of filter weights are pre-computed and stored in two separate LUT. Notably, the least significant bits (LSBs) of registers form the address bits for filter weight LUT (W-LUT). Based on the combination of input samples bit-slices $x_{k,B-1-j}$ with $k \in [0, N-1]$, any combination of filter weights partial product can be accessed from LUT which after undergoes shift-accumulation.

So far the mathematical description of proposed scheme for filtering operation has been carried out. Now, consider the system level diagram of proposed design, as shown in Fig. 2. It comprises of two LUTs namely, weights updating LUT (W-LUT) and input samples updating LUT (X-LUT) which store their OBC combinations. The contents of W-LUT has to be updated from time-to-time. To do that, the proposed algorithm employed X-LUT which stores the OBC combinations of input

samples. Unlike weight adaptation of MAC based LMS filter (3), the weight adaptation of proposed scheme is performed at the contents of W-LUT using the contents of X-LUT, according to

$$\sum_{k=0}^{N-1} a_k^r w_k(n+1) = \sum_{k=0}^{N-1} a_k^r w_k(n) + \mu e(n) \sum_{k=0}^{N-1} a_k^r x(n-k)$$

(11)

where a_k^r is the kth bit in the N-bit representation of the address r. Mathematically,

$$r = \sum_{k=0}^{N-1} a_k^r 2^k \quad (12)$$

That is, OBC combinations of input samples are stored in X-LUT at time-instant n which later used to update the contents of W-LUT. By noting (7), the combined term $\mu e(n)$ can be quantized in powers of two [3] to simplify the multiplication by shifting operation.

III. Proposed Architecture

In order to carry out the necessary tasks required for filtering and weight updating operations. We have to understand the filtering operation of proposed design. Initially, all the input samples are stored in register bank with least-significant-bits (LSBs) of each register forms the address lines of W-LUT. The successive reading of contents from W-LUT and followed by shift-accumulation (SA) will produce the output, according to (10). The number of times shift-accumulation is performed depends on the wordlength of input samples. Since the term $c_{initial}$ has been taken care by SA unit in every first accumulation clock cycle. This is in accordance with (10) which requires a 2-to-1 multiplexer, as shown in Fig. 1(b). Note that the SA operation is performed in parallel with X-LUT update for either W or 2^{N-1} clock cycles, where W is wordlength of input samples. The output so obtained is subtracted from $d(n)$, according to (2). After that, the contents of X-LUT is updated whose output is multiplied with the product $\mu e(n)$. The terms $\mu e(n) \sum_{k=0}^{N-1} a_k^r x(n-k)$ will be added to the corresponding contents of W-LUT. In the proposed implementation, both the

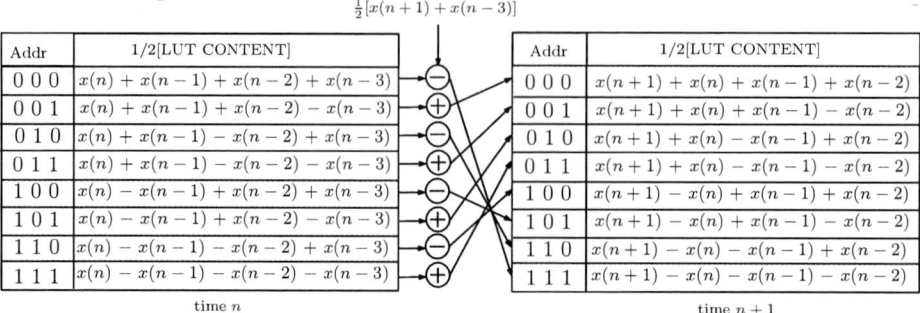

$\frac{1}{2}[x(n+1) + x(n-3)]$

Addr	1/2[LUT CONTENT]
0 0 0	$x(n) + x(n-1) + x(n-2) + x(n-3)$
0 0 1	$x(n) + x(n-1) + x(n-2) - x(n-3)$
0 1 0	$x(n) + x(n-1) - x(n-2) + x(n-3)$
0 1 1	$x(n) + x(n-1) - x(n-2) - x(n-3)$
1 0 0	$x(n) - x(n-1) + x(n-2) + x(n-3)$
1 0 1	$x(n) - x(n-1) + x(n-2) - x(n-3)$
1 1 0	$x(n) - x(n-1) - x(n-2) + x(n-3)$
1 1 1	$x(n) - x(n-1) - x(n-2) - x(n-3)$

time n

Addr	1/2[LUT CONTENT]
0 0 0	$x(n+1) + x(n) + x(n-1) + x(n-2)$
0 0 1	$x(n+1) + x(n) + x(n-1) - x(n-2)$
0 1 0	$x(n+1) + x(n) - x(n-1) + x(n-2)$
0 1 1	$x(n+1) + x(n) - x(n-1) - x(n-2)$
1 0 0	$x(n+1) - x(n) + x(n-1) + x(n-2)$
1 0 1	$x(n+1) - x(n) + x(n-1) - x(n-2)$
1 1 0	$x(n+1) - x(n) - x(n-1) + x(n-2)$
1 1 1	$x(n+1) - x(n) - x(n-1) - x(n-2)$

time $n + 1$

Fig. 4.   Proposed LUT update scheme for 4th of DA based adaptive filter.

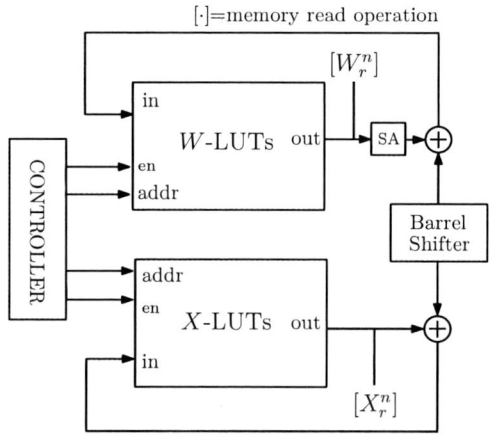

Fig. 5.   Detailed diagram of proposed architecture for DA based adaptive filter.

1: **loop**
   $y(n) = \sum_{j=0}^{W-1} c_{W-1-k} 2^{-k} + c_{initial} 2^{-(W-1)}$
2: **for** $r = 0$ to $2^N - 1$ **do**
   $c_{W-1-j} \leftarrow w_{k,W-1-j}(n) x_i(n)$
3:     **if** $r \bmod(2^{N-1}) == 0$ **then**
        $X_r(n+1) \leftarrow S + X_{2r+1}(n)$
4:     **else**
        $X_r(n+1) \leftarrow S - X_{2(2^N-1-r)}(n)$
5:     **end if**
6:     **end for**
7:     $e(n) \leftarrow d(n) - y(n)$
8:     **for** $r = 0$ to $2^{N-1} - 1$ **do**
       $W_r(n+1) = W_r(n) + \mu e(n) X_r(n)$
9:     **end for**
10:    **return** $y(n)$
11:    $n \leftarrow n + 1$
12: **end loop**

Fig. 6.   Algorithm explaining proposed DA based adaptive filter.

LUTs store similar OBC combination at time instant n, as shown in Fig. 3. Hence, the proposed design does not require external register and decomposition circuitry for LUTs, as in case of [6]. Thereby, it results in significant reduction of area and power consumptions. This is because when two LUTs operated in parallel, it comparatively consumes large power due to higher activity over a single non-decomposed LUT architecture for a given algorithm [2].

The X-LUT update scheme in the proposed implementation from time n to $n+1$ is explained as follows. The external term $x(n+1) + x(n-3)$ is used to map the contents of X-LUT from time instant n to the time instant $n + 1$. The contents of LUT for the next iteration must have different address locations. This can be understood with the following example: Consider a proposed 4th order filter with X-LUT update scheme, as shown in Fig. 4. At time instant n, the content at 0th address location of X-LUT is $\frac{1}{2}[x(n)+x(n-1)+x(n-2)+x(n-3)]$. This can be updated by subtracting from the external term $1/2[x(n+1) + x(n-3)]$ which would give $\frac{1}{2}[x(n+1) - x(n) - x(n-1) - x(n-2)]$. Interestingly, the term does not involve oldest sample $x(n-3)$. Hence, in similar manner, all the address locations of X-LUT are updated with the same external term $1/2[x(n+1) + x(n-3)]$. Mathematically, the contents of X-LUT at r^{th} address location at time instant $n+1$

$X_r(n+1)$ can be given as

$$X_r(n+1) = S + X_{2r+1}(n) \quad \text{with } r|r < 2^{N-1}$$
$$X_r(n+1) = S - X_{2(2^N-1-r)}(n) \quad \text{elsewhere} \quad (13)$$

where $S = \frac{1}{2}[x(n+1) + x(n-3)]$ denotes the external term to update the X-LUT contents. Once the process of X-LUT update over, the contents of updated X-LUT would be used to update the contents of W-LUT. This can be accomplished by an error signal $e(n)$, as computed in (2) followed by scaling with μ. Note that when filtering operation is completed using W-LUT and then again X-LUT is updated from X-LUT(n) to X-LUT$(n+1)$. Thus, one step of the filtering operation and adaptation at time n is completed. A similar explanation can be given for different time instances for each new sample.

The multiplication required in (11) can be realized by using a custom hardware multiplier. But, that requires significant area on the chip. In order to simplify this multiplication, we consider an approximation for the term $\mu e(n)$. It states that only most significant bit (MSB) of error signal has been taken into account. Thus, the area due to $\mu e(n)$ is reduced since it replaced the hardware multiplier by a simple barrel shifter,

TABLE I
Time and Hardware Complexity of Various Existing Schemes.

Design	Throughput	Adders (per cycle)	Registers	SH	Memory
DA$_0$ [3]	$1/[m_0(T_R + (k-1).T_M + T_A)]$	$m.(2^{k-1} + 2^k) + m.B - 1$	$m.(1+k) + 2$	m	$2m.(2^k - 1)$
DA$_1$ [4]	$1/[m_1(T_R + T_A)]$	$m.(2^{k-1} + k) + m.B - 1$	$m.(2 + 2k) + 1$	$m.k$	$m.(2^k - 1)$
DA$_2$ [6]	$1/[m_2(T_R + (k+1).T_M + 2T_A)]$	$m.(2^{k/2+1} + 2^{k/2+2} + 2) + m.B + 1$	$m.(4+k) + 2$	m	$m.2^k$
Proposed	$1/[m_0(T_R + (k-1)T_M + T_A)]$	$m.(2^{k-1} + 2^k) + m.B + 1$	$m.(1+k) + 2$	m	$m.2^k$

$N = m\times k$; B, W are wordlengths of input samples and filter weights; $m_0 = 2^k + \max(W, 2^{k-1}) + log_2 m$, $m_1 = 2^{k-1} + log_2 m + B + 1$, $m_2 = 2^{k/2} + \max(W, 2^{(k/2)-1}) + log_2 m + 1$, T_R = LUT access time, T_M = 2-to-1 multiplexer delay and T_A = adder delay. Note that DA$_2$ [6] scheme requires extra $2mk$ 2-to multiplexers. The design in [7] has been excluded since it is based on pipelined LMS algorithm.

as shown in Fig. 5. In other words, the combined product of X-LUT contents and $\mu e(n)$ is approximated as a right shift version of the X-LUT contents. By doing so, the throughput of proposed filter does not degrade, but it slightly degrades the convergence. The overall operation of proposed filter is explained with proposed algorithm in Fig. 6.

IV. Results and Discussions

For the sake of simplicity, we refer the designs in [3], [4] and [6] as DA$_0$, DA$_1$, DA$_2$ respectively. In addition, it is also assumed that each design has m sub-filter units order k such that $N = m \times k$ (N is a composite number). The expressions of throughput and hardware complexities of different designs are listed in Table I. The LUT complexity of DA$_0$, DA$_1$ and DA$_2$ designs are $m.(2^{k+1} - 2)$, $m.(2^k - 1)$, and $m.2^k$ respectively. While the proposed design has same LUT complexity as that of DA$_2$ scheme. But, DA$_1$ and DA$_2$ require different weight adaptation schemes. It may be noted that the proposed design is based on the OBC combinations of input samples and filter weights as similar to DA$_2$ scheme. However, the proposed design does not require decomposed LUT, as in case of [6]. This require lesser hardware complexity over the existing designs. For better clarity, we have shown explicit plots for adders (in Fig. 8) and registers (in Fig. 9) as required for the implementation of different designs. It can be noted that the proposed design provides 7.5 % less number of adders 20 % less registers for 32nd filter order respectively. Importantly, the proposed design does not require multiplexers for decomposing the LUT into two small LUTs, as in the case of DA$_2$ scheme. In addition, the proposed design also not requires an extra adder during the update of LUT, thus sampling period is reduced for the proposed design has been reduced over the DA$_2$ scheme.

Throughput of an adaptive filter is defined as the ratio of clock rate to the number of clock cycles required in processing the input sample. Mathematically,

$$\text{Throughput=Clock rate/Number of clock cycles} \qquad (14)$$

where Clock rate=1/Critical path. As derived, the critical path of proposed design is reduced by $T_A + 2T_M$ time units over the DA$_2$ scheme. This will eventually gives higher sampling rate which can be used to reduce the power consumption [3]. To reduce the critical path further, we can employ 3:2 compressor (or CSFA) followed by an carry propagation adder

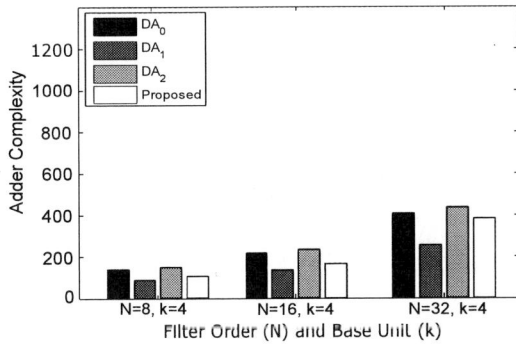

Fig. 7. Comparison of number of adders for proposed and existing designs.

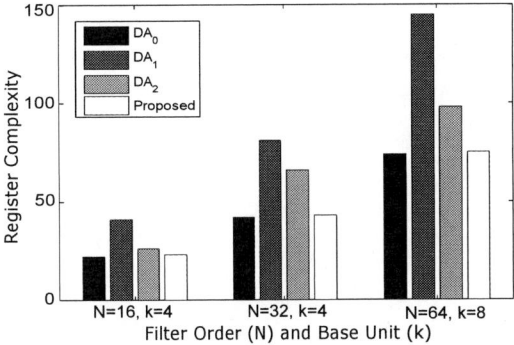

Fig. 8. Comparison of number of registers for proposed and existing designs.

[11]. Notably, the proposed designs have same number of clock cycles as that of DA$_0$ scheme. In order to verify the validity of proposed design, we performed the simulation in verilog. Subsequently, we carried-out application- specific-integrated-circuit (ASIC) synthesis to estimate area, power and throughput of the design using UMC 180 nm CMOS library by Cadence RTL Compiler for $N = 16$ and 32. The wordlength of input samples and filter weights were taken to be 8-bits. The estimated area and power consumptions for the proposed and existing designs are listed in Table II. From the listed results, it is clear that the area figures of proposed design are significantly reduced over the DA$_2$ scheme, especially for large k and N. This is due to fact that the proposed design exploits OBC combinations of filter weights and input samples, as similar to DA$_2$ scheme. In addition, the proposed design does not require multiplexers for the decomposition of LUT into

978-1-5386-3693-0/18 $31.00 © 2018 IEEE

TABLE II
COMPARISON OF AREA, POWER AND THROUGHPUT
FOR $W, B = 8$ AND $N = 16$ AND 32

Design	Area (mm²)		Power (mW)		Throughput (per μsec)	
	$L = 16$	$L = 32$	$L = 16$	$L = 32$	$L = 16$	$L = 32$
DA$_0$-ADF [3]	0.191	0.376	119.76	245.69	18.14	17.52
DA$_1$-ADF [4]	0.156	0.308	67.92	140.75	30.96	29.41
DA$_2$-ADF [6]	0.125	0.247	52.98	109.78	28.36	26.59
Proposed	0.102	0.198	44.13	87.23	20.54	19.67

two smaller LUTs, hence additional savings in area and power are achieved. For example, an 32nd order proposed filter with 4th base order units, the proposed design offers 19.83 % less area and 20.54 % less power over the DA$_2$ design for 16th order filter. The savings become even more for large order filter with large base order units. The power consumption of proposed design has lesser values over the DA$_2$ design since no decomposition of LUTs are exploited. Hence, the power consumption would have even lesser values over the DA$_2$ design. In addition, the OBC combinations of input samples and filter weights have less LUT requirements over DA$_0$ and DA$_1$ designs.

We have also performed field-programmable-gate-array (FPGA) synthesis on Altera Cyclone III EP3C55F484C6 at 100 MHz for 32nd filter order with 4th order base unit. And, the corresponding results in terms of slice LUT and flip-flop (FF) are shown in Fig. 10 and Fig. 11 respectively. Since the proposed design is based on OBC scheme, hence it offers significant savings in number of sliced LUT and FF over the existing designs. For example, an 32nd order filter, the savings in the number of sliced LUT and FF are 16.67 % and 19.04 % respectively over the DA$_2$ scheme.

V. CONCLUSION

In this paper, a new area and power efficient design for DA based LMS adaptive filter has been presented. The proposed approach is based on storing OBC combinations of input samples and filter weights in two separate LUTs. In the proposed implementation, the recent sample has been stored in LUT due to that decomposition of LUT is not possible, unlike the case of [6]. Thus, the savings in area and power are significant due to less hardware complexity and non-concurrent LUT update scheme over [6]. From synthesis results, it is shown that the proposed design with 32nd order occupies nearly 19.83 % less area and consumes 20.54 % less power; utilizes 16.67 % and 19.04 % less number of LUT and FF respectively over the best existing scheme.

REFERENCES

[1] S. Haykin, *Adaptive Filter Theory (3rd Ed.)*. Upper Saddle River, NJ, USA: Prentice-Hall, Inc., 1996.

[2] A. Croisier, D. Esteban, M. Levilion, and V. Riso, "Digital filter for pcm encoded signals," Dec. 4 1973, uS Patent 3,777,130.

[3] D. J. Allred, H. Yoo, V. Krishnan, W. Huang, and D. V. Anderson, "LMS adaptive filters using distributed arithmetic for high throughput," *Circuits and Systems I: Regular Papers, IEEE Transactions on*, vol. 52, no. 7, pp. 1327–1337, 2005.

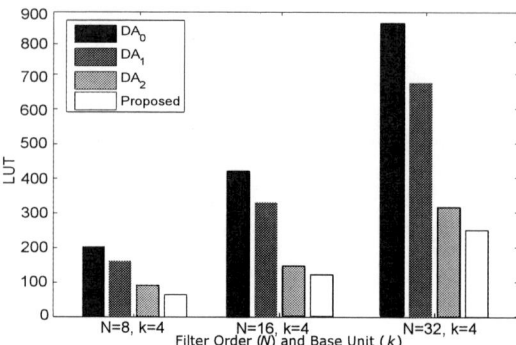

Fig. 9. Comparison of number of slice LUTs between proposed and existing designs.

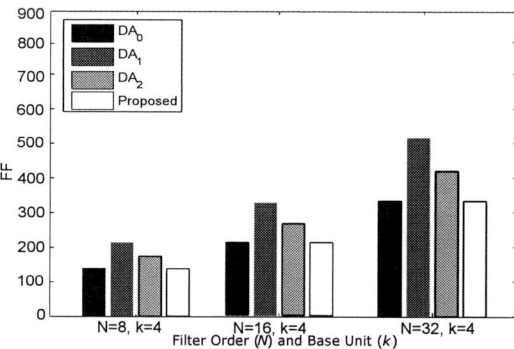

Fig. 10. Comparison of number of flip-flops (FF) between proposed and existing designs.

[4] R. Guo and L. S. DeBrunner, "Two high-performance adaptive filter implementation schemes using distributed arithmetic," *Circuits and Systems II: Express Briefs, IEEE Transactions on*, vol. 58, no. 9, pp. 600–604, 2011.

[5] Guo, Rui and DeBrunner, Linda S, "A novel adaptive filter implementation scheme using distributed arithmetic," in *Signals, Systems and Computers (ASILOMAR), 2011 Conference Record of the Forty Fifth Asilomar Conference on*. IEEE, 2011, pp. 160–164.

[6] M. S. Prakash and R. A. Shaik, "Low-area and high-throughput architecture for an adaptive filter using distributed arithmetic," *Circuits and Systems II: Express Briefs, IEEE Transactions on*, vol. 60, no. 11, pp. 781–785, 2013.

[7] S. Y. Park and P. K. Meher, "Low-power, high-throughput, and low-area adaptive fir filter based on distributed arithmetic," *Circuits and Systems II: Express Briefs, IEEE Transactions on*, vol. 60, no. 6, pp. 346–350, 2013.

[8] P. K. Meher and S. Y. Park, "High-throughput pipelined realization of adaptive fir filter based on distributed arithmetic," in *VLSI and System-on-Chip (VLSI-SoC), 2011 IEEE/IFIP 19th International Conference on*. IEEE, 2011, pp. 428–433.

[9] B. K. Mohanty, P. K. Meher, and S. K. Patel, "LUT optimization for distributed arithmetic-based block least mean square adaptive filter," *IEEE Transactions on Very Large Scale Integration (VLSI) Systems*, vol. 24, no. 5, pp. 1926–1935, 2016.

[10] B. K. Mohanty and P. K. Meher, "A high-performance energy-efficient architecture for fir adaptive filter based on new distributed arithmetic formulation of block lms algorithm," *IEEE transactions on signal processing*, vol. 61, no. 4, pp. 921–932, 2013.

[11] K. K. Parhi, *VLSI digital signal processing systems: design and implementation*. John Wiley & Sons, 2007.

978-1-5386-3693-0/18 $31.00 © 2018 IEEE

2018 31th International Conference on VLSI Design and 2018 17th International Conference on Embedded Systems

Novel High speed Vedic Multiplier proposal incorporating Adder based on Quaternary Signed Digit number system

Preyesh Dalmia, Vikas, Abhinav Parashar, Akshi Tomar and Dr. Neeta Pandey

Dept. of Electronics and Communication Engineering
Delhi Technological University (formerly Delhi College of Engineering)
New Delhi, India
{preyeshdalmia, vikas.dce2016, abhinavparashar.1810, akshitomar274}@gmail.com, neetapandey@dce.ac.in

Abstract— This paper presents a high-speed Vedic multiplier based on the Urdhva Tiryagbhyam sutra of Vedic mathematics that incorporates a novel adder based on Quaternary Signed digit number system. Three operations are inherent in multiplication: partial products generation, partial products reduction and addition. A fast adder architecture therefore greatly enhances the speed of the overall process. A Quaternary logic adder architecture is proposed that works on a hybrid of binary and quaternary number systems. A given binary string is first divided into quaternary digits of 2 bits each followed by parallel addition reducing the carry propagation delay. The design doesn't require a radix conversion module as the sum is directly generated in binary using the novel concept of an adjusting bit. The proposed multiplier design is compared with a Vedic multiplier based on multi voltage or multi value logic [MVL], Vedic Multiplier that incorporates a QSD adder with a conversion module for quaternary to binary conversion, Vedic multiplier that uses Carry Select Adder and a commonly used fast multiplication mechanism such as Booth multiplier. All these designs have been developed using Verilog HDL and synthesized by Synopsys Design Compiler. They have been realized using the open source NAN gate 15nm technology library. The proposal shows a maximum of 88.75% speed improvement with respect to Multi Value logic based 128x128 Vedic multiplier while the minimum is 17.47%.

Keywords-Multiplier; Quaternary Signed Digit adder [QSD]; Urdhva Tiryagbhyam; Vedic Mathematics

I. INTRODUCTION

One of the primary features that help us determine the computational power of a processor is the speed of its arithmetic unit. An important function of an arithmetic block is multiplication because, in most mathematical computations, it forms the bulk of the execution time. Thus, the development of a fast multiplier has been a key research area for a long time.

Some of the important algorithms proposed for fast multiplication in literature are Array, Booth and Wallace multipliers [1]-[5]. Vedic Mathematics [6, 7] is a methodology of arithmetic rules that allows for more efficient implementations regarding speed. Multiplication in this methodology consists of three steps: generation of partial products, reduction of partial products, and finally carry-propagate addition. Multiplier design based on Vedic mathematics has many advantages as the partial products and sums are generated in one step, which reduces the carry propagation from LSB to MSB. This feature helps in scaling the design for larger inputs without proportionally increasing

the propagation delay as all smaller blocks of the design work concurrently. References [8], [9] and [11] compared Vedic Multiplier with other multiplier architectures namely Booth, Array and Wallace on the basis of delay and power consumption. Vedic multiplier showed improvements in both the parameters over other architectures. Thus, many implementations of multiplication algorithms based on Vedic sutras have been reported in literature [10]-[12]. Vedic multiplier schemes proposed in literature are based on Urdhva Tiryagbhyam and Nikhilam sutras of Vedic Mathematics. As Nikhilam sutra is only efficient for inputs that are close to the power of 10, in this paper a design to perform high-speed multiplication based on the Urdhva Tiryagbhyam sutra of Vedic Mathematics which is generalized method for all numbers, has been presented.

The final step, carry-propagate addition, requires a fast adder scheme because it forms a part of the critical path. A variety of adder schemes have been proposed in literature to optimize the performance of Vedic multiplier [13]. Adder based on QSD shows an improvement in speed over other state of the art adders [14, 15]. Earlier implementations of QSD adder were based on Multi Voltage or Multi Value Logic (MVL) [16]. The difficulty in application of quaternary addition outside MVL (Multi Voltage logic) is that, the adder is only a small unit of the design whose outputs will needed to be converted back to binary for further processing. However, use of a conversion module undermines the advantages gained in speed by using QSD. In this paper, a novel implementation of an adder based on QSD is proposed, which reduces the carry propagation delay in the design by making use of carry free arithmetic. The proposed adder design works on a hybrid of binary and quaternary number systems wherein the sum is directly generated in binary using the concept of an adjusting bit, eliminating the conversion module. The design can be scaled to larger bit implementations such as 32, 64, 128 or more with minimal increase in propagation delay owing to the parallelism prevalent in the design. We have compared our design with a Vedic multiplier based on MVL logic that uses a ripple carry adder [16], Vedic Multiplier that incorporates a QSD adder and a conversion module for quaternary to binary conversion, Vedic multiplier that uses state of the art fast adder scheme such as Carry select adder [17] and a commonly used fast multiplication mechanism such as Booth multiplier [18], to prove the feasibility of our design across important comparison points.

This paper is organized as follows. Section II describes the Basic Terminology associated with our design. Section III describes the Proposed Multiplier architecture based on Vedic

978-1-5386-3693-0/18 $31.00 © 2018 IEEE

289

multiplication and Quaternary addition. Section IV comprises of Result in which device utilization summary and computational path delay obtained for the proposed Vedic multiplier (after synthesis) is discussed and Section V consists of Conclusion.

II. BASIC TERMINOLOGY

A. Urdhva Tiryagbhyam (UT) Sutra

The UT sutra is an ancient Vedic Mathematics sutra that can be used for multiplication of two numbers in any number system. It is based on "Vertical and Crosswise" multiplication. A 2x2 multiplier based on UT sutra is depicted in Fig. 1 and Fig. 2, where X and Y represent inputs while Z corresponds to output. Stepwise procedure is outlined below.

Step1: Vertical Multiplication: The least significant digits of the multiplicand and the multiplier are multiplied, as in (1).

$$Z0 = X0.Y0 \tag{1}$$

Step2: Crosswise Multiplication and Addition: Z1, in (2), is obtained by cross multiplying X1 and Y0, and Y1 and X0 and subsequently adding the two products. In this stage a carry C1, as in (3), might be generated, that is propagated to the next step.

$$Z1 = (X0.Y1) \oplus (X1.Y0) \tag{2}$$
$$C1 = X0.X1.Y0.Y1 \tag{3}$$

Step3: Vertical Multiplication and Addition: The most significant digits of the multiplicand and the multiplier are multiplied, and the product is added with the carry of the previous step to obtain Z3 and Z2, as in (4) and (5) respectively.

$$Z2 = (X1.Y1) \oplus C1 \tag{4}$$
$$Z3 = X0.X1.Y0.Y1 \tag{5}$$

The final result is concatenation of Z3, Z2, Z1 and Z0.

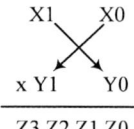

Fig. 1. Vertical and Crosswise multiplication

The logic circuit for 2x2 UT multiplier is shown Fig. 2.

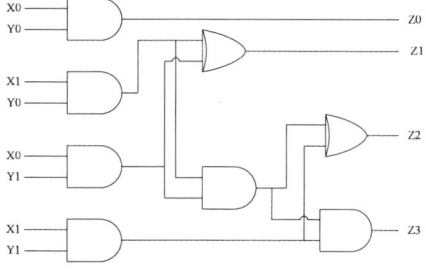

Fig. 2. 2x2 UT multiplier

B. Quaternary Signed Digit (QSD number system)

The QSD is a radix-4 number system that provides the benefit of faster arithmetic calculations over binary computation, as it eliminates rippling of carry during addition. Every number in QSD can be represented using digits from the set {-3,-2,-1, 0, 1, 2, 3}. Being a higher radix number system it utilizes less number of gates and hence saves on time and reduces circuit complexity. The stages involved in addition of two numbers in QSD are:

Stage1: Generation of intermediate carry and sum: When two digits are added in QSD number system, the resulting sum ranges between -6 to +6. Numbers with magnitude higher than 3 are represented by multiple digits with least significant digit representing sum and the next digit corresponds to carry. Also, every number in QSD can have multiple representations [14, 15]. The representation is chosen such that the magnitude of sum digit is 2 or less than 2 and the magnitude of carry digit is 1 or less than 1, the reason for which is explained in the next stage.

Stage2: The intermediate sum and carry have a limit fixed on their magnitude because this allows carry free addition in the second step. The result can be obtained directly by adding the sum digit with the carry of the lower significant digit [14, 15].

III. PROPOSED DESIGN

A. 4x4 Multiplier

Block diagram of a 4x4 multiplier is shown in Fig. 3. In this multiplier, four 2x2 multipliers are arranged systematically. Each multiplier accepts four input bits; two bits from multiplicand and other two bits from multiplier. Addition of partial products is done using two four bit Quaternary adders, a two-bit adder and a half adder. The final result is obtained by concatenating the least significant two bits of the first multiplier, four sum bits of the second four-bit Quaternary adder and the sum bits of two-bit adder.

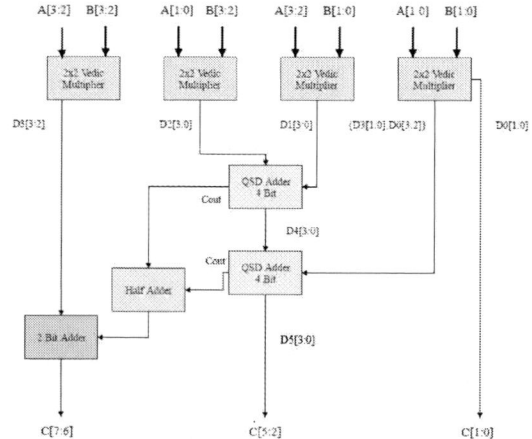

Fig. 3. Proposed 4x4 Multiplier

Table I shows all intermediate and final results involved in the multiplication process of two binary numbers, $A = (1111)_2$ and $B = (1001)_2$.

The data flow in the proposed 4x4 multiplier is given below:

1) A[1:0] and B[1:0], A[3:2] and B[1:0], A[1:0] and B[3:2], and A[3:2] and B[3:2] are multiplied by 2x2 Vedic multipliers, giving output D0[3:0], D1[3:0], D2[3:0] and D3[3:0] respectively.

978-1-5386-3693-0/18 $31.00 © 2018 IEEE

2) D1 [3:0] and D2[3:0] are added by the proposed 4 bit QSD adder, giving D4[3:0] and a carry out as the outputs.

3) D4[3:0] and {D3[1:0], D0[3:2]} are added by the second 4 bit QSD adder, giving D5[3:0] and a carry out as the outputs.

4) The half adder is used to add the carry outs of the QSD adders. The output obtained is fed to the 2 Bit Adder along with D3[3:2].

5) The result, C, in binary is obtained by concatenation of output of 2 Bit Adder, D5[3:0] and D0[1:0].

The proposed design can be extended to multiply both negative and positive integers by an addition of a sign bit in both inputs. An XOR logic can then be used to compute the sign bit of the final output. The multiplication of the magnitudes will proceed simultaneously in a similar manner to the example described above.

TABLE I. MULTIPLICATION RESULT OF TWO 4 BIT BINARY NUMBERS USING THE PROPOSED DESIGN

	Binary equivalent	Decimal equivalent	Explanation
A	$(1111)_2$	15	Input 1
B	$(1001)_2$	9	Input 2
D0	$(0011)_2$	3	Output of 2x2 Vedic Multiplier 1
D1	$(0011)_2$	3	Output of 2x2 Vedic Multiplier 2
D2	$(0110)_2$	6	Output of 2x2 Vedic Multiplier 3
D3	$(0110)_2$	6	Output of 2x2 Vedic Multiplier 4
D4	$(01001)_2$	9	Output of 4 bit QSD adder 1 (D1+D2)
D5	$(10001)_2$	17	Output of 4 bit QSD adder 2 (D4 +{D3[1:0],D0[3:2]})
C[1:0]	$(11)_2$	3	D0[1:0]
C[5:2]	$(0001)_2$	1	D5[3:0]
C[7:6]	$(10)_2$	2	Output of 2 Bit Adder (D3[3:2]+D4[4]+D5[4])
C[7:0]	$(10000111)_2$	135	Final Result

B. 32x32 multiplier

The 4x4 multiplier design can be scaled to multiply larger numbers as shown in Fig. 4, where the design is scaled up for a 32 bit multiplier.

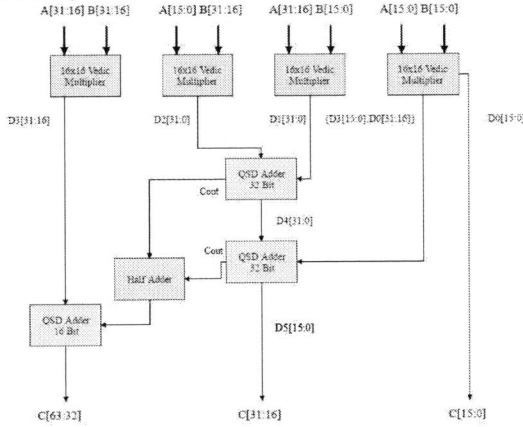

Fig. 4. Proposed 32x32 Multiplier

C. Proposed adder design based on QSD

In this paper, a novel idea of an adder, based on QSD (Quaternary Signed Digit) is proposed. The algorithm for the proposed adder uses a hybrid of quaternary and binary number systems. The outputs from smaller multipliers are obtained as binary strings. Inside the addition module, this string is broken into quaternary digits of two bits each. Addition using QSD allows us to reduce the carry propagation delay by making use of carry free arithmetic i.e. the carry doesn't ripple past the subsequent quaternary digit. Especially for higher bit input strings this method is extremely efficient.

The difficulty in application of quaternary addition outside MVL (Multi Voltage logic) is that the least significant 2 bits of the binary representation of the quaternary digits can't be directly concatenated to form an output binary string for every case as depicted in Table II. Each string would have to be read individually and a conversion module that converts quaternary to binary would have to be employed. To overcome this limitation, the concept of an adjusting bit has been introduced.

TABLE II. CONVERSION OF A QUATERNARY NUMBER TO BINARY NUMBER SYSTEM

Quaternary number A	$2\ 1 \rightarrow 010\_001$	Quaternary number B	$2\ \bar{1} \rightarrow 010\_111$
Binary equivalent of A	1001	Incorrect Binary equivalent of B	1011
Decimal equivalent of A	9	Incorrect Decimal equivalent of B	11

The Intermediate sum lies in the range [0, 6], as the operands are unsigned numbers. From [16], for quaternary addition to be carry free beyond the first stage, the intermediate sum can't be greater than 2. To ensure this stipulation holds true, the $(1\bar{1})_4$ representation of 3 needs to be chosen while adding. However, this represents a blocking case when converting the final output string back into binary as it prohibits us from simply concatenating the lower two bits of quaternary output strings to get the binary equivalent.

For addition of unsigned numbers, if the $(03)_4$ representation would have been used, direct concatenation of results could have been possible. But, then that wouldn't have always been carry free after the initial stage. Thus, the concept of an adjusting bit has been devised to solve the dilemma of which representation of 3 to use, such that both carry free addition and concatenation of output string bits to get the final output can be realized in the same design.

The solution to the problem described above, is that the $(03)_4$ representation of 3 is required to be taken instead of the $(1\bar{1})_4$ representation in some cases. But, determining when such a change is required before proceeding with the addition will increase the delay of the design and be counter-productive. Thus, the $(1\bar{1})_4$ representation of 3 is always selected in stage 1, to satisfy necessary conditions for carry free arithmetic. While necessary adjustments are made in stage 2 if $(03)_4$ representation was to be taken, the need for such an adjustment is determined via an adjusting bit.

OBSERVATION 1: In both quaternary representations of 3, $(03)_4$ or $(1\bar{1})_4$, the two least significant bits of the least significant digit are 11. Thus, regardless of which representation was supposed to be taken, the lower two bits of the intermediate sum will remain same and these are the two-bit positions that would be concatenated in the end.

The problem of incorrect representation would come under certain cases. To better understand these cases an example is described. The example uses two numbers as inputs represented using three quaternary digits each:

Input A= $(X_3X_2X_1)_4 = (A_8A_7A_6A_5A_4A_3A_2A_1A_0)_2 = (030)_4$
Input B = $(Y_3Y_2Y_1)_4 = (B_8B_7B_6B_5B_4B_3B_2B_1B_0)_2 = (001)_4$

The Base case: For addition of X_2 and Y_2, if the intermediate sum comes out to be 3, as stated above $(1\bar{1})_4$ representation will be chosen in stage 1. The Intermediate sum for this digit addition becomes $\bar{1}$ or $(111)_2$. If then the intermediate carry from the addition of X_1 and Y_1 is 0, the final output after stage 2 for this addition would be $\bar{1}$. The intermediate carry that will be added to addition of X_3 and Y_3 would be 1. The binary output thus received after the concatenation of lower two bits will be wrong, as shown in Table III.

As established above, this problem wouldn't have been there if the $(03)_4$ representation of 3 would have been used. According to the findings of observation 1, the intermediate carry from addition of X_2 and Y_2 needs to be negated for the correct result because for $(03)_4$ there would have been no carry. This negation will be done by the adjusting bit.

TABLE III. EXAMPLE OF QUATERNARY ADDITION USING ORIGINAL LOGIC

$X_3X_2X_1$	0 3 0 → 000_011_000	Input A
$Y_3Y_2Y_1$	0 0 1 → 000_000_001	Input B
U1	0 $\bar{1}$ 1 → 000_111_111	Stage 1 output (Intermediate sum)
U2	0 1 0	Stage 1 output (Intermediate carry)
R	1 $\bar{1}$ 1 → 001_111_001	Result (Before concatenation)
R'	$(01\ 11\ 01)_2$	Incorrect Result (After concatenation)

Mathematically this can be written as:

Final output = Intermediate sum + Intermediate carry – Adjusting bit.

Thus, adjusting bit can be said to be 1 when $(S_{n-1}.\bar{c})$ is true where S_{n-1} and $\bar{c}$ are defined as:

S_{n-1}: True if n-1th intermediate sum digit is 3.
$\bar{c}$: True if there is no carry from n-2th digits sum.

Secondly, another special case could arise when the intermediate sum for addition of X_2 and Y_2 and X_1 and Y_1 are both 3. For example if A = $(030)_4$ and B = $(003)_4$. Now as per previously devised logic the addition would have proceeded as in Table IV.

Thus, the final result as shown in Table IV, would have been $(01\ 11\ 11)_2$ which is incorrect. The intermediate carry from the addition of X_2 and Y_2 hasn't been negated while carry from addition of X_1 and Y_1 has. This is because

intermediate carry from X_1 and Y_1 is taken as 1 while calculating the adjusting bit for X_3 and Y_3 While an adjustment is made to it later to negate it to 0. This adjustment hasn't been factored into the formula. Thus, the modified and complete formula for adjusting bit becomes as in (10).

TABLE IV. EXAMPLE OF QUATERNARY ADDITION USING INITIAL MODIFIED LOGIC

$X_3X_2X_1$	0 3 0 → 000_011_000	Input A
$Y_3Y_2Y_1$	0 0 3 → 000_000_011	Input B
U1	0 $\bar{1}$ 1 → 000_111_111	Stage 1 output (Intermediate sum)
U2	0 1 1	Stage 1 output (Intermediate carry)
A	0 1 0	Adjusting Bit
R	1 $\bar{1}$ 1 → 001_111_111	Result (Before concatenation)
R'	$(01\ 11\ 11)_2$	Incorrect Result (After concatenation)

$$\text{Adjusting bit} = S_{n-1}.(S_{n-2}+\bar{c}) \tag{10}$$

Where S_{n-2} is true if n-2th intermediate sum digit is 3. This formula can cover the problem of n consecutive 3's in a similar manner.

The adjusting bit can be predicted based on the initial inputs to the adders itself. It can be computed in parallel with Stage 1. Thus, effect on delay of the adder is minimal. The above example is revaluated with the modified formula:

Input A= $(X_3X_2X_1)_4 = (A_8A_7A_6A_5A_4A_3A_2A_1A_0)_2 = (030)_4$
Input B = $(Y_3Y_2Y_1)_4 = (B_8B_7B_6B_5B_4B_3B_2B_1B_0)_2 = (003)_4$

Adjusting Bit for addition of X_n and Y_n is $S_{n-1}.(S_{n-2}+\bar{c})$. As can be seen from the flow of data shown in Table V. The modified formula gives the correct binary output after concatenation.

The proposed adder works in two stages, as shown in Fig. 5.

1) In the first stage, as in Fig. 5(a), every individual digit at the same position in the quaternary representation of two n-bit numbers A and B is added using a 2 Bit Adder to generate a sum. This sum lies in the range [0, 6]. From the sum obtained from the adder, the intermediate sum and intermediate carry for the next stage are calculated in parallel using 2x1 multiplexers. The logic for the selection of the representation of sum and carry has been explained in [16]. The adjusting bit is also computed in parallel with the addition process. The input to the adjusting bit calculation block for every quaternary digit addition are the previous two quaternary digits of A and B signified by [n-2: n-5].

2) Second stage has two modules as shown in Fig. 5(b). One is a one-bit module that performs the computation (A+B-C). In this case A would be LSB of intermediate sum, B would be carry from the previous quaternary digit addition and C would be the adjusting bit. The other module will be a half adder which will add the carry from the (A+B-C) module and the bit to the left of the least significant bit of the intermediate sum. As for the final concatenation, the sign bit would not be

used owing to the adjustments proposed in the design. Thus, its final value is not computed.

TABLE V.   EXAMPLE OF QUATERNARY ADDITION USING PROPOSED LOGIC

A	$(1100)_2$	Input 1
B	$(0011)_2$	Input 2
Q1	11_00 → 3 0	Quaternary representation of Binary number A
Q2	00_11 → 0 3	Quaternary representation of Binary number B
X_2X_1	3 0 → 011_000	Input A
Y_2Y_1	0 3 → 000_011	Input B
U1	$\bar{1}\,\bar{1}$ → 111_111	Stage 1 output (Intermediate sum)
U2	1 1	Stage 1 output (Intermediate carry)
S2	1	2nd intermediate sum digit is 3
S1	1	1st intermediate sum digit is 3
S0	0	0th digits do not exist
C1	1	Carry from sum of 2nd digits is 1
C0	1	Carry from sum of 1st digits is 1
A2	$S2.(S1+\overline{C1}) = 1$	2nd Adjusting Bit
A1	$S1.(S0 + \overline{C0}) = 0$	1st Adjusting Bit
A	1 0	Stage 1 output (Adjusting Bit)
U3	001_111_111	Stage 2 output (Before Adjusting Bit logic)
U4	000_111_111	Stage 2 output (After Adjusting Bit logic)
R	$(1111)_2$	Result after concatenation

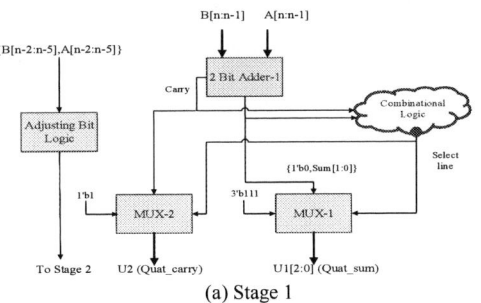

(a) Stage 1

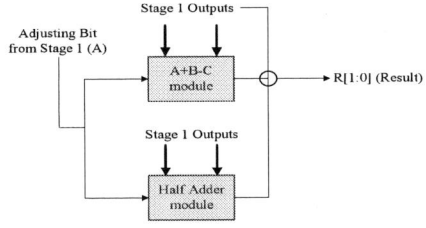

(b) Stage 2

Fig. 5. Proposed Adder

IV. RESULTS

In this section, we present a comparison between proposed design of multiplier and existing architectures namely Vedic multiplier based on MVL logic that uses a ripple carry adder [16], Vedic Multiplier that incorporates a QSD adder with a conversion module for quaternary to binary conversion, Vedic multiplier based on a different fast adder scheme such as Carry select adder [17] and a known fast multiplication scheme such as Booth multiplier [18]. These four architectures were chosen and implemented to verify the viability of proposed design across all domains it's pertinent to. All architectures are described using Verilog HDL and all the possible states including corner cases for digit by digit multiplication blocks are verified using simulation with Xilinx ISim simulator. The design synthesis has been carried out using Synopsys Design Compiler, using the open-source NAN gate 15nm technology library[19]. Table VI shows that proposed design has made substantial improvements in terms of speed over the existing designs. The total delay of 128x128 Multiplier based on Proposed Design comes out to be 578.85 ps which is 88.75% faster than booth multiplier,71.35% faster than MVL multiplier based Multiplier, 17.47% faster than Carry select adder based Multiplier and 51.69% faster when compared with QSD Adder based Vedic Multiplier using conversion module.

Proposed 128x128 design has 7.7% lower implementation area then CSA based Vedic multiplier but shows an increase in area over other three designs, as shown in Table VII, it can be considered as a tradeoff for the substantial improvement in speed over those designs. As shown in Table VIII, for 16 input bit value the proposed design consumes the lowest power amongst the designs compared. Whereas, for the larger input sizes, the power consumed by proposed designs is 25.14% and 20.64% more than the lowest recorded power amongst the designs compared for 64 bit and 128 bit respectively.

TABLE VI.   COMPARISON OF PROPOSED DESIGN WITH OTHER MULTIPLIER ARCHITECTURES ON THE BASIS OF TOTAL DELAY

Type Of Multiplier	Delay (ps)			
	16 bit	32 bit	64 bit	128 bit
Proposed Design	266.13	422.65	501.68	578.85
QSD Adder based Vedic Multiplier with conversion module	308.27	506.96	878.25	1198.21
CSA based Binary Vedic Multiplier	362.61	484.88	595.11	701.43
MVL Multiplier	431.18	949.25	1763.98	2020.71
Booth Multiplier	637.57	1259	2604	5148.56

TABLE VII.   COMPARISON OF PROPOSED DESIGN WITH OTHER MULTIPLIER ARCHITECTURES ON THE BASIS OF AREA

Type Of Multiplier	Area (No. of Cells)			
	16 bit	32 bit	64 bit	128 bit
Proposed Design	768	3475	14440.6	58842.6
QSD Adder based Vedic Multiplier with conversion module	678.9	2660.8	12303	50181.9
CSA based Binary Vedic Multiplier	938	3884.2	15801	63767
MVL Multiplier	432	1765.1	7212	29464.2
Booth Multiplier	605	2332	8977	35987

978-1-5386-3693-0/18 $31.00 © 2018 IEEE

TABLE VIII.   COMPARISON OF PROPOSED DESIGN WITH OTHER
MULTIPLIER ARCHITECTURES ON THE BASIS OF POWER

Type Of Multiplier	Power (mW)			
	16 bit	32 bit	64 bit	128 bit
Proposed Design	737.72	4446.7	24132	99630
QSDA based Vedic Multiplier with conversion module	774.35	4066.3	20958	87176
CSA based Binary Vedic Multiplier	1024.4	4558.7	19284	82580
MVL Multiplier	912.5	4722.2	21189	96478
Booth Multiplier	800.62	4627.1	19659	86556

V. CONCLUSION

It can be concluded that the design when scaled to higher bits only shows a marginal rise in delay due to its core strengths. Firstly, the parallelism involved in its partial product generation. Secondly, reduction of carry propagation delay in the novel adder it incorporates. Due to the use of QSD, the design is able to incorporate carry free arithmetic while eliminating radix conversion module speed overhead by integrating concept of adjusting bit logic in its architecture.

The proposed design showed an increase in implementation area over some designs due to increased parallelism even in finer nuances of the architecture. The proposed design is targeted towards digital systems requiring high throughput and low latency at the cost of area overhead. For example, in a DSP system, operations such as Fast Fourier Transform, Convolution, Filtering and Discrete Wavelet transform etc. Multipliers play a key role in determining the speed of the system. Similarly, this architecture would be a good candidate to be implemented as a large part of systems like DCT, Central Processing Unit (CPU), MAC (Multiply and Accumulate) Unit, Image Processors where high-speed multiplications are critical to the performance of the system.

It can also be observed that despite the objective of decreasing the delay, the proposed design performs better than most designs compared in terms of power for lower input bit sizes [16 and 32 bit]. Although it consumes more power than other designs higher input bit sizes [64 and 128 bit], it is justifiable when factored in with advantages gained in speed for higher input bits.

REFERENCES

[1] M. Rabaey, A. Chandrakasan, and B. Nikolic, "Digital Integrated Circuits A Design Perspective," PHI, 2003.

[2] B. Pahrami, "Computer Arithmetic and Hardware Design," New York, Oxford University Press, 2000.

[3] M. Ercegovac, and T. Lang, "Digital Arithmetic, San Francisco, Morgan Kaufmann," 2004.

[4] C S Wallace, "A Suggestion for a Fast multiplier", IEEE Transactions on Electronic Computers, Vol. EC-13, Issue 1, pp. 14-17, 1964.

[5] K. Choi and M. Song, "Design of a high performance 32x32-bit multiplier with a novel sign select booth Encoder," in IEEE International Symposium on Circuits and Systems, Volume 2, 2001, pp. 701-704.

[6] J. Swami S. B. K.Tirthaji Maharaja, "Vedic Mathematics: Sixteen Simple Mathematical Formulae from the Veda," Delhi, 1965.

[7] S. N. A and K. N, "Implementation of Power Efficient Vedic Multiplier," International Journal of Computer Applications (0975 – 8887), Vol. 43– No.16, 2012, pp. 21-24.

[8] S. Vaidya and D. Dandekar, "Delay-Power Performance Comparison of Multipliers In VLSI Circuit Design," International Journal of Computer Networks & Communications (IJCNC), Vol.2, No.4, July 2-010.

[9] H. Thapliyal and M. B. Srinivas, "High Speed Efficient N X N Bit Parallel Hierarchical Overlay Multiplier Architecture Based On Ancient Indian Vedic Mathematics", Enformatika (Transactions on Engineering, Computing and Technology),Vol. 2, Dec 2004, pp. 225-228.

[10] H. D. Tiwari, G. Gankhuyag, C. M. Kim and Y. B. Cho, "Multiplier design based on ancient Indian Vedic Mathematics," International SoC Design Conference, 2008, pp. 65-68.

[11] D. Jaina, K. Sethi and R. Panda, "Vedic Mathematics based multiply accumulate Unit," IEEE International Conference on Computational Intelligence and Communication Systems.2011, pp. 754-757.

[12] S. Jinesh, P. Ramesh and J. Thomas, "Implementation of 64 bit high speed multiplier for DSP application- based on Vedic mathematics," in IEEE TENCON,2015, pp. 1-5.

[13] J. Thomas, R Pushpangadan, S Jinesh, "Comparative study of performance of vedic multiplier on the basis of adders used," IEEE-WIECON, 2015, pp. 325-328.

[14] R. Rani, L. K. Singh and N. Sharma, "FPGA Implementation of Fast Adders using Quaternary Signed Digit Number System," 2009 International Conference on Emerging Trends in Electronic and Photonic Devices & Systems (ELECTRO-2009), pp. 132 - 135.

[15] Nagamani A. N, Nishchai S, "Quaternary High Performance Arithmetic Logic Unit Design", 14th Euromicro Conference on Digital System Design 2011 IEEE.

[16] A. S. Shende, M. A. Gaikwad and D. R. Dandekar, "Design of efficient 4X4 Quaternary Vedic Multiplier Using Current Mode Multi Valued Logic," Int. J. on recent Trends in Engineering and Technology,Vol 10, No 2, Jan. 2014, pp. 59-69.

[17] G. R. Gokhale, S. R. Gokhale, "Design of area and delay efficient Vedic multiplier using Carry Select Adder," International Conference on Information Processing (ICIP), 2015, pp. 295 – 300.

[18] S. Kim and K. Cho, "Design of High-speed Modified Booth Multipliers Operating at GHz Ranges," World Academy of Science, Engineering and Technology 61, 2010, pp. 1-4.

[19] M. Martins, J. M. Matos, R. P. Ribas, A. Reis, G. Schlinker, L. Rech and J. Michelsen, "Open Cell Library in 15nm FreePDK Technology", In Proceedings of the International Symposium on Physical Design (ISPD), 2015.

2018 31th International Conference on VLSI Design and 2018 17th International Conference on Embedded Systems

Design Space Exploration of an Execution-Driven Functional Simulation Methodology

Ipsita Biswas Mahapatra[1], Utkarsh Agarwal[2], Chandrashekhar Azad[3] and S. K. Nandy[4]

[1]Centre for Nano Science and Engineering, Indian Institute of Science, Bangalore, India 560012
[2]Adobe Systems, Bangalore, India 560012, [3]Bhagalpur College of Engineering, Bhagalpur, India 813210
[4]Deparment of Computational and Data Sciences, Indian Institute of Science, Bangalore, India 560012
Email: ipsita.mahapatra@gmail.com, utkarsh.ag2@gmail.com, cazad3011@gmail.com, nandy@iisc.ac.in

Abstract—Exploration of an efficient functional simulation methodology that has the capability to encounter conflicting conditions such as: maximizing hardware occupancy using efficient partitioning and mapping algorithms and minimizing inter hardware communication using optimized hardware dimensions, is very important. In this paper, we explore the design space of an execution-driven functional simulation methodology named EX-DRIVE. It performs functional simulation of a design under test (DUT) without the need for hardware synthesis and implementation of the DUT, offering significant improvement in functional simulation time. To realize this methodology we use a Network of Interconnected HyperCells (NIHC) as the meta platform. We explore the design space of EX-DRIVE for various dimensions of NIHC fabric and different partitioning and mapping algorithms. For this study we investigate six different hardware dimensions having a fixed hardware capacity and three partitioning and mapping algorithms: a Discrete Particle Swarm Optimization based algorithm (DPSO), a heuristic and a convex algorithm. We find that, for a fixed hardware capacity, the heuristic and convex algorithm proves to be more efficient for large and densely connected DUTs whereas the DPSO based algorithm proves to be more efficient for smaller and sparsely connected data flow graphs. The proposed algorithms are generic enough to be applied to any coarse grained re-configurable array assisted functional simulation platform.

Keywords-VLSI; Functional simulation; Execution-driven simulation; Re-configurable hardware

I. INTRODUCTION

The enormous increase in complexity of recent Very Large Scale Integration (VLSI) circuits has only been possible due to efficient Design Space Exploration (DSE) of functional simulation systems. Hence, exploration of an optimized system that has the capability to encounter conflicting conditions such as: maximizing hardware occupancy using efficient partitioning and mapping algorithms and minimizing inter hardware communication using optimized hardware dimensions, is extremely significant. In this paper, we present the DSE of our in-house functional simulation system, named EX-DRIVE, for metrics that are important from design point of view.

EX-DRIVE (EXecution-DRIVEn functional simulation), an execution-driven functional simulation system, performs "execution" of a new or revised DUT of a digital design, without synthesizing and implementing the DUT onto a simulation platform. Traditionally, in simulation acceleration, emulation and FPGA prototype, the DUT is mapped through hardware

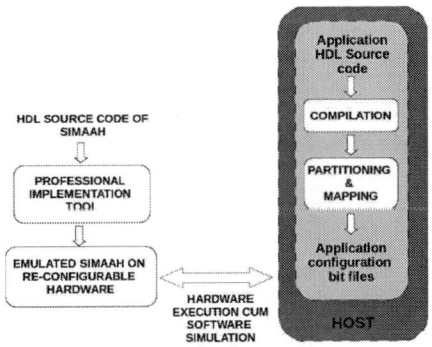

Fig. 1. Proposed approach for EX-DRIVE

synthesis (HS) and implementation, to a FPGA or an ASIC or a processor for performing its functional simulation. In our approach, the mapping of DUT is achieved during the compilation of the DUT for EX-DRIVE. The DUT is in the form of a Data Flow Graph (DFG). For functional simulation of a DUT, we configure a target hardware as per the DFG of the DUT, using a partitioning and mapping algorithm (Figure 1). The interconnection and functional units of the target hardware gets configured as per the edges and nodes of the DFG.

To realize the proposed methodology we use a Network of Interconnected HyperCells (NIHC) as the meta platform. A HyperCell (HC) [1][2] is an array of highly parallel hardware structural units, comprising of compute units (CU), crossbar switches and memory units (Figure 2). For every new or revised DUT considered for functional simulation, we reconfigure the CUs for functionality and the interconnects for connectivity, thereby executing the register transfer operations of the DUT on the array of interconnected HCs.

In our framework, dimension of NIHC fabric and the adopted partitioning and mapping algorithm are two important metrics which has a direct impact on the performance of EX-DRIVE. Hence it becomes inevitable to determine the most optimized dimension of NIHC fabric and the most efficient partitioning and mapping algorithm that will yield the best performance for EX-DRIVE. This calls for a proficient DSE of EX-DRIVE by which we can come up with the

978-1-5386-3693-0/18 $31.00 © 2018 IEEE

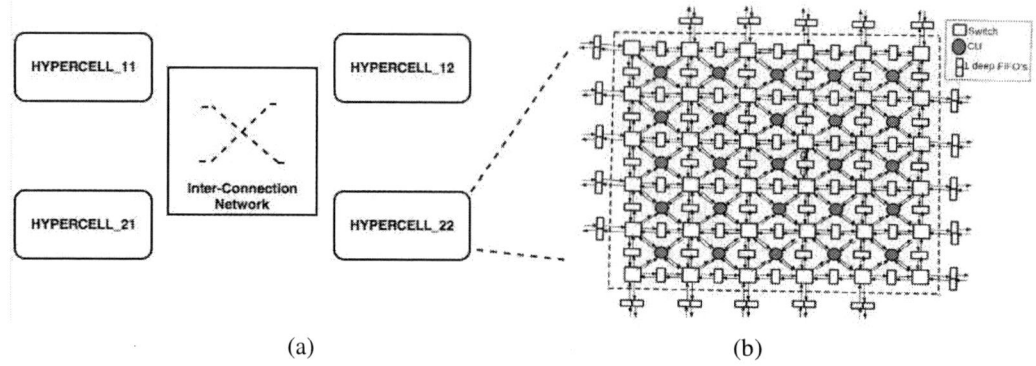

Fig. 2. (a) NIHC architecture, (b) HyperCell

most suitable HC and CU network dimension and the most efficient partitioning and mapping algorithm. Palermo et al. [3][4] proposed to find approximate pareto points over a design space as a method for proficient DSE. Pareto efficiency represents a state where it is difficult to enhance quality of a design objective without making no less than one other design target more worse off. We make use of pareto efficiency and investigate six different hardware dimensions having a fixed hardware capacity and three partitioning and mapping algorithms: DPSO, heuristic and convex algorithm.

II. EXPLORING PARTITIONING AND MAPPING ALGORITHMS

In the past few years, various functional simulation flows has been proposed and developed along with their associated partitioning and mapping techniques. These techniques belong to two main categories: modulo scheduling [5] and scheduling a single loop iteration thereby exploiting inter-iteration parallelism of an individual loop [6]. In our framework, NIHC architecture is neither fixed like ASICs nor variable like FPGAs. Hence, the existing partitioning and mapping algorithms can be employed in our work, but the fact is, it has to be customized for our functional simulation system. In this section, we present three algorithms: a Discrete Particle Swarm Optimization based partitioning and mapping algorithm (DPSO-PA), an effective partitioning & mapping heuristic and a convex algorithm, because the heuristic and convex algorithm appears to be more efficient for large and densely connected DFGs whereas the DPSO-PA proves to be more efficient for smaller and sparsely connected DFGs.

All the proposed algorithms partition DFG representation of structural verilog source code. Verilog offers design hierarchy information through modules and their instances. In this work, we exploit this information and combine it with the proposed algorithms. Hence, we partition modules/instances instead of gates. The proposed algorithms perform allocation and binding of variables and operations of DUTs into a simulation platform. The scheduling step remains implicit in our approach.

The proposed algorithms can be applied to any coarse grained reconfigurable array (CGRA) assisted functional simu-

lation platform whose internal architecture comprise a two dimensional $r \times c$ matrix, which in turn consists of k ($k = r \times c$) programmable devices. In order to explore the design space of EX-DRIVE for the proposed partitioning and mapping algorithms, we adopt the following dimension of NIHC fabric: a 3×3 array of HC and each HC is a 5×5 array of CUs.

A. DPSO based partitioning and mapping

PSO [7][8], is an algorithm for solving continuous optimization problems. Hence, it is necessary to adapt PSO to the discrete problem space. In $G(V, E)$, nodes are numbered from 1 to N. Each element of E is defined as (i, j), such that $i \epsilon \{1, ..., N\}$ and $j \epsilon \{1, ..., N\}$. As we are looking for mapping of $G(V, E)$, we divide $G(V, E)$ into smaller multiple subgraphs of the form $G'(V', E')$, where $V' \subset V$ and $E' \subset E$. Each subgraphs consist of, approximately, m number of nodes where, $m < N$. Such a subgraph is called a "Partition" of $G(V, E)$. We allocate the nodes of an individual partition set to the CUs of a HC. This forms a "Mapping" of $G'(V', E')$ and manifests as a particle in the DPSO-PA formulation. DPSO-PA is iteratively applied on all such $G'(V', E')$ which collectively constitutes $G(V, E)$. The position of a particles is a one dimensional vector which denotes the CU number of the selected HC to which a nodes of $G(V, E)$ has been mapped. Hence the search space is defined as the finite set of all possible "Mappings".

First step: Considering the above relationship between DPSO problem space and DFG partitioning and mapping problem space, let the swarm, in DPSO-PA, consists of a group of particles $(M_1, M_2, ..., M_n)$. First an initial set of particles M_i is created randomly in order to form a population. Particle's positions are updated in a fixed number of iterations, and the position of the best particle is searched using a fitness function, F. The value returned by F represents the total time taken by the corresponding HC to execute the functionalities of all the nodes mapped onto it. Lower the execution time higher will be the value of F.

Second step: The k_{th} particle represents a N dimensional vector X_k, where ($k = 1, 2, ..., n$). It means that the k_{th} particle with 'flying' velocity $V_k = (V_k^1, V_k^2, ..., V_k^D)$, is located at $X_k = (X_k^1, X_k^2, ..., X_k^D)$, where $(k = 1, 2, ..., n)$ in the searching space. Particle's positions are updated in a fixed number of iterations i, and the position of the best particle is searched using F. i accounts for the number of iterations for which the DPSO-PA optimization procedure is carried out. We have chosen i to be 1000 because the best value of $G_{bestposition}$ is achieved at this value. There is a velocity associated with each particle. It is a swap operator that updates the position of the particle at every iteration using the following equation:

$$V_k^{i+1} = c_1 \times V_k^i + rand \times c_2 \times (P^i_{k\,bestposition}) + rand \times c_3 \times (G^i_{bestposition})$$

Here, P is a particle. c_1, c_2 and c_3 are constants called acceleration coefficients. We have chosen $c_1 = c_2 = c_3 = 2$. $rand$ is a function which is used to generate random numbers in between the range $[0, 1]$. V_k^{i+1} and V_k^i are the velocities of particle k at $(i+1)th$ and $(i)th$ iterations. $P^i_{k\,bestposition}$ is the best position attained by particle k at $(i)th$ iteration and $G^i_{bestposition}$ is the global best position of the $(i)th$ iteration. The $P_{bestposition}$ is the particle's best position at which the best fitness_value (F) has been accomplished so far. The $G_{bestposition}$ is the global best position over all the particles, recorded so far, which is stored as the best F of that iteration. The position of a particle updates itself in the following way:

$$X_k^{i+1} = X_k^i + V_k^{i+1}$$

Here, X_k^{i+1} and X_k^i are the positions of particle k at $(i+1)th$ and $(i)th$ iterations. V_k^{i+1} is the velocity of particle k at $(i+1)th$ iteration.

If the updated position of a particle is better than its own $P_{bestposition}$ and $G_{bestposition}$ of that population, then both the $P_{bestposition}$ and $G_{bestposition}$ gets updated. The convergence factor lies in the fact that with each iteration we get better or similar $P_{bestposition}$ and $G_{bestposition}$. The final position information, associated with the particle displaying $G_{bestposition}$, is used for mapping the nodes of the current partition set onto the CUs of the corresponding HC. The final output of the DPSO-PA is the address of the CU and the HC, associated with each node of $G(V, E)$.

B. Heuristic based mapping

The proposed execution-driven functional simulation of a DUT is achieved through allocation and binding of hardware units of the DUT onto NIHC. The partitioning and scheduling steps are implicit in this method.

Mapping phase: Algorithm 1 explains the proposed mapping heuristic. In this work, a function called "find_node" is used to search the root nodes. Root nodes are the nodes which donot have any children. These nodes serve as input to the mapping process. A combination of forward and backward BFS algorithm is used to map the nodes of $G(V, E)$ onto a HC. The mapping process stops with a break operator. This scenario arises when no more hardware units are available for mapping the root nodes returned by the "find_node" function.

Algorithm 1 "Map" function

---

function Map

Call function Check, Find_Node

for *i:=0; i<number of partition sets; i++* **do**

Create a new partition set to place the nodes

while *number of nodes in the current partition set < 20* **do**

Apply Backward BFS on the present node

Place all non allocated ancestors of the present node into the current partition set

Apply Backward BFS again to allocate non allocated ancestors into the current partition set

When all the connected components of the graph gets completely mapped onto current HC

Call "Find_Node" function

end

The mapping process also stops when the "map" function returns 0. This case arises when all the CUs adjacent to the previously mapped CUs are occupied. In such a situation, though there are other CUs available for mapping, the heuristic will not be able to supply a CU for mapping the nodes under consideration.

Routing phase: Routing is done using "minimum hop" criterion. "Hop" is a numerical value which represents the number of interconnection units used while establishing connections between the consumed CUs of a HC. Hence, the minimum hop path, involving the least number of interconnection units, is the shortest path between two CUs among which connection has to be established.

C. Convex partitioning and mapping algorithm

In this section, we perform graph partitioning using graph covering algorithm. The proposed algorithm partitions a given application DFG into two or more sub graphs such that the total weight of edges interconnecting these sub graphs are minimized and then check for convexity of the partition sets thus produced. While doing this we also maintain a given balance constraint among the sizes of the partition sets. In this section, we also present an algorithm for minimizing the critical path length of a partition set which violates convexity (i.e non convex partition set) and for calculating the longest depth of each partition set whether convex or non convex. Our proposed algorithm for graph partitioning is based on FMS algorithm [9]. In FMS algorithm, balance criterion for size of partition sets states that size of each partition set should be in a close integral range. But, we adopt a balance criteria which allows the size of each partition set within the range of $[0, n']$, where n' is the total number of nodes present in a resource graph $R(V', E')$.

The proposed algorithm comprises two main parts. The first part performs graph partitioning while the second part deals with the detection of violation of convexity and calculation of the longest depth of partition set Π of $G(V, E)$. The following steps explain the algorithm in details.

1) We randomly partition $G(V, E)$ into direct k way partition $\Pi = (P_1, \ldots, P_k)$, while maintaining the balance constraints of each partition set P_i (i.e. $L(P_i) \leq w(P_i) \leq U(P_i)$, $L(P_i) = 0$ and $U(P_i) = n'$). Here, $L(P_i)$ and $U(P_i)$ are integer lower and upper bounds on the size of partition set P_i, respectively. $w(P_i)$ is the size of each partition set P_i. We take upper bound $U(P_i)$ of partition set P_i equal to n' and lower bound $L(P_i)$ of partition set P_i equal to 0, because each partition set should be covered by the resource graph $R(V', E')$.

2) We start with an initial partition set $\Pi = (P_1, \ldots, P_k)$. We calculate the initial cost of each node and assume that the final cut size of graph $G(V, E)$ is equal to the initial cut size. At each iteration, during a pass, we consider all possible moves of each unlocked nodes, belonging to its own partition set, to any of the other $(k - 1)$ partition sets. We choose the move having the best maximum move gain and update the node gain and move gain of all effected nodes. We carry out such passes until no improvement in cut size is obtained.

3) For each i, $i = i$ to k, we calculate every path p between an input node, $v_i \epsilon V$ and an output node, $v_o \epsilon V$ of partition set P_i. We check whether any node of path p belongs to P_t where $P_i \neq P_t$. If such a node is found then we add partition set P_i to non convex partition set $P_{nc} \epsilon \Pi$ else we add it to convex partition set $P_c \epsilon \Pi$. We then choose the following parameters: longest path L_i among every path p of partition set P_i, longest depth L_{nc} of non convex partition set P_{nc} among the longest depth L_i of each partition set $P_i \epsilon P_{nc}$, $1 \leq i \leq k$ and longest depth L_c of convex partition set P_c among the longest depth L_i of each partition set $P_i \epsilon P_c$, $1 \leq i \leq k$.

4) We increment the value of iteration (itr) by 1. Steps 1, 2, 3 and 4 are repeated till (itr) becomes maximum or $L_{nc} \leq L_c$.

III. Exploring various sizes of NIHC fabric

In this section we will explore the design space of EX-DRIVE for various dimensions of NIHC fabric. A NIHC fabric comprises a 3×3 array of HC and each HC is a 5×5 array of CUs. Let us consider that the HC fabric, represented as M, has dimensions $x \times y$. Similarly, the CU fabric, represented as N, has dimensions $p \times q$. For any fixed hardware capacity C, where, $C = M \times N$, we need to find out an unique combination of xy and pq which will produce the best performance of EX-DRIVE.

Now, let us consider a fixed hardware capacity of 225. This implies that there are 225 CUs available on NIHC fabric, for mapping the nodes of a DFG of any test case. Hence, following are a set of dimensions that will allow all the 225 CUs to be used in mapping:

- Hypercell network: 3×1, 5×5, 3×3, 9×1
- CU network: 3×25, 5×15, 3×15, 5×9, 5×5

Here, 5×5 and 3×3 are symmetrical networks of HC whereas the others are unsymmetrical networks of HC. Hence, by exploring both the options of symmetrical and unsymmetrical

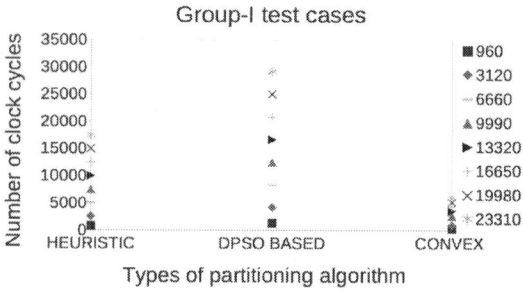

(a)

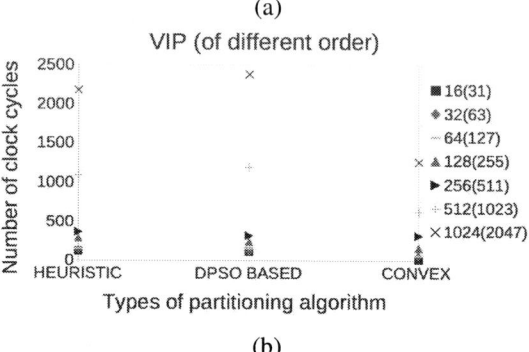

(b)

Fig. 3. Number of CC required to execute Group-I test cases & VIP while adopting each algorithm.

networks of HC, we can prove that EX-DRIVE works for both the types of NIHC fabric.

IV. Results and discussions

In order to prove the effectiveness and viability of the proposed functional simulation flow, we have performed experiments on 32 different test cases. These test cases can be classified into three major categories (Table 1):

- Group I - Large dis-connected DFGs (with number of nodes varying from 1000 - 23000) of hypothetical test cases
- Group II - Large connected DFGs (with number of nodes varying from 30 - 8000) of mathematical applications
- Group III - small DFGs (with number of nodes varying from 18 - 300) of benchmark circuits from [10]

The partitioning and mapping algorithms are developed in C11 (GNU C Compiler - gcc 4.8.2). It has been implemented on an Intel(R) Core(TM) CPU (i7-4770) with a 3.40 GHz processor and 32.00 GB of RAM, running on the CentOS 6.5 operating system. To perform a comparative study, we compute the number of clock cycles (CC) required by EX-DRIVE to execute the test cases while adopting each fabric dimension and algorithm. For this, we perform logic synthesis of NIHC fabric on a VIRTEX6 XCVLX760T device.

For exploring the proposed algorithms, we configure NIHC fabric for the test cases using the proposed algorithms. We then perform their individual functional simulation using Xilinx ISIM software, on a Linux server (Intel Xeon E3-1230 V2 @ 3.30GHz, 32.00 GB RAM, Cent OS 6.6). The numbers thus computed for Group-I and Group-II test cases has been

TABLE I

TEST CASES AND NUMBER OF ASSOCIATED NODES (VIP - VECTOR INNER PRODUCT, VMP - VECTOR MATRIX PRODUCT, MMP - MATRIX MATRIX PRODUCT)

Group I - Large disconnected DFGs		Group II - Large connected DFGs				Group III - Small DFGs	
Hypothetical test cases	Number of nodes	Mathematical applications	Number of nodes	Mathematical applications	Number of nodes	Benchmark circuits	Number of nodes
DFG_1	999	VIP (n=16)	31	VMP (n=4)	28	Mesa horner bezier	18
DFG_2	3330	VIP (n=32)	63	VMP (n=8)	120	Auto regression filter	28
DFG_3	6660	VIP (n=64)	127	VMP (n=16)	496	Mpeg motion vector	32
DFG_4	9990	VIP (n=128)	255	VMP (n=32)	2016	Mesa feedback points	53
DFG_5	13320	VIP (n=256)	511	MMP (n=2)	12	Cosine 2	82
DFG_6	16650	VIP (n=512)	1023	MMP (n=4)	112	Mesa interpolate aux	108
DFG_7	19980	VIP (n=1024)	2047	MMP (n=8)	960	Mesa smooth triangle	197
DFG_8	23310	VMP (n=2)	6	MMP (n=16)	7936	Mesa invert matrix	333

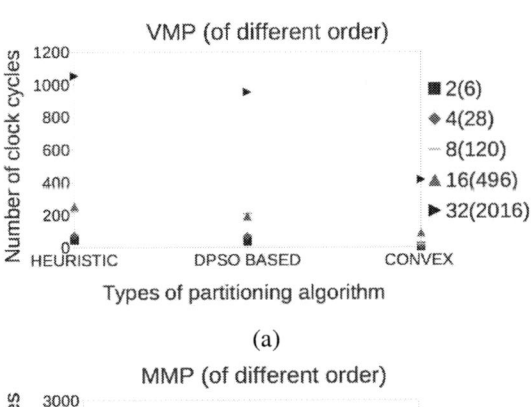

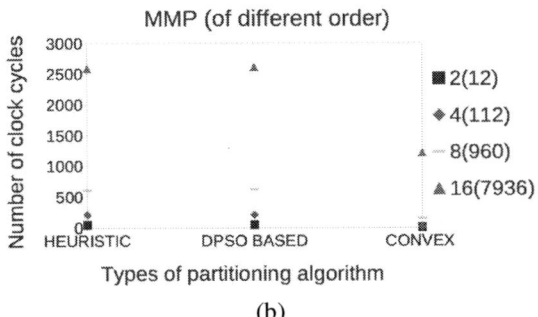

Fig. 4. Number of CC required to execute VMP & MMP while adopting each algorithm.

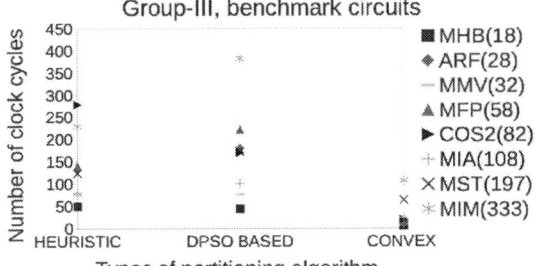

Fig. 5. Number of CC required to execute Group-III benchmarks while adopting each algorithm.

shown in Figure 3, 4 & 5. The legend of the plots depicts the number of nodes of the DFGs of the respective test cases. As can be seen in Figure 3(a), minimum number of CCs gets consumed when convex scheme is adopted and DPSO-PA scheme consumes the maximum CCs. For the Group-II test cases (Figure 3(b) & 4), smaller test cases donot exhibit much variation in the number of CCs. But as the test cases grow in size, best performance in terms of number of CCs is again achieved by convex scheme. Among DPSO-PA and heuristic, with increasing size of test cases, number of CCs consumed increases rapidly in the case of DPSO-PA. This trend continues to prevail even for VMP and MMP respectively. But in these two cases, what we notice is that, the performance of heuristic degrades gradually for higher dimensions of test cases. This is due to the increase in number of interconnections as the DFGs of higher dimensions are more densely populated. Figure 5 shows the plots for very small DFGs i.e, the Group III benchmarks. Here also, the above trend prevails. As the test cases grows in size, number of CCs consumed increases rapidly in the case of DPSO-PA, whereas, the number of CCs decreases gradually while adopting convex algorithm. Hence, in general, we can infer that, given any type of DFGs i.e, small or big, simple or complex, the convex partitioning and mapping algorithm performs better as compared to the DPSO-PA and heuristic based schemes. Among heuristic and DPSO-PA, for very small test cases with lesser interconnections, DPSO-PA performs better than heuristic. But as the test cases grows large and complex in size, the heuristic offers better performance.

For exploring NIHC fabric size, we compute the number of CCs required to execute the test cases by all the NIHC fabric dimensions considered above, while adopting the convex algorithm. The numbers computed for Group-I and Group-II test cases has been shown in Figure 6 & 7. The legend of the plots depicts the number of nodes of the DFGs. As can be seen in Figure 6(a), except for the smallest test case, all other test cases gets executed in least number of CCs when we use the 3×3 HC fabric comprising of 5×5 CU network. For VIP (Figure 6(b)) also, minimum number of CC gets

978-1-5386-3693-0/18 $31.00 © 2018 IEEE

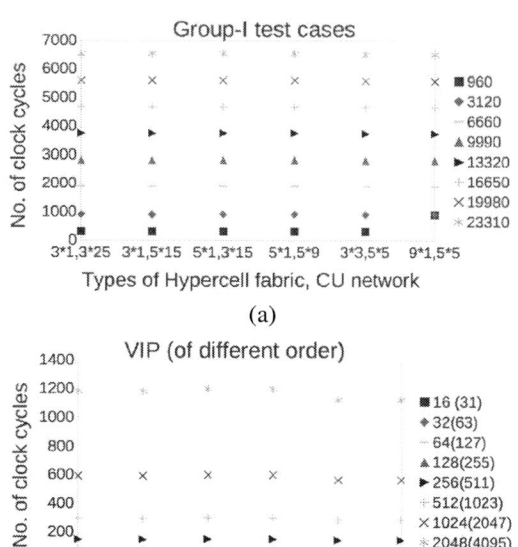

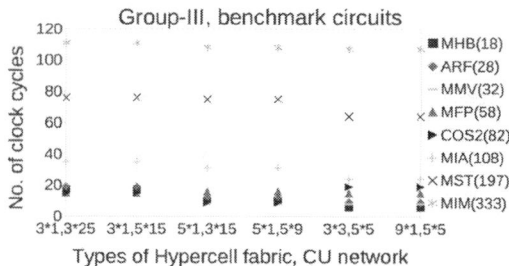

Fig. 8. Number of CC required to execute Group-III benchmarks for each fabric size.

(a)

(b)

Fig. 6. Number of CC required to execute Group-I test cases & VIP for each fabric size.

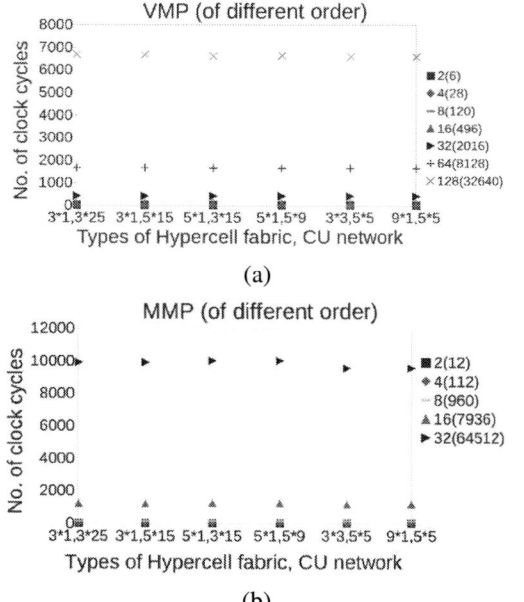

(a)

(b)

Fig. 7. Number of CC required to execute VMP & MMP for each fabric size.

consumed while using the 3×3, 5×5 configuration of NIHC fabric. For VMP and MMP (Figure 7) test cases also, the trend remains the same. But for very small DFGs i.e, the Group-III benchmark circuits (Figure 8), the trend gets violated. For very small DFGs, minimum CCs gets consumed while adopting the 5×1, 3×15 configuration of NIHC fabric. But as we consider larger DFGs, the best performance in terms of number of CCs is again achieved by the 3×3, 5×5 configuration. Hence, in general we can infer that 3×3 HC fabric comprising of 5×5 CU network provides the best performance for EX-DRIVE.

V. CONCLUSION

In this paper, we have explored the design space of EX-DRIVE for various dimensions of NIHC fabric and different partitioning and mapping algorithms. For this study we have presented six different configurations of NIHC fabric and three partitioning and mapping schemes: DPSO-PA, an effective heuristic and a convex algorithm. We have found that, for a fixed hardware capacity, the heuristic and convex algorithm proves to be more efficient for large and densely connected DUTs whereas the DPSO-PA proves to be more efficient for smaller and sparsely connected DFGs. The proposed algorithms could also be applied to any CGRA assisted functional simulation platform.

REFERENCES

[1] I. B. Mahapatra et al., "SIMAAH: RTL simulation accelerator for complex SoC's", IEEE CONECCT 2015, India.

[2] S. Das et al., "A framework for post-silicon realization of arbitrary instruction extensions on reconfigurable data-paths", J. Sys. Architec., Vol. 60, pp. 592-614, 2014.

[3] G Palermo et al., "Multi-objective design space exploration of embedded systems", J. Embedded Comput., 1(3):305– 316, August 2005.

[4] G. Agosta et al., "Multi-objective co- exploration of source code transformations and design space architectures for low-power embedded systems", ACM SAC 2004, pp. 891–896.

[5] B. Mei, "Exploiting loop-level parallelism on coarse-grained reconfigurable architectures using modulo scheduling", DATE 2003.

[6] J. A. Fisher, "Trace scheduling: a technique for global microcode compaction", IEEE Transactions on Computers C-30, pp. 478-490, 1981.

[7] Eberhart, et al., "Particle Swarm Optimization", IEEE Transactions on Evolutionary Computation, pp. 201-203, 2004.

[8] J. Kennedy, et al., "Particle Swarm Optimization," Swarm Intelligence, Vol. 1, pp. 33-57, 2007.

[9] L. A. Sanchis, "Multiple-way network partitioning," IEEE Trans. Comput., vol. 38, no. 1, pp. 62–81, Jan. 1989.

[10] Available Online: http://express.ece.ucsb.edu/benchmark.

978-1-5386-3693-0/18 $31.00 © 2018 IEEE

2018 31th International Conference on VLSI Design and 2018 17th International Conference on Embedded Systems

A Practical Methodology to Compress Technology Libraries using Recursive Polynomial Representation

Sneh Saurabh
Indraprastha Institute of Information Technology Delhi, India
Email: sneh@iiitd.ac.in

Priyanka Mittal
Indraprastha Institute of Information Technology Delhi, India
Email: priyanka16103@iiitd.ac.in

Abstract—With the advancement in technology, the libraries used for storing timing, power and other related information of cells have become voluminous. As a result, run-time of loading libraries, that is often governed by I/O or network bottlenecks, has become unacceptably high. Traditionally, this problem is tackled by compressing technology libraries using gzip or other lossless compression technique. In this paper, we propose a practical methodology to compress technology library with a high compression ratio and tolerable errors. The compressed representation is based on representing the discrete functions of the library as suitable polynomials defined recursively with respect to all the independent variables of that function. Our implementation of the proposed methodology shows that the compression ratio of $8 - 10\times$ can be achieved for realistic error measures.

Index Terms—integrated circuits (IC), technology library

I. INTRODUCTION

Technology libraries of standard cells and macros are integral part of digital implementation and sign-off flows [1], [2]. They provide a high-level view of the cells to the tools employed for synthesis, timing and power optimization etc. by abstracting out transistor and layout details of the cells. However, with the advancement in the technology nodes, because of larger non-linear distortion of signals, increased susceptibility to process-induced variations, greater impact of noises etc. technology libraries have now become enormous [3]–[7]. Additionally, at advanced processes, the number of process-corners/modes combination increases dramatically leading to a requirement to load a large number of libraries. As a result, run-time required to load the data contained in the technology libraries has become unacceptably high.

In a widely used work model, large technology libraries, that often occupy ten/hundred GB of disk space, are stored on a fast and reliable central file server and are accessed by multiple designers from different locations whenever required. An important bottleneck for the designers in this work model is the large run-time required in accessing the libraries from the central file server. This run-time is often governed by the delay in accessing the file over the network rather than by the run-time spent in processing the information which are easily scaled down by multi-threading/multi-processing techniques. Therefore, this problem is traditionally addressed by keeping a compressed (generally gzipped) version of the

technology libraries at the central location and decompressing them only at the individual designers end. However, traditional lossless compression techniques generally offer only $2 - 3\times$ compression ratio and fail to fully address the above mentioned problem. Therefore, a technique to compress technology libraries with a high compression ratio, and possibly with some small tolerable error, say less than 5% of relative error, is desirable. The methodology described in this paper addresses this problem by proposing a novel compressed representation of the information contained in a technology library, devising an algorithm to derive the compressed representation and demonstrating that a high compression ratio with tolerable errors is achievable using the proposed technique.

Technology libraries are typically written in Liberty format and are plain ASCII files [1]. They are generated by characterizing the behavior of the library cells under wide ranging circuit and environmental conditions. In the past, a few techniques of compressing technology libraries have been proposed [8]–[12]. A table compression technique based on dynamic programming was proposed in [8]. The table size was reduced by removing oversampling points in the linear regions of the data and failed to achieve a high compression ratio due to non-linear regions [8]. A multiple-variable based piece-wise polynomials that optimally fitted the functions in the technology libraries was proposed in [9]. The Delay and Power Calculation Language (DPCL) was devised to model these piece-wise polynomials and was largely not accepted by the designers possibly due to its unwieldy complexity. There are other information specific compression techniques that have been proposed for technology libraries [10]–[12].

In this paper, we propose a technique that can be employed to compress general multi-dimensional tables of the technology libraries. The technique is based on representing functions specified at discrete points in the library as suitable polynomials of one independent variable and exploits the distinct dependency of the function on that independent variable. The process is recursively repeated on the coefficients of the polynomials using all other independent variables until a compressed representation is obtained. The degree of polynomials is the handle that controls the trade-off between the compression ratio and the accuracy of the compressed representation.

978-1-5386-3693-0/18 $31.00 © 2018 IEEE

301

It is important to note that, in this paper, we are not proposing to directly use the compressed analytical representation by applications such as Static Timing Analysis (STA). The compressed representation is decompressed to the table-based representation, before being used by the applications. Therefore, proposed methodology is not replacing table-based representation in contrast to the earlier works [13], [14]. During deriving the compressed representation it is ensured that the behavior is monotonic at the *sampled* points and the behavior of the polynomials at *non-sampled points* has no consequence for the applications.

The rest of our paper is organized as follows. In section II, we describe the compressed representation of the proposed methodology and in section III we present an algorithm that derives the compressed representation. In section IV we demonstrate the results obtained using the proposed technique and in section V we make concluding remarks.

II. COMPRESSION METHODOLOGY

A. Discrete Functions in Technology Libraries

The information contained in a technology library is organized as functions of independent variables [1]. These functions are generally dependent on one, two or three variables. For instance, the delay of a timing arc for a pin could be a function of the slew of the input signal and the load driven by the pin. Additionally, these functions are defined at only certain specific values of independent variables. The specific values of independent variables are known as the characterization points of the libraries. Depending on the application of a given technology library, the functions are interpolated at points that are not characterized in the library. The range of independent variables are chosen such that extrapolation is never required.

An important point to note is that the bulk of disk space, typically more than 75% of the technology library, is consumed by the above mentioned discrete functions. Therefore, in this work, we have focused on deriving a compressed representation of the discrete functions in the technology library. Moreover, it is observed that the functions of two independent variables generally occupy a large volume of the technology library. Therefore, in this paper, we demonstrate the proposed methodology on functions of two variables. However, the proposed compression technique can easily be extended to functions of higher or lower dimensions.

P \ Q	q_1	q_2	q_3	q_4	q_5	q_6	q_7
p_1	y_{11}	y_{12}				y_{16}	y_{17}
p_2	...	...	...	...	...	...	...
p_3	...	...	...	...	...	...	...
p_4	...	...	...	...	...	...	...
p_5	...	...	...	...	...	...	...
p_6	...	...	...	...	...	...	...
p_7	y_{71}	y_{72}	...	...	...	y_{76}	y_{77}

TABLE I. An example illustrating a discrete function Y defined for M sampling points of P and N sampling points of Q, where, $M = 7, N = 7$

A discrete function of two dimension can be represented as a two dimensional table as shown in Table I. The independent variables p and q are defined at M and N discrete points, respectively, and can be represented as an array:

$$P = \{p_1, p_2, p_3, ..., p_M\} \tag{1}$$

$$Q = \{q_1, q_2, q_3, ..., q_N\}. \tag{2}$$

The given two dimensional function can be represented as:

$$y_{ij} = f(p_i, q_j) \tag{3}$$

where $1 \leq i \leq M$ and $1 \leq j \leq N$. The two-dimensional discrete function can naturally be represented in a form of a matrix as:

$$Y = \begin{bmatrix} y_{11} & y_{12} & \cdots & y_{1N} \\ y_{21} & y_{22} & \cdots & y_{2N} \\ \cdots & \cdots & \cdots & \cdots \\ y_{M1} & y_{M2} & \cdots & y_{MN} \end{bmatrix} \tag{4}$$

In this paper, the i-th row of the matrix Y is represented as Y_{i*} and the j-th column of the matrix is represented as Y_{*j}.

B. Representing Discrete Functions using Polynomial(s)

In the proposed methodology, a discrete two dimensional table Y is represented compactly in the form of arrays of suitable polynomials. Firstly, all the columns of the matrix Y are represented as an array of polynomials: a polynomial being defined by the list of its coefficients. This is followed by representing the rows of coefficients of the polynomial as an array of polynomial with another smaller set of coefficients. These reduced set of coefficients of the polynomial are stored in the compressed technology library. In the following paragraphs, the compressed representation of the discrete function is described in detail.

Each column j of the matrix Y, i.e. Y_{*j} can be represented in the form of a polynomial of degree D_C:

$$y_{ij} = \sum_{k=0}^{D_C} a_{kj} * p_i^k + \epsilon_{ij} \tag{5}$$

where a_{kj} are the coefficients of the polynomial that needs to be determined and ϵ_{ij} is the error in the representation of the function in the library by the polynomial. In matrix notation, this can be represented as:

$$Y_{*j} = \begin{bmatrix} y_{1j} \\ y_{2j} \\ \cdots \\ y_{Mj} \end{bmatrix} = \begin{bmatrix} 1 & p_1 & p_1^2 & \cdots & p_1^{D_c} \\ 1 & p_2 & p_2^2 & \cdots & p_2^{D_c} \\ \cdots & \cdots & \cdots & \cdots & \cdots \\ 1 & p_M & p_m^2 & \cdots & p_M^{D_c} \end{bmatrix} \begin{bmatrix} a_{0j} \\ a_{1j} \\ \cdots \\ a_{D_C j} \end{bmatrix} + \begin{bmatrix} \epsilon_{1j} \\ \epsilon_{2j} \\ \cdots \\ \epsilon_{Mj} \end{bmatrix} \tag{6}$$

Equivalently,

$$Y_{*j} = P_V A_{*j} + E_j \tag{7}$$

where P_V is the Vandermonde matrix for P and is computed as:

$$P_V = \begin{bmatrix} 1 & p_1 & p_1^2 & \dots & p_1^{D_C} \\ 1 & p_2 & p_2^2 & \dots & p_2^{D_C} \\ \dots & \dots & \dots & \dots & \dots \\ 1 & p_M & p_M^2 & \dots & p_M^{D_C} \end{bmatrix}, \qquad (8)$$

A_{*j} is the coefficient matrix for the j-th column of Y and is given as:

$$A_{*j} = \begin{bmatrix} a_{0j} \\ a_{1j} \\ \dots \\ a_{D_C j} \end{bmatrix}, \qquad (9)$$

and $E_j = \begin{bmatrix} \epsilon_{1j} \\ \epsilon_{2j} \\ \dots \\ \epsilon_{Mj} \end{bmatrix}$ is the matrix containing errors in the

polynomial representation of the function for the j-th column. The A_{*j} matrix is found for all columns of Y and the complete coefficient matrix A can be written as:

$$A = \begin{bmatrix} a_{01} & a_{02} & \dots & a_{0N} \\ a_{11} & a_{12} & \dots & a_{1N} \\ \dots & \dots & \dots & \dots \\ a_{D_C 1} & a_{D_C 2} & \dots & a_{D_C N} \end{bmatrix} \qquad (10)$$

It is evident from equations (4) and (10), that the the coefficient matrix A represents Y using $\{(D_C+1)\times N\}$ points instead of $(M \times N)$ points. A compression can be obtained if:

$$(D_C + 1) < M \qquad (11)$$

The coefficient matrix can be further represented in terms of polynomials row-wise or in terms of the other independent variable, viz. Q. Each row k of the coefficient matrix A, i.e. A_{k*} can be represented in terms of a polynomial of degree D_R in q_j:

$$a_{kj} = \sum_{l=0}^{D_R} b_{kl} * q_j^l + \epsilon_{kl} \qquad (12)$$

where b_{kl} are the coefficients of the polynomial that needs to be determined and ϵ_{kl} is the error in the polynomial representation. In matrix notation it can be written as:

$$A_{k*}^T = Q_V B_{k*}^T + E_k^T \qquad (13)$$

where Q_V is the Vandermonde matrix for Q:

$$Q_V = \begin{bmatrix} 1 & q_1 & q_1^2 & \dots & q_1^{D_R} \\ 1 & q_2 & q_2^2 & \dots & q_2^{D_R} \\ \dots & \dots & \dots & \dots & \dots \\ 1 & q_N & q_N^2 & \dots & q_N^{D_R} \end{bmatrix}, \qquad (14)$$

B_{k*} is the coefficient matrix for the k-th row of A and can be represented as:

$$B_{k*} = \begin{bmatrix} b_{k0} & b_{k1} & \dots & b_{kD_R} \end{bmatrix}, \qquad (15)$$

and $E_k = \begin{bmatrix} \epsilon_{k0} & \epsilon_{k1} & \dots & \epsilon_{kD_R} \end{bmatrix}$ is the matrix containing errors in the polynomial representation of the k-th row of A.

The B_{k*} matrix is found for all rows of A and the complete coefficient matrix B can be written as:

$$B = \begin{bmatrix} b_{00} & b_{01} & \dots & b_{0D_R} \\ b_{10} & b_{11} & \dots & b_{1D_R} \\ \dots & \dots & \dots & \dots \\ b_{D_C 0} & b_{D_C 1} & \dots & b_{D_C D_R} \end{bmatrix} \qquad (16)$$

The coefficient matrix B is stored as a compressed representation of Y. For functions of more than two independent variables, the compressed polynomial representation can be recursively derived using B and other independent variables.

It is evident from equations (4) and (16), that the the coefficient matrix B represents Y using $\{(D_C+1)\times(D_R+1)\}$ points instead of $(M \times N)$ points. A compression can be obtained if:

$$(D_C + 1) \times (D_R + 1) < (M \times N) \qquad (17)$$

The functional compression ratio of the polynomial representation can be defined as:

$$Functional\ Compression\ Ratio = \frac{(M \times N)}{(D_C + 1) \times (D_R + 1)} \qquad (18)$$

It should be noted that, to further compress the technology libraries, the lossless compression utilities that are traditionally employed (such as gzip) can be further applied on the files compressed using the proposed technique.

C. Practicality of the Compression Methodology

The proposed methodology is found to yield a high compression ratio with reasonable accuracy on several technology libraries. The root cause appears to be that the condition defined by (17) is found to be satisfied for most of the two-dimensional functions in the technology libraries for reasonable error tolerances. This is possible because, quite often, two-dimensional functions in technology libraries do not strongly depend on one of the two variables P or Q and sometimes on both the variables P and Q. Since, the data in a technology library is represented in a generalized manner, the distinct dependency of the function on the independent variables is not effectively exploited in representing functions in a technology library. As a result, there are several redundant data in a technology library which can be removed to yield high compression ratio. For instance, in a technology library, the output slew at a pin is generally represented as a function of input slew and the output load at that pin. However, for cells consisting of multiple stages, the effect of the input slew dies down after a couple of stages and the output slew becomes independent of the input slew. In the generalized representation, the output slew is still represented as a matrix Y with the same values across a column yielding redundant data.

In the proposed methodology, compression is achieved by representing each functions in a flexible manner and by exploiting the distinct dependency of each function on the independent variables P and Q. For Y that is not dependent on P or Q in a complicated manner, polynomial representations with a lower degree (e.g. zero, first or second degree) are sufficient to obtain a reasonable accuracy. For the example described above, since the data is the same across a column, it can be represented by a polynomial of zero degree and B degenerates to a row matrix. However, for Y that is dependent on P and Q in a complex manner, a higher degree polynomial is required in their representation and a low compression ratio is obtained for that particular Y. However, practically, this situation is encountered in very small percentage of all functions in the technology library, and, overall, the proposed methodology exhibits a high compression ratio.

D. Restoring Discrete Functions using Polynomial Representation

Restoring the discrete function Y from the coefficient matrix B is straight forward. Firstly, the coefficient matrix A is restored row-wise from B. The k-th row of A is restored using the k-th row of B, i.e. B_{k*} as follows:

$$\hat{A}_{k*}^T = Q_V B_{k*}^T \qquad (19)$$

A ˆ (hat) over A in the above equation denotes that, in general, the restored values could be different from the original values in A due to the error terms. This is followed by restoring Y column-wise from the coefficient matrix $\hat{A}$. The j-th column of Y is restored using the j-th column of $\hat{A}$, i.e. $\hat{A}_{*j}$ as follows:

$$\hat{Y}_{*j} = P_V \hat{A}_{*j} \qquad (20)$$

E. Generating the Polynomial(s) of the Compressed Representation

A polynomial fitting a given discrete function can be found in several ways. The coefficient matrices A and B have to be determined such that the absolute value of the errors in (5) and (12) are minimized.

In this work, we have employed Least Square Error (LSE) method to find the best polynomial fitting a given discrete function [15], [16]. For the relation (5), the square of residual of fitting the j-th column of Y is defined as:

$$R_j^2 = \sum_{i=1}^{M} (y_{ij} - \sum_{k=0}^{D_C} a_{kj} * p_i^k)^2 \qquad (21)$$

The residual is a measure of the error of the fitting polynomial and is minimized in the LSE method. Using (7) and the LSE optimization criteria, the coefficient matrix A can be computed column-wise as:

$$A_{*j} = \{P_V^T P_V\}^{-1} P_V^T Y_{*j} \qquad (22)$$

Since only distinct characterization points are specified in technology libraries, the elements in P are distinct. Additionally, since P_V is a Vandermonde matrix of P, P_V has linearly independent columns. Therefore, when $(D_C < M)$, a unique solution for A_{*j} can be computed using (22). Similarly, using (13) and the LSE optimization criteria, the coefficient matrix B can be computed row-wise as:

$$B_{k*}^T = \{Q_V^T Q_V\}^{-1} Q_V^T A_{k*}^T \qquad (23)$$

F. Quantifying the quality of the Compressed Representation

The effectiveness of the proposed compression methodology is quantified using compression ratio which can be defined as:

$$Compression\ Ratio = \frac{Size\ of\ the\ original\ file}{Size\ of\ the\ compressed\ file} \qquad (24)$$

It should be noted that the proposed compression technique is a lossy compression technique. Therefore, the inaccuracies of the restored functions needs to be quantified carefully. In this paper, we have quantified the inaccuracy of the compressed file using the following three error measures:

1) Root Mean Square Error (E_{RMS}): It is defined as:

$$E_{RMS} = \sqrt{\frac{1}{MN} \sum_{i=0}^{M} \sum_{j=0}^{N} (y_{ij} - \hat{y}_{ij})^2} \qquad (25)$$

where, y_{ij} represents the actual value in the technology library as in (4) and $\hat{y}_{ij}$ represents the restored value as computed using (20).

2) Maximum Error (E_{MAX}): It is defined as

$$E_{MAX} = \max_{\forall_i \leq M, \forall_j \leq N} | y_{ij} - \hat{y}_{ij} | \qquad (26)$$

3) Maximum Relative Error ($E_{MAX-REL}$): For $y_{ij} \neq 0$, it is defined as

$$E_{MAX-REL} = \max_{\forall_i \leq M, \forall_j \leq N} | \frac{y_{ij} - \hat{y}_{ij}}{y_{ij}} | \qquad (27)$$

III. ALGORITHM

The compression of technology library starts by reading a library file and extracting independent variables and discrete functions. These functions are passed one-by-one to the top-level compression routine $COMPRESS\_FUNCTION$ shown in Algorithm-1.

Algorithm 1: COMPRESS_FUNCTION

Input : $Y, P, Q,$
$\qquad\qquad Tolerance = \{E_{RMS}, E_{MAX}, E_{MAX-REL}\}$
Output: $Success$ or $Failure$

1: $M = size\ of\ P$
2: $N = size\ of\ Q$
3: $(A, D_C) = COMPRESS\_C\_WISE(Y, P, M, N)$
4: $(B, D_R) = COMPRESS\_R\_WISE(A, Q, N, D_C)$
5: $\hat{Y} = RESTORE\_FUNCTION(B, P, Q)$
6: **Compute** $Error = \{E_{RMS}, E_{MAX}, E_{MAX-REL}\}$
    using Y and $\hat{Y}$ employing (25), (26) and (27)
7: **if** $Error \leq Tolerance$ **then**
8:     **Store** B as a compressed representation of Y
9:     **return** $Success$
10: **else**
11:     **return** $Failure$
12: **end if**

The routine $COMPRESS\_FUNCTION$ takes as input the $M \times N$ matrix Y and the independent variables

P and Q of size M and N, respectively. Additionally, $COMPRESS\_FUNCTION$ takes $Tolerance$ as a parameter from the user which is defined in terms of error measures described in the previous section. For a given discrete function, if the $COMPRESS\_FUNCTION$ is able to compress the function within the $Tolerance$ limit, a compressed representation of the function is stored, else the original function is stored in the compressed library.

Algorithm 2: COMPRESS_C_WISE

Input : $Y, P, M, N, Tolerance$
Output: A, D_C

1: **for** $D_C \leftarrow 0$ to $(M - 1)$ **do**
2:    $SatisfyErrCriteria = 1$
3:    **Initialize** A of size $\{(D_C + 1) \times N\}$ with all 0
4:    **for** $j \leftarrow 1$ to N **do**
5:       $(CoeffColJ, Error) = POLY\_FIT(Y_{*j}, P, D_C)$
6:       **if** $Error > Tolerance$ **then**
7:          $SatisfyErrCriteria = 0$
8:          **break inner for-loop**
9:       **end if**
10:       **Update** j-th column of A with $CoeffColJ$
11:    **end for**
12:    **if** $SatisfyErrCriteria = 1$ **then**
13:       **break outer for-loop**
14:    **end if**
15: **end for**
16: **return** (A, D_C)

The $COMPRESS\_FUNCTION$ first compresses the function Y column-wise using $COMPRESS\_C\_WISE$ shown in Algorithm-2. It tries out different degrees of polynomials starting from 0 to $M-1$ that could satisfy the $Tolerance$ criteria specified by the user. Whenever, the tolerance criteria is met for all the columns of Y for the lowest degree of polynomial, the corresponding degree D_C and the coefficient matrix A is accepted.

Algorithm 3: POLY_FIT

Input : Y_{*j}, P, D_C
Output: $A_j, Error$

1: **Compute** P_V using P and D_C employing (8)
2: **Compute** A_{*j} using P_V and Y_{*j} employing (22)
3: **Compute** $\hat{Y}_{*j}$ using A_{*j} and P_V employing (20)
4: **Compute** $Error$ using Y_{*j} and $\hat{Y}_{*j}$ employing (25), (26) and (27)
5: **return** $(A_{*j}, Error)$

To find the polynomial of a given degree, the routine $POLY\_FIT$ shown in Algorithm-3 is used. It takes an array of sampling points with their corresponding values and returns a polynomial of the specified degree. In our work, we have used Gaussian Elimination Method to solve the matrix inversion problem encountered in computing the fitting polynomial,

though other efficient algorithms of matrix inversion can also be used effectively. The routine $POLY\_FIT$ also computes the values of the fitted polynomial at the sampling points and returns the error measures.

After compressing the function column-wise, the routine $COMPRESS\_FUNCTION$ compresses the coefficients of A row-wise using $COMPRESS\_R\_WISE$. This routine is not shown in the paper since it is similar to $COMPRESS\_C\_WISE$.

Algorithm 4: RESTORE_FUNCTION

Input : B, P, Q
Output: $\hat{Y}$

1: **Compute** $\hat{A}$ using B and Q employing (19)
2: **Compute** $\hat{Y}_{*j}$ using $\hat{A}$ and P employing (20)
3: **return** $\hat{Y}$

Finally, the routine $COMPRESS\_FUNCTION$, restores the original function using B and the independent variables employing $RESTORE\_FUNCTION$ shown in Algorithm-4. It computes the error measure of the compressed representation and if the error measures are acceptable then the compressed representation of the function are stored in the technology library.

IV. RESULTS

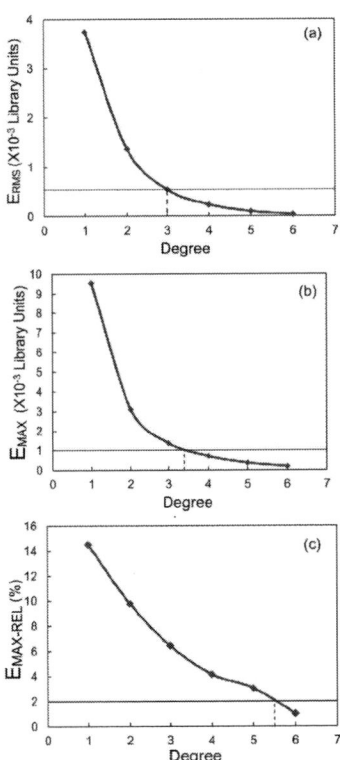

Fig. 1. Trade-off between the degree of polynomial and error measures (a) Root Mean Square Error (b) Maximum Error (c) Maximum Relative Error

	Original Size (KB)	Compression Ratio for $E_{RMS} = 0.004$ (library units)	Compression Ratio for $E_{RMS} = 0.001$ (library units)	Compression Ratio for $E_{MAX} = 0.006$ (library units)	Compression Ratio for $E_{MAX-REL} = 5\%$
Library-1	11	12.6	7.1	12.6	7.1
Library-2	0.11	13.9	8.1	13.9	8.1
Library-3	4.4	9.1	7.1	9.1	7.1
Library-4	6.5	11.8	8.7	6.2	8.7
Library-5	10.9	6.8	4.4	6.8	4.4

TABLE II. Compression Ratio of different library files for different tolerances

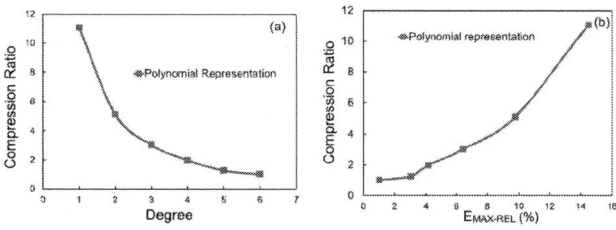

Fig. 2. Dependency of Compression Ratio on (a) Degree of modeled polynomial (b) Maximum Relative Error

The proposed technique was implemented and tested on a 2.37 GHz 64-bit Linux machine with 8 GB memory. Firstly, a typical library was compressed using different degree of polynomials related to all the independent variables and the error measures were quantified for different representations, as shown in Fig. 1(a)-(c). The results indicate that the error measures decrease as the degree of polynomials increase.

Then, a study was carried out to investigate the relationship between the compression ratio and the size of polynomials and the error measures. Fig. 2(a) and (b) show that the compression ratio can be increased by decreasing the degree of the polynomial representation or by increasing the tolerance of error.

Table-II shows the compression ratio of different libraries for "65 nm" technology for different tolerance settings. The compression was obtained using polynomial representation followed by traditional *gzip*. For an allowed RMS Error of 0.001 library units, a compression ratio of around $8\times$ is obtained. For all the libraries that we have benchmarked, the library timing unit was "ns". Therefore, for an RMS error of less than 1 *ps*, compression ratio of around $8\times$ is obtained. For a maximum error of 6 *ps*, around $10\times$ compression can be obtained and for a maximum relative tolerance of 5% a compression ratio of around $8\times$ is obtained.

V. CONCLUSION

In this paper, we have proposed a practical compression methodology and have shown that a high compression ratio can be obtained with a tolerable error. However, the impact of the errors (for example 5% maximum relative error) on the tools for timing/power analysis etc. needs to be evaluated carefully. Additionally, the effectiveness of the proposed technique needs to be evaluated on tables of advanced process nodes, though we believe that the tables of "65 nm" technology that we have used had sufficient complexity for benchmarking.

Moreover, in this work, we have not specifically shown the decompression run times, we have found it to be negligible compared to other bottleneck functions in reading a library.

REFERENCES

[1] Synopsys Inc., "Liberty library modeling," https://www.synopsys.com/community/interoperability-programs/tap-in.html.

[2] J. Bhasker and R. Chadha, *Static timing analysis for nanometer designs: A practical approach.* Springer Science & Business Media, 2009.

[3] R. Trihy, "Addressing library creation challenges from recent liberty extensions," in *Proceedings of the 45th annual Design Automation Conference.* ACM, 2008, pp. 474–479.

[4] I. Keller, K. H. Tam, and V. Kariat, "Challenges in gate level modeling for delay and si at 65nm and below," in *Design Automation Conference, 2008. DAC 2008. 45th ACM/IEEE.* IEEE, 2008, pp. 468–473.

[5] V. Kariat, "ECSM 2.0 static modeling format standard," 2010.

[6] L. Brusamarello, G. I. Wirth, P. Roussel, and M. Miranda, "Fast and accurate statistical characterization of standard cell libraries," *Microelectronics Reliability*, vol. 51, no. 12, pp. 2341–2350, 2011.

[7] Q. Xie, X. Lin, Y. Wang, M. J. Dousti, A. Shafaei, M. Ghasemi-Gol, and M. Pedram, "5nm FinFET standard cell library optimization and circuit synthesis in near-and super-threshold voltage regimes," in *VLSI (ISVLSI), 2014 IEEE Computer Society Annual Symposium on.* IEEE, 2014, pp. 424–429.

[8] J. F. Croix and D. F. Wong, "A fast and accurate technique to optimize characterization tables for logic synthesis," in *Proceedings of the 34th Design Automation Conference*, June 1997, pp. 337–340.

[9] G. Wang and R. Gopisetty, "Efficient generation of timing and power polynomial models from lookup tables for SoC designs," in *Twelfth Annual IEEE International ASIC/SOC Conference (Cat. No.99TH8454)*, 1999, pp. 216–220.

[10] S. Hatami, P. Feldmann, S. Abbaspour, and M. Pedram, "Efficient compression and handling of current source model library waveforms," in *Design, Automation & Test in Europe Conference & Exhibition, 2009. DATE'09.* IEEE, 2009, pp. 1178–1183.

[11] S. Saurabh, N. Kumar, and I. Keller, "Method and apparatus for comprehension of common path pessimism during timing model extraction," Jan. 20 2015, US Patent 8,938,703.

[12] S. Saurabh and N. Kumar, "Method and apparatus for efficient generation of compact waveform-based timing models," Aug. 8 2017, US Patent 9,727,676.

[13] A. Goel and S. Vrudhula, "Statistical waveform and current source based standard cell models for accurate timing analysis," in *2008 45th ACM/IEEE Design Automation Conference*, June 2008, pp. 227–230.

[14] T. Liu, C. C. Chen, and L. Milor, "Accurate standard cell characterization and statistical timing analysis using multivariate adaptive regression splines," in *Sixteenth International Symposium on Quality Electronic Design*, March 2015, pp. 272–279.

[15] W. H. Press, *The art of scientific computing.* Cambridge university press, 1992.

[16] Å. Björck, *Numerical methods for least squares problems.* SIAM, 1996.

CLRFrame: An Analysis Framework for Designing Cross-Layer Reliability in Embedded Systems

Siva Satyendra Sahoo, Bharadwaj Veeravalli
Department of ECE
National University of Singapore
Singapore
satyendra@u.nus.edu, elebv@nus.edu.sg

Akash Kumar
Center for Advancing Electronics Dresden
Technische Universität Dresden
Dresden, Germany
akash.kumar@tu-dresden.de

Abstract—Continued transistor scaling and increasing power density have led to considerable increase in fault rates in silicon nanotechnology-based real-time systems. Instead of fixing everything at the hardware layer, cross-layer fault tolerance techniques present a more cost-efficient methodology for adapting to such increased fault rates. The effectiveness (*Coverage, Fault-Masking*, and *Recovery*) and overheads (*Execution time, Energy* and *Cost*) of each fault-tolerance technique vary with the layer and the frequency at which it is implemented. Therefore, appropriate modeling of fault-mitigation methods is necessary for efficient cross-layer design space exploration (DSE). To this end, we propose a first-order framework for analyzing the effects of implementing fault-tolerance across multiple layers. We also propose a Markov-chain based methodology for analytical modeling of fault-mitigation methods and their inter-layer interaction. As a case-study, we model some generic fault-mitigation methods and provide detailed modeling of typical application execution involving fault-mitigation at different layers.

Keywords-Cross-layer Resilience, Real-time systems, Fault-Tolerance

I. INTRODUCTION

Technology scaling and architectural innovations have been the driving force behind the increasing ubiquity of embedded systems. However, these approaches have also led to significant increase in Soft Error Rate (SER) of logic circuits [1]. While the logical masking effect remains unaltered, electrical masking effect has reduced due to scaling down transistor size and supply voltage [2]. Similarly, deeper pipelines used for enabling higher clock speed have resulted in the reduction of latching-window masking, leading to even higher SER in microprocessors. Therefore, extracting increasing usable performance out of embedded systems requires building resilient systems out of increasingly unreliable hardware [3]. Traditional approaches to fault-aware design adopted a phenomenon-based approach that focuses on mitigating all physical faults at the hardware layer. However, such an approach is becoming increasing infeasible due to the increasing complexity, higher SER and tighter cost and energy constraints of embedded systems.

In contrast, cross-layer design approach involves distributing fault-mitigation activities to several layers of the system stack [4]. This entails utilizing the information and capabilities of each layer to provide adequate overall system resilience. Such an approach can reduce fault-mitigation effort at hardware layer leading to more cost-effective designs. Designing an effective cross-layer resilience strategy requires formulating efficient methods for cross-layer design

space exploration (DSE)– both at design-time and run-time for finding the right *selection* and *configuration* of fault-mitigation methods that should be implemented at each layer to meet the system-level goals and constraints. Given the increasing design space due to the choices of fault-mitigation methods and configurations at each layer, application complexity, hardware variations etc., non-traditional DSE methods such as *Evolutionary Computing* and *Randomized Algorithms* are being increasingly used in related *Electronic Design Automation* (EDA). The effectiveness of such methods hinges on the efficient estimation of performance metrics and associated overheads for a set of design decisions. For example, in *Genetic Algorithms* (GA) the computation time and accuracy of evaluating the *fitness* of each *individual* of a *generation* can have a significant effect on the overall DSE performance. Therefore, efficient evaluation of design decisions from early stages of DSE can lead to the faster design of cost-effective error-resilient systems. To this end, we propose a methodology for designing cross-layer resilience for embedded systems.

Contributions: Our contributions can be listed as below:

- We propose a first-order framework for comparing different cross-layer designs. The framework includes the effect of implicit fault-masking at different layers along with explicitly implemented fault-mitigation methods to estimate system-level performance metrics.
- We propose a Markov-chain based methodology for analytical modeling of fault-mitigation methods. We demonstrate the usage of this model by modeling some application-agnostic fault-mitigation methods.
- We extend our proposed modeling methodology to account for inter-layer interactions of fault-mitigation implemented at different layers of the system stack.

The rest of the paper is organized as follows: In Section II we provide a brief background of cross-layer reliability and survey some state-of-the-art approaches to designing cross-layer reliability. The proposed framework for cross-layer reliability analysis is discussed in Section III. In Section IV we provide details of Markov-chain based analytical modeling methodology. We conclude the paper in Section V by providing the scope for future work.

II. BACKGROUND

Traditional single-layer reliability approaches focus on mitigating all physical faults at the circuit or the hardware layer. A phenomenon-based approach is usually used, i.e., each fault mechanism (NBTI, EM, Single Event Upsets

(SEU) etc.) is mitigated separately to provide an error-free hardware platform. Although it provides a convenient abstraction to the software developer, the rising costs – *area* and *power* – for mitigating the effects of increasing fault rates can make such an approach infeasible for many applications. Therefore, fault-mitigation needs to be implemented at multiple layers of the system stack. In [5], the authors provide a brief survey of such techniques at different abstraction layers. Designing cross-layer reliability involves finding the right *selection* and *configuration* of fault-mitigation methods that should be implemented at each layer to meet the system-level goals and constraints.

A. Cross-layer Reliability

In contrast to the single-layer phenomenon-based design approach, the cross-layer approach provides a more application-specific and cost-efficient method for reliability-aware system design. Since the fault-mitigation activities are not limited the hardware layer, an appropriate combination of methods that meets the design goals and constraints can be implemented. As discussed in [6], implementing separate fault tolerance stages at different layers can result in reduced power and area overheads. For example, TMR, which provides complete protection from single errors, has more than 200% area and power overheads. However, error/fault detection by Dual Modular Redundancy (DMR) usually has less (100%) overhead. Therefore, an implementation that uses DMR-based hardware error detection and software recovery methods can reduce overheads. Further, distributing fault tolerance tasks to higher layers enable the designer to take advantage of the masking effects of more layers [7]. In [4], the authors outline a methodology for implementing cross-layer resilience. Additional subsystems *Error handler routine, Resource Map, Hardware Configuration Routine,* and *Task Scheduler* in the operating system are used to trigger the appropriate fault-tolerant technique at the appropriate layer. In [8], the authors proposed new techniques – *Error-aware placement* and *Failure prediction* – for globally-optimized cross-layer resilience. Similarly, in [9], the authors propose various cross-layer techniques – from *microarchitecture* to *application* level – for both general purpose processor-based and reconfigurable processor-based embedded systems. In all methods presented, every layer takes advantage of the information available at its adjacent layers. In [6], the authors present a cross-layer approach providing resilience in multimedia applications. Specifically, the proposed method uses hardware layer for error detection, middleware for Drop and Forward recovery and application layer for error resilient application design. In [10], the authors provide a heuristic-based methodology for combining several hardware and software techniques – *Circuit-level hardening, logic-level parity checking, microarchitectural recovery,* and *ABFT* – to provide soft-error tolerance in processor cores.
Most of the state-of-the-art cross-layer reliability techniques lack a holistic approach and do not consider all reliability metrics – *Functional, Timing* and *Lifetime*. For instance, the approach described in [10] involves maximizing the fault-

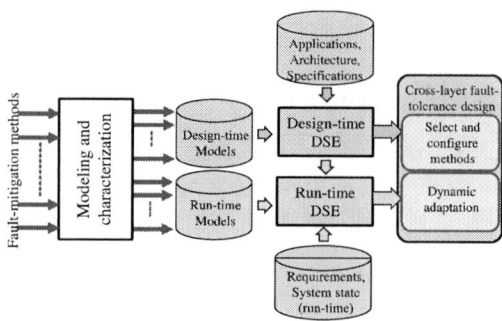

Figure 1: Cross-layer reliability design methodology

mitigation by software layers. Usually, software mitigation of hardware faults – based on *temporal redundancy* – incurs lesser area/power overheads. However, the increased execution time can lead to faster aging in the long run. Therefore, systems that have design constraints of system lifetime have to use additional processing units. This can offset some of the area/cost advantages. In [11], the authors show the adverse effects of increasing checkpoints, a temporal redundancy-based method, on permanent fault tolerance. Further, barring a few application areas that require high reliability in all three measures, most applications, especially soft real-time systems, can tolerate some degradation in each or all of the metrics. We envision a more application-specific and holistic methodology, as depicted in Figure 1, for incorporating cross-layer reliability into system design. Fault-mitigation methods need to be modeled for their characterization. Such models, along with a cross-layer DSE framework, can be used during design-time to determine the feasibility of the method for a particular application, estimate the range of variations in system state that the method can handle, compare different methods and determine the appropriate configuration of various methods at different layers from an early stage in the design phase. Similarly, run-time models of such mitigation methods can be used to adapt the methods for changing system state or switch to other fault-mitigation methods.

III. CROSS-LAYER FAULT-MITIGATION FRAMEWORK

Cross-layer fault-tolerance entails distributing fault mitigation activities among different layers, in order to achieve system-level objectives within the constraints. Fault mitigation can be broadly divided into two stages – *Detection/Validation* and *Tolerance*. Detection methods can be characterized by their latency (T_{FD}), coverage (Cov_{FD}) and the associated power and area overheads (Ov_{FD}). Similarly, a tolerance technique can be characterized by reconfiguration latency (T_{FT}), and associated overheads (Ov_{FT}). For instance, Dual Modular Redundancy (DMR) in hardware usually has a latency equal to that of the comparator, 100% power and area overheads and a coverage equivalent to *1 - (fraction of faults affecting both modules simultaneously)*. Similarly, Triple Modular Redundancy (TMR) in hardware has a reconfiguration latency equivalent to the latency of the voter module, and 200% power and area overheads. Each fault mitigation method, a combination of detection and

978-1-5386-3693-0/18 $31.00 © 2018 IEEE

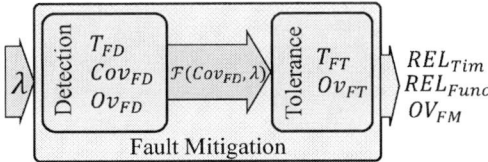

Figure 2: Characterization of fault-mitigation methods

tolerance methods can be characterized as shown in Figure 2. Masking factor (MF_{FM}) is the ratio of error rate with fault mitigation (λ') to that without (λ). CDF_{FM} denotes the cumulative distribution function (CDF) of execution latency and Ov_{FM} denotes overheads in area and power. These three quantities, as functions of λ and other characteristics shown inside the Fault Mitigation block in Figure 2, can be used to estimate the performance metrics of the method and also form the interface for plugging the method into the overall cross-layer analysis framework.

With a cross-layer approach, the convenient abstractions of single-layer design cannot be used. Designing for system-level fault-tolerance must incorporate effects of fault mitigation at all layers. Hence, appropriate resilience interfaces are required for effective transfer and usage of inter-layer information. Figure 3 shows a framework with such interfaces for cross-layer fault-tolerance design. Functional reliability, the rate of errors during application execution, can be determined by considering the masking factor, MF_X, of each layer (X). It consists of implicit fault masking at that layer, MF_{X0} and the masking effect due to fault mitigation methods implemented at that layer. Assuming all detected errors/faults are mitigated, the product of error rate at X, i.e. λ_X, and the coverage of detection methods at layers below X, i.e. Cov_X denotes the masking effect due to detection at layers below X. Similarly, $(1 - MF'_{X0}).Cov_X$ is the masking due to detection methods used at layer X. MF'_{X0} is the net effect of MF_{X0} and $\lambda_X.Cov_X$. Please note that simple proportionality relations have been used among various quantities only as an illustration. Actual implementation will involve more complex method-specific relations among various quantities. Timing reliability of the system, quantified by the probability of execution completion before a deadline, can be obtained from the CDF of execution latency (CDF_X) due to mitigation methods implemented at each layer X. Similarly, the overhead information is essential for computing overall system performance and satisfying the design constraints. The average execution time, $T_{avg(X)}$, can be used to quantify metrics such as throughput, energy consumption, system lifetime etc. Similarly, area and power overheads information can be used to satisfy system constraints. The three interfaces described here provide methods to collect appropriate information for cross-layer analysis. The structures necessary for carrying the information vary with each layer. For instance, while λ_X provides a metric to quantify the effect of fault-mitigation at one layer on all upper layers, the granularity of structures for capturing the effect of λ_X depends on the designer and the intended application. In [12], Instruction Vulnerability Index and Function Vulnerability Index were used to incorporate

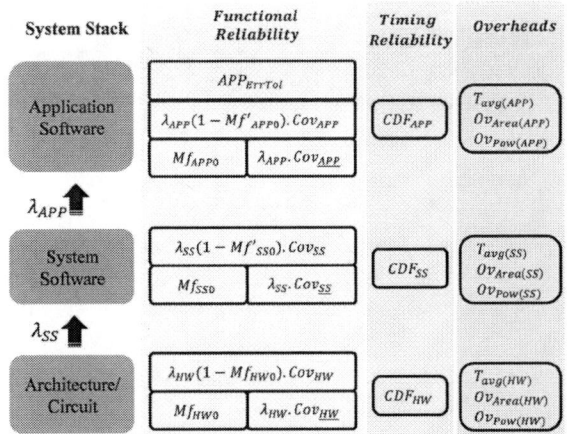

Figure 3: Framework for cross-layer fault-tolerance analysis

the effect of λ_X and implicit masking effects (MF_{X0}) at the architecture and software layers. Similarly, structures suitable for each interface can be used to enable inter-layer information.

IV. Modeling Cross-layer Fault-Mitigation

A. Modeling Fault-Mitigation methods

Designing cross-layer reliability involves implementing various fault-mitigation methods at different layers of the system stack. To select appropriate methods and their configurations for each layer, the effect of implementing each method needs to be estimated. To this end, we use analytical models for computing the following performance metrics:

- **Functional Reliability:** It refers to the probability of getting functionally correct outputs in the presence of faults. As shown in Figure 2, the functional reliability is a function of the coverage of the detection method implemented and the fault/error rate.

- **Timing Reliability:** It refers to the probability of the execution completing within the specified deadline. Timing reliability can be determined from the cumulative distribution function of the execution time.

- **Overheads:** We limit our analysis to the determine the following overheads – cost (quantified by the physical area of implementation), power, energy and average execution time T_{avg}. Lifetime reliability can be estimated as a function of T_{avg}.

We model each application-agnostic fault-mitigation method discussed in this article as an absorbing Markov-chain. This enables the computation of T_{avg} and the cumulative distribution function (CDF) of the execution time analytically. For an absorbing Markov-chain, the transition matrix P can be represented as $P = \begin{bmatrix} I & 0 \\ R & Q \end{bmatrix}$, where I is the identity sub-matrix, R is the transition sub-matrix from transient to absorbing states and Q is a $n \times n$ transition sub-matrix for transient states only [13].

The fundamental matrix F can be obtained as $F = (I - Q)^{-1}$. Each entry $F_{i,j}$ represents the expected number of periods that the chain spends in transient state j given

978-1-5386-3693-0/18 $31.00 © 2018 IEEE

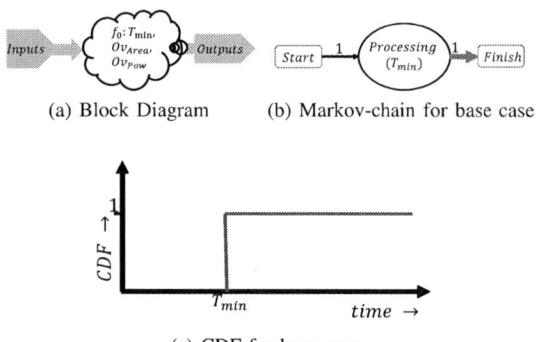

(a) Block Diagram   (b) Markov-chain for base case

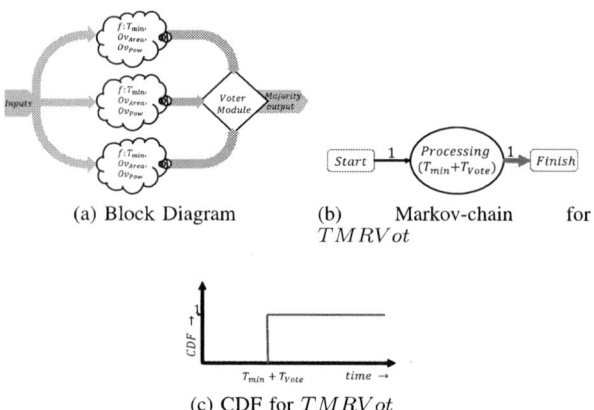

(a) Block Diagram   (b) Markov-chain for $TMRVot$

(c) CDF for base case

Figure 4: Base Case: No Fault Mitigation

(c) CDF for $TMRVot$

Figure 5: $TMRVot$: TMR with Voting

that the chain began in transient state i. Further, the average execution time can be computed as $T_{avg} = T_{exec} \times (e_1^T F)$, where $e_1 = [1, 0, ...0]$ and T_{exec} is the state-residence time of the states in the Markov-chain.

Fault/Error Model: We assume the fault/error-mitigation methods discussed in this article are for mitigating the effects of operational transient faults and resulting errors and failures in the datapath, i.e. computational faults/errors. We assume the occurrence of faults/errors follows a Poisson distribution, λ being the fault/error-rate per module. Our analysis assumes that multiple faults/errors can occur during a module's execution. In the current article, we provide the modeling details of 2 methods in comparison to the Base case of not implementing any fault-mitigation.

1) Base Case (No Fault-Mitigation): We define the base case as– *Processing without implementing any fault-mitigation method.* The block diagram for the base case is shown in Figure 4a. The base case forms the reference point for modeling and comparing the various fault-mitigation methods. Also, the notations used in describing the base case will be followed for the rest of the methods modeled in the current article.

Functional Reliability: The probability of a fault-induced error occurring during processing is shown in Eq. 1. We assume no implicit masking of errors.

$$
\begin{aligned}
&Prob(Error\ without\ Fault - Mitigation) \\
&= 1 - Prob(No\ faults\ during\ processing) \\
&= (1 - e^{-\lambda T_{min}})
\end{aligned}
\tag{1}
$$

Timing Reliability: The Markov-chain for the base case is shown in Figure 4b. $Finish$ signifies the absorbing state in the chain. The state-residence time is shown in the state in parenthesis. The CDF of the execution time is shown in Figure 4c.

Overheads: Since the base case does not involve any form of spatial or temporal redundancy, there are no extra overheads incurred. The resulting costs are shown in Eq. 2

$$
\begin{aligned}
Average\ Execution\ time &: T_{avg} = T_{min} \\
Area\ cost &: Ov_{Area} \\
Power\ cost &: Ov_{Pow} \\
Energy\ Consumption &: Ov_{Energy} = T_{avg} \times Ov_{Pow}
\end{aligned}
\tag{2}
$$

2) Triple Modular Redundancy with Voting (TMRVot): TMR is one of the most commonly used forms of spatial redundancy. TMR can mask a large proportion of faults/errors, at high area and power overheads. Figure 5a shows the block diagram for a TMR implementation. We assume the voter module to compute the majority of the 3 modules' outputs as the final output. T_{min} is the minimum execution time of each module. Similarly, Ov_{Area} and Ov_{Pow} are the area and power costs of implementing each module.

Functional Reliability: In TMR, the fault-masking is achieved by taking the majority of the outputs. Hence, no separate fault-detection method needs to be used. So, the coverage is equivalent to the coverage of the implicit fault-masking. The output can have a fault-induced errors in the case – A fault/error occurring in at least 2 of the three modules. So, the probability of a computational error can be estimated as shown in Eq. 3.

$$
\begin{aligned}
Prob(Error) &= 3 \times Prob(Error\ in\ any\ 2\ modules\ only) \\
&+ Prob(Error\ in\ all\ three\ modules) \\
&= 3(1 - e^{-\lambda T_{min}})^2 e^{-\lambda T_{min}} + (1 - e^{-\lambda T_{min}})^3 \\
&= 1 - 3e^{-2\lambda T_{min}} + 2e^{-3\lambda T_{min}}
\end{aligned}
\tag{3}
$$

The probability of computational error in the absence of any fault-mitigation implementation and the resulting *Masking Factor* is shown in Eq. 4.

$$
\begin{aligned}
&Prob(Error\ without\ Fault - Mitigation) \\
&= Prob(Atleast\ one\ error\ during\ application\ execution) \\
&= (1 - e^{-\lambda T_{min}}) \\
&Masking\ factor : MF_{TMRVot} \\
&= \frac{Prob(Error\ without\ Fault - Mitigation)}{Prob(Error\ with\ Fault - Mitigation)} \\
&= (1 - e^{-\lambda T_{min}})/(1 - 3e^{-2\lambda T_{min}} + 2e^{-3\lambda T_{min}})
\end{aligned}
\tag{4}
$$

Timing Reliability: TMR does not involve any form of temporal redundancy. The resulting Markov-chain is shown in Figure 5b. $Finish$ signifies the absorbing state in the chain. The state-residence time includes the time taken by the $Voter\ Module$. The CDF of the execution time is shown in Figure 5c.

Overheads: Since TMR involves 3 parallel modules, the

978-1-5386-3693-0/18 $31.00 © 2018 IEEE   310

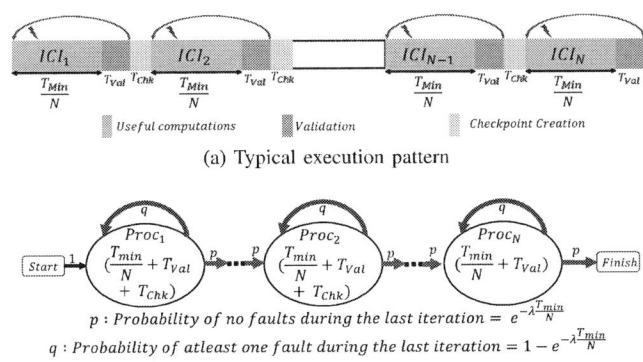

(a) Typical execution pattern

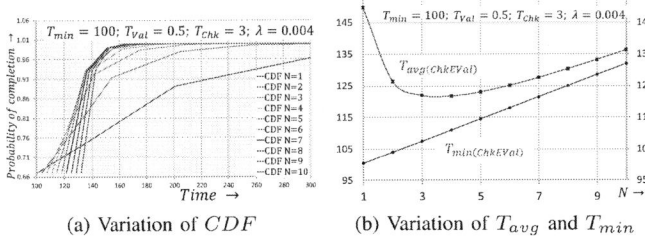

(a) Variation of CDF (b) Variation of T_{avg} and T_{min}

Figure 8: $ChkEVal$: Variation of performance metrics with number of ICIs

(b) Markov-chain for $ChkEVal$

Figure 6: Checkpointing and rollback recovery with N ICIs

States	$Proc_1$	$Proc_2$	....	$Proc_i$	.....	$Proc_N$	$Finish$
$Proc_1$	q	p	....	0	....	0	0
$Proc_2$	0	q	....	0	....	0	0
:	:	:	...	:		:	:
$Proc_i$	0	0	....	q	....	0	0
:	:	:		:	·.	:	:
$Proc_N$	0	0	....	0	....	q	p
$Finish$	0	0	....	0	.....	0	1

Figure 7: Transition Matrix for Markov-chain of $ChkEVal$

area and power costs incur around 200% increase over the base case. The resulting overheads are shown in Figure 5.

$$Average\ Execution\ time: T_{avg(TMRVot)} = T_{min} + T_{Vote}$$
$$Area\ cost: Ov_{Area(TMRVot)} = 3 \times Ov_{Area}$$
$$Power\ cost: Ov_{Pow(TMRVot)} = 3 \times Ov_{Pow} \quad (5)$$
$$Energy\ Cost: $$
$$= T_{avg(TMRVot)} \times Ov_{Pow(TMRVot)}$$

3) Checkpointing and Rollback Recovery with end-Validation (ChkEVal): Checkpointing with rollback recovery is one of the more prevalent methods used for fault-mitigation. It involves saving the program state at regular intervals and restoring the saved state in the event of a fault/error. The timing overheads associated with checkpoint creation and validation determine the effectiveness of this method. The typical execution pattern of processing with N inter-checkpoint intervals (ICIs) is shown in Figure 6a. Please note that our model assumes validating the computation results at the end of each ICI. Further, we assume fault-free validation and checkpoint creation. T_{Val} and T_{Chk} denote the time spent on performing each validation and creating each checkpoint respectively.

Functional Reliability: The fault-mitigation coverage is equal to the coverage of the validation method implemented.

Timing Reliability: The Markov-chain for $ChkEVal$ is shown in Figure 6b and its corresponding transition matrix ($P_{ChkEVal}$) is shown in Figure 7. The CDF and T_{avg} is a function of the number of ICIs, N. Figure 8a shows the varying CDF with increasing N for a test-case.

Overheads: The average time to completion can be obtained from the absorbing Markov-chain's transition matrix. Eq. 6 shows the overheads associated with $ChkEVal$. $F_{ChkEVal}$ corresponds to the *Fundamental Matrix* derived from $P_{ChkEVal}$. Further, Figure 8b shows the variation of $T_{avg(ChkEVal)}$ with increasing N.

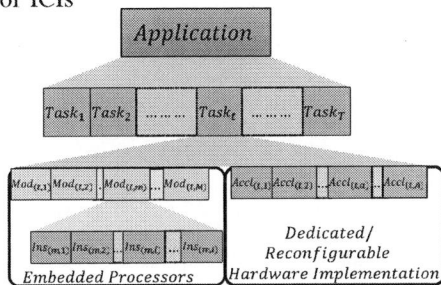

Figure 9: Multi-layer view of a typical hierarchy in application execution

$$Average\ Execution\ time: T_{avg(ChkEVal)}$$
$$= \left(\frac{T_{min}}{N} + T_{Val} + T_{Chk} \right) \times (e_1^T F_{ChkEVal}) - T_{Chk}$$
$$Area\ cost: Ov_{Area(ChkEVal)} = Ov_{Area} \quad (6)$$
$$Power\ cost: Ov_{Pow(ChkEVal)} = Ov_{Pow}$$
$$Energy\ Cost: Ov_{Energy(ChkEVal)}$$
$$= T_{avg(ChkEVal)} \times Ov_{Pow(ChkEVal)}$$

B. Modeling Inter-layer interaction of Fault-Mitigation methods

The typical hierarchy involved in an application's execution on a heterogeneous hardware platform can be represented is shown in Figure 9. Any application can be represented as a set of tasks – $Task_t$, where $t \in \{1, 2..., T\}$. Further, each task $Task_t$ can be modeled as composed of a set of modules (or function calls) – $Mod_{(t,m)}$, where $m \in \{1, 2..., M\}$ – that is executed on instruction-set based embedded processors as a sequence of instructions – $Ins_{(m,i)}$, where $i \in \{1, 2..., I\}$. Similarly, to satisfy the performance requirements of the application, some portion of the task may need to be processed on accelerators – $Accl_{(t,a)}$, where $a \in \{1, 2..., A\}$ – implemented on dedicated hardware and/or dynamically reconfigurable fabric. Please note that, as shown in Figure 9, we model the application, tasks, and modules as linear sequences of their constituent tasks, modules and accelerators, and instructions respectively. For the current article, we have not considered typically occurring branching and dependencies among the various tasks, modules, accelerators and instructions. However, our modeling methodology can be extended to consider such dependencies and is planned for future research.

Due to the randomness involved in the occurrence of faults

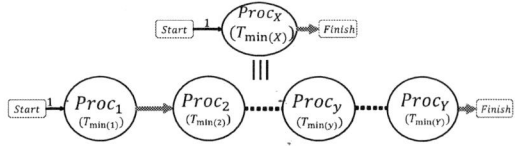

Figure 10: Modeling a process having sub-processes as an absorbing Markov-chain

and their mitigation, the execution of each application (task, module) can be modeled as a stochastic process comprising of a set of independent sub-processes. As shown in Eq. 7, each of the processes and sub-processes can be characterized by the set of parameters discussed in Section IV-A. We discus each of the parameters below.

$$
\begin{aligned}
& X(RelF_{(X)}, T_{min(X)}, CDF_{(X)}, T_{avg(X)}, \\
& \quad Ov_{Area(X)}, Ov_{Pow(X)}, Ov_{Energy(X)}) \\
& X \epsilon \{Ins_{(m,i)}, Mod_{(t,m)}, Accl_{(t,a)}, Task_t, Application\}
\end{aligned} \tag{7}
$$

Functional Reliability: $RelF_{(X)}$ refers to the probability of a fault-induced computational error occurring during the process X. It should be noted that the assumption of independence, w.r.t. functional reliability, of the sub-processes of a process holds iff the probability of an error occurring in one sub-process does not affect the error occurrence in any other. In the current article, we consider only fault-induced errors. Hence, $RelF_{(X)}$ can be defined as shown in Eq. 8.

$$
RelF_{(X)} = \begin{cases} if\ sub-processes\ exist,\ \max\limits_{Y} RelF_{(y)}, \\ where,\ Y: set\ of\ sub-processes\ of\ X \\ else,\ Probability\ of\ error\ with\ fault-mitigation \end{cases} \tag{8}
$$

Timing Reliability: The Markov-chain model used in Section IV-A can be extended to a combination of sub-processes as shown in Figure 10. Further, the estimation methods on an absorbing Markov-chain discussed earlier can be used to find the CDF of the overall execution time. However, a faster method to estimate the CDF is by taking the product of the CDFs of the sub-processes. Assuming independence of the sub-processes, the probability distribution function (pdf) of the complete process is the convolution of the pdfs of the sub-processes. Hence, the CDF of the process is the product of the CDFs of the sub-processes. So, $CDF_{(X)} = \prod\limits_{y \epsilon Y} CDF_{(y)}$. Similarly, $T_{min(X)}$, the minimum execution time of process X is the sum of the minimum execution times of the sub-processes: $T_{min(X)} = \sum\limits_{y \epsilon Y} T_{min(y)}$

Overheads: As mentioned before, the $T_{avg(X)}$ can be estimated from the expanded Markov-chain based representation of the process. However, under the independence assumption, the overheads can be computed as shown in Eq. 9. $Ov_{Area(X)}$ is a determined by the mapping of tasks/modules/accelerators. Similarly, $Ov_{Pow(X)}$ is a function of the scheduling of the tasks and fault-mitigation methods implemented.

$$
T_{avg(X)} = \sum\limits_{y \epsilon Y} T_{avg(y)}; \quad Ov_{Energy(X)} = \sum\limits_{y \epsilon Y} Ov_{Energy(y)} \tag{9}
$$

V. CONCLUSION

With increasing susceptibility of hardware to physical faults, a comprehensive fault-aware cross-layer design approach is necessary. To this end, a cross-layer analysis framework is proposed. The proposed framework considers the effect of using different fault-detection and fault-tolerance methods, their coverage and the layer of implementation to determine the overall system-level effect. While the proposed framework is used to estimate the effect of some combination of fault-mitigation methods, each fault-mitigation method needs to be modeled appropriately to fit into the framework. To this end, a Markov-chain based analytical modeling methodology to model methods and their cross-layer interaction was proposed. Further research is required towards integrating the effects of dependencies in tasks and non-linear, non-sequential execution of tasks on real hardware platforms.

ACKNOWLEDGMENT

This work is supported in part by the German Research Foundation (DFG) within the Cluster of Excellence "Center for Advancing Electronics Dresden" (cfaed) at the Technische Universität Dresden.

REFERENCES

[1] A. Geist, "Supercomputing's monster in the closet," *IEEE Spectrum*, March 2016.

[2] P. Shivakumar, M. Kistler, S. W. Keckler, D. Burger, and L. Alvisi, "Modeling the effect of technology trends on the soft error rate of combinational logic," in *Dependable Systems and Networks*, 2002.

[3] S. Borkar, "Designing reliable systems from unreliable components: the challenges of transistor variability and degradation," *Micro, IEEE*, 2005.

[4] N. P. Carter, H. Naeimi, and D. S. Gardner, "Design techniques for cross-layer resilience," in *DATE*, 2010.

[5] S. S. Sahoo, B. Veeravalli, and A. Kumar, "Cross-layer fault-tolerant design of real-time systems," in *DFTS*, 2016.

[6] K. Lee, A. Shrivastava, M. Kim, N. Dutt, and N. Venkatasubramanian, "Mitigating the impact of hardware defects on multimedia applications: a cross-layer approach," in *Proceedings of the 16th ACM international conference on Multimedia*, 2008.

[7] T. Santini, P. Rech, A. Sartor, U. B. Corrêa, L. Carro, and F. R. Wagner, "Evaluation of failures masking across the software stack," *MEDIAN*, 2015.

[8] L. Leem, H. Cho, H. H. Lee, Y. M. Kim, Y. Li, and S. Mitra, "Cross-layer error resilience for robust systems," in *ICCAD*, 2010.

[9] J. Henkel, L. Bauer, H. Zhang, S. Rehman, and M. Shafique, "Multi-layer dependability: From microarchitecture to application level," in *DAC*, 2014.

[10] E. Cheng, S. Mirkhani, L. G. Szafaryn, C.-Y. Cher, H. Cho, K. Skadron, M. R. Stan, K. Lilja, J. A. Abraham, P. Bose, and S. Mitra, "CLEAR: Cross-Layer Exploration for Architecting Resilience - Combining Hardware and Software Techniques to Tolerate Soft Errors in Processor Cores," ser. DAC, 2016.

[11] A. Das, A. Kumar, and B. Veeravalli, "Aging-aware hardware-software task partitioning for reliable reconfigurable multiprocessor systems," in *CASES*, 2013.

[12] S. Rehman, M. Shafique, F. Kriebel, and J. Henkel, "Reliable software for unreliable hardware: embedded code generation aiming at reliability," in *ISSS+CODES*, 2011.

[13] J. G. Kemeny, J. L. Snell, and G. L. Thompson, *Introduction to Finite Mathematics*. Prentice Hall Inc, 1974.

Computing Fréchet Distance Metric based L-Shape Tile Decomposition for E-Beam Lithography

Arindam Sinharay[1], Pranab Roy[2] and Hafizur Rahaman[3]

Indian Institute of Engineering Science & Technology, Shibpur, Howrah, India

[1]arindam.sinharay@gmail.com, [2]ronmarine14@yahoo.co.in,
[3]rahaman_h@yahoo.co.in

Abstract— A Basic step in mask data preparation is layout decomposition. This is also a fundamental step in e-beam lithography (EBL) writing. To achieve better throughput in EBL - shape writing technique has been adopted. It is termed as L-shape fracturing by keeping similarity with rectangular fracturing. While implementing this new technique, a very thin/narrow feature is being generated termed as sliver. To achieve better manufacturability, proper methodology needs to be developed for minimization of sliver. This paper proposes a framework for L-shape fracturing with map matching with respect to Fréchet distance metrics. The framework starts with finding out concave vertices for a given layout as input. It then proceeds to next level by computing most densely populated concave points with a goal to form balanced partition of concave points of the given layout. In the subsequent steps, Fréchet distance based layout fracturing is finally accomplished. In this paper a heuristic function is proposed for sliver minimization as well as generation of L-shape features.

Index Terms— *layout decomposition, L-shape fracturing, sliver minimization mechanism, Fréchet distance metrics, e-beam lithography.*

I. Introduction *(Heading 1)*

E-BEAM Lithography (EBL) [2] is variedly adopted in the mask manufacturing. Because it controls the critical dimension (CD) and it intervenes the fidelity of printed image on wafer. As EBL is capable to generate accurate pattern, it is used in sub 22nm logic nodes [3]. It is well adopted in DPL/MPL [4][5] technology.

While writing image by EBL, a mandatory intermediate step is layout fracturing. There the input layout is being fractured into multiple non-overlapping rectangles. In the successive steps EBL writing machine prepares layout onto – wafer or mask, where decomposed rectangles are shot by variable shape beam (VSB) [6]. With the lowering of minimum feature size, a gradual increase in fractured rectangle number has been noticed. Very complex optical proximity correction (OPC) involves in creation of huge data volume and longer duration of writing time. As larger data volume and longer writing time directly proportional to manufacturing cost, high throughput is mostly targeted. This makes a bottleneck condition in EBL technology.

To solve the above mentioned manufacturing problem, a number of researchers proposed different optimization techniques [7], [8], [9]. These methods have demanded a substantial reduction in EBL writing time .Among all these all these methods L-SHAPE layout decomposition is very straight forward, effective, geometrical computing technique [7], [8]. It is capable of reduction mask generation, and also enhances the throughput. The overall lithographic cost has been bounded to reduce by the application of this technique.

VSB shots are generated for conventional EBL [12] when rectangular fracturing is considered. The demand of time is to develop novel fracturing algorithm based technique to generate L-shape in fractured layout. Theoretically, 50% of time and cost optimization in EBL writing can be achieved by adapting L-Shape fracturing technique.

We distinguish the differences of layout fracturing problem from well-studied orthogonal polygon decomposition problem [1] in a geometrical computation domain. In this application domain, CD control and yield optimization techniques have to be displayed .each of the VSB shots should be greater than a predefined and fixed threshold value, δ. A shot of width < δ, causes sliver generation. Optimization of Sliver [9] is an important objective to be achieved in layout fracturing.

Several studies have been made on controversial rectangular fracturing [10] [11] [12]. While in L-shape based decomposition, heuristics functions are directed towards horizontal decomposition only. This may result in inefficient sliver optimization and cannot be deployed in L-shape fracturing problem. Integer Linear Programming (ILP) based problem formulation have been proposed by Kahng et al.[10] matching based speed up technique for cost & time optimization has been discussed in[9][10]. Heuristic function driven rectangular shots and sliver optimization was proposed by Ma et al [12]. We noticed that L-shape fracturing is comparatively new application oriented approach of VSB technology. We are inspired by the invention off Lopez and Meheta [15]. They have presented L-shape fracturing for circuit component partitioning. We implemented these concepts in L-shape fracturing for VSB writing.

This paper identifies L-shape fracturing problem in EBL domain for VSB writing. This geometric fracturing problem upon orthogonal polygons has been studied. Then Fréchet distance based orthogonal polygon fracturing is done recursively to create L-shape and R-shape features.

The rest of the paper is organized as follows: - basic definition and problem discussion is given in section 2. Associated grid bounded rectangle of layout computation is elaborated in section 3. Section 4 illustrates the concave point finding methodology. Computation of closed curve has been presented in the section 5. In section 6, Frechet distance has been computed. Section 7 elaborates the layout decomposition process. Section 8 represents the results and

978-1-5386-3693-0/18 $31.00 © 2018 IEEE

discussions, followed by conclusion in Section 9. Type Style and Fonts

II. EASE OF USE

We would like to illustrate some basic notations and definitions for better understanding of the problem formulation. In this paper we only consider orthogonal fatty polygons. Hence the terms polygon and orthogonal polygon and orthogonal fatty polygons are used interchangeably.

Definition 1 (Orthogonal polygon [1]). *An orthogonal polygon is a polygon with axis aligned edges.*

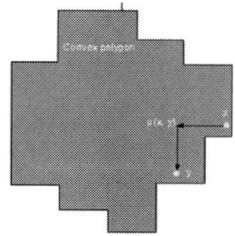

Fig. 1. (a) Orthogonal polygon; (b) Convex Polygon

Hence, edges alternate between horizontal and vertical axis. Moreover, the edges will coincide orthogonally i.e. the internal angle will be 90° or 270°. Orthogonal polygon shown in fig 1(a).

Definition 2 (Convex Polygon). *If for any two arbitrary points x and y inside the input polygon P, there exist a rectilinear path μ(x, y) with not more than single bend such that μ(x, y) connects x and y lies entirely inside P, than the polygon P is called convex polygon(fig2 (b)).*

Let P be an orthogonal polygon.

The bounding rectangle or minimum super scribing rectangle of P together with the extensions of its vertical and horizontal edges define an associate grid of P. With this associate grid of P optimal approximation of polygon can be done.

We can traverse the polygon clockwise or counterclockwise.

Definition 3 (concave vertex). *When the internal angle subtends at a corner point of the orthogonal layout polygon is 270, and then the corner point defined is concave point.*

Let α be the number of concave vertices (fig. 2(a)) in the layout polygon and total number of vertices in L_P be v, then by [16] we can say that there exists a relationship between α and v given by,

$$v = 2\alpha + 4$$

If α is odd number in a given L_P, then we call L_P is odd polygon; otherwise it is called even polygon.

Definition 4 (Cut). *A cut of L_P defined to be a horizontal or vertical line segment started at one corner point and terminated on another corner point or boundary of the polygon (as in fig. 2(b)).*

Our analysis will emphasize on the cuts needed to fracture the layout L_P. We have identified two types of possible cuts with respect to L_P as discussed below.

Definition 5 (Anchored cut [14]). *When a cut of L_P has at least one end point as corner point of the L_P, then it termed as anchored cut.*

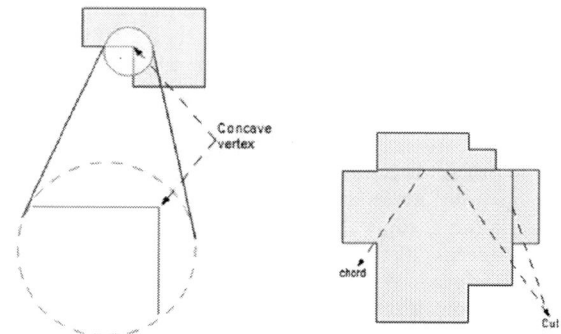

Fig. 2. (a) Concave vertex; (b) Example of Cut

There is special case of anchored cut, i.e when two ends of the cut arc corner points of L_P. In that case, we define it as chord.

Definition 6 (Floating cuts). *When a cut has none of its end points at a corner point of L_P, then the cut is floating.*

A floating cut is strictly interior to a given L_P. It can float (or move) with in a finite range of horizontal or vertical length or distance without altering its length. Our concern will be more focused on the anchored cuts and its special case, chord. We notice that in certain case, anchored cut ends up on midline [14] grid, which can be computed in polynomial time.

If a cut of or chord of a given polygon partition the set of concave vertices into two set of concave vertices having one or either of the set(s) odd number of concave vertices then the cut (or chord) is considered as odd cut(or odd chord)[15].

Definition 7 (R-shape). *An R-shape is polygon shaped in the form of rectangle.*

It is the polygon having least number of vertices. It is also the convex polygon of orthogonal type consisting of least number of vertices as well as a convex hull. The sides of the R-shapes are parallel to the polygonal edges.

Definition 8 (L-shape). *A polygon which looks like the letter 'L' is defined as L shape polygon.*

An L-shape polygon is least possible convex polygon but it is not a convex hull. It can be constructed with two R-shapes. In other word, an L-shape can be fractured into two R-shapes. To examine whether a polygon is L-shaped or not, we have to check the number of vertices. In this case it must have 6 vertices. Another properties of L-shape is that the existence of only one concave point.

978-1-5386-3693-0/18 $31.00 © 2018 IEEE   314

Definition 9 (Sliver length). *If the width of the bounding box of L-shape or R-shape is less than or equals to predefined number < δ, than it is called the length of sliver of the bounding box. Otherwise, the sliver length is 0.*

Problem (**L-shape based layout decomposition**). *Given an input layout in the form of orthogonal polygon, our objective is to fracture or decompose the orthogonal layout polygon into a set of L-shapes and/ or R-shapes to minimize number of shots. The fracturing should be initiated from the Frechet distance concave vertex. Meanwhile the sliver length of fractured shots should be minimized simultaneously.*

III. Prepare Your Paper Before Styling

Let L_P be the given orthogonal layout polygon (in fig 3(a)). An associated grid of L_P is the bounding rectangle of L_P in combination with the extension of its horizontal and vertical edges. Each edge of L_P denotes the grid segment.

Figure 3(b) shows the L_P, its associated grid and associated grid bounded rectangle. The following lemma gives us the fast approximation of L_P area.

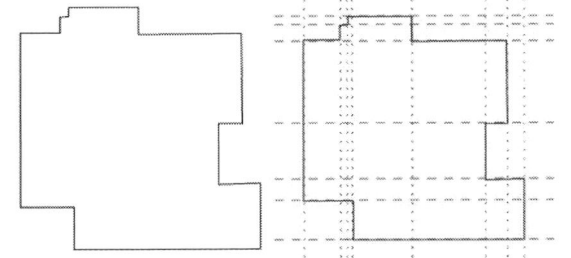

Fig. 3. (a)Given Layout Polygon; (b) Associated Grid of Layout Polygon

Lemma 1. *Let L_P be the given orthogonal polygon with n corner points. We assume RL_P be a rectangle. RL_P is the fast approximation of L_P if RL_P drawn on associated grid and none of the four sides intersects L_P.*

According to Lemma, there exists a fast approximating rectangle, RL_P which is on the associated grid of L_P(as in fig 4). As no side of RL_P is going to intersect L_P, there can be two probable cases,

Case 1. No point of LP coincides with the sides of RL_P.

Case 2. There are points (at least two corner points or multiple of 2 corner points) which coincide with a side or more sides of RL_P.

If we consider the case (2), then for exact upper approximation of 4 sides (which coincides with corner points of L_P), we get RL_P which fast approximate the area of L_P.

A. Complexity of lemma 1

Let there be n corner points of L_P. So there must be n number of edges of L_P which can be drawn in O (n) time. This implies that the associated grid of L_P can be computed in O (n) time. Now we can find four associated grid lines, which do not intersect the L_P and even of corner points lies on associated grid lines. As the max number of associated grid, lines are n, so it can be computed in O (n) time also. Therefore, the construction of RL_P can be done in constant time K with four associated grid lines. Therefore, we conclude that the computational complexity of this algorithm is O (n).

IV. Using the Template

Before going into the procedure of clustering concave corner points of layout polygon it is earnestly desired to find out concave corner points. By definition (3), we have developed the method for finding out concave vertices for a given layout polygon.

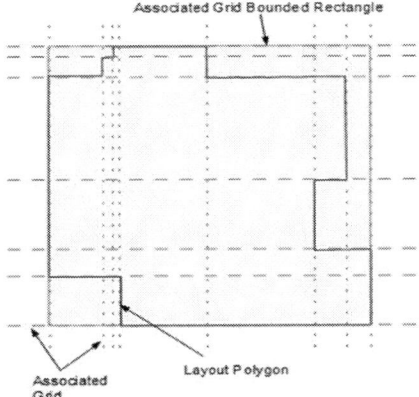

Fig. 4. Associated Grid Bounded Rectangle of Layout Polygon.

Let there be n number of corner points. Therefore, 2n numbers of data (one for x-coordinate and one for y-coordinate) have been taken in a form of circular Queue data structure. For examining the characteristic of ith corner point, we fetch out the duplet data of $(i+1)^{th}$ point and $(i-1)^{th}$ point. We have noticed that for i^{th} point to be a concave point, one of the following four conditions to be satisfied.

i) $\left(x^+ y^0 \wedge x^0 y^+ \right)$

ii) $\left(x^- y^0 \wedge x^0 y^- \right)$

iii) $\left(x^0 y^+ \wedge x^- y^0 \right)$

iv) $\left(x^0 y^+ \wedge x^+ y^0 \right)$

where

+ indicates increase in value
− indicates decrease in value
0 indicates no change in value
∧ indicates both conditions to be satisfied

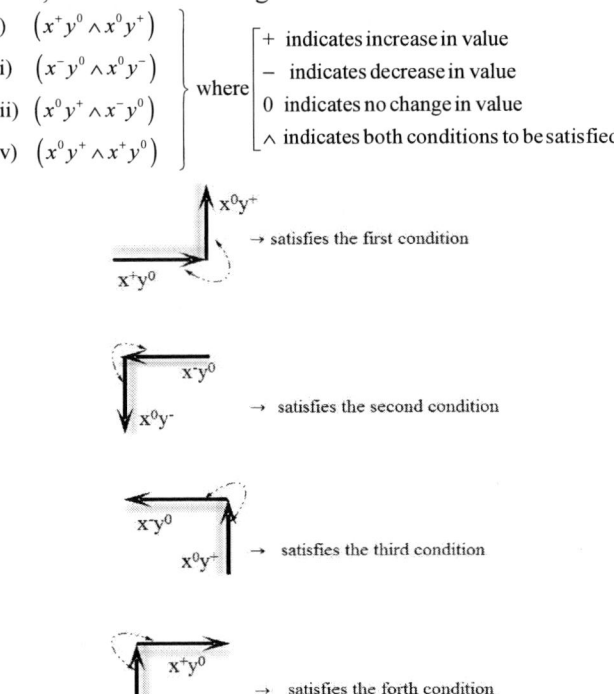

Fig. 5. Different Types of Concave Corners

In the fig. 5, Arrows indicates the directions of x and y coordinates to and from the considered point. Shadow in the above figure signifies the outer regions of the layout polygon (fig 6).

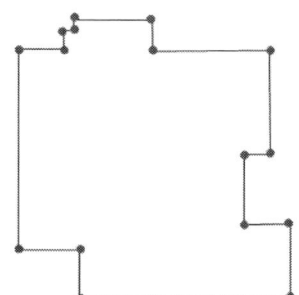

Fig. 6. Convex and Concave Corner Points of Layout Polygon

Following the above description, depending on the algorithm below, we have found out all the concave points of the layout polygon. Then we place a pointer to each concave point (coordinates of each point) and store them in a array of pointer.

Once the corner points have been computed the next job is to find the point of initiation vertex counting. The point of initiation can be computed as follows:-

➤ Find the minimum x-coordinate of vertices.

➤ Find the minimum y-coordinate of the vertex within the set of minimum x-coordinate.

For 'n' number of vertices present in the given orthogonal layout polygon, 'n' number of x-coordinates are present. This n number of data can be efficiently sort in $O(n \log n)$ time.

Proposition: - *Except for L-shape orthogonal polygon, any orthogonal polygon of 2n vertices can have at most n-number of corner points. Among these corner points, at most n-number of corner points can have minimum value of x-coordinates. For L-shape tile, it is 2 corner points having minimum x-coordinates values. This is same for rectangular shaped tile also.*

Continuing with this discussion, it is clear that among the set of vertices having minimum value(s) of x-coordinate(s), minimum value of y-coordinate can be found in $O(n \log n)$ time. The following lemma is obvious from this analysis.

Lemma: - the point of initiation can be found in $O(n \log n)$ time.

V. COMPUTING CLOSED CURVE FROM LAYOUT POLYGONAL EDGES

If two curves share, common starting and ending points, then that can be defined as closed curves. For example, the parameterized curve ($\cos \varphi, \sin \varphi$), where $\varphi \in [0, 2\pi]$.

In the present context, from the point of initiation, vertex processing has been started in both directions, which are orthogonal to each other. This can be seen as visiting the vertices of layout polygon on clockwise and counter clockwise direction. These two traversals end at the common point where they meet. As the orthogonal polygon always have even number of vertices, two polygonal curves

will have equal number of vertices. These two curves share the starting and ending points in their common.

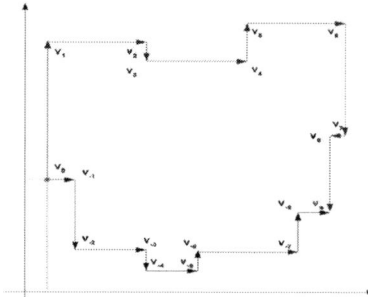

Fig. 7. Initial Layout with corner points

Following observations have been pointed out and they can be put down together as followings.

➤ These two polygonal curves are closed by construction.

➤ Each of the curves is formed with exactly half the edges (in terms of number of edges) of the layout polygon.

➤ Both the curves are orthogonal.

In figure 7, the closed curves are shown. C_1 has been constructed by the following sequence of vertices $\langle v_0, v_1, v_2, v_3, v_4, v_5, v_6, v_7, v_8, v_9 \rangle$.

Similarly $\langle v_0, v_{-1}, v_{-2}, v_{-3}, v_{-4}, v_{-5}, v_{-6}, v_{-7}, v_{-8}, v_{-9} \rangle$ makes the C_2. From layout polygon in figure C, It is clear that $v_9 = v_{-9}$.

The closed curves C_1 and C_2 can be formed by two linear arrays. These two linear arrays has have starting and ending data I their common. It can be seen that C_1 and C_2 are formed with edges. Therefore, parameterization is needed.

VI. DEFINING PARAMETERIZATION OF C1 AND C2

Let $C_i : [0, n] \rightarrow F$ be a polygonal curve. The sequence $\langle v_0, v_1, ..., v_n \rangle$ of end points of the line segments of C is denoted by $\sigma(C_i)$. Let C_1 and C_2 be polygonal curves and $\sigma(C_1) = \langle v_0, v_1, ..., v_n \rangle$ and $\sigma(C_2) = \langle v_0, v_{-1}, ..., v_n \rangle$ the corresponding sequence (in fig 8(a), C_1 and C_2).

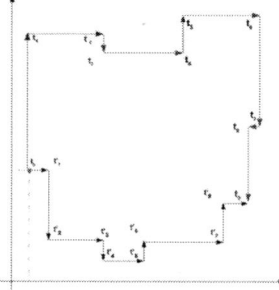 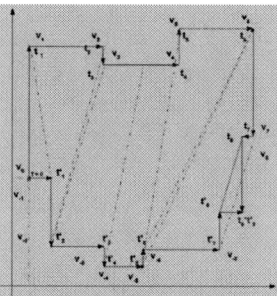

Fig. 8. (a)C_1 in blue color, clockwise and C_2 in black color, and anti-clockwise; (b) Coupling between C_1 and C_2

In the present algorithm, C_1 is parameterized by the function $\alpha(t)$ and C_2 is parameterized by the function $\beta(t)$. For example, in case of C_1, $\alpha_1(t) = (t_0 \rightarrow t_1)$ represented by the edge $\langle v_0, v_1 \rangle$. In the same way, other line segments of the C_1 are parameterized.. For the closed curve C_2, $\beta_1(t') = (t_0 \rightarrow t_1)$ is represented by the edge $\langle v_0, v_{-1} \rangle$. Figure 8(a) shows the two closed curves along with

parameterizes line segments those have constructed the two curves C_1 and C_2 and also their corresponding edges in the layout polygon.

VII. COMPUTING DISCRETE FRÉCHET DISTANCE

To compute the Discrete Fréchet Distance between C_1 and C_2, we need to establish a coupling between C_1 and C_2. Let a coupling L between C_1 and C_2 is a sequence $\left(v_{a_1}, v_{b_1}\right), \left(v_{a_2}, v_{b_2}\right), \ldots, \left(v_{a_m}, v_{b_m}\right)$ of distinct pairs from $\sigma(C_1) \times \sigma(C_2)$ such that $a_1 = t_p, b_1 = t_p, a_m = t_p = t_s, b_m = t_p = t_s$ and for all $i = 0, 1, \ldots, t_p$, we have $\left(a_{i+1} = a_i + 1\right), \left(b_{i+1} = b_i + 1\right)$. The coupling has shown in figure 8(b).

Thus the coupling has to follow the order of the points of the C_1 and C_2. The length $\|L\|$ of the coupling L is the length of the longest distance in L, that is, $\|L\| = \max_{i=1,\ldots,m} d\left(v_{a_i}, v_{b_i}\right)$. Therefore, for the polygonal curve C_1 and C_2, their Discrete Fréchet Distance is defined by Frechet Distance metric based L-shape tile decomposition fracturing Frechet Distance metric based L-shape tile decomposition fracturing (in fig 8(b)),

$$d_{dF}(C_1, C_2) = \min \{\|L\| \mid L \text{ is a coupling between } C_1 \text{ and } C_2.\}$$

It is clear from figure 9, $t_8 t_8$ is the Frechet distance between C_1 and C_2. Now, we get the t_8 as the point of initiation of the cut. A vertical cut in the positive direction of Y-axis will (1) imbalance the two resulting orthogonal polygon (2) one resulting polygon will be Rectangle and (3) there will generate a sliver. A horizontal cut will not generate a terminating R-shape in any one of the decomposed sub problem and two decomposed polygons will be balance in terms of area and numbers of concave corner points. Therefore, in this case, a horizontal cut will be more appropriate. We note that this deciding result varies on the above discussed deciding factors.

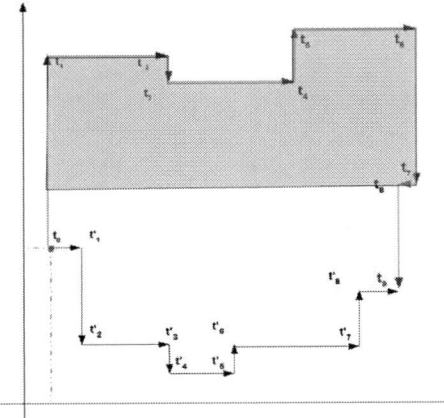

Fig. 9. Horizontal cut through t_8

The above discussed coupling measure can be efficiently computed by dynamic programming. It needs only two simple data structures such as arrays. In figure 10(a) and

10(b), decomposition of first sub problem and final decomposition of layout polygon have been shown respectively.

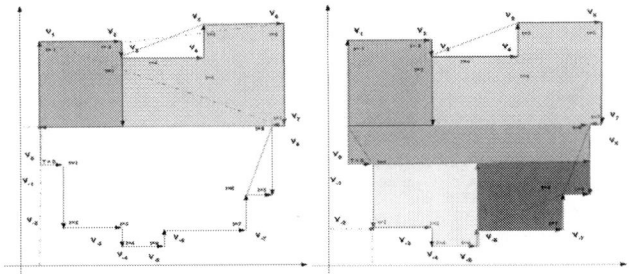

Fig. 10. (a) Decomposing first sub problem;
(b) Final decomposition of L_P

VIII. EXPERIMENTAL RESULT

Experiments for the proposed Frechet Distance metric based L-shape tile decomposition fracturing algorithm have been carried out through MATLAB 2014a computational geometry toolbox. The experiments are carried out on Intel core i3-2330M processor machine on Windows 2007 operating system with 4GB DDR3 RAM. ISCAS 85&89 benchmarks are scaled downed to 28nm logic node followed by accurate lithographic simulation performed to Metal 1 layers. Table 1 shows the results of experiments driven by the developed algorithms discussed in this paper. The approaches in [13] and [16] have been carried over in present experimental environment and the results are compared with the newly developed algorithm under consideration. All fracturing methods are evaluated with the consideration of sliver length 5nm. For each method "shots", "sliver" denotes shot numbers, total sliver length.

It is interesting to note that for all the benchmarks considered, our proposed method performs better than [13], and are better than the results reported in [16] for most of the benchmarks.

IX. CONCLUSION

In this paper, we propose an algorithm for EBL with the new Frechet Distance metric based L-shape tile decomposition fracturing for sliver minimization. This is a novel approach for integrating Frechet distance metric. This has shown best metric for sliver minimization. This approach claims to be the first algorithmic endeavour with the Frechet metric in EBL L-shape decomposition with sliver minimization. The current researchers carried out further experiments and analysis on this field of layout fracturing using above metric. Due to space constraint, we omit some fundamental discussions, detailed experimental analysis and few more snapshots.

REFERENCES

[1] Joseph O'Rourke, Orthogonal Polygon, Chapter 2, Art Gallery Theorems and Algorithms, Oxford University Press, 1987, pp.31-80.

[2] L. Pain, M. Jurdit, J. Todeschini, S. Manakli, B. Icard, B. Minghetti, G. Bervin, A. Beverina, F. Leverd, M. Broekaart, P. Gouraud, V. D. Jonghe, P. Brun, S. Denorme, F. Boeuf, V.Wang, and D. Henry, "Electron beam direct write lithography flexibility for ASIC

manufacturing: an opportunity for cost reduction," in Proc.of SPIE, vol. 5751, 2005.

[3] Y. Arisawa, H. Aoyama, T. Uno, and T. Tanaka, "EUV flare correction for the half-pitch 22nm node," in Proc. of SPIE, vol. 7636, 2010.

[4] A. B. Kahng, C.-H. Park, X. Xu, and H. Yao, "Layout decomposition for double patterning lithography," in IEEE/ACM International Conference on Computer-Aided Design (ICCAD), 2008, pp. 465–472.

[5] B. Yu, K. Yuan, B. Zhang, D. Ding, and D. Z. Pan, "Layout decomposition for triple patterning lithography," in IEEE/ACM International Conference on Computer-Aided Design (ICCAD), 2011, pp. 1–8.

[6] S. Jiang, X. Ma, and A. Zakhor, "A recursive cost-based approach to fracturing," in Proc. of SPIE, vol. 7973, 2011.

[7] E. Sahouria and A. Bowhill, "Generalization of shot definition for variable shaped e-beam machines for write time reduction," in Proc.of SPIE, vol. 7823, 2010.

[8] A. Elayat, T. Lin, E. Sahouria, and S. F. Schulze, "Assessment and comparison of different approaches for mask write time reduction," in Proc. of SPIE, vol. 8166, 2011.

[9] K. Yuan, B. Yu, and D. Z. Pan, "E-Beam lithography stencil planning and optimization with overlapped characters," IEEE Transactions on

Computer-Aided Design of Integrated Circuits and Systems (TCAD), vol. 31, no. 2, pp. 167–179, Feb. 2012.

[10] A. B. Kahng, X. Xu, and A. Zelikovsky, "Yield-and cost-driven fracturing for variable shaped-beam mask writing," in Proc. of SPIE, vol.5567, 2004.

[11] ——, "Fast yield-driven fracture for variable shaped-beam mask writing," in Proc. of SPIE, vol. 6283, 2006.

[12] B. Dillon and T. Norris, "Case study: The impact of vsb fracturing," in Proc. of SPIE, vol. 7028, 2008.

[13] X. Ma, S. Jiang, and A. Zakhor, "A cost-driven fracture heuristics to minimize sliver length," in Proc. of SPIE, vol. 7973, 2011.

[14] J.O. Rourke, G.Tewari, "The structure of optimal partitions of orthogonal polygons into fat rectangles" , 14th Canadian Conference on Computational Geometry - CCCG02, Volume 28, Issue 1, May 2004, pp. 49–71.

[15] M. A. Lopez and D. P. Mehta, "Efficient decomposition of polygons into L-shapes with application to VLSI layouts," ACM Transactions on Design Automation of Electronic Systems (TODAES), vol. 1, no. 3, pp. 371–395, Jul. 1996.

[16] B.Yu, J-R.Gao and, D.Z.Pan, "L-Shape Based Layout Fracturing for E-Beam Lithography", Design Automation Conference (ASP-DAC), 2013 18th Asia and South Pacific, Yokohama, pp.249-254.

TABLE 1: RESULTS AND COMPARISON

CIRCUITS	POLYGON	[13]		[16]		FRECHET ALGORITHM		COMPARISON(BETTER THAN)	
		SHOT#	SLIVER (MM)	SHOT#	SLIVER (MM)	SHOT#	SLIVER (MM)	SHOT#	SLIVER
C432	1109	6898	48.3	4214	7.4	3158	4.81	[13], [16]	[13], [16]
C499	2216	13397	96	8112	11.8	6950	9.95	[13], [16]	[13], [16]
C1355	3262	23283	185.2	13936	24.8	12559	16.54	[13], [16]	[13], [16]
C2670	7933	56619	525.4	34102	114.8	30361	59.78	[13], [16]	[13], [16]
C7552	21253	151643	1334.6	91157	290.7	85752	118.49	[13], [16]	[13], [16]
S1488	4611	37126	303.7	22099	31.6	20259	33.25	[13], [16]	[13], [16]
S38417	67696	454307	4040.2	275054	729	248517	429.10	[13], [16]	[13], [16]
S35932	26267	163956	1470.4	100629	284	75365	149.81	[13], [16]	[13], [16]
S38584	168319	1096363	10045.2	666906	1801.7	620785	906.77	[13], [16]	[13], [16]

2018 31th International Conference on VLSI Design and 2018 17th International Conference on Embedded Systems

2nd Order Sallen Key Switched Capacitor LPF with N-type Transistors

Kamal Chapagai, Pydi Bahubalindruni and Nishtha
Electronics and Communications Department
Indrapastha Institute of Information Technology
110020 - Okhla Phase 3, Delhi
Email: kamal16130@iiitd.ac.in

Abstract—**This paper presents active low pass RC and switched capacitor filters with only N-type transistors. Two different high-gain OpAmp configurations with positive feedback are explored in this paper: One circuit based on cascaded inverters and the other based on capacitor bootstrapping. While the cascaded inverter configuration gives a gain of more than 50 dB and bandwidth of 48 MHz with 210 μw of power consumption, OpAmp with capacitor bootstrapping gives a gain around 40 dB and bandwidth of 2 MHz consuming 70 μw power. Circuit simulations were carried out in standard 180 nm technology. With these OpAmps, RC and SC filter implementations are presented. SC filter using OpAmp with cascaded inverter gives a THD of 5.36 % with 946.6 μw of power consumption, while SC filter with capacitive bootstrapped OpAmp give THD of 7.47 % with 351.7 μw of power consumption.**

Index Terms—**N-type transistors, switched capacitor, filter, OpAmp, CMFB, bootstrapping, positive feedback.**

I. INTRODUCTION

Switched capacitor circuits are important building blocks in signal processing systems. Their main advantages are low power consumption, relative lower chip area due to absence of resistors and high accuracy since the performance is function of ratio of capacitance instead of isolated capacitance, which is typically insensitive to the variations in processing conditions.

Switched capacitor filters with low-power consumption and high linearity have been reported in CMOS technology [1],[2]. However, many emerging technologies, namely amorphous indium-gallium-zinc oxide (a-IGZO) and amorphous hydrogenated-silicon (a-Si:H) thin-film transistors (TFTs), that allow flexible electronics for potential real world applications (wearable technology, smart packaging) lacks stable reproducible complementary (P-type) device. Circuit design with these technologies has to be carried out only with N-type transistors [3].

Filters are important functional blocks in afore mentioned applications, however, there is very limited work done in this area[4], which is confined to RC implementation.

Design of circuits using only N-type transistors impose several restrictions as mentioned below:

- In switched capacitor circuits, switches are important functional elements. It is challenging to implement high performance switch since it is not possible to implement transmission gate. All switches need to be implemented

with only N-type transistors, which typically shows non-linear on-resistance with respect to the input signal variations. This will impact the overall circuit performance.
- Due to the use of NMOS transistor as load, the overall gain of the amplifier (common source amplifier with N-type transistor as load) is very low, requiring gain enhancement techniques.

Operational amplifiers are inevitable building blocks in switch capacitor filters. It is challenging to design high-gain OpAmp only with N-type transistors. An internally compensated, process independent operational amplifier was presented in [5] consisting of input stage, differential to single ended conversion stage, cascode stage to reduce miller effect and output stage. With 26 transistors the gain presented was 50 dB, UGB of 5 MHz and power consumption of 150 mW.

Another design was reported in [6] using 25 NMOS transistors, showing a gain of 52 dB and GBW of 1 MHz. A high gain common source amplifier with bootstrapped load topology was presented using only N-type transistors in 0.35 μm CMOS technology [7]. It has a gain of 50 dB with bandwidth of 1 MHz and a power consumption of 109 μW.

An implementation technique for a common source amplifier with positive feedback for a-Si:H TFTs, containing only N-type transistor was presented [8]. The gain presented for a common source amplifier with positive feedback was around 10 dB. Using the same principle, an Operational Amplifier and a differential difference Amplifier (DDA) in a-Si:H TFT with a gain and GBW of 43 dB, 21.6 KHz and 40 dB, 5 KHz respectively were presented [3],[4].

In a-GIZO TFTs, pseudo-CMOS and positive feeback based operational amplifiers were presented in [9],[10] to give a gain around 22.5 dB. However, a common source amplifier with bootstrapped load presented in [11] has a gain of 34 dB. Nevertheless, the design of filter circuits with only N-type transistors is scarce (NMOS, a-Si:H TFT or IGZO TFT).

For the first time, active low-pass switch capacitor filter is designed and implemented using only N-type transistors. This design can be directly adapted to various TFT technologies that are lacking complementary device. This paper also presents a comparative study of OpAmps with positive feedback and Operational amplifier with bootstrapped load in terms of gain, bandwidth, power and complexity under same

978-1-5386-3693-0/18 $31.00 © 2018 IEEE

bias condition. With the designed OpAmp, an implementation of sallen key RC and switched capacitor low pass filters are presented.

The rest of the paper is organized as follows. Section II presents techniques for gain enhancement in OpAmps and design considerations of RC and SC filters. Section III presents the simulation results of the OpAmps and the filters and finally section IV draws conclusions.

II. GAIN ENHANCEMENT TECHNIQUES

A. Gain enhancement techniques for Common source Amplifiers

For a common source amplifier, the net impedance offered by the load transistor translates to the overall gain of the amplifier. Most simplest forms are:

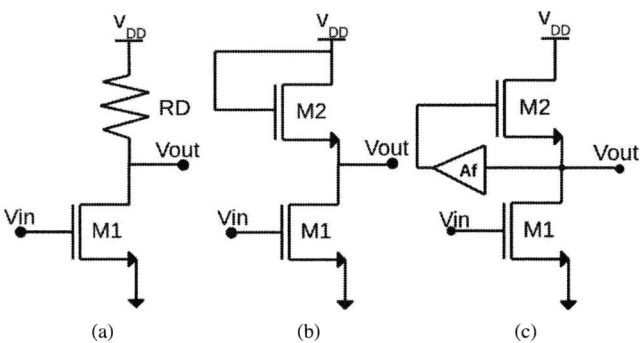

Fig. 1. Common source amplifier with only N-type transistor(s) (a) resistive load. (b) diode connected load. (c) positive feedback.

1) Common source with resistive load: Assuming that the channel length modulation is neglected, a common source amplifier with resistive load (see Fig. 1(a)) gives a gain of:

$$A_{v1} = -gm1 * R_D. \tag{1}$$

The gain of the amplifier is limited by the value of load resistor R_D and transconductance of the driving transistor $gm1$, which is directly proportional to the aspect ratio of the driving transistor. Large load resistor increases the power consumption while large aspect ratio of driving transistor increases parasitic capacitance and area.

2) Common source with diode connected transistor: A common source amplifier with diode connected load (see Fig. 1(b)) gives a gain of:

$$A_{v2} = -\frac{gm1}{gm2 + gds_1 + gds_2} \tag{2}$$

For a technology, if the transconductance is considered much larger i.e. $gm >> gds$, the gain arrives to the ratio of the root of the aspect ratios of the two transistors as :

$$A_{v2} = -\sqrt{\frac{(W/L)_1}{(W/L)_2}} = -\frac{W_1}{W_2} \tag{3}$$

For the same channel length, gain is proportional to the widths. For large gain the width of driving transistor is

significantly larger than the load transistor. Large width again gives rise to large parasitics and limited bandwidth due to miller capacitance.

3) Common source with feedback: A common source amplifier with feedback (see Fig. 1(c)) gives a gain of:

$$A_{v3} = -\frac{gm_1}{(1 - A_f)gm_2 + gds_1 + gds_2} \tag{4}$$

The basic idea here is to make the gain of the feedback network close to 1 such that the overall gain of the amplifier resembles the gain given by CMOS topology:

$$A_{v3} = -\frac{gm_1}{gds_1 + gds_2} \tag{5}$$

B. Gain enhancement techniques with Feedback

Two circuits have been used for gain enhancement with feedback technique:

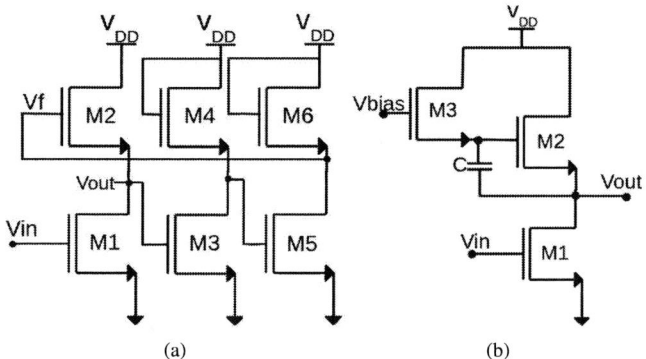

Fig. 2. Gain enhancement with positive feedback (a) Cascade of two inverters (method 1). (b) Capacitive bootstrapping (method 2).

1) Gain enhancement with cascade of inverters (method I): A gain enhancement topology with cascade of two inverters is shown in Fig. 2(a). It uses a two-stage common source feedback network giving a feedback gain equivalent to:

$$Af = \frac{gm_3 * gm_5}{gm_4 * gm_6}. \tag{6}$$

The feedback gain can be made close to 1 by carefully choosing the aspect ratio of the feedback transistors and the biasing currents. This principle is used in a differential Amplifier (OpAmp1) as in Fig. 3(a). The OpAmp is a self compensated and self biased circuit using common-mode-feedback (CMFB). Transistors M1-M4 form the first stage differential amplifier. It has an over all gain as in equation (4). The transistors M5 and M6 are the self biasing mirrored transistors with $I_5 = I_6$. The biasing currents $I_5 = I_6$ is direct function of $V_{CM1} = V_{GS6} = V_{GS5}$. V_{CM} is proportional to the common-mode voltage at node A and A', aspect ratios of common-mode sensing and mirror transistors. The common mode sensing (not shown on schematic) is formed by two common drain transistors with shorted source. The gate-drain voltage of mirrored biasing transistors is the source voltage of sensing transistor. This gives rise to a negative feedback

978-1-5386-3693-0/18 $31.00 © 2018 IEEE   320

system such that any change in output signal variations will be compensated by the biasing network. The internal transistors M7-M10 form a positive feedback network while M11 and M12 are the biasing transistors for the feedback network. The common-mode feedback based biasing and compensation of this feedback circuit uses same theory as in the first stage. From the small signal analysis we can find the gain of the two stages as:

$$StageI : A_1 = \frac{gm_1}{gm_3(1 - A_f) + g_{ds1} + g_{ds3}} \quad (7)$$
$$= \frac{gm_2}{gm_4(1 - A_f) + g_{ds2} + g_{ds4}}$$

$$FeedbackStage : A_f = \frac{gm_7}{gm_9 + g_{ds7} + g_{ds9}} \quad (8)$$
$$= \frac{gm_8}{gm_{10} + g_{ds8} + g_{ds10}}$$

Transistors M13-M16 form the differential to single ended conversion. There are two paths in the circuit, one leading from M13 as source follower with M15 and M16 as load through M16, which is a common source amplifier (inverting) with M14 as load. The second path is through M14 which is a source follower with M16 and M15 as load. The net voltage at the output node are in phase. The total gain of the circuit as given by [5] is:

$$A_2 = \frac{gm_{14}}{2(gd_{16} + gm_{14})} \left[1 + \frac{gm_{13}gm_{16}}{gm_{14}(gm_{13} + gm_{15})} \right] \quad (9)$$

2) Gain enhancement with capacitive bootstrapping (method II): A gain enhancement technique with capacitive bootstrapping is shown in Fig.2(b). The bootstrapping consists of the capacitor C and transistor M3. The basic idea is to increase the equivalent active load value to a point where the global gain gets very close to the gain achieved by standard CMOS based single stage common source amplifier [11]. For proper operation, the transistor M1 and M2 will operate in saturation region while transistor M3 is cutoff. The gain of this circuit is still given by the equation:

$$A_v = \frac{gm_1}{gm_2(1 - A_f) + g_{ds1} + g_{ds2}} \quad (10)$$

where the path through the capacitor C and the transistor M3 forms the positive feedpath and the gain of the path as given by [11] is:

$$A_f = \frac{C}{C + C_{eq}} \quad (11)$$

where C_{eq} is the parasitic capacitance of the feedback transistor given by:

$$C_{eq} = C_{gs3} + C_{gd2} \quad (12)$$

Therefore, the overall gain of the amplifier is directly dependent on the value of the feedback capacitor and the aspect

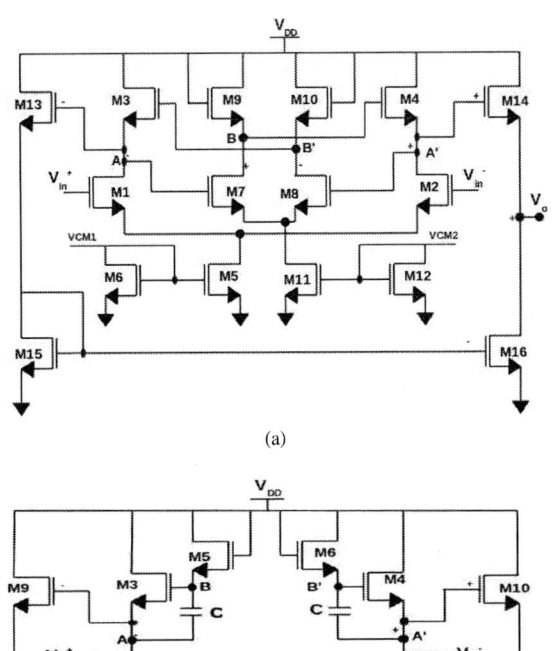

(a)

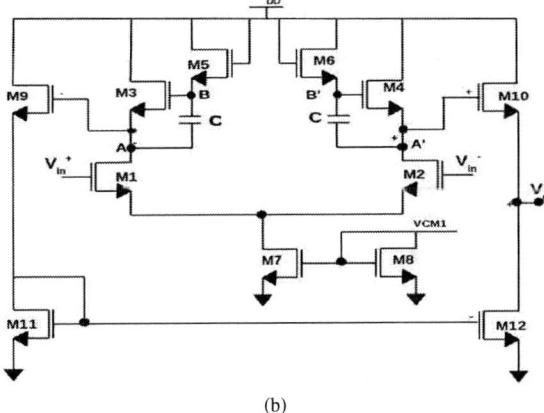

(b)

Fig. 3. OpAmp with gain enhancement (positive feedback) (a) OpAmp with cascaded inverter (OpAmp1). (b) OpAmp with Capacitive Bootstrapping (OpAmp2).

ratio of the feedback transistor. One of the uniqueness of the amplifier with bootstrapped load is that the response shows a bandlimited/bandpass response. i.e. it has both lower and higher cutoff frequency and the lower cutoff frequency value can be controlled by the aspect ratio of biasing transistors (M5/M6) and the bootstrapping capacitance (C).

Using the same principle, an operational amplifier (OpAmp2) is designed as shown in Fig. 3(b). The operation of the OpAmp can be derived from that of Fig. 2(b) and Fig. 3(a). The inner loop consisting of M1-M6 form the driving and load transistors for the first gain stage with an overall gain given by equation (12)-(14). The outer loop trasistors M9-M12 form the differential to single ended conversion, giving the gain as per equation (11). The overall gain is now:

$$A_v = \frac{gm_1}{gm_2(1 - A_f) + gds_1 + gds_2} \cdot \frac{gm_{10}}{2(gds_{12} + gm_{10})} \quad (13)$$
$$\cdot \left[1 + \frac{gm_9gm_{12}}{gm_{10}(gm_9 + gm_{11})} \right]$$

978-1-5386-3693-0/18 $31.00 © 2018 IEEE

III. RC AND SC FILTER DESIGN

With these two high-gain OpAmps (OpAmp1 and OpAmp2) using only N-type transistors, active low pass Sallen Key RC filter and SC filter were implemented. The circuit presented in Fig. 4(a) is a standard second order Sallen key LPF with transfer function of the form given in [12] as:

$$A(s) = \frac{1}{1 + as + bs^2} \qquad (14)$$

Where :

$$a = \omega_c C_1 (R_1 + R_2) \qquad (15)$$
$$b = \omega_c^2 C_1 C_2 R_1 R_2) \qquad (16)$$
$$\omega_c = \frac{1}{RC} \qquad (17)$$

With consideration that $\mathbf{R_1} = R_2 = R$ and $\mathbf{C_1} = C_2 = C$ and unity gain configuration, The value of $\mathbf{Q} = 0.5$. By choosing a suitable value of C the values of R can be determined for a given cutoff frequency ω_c.

Usage of resistors in circuit increases the chip area, inaccuracy and power consumption. Hence, switched capacitor filters are prefered in integrated circuits. A stray-insensitive switched capacitor (see Fig. 4(b)) filter is designed by direct replacement of switched capacitor equivalent of resistor [13]. The transformation equation is given by $R = T/C$.

IV. CIRCUIT SIMULATION AND RESULTS

All the circuit simulations are carried out in standard 180 nm technology using a power supply of 1.8 V. The simulation response for circuit in Fig. 2(b) is presented in Fig. 5(a) and Fig. 5(b) with respect to different values of C and aspect ratios of biasing transistor (M3).

For both the OpAmp topologies (Fig. 3(a),3(b)), the first gain stage with feedback constitute the dominant gain stage as well as the dominant pole. Therefore, the design is aimed at optimizing the gain at the first stage and phase at the second. The second stage is a buffer circuit with almost unity gain, leading to ineffective use of frequency compensation between the two stages. Hence, large aspect ratio of the second stage is used for better phase response. The aspect ratios of the transistors are shown in Table. I.

TABLE I
TRANSISTORS SIZING IN OPAMPS

Transistor/Topology	OpAmp1	OpAmp2	Remarks
M1,2 (μm)	4	10	-
M3,4 (μm)	0.6	2	-
M5,6 (μm)	2	3	-
M7,8 (μm)	8	3	-
M9,10 (μm)	5.8	3	-
M11,12 (μm)	2	3*	*CM Sensing
M13,14 (μm)	15	C=10p	-
M15 (μm)	2	-	-
M16 (μm)	20	-	-
M17,18 (μm)	4**	-	**CM Sensing
M19,20 (μm)	1**	-	**CM Sensing

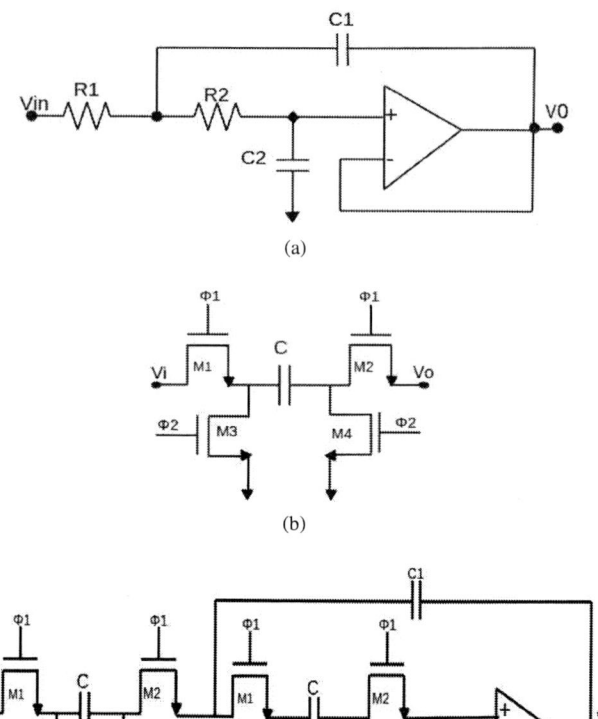

Fig. 4. Active Sallen Key filters based on (a) RC implementation. (b) Resistors implemented with switched capacitor equivalent. (c) SC implementation.

The response of the OpAmps are shown in Fig. 6(a) and Fig. 6(b). The OpAmps are designed for a gain of 40 dB and optimized for gain-bandwidth. The results of the OpAmps are shown in Table. II. Fig. 7 shows the open loop large signal transient response of the OpAmp.

TABLE II
PARAMETER OF OPAMPS

Transistor/Topology	OpAmp1	OpAmp2	Previous Work [4]
Technology node	CMOS 180	CMOS 180	a-Si TFT
Vdd (v)	1.8	1.8	25
Gain (dB)	51.43	41.3	40
GBW (MHz)	48.74	61.56	5.8K
Phase (deg)	86	-	76
Power (mW)	0.209	0.071	3.465
CMRR (dB)	99.89	80.34	-
PSRR (dB)	66.24	88.37	-
Load	C = 1pF	C = 1pF	15pF,1M Ω

Using these OpAmps designed above, Sallen Key unity gain, RC and SC Low pass filters were designed and implemented. The resistance $R = 159 \, MOhm$ and $C = 1 \, pF$

978-1-5386-3693-0/18 $31.00 © 2018 IEEE

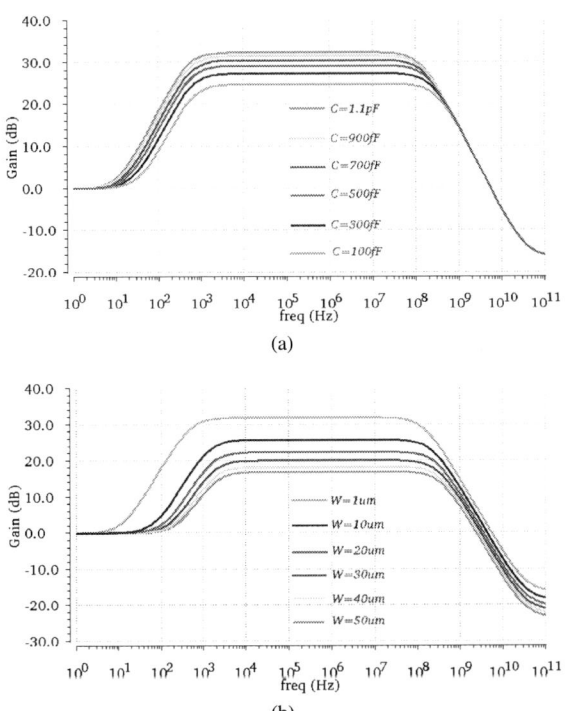

(a)

(b)

Fig. 5. Frequency response (a) Frequency response for various C (W3 = 1 μ m). (b) Frequency response for various W3 (C = 1 pF).

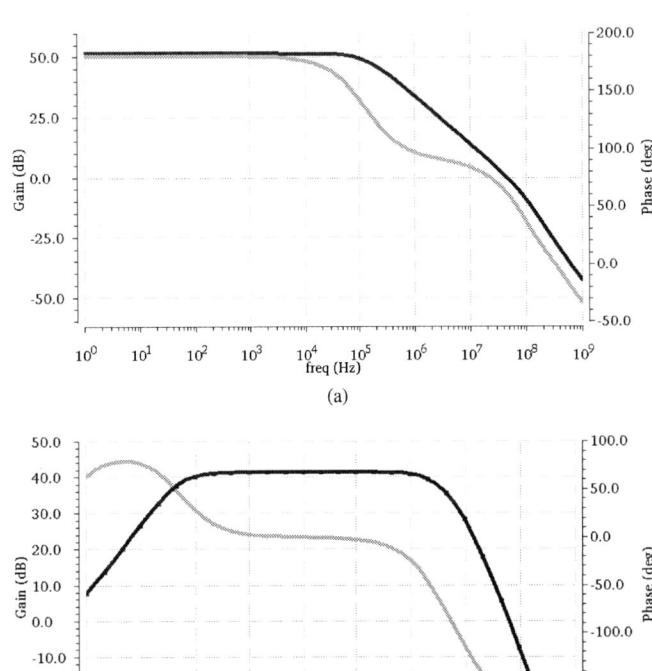

(a)

(b)

Fig. 6. Frequency response of OpAmps with positive feedback (a) OpAmp1 (b) OpAmp2.

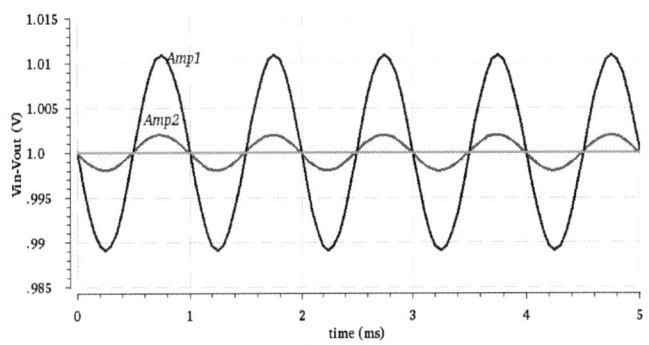

Fig. 7. Transient response of OpAmps.

for a cutoff frequency of 1 kHz. The response and parameters are shown in Fig. 8. and Table. III respectively. A switched capacitor filter was also designed and implemented. With a clock frequency of 100 kHz, the capacitor equivalent of resistor R = 159 MOhm is C_{eq} = 62.89 fF. PSS and PAC analysis of the circuits give responses as in Fig. 9 and parameters as in Table. III.

From Fig. 8(b) and 9(b), it can be seen that there is a dependance of gain to frequency at very low frequency for the filter. It is due to the bandpass effect of OpAmp2. The response of the filters in Fig. 8 shows that, the gain rolloff in filter with OpAmp2 is decreased when compared to filter with OpAmp1. However, there is significant decrease in the power consumption. Switching from RC filter to switched capacitor filter have shown improvement in terms of input and output noise due to absence of resistor in the SC circuit. However, the RC network is still superior in terms of the THD.

ACKNOWLEDGMENT

This work is supported by early career research grant with project ref. ECR/2017/000931.

V. CONCLUSION

Two configurations of gain enhancements are explored only with N-type transistors. One based on positive feedback and the other based on bootstrapped feedback. Using these techniques, high gain, low power OpAmps are presented. The OpAmps with bootstrapped load has decreased power consumption for same gain but cannot operate at DC. Using these two configurations of OpAmp, design and implementation of Sallen-Key Low pass Filter is presented in RC network and switched capacitor network for a 180 nm CMOS technology. These designs can be directly adaptable in various emerging TFT technologies that allows low-cost flexible electronics to implement various real-world applications.

REFERENCES

[1] C. Shi, Y. Wu, and M. Ismail, "A NOVEL LOW-POWER HIGH-LINEARITY CMOS FILTER," Proc. 43rd IEEE Midwest Symp. on circuits and Systems, Aug. 2000.

978-1-5386-3693-0/18 $31.00 © 2018 IEEE

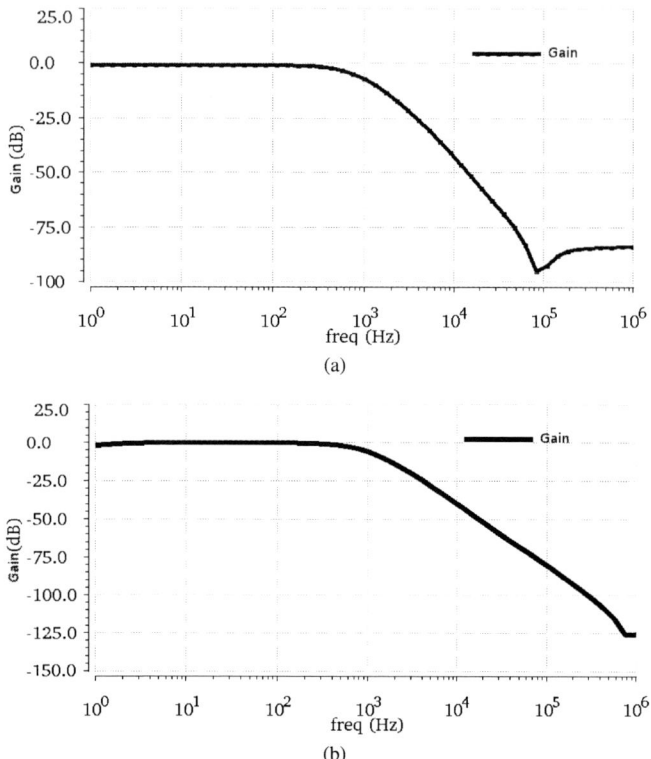

Fig. 8. Frequency response of LPFs (a) RC LPF with OpAmp1. (b) RC LPF with OpAmp2.

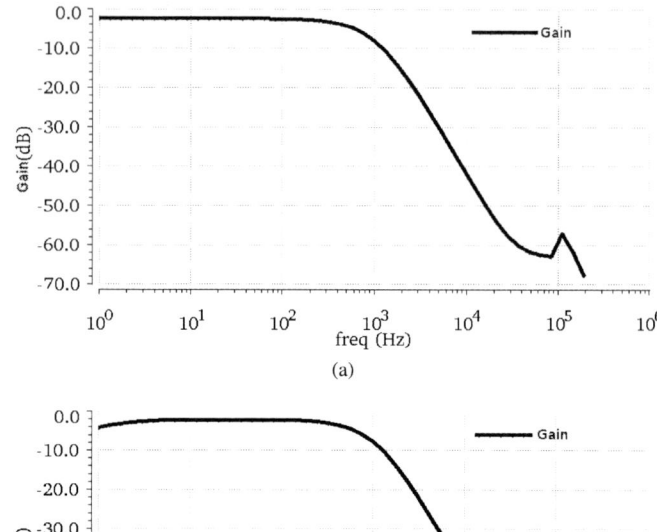

Fig. 9. Frequency response of SC LPF (a) SC LPF with OpAmp1. (b) SC LPF with OpAmp2.

TABLE III
OUTCOME OF FILTERS AND COMPARISON

Parameter /Topology	RC LPF with OpAmp1	RC LPF with OpAmp2	SC LPF with OpAmp1	SC LPF with OpAmp2	Previous Work [4]
Tech Node	CMOS 180n	CMOS 180n	CMOS 180n	CMOS 180n	a-Si 8μ
Topology	RC Bi-quad	RC Bi-quad	SC Biquad	SC Biquad	RC Biquad
Gain (dB)	-0.8	-1.98	-3.13	-3.88	-0.12
f_{-3dB} (Hz)	611.12	856.5	600	940.6	150
f_{-20dB} (Khz)	2.89	3.347	2.56	3.53	-
$V_{Noiseout}$ ($fv/\sqrt{hz}$)	32.035	14.46	21.02	3.945	-
$V_{Noisein}$ ($\mu v/\sqrt{hz}$)	358.14	97.41	61.27	1.56	-
THD (%)	4.87	4.99	5.365	7.47	-28.48dB
Power (μw)	413.7	68.29	296.6	67.48	3.7mw

[2] J. Crols and M. Steyaert, "Switched-Opamp: An approach to Realize Full CMOS Switched-Capacitor Circuits at Very Low Power Supply Voltages," IEEE journal of Solid-state circuits, vol. 29, no. 8, Aug. 1994.

[3] Y. C. Tarn, P. C. Ku, H. H. Hsien and L. H. Lu, "An Amorphous-Silicon Operational Amplifier and its Application to a 4-Bit Digital-to-Analog Converter," IEEE Journal of Solid state circuits, vol. 45, no. 5, May 2010.

[4] C. H. Wu, H. H. Hsieh, P. C. Ku, and L. H. Lu, "A Differential Sallen-Key Low Pass Filter in Amorphous-Silicon Technology," IEEE Journal of Display Technology, vol. 6, no. 6, Jun. 2010.

[5] Y. P. Tsividis and P. R. Gray, "An Integrated NMOS Operational Amplifier with Internal Compensation," IEEE journal of Solid-state circuits, vol. sc.

11, no. 6, Dec. 1976.

[6] P. U. Calzolari, G. Masetti and M. Severi, "Integrated NMOS Operational Amplifier for CCD Transversal filters," Electronics Letters, vol. 15, no. 1, pp.29-31, Jan. 1979.

[7] P. Bahubalindruni, V. G. Tavares, P. Barquinha, R. Martins and E. Fortunato, "High-Gain Topologies for Transparent Electronics," Journal of Display Technology, vol. 6, no. 6, Jun. 2010.

[8] S. Sambandan, "High Gain Amplifiers with Amorphous-Silicon Thin-Film Transistors," IEEE Electron Device Letters, vol. 29, no. 8, Aug. 2008.

[9] K. Ishida, R. Shabanpour, L. Petti and N. S. Munzenrieder, "22.5dB Open-Loop Gain, 31KHz GBW Pseudo-CMOS Based Operational Amplifier with a-IGZO TFTs on a Flexible Film," IEEE Asian Solid-State Circuits Conference (A-SSCC), pp.313-316, Nov. 2014.

[10] R. Shabanpour, K. Ishida, T. Meister and N. Munzenrieder, "A 70deg Phase Margin OPAMP with Positive Feedback in Flexible a-GIZO TFT Technology," IEEE Asian Solid-State Circuits Conference, Nov. 2014.

[11] P. Bahubalindruni, B. Silva, V. G. Tavares, P. Barquinha, R. Martins and E. Fortunato, "Analog circuits with high-gain topologies using a-GIZO TFTs on Glass," Journal of Display Technology, vol. 11, no. 6, Jun. 2015.

[12] R. P. Sallen and E. L. Key, "A Practical method of designing RC active filter," IRE Transaction-Circuit Theory, vol. CT-2, pp. 74-85 Mar. 1955.

[13] R. Gregorian, K. W. Matin and G. C. Themes, "Switched-Capacitor Circuit Design," Proceedings of the IEEE, vol. 71, no.8, Aug. 1983.

2018 31th International Conference on VLSI Design and 2018 17th International Conference on Embedded Systems

Pseudo-continuous Output Switched-Capacitor Amplifier for Rail-to-Rail Current Sensing Application

Anjali Gopinath, Ravi Kumar Adusumalli, Rohit Ranganathan, Arya S
ams Semiconductors India Pvt. Ltd., Hyderabad, INDIA
e-mail:RaviKumar.Adusumalli@ams.com

Abstract—**This paper proposes a switched-capacitor amplifier capable of generating a constant output voltage in both the phases of the clock, even in the presence of op-amp offset. Thus it has the advantage that the subsequent A/D stage can operate with at least double the data rate. The architecture is implemented in AMS 0.35um CMOS process and operates on a supply of 5V.**

Keywords—Switched-capacitor, current sensing, multi-phase output, continuous output, high-side, low-side.

I. Introduction

Accurate calculation of State-of-Charge (SOC) and State-of-Health (SOH) of the battery requires almost simultaneous measurement of battery voltage, current and temperature. To meet the high-precision and low-noise requirements of the battery measurement application, we operate the sigma-delta modulator (SDM) with a high over-sampling ratio (OSR). Conventional switched- capacitor(SC) amplifiers have the data available only during one phase of operation. This limits the speed of operation and the data-rate of the SDM ADCs using switched-capacitor front ends. Additionally, the voltage drop across the low-side current sensing shunt resistor can severely limit the performance of some of the applications like mobile and IoT. The high-side current sensing approach provides more flexibility for the position of the shunt-resistor. Most circuits that can operate for both low-side and high-side measurement either require a voltage level-shifter before the SC-amplifier or wide input range op-amp for a rail-to-rail input common-mode operation .

In the proposed architecture we present a pseudo-continuous output SC amplifier that provides output in both the phases of the amplifier clock. This at least doubles the sampling frequency of the SDM, thereby effectively making the data-rate twice. The SC amplifier samples the signal over its own common-mode as in [1]. This eliminates the need for either a level-shifter or an op-amp with rail-to-rail input requirements allowing the use of the same SC amplifier design for both high-side and low-side current-sensing applications.

Section II summarizes some of the existing SC architectures supporting multi-phase output generation or continuous output generation. Section III explains the proposed architecture. Section IV and V discusses the simulation results and conclusion.

II. Existing Continuous Output Switched-Capacitor Amplifier Architectures

In conventional SC amplifiers with two phase operation - sampling and amplification, the output is taken to be valid during the amplification phase [2]. However there are architectures which are capable of generating valid outputs in both the phases. The basic clocking scheme of a system using conventional SC amplifier and an SC amplifier with continuous multi-phase output is compared in Fig. 1. The dotted arrows indicate the phases where output is valid. If a conventional SC amplifier is used, the SDM sampling frequency cannot be more than the operating frequency of SC amplifier. However with multi-phase output SC architecture the operating frequency of the SDM amplifier can be at least twice that of the SC amplifier. SC amplifier architectures which are capable of generating valid outputs in both the phases of operation are discussed in [3] and [4].

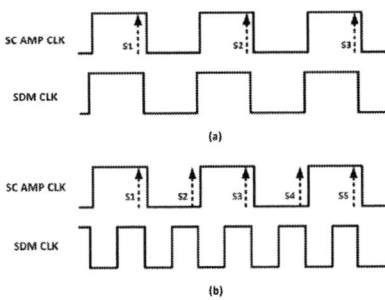

Fig. 1. Clocking scheme (a) Conventional SC amplifier (b) Multi-phase output SC amplifier

In the gain and hold SC architecture discussed in [3] the output remains the same in both sampling and amplification phases. However in the presence of op-amp offset the differential output between two consecutive phases differ by offset amount. In such cases even chopping technique cannot be helpful in removing the offset. As a result the SC amplifier output in the amplification phase can only be given to sigma-delta ADC. This limits the architecture from using in applications requiring high speed operation of sigma-delta ADC.

The multi-path SC architecture discussed in [4] is capable of generating continuous outputs. But it happens at the expense

978-1-5386-3693-0/18 $31.00 © 2018 IEEE   325

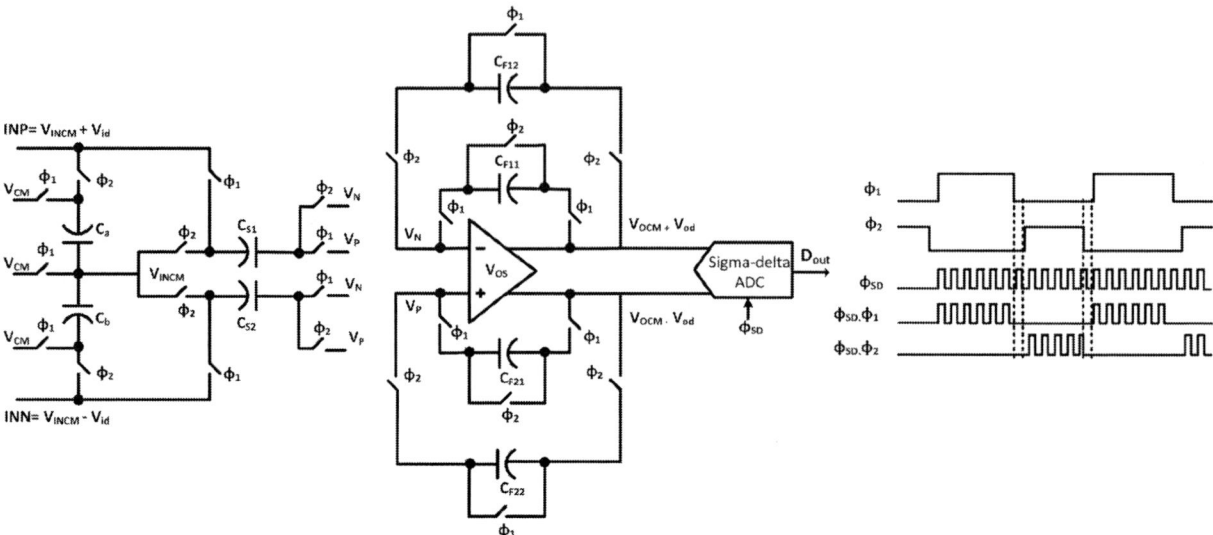

Fig. 2.   Proposed SC architecture with clock scheme

of area and power. Also the op-amp in this SC architecture resets during one of the phases making the bandwidth and slew requirements of the op-amp quite high.

III.   PROPOSED SC AMPLIFIER ARCHITECTURE

The proposed SC architecture is shown in Fig. 2. In this architecture during the ϕ_1 phase input is sampled on the capacitors C_{S1} and C_{S2} and the corresponding charge is transferred to the output during the same phase using capacitors C_{F11} and C_{F21}. During the ϕ_2 phase, the average of input signal common-mode voltage (V_{INCM}) is forced on to the bottom plate of capacitors C_{S1} and C_{S2} and charge transfer happens through capacitors C_{F12} and C_{F22} while the capacitors C_{F11} and C_{F21} get reset. During ϕ_1 phase, the bottom plates of capacitors C_{S1} and C_{S2} are connected to inputs $V_{INCM} + V_{id}$ and $V_{INCM} - V_{id}$. As a result the virtual node V_N tend to increase and V_P tend to decrease. In order to prevent this the corresponding output nodes become $V_{OCM} - V_{id}$ and $V_{OCM} + V_{id}$ so that the swinging of virtual nodes get cancelled. Thus an output which is out of phase when compared to the input is obtained during this phase. During ϕ_2 phase, V_{INCM} is forced on to the bottom plates of capacitors C_{S1} and C_{S2}. In order to maintain the charge in the input capacitors as in the previous phase the virtual node V_N tends to decrease and V_P tend to increase. In order to nullify this effect the corresponding op-amp output nodes become $V_{OCM} + V_{od}$ and $V_{OCM} - V_{od}$. Thus the output obtained would be inverted when compared to the output in the ϕ_1 phase. The additional switches introduced at the virtual node as shown in Fig. 2 prevents this phase change from ϕ_1 to ϕ_2 phase.

From [5], if the average of inputs is forced to the bottom plates of the sampling capacitors during ϕ_2 phase then the op-amp input terminals could be fixed at a particular voltage (V_{CM}) in both the phases irrespective of the input common-mode voltage. This in turn helps in supporting rail-to-rail input common-mode variations without having to use a rail-to-rail

op-amp. The proposed technique thus incorporates the idea of shifting the input common-mode voltage to a fixed voltage level [6].

The clocking scheme used here is such that the SDM samples the SC amplifier output multiple times during the phase where output is valid. Thus for a fixed frequency of SC amplifier, the SDM data rate can be more than twice that of SC amplifier frequency. Similar clocking scheme is used in [4].

A.   Expression for op-amp output voltage

The charge associated with the sampling and feedback capacitors during ϕ_1 phase is given by equations (1)-(6).

$$Q_{CS1}(\phi_1) = C_{S1}(V_{INCM} + V_{id} - V_X(\phi_1) + \frac{V_{OS}}{2}) \quad (1)$$

$$Q_{CF21}(\phi_1) = C_{F21}(V_X(\phi_1) - \frac{V_{OS}}{2} - V_{OCM} + V_{od}(\phi_1)) \quad (2)$$

$$Q_{CS2}(\phi_1) = C_{S2}(V_{INCM} - V_{id} - V_X(\phi_1) - \frac{V_{OS}}{2}) \quad (3)$$

$$Q_{CF11}(\phi_1) = C_{F11}(V_X(\phi_1) + \frac{V_{OS}}{2} - V_{OCM} - V_{od}(\phi_1)) \quad (4)$$

$$Q_{CF12}(\phi_1) = 0 \quad (5)$$

$$Q_{CF22}(\phi_1) = 0 \quad (6)$$

During ϕ_2 phase,

$$Q_{CS1}(\phi_2) = C_{S1}(V_{INCM} - V_X(\phi_2) - \frac{V_{OS}}{2}) \quad (7)$$

$$Q_{CF21}(\phi_2) = 0 \quad (8)$$

$$Q_{CS2}(\phi_2) = C_{S2}(V_{INCM} - V_X(\phi_2) + \frac{V_{OS}}{2}) \quad (9)$$

$$Q_{CF11}(\phi_2) = 0 \quad (10)$$

978-1-5386-3693-0/18 $31.00 © 2018 IEEE   326

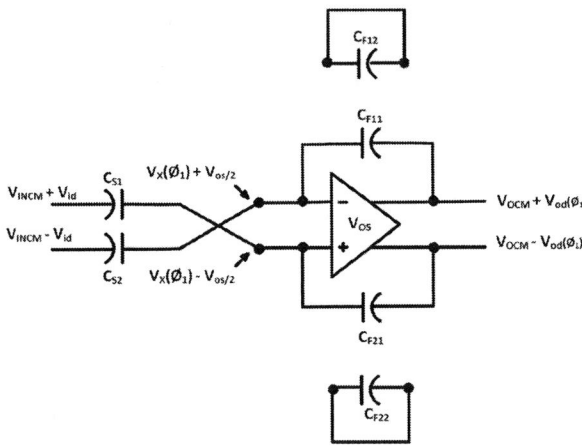

Fig. 3. ϕ_1 phase of operation

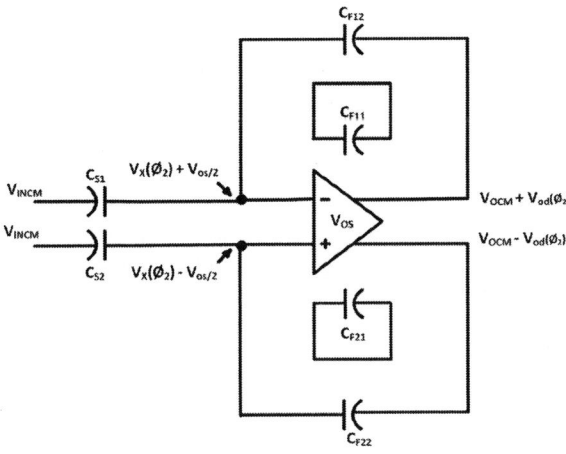

Fig. 4. ϕ_2 phase of operation

$$Q_{CF12}(\phi_2) = C_{F12}\left(V_X(\phi_2) + \frac{V_{OS}}{2} - V_{OCM} - V_{od}(\phi_2)\right) \quad (11)$$

$$Q_{CF22}(\phi_2) = C_{F22}\left(V_X(\phi_2) - \frac{V_{OS}}{2} - V_{OCM} + V_{od}(\phi_2)\right) \quad (12)$$

From ϕ_1 to ϕ_2 phase, according to charge conservation principle,

$$\Delta Q_{S1} = \Delta Q_{CF12} \quad (13)$$

$$\Delta Q_{S2} = \Delta Q_{CF22}. \quad (14)$$

Also $V_X(\phi_1) = V_X(\phi_2) = V_{CM}$ and the CMFB loop ensures that op-amp output common-mode voltage (V_{OCM}) is fixed at V_{CM}. Subtracting (14) from (13), assuming $C_{S1} = C_{S2} = C_S$ and $C_{F12} = C_{F22} = C_F$

$$V_{od}(\phi_2) = \frac{C_S}{C_F}V_{id} + \frac{C_S}{C_F}V_{OS} + \frac{V_{OS}}{2} \quad (15)$$

Similarly from ϕ_2 to ϕ_1 phase, according to charge conservation principle,

$$\Delta Q_{S1} = \Delta Q_{CF21} \quad (16)$$

$$\Delta Q_{S2} = \Delta Q_{CF11}. \quad (17)$$

Upon solving the equations as in the previous case we get,

$$V_{od}(\phi_1) = \frac{C_S}{C_F}V_{id} + \frac{C_S}{C_F}V_{OS} + \frac{V_{OS}}{2} \quad (18)$$

Thus from (15) and (18) it is clear that the op-amp output in both the phases remains constant. As the SC amplifier offset remains the same in both the phases system chopping can be used to remove it.

IV. SIMULATION RESULTS

The performance of the proposed architecture has been compared with the existing designs by simulating in CMOS $0.35\mu m$ process. The op-amp architecture implemented is folded cascode with PMOS input pairs in order to ensure high enough gain for meeting the linearity requirement. The op-amp has an open-loop gain of 128dB which varies across corners by +/-10dB. The operating frequency of the SC amplifier and SDM is chosen to be 250KHz and 1MHz respectively.

In the circuit implementation, the capacitance values were selected such that the gain turns out to be 1 for simplicity of analysis ($C_S = C_F = 2pF$). For a differential input of 1V, with an op-amp intentional offset (V_{OS}) of 10mV, according to equations (15) and (18), the offset at the output would be $1.5V_{OS}$. Thus if output is taken differentially it deviates from input by $3V_{OS}$ which is 30mV. This is shown in Fig. 5.

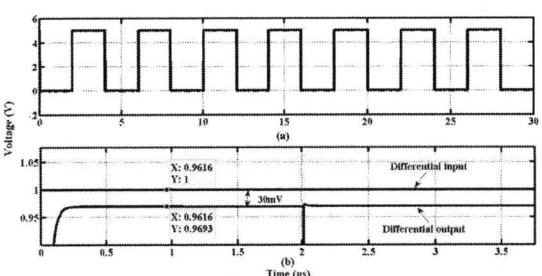

Fig. 5. (a) Clock Input (b) Differential output obtained for a 1V differential input with op-amp input offset of 10mV

The proposed architecture is compared with the prior-art [3],[4] based on the differential output voltages (V_{od}) and the op-amp input terminal voltages ($V_N = V_P = V_X$) obtained corresponding to inputs near to rails. This is summarized in Table I and Table II. The simulations were performed by introducing an intentional offset V_{OS} at the op-amp input.

From Table I it is clear that for the proposed architecture and multi-path SC architecture the differential output remains constant in both the phases. However in the gain and hold SC architecture the op-amp differential output voltages vary from ϕ_1 to ϕ_2 phase by offset amount. Thus even chopping technique cannot be useful in this case. Table II shows that the op-amp input terminal voltages in the proposed architecture remain constant at a particular voltage ($V_{CM} = 2.5V$) irrespective of the value of input common-mode voltage. However in [3] and [4] the op-amp input terminal voltages vary with varying input common-mode voltages thus requiring a rail-to-rail op-amp to support rail-to-rail input variations.

978-1-5386-3693-0/18 $31.00 © 2018 IEEE   327

TABLE I.   COMPARISON OF AMPLIFIER DIFFERENTIAL OUTPUT

	$INN = 0.5V, INP = 1.5V, V_{OS} = 10mV$		
	Gain and hold [3]	Multi-path [4]	Proposed
$V_{od}(\phi_1)$ (V)	0.989	0.999	0.969
$V_{od}(\phi_2)$ (V)	0.999	0.999	0.969
$\frac{\Delta V_{od}}{2}$ (V)	0.005	No chop needed	0
	$INN = 3.5V, INP = 4.5V, V_{OS} = 10mV$		
	Gain and hold [3]	Multi-path [4]	Proposed
$V_{od}(\phi_1)$ (V)	0.989	0.999	0.969
$V_{od}(\phi_2)$ (V)	0.999	0.999	0.969
$\frac{\Delta V_{od}}{2}$ (V)	0.005	No chop needed	0

TABLE II.   COMPARISON OF OP-AMP INPUT TERMINAL VOLTAGE

	$INN = 0.5V, INP = 1.5V, V_{OS} = 0mV$		
	Gain and hold [3]	Multi-path [4]	Proposed
$V_X(\phi_1)$ (V)	2.50	2.50	2.50
$V_X(\phi_2)$ (V)	3.25	4.00	2.50
	$INN = 3.5V, INP = 4.5V, V_{OS} = 0mV$		
	Gain and hold [3]	Multi-path [4]	Proposed
$V_X(\phi_1)$ (V)	2.50	2.50	2.50
$V_X(\phi_2)$ (V)	1.75	1.00	2.50

A comparison of the three architectures based on various performance parameters is shown in Table III. The three architectures were implemented and simulated using the same op-amp (folded cascode) and by using similar sampling and feedback capacitance values. From the table it can be inferred

TABLE III.   PERFORMANCE COMPARISON

Parameters	Gain and hold	Multi-path	Proposed
Power (mW)	0.45	0.90	0.45
Area (mm^2)	0.124	0.201	0.139
SDM Data Rate	1	2	2
Output	Not continuous	Continuous	Continuous

that the proposed architecture uses lesser power and area compared to the multi-path architecture (current consumption in multi-path architecture is half as only one amplifier is considered corresponding to each path for analysis). Though the proposed architecture has higher area when compared to the gain and hold architecture, it can improve the SDM operating frequency at least by twice as the output remains constant in both the phases of the clock. This is with the assumption that the SC amplifiers in all the three architectures have the same frequency of operation. The data rate shown in table is normalized to the gain and hold architecture. The SFDR plot for the proposed architecture for a single-ended input of $200mV_{pk-pk}$ over an input common-mode of 5V and 0V is shown in Fig. 6 and Fig. 7 respectively. The proposed technique ensures high linearity over rail-to-rail input common-modes.

V.   CONCLUSION

The proposed architecture discusses a switched-capacitor amplifier whose output remains constant in both the phases of operation. This allows the SDM following the SC amplifier to function at minimum double the frequency thereby effectively increasing the throughput of the system. Being able to generate

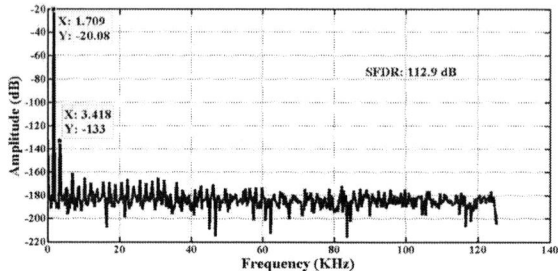

Fig. 6.   SFDR plot for input common-mode of 5V

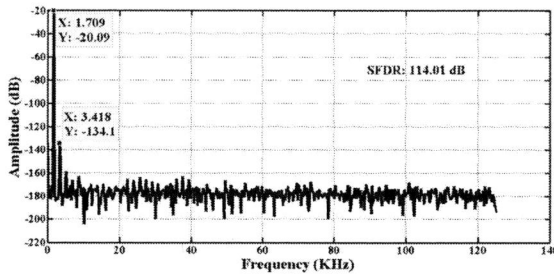

Fig. 7.   SFDR plot for input common-mode of 0V

pseudo-continuous output also helps to make use of system chopping to effectively remove the op-amp offset. The proposed technique has lower power for a given data rate and is an area optimized solution with respect to conventional switched-capacitor architectures.

VI.   ACKNOWLEDGMENT

We extend our sincere thanks to the Design team in ams Semiconductors India Pvt. Ltd for supporting us during this work.

REFERENCES

[1] Anjali Gopinath, Ravi Kumar Adusumalli, Veeresh Babu Vulligaddala, M.B. Srinivas, "A Switched-Capacitor Amplifier with True Rail-to-Rail Input Range without Using a Rail-to-Rail Op-Amp," *VLSID*, pp. 329-334, 2017.

[2] R. Gregorian, "High-resolution switched-capacitor D/A converter," *Microelectron. J.*, no. 12, pp. 10-13, 1981.

[3] Schoenberg, B. J. Hosticka, F. V. Schnatz, "A CMOS Readout Amplifier for Instrumentation Application," *IEEE J. Solid-State Circuits*, vol. 26, no. 7, 1991.

[4] T.Rahul, Bibhudatta Sahoo, S. Arya, S.J. Parvathy, Veeresh Babu Vulligaddala, "A Wide Dynamic-Range Low-Power Signal Conditioning Circuit for Low-Side Current Sensing Application," *VLSID*, pp. 265-270, 2015.

[5] Andre Luis Vilas Boas, Andre L. R. Mansano, Alfredo Olmos, Fabio de Lacerda, "Switched-Capacitor Amplifier Circuit," U.S. Patent 8,198,937 B1, Jun.12, 2012.

[6] Evgeny V. Ivanov, "Switched-capacitor level-shifting technique with sampling noise reduction for rail-rail input range instrumentation amplifiers," *IEEE Trans. Circuits Syst. I, Reg. Papers*, vol. 59, no. 12, pp. 2867-2880, 2012.

978-1-5386-3693-0/18 $31.00 © 2018 IEEE

2018 31th International Conference on VLSI Design and 2018 17th International Conference on Embedded Systems

Design Considerations of a Sub-50 µW Receiver Front-end for Implantable Devices in MedRadio Band

Gregory Chang, Shovan Maity, Baibhab Chatterjee and Shreyas Sen, *Member, IEEE*

School of Electrical and Computer Engineering, Purdue University, West Lafayette, Indiana – 47907, USA

E-mail: *{changg, maity, bchatte, shreyas}@purdue.edu*

Abstract—**Emerging health-monitor applications, such as information transmission through multi-channel neural implants, image and video communication from inside the body etc., calls for ultra-low *active* power (<50µW) high data-rate, energy-scalable, highly energy-efficient (pJ/bit) radios. Previous literature has strongly focused on low *average* power duty-cycled radios or low-power but low-date radios. In this paper, we investigate power performance trade-off of each front-end component in a conventional radio including active matching, down-conversion and RF/IF amplification and prioritize them based on highest performance/energy metric. The analysis reveals 50Ω active matching and RF gain is prohibitive for 50µW power-budget. A mixer-first architecture with an N-path mixer and a self-biased inverter based baseband LNA, designed in TSMC 65nm technology show that sub 50µW performance can be achieved up to 10Mbps (< 5pJ/b) with OOK modulation.**

Keywords—**MedRadio, Low-power, N-path Mixer, Receiver**

I. INTRODUCTION

IMPLANTABLE devices for healthcare monitoring applications often require to operate at extremely low power. Applications such as multichannel neural implants [1] and ingestible image/video transceivers [2] demand high rate communication. Past works in Medical Device Radio Communication (MedRadio) spectrum around 400 MHz had offered promising opportunities in health care monitoring through extensive use of medical telemetry at low-data-rate [3]. Yet, low-power high data-rate applications continues to be a challenge as transceiver system are size constrained and energy-sparse. For implantable use, to avoid frequent surgeries, batteries must sustain device operation for years; hence, RF transceivers must consume low power during communication. Other requirements such as small form factor, channel selectivity and interference rejection capability still remain crucial [4]. These stringent requirement poses the need to 1re-scrutinize receiver architectural performance trade-offs under the given communication standards and silicon process technology limits. In [5], [6], it is shown that a high data-rate RF receiver typically consumes 1 mW@ 1 Mbps, translating to 1nJ/b energy-efficiency. Hence, improvement in energy-efficiency of low-power high data-rate wireless communication system is the top priority. While recent efforts in pJ/b BAN also include human body communication [7]–[9], we will focus on standard wireless radio in this paper.

In terms of standards, short-range transmission on 401-406 MHz MedRadio band is suitable for the design space of low power operation. Since communication range rarely exceeds 3 m in a BAN, sensitivity requirements are relaxed [10]. The free-space line-of-sight path loss (FSPL) at 403.5 MHz can be calculated using Frii's formula [11], as shown in [12], while keeping the maximum effective isotropic radiated power

(EIRP) at -16 dBm for transmitter as per MedRadio standards, the minimum sensitivity required at the receiver is about -64 dBm, which enables lower power operation. Spectrally efficient but energy inefficient modulation schemes are not desirable in such applications. Higher order modulation schemes require higher E_b/N_0 and complex decoding, hence higher power. With this constraint, BPSK or OOK is most suitable, while OOK allows simpler and lower power receiver implementation.

The goal of this paper is to explore the performance trade-offs of an Ultra Low Power (ULP) receiver front-end and to prompt for an architecture suitable for sub-50 µW receiver implementation in the TSMC 65nm technology. Here we present 1) the energy-cost trade-off trends of each front-end component such as data bandwidth and sensitivity trades-off with power, 2) justification that mixer-first is a desirable architecture as energy-cost of active matching to 50 ohms is high and 3) with only baseband amplification 3.4pJ/bit energy efficiency is achieved in simulation with OOK transmission.

II. POWER VS. OPERABLE FREQUENCY TRADE OFF LIMITATION

A. Low Noise Amplifier (LNA) Design Choice

Conventional LNA often employs inductive source degeneration. This structure is known for improved noise performance and provides passive matching. However, for limited power budget operation, this structure suffers from low transconductance (g_m), impacting noise and gain performance. In addition, a larger inductor would be required for proper matching [13]. In many ULP systems, the popular choice of LNA structure is the inverter-based resistive feedback topology [13]-[14]. As shown in Figure 1, signal is also fed to the active load such that bias current is reused but the gain is boosted by the active load's g_m. It had also been shown in [13] that due to increased g_m, such topology has a better noise figure performance. In [15]-[16], low power consumption was achieved through Ultra-Low-Voltage (ULV) designs at V_{DD}=0.4V and 0.5V, respectively. However, the LNA power consumption is about 100µW in both designs.

To evaluate power and operable bandwidth trade-off of the LNA shown in Figure 1, the design was simulated in TSMC 65nm technology. The gate biasing of the NMOS is implemented by the decoupling capacitor C_{PN}, separating bias DC levels from both the signal input and LNA output. The mobility ratio of electrons and holes in 65 nm technology is taken into consideration to set the widths of the PMOS and NMOS transistors, so that node RF_{OUT} will be biased at roughly V_{DD} / 2. It is desired to have power consumption of the LNA

978-1-5386-3693-0/18 $31.00 © 2018 IEEE

329

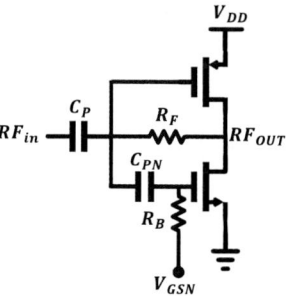

Figure 1: RF/BB Scalable LNA design with NMOS Gate Control

below 30 µW for our target applications.

1) Energy Cost of Active Matching

Matching is an important factor in RF system. As shown in [17], active matching can be employed by using a shunt feedback. However, extra power is required for such shunt feedback. For the design choice of Figure 1, matching is realized through the feedback resistor R_f. It can be shown analytically that in order to match to 50 Ω, high power is required as it would be limited by technology's $g_m r_o$. The input impedance shown in Eq. 2, which is simplified to Eq. 3.

$$Z_{in} = \frac{R_f}{1+g_m r_o} + \frac{r_o}{1+g_m r_o} \quad (2)$$

$$Z_{in} = \frac{1}{1+k}\left(R_f + \frac{k}{g_m}\right) \quad (3)$$

Where $k = g_m r_o$, $g_m = g_{m,n} + g_{m,p}$, and $r_o = r_{o,p}||r_{o,n}$. Given the k value for a particular technology, input impedance matching can be achieved from (3) by choosing appropriate value of R_f and transistor width which controls k/g_m under a given gate bias voltage. In [14] and [18], implementation is done through 180 nm and 130 nm technologies where $g_m r_o$ is

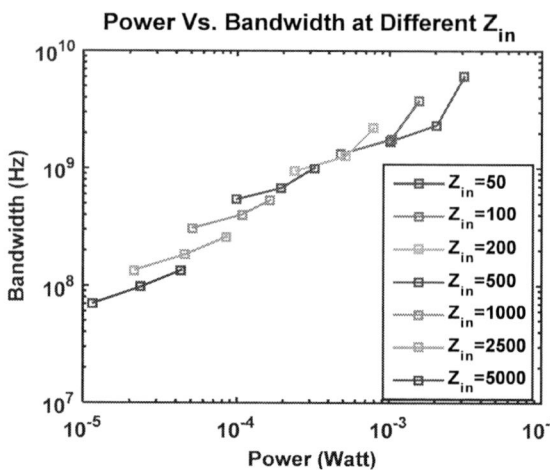

Figure 3: Power vs Bandwidth of LNA at different Z_{in} configuration. High active power is needed for 50 ohms matching and entails excess bandwidth. For µW operations, active matching need to be at kilo-ohms level and utilize passive matching network, avoiding active matching to 50 ohms at RF. The three points along each segment is derived from the three biasing condition V_{gs} shown in Figure 2.

higher than to 65 nm. When biasing at optimum $\frac{g_m f_t}{I_d}$, k is around 10~11 for 65 nm. Hence when matching to 50 Ω, it could easily require double the power as compared to 180 nm, as it requires $\left(R_f + \frac{k}{g_m}\right) = 550$. Although value of R_f does not directly impact power, large transistors that supports high current must be used in order to reduce the second term. With this principle, the LNA is shown to perform at various input impedance and gate bias. For low resistance matching such as 50 Ω, the PMOS is easily scaled to 600µm and NMOS at 300µm so that k/g_m term is less than 300. Figure 2 shows the power, bandwidth, R_f required and noise figure achieved at various input impedance. When biasing at higher $V_{GS,N}$, higher bandwidth is achieved from the same Z_{in}, but more power will then be consumed. This leads to an interesting observation between power and bandwidth trade off at different Z_{in} matching schemes. In order to achieve low impedance, more power is consumed. Low impedance also exhibits excess bandwidth for a 400MHz band receiver.

Performance tunability is largely achieved by controlling the gate bias. Derived from Figure 2, Figure 3 shows that matching to 50 Ω easily requires at least 1 mW power and will obtain GHz bandwidth. For a MedRadio transceiver design, such LNA would prove to be overly power-hungry. Further reduction in overhead bandwidth can be achieved through deep sub-threshold gate bias below 0.45V, but this would require unpractically large transistor device and can exceed technology allowed maximum for 65nm, hence only three gate biasing choices were investigated in Figure 2 and Figure 3 .

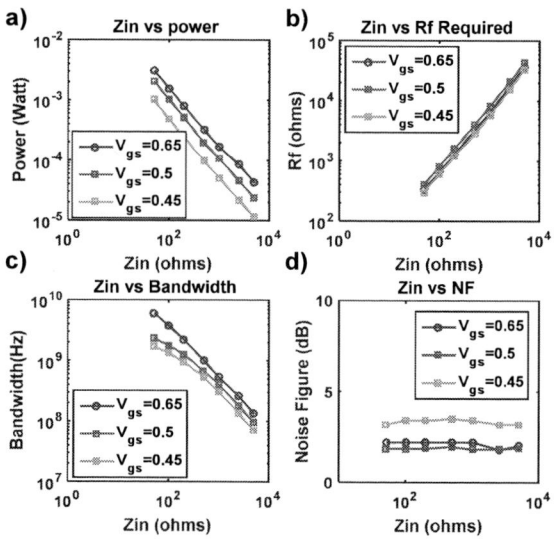

Figure 2: a) Power required for a given Z_{in}, showing that at different LNA gate bias, increasing power was required for lower impedance matching. b) Feedback resisor varies proportionally with given Z_{in} as accordance to equation (2). c) Bandwidth increases as Z_{in} is lowered, leads to an interesting observation between power and bandwidth trade off at different Z_{in}. d) Noise figure of the LNA under any Z_{in} in general does not vary under matched condition.

2) Energy Cost of BW

The power consumption for the LNA varies directly to its bandwidth, as most of the power consumption results from the quiescent current used to bias the circuit. A higher quiescent current alleviates the short-channel resistance r_o, lowering any RC combination at output node for voltage-voltage mode

978-1-5386-3693-0/18 $31.00 © 2018 IEEE   330

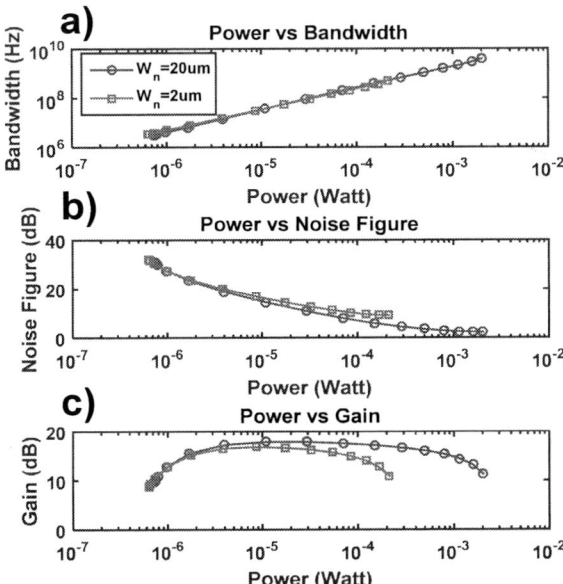

Figure 4: a) LNA's power varies proportionally with bandwidth, where large size LNA can achieve higher bandwidth range. b) Noise figure improves as more power is consumed. c) The flat-band gain increases initially and then saturates with increasing power consumption

operation, expanding the bandwidth closer to the technology allowed f_t.

As active matching had proved to be power hungry, it motivates LNA for baseband usage rather than RF usage. For BB-LNA, power and bandwidth trade-off trend still exits. As shown in Figure 4 power adjustment can be done through sizing and gate voltage tuning. If bandwidth is limited to MHz range, sub-50 µW implementation can be achieved. This illustrates an important bandwidth-power trade off and shows that it is the desirable design for baseband LNAs.

B. Frequency Down conversion with N-path Passive Mixer

Frequency translation are another crucial component in power budgeting a receiver. As described in [19], N-path filtering with passive mixers were first proposed as RF frequency bandpass filter. This technique is now often used for narrow band down-conversion by utilizing the frequency translation capability. For such down-conversion design, N parallel paths of passive switches are connected to one input and each path is loaded with a capacitor. To down-convert properly, a non-overlapping LO (NLO) is used to control the mixing operation, such that the baseband capacitor is connected to only one path at a time. The power consumed to operate such device varies with both number of path used, size of switches and operation frequency.

In this implementation, NLO is produced from a divider circuit functioning as the phase generator from a LO source. For front-end power budgeting, LO generation is not considered but phase generation is. For this purpose, two divider architectures shown in Figure 6 were investigated : a conventional flip-flop based divider and a circular divider presented in [20]. Both divider generates 4-phase LO that are 90 degrees apart. These signals were fed to AND gates to produce 25% duty cycle non-

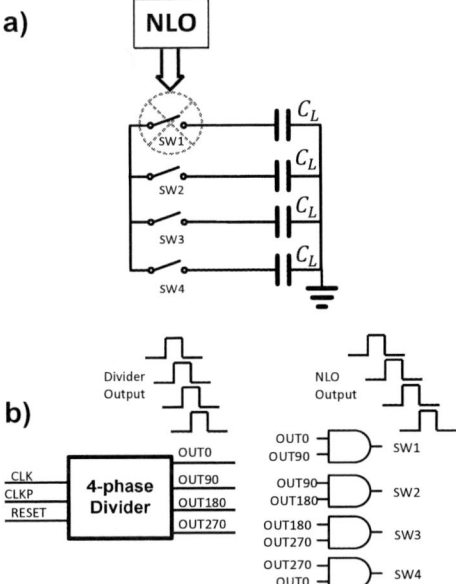

Figure 5: a) 4-Path switches perform passive downconversion controlled by NLO, switch is implemented with NMOS. b) NLO phase generation circuit. Divider generates 90 degree apart 4 phase signl, then fed to AND gates to convert into non-overlapping 4-phase signals.

overlapping LO as shown in Figure 5b. Figure 7 shows the power consumption comparison between the circular and flip-flop based divider architectures. It can be seen that at frequency of interest for RF receiver the circular divider consumes less power and hence is used for designing this receiver.

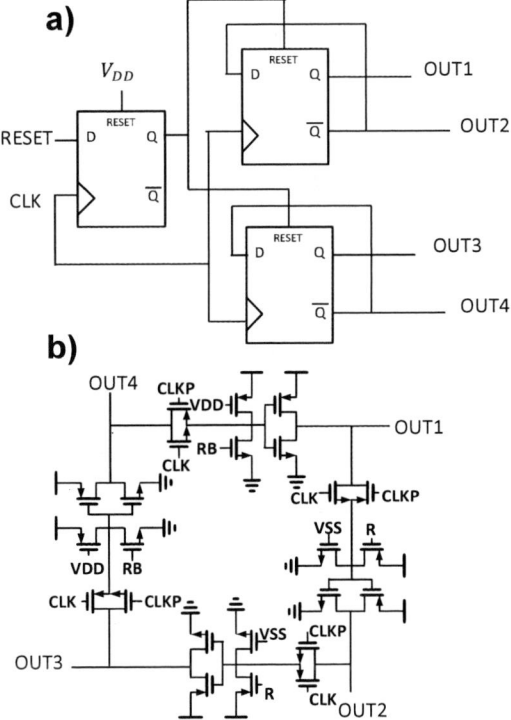

Figure 6: a) Flip-flop based divider. b) Circular divider of [20] . Out1, out2, out3 and out4 are 4-phase 90 degree offset signal

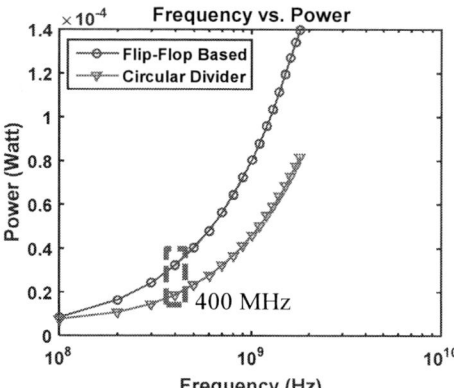

Figure 7: 2 dividers structure performance compared. Circular divider of [20] showed better power performance at 400MHz

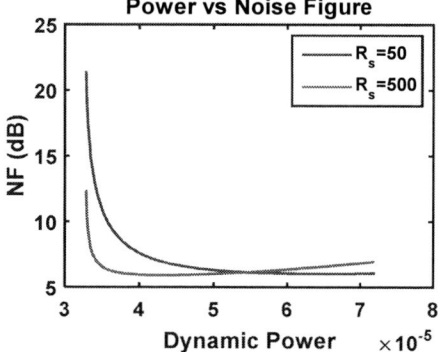

Figure 8: Dynamic power of 4-path mixer vs noise figure achieved under 2 different source impedance. This variation shows that to improve mixer-first's performance it is crucial to transform source impedance from 50 to larger values.

The power consumption of the circular divider along with the passive mixer is simulated with supply voltage of 1V and a load capacitance of 5pF. Figure 8 shows noise figure tradeoff with dynamic power of the mixer. The divider and MOS switches were simulated with a fixed supply voltage and the variation in dynamic power is obtained through variation in MOSFET size, as the gate capacitance of the switches act as extra capacitive load. From Figure 8 , it can be seen that increasing dynamic power beyond a certain point does not improve noise figure. Simulation results show that the optimum point of operation corresponds to switch size of 10μm. Figure 8 also suggests larger source impedance improves noise performance. From [19], the noise figure can be simplified to Eq. (4) and that minimizing switch resistance R_{sw} and maximizing source resistance R_s would optimize noise performance. However, R_{sw} is inversely proportional to transistor width, which varies with power consumption proportionally. From power budgeting prospective, increasing R_s should be prioritize over reducing R_{sw}.

$$NF = \frac{\pi^2}{4} \frac{1+\frac{R_{sw}}{R_s}}{1-\frac{R_{sw}}{R_s}} \quad (4)$$

Since power consumption in a digital circuit is proportional to its frequency and the band of operation of this receiver is 400MHz, relatively low compared to the 2.4GHz band, a digital mixer is ideal for this implementation. Although the NLO had

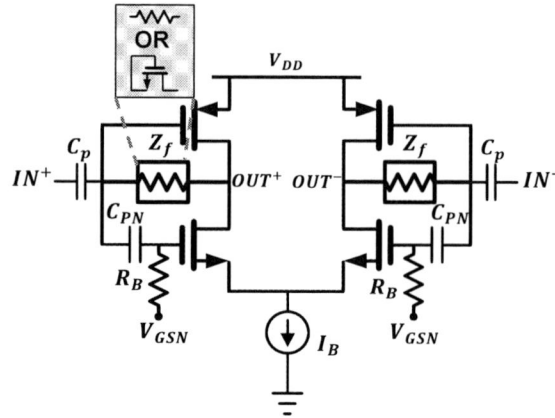

Figure 9: Differential Baseband LNA employing resistive feedback or LNA employing off-MOS feedback if bandwidth operation less than 1MHz. Outputs were loaded with 150fF capacitor to account for loading by latter stages of systems

4 25% duty cycle, 2 of the phases (0 and 180) were chosen to use for the down-conversion since OOK modulation does not utilize quadrature-phase to transmit separate message. In addition, 25% duty cycle clock was chosen, since it provides better selectivity compared to two phases with 50% duty cycle.

III. RECEIVER ARCHITECTURE AND SIMULATED PERFORMANCE

A. Complete Receiver

With the system component's power and operable frequency range closely examined, it can be easily concluded that in order to avoid overhead bandwidth, optimal design should be implemented with passive matching instead of active matching, and baseband amplification should be strictly control to only support up to the bandwidth of the allocated transmission band.

For this front-end implementation, a mixer first architecture was adopted. The circular divider presented in [20] was used to construct the front-end. As mentioned, 2 path passive mixer was used instead of 4. Since passive mixer splits signal differentially, the LNA is designed into differential form as shown in Figure 9, while using the same biasing condition as depicted in section II A. For LNA operation bandwidth lower than 1MHz, off-transistors were used instead of feedback resistor as shown in Figure 9. The LNA outputs were loaded with 150fF capacitor to account for loading by latter stages of systems.

Operating under OOK scheme with DDR transmission signal input, the receiver is designed to be zero-IF, hence the LO frequency is equal to the carrier frequency. The receiver front-end is shown in the red dashed box in Figure 10. The system utilizes a 50 ohm off-chip resistor in the front end and uses off-chip matching network to perform step-up impedance transformation seen by the mixer, improving noise performance according to Eq. (4). While using high resistance at the mixer switch's input in conjunction with step-down matching is theoretically equivalent, large on-chip resistor are less desirable. The inductor L_m used was 180nH and C_m was 2pF, simulated along with the front-end but envisioned to be implemented off-chip.

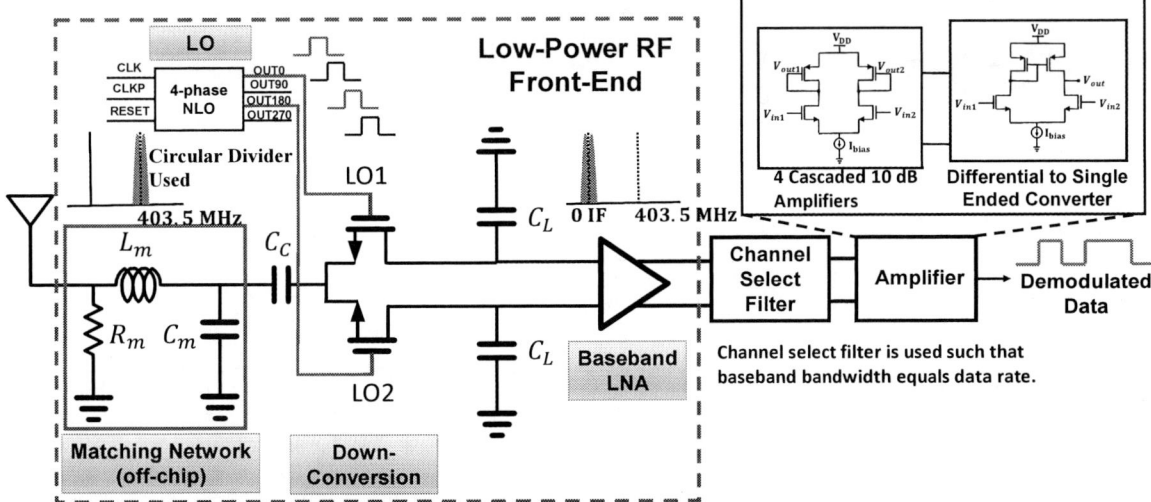

Figure 10: Low Power Front-end System. Components in the red dashed box is the simulation scope of this paper. 50 ohm R_m matches to antenna and the step-up transformation seen from mixer's input improves noise performance and avoids large on-chip resistor. 2-path mixer was used as OOK doesn't utilize signal's phase to transmit information. Channel select filter and 4 cascaded 10 dB gain amplifier were used to test the functionality.

The bandwidth of LNA is set to the spectrical occupancy of a desired data rate. For MedRadio compliant, ideally the LNA would be limited to 300 kHz bandwidth. High data rate mode was also simulated as motivated in section I.

B. Simulation Results

The receiver front-end of Figure 10 was simulated with TSMC 65nm in SPECTRE. OOK signals inputs were also fed as input to the front-end. To test the functionality, channel select filter and 4 cascaded 10-dB gain amplifiers were used, providing enough gain to verify the demodulated data, but are not in the simulation scope of this paper. Power consumption and sensitivity were estimated with PSS and PNOISE simulation tools. 8bit-10bit coding was used such that majority of baseband information's energy is located at the Nyquist frequency. Noise figure was simulated and measured at this frequency. The theoretical sensitivity then is estimated through equation (5)

$$-174 + 10Log_{10}(BW) + SNR_o + NF \quad (5)$$

The envisioned channel selection filter would pass only the desirable bandwidth hence *BW* was taken as the bandwidth of the channel where the data occupies, making the two equivalent. The SNR_o term represents the signal to noise power ratio for a given BER under a modulation scheme. While modulation BER performance is given in E_b/N_0 form, it can be converted to SNR power ratio as shown in (6)[21].

$$\frac{S}{N} = \frac{R \times E_b}{BW_{data} \times N_0} \quad (6)$$

E_b is energy per bit, N_0 is noise spectral density and R is data rate. For the front-end aiming to utilize OOK, the data rate is the same as data occupied baseband bandwidth with DDR. As shown in [22] and with Eq. 6, OOK with BER of 10^{-3} requires about 13dB SNR. Figure 11 shows the possible deign point simulated for the mixer-first receiver front-end. The theoretical sensitivities were estimated at different system bandwidth operation for optimal bandwidth utilization.

For current standard compliant receiver, the front-end is also adopted with LNA operated with 1MHz bandwidth for optimal sensitivity. Motivated by the fact that in Figure 12, system

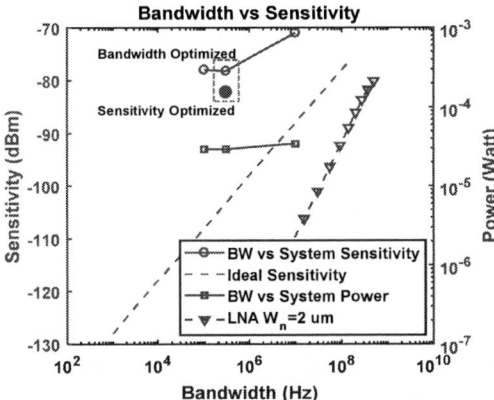

Figure 11: Bandwidth vs Sensitivity and Power required. Ideal sensitivity tendency on bandwidth and LNA power variation is shown for comparison. BW and sensitivity optimal design are shown separately as annotated. Standard compliant performance shown in red dashed box

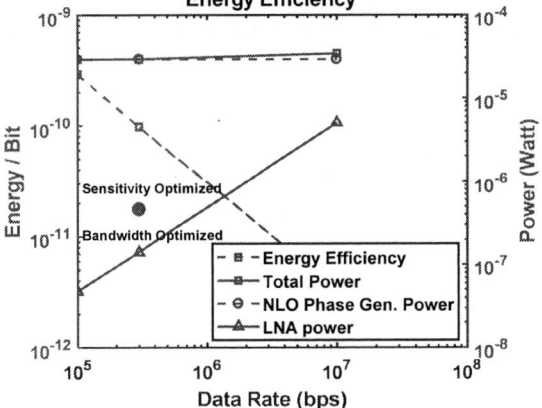

Figure 12. Bandwidth vs Energy per Bits. Data rate reduction does not improve energy efficient infinitely as power is limited by LO synthesis. Phase gen. power exhibits no data rate dependency yet total power is limited.

TABLE I : FRONT END PERFORMANCE COMPARISON

Symbol	[24]	[25]		[18]	[23]	This work (Simulation)	
Receiver Purpose	MICS Transceiver	WuRx		MICS Front-end	MICS Front-end	MedRadio Front-end	
Architecture	RF LNA	Mixer-First		RF LNA	RF LNA	Mixer-First	
Modulation	OOK	OOK		-	FSK	OOK	
Supply Voltage	1.8	0.75 V		1V	1.8 V	1 V	
Frequency	400MHz	2.45 GHz		401-457MHz	402-405MHz	402-405MHz	
Power	3.4 mW	50 µW		370µW	1.3mW	29.2µW	34 µW
BW/Data Rate	2Mbps	250kbps	650kbps	1Mbps	200kbps	300kbps*	10Mbps*
Energy Efficiency	1.71 nJ/b	200pJ/b	77pJ/b	370pJ/b	6.5nJ/b	97pJ/b	3.4pJ/b
Sensitivity (dBm) 10^-3 BER	-80.2	-88	-71	-96(if were to use OOK)	-96.8	-83#	-70#
Technology	0.18 µm	65 nm		0.18 µm	0.18 µm	65 nm	

*OOK DDR signaling potentially supporting up to 300kbps data rate under 300 kHz bandwidth limitation, or 10Mbps under 10MHz bandwidth, here is used to estimate the sensitivity and energy efficient, # No implementation loss included and should be added into design consideration

power ceases to reduce at 300kbps, but noise figure from LNA continues to degrade. This translates to -83dBm sensitivity as shown in the red dashed box in Figure 11, consuming about 29.2 µW. The sensitivity for optimal bandwidth setting, LNA having 300kHz bandwidth, were also shown in Figure 11. In Table 1, the 10Mbps results are compared with other literature reported in [23], [24], [25]. The energy efficiencies were estimated under the desired data rate.

In Figure 12, one can observe that although decreasing bandwidth at circuit level reduces power, at system level, it does not translate to better energy efficiency. This is shown with the power break-down of the system simultaneously. When trying to reduce power through bandwidth adjustment, low bandwidth operation is eventually limited by the LO phase generation circuit's power hence the energy/bit does not improve after certain operation data-rate.

IV. CONCLUSION

Implantable healthcare monitoring calls for ultra-low active power (<50µW) high data-rate, energy-scalable, highly energy-efficient (pJ/bit) radios. Investigation of power performance trade-off of each front-end component in a conventional radio reveals 50Ω active matching and RF gain is prohibitive for 50µW power-budget. Receiver front-end designed with N-path mixer and baseband LNA in TSMC 65nm technology achieved sensitivity -83 dBm sensitivity and <100 pJ/b for MedRadio compliant data rate and -70 dBm sensitivity with <5pJ/b at 10Mbps. Since LO synthesis power dominated the total power, future work would focus on energy/performance scalable LO and high-impedance interfaces.

V. ACKNOWLEDGEMENT

The work is supported by Semiconductor Research Corporation (SRC) Grant No. 2720.001 and National Science Foundation (NSF) CRII Award, CNS Grant No. 1657455.

VI. REFERENCES

[1] C. M. Lopez et al., "An Implantable 455-Active-Electrode 52-Channel CMOS Neural Probe," IEEE JSSC 2014

[2] "PillCam Capsule Endoscopy - Given Imaging." [Online]. Available: http://www.givenimaging.com/en-int/Innovative-Solutions/Capsule-Endoscopy/Pages/default.aspx. [Accessed: 31-Jul-2017].

[3] D. Panescu, "Emerging Technologies [wireless communication systems for implantable medical devices]," IEEE Eng. Med. Biol. Mag 2008.

[4] P. D. Bradley, "An ultra low power, high performance Medical Implant Communication System (MICS) transceiver for implantable devices," in BioCAS 2006

[5] S. Sen, "Invited: Context-aware energy-efficient communication for IoT sensor nodes," in DAC 2016

[6] S. Sen, et al., "TRIFECTA: Security, Energy-Efficiency, and Communication Capacity Comparison for Wireless IoT Devices," IEEE Internet Comput., In Press.

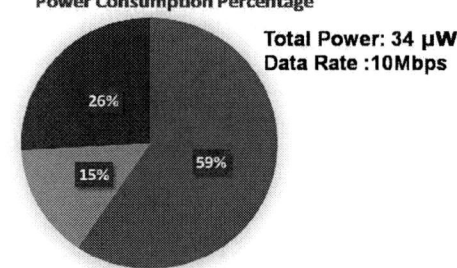

Power Consumption Percentage

Total Power: 34 µW
Data Rate :10Mbps

■ NLO Phase Generation-AND gates
■ LNA
■ NLO Phase Generation-Circular Divider

Figure 13. Power distribution amongst key front-end components. LO synthesis components occupy the majority of the power.

[7] S. Sen, "SocialHBC: Social Networking and Secure Authentication using Interference-Robust Human Body Communication," in IEEE ISLPED 2016.

[8] S. Maity et al., "Adaptive interference rejection in Human Body Communication using variable duty cycle integrating DDR receiver," in DATE), 2017

[9] S. Maity et al., "Secure Human-Internet using dynamic Human Body Communication," in 2017 IEEE/ACM ISLPED 2017

[10] "Medical Device Radiocommunications Service (MedRadio)," Federal Communications Commission, 09-Dec-2011. [Online]. Available: https://www.fcc.gov/wireless/bureau-divisions/broadband-division/medical-device-radiocommunications-service-medradio.

[11] H. T. Friis, "A Note on a Simple Transmission Formula," Proc. IRE, vol. 34, no. 5, pp. 254–256, May 1946.

[12] A. J. Johansson, "Performance of a radio link between a base station and a medical implant utilising the MICS standard," in The 26th Annual International Conference of the IEEE Engineering in Medicine and Biology Society, 2004

[13] H. K. Cha, et al., "A CMOS MedRadio Receiver RF Front-End With a Complementary Current-Reuse LNA," TMTT 2011

[14] T. Taris, et al. "A 60µW LNA for 2.4 GHz wireless sensors network applications," in RFIC 2011

[15] M. Parvizi, et al., "A 0.4V ultra low-power UWB CMOS LNA employing noise cancellation," ISCAS2013

[16] M. Parvizi, et al., "Short Channel Output Conductance Enhancement Through Forward Body Biasing to Realize a 0.5 V 250 µW 0.6-4.2 GHz Current-Reuse CMOS LNA," IEEE JSSC 2016

[17] M. Parvizi, et al., "An Ultra-Low-Power Wideband Inductorless CMOS LNA With Tunable Active Shunt-Feedback," TMTT 2016.

[18] C. Choi, et al., "A 370µW CMOS MedRadio Receiver Front-End With Inverter-Based Complementary Switching Mixer," MWCL 2016

[19] C. Salazar, et al., "A 2.4 GHz Interferer-Resilient Wake-Up Receiver Using A Dual-IF Multi-Stage N-Path Architecture," IEEE JSSC 2016

[20] A. Ba et al., "A 1.3 nJ/b IEEE 802.11ah Fully-Digital Polar Transmitter for IoT Applications," IEEE JSSC 2016

[21] D. Terlep, "Receiver Sensitivity Equation for Spread Spectrum Systems." Maxim Integrated Products, Inc, 28-Jun-2002.

[22] Q. Tang et al., "BER performance analysis of an on-off keying based minimum energy coding for energy constrained wireless sensor applications," ICC 2005

[23] H. Cruz, et al. "A 1.3 mW low-IF, current-reuse, and current-bleeding RF front-end for the MICS band with sensitivity of -97 dbm," IEEE TCASI 2015

[24] Y.-H. Liu, et al. "A low-power asymmetrical MICS wireless interface and transceiver design for medical imaging," in BioCAS 2006.

[25] C. Bryant et al., "A 2.45GHz, 50uW wake-up receiver front-end with -88dBm sensitivity and 250kbps data rate," in ESSCIRC 2014

978-1-5386-3693-0/18 $31.00 © 2018 IEEE

2018 31th International Conference on VLSI Design and 2018 17th International Conference on Embedded Systems

A 0.6 mW 1.6 dB Noise Figure Inductorless Shunt Feedback Wideband LNA With G_m Enhancement and Current Reuse in 65 nm CMOS

Narendra Nath Ghosh[*], Prakash Kumar Lenka[†], SriHarsa Vardan G[‡] and Ashudeb Dutta[§]

Department of Electrical Engineering, Indian Institute of Technology Hyderabad

Email: [*]narenbcrec@gmail.com, [†]ee15mtech11028@iith.ac.in [‡]ee15resch11008@iith.ac.in, [§]asudeb_dutta@iith.ac.in

Abstract—This paper presents the design of an inductorless low power differential Low Noise Amplifier (LNA) for multi-standard radio applications between 0.2-3.2 GHz. Conventionally, the low power shunt-feedback based LNA noise performance suffers due to the low intrinsic g_m of MOS transistors. A single stage differential shunt-feedback LNA which incorporates both, G_m boosting and current-reuse technique is proposed to overcome the noise performance degradation in low power designs. A detailed analysis of the conventional inductorless shunt-feedback based LNA along with the proposed technique is provided. It provides a good trade-off between different performance parameters after sizing and biasing optimization under ultra low power design constraint. The proposed technique is implemented in 65 nm CMOS technology and occupies an active area of 0.25 mm^2. It exhibits a power gain of 13.5 dB with 1.6 dB NF while dissipating only 0.6 mW power.

Index Terms - Ultra low power(ULP), Wireless Sensor Network (WSN), Ultra Low Voltage(ULV), Noise Figure(NF), Shunt Feedback(SF), Ultra wideband (UWB).

I. INTRODUCTION

Proliferation of portable wireless devices and wireless sensor networks (WSN) necessary for contemporary consumer electronics and applications involving internet of things (IOT) , calls for the design of ultra-low-power(ULP) RF front-end circuits. These applications demand long battery lifetime and portability. The low-noise amplifier(LNA) is first block in RF receiver chain. LNA has to provide a wide-band matching, low noise, high gain and moderate linearity. Conventionally, LNA designs require to consume more power to achieve low noise figure(NF) and linearity, along with other performance parameters. To achieve very low NF and high gain, while maintaining gain flatness throughout the wide RF bandwidth, is a challenging task in ULP designs [1], [2]. This has accentuated research in the field of wide-band ULP LNA design(<1 mW) [3]–[16].

A common gate LNA design using self body bias technique to enhance G_m, and the present state-of-the-art use inductors for wide-band input matching and maintain gain flatness at higher frequencies. But this technique achieves low gain (e.g. 7.9 dB) and occupies a large chip area [3]. [4] presents a good input matching circuit trough reduction in the input impedance by enhancing the loop gain for a shunt feedback topology. But this method is able to achieve only 10 dB power gain and 5.5 dB NF, although power consumption is low

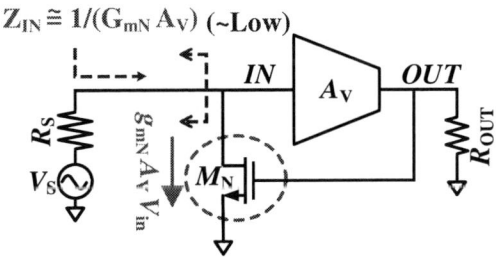

a : Shunt feedback LNA topology

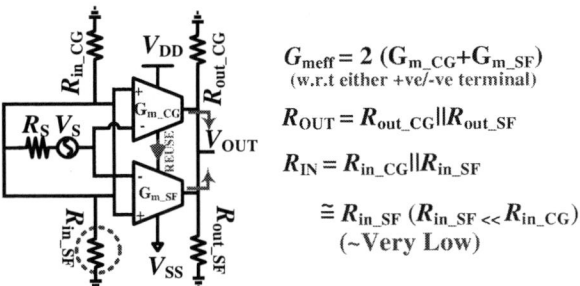

b : G_m boosted current-reuse shunt feedback LNA topology

Fig. 1 *Conceptual block diagram of proposed design*

[4]. The current-reuse and noise cancellation technique in [5] shows a very ULP LNA providing high frequency matching and gain flatness. However the use of inductor makes it bulky. The shunt feedback and current-reuse LNA presented in [6] provides ails from low bandwidth, even while using the G_m boosting technique. An inductor-less CG LNA architecture involving G_m boosting by two stage is discussed in [7]. This technique improves both, the input matching and G_m of the circuit. [8] reports a similar kind of ULP design, (0.4mWwith a supply voltage of 0.8V), employing G_m boosting technique. It introduces a feed-forward noise cancellation technique which leads in reduction in noise contribution in the main circuit. It has the drawback of requiring a off-chip balun, hence increasing the cost of system integration and low BW. The forward body biasing scheme tuning the feedback coefficient and complementary input characteristics, resulting in partial distortion cancellation is introduced in [9]. This

978-1-5386-3693-0/18 $31.00 © 2018 IEEE   335

paper manages to achieve 12dB gain and 4.9dB NF over 0.1 - 2.2 GHz bandwidth with -10dBm IIP3 while consuming only 400µA from a 1 V supply. A ULP wide-band LNA using series inductive peaking in feedback loop and current-reuse scheme is shown in [10]. The design achieves 12.6dB gain, 0.1-7GHz bandwidth, 5.5dB NF and -9dBm IIP3 using only 0.75 mW. However it is not an area efficient design. [11] achieves a low power input matching by using ULP shunt feedback architecture. It also involves inductive G_m boosting to increase BW without consuming additional power. The LNA achives 14dB gain, 4dB NF, -10dBm IIP3 and 0.6 to 4.2GHz bandwidth consuming only 50µA from 0.5V supply. Hence, it suffers form a degraded NF.

Integrating all the beneficial design techniques from all the above mentioned references, the proposed work focuses on the design of a inductorless LNA, targeting low noise figure with ultra low power consumption. A G_m boosting with current-reuse structure is utilized to obtain low noise figure with low power consumption. Moreover, the proposed topology also incorporates a partial noise cancellation mechanism. Transistor level sizing and biasing is done for optimum power consumption keeping in mind its proper integration in RF receiver chain as an end product.

The paper is organized as follows: Section II discusses about the issues, challenges, choice of design parameters and region of operation pertinent to the design of ULP LNAs. An intuitive analysis of the active shunt feedback topology is provided in Section III. Section IV includes the optimization and detailed analysis of gain, BW, linearity, noise and stability of the proposed LNA topology. Finally, the simulation results are presented in Section V.

II. ULP DESIGN ISSUES

A low voltage and low power design topology is very much desirable in applications such as WSNs and systems operated on energy scavenged from almost all resources(environmental or piezoelectric). Since the energy scavenged from these sources is very low, low power consumption is mandatory for remote IoT application. In low power operation, a MOS transistor has several impacts on its characteristics as intrinsic gain, transit frequency, NF and linearity.

A. Transconductance Efficiency (Optimization of g_m/I_D)

The deep weak inversion(WI) region provides maximum transconductance efficiency. It reaches half of maximum value in moderate inversion(MI) region and in strong inversion(SI) region its value is quite low [10]. On the other hand, the power consumption is minimum in WI region but linearity is poor. Further, linearity is better in SI region but power consumption is huge. So, MI is better choice in terms of low power consumption and moderate linearity. Keeping the ratio of transconductance to dc drain current(g_m/I_D) in between 8-12 V^{-1} (g_m/I_D of the MI region is between 5-20 V^{-1} [10]) is the good choice for transconductance efficiency of the input MOS.

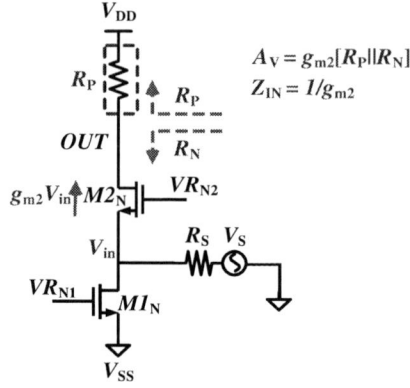

a : *Common-gate LNA*

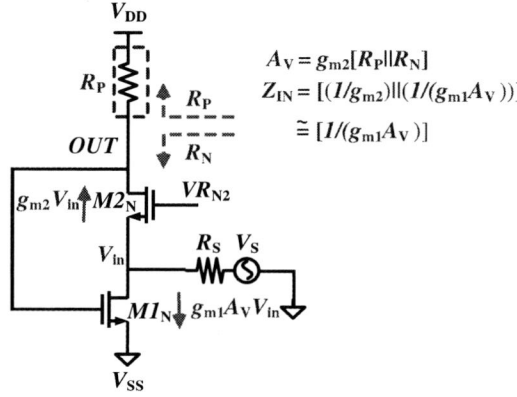

b : *Shunt feedback common-gate LNA*

Fig. 2 *Conventional CG and SF common-gate based LNA*

B. Intrinsic voltage gain and Transit Frequency

The intrinsic voltage gain of MOS transistor depends upon its region of operation. Intrinsic gain, i.e. g_m/g_{ds},has a high value in WI region (>10 V^{-1}) , moderate in MI region g_m/g_{ds} (between 6-10 V^{-1}) and low in SI region (<6 V^{-1}). Transit frequency(f_T) is another important parameter for RF LNA design. f_T is very low and high in WI region and MI region respectively. The f_T is directly proportional to g_m and inversely proportional to parasitic capacitances ($C_{gs} + C_{gd}$). For low power design, usually g_m is very less and parasitic capacitances are directly proportional to the area of MOS. So, smaller device area will provide us more f_T. Shunt feedback technique allows us to achieve more f_T along with both, good wideband input matching and low power consumption.

III. LNA CIRCUIT DESCRIPTION

In this section, we elaborate step-by-step evolution of the low-power design techniques for the proposed ULP LNA. Prior to the design of the LNA, lowering of the input impedance while consuming very low power by active shunt feedback is analytically justified. It is followed by an study on the effect of g_m boosting and current-reuse on various

978-1-5386-3693-0/18 $31.00 © 2018 IEEE   336

performance parameters (e.g. gain, noise figure, etc.) for the proposed LNA.

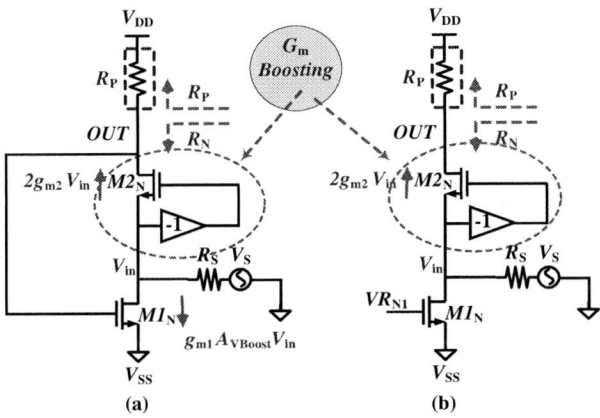

a : G_m boosting in CG and SFB

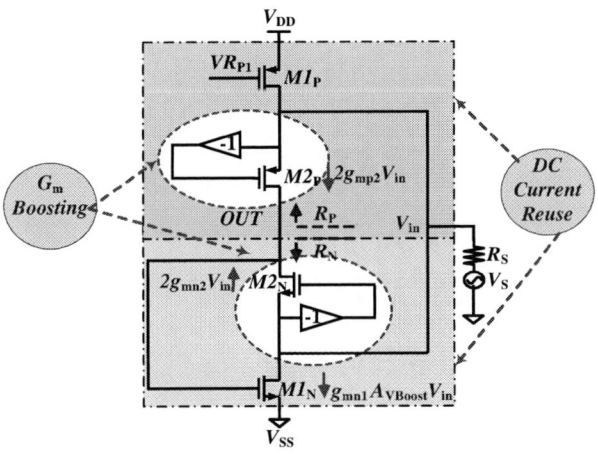

b : G_m boost + Current reuse

Fig. 3 G_m boosting method and current-reuse

A. Active Shunt Feedback Topology

Low power LNA design basically has been categorized into two types of topologies viz. CS and CG. The CS topology with inductive degeneration occupies a huge inductor area. Although resistive shunt feedback topology on CS gives good matching along with gain/NF, but these designs involve higher power consumption. Another well-known approach is the popular CG LNA, which provides better wideband input matching beside reverse isolation w.r.t. CS LNAs. Although NF of CG LNAs are quite high, their major drawback is the system g_m. The system g_m is determined by the input matching criteria at the cost of higher power consumption.

Fig.1a shows the conceptual shunt feedback topology and illustrates the real input matching provided by the feedback. The figure shows that the output of a non-inverting amplifier block (having gain A_V) connected to the gate of a MOS(M_N),

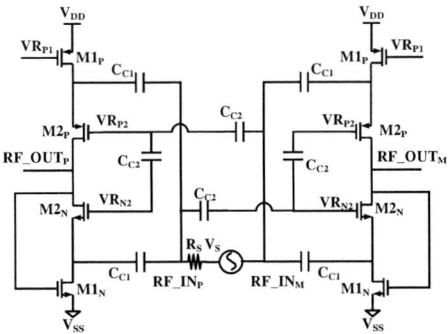

Fig. 4 *Schematic of proposed LNA(buffer not shown)*

whose drain is also connected to the input of the same amplifier. This leads to a huge signal current flow ($A_V \times g_{mN}V_{in}$) through M_N from drain to source terminal.

So, the input impedance of complete block is drastically lowered down to $Z_{IN} \cong 1/[g_{mN}A_V]$ without depending upon the input impedance of A_V(CG comparatively high)block. So, A_V block which can now be designed in common-gate structure, whose input transistor can take any value of g_m or bias current. It allows us to design LNA in low power (required power becomes A_V times lower than the conventional CG LNA (Fig.2a)) and also provides positive A_V.

Hence, wideband LNA in shunt feedback topology designed for low power consumption gives lower g_m. Moreover, for low Noise Figure(NF) and wide frequency coverage, a higher g_m is required. So, the G_m boosting technique is introduced to fulfill the above requirement. As shown in Fig.3a, the structure gives twice g_m by providing the signal and its out of phase (with 180°phase shift) component to gate and source of input transistor respectively. Fig.3b shows g_m boosting as well as current-reuse by using a PMOS transistor based common gate amplifier which gives extra g_m to the circuit, resulting in a high output impedance. Z_{IN} of the circuit is dominated by the $Z_{IN,SF}$, which is much lesser than the $Z_{IN,CG}$. Conceptual block diagram of proposed LNA is shown in Fig.1b.

IV. DESIGN OF ULP WIDE-BAND LNA

A. Transconductance

Designing LNA in low P_{DC} entails low g_m values. As shown in Fig. 1a, in common gate LNA, only transistor $M2_N$ provides $G_{m,CG}(\cong g_{m2,N} + g_{mb2,N})$. Similarly, a shunt-feedback common gate LNA provides an equal transconductance $G_{m,SF}(\cong g_{m2,N} + g_{mb2,N})$. But due to low power and low $G_{m,SF}$, this topology is subjected to the poor NF and gain performance. G_m boosting provides an efficient solution to this problem. A simple G_m boosting LNA provides double effective transconductance $G_{m,Boost}(\cong 2(g_{m2,N} + g_{mb2,N}))$ than the standard shunt-feedback common gate LNA.

Finally, the proposed ULP wideband LNA provides transconductance $G_{m,tot}$ due to both, the common gate boosted G_m and shunt-feedback common gate boosted G_m structure. Because of the current-reuse technique, the LNA

978-1-5386-3693-0/18 $31.00 © 2018 IEEE   337

avoids multi-stage structures, making it more stable and less power hungry.

$$\begin{aligned} G_{m,tot} &\cong 2[(g_{m2,N} + g_{mb2,N}) + (g_{m2,P} + g_{mb2,P})] \\ &\cong [G_{m,Boost,N} + G_{m,Boost,P}] \end{aligned} \quad (1)$$

B. Output Impedance

For the proposed ULP Wide-band LNA, the output impedance $Z_{out,tot}$ which consists of parallel combination of $Z_{out,SF}$ and $Z_{out,CG}$ (taking into the effect of all parasitics and output capacitance(C_{out})) is

$$\begin{aligned} Z_{out,tot} &\cong \left(\frac{r_{o2,N}(1 + g_{m2,N}Z_s)}{1 + g_{m1,N}g_{m2,N}Z_S} \right) \Big\| \\ & \left(r_{o2,P} + (Z_S \parallel r_{o1,P}) + \right. \\ & \left. r_{o2,P}g_{m2,P}(Z_S \parallel r_{o1,P}) \right) \Big\| \frac{1}{sC_{out}} \end{aligned} \quad (2)$$

where, $Z_S = R_S \parallel \frac{1}{s(C_{P1}+C_{P2})}$ and $C_{P1} = C_{gs2,N}+C_{gd1,N}+C_{sb2,N}+C_{db1,N}$ and $C_{P2} = C_{gs2,P}+C_{gd1,P}+C_{sb2,P}+C_{db1,P}$

C. Voltage Gain & Bandwidth

The proposed structure takes the advantage of the G_m boosting architecture. Hence it exhibits greater voltage gain compared to the conventional common gate and SFB circuits, designed with same power dissipation.The voltage gain is expression as

$$\begin{aligned} A_{v,tot} &\cong G_{m,tot} \times Z_{out,tot} \\ &\cong \frac{G_{m,tot}}{g_{m1,N}} \left[1 + \frac{1}{g_{m2,N}Z_s} \right] (approx.) \end{aligned} \quad (3)$$

To achieve a high dc gain, the $(g_{m1,N}g_{m2,N}Z_s)$ component should be much lesser than unity, as is evident from eq.(3). It can be seen that there are three poles in the circuit, out of which the second dominant pole limits the bandwidth of LNA. The first pole, ω_{p1}, is due to large bypass capacitors and input impedance as well as biasing circuit impedance. The second pole, ω_{p2}, appears due to the output load capacitance (buffer has been included) and the output impedance.

$$\omega_{p2}(dominant) = \frac{1}{C_{out}R_{out}}$$

D. Input Matching

For input impedance matching, R_{in} has to be equal to $R_S(50\Omega)$, which requires a high effective transconductance, hence resulting in a large power consumption in conventional LNAs. The input impedance of a common gate LNA, $Z_{in,CG}, \cong \frac{1}{G_{m,CG}}$ and for shunt-feedback common gate LNA, $Z_{in,SF}$, can be written as

$$\begin{aligned} Z_{in,SF} &\cong \frac{1}{g_{m1,N}[(G_{m,SF})(R_P \parallel Z_{out,SF})]} \Big\| \frac{1}{G_{m,SF}} \\ &\cong \frac{1}{G_{m,SF}(1 + g_{m1,N}Z_{out,SF})} \end{aligned}$$

Similarly, for the proposed structure input impedance(4) equals to

$$Z_{in,tot} \cong \frac{1}{g_{m1}[(G_{m,tot})(Z_{out,tot})]} \Big\| \frac{1}{G_{m,Boost,N}} \Big\| \frac{1}{\frac{1}{G_{m,Boost,P}}}$$

$$Z_{in,tot} \cong \frac{1}{G_{m,tot}(1 + g_{m1,N}Z_{out,tot})} \quad (4)$$

Since, $G_{m,tot}$ of proposed structure is higher, the input resistance becomes substantially lower, which in turn enables the designer to use less power for 50Ω impedance matching.

E. Noise Calculation and Cancellation

The noise analysis is done using small signal analysis and taking into effect the noise current of each transistor as shown in Fig.5. Each transistor's noise contribution to the output is also tabulated.

As shown in the Table I, the noise current contribution

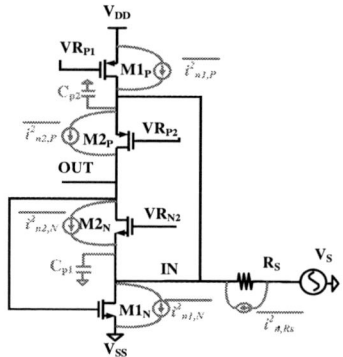

Fig. 5 *Equivalent noise circuit model of LNA*

of $M2_N$ and $M2_P$ can be partially canceled by choosing $g_{m2,N}$ and $g_{m2,P}$ optimally. Finally the noise figure(NF) is calculated as in Eq.(5). As the proposed structure exhibits a higher effective transconductance($G_{m,tot}$), the achieved NF is significantly low.

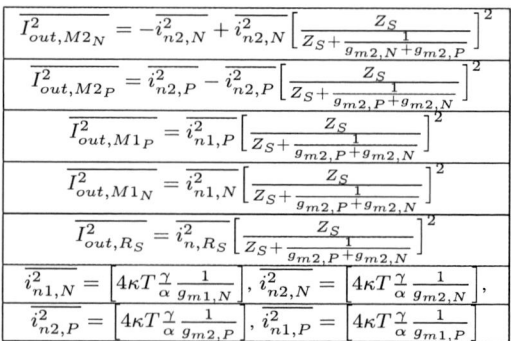

TABLE I: Noise Contributions

$$NF = 1 + \frac{\overline{I^2_{out,M2_N}} + \overline{I^2_{out,M1_N}} + \overline{I^2_{out,M1_P}} + \overline{I^2_{out,M2_P}}}{G^2_{m,tot} \times 4\kappa T R_S} \quad (5)$$

TABLE II: Designed Operating Point Parameters

$g_{m1,N}$	$g_{m2,N}$	$g_{m1,P}$	$g_{m2,P}$	$I_{D,half}$
1.56 mS	3.26mS	2.05 mS	2.75 mS	250 µA

$g_{ds1,N}$	$g_{ds,N}$	$g_{ds1,P}$	$g_{ds2,P}$	
0.45 mS	0.43 mS	0.32 mS	0.31 mS	

F. Linearity

A weak static non-linear amplifier can be approximated by a polynomial $v_{out} = \alpha_1 v_{in} + \alpha_2 v^2_{in} + \alpha_3 v^3_{in}$. In the proposed LNA $\alpha_1, \alpha_2, \alpha_3$ is obtained by taking the first, second and third derivative of I_D w.r.t. V_{in} respectively.The expressions for α_1, α_3 and A_{IIP3} can be summarized as $\alpha_1 = -\frac{G_{m,tot}}{1+G_{m,tot}R_s}$, $\alpha_3 = \frac{1}{6} \frac{12K^2 R_S}{(1+1+G_{m,tot}R_s)}$

$$A_{IIP3} = \sqrt{\frac{4}{3}\frac{\alpha_1}{\alpha_3}}$$
$$= \sqrt{\frac{2G_{m,tot}}{3R_S}\frac{(1+G_{m,tot}R_S)^2}{K}}$$

where $K = (1/2)\mu_n C_{ox}(W/L)$ and R_S is the source impedance.

Since this structure utilizes the benefit of the G_m boosting technique, the IIP3 performance is improved compared to the conventional CG LNA designed with an equal power consumption. The improvement in the IIP3 performance can be attributed to the lesser voltage swing at the input node(as the low input impedance is obtained for lower current).

G. Stability

Using k-factor as an indicator, the stability of the LNA is investigated. The proposed LNA exhibits an unconditional stability as the k-factor>1 for the frequency band of interest as shown in Fig.6f. The k-factor is expressed as $k = \frac{1-|S_{11}|^2-|S_{22}|^2+|S_{11}S_{22}-S_{12}S_{21}|^2}{2|S_{12}S_{21}|}$.

V. SIMULATION RESULTS

The LNA(fully-differential) is designed in 65 nm low-power UMC-CMOS process. The circuit draws 500µA from a 1.2 V power supply. A buffer is also incorporated to isolate the LNA outputs from the testing apparatus, which consumes 5 mA from same power supply. The circuit was designed using the Cadence design environment and simulated using Spectre-RF. Fig.6a shows S11 for input matching. Further, the extracted result of S11 is below -10 dB from 250 MHz to 3.2 GHz. As shown in Fig.6b the extracted power gain (S21), with the buffer, achieved at maximum of 800 MHz is 13.5 dB and remains at a 3-dB flatness from 250 MHz to 3.2 GHz. The minimum noise figure of the LNA shown in Fig.6c is 1.63 dB at 1.2 GHz frequency and average NF

TABLE III: Comparison With Published Low Power LNA

Specification/Reference	This Work	[17]	[18]	[19]	[20]	[11]	[9]
Inductor?	No	No	No	No	No	Yes	No
Differential?	Yes	Yes	Yes	No	Yes	No	No
3 -dB BW(GHz)	0.25 ~3.2	0.1 ~4.3	0.4 ~1	0.1 ~2.1	0.1 ~2	0.6 ~4.2	0.1 ~2.2
Power(mW)	0.6	2	0.2	7.06	21.3	0.25	0.4
Supply(V)	1.2	1.2	1	1.3	2.2	0.5	1
Gain max(dB)	23.76/13.5^{PG}	21.2	18	21.2	17.5	14	12.3
Min NF(dB)	1.6~2	2.8	4.2	2	2.9	4	4.9
IIP3 (dBm)	-10.7	-7.7	-21	12.4	10.6	-10	-11.5
CP_{1dB}(dBm)	-18.18	-	-	-	-	-20	-
Tech(nm)	65	65	180	130	180	130	130
Area(mm^2)	0.25	0.05	0.27	0.007	0.63	0.39	0.005

*PG:Power Gain

is below 2dB across the entire band. Fig.6e highlights the 1-dB compression point which is -18.18 dBm. A two-tone test is applied to determine the IIP3 of the circuit. Fig.6d shows the IIP3 -10.7 dBm at 1 GHz frequency with two tone spacing of 10 MHz.

VI. CONCLUSION

An ULP, low supply voltage wideband CMOS LNA is proposed and designed based on the gm/Id biasing methodology. The current-reuse scheme (to lower the power consumption) along with g_m boosting structure (to reduce noise figure below 2 dB) is employed in this LNA implementation. The LNA operates over a bandwidth of 0.25-3.2GHz , achieves a power gain of 13.5dB , noise figure of 1.6dB and worst case IIP3 of -10.7dBm while consuming 0.6mW from 1.2Vsupply voltage. The low power consumption of the proposed LNA makes it an attractive option for portable wireless receivers. The low noise figure of this LNA impacts positively on the subsequent stages of receiver chain.

REFERENCES

[1] S. Solda, M. Caruso, A. Bevilacqua, A. Gerosa, D. Vogrig, and A. Neviani, "A 5 Mb/s UWB-IR Transceiver Front-End for Wireless Sensor Networks in 0.13µm CMOS," *IEEE Journal of Solid-State Circuits*, vol. 46, no. 7, pp. 1636–1647, July 2011.

[2] M. Crepaldi, C. Li, J. R. Fernandes, and P. R. Kinget, "An Ultra-Wideband Impulse-Radio Transceiver Chipset Using Synchronized-OOK Modulation," *IEEE Journal of Solid-State Circuits*, vol. 46, no. 10, pp. 2284–2299, Oct 2011.

[3] J. F. Chang and Y. S. Lin, "0.99 mW 3-10 GHz common-gate CMOS UWB LNA using T-match input network and self-body-bias technique," *Electronics Letters*, vol. 47, no. 11, pp. 658–659, May 2011.

[4] K. Allidina and M. N. El-Gamal, "A 1V CMOS LNA for low power ultra-wideband systems," in *2008 15th IEEE International Conference on Electronics, Circuits and Systems*, Aug 2008, pp. 165–168.

[5] M. Parvizi, K. Allidina, F. Nabki, and M. El-Gamal, "A 0.4V ultra low-power UWB CMOS LNA employing noise cancellation," in *2013 IEEE International Symposium on Circuits and Systems (ISCAS2013)*, May 2013, pp. 2369–2372.

[6] S. B. T. Wang, A. M. Niknejad, and R. W. Brodersen, "Design of a Sub-mW 960-MHz UWB CMOS LNA," *IEEE Journal of Solid-State Circuits*, vol. 41, no. 11, pp. 2449–2456, Nov 2006.

[7] F. Belmas, F. Hameau, and J. M. Fournier, "A Low Power Inductorless LNA With Double G_m Enhancement in 130 nm CMOS," *IEEE Journal of Solid-State Circuits*, vol. 47, no. 5, pp. 1094–1103, May 2012.

[8] Z. Li, L. Sun, and L. Huang, "0.4 mW wideband LNA with double Gm enhancement and feed-forward noise cancellation," *Electronics Letters*, vol. 50, no. 5, pp. 400–401, Feb 2014.

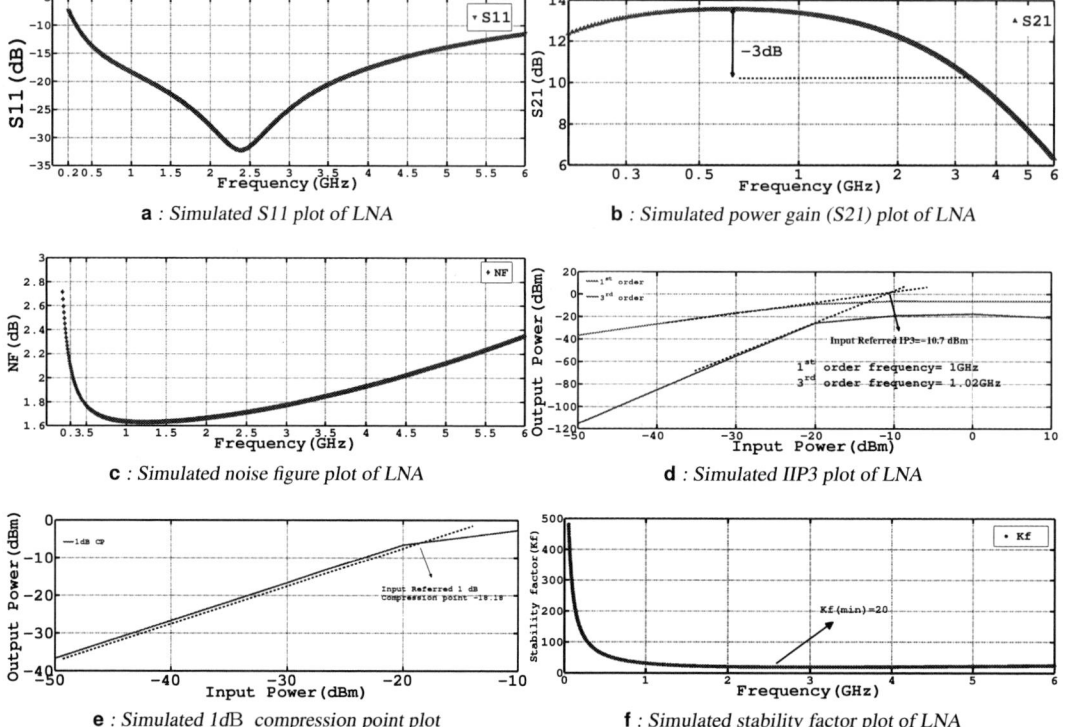

a : *Simulated S11 plot of LNA*

b : *Simulated power gain (S21) plot of LNA*

c : *Simulated noise figure plot of LNA*

d : *Simulated IIP3 plot of LNA*

e : *Simulated 1dB compression point plot*

f : *Simulated stability factor plot of LNA*

Fig. 6 *Simulated performance of proposed LNA*

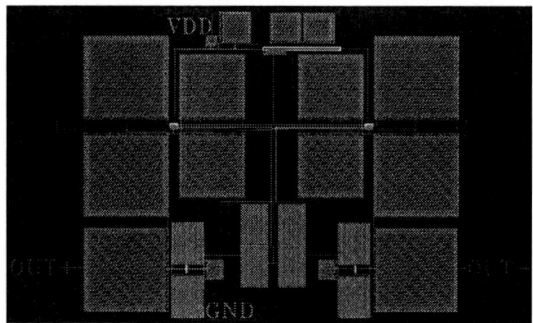

Fig. 7 *Layout diagram of LNA*

[9] M. Parvizi, K. Allidina, and M. N. El-Gamal, "An Ultra-Low-Power Wideband Inductorless CMOS LNA With Tunable Active Shunt-Feedback," *IEEE Transactions on Microwave Theory and Techniques*, vol. 64, no. 6, pp. 1843–1853, June 2016.

[10] ——, "A Sub-mW, Ultra-Low-Voltage, Wideband Low-Noise Amplifier Design Technique," *IEEE Transactions on Very Large Scale Integration (VLSI) Systems*, vol. 23, no. 6, pp. 1111–1122, June 2015.

[11] ——, "Short Channel Output Conductance Enhancement Through Forward Body Biasing to Realize a 0.5 V 250 μW 0.6 - 4.2 GHz Current-Reuse CMOS LNA," *IEEE Journal of Solid-State Circuits*, vol. 51, no. 3, pp. 574–586, March 2016.

[12] H. Wang, L. Zhang, and Z. Yu, "A Wideband Inductorless LNA With Local Feedback and Noise Cancelling for Low-Power Low-Voltage Applications," *IEEE Transactions on Circuits and Systems I: Regular Papers*, vol. 57, no. 8, pp. 1993–2005, Aug 2010.

[13] J. Kim, S. Hoyos, and J. Silva-Martinez, "Wideband Common-Gate CMOS LNA Employing Dual Negative Feedback With Simultaneous

Noise, Gain, and Bandwidth Optimization," *IEEE Transactions on Microwave Theory and Techniques*, vol. 58, no. 9, pp. 2340–2351, Sept 2010.

[14] S. C. Blaakmeer, E. A. M. Klumperink, D. M. W. Leenaerts, and B. Nauta, "Wideband Balun-LNA With Simultaneous Output Balancing, Noise-Canceling and Distortion-Canceling," *IEEE Journal of Solid-State Circuits*, vol. 43, no. 6, pp. 1341–1350, June 2008.

[15] ——, "The Blixer, a Wideband Balun-LNA-I/Q-Mixer Topology," *IEEE Journal of Solid-State Circuits*, vol. 43, no. 12, pp. 2706–2715, Dec 2008.

[16] M. Parvizi and A. Nabavi, "Improved derivative superposition scheme for simultaneous second- and third-order distortion cancellation in LNAs," *Electronics Letters*, vol. 45, no. 25, pp. 1323–1325, December 2009.

[17] Z. Pan, C. Qin, Z. Ye, and Y. Wang, "A Low Power Inductorless Wideband LNA With Gm Enhancement and Noise Cancellation," *IEEE Microwave and Wireless Components Letters*, vol. 27, no. 1, pp. 58–60, Jan 2017.

[18] H. J. Liu and Z. F. Zhang, "An Ultra-Low Power CMOS LNA for WPAN Applications," *IEEE Microwave and Wireless Components Letters*, vol. 27, no. 2, pp. 174–176, Feb 2017.

[19] M. D. Souza, A. Mariano, and T. Taris, "Reconfigurable Inductorless Wideband CMOS LNA for Wireless Communications," *IEEE Transactions on Circuits and Systems I: Regular Papers*, vol. 64, no. 3, pp. 675–685, March 2017.

[20] B. Guo, J. Chen, L. Li, H. Jin, and G. Yang, "A Wideband Noise-Canceling CMOS LNA with Enhanced Linearity by Using Complementary nMOS and pMOS Configurations," *IEEE Journal of Solid-State Circuits*, vol. 52, no. 5, pp. 1331–1344, May 2017.

An NMOS Low Drop-out Voltage Regulator with -17dB Wide-band Power Supply Rejection for SerDes in 22FDX

Nitin Bansal
Invecas Technologies Pvt. Ltd.
Noida, India
nitin.bansal@invecas.com

Rahul Gupta
Invecas Technologies Pvt. Ltd.
Noida, India
rahul.gupta@invecas.com

Abstract— **A fully on-chip Low Drop Out (LDO) regulator that uses an NMOS output transistor, designed in 22FDX technology, is presented. Proposed LDO capitalizes on technological advantages of back gate bias available in 22FDX to reduce the drop out voltage for regulator and achieve wide band power supply rejection. Charge-pump is used to apply forward back-bias on the NMOS pass device to reduce its threshold voltage down to 0V and enable low drop out regulation. The LDO generates 0.91V output from 1.35V input supply with load current capacity of 15mA. Proposed LDO provides wide band power supply rejection achieving worst case PSR of -17dB across 1Hz-10GHz frequency spectrum essential for SerDes applications. Unity gain bandwidth of >15MHz is achieved along with fast transient response of 63mV peak-to -peak for a current step of 14mA in 100ns.**

Keywords- CMOS, voltage regulators, linear voltage regulators, LDO, PSRR, FDSOI, wide band power supply rejection , fast transient.

I. INTRODUCTION

As the power dissipation is becoming the key parameter of optimization in IC design, the low voltage design is becoming important. For SerDes applications it is desired to reduce the operating analog supply to minimize the power consumption for transmitter and receiver. As the serial links transmit or receive data at frequencies in the range of GHz and have tight phase noise and clock jitter specifications, they need clean power supply for Voltage Controlled Oscillators, Phase Locked Loops and Clock Data Recovery circuits. Self-induced noise from different sub-blocks of SerDes may also induce jitter and affect the eye clearance. Thus it is required to have internal LDOs which have high wide-band power supply rejection.

PMOS based regulator is the natural choice when it comes to low drop out application since the voltage headroom is more relaxed than NMOS regulator. PMOS regulator, however, suffer from limitations of low PSRR across frequency spectrum and slow transient response. These limitations can be overcome by placing large capacitance at the output, typically of the order of 1uF, and making the external pole as the dominant pole [1-2]. However, this technique places overhead on the overall pin count and die area. Attempts have also been made to improve PSR without external capacitance usage. [3] uses a PSR enhancer circuit to couple the noise to the gate of PMOS to achieve improved PSR. [4] uses a Tri-loop LDO technique to achieve a worst case PSR of -12dB@5MHz. [5] proposes a power-supply ripple injection filter to achieve a PSR of -15dB up-to 15MHz. However, the PSR tends to degrade at higher frequencies for these architectures. Beside the techniques described in [3-5] tend to become complex to implement and difficult to satisfy all corner conditions in addition to area/power overheads.

The limitations described above make the designer to choose NMOS device as the pass element to achieve high PSRR and good load transient response. Low output impedance of the NMOS helps in pushing the pole associated at the output node at higher frequency and thus achieves high bandwidth i.e. good transient response and high PSRR for all frequency range. However, these type of regulators are not suitable for low voltage application as large voltage headroom is needed to drive NMOS. Architectures have been proposed to reduce the voltage headroom limitation [6-10], discussed in following section. Here, we propose a low voltage drop-out, fully integrated capacitor-less, NMOS based linear regulator for high speed SerDes applications. We exploit the back gate bias option available in FDSOI technology to reduce the operative voltage headroom and utilize NMOS as the pass element.

Rest of the paper is organized as follows. Section II reviews some existing NMOS LDO architectures. Section III provides an overview of 22FDX technology and in section IV we describe the design architecture and circuit implementation. It is followed by simulation results in section V along with comparison to other architectures. In section VI we present the conclusion.

II. PREVIOUS WORK ON NMOS VOLTAGE REGULATORS

There has been many paper presented on NMOS regulator which address the issue of low operating supply voltage. Since the fundamental limitation associated with NMOS regulator is voltage headroom, different techniques have been proposed to overcome this. However, they all rely on generating a higher voltage at the gate of NMOS to generate sufficient overdrive. [6-7] propose usage of charge-pump in the regulation loop to boost the gate drive voltage. [8-9] describe switched floating capacitor based gate overdrive architecture to drive the NMOS and achieve low drop-out. Charge-pump based or switched floating capacitor technique are useful when the application is

978-1-5386-3693-0/18 $31.00 © 2018 IEEE

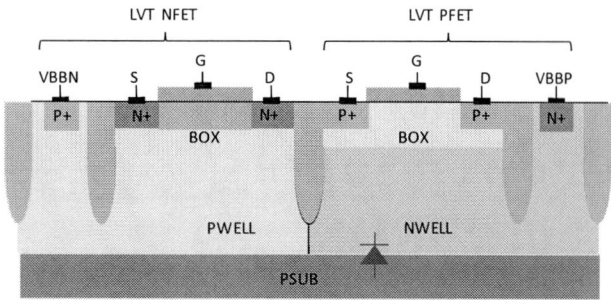

Fig. 1a. UTTB FDSOI device structure

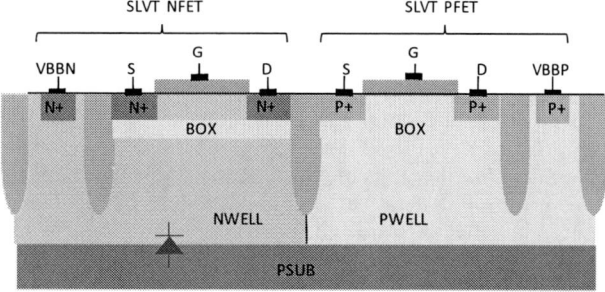

Fig. 1b. UTTB FDSOI SLVT device structure

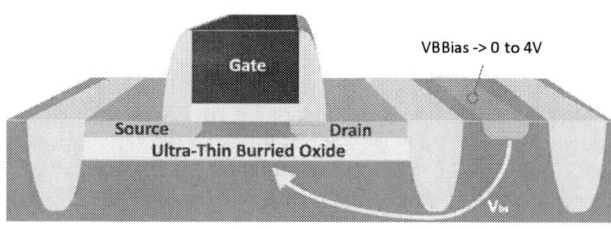

Fig. 1c. Back Gate bias in UTTB-FDSOI

to supply the digital blocks because placing the charge-pump inside the control loop will generate noise at the gate of pass device. As the NMOS pass devices is just acting as a buffer, The noise at gate will be reflected at the output voltage of regulator and will degrade the integrity of DC voltage. Large capacitance or an RC filter can be placed at the charge-pump output to filter the noise but it will slow down the control loop and will degrade the transient performance of regulator. Secondary regulation loops are often introduced in these methods to provide faster regulation as the main loop tends to be slower due to introduction of charge-pump. [10] describes a floating gate NMOS based regulator where charge is pre-programmed on the floating gate to modify the threshold voltage of the NMOs pass element.

[11] describes a sub-unity gain positive feedback based NMOS regulator which uses NMOS output stage providing a worst case PSR of -17dB across 1Hz-10GHz frequency spectrum, however it does not address the voltage headroom issue.

Another possible implementation could be to boost the input supply of regulator and thus remove the headroom overhead. This approach is rarely used because the pass device will load the boosted supply coming from the charge-pump. Load current driving requirement will constrain the charge-pump design for large switch sizes and capacitors which in turn will blow up the area.

We propose a regulator architecture where in the technological advantage of 22FDX process is used to eliminate the overheads associated with the usage of boosted supply at the gate or at the supply of pass device.

III. 22FDX TECHNOLOGY OVERVIEW

The evolution of FDSOI [Fully Depleted Silicon On Insulator] technology has ensured continuation of Moore's law[12]. Ultra-thin body created by introduction of ultra-thin Buried Oxide (BOX) provided a way to overcome undesired shrinking dimensions' effect of bulk technologies, Fig. 1a. It enabled advantages in terms of reduced bulk leakage, reduced source drain capacitances and no possibility of latch up. The body of transistor is made independent of the channel and thus can be 'flipped' i.e. NMOS transistor can be fabricated over NWell and PMOS devices can be fabricated over Pwell. These devices have lower threshold voltages compared to regular devices and are called Super Low Voltage Threshold (SLVT) devices, Fig 1b. Ultra-Thin Body and Buried Oxide (UTTB) provides another terminal to alter and control the performance of MOS devices. The BOX can be used as a back gate to dynamically increase or decrease the threshold voltage of devices with suitable biasing [13-15], Fig. 1c.

The junctions associated with bulk is through BOX and p-substrate. The thickness of BOX is 200Å and can sustain voltages higher than safe operating voltages of the device. Also the parasitic diode formed between Nwell and the p-substrate, of SLVT NMOS device is low doped and has breakdown voltages of around 7v and higher. Thus the bulk can be biased at a potential much higher than normal operating voltages of the devices.

IV. DESIGN ARCHITECTURE AND CIRCUIT IMPLEMENTATION

The architecture of the proposed design is shown in Fig. 2. The design is done using 1.8V devices available in the 22FDX technology. An error amplifier drives the NMOS pass

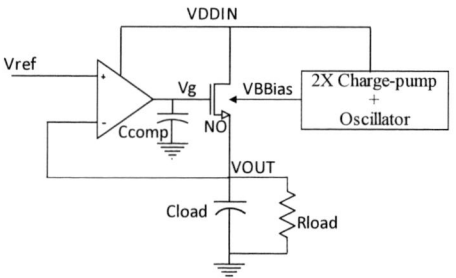

Fig. 2. NMOS LDO architecture

978-1-5386-3693-0/18 $31.00 © 2018 IEEE   342

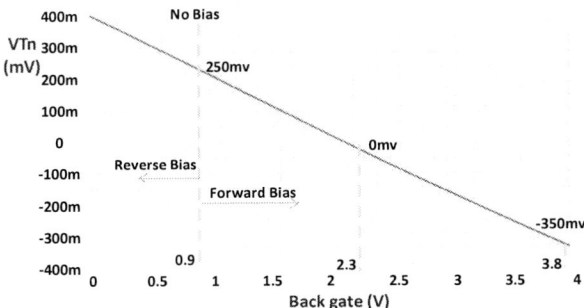

Fig. 3. Threshold voltage variation with Back Gate Bias for NMOS SLVT device in 22FDX (Vsource = 0.9V)

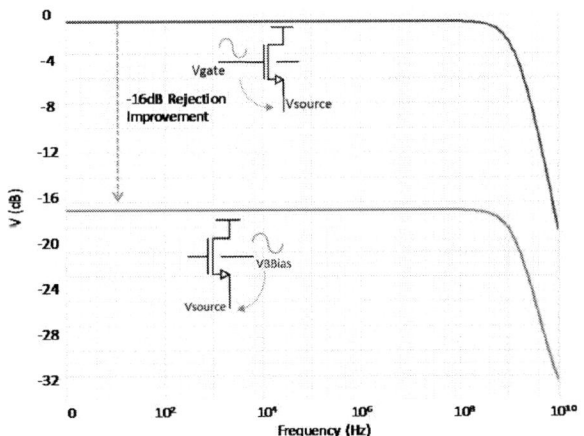

Fig. 4. Noise rejection comparison from gate and back gate for NMOS source follower

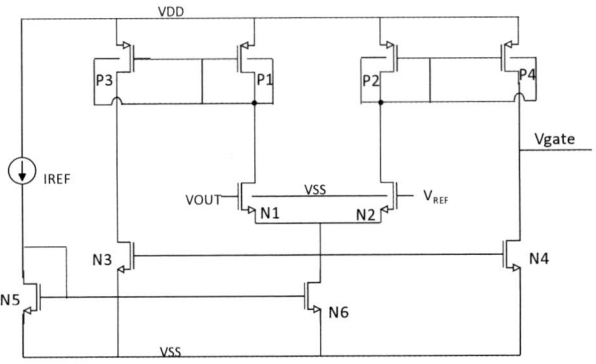

Fig. 5. Schematic of push-pull error amplifier

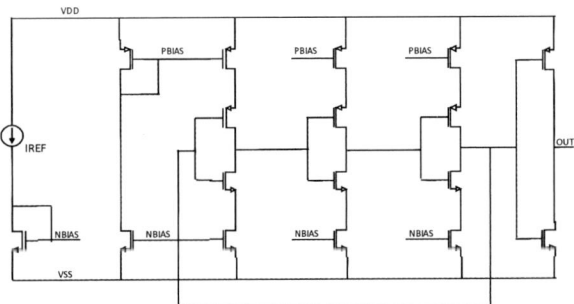

Fig.6. Schematic of a 4uA current controlled oscillator for Generating ~1MHz charge-pump clock

transistor to provide the regulated voltage and load current. Since the pass device is of NMOS type the min supply which is needed for the loop to regulate is:

$$VDD_{MIN} = V_{OUTMAX} + VT_{N0} + V_{OVN0} + V_{OVEA}$$

Where VT_{NO} is the threshold voltage of pass transistor, V_{OVNO} is the overdrive voltage needed to drive the load current through the NMOS and V_{OVEA} is the overdrive voltage of the output stage of error amplifier to ensure that it remains in saturation and provides sufficient gain.

In order to design NMOS regulator at minimum possible supply, the overdrive voltages needed for N0 and Error Amplifier's output stage can be optimized by increasing the W/L ratio. V_{OVN0} of ~ 300mV for 15mA load current and V_{OVEA} of ~100mV may be achieved but VT_{N0} is not the design parameter which can be controlled. With VT_{N0} of ~250mV and V_{OUTMAX} of 0.91V, the minimum operating supply turns out to be ~1.56V thus limiting the NMOS pass transistor design. Here the back gate bias advantage associated with FDSOI can be used to reduce the VT of pass transistor. Fig. 3, shows the behavior of transistor VT with back-bias. At VBBias lower than V_{OUT}, the transistor is in Reverse back bias. With VBBias higher than V_{OUT} the transistor sees Forward Back-Bias(FBB) and its VT reduces. With ~2.3v FBB the VT reduces to Zero

and becomes negative for higher FBB. VT reduction helps in gaining more headroom for LDO operation. A charge-pump is used to generate 2X VDDIN voltage for back-bias.

Fig. 4 presents a comparison of noise rejection when noise source is applied at gate or back gate of the NMOS SLVT device in 22FDX technology. While the noise present at gate is almost completely transferred to source, the back gate terminal provides a rejection of 16dB. The result makes it evident that it is better to use charge-pump to generate higher back-bias and reduce VT instead of raising gate voltage with charge-pump.

One issue associated with Zero or negative VT of MOS achieved with FBB, is the increase in leakage of device. However, in this case as the source voltage is around 0.9v, the gate voltage can be forced at 0.9v or lower level by the error amplifier in the loop to keep the leak current under control. At no-current load conditions ability of gate voltage to go below the source voltage becomes important as otherwise a large leak current flow may affect the voltage regulation accuracy.

To ensure high voltage swing at the output of error amplifier and be sufficient to drive the NMOS for min /max load, the push-pull error amplifier topology is chosen. The output voltage can swing between (VDD-V_{OVP}) to V_{OVN}, Fig. 5. To optimize the design back-bias is used in the amplifier with voltages available in the design. It helps in reducing devices' area, improve Gm and operating margins for the amplifier [16].

978-1-5386-3693-0/18 $31.00 © 2018 IEEE

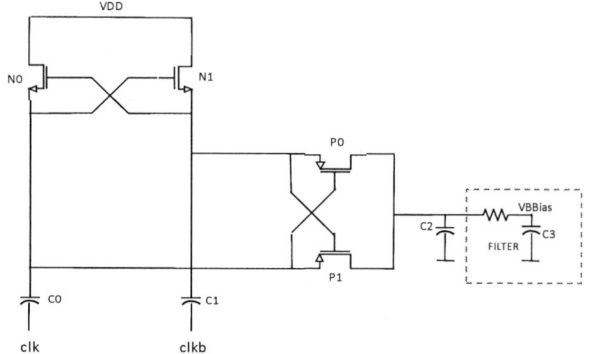

Fig. 7. Schematic of voltage doubler charge-pump along with a passive RC-filter

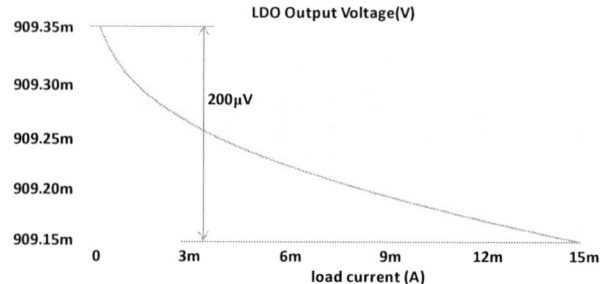

Fig. 8. DC load regulation curve of proposed LDO

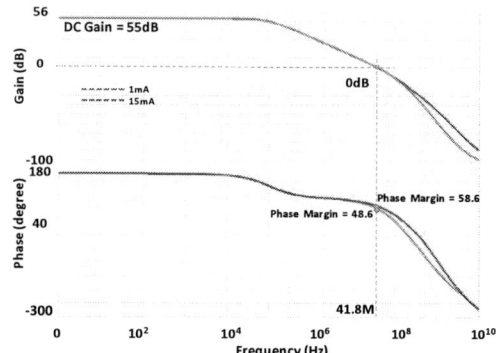

Fig. 9. Stability curve of proposed LDO

The open loop output impedance of the regulator is $1/gm_{N0}$. Gm of the NMOS pass device will be very high as the transistor sizes are large so as to minimize the overdrive and drive 15mA load current. It is easier to make the dominant pole at the output of error amplifier. This makes the compensation of the regulator quite simple as the regulator can be stabilized by placing a capacitance at the output of error amplifier such that the non-dominant poles lie outside the UGB.

A low power and low area current controlled ring oscillator is used to generate clock for the charge-pump of the order of 1MHz, Fig 6. Non overlapping clocks are used for charge-pump to boost the supply to 2X. As the output of charge-pump is driving the back gate of pass device and needs no dynamic control there will not be any load current drawn from charge-pump. This enables usage of small enough transistor switches and capacitance values which makes it low area. Conventional architecture [17] is used for the charge-pump as shown in Fig. 7. The startup time achieved with the charge-pump is 50us.

It is important to consider the noise contributors in the design for effective PSRR analysis. There are 3 Noise components present in the LDO: input supply, charge-pump and band gap reference.

A. Noise from Input Supply Voltage

At lower frequencies, the supply noise will be rejected by the loop gain and the attenuated component will appear at the output. After the loop gain reaches unity the open loop rejection will decide the noise component at the output. Intuitively we can see that the noise path from supply to output is through gate of NMOS. From the gate to output the NMOS gain is approximately unity, so to improve the PSRR at high frequency the noise at the gate of NMOS need to be minimized. The compensation scheme we have used makes sure that at noise frequencies greater than UGB the compensation cap is already acting as a short to the ground and thus achieve high PSRR even at higher frequency.

B. Noise from Charge-pump

There are two sources of charge-pump noise at the back gate. First component is the switching noise of the charge-

pump itself and second component is the supply noise coupled at the output of charge-pump i.e. PSR of the charge-pump. The worst case switching noise generated by the charge-pump is of the order of 10mV and is filtered by a passive RC filter. The cutoff frequency of the filter is set equal to the regulation loop bandwidth because the noise at the Back-gate will be corrected by the loop till the UGB and thereafter the filter should start working to reject the switching noise of charge-pump. The worst case noise ripple achieved at the back-gate is 0.6mV-pp after filter.

The supply noise rejection of the charge-pump is analyzed by giving transient noise tones of different frequency at the supply of charge-pump. The charge-pump itself provides at-least 20dB reduction in supply noise which is further attenuated by the RC filter placed at the output of charge-pump.

C. Noise from Bandgap

The regulator is placed in the unity gain configuration so, the noise at reference voltage will be reflected at the output as it is. Normally the band-gap PSR is optimized as part of band-gap design and is not covered here.

V. SIMULATION RESULTS AND COMPARISON

The LDO proposed in this paper has been characterized for all possible process corners, temperature range of -40C to 125C and supply variation of 1.35V to 1.98V. A load capacitance of 30pF is integrated within the regulator.

978-1-5386-3693-0/18 $31.00 © 2018 IEEE

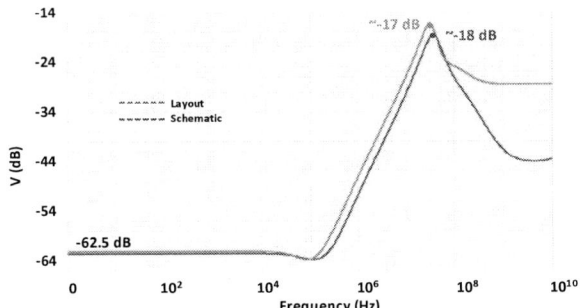

Fig. 10. Worst case PSR curve of proposed LDO

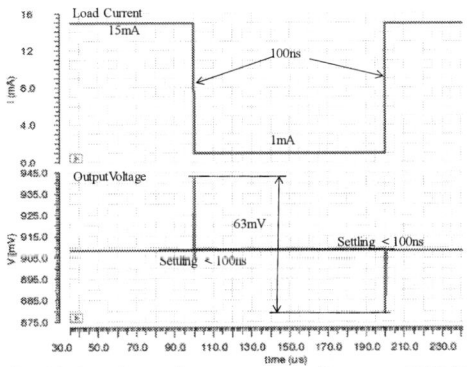

Fig. 11. Load transient response of proposed LDO

The worst case DC load regulation achieved with the proposed LDO is less than 0.2mV-pp, Fig. 8. The proposed regulator achieves the worst case phase margin of 45 degrees across all resistive and current loads, Fig. 9. At minimum load current the output impedance increases which bring the non-dominant poles closer to UGB. So, minimum load is the worst case for the phase margin. The worst case PSR achieved is -17dB across all frequency, load and PVTs, Fig. 10.

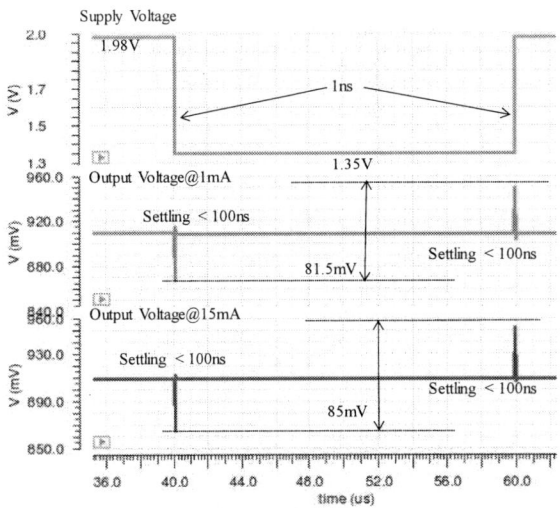

Fig. 12. Supply voltage step response of proposed voltage regulator

Load transient performance is evaluated by giving a current step of 1mA to 15mA in 100ns at the regulator output. The worst case deviation in the output voltage is 63mV-pp. The transient output voltage and current waveform are shown in Fig. 11.

Line regulation simulation is performed by stepping the input supply voltage from 1.35V to 1.98V and checking the output voltage variation. The peak-to-peak deviation in output voltage is <85mV, Fig. 12. The DC line regulation is less than 1mV-pp.

The overall area of the regulator is 0.017mm² including the 30pF capacitive load, oscillator and charge-pump, Fig 13. The regulator consumes around 190uA of bias current while the current consumption in oscillator and charge-pump is less than 10uA. Results are summarized in Table I and comparison with

Table 1 Results of proposed LDO and comparison

	[3]	[4]	[5]	[7]	[8]	This work *
Year	2014	2015	2012	2012	2009	**2017**
Technology (nm)	180	0.065	0.13	0.065	130	**22FDX**
Pass device	PMOS	PMOS	PMOS	NMOS	NMOS	**NMOS**
V_{IN}, or V_{SUPPLY} (V)	1.8	1.15-1.4	2.07-5.5	2.0-5.5	1.6	**1.35-2.0**
V_{OUT} (V)	1.6	0.95v-1.2	1.3	1.3	1.5	**0.77 – 0.91**
Imax (mA)	50	50	50	200	50	**15**
Line Reg. (mV/V)	NA	37	8.1	NA	NA	**0.5**
Load-Reg. Time, T_R ($C_{OUT}\Delta V_{OUT}/I_{MAX}$, ns)	0.195	1.15	0.022	NA	NA	**0.135**
DC Load Reg. (mV)	NA	1.1	0.055	NA	NA	**0.4**
PSR up-to 10MHz (dB),min	-37	-12	-15dB	-50dB@1MHZ	-57dB@1KHz	**-35**
Min PSR (1Hz-10GHz) (dB)	NA	-12@5MHz	NA	NA	NA	**-17@40MHz**
Active Area (mm²)	0.14 w/o cap	0.026	0.018	0.21	0.018	**0.02**
ΔI/ΔT at input (mA/ns)	50/100	10/0.2	50/200	100/10	50/200	**14/100**
T_{SETTLE} (for load reg., µs/mA)	0.6/50	NA	0.4/50	NA	0.6/50	**0.1/14**
C_{OUT} (pF)	128	140	20	4400	70	**30**
FOM ($C_{OUT}\Delta V_{OUT}I_Q/I_{MAX}^2$, ps)	0.21	5.74	0.017	23	NA	**1.8**

* Simulation Results

978-1-5386-3693-0/18 $31.00 © 2018 IEEE   345

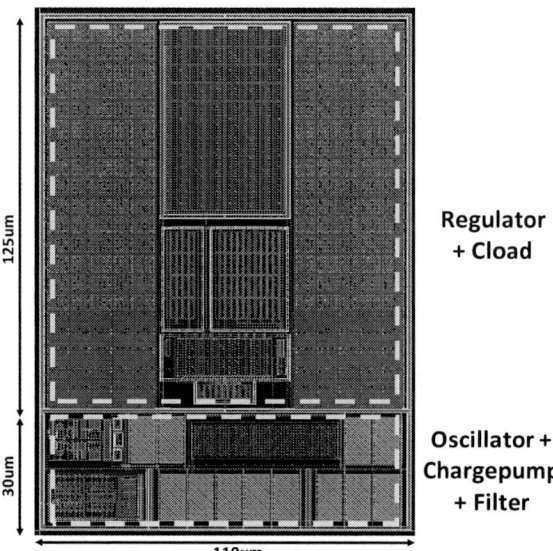

125um | **30um** | **110um**

Regulator + Cload

Oscillator + Chargepump + Filter

Fig. 13. Layout of proposed LDO in 22FDX

other architectures is also presented. The regulator achieves best-in-class wide band -17dB PSR rejection using NMOS pass device. The regulator achieves an FOM [18] of 1.8 (smaller the FOM number better it is). As can be seen the proposed regulator provides good performance with a simple and low cost architecture. Unlike other charge-pump based NMOS regulators, the proposed design injects no switching noise at the output.

VI. CONCLUSION

Low drop out NMOS regulator is presented for low voltage high speed applications. Back-bias usage in FDSOI enables the low voltage NMOS pass transistor based LDO design achieving wide band PSRR and fast load transient response. Simulation results presented confirms the performance advantages of the presented architecture.

REFERENCES

[1] S. Yeung, J. Guo, and K.N. Leung, "25mA LDO with -63dB PSRR at 30MHz for WiMax," *Electronic Letters*, vol. 46, Issue 15, pp. 1080–1081, July 2010.

[2] J. Guo, and K.N. Leung, "A 25mA CMOS LDO with -85dB PSRR at 2.5MHz," *IEEE Asian Solid-State Circuits Conference* pp. 381-384, Nov. 2013.

[3] C. J. Park, M Onabajo, J Silva-Martinez, "External Capacitor-less low drop-out regulator with 25 dB superior power supply rejection in the 0.4-4MHz range," *IEEE J. Solid-State Circuits*, vol. 49, no. 2, pp. 486-501, Feb. 2014.

[4] Y. Lu, Y. Wang, Q. Pan, W Ki, C.P. Yue, "A fully integrated low-dropout regulator with full spectrum power supply rejection," *IEEE Transactions on Circuits and Systems I*, vol. 62, Issue 3, pp. 707-716, Jan. 2015.

[5] E.N.Y. Ho, P.K.T. Mok, "Wide-loading-range fully integrated LDR with a power-supply ripple injection filter", *IEEE Transactions on Circuits and Systems II*, vol. 59, Issue 6, pp. 356-360, May 2012.

[6] G. Nebel, T. Baglin, I.S. Sebastian, H. Sedlak, U. Weder, "A very low drop out voltage regulator using an NMOS output transistor", *IEEE Int. Symposium on Circuits and Systems*, pp. 3857-3860, Vol. 4 , May 2005.

[7] T. Jackum, W. Pribyl, F. Praemassing, G. Maderbacher, and R. Riederer, "Capacitor-less LVR for a 32-Bit automotive microcontroller SoC in 65nm CMOS", *IEEE European Solid-State Circuit Conf.*, pp. 329 – 332, 2012.

[8] D. Camacho, P. Gui, P. Moreira, "An NMOS low dropout voltage regulator with switched floating capacitor gate overdrive," *IEEE International Midwest Symposium on Circuits and Systems*, pp. 808-811, Aug 2009.

[9] V. Molata, V. Kote, J. Jakovenko, "Capacitor-less linear regulator with NMOS power transistor", *Electroscope*, Rocnik 2013.

[10] H.X. Roman, G.J. Serrano, "A 1uA linear regulator with programmable NMOS pass device," *IEEE Dallas Circuits and Systems Conference*, pp. 1-4, Oct. 2015.

[11] S.K. Singh, N. Bansal, "Any capacitor stable LVR using sub-unity gain positive feedback loop in 65nm CMOS," *International Conference on VLSI Design*, pp. 260-264, Jan. 2015.

[12] "Viewpoint: FD-SOI supports Moore's Law," http://www.eetasia.com/articleLogin.do?artId=8800695818&fromWhere =/ART_8800695818_480200_NT_1f50a2f3_2.HTM&catId=480200&n ewsType=NT&pageNo=2&encode=1f50a2f3, Mar 2014.

[13] Mark Lundstrom, "Overview of SOI Technology," https://nanohub.org/resources/2038/download/2006.11.10-ece612-l28.pdf, 2006.

[14] T. Skotnicki, S. Monfray, "UTTB FDSOI: evolution and opportunities," *IEEE European Solid State Device Research Conference*, pp 76-79, Sept. 2015.

[15] P. Magarshack, P Flatresse, G. Cesana, "UTTB FD-SOI: A process/design symbiosis for breakthrough energy-efficiency", *IEEE Design, Automation & Test in Europe,* Mar 2013.

[16] S.K. Singh, G.D. Kanungo, "Ultra-fast cap-less LDO for dual lane USB in 28FDSOI", *IEEE VLSI Design Conference*, pp. 254-259, Jan 2015.

[17] P. Favrat, P. Deval, M.J. Declercq, "A high efficiency CMOS voltage doubler," *IEEE J. Solid-State Circuits*, vol. 33, no. 3, pp. 410-416, Mar. 1998.

[18] P. Hazucha, T. Karnik, B. A. Bloechel, C. Parsons, D. Finan and S. Borkar, "Area-efficient linear regulator with ultra-fast load regulation," *IEEE J. Solid-State Circuits*, vol. 40, no. 4, pp. 933–940, Apr. 2005.

978-1-5386-3693-0/18 $31.00 © 2018 IEEE

Single Inductor Dual Output Buck Converter for Low Power Applications and its Stability Analysis

Sowmya Sankaranarayanan*, Kulkarni Chaitali Vinod*, Aswanth Sreekumar*, Tonse Laxminidhi*,
Vipul Singhal[†] and Rajat Chauhan[†]

*Department of E&C Engineering, N.I.T.K. Surathkal, Mangalore, India

Email: s.sowmya2510@gmail.com, chaitalikulkarni.13@gmail.com, sk.aswanth@gmail.com, laxminidhi_t@yahoo.com

[†]Kilby Labs, Texas Instruments, Bangalore, India

Email: vipul@ti.com, rajat.chauhan@ti.com

Abstract—The applications like sensor nodes and wearables, which run on coin/button cell and/or harvested energy source need small form factor and very low power consumption. A single inductor multiple output (SIMO) converter provides saving on inductor count and hence becomes a right choice for such applications. This paper presents a single inductor dual output (SIDO) buck converter targeting light load applications. The architecture uses discontinuous conduction mode (DCM) with pulse frequency modulation (PFM) control and the switching scheme ensures almost zero cross-regulation. The proposed converter is simulated in 180 nm CMOS technology showing zero cross-regulation. An efficiency of above 88% is achieved considering inductor and package losses in load range of micro-amperes to a few milli-amperes. This paper also presents a detailed stability analysis and model for the selected SIMO architecture along with some interesting observations and inferences derived from this analysis.

Index Terms—Low power, DC-DC converter, buck, SIMO, stability, DCM, PFM, IAC

I. INTRODUCTION

Most of the ultra-low power systems, for example sensor nodes and wearable products, running on coin/button cell and/or harvested energy source utilize duty-cycling of the active mode and spend majority of the time in sleep mode. These systems have sleep mode current consumption in few micro-amperes, while in active mode, current consumption can be in few tens of milli-amperes (mostly limited by supply source). Along with low power, due to the small form-factor, it is also desirable to keep component count as low as possible for these systems. Power management for these systems needs to provide high efficiency at such low load currents (micro-amperes to milli-amperes range) with minimal number of board components.

This paper discusses selecting appropriate DC-DC converter architecture for such low load current range and uses it to build a zero cross-regulation SIMO DC-DC buck converter to save on inductor count. This paper also analyses the stability of the selected architecture along with validation of the model through simulations which provide interesting inferences.

II. BACKGROUND

SIMO buck converters provide a more compact solution for applications which demand small form-factor. Many different

control and feedback schemes have been proposed [1]–[3]. Typically for the low power applications, the architecture that fits well is PFM control, DCM operation buck converter. Compared to pulse width modulation (PWM), PFM control-loop architecture provides better efficiency at low load currents. Also the DCM operation allows switching between the SIMO outputs during the zero inductor current period. This ensures no interaction between the multiple outputs and hence near to zero cross-regulation. This architecture, its operation and simulation results are discussed in detail in Section III. Ensuring the stability of SIMO loops is very important to guarantee small ripple and regulation of the output voltage especially during load and line transients. Though there has been literature published on DC-DC stability models, most of them discuss continuous conduction mode (CCM) or PWM control architectures [4]–[6]. There are very few publications on PFM control DCM buck converters [7].

Advancing on the existing work, this paper provides more detailed analysis, specifically for the feed-forward path and also provides more intuitive understanding of the path. For the selected architecture, this paper goes on to develop a stability model, provide stability analysis and verify the model accuracy against SPICE simulation. The stability for a PFM control-loop DCM SISO buck converter is analyzed using the injected absorbed current (IAC) method [8], which is then extended to a SIMO buck converter.

III. PROPOSED SIMO ARCHITECTURE

In CCM operation, the inductor current remains above zero throughout the switching period, while in DCM, the inductor current reaches zero and stays zero for a portion of the switching cycle as shown in Fig. 1.

Thus, the average inductor current in a cycle in DCM is generally lower than in CCM, making DCM the preferred mode of operation for light loads. Using PFM control during DCM operation would cause the switching frequency to reduce with the load, especially under light load conditions. This reduces the dynamic power at light loads resulting in higher conversion efficiency.

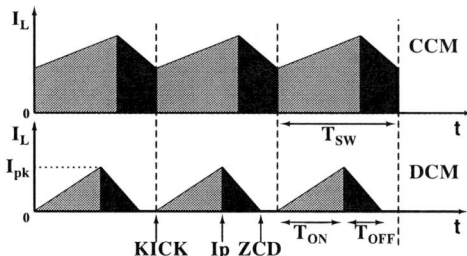

Fig. 1: SISO buck inductor current in CCM and DCM

A. Operation of SISO Buck Converter with Proposed Control Scheme

The block diagram of the sampled loop SISO buck converter operating in DCM with PFM control is shown in Fig. 2. The output voltage V_{out} is compared with the reference voltage $V_{out,ref}$ and sampled at the rising edge of the clock. The charge cycle is initiated only if the output voltage falls below the reference $V_{out,ref}$. Hence it is a need-based charging scheme and the converter does not initiate a charge cycle unless the output demands it. The converter operates in DCM for the entire range of loads and uses PFM control as the charge cycles come at variable intervals of time, depending on the load current. The detailed working is as follows:

By sensing the output voltage and inductor current, we get the signals KICK, Ip and ZCD at the points indicated in Fig. 1.

- KICK signal goes high when the output voltage V_{out} falls below the desired reference voltage $V_{out,ref}$ and starts the charge cycle.
- Ip signal goes high when the inductor current reaches the set current reference value I_{pk}.
- ZCD (zero crossing detect) signal goes high when the inductor current returns to zero.

These signals control the powerFET (pull-up and pull-down) switches that energize and de-energize the inductor during a charge cycle. The sequence of operations (carried out by the Set-Reset Logic block) during one energize/de-energize cycle is as follows:

- KICK makes PC go low, turning on the pull-up (PFET) switch to energize the inductor.
- Ip makes PC and NC go high, turning off the pull-up switch and turning on the pull-down (NFET) switch, de-energizing the inductor.
- ZCD makes NC go low, turning off the pull-down switch and completing the cycle.

Note that while DC-DC switching frequency varies with the load, sampling clock (Clk) is a free running clock with frequency set as per maximum supported load. In this scheme the switching frequency is the same as the KICK frequency. The peak value of the inductor current is maintained constant (equal to the reference I_{pk}) in the SISO operation and in the stability analysis of the sampled loop SISO buck converter described in Section IV.

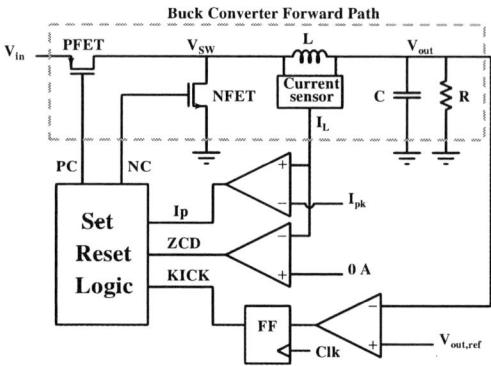

Fig. 2: Block diagram of the SISO buck converter

B. SIMO Buck Converter

SIMO converters can be viewed as circuit extrapolations of corresponding SISO converters [9]. Since all the outputs share one common inductor, variations in one output might affect the others, referred to as cross-regulation. Fig. 3 shows the block diagram of a single inductor dual output (SIDO) buck converter with PFM control.

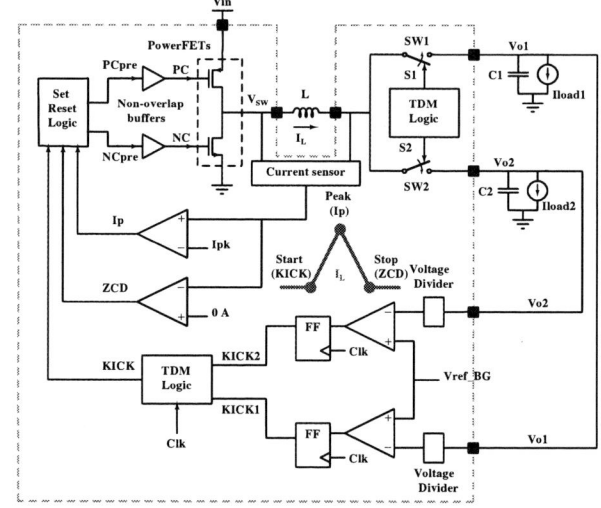

Fig. 3: Block diagram of SIDO buck converter circuit

The SIDO buck converter in Fig. 3 is implemented and verified using SPICE simulator in 180 nm technology. Since maximum current demands in light load applications are within few tens of milli-amperes, we pick the scheme shown in Fig. 4 to multiplex the inductor current between the outputs. In this scheme, the switching between the two outputs happens only when inductor current is zero. This makes the outputs and their feedback loops independent, with no interaction with each other. As a result, cross-regulation is eliminated and each of the outputs can be treated as two separate SISO converters sharing a common inductor. We refer to this configuration as

978-1-5386-3693-0/18 $31.00 © 2018 IEEE   348

time division multiplexed SIDO (TDM SIDO) buck converter. The switching between the two outputs happens on demand basis which means that for equal load on both outputs, they are served in alternate cycles while for unequal loads, the one with the higher load is served more often than the other. This allocation is tracked automatically for varying load current by TDM Logic. SIDO was designed to support a total load current ($I_{load1} + I_{load2}$) of 50 mA supporting a full range of load combinations i.e. "0 mA+50 mA" to "25 mA+25 mA" to "50 mA+0 mA".

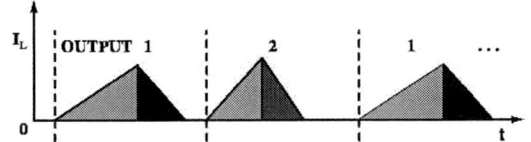

Fig. 4: SIDO buck converter inductor current in DCM

KICK1 and KICK2 signals are combined into a single KICK to control the powerFETs and also generate control signals for output switches SW1 and SW2. As proof of concept, a SIDO buck converter is designed for $V_{in} = 3.6$ V, $V_{o1} = 1.8$ V, $V_{o2} = 1.2$ V, $L = 4.7\,\mu$H, $C = 10\,\mu$F, $I_{pk} = 130$ mA and Clock (Clk) Frequency = 1 MHz. Fig. 5 shows performance of the converter for V_{in} step with $I_{load1} = I_{load2} = 1$ mA. Fig. 6 shows the response when I_{load1} is stepped from 0 to 15 mA and $I_{load2} = 15$ mA. Observe that there is no effect of V_{o1} on V_{o2} and vice-versa. This proves that, with the proposed switching scheme, the proposed architecture results in zero cross-regulation.

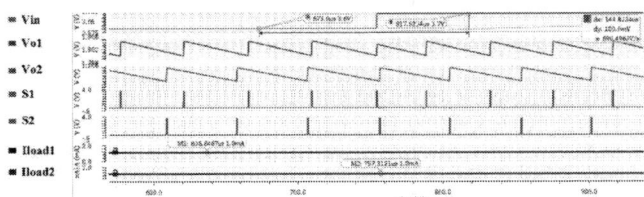

Fig. 5: SIDO buck converter response for V_{in} step

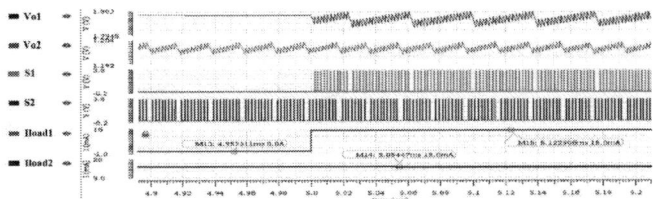

Fig. 6: SIDO buck converter response for load step at one output

Fig. 7 shows efficiency versus total load current ($I_{load1} + I_{load2}$). It includes losses due to inductor and package.

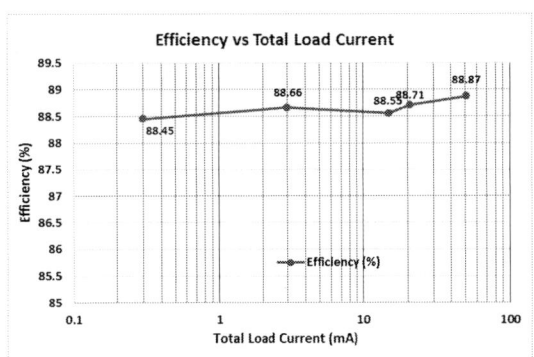

Fig. 7: SIDO efficiency versus total load current

The proposed control scheme can be extended to the design of a SIMO converter with N outputs. In this case, the maximum rate at which charge can be delivered to each output is $\frac{1}{N}$ times the sampling clock frequency. As a result, the voltage ripple at the outputs of the SIMO converter will increase with number of outputs for a given clock frequency. This ripple can be brought down by increasing the clock frequency.

IV. MODELING OF SISO BUCK CONVERTER

The stability of the buck converter can be determined from the phase and gain margins of the open loop system. To obtain these margins, a model of the buck converter must be obtained. To obtain a model for the SIMO buck converter, a SISO buck converter is modeled first. The SISO buck converter model can be easily extended to the SIMO buck converter, as will be explained in the following sections.

A. Modeling of Forward Path

There are two popular methods for modeling switched power converters- state space averaging (SSA) [10] and injected absorbed current (IAC) method. For DCM operation, the SSA method requires an additional constraint to account for the discontinuity in the inductor current. Both the methods result in the same transfer function [8]. For ease of algebra, the IAC method is chosen and described in this paper. The SISO PFM buck converter shown in Fig. 2 is modeled as represented in the block diagram shown in Fig. 8. The forward path has two inputs, input voltage V_{in} and KICK frequency F, and a single output V_{out}. The feedback path is modeled with V_{out} as the input and F as its output, thus closing the loop.

V_{in}, F, V_{out} are the nominal values of the operating point over which the circuit is to be linearized. From the geometry of the DCM inductor current waveform shown in Fig. 1 the following steady state equations are obtained:

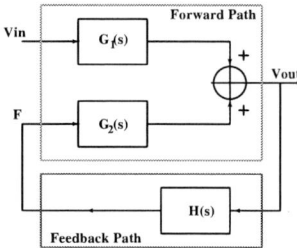

Fig. 8: Block diagram of SISO PFM buck converter

$$I_{L,avg} = \frac{I_{pk}(T_{ON} + T_{OFF})}{2T_{sw}}$$

$$T_{ON} = \frac{I_{pk}L}{V_{in} - V_{out}}$$

$$T_{OFF} = \frac{I_{pk}L}{V_{out}}$$

$$I_{L,avg} = \frac{I_{pk}{}^2 L V_{in} F}{2V_{out}(V_{in} - V_{out})}$$

$$F = \frac{1}{T_{sw}}$$

Taking partial derivatives, linearizing and taking the Laplace transform we get,

$$I_L(s) = \left(\frac{I_{pk}{}^2 L V_{in}}{2V_{out}(V_{in} - V_{out})}\right) F(s)$$
$$- \left(\frac{I_{pk}{}^2 L F}{2(V_{in} - V_{out})^2}\right) V_{in}(s)$$
$$- \left(\frac{I_{pk}{}^2 L F V_{in}(V_{in} - 2V_{out})}{2V_{out}{}^2(V_{in} - V_{out})^2}\right) V_{out}(s)$$

For a resistive load, $I_L(s) = V_{out}(s)\left[\frac{1}{R} + sC\right]$ we get,

$$V_{out}(s)\left[\frac{1}{R} + sC\right] = \left(\frac{I_{pk}{}^2 L V_{in}}{2V_{out}(V_{in} - V_{out})}\right) F(s)$$
$$- \left(\frac{I_{pk}{}^2 L F}{2(V_{in} - V_{out})^2}\right) V_{in}(s)$$
$$- \left(\frac{I_{pk}{}^2 L F V_{in}(V_{in} - 2V_{out})}{2V_{out}{}^2(V_{in} - V_{out})^2}\right) V_{out}(s)$$

Using the relation $I_{L,avg} = \frac{I_{pk}{}^2 L V_{in} F}{2V_{out}(V_{in} - V_{out})} = \frac{V_{out}}{R}$

$$V_{out}(s)\left[\frac{1}{R}\left(\frac{2V_{in} - 3V_{out}}{V_{in} - V_{out}}\right) + sC\right] = \left(\frac{V_{out}}{RF}\right) F(s)$$
$$- \left(\frac{V_{out}{}^2}{R V_{in}(V_{in} - V_{out})}\right) V_{in}(s)$$

By setting $F(s) = 0$, we get $\frac{V_{out}(s)}{V_{in}(s)}$ and by setting $V_{in}(s) = 0$ we get $\frac{V_{out}(s)}{F(s)}$.
Thus the transfer functions are:

$$\frac{V_{out}(s)}{V_{in}(s)} = \frac{-\frac{V_{out}{}^2}{V_{in}(2V_{in} - 3V_{out})}}{s\frac{RC(V_{in} - V_{out})}{(2V_{in} - 3V_{out})} + 1} = G_1 \tag{1}$$

$$\frac{V_{out}(s)}{F(s)} = \frac{\frac{V_{out}(V_{in} - V_{out})}{F(2V_{in} - 3V_{out})}}{s\frac{RC(V_{in} - V_{out})}{(2V_{in} - 3V_{out})} + 1} = G_2 \tag{2}$$

It is observed from (1) and (2) that DCM forward path with constant I_{pk} control is a first order system. We also observe a pole at $s = -\frac{(2V_{in} - 3V_{out})}{RC(V_{in} - V_{out})}$. When $V_{out} > \frac{2V_{in}}{3}$, the pole moves to the right half of the s-plane and causes the forward path to become unstable, as mentioned in [7]. The derivation is repeated assuming current source load (I_{load}) instead of a resistor. The transfer function takes the form as given in (3).

$$\frac{V_{out}(s)}{V_{in}(s)} = \frac{-\frac{V_{out}{}^2}{V_{in}(V_{in} - 2V_{out})}}{\frac{V_{out}C(V_{in} - V_{out})}{I_{load}(V_{in} - 2V_{out})}s + 1} \tag{3}$$

We observe that with current source load, the pole is at $s = \frac{-I_{load}(V_{in} - 2V_{out})}{CV_{out}(V_{in} - V_{out})}$. For this case, the pole moves to the right half of the s-plane when $V_{out} > \frac{V_{in}}{2}$. Following are interesting observations from the forward path transfer functions, followed by the MATLAB simulations of the transfer functions and validation against circuit SPICE simulations.

1) From (1) and (2) we can say that the forward path model resembles a simple first order RC circuit with transfer function of the form $\frac{A}{1 + \frac{s}{\omega_p}}$, where $\omega_p = \frac{1}{R_{eq}C}$. For a resistive load, $R_{eq} = \frac{R(V_{in} - V_{out})}{(2V_{in} - 3V_{out})}$.

2) For the positive equivalent resistance (R_{eq}) i.e. $V_{out} < \frac{2V_{in}}{3}$, the resistance drains charge out of the capacitor and the charge does not build up as can be seen from Fig. 9-12, and the corresponding values are tabulated in TABLE I and TABLE II.

3) For $V_{out} > \frac{2V_{in}}{3}$, R_{eq} is negative. This means that charge is continuously pumped into the capacitor. The voltage across the capacitor rises exponentially to V_{in}, thus leading to instability. The instability indicated by the derived model is verified by setting V_{out} accordingly in SPICE simulations. Due to the inversion in the $\frac{V_{out}(s)}{V_{in}(s)}$ transfer function, when V_{in} is stepped down, V_{out} rises continuously and saturates at V_{in}. This is shown in Fig. 13, where $V_{out} = 2.4\,\text{V} = \frac{2V_{in}}{3}$ and V_{in} is stepped from 3.6 V to 3.55 V. If V_{in} is stepped up, V_{out} decreases and tries to exponentially decay down to zero. Eventually, V_{out} falls below $\frac{2V_{in}}{3}$, hence entering the stable region and settles at the new steady state. This is shown in Fig. 14, where $V_{out} = 3\,\text{V} > \frac{2V_{in}}{3}$ and V_{in} is stepped from 3.6 V to 3.65 V. In both these cases the converter in open loop is unstable since V_{out} is drifting away from its desired level by a large amount due to a small perturbation.

4) Assuming a current source load, $R_{eq} = \frac{V_{out}(V_{in} - V_{out})}{I_{load}(V_{in} - 2V_{out})}$. Negative equivalent resistance (R_{eq}) arises for $V_{out} > \frac{V_{in}}{2}$. Which means, forward path with a resistive load is stable for a wider range of V_{out} than with a current source load. Since it is a first order system, the only source of instability is accumulation (or loss) of charge causing the voltage to rise (or fall) continuously and driving the system to saturation. Current drawn by a

978-1-5386-3693-0/18 $31.00 © 2018 IEEE

resistive load is a function of V_{out}. For rising V_{out}, current (charge) drawn by the resistive load increases, thus draining charge accumulated on the V_{out} node. In case of current source load, the current (charge) drawn remains constant and so charge accumulation cannot be counteracted. Hence, the range of voltages for stable operation for the two loads is justified.

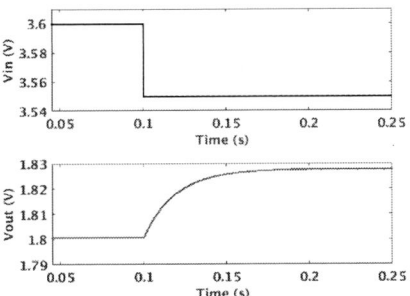

Fig. 9: SPICE simulation of forward path for V_{in} step

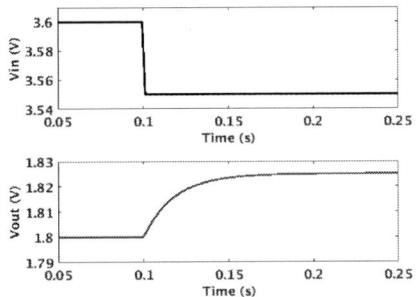

Fig. 10: Response of forward path transfer function in MAT-LAB for V_{in} step

TABLE I: Observations from transient simulations for V_{in} step in MATLAB and SPICE simulator

$V_{out}(V)$	DCgain		Rise time (ms)	
	MATLAB	SPICE	MATLAB	SPICE
1.2	-0.11	-0.112	17.75	17.2
1.6	-0.29	-0.31	29.3	29.5
1.8	-0.5	-0.53	39.73	42.35

TABLE II: Observations from transient simulations for F step in MATLAB and SPICE simulator

$V_{out}(V)$	DCgain		Fall time (ms)	
	MATLAB	SPICE	MATLAB	SPICE
1.2	3.972×10^{-5}	3.979×10^{-5}	17.6	17.4
1.6	5.972×10^{-5}	5.862×10^{-5}	29.3	29.0
1.8	7.943×10^{-5}	7.384×10^{-5}	39.6	35.3

B. Modeling of Feedback Path

The feedback path (H) consists of a continuous comparator followed by a latch, which sample the output at the rising edge

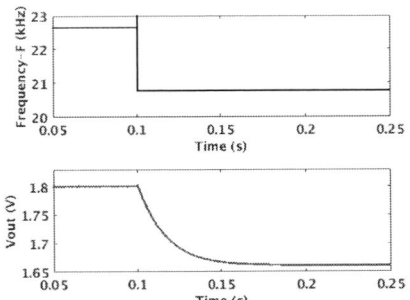

Fig. 11: SPICE simulation of forward path for F step

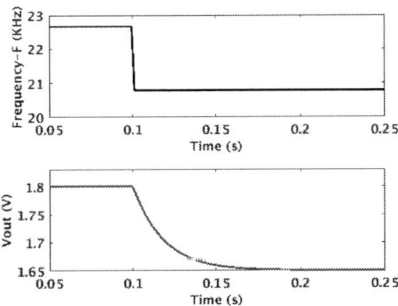

Fig. 12: Response of forward path transfer function in MAT-LAB for F step

of the clock. The clock frequency is always set much larger than the required powerFET switching frequency. Also the delay (lag) from the comparator and latch are comparatively very small. This means that for any small change in V_{out}, there is an instantaneous and large change in KICK frequency (F). Hence the feedback path can be safely assumed to be continuous, having a very large gain and bandwidth. Therefore the feedback path is modeled as a block with very high gain and no poles. For simulation purposes, the feedback path model is taken as a constant scaling factor $H = 10^{10}$.

The pole-zero plot of the closed loop system is shown in Fig. 15. Due to the large feedback factor, the location of the closed

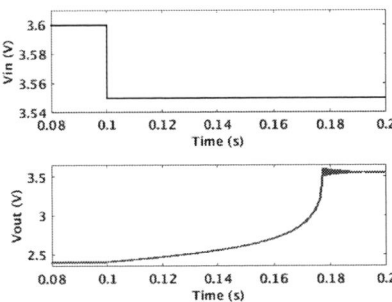

Fig. 13: SPICE simulation of forward path for V_{in} step down at $V_{out} = 2.4\,V$

978-1-5386-3693-0/18 $31.00 © 2018 IEEE

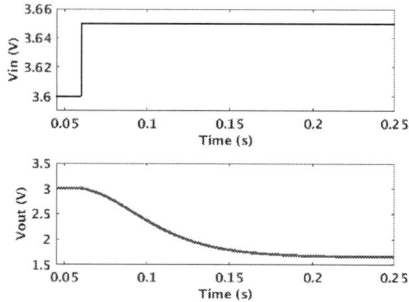

Fig. 14: SPICE simulation of forward path for V_{in} step up at $V_{out} = 3\,\text{V}$

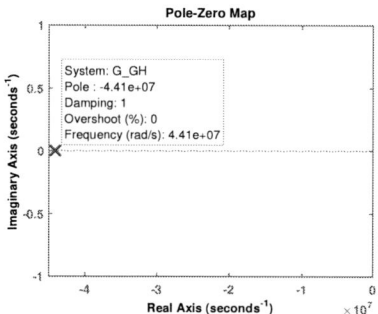

Fig. 15: Pole-zero plot of closed loop buck converter

loop pole is fixed irrespective of the load conditions. From SPICE simulations, we observe that the transient response of the closed loop SISO buck converter with PFM control is the same for all supported load currents.

The SISO buck converter forward path model has a first order transfer function whose gain starts from below $0\,\text{dB}$ and the gain and phase margins are infinite for $V_{out} < \frac{2V_{in}}{3}$. For $V_{out} > \frac{2V_{in}}{3}$, the pole moves to the right half of s-plane, thus making the forward path unstable.

Open loop pole $(\omega_0) = \frac{-(2V_{in} - 3V_{out})}{RC(V_{in} - V_{out})}$,

Gain $(A) = \frac{-V_{out}^2}{V_{in}(2V_{in} - 3V_{out})}$,

When the loop is closed with very large feedback factor ($H = \infty$ ideally), the transfer function reduces to $\frac{G}{1 + GH} = \frac{1}{H}$. Setting H to infinity makes the transfer function go to 0. Thus the output remains unaffected by perturbations in the input as can be seen in Fig. 5 and the closed loop SISO buck converter is stable under all operating conditions.

V. SIMO BUCK CONVERTER STABILITY

The SISO buck converter stability model can be extended to the SIDO buck converter architecture described in this paper. The switching between the SIDO's two outputs happens only when inductor current is zero, ensuring non-overlap of output switches. This makes the two outputs and their feedback loops independent, with no interaction with each other. The outputs can be treated as two independent SISO converters sharing a common inductor. Each output is given a charge cycle at a rate proportional to the load it needs to support. Since the SISO buck converter is stable, the discussed SIDO buck converter is also concluded to be stable.

VI. CONCLUSION

The TDM SIMO approach was chosen to design a SIDO buck converter operating in DCM as this eliminates cross-regulation and enables highly asymmetrical loads to be supported at the outputs. The SIMO buck converter switches between its multiple outputs during the zero inductor current time. So, the SIMO buck converter can be analyzed as multiple SISO buck converters that are independent of each other. A SISO DC-DC buck converter operating in DCM PFM has been modeled. This model is also verified using MATLAB and SPICE simulations by comparing DC gain and rise/fall time. It is proved that closed loop SISO buck converter is stable under all operating conditions. By extension of the SISO buck converter stability derivation, the SIMO buck converter is also concluded to be stable.

ACKNOWLEDGMENTS

The authors would like to thank Mahesh Mehendale from Kilby Labs at Texas Instruments for his guidance and crucial feedbacks during the course of this project. The authors would also like to thank the Ministry of Electronics & Information Technology (MeitY), Government of India, for providing EDA tools through SMDP-C2SD Project.

REFERENCES

[1] H. Eachempatti, S. Ganta, J. Silva-Martínez, and H. Martínez-Garcia, "SIDO buck converter with independent outputs," in *53rd IEEE International Midwest Symposium on Circuits and Systems (MWSCAS)*, pp. 37–40, 2010.

[2] M. Belloni, E. Bonizzoni, and F. Maloberti, "On the design of single-inductor multiple-output DC-DC buck converters," in *IEEE International Symposium on Circuits and Systems (ISCAS)*, pp. 3049–3052, 2008.

[3] P. Patra, J. Ghosh, and A. Patra, "Control scheme for reduced cross-regulation in single-inductor multiple-output DC-DC converters," *IEEE Transactions on Industrial Electronics*, vol. 60, no. 11, pp. 5095–5104, 2013.

[4] W.-H. Chang, J.-H. Wang, and C.-H. Tsai, "A peak-current controlled single-inductor dual-output DC-DC buck converter with a time-multiplexing scheme," in *IEEE International Symposium on VLSI Design Automation and Test (VLSI-DAT)*, pp. 331–334, 2010.

[5] Y. Wang, J. Xu, and G. Zhou, "A cross regulation analysis for single-inductor dual-output CCM buck converters," *Journal of Power Electronics*, vol. 16, no. 5, pp. 1802–1812, 2016.

[6] D. Ma, W.-H. Ki, C.-Y. Tsui, and P. K. Mok, "Single-inductor multiple-output switching converters with time-multiplexing control in discontinuous conduction mode," *IEEE Journal of Solid-State Circuits*, vol. 38, no. 1, pp. 89–100, 2003.

[7] B. Arbetter, R. Erickson, and D. Maksimovic, "DC-DC converter design for battery-operated systems," in *26th Annual IEEE Power Electronics Specialists Conference, PESC'95 Record*, vol. 1, pp. 103–109, 1995.

[8] A. Kislovski, *Dynamic Analysis of Switching-Mode DC/DC Converters*. Springer Science & Business Media, 2012.

[9] D. Kwon and G. A. Rincón-Mora, "Single-inductor multiple-output (SIMO) switching DC-DC converters," *IEEE Transactions on Circuits and Systems II: Express Briefs*, vol. 56, no. 8, pp. 614–618, 2009.

[10] L. Umanand, *Power Electronics: Essentials and Applications*. New Delhi, India: Wiley India, 2009.

A 0.29ps FOM Fast Transient Any Cap Stable LVR in 28FDSOI

Nitin Bansal
Invecas Technologies Pvt. Ltd.
Noida, India
nitin.bansal@invecas.com

Saurabh Kumar Singh, Hemant
Shukla
Cirrus Logic, UK

Madhvi Sharma
STMicroelectronics Pvt. Ltd.
Noida, India
madhvi.sharma@st.com

Abstract— A fully on-chip, fast load transient linear voltage regulator (LVR) based on sub-unity gain positive feedback loop utilizing a current subtractor loop and a transient enhancement loop is presented. Current subtractor loop is intended to improve DC load regulation and reduces quiescent current requirement of the LVR while transient enhancement loop provides better load transient performance. The LVR is implemented in 28FDSOI and generates 1.05V output from 1.3-1.98V primary input while using another 2.5-3.6V supply to power regulation loop. Load transient measurement results show that output exhibits a drop of only 65mV without any undershoot or overshoot for fast changing load current step of 75mA in 60ns and is stable with a wide range of capacitive loads from 100pF to 1µF. The LVR achieves an FOM of 0.29 along with wide band power supply rejection of -22dB@100MHz and supports convenient pass transistor distribution across the chip for uniform thermal dissipation.

Keywords— *CMOS, Linear Voltage Regulator, Capacitor-less regulator, NMOS regulator, Any Cap Stable regulator , 28FDSOI, Device Reliability.*

I. INTRODUCTION

Fully on-chip, capacitors less LVR implementations are crucial to achieving a higher level of integration for SOC as they don't require any off-chip components. The main challenge in capacitor-less LVR has been to achieve a fast load transient response as the output capacitance is limited by available on-chip capacitance.

Capacitor-less LVR circuits use various transient enhancement circuits or techniques to improve the load transient performance. Guo and Leung [1] describe a capacitive coupling based spike detection method to reduce the overshoots and undershoots. Jackum *et al* [2] describe an NMOS LVR with fast assist loop which enforces a tradeoff between power supply rejection (PSR) and transient performance. Ming *et al* [3] employ common gate error amplifier to increase slew rate with capacitive coupled direct dynamic charging technique to enhance load transient performance. Jackum *et al* [4] describe a fast LVR using NMOS pass element. This LVR combines the fast response provided by a fast regulator loop and good DC performance achieved through a slow regulation loop. Guo and Leung [5] describe a high PSR LDO which require an external capacitance of 4.7µF and thus limits the usage. Lu *et al* [6] describe fast LVR structure which contains tri-loop error

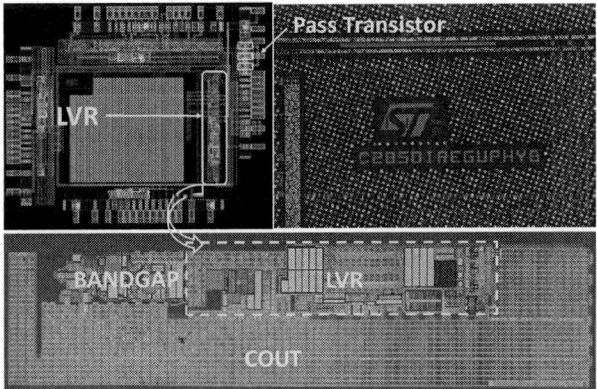

Fig 1. Chip Micrograph of proposed LVR

amplifier. Structures described in [1], [2], [3], [4], [6] make the internal pole dominant and place an upper limit restriction on output load capacitor to ensure stability. For digital applications, it is an indispensable practice to make robust power/ground network by maximizing decap insertion and distribute them efficiently to take care of high simultaneous switching currents. Thus, it is desirable to have wide range load capacitance stable LVR structure. For example, [1] and [3] place an upper limit of 50pF and 100pF on load capacitor respectively, which could be quite low and pose a limitation for digital applications. Modular and distributed structure of LVR helps in managing higher thermal dissipation of power inefficient LVRs better. Sajal [8] describes a distributed topology which places multiple regulators in parallel ensuring overall stability and higher load current.

This work focuses on NMOS based capacitor-less LVR employing a fast current mode regulation loop, capacitively coupled transient enhancement and current subtraction techniques for better transient response and load regulation. It achieves a peak-to-peak voltage spread of 65mV for a load current step of 75mA in 60ns.The proposed techniques do not impose any load capacitance constraints for stability and provide good PSR of -22dB@100MHz. LVR is also modular in nature to support distributed implementation

The design was fabricated in STMicroelectronics' 28FDSOI technology (Fig. 1). The circuit is presented in Section II; measurement results are shown in section III; a

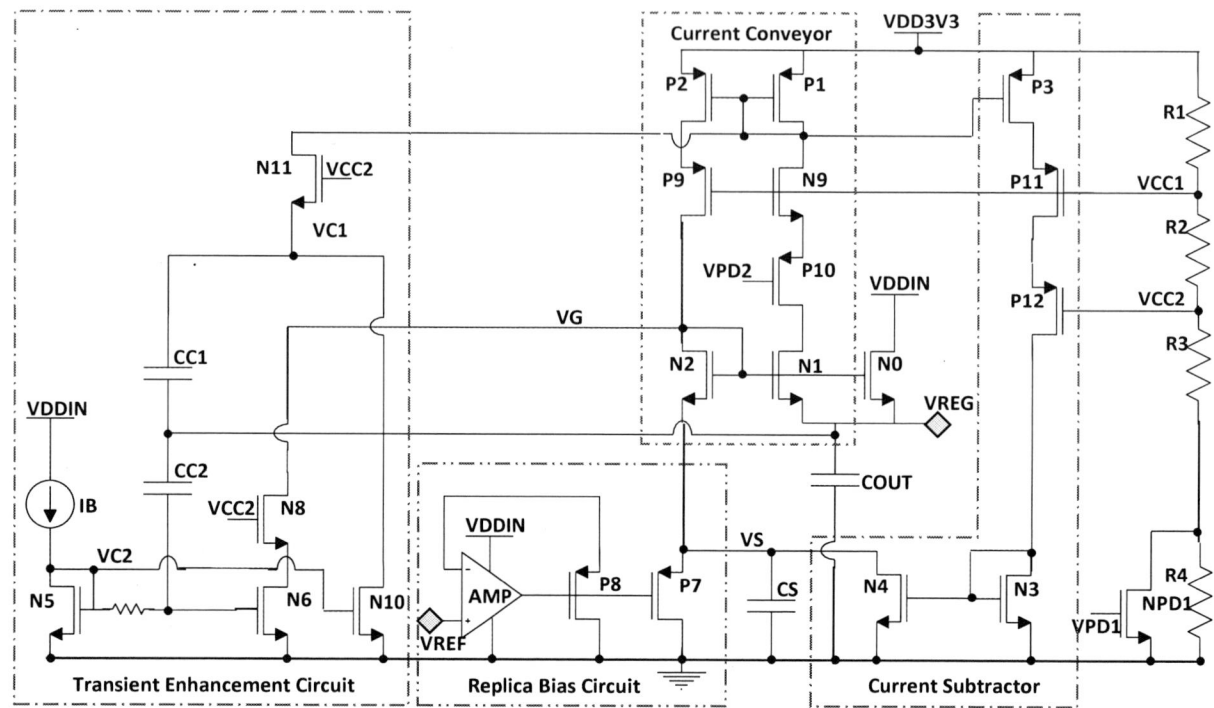

Fig 2. Simplified topology of proposed LVR

performance comparison is done in section IV and conclusions are drawn in section V.

II. CIRCUIT DESCRIPTION OF PROPOSED LVR

The regulator was developed to support digital part of serial interfaces such as USB, Display Port and HDMI thus 3.3V supply was available on-chip. This enabled selection of NMOS pass transistor as it offers better area efficiency.

A simplified circuit description of proposed LVR is shown in Fig 2. Pass transistor N0 is connected between 1.8V input supply VDDIN, which can go down to 1.3V, and LVR output VREG. N1, N2, P1 and P2 form a first generation current conveyor like circuit powered by the 3.3V supply VDD3V3. The current conveyor acts as the main regulation loop having sub-unity gain positive feedback. Node VS acts as the reference voltage for the LVR loop which is supplied by PMOS source follower P7. P7 is biased through replica loop containing AMP and P8 supplied by VDDIN [7]. A 100pF on-chip capacitor COUT represents various on-chip decoupling capacitors and load capacitor offered by circuit being driven by the LVR. MOS transistors used in this circuit implementation have a maximum operating voltage of 1.98V. Since part of the circuit operate at 3.3V supply, device reliability needs to be ensured.MOS transistors (P9-P12 and N8-N9) are inserted at various nodes forming protective cascode structures. The protection MOS are biased through a resistive divider comprising of R1-R4 which consume less than 3% of overall bias current. NPD1 and P10 ensure circuit reliability during power down mode by reducing the gate voltages of protective MOS transistors.These protective devices do not affect the

functioning of design and are ignored in the following description for simplicity.

A. Current Subtractor circuit

Node VS should remain constant across all load conditions hence, should be a low impedance node. Current subtractor accomplishes this by keeping the current through P7 constant over varying current load conditions. The principle here is to take out the current added by regulation loop (N1, P1, P2,N2) with increasing load current through a subtractor loop (P3, N3, N4) so that the operating current of current conveyor increases adaptively with load current but the net current in source follower P7 remains constant. Capacitor CS of 1pF helps to reduce impedance at high frequencies. Maintaining low impedance at VS without current subtractor circuit would take large current through source follower P7. Hence, current subtractor also helps in reducing quiescent current of LVR. To ensure stability net current taken out by N4 is little less than the current added to P7 by the regulation loop.

B. Transient Enhancement circuit

Transient enhancement circuit provides a fast regulation path from VREG to VG node through N6, N11 and capacitors CC2 and CC1 respectively, for rapidly changing load currents. Through capacitive coupling, a current is pushed or pulled out of current conveyor to move VG node in response to varying current load conditions. Fig 3 shows the simulated open loop bode plots for transient enhancement circuit indicating phase margin in excess of 35° for different load current conditions. For higher capacitive load of 100nF and 1uF the gain may not

978-1-5386-3693-0/18 $31.00 © 2018 IEEE   354

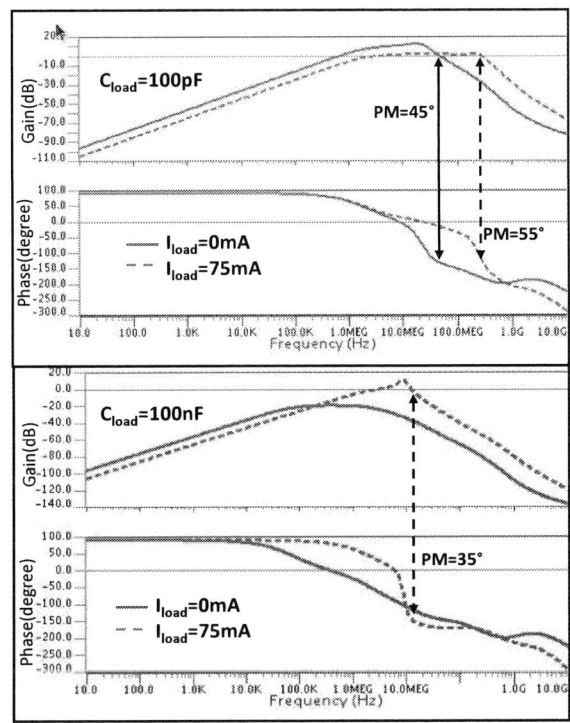

Fig 3. Simulated Bode Plot of transient enhancement circuit for 100pF and 100nF load capacitance

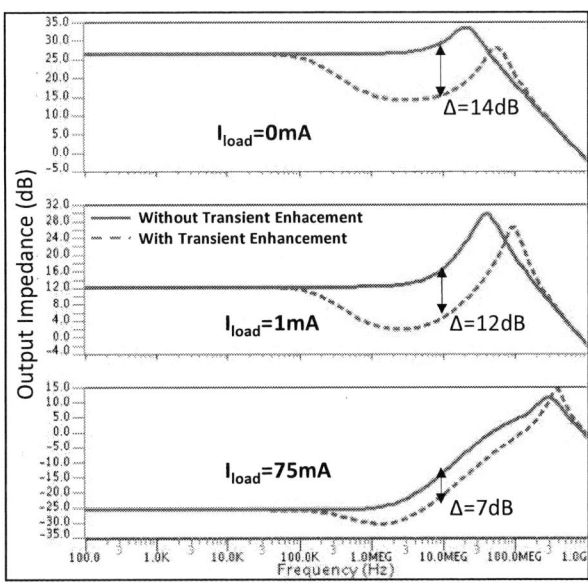

Fig 4. Simulated Output Impedance curves for Cload=100pF with and without transient enhancement circuit

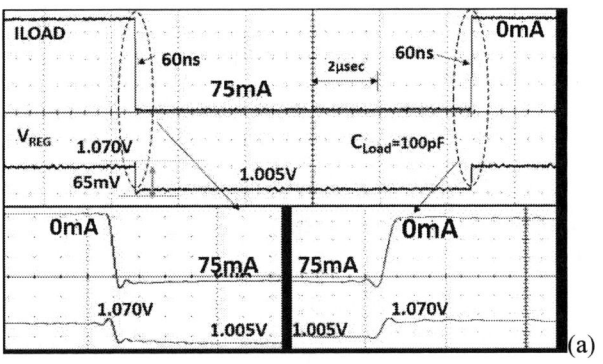

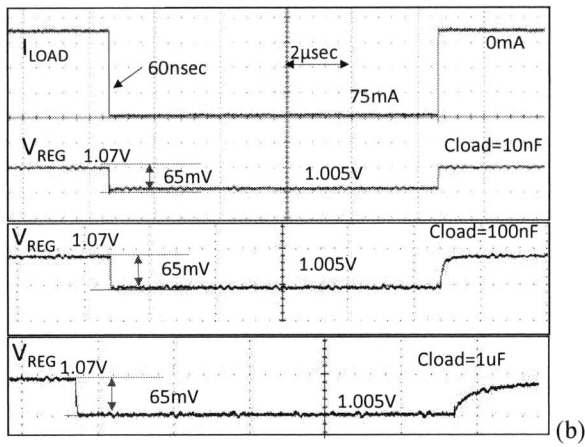

Fig 5. Measured load transient response of LVR with (a) 100pF (b) 10nF, 100nF and 1μF load capacitance

cross above 0dB thus the system would be unconditionally stable. Fig. 4 shows the output impedance curves for the regulator. Output impedance reduction of 14dB is achieved for lower load currents and upto 7dB for higher load currents at 10MHz frequency by using transient enhancement circuit. In addition, it also results in increase of output impedance peak frequency by upto 60MHz thus achieving faster load transient performance.

III. MEASUREMENT RESULTS

Pass transistor N0 was distributed in three identical blocks each occupying 0.0013mm². Each pass transistor can supply 25mA and can be placed in distributed manner around the chip. Rest of LVR occupies 0.011mm². The total area of the LVR implementation is thus 0.0149mm² including coupling capacitors CC1 and CC2 which are realized through the metal-metal fringe capacitor. The reference voltage and the bias current are provided by an on-chip Bandgap (area and current consumption of Bandgap are not included in the results).

LVR has been tested on three temperatures of -40°C, 25°C and 125°C, with VDDIN supply range from 1.3V to 1.98V and VDD3V3 supply range from 2.5V to 3.6V. Fig. 5(a) shows load transient response when load current changes between 0 to 75mA with an edge time of 60nsec. Peak to peak voltage variation is 65mV with 100pF on-chip capacitance. It

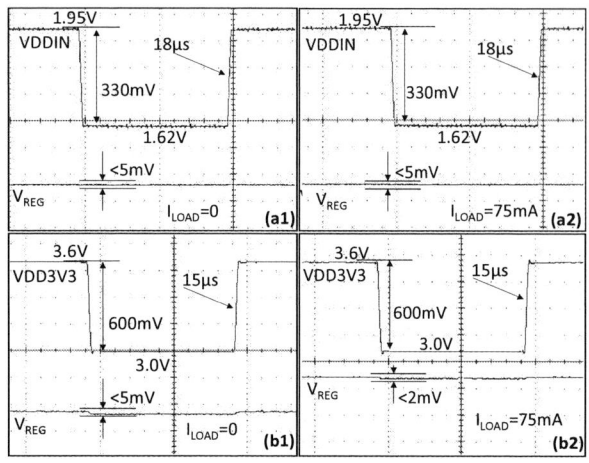

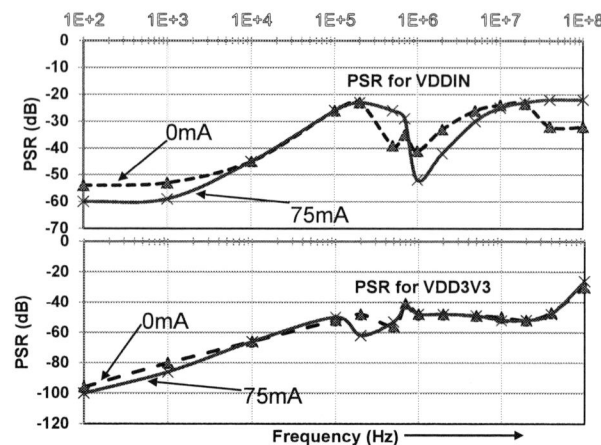

Fig 6. Measured line transient response (a) VDDIN change between 1.62V and 1.95V in 18μs (b) VDD3V3 change between 3.0V and 3.6V in 15μs for No load and Full load conditions

Fig 7. Measured PSR for VDDIN and VDD3V3 supplies

can be seen that LVR settles without any undershoot or overshoot. Fig. 5(b) shows the load transient performance of the LVR with 10nF, 100nF and 1μF off-chip output capacitances which demonstrate the ability of the proposed LVR to be stable with any capacitive load. Fig. 6(a1, a2) show measured line transient response of 1.8V input supply. For a supply jump of 330mV, LVR output peak to peak variation is <5mV with 100pF load capacitance. Fig. 6(b1, b2) show similar measurement for VDD3V3 input. Measured PSR curves with respect to VDDIN and VDD3V3 supplies are shown in Fig. 7 for two different load currents of 0mA and 75mA and 100pF load capacitance.

IV. PERFORMANCE COMPARISON

Performance comparison between the proposed regulator and existing LVRs published in the literature is shown in

Table I. FOM as described in [9] is considered to compare the load transient performance among all LVR. A smaller FOM indicates better LVR performance. The FOM of the proposed LVR is the lowest among all compared designs. All other structures place an upper limit for the capacitive load however the proposed LVR places no upper limit on output capacitance which can enable better decoupling capacitor placement in digital cores for local current distribution. The PSR claimed in [6] has a peak of -11dB@7MHz, whereas the proposed LVR has PSR better than -22dB for a wide frequency range of 100Hz to 100MHz. There is a substantial performance advantage in all parameters compared to [7]. In comparison to [8], the proposed LVR requires only the pass transistor to be distributed around the I/O ring which is more convenient for distributing thermal power.

V. CONCLUSION

A fully on-chip capacitor-less LVR with fast load transient is presented. LVR is based on sub-unity gain positive feedback

TABLE I. RESULTS OF PROPOSED LVR AND COMPARISON

	[1]	[2]	[3]	[4]	[5]	[6]	[7]	**This work**
Publication and Year	JSSC 2010	ISCAS 2011	IEEE Tran. 2012	ESSCIRC 2012	ASSC 2013	IEEE Tran 2015	VLSID 2015	**2017**
Technology	90nm	65nm	0.35μ	65nm	0.18um	65nm	65nm	**28FDSOI**
V_{IN}, (V)	0.75-1.2	1.8/3.6	2.5-4	2.07-5.5	1.8	1.2	1.6/3.6	**1.3/2.5**
V_{OUT} (V)	0.5-1.0	1.2	2.35	1.3	1.2	1	1.25	**1.07**
Imax (mA)	100	200	100	200	25	10	200	**75**
I_Q (μA)	8	132	7@I_{LOAD}=0	176	15	50	470@ I_{LOAD}=0	**250@I_{LOAD}=0**
Active Area (mm^2)	0.019	0.08	0.064	0.21	0.08	0.023	0.18	**0.0149**
Active Area/Imax (μm^2 /mA)	190	400	640	1050	1680	2300	900	**198**
ΔV(mV)	114	120	455	184	70	130	158	**65**
ΔI/ΔT at Vout (mA/ns)	3-100/100 =0.97	1-150/100 =1.49	0.05-100/500 =0.2	1-150/16 =9.31	1-25/40 =0.6	0-10/0.2 =50	0-115/140 =0.82	**0-75/60 =1.25**
C_{OUT} (pF)	50	150	100	4400	4.7 x 10^6	140	Any cap stable	**Any cap stable**
PSR	0 @1MHz	-34dB/-24dB @1MHz	-	-50dB @1Mhz	-55dB @5MHz	-11dB @7MHz	-16dB @7MHz	**-30dB /-50dB @7MHz -22dB/-22dB @100MHz**
FOM* (ps) [9]	1.71$	0.9$	10	23$	NA	3.01	5.61	**0.29**

*FOM= (C_{OUT} X ΔV / I_{MAX}) X (I_Q/ I_{MAX})

$ Minimum load current required by design is added to I_Q for FOM calculation

regulation loop and uses NMOS pass transistor. Current subtractor and transient enhancement techniques are described to achieve a FOM of 0.29ps. Measurement results show insignificant undershoot or overshoot for current steps of 75mA in 60nsec.

REFERENCES

[1] Jianping Guo and Ka Nang Leung, "A 6-µW chip-crea-efficient output-capacitorless LDO in 90-nm CMOS technology", *IEEE J. Solid-State Circuits*, vol. 45, no. 9, pp. 1896-1905, Sep. 2010.

[2] Thomas Jackum, Gerhard Maderbacher , Wolfgang Pribyl , Roman Riederer " Fast transient response capacitor-free linear voltage Regulator in 65nm CMOS," *ISCAS* , pp. 905-908, 2011

[3] Xin Ming, Qiang Li, Ze-kun Zhou and Bo Zhang, "An ultrafast adaptively biased capacitorless LDO with dynamic charging control, " *IEEE Tran. Circuits, Syst. II*, vol. 59, no. 1, pp. 40-44, Jan. 2012.

[4] Thomas Jackum, Wolfgang Pribyl, Frank Praemassing, Gerhard Maderbacher, and Roman Riederer, "Capacitor-less LVR for a 32-Bit automotive microcontroller SoC in 65nm CMOS", *IEEE European Solid-State Circuit Conference*, pp. 329 – 332, 2012.

[5] Jianping Guo, Ka Nang Leung, "A 25mA CMOS LDO with -85dB PSRR at 2.5MHz", *Asian Solid State Circuits Conference*, pp 381-384, 2013

[6] Yan Lu, Yipeng Wang, Quan Pan, Wing-Hung Ki, C. Patrick Yue, "A fully Integrated low-dropout regulator with full-spectrum power supply rejection," *IEEE Tran. Circuits, Syst. I*, vol. 62, no. 3, Mar 2015

[7] Saurabh Kumar Singh, Nitin Bansal, "Any capacitor stable LVR using sub-unity gain positive feedback loop in 65nm CMOS," *International Conference on VLSI Design*, pp. 260-264, Jan. 2015.

[8] Sajal Kumar Mandal, "A distributed regulated power conversion topology to avoid thermal talks with core loads," *International Conference on Electronics, Circuits and Systems*, pp 970-973, 2008.

[9] Peter Hazucha, Tanay Karnik, Bradley A. Bloechel, Colleen Parsons, David Finan and Shekhar Borkar, "Area-efficient linear regulator with ultra-fast load regulation," *IEEE J. Solid-State Circuits*, vol. 40, no. 4, pp. 933–940, Apr. 2005

2018 31th International Conference on VLSI Design and 2018 17th International Conference on Embedded Systems

A High Performance Gated Voltage Level Translator with Integrated Multiplexer

Dharshak B S and Rahul M. Rao

Email: dharshs1@in.ibm.com

IBM India Private Limited, Bangalore-560045, India

Abstract— **Multiple supply voltages are commonly used in designs to enable better power performance through dedicated control of the supply voltage of the various functional units. In multiple supply voltage designs, circuits are partitioned into voltage islands that operate at their optimum supply voltages which necessitates the use of voltage level translators between them. This paper presents a high performance voltage level translator design aimed at minimizing insertion penalty by minimizing logic contention and thereby improving latency. In addition, the proposed voltage level translator design has an integrated logic multiplexer function built in through an enable signal. Simulation results of the proposed voltage level translator in comparison with the conventional voltage level translator shows upto 42% delay reduction, combined with a power benefit upto 15%, for supply voltage ranging from near-threshold to above-threshold levels.**

Keywords—Level translator, Near-threshold, Above-threshold, Multiple supply voltages, Logic contention, Latency, Multiplexing.

I. INTRODUCTION

Multiple Supply Voltage Domain (MSVD) is a popular design technique in today's microprocessors and System-on-Chips (SoC) [1]. These designs involve partitioning the microprocessors and SoCs into separate supply voltage domains or voltage islands, wherein each island operates at an optimum supply voltage to meet its timing demands. This enables operation of timing critical regions at higher voltage supply, while running timing non-critical sections at lower supply voltage. Such configurations require a voltage level translator to move signals from the lower supply voltage to the higher voltage supply, without creating excessive shoot through currents. These voltage level translators need to operate over a wide voltage range from near-threshold to nominal or above-threshold depending on the application range and the optimal operating conditions for the various units.

Many voltage level shifters have been discussed in the past, except for the absence of a gated voltage level shifter to control its operation. [2] presents a conventional level shifter using multi-threshold design technique to reduce contention. Unlike the standard level converters based on feedback, the new circuits in [3] avoid feedback and employ multi-threshold transistors in order to reduce contention, which does not fully eliminate standby currents. A subthreshold level converter is demonstrated in [4] by using a cascaded two-stage design to convert signals from subthreshold up to the nominal supply voltage, which while good from a power efficiency point of view, is unsuitable for high performance designs. [5] introduces a combination of NMOS-diode current limiters and multi-threshold CMOS technique to reduce the current contention in the conventional level shifter topology. [6] discusses another multi-threshold CMOS technique for voltage level conversion. The pass-transistor topology described in [7] contains small number of transistors and is a promising level translator to minimize delay and area penalty. However, the existence of pass transistors in [7] causes a reverse current flow originating from the high voltage supply of the level translator and passes through the low voltage supply of the level translator as well as the power supply of the previous logic stage, which is undesirable. Voltage level shifters based on current mirror technique are discussed in [8, 9 and 10]. Some of the above configurations also include some devices that are always *on*, that suffer from device reliability due to aging, especially at higher voltage and temperature operational conditions of high performance designs. A different approach using level translation at flip-flops is presented in [12]. However, this restricts voltage translation to sequential boundaries, thereby restricting design choices.

In this work, a modified voltage level translator that aims to minimize level translator insertion delay, primarily by reducing logic contention is presented. The remainder of the paper is organized as follows: Section II explains the motivation behind the proposed high performance voltage level translator. Section III illustrates the key idea of the proposed voltage level translator to reduce logic contention. Section IV presents the applications of proposed level translator. The simulation results are discussed in section V. Concluding remarks are presented in section VI.

II. PRIOR ART AND MOTIVATION

Fig. 1 shows the conventional voltage level translator which uses cross-coupled PMOS devices to achieve full-swing conversion from input voltage VDIN to output voltage VDOUT. The conventional level translator design shown in Fig.1 is used as a reference for comparison with the proposed level translator design. VDIN and VDOUT are the input and output supply voltages of the level translator respectively, and VSS is the ground connection of the level translator.

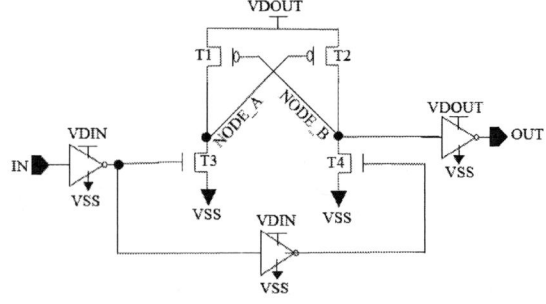

Fig. 1. Conventional Voltage Level Translator

978-1-5386-3693-0/18 $31.00 © 2018 IEEE   358

The PMOS transistors (T1 and T2) act as a cross-coupled load. When the input signal 'IN' is low (logic 0), NMOS transistor T3 is turned *on*, which drives node 'NODE_A' low. Additionally, NMOS transistor T4 is turned *off* and PMOS transistor T2 is turned *on*, due to which node 'NODE_B' is pulled high (logic 1) to VDOUT. Thus, the output signal $\overline{\text{OUT}}$ becomes low. The operation reverses when the input signal 'IN' is switched to high. This conventional voltage level translator has large delay, because it suffers from contention between the pull-down transistors (T3 and T4) and the pull-up transistors (T1 and T2). This paper proposes an approach in which the contention is reduced to gain better performance in delay.

III. PROPOSED VOLTAGE LEVEL TRANSLATOR

The proposed voltage level translator is shown in Fig. 2. The level translator shown has an integrated enable (EN) signal which acts as a control signal with the circuit functioning correctly, when EN='0' and EN_MUX='0'. ENB is the complement of EN. VDIN and VDOUT are the input and output supply voltages of the proposed level translator respectively, and VSS is the ground connection of the proposed level translator.

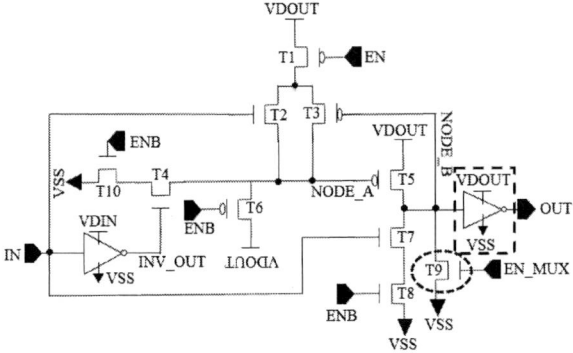

Fig. 2. Proposed Voltage Level Translator

With reference to Fig. 2, assuming that initially the input 'IN' is low, the NMOS transistors T2 and T7 are *off*; the node 'INV_OUT' is high which turns *on* the NMOS transistor T4 and pulls the contention node 'NODE_A' low. Hence, the PMOS transistor T5 is turned *on* and the contention node 'NODE_B' is pulled high to VDOUT voltage, which cuts-off the PMOS transistor T3. Thus the output 'OUT' is low. Now when the input signal 'IN' makes a low to high transition, the NMOS transistors T2 and T7 are turned *on*; the node 'INV_OUT' goes low turning *off* the NMOS transistor T4. Turning *on* the NMOS transistor T2 pulls the contention node 'NODE_A' to VDIN – Vth *(where Vth being the threshold voltage of T2)*. Thus, pulling the contention node 'NODE_A' initially to an intermediate voltage state *(rather than leaving it at its initial VSS, i.e. logic 0 state)* helps the PMOS transistor T3 in pulling 'NODE_A' speedily to VDOUT *(with the assistance of 'NODE_B' going low by the turning 'on' of the NMOS transistor T7)* and turn *off* the PMOS transistor T5 and win the fight against T5. Hence, the 'OUT' node goes high to VDOUT. Subsequently, the contention between the PMOS transistors T3 and T5 during the high to low transition of 'OUT' is overcome by making T3 a little weaker in strength as compared

to T5. This facilitates T5 to easily win the fight against T3 *(with the aid of 'NODE_A' going low by turning 'on' of the NMOS transistor T4, driven by 'INV_OUT' going high, along with the NMOS transistor T7 being turned 'off')* and pulls 'NODE_B' to VDOUT; thereby transitioning 'OUT' to VSS. The strength of T3 can slightly be compromised with that of T5, as T3 is being aided by T2 during the low to high transition of 'OUT'.

The Transistors T6 and T9 are sized to be of minimum width as they are functional only while the level translator is turned *off*. Also, these transistors could be of high threshold voltage (HVT) as they do not play a role during the normal operation of the level translator. The rest of the circuit are implemented using regular threshold voltage transistors (RVT) to meet the performance requirements *(though they could be designed with HVT if a different power performance point is desired)*. The transistors T7 and T8 are sized such that the equivalent pull down strength is same as the pull down strength of transistors T4 and T10 together. The input inverter is sized depending on the input slew range to correct the slew of the incoming signal. The pull down to pull up network sizing are in the ratio of approximately 2:1.

IV. APPLICATIONS OF PROPOSED VOLTAGE LEVEL TRANSLATOR

A. Power Gating

The enable signals EN, ENB and EN_MUX in the proposed level translator could effectively be used as fence and partial power gate signals, while the level translator is not being employed for operation.

The normal operation of the proposed level translator requires EN=EN_MUX='0' and ENB='1'; flipping these signals make the level translator non-operational. From Fig. 2, when EN='1' and ENB='0' the transistors T1, T8 and T10 are turned *off* detaching the rest of the circuit from the supplies – VDOUT and VSS. Making EN='1' and ENB='0' turns *on* transistor T6 which in turn shuts *off* transistor T5, thus separating it from the VDOUT supply. The transistor T9 is turned *on* with EN_MUX='1', pulling the 'OUT' to VDOUT. Table I below shows the truth table of the proposed level translator.

TABLE I.   TRUTH TABLE OF PROPOSED LEVEL TRANSLATOR

EN *(VDOUT)*	ENB *(VDOUT)*	EN_MUX *(VDOUT)*	OUT *(VDOUT)*
Logic 0	Logic 1	Logic 0	IN
Logic 1	Logic 0	Logic 1	Logic 1
Logic 1	Logic 0	Logic 0	Z

The presence of enable signals in the proposed level translator entitles power saving during non-operating conditions, while the conventional voltage level translator does not support this application.

B. Logic Multiplexing

A pair of proposed voltage level translators can be employed to voltage level translate one of the two signals, originating from different voltage domains VDIN1 and VDIN2, to a common VDOUT voltage domain; by deploying it to function as a multiplexer as shown in Fig. 3. This is enabled through the use of EN in conjunction with the EN_MUX. As shown in Table I, when EN_MUX is a logic low, while EN is logic high, the level translator output is at a high impedance state. Thus, in Fig. 3, the logic values of the enable signals EN and ENB activates one of the two level translators either to pull up or pull down the 'OUT' node depending on the corresponding input, with the other level translator being in high impedance and low power state as shown in Table II.

TABLE II. TRUTH TABLE OF INTEGRATED MULTIPLEXER

EN_MUX = 0		
EN *(VDOUT)*	**ENB** *(VDOUT)*	**OUT** *(VDOUT)*
Logic 0	Logic 1	IN1
Logic 1	Logic 0	IN2

As shown in Fig. 3, the multiplexing function is implemented at 'NODE_B' of the proposed level translator by excluding the inverter in the dotted box, from Fig. 2, and placing a single inverter of the same drive strength, as in Fig. 2, at the common output.

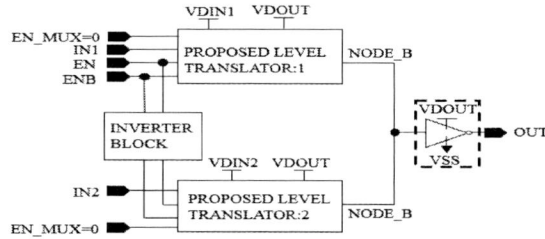

Fig. 3. Multiplexing Application of Proposed Voltage Level Translator

The multiplexing application, using the conventional level translator, could be attained by including a simple 2:1 multiplexer before the level translator. This requires the condition that, both IN1 and IN2 signals are bound to the same voltage domain VDIN. Nevertheless, two signals tied to different voltage domains VDIN1 and VDIN2 can also be multiplexed using the conventional level translator by appending a 2:1 multiplexer after the output of the level translators. However, these two implementations are achieved at the cost of additional delay and power contributed by the extrinsic multiplexer.

V. RESULTS

Simulations are performed for both the discussed level translators in an advanced FinFET Technology at nominal process conditions with input and output voltages (VDIN and VDOUT respectively) ranging from 0.6V to 1.1V using spectre circuit simulator [14]. The simulations are done using identical load FO6 and input slew conditions for both the discussed level translators.

Table III shows the delay improvement (in %) of the proposed level translator as compared to the conventional level translator for their operating voltage range. The "*rise_delay*" is the delay from rising edge of 'IN' to rising edge of 'OUT', similarly "*fall_delay*" is the delay of the level translator in transitioning a falling edge from 'IN' to 'OUT'. The numbers within '( )' in table III refer to the rise delay improvement, while the numbers within '[ ]' refer to the fall delay improvement. The delays measured are from the 'IN' pin to the 'OUT' pin, in the discussed level translators.

TABLE III. RISE/FALL DELAY IMPROVEMENT OF PROPOSED LEVEL TRANSLATOR (IN % COMPARED TO CONVENTIONAL LEVEL TRANSLATOR)

		VDIN					
		0.6V	**0.7V**	**0.8V**	**0.9V**	**1.0V**	**1.1V**
VDOUT	**0.6V**	(29.34) [12.51]	(26.23) [8.23]	(25.65) [5.79]	(26.15) [3.94]	(27.32) [2.29]	(28.78) [0.71]
	0.7V	(29.91) [18.42]	(25.32) [12.42]	(24.87) [9.11]	(26.20) [6.82]	(28.07) [4.93]	(30.28) [3.20]
	0.8V	(30.92) [24.22]	(24.31) [16.39]	(22.99) [12.12]	(24.29) [9.30]	(26.62) [7.12]	(29.27) [5.25]
	0.9V	(31.93) [30.22]	(23.66) [20.19]	(21.11) [14.98]	(21.79) [11.59]	(24.06) [9.08]	(26.93) [7.02]
	1.0V	(32.42) [36.39]	(23.29) [23.87]	(19.53) [17.74]	(19.36) [13.79]	(22.09) [10.92]	(23.97) [8.64]
	1.1V	(32.17) [42.34]	(22.96) [27.51]	(18.26) [20.40]	(17.22) [15.92]	(18.47) [12.70]	(20.90) [10.19]

The proposed level translator shows a speedup of upto 42% as compared to the conventional level translator; thereby reducing the insertion penalty. This speedup (of > 10%) is fairly consistent across process corners as well, as shown in Fig. 4 which plots the level translator insertion delay of the proposed voltage level translator under different process conditions – TT:nominal, FF:fast-fast, SS:slow-slow, FS:fast-slow and SF:slow-fast. The delays of the proposed level translator are normalized with respect to the conventional level translator. The dotted line represents the delay of the conventional level translator.

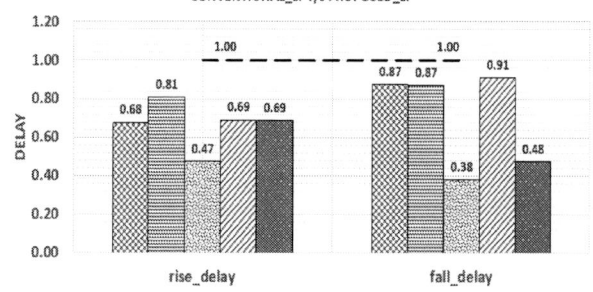

Fig. 4. Worst Delay Comparison across Process Corners

The improved performance also comes with a reduction in the power consumption of the proposed level translator. Table IV shows the percentage reduction in the AC (dynamic) power consumption of the proposed level translator during its operation, as compared to the conventional level translator.

TABLE IV. AC POWER SAVINGS OF PROPOSED LEVEL TRANSLATOR (IN % COMPARED TO CONVENTIONAL LEVEL TRANSLATOR)

		VDIN					
		0.6V	**0.7V**	**0.8V**	**0.9V**	**1.0V**	**1.1V**
VDOUT	**0.6V**	11.74	11.56	11.55	12.12	13.43	15.19
	0.7V	10.36	9.86	9.48	9.32	9.74	10.89
	0.8V	9.39	8.61	8.07	7.54	7.35	7.75
	0.9V	8.74	7.55	7.00	6.31	5.74	5.60
	1.0V	8.37	6.60	6.07	5.39	4.61	4.09
	1.1V	8.22	5.78	5.19	4.60	3.78	3.02

A marginal DC (static) power cost is observed with the proposed level translator as shown in table V. The negative numbers show worse power consumption in the proposed level translator as compared to the conventional level translator.

TABLE V.   DC POWER SAVINGS OF PROPOSED LEVEL TRANSLATOR (IN % COMPARED TO CONVENTIONAL LEVEL TRANSLATOR)

		VDIN					
		0.6V	0.7V	0.8V	0.9V	1.0V	1.1V
VDOUT	**0.6V**	-8.90	-2.29	4.39	10.98	17.36	23.41
	0.7V	-15.05	-9.10	-2.88	3.47	9.82	16.03
	0.8V	-20.12	-14.90	-9.28	-3.36	2.73	8.87
	0.9V	-24.26	-19.76	-14.79	-9.43	-3.76	2.11
	1.0V	-27.64	-23.81	-19.49	-14.72	-9.56	-4.10
	1.1V	-30.38	-27.14	-23.43	-19.26	-14.65	-9.67

In several applications, the voltages at the input and output side of a level translator may vary depending on the performance requirements of the various functional units which may be operated under dynamic voltage and frequency scaling scenarios. Hence, the power consumption of the level translator when going from high to low is as important as when going from low to high, in addition to the scenario where both the input and the output voltage levels are identical.

As discussed in the earlier section, one application of the proposed level translator is that, it effectively can be used to power gate while the level translator is not being employed for operation. Figures 5 and 6 show the power savings obtained by power gating the proposed level translator *(EN='1', ENB='0' and EN_MUX='1')*; for the conditions when 'IN' is switching, and is static respectively. This is done by comparing the power consumption of the proposed level translator, in power gating mode, with that of the un-gated conventional level translator; for the same 'IN' conditions. Power savings in the range of 60 to 95% and 25 to 40%, while 'IN' switching and is static respectively, is benefited while the level translator is non-operational.

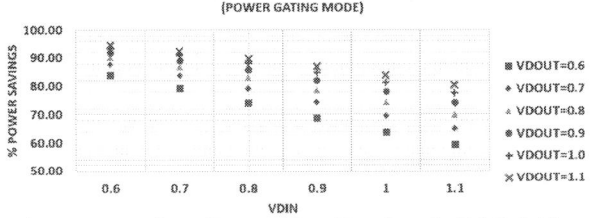

Fig. 5. Power Savings of Proposed Level Translator for 'IN' Switching

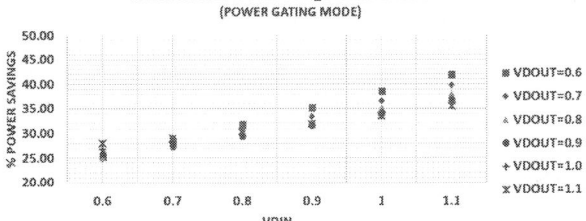

Fig. 6. Power Savings of Proposed Level Translator for 'IN' Static

CONCLUSIONS

This paper proposes a voltage level translator for high speed applications with increased performance by reducing the logic contention which facilitates lower delays in transmitting both rising and falling edges from input to output of the level translator. The integration of enable signal into the level translator al-

lows controlling its functionality and entitles it for power saving and logic multiplexing applications. Around 42% improvement in speed is observed in the proposed voltage level translator, and a minimum of 10% across process corners. In addition, upto 15% reduction in the power consumption of the proposed level translator is noticed. A significant power savings is also observed while the proposed level translator is deployed in power gating mode, as compared to the un-gated conventional level translator.

ACKNOWLEDGEMENTS

The authors acknowledge the various technical discussions with colleagues at IBM that helped improve this work.

REFERENCES

[1] Volkan Kursun and Eby G. Friedman, "*Multi-voltage CMOS Circuit Design*", ISBN:978-0-470-01023-5, Wiley Publications

[2] Marco Lanuzza, Pasquale Corsonello and Stefania Perri, "*Low Power Level Shifter for Multi-Supply Voltage Designs*", IEEE Transactions on Circuits and Systems II: Express Briefs, DOI: 10.1109/TCSII.2012.2231037, Volume:59, Issue:12, Page(s): 922-926, December-2012

[3] Sherif A. Tawfik and Volkan Kursun, "*Low Power and High Speed Multi Threshold Voltage Interface Circuits*", IEEE Transactions on Very Large Scale Integration Systems, DOI: 10.1109/TVLSI.2008.2006793, Volume:17, Issue:5, Page(s):638-645, April-2009

[4] Stuary N. Wooters, Benton H. Calhoun and Travis N. Blalock, "*An Energy-Efficient Sub-threshold Level Converter in 130nm CMOS*", IEEE Transactions on Circuits and Systems II: Express Briefs, DOI: 10.1109/TCSII.2010.2043471, vol:57, no:4, Page(s):290-294, April-2010

[5] Wenfeng Zhao, et al, "*A 65nm 25.1ns 30.7fJ Robust Sub-threshold Level Shifter with Wide conversion Range*", IEEE Transactions on Circuits and Systems II: Express Briefs, DOI:10.1109/TCSII.2015.2406354, Volume:62, Issue:7, Page(s):671-675, July-2015

[6] Marco Lanuzza, Pasquale Corsonello and Stefania Perri, "*Fast and Wide Range Voltage Conversion in Multisupply Voltage Designs*", IEEE Transactions on Very Large Scale Integration Systems, DOI:10.1109/TVLSI.2014.2308400, vol:23, no:2, p:388-391, Feb-2015

[7] Sarvesh H. Kulkarni and Dennis Sylvester, "*High Performance Level Conversion for Dual VDD Design*", IEEE Transactions on Very Large Scale Integration Systems, vol:12, no:9, pp(s):926-936, Sept-2004

[8] Shien Chun Luo, et al, "*A Wide Range Level Shifter Using a Modified Wilson Current Mirror Hybrid Buffer*", IEEE Transactions on Circuits and Systems I: Regular Papers, DOI:10.1109/TCSI.2013.2295015, vol:61, no:6, pp(s):1656-1665, June-2014

[9] Sven Lutkemeier and Ulrich Ruckert, "*A Subthreshold to Above-Threshold Level Shifter Comprising a Wilson Current Mirror*", IEEE Transactions on Circuits and Systems II: Express Briefs, DOI:10.1109/TCSII.2010.2056110, vol:57, no:9, pp(s):721-724, Sept-2010

[10] Jun Zhou, et al, "*An Ultra-Low Voltage Level Shifter using Revised Wilson Current Mirror for Fast and Energy Efficient Wide Range Voltage Conversion from Sub-threshold to I/O Voltage*", IEEE Transactions on Circuits and Systems I: Regular Papers, DOI:10.1109/TCSI.2014.2380691, Volume:62, Issue:3, Page(s):697-706, March-2015

[11] S. Rasool Hosseini, Mehdi Saberi and Reza Lotfi, "*A Low Power Subthreshold to Above-threshold Voltage Level Shifter*", IEEE Transactions on Circuits and Systems II: Express Briefs, vol:61, no:10, pp(s):753-757, Oct-2014

[12] Fujio Ishihara, Farhana Sheikh and Borivoje Nikolic, "*Level Conversion for Dual Supply Systems*", IEEE Transactions on Very Large Scale Integration Systems, DOI:10.1109/TVLSI.2003.821548, Volume:12, Issue:2, Page(s):185-195, February-2004

[13] Marco Lanuzza, et al, "*An Ultralow Voltage Energy Efficient Level Shifter*", IEEE Transactions on Circuits and Systems II: Express Briefs, DOI:10.1109/TCSII.2016.2538724, Volume:64, Issue:1, Page(s):61-65, January-2017

[14] https://en.wikipedia.org/wiki/Spectre_Circuit_Simulator

2018 31th International Conference on VLSI Design and 2018 17th International Conference on Embedded Systems

Accelerating Hash Computations through Efficient Instruction-set Customisation

Mayuran Sivanesan*, Anupam Chattopadhyay[†] and Ronak Bajaj[§]

School of Computer Science and Engineering, Nanyang Technological University, Singapore

Email: *mayuran001@e.ntu.edu.sg, [†]anupam@ntu.edu.sg, [§]ronak1@ntu.edu.sg

Abstract—Security has become a part of our everyday life due to the breakthrough technologies of wireless communication, e-commerce and digital content delivery. For ensuring authenticity of the transactions, cryptographic hash algorithms, such as SHA-1/2/3 are standardised in different protocols. In this paper, we undertake a detailed instruction-set customization study for accelerating SHA algorithms on a commercially available customisable processor, Tensilica Xtensa core. We have explored the bottlenecks of implementations on Tensilica Xtensa and designed efficient custom instructions, resulting in 30% improvement in performance compared to implementations using generic instructions. Furthermore, we highlight the usage of SHA-256 as a widely used cryptographic hash function for blockchain and cryptocurrency mining.

I. INTRODUCTION

In wireless devices and embedded systems, cryptographic algorithms are being widely used for communication and data storage devices. Cryptographic algorithms also play a major role in verification of integrity of data, as it moves through many layers and communication channels, where data could be corrupted by some external agent. For Internet-of-Things (IoT) applications, cryptographic hash algorithms are required to provide high speed and near-realtime performance, which can be achieved using hardware based implementation on Application Specific Integrated Circuits (ASIC), Field Programming Gate Arrays(FPGA), or General Purpose Graphical Processing Units(GPGPU). While securing information, hash generated using a standard algorithm is generally appended with the message. At destination of data or at the time of message retrieval, hash for the message is recalculated using the same mechanism as at source of the data and is compared with the original hash appended with the message to determine the authenticity of the data received.

In the last few decades, several versions of Secure Hash Algorithm (SHA) has been released which are primarily based on the Message-Digest 4(MD4) algorithm, developed by Ronald Rivest [1]. As initial algorithms were developed with the focus on executing them on general purpose machines, MD4 used 32-bit variables, and software friendly boolean operations such as AND, OR, XOR, and NOT. However, with the advancements in general purpose computing as well as with the availability of various hardware platforms, more complexities have been added to these algorithms to ensure a more secure transmission and authentication.

Lately, application of SHA-256 in blockchain, specifically bitcoin mining has also gained a lot of traction. As the computational requirements for bitcoin mining keep increasing, it has moved from using general purpose Central Processing Units(CPU), to GPUs, FPGAs, and then ASICs developed specifically for bitcoin mining. Bitcoin mining is a trial and error procedure because of the "proof of work", which puts restrictions on the maximum value of hash calculated. As reward for successfully creating a hash is significant (25 bitcoins), its mining has been very competitive, and requires serious investments in ASICs, with high power usage. CPUs can also be used for bitcoin mining, but they are relatively slow compared to other dedicated hardware options available. However, it can be significantly accelerated with an optimized implementation of SHA-256 integrated to the generic CPU, as SHA-256 is the primary computational block of bitcoin mining, which is computed repetitively with different inputs.

In this paper, we have explored the implementations of SHA-1, SHA-256, and SHA-3 on Tensilica Xtensa. We first identify the bottlenecks of algorithms by analyzing different datapaths, memory access patterns and ways to overcome them using lookup table integration and microcode extension. In order to improve the performance of SHA algorithms, we then extend the Xtensa core additional custom instruction to improve execution speed, memory, area, and power. We also present a case study of bitcoin mining application implemented on Xtensa using the proposed custom instructions. We have integrated bitcoin mining into the Tensilica Xtensa as an "opportunistic bitcoin miner", i.e., main Tensilica Xtensa core detects when the CPU is idle, and automatically switches to bitcoin mining mode, utilising the extended instruction set for efficient implementation of SHA-256. We present optimizations for SHA-1 as SHA-256 is primarily based on the similar concepts to SHA-1. As SHA-3 is the latest algorithm for SHA, and can potentially replace SHA-256 over the time, we have improved the performance of SHA-3 as well.

Remainder of the paper is organized as follows. In Section II, we discuss different SHA algorithms in detail. Section III discuss the prior works on SHA algorithms on different platforms. Analysis and identification of bottlenecks of SHA algorithms is presented in Section IV with extension of instruction set discussed in Section V. Section VI presents results and analysis of the implementation of SHA algorithms using the extended instruction set on Xtensa. Case study on bitcoin mining is also presented in this section. Section VII concludes our work.

II. SECURE HASH ALGORITHMS

A. SHA-1

SHA-1 produces a 160-bit (20 byte) hash values for a message which can be from 160-bit to 2^{64}-bit long. Preprocessing and Hash computation are the two major steps

978-1-5386-3693-0/18 $31.00 © 2018 IEEE

362

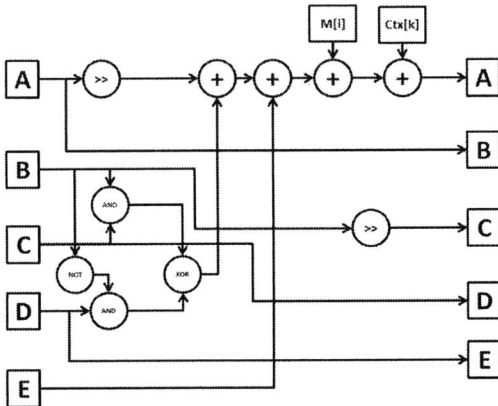

Fig. 1: *Data Flow Graph of one of the first 20 iteration of the SHA-1 compression function.*

involves to generate the final hash value in SHA-1. In pre-processing step, input message is padded and divided into multiples of 512-bit data lengths. After appending 1 to the input message, rest of the message is padded with zeros, and in the end, 64 bit of total length of padded message is added. Each 512 bit data is then divided into 16 words of 32-bit each and hash value is initialized with five pre-defined words of 32-bit each.

In the hash computation step, to compute the final hash value, SHA-1 performs 80 rounds of computations divided in 4 stages of 20 rounds each. Each stage uses same set of basic functions with different internal architecture and different constant value for each stage to compute the output of each round. Five working registers of 32 bit each are updated at the end of each round. After completing 80 rounds of computations, final values in the registers are added with their initial values to compute the final hash value. Fig. 1 shows the dataflow diagram of first stage of the four stages. Note that standard boolean operations such as shift, XOR, AND, and arithmetic operation addition are used for hash computations. SHA-1 cannot be fully parallelized as input of each round is dependent on the output of previous round.

B. SHA-256

In 2001, three variants of SHA-1, SHA-256, SHA-384 and SHA-512 were introduced with message digest lengths of 256, 384 and 512 respectively. SHA-256 is implemented with the padding step same as of SHA-1 whereas in hash computation step it uses 8 working registers instead of 5 used in SHA-1 and the hash value is computed using 64 round with single stage of function and finally the output is added, similar to SHA-1. Fig. 2 explains one step of the state update transformation of SHA-256 [2]. As shown in Fig. 2, SHA-256 uses ADDITION, AND, and SHIFT operators for updating hash values. As in case with SHA-1, hash values computations cannot be fully parallelized due to the inter data dependency between the each iteration.

C. SHA-3

SHA-256 was developed on similar lines to SHA-1 while increasing the complexity of the algorithm. However, for SHA-3, an altogether new approach has been used. SHA-3 uses the

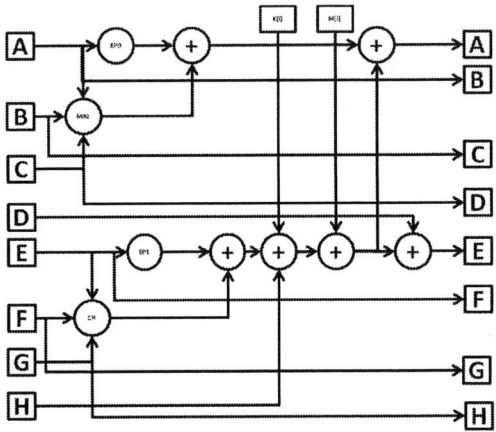

Fig. 2: *One iteration of the state update transformation of SHA-256.*

TABLE I: *Summary of Area and power consumption of SHA implementation from published ASIC/FPGA Implementations.*

Platform	Algorithm	Area (in Slices)	Power (in mW)
Artix-7 FPGA(192MHz)	SHA-3	982	612
Xilinx Spartan3E (81.10MHz)	Keccak-256	3036	263
Xilinx Spartan3E (73.05MHz)	Keccak-512	2785	301
VIRTEX II Pro FPGA	SHA-256	755	N/A
ASIC (819MHz)	SHA-256	140644	N/A

sponge construction, in which, data is absorbed in the sponge and the results is squeezed out. The final hash is produced after 24 permutation rounds in Keccack with the invocation of 5 modules knowns as θ (theta), ρ (rho), ϕ (phi), χ (chi), and ι (iota). In the pre-processing step of SHA-3, input string are converted to state arrays. θ **step:** In this step, the parity of the pre-processed data of 5×5 array of 64-bit array is calculated using XOR operation and the results are stored. ρ **and** ϕ **step:** 64-bit elements are rotated by a triangular number by excluding the first element in the step ρ. Then the permutation follows by fixed assignment are executed by ϕ module. χ **step:** This module is adds a non-linear aspects to previous round. ι **step:** This module is responsible for breaking any symmetry caused by other modules in previous stages.

III. RELATED WORK

Current trend of processing units that are used in embedded system varies among ASICs, FPGAs, GPGPUs, Digital Signal Processors(DSP) and Single-chip heterogeneous computers, to improve the performance, area, and power of the system. ASICs are best fit for low-power systems, but these are less flexible and have a long time-to-market. FPGAs are useful in development/prototyping phase, since the cost of these devices is generally high for large scale deployment. GPGPUs are applicable for Single Instruction Multiple Data(SIMD) applications. Application Specific Instruction set Processors(ASIP) provide a trade-off between fully configurable devices like FPGAs and hard wired ASICs.

The two major categories of ASIPs are, Instruction Set Architecture(ISA) based ASIPs and Programmable hardware based ASIPs. Specialized resources are used in ISA based ASIPs where as FPGAs are used in programmable hardware based ASIPs to improve the performance of the application. However, hybrid of these have also started coming to the market. Many strategies are used to improve the performance

978-1-5386-3693-0/18 $31.00 © 2018 IEEE

of ISA based ASIPs. Some of these are: implementing parallel processing modules using multiple Processing Elements(PE), special purpose hardware such as co-processors or functional units, memory architectures for multi-threading and task specific memories, on-chip communication mechanism to connect co-processor memory and peripheral support. PEs can be pipelined or designed as symmetric architectures, based on the nature of the data processing required for the application.

Programmable hardware based ASIPs can be categorized according to the time required to program the ASIP and the granularity of the programmability. Steps of mapping a design onto these ASIPs can be divided into four major phases. These are: mapping of target application for processor architecture (an iterative process until design specifications are met), translating the specified architecture into synthesizable form, designing software layer and system integration, and final step is testing of the system. Architectural Descriptive Languages (ADLs) play a major role in designing ASIPs, which can be categorized into Instruction Set centric, architecture centric, or mixed type. Our work is primarily focused on Tensilica Xtensa, which falls into the category of 'mixed' type ADLs same as LISA. Mixed type of ADLs can support end-to-end process of compiling application with new instructions, to simulations and validation for comparing various metrics like area, power, or performance.

In recent years, FPGAs have been used to benchmark and improve the throughput of many algorithms. FPGA implementation of SHA-3 on Artix-7 was presented in [3], which is a more suitable option for IoT devices, as it consumes low power (612 mW), less area footprint (982 slices), with operating frequency of 192 MHz. In [3], all SHA-3 candidates (JH, BLAKE, GROSEL, KECCAK, and SKEIN) are synthesized for Xilinx Virtex-5, -6, and -7, providing area and throughput comparisons. KECCAK was identified as an optimum candidate for hardware implementation based on study by [3].

In [4], SHA-3 was implemented on PIC24 core developed on LISA. The description was automatically translated to Register Transfer Level(RTL) (VHDL) and synthesized using Synopsys, and execution cycles were calculated. Then Instruction Set Extension(ISE) was implemented and performance was improved by 172%, with 10% of extra core area and 70% reduction in memory consumption.

ASIC implementations of different SHA-3 candidates has been implemented in [5], based on SASEBO-R board. Interfacing between PC and SHA-3 ASIC implementation is done via USB interface, enabling the communication to C# based software tester through FPGA. Based on the study done in [5], Keccak was identified as the best candidate for hardware implementation, with less resource required and high energy efficiency. The work in [6] presents a generic framework for analyzing and estimating the performance of SHA-3, based on cell broadband engines and NVIDIA GPUs. Cell broadband engines are based on Synergistic Processing Elements, consisting 8 Synergistic Processing Units (SPUs), with 128 registers, each 128 bits wide. To make code more suitable for execution in an identical way on cell, authors have reduced the use of branches to process all four input strings. In case of NVIDIA, using PTX compact and non-divergent kernels allows for execution of even more simultaneous threads.

TABLE II: *Functions of SHA consuming most execution cycle.*

Function Name	Total Cycle count in %
SHA-1 Functions	
sha1_transform()	76.76
sha1_update()	20.15
SHA-256 Functions	
sha256_transform()	83.30
SHA-3 Functions	
sha3_round()	42.32

Unrolling the compression functions, thus eliminating the branch instructions increased the performance for NVIDIA GPUs. Internal states and input message blocks were stored inside the shared memory. SHA-3 Keccak implementation on NVIDIA's GTX 295 GPU was reported in [7], parallelized using tables implementing modulo operations and 25 threads in parallel, with state variables stored in the shared memory. In this report, tree mode with inter-leaving of processing was used to perform parallel and independent hashing.

Detailed power and performance of SHA-3 candidates were analyzed and reported in [8] based on Xilinx Spartan3E FPGA. The two different version of Keccak, Keccak-512 and Keccak-256 were implemented. Keccak-256 consumed maximum power of 263 mW for speed optimized settings, running at 81.10 MHz, with 3036 slices used. But it was improved to 222.4 mW with area optimization settings with 3037 slices at 54.64 MHz. On the other hand Keccak-512, with high speed configuration, consumed 301 mW in 73.05 MHz operating frequency. This power consumption was reduced to 226 mW by configuring the tool to optimize the area, running at 52.38 MHz. In both these cases, slice utilization remained constant value (2785). Area and power consumption of these implementations are summarized in Table I.

[9]is discussing various methods to achieve higher throughput in Xilinx Virtex-6 FPGA platform such as unfolding, preprocessing and pipelining. Xilinx Fast Simplex Link(FSL) stream bus is used by the authors to create an interface between Microblaze soft CPU and SHA-1 core in this work.

IV. IDENTIFICATION OF INSTRUCTION SET EXTENSION

Various algorithms to improve performance based on identifying computation patterns using data/control flow graph of an application have been proposed in the literature. However, instruction extension and I/O bandwidths are limited by the register files of a processor unit. In this work, we have designed instructions and application-specific functional units (AFUs) to speed up the states consuming high number of cycles. In addition to special instruction, we also utilize the high I/O bandwidth of newly implemented 128-bit register files. In order to identify the extensions to instruction set required to improve the performance, we follow these three steps:

1) **Profiling:** The application is profiled using inbuilt Xtensa profiler. This provides the number of clock cycles required for different functions, number of times a function is called, number of committed instructions, and number of load, store, and pipeline interlocks.

2) **Identification:** After profiling the application, we identified the functions/sub-functions of the application limiting the throughput.

3) **Memory access:** Although width of data bus of traditional D570_T CPU is 32 bits, we have observed that consecutive memory accesses are required to improve the performance of load/store instructions.

Table II shows the cycle count of main functions (in %age) of SHA. As these functions are bottleneck to high performance, our work is primarily focused on improving these functions. Fifth chapter of "Processor Description Languages" book [10] explains the ASIP design methodology using LISA with a case study of retinex-like image and video processing. LISA supports automatic generation of synthesizable RTL, compiler, assembler, simulator and etc those are necessary to analyze the performance of the target application. Simulator of LISA can be used to study data-flow constraints, latency, area constraints and architectural constraints. LISA enables the designer to configure operation of the custom ISA, pipeline stages, instruction decoder capabilities and clock domains to achieve the optimal RTL design of the processor core.

V. INSTRUCTION SET EXTENSION USING TENSILICA XTENSA

Fig. 3 illustrates the Tensilica Developer toolkit which consists of compiler, instruction set simulator (ISS), profiler, debugger and other tools required to analyze high-performance applications specific to Xtensa core. Designer normally configure, create custom instruction and simulates using ISS and measures the performance using profiler until the requirements are satisfied. Xtensa Processor Generator(XPG) from Tensilica generates the bitstrem for Xilinx KC705 development board [11], which consists of Kintex-7 XC7K325T-2FFG900C FPGA, 1 GB DDR3 memory, 200 MHz LVDS oscillator and communication interfaces such as I2C and Ethernet. The XC7K325T FPGA is incorporated with 326,080 logic cells, 50,950 slices, 4,000Kb of maximum distributed RAM, 840 DSP slices, one PCIe interface and 500 user I/Os. Kintex-7 series FPGA slice contains four look up tables and eight flip-flops and DSP slice contains a pre-adder, a 25 x 18 multiplier, an adder, and an accumulator. The On-Chip Debug (OCD) can be performed using the USB interface and OnCE JATG probe those were fixed in the daughter board.

Tensilica provides various techniques to implement and improve any given application. Following are some of the performance improvement techniques that can be configured in XPG. 3-Way FLIX with 64 instruction length and ConnX Vectra LX DSP Engineare configurable co-processors are available in Xtensa. Xtensa C-library provides fast and reentrant execution support for multi-threaded environment with windowed register file to improve the performance and code size. Real Time Operating System(RTOS) compatibility option including alternate reset base option can be used for supporting RTOS features. Power consumption can be reduced by global clock gating and functional clock gating in the clock tree. Power consumption while executing WAITI instructions could be minimized by power down the core area and waits for interrupt. MUL16, MUL32, 32-bit integer divider, MAC16 for DSP options and CLAMPS for saturation arithmetic are the selectable and configurable instruction in XPG. 1-bit boolean registers, floating point registers, MAC16 data registers and custom DSP registers are available in this platform with

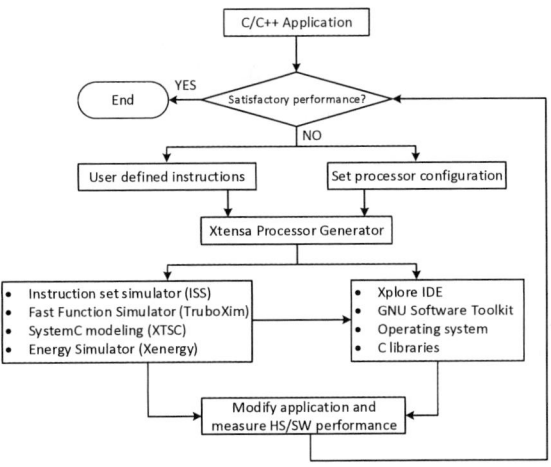

Fig. 3: *Design flow in Xtensa Xplorer: Automatic Methodology of generation of customized processor and matching software tools.*

automatically enabled load store and move instruction. Queue interface developed in Tensilica Instruction Extension(TIE) language is more suitable for first-in first-out implementation to communicate with the RTL developed blocks. High throughput can be achieved using this communication since custom instructions can be designed to access this interface via processor data-path. After analyzing the performance bottlenecks in Section IV, following techniques have been used to improve the performance of SHA algorithms.

- **Register file:** Caches also can be configured based on the cache ways, size, line width and banking. We increased the word length of data memory and cache to 128 bits, so that 4 word can be read simultaneously. This has been implemented by including a new register file REG_SIMD, which is then used by most of the extended instructions to fetch 128 bit data in SIMD fashion. Addition of this register file increased the number of slices utilized by 6292. Due to the addition of this new register file, Tensilica Xtensa automatically created LOAD, STORE, and MOVE instructions for the register file, utilizing 308 slices and 8 slices for LOAD/STORE and MOVE respectively.

- **TIE Tables:** SHA algorithms uses pre-defines tables. In the proposed extended ISA of Tensilica Xtensa, instead of storing these constants in a memory and accessing using LOAD/STORE instructions, we have implemented these tables as hardware constants, which can be accessed without any memory interaction. As these constants are fixed for the algorithms and are used in each iteration, it reduces the code size significantly and improves the performance of final implementations.

- **Extension of Computational Units through Microcoded Multi-Cycle Instructions:** We implemented some of the multi-cycle and multi data instructions, by introducing AFUs with special data path using TIE language. Based on the algorithms, extended ISE play a major role to minimize the extra load/store and memory read/write. Resource utilisation and power of these in-

Type	Description
Committed instructions (CI)	The total number of instruction executed will be provided by this value.
Taken branches (TB)	This value indicates the total number of branch instruction executed.
Pipeline interlocks (PI)	This value is an indicator of total pipeline interlocks to eliminate data and control hazards.
Loads (LD)	Number of load executed.
Stores (ST)	Number of store executed.
Total Cycles (TC)	Total number of execution cycle

TABLE III: *Parameter considered for performance improvement.*

structions is discussed in more detail in Section VI and Table IV.

VI. RESULTS

Table III lists the parameters we have used to compare the performance provided by cycle accurate simulation from Xtensa Xplorer, with 764 MHz of core speed and estimated target geometry of 28nm HPM process. Branch delays and interlocks are best indicators of CPU cycle wastage of an application. Using the load/store values, the memory intensive process is optimized and improved by wider data buses. Speedup due to extended instruction set is calculated as,

$$Speedup\ (in\%) = \left(\frac{TC_{Before} - TC_{After}}{TC_{Before}} \right) \times 100 \quad (1)$$

In [12], authors discussed about re-usability factor and extra cost factor, which are more related to measure the flexibility of an ASIP. Embedded system development has been targeted for various application domain such as automotive, multimedia and security. So, the existing instruction of the base processor can be considered as inter-domain ISAs and it is necessary to measure the design efforts to design intra-domain custom instruction. Re-usability factor and extra cost factor are calculated as following:

$$Reusability\ Factor = \left(\frac{Instruction_{BaseISA}}{Instruction_{mASIP}} \right) \times 100 \quad (2)$$

$$Extra\ Cost\ Factor = \left(\frac{ISA_{App} - ISA_{BaseISA}}{ISA_{mASIP}} \right) \times 100 \quad (3)$$

Considering the extended instruction set discussed in Section V, re-usability and extra cost factors are 97.9% and 2.1% respectively. This proves that the newly implemented ASIP to speed up SHA algorithms can be easily developed by adding 2.1% of extra instructions and it can still provide 97.9% of support for the intra domain applications which was supported by its base processor.

Table IV shows the resources (number of slices) and power required for each of the extended instructions added to improve performance of SHA algorithms. We have primarily improved the hash computation, which is computed repetitively for each round. As tabulated in Table IV SHA-256 implementation consumes 1557, 3036 and 755 slices on, Tensilica platform, Xilinx Spartan 3E and VIRTEX II FPGA respectively. 1610 Slices were estimated for our design of SHA-3 but 982 slices were used to implement SHA-3 on Artix-7. Our design consumes 1903 slices to implement SHA-1 but 1649 slices were utilized by the work done in [9].

Instruction Name & Description	Area (in Slices)	Power (in pJ/cycle)
Instructions for SHA-1		
SHA1_SHIFT: Rearranges the 4 bytes into reverse order.	16	0.063
SHA1_5SHIFT_1: Perform the MD4 stage 1 shift operation of 5 32-bit working registers.	619	17.277
SHA1_5SHIFT_2: Perform the MD4 stage 2 and 4 shift operation of 5 32 bits working registers.	591	16.999
SHA1_5SHIFT_3: Perform the MD4 stage 3 shift operation of 5 32 bits working registers.	677	18.017
Instructions for SHA-256		
SHA1_SHIFT: Rearranges the 4 bytes into reverse order.	16	0.063
SHA256_SIG0 : ROTR7(x) $\oplus$ ROTR18(x) $\oplus$ SHR3(x)	91	6.526
SHA256_SIG1 : ROTR17(x) $\oplus$ ROTR19(x) $\oplus$ SHR10(x)	81	6.526
SHA256_4SHIFT_1 : Shift operation of 4 32 bits working registers.	853	18.698
SHA256_4SHIFT_2 : Shift operation of 4 32 bits working registers.	516	22.664
Instructions for SHA-3		
ROTLEFT64: Shifts left by 1, right by 63 and do logical OR.	8	7.552
SHA3_RPSHIFT: Shifts left and right based on ρ and ϕ step then do logical OR.	1602	10.188

TABLE IV: *Instruction that were implemented to speedup SHA-1, SHA-256 and SHA-3 with their area and power consumption. ROTRn: Rotate right for n bits; SHRn: Shift right for n bits.*

Fig. 4 shows the performance improvements for the parameters shown in Table III for all three SHA algorithms. As shown in the Fig. 4a, total execution cycles were reduced by 29.58% to perform SHA-1 by introducing four instruction with 1610 additional slices.

The processing speed of SHA-256 was improved by introducing extra instructions shown in Table IV. In addition to hashing functions for SHA-256, two extra instructions to calculate sigma values are also implemented. As shown in the figure 4b, the total execution cycles were reduced by 52.33% to perform SHA-256 by introducing addition four instruction and 1,557 additional slices were consumed. Specially, pipeline interlocks and number of load instruction were decreased by introducing ISE. Moreover, by minimizing committed instructions, program memory area also reduced by 53.06%.

Performance of SHA-3 (Keccak) also improved significantly due to the instructions shown in Table IV. Specially, shifting was significantly improved in ρ and ϕ steps by introducing a hardware table. As shown in the Fig. 4c, the total execution cycles were reduced by 10.28% to perform SHA-3 by introducing additional two instruction.

A. Case Study: Bitcoin Mining

Bitcoin is a currency with distributed ledger system that can be traded with traditional government-controlled centralized currencies. Hardware implementations for bitcoin mining varying from using a generic CPU to dedicated ASICs are commercially available, with different optimizations like pipelining, delay balancing, loop unrolling, different type of adders and more. Transactions of bitcoin are populated to miners who confirms the transactions and generate a hash for new transactions which is then added to ledger, known as blockchain. For each set of transactions, miners do not modify the transaction data, but use another random piece of data called a 'nonce' with transaction data to generate final hash. This process is repeated with different nonce until the final output hash confirms with the bitcoin protocol. The first miner to generate the hash satisfying the protocol is awarded 25 bitcoins and appends the generated hash to blockchain. Fig. 5

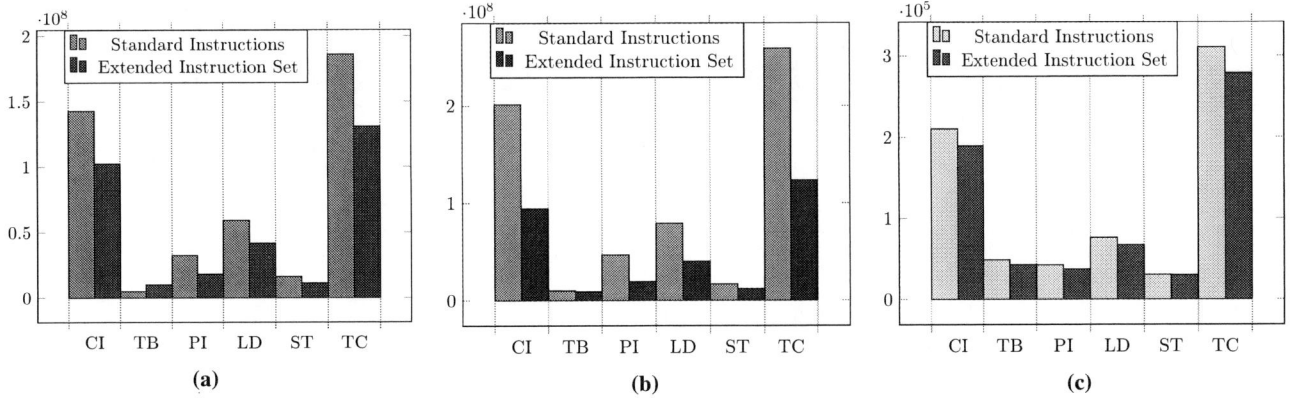

Fig. 4: *Performance comparison of (a) SHA-1 (b) SHA-256 (c) SHA-3.*

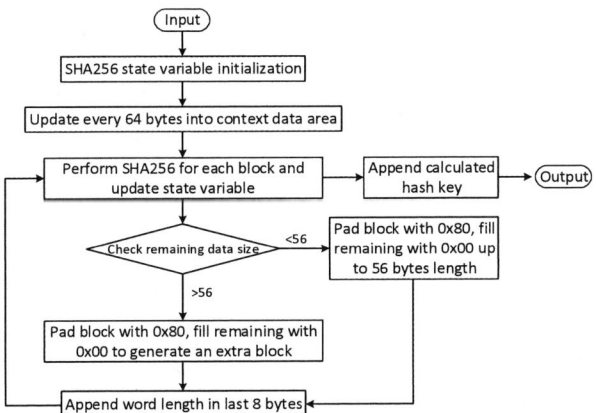

Fig. 5: *Data Flow Graph of Bitcoin mining case study.*

depicts the implementation of SHA-256 in bitcoin mining process for single 512-bit block.

We have integrated bitcoin miner implemented using the extended instruction set discussed in this paper as an "opportunistic bitcoin miner", i.e., the Tensilica Xtensa core switches to mining mode when it detects the CPU is idle and can be used for bitcoin mining without affecting the performance of any programs run by the user, as bitcoin mining is computationally intensive. A similar idea was commercially launched by 21 Inc [13], providing a marketplace where any computer can be used for bitcoin mining through HTTP requests after downloading a software provided by the company. Using the conventional instruction set of Tensilica Xtensa, bitcoin miner can compute 13053 Hash/sec. Using the extended instruction set, we are being able to improve the performance to 20437 Hash/sec, an increase of 57%. Many bitcoin miners have been reported in academia as well as commercially available, and can reach high number of hashes/sec. Dedicated ASIC based bitcoin miner can reach computation speed of 1.6 GHash/sec [14]. Here, note that these are dedicated hardware specifically designed for the purpose of mining and have multiple instance of SHA-256. In our case, the bitcoin miner is an opportunistic one, and furthermore, can be adapted to fit different mining protocols as well as other applications involving block chain.

VII. CONCLUSION AND DISCUSSION

This paper presents implementations of SHA-1, SHA-256, and SHA-3 algorithms on Tensilica Xtensa platform with custom ISE. We have shown how to identify the bottleneck operations and then instruction set extensions, resulting in significant improvements in total number of cycles, with minimal load/store instructions. On an average, we are able to achieve 30% performance improvements. We have also presented a case study of bitcoin miner, which utilises the optimized SHA-256 implementations, and is integrated with the Tensilica Xtensa such that it can automatically switch to miner mode when the CPU is idle, computing 20437 Hash/sec.

REFERENCES

[1] R. L. Rivest, "The MD4 message digest algorithm," Internet RFC 1320.
[2] C. Paar and J. Pelzl, *Understanding Cryptography: A Textbook for Students and Practitioners.* Springer-Verlag New York Inc, 2010.
[3] H. T. Y. Jararweh, L. Tawalbeh and A. Mohd, "Hardware performance evaluation of SHA-3 candidate algorithms," *Journal of Information Security*, vol. 3, no. 2, pp. 69–76, October 2012.
[4] J. H. F. Constantin, A. P. Burg, and F. K. Gurkaynak, "Instruction set extensions for cryptographic hash functions on a microcontroller architecture," in *Application-Specific Systems, Architectures and Processors (ASAP), 2012 IEEE 23rd International Conference on*, July 2012, pp. 117–124.
[5] X. Guo, M. Srivistav, S. Huang, D. Ganta, M. Henry, L. Nazhandali, and P. Schaumont, "Performance Evaluation of Cryptographic Hardware and Software – Performance Evaluation of SHA-3 Candidates in ASIC and FPGA," November 2010, http://rijndael.ece.vt.edu/sha3/.
[6] J. W. Bos and D. S. ., "Performance analysis of the sha-3 candidates on exotic multi-core architectures."
[7] G. H. Pierre-Louis Cayrel and M. Schneider., "Gpu implementation of the keccak hash function family."
[8] A. Jaffar and C. J. Martinez, "Detail power analysis of the sha-3 hashing algorithm candidates on xilinx spartan-3e," *International Journal of Computer and Electrical Engineering*, vol. 5, no. 4, pp. 410–413, 2013.
[9] J. woon Kim, H. ung Lee, and Y. Won, "Design for high throughput sha-1 hash function on fpga," in *2012 Fourth International Conference on Ubiquitous and Future Networks (ICUFN)*, July 2012, pp. 403–404.
[10] A. Chattopadhyay, H. Meyr, and R. Leupers, *LISA: A Uniform ADL for Embedded Processor Modelling, Implementation and Software Toolsuite Generation.* Morgan Kaufmann, June 2008, ch. 5, pp. 95–130.
[11] "KC705 Evaluation Board," https://www.xilinx.com/products/boards-and-kits/ek-k7-kc705-g.html.
[12] R. Ragel, S. Radhakrishnan, and A. Ambrose, "Instruction-set selection for multi-application based asip design: An instruction-level study," in *Information and Automation for Sustainability (ICIAfS), 2012 IEEE 6th International Conference on*, Sept 2012, pp. 141–146.
[13] "[Online] 21 Inc," https://21.co/.
[14] "[Online] Ant Miner U1," https://www.bitmaintech.com/productDetail.htm?pid=00020140107162953571 7ycyGoWo06FE.

978-1-5386-3693-0/18 $31.00 © 2018 IEEE

Lightweight Forth Programmable NoCs

Vinay B. Y. Kumar, Deval Shah, Mandar Datar, and Sachin B. Patkar
Department of Electrical Engineering
Indian Institute of Technology Bombay, India
Email:{vinayby, deval281shah, mandardatar, patkar}@ee.iitb.ac.in

Abstract—**Designing around packet-switching networks-on-chip (NoCs) has been an increasingly popular architecture paradigm for distributed on-chip computing. The availability of highly configurable and synthesizable NoC generators, such as CONNECT, has freed designers to focus instead on application mapping and other aspects of this design paradigm. In conventional NoCs, routers throughout the network are all uniformly provisioned with dedicated resources (e.g. crossbar switches, virtual-channels, arbiters) so as to route data as quickly as possible while keeping average message deliver times low. Many applications however, may not be sensitive to these metrics, and may otherwise have needs that may not be readily supported, or may present opportunities to simplify routers. We present a configurable synthesizable NoC generator with programmable routers built around lightweight Forth cores. The use of programmable routers in onchip networks opens up interesting possibilities for applications, and this notion, at least in the lightweight category, has never been explored. For illustration, support for features such as broadcast and multicast, adaptive routing, and systolic array computing have been implemented through programming the routers. The approach has been evaluated through a comparison in terms of network performance characterization and resource overhead.**

I. INTRODUCTION

Packet-switching networks-on-chip (NoCs) are increasingly seen as convenient and reliable substrates to support on-chip distributed computing. Among other things, designing around NoCs reduces verification effort, enforces modular design, encourages reuse. There have also been calls for availability of components of NoCs as embedded hard resources on FPGAs [1], [2]. While a lot is happening towards design of NoCs themselves, availability of cycle accurate NoC simulators such as BookSim [3], and highly configurable synthesizable NoC generators such as CONNECT [4] has freed application designers to focus on other aspects of evaluating mapping their applications to NoCs.

The design objectives of routers in typical NoCs are to route data as soon as possible (routing latency, crossbar switches) while also minimizing the average message delivery times (through routing algorithms, use of virtual channels). Routers are then provisioned with dedicated hardware resources, typically uniformly, throughout the NoC. Many applications, however, may not be sensitive to or need these aspects of router design. For instance, pure systolic algorithms, on one extreme, need very little from the routers (as routing may just be `send right` and latency is not important) and applications such as vision/signal processing pipelines do not need low-latency routing. Some applications may not need such routers everywhere on the network, while, some applications may have

special needs such as collective communication capability such as support for broadcast or multicast.

Although it is conceivable that this extra level of customizability could potentially be provided, at the RTL, as a part of a configurable-NoC generator such as CONNECT [4], we propose a more flexible alternate approach. We present a configurable NoC generator (much like [4]), with programmable routers built around lightweight Forth [5], [6] cores. The routers in the generated Forth NoCs (FNoCs, henceforth) are programmed with Forth [5], an interesting stack-oriented concatenative programming language. A Forth core is a lightweight stack-based processor, the language easy to learn and expressive, and the compiler elementary with no explicit grammar (more in Sec. II). Programmable routers have been gainfully used in traditional (off-chip) communication networks, but have not been explored (except [7]) for use in onchip networks. The work in [7] proposes programmable routers in the case of MPSoC based NoCs using conventional processor cores.

A part of the motivation for the present work was to supplement an in-house application-level design automation flow (early work [8]) where applications described in coordination-model semantics are compiled and mapped to user specified NoC configurations, targeting single or multiple FPGA platforms. CoRAM [9] is another application-level synthesis flow around NoCs where programmable routers could add value. Using Forth was preceded by the use of stack cores in [10].

The proposed approach—with only marginally extra cost in terms of resources and latency compared to conventional NoCs—the configurability during generation together with runtime programmability in FNoCs opens up interesting possibilities such as listed below. The items marked I have been implemented for illustration.

- (I) user specified, runtime updatable, routing functions (e.g. X-Y, lookup-table, pure/quasi systolic)
- (I) dynamic routing decisions (e.g. turn model)
- collective operations, e.g. broadcast/multicast/barrier (I)
- computing in transit on the network (e.g. reductions)
- vehicle for NoC emulation (ref. BookSim [3])
- programmed buffer management (e.g. [11])
- bridging to an IoT messaging system

A. Organization

The content is organized as follows. Section 2 introduces aspects of the forth environment. Section 3 discusses relevant aspects of conventional NoCs and how programmable NoCs

978-1-5386-3693-0/18 $31.00 © 2018 IEEE 368

could add value. The Forth programmable router design and features implemented for illustration are discussed in Section 4. Performance characterization and evaluation of FNoCs is presented in Section 5.

II. FORTH ENVIRONMENT

Forth [5] is a interesting stack-oriented concatenative programming language due to Charles Moore [12]. Forth is considered extremely suitable for embedded IoT applications and it has always had a niche in embedded systems where it has been used to compactly write anything from applications down to device drivers, hardware interfaces, and controllers. For instance, J1 [6] core, written in 200 lines of Verilog, runs a camera streaming application where the mac-controller up to the tcp-ip stack were all written in forth, fitting in 8KB.

Forth environment is a tight knit of the forth program, compiler and the processor; a tight knit in the sense that a typical forth programmer is necessarily well aware of all these components. Forth has no explicit grammar (sequence of numbers and symbols separated by whitespace) and that makes its compiler elementary . This makes writing forth programs easy and allows more freedom in expressing programs compared to traditional languages. Forth encourages bottom up thinking/coding, which can be a learning curve, but is ideal for low-level interfacing, and this naturally leads to incremental program verification and compact applications. A Forth core running a forth program can be more 'application specific' than an application specific program, say in C, running on a traditional general purpose processor. For its intended role in this work, as a lightweight processor in a packet-switching router, the choice of Forth environment was a natural one. The components of the forth environment are discussed next in a limited sense.

A. Language and syntax

Forth uses reverse polish notation and reflects stack architecture in its syntax. A forth program consists of a sequence of numbers (literals, constants or variables) and words (subroutines). These are separated by whitespace. For example, the program '1 4 squared +' results in 17 on the top of the stack (T), and the word squared is defined as follows.

```
: squared ( n -- n * n )
  dup (duplicates top of the stack (T))
  *   (execute operator *) ;
```

Notes: Text within (and) are comments. New word definitions start with a ':' followed by the word name and end with a ';'. A special comment style is used to show the contents on stack (before -- and after) execution of a word. E.g. (n -- n*n) for squared. Also, because of the semantic significance of whitespace, identifier names can have and be special characters (e.g. @, !, io!, io@ are valid word names).

B. Processor

There are many flavors of Forth cores (processors) and corresponding language variants. We extend the Forth core

J1 [6] adding support for (hardcoded) interrupt service routines (ISRs). Specifications summary: 32bit data, 16bit instruction, 13bit PC; 32 deep data stack and return address stack; ISRs; peripheral input/output data read/write operations; and single cycle instructions and single cycle subroutine calls.

C. Compiler

As mentioned earlier a forth program consists of numbers and words separated by whitespaces where each word is defined in terms of smaller words until we reach the basic words of the language, say in a file basewords.fs. The basic words are further defined in terms of machine words, in a file crossforth.fs. It is effectively a preorder traversal starting from squared.fs. Gforth generates the final machine code with: gforth crosswords.fs basewords.fs squared.fs

Compared to conventional processors and corresponding compiler toolchains, it is trivial to extend the Forth core (hardware) to add a new instruction or even peripherals (here, as io peripherals). This just involves adding 1 or 2 lines in the files crossforth, basewords. Here is an illustration to compute, say, e^{ax^2+b}, with e^x as an RTL IP and attached as a memory mapped peripheral to the Forth core.

```
: exponent_ax2_plus_b (b a x -- e^(ax^2+b) )
  squared * + exponent ;
: exponent (a -- e^a) (an I/O peripheral)
  get_oaddr_exponent (a -- a oaddr)
  io! (a oaddr -- empty) (io[oaddr] <= a)
  get_iaddr_exponent (empty -- iaddr)
  io@ (iaddr -- e^a) ;
```

III. NOC ARCHITECTURE

User processing elements or cores or nodes are attached to routers which are in turn connected in a given or chosen topology. A conventional router microarchitecture, shown in Fig. 1, is composed of input buffers, arbiters, routing unit, and crossbar switch. A router has ports that connect to other routers or nodes. The links, often bidirectional, can transfer phit bits (for physical digits) of data per cycle per direction. Messages over the network are exchanged in terms of application level packets. A source node splits a packet to be transmitted into smaller units called flits (for flow control digits) and enqueues them into its local router, the destination node then reassembles these flits. Other than the payload data, each flit carries header information as required by the router and as necessary for the routing strategy. Also, a flow control mechanism is used to safely transfer flits in and out of routers. The phit and flit widths are typically equal except for instance when a link crosses chip boundaries (e.g. using high-speed serial links). Flit widths can be particularly wide on FPGAs [4]. The flits may arrive out of order (depending on router configuration) and is something the nodes may have to account for.

A. Expectations

In a typical NoC all routers are identical in terms of design and configuration. And as with CONNECT [4] a number of (supported) microarchitectural options for routers (along with

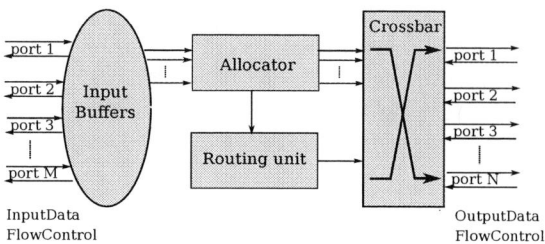

Fig. 1: Conventional router architecture

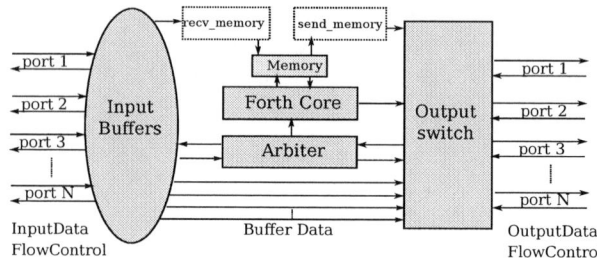

Fig. 3: Components of the forth programmable router

network topology) can be chosen 'at the time of generating a network' for an application. Many applications can benefit from the extra flexibility due to programmable routers. For instance, applications such as signal processing pipelines are not routing latency sensitive; resources such as virtual channels may not be necessary at all routers; algorithms using pure or even quasi systolic communication may have simpler yet specific routing needs; and many scenarios in general benefit from dynamic and adaptive routing strategies (but often considered an expensive to implement in hardware).

Towards this, one should expect to incur overheads, however, to make trade-off worthwhile, it is desirable that a programmable router (PR) meet the following three expectations: 1. resource usage should be minimal compared to a traditional router (TR); 2. routing latency of a PR must be close to TRs for conventional routing schemes; 3. likewise, for network throughput; 4. compiled programs should be compact.

IV. FORTH PROGRAMMABLE ROUTER

The forth system, being very lightweight in relative terms, is a natural choice around which to build a programmable router. It meets the basic expectations listed in Sec III-A. A basic Forth core (details in Sec II-B) has been extended to support routing and I/O operations. The arbiter component of the router is hardwired, and Forth core can be programmed to allocate an output port for a given destination node, and more.

A. Interfaces

To encourage (drop-in-) evaluation by other users, router port interfaces (Fig. 2) match CONNECT [4] generated NoC interfaces with peek-flow flow control. The flit format also matches that of [4], however, for native applications exploiting programmability, the flit format is in users control.

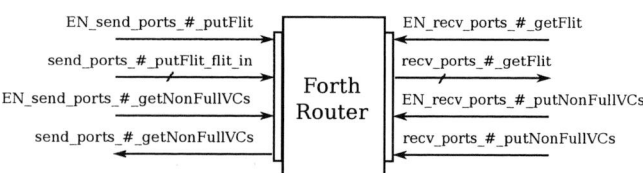

Fig. 2: Forth router send and recv port interfaces

B. Architecture

The components of the forth programmable router are shown in Fig. 3 and those of a conventional router in Fig. 1.

Router uses a three-stage pipeline. First stage consists of buffering of flits in the input FIFOs and arbiter decision. In the second stage, routing decision is taken by the Forth core according to the user supplied routing function. This is also the place to execute user programs to 'work with the incoming packets' before routing them out. In the third stage of pipeline, the flits of the packet being routed, are sent to the appropriate output port according to the routing decision respecting the flow control constraints.

1) Input buffers and output switch: There is one input buffer per input port, and buffer depths are configurable. Explicit virtual channels have been deliberately been avoided for now as more could be done later through programmed buffer management. The input buffer enqueues valid incoming flits. Flow control status per input buffer is sent on upstream links. Arbiter selects input buffers with flit data for further action through the Forth core. Based on the routing decisions taken by the Forth core, the output switch sends flits from input buffers (or memory) to appropriate output port if allowed by the flow control state of the corresponding downstream link.

2) Forth core: The Forth core serves as the heart of the router. The core has configurable amount (usually small) of data and program memory as BRAMs which can be updated at runtime (as messages on the NoC) or after bitstream generation (using Xilinx/data2mem). It can be considered a programmable FSM controlling other datapath components in the router. A request forwarded from the arbiter interrupts the Forth core. Depending on the type of the request associated with the incoming packet, the corresponding interrupt service is triggered. Apart from routing, other requests could be for broadcast, multicast, etc. For a basic or default routing request, the Forth core gets the destination node id and the corresponding routing ISR is shown below.

```
: interrupt_service_routine intrhand @ route ;
```

Here, intrhand is custom word which stores destination address on the top of the stack (TOS). With lookup based routing, we have stored the output port id for each destination in the data memory. The word @ (here and in other examples later) reads this output port id from memory and stores it to TOS. The word route signals the particular output port to route the packet. Apart from these ISRs, operations such as computing on the packets before forwarding can also be done. The incoming packets could be unmarshalled and stored to data memory before they are sent back on the network.

3) Arbiter: As the router can serve only one input buffer at a time, the arbiter serves as the link between input buffers and the Forth core. Arbiter takes requests from input buffer and forward one of these requests in a round robin fashion. Once the request is served by the Forth core, arbiter again starts forwarding a new request.

C. Features

1) Broadcast: When broadcast is not natively supported by the routers of an NoC, it is done through multiple point-to-point transfers. For applications with such traffic, this is inefficient. This functionality can be achieved through an appropriate ISR in the Forth core without needing any changes in hardware. For the sample implementation of this ISR, broadcast destination is N with an NoC with nodes 0 through $N - 1$. The first flit specifies the tag and size of the packet.

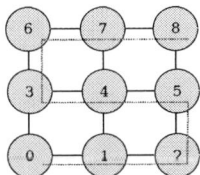

Fig. 4: Tree path chosen for 3×3 mesh (likewise for 4×4)

The simple tree-based approach is used for broadcasting where a tree connecting all nodes (Fig. 4) of the NoC is stored in the data memory of all Forth cores. When the core is interrupted with a broadcast request, it checks the incoming direction and based on the tree-path it decides the output port id. While sending the packet through this output port, it is also stored in Forth data memory temporarily, which is later sent to the attached processing node/s. Approaches that are more appropriate for the chosen topology may likewise be programmed.

2) Multicast: Multicast is a communication pattern in which a single source sends a packet to multiple destinations. This can be implemented again by multiple point to point communications, but if the same can be done with support from routers, it can reduce the traffic on the network. This was implemented along the lines discussed in the previous section. Broadcast is a special case of multicast. Here, the multicast group information is encoded in the second flit. Each Forth core receives the packet, replicates and forwards it, while sending it to its local node if it falls in the multicast group.

3) Deflection routing: Whenever Forth core gets a request from any packet, it decides the output port according to the destination of the packet. Now in heavy traffic, the particular output port might not be free. In such a case the Forth core is programmed to assign different output port to the packet and deflects it. This prevents unnecessary waiting time for a packet. For evaluation, we have considered minimal path only for deflection. Different approaches for such deflection based routing exist. A turn based model for non-minimal path is also discussed in [13]. Such algorithms can be programmed in, instead of adding extra hardware. For the ISR below, for a

3×3 mesh, the deflections on each router are stored next to the table for regular routing (hence the +9).

```
: interrupt_service_routine_deflection
  intrhand ( empty -- actual_dest)
  dup@ ( actual_dest -- actual_dest routedecision)
  duproute (actual_dest routedecision --
                 actual_dest buffer_status)
  if ( actual_dest buffer_status -- actual_dest)
    9 + @ ( actual_dest -- new_route_decision )
    route
  else then ;
```

V. EVALUATION

We evaluate FNoCs in terms of the typical NoC evaluation metrics such as resource utilization, load-delay characterization, and operating frequency. Programmed features such as broadcast and deflection routing are also evaluated. Two case examples with contrasting scenarios have also been implemented. A comparison is drawn between FNoCs (using labels FNoC_bfc and FNoC_bfc+ where bfc stands for basic Forth core) and appropriately equivalently provisioned CONNECT generated NoCs (using the label CONNECT_IQ) with the following parameters: simple input queued router, peek flow control, buffer depth of 16 flits, separable input-first round robin allocator. (Compared to bfc, the bfc+ supports more instructions (arith/logic) beyond necessary for routing, and has higher instruction (16 KB) and data (16 KB) memory sizes compared to 512 B and 2 KB respectively in the bfc version.)

A. Latency (zero-load); Design frequency

With its 3 stage pipeline, the forth router has a latency of 3 cycles per hop. In Fig. 4, the latency under zero-load for a flit from PE 0 to PE 8, at 5 hops, will be 15 cycles.

Table I shows a comparison of achievable design frequencies targeting Xilinx VC709 board (to relate to [4]) with different topologies and flit widths of 32, 64 and 128 bits.

TABLE I: Maximum frequency comparisons between Forth and CONNECT NoCs for flitwidths 32, 64 and 128 bits

Topology	2×2 Mesh			3×3 Mesh		
NoC	FMax in MHz			FMax in MHz		
	32 bit	64 bit	128 bit	32 bit	64 bit	128 bit
FNoC_bfc+	142	139	160	139	144	138
FNoC_bfc	160	157	162	145	126	145
CONNECT_IQ	164	166	167	114	124	115

B. Network Load-Delay Characterization

For the load-delay characterization of FNoCs two kinds of traffic patterns (uniform random, and unbalanced random) are used along with other parameters chosen as in [4]. For uniform random traffic, each node sends packets to any node with equal probability. For unbalanced random traffic, each node injects packets such that 90% of the them are addressed to neighbors and the rest 10% are addressed to uniformly randomly chosen nodes. (The unbalanced traffic pattern represents heavy local communications.) The packet size (in terms of number of flits) is chosen to be half of the buffer depth, and the simulations

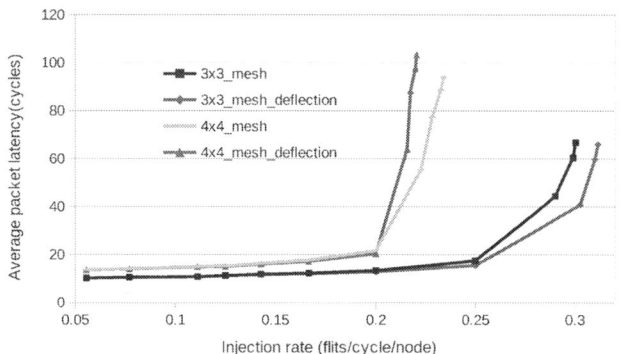

Fig. 5: Load-Delay plots: FNoCs/uniform random traffic

were run for 80000 cycles. Improvements due to routers programmed for deflection routing are also discussed.

For uniform random traffic (Fig. 5), as is expected, the 4×4 mesh network gets saturated for lower injection rates compared to 3×3 network because it has more nodes and hence more packets on the network for the same injection rate. Also programming for deflection routing does not give a significant performance gain for uniform traffic as all ports are near saturation.

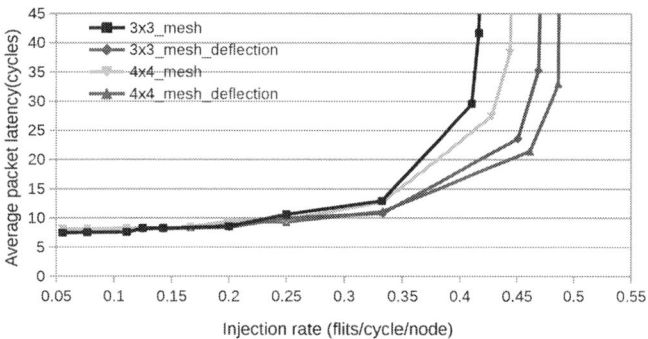

Fig. 6: Load-Delay plots: FNoCs/unbalanced random traffic

As unbalanced traffic (Fig. 6) has more local communication, networks gets saturated at a higher injection rate compared to balanced traffic. Here for all the topologies, programming for deflection results in noticeable improvement in the performance. Unbalanced traffic pattern has lower non-local traffic, so not all the output ports are saturated. In such a case, deflection results in improvement in latency as well as maximum injection rate.

Performance of double ring topology with 9 nodes for uniform and unbalanced random traffic and fat tree (as in [4]) topology with 8 nodes were analyzed as shown in Fig. 7. Unbalanced traffic pattern results in saturation at higher injection rate because of lower non-local communication. Fig. 8 compares load-delay curves with 3×3 mesh CONNECT NoCs.

C. Broadcast performance

Here, the routers are programmed with a tree based broadcast algorithm. For a 3×3 mesh topology the tree is shown

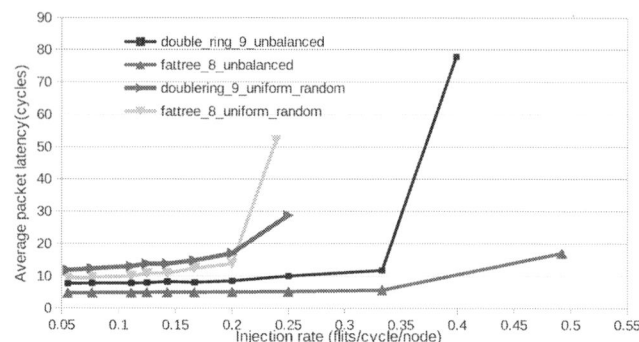

Fig. 7: Load-Delay plots: doublering 9 nodes, fat tree 8 nodes

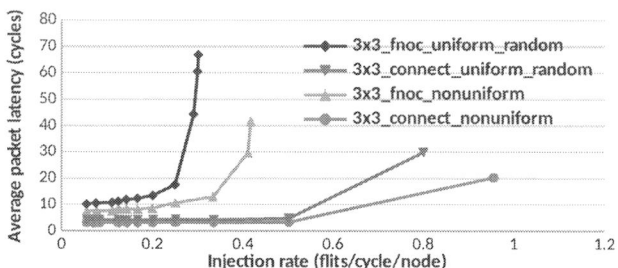

Fig. 8: Load-Delay plots for 3×3 mesh FNoC vs. CONNECT

in Fig. 4. Table II lists broadcast latencies for 8×8 mesh FNoC and CONNECT for different packet sizes and source nodes. As it should be, broadcasts along a tree path is better than with point-to-point messaging. Barrier synchronization operation also likewise benefits.

TABLE II: Broadcasts on 8×8 mesh: FNoC vs. CONNECT

Broadcasting node	node-0		node-32	
Packet Length (flits)	10	50	10	50
Latency (cycles) FNoC	330	370	229	309
Latency (cycles) CONNECT_IQ	655	3215	651	3211

D. Resource utilization

This section shows that, as is desirable, the resource utilization of programmable FNoCs is comparable or is only marginally more than a conventional NoCs. The utilization in Table III shown in terms of slice LUTs when targeting VC709. The FNoC_bfc version fares competitively compared to CONNECT.

E. Small example applications

The following two toy applications, of contrasting characteristics, are discussed as an illustration of programmability of routers. Both of these cases use the native routing Forth cores for computation and routing. (Note that an FNoC with additional threads (with additional data-stacks) and a floating-point unit in each Forth core can be a good work-horse for more serious computing.)

978-1-5386-3693-0/18 $31.00 © 2018 IEEE

TABLE III: Resource utilization on FPGA (VC709) in terms of slice LUTs (out of 433,200 available)

NoC version	3 Ports			4 Ports			5 Ports		
	32bit	64bit	128bit	32bit	64bit	128bit	32bit	64bit	128bit
FNoC_bfc+	1522	1761	2241	1810	2061	2830	2139	2484	3586
FNoC_bfc	980	1388	2058	1284	1448	2176	1492	1905	2980
CONNECT_IQ	403	594	1253	808	964	1520	1146	1565	2588

1) Systolic polynomial evaluation: Polynomial evaluation through Horner's rule is a classic example [14] for systolic array based computing. $p(x0) = \sum_{k=0}^{m} c_k x^k$ can be rewritten as an alternating sequence of multiplication and addition as $p(x) = (((c_m x + c_{m-1})x + c_{m-2})x + ... + c_1)x + c_0$. The systolic array arrangement for this form is shown in Fig. 9, where each node takes data from an upstream node computes, and forwards to the downstream node: $x_{out} = x_{in}, y_{out} = x_{in}c_m + y_{in}$. With a corresponding linear FNoC, the routers are programmed as shown below with just a few lines added to the routing ISR. A toy polynomial evaluation was implemented using the routers (for $m = 6$ and 6 routers). For the given code, each hop now takes 16 cycles. Note that the flit format is automatically simplified.

```
: interrupt_service_routine
  intrhand io@ io@ io@ calculate
  io! (send left) ;
: calculate (x y -- x x*y+c_0)
  over     (x y -- x y x)
  * c_0 +  (x y x -- x xy+c_0)
           (c_0 defined as a constant)
  ny0 ! nx0 ! ; (ny0, nx0 are variables)
```

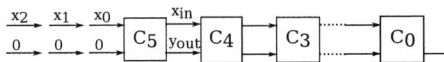

Fig. 9: Systolic array model for polynomial function evaluation

2) K-means clustering: Distributed K-means is an iterative process starting with a K random vectors as centroids of K clusters (constituting a total of N vectors). At each step of the iteration, cluster memberships of vectors (in terms of distance to centroids) are decided followed by updating the centroids with the means of the new clusters. On the network, this results in concurrent broadcasts. Best done with dedicated nodes, the following is a toy implementation with 1D vectors and $K = 3$ executing natively on a ring of 3 routers.

At the beginning 3 nodes hold 3 vectors together with their cluster membership information, and each node elects and broadcasts their respective centroids. With respect to the received centroids, each node calculates the new cluster memberships of its vectors, and broadcasts local partial sums for each cluster-id and the number of vector members with that id. After receiving the partial sums and counts from all nodes, each node calculates the centroid of respective cluster followed by the broadcast phase. The total communication time using FNoC broadcast feature takes 64 cycles whereas with multiple unicasts, takes 80 cycles. Overall calculation time per iteration is 301 cycles and total 2 iterations are required.

VI. CONCLUSION

We make a novel case for and explore the use of lightweight programmable routers (LPRs) in NoCs that we build around the versatile Forth environment. These NoCs with Forth LPRs are generated much like the popular CONNECT generator tool (with interface backward compatibility for convenient evaluation). We show that the resource usage and latencies of these LPRs is only marginally higher in comparison to traditional routers and load-latency characterization is also shown to be within acceptable ranges. That this cost can be traded for interesting possibilities and scenarios as listed in the introduction is notable, and are targets for follow-up work—in particular: 'computation in transit on the network'.

VII. ACKNOWLEDGEMENTS

We thank Nishant Kathpal for help with reviewing some of the results. This work was supported by Dr. Suhas Pai through IIT Bombay Heritage Fund.

REFERENCES

[1] N. Kapre and J. Gray, "Hoplite: Building austere overlay NoCs for FPGAs," *25th International Conference on Field Programmable Logic and Applications, FPL 2015*, no. 1, 2015.

[2] M. S. Abdelfattah and V. Betz, "The case for embedded networks on chip on field-programmable gate arrays," *IEEE Micro*, vol. 34, 2014.

[3] N. Jiang, J. Balfour, D. U. Becker, B. Towles, W. J. Dally, G. Michelogiannakis, and J. Kim, "A detailed and flexible cycle-accurate network-on-chip simulator," in *2013 IEEE International Symposium on Performance Analysis of Systems and Software (ISPASS)*, 2013, pp. 86–96.

[4] M. K. Papamichael and J. C. Hoe, "CONNECT: Re-examining conventional wisdom for designing NoCs in the context of FPGAs," in *Proceedings of the ACM/SIGDA international symposium on Field Programmable Gate Arrays*. ACM, 2012, pp. 37–46.

[5] https://www.forth.com/resources/forth-programming-language/

[6] J. Bowman, "J1: a small Forth CPU core for FPGAs," 2010, pp. 43–46. http://www.excamera.com/sphinx/fpga-j1.html

[7] H. C. Freitas, T. G. S. Santos, and P. O. A. Navaux, "Design of programmable noc router architecture on fpga for multi-cluster nocs," *Electronics Letters*, vol. 44, no. 16, pp. 969–971, July 2008.

[8] V. B. Y. Kumar, P. Engineer, M. Datar, Y. Turakhia, S. Agarwal, S. Diwale, and S. B. Patkar, "Framework for application mapping over packet-switched network of fpgas: Case studies," *CoRR*, vol. 1508.06823, 2015.

[9] E. S. Chung, J. C. Hoe, and K. Mai, "Coram: An in-fabric memory architecture for fpga-based computing," in *Proceedings of the 19th ACM/SIGDA International Symposium on Field Programmable Gate Arrays*, ser. FPGA '11. ACM, 2011.

[10] V. B. Y. Kumar, K. Dhiman, M. Datar, A. Pacharne, H. Narayanan, and S. B. Patkar, "Relaxation based circuit simulation acceleration over cpu-fpga," in *2016 29th International Conference on VLSI Design (VLSID)*.

[11] D. U. Becker, "Efficient microarchitecture for network-on-chip routers," Ph.D. dissertation, STANFORD UNIVERSITY, 2012.

[12] C. H. Moore, "Programming a problem-oriented language," 1970, 2011. www.forth.org/POL.pdf

[13] W.-C. Tsai, K.-C. Chu, Y.-H. Hu, and S.-J. Chen, "Non-minimal, turn-model based noc routing," *Microprocessors and Microsystems*, 2013, embedded Multicore Systems: Architecture, Performance & Application.

[14] M. S. Lam, *A Systolic Array Optimizing Compiler*. Kluwer Academic Publishers, 1989.

978-1-5386-3693-0/18 $31.00 © 2018 IEEE

2018 31th International Conference on VLSI Design and 2018 17th International Conference on Embedded Systems

An Adaptive Deflection Router
with Dual Injection and Ejection Units for Mesh NoCs

John Jose, Abhijit Das

MARS Research Lab, Dept. of CSE, Indian Institute of Technology Guwahati, India
{johnjose, abhijit.das}@iitg.ernet.in

Abstract—With the ever increasing core counts in Chip Multi-Processors (CMPs), Network-on-Chip (NoC) has emerged as a preferred framework for communication among various chip components. Among other factors, energy efficiency and congestion management play a vital role in identifying an efficient NoC design. Thus, NoCs with side-buffered deflection routers have gained popularity; mainly because of their simplicity in router design, low energy consumption and better load balancing capacity. Standing on the shoulders of existing state of the art, this paper proposes ADIEU; An Adaptive Deflection Router With Dual Injection and Ejection Units. This router has dual injection units and a minimal set of side-buffers to make adaptive routing decisions. Experimental results on the proposed microarchitecture using both real and synthetic workloads shows reduced average latency, buffer occupancy and deflection rate of flits when compared with the existing side-buffered deflection routers without any change in the critical path delay.

Keywords-Congestion, buffer-occupancy, side-buffer, starvation

I. INTRODUCTION

Advancing VLSI technology by decreasing feature sizes and shortening wire widths unmasked the constraints of the traditional bus based on-chip interconnection systems. Furthermore, this technology scaling has also increased the performance gap between computation and communication efficiency in modern SoCs. Apart from high speed computing cores, efficient and reliable communication is also essential for achieving high performance in multi-core systems. Network-on-Chip (NoC) is now an established framework that can efficiently support the integration of a massive number of cores on a chip by decoupling the on-chip computation and communication infrastructure, thus overcoming the scalability issues in conventional buses [1].

Input buffered routers dominated initial NoC designs due to their simple wormhole switching [1] and high load handling capacity. However, they consume a significant portion of chip power due to the presence of buffers. Studies show that approximately 30% to 40% of chip power is consumed by the NoC [2][3]. Thus, recent router designs are focusing on a short critical path and minimal buffer footprint [4][5].

Buffer-less routers are proposed as an energy efficient alternative to the traditional input buffered routers. Our simulations (Section V describes the experimental setup)

with real workloads on mesh NoCs with input buffered routers show that for low injection rate applications, in 90% cases, less than 25% of the buffers are being occupied, thereby exposing the over provisioning of buffers in routers. For low to medium injection rate applications, buffer-less NoC router is an optimal design choice [6][7].

Deflection routing is the most commonly used approach in buffer-less routers. When two flits (packets are broken down into multiple flits) that want to have the same output port reach a buffer-less deflection router, only one gets the requested port, and the other flit is deflected through an undesired port. The deflected flits eventually reach the destination by proper livelock prevention mechanisms. Buffer-less routers may not be a good design choice for high injection rate applications as the flits experience high deflection rate. Buffer-less routers that are equipped with side-buffers can accommodate some deflected flits thereby reducing the deflection rate [5][6].

In this paper, we address few critical limitations of the existing state-of-the-art side-buffered deflection routers through a proposed energy efficient **A**daptive Deflection Router with **D**ual **I**njection and **E**jection **U**nits (ADIEU) that effectively handles network congestion leading to the reduction in average latency, buffer occupancy and deflection rate of flits.

II. BUFFER-LESS ROUTERS: RELATED WORK

As buffers in the NoC routers are power hungry and buffer management circuits are complex, buffer-less routers are gaining popularity on large mesh networks. Few works have also exploited a hybrid approach that uses a conventional buffered router with a provision to switch to buffer-less mode under low network load by using power gating techniques [8][9].

In buffer-less deflection routers, storage of flits happens only in pipeline registers. Buffering of flits that fail in getting the desired port is replaced by the concept of deflecting the flits [10] to non-productive ports. To avoid fragmentation, deflection routers employ flit level routing. In every cycle, maximum of four flits each can enter or leave a router. The incoming flits enter a routing unit, which computes the desired output port for the flit. After routing, if there is any flit destined to the local core, it is ejected. The port

978-1-5386-3693-0/18 $31.00 © 2018 IEEE   374

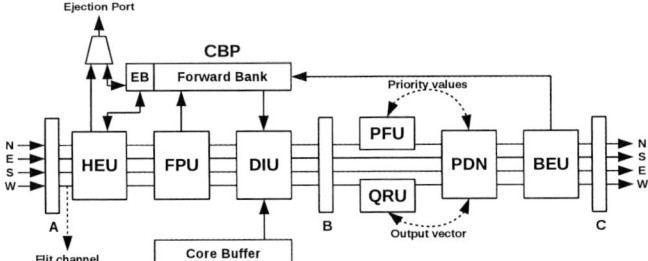

Fig. 1. Router pipeline for DeBAR. HEU-Hybrid Ejection Unit, FPU-Flit Preemption Unit, DIU-Dual Injection Unit, PFU-Priority Fixer Unit, QRU-Quadrant Routing Unit, PDN-Permutation Deflection Network, BEU-Buffer Ejection Unit, CBP-Central Buffer Pool.

allocator assigns output ports to all the flits present in the router. Flits which get the same output port as requested are called productively assigned flits, and the others are called deflected flits. BLESS [11] uses a crossbar with sequential output port allocation unit which increases the router critical path. This allocation unit is replaced by a Permutation Deflection Network (PDN) in CHIPPER [4], which considerably reduces the critical path delay at the expense of increased deflection rate.

Buffer-less deflection router suffers from performance degradation at high injection rate due to high deflection rate of flits. To address this issue, MinBD [5], DeBAR [6] and SLIDER [12] use a minimal set of side-buffers. Entry path to these side-buffers is kept after the PDN unit to accommodate a fraction of the deflected flits, thereby reducing the mis-routed flit traffic in the network. Minimally buffered routers outperform input buffered routers in low injection traffic and buffer-less defection routers in high injection traffic.

DeBAR is the best available deflection router that proposed an effective solution by combining the merits of buffered and buffer-less routing. It has also successfully addressed the limitations of BLESS, CHIPPER and MinBD.

III. MOTIVATION

DeBAR is a 2-stage deflection router that uses a Central Buffer Pool (CBP) to accommodate a fraction of misrouted flits. The block diagram of DeBAR is shown in Fig. 1 where A, B, and C are the pipeline registers. Four internal flit channels carry input flits through various units of the router pipeline. The core-buffer contains newly created flits from the local processing core. We identify three performance limitations in the DeBAR design that motivated us for the proposed work. We analyse the cause for each of these limitations and suggest suitable cost-effective solutions.

A. Starvation of Side-Buffered Flits Due to Ineffective Priority Scheme

In DeBAR, flit priority is calculated based on the hops-to-destination of the flit from current router. The flit with least hops-to-destination is given the highest priority during port allocation. Because of port conflicts, flits with low priority may be allocated non-productive output ports, leading to subsequent buffering in CBP by BEU. As the priority of flits buffered in CBP is not changing when such flits are re-injected into the network (because hops-to-destination of those flits are not changed), there is a high chance that they can be buffered again in CBP due to port conflicts. This leads to starvation of such flits and increases the average flit latency. Close to saturation load, for uniform traffic we identify 27% of such starvation cases (flits from CBP get back to CBP again due to low priority) upon using DeBAR for an 8×8 mesh NoC system. We propose that the flits that are buffered in CBP should get a higher priority when they are re-injected to ensure that they make forward progress.

B. Output Channel Wastage

In DeBAR, at saturation load, we observe that in 35% of cases at least one of the output ports of a router is idle while flits are waiting in CBP or core-buffer with those unused ports as their productive ports. This is due to the forwarding of misrouted flits to CBP after PDN by BEU. Flits waiting in CBP / core-buffer may not be able to inject to DIU if all four internal flit channels are busy. However, after PDN, due to side-buffering of a deflected flit in CBP the port already assigned to such a flit by PDN will be idle. Presence of such idle output channel is a wastage of resource. If an idle channel can be assigned to a flit that is waiting in either CBP or core-buffer, this channel wastage can be reduced.

C. Sequential Positioning of Independent Operations

HEU in DeBAR is functional only if there is a flit ejection. FPU aims at creating an idle slot in the internal flit channels of the router pipeline so that flits from core-buffer and CBP can be injected into these slots when they reach their starvation thresholds. A flit removed by HEU for ejection creates an idle slot in the internal flit channel. If HEU ejects a flit, FPU does not perform any flit removal as HEU has already created an idle slot. If all flit channels are busy and none of them are ejection flits, then HEU can be idle, and FPU has to pre-empt one flit. This means that only one of these two units are operational in a given cycle. Hence these two units can be combined to form a single unit to reduce hardware cost and critical path delay.

Experimental analysis on real and synthetic workloads have confirmed that the above-identified limitations of DeBAR create a critical performance bottleneck. In the proposed work, we modify the existing priority scheme to reduce the starvation of side-buffered flits. We also provide one more injection unit late in the pipeline thereby reducing the side-buffer/core-buffer occupancy of flits. This additional injection unit can also reduce channel wastage.

IV. ADIEU ARCHITECTURE

The basic working of ADIEU is similar to that of DeBAR except for few additional units that improve performance.

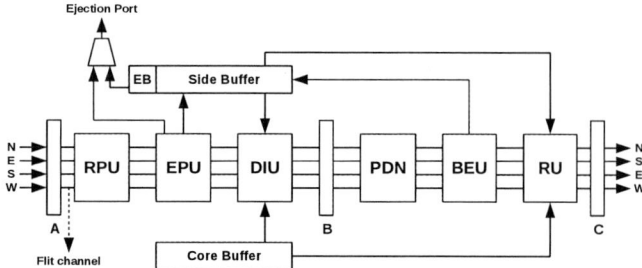

Fig. 2. Router pipeline of ADIEU. RPU-Routing and Priority unit, EPU-Ejection and Pre-emption unit, DIU-Dual Injection Unit, PDN-Permutation Deflection Network, BEU-Buffer Ejection Unit, RU-Re-injection Unit, EB-Ejection Bank.

Fig. 2 shows the block diagram of ADIEU. Like DeBAR, here also input flits are stored in a pipeline register, and a fraction of deflected flits are stored in side-buffers. ADIEU differs from DeBAR in the following aspects:

- The priority scheme is modified such that the re-injected flits to the DIU will get the highest priority and will not be side-buffered again on the same router. This reduces the side-buffer occupancy of flits thereby addressing the issue mentioned in Section III-A.
- A Re-injection Unit (RU) is included as the last unit in ADIEU pipeline to give chance for injecting flits that are waiting in the core/side-buffer. These flits are assigned productive output ports in idle output channels to address the issue mentioned in Section III-B.
- HEU and FPU of DeBAR are combined into a single unit called Ejection and Pre-emption Unit (EPU) to address the issue mentioned in Section III-C.
- Route and priority computations are done in the first stage of the pipeline (at RPU) to accommodate the new RU in the second stage.

The internal architecture and working of various units in ADIEU are discussed below.

A. Routing and Priority Unit (RPU)

RPU reads the destination information of all the incoming flits from pipeline register A. Based on the destination address of a flit, the desired output port is identified. We use dimension order routing algorithm [1] for identification of a productive output port. By this routing operation, the locally destined flits (ejection flits) are also identified. For the ejection flits, RPU sets an ejection flag in the flit header. Similarly, from the destination address of a flit, the hops-to-destination value is computed, which is considered as the priority of that flit. The 2-bit priority value (similar to the one as in DeBAR) is stored in the flit header itself.

B. Ejection and Pre-emption Unit (EPU)

EPU can act as an ejection unit or as a pre-emption unit based on the value of the ejection flag (already set/reset

by RPU) in the incoming flits. EPU consists of an ejection flag checking circuit and two parallel combinational blocks; one for ejection unit and other for pre-emption unit. If the ejection flag is set, EPU forwards the flit from router pipeline to the ejection port. Similar to DeBAR, EPU performs at most two flit ejections/cycle with the help of single ejection port and the Ejection Bank (EB) in the side-buffer.

If the ejection flag is not set, EPU acts like a flit pre-emption unit. It checks whether all internal flit channels are occupied and whether the starvation threshold is crossed or not. If so, EPU will pre-empt a flit from the router pipeline to the side-buffer. Similar to DeBAR, the starvation of flits waiting in the buffers are addressed by fixing a threshold to Re-Inject Interval (RII) for side-buffer and Core Inject Interval (CII) for core-buffer. By this flit pre-emption, EPU makes a free channel for buffer injection. We consider the threshold value of CII and RII as 2 cycles each.

C. Dual Injection Unit (DIU)

The basic working of DIU in ADIEU is same as that in DeBAR except for the priority variation of re-injected side-buffered flits. In DeBAR the priority of flits does not change even if they are re-injected from the CBP. We see that this could lead to unnecessary penalisation of flits entering the side-buffer. In ADIEU, the re-injected flits from the side-buffer are assigned the highest priority to ensure that the side-buffered flits are not penalised again on the same router. At the end of the first cycle, all the flits reach pipeline register B.

D. Permutation Deflection Network (PDN) and Buffer Ejection Unit (BEU)

PDN and BEU in ADIEU are same as that in DeBAR. PDN is a two-stage arbitration circuit that performs parallel allocation of output ports. Fig. 3 shows how the arbiter blocks (A, B, C, and D) are arranged to form a PDN. For each arbitration stage, the priority level and the desired output port of incoming flits (given by RPU) are used for determining the actual output port. The highest priority flit always gets its productive port. Other flits may or may not get a productive port depending on current port conflicts. From among the flits coming out from PDN, BEU selects at most one flit that is assigned a non-productive port for storing into the side-buffer. This side-buffering reduces average deflection rate, thereby bringing down the unwanted flit movements in the network.

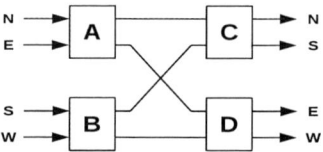

Fig. 3. Permutation Deflection Network (PDN).

Percentage Miss Rate	Benchmarks
Low MPKI (less than 5)	calculix, gobmk, gromacs, h264ref
Medium MPKI (between 5 and 25)	bwaves, bzip2, gamess, gcc
High MPKI (greater than 25)	hmmer, lbm, mcf, leslie3d

Table I. Classification of benchmarks based on cache MPKIs

Mix #	SPEC CPU 2006 Benchmarks					
M1	calculix(16)	gobmk(16)	gromacs(16)	h264ref(16)		
M2	bwaves(16)	bzip2(16)	gamess(16)	gcc(16)		
M3	hmmer(16)	lbm(16)	mcf(16)	leslie3d(16)		
M4	calculix(16)	gobmk(16)	gamess(16)	gcc(16)		
M5	bwaves(16)	bzip2(16)	mcf(16)	leslie3d(16)		
M6	hmmer(16)	lbm(16)	gromacs(16)	h264ref(16)		
M7	calculix(10)	gromacs(10)	bwaves(10)	gamess(10)	hmmer(12)	mcf(12)

Table II. Various workload mixes

L1 cache miss creates a 1 flit request packet to the respective core where the shared distributed L2 cache is mapped. Then, the respective core responds with a 4 flit reply packet.

E. Re-injection Unit (RU)

We observe that only in less than 10% cases, all output ports of DeBAR are full. Under such conditions, to exploit the slot wastage, the newly added RU search among the buffered flits (which are there in side/core-buffers) to find if their desired output ports match with any idle output channels. If found, RU assigns respective idle output channels to each such flits. As in the case of DIU, side-buffer and core-buffer re-injections are given alternate priority in odd and even cycles, respectively to ensure fairness.

V. EXPERIMENTAL METHODOLOGY

We use a cycle-accurate simulator, BookSim 2.0 [13] for the NoC simulation. We modify BookSim to model two-cycle deflection router microarchitectures of MinBD, DeBAR and ADIEU for an 8×8 mesh network. We consider flits with necessary header information to facilitate independent routing as practised in standard deflection routers [5]. Necessary reassembly mechanism is employed for handling out-of-order delivery of flits. The flit channel is 140-bit wide: 128-bit data field and a 12-bit header field. We first consider synthetic workloads for the evaluation of our proposed router design. Average latency, buffer occupancy and deflection rate of flits are collected for each traffic pattern with injection rate varying from zero to saturation.

To evaluate our design with real workloads, SPEC CPU 2006 benchmarks are used, which are classified according to their Misses Per Kilo Instructions (MPKI) on a 64KB L1 cache as shown in Table I. This is to classify the applications to different network injection intensity groups. Based on this network injection intensity, we create 7 workload mixes (M_is) consisting of SPEC CPU 2006 benchmarks as shown in Table II. Consider mix 1 (M_1); where out of 64 cores that we model, 16 cores run *calculix*, 16 cores run *gobmk*, 16 cores run *gromacs* and last 16 cores run *h264ref* benchmark. Similarly, other workload mixes (M_2 - M_7) can also be described.

We run 64 application instances of the respective workload mixes (as mentioned above) in gem5 simulator [14], which models a 64-core CMP setup with CPU cores and 2 levels of cache hierarchy. Each core consists of an out-of-order x86 processing unit with a 64KB, 4-way associative, 32B block, dual ported, unified, private L1 cache and a 32MB, 16-way associative, 64B block, shared distributed L2 cache (i.e. 512KB/core). We create a request packet for each L1 cache miss and feed it to BookSim to model the NoC traffic. Network statistics are collected and analysed. Each

VI. EXPERIMENTAL ANALYSIS

We compare the performance of ADIEU with both De-BAR and MinBD routers, as they are considered the best in the available literature. An analysis is done on average latency, buffer occupancy and deflection rate for both synthetic and real workloads.

A. Effect on Average Flit Latency

Fig. 4 shows a set of injection rate vs average flit latency graphs for MinBD, DeBAR and the proposed ADIEU routers using synthetic traffic patterns. We can see that across all traffic patterns ADIEU shows either same or lower average flit latency than MinBD and DeBAR. Also across all traffic patterns, ADIEU saturates later than MinBD and DeBAR. This makes our proposed ADIEU a better design choice for high injection rate applications.

Fig. 7 shows percentage reduction in average flit latency of DeBAR and ADIEU with respect to MinBD for various SPEC CPU 2006 benchmark mixes. We can see that for all the mixes ADIEU shows a reduction in average flit latency than DeBAR. Significant reduction in latency can be seen for high injection rate mixes like M_3 and M_5.

B. Effect on Average Buffer Occupancy

Buffer occupancy of a flit in side-buffered deflection routers refers to the number of cycles spent by a flit in side-buffers in its entire lifetime. It gives the waiting time of the flit in the side-buffers until it gets re-injected into the router pipeline. Average buffer occupancy, B_{occ} is given by,

$$B_{occ} = \frac{\sum_{i=1}^{N} b_i}{N} \qquad (1)$$

where b_i is the total number of cycles a flit stays in side-buffers of all routers in its path to destination and N is the total number of injected flits.

In DeBAR, to reduce the deflection rate of flits, one of the flits that are assigned a non-productive port is moved to the side-buffer. If a buffered flit gets delayed in re-injecting into the router pipeline, it can increase average buffer occupancy. An increase in either deflection rate or buffer occupancy can lead to increase in the overall flit latency. Since we give the highest priority to the re-injected flits from side-buffer,

978-1-5386-3693-0/18 $31.00 © 2018 IEEE

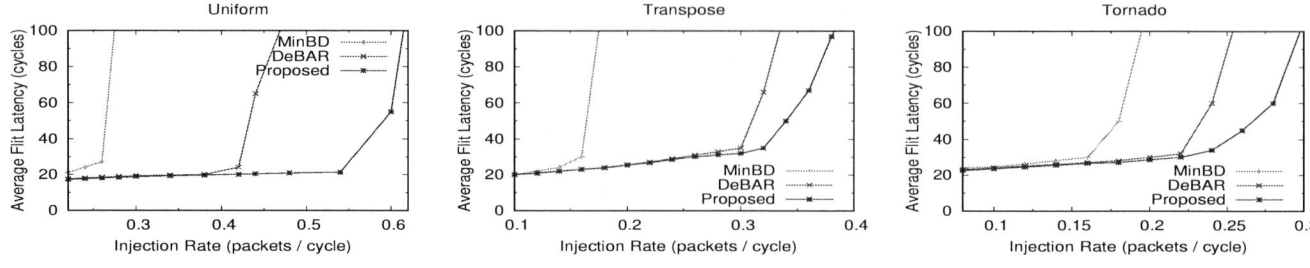

Fig. 4. Comparison of average flit latency for various synthetic traffic patterns in 8×8 mesh network.

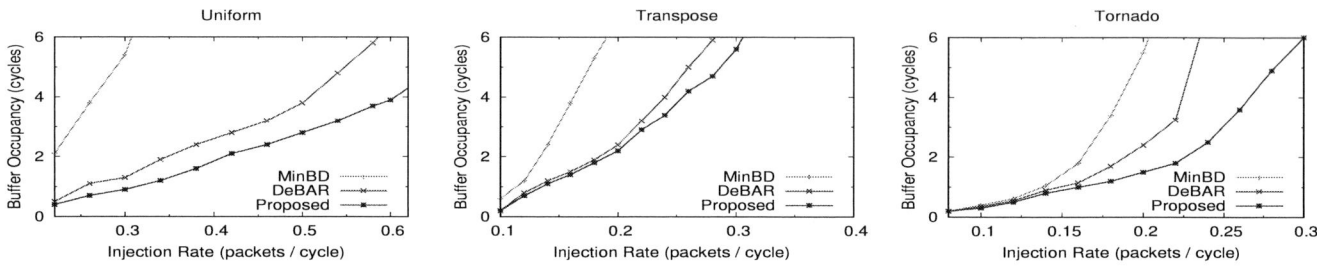

Fig. 5. Comparison of average flit buffer occupany for various synthetic traffic patterns in 8×8 mesh network.

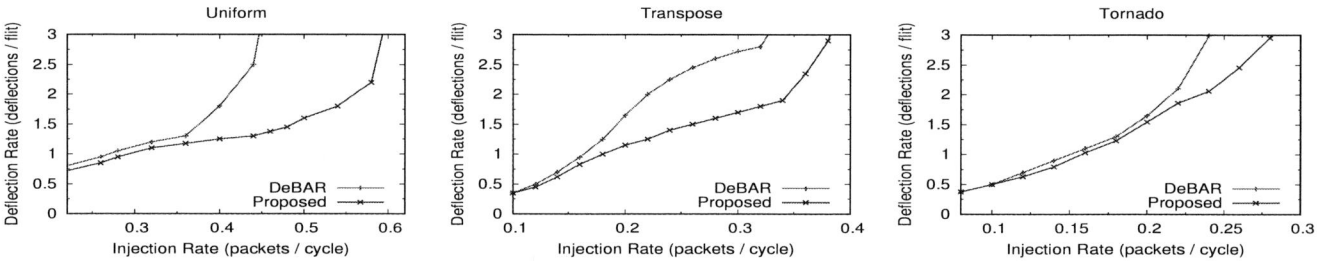

Fig. 6. Comparison of average flit deflection rate for various synthetic traffic patterns in 8×8 mesh network.

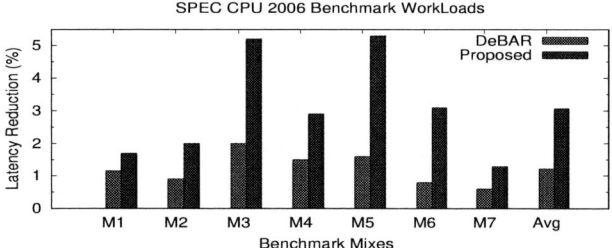

Fig. 7. Percentage reduction in average flit latency.

Fig. 8. Reduction in average flit buffer occupancy.

they will always get their desired output ports. Thus the re-injected flits will not go to side-buffer on the same router or get deflected away. Hence using ADIEU, we expect an overall reduction in the average buffer occupancy of flits.

Fig. 5 shows the buffer occupancy comparison for MinBD, DeBAR and ADIEU designs. At higher injection rate, ADIEU design has significantly lower buffer occupancy in all traffic patterns. This result shows the drawback in DeBAR design due to starvation of flits in side-buffer. By giving highest priority to the re-injected flits, we avoid this starvation scenario, thereby reducing flit latency.

Fig. 8 shows the average buffer occupancy of DeBAR

and ADIEU designs for SPEC CPU 2006 benchmark mixes. Since the respective values for MinBD is very high, we are not plotting them in Fig. 6, Fig. 8 and Fig. 9. Here we focus more on how much improvement we attain with respect to DeBAR design. We can see that for all the mixes ADIEU shows a reduction in buffer occupancy of flits than DeBAR. A significant difference in buffer occupancy can be seen at high injection rate mixes M_3 and M_5.

C. Effect on Deflection Rate

Deflection rate is defined as the number of non-productive hops a flit takes on an average to reach its destination. Deflection rate comparison between DeBAR and ADIEU

978-1-5386-3693-0/18 $31.00 © 2018 IEEE

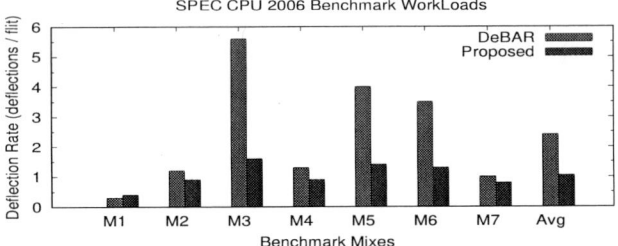

Fig. 9. Reduction in average flit deflection rate.

for synthetic traffic is shown in Fig. 6. It is clear that for all the traffic patterns, ADIEU achieves lower deflection rate as compared to DeBAR. The difference is more evident at higher injection rates. This is due to the better priority scheme used in ADIEU and the re-injection unit that increases the chance for flits to get their desired output port.

Fig. 9 shows the deflection rate of DeBAR and ADIEU designs for SPEC CPU 2006 benchmark mixes. We can see that for all the mixes (except M_1) ADIEU shows a reduction in deflection rate of flits than DeBAR. Lower the deflection rate; lower will be the network activity and hence lower dynamic power dissipation through the links.

Our simulations show that by using ADIEU router, there is a reduction of 11.5% in dynamic power with respect to DeBAR due to lower buffer occupancy and lower deflection rate.

D. Effect on Router Critical Path, Area, and Power

We implement DeBAR and ADIEU in Verilog and synthesise using Synopsys Design Compiler with 65nm cell library to obtain timing delay. In ADIEU, due to the removal of routing and priority units, and the addition of RU, the latency of stage 2 is unchanged with respect to DeBAR. The latency of stage 2 dominates over stage 1 in both DeBAR and ADIEU. We experimentally find that ADIEU can be operated at the same frequency as that of DeBAR.

We compute the area and power estimates of DeBAR and ADIEU using Orion 2.0 [15]. We assume 65nm technology for a NoC operating at 1GHz frequency with an inter-router link delay of 1 cycle. Due to the presence of RU and additional circuits in ADIEU, we incur an area overhead of 2.5% and a static power overhead of 3.8% with respect to DeBAR. Nevertheless, the performance gained with ADIEU is much more significant than this negligible overhead.

VII. CONCLUSION

By identifying the performance limitations in existing baseline models including MinBD and DeBAR, we proposed ADIEU, an adaptive deflection router microarchitecture with minimal side-buffering. ADIEUs superior design is based on enhancements proposed in primitive DeBAR design to improve overall system performance. The modification in priority scheme and the inclusion of an extra re-injection logic facilitate all possible opportunities for idle flits to

move out of the router. ADIEU microarchitecture stands above both MinBD and DeBAR, in terms of better overall average latency, buffer occupancy and deflection rate. All these enhancements and optimisations make ADIEU an ideal implementation choice for minimally buffered NoC routers.

ACKNOWLEDGMENT

This research is supported in part by Department of Science and Technology (DST), Government of India vide project grant ECR/2016/000212. The authors would like to thank R&D Section, IIT Guwahati for the SuG grant given to do this work.

REFERENCES

[1] W. J. Dally and B. P. Towels, *Principles and Practices of Interconnection Networks*. Morgan Kaufmann, 2004.

[2] Y. Hoskote *et al.*, "A 5-GHz Mesh Interconnect for a Teraflops Processor," *IEEE Micro*, vol. 27, no. 5, pp. 51–61, 2007.

[3] M. B. Taylor *et al.*, "Evaluation of the Raw Microprocessor: An Exposed-Wire-Delay Architecture for ILP and Streams," in *International Symposium on Computer Architecture (ISCA)*, pp. 2–13, 2004.

[4] C. Fallin *et al.*, "CHIPPER: A Low-complexity Bufferless Deflection Router," in *International Symposium on High Performance Computer Architecture (HPCA)*, pp. 144–155, 2011.

[5] C. Fallin *et al.*, "MinBD: Minimally-Buffered Deflection Routing for Energy-Efficient Interconnect," in *International Symposium on Networks-on-Chip (NOCS)*, pp. 1–10, 2012.

[6] J. Jose *et al.*, "DeBAR: Deflection Based Adaptive Router With Minimal Buffering," in *Design, Automation and Test in Europe (DATE)*, pp. 1583–1588, 2013.

[7] R. James *et al.*, "Smart Port Allocation for Adaptive NoC Routers," in *International Conference on VLSI Design (VLSID)*, pp. 475–480, 2015.

[8] S. A. R. Jafri *et al.*, "Adaptive Flow Control for Robust Performance and Energy," in *International Symposium on Microarchitecture (MICRO)*, pp. 433–444, 2010.

[9] G. Kim *et al.*, "FlexiBuffer: Reducing Leakage Power in On-Chip Network Routers," in *Design Automation Conference (DAC)*, pp. 936–941, 2011.

[10] E. Nilsson *et al.*, "Load Distribution with the Proximity Congestion Awareness in a Network on Chip," in *Design, Automation and Test in Europe (DATE)*, pp. 1126–1127, 2003.

[11] T. Moscibroda and O. Mutlu, "A Case for Bufferless Routing in On-Chip Networks," in *International Symposium on Computer Architecture (ISCA)*, pp. 196–207, 2009.

[12] B. Nayak *et al.*, "SLIDER: Smart Late Injection DEflection Router for Mesh NoCs," in *International Conference on Computer Design (ICCD)*, pp. 377–383, 2013.

[13] N. Jiang *et al.*, "A Detailed and Flexible Cycle-Accurate Network-on-Chip Simulator," in *International Symposium on Performance Analysis of Systems and Software (ISPASS)*, pp. 86–96, 2013.

[14] N. Binkert *et al.*, "The gem5 Simulator," *SIGARCH Computer Architecture News (CAN)*, vol. 39, no. 2, pp. 1–7, 2011.

[15] A. B. Kahng *et al.*, "ORION 2.0: A Fast and Accurate NoC Power and Area Model for Early-Stage Design Space Exploration," in *Design, Automation Test in Europe (DATE)*, pp. 423–428, 2009.

978-1-5386-3693-0/18 $31.00 © 2018 IEEE

2018 31th International Conference on VLSI Design and 2018 17th International Conference on Embedded Systems

Towards Near Data Processing of Convolutional Neural Networks

Palash Das, Shivam Lakhotia, Prabodh Shetty and Hemangee K. Kapoor

Indian Institute of Technology, Guwahati, Assam 781039

Email: {palash.das, s.lakhotia, p.shetty, hemangee} @iitg.ernet.in

Abstract—The gap between the processing speed of the CPU and the access speed of the memory is becoming a bottleneck for many data intensive applications. This gap can be reduced if the computation can be taken near to the data. Recent advancement in memory technology has made it feasible to have 3D stacked memory along with the capability of having an integrated logic layer, thus making near data processing (NDP) feasible.

Convolutional Neural Networks (CNNs) are used in a wide range of applications such as image processing, video analysis, natural language processing, etc. They are data intensive and have highly parallel computations. If we can perform them near the memory, we can achieve higher throughput. This paper proposes a CNN Logic Unit (CLU) as a hardware implementation in the logic layer associated with 3D stacked memory, e.g. the Hybrid Memory Cube (HMC) implementing the concept of NDP. Full system simulation results show that the method is very promising as it gives several times improvement for CNN operations when performed near memory as compared to in conventional CPU based systems with DRAM. We get an improvement of 76x in performance while energy consumption is reduced by 55x.

Keywords—Hybrid Memory Cube, Convolution Neural Networks, Near Data Processing, 3D-Stacked Memory, CNN Logic Unit (CLU), SRAM cache.

I. INTRODUCTION

The shift in trend from CPU-bound application to data-bound applications has resulted in the development of in-memory data systems which certainly deliver better throughput compared to disk based systems. However, this still does not completely solve the problem of the *memory-wall* as the increase in memory access speed is relatively very slow compared to the increase in the computation speed of the CPU. For these data-bound applications the ever rising cost of moving the data from the main memory to the processor through the entire memory hierarchy is still a problem that needs to be addressed. A very simple solution to reduce this cost is to remove the data movement altogether; and one of the ways to do this is by using Near Data Processing (NDP).

NDP is an approach where a computational hardware unit is placed close to the data. This hardware performs a specific set of simple operations. NDP is not a new approach. There have been many NDP prototypes since 1990s [1], [2], [3], [4] and they have proven that NDP can break the *memory-wall*. But it could not be commercially adopted because incorporating memory elements with logic on a single chip in turn makes higher per bit cost to implement memory. Moreover, it also demands newer programming models. However, with the

emergence of 3D-die stacking with tight integration of logic and memory by using Through-Silicon-Vias (TSVs) has again resurged processing-in-memory (PIM) after a dormant decade.

The aim of this paper is to present a design that can use NDP for Convolutional Neural Networks (CNNs). CNNs are widely used in machine learning, data analysis in back-end data centers, image recognition, video analysis, natural language processing, drug discovery etc. CNNs also have high degree of compute parallelism and they deal with huge data sets which make them an ideal candidate for NDP. In this paper, we present the design of a dedicated CNN Logic Unit (CLU) that performs convolutions in-memory. We have chosen Hybrid Memory Cube (HMC) as our memory unit. We divide the data and store it in separate vaults of the HMC and simultaneously perform convolutions using the CLUs on each vault.

This paper makes the following contributions:

- We use NDP to improve the performance of CNNs. We present the design of a CLU which performs convolutions in the memory.
- Data parallelism is exploited by performing simultaneous computations in all the vaults of the HMC.
- We get 76 times performance improvement and 55 times reduction in energy consumption on an average.

II. BACKGROUND AND MOTIVATION

A. Related Work

2D Process-in-memory: Researchers have put up a lot of effort from 1995 to 2005 in integrating DRAM with simple logic. Although some of them did give good results, their cost of manufacturing was not feasible enough to have widespread use. Most of them can be classified into two types, integrating an accelerator on top of DRAM chip or changing the DRAM itself for calculating certain operations. JAFAR [5], Buffered Compares [6], NDA [7] are some of the 2D PIM approaches. Whereas, Fast Bulk bitwise AND and OR [8] belongs to the later. JAFAR was used for the select operation in the databases. Whenever host processor sends select request, JAFAR would send only qualifying data to the processor thus reducing traffic between processor and DRAM. Buffered Compares also followed similar logic for key-value pair evaluation. NDA stacks Coarse-Grained Reconfigurable Architecture (CGRA) units on top of DRAM modules. These CGRAs were used for computation near DRAM. NDA would then send the evaluated data back to the processor. Fast bulk bitwise AND and OR

978-1-5386-3693-0/18 $31.00 © 2018 IEEE   380

operations modified DRAM itself. They connected 3 cells directly to a bit-line in such a way, as to provide AND and OR operation depending on the value of one of these cells.

NDP on 3D stacked Memory: 3D stacking of multiple layers using through-silicon-vias (TSVs) reduced energy consumption and increased performance of the system. Generally, 3D DRAM consists of DRAM layers stacked on top of each other and one logic layer at the bottom of these layers. These layers interact with each other using TSVs. Micron's Hybrid Memory Cube (HMC) [9] and JEDEC's High Memory Bandwidth DRAM [10] are popular 3D DRAM structures. Some used it for highly parallel map-reduce operations [11] while others used it for data re-organization [12]. In HRL [13], a Heterogeneous Reconfigurable Logic has been proposed as a near data processing element. While executing MapReduce, graph processing, and deep neural networks, HRL has been proven better than both fine-grained (FPGA) and coarse-grained (CGRA) reconfigurable logic in terms of performance and power. However, HRL is lagging behind of custom accelerators in terms of performance. A Logic-in-Memory accelerator [14] has also been proposed which is tightly integrated with 3D die stacked DRAM device which is quite similar to HMC. Significant performance gain and power efficiency has been achieved with respect to CPU based system while implementing target application like SpGEMM, 2D FFT. PIM-Enabled Instructions [15] have also employed special hardware for NDP operations. However there is no mention about CNNs. But our proposed design is focused on CNN implementation, resulting in simpler design to fit close to the main memory within its limited power and area budget. CNNs being widely used in sophisticated devices it is flexible which motivated us to implement it as NDP. We use HMC for convolution, however, with modification of I/O management, one can integrate CNN for other 3-D memory technologies.

B. Hybrid Memory Cube

HMC is made up of 4 to 8 dies of DRAM cell arrays (DRAM layers). Each of these layers is connected with TSVs for transferring data between the layers. It also consists of a single logic layer that lies at the bottom of the DRAM layers. Each of these layers is divided into 16 parts. A stack of these partitions is called a vault. Processing in each of these vaults is independent of the other vaults. HMC implements the vault controller in the logic layer as well as the SerDes links for off-stack communication. A conceptual view is represented in Figure 1 [16]. Prior/inbuilt existence of logic layer with ready I/O links provided by HMC make them an attractive option for NDP.

C. Convolutional Neural Nets

Convolutional Neural Nets (CNNs) are a type of feed-forward neural nets. It consists of two types of matrices: data matrix and filter matrix. Dimensions of the filter matrices are lower than that of the data matrix.

The filter matrix traverses whole of the data matrix for convolution. At each position, CNN evaluates the dot product

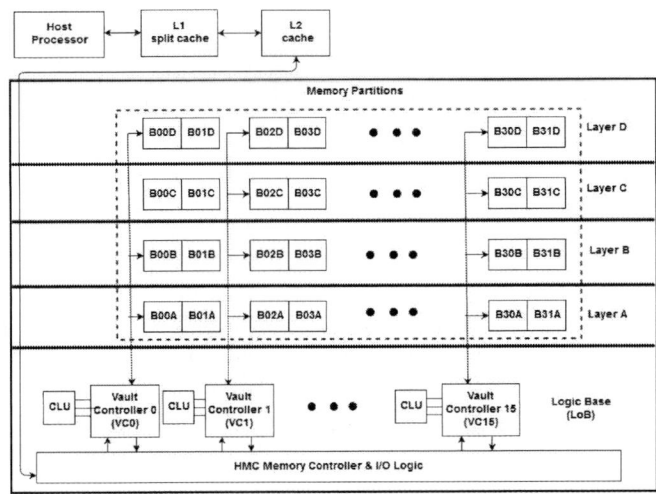

Fig. 1. Architecture of our proposed NDP system showing the CLU integrated with HMC logic layer.

between the data matrix and the filter matrix. After one convolution we arrive at a single layer. We keep doing this for various filter matrices and thus arrive at multiple layers. These layers are again sent to the CNN and again evaluated with a different set of filter matrices. We keep forwarding these layers until we reach the final layer of CNN where we evaluate the loss function. This initiates the back-propagation stage of the CNN. From the algorithm described earlier, it can be observed that CNNs use convolution a lot of time. Convolution is a dot product of two matrices at various filter positions. This fact alone can be used to parallelize convolution.

- To parallelize the operation, we divided the data matrix into 16 parts (as there are 16 vaults in our selected HMC). Each vault would then process on its corresponding data. However, in case of a different number of vaults, the data can be partitioned accordingly.
- Another thing to be observed is that convolution is highly data intensive. If all of these layers can be evaluated in DRAM itself, this would reduce the traffic between the processor and the DRAM.

Thus we arrived at the idea of using HMC for convolution. Figure 2 shows ConvNet of a basic CNN. This is the model

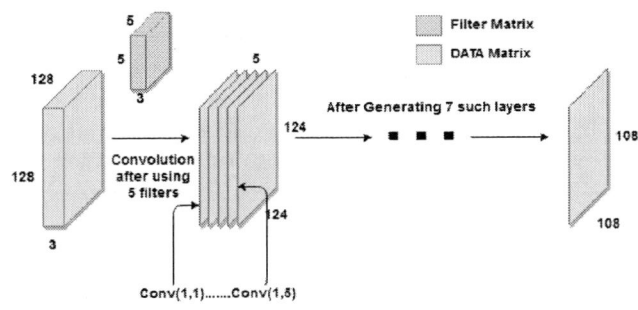

Fig. 2. Convolutional Neural Network.

that we will be using to evaluate our proposed model against a CPU based baseline model.

III. System Architecture and Operational Steps

Our design philosophy was to take advantage of two major factors: first, harness the parallelism present in performing convolution and second, reduce the data movement overhead by placing the CLU as close to the data as possible. Parallelism can be harnessed by using a CLU per vault of the HMC. This leads to the requirement of communication between the vaults as each vault needs the data computed by the next vault. There have been many implementations of NDP where the logic unit has been placed outside the DRAM. However to reduce data movement further we place the CLU in the logic layer of the HMC. In the following sub-section we describe the architecture of the CLU.

We have designed the full system architecture as shown in Figure 1 which is highly efficient in terms of power and energy. In our system, we have a host processor with one core which is connected with L1 split cache which in turn is connected with a L2 cache. This L2 cache is the last level cache connected with the host controller of the HMC.

We use HMC like structure that consists of 4-Gbit layers which sum up to 2GB of memory in total (4 layers: layer A, layer B, layer C and layer D). Each layer has 16 vaults and each vault consists of 2 banks (named as B00A, B01A,...) per layer.

All 16 vault controllers (named as VC0, VC1,.., VC15) for each vault are present in the LoB (Logic base die in HMC). CLU is the logic unit attached with the vault controller as shown in Figure 1. The function of the CLU is to perform CONV operation (Convolution) in the convolution layer of CNNs. The application specific design of CLU improves the overall system performance significantly compared to baseline with reduced energy consumption.

Figure 3 depicts the block-level design of the CLU. It consists of the CLU controller that drives the entire module to execute the CONV operation. Various registers are used to hold the temporary values of the operation. After computation the final result is sent back to the memory die by the vault controller. To maintain the accuracy of data, IEEE-754 format has been used. We have used one IEEE-754 double precision floating point multiplier and one IEEE-754 double precision floating point adder for the entire CONV operations. We have also implemented a small SRAM memory to load the filter elements. There are some trade-offs regarding the use of SRAM in the design:

- We can load all the dimensions of the filter in the SRAM. In this case, SRAM size will be a little higher which will increase the power requirements of the system.
- We can load only one dimension of each filter in the SRAM. In this case, SRAM size will be small, power requirements will be less, but for data intensive CNN applications, DRAM accesses will increase.

Filter sizes are generally small in real implementations. We are loading only one filter at a time and performing the entire

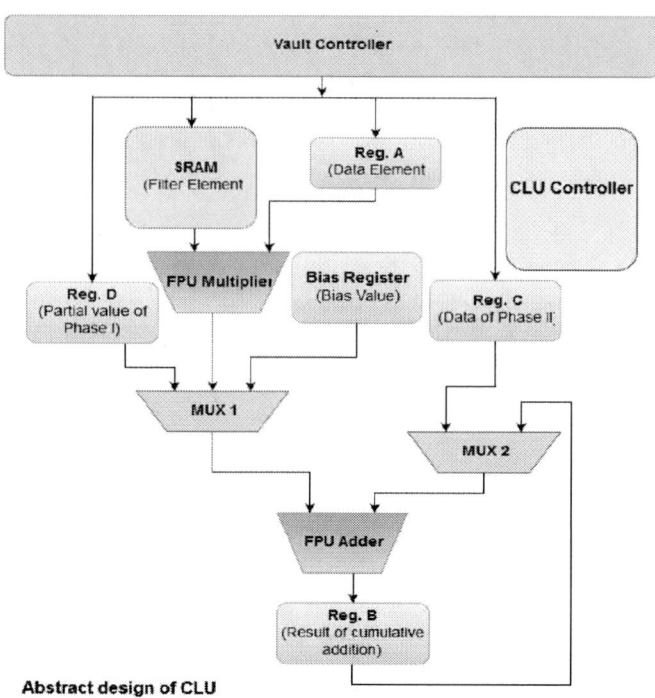

Abstract design of CLU

Fig. 3. Abstract data path of CLU.

convolution operation using it. As a result, we have chosen the first trade off in our proposed CLU design methodology.

A. Operational Steps

As shown in Figure 4, all the dimensions of the data are split horizontally and stored in different vaults. This will be done by the host processor during the load time of the data in the main memory. The host processor can then send a special start signal to the CLU unit to process the data. As the data is distributed across all the vaults, all the CLUs will be able to process the data simultaneously. This vault level parallelism improves the

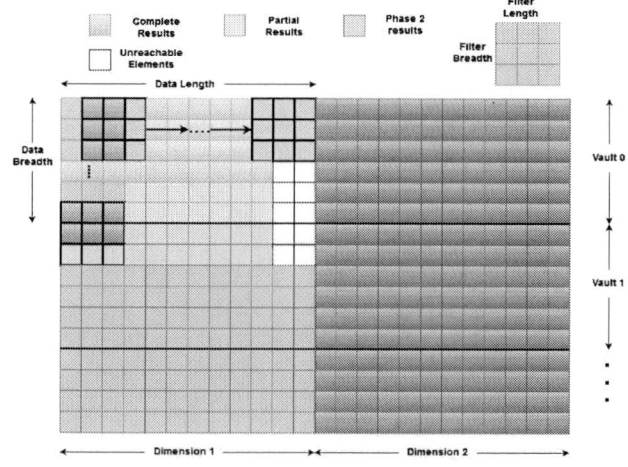

Fig. 4. Data distribution and processing of data across the vaults.

performance significantly. The CLU that we have implemented goes through three main phases: I, II and III, described next. Details are given in Algorithm 1, 2 and 3 respectively.

1) Phase I: As shown in Figure 3-

- All the filter elements are loaded into the SRAM by the CLU controller with the help of vault controller from the memory die of HMC.
- CLU controller also brings one single image data element from DRAM bank of HMC to register-A with the help of the vault controller.
- FPU multiplier multiplies these two and adds the answer to the cumulative value in register B.
- If it is dimension-1, then a bias value is added by the CLU controller with this result available in B.
- This result is then written back to the memory bank with the help of the vault controller.

This process continues in all vaults and results into a snapshot of data as shown in the Figure 4. This gives some complete results and some partial results. Partial results are generated due to lack of all data, as they are distributed in other vaults for parallel processing and we do not keep replicas for space efficiency.

There will be $dimension*(filterbreadth-1)*datalength$ number of partial results in each vault other than the last vault.

Algorithm 1 Phase 1

Input: Data matrices and filter matrices
Output: One layer of ConvNet with few partial results

1: $d \leftarrow 0$
2: **while** $d < dimension$ **do**
3:    $initializations$
4:    $f_i \leftarrow 0$
5:    **while** $f_i < datalength$ **do**
6:       $f_j \leftarrow 0$
7:       **while** $f_j < (data\_length - filter\_length) + 1$ **do**
8:          **if** $d == 0$ **then**
9:             $B \leftarrow B + bias$
10:          $i \leftarrow f_i$
11:          **if** $(f_i + filter\_breadth) \leq data\_breadth$ **then**
12:             $i\_range \leftarrow f_i + filter\_breadth$
13:          **else**
14:             $i\_range \leftarrow data\_breadth$
15:          **while** $i < i\_range$ **do**
16:             $j \leftarrow f_j$
17:             **while** $j < f_j + filter\_length$ **do**
18:                $Multiply$
19:                $MUX1 \leftarrow 1$
20:                $MUX2 \leftarrow 1$
21:                Add
22:                $j \leftarrow j + 1$
23:             $i \leftarrow i + 1$
24:          Store(B)
25:          $f_j \leftarrow f_j + 1$
26:       $f_i \leftarrow f_i + 1$
27:    $d \leftarrow d + 1$

---

2) Phase II: To achieve data parallelism, we have divided the data among various vaults. Due to this, during last few row-wise shift operations, the filters are not able to cover all rows in a given partition of data. In other words, the filters need to be applied on set of rows located on two different vaults. In phase II, we compute partial results for such rows and store the results locally, which will be used in phase III. In particular,

there will be $dimension*(filterbreadth-1)*datalength$ number of data (indicated as first two light blue rows in the vault 1 of Figure 4) that will be computed in phase II.

Algorithm 2 Phase 2

Input: Data matrices and filter matrices
Output: Convoluted partial Data for sending to the previous vault (using NoC) to make the partial results of phase I, complete

1: $d \leftarrow 0$
2: **while** $d < dimension$ **do**
3:    $f_i \leftarrow 0$
4:    **while** $f_i < filter\_breadth - 1$ **do**
5:       $f_j \leftarrow 0$
6:       **while** $f_j < data\_length - filter\_length + 1$ **do**
7:          $i \leftarrow 0$
8:          **while** $i \leq f_i$ **do**
9:             $j \leftarrow f_j$
10:            **while** $j < f_j + filter\_length$ **do**
11:               $Multiply$
12:               $MUX1 \leftarrow 1$
13:               $MUX2 \leftarrow 2$
14:               Add
15:               $j \leftarrow j + 1$
16:             $i \leftarrow i + 1$
17:          $Store(B)$
18:          $f_j \leftarrow f_j + 1$
19:       $f_i \leftarrow f_i + 1$
20:    $d \leftarrow d + 1$

---

3) Phase III: For the rows having incomplete results due to partitioned data, the final value is computed by addition of the two partial results. One partial value is available on the local vault while the second needs to be fetched from another vault using inter-vault links of LoB. Phase III does this task and keeps the final value at its appropriate position in its local vault. CLU controller first loads the partial value in the register D (as shown in Figure 3) and then it loads the computed value of phase II from the adjacent vault to register C (as shown in Figure 3). These values are then added one by one using the FPU adder and stored in register B. The partial values are replaced in the memory when the results from register B are written back to the corresponding memory locations to make it complete.

Algorithm 3 Phase 3

Input: Partial results of phase I and partial data computed in phase II
Output: One layer of ConvNet with all complete results

1: $Initializations$
2: $i \leftarrow 0$
3: **while** $i < filter\_breadth - 1$ **do**
4:    $j \leftarrow 0$
5:    **while** $j < data\_length - filter\_length + 1$ **do**
6:       $MUX1 \leftarrow 0$
7:       $MUX2 \leftarrow 0$
8:       Add
9:       $Store(B)$
10:       $j \leftarrow j + 1$
11:    $i \leftarrow i + 1$

At the end of phase III, we eventually get the results of one CONV operation made by one filter. Same steps are repeated for individual filters to form the next convolution layers of the entire ConvNet.

TABLE I
SPECIFICATION OF THE SIMULATION SETUP.

Host Processor [GEM5]	
Core	Single core, x86-64, Simple timing CPU (GEM5), 2GHz
L1 i-cache	64 KB
L1 d-cache	64 KB
L2 cache	256 KB
Main Memory	2 GB
Proposed CLU for NDP	
CLU	16 in number, 550MHz and each has 2 FPU units
Cache	1 SRAM cache, 1 KB in capacity
DDR3_1600_x64	
Timing Parameters	tCK = 1.25ns, tRCD = 13.75ns, tCL = 13.75ns, tRP = 13.75ns, tRAS = 35ns, tRRD = 6ns, tWR = 15ns, tCS = 2.5ns
3D Memory Stack (HMC)	
Timing Parameters [17]	tRP =13.75ns, ttCCD=5ns, tRCD=13.75ns, tCL=13.75ns, tWR=15ns, tRAS=27.5ns, tCK = 0.8ns
Energy [17] [16]	3.7 pj/bit for DRAM read, 6.78 pj/bit for, SerDes hop
Power[18]	11.08 W
Logic Die	90nm, dimension 27 X 27 mm^2

IV. EXPERIMENTAL EVALUATION

A. Conventional CPU-based baseline:

To model the execution time of CNN computation in the processor, we used Gem5. It is a cycle accurate simulator which is widely used for full system simulation. We use McPAT 1.3 for modeling energy consumption of the core and CACTI 6.5 for computing energy values of DRAM. The processor model has L1, L2 in the memory hierarchy. Here to perform the convolutions the image and filter data are fetched to the processor from the memory via the cache hierarchy. Subsequently, the results travel through the cache line up to the memory.

B. NDP hardware:

The CLU hardware logic is designed and synthesized using Verilog and Cadence. For synthesizing we use the UMC 90nm technology at the operating frequency of 550MHz and supply voltage of 1V. The required parameters are taken from [9], [19], [18], [20].

C. Results

Specification parameters of the processor based system and proposed NDP+HMC based system are shown in Table I and the input parameters of convolution operations are given in Table II.

Performance improvement: Simulation time for one CONV of our baseline has been calculated by Gem5. We have assumed the data to be in-memory and loaded by the

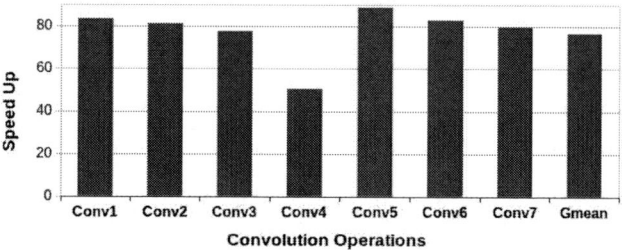

Fig. 5. Speedup in execution time normalized w.r.t. baseline CPU based model.

host processor. Timing for CLU was calculated using Cadence tool-set. We achieved a speedup of 76 times on average compared to the CPU-based baseline model. Table III shows the computation time for baseline and CLU core for various convolutions. From the experimental results, it can be seen that the speedup in CONV4 is little less with respect to the other CONV operations. The speedup depends on the performance of the baseline CPU based system. This performance of CPU based system is somewhat dependent on the locality of references which is data dependent.

Energy Savings: For energy computations of the baseline model, we used the tool McPAT and CACTI. Parameters for McPAT were obtained from Gem5. Energy computations of CLU cores are measured using Genus (cadence tool set). We achieved reduction in energy consumption by around 55 times

TABLE II
CONVOLUTION DETAILS

Convolution	Input Data dimensions	Filter dimensions
CONV1	128x128x3	5x5x3
CONV2	124x124x3	5x5x3
CONV3	120x120x3	4x4x3
CONV4	117x117x4	4x4x4
CONV5	114x114x5	3x3x5
CONV6	112x112x3	5x5x3
CONV7	108x108x3	5x5x3

TABLE III
PERFORMANCE COMPARISON

Convolution	Baseline (in sec)	CLU (in sec)	Speed up (times)
CONV1	1.1490	0.0138	83.4
CONV2	1.0808	0.0133	81.1
CONV3	0.6643	0.0086	77.6
CONV4	0.8320	0.0165	50.4
CONV5	0.8080	0.0091	88.7
CONV6	0.8730	0.0105	82.8
CONV7	0.8090	0.0102	79.7

978-1-5386-3693-0/18 $31.00 © 2018 IEEE

TABLE IV
ENERGY COMPARISON

Convolution	Baseline (in J)	CLU (in J)	Reduction (times)
CONV1	11.0619	0.1848	59.9
CONV2	10.4054	0.1779	58.5
CONV3	6.3960	0.1148	55.7
CONV4	8.0103	0.2057	38.9
CONV5	7.7793	0.1180	65.9
CONV6	8.4049	0.1415	59.4
CONV7	7.7888	0.1354	57.5

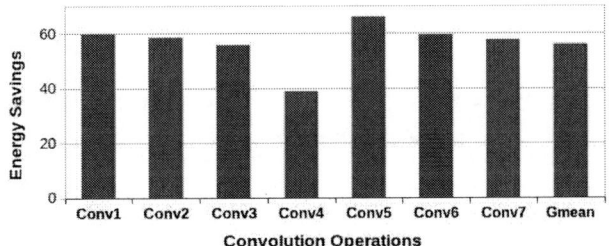

Fig. 6. Reduction in energy consumption normalized w.r.t. baseline CPU based model.

with respect to baseline model. Table IV shows comparison of energy between the baseline and proposed CLU core based system for various convolutions.

We are using NDP and the logic units are placed on the logic die itself which is making the energy of transferring bits, almost negligible as compared to the conventional system where it incurs a big cost in it. The logic unit that we have used is not a full-fledged processor as used in many literatures [21]. Rather it is a dedicated small unit which only does one task and uses only the necessary elements.

Area Analysis: We have used post synthesis area analysis after getting the netlist from Genus and used Innovus for this task. Our implementation of CLU at 90nm has been placed in 0.78mm x 0.78mm square box layout which introduces the area overhead of $0.61mm^2$. Area overhead of all the CLUs is only 1.34% of the logic die area which is negligible.

V. CONCLUSION

CNNs have application in variety of areas and these being data parallel show scope for near data processing. In this paper, we have designed a dedicated hardware logic unit (CLU) and integrated it with HMC's logic layer. The proximity of the CLU to the data in HMC vaults improves performance. We have also proposed to parallelize the computation across vaults and achieved a speedup of 76x over baseline CPU-based system. The energy saving of 55x were obtained by avoiding expensive data transfer to the CPU across memory hierarchy.

We are planning to improve this architecture along several directions: increasing the clock frequency, implementing other layers of CNNs by upgrading the design of CLU, technology scaling, and comparing throughputs with other hardware like GPU, testing throughputs of the proposed system for the common neural networks like AlexNet.

The results from this work show promise of using NDP for data parallel applications like CNNs by use of dedicated hardware instead of complete processor.

REFERENCES

[1] Peter M Kogge. Execube-a new architecture for scaleable mpps. In *Parallel Processing, 1994. Vol. 1. ICPP 1994. International Conference on*, volume 1, pages 77–84. IEEE, 1994.

[2] Yi Kang, Wei Huang, Seung-Moon Yoo, Diana Keen, Zhenzhou Ge, Vinh Lam, Pratap Pattnaik, and Josep Torrellas. Flexram: Toward an advanced intelligent memory system. In *Computer Design (ICCD), 2012 IEEE 30th International Conference on*, pages 5–14. IEEE, 2012.

[3] Junwhan Ahn, Sungpack Hong, Sungjoo Yoo, Onur Mutlu, and Kiyoung Choi. A scalable processing-in-memory accelerator for parallel graph processing. In *Computer Architecture (ISCA), 2015 ACM/IEEE 42nd Annual International Symposium on*, pages 105–117. IEEE, 2015.

[4] Michael Schaffner, Frank K Gürkaynak, Aljoscha Smolic, and Luca Benini. Dram or no-dram? exploring linear solver architectures for image domain warping in 28 nm cmos. In *Design, Automation & Test in Europe Conference & Exhibition (DATE), 2015*, pages 707–712. IEEE, 2015.

[5] Sam Likun Xi, Oreoluwa Babarinsa, Manos Athanassoulis, and Stratos Idreos. Beyond the wall: near-data processing for databases. In *Proceedings of the 11th International Workshop on Data Management on New Hardware*, page 2. ACM, 2015.

[6] Jinho Lee, Jung Ho Ahn, and Kiyoung Choi. Buffered compares: Excavating the hidden parallelism inside dram architectures with lightweight logic. In *Design, Automation & Test in Europe Conference & Exhibition (DATE), 2016*, pages 1243–1248. IEEE, 2016.

[7] Amin Farmahini-Farahani, Jung Ho Ahn, Katherine Morrow, and Nam Sung Kim. Nda: Near-dram acceleration architecture leveraging commodity dram devices and standard memory modules. In *High Performance Computer Architecture (HPCA), 2015 IEEE 21st International Symposium on*, pages 283–295. IEEE, 2015.

[8] Vivek Seshadri, Kevin Hsieh, Amirali Boroum, Donghyuk Lee, Michael A Kozuch, Onur Mutlu, Phillip B Gibbons, and Todd C Mowry. Fast bulk bitwise and and or in dram. *IEEE Computer Architecture Letters*, 14(2):127–131, 2015.

[9] Hybrid Memory Cube Consortium et al. Hybrid memory cube specification 1.0. *Last Revision Jan*, 2013.

[10] JEDEC Standard. High bandwidth memory (hbm) dram. *JESD235*, 2013.

[11] Seth H Pugsley, Jeffrey Jestes, Huihui Zhang, Rajeev Balasubramonian, Vijayalakshmi Srinivasan, Alper Buyuktosunoglu, Al Davis, and Feifei Li. Ndc: Analyzing the impact of 3d-stacked memory+ logic devices on mapreduce workloads. In *Performance Analysis of Systems and Software (ISPASS), 2014 IEEE International Symposium on*, pages 190–200. IEEE, 2014.

[12] Berkin Akin, Franz Franchetti, and James C Hoe. Data reorganization in memory using 3d-stacked dram. In *ACM SIGARCH Computer Architecture News*, volume 43, pages 131–143. ACM, 2015.

[13] Mingyu Gao and Christos Kozyrakis. Hrl: efficient and flexible reconfigurable logic for near-data processing. In *High Performance Computer Architecture (HPCA), 2016 IEEE International Symposium on*, pages 126–137. Ieee, 2016.

[14] Qiuling Zhu, Berkin Akin, H Ekin Sumbul, Fazle Sadi, James C Hoe, Larry Pileggi, and Franz Franchetti. A 3d-stacked logic-in-memory accelerator for application-specific data intensive computing. In *3D Systems Integration Conference (3DIC), 2013 IEEE International*, pages 1–7. IEEE, 2013.

[15] Junwhan Ahn, Sungjoo Yoo, Onur Mutlu, and Kiyoung Choi. Pim-enabled instructions: A low-overhead, locality-aware processing-in-memory architecture. In *Computer Architecture (ISCA), 2015 ACM/IEEE 42nd Annual International Symposium on*, pages 336–348. IEEE, 2015.

[16] Joe Jeddeloh and Brent Keeth. Hybrid memory cube new dram architecture increases density and performance. In *VLSI Technology (VLSIT), 2012 Symposium on*, pages 87–88. IEEE, 2012.

[17] Erfan Azarkhish, Davide Rossi, Igor Loi, and Luca Benini. High performance axi-4.0 based interconnect for extensible smart memory cubes. In *Design, Automation & Test in Europe Conference & Exhibition (DATE), 2015*, pages 1317–1322. IEEE, 2015.

[18] J Thomas Pawlowski. Hybrid memory cube (hmc). In *Hot Chips 23 Symposium (HCS), 2011 IEEE*, pages 1–24. IEEE, 2011.

[19] Dong Uk Lee, Kyung Whan Kim, Kwan Weon Kim, Hongjung Kim, Ju Young Kim, Young Jun Park, Jae Hwan Kim, Dae Suk Kim, Heat Bit Park, Jin Wook Shin, et al. 25.2 a 1.2 v 8gb 8-channel 128gb/s high-bandwidth memory (hbm) stacked dram with effective microbump i/o test methods using 29nm process and tsv. In *Solid-State Circuits Conference Digest of Technical Papers (ISSCC), 2014 IEEE International*, pages 432–433. IEEE, 2014.

[20] Christian Weis, Norbert Wehn, Loi Igor, and Luca Benini. Design space exploration for 3d-stacked drams. In *Design, Automation & Test in Europe Conference & Exhibition (DATE), 2011*, pages 1–6. IEEE, 2011.

[21] Mingyu Gao, Grant Ayers, and Christos Kozyrakis. Practical near-data processing for in-memory analytics frameworks. In *Parallel Architecture and Compilation (PACT), 2015 International Conference on*, pages 113–124. IEEE, 2015.

978-1-5386-3693-0/18 $31.00 © 2018 IEEE

2018 31th International Conference on VLSI Design and 2018 17th International Conference on Embedded Systems

PPU: Privacy-Aware Purchasing Unit for Residential Customers in Smart Electric Grids

Amrita Roy Chowdhury* and Parameswaran Ramanathan*
University of Wisconsin-Madison
amrita@cs.wisc.edu,parmesh@ece.wisc.edu

Abstract— **Retail energy markets where residential customers with storage devices can purchase energy ahead of time are beneficial to both customers and energy suppliers. The customers can maximize their utility by making strategic purchase decisions while suppliers can leverage on the early purchase decisions to develop a better cost-effective plan to serve the demands. A crucial observation is that the flexibility of storing energy in a battery separates the time of purchase and time of use for the stored energy. This can be exploited to address a rising privacy concern amongst the smart grid users which is maintaining the confidentiality of the consumption profile of the customers. In this paper, we propose an embedded unit that implements an energy purchase strategy which integrates ahead of time purchases with the privacy concern of the customers in a cost effective fashion. The embedded unit called PPU supports a quantifiable trade-off between the privacy and the utility enjoyed by the residential customers. Simulations results characterizing the performance of the proposed PPU are included in the paper. We outline an 'ideal' attack and show that PPU can still provide a privacy guarantee of 90%.**

Index Terms— **Real Time Pricing; Energy Storage; Purchase Ahead; Smart Grid; Correlation; Privacy-Aware;**

I. Introduction

Energy markets are changing rapidly and electric grids are undergoing an infrastructural overhaul. With the progressive integration of smart embedded systems capable of advanced metering and bidirectional communication, the roll-out of smart grids has accelerated world wide. This paper is based on the following three premises about the emerging smart electric grid: (i) real-time pricing is an integral part of the residential electricity demand management, (ii) energy storage devices are cheap and readily available to the residential customers, and (iii) energy suppliers offer future prices to the residential customers, allowing them to purchase energy in advance of delivery. Given these aforementioned premises, the paper proposes an embedded energy purchasing unit that implements a one-slot purchase ahead strategy to meet the energy demands of a residential customer while alleviating some of his/her key privacy concerns.

Real-time pricing is a widely-used technique where the energy supplier changes the price of energy for residential customers to manage the demand for its limited energy supply. The rationale is to incentivize the customers to shift their demands to times when the energy supply is abundant or 'off-peak'hours by quoting lower prices for those intervals.

*This work is supported in part by the National Science Foundation grant CNS1329452.

This way the supplier can balance the customers' demand profile and the real-time variation in production capacity.

Energy storage devices (like lithium-ion battery manufactured by Tesla, Samsung etc) are increasingly available for residential customers in many countries including India, USA and China. Such devices offer the customers an additional flexibility; they can purchase energy and store it for later use. Thus, while the energy supplier can indirectly control the time when the customers make purchase decisions by dynamically varying the price, the customers can still consume that energy at their convenience. In other words, energy purchases can now occur at times different from the time when it is consumed. This separation between the time of purchase and time of use can be leveraged upon by the customers to shield their consumption profile from their energy suppliers. Consequently the smart grid customers can now alleviate an increasing privacy concern of revealing sensitive information about their energy consumption. For instance, the energy purchase informations can act as a side-channel divulging personal details about customer behavior like when they are away from home, have they turned on their television etc.

Another characteristic of purchase decisions is that they have strong temporal correlations, i.e., in addition to affecting the immediate benefit, a customer's current decision will also influence his/her demand and benefit in the future. Without future pricing, customers are forced to make suboptimal purchase decisions based on the instantaneous price and demand with no opportunity for accounting for the possibility of lower prices/higher demand in the near future. Currently countries like USA, New Zealand and Singapore have energy future markets. The shortcomings of the myopic optimization approach which does not use future market, can be overcome by a purchase ahead strategy. At the same time, purchases made by the customers ahead of delivery give energy suppliers more time to plan and schedule generation and procurement of energy.

In this paper, we have combined all of the three aforementioned premises to propose a design for a smart embedded unit capable of an energy purchase strategy that maximizes a customer's satisfaction from his/her stored energy management while addressing privacy constraints. We name this smart embedded unit as 'Privacy-Aware Purchasing Unit' and denote it by the abbreviation **PPU** for the rest of the paper.

Some of the recent works on optimal energy purchase include formulating it as a multi-objective optimization

978-1-5386-3693-0/18 $31.00 © 2018 IEEE

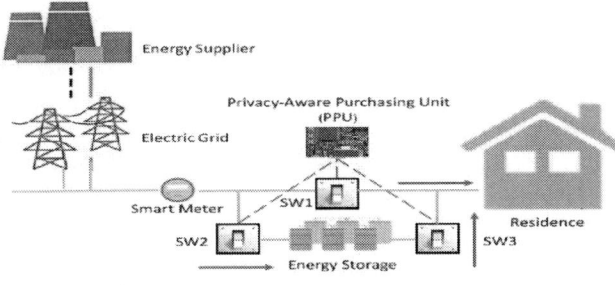

Fig. 1. System Setting

problem [2], a Markov decision process [12] and a linear programming model [11]. Existing literature for ensuring privacy of customers' consumption profile can be broadly divided into two approaches - data aggregation and data anonymization. Data aggregation may occur along three dimensions- space, time or precision. Space based data aggregation [8], [9] incorporate cryptographic protocols using gateways where the consumption data of a neighborhood is aggregated before communicating it to the energy supplier. Examples of data aggregation across time include the works [13], [7], [10] which use re-chargeable batteries to hide the fine-grained power consumption trace by charging and discharging operations. Papers [14], [6] demonstrate aggregation along precision where a random noise is purposely introduced to distort customers' power consumption data. Works using data anonymization include a linkable anonymous credential protocol [4] and a solution based on assigning each smart meter two different identifiers [5]. In a nut shell, all data aggregation based approaches are aimed at masking the fine grained consumption details from the energy supplier either by adding explicit noise or via systematic battery charge/discharge operation or even homomorphic encryption based neighborhood level data summing. This is however detrimental to the operation value of the smart grid system as, in order to fully exploit the advantages of real-time pricing, the energy supplier needs complete knowledge of the customers' purchase decisions. The proposed **PPU** is superior to the previous works in this respect as the supplier has access to the undistorted fine grained information about every individual purchase decision of the customer. The data anonymization based approaches also have their share of problems. For instance, [5] requires third party escrow services for authenticated anonymous smart meter reading. Our **PPU** approach requires no such third party entity.

Unlike the aforementioned works, **PPU** makes optimal ahead of time purchases in tune with the real-time price fluctuations that adequately addresses the privacy concerns of a customer. **PPU** integrates the customer's privacy concerns directly with the purchase decisions it makes and is thus completely orthogonal to existing works. Another salient point is that **PPU** supports a simple customer tunable parameter to trade-off privacy with customer's utility. It is important to note that, in this architecture any customer motivated to uphold his/her privacy can use **PPU** completely

independently of all other entities in the smart grid system, i.e., the deployment of the proposal requires no consensus amongst the different components of the grid.

The rest of the paper is organized as follows. Section II describes the system model and the problem formulation. The proposed design of the purchase ahead strategy for **PPU** is explained in Section III followed by extensive evaluations in Section IV. Finally we conclude the paper in Section V.

II. System Model

A. Notations

The different parameters in the model are :
$U(t)$=*Utility of the customer at time slot t*
$\hat{U}(t)$=*Expected utility of the customer at time slot t*
$B(t)$ = *Battery content at time slot t after delivery of total requested energy*
$C(t)$=*Actual consumption at time slot t*
$\eta(t)$=*Units purchased at time t for delivery at time slot t+1*
$\mu(t)$=*Units purchased at time t for delivery at time slot t*
$P_\mu(t)$=*Unit price at time t for delivery in time slot t*
$P_\eta(t)$=*Unit price at time t for delivery in time slot t+1*
$s(t)$=*State of energy consumption*
$\hat{P}_\mu(t)$=*Expected value of $P_\mu(t)$*
$q_L(s(t))$=*Prob(s(t + 1) =low)*
$q_H(s(t))$=*Prob(s(t + 1) =high)*

B. System Setting

Our work focuses on the residential market of the smart grid which comprises of many electricity customers, each with their own time-varying power demands. The customers who are concerned with their privacy are equipped with an energy storage device, a smart meter for tracking the energy usage and the proposed **PPU**. We consider a discrete time model where the system operates in slotted time. The duration of a slot depends on the time resolution at which the transactions are made (15-minutes, 1-hour, 1-day etc.). In addition to the 'spot price', i.e., unit cost to purchase energy in real-time, the supplier also announces one-slot ahead prices, i.e., unit price of energy to be delivered in the immediate next time slot. Specifically at each time interval, **PPU** makes two purchase decisions namely, $\mu(t)$ at unit price $P_\mu(t)$ for the current slot and $\eta(t)$ at unit price $P_\eta(t)$ to be delivered in the next slot. At time slot t, the supplier delivers the total requested amount for the particular slot, i.e., $\eta(t-1) + \mu(t)$. As shown in Fig. 1, **PPU** implements its purchase decision by controlling three switches: (i) SW1 controls the energy delivered directly from the grid for consumption (ii) SW2 switch controls the energy drawn off the grid for storage (iii) SW3 switch controls the energy drawn from storage to meet customer's demand.

Consumption Model: We assume that **PPU** has full knowledge of the consumption $C(t)$ only for the current time slot, i.e., t. For all future slots $t' > t$, only the stochastics of consumption, $C(t')$ are known. We model a customer's demand stochastics as a finite state Markov-modulated process where its 'state', $s(t)$ captures the temporal fluctuation in the demand and evolve with a transition probability

$\text{Prob}\big(s(t+1)|s(t)\big)$. For simplicity, we assume that there are only two states of consumption, *high* and *low*. Within a given state $s(t)$, the stochastics of energy consumption are assumed to follow an uniform distribution, i.e., $C(t)$ is uniformly distributed between $[0,C_L]$ and $[C_L,C_H]$ for $s(t) = low$ and $s(t) = high$ respectively. Note that only the 'state', $s(t)$ is binary, the actual consumption remains fine grained.

Energy Storage Model: In each time slot if the stored energy is less than the demand then atleast the deficit is bought on the spot, at spot price $P_\mu(t)$. Additionally, the customer makes a purchase ahead decision $\eta(t)$ at $P_\eta(t)$ for delivery in the next time slot. The surplus energy, if any is saved in the battery. In practice, there are physical constraints like batteries have a finite capacity.

$$0 \le B(t) \le B_{max} \qquad (1)$$

Likewise, there may be bounds on the total amount a customer may purchase at a given time.

$$0 \le \eta(t) \le \eta_{max}, 0 \le \mu(t) \le \mu_{max} \qquad (2)$$

Purchase Decision Model: We model the customers objective in making purchase decision at time slot t for time slot $t+1$ through an utility function

$$U(t+1) = \alpha \log \big(1 + B(t+1)\big) - \eta(t)P_\eta(t) \\ - \mu(t+1)P_\mu(t+1) \qquad (3)$$

where α is a weighting coefficient. The first term models the increasing customer satisfaction when more energy is stored in the battery. The second term denotes the cost of purchasing ahead of time and the final term accounts for the extra cost to cover the energy deficit in real-time by making a spot purchase. Such a logarithmic function models a decreasing marginal utility with increasing stored energy and is used often in microeconomics. **PPU**'s buying decision aim at maximizing this utility subject to certain privacy constraints. The spot purchase decision at time slot t, $\mu(t)$ strives to maximize the utility of the current time slot, $U(t)$ while the ahead of time purchase decision, $\eta(t)$ attempts to maximize the expected utility of the next slot, $\hat{U}(t+1)$. The strategies will be discussed in Section III.

Privacy Model: Even if the all communications of the smart meters are protected via standard cryptographic tools to thwart traffic analysis attacks, the exact quantitative information about every purchase has to be conveyed to the energy supplier. This is the least amount of information that the supplier needs to know in order to deliver the desired amount of energy to the customer. Hence the attacker is not necessarily a third party entity snooping on the transaction, in fact the energy supplier itself can sell the data to a third party entity. In this paper, we strive to address the privacy challenge, that given the knowledge of purchase history and the current purchase decisions, an attacker should not be able to estimate a customer's current energy consumption. In most situations, an attacker is satisfied by an estimation the customer's state of consumption, i.e., the exact quantitative value of the energy consumption at a given time might not be

needed. For example, to guesstimate whether a customer is at home, it may suffice to know if the state of the consumption at the concerned time is high or low.

It is important to note that in any buying strategy involving customer's welfare maximization, the optimization problem formulation is bound to be a function of the customer's consumption. Consequently there is a strong dependency between the units of electricity to be purchased, obtained as the solution of this function, and the consumption statistics. For example if the probable consumption of a certain time slot is high, then the energy purchase for the concerned time slot will also be sufficiently high. This statistical relationship between the purchase decisions and the consumption can be characterized mathematically through their correlation. This correlation opens a privacy vulnerability to an attacker who can exploit this predictive relationship to estimate the customer's consumption pattern. We model the privacy constraints via this correlation and **PPU** makes optimal purchase decisions with an objective of reducing this correlation between recent consumption and purchase history.

III. PRIVACY-AWARE PURCHASING UNIT (**PPU**)

Algorithm for PPU at each time slot t

1) **PPU** receives spot price, $P_\mu(t)$ and purchase ahead price, $P_\eta(t)$ from the supplier and exact consumption, $C(t)$ from the customer for the current slot t.
2) **PPU** makes spot purchase decision for the current slot, i.e., $\mu(t)$
3) **PPU** makes purchase ahead decision for energy to be delivered in the next time slot, i.e., $\eta(t)$
4) **PPU** receives total requested $\eta(t-1) + \mu(t)$ units of energy from the supplier.

Hence $B(t)$, i.e., the content of the battery at time slot t after the delivery of the total amount of requested energy bought both on the spot at t and ahead of time at $t-1$, is given by

$$B(t) = B(t-1) - C(t-1) + \eta(t-1) + \mu(t) \qquad (4)$$

1) Spot Purchase Decision: Considering non-deferable load, a customer is forced to make a spot purchase when the amount of energy stored in the battery is inadequate to run the scheduled load. Keeping this in mind, at each time slot **PPU** makes spot purchases $\mu(t)$ in a two step process.

Step 1: Calculate optimum spot purchase
$$U^*(t) = \max_{\mu(t)}\big\{\alpha \log\big(1 + B(t)\big) - \mu(t)P_\mu(t)\big\} \qquad (5)$$
subj to: $0 \le \mu(t) \le \mu_{max}$ * Maximum Purchase constraint *\
Step 2: Minimize correlation with spot purchase
$$\arg\min_{\mu(t)}\{\text{SpotCorr}_k(t)\} \qquad (6)$$
subj to: $U(t) \ge U^*(t)(1-\delta), \delta \in [0,1)$
$\mu(t) \ge \max\{C(t) - B(t-1) + C(t-1) - \eta(t-1), 0\}$
$\mu(t) \le \min\{\mu_{max}, B_{max} - (B(t-1) - C(t-1) + \eta(t-1) - C(t))\}$
where $\text{SpotCorr}_k(t) = \dfrac{\frac{1}{k}\sum_{j=0}^{k-1}\big(\mu(t-j)-\bar{\mu}\big)\big(C(t+i-j)-\bar{C}\big)}{\sigma(\mu).\sigma(C)}$

In **Step 1**, **PPU** makes an optimal spot purchase based on maximizing the actual utility of the current slot $U(t)$ while **Step 2** models our privacy constraint. δ is selected by the customer and denotes the slack allowed on obtained utility w.r.t $U^*(t)$. k denotes the population size for $\mu(t)$ and $C(t)$ including the current time slot. $t+i$ denotes the time slot whose consumption $\mu(t)$ should be decorrelated with. $\bar{\mu}$ and $\sigma(\mu)$ represent the mean and standard deviation of the last k spot purchases respectively. $\bar{C}$ and $\sigma(C)$ are the corresponding values for consumption.

2) Purchase ahead decision for the next slot: Purchase ahead decision at time slot t, $\eta(t)$ is made by maximizing the expected utility of the next time slot $t+1$. The expectation is over the possible values of consumption at time slot $t+1$, $C(t+1)$ because it determines the spot purchase in the next slot $\mu(t+1)$. Also since the spot price of the next slot is unknown, we use its expected value $\hat{P}_\mu(t+1)$. Specifically we decide on $\eta(t)$ in the following two steps

Step 1: Calculate optimal ahead of time purchase
$$\hat{U}^*(t+1) = \max_{\eta(t)} \Big\{ \mathop{\mathbb{E}}_{C(t+1)} \big(\alpha \log(1+B(t+1)) - \eta(t)P_\eta(t) \\ -\mu(t+1)\hat{P}_\mu(t+1) \big) \Big\} \tag{7}$$
subj to: $C(t+1) \le B(t+1) \le B_{max}$
$0 \le \mu(t+1) \le \mu_{max}$
$0 \le \eta_t(t) \le min\Big\{ \eta_{max}, B_{max} - \big(B(t)-C(t)\big) \Big\}$

Step 2: Minimize correlation with ahead of time purchase
$$\arg\min_{\eta(t)} \{ \text{AheadCorr}_k(t) \} \tag{8}$$
subj to: $\hat{U}(t+1) \ge \hat{U}^*(t+1)(1-\delta), \delta \in [0,1)$
$0 \le \eta_t(t) \le min\{ \eta_{max}, B_{max} - (B(t)-C(t)) \}$
where $\text{AheadCorr}_k(t) = \frac{\frac{1}{k}\sum_{j=0}^{k-1}\big(\eta(t-j)-\bar{\eta}\big)\big(C(t+i-j)-\bar{C}\big)}{\sigma(\eta).\sigma(C)}$

$\bar{\eta}$ and $\sigma(\eta)$ denote the mean and variance of the last k ahead of time purchases respectively. Considering a two-state Markov chain model for the consumption as discussed, the optimal solution for **Step 1** turns out to be

$$\eta_t^*(t+1) = max\{0, B^*(t+1)+C(t)-B(t)\} \tag{9}$$

where the optimal target value for battery in the next slot, $B^*(t+1)$ is given by

$$B^*(t+1) = \begin{cases} \frac{-b-\sqrt{b^2-4ac}}{2a} & \text{if } 0 \le B^*(t+1) < C_L \\ \quad a = -q_L\hat{P}_\mu \\ \quad b = C_L(\hat{P}_\mu - P_\eta) - q_L\hat{P}_\mu \\ \quad c = C_L(\alpha+\hat{P}_\mu - P_\eta) \\ \frac{-b-\sqrt{b^2-4ac}}{2a} & \text{if } C_L \le B^*(t+1) < C_H \\ \quad a = -q_H\hat{P}_\mu \\ \quad b = -P_\eta(C_H-C_L)+q_H\hat{P}_\mu(C_H-1) \\ \quad c = (C_H-C_L)(\alpha-P_\eta)+q_H\hat{P}_\mu C_H \\ \frac{\alpha}{P}-1 & \text{if } C_H \le B^*(t+1) \\ \quad \text{where } q_L = q_L(s(t)), q_H = q_H(s(t)), \\ \quad \hat{P}_\mu = \hat{P}_\mu(t+1), P_\eta = P_\eta(t) \end{cases} \tag{10}$$

It is important to note that the slack parameter δ is completely independent of other parameters and captures the customer acceptable privacy/utility trade-off. It is so because higher the value of δ, the greater is the flexibility in obtaining lower correlation (hence better privacy) but at the cost of reduced resulting utility and vice-versa.

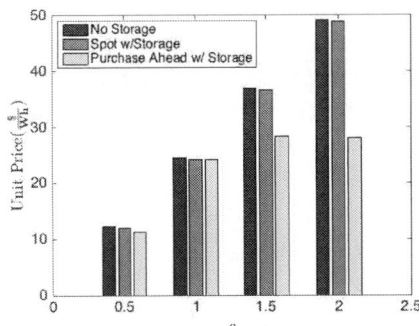

Fig. 2.　Cost-effectiveness of **PPU**

3) Discussion on privacy parameters: The customer is free to choose δ, which denotes the amount of slack to be allowed on the obtained utility as a trade-off for privacy, according to his/her level of satisfaction. For selecting the time slot $t+i$, i.e., the slot whose consumption $\mu(t)$ and $\eta(t)$ should be decorrelated with, we see that current time slot t is an obvious choice. It is so because from the utility function formulation, it is easy to see that any purchase decision is most strongly tied to the current consumption, $C(t)$. Another option can be decorrelation w.r.t to the consumption of the previous slot $C(t-1)$ because the policy being a one-slot purchase ahead one, $C(t-1)$ can affect $\mu(t)$ and $\eta(t)$. However, since **PPU** only has the stochastics for the future demands, a constraint based on expected consumption $C(t+1)$ might not be useful. The population size, k decides the relative weightage of recent and past history, which involves balancing the stability and agility. A large k produces a smooth, stable constraint whereas a small k adapts quickly with recent history. It is important to note that **PPU** is not bound to any specific values of the three parameters and the customer can use multiple combinations of them. For eg, if the customer wants to decorrelate for n different population sizes given by the set $\mathbb{K}$, we can tweak the privacy constraint (8) to

$$\arg\min_{\eta(t)} \Big\{ \arg\max_{k \in \mathbb{K}} \big\{ \text{AheadCorr}_k(t) \big\} \Big\}, \tag{11}$$

Another salient aspect is that **PPU** is in fact completely independent of the privacy constraints. If the customer is not concerned about privacy, he/she can simply set slack parameter $\delta = 0$ and incur no cost penalty due to privacy.

IV. Evaluation

In this section, we provide simulations to illustrate the performance of the proposed energy purchase strategy implemented by **PPU**. The consumption trace was collected from residential customers in Madison, Wisconsin, USA by researchers from WEMPEC (Wisconsin Electric Machines and Power Electronics Consortium) for a full year from September 15, 2015 to September 13, 2016 [3]. In this paper, we use the data for the month of November, 2015 and compute the values for the parameters of the two-state Markov chain namely C_L, C_H and the transition probabilities $q_L(s(t))$ and $q_H(s(t))$ from it. We use the Locational Marginal Price (LMP) from Midcontinent Independent System Operator (MISO) [1] as our purchase ahead price $P_\eta(t)$. We assume $P_\eta(t)$ to also follow a simple two-state Markov chain where $P_\eta(t)$ is uniformly distributed between $[P_\eta^{Lmin}, P_\eta^{Lmax}]$ and $[P_\eta^{Hmin}, P_\eta^{Hmax}]$ for the two states, $s_p(t) = low$ and $s_p(t) = high$ respectively. All the aforementioned parameters are computed from the trace data and are used only for generating the purchase decisions made by **PPU**. The raw trace data itself is used for the actual consumptions of the customers. As for the spot price, $P_\mu(t)$ we compute it by assuming $P_\mu(t) = \rho.P_\eta(t)$. Population size k is chosen to be 5 for all the simulations. We also define a new evaluation parameter where E_t is the expectation over time

$$\text{AvgMaxCorr} = E_t\left(max\{\text{SpotCorr}_k(t), \text{AheadCorr}_k(t)\}\right)$$

For evaluating **PPU**, we present four simulation results.

1) Cost-effectiveness of PPU: Fig. 2 illustrates the cost-effectiveness of **PPU**. It shows the unit price incurred by the customer with a battery of $30Wh$ over a run of 10,000 time slots across varying values of ρ for three different cases. As mentioned above, ρ is the ratio of spot price to ahead price. $\rho > 1$ implies higher spot price hence intuitively the majority of the purchases should be shifted ahead of time and vice versa. The ahead price $P_\eta(t)$ is kept the same for all three cases. The first scenario is that of no storage device which is the baseline case, where in each slot purchases exactly equal to the consumption $C(t)$ are made at the spot rate, $P_\mu(t)$. Next we consider the case of a customer with a storage device who only buys on the spot based on equation (5). Finally we consider the **PPU** purchase ahead strategy. Note that here we showcase the performance of the bare purchase strategies without addressing the privacy concern, i.e, $\delta = 0$. We can make threefold observations from Fig. 2.

1) **PPU** significantly reduces the price incurred, 33% for $\rho = 1.5$ and 50% for $\rho = 2$.
2) The cost for the 'No storage'case and only spot purchase with storage is almost the same. This means, storage device adds negligible cost efficiency on its own and a purchase ahead strategy is needed to fully exploit the flexibility of the battery. It is so because otherwise, the customers can only make sub-optimal decisions based on the current price and demand without accounting for the possibility of lower prices/higher demand in the near future.
3) The cost incurred for the purchase ahead strategy increase only negligibly with increasing value of ρ from 1 to 2. A correctly formulated purchase ahead strategy should be able to make judicious decisions for the ahead time slot such that the spot purchases become negligible. **PPU** does just that, shifting almost all the purchases one slot ahead. Since the ahead price does not change with ρ, we see negligible resulting cost penalty. One additional point to be made is that even when $\rho = 1$, i.e., the ahead and spot prices are identical, it is still desirable to buy ahead of time.

The reason behind this is that, the energy suppliers get adequate time to chalk out a suitable production plan to meet the projected demand and the customers are increasingly satisfied because of the guaranteed service (in the form of agreed energy units) in the future.

It is also to be observed that **PPU** is sensitive to the case when spot price is less than the ahead price ($\rho = 0.5$); it makes spot purchases only and results in no additional overhead due to the pricey time ahead purchases.

2) Variation of system variables with design parameter δ: In Fig. 2 we demonstrate the variation of customer's obtained utility, incurred price and resulting AvgMaxCorr with the slack parameter δ. In other words it shows the privacy/utility (or privacy/cost) trade-off associated with δ. The battery size considered is $30Wh$ and $\rho = 2$. The values of utility and price are normalized to that of the base case which is $\delta = 0$ (no privacy). As expected the utility decreases with increasing δ, it drops by 23% with a slack of 25%. Quite intuitively, price traces out an opposite trajectory registering an 24% increment at a slack of 25%. AvgMaxCorr shows a dramatic drop in its value from 0.9521 to 0.2331 even with a slack of just 5%. From Fig. 2 we conclude that $\delta = 0.1$ may be a good value for the allowed slack as we get AvgMaxCorr=0.22 with just 7% reduction in the obtained utility.

3) Variation in AvgMaxCorr with battery capacity: Intuitively increasing the capacity of the battery should add value to the system, as the system now has a larger window for making the purchase decisions and greater opportunity for minimizing the correlation. This trend is demonstrated in Fig. 3 as we see AvgMaxCorr fall at an decreasing rate saturating when battery is $35Wh$. For instance, a battery of size $30Wh$ and $\delta = 0.1$ can reduce AvgMaxCorr from 0.97 to 0.2. However, increasing battery size has no such observable effect for the no privacy base case.

4) Privacy achieved from PPU: To test the robustness of **PPU** we show its resilience to an ideal attack in Fig 5.
Ideal Attack Model: The goal of the attacker is to estimate the state of consumption. Thus the success of an attacker can be denoted mathematically as $Pr\left(s' = s(t)\big|H_k\right)$ where s' is the estimated state of the current consumption and H_k is the history of last k buying decisions including that of the current time slot. We construct an ideal attacker by training a Support Vector Machine (SVM) with a feature set constructed from actual data of both the ahead and spot purchases and the correct state of consumption. This represents the best possible scenario for the attacker information theoretically as it has unbridled access to all the data. Any other attack constructed can only use a subset of this information. The total dataset size is 10,000 time slots and the training to test size ratio is 80:20. An ideal attack like this is infeasible in the practice because the actual consumption state data used for the training is a private information of the customer and is not available to any attacker. In other words the training data will always be erroneous in practice. Additionally estimating just the state of the consumption instead of the quantitative consumed units and the fact that there are only two states

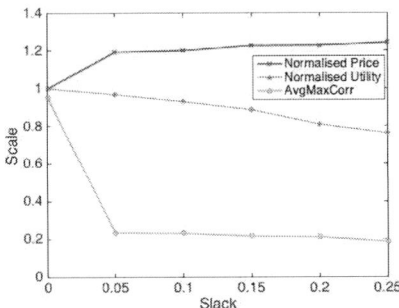

Fig. 3. Utility and Cost trade-off with AvgMaxCorr via slack parameter δ

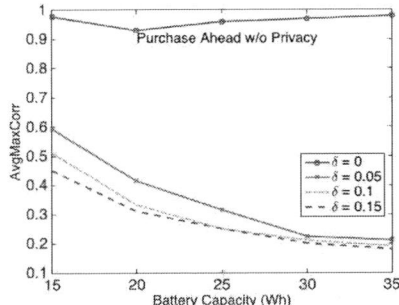

Fig. 4. Variation of AvgMaxCorr with battery size

in the assumed system represents the easiest case from the attacker's perspective. We define our privacy parameter as

$$Privacy = 1 - Pr\left(s' = s(t)\Big|H_k\right) = 2\frac{\#IncorrectResults}{SizeOfTestSet}$$

The factor of 2 comes from the fact that, since the system has only two states, inaccuracy % can be 50% at maximum. In Fig. 5 we analyze both ahead of time purchase and only spot purchase along with their respective privacy aware versions. δ is 0.1 for both the privacy aware cases and $\rho = 2$ in all the four cases. We can see that **PPU** achieves a privacy of 90% with a battery capacity of $30Wh$ even under the ideal attack. Quite expectedly due to the high correlations between the purchases and the respective consumptions, both of the no privacy base cases, i.e., $\delta = 0$ provide negligible privacy. This shows that neither storage capacity nor ahead of time purchase alone can conceal information about the customer's consumption state. An interesting point is that for a battery size of $30Wh$, the privacy-aware spot purchase only strategy achieves a privacy of just 44%, however an extra 46% privacy is obtained if we make privacy-aware purchase ahead decisions as well. This shows the divide between the privacy gained solely due to the storage and the privacy gained from purchases made ahead of time. The latter increments the privacy two-fold due to the increased separation between the time of purchase and time of use.

V. CONCLUSION

We propose a privacy-aware purchasing unit **PPU** that can purchase energy ahead of time for storage device backed

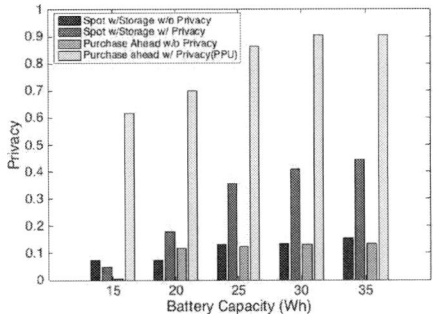

Fig. 5. Evaluation of the privacy metric

customers and provides a privacy of 90% even under an ideal attack. As a future work we propose to extend **PPU** to make multi-slot ahead of time purchases so that it can optimize the purchases better, over multiple time slots. Integrating real time scheduling with purchase is another research prospect.

REFERENCES

[1] *Locational Margin Price*. https://www.misoenergy.org/MarketsOperations/Prices/Pages/Prices.aspx.

[2] A. Anvari-Moghaddam, H. Monsef, and A. Rahimi-Kian, "Optimal smart home energy management considering energy saving and a comfortable lifestyle," *IEEE Transactions on Smart Grid*, vol. 6, no. 1, pp. 324–332, Jan 2015.

[3] A. Brooks, A. Manur, and G. Venkataramanan, "Energy modeling of aggregated community scale residential microgrids," In *Proceeding of International Conference on Sustainable Green Buildings and Communities*, Dec 2016.

[4] F. Diao, F. Zhang, and X. Cheng, "A privacy-preserving smart metering scheme using linkable a anonymous credential," *IEEE Transactions on Smart Grid*, vol. 6, no. 1, pp. 461–467, Jan 2015.

[5] C. Efthymiou and G. Kalogridis, "Smart grid privacy via anonymization of smart metering data," In *Proceedings of International Conference on Smart Grid Communications*, pp. 238–243, Oct 2010.

[6] X. He, X. Zhang, and C. C. J. Kuo, "A distortion-based approach to privacy-preserving metering in smart grids," *IEEE Access*, vol. 1, pp. 67–78, 2013.

[7] G. Kalogridis, C. Efthymiou, S. Z. Denic, T. A. Lewis, and R. Cepeda, "Privacy for smart meters: Towards undetectable appliance load signatures," In *Proceedings of International Conference on Smart Grid Communications*, pp. 232–237, Oct 2010.

[8] F. Li, B. Luo, and P. Liu, "Secure information aggregation for smart grid using homomorphic encryption," In *Proceedings of International Conference on Smart Grid Communication*, pp. 327–332, Oct 2010.

[9] H. Li, X. Liang, R. Lu, X. L. H. Yang, and X. Shen, "Eppdr: An efficient privacy-preserving demand response scheme with adaptive key evolution in smart grid," *IEEE Transactions on Parallel and Distributed Systems*, vol. 25, no. 8, pp. 2053–2064, Aug 2014.

[10] S. McLaughlin, P. McDaniel, and W. Aiello, "Protecting consumer privacy from electric load monitoring," In *Proceedings of Conference on Computer and Communications Security*, pp. 87–98, 2011.

[11] J. Rajasekharan and V. Koivunen, "Optimal energy consumption model for smart grid households with energy storage," *IEEE Journal of Selected Topics in Signal Processing*, vol. 8, no. 6, pp. 1154–1166, Dec 2014.

[12] Y. Zhang and M. van der Schaar, "Structure-aware stochastic storage management in smart grids," *IEEE Journal of Selected Topics in Signal Processing*, vol. 8, no. 6, pp. 1098–1110, Dec 2014.

[13] Z. Zhang, Z. Qin, L. Zhu, J. Weng, and K. Ren, "Cost-friendly differential privacy for smart meters: Exploiting the dual roles of the noise," *IEEE Transactions on Smart Grid*, vol. 8, no. 2, pp. 619–626, 2017.

[14] J. Zhao, T. Jung, Y. Wang, and X. Li, "Achieving differential privacy of data disclosure in the smart grid," In *Proceedings of International Conference on Computer Communications*, pp. 504–512, April 2014.

2018 31th International Conference on VLSI Design and 2018 17th International Conference on Embedded Systems

A High-performance and Area-efficient VLSI Architecture for the PRESENT Lightweight Cipher

Jai Gopal Pandey, *Senior Member*, *IEEE*, Tarun Goel*, Abhijit Karmakar

CSIR - Central Electronics Engineering Research Institute (CEERI)
*Academy of Scientific & Innovative Research (AcSIR), CSIR-CEERI Campus
Pilani-333031, India
jai@ceeri.res.in, tarungoel.com@gmail.com, abhijit@ceeri.res.in

Abstract— **Security and privacy are of prime concern in the emerging internet of things (IoT) and cyber-physical systems (CPS) based applications. Lightweight cryptography plays an essential role in securing the data in this emerging pervasive computing environments. In this paper, we propose a high-performance and area-efficient VLSI architecture with 64-bit datapath for the PRESENT block cipher. The proposed architecture performs an integrated encryption/decryption operation for both 80-bit and 128-bit key lengths. The architecture is synthesized for the Virtex-5 XC5VLX110T FPGA device, available on the Xilinx ML-505 platform. It has been observed that the proposed architecture utilizes 0.73% and 0.87% of FPGA slices for 80-bit and 128-bit key lengths, respectively. A throughput of 410 Mbps and power consumption is about 16 mW for both the key lengths.**

Keywords— *Lightweight cryptography; PRESENT block cipher; Integrated encryption/decryption; VLSI architecture; FPGAs.*

I. INTRODUCTION

The rapidly-growing area of Internet of things (IoT) and cyber-physical systems (CPS) is based on an ecosystem which eventually relies on billions of tiny interconnected computing-devices [1], [2]. The ability of selective computing, sensing, control and communication, makes these ever-ready devices effective, efficient and intelligent. Some of the day-to-day applications around these devices include car-locks, e-cash cards, electronic gadgets, digital lockers, secure communication, and many more. These tiny devices and their network create a wide-spread pervasive-computing infrastructure in emerging applications. Ever-increasing applications of these devices create an extensive demand of smart computing system and their energy-efficient field deployment. Besides design goals, security and privacy are the prime aspects of this IoT-based CPS infrastructure. Here, the field of cryptography and related ciphers provide a mechanism by which data can be efficiently secured. To secure the transmitted data through an electronic system, a variety of ciphers have been used for years. The deployment trend of ciphers in electronic systems is shown in Fig. 1.

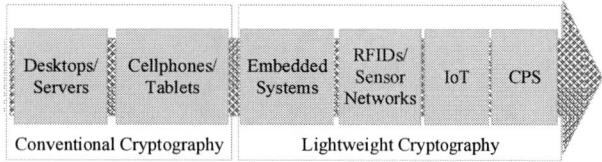

Fig. 1. Deployment trend of ciphers in electronic systems.

This research work has been carried out under the *SMDP-C2SD* project, sponsored by *MeitY, Govt. of India.*

As shown in the Fig. 1, the majority of the conventional cryptographic algorithms have been developed around desktop/server centric environments. Therefore, many of these cryptographic algorithms are generally unsuitable for implementation in constrained devices which are used in the modern-age applications. In many conventional cryptographic standards, the trade-offs between security, performance and resource requirements were optimized for desktop and server environments. This makes the implementation of conventional ciphers difficult in resource-constrained applications, and their performance may not be acceptable. The shift from desktop-based applications to small-device centric applications bring a wide range of security and privacy concerns. Lightweight cryptography provides a solution tailored for resource-constrained devices and their efficient VLSI implementations [3].

Recently, national institute of standards and technology (NIST) provided a report containing an overview of lightweight cryptography and an outline of NIST's plan for standardizing the lightweight cryptographic algorithms [4]. Further, a detailed taxonomy of the lightweight block ciphers can be found in [3] and [5]. Systematic surveys of lightweight-cryptography ciphers and their software and hardware implementations with detailed description and related discussions can be found in [3], [5] and [6]. Here, it has been emphasized that efficient implementation of the ciphers are closely dependent on the selection of appropriate architecture, as they result in low implementation complexity and high-performance in actual realizations. To propose a new architecture for the lightweight cryptography, there is always trade-offs between the three prime objectives i.e. security, cost and performance, which is shown in Fig. 2 [7].

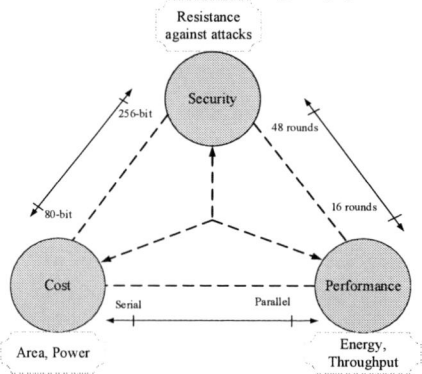

Fig. 2. Architectural trade-offs between security, cost and performance. Adapted from [7].

978-1-5386-3693-0/18 $31.00 © 2018 IEEE

In the context of lightweight cryptography, ISO/IEC 29192-2 has standardized a symmetric block cipher algorithm in the year 2012, which is known as the PRESENT cipher [8]. The algorithm provides an adequate level of data security alongwith hardware-oriented performance attributes, which makes it a prominent choice for developing most of the secure and lightweight applications [5] and [9].

In this paper, we propose a high-performance and area-efficient VLSI architecture for the PRESENT block cipher that completely integrates both encryption and decryption engines. The architecture has been implemented in the Xilinx Virtex-5 XC5VLX110T FPGA device [10]. The experimental results of the implementation show that the proposed architecture consumes a number of 126 slices for the 80-bit key and 150 slices for the 128-bit key lengths. The architecture is capable of running at a clock frequency of more than 210 MHz. The dynamic power dissipation is about 16 mW when the architecture has been operated with around two thousand random input vectors.

Rest of this paper is organized as follows: In Section II, an overview of the PRESENT algorithm is given. Section III is used to discuss the existing implementations of the PRESENT block cipher. An integrated architecture for the PRESENT block cipher alongwith a detailed description of various basic building blocks is proposed in Section IV. Section V is used to provide experimental results alongwith a comparison with an existing architecture from literature. Finally, conclusions are drawn in Section VI.

II. THE PRESENT ALGORITHM

The PRESENT algorithm works on a block size of 64-bit and supports two key length variants of 80-bit and 128-bit. The principle of PRESENT cipher is based on the concept of substitution and permutation network (SPN). There are total 31 rounds and each round consists of an XOR operation, which is required to introduce a round key K_i, for $0 \le i \le 31$, where K_{31} is used for post-whitening operation [9]. A non-linear substitution layer operation is performed in each round and this layer consists of a 4-bit S-box which is applied 16-times in parallel. In addition to that, there is a linear bitwise permutation layer. These operations are described below:

A. The Key Schedule

The cipher requires a unique round key (K_i) in each round, where the input key is stored in a key register $K(k_{79}k_{78}...k_0)$ for 80-bit key or $K(k_{127}k_{126}...k_0)$ for 128-bit key length.

B. Add Roundkey Operation (AddRoundKey)

With the current state $b_{63}...b_0$ and for the given leftmost 64-bit of the round key $K_i = k_{63}^i ... k_0^i$ (or $K_i = k_{127}^i ... k_{64}^i$); $0 \le i \le 31$, the *AddRoundKey* operation is defined as $b_j = b_j \oplus k_j^i$ for $0 \le j \le 63$ [9].

C. The S-box and Inverse S-box

The PRESENT algorithm requires a 4-bit-to-4-bit S-box (S) as $\mathbb{F}_2^4 \to \mathbb{F}_2^4$ [9]. The 64-bit current state $b_{63}...b_0$ is taken

as sixteen 4-bit words $w_{15}...w_0$, where $w_i = b_{4\times i+3} \| b_{4\times i+2} \| b_{4\times i+1} \| b_{4\times i}$ for $0 \le i \le 15$. The output $S[w_i]$ provides the updated state values as per [9]. The S-boxes are used in each of the rounds and in the key scheduling operation. For inverse S-box, the S-box does not satisfy $S(S(x)) = x, x \in \mathbb{F}_2^4$; thus, same S-box cannot be used in encryption and decryption both. For the PRESENT cipher, a relation between the S-box and inverse S-box is given by expression, $S^{-1}(S(x)) = x, x \in \mathbb{F}_2^4$.

D. The Bit Permutation (p-layer) and Inverse Bit Permutation (inv-p-layer) Operations

The bit permutation layer is used to move bit i of the state to bit position $P(i)$ [9] and it is given by the following expression,

$$P(i) = \begin{cases} i.16 \bmod 63, & i \hat{I} \{0,...,62\} \\ 63, & i = 63 \end{cases} \tag{1}$$

Similarly, the inverse bit permutation operation is defined by the below equation,

$$P^{-1} \begin{cases} i.16 \bmod 63, & i \in \{0,...,62\} \\ 63, & i = 63 \end{cases} = i \tag{2}$$

In the following section, some of the work related to the implementations of the PRESENT cipher is provided.

III. RELATED WORK

An architectural design space exploration for encryption and decryption operations can be found in [11]. In this Spartan-III FPGA device based implementation, the encryption and decryption operations require a total of 373 slices and 423 slices for the 80-bit and 128-bit key lengths, respectively [11]. In another implementation, a cryptographic co-processor with encryption and decryption capabilities has been provided by [12]. In this paper the architectural exploration for serial, iterative and parallel variants of the cipher has been provided.

For encryption operation, an FPGA-based implementation of the PRESENT cipher that uses 117 slices of the Xilinx Spartan-3 XC3S50 FPGA device has been reported in [13]. Here, a throughput of 28.46 Mbps at a maximum frequency of 114 MHz has been obtained. Two different RAM-based implementations of the cipher have been provided in [14]. In the first implementation, the substitution box of the cipher has been realized into the FPGA slices. In the second implementation, it has been integrated into the RAM. In these implementations, the first design occupies 83 slices and the second design consumes 85 slices of Xilinx Spartan XC3S50 device. These realizations produce a throughput of 6.03 Kbps and 5.13 Kbps at 100 KHz system clock respectively.

Related to the encryption-only operation, one such implementation of the PRESENT cipher with 8-bit datapath has been given in [15]. This design utilizes 62 slices of the Xilinx Virtex-5 XC5VLX50 device and provides a latency of 295 clock cycles, a throughput of 51.32 Mbps at the maximum frequency of 236.574 MHz. Another implementation using a 64-bit datapath has been reported in [16] that consumes 74 slices of the Xilinx Spartan-6 XC6SLX16-3CSG324C FPGA device. At a maximum clock frequency of 221.63 MHz and with 33 clock latency, a throughput of 221.63 Mbps has been obtained. Similar to this a 64-bit datapath based architecture has been synthesized on 87 slices of the Xilinx Virtex-5 XC5VLX50 FPGA device [17]. Here a latency of 47 clocks, maximum frequency of 221.64 MHz and a throughput of 341.64 Mbps have been reported.

As evident, most of the authors have provided an architecture for encryption-only operation and they have assumed that the decryption operation works opposite to that of the encryption operation and hence would require roughly same logic complexities and hardware resources. However, in our opinion, the decryption operation is a bit complex in comparison to the encryption operation. This is due to the fact that to start the decryption operation, first, we need the last round key that is generated from the key scheduling operation. In connection to this, one of the implementations in which both encryption and decryption operations have been tackled [11], assumes that last round key is available at the starting of the decryption operation. They have considered that the key is static in nature throughout all of the encryption and decryption operations for all the input blocks. In another implementation by [17], for PRESENT cipher context, it has been given that the decryption operation requires almost same area as of encryption when implemented separately. In the following section, an integrated encryption and decryption architecture is proposed and described in detail.

IV. AN INTEGRATED VLSI ARCHITECTURE FOR THE PRESENT LIGHTWEIGHT BLOCK CIPHER

The proposed architecture for an integrated encryption and decryption of PRESENT block cipher is shown in Fig. 3. Here, an iterative type of architecture is considered for saving the resources and computation time. To implement the PRESENT block cipher a 64-bit datapath is chosen, mainly to implement the permutation operation efficiently.

In the proposed architecture there are three main components which are encryption/decryption engine, key scheduling unit and controller. There is a 1-bit input signal 'enc_dec' which is used to select the encryption or decryption operations. If 'enc_dec' is at logic '1' level then encryption operation is performed, else the decryption operation is executed. An up-down counter facilitates the integrated encryption/decryption operation. The main building blocks of the proposed architecture have been arranged in different subsections which are described below.

A. Datapath of the Integrated PRESENT Architecture

As shown in the Fig. 3, the architecture consists of a set of registers, multiplexers and XOR gates. The bit permutation (1) and inverse bit permutation operations (2) are simple bit-transposition, which require only simple wirings. There is one 64-bit multiplexer which is required to switch the state between the load phase (*Input*) and the intermediate state. Next, the multiplexer passes the state to a 64-bit state register. This register is used to store the intermediate state and passes it to the 64-bit XOR gate. This gate performs the XOR operation of intermediate state coming from the state register with 64-bit round key coming from the key scheduling unit. In the architecture, both the S-box (S) and inverse S-box (S^{-1}) are realized by the area-optimized combinational logic implementation. To differentiate between the encryption and decryption operations, two 64-bit multiplexers are deployed in the datapath. In the proposed architecture, the inputs and outputs are registered. The output register is added to synchronize the output with the last round.

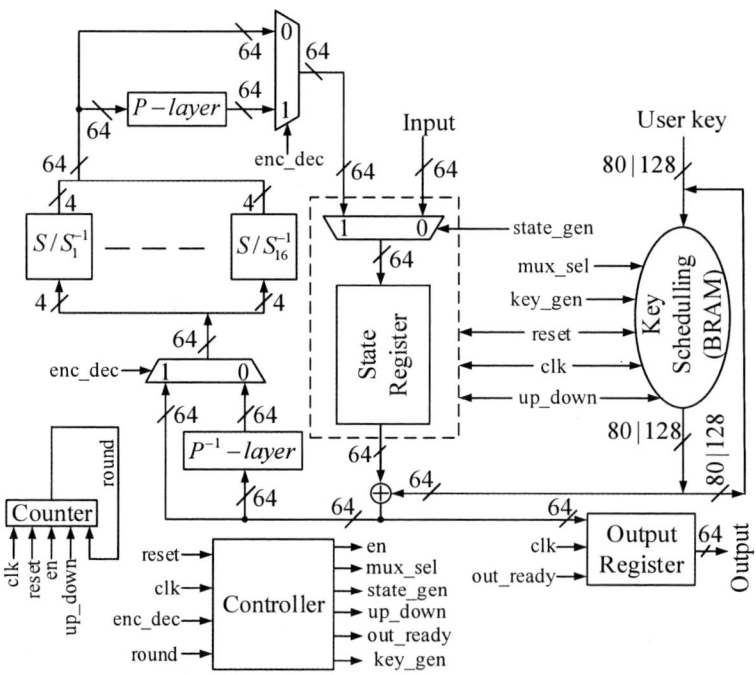

Fig. 3. A VLSI architecture of the PRESENT block cipher for an integrated encryption/decryption operation.

978-1-5386-3693-0/18 $31.00 © 2018 IEEE   394

As per Fig. 3, a total of 33 clock cycles are required for the encryption operation to get the ciphertext. Out of 33 clock cycles, the first cycle is used to load the user key and plaintext. In next 31 clock cycles, 31 intermediate states are computed for each of the rounds and finally in the last clock cycle output is available in the output register. Whereas, in the decryption operation, total 64 clock cycles are required to obtain the first block of decrypted output. Out of these 64 clock cycles, the 31 clock cycles are required to compute the last round key and 33 clock cycles are required to perform the decryption operation. Here, the computed keys have been simultaneously stored in a block RAM (BRAM) so that there is no need to compute the last round key for other blocks of input. Thus, only 33 clock cycles are required to decrypt the remaining blocks of ciphertext. The advantage of using the integrated architecture is that there are some resources which can be used in both encryption and decryption operations. The detailed description of the key scheduling operation is detailed below.

B. Architecture for the Key Scheduling Operation

Key scheduling unit works in the storage mode. Here, computation of the round keys is performed only for the first block of data and the computed round keys are stored simultaneously in the BRAM. This computation mode offers a reduced number of clock cycles for the decryption operation. The key storage mode is also beneficial for processing a large chunk of data which contains multiple blocks that has to be encrypted or decrypted with the same key. A detailed description of the round key computation for both 80-bit and 128-bit key lengths is given in [9]. In Fig. 3, the key scheduling unit is shown in the datapath of the integrated encryption/decryption architecture. A detailed architecture of the same is given in Fig. 4, for both key-lengths of 80-bit and 128-bit.

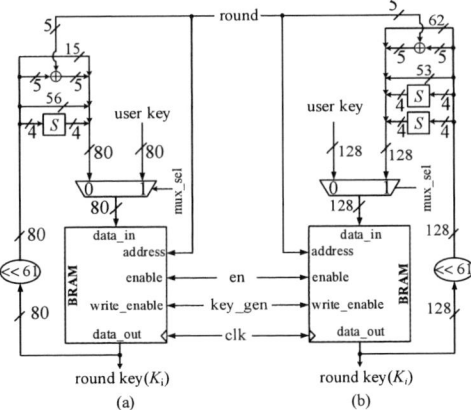

Fig. 4. The key-scheduling process in the PRESENT cipher (a) with the 80-bit input key (b) with the 128-bit input key.

There are three steps in the key scheduling process. First, the intermediate key is left-rotated by 61 bits. Then, the first leftmost 4 bits are passed through one S-box for 80-bit key length and two S-boxes for 128-bit key length, as shown in Fig. 4 (a) and (b), respectively. Finally, in the third step, 5 bits of the intermediate key are XORed with the counter value and after that, the 80-bit (or 128-bit) key is written to the data_in

port of the BRAM. The intermediate key is read out from the data_out port of the BRAM and first leftmost 64 bits of intermediate key i.e., round key is XORed with the intermediate state of that particular round. There are three control signals, 'en', 'key_gen' and 'mux_sel' generated by the controller for monitoring the key scheduling process, which are shown in Fig. 4. Alongwith the BRAM, one 80-bit or 128-bit multiplexer is also used. The 'mux_sel' signal is used to switch between the load phase (user key) and intermediate keys. The 'key_gen' signal is given to the write_enable port of the BRAM and it enables the BRAM for updating its content. The controller that controls the datapath and the key generation unit is described below.

C. Controller for the Encryption/Decryption Operations

The controller for the integrated encryption/decryption operation of the PRESENT cipher is given in Fig. 5. To make the proposed architecture (as in Fig. 3), capable of performing the encryption and decryption operations for the multiple blocks with the same user key, an appropriate controlling mechanism is required. The controller, thus, is designed to provide various control signals for monitoring the key generation unit and for controlling the encryption/decryption engine. There are total nine states in the controller, which are used for performing the encryption and decryption operations. Out of these nine states, two states i.e. Start (S_0) and the Reg_out (S_{OUT}) are common for both the encryption and decryption. In the first state, S_0, the counter is enabled through 'en' signal and works as an up counter by setting the 'up_down' signal to logic '1'. In this state, input and the user keys are loaded and the key scheduling unit starts computation through 'key_gen' signal which is at logic '1'. In the next clock cycle, the state switches to either S_{1E} state or S_{1D} state for enc_dec='1' or '0', respectively.

For controlling the encryption operation given in Fig. 5, in the state S_{1E}, multiplexers are switched as 'mux_sel' signal for the key scheduling unit gets logic '1' and the signal 'state_gen' for the datapath gets high so the encryption of the first block of plaintext begins in this state. Simultaneously, all the generated round keys are stored in the BRAM. The state remains in S_{1E}, until the counter value reaches 31. Then, the state switches to S_{OUT} state, where the counter is disabled through en='0' and 'out_ready' signal becomes logic '1'. In this state, the key generation unit finishes its computation operation and it is only used for fetching the stored round keys. So, here, the 'key_gen' signal is kept at logic '0'. In the next clock cycle, the first block of ciphertext is available through the output register.

Further, for the encryption of other blocks of plaintexts, the state switches to the S_{2E} state. In this state, other blocks of plaintext are loaded to the state register. This state is very similar to the initial state S_0, except that now there is no need to compute the round keys as they have already been stored in the BRAM. So, here, 'key_gen' signal is at logic '0'. After that, the state switches to the S_{3E} state. The state S_{3E} is similar to the state S_{1E} except here the 'key_gen' signal changes to logic '0'. When the counter value reaches at 31, the state switches to the output state S_{OUT}. This cycle continues till all the blocks of plaintexts are encrypted.

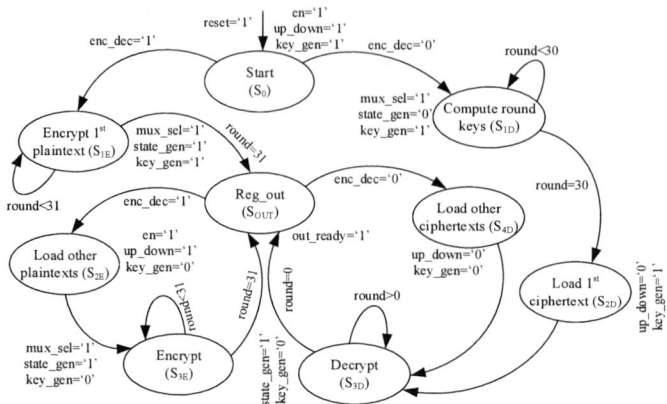

Fig. 5. Controller for the integrated encryption and decryption operations.

In the processing of the decryption operation, which is shown in Fig. 5, the state switches at S_{1D} from the initial state S_0. In the state S_{1D}, all the round keys are computed and stored in the BRAM. In this state, the decryption process does not start as the '*state_gen*' signal is at logic '0'. When the counter value reaches to 30, the state is switched to S_{2D} state. Here, the first block of ciphertext is loaded to the state register and now the counter starts to work as down counter by switching the '*up_down*' signal to logic '0'. Further, the state is switched to S_{3D} state, where the decryption process starts by assigning the '*state_gen*' signal to logic '1'. Further, the key generation unit completes its computation operation with the '*key_gen*' signal becoming logic '0'. As counter value becomes 0, the state switches to the final state S_{OUT} and the first block of decrypted ciphertext is available through the output register as '*out_ready*' signal becomes logic '1'. Next, to decrypt the remaining blocks of ciphertext the state does not go back to the initial state because pre-computed round keys are available in the BRAM. So, the state moves to S_{4D} state, which is very similar to the state S_{2D} except that '*key_gen*' signal is at logic '0'. In this state, other blocks of ciphertext are loaded into the state register. To perform the decryption operation for the rest blocks, the state switches to the S_{3D} state. This cycle continues until all the blocks of ciphertext are decrypted.

V. EXPERIMENTAL RESULTS FOR AN FPGA DEVICE

The proposed architecture is implemented in the VHDL language and synthesized using Xilinx Design Suite 14.7 for the Xilinx Virtex-5 XC5VLX110T-1-FF1136 FPGA device on Xilinx ML-505 platform. The synthesis process for the implementation has been configured with design goal as *balanced* and strategy as *Xilinx default*. The FPGA device utilization summary of the proposed architecture is given in Table I.

As per the table, the architecture with an 80-bit key (*PRE_80*) is consuming only 0.73% slices while the proposed architecture with a 128-bit key (*PRE_128)* is consuming 0.87% slices. The *PRE_80* architecture consumes 32×80 bit size block memory, whereas, the *PRE_128* architecture needs 32×128 bit size of memory to store the intermediate keys.

TABLE I. DEVICE UTILIZATION SUMMARY ON THE XILINX VIRTEX-5 XC5VLX110T FPGA DEVICE.

Elements	Available Resources	Resource Utilization	Resource Utilization
		PRE_80	*PRE_128*
LUTs	69120	348	396
Registers	69120	137	137
Total Slices	17280	126	150

The performance of the design is evaluated in terms of power dissipation, latency, maximum frequency and throughput. The performance of the proposed architecture is given in Table II.

TABLE II. PERFORMANCE ON THE XILINX VIRTEX-5 XC5VLX110T FPGA DEVICE.

Elements	Resource Utilization	Resource Utilization
	PRE_80	*PRE_128*
Latency	33	33
Max. frequency (MHz)	215.42	212.13
Throughput (Mbps)	417.79	411.41
Efficiency (Mbps/#Slices)	3.32	2.74
Power (mW)	16.59	16.80

As given in Table II, the architecture consumes around 16 mW power at 215 MHz frequency. Both the operations have a latency of 33 clock cycles. However, the decryption operation requires an additional 33 cycles for round key generation for the first block of ciphertext. Throughput of the design is around 410 Mbps for both the key lengths, that is computed for 64-bit datapath by expression, $throughput = (max.freq. \times total\ no.\ of\ bits)/latency$.

To compare the proposed design with an existing design available in the literature, the selected design metrics are: slice LUTs, registers and a total number of consumed slices. To perform a comparison at the architectural-level, the proposed integrated architecture is tuned to match the architectural capability of [11]. Therefore, for comparison, the key scheduling unit is implemented using *on-the-fly* mode rather

978-1-5386-3693-0/18 $31.00 © 2018 IEEE

than storing the computed keys in the BRAM. An architectural-level comparison between the proposed design and the design of [11] is given below.

A. Architectural-level Comparison

The architecture presented in [11] is one of a few established ones that provides decryption operation for the FPGA. This architecture has been implementation on the Xilinx Spartan-IIIXC3S400 FPGA device. Thus, to perform a fair comparison of utilized device resources, we have targeted the same FPGA device and equal speed grade. Similar to [11], we also kept the synthesis tool project design goals and strategies with synthesis and place and route (PnR) effort properties as *high* and the PnR extra effort at *continue on impossible*. The implementation has been performed for both 80-bit key length (*PRE_80*) and 128-bit key length (*PRE_128*). The synthesis results for both the architectures are compared and shown in Fig. 6.

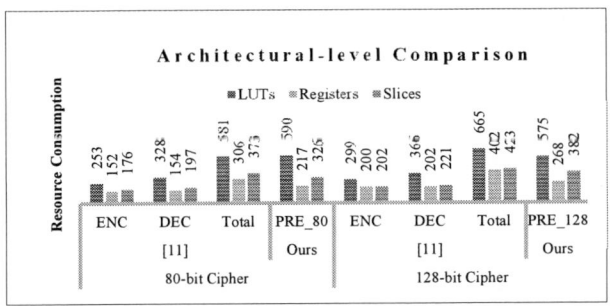

Fig. 6. Architectural-level comparison between the architecture of [11] and the proposed architecture using Xilinx Spartan-III XC3S400 FPGA.

All the data presented in Fig. 6, are from the post place and route (*PnR*) report. It can be observed from the above figure that, in comparison to architecture [11], the proposed architecture with 80-bit key length (*PRE_80*) requires 12.6% lower FPGA slices and with 128-bit key length (*PRE_128*) consumes 9.7% lesser slices. By this, we can say that the proposed integrated architecture is capable of performing both the encryption (ENC) and decryption (DEC) by the same set of hardware, which is an essential requirement in any practical lightweight cipher-based system. Also, the integrated architecture consumes lesser slices in comparison to two separate modules for performing encryption and decryption. It can be noted that our design requires an extra clock cycle in comparison with [11] to perform the operations as we have considered the registered output.

VI. CONCLUSION

An integrated VLSI architecture for PRESENT lightweight block cipher is presented. The architecture supports both the encryption and decryption operations with 80-bit and 128-bit key lengths. The design is modeled in the VHDL language and synthesized in Xilinx Virtex-5 XC5VLX110T-1-FF1136 FPGA device on ML-505 platform. The architecture utilizes 0.73% and 0.87% of FPGA slices for 80-bit and 128-bit key length, respectively. The throughput of the design is around 410 Mbps and power consumption is around 16 mW for both the key lengths. The proposed architecture is area-efficient with high-performance capability for providing an adequate level of security under the resource-constrained environment for IoT and CPS applications.

REFERENCES

[1] E. A. Lee and S. A. Seshia, Introduction to Embedded Systems, A Cyber-Physical Systems Approach, 2011.

[2] T. Xu, J. B. Wendt and M. Potkonjak, "Security of IoT systems: Design challenges and opportunities," in IEEE/ACM Int'l Conf. on Comp.-Aided Design, San Jose, Califo., pp. 417-423, 03 Nov. 2014.

[3] A. Biryukov and L. Perrin, "Lightweight Block Ciphers," [Online]: https://www.cryptolux.org/index.php/Lightweight_Block_Ciphers. [Accessed 06 Jan. 2017].

[4] K. McKay, L. E. Bassham, M. S. Turan and N. W. Mouha, "NISTIR 8114 - Report on Lightweight Cryptography," National Institute of Standards and Technology (NIST), Gaithersburg, March 2017.

[5] B. J. Mohd, T. Hayajneh and A. V. Vasilakos, "A survey on lightweight block ciphers for low-resource devices: Comparative study and open issues," Jour. of Network and Computer Appl., vol. 58, pp. 73-93, 2015.

[6] T. Eisenbarth, S. Kumar, C. Paar, A. Poschmann and L. Uhsadel, "A survey of lightweight-cryptography implementations," IEEE Design & Test of Computers, vol. 24, no. 6, pp. 522-533, 2007.

[7] A. Y. Poschmann, "Lightweight cryptography: cryptographic engineering for a pervasive world," Ph. D. Thesis, Ruhr-University, Bochum, Germany, 2009.

[8] "Information technology – Security techniques – Part 2: Block ciphers," Jan. 2012.

[9] A. Bogdanov, L. R. Knudsen, G. Leander, C. Paar, A. Poschmann, M. J. B. Robshaw, Y. Seurin and C. Vikkelsoe, "PRESENT: An ultra-lightweight block cipher," in Int'l Workshop on Cryptographic Hardware and Embedded Systems, Vienna, Austria, Springer, pp. 450-466, 2007.

[10] Xilinx Inc., "Virtex-5 FPGA user guide UG190 (v5.4)," www.xilinx.com, 2012.

[11] M. Sbeiti, S. Michael, A. Poschmann and C. Paar, "Design space exploration of present implementations for FPGAS," in 5th Sout. Conf. on Prog. Logic, Sao Carlos, Brazil, pp. 141-145, 1-3 April 2009.

[12] C. Rolfes, A. Poschmann, G. Leander and C. Paar, "Ultra-lightweight implementations for smart devices–security for 1000 gate equivalents," in Int'l Conf. on Smart Card Research and Advanced Applications, London, UK, pp. 89-103, 08-11 Sep. 2008.

[13] P. Yalla and J. P. Kaps, "Lightweight cryptography for FPGAs," in IEEE Int'l Conf. on Reconfigurable Computing and FPGAs (ReConFig'09), Cancun, Mexico, pp. 225-230, 09 Dec. 2009.

[14] E. B. Kavun and T. Yalcin, "RAM-based ultra-lightweight FPGA implementation of PRESENT," in Int'l Conf. on Reconf. Computing and FPGAs (ReConFig'11), Cancum, Mexico, pp. 280-285, 30 Nov-02 Dec 2011.

[15] J. J. Tay, M. L. D. Wong, M. M. Wong, C. Zhang and I. Hijazin, "Compact FPGA implementation of PRESENT with Boolean S-Box," in 6th Asia Symp. on Quality Elect. Design (ASQED), Kula Lumpur, Malaysia, pp. 144-148, 04 Aug. 2015.

[16] C. A. Lara-Nino, M. Morales-Sandoval and A. Diaz-Perez, "Novel FPGA-based low-cost hardware architecture for the PRESENT block cipher," in 2016 Euromicro Conf. Digital System Design (DSD), Limassol, Cyprus, pp. 646-650, 31 Aug. 2016.

[17] N. Hanley and M. O'Neill, "Hardware Comparison of the ISO/IEC 29192-2 Block Ciphers," in *I*EEE Computer Society Annual Symposium on VLSI (ISVLSI), Amherst, MA, USA, pp. 57-62, 19-21 Aug. 2012.

2018 31th International Conference on VLSI Design and 2018 17th International Conference on Embedded Systems

Security Vulnerabilities of Unmanned Aerial Vehicles and Countermeasures: An Experimental Study

Vishal Dey*, Vikramkumar Pudi[†], Anupam Chattopadhyay[†] and Yuval Elovici[‡]

* Department of Computer Science and Technology, Indian Institute of Engineering Science and Technology, Shibpur, India
[†] School of Computer Science and Engineering, Nanyang Technological University, Singapore
[‡] Telecom Innovation Laboratories, Ben-Gurion University of the Negev, Israel

Abstract—Drones present a novel airborne platform for new commercial and consumer tasks. Due to their increasing use, security analysis has become necessary. In this paper, we present a collection of fatal attacks, minor observations and associated security vulnerabilities for DJI Phantom 4 Pro(P4P) and Parrot Bebop 2 drones. We also propose some countermeasures to bolster the security of the analyzed drones. In addition to this, we provided a brief overview of the system and embedded architecture of these vehicles. Although DJI has tried to improve security by introducing radio communication and partially encrypted firmware, the P4P still remains vulnerable to GPS spoofing, jamming and other vulnerabilities through DJI-SDK has been exposed in our experimental study. In contrast, Bebop drones are more vulnerable to wireless attacks and we have experimentally validated this for Bebop 2 drone.

I. INTRODUCTION

The demand for unmanned aerial vehicle (also known as drone) technology is rapidly growing [1]. Today, drones are increasingly used not only in military applications, but also for civilian tasks [2], such as delivery services, traffic surveillance, agriculture and farming, aerial surveys, and a variety of tasks that are too dangerous or remote for humans [3]–[5]. It won't be long before drones can also be used in disaster relief, expanding Internet access, and more [4]. Recently, The German logistics company, DHL, and Amazon announced the launching of a new drone-based delivery service. An Australian textbook rental company, Zookal, has already started using drones to deliver books. Drones have become part of Internet-of-Things (IoT) as well, given the increased use of drones in commercial areas like, agricultural and surveillance, in which vast amounts of visual data are collected from the air and passed to the cloud for analysis using the Internet.

In future, drones ought to replace connected sensors at rest in the IoT, as drones are 1) deployable in different locations, 2) capable of carrying flexible payloads, and 3) reprogrammable in mission. Recently Broadcom introduced the WICED Sense Development Kit, an all-in-one IoT prototyping kit for drones. The Erle Robotics company offers Erle-Brain operating systems capable of connecting drones with IoT. In a few years, Internet connected drones will be widely used in commercial and civil tasks. Because IoT drones are directly accessible over the Internet, require GPS signals and communicates wirelessly, they cause a serious threat to the security of

the drone platform. For example, a drone-hijacking program called SkyJack was recently engineered to autonomously seek out, hack, and wirelessly take over other drones within Wi-Fi distance and turn them into zombie drones under the control of attackers. Thus, it is critical to address the security of drones connected to the Internet.

A. Motivation

Future smart cities will be swarming with drones for different commercial and personal purposes in the hope that it will improve the lives of humans [6]. Drones are harmless when used properly and can be extremely useful, for example, for taking beautiful aerial photographs and cinematography. Although when misappropriated, drones can lead directly to invasions of privacy, concerns with aircraft safety, and even result in personal injury.

An armed drone used for military purposes can lead to catastrophic results if it gets hacked and lands in the hands of terrorists. Small drones like those used by civilians are even easier to attack; these drones use a wireless transmission protocol for navigation (radio waves or WiFi) and telemetry, and the navigation signals associated with this are easy to manipulate in order to hack the drone. Attackers may attack drones for a variety of reasons: to obtain objectionable media, valuable personal information or data. Companies like Amazon and, Dominos have planned to deliver items using drones, which, if hijacked will not reach the target customer. Given increased popularity and numbers, drones pose an immediate security concern, thus making a security analysis mandatory.

B. Security Issues in Unmanned Aerial Vehicles (UAVs)

We survey the vulnerabilities found in Phantom and Bebop drones, including drone hacks based on unprotected WiFi, access to drone configuration files and, changes to settings in flight, and GPS attacks.
WiFi insecurity: The Parrot drones allow multiple connections through WiFi, which allows an attacker to tamper with the system and hack the drone.
SkyJack: The SkyJack [7] developed by Samy Kamkar uses a Raspberry Pi and aircrack-ng [8] mounted on a WiFi-based drone, e.g. Phantom 2 Vision or Parrot drones, enabling it to fly, around and get into the network of nearby drones. The

978-1-5386-3693-0/18 $31.00 © 2018 IEEE

398

SkyJack disconnects the controller by changing the SSID and, immediately reconnects and is then able to transmit commands through the malicious drone to the hijacked drone, enabling it to take control of other drone in flight using a customized script, hereby forming zombie drones. The DJI Phantom 4 Pro is not vulnerable to this attack whereas the Bebop2 drone is.

GPS based attacks: Drones uses GPS signals for navigation. The GPS signals used by civilian drones are not encrypted. Thus, drones are vulnerable to the GPS attacks. The necessary conditions to be met for successful spoofing and different capture and post-capture control techniques are demonstrated in [9].

Maldrone: Maldrone [10] sets up a proxy serial port, intercepts flight commands from the controller, and redirects actual serial port communication to fake ports, while forwarding the hijackers commands to the drone. While the Bebop 2 will be affected, this attack is not applicable for Phantom 4 Pro.

In our work, we performed some known experiments like GPS Spoofing, wireless attacks and also exposed new vulnerabilities associated with DJI P4P and Parrot Bebop 2 drones like cracking DJI-SDK, reverse-engineering firmware. Most of the published wireless attacks desribed above are reproduced for Parrot Bebop 2 which proved succesful and the attacks for this drone as presented in Section III are derived from previous attacks. The works in [11], several hackers in [12] in DEFCON mainly demonstrated attacks on the DJI Phantom 3 Standard drone. We found that most of the attacks demonstrated are not applicable to the DJI Phantom 4 Pro. Based on our experiments, we found that GPS spoofing is the most prominent attack against the DJI P4P, but it does not affect the Bebop 2 drone.

The rest of paper is organized as follows. Section-II provides a detailed analysis of the architecture, internal operations and governing system of the P4P and Bebop 2 drones. In Section III we, describe the experiments conducted on Phantom 4 and Bebop 2 drones to identiy security vulnerabilities: reverse engineering drone firmware, removing authentication from DJI SDK, a GPS spoofing attack on P4P, and wireless attacks. A comparison of the two drones in terms of user-friendliness and security issues is presented in Section IV, proposed countermeasures against the security threats are presented in Section V. In section VI, we conclude and discuss future work.

II. DJI PHANTHOM 4 AND PARROT BEBOP 2 DRONES ARCHITECTURE AND OPERATION OVERVIEW

In this section, we provide a detailed description of the internal architectures, internal operations, and operating system of DJI Phantom 4 Pro and Parrot Bebop 2 drones.

A. DJI Phantom 4 Pro

A brief overview of the DJI Phantom 4 Pro drone system and its components follows.

The DJI Phantom 4 Pro drone consists of a remote controller, mobile device and drone components as shown in Fig. 1. The mobile device uses the DJI GO 4 app which acts as a user-friendly interface to control the drone and displays the

(a) Controller with phone (b) Phantom 4 Pro

Fig. 1: DJI Phantom 4 Pro with controller

live video feeds. It has many advanced features such as RTH (return-to-home), Waypoint mission and other intelligent flight modes including TapFly (to fly in the designated position without using the remote controller), ActivTrack (to mark/track a moving object on the mobile screen, e.g., tracking a car in a race), and Draw (for Waypoint missions).

The DJI Phantom 4 Pro operates using the Rockchip ARM-processor and Unix-based operating system, BusyBox. It comes with a powerful remote controller that has Lightbridge technology. Lightbridge is a one way stream of data without two way handshaking, minimizing video latency while maximizing flight distance. As a result, latencies with Lightbridge are more consistent at a distance versus standard WiFi. Lightbride is much faster at re-establishing a connection when it reaches its distance limit. The GL300 main board is the most important hardware component, supporting the processor, Ambarella, CPU, FPGA Altera Cyclone V, microSD slots, radio transceiver modules, etc.

The OFDM (orthogonal frequency division multiplexing) Receiver module is responsible for pairing with RC and communicating with it. The OFDM receiver implements the Lightbridge technology, and is also a middleman between the gimbal and flight controller. The control signals and video is transmitted over 2.4 GHz using digital transmission. Ambarella makes a video stream to the davinchi module via HDMI (or USB) and at the same time writes a video stream to microSD (using video compression H.264 or H.265). Then DaVinchi sends video to the Encryption module, after that encrypted stream goes to RF transceiver (AD9364).

B. Parrot Bebop 2

The Parrot Bebop 2, an easy-to-fly user-friendly drone, is shown in Fig. 2. The Bebop 2 drone setup requires a drone and mobile device. The mobile device uses "FreeFlight Pro," an official app, for control and communication of the drone. The mobile device binds to the network and, establishes a connection with the drone via Wi-Fi. It sends drone commands and receives drone data and other sensor readings and sends video packets over UDP. The drone acts as an 802.11 Wi-Fi hotspot for 2.400 GHz - 2.483 GHz and establishes a Bebop network once powered on. Both the controller and the drone maintain their open access points (APs) active.

Bebop 2 uses a P7 dual core CPU, a quad-core GPU, and 8 GB of memory, and the operating system is a striped down version of Unix: BusyBox. The Bebop memory is accessed

978-1-5386-3693-0/18 $31.00 © 2018 IEEE

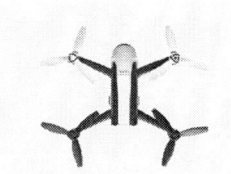

Fig. 2: Parrot Bebop 2 drone

through its WiFi connection and the use of communication protocols such as Telnet and FTP. The onboard flight controller controls the drone by combining input from several sensors, including a *GPS chip* (Furuno GN-87F GPS chip or u-blox Neo 8M), a *vertical camera* (MT9V117 camera) used for adjusting of the Bebop drone's orientation and XY position, and other sensors (e.g. barometer, accelerometer, and magnetometer etc.) for assistance.

III. ATTACKS PERFORMED

A. DJI Phantom 4 Pro

1) Cracking DJI SDK: DJI provides mobile and on-board SDK which are available for download from [13]. This attack is based on the removal of authentication between the DJI app and the DJI server. DJI offers an SDK to build customized Android and iOS apps. This enables to modification of the SDK by decompiling and patching portions of code. The first use of DJI Go or any customized app using DJI-SDK, requires one time registration and activation.

We decompiled the library jar file dji-sdk using [14] and edited the code using [15]. We found that the following functions largely handle this authentication: registerApp, isSD-KActivated, shouldAllowAccessSDK and hasSDKRegistered. startConnectionToProduct(). These functions calls shouldAl-lowAccessSDK(), checks whether the user will be given access to SDK. The hasSDKRegistered() function registered with a valid AppID(key) against a registered user, which is obtained after registration of an app. This key is pasted as an AppID in the main Manifest.xml file of the customized app to activate DJI-SDK.

We patched two functions(shouldAllowAccessSDK() and hasSDKRegistered()) to return true always without any further permission checking, thus effectively removing authentication. Therefore even a wrong key or null string used as the AppID would not result in an error that would prevent the app from opening. This authentication is crucial in preventing malicious users from taking control of a drone away from an authentic user. We built custom apps for controlling the camera and, taking photos and, videos, and a playback manager, using the cracked SDK. The custom apps were successfully authenticated and, worked perfectly fine.

2) Reverse Engineering Firmware: The binary of the DJI Phantom 4 Pro firmware is encrypted and digitally signed. We unpacked firmware using RKflash [16] : a flashing tool for Rockchip based systems, and obtained standard images. The image *embedded-update.img* was further extracted into standard image modules: *boot.img, recovery.img, system.img.* We further unpacked into standard Android partitions using different Android image flash tools. Although we were able to unpack the binary, it was not possible to reflash DJI Phantom 4 Pro with modified binary. This is likely because the binary is not completely decrypted or unpacked by standard tools.

3) GPS Spoofing: DJI Phantom drones are GPS based drones. GPS enables a drones navigation, and due to the lack of encryption of the civilian GPS signals, they can be easily spoofed. The basic idea in GPS spoofing is the transmission of fake GPS coordinates to the flight controller of the drone. This will enable an attacker to hijack the drone and place it in the complete control of the attacker. To spoof a GPS receiver, a transmitter is used to transmit false GPS signals, forcing the drone GPS to synchronize with the attackers signals, including spoofed information regarding the ephemeris and almanac data. If the GPS receiver switches from legitimate GPS signals to false GPS signals, causing the drone to veer off its path and sending the drone in a direction specified by the attacker.

Though DJI improved the navigation by utilizing inertial measurement unit (IMU) in its drones, instead of relying only on GPS, we still managed to spoof the GPS receiver using the LabSat3 GPS simulator [17]. DJI Phantom 4 Pro comes with the u-blox NEO 8M GPS receiver which has the ability to detect jamming with the help of J/N monitoring. J/N monitoring triggers an alarm if the received power is more than the triggering threshold, preventing a jam-and-then-spoof kind of attack. We performed the spoofing attack inside a screened environment (shielded room) in the presence of weak real GPS signals. It was difficult to conduct the experiment outdoors due to the presence of strong legitimate GPS signals.

Experimental Procedure: First, real GPS signals are recorded using the active antenna RLACS198 provided with the LabSat kit. We can also generate fake GPS signals using SatGen software provided with LabSat. The location of the drone was successfully spoofed by LabSat. The drone was placed inside the HESL lab at Nanyang Technological University (NTU). Although the DJI Phantom 4 Pro has a u-blox GPS module with J/N monitoring, it failed to detect spoofing in our case, because the power of transmitted signal was very low. The location, altitude, and speed are provided by the DJI Go 4 app. In our experiments several kinds of hacks were performed: force landing a drone spoofing the location so as to misdirect the drone to NFZ, flying a drone in a NFZ, and a height hack. In figure 3b, the drone is seen to be located at NTU, the spoofed location shows the drone to be in Changi South Avenue near SUTD in figure 3c, which is a no-fly-zone. Once spoofed, the drone could perform an emergency landing receiving NFZ coordinates from fake GPS signals. The movement of the drone could be faked by replaying a pre-recorded GPS data using LabSat as in figure 3d.

B. Parrot Bebop 2

Most of the attacks on Parrot Bebop 2 drone conducted in our experiments are targeted at the open WiFi. Parrot manufacturers introduced the feature of adding the WPA/WPA2

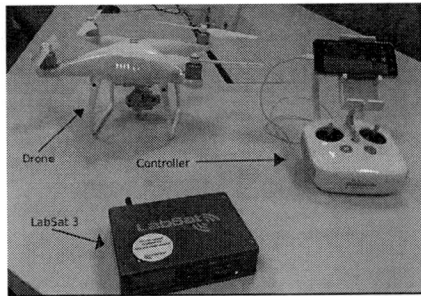

(a) GPS Spoofing Setup for DJI Phantom Drone

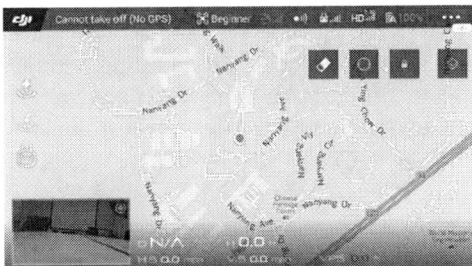

(b) Original Location of Drone (NTU, Singapore)

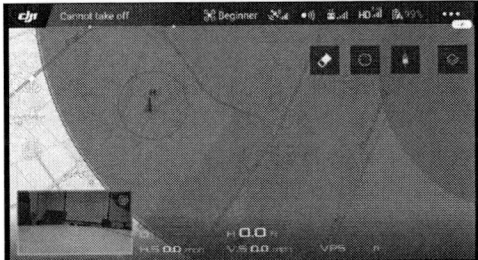

(c) No Fly Zone near SUTD, Changi Airport

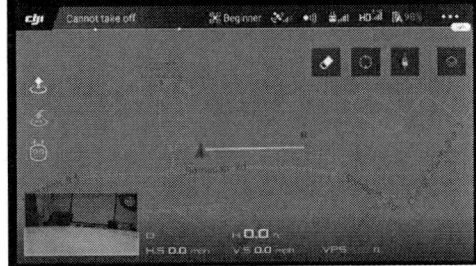

(d) Drone Spoofed Location Movement

Fig. 3: DJI Phantom Drone GPS Spoofing Experimental Setup and Drone GPS Spoofed Locations

password to the Bebop's WiFi which obviously decreases range and speed. Nonetheless, it remains vulnerable to WiFi attacks. Weak password can easily be cracked by tools like [8].

Network Mapping: The Nmap utility shows the drones ports and the different services running on these ports utilizing it. Based on the Nmap scan of the drone's IP, we identified three open ports running the following services FTP, Telnet, and Tor-SOCKS.

1) Open WiFi: The Bebop 2 drone uses an IEEE Wi-Fi radio (802.11ac), and the IP address of the drone is 192.168.42.1. An unencrypted open Wi-Fi allows any one to connect the Bebop 2 drone and hijack the drone. Since Bebop2 allows for multiple clients to connect to the network, the problem is more critical i.e. more than one user can control the drone and there is no way of validating the original owner.

2) Deauthenticating Owner: We performed the deauthentication attack on the drone using method presented in [8]. This attack consists of capturing packets of the Bebop network without even connecting to drone, thereby disconnecting the authenticated owner. Initially the wireless network is scanned in monitor mode using airmon-ng. After the Bebop network is found, it snoops into the network using airodump-ng capturing all the packets from that network only. When the list of clients is available, a deauthentication attack is performed by executing aireplay-ng, which sends de-authentication packets continuously to the clients, disconnecting them until the program executes. In the meantime, the hacker is able to get connected and as the only client connected, can hijack the drone.

In our experiments, the deauthentication packets are sent continuously from the attacker's laptop to the connected client whose MAC address is 48:88:CA:62:9B:A3 where the drone's BSSID is A0:14:3D:C1:F4:2B. The entire network can be jammed by sending deauthentication packets continuously, preventing anyone from connecting to the network. Aireplay-ng sends 128 deauthenticate packets: 64 packets to drone BSSID and 64 packets to the client.

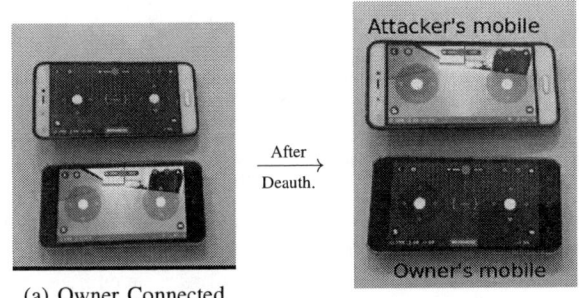

(a) Owner Connected

(b) Owner disconnected

Fig. 4: Deauthenticating owner's mobile

In Figure 4, the mobile phone belonging to the owner (bottom) remains disconnected as long as aireplay-ng continues to send directed deauthenticate packets. The attacker takes advantage of this to connect his/her own mobile phone(above) and can now use it to hijack the drone.

3) Open Telnet: Once connected to the Bebop 2's Wi-Fi, the user can Telnet to it. The drone's IP address is 192.168.42.1. Since multiple users can connect to the drone's Wi-Fi, just attempting to connect to the network, generates multiple requests in a short span of time, which may cause

978-1-5386-3693-0/18 $31.00 © 2018 IEEE

a **DoS** attack, resulting in an interruption of the original commands. Even by telnetting to it, the user obtains root access to the entire file system. By telnetting into the drone, we were able to kill the main process (*dragon-prog* the main Bebop process governing the whole system of flight control), which would stop motor arms immediately, causing the drone to fall to the ground. We also found a list of other interesting processes running by *ps* command of UNIX. We also found important shell scripts */usr/bin/ardrone3_shutdown.sh* and */usr/bin/DragonStarter.sh* We observed that executing *ardrone3_shutdown.sh* also stopped the main process shutting down the drone immediately.

4) Root Access to File System: Since the release of the Parrot AR.Drone 3.0, most of the files are read-only, except the /data directory which is the FTP media directory. Though the files are read-only by default, the attacker can override this by remounting the entire file system with

$$\# \ mount \ -o \ remount, \ rw \ /$$

5) Open FTP Port: Without any authentication for FTP, attackers may connect to the WiFi and login to the FTP to steal personal information including media, flight data, etc. Based on the factory settings, FTP access does not allow access to the entire files system, and only allows access to the FTP directory */data/ftp/internal_000*, which is where images and videos are recorded, along with other useful files (e.g. pud files). The limited FTP access can be taken over by the attacker by editing the file */etc/inetd.conf.* After the following lines,

21 stream tcp nowait root ftpd wSS /data/ftp
51 stream tcp nowait root ftpd wSS /update
61 stream tcp nowait root ftpd wSS

we need to add the following line to obtain access to the entire file system:

71 stream tcp nowait root ftpd ftpd SS /

6) Snooping into the WiFi and Packet Capture: We used Wireshark to capture packets during the connection establishment and throughout the entire flight. The flight commands and video are passed as UDP packets, while the initial connection setup is done by TCP.

7) Additional Vulnerabilities:

a) Reversing firmware: The Bebop 2 firmware is completely reversible and can be easily modified and the new binary can be reflashed. Reversing firmware gives insight into the drone workflow and file system, which makes it easier for the attacker to exploit vulnerabilities and target specific locations.

b) Modifying Files: The entire file system can be accessed using Telnet after being connected to the Bebop's Wi-Fi. The incorrect modification of configuration files/scripts such as dragon.conf to crash the software and hamper normal drone operation. Modifying the /etc/passwd file may brick the drone as changing some passwords may limit the owner's ability to telnet into it, or connect to the drone.

IV. COMPARISON OF TWO DRONES

From our experiments and keen observations made after analyzing the security aspects of both the Phantom 4 Pro and Bebop 2 drones, we found that Phantom 4 Pro drone is more secure than Bebop 2 drone. Phantom 4 Pro is only vulnerable to GPS spoofing and interception/jamming of radio signals using SDR. Bebop 2 is prone to various wireless attacks and can be easily hijacked. While SDR equipment and/or GPS spoofers can be difficult to acquire, due to limited availability and cost, several wireless tools and network scanner, analyzers are readily available online at no cost. In addition, it appears that DJI has tried to improve the security of its drones by introducing communication and video feed transmission via radio signals and DJI claims to have an on-board FPGA video encryption algorithm which can be easily verified by scanning transmitted radio signals using SDR or tapping the output of FPGA before it goes to the transceiver. DJI also partially encrypts the firmware, and the binary is digitally signed, although it can be decrypted to some extent.

V. PROPOSED COUNTERMEASURES AGAINST THE DRONE SECURITY VULNERABILITIES

In this section, we propose some methods/actions that can be implemented to increase the security of drones. Drones will remain vulnerable to GPS spoofing unless receivers gain the capability of detecting spoofing (e.g. SAASM used by military) thus, there is a need for developing anti-spoofing and anti-jamming receivers. While there has been a lot of promising work and methods proposed for the detection and avoidance of civilian GPS anti-spoofing and anti-jamming in the literature similar to that in [18]. Those methods can be broadly classified into cryptographic (spread spectrum, dual receiver correlation), and non-cryptographic (antenna array). These methods are either difficult to implement or require costly hardware. Therefore we propose some simpler software-based techniques for spoof detection.

Checking latency: The motion speed can be validated for a change in location in just a brief instant, (e.g. an individual cannot travel and change his/her location from New York to Beijing in a few seconds). The changes of coordinates can be recorded and validated given the time it takes to change the coordinates.

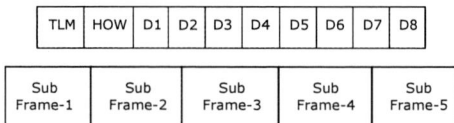

Fig. 5: GPS frame structure

Checking GPS Subframe Data: The GPS frame starts with a telemetry word used as preamble, which provides details of the satellite. Handover word (HOW) provides the GPS system time. HOW is followed by eight data words with parity bits. Subframe-1 carries information to correct the satellite clock. Subframe-2 and subframe-3 provide ephemeris data for the satellites. Subframe-4 and subframe-5 send the almanac data. GPS subframe data can be validated easily by checking each of the subframes. Fake GPS data has been found to have incorrect subframe data, thus it is possible to detect spoofing. As a countermeasure that can be used to easily detect GPS spoofing, we propose using an onboard Raspberry Pi on the drone to validate time-motion or GPS subframe data. This adds an extra overhead for checking GPS data before passing to the flight controller which affects the latency of the system.

In addition to GPS anti-spoofing and anti-jamming, we recommend some modifications which can improve the security of drone.

- Using encryption/packer to protect library files
- Using an obfuscator to prevent decompiling and, reverse engineering of the firmware
- Improve SDK authentication, (although this is, now being done between the app and server, the drone must also be included in the one-time authentication)
- Encrypt the entire firmware binary, and store, the encryption key in the hardware

These above mentioned strategies are proposed to be implemented by the manufacturer and we left these as recommendations. In contrast, many of the attacks described in section III-B for Bebop 2 are possible only due to open WiFi and open ports. In addition, FTP need not be activated inflight in order to prevent theft of flight data and media. The backing up of Bebop 2 drone data offline provides security for personal data. For the Bebop 2, and other Wi-Fi based UAVs, we propose the following countermeasures:

a) Adding WPA security: The parrot Bebop 2 developers added WPA/WPA2 security to the WiFi using Freeflight Pro app. Even though it is possible to crack the Wi-Fi password using tools like Aircrack-ng. Still adding WPA to Wi-Fi improves the security of the parrot Bebop 2 drone. This security improvement may cause decrease in range and transmission speed.

b) Adding Telnet Password: Bebop 2 should not allow root access to the file system once connected and telnetted. This can be managed by adding a Telnet password modifying /etc/passwd file.

c) MAC-Filtering and Hidden SSID: These can be achieved by installing bcmwl program in the drone. bcmwl controls all of the wireless wi-fi connectivity of the drone. Any changes made can be temporary or permanent. In order to reversibly hide the Wi-Fi SSID, the user can open a Telnet session and enter the following command:

```
# bcmwl closed 1
```

This makes it more difficult for others to connect to the drone, since the Wi-Fi network will not appear in the list of available networks. To activate MAC address filtering, enter the following command

```
# bcmwl mac MA:CA:DD:RE:SS:01 MA:CA:DD:
RE:SS:02 ...
```

where all the possible MAC addresses are provided by the owner followed by:

```
# bcmwl macmode 2
```

which sets MAC filtering to only accept connections from the previous list of devices. To permanently introduce these changes, the /sbin/broadcom_setup.sh file must be modified.

VI. CONCLUSION AND FUTURE WORK

In this paper, we have investigated security vulnerabilities associated with DJI Phantom 4 Pro and Parrot Bebop 2 drones, and proposed some countermeasures against these security vulnerabilities. The P4P is one of the most secure and robust drones available in the commercial market. Although DJI attempted to develop the P4P, so that it is less vulnerable than its predecessors, it still needs further work and as well as comprehensive security analysis. In contrast, the Parrot manufacturers did not include any security measures for the Bebop drones, and Bebop 2 is too risky for any personal or commercial use. Though we performed our experiments on two drones, most of the attacks will be applicable to drones of similar architecture and communication protocols. In future, we are planning to analyze the P4P drone communication signals using the SDR equipment, such as HackRF and BladeRF, and live video transmission.

REFERENCES

[1] B. Meyerson, "Top 10 emerging technologies of 2015," in *The World Economic Forum Blog, Agenda.* https://agenda.weforum.org/2015/03/top-10-emerging-technologies-of-2015-2, 2015.

[2] E. Vattapparamban, İ. Güvenç, A. İ. Yurekli, K. Akkaya, and S. Uluağaç, "Drones for smart cities: Issues in cybersecurity, privacy, and public safety," in *Wireless Communications and Mobile Computing Conference (IWCMC), 2016 International.* IEEE, 2016, pp. 216–221.

[3] M. A. Ruiz Estrada, "How unmanned aerial vehicles–uavs–(or drones) can help in case of natural disasters response and humanitarian relief aid?" 2017.

[4] D. Williams, "Ten ways drone technology is shaping our future," 2015.

[5] M. Corcoran, "Drone journalism: Newsgathering applications of unmanned aerial vehicles (uavs) in covering conflict, civil unrest and disaster."

[6] K. Nonami, "Prospect and recent research & development for civil use autonomous unmanned aircraft as uav and mav," *Journal of system Design and Dynamics*, vol. 1, no. 2, pp. 120–128, 2007.

[7] S. Kamkar, "Skyjack: autonomous drone hacking," *Available: http://samy.pl/skyjack*, 2013.

[8] "Aircrack-ng: tools to access wifi network security," *Available: http://www.aircrack-ng.org.*

[9] A. J. Kerns, D. P. Shepard, J. A. Bhatti, and T. E. Humphreys, "Unmanned aircraft capture and control via gps spoofing," *Journal of Field Robotics*, vol. 31, no. 4, pp. 617–636, 2014. [Online]. Available: http://dx.doi.org/10.1002/rob.21513

[10] "Maldrone: Backdoor for drones," *Available: http://garage4hackers.com/entry.php?b=3105*, 2015.

[11] F. Trujano, B. Chan, G. Beams, and R. Rivera, "Security analysis of dji phantom 3 standard," *Available: https://courses.csail.mit.edu/6.857/2016/files/9.pdf*, 2016.

[12] L. Huang and K. Yang, "Gps spoofing: Low-cost gps simulator," presented in DEFCON 23 hacking conference.

[13] "Dji sdk for developers," *Available: https://developer.dji.com.*

[14] "Jd-gui: standalone java decompiler," *Available: http://jd.benow.ca.*

[15] "Java bytecode editor," *Available: http://set.ee/jbe.*

[16] "rkflashtool: tools for flashing rockchip devices," *Available: https://github.com/neo-technologies/rkflashtool.*

[17] "Labsat 3: A multi-constellation global navigation satellite simulator," *Available: https://www.labsat.co.uk/index.php/en/products/labsat-3.*

[18] K. Wesson, D. Shepard, and T. Humphreys, "Straight talk on anti-spoofing," *GPS World*, 2012.

AMS-Miner: Mining AMS Assertions using Interval Arithmetic

Antonio Anastasio Bruto da Costa, Shriya Dharade, Sudipa Mandal and Pallab Dasgupta

Dept. of Computer Science and Engg., Indian Institute of Technology Kharagpur, West Bengal, India 721302

Email: bruto@cse.iitkgp.ernet.in, shriya.dharade27@gmail.com, sudipa.mandal@cse.iitkgp.ernet.in, pallab@cse.iitkgp.ernet.in

Abstract—We present AMS-Miner, a methodology for generating Analog Mixed-Signal (AMS) Assertions automatically. The proposed methodology uses predicates over real variables (PORVs) to map real-valued signals obtained from timestamped traces of the AMS design to Boolean signals. Dense time for AMS designs is dealt as intervals of time. This mapping yields a set of time intervals wherein each PORV is true. Interval arithmetic coupled with decision tree learning are used to operate on time-intervals in the AMS-data space to extract assertions in the form of cause-effect patterns, as timed sequences of PORVs. This article, for the first time explores assertion mining for AMS and builds the theory for interval based assertion mining for AMS traces.

Index Terms—Analog and Mixed-Signal, Assertions, Simulation, Data Mining

I. INTRODUCTION

The basis of formal analysis of systems, is in having the ability to formally specify how the system must behave. This specification is captured in the form of *assertions*. Assertions have been at the heart of testing and verification for both fully digital and analog mixed-signal (AMS) designs. SystemVerilog Assertions (SVA) [1] has been the champion of assertion specification languages for digital designs for its richness of expressions, and has formed the basis of assertion monitoring through simulation and formal verification in off-the-shelf CAD tools. Assertion based validation has become the foundation for design validation across the abstraction hierarchy [2], and the most coveted factor in guaranteeing the success of assertion based verification is in compiling a complete and correct collection of assertions [3].

Assertion writing is, for the most part, a manual effort. In the digital design domain, studies in automatically generating assertions using tools such as GoldMine [4], [5] have the potential for dramatically reducing the manual effort involved in writing assertions for digital hardware. The pain-point emphasized by Ref. [5] is in the manual expression of temporal assertions that span multiple clock cycles, with interesting yet intricate relationships which are tedious and complex for a human being to reason with. The authors therefore suggest the use of data mining techniques to automatically extract complex temporal assertions, in the form of Linear Temporal Logic (LTL) like formulae from traces of digital designs.

The conundrum of assertion specification intensifies for an AMS design due to elaborate ways in which analog artifacts influence each other, with the addition of real valued signals and dense time over which signal changes are observed. Testing of AMS designs is a tedious and time-consuming task. Past work has focused on using assertions based on

AMS extensions of LTL for monitoring AMS designs through simulation [6], [7], [8]. These extensions introduce predicates over real variables (PORVs) [6], analysis over dense time, the notion of events and the use of real-valued local variables in the assertion framework [7]. An SVA-like language for analyzing AMS designs was introduced in [9], in the Feature Indented Assertion language (FIA). FIA, used for expressing *features*, associates a computable function to the match of an SVA-like sequence expression. Features in FIA are real-valued attributes overlayed on assertions and can express complex AMS attributes, such as rise-time, overshoot, settling time, etc. The formal analysis of features for AMS designs has also been explored in [10].

The use of sequence expressions is common place when writing assertions in the digital domain. Digital circuit designers are familiar with writing SystemVerilog Assertion (SVA) properties for verifying their designs. Sequence expressions for AMS have been explored extensively in [9], [10], [11]. These use-cases extend SVA properties with the use of dense time, PORVs and events over PORVs.

In the digital domain, assertion mining has been extensively studied and mature tools have resulted from these studies. Goldmine [4], one of the more popular tools for assertion mining for digital designs, presents assertions in the form of clocked temporal sequences of Boolean signals. For instance for Boolean signals x, y and z, Goldmine may present an assertion of the form "x ##2 y => ##3 z". Here ##2 and ##3 indicate a clock cycle separation of two and three clock cycles respectively.

For AMS, a sequence expression takes the following form:

(X>20) ##[2:3.2] (Y<10) |=> ##[1.3:1.9] (Z>2)

In this sequence, (X>20), (Y<10) and (Z>2) are predicates over the real variables X,Y and Z; and ##[2:3.2] and ##[1.3:1.9] indicate dense time intervals during which the predicate that follows may become true. Simply extending present techniques for digital designs to mine assertions for AMS designs would be a wasteful effort. Analog signals in an AMS design are related to the rest of the design in time that is *real* valued, not clocked. Since the input to digital assertion miners is a clocked trace, for an AMS design, a sampling clock of sufficiently high frequency f, together with an appropriate extension of present techniques would be required to generate digital-like assertions for the AMS design. In addition to the potential dramatic increase in simulation time, and the shear size of the resulting trace, as per Ref. [4] to mine temporal assertions with a temporal length of $\mathcal{T}$, an input dataset with

978-1-5386-3693-0/18 $31.00 © 2018 IEEE

404

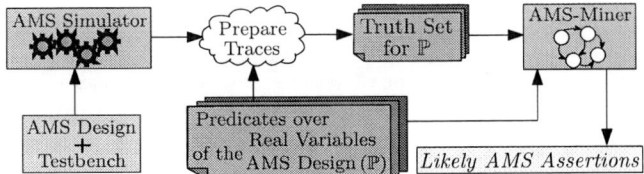

Fig. 1: The AMS-Miner Methodology

each row of size M would require each row to expand by $M \times \mathcal{T} \times f$, thereby exacerbating scalability issues. Moreover, the representation of the continuous truth of a predicate is lost through sampling. Assertions generated for the design would try to relate the truth of predicates over cycles of the sampling clock, yielding a large number of spurious assertions, rendering expressing properties over dense time extremely tedious. Furthermore, the semantics of real-time extensions of LTL may not be suitable for expressing AMS assertions. Instead, it is preferable to mine assertions for AMS designs using suitable and scalable data abstractions for the trace data, and thereafter use algorithms, specifically tuned to mine cause-effect relationships in the AMS design, to mine likely assertions as sequences over dense time.

In this article, we introduce AMS-Miner, that automatically mines AMS assertions from signal event traces of AMS designs as sequence expressions similar to those used in FIA [9], [10]. An outline of the proposed methodology is shown in Figure 1. On the basis of PORVs of interest, traces of the AMS design are prepared to indicate the time intervals in which each PORV is true, the truth set. This along with the list of PORVs are fed to AMS-Miner, which generates likely AMS assertions for the design. The extensions of assertion mining to AMS designs impacts the analysis of real-world systems allowing interesting cause-effect patterns to be learned from traces of these systems.

II. AMS SEQUENCE ASSERTION MINING METHODOLOGY

The AMS assertion miner methodology developed in this article learns assertions as sequence expressions that have the following form:

$$P_1 \; \#\#[a_1, b_1] \; P_2 \; \#\#[a_2, b_2] \; \dots \; \#\#[a_{N-2}, b_{N-2}] \; P_{N-1}$$
$$|\!=\!> \; \#\#[a_{N-1}, b_{N-1}] \; P_N$$

$P_1, P_2, ..., P_N$ are PORVs and $a_i, b_i \in \mathbb{R}_{\geq 0}$, $a_i \leq b_i$, $1 \leq i < N$. Assertion mining begins with a set of truth intervals for each PORV P_i, $1 \leq i \leq n$. The consequent PORV P_N is chosen apriori. Statistical measures used in the algorithm are computed with respect to P_N.

Table I presents a collection of notations used throughout this article.

Definition 1. Time Interval: *A time interval I is a non-empty convex subset of $\mathbb{R}_{\geq 0}$ expressed as $[a : b]$, $(a : c)$, $(a : c]$, $[a : c)$; where $a, b, c \in \mathbb{R}_{\geq 0}$ and $b \geq a$, $c > a$. $l(I)$ and $r(I)$ are used to denote the left and right ends of interval I.*

The notation $I \oplus [c, d]$, where $c, d \in \mathbb{R}_{\geq 0}$, $c \leq d$, denotes the Minkowski sum, and is defined as follows:

$$I \oplus [c : d] = [l(I) + c : r(I) + d]$$

Notation	Description
P	A Predicate over Real Variables (PORV)
$\mathbb{P}$	The set of all PORVs under consideration, $|\mathbb{P}| = n$
$\mathcal{T}$	Simulation trace
I_P	An interval $[a : b]$ during which P is true, $a, b \in \mathbb{R}^+, a \leq b$
$\mathcal{I}_\tau(P)$	Set of all truth intervals of P in trace τ
$\mathcal{A}$	A list of non-conflicting truth assignments to PORVs of the form $\{\langle P_i, b\rangle, \langle P_j, b\rangle, ..., \langle P_k, b\rangle | b \in \{\bot, \top\}\}$
$\tau_\mathcal{A}$	The sub-trace of τ under the assignments $\mathcal{A}$
$\mathcal{I}_{\tau_\mathcal{A}}(P)$	Set of all truth intervals of P in trace τ constrained by the assignments $\mathcal{A}$
$\mathbb{I}_\tau(\mathbb{P})$	The set of sets of truth intervals, $\mathbb{I}_\tau(\mathbb{P}) = \{\mathcal{I}_\tau(P), P \in \mathbb{P}\}$
$|I_P|$	$b - a$, where $I_P = [a, b)$, $a, b \in \mathbb{R}^+, a \leq b$
$|\mathcal{I}_\tau(P)|$	$\Sigma|I|, I \in \mathcal{I}_\tau(P)$. Refer to Equation 1.
$|\mathbb{I}_\tau(\mathbb{P})|$	Length of time over which $\mathbb{I}_\tau$ is defined i.e. length of τ.
$|\mathbb{I}_{\tau_\mathcal{A}}(\mathbb{P})|$	Length of time over which $\mathbb{I}_\tau(\mathbb{P})$ is defined constrained by the assignments $\mathcal{A}$. Refer to Definition 4.

TABLE I: Notations Used

The notation $I \ominus [c, d]$ denotes the Minkowski difference, and is defined as follows:

$$I \ominus [c : d] = [l(I) - d : r(I) - c]$$

For an interval I, and for $t \in \mathbb{R}_{\geq 0}$, $(I + t)$ and $(I - t)$ respectively denote the intervals $\{t' + t | t' \in I\}$ and $\{t' - t | t' \in I, t' - t \geq 0\}$. □

It may be noted that in this article, we deal with intervals of truth. Hence all intervals are of the form $[a, b)$.

Definition 2. Interval Set: *The Interval Set for the PORV P, $\mathcal{I}_\tau(P = b)$, $b \in \{\top, \bot\}$, is the set of all time intervals, $[c, d)$; $c, d \in \mathbb{R}_{\geq 0}$; $c < d$, in trace τ where P evaluates to b.* □

We use $\mathcal{I}_\tau(P)$ to refer to $\mathcal{I}_\tau(P = \top)$. The complement interval set, defining when the PORV is false, that is $\mathcal{I}_\tau(P = \bot)$ is represented as $\bar{\mathcal{I}}_\tau(P)$.

We define the length of the interval set $\mathcal{I}_\tau(P)$, denoted $|\mathcal{I}_\tau(P)|$, as follows:

$$|\mathcal{I}_\tau(P)| = \Sigma_{\forall I = [a,b) \in \mathcal{I}_\tau(P)}(b - a) \qquad (1)$$

We use the notation $\mathbb{P}$ to denote the set of all PORVs under consideration. Traces of digital circuits have the advantage of being represented as a truth table with an underlying clock, i.e. rows of the truth table represent the state of the circuit at successive clock cycles. When we make the switch to analog, the notion of a clock and clock boundaries dissolve into the expanse of dense real time. Additionally, the truth of a predicate is evaluated over an interval of time. The representation of a trace for an AMS circuit must include analog attributes to ensure that results are meaningful, making the format of a truth table ineffective. In this article the preferred form of representation is intervals of truth for a predicate. Expressing the truth of predicates as intervals of truth incorporates the notion of dense time and provides us the structure we require for processing truth data usefully. A trace is therefore viewed from the perspective of a *truth set*, given in the definition below.

Definition 3. Truth Set: *The Truth Set for $\mathbb{P}$, $\mathbb{I}_\tau(\mathbb{P})$, is the set of all Interval Sets for the trace τ of all PORVs $P \in \mathbb{P}$. $\mathbb{I}_\tau = \{\mathcal{I}_\tau(P) | P \in \mathbb{P}\}$.* □

For instance the truth set for the trace describing the truth

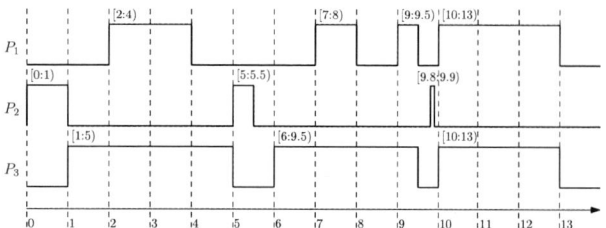

Fig. 2: Truth waveforms for predicates P_1, P_2 and P_3.

of predicates $\mathbb{P} = \{P_1, P_2, P_3\}$, in Figure 2, is $\mathbb{I}_\tau(\mathbb{P}) = \{\{[2 : 4), [7 : 8), [9 : 9.5), [10 : 13)\}; \{[0 : 1), [5 : 5.5), [9.8 : 9.9)\}; \{[1 : 5), [6 : 9.5), [10 : 13)\}\}$.

While evaluating the truths of PORVs, it is relevant to view all segments of the trace for which a PORV P is true. For trace $\mathcal{T}$, we define an assignment list $\mathcal{A}$, a set of non-conflicting truth assignments to PORVs. For instance, for PORVs P_1, P_2, P_3; $\mathcal{A} = \{P_1 = \top, P_2 = \top\}$, $\mathcal{A} = \{P_1 = \bot, P_3 = \top\}$, $\mathcal{A} = \{P_1 = \top, P_2 = \bot, P_3 = \top\}$ are consistent assignment lists. However $\mathcal{A} = \{P_1 = \top, P_1 = \bot\}$ is inconsistent as it assigns both truth values $\top$ and $\bot$ to P_1. The sub-trace (containing possibly disconnected trace segments) adhering to the assignment list $\mathcal{A}$ is denoted $\mathcal{T}_\mathcal{A}$. It is understood that the predicates $P_1 \equiv (x > 5)$ and $P_2 \equiv (x < 3)$ are together unsatisfiable, and we rely on the correctness of the input data to assert that P_1 and P_2 are never true simultaneously. Additionally, techniques for satisfiability checking, suggested in [12], may be used during the assertion mining process to ensure that an unsatisfiable combination of predicates are eliminated from consideration.

When looking at a trace through the lens of a PORVs truth, for an assignment list $\mathcal{A}$, we compute the constrained sub-truth set as follows:

$$\mathcal{I}_{\tau_\mathcal{A}}(P) = \bigcup_{\substack{I \in \mathcal{I}_\tau(P) \\ I' \in \mathcal{I}_\tau(Q), Q \in \mathcal{A}}} (I \cap I') \tag{2}$$

Definition 4. *Length of a Truth Set: The length of a Truth Set $\mathbb{I}_{\tau_\mathcal{A}}(\mathbb{P})$, under PORV truth assignments $\mathcal{A}$, is recursively defined as follows:*

$$
\begin{aligned}
|\mathbb{I}_{\tau_\mathcal{A}}(\mathbb{P})| &= \quad Length\ of\ \tau & if\ \mathcal{A} = \phi \\
&= \quad |\mathcal{I}_\tau(P)| & if\ \mathcal{A} = \langle P, \top \rangle \\
&= \quad |\mathbb{I}_\tau(\mathbb{P})| - |\mathcal{I}_\tau(P)| & if\ \mathcal{A} = \langle P, \bot \rangle \\
&= \quad |\mathcal{I}_{\tau_{\mathcal{A}/\{P=\top\}}}(P)| & if\ |\mathcal{A}| > 1\ \wedge \\
& & \langle P, \top \rangle \in \mathcal{A} \\
&= \quad |\mathbb{I}_{\tau_{\mathcal{A}/\{P=\top\}}}(\mathbb{P})| - |\mathcal{I}_{\tau_{\mathcal{A}/\{P=\top\}}}(P)| & if\ |\mathcal{A}| > 1\ \wedge \\
& & \langle P, \bot \rangle \in \mathcal{A}
\end{aligned}
$$

$\square$

$|\mathcal{I}_{\tau_{\mathcal{A}/\{P=\top\}}}(P)|$ can be computed using Equations 1 and 2.

The computation of statistical functions of mean and error, used by the assertion miner are detailed in the following sections.

A. Computation of Mean

In the digital domain, a trace of the digital design can be represented as a truth table. Statistical counts for a Boolean variable x are viewed in terms of the number of rows of the table in which x is either $\top$ or $\bot$. Therefore, in a table of K rows, where J is the quantum of rows in the trace that are witness to x's truth, i.e. state x is $\top$, the mean is $\frac{J}{K}$. However, for AMS designs, time is dense and a truth table is no longer a viable form of representation for a trace of the AMS design. The use of time intervals that represent the truth of a predicate allows us to cope with dense time. This also requires us to adapt the definition of mean to handle intervals of truth. Equations 1 and 2 together, define the quantum of time for which a PORV P is true. The notion of a truth table, adapted to time intervals is captured in the definition of interval sets in Definition 4. Mean, computed with respect to PORV P, is the quantum of time that P is true in the current view of the trace (possibly constrained by the truth of other predicates) and is therefore computed as follows:

$$\mu_{\tau_\mathcal{A}}(P) = \frac{|\mathcal{I}_{\tau_\mathcal{A}}(P)|}{|\mathbb{I}_{\tau_\mathcal{A}}(\mathbb{P})|} \tag{3}$$

B. Computation of Error

The *error* is a measure of deviation from the mean. From the perspective on Boolean valued predicates, we wish to identify which predicates contribute most to reducing the deviation (error). Therefore, intuitively error measures the quantum of time for which the consequent is true and the quantum of time the consequent is false offset by the mean. Since the domain of values for a predicate is $\{\top, \bot\}$, the error function is expressed as follows:

$$
\begin{aligned}
\epsilon_{\tau_\mathcal{A}}(P) &= \frac{|1 - \mu_{\tau_\mathcal{A}}(P)| \times |\mathcal{I}_{\tau_\mathcal{A}}(P)| + |0 - \mu_{\tau_\mathcal{A}}(P)| \times (|\mathbb{I}_{\tau_\mathcal{A}}(\mathbb{P})| - |\mathcal{I}_{\tau_\mathcal{A}}(P)|)}{|\mathbb{I}_{\tau_\mathcal{A}}(\mathbb{P})|} \\
&= \frac{|\mathcal{I}_{\tau_\mathcal{A}}(P)| - 2 \times \mu_{\tau_\mathcal{A}}(P) \times |\mathcal{I}_{\tau_\mathcal{A}}(P)| + \mu_{\tau_\mathcal{A}}(P) \times |\mathbb{I}_{\tau_\mathcal{A}}(\mathbb{P})|}{|\mathbb{I}_{\tau_\mathcal{A}}(\mathbb{P})|} \\
&= \frac{(1 - 2 \times \mu_{\tau_\mathcal{A}}(P)) \times |\mathcal{I}_{\tau_\mathcal{A}}(P)|}{|\mathbb{I}_{\tau_\mathcal{A}}(\mathbb{P})|} + \mu_{\tau_\mathcal{A}}(P) \\
&= (1 - 2 \times \mu_{\tau_\mathcal{A}}(P)) \times \mu_{\tau_\mathcal{A}}(P) + \mu_{\tau_\mathcal{A}}(P) \\
&= 2 \times \mu_{\tau_\mathcal{A}}(P) \times (1 - \mu_{\tau_\mathcal{A}}(P))
\end{aligned}
$$

Therefore the computation for error contributed by constraints $\mathcal{A}$ to the consequent PORV P is computed as follows:

$$\epsilon_{\tau_\mathcal{A}}(P) = 2 \times \mu_{\tau_\mathcal{A}}(P) \times (1 - \mu_{\tau_\mathcal{A}}(P)) \tag{4}$$

The assertion mining algorithms presented in this article use the statistical measures of Mean and Error of Sections II-A, II-B to chose which predicates and time intervals between predicates best contribute to the consequents truth.

C. AMS-Miner

AMS-Miner uses data mining techniques inspired by the decision tree learning algorithms in [13], [14], [15] to identify correlations between characteristic PORVs and a consequent PORV. The consequent PORV of the assertions generated may be in either its positive or negative literal form, thereby ascertaining the causes for both, its truth and falsification. For an assertion $A : S \Rightarrow T$, statistical measures of *support* and *confidence* help validate the correlation between the antecedent S and consequent T. The support of A determines the quantum of time in the truth set for which T is true given that S is true. The confidence of A is an estimate of the conditional probability, $Prob(T|S)$. AMS-Miner requires that all assertions generated by it have 100% confidence, if not then S correlates both with true and false incidences of T.

978-1-5386-3693-0/18 $31.00 © 2018 IEEE

Algorithm 1: `AMS-Miner`: Assertion Mining for AMS Design Traces

Input: Truth Set $\mathbb{I}_\tau(\mathbb{P})$ for trace τ, PORV List $\mathbb{P}$, Consequent PORV P_c, Assignment List $\mathcal{A}$.

Output: Assertions $\mathcal{R}$, structured as the decision tree.

1   $m \leftarrow \mu_{\tau_\mathcal{A}}(P_c); \ e \leftarrow \epsilon_{\tau_\mathcal{A}}(P_c);$
2   **if** $e = 0$ **then** $\mathcal{R} \leftarrow \mathcal{R} \cup (\mathcal{A} \Rightarrow P_c\langle m \rangle);$ **return** ;
3   $P_{best} \leftarrow \phi; \ g_{best} \leftarrow -\infty;$
4   **forall** $P \in \mathbb{P}$ **do**
5     $g \leftarrow e - \epsilon_{\tau_{\mathcal{A} \cup \langle P, \top \rangle}}(P_c) - \epsilon_{\tau_{\mathcal{A} \cup \langle P, \bot \rangle}}(P_c);$
6   **if** $g > g_{best}$ **then** $P_{best} \leftarrow P; \ g_{best} \leftarrow g$;
7   `AMS-Miner`$(\mathbb{I}_{\tau_{\mathcal{A} \langle P_{best}, \top \rangle}}(\mathbb{P}/P_{best}), \mathbb{P}/P_{best}, \mathcal{A} \cup \langle P_{best}, \top \rangle);$
8   `AMS-Miner`$(\mathbb{I}_{\tau_{\mathcal{A} \langle P_{best}, \bot \rangle}}(\mathbb{P}/P_{best}), \mathbb{P}/P_{best}, \mathcal{A} \cup \langle P_{best}, \bot \rangle);$

The input to AMS-Miner is a set of dense time intervals during which the PORVs truth is observed, for each PORV in the list $\mathbb{P}$. For simulatable AMS designs these truth intervals can be measured accurately with specified error and time tolerances supported by standard simulators. Ref. [8] discusses this further, and delves into how error tolerances figure in the sampling of signals at cross events, that in turn determines the truth interval boundaries that we compute.

The AMS-Miner algorithm presented as Algorithm 1, implicitly uses a measure of *gain*, where the gain of choosing a PORV P at a node in the decision tree is an indication of the reduction in the quantum of error between that node and the child nodes produced if P is used to partition the truth set.

We initially present AMS-Miner as an algorithm that, on its own, does not take into consideration the temporal sequences detailed earlier. In Section III, we detail the mechanism for transforming the input truth set to accommodate temporal artifacts such that, the transformed truth set when provided as input to a variant of AMS-Miner generates rules of the following form (as shown earlier):

$$P_1 \ \#\#[a_1, b_1] \ P_2 \ \#\#[a_2, b_2] \ \ldots \ \#\#[a_{N-2}, b_{N-2}] \ P_{N-1}$$
$$|=> \ \#\#[a_{N-1}, b_{N-1}] \ P_N$$

D. An Example run through AMS-Miner

The semantics of sequence expressions allows for both, *immediate causality* and *future causality* to be asserted. Immediate causality, expressed as $S |=> R$, is observed when the truth of the sequence S at time t causes the consequent R to be true at time t. Future causality, expressed by the assertion $S |=>\#\#[a:b] R$ relates the truth of the sequence S at time t with the truth of R at time $t' \in [t + a, t + b]$. In this section, we describe how AMS-Miner generates assertions that carry the semantics of immediate causality. We explore a run of Algorithm 1 on the interval sets listed in Table II. In Section III we extend AMS-Miner to compute sequence expressions that describe future causal relationships.

The length of the trace $\mathcal{T}$ is 13. For simplicity, in this article, we cannot describe how all temporal combinations are tested, and hence to introduce and explain the algorithm we show how we would test for the interval sets $\mathcal{I}_\tau(P_1)$ and $\mathcal{I}_\tau(P_2)$. The decision tree produced by Algorithm 1 is shown in Figure 3.

At the root node (lines 1 and 2 of Algorithm 1) the $\langle mean, error \rangle$ tuple is calculated for the PORV P_3 to be

Interval Set	Time Intervals
$\mathcal{I}_\tau(P_1)$	$\{[2:4], [7:8), [9:9.5), [10:13]\}$
$\mathcal{I}_\tau(P_2)$	$\{[0:1), [5:5.5), [9.8:9.9)\}$
$\mathcal{I}_\tau(P_3)$	$\{[1:5), [6:9.5), [10:13]\}$

TABLE II: Interval Sets for $\mathbb{I}_\tau(\mathbb{P})$.

$\langle 0.81, 0.31 \rangle$. For each PORV, the gain (at line 8) is computed. For PORV P_1, choosing P_1 for the partitioning step, causes the interval sets to be constrained as follows:

Choosing $\begin{cases} \mathcal{I}_{\tau_{\langle P_1, \top \rangle}}(P_1) = \{[2:4], [7:8), [9:9.5), [10:13]\} \\ \mathcal{I}_{\tau_{\langle P_1, \top \rangle}}(P_3) = \{[2:4], [7:8), [9:9.5), [10:13]\} \end{cases}$
$P_1 = \top$

Choosing $\begin{cases} \mathcal{I}_{\tau_{\langle P_1, \bot \rangle}}(\neg P_1) = \{[0:2), [4:7), [8:9), [9.5:10)\} \\ \mathcal{I}_{\tau_{\langle P_1, \bot \rangle}}(P_3) = \{[1:2), [4:5), [6:7), [8:9)\} \end{cases}$
$P_1 = \bot$

Choosing P_2 for the partition produces the following constrained lists:

Choosing $\begin{cases} \mathcal{I}_{\tau_{\langle P_2, \top \rangle}}(P_2) = \{[0:1), [5:5.5), [9.8:9.9)\} \\ \mathcal{I}_{\tau_{\langle P_2, \top \rangle}}(P_3) = \phi \end{cases}$
$P_2 = \top$

Choosing $\begin{cases} \mathcal{I}_{\tau_{\langle P_2, \bot \rangle}}(\neg P_2) = \{[1:5), [5.5:9.8), [9.9:13]\} \\ \mathcal{I}_{\tau_{\langle P_2, \bot \rangle}}(P_3) = \{[1:5), [6:9.5), [10:13]\} \end{cases}$
$P_2 = \bot$

When choosing P_1 for the partitioning, the $\langle mean, error \rangle$ are $\langle 1, 0 \rangle$ when P_1 is true and $\langle 0.62, 0.47 \rangle$ when false. The gain computed for the choice of P_1 is $0.31 - 0 - 0.47 = -0.16$. Similarly choosing P_2 for the partitioning step, $\langle mean, error \rangle$ are $\langle 0, 0 \rangle$ when P_2 is true and $\langle 0.91, 0.16 \rangle$ when P_2 is false, resulting in a gain of $0.31 - 0 - 0.16 = 0.15$. The gain from choosing P_2 is larger (indicating a better correlation with P_3), hence, at line 10 in Algorithm 1 the predicate chosen in P_2. Every path in the decision tree represents an assertion. In the decision tree (Figure 3), the child node corresponding to the assignment $\langle P_2, \top \rangle$ results in a zero error node indicating that an assertion is generated here. Since the mean at the node is zero, this indicates a correlation with the negation of the consequent P_3. Therefore the assertion generated at this node is $P_2 |=> \neg P_3$.

Moving on, at the child node where P_2 is false, since P_1 is the only PORV left for splitting, in a manner similar to the one discuss earlier, the data space is split on the truth of P_1, as follows:

Choosing $\begin{cases} \mathcal{I}_{\tau_{\{\langle P_2, \bot \rangle \atop \langle P_1, \top \rangle\}}}(P_1) = \{[2:4], [7:8), [9:9.5), [10:13]\} \\ \mathcal{I}_{\tau_{\{\langle P_2, \bot \rangle \atop \langle P_1, \top \rangle\}}}(P_3) = \{[2:4], [7:8), [9:9.5), [10:13]\} \end{cases}$
$P_1 = \top$

Choosing $\begin{cases} \mathcal{I}_{\tau_{\{\langle P_2, \bot \rangle \atop \langle P_1, \bot \rangle\}}}(\neg P_1) = \{[1:2), [5.5:7), [8:9), [9.9:10)\} \\ \mathcal{I}_{\tau_{\{\langle P_2, \bot \rangle \atop \langle P_1, \bot \rangle\}}}(P_3) = \{[1:2), [6:7), [8:9)\} \end{cases}$
$P_1 = \bot$

The values of $\langle mean, error \rangle$ under the assignments $\{\langle P_2, \bot \rangle \langle P_1, \top \rangle\}$ and $\{\langle P_2, \bot \rangle \langle P_1, \bot \rangle\}$ are respectively, $\langle 1, 0 \rangle$ and $\langle 0.83, 0.28 \rangle$. In the decision tree, on splitting, one child node has a non-zero error, indicating that the correlation for the assignments at that node are still inconsistent with the consequent. Therefore at this stage, since we have no further predicates to refine the data space, the node is discarded. The other child node with the assignment $\langle P_1, \top \rangle, \langle P_2, \bot \rangle$ results in a zero error node, with a mean of one, asserting a correlation

978-1-5386-3693-0/18 $31.00 © 2018 IEEE   407

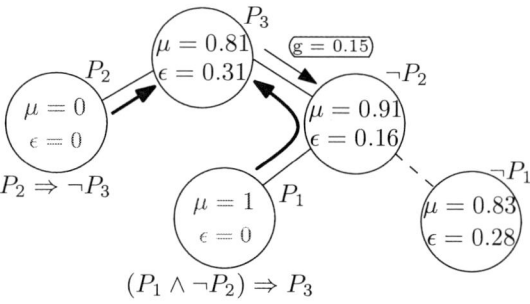

Fig. 3: Decision tree generated by AMS-Miner for Interval Sets of Table II.

with the positive occurrence of the consequent P_3. Therefore the assertion generated at this node is $(P_1 \wedge \neg P_2)$ |=> P_3.

III. TEMPORAL SEQUENCE EXPRESSIONS

The primary challenge faced in assertion writing is in handling assertions that span over time. The assertion mining algorithm, presented in Algorithm 1, produces assertions of the form $X \wedge Y$ |=> Z, however these are void of the temporal dimension that is crucial to understanding the more involved cause-effect relationship between artifacts in the design. To infuse the notion of timed sequences into assertions generated by the mining algorithm, we must identify relationships between the consequent and its possible causes over time. We resolve the issue of identifying exact points of time that reconcile the notion of temporal influence by taking advantage of the non-determinism that results from the use of the sequence delay operator ##[a:b]. The semantics of the delay operator in the sequence expression X ##[a:b] Y, requires Y to be true at any point within the time interval [a:b] after X is true. An implication of the form X |=> ##[a:b] Y asserts that whenever X is true, Y becomes true within the time interval [a:b] of X being true. Since X can be true over a continuum of time, the assertion requires that for every time point in the continuum of the truth of X, a corresponding time point exists within [a:b] during which Y is true. This entails looking at the truth intervals of Y and identifying the influence of Y in the past upto a bound k, i.e. to answer the question, "Does X's truth during the interval [2:3.1] have an influence on the truth of Y within the interval [5:7] with a separation of [0.5:1.2]?", we must project the consequents truth interval back by computing [5:7] $\ominus$ [0.5:1.2] = [3.8:6.5] and determine if [2:3.1] falls within this window. In this case, since [2:3.1] $\cap$ [3.8:6.5] = ϕ, there is no influence of the interval on the consequents truth.

To develop temporal assertions, we transform the truth set $\mathbb{I}_\tau(\mathbb{P})$ by representing the truth of the consequent PORVs influence at past intervals with time t offset by 0, k, $2k$, $3k$, ..., $(N-1)k$ in the past. Additionally, to enable wider correlations to be identified, we use a temporal shift of [0:i×k]. We denote the consequent P_c shifted by i×k as P_c^i. For PORVs X and Y in the sequence expression X |=> ##[a:b] Y, if the sequence associates Y, at time t, with X with a temporal shift of ##[0:k], this would mean that Y is associated with X being true, at any time within the window [t:t]$\ominus$[0:k]=[t-k:t].

Algorithm 2: AMS-Miner^T: Assertion Mining Temporal Assertions for AMS Design Traces

Input: Truth Set $\mathbb{I}_\tau(\mathbb{P})$ for trace τ, PORV List $\mathbb{P}$, Consequent PORV P_c, Assignment List $\mathcal{A}$, Sequence Length N, Sequence Position for the split T.
Output: Assertions $\mathcal{R}$, structured as the decision tree.

1   $M, E \leftarrow$ Lists of length N.
2   **for** $0 \leq i < N$ **do** $M[i] \leftarrow \mu_{\tau_\mathcal{A}}(P_c^i)$; $E[i] \leftarrow \epsilon_{\tau_\mathcal{A}}(P_c^i)$;
3   **if** $e = 0$ **then** $\mathcal{R} \leftarrow \mathcal{R} \cup (\mathcal{A} \Rightarrow P_c\langle M[T]\rangle)$; **return** ;
4   $P_{best} \leftarrow \phi$; $T_{best} \leftarrow \phi$; $g_{best} \leftarrow -\infty$;
5   **for** $0 \leq i < N$ **do**
6   **forall** $P \in \mathbb{P}$ **do**
7   **if** $i==0$ **then**
8       $g \leftarrow E[i] - \epsilon_{\tau_{\mathcal{A}\cup\langle P,\top\rangle}}(P_c^i) - \epsilon_{\tau_{\mathcal{A}\cup\langle P,\bot\rangle}}(P_c^i)$;
9   **else** $g \leftarrow max(E[i] - \epsilon_{\tau_{\mathcal{A}\cup\langle P,\top\rangle}}(P_c^i)$,
10          $E[i] - \epsilon_{\tau_{\mathcal{A}\cup\langle P,\bot\rangle}}(P_c^i))$;
11   **if** $g > g_{best}$ **then** $P_{best} \leftarrow P$; $T_{best} \leftarrow i$; $g_{best} \leftarrow g$;
12   AMS-Miner^T($\mathbb{I}_{\tau_{\mathcal{A}\cup\langle P_{best},\top\rangle}}(\mathbb{P}/P_{best})$, $\mathbb{P}/P_{best}$, $\mathcal{A} \cup \langle P_{best}, \top\rangle$, N, T);
13   AMS-Miner^T($\mathbb{I}_{\tau_{\mathcal{A}\cup\langle P_{best},\bot\rangle}}(\mathbb{P}/P_{best})$, $\mathbb{P}/P_{best}$, $\mathcal{A} \cup \langle P_{best}, \bot\rangle$, N, T);

This allows us to find PORVs, temporally further from the consequent PORV, at the initial levels of the decision tree, thus giving us the potential to better partition the data space on the basis of the more dominant correlations in the trace. The user specifies k, the maximum interval separating two PORVs in the sequence, and N, the maximum length of a sequence. The truth set is augmented to $\hat{\mathbb{I}}_\tau(\mathbb{P})$ as follows:

$$\hat{\mathbb{I}}_\tau(\mathbb{P}) = \mathbb{I}_\tau(\mathbb{P}) \cup \left(\bigcup_{1 \leq i < N} \mathcal{I}_\tau(P_c) \ominus [0 : i \times k] \right) \quad (5)$$

where,

$$\mathcal{I}_\tau(P) \ominus [a : b] = \bigcup_{I \in \mathcal{I}_\tau(P)} I \ominus [a : b] \quad (6)$$

We provide the augmented truth set $\hat{\mathbb{I}}_\tau(\mathbb{P})$ of Equation 5 as input to Algorithm 2. Additionally, the computation of gain must be adapted to account for the fact that, due to the non-determinism that is integral to the semantics of future causality in a sequence expression, a predicates influence of truth and falsity can overlap, that is, at a time point t in the trace, t can cause the predicate to be either true or false. Consider the backward influence of P_1 in the interval [0:2] as follows:

$$\mathcal{I}_\tau(P_1)\ominus \text{[0:2]} = \{\text{[0:4]},\text{[5:13]}\}$$
$$\mathcal{I}_\tau(\neg P_1)\ominus \text{[0:2]} = \{\text{[0:10]}\}$$

To account for this overlap, we use the best of the gains computed for the true and false interpretations of the consequent. The gain for PORV P is computed as follows:

$$gain = max(\epsilon_{\tau_\mathcal{A}} - \epsilon_{\tau_{\mathcal{A}\cup\langle P,\top\rangle}}(P_c), \epsilon_{\tau_\mathcal{A}} - \epsilon_{\tau_{\mathcal{A}\cup\langle P,\bot\rangle}}(P_c))$$

The theory for assertion generation extends naturally to sequences over time, resulting in assertions of the form X ##[a:b] $\wedge$ Y |=> ##[c:d] Z. Such an assertion in English is interpreted as "If X and Y respectively occurred within the past time windows of [a+c:b+d] and [c:d] before a time point t, then Z is true at t". At the stage of the leaf nodes, where assertions are generated, PORV truth associations that

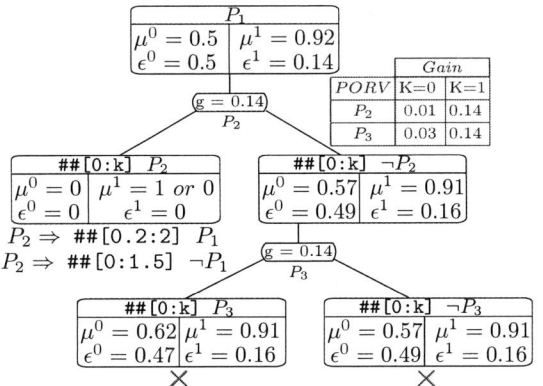

Fig. 4: AMS-Miner^T decision tree for Truth Set of Table II. K is the position of temporal influence from the consequent.

correlate with the consequents truth, along with the temporal shift made to each PORV are available. Additionally, the truth intervals for each component PORV at the leaf node are also known. We then use the Minkowski difference to generate the time intervals separating the PORVs to form a sequence expression like the one shown earlier.

Figure 4 describes a run of Algorithm 2 on the truth set of Table II, with the consequent being P_1. Table II describes a very interesting relationship between the truth of P_2 and the truth of P_1, i.e. P_2 becomes true a few time units earlier than P_1. Additionally, when P_2 is true, P_1 is false for a period of time before it becomes true. We see how AMS-Miner^T is able to extract these patterns from the trace.

In Figure 4, μ^K and ϵ^K are the mean and error computed with respect to the consequents truth, taken K steps into the past (as discussed earlier), and are indicative of the future causality of the assertion at the node with the consequents truth. The value of the time quantum used for shifting truth intervals is $k = 2$. As shown in Figure 4, the best gain at the root node is for P_2 associated with the future causality of P_1. Therefore a split with respect to P_2 is performed, yielding one child node for which the error is zero and future causality on the consequent is associated with both, when the consequent is true and when it is false. The truth sets when P_2 is used a splitting point (with respect to P_1 at a distance of `##[0:2]`) are as follows:

$$\text{Choosing} \left\{ \begin{array}{l} \mathcal{I}_{\tau_{\langle P_2, \top \rangle}}(P_1^1) = \{[0:1], [5:5.5], [9.8:9.9]\} \\ \mathcal{I}_{\tau_{\langle P_2, \top \rangle}}(\neg P_1^1) = \{[0:1], [5:5.5], [9.8:9.9]\} \end{array} \right.$$

$$\text{Choosing} \left\{ \begin{array}{l} \mathcal{I}_{\tau_{\langle P_2, \bot \rangle}}(P_1^1) = \{[1:4], [5:9.8], [9.9:13]\} \\ \mathcal{I}_{\tau_{\langle P_2, \bot \rangle}}(\neg P_1^1) = \{[1:5], [5:9.8], [9.9:10]\} \end{array} \right.$$

At the node for $P_2 = \top$, using the Minkowski operations, two assertions are generated, namely P_2 `|=> ##[0.2:2]` P_1 and P_2 `|=> ##[0:1.5]` $\neg P_1$. Exploring the branch when $P_2 = \bot$ does not yield assertions, as indicated in the figure.

The timing information from the assertions generated by AMS-Miner^T is especially useful. For instance, if P_2 were the *enable* for a Low dropout regulator (LDO), and P_1 is the PORV ($v > 4.2$), where v is the output voltage and 4.2 is the rated voltage for the LDO; the assertions generated indicate that evidence shows that the LDO's output is under the rated voltage for at most 1.5 time units and reaches the rated voltage between 0.2 and 2 time units after the *enable* is asserted. The time window `[0.2:1.5]` is fuzzy, i.e. P_1 may or may not be true during this time window.

IV. CONCLUSION

In this work we present AMS-Miner^T, a methodology to automatically extract AMS assertions in the form of sequence expressions expressing cause-effect relationships between signals in an AMS design. Our methodology uses time intervals to handle dense time and predicates over real variables to handle real-valued analog signals, and builds on techniques of data mining using decision tree learning to produce complex temporal assertions. The methodology for mining AMS assertions was demonstrated on an example. In the future, we plan on building AMS-Miner^T into a CAD tool and have it tested on complex designs from AMS circuit families, such as Low-Dropout Regulators, PLLs, Buck Regulators; from the control domain, as well as other complex integrated systems. We believe that AMS-Miner^T is an important step towards increasing the productivity of AMS verification engineers and in revealing previously unknown, desirable or undesirable, patterns of behaviour, enabling early action, and minimizing human resource effort and cost.

REFERENCES

[1] *1800-2012 - IEEE Standard for SystemVerilog*, IEEE Std. [Online]. Available: http://standards.ieee.org/findstds/standard/1800-2012.html

[2] A. Gupta, "Assertion-based Verification Turns the Corner," *IEEE Design & Test*, vol. 19, no. 4, pp. 131–132, Jul. 2002.

[3] P. Dasgupta, *A Roadmap for Formal Property Verification*. Secaucus, NJ, USA: Springer-Verlag New York, Inc., 2006.

[4] S. Vasudevan *et al.*, "GoldMine: Automatic assertion generation using data mining and static analysis," in *DATE*, March 2010, pp. 626–629.

[5] S. Hertz, D. Sheridan, and S. Vasudevan, "Mining hardware assertions with guidance from static analysis," *IEEE TCAD*, vol. 32, no. 6, pp. 952–965, Jun. 2013.

[6] O. Maler and D. Nickovic, "Monitoring temporal properties of continuous signals," in *FORMATS-FTRTFT*, ser. LNCS, vol. 3253. Springer, 2004, pp. 152–166.

[7] S. Mukherjee and P. Dasgupta, "Incorporating Local Variables in Mixed-Signal Assertions." in *IEEE International Conference TENCON*, 2009.

[8] S. Mukherjee, P. Dasgupta *et al.*, "Synchronizing ams assertions with ams simulation: From theory to practice," *ACM TODAES*, vol. 17(4), pp. 38:1–38:25, 2012.

[9] A. Ain *et al.*, "Feature Indented Assertions for Analog and Mixed-Signal Validation," *IEEE TCAD*, vol. 35, no. 11, pp. 1928–1941, Nov 2016.

[10] A. A. B. da Costa, P. Dasgupta, and G. Frehse, "Formal Feature Analysis of Hybrid Automata," in *Proc. of MEMOCODE*, 2016.

[11] A. A. B. da Costa and P. Dasgupta, "Formal interpretation of assertion-based features on AMS designs," *IEEE Design & Test*, vol. 32, no. 1, pp. 9–17, 2015.

[12] S. Mukherjee, P. Dasgupta, and S. Mukhopadhyay, "Auxiliary specifications for context-sensitive monitoring of ams assertions," *IEEE TCAD*, vol. 30(10), pp. 1446–1457, 2011.

[13] L. A. Breslow and D. W. Aha, "Simplifying decision trees: A survey," *Knowl. Eng. Rev.*, vol. 12, no. 1, pp. 1–40, Jan. 1997.

[14] J. R. Quinlan, "Induction of decision trees," *Mach. Learn.*, vol. 1, no. 1, pp. 81–106, Mar. 1986.

[15] J. R. Quinlan, *C4.5: Programs for Machine Learning*. San Francisco, CA, USA: Morgan Kaufmann Publishers Inc., 1993.

2018 31th International Conference on VLSI Design and 2018 17th International Conference on Embedded Systems

ELURA: A Methodology for Post-silicon Gate-level Error Localization Using Regression Analysis

Ankit Jindal[1], Binod Kumar[1], Kanad Basu[2] and Masahiro Fujita[3]

Indian Institute of Technology Bombay, India[1], New York University, USA[2], University of Tokyo, Japan[3]

Email: {ankitjindal, binodkumar}@ee.iitb.ac.in[1], kb150@nyu.edu[2], fujita@ee.t.u-tokyo.ac.jp[3]

Abstract—Limited observability of internal design states exacerbates the problem of error localization at post-silicon stage. With the assistance of system-level observed features, errors can be localized at higher design abstraction level during post-silicon validation. However, localizing error(s) at the gate-level becomes very difficult due to the tremendous design complexity of modern circuits. In this work, we propose a methodology for localizing error to a small portion of the gate-level design description based on a regression analysis of circuit responses to synthetically injected errors. The model learns the flip-flop states upon error introduction and given an actual erroneous netlist, provides ranking of infected flip-flops. The error localization analysis is done by testing with states of the flip-flops which are traced through trace buffer mechanism. On-chip trace buffers are capable of overcoming the obstacle of diminished observability, which can be further enhanced with the help of state restoration technique. We demonstrate that state restoration significantly improves error localization. Experimental results on benchmark circuits justify the efficacy of the proposed methodology on two gate-level error models which represent logical error manifestations.

Index Terms—Post-silicon validation, Trace signal selection, Error Localization, Error detection, Design topology.

I. INTRODUCTION

Post-silicon validation and debug have attained a position of utmost importance in the modern design implementation cycle [1]. The verification efforts at the pre-silicon stage are not capable of exposing some design bugs which ultimately end up slipping to the first silicon as simulation is many orders of magnitude slower than actual chip execution. In addition to that, due to inaccurate abstraction of simulation models and process variations, there are chances that many electrical bugs appear in the first released silicon. An efficient bug localization process at the post-silicon stage helps in avoiding re-spins of the design. However, in the present era of designs with tremendous complexity, the limited accessibility of internal signals of design is the main obstacle during post-silicon validation and debugging.

Abramovici et al. [2] proposed employing few trace buffers as a solution to tackle the hindrance of restricted post-silicon visibility. A subset of internal signals of the design need to be selected for tracing. These signal states can be used off-line to analyze the root-cause of observed malfunctioning/erroneous behavior. Since area overhead minimization forces us to use limited number of such buffers, the observability can be extended to a higher extent through use of state restoration technique (i.e., deriving untraced signal values

with the assistance of traced signal bits) [3]–[5]. Different signal selection methodologies achieve maximization of state restorability through various means. However, very few signal selection approaches attempt to link state restoration with either error detection or localization. Basu et al. [6] proposed trace signal selection in a dynamic manner based on location of some error-prone design zones. They ranked flip-flops in term of their utility through computation of conditional error propagation probabilities of signal transitions from gate inputs to outputs on paths between various flip-flops. Kumar et al. [7] have shown results for some design errors on ISCAS'89 benchmark circuits that state restoration does not explicitly target error detection. They assigned ranks to flip-flops through simulations based on pre-determined list of netlist based errors. However, error localization using circuit responses from trace buffers have not been investigated by any of the signal selection approaches.

This paper proposes an error localization methodology at the gate-level to assist the debug engineer in analyzing the root-cause of failures at the post-silicon stage. We synthetically inject errors into the netlist and based on the circuit response (i.e., signal states of flip-flops of the design), a regression model is developed for analysis. One of the intended bug injection methods is random gate replacement technique [7], [8] which acts as a representative of many logic bugs. Considering the large number of gates in modern designs, replacement of each gate with a random one would turn out to be highly exhaustive. Therefore, we iteratively perform error injection into very small portions of the design (netlist) and then apply the regression analysis for building the model. The localization experiments are performed by utilizing responses (i.e., flip-flop state values) gathered from the on-chip trace buffers. We obtain a list of flip-flops (corresponding to erroneous netlist regions) infected by the error (injected during testing) which are ranked based on a pre-defined scoring mechanism.

The remainder of the paper is organized as follows. Section II summarizes related work on post-silicon error localization techniques at different abstraction levels of design. Section III presents our proposed methodology and different terminologies associated with it. Section IV explains our testing methodology for a given erroneous design and analyses the results obtained from our experiments and some inferences are discussed. Section V finally concludes the paper with directions on some future research.

978-1-5386-3693-0/18 $31.00 © 2018 IEEE   410

II. Previous Work

Park et al. [9] proposed IFRA (Instruction Footprint Recording and Analysis) for debugging processor systems for errors with a block-level spatial localization by the use of low-cost hardware recorders. However, such an architectural block in a processor typically contain thousands of gates which makes it very difficult to debug at the gate-level. Furthermore, IFRA assists in localizing the bugs only when errors are detected on-chip within about a thousand cycles. DeOrio et al. [10] proposed a post-silicon bug diagnosis methodology based on data collection from failing tests and then applying clustering technique to form signal groups. Since during post-silicon validation, a large number of tests are applied, the amount of logs collected is also very large. Therefore, machine learning techniques like clustering, regression analysis can be deployed to extract hints for bug/error localization from the obtained test response logs. Mammo et al. [11] have proposed an automatic mismatch diagnosis approach for exactly identifying the buggy unit(i.e., processor block) with the assistance of hints from instruction set simulation in addition to the normal test execution on DUV (design-under-verification). Similar to [10], Bertacco et al. [1] proposed a mechanism to detect root-cause signals by identifying "outliers" signals from post-silicon tests and iteratively selecting signals to be monitored. Khudia et al. [12] have analyzed bugs that manifest inconsistently over repeated executions of the same test. They classified the internal signals into *passing groups* and *failing groups* for bug diagnosis upon test executions. A methodology to bridge pre-silicon and post-silicon verification by on-chip detection of protocols (which have been extracted during pre-silicon) and then performing an off-line software diagnosis for localization at architectural block level and the corresponding buggy signal along with the detection cycle [13]. With the help of information obtained from trace-buffers, a technique for post-silicon fault localization has been proposed by Zhu et al. [14] using invariant identification with a SAT formulation.

III. Proposed Methodology

We propose a methodology, *ELURA* (Error localization using regression analysis) which assists in localizing the infected flip-flop(s), which in turn hint towards buggy gate(s) in a erroneous design. The proposed methodology includes two phases: *training* and *testing*. The *training* phase includes building a model which consists of a relationship between smaller region of circuit (in which bugs are injected synthetically) and the corresponding circuit responses which, in turn are to be utilized for identifying the buggy gates given an erroneous netlist. The model is developed through a large number of iterations, where in each iteration one error is injected into the design description and the complete error signature (state of all the flip-flops of the design for a fixed number of cycles) are used for *training*. During *testing* phase, for localizing the error location in a buggy netlist based on the partial bug signature (state of certain number of flip-flops of the design) obtained through on-chip trace buffers. While the localization ability of the proposed model depends on the training of the model, error localization to a small portion of the design depends on the post-silicon observability available through the trace buffers. The proposed methodology makes use of the advantage of indirect expansion of visibility through state restoration technique [3]–[5]. Since restoration of untraced signal states varies with the trace signal selection methodology, variation in localization results is expected with each technique up to some extent. Table I introduces the terminology for various parameters to be used.

TABLE I
TERMS AND THEIR MEANING

Term	Meaning
M	training model
TBw	width of trace buffer
D_{tb}	depth of trace buffer
ffZ_k	any *flip-flop zone* of the circuit
K	no. of iterations of training
Sle_i	signature used for training
Cy	no. of cycles in each iteration of training
n	number of flip-flops in netlist
FF_j	error signature of jth flip-flop for training
Ste	signature used for testing
$rank_j$	*rank* of jth flip-flop
ff_j	error signature of jth flip-flop for testing

A. Finding smaller zones in circuit

The first step in the *training* phase is the introduction of bug/error in smaller portion of the netlist for the purpose of training. The circuit description is divided into a fixed partitions, termed as flip-flop zones (ffZ) meaning a collection of gates connected to one particular flip-flop. Under this step, a particular ffZ can have gates connected in certain levels (generally one to six). If the circuit has n flip-flops, there can be a maximum of n flip-flop zones. The idea behind ffZ finding approach is that once we are able to locate any ffZ, the individual erroneous gate(s) can be identified with the help of that information. Figure 1 illustrates ffZ identified from a design netlist. The portion outlined (with solid line) comprising of one flip-flop and six gates is ffZ corresponding to *FF1*. The other region (outlined with dashed line) depicts ffZ corresponding to second flip-flop (FF2).

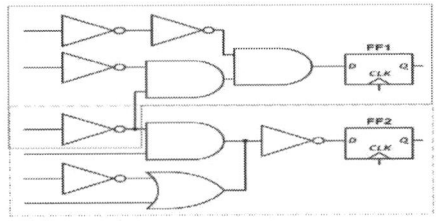

Fig. 1. Circuit for illustrating ffZ identification

The error propagation to a flip-flop depends on the conditional probabilities of signal transitions at different inputs of gates following on the path leading to it. Therefore, for developing M, we take one ffZ and inject error into it and then obtain a bug(error) signature. During the next iteration, we select (in random fashion) another ffZ and repeat the same procedure. Note that it is possible that some gate(s) may be common between two flip-flop zones of the design. During testing phase of the proposed methodology, error localization up to either one ffZ or a small list of ffZ is achieved with the help of statistical analysis (regression).

B. Error injection in smaller zones

At each iteration of training, we inject error by the method of random gate replacement, where one gate is randomly replaced by another [7], [8]. This model resembles manifestations of many kind of mutations at RTL after the synthesis step. For some cases, this model can serve as a representative of electrical errors also. For example, the *OR* gate from which a flip-flop in design derives its input gets replaced by *XOR* gate and in some cycle the content of this flip-flop is "1". When both inputs to the new gate become "1", its outputs become "0" causing a bit-flip at the concerned flip-flop.

Note that any other error-model can also be utilized for the purpose of generating buggy signatures to develop the regression model (M). We introduce random error to remove any bias towards any specific region in the circuit during *training phase*. Figure 2 shows an example ffZ where one *AND* gate (shown in the left side) is replaced by one *OR* gate (shown in red color in the right side).

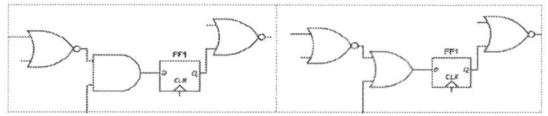

Fig. 2. Circuit for illustrating error injection

It is likely that during any iteration of the *training* phase, error is not injected due to nature of the selected ffZ in this iteration. For instance- if ffZ contains only "NOT" gates, error injection can not proceed. Another likely scenario is that during some iterations (out of total K) of training, same ffZ is selected for error injection since this selection is done in random manner. However if during $(i+1)^{th}$ iteration, selected ffZ is same as that of ith iteration, the error injected is likely to be different giving a different error signature.

The error signature generated during each iteration of this phase is related with the infected flip-flop zone(ffZ) and over a large number of iterations, a regression model is developed between the two parameters. Thus, error signature, Sle_i (in each iteration) leads to a training vector, Trv_i in the model M which can be related with the corresponding ffZ.

C. Developing the regression model

For analyzing regression between the injected error (in ffZ) and respective FF_i (error signature), the procedure described previously is iterated K times (depending on number of flip-flops of the netlist). The *error signature* consists of states of all flips of design for Cy cycles. Two possible methods to compute error signature (FF_i) are given as follows:

- *L1*: Recording only the erroneous states of all the flip-flops, assuming no golden (error-free) signature.
- *L2*: Assuming that a golden signature is available, computing XOR values of the golden and erroneous signature.

Corresponding to these two training methods, difference is expected between error localization ability of the resultant model. The steps of developing the regression model are summarized in the $Training$ procedure.

```
1:  procedure Training(In:Netlist,K,choice,n; Out:M)
2:      M ← ∅
3:      goldensignature ← state of n flip-flops for Cy cycles
4:      for i=1 to K do
5:          find flip-flop zones from Netlist
6:          S(ffZ) ← flip-flop zones from Netlist
7:          ffZ_k ← randomly selected one ffZ from S(ffZ)
8:          inject random error in ffZ_k
9:          for each of n flip-flops for Cy cycles do
10:             FF_jp ← state of j^th flip-flop in p^th cycle
11:             errorsignature ← state of all FF_jp's
12:             if choice = L1 then
13:                 I ← {ffZ_k,errorsignature}
14:             else if choice = L2 then
15:                 XOR(goldensignature,errorsignature)
16:                 I ← {ffZ_k,XORsignature}
17:             end if
18:         end for
19:         M ← I ∪ M
20:         S(ffZ)← ∅, I ← ∅
21:     end for
22: end procedure
```

Each training vector (Trv_i) comprises of a matrix of $nxCy$ state bits (flip-flop responses). For K iterations of training, M contains K labels where label ffZ_1 correspond to Trv_1, label ffZ_2 correspond to Trv_2 and so on.

D. Evaluating the regression model

1) Defining Objective function: During testing phase, signature (Ste) generated from trace buffers is the testing vector, Ttv. From this testing vector, an *Objective function($Objfn$)* is computed for each label (ffZ) which denotes the deviation of each flip-flop , ff_j (of Ttv) from M. For $L1$ training method, $Objfn$ is calculated as per Equation 1, where FF_j denotes corresponding flip-flops in the training vector(s). For each ffZ, *Objective function* is depicted by $Objfn_{L1}(ffZ)$ when $L1$ training method is adopted.

For the second training method, *Objective function* is given by Equation 2, where the only difference is that FF_j and ff_j are replaced by XOR values of their golden and erroneous signature values. In both the equations, p denotes the cycle of jth flip-flop. This factor measures distance as the cycle-by-cycle computation($Objfn$) of all flip-flops (beginning from the first flip-flop) proceeds.

$$Objfn_{L1}(ffZ) = \sum_{j=1}^{j=n} \sum_{p=1}^{p=Cy} (FF_{jp} - ff_{jp})^2 \qquad (1)$$

$$Objfn_{L2}(ffZ) = \sum_{j=1}^{j=n} \sum_{p=1}^{p=Cy} (FF'_{jp} - ff'_{jp})^2 \qquad (2)$$

978-1-5386-3693-0/18 $31.00 © 2018 IEEE

where, we have

- FF_j' given by XOR(golden FF_j, erroneous FF_j)
- ff_j' given by XOR(golden ff_j, erroneous ff_j)

For an effective localization, $Objfn$ must be minimized justifying that localization happens based on the model (M) if Ttv is similar to any one of the Trv_i. Thus, $Objfn$ captures the dissimilarity between all Trv and Ttv for each ffZ. In above Equation, FF_j' (training vector components) and ff_j' (testing vector components) are compared only in those clock cycles where there is a difference between the two.

It is worth to note that the model M contains K labels corresponding to K iterations, where each label correspond to a ffZ (which consists of one flip-flop with some gates). From the perspective of a regression model, the label which corresponds to best error localization in M is represented by Equation 3 where $a_1, a_2 \ldots\ldots a_K$ are constants and ffZ_1, $ffZ_2 \ldots\ldots ffZ_K$ are the corresponding labels (ffZ).

$$ffZ_k = a_1 ffZ_1 + ..a_i ffZ_i + .. + a_K ffZ_K \quad (3)$$

The actual erroneous ffZ_k is any one ffZ_i (the constant a_i corresponding to that is 1, while all others are 0). This ffZ_k corresponds to the minimum value of either $Objfn_{L1}$ or $Objfn_{L2}$ as per the respective training method ($L1$ or $L2$).

2) Analysis of profitability of training methods: For $L1$ method, flip-flop state values in testing vector are to be compared with those in training vectors of M, while in $L2$ only XOR values of golden and erroneous in testing vector are compared with XOR values in training vector. Thus, in $L2$ we have a refined data ($Objfn$ values) for the purpose of comparison. This is because we have comparison points ($Objfn$ values) in $L2$ case only when bit values of XOR between golden and erroneous signature are "1".

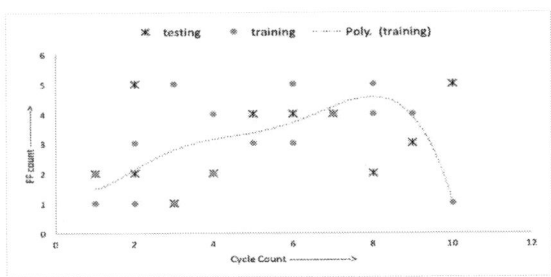

Fig. 3. Testing and training methodology illustration as per $L1$

For the purpose of illustration, let us assume a design with small number of flip-flops (six) and we wish to test the training (corresponding to few clock cycles) of M with a training vector. Figure 3 illustrates $L1$ case. The signatures here depict flip-flops states whenever these are "1" for respective clock cycle. Note that training signatures fit into a curve (represented by poly.(training)) and there are flip-flop responses (of testing) for which distance ($Objfn$ value) is to be measured from the fitted curve to measure the dissimilarity. Thus, the distance between test points (i.e., response values) and the training curve needs to be minimized for optimizing $Objfn$. Note that

this distance ($Objfn$ value) is large in case of $L1$ while it is substantially reduced for $L2$ case.

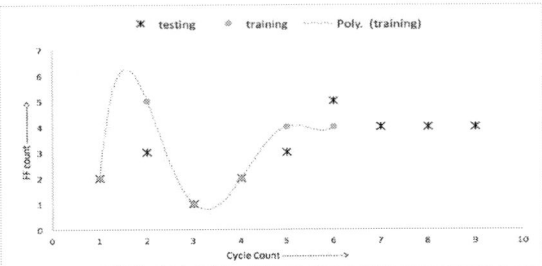

Fig. 4. Testing and training methodology illustration as per $L2$

The scenario for $L2$ case is illustrated in Figure 4 where we have done curve-fitting (represented by poly.(training)) for the XOR signature (used in training) in those cycles i.e., when the erroneous and golden response bits do not match and the resultant XOR bits are "1". Thus, we have $Objfn$ values in this case for those clock cycles, where the injected error has effected the corresponding flip-flop (or, ffZ).

The computation of $Objfn$ (in Equation 1 and 2) proceeds differently with the post-silicon observability available through on-chip trace buffers. This is because the amount of known bits (0 or 1) and don't care bits (X) varies whether restoration technique is applied (or not). After utilizing restoration, $ObservationWindow(ObsW)$ which denotes the number of flip-flop states visible/observable changes. To account for this variation, a factor ($1/ObsW$) needs to be introduced in $Objfn$ computation. Note that for restoration, we can estimate $ObsW$ as *restored and traced states* (rts), calculated by sum of *traced states* ($=TBw*D_{tb}$) and *restored states* whereas for non-restoration, we can estimate $ObsW$ as *traced states* (ts) only , computed by $TBw*D_{tb}$. It is obvious that $ts < rts$ whenever we achieve non-zero restoration using the traced signal states.

$$Objfn = \frac{1}{ts} \sum_{j=1}^{j=n} \sum_{p=1}^{p=Cy} (FF_{jp} - ff_{jp})^2 \quad (4)$$

$$Objfn = \frac{1}{rts} \sum_{j=1}^{j=n} \sum_{p=1}^{p=Cy} (FF_{jp} - ff_{jp})^2 \quad (5)$$

For the same injected error (of random nature), if we do not apply restoration, number of state bits in the test vector error signature are less. However, in case of restoration (obtained from trace signals selected by any restorability maximization technique [3]–[5]) number of unknown bits are reduced. It is obvious from the above expressions that in a scenario when we get multiple suspects, we obtain lesser $Objfn$ with restorability technique as compared to non-restoration.

IV. EXPERIMENTAL FORMULATION AND RESULT

A. Experimental Setup

We have chosen seven benchmark circuits (ISCAS-89, ITC-99 and OpenCore) for testing the localization efficiency of the proposed regression analysis model. For building the regression model, we selected value of K (which can be variable for each benchmark) between 1000 to 5000. Note that random input test vectors have been used during the *testing* phase. For the *training* phase also, random inputs were applied.

TABLE II
CHARACTERISTIC OF BENCHMARKS

Sl.No.	Name	No. of FF's	No. of gates
1	$b17$	864	13988
2	$b18$	3320	114621
3	$p16c5x$	739	4107
4	$softUSB\_navre$	317	4718
5	usb	1761	10650
6	$s38417$	1636	22179
7	$s38584$	1426	19253

For experiments, TBw is chosen as 32 and D_{tb} as 1024 for trace-buffers. Thus, for testing an erroneous design netlist, we have the *error signature* (comprising of these 32 flip-flops for 1024 cycles (Cy)). We inject error into the design netlist by 2 methods [7], [8] separately (during each experiment):

- *e1*: random gate replacement
- *e2*: wire exchange between inputs to different gates, selected in a randomly manner

B. Formulation for identifying false positives

During *testing* phase, based on the *error signature*, we obtain the suspect ffZ (i.e., corresponding flip-flops in ffZ). One important issue is to verify the suspect ffZ obtained from M. As is possible with any statistical process, there are chances of *false positives* i.e., cases when some ffZ are not affected but during testing, regression model M gives these *flip-flop zones* as suspects(S). To verify that no *false positives* are provided during *testing phase*, we perform a topological connection based analysis of the injected error location($error_i$) and identify the flip-flops (i.e., few ffZ) that can be infected by $error_i$. Through this topological analysis, we obtain a parameter $n(S_{con})$, whereas from the regression model M, we obtain another parameter $n(S)$.

- $n(S_{con})$: number of flip-flops (or, ffZ regions) directly connected to the location of $error_i$ (in the forward logical-cone).
- $n(S)$: number of suspects (ffZ) obtained during *testing*.

For efficient localization, we want $n(S)$ to have the least possible value (i.e., 1). However during testing, when we have multiple suspect ffZ, these probable ffZ are ordered by $rank_j$ as per their respective $Objfn$ values. The regression model, M must abide by the following two conditions while being evaluated on any testing vector for error localization:

- $n(S_{con}) \geq n(S)$
- $n(S) \subseteq n(S_{con})$, for occurrence of no *false positives*

C. Defining error localization metric

We repeat the testing experiment a number of times (denoted by $N_{testing}$) and define the following parameter to measure

the quality of error localization. Lower the value of *elmetric*, better is the error localization.

$$elmetric = \frac{\sum_{j=1}^{j=N_{testing}} n(S)}{N_{testing}} \qquad (6)$$

It is worth to note that in every iteration of *testing* phase, a separate error is introduced for the purpose of testing. The error injection is purely random in nature. We choose $N_{testing}$ as 100 in our experiments. During our experiments we observed $n(S)$ as 1 in many iterations for all the benchmark circuits, signifying that the proposed methodology is able to identify the actual suspect ffZ. It is worth to note that the values of $n(S_{con})$ vary from a range of 1X-6X to that of $n(S)$. For all the testing experiments, we obtained that $n(S) \subseteq n(S_{con})$ depicting that the proposed regression analysis avoids the occurrence of *false positives*.

D. Results on elmetric computation and discussions

1) Comparative evaluation of training methods for e1: Figure 5 depicts variation in the proposed evaluation parameter, *elmetric* for both the training methods ($L1$ and $L2$) when injected error is of $e1$ type. As explained in Section III-D2, the XOR based training model is expected to assist the localization in a better way. As it is clear from Figure 5, $n(S)$ has values greater than 1. Under these cases, the suspects are ranked ($rank_j$) by their $Objfn$ values. The lesser $Objfn$ values (more closeness between M and testing vector), higher the chances of it being the actual infected (i.e., suspect) ffZ.

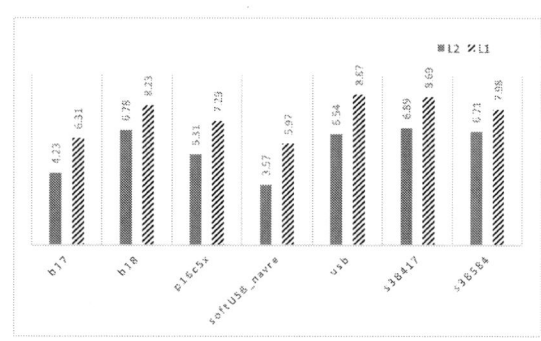

Fig. 5. Comparison of *elmetric* between $L1$ and $L2$ ($e1$)

2) Comparative evaluation of restoration and non-restoration methods for e1: To compare *elmetric* for the cases of only-traced (i.e., not restored) post-silicon observability and traced (and restored) observability, we utilize restoration technique of [4]. An increase in ease of localization is obtained by using state restoration technique as we observe lesser $n(S)$ values with this technique. This observation justifies the proposals of concept of maximization of restoration suggested by authors in restorability techniques such as [3]–[5].

3) Comparative evaluation of training methods for e2: When the injected error is of $e2$ type, the training methods ($L1$ and $L2$) are compared in Figure 7. We observed that localization of wire-exchange error is poor than the gate-replacement error. We figured out that our error injection mechanism fails to create an erroneous signature during few

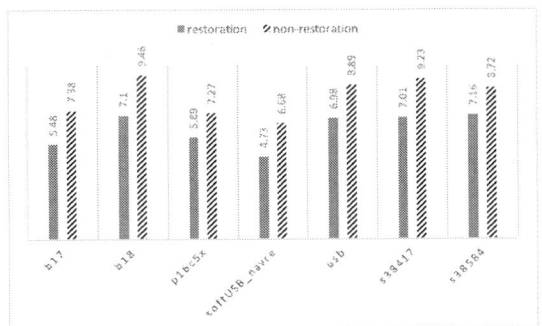

Fig. 6. Comparison of *elmetric* between *res.* and *non-res.* (e1)

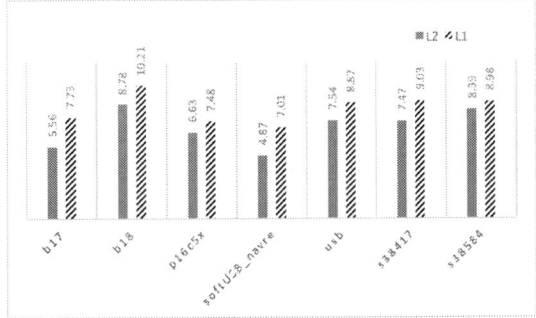

Fig. 7. Comparison of *elmetric* between $L1$ and $L2$ (e2)

training iterations which may have led to poor training in M. However, localization still succeeds in many iterations (during testing phase) and we obtain $n(S)$ as 1 (best value of *elmetric*).

4) Comparative evaluation of restorability maximization methods for error localization: As explained in Section III-D2, error localization varies with the amount of restored state values. We analyzed on 4 benchmark circuits with three restorability maximization techniques [3]–[5]. Since restoration is

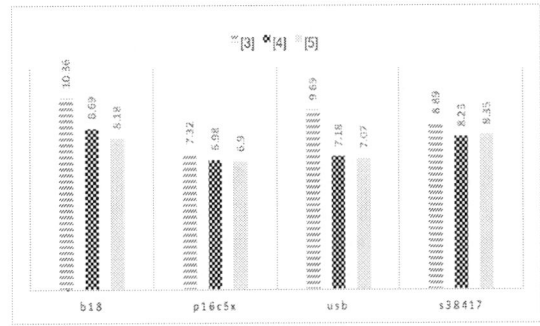

Fig. 8. Comparison of *elmetric* between *restoration* methods (e1)

lesser with [3] out of these three techniques, corresponding *elmetric* has a higher value for all the benchmark circuits with this approach. The other two approaches provide comparable *elmetric* values.

V. CONCLUSION

This paper proposed a methodology for error localization to a smaller region of the circuit under the limited post-silicon visibility. Using regression analysis, a model was developed by

training for a large number of iterations which was used to test an erroneous design to obtain probable suspects with a ranking scheme. The proposed technique localized to one suspect during the experiments in many cases. The proposed methodology can be improved by various means like employing a suitable technique for training a large number of features during the training phase. Increasing number of features during training can help localize to the particular erroneous gate instead of a smaller region of the circuit. Identifying such training features is expected to play an important role in error localization. The definition of the objective function may also be refined to achieve a better ranking of suspects. One such refinement can be to define a scaling factor which can account for the selection of flip-flops as trace signals. This can also assist in shaping the characteristics of the regression model in a better way. We intend to evaluate the efficacy of the proposed methodology on other kind of error models like bit-flips.

REFERENCES

[1] V. Bertacco and W. Bonkowski, "Ithelps: Iterative high-accuracy error localization in post-silicon," in *33rd IEEE International Conference on Computer Design, ICCD 2015, New York City, NY, USA, October 18-21, 2015*, 2015, pp. 196–199.

[2] M. Abramovici, P. Bradley, K. Dwarakanath, P. Levin, G. Memmi, and D. Miller, "A reconfigurable design-for-debug infrastructure for socs," in *2006 43rd ACM/IEEE Design Automation Conference*, 2006, pp. 7–12.

[3] K. Basu and P. Mishra, "Rats: Restoration-aware trace signal selection for post-silicon validation," *IEEE Transactions on Very Large Scale Integration (VLSI) Systems*, vol. 21, no. 4, pp. 605–613, April 2013.

[4] K. Rahmani, P. Mishra, and S. Ray, "Efficient trace signal selection using augmentation and ilp techniques," in *Fifteenth International Symposium on Quality Electronic Design*, March 2014, pp. 148–155.

[5] M. Li and A. Davoodi, "A hybrid approach for fast and accurate trace signal selection for post-silicon debug," in *Design, Automation Test in Europe Conference Exhibition (DATE), 2013*, March 2013, pp. 485–490.

[6] K. Basu, P. Mishra, P. Patra, A. Nahir, and A. Adir, "Dynamic selection of trace signals for post-silicon debug," in *14th International Workshop on Microprocessor Test and Verification*, Dec 2013, pp. 62–67.

[7] B. Kumar, A. Jindal, M. Fujita, and V. Singh, "A methodology for trace signal selection to improve error detection in post silicon validation," in *2017 30th International Conference on VLSI Design(VLSID'17)*, Hyderabad, India, Jan 2017.

[8] A. G. Veneris and I. N. Hajj, "Design error diagnosis and correction in vlsi digital circuits," in *Proceedings of 40th Midwest Symposium on Circuits and Systems.*, vol. 2, Aug 1997, pp. 1005–1008 vol.2.

[9] S. B. Park and S. Mitra, "Ifra: Instruction footprint recording and analysis for post-silicon bug localization in processors," in *Design Automation Conference. DAC 2008. 45th ACM/IEEE*, pp. 373–378.

[10] A. DeOrio, Q. Li, M. Burgess, and V. Bertacco, "Machine learning-based anomaly detection for post-silicon bug diagnosis," in *Design, Automation and Test in Europe, DATE 13, Grenoble, France, March 18-22, 2013*, 2013, pp. 491–496.

[11] B. Mammo, M. Furia, V. Bertacco, S. A. Mahlke, and D. S. Khudia, "Bugmd: automatic mismatch diagnosis for bug triaging," in *Proceedings of the 35th International Conference on Computer-Aided Design, ICCAD 2016, Austin, TX, USA, November 7-10, 2016*, 2016, p. 117.

[12] A. DeOrio, D. S. Khudia, and V. Bertacco, "Post-silicon bug diagnosis with inconsistent executions," in *2011 IEEE/ACM International Conference on Computer-Aided Design (ICCAD)*, Nov 2011, pp. 755–761.

[13] A. DeOrio, J. Li, and V. Bertacco, "Bridging pre- and post-silicon debugging with biped," in *2012 IEEE/ACM International Conference on Computer-Aided Design (ICCAD)*, Nov 2012, pp. 95–100.

[14] C. S. Zhu, G. Weissenbacher, D. Sethi, and S. Malik, "Sat-based techniques for determining backbones for post-silicon fault localisation," in *High Level Design Validation and Test Workshop (HLDVT), 2011 IEEE International*, Nov 2011, pp. 84–91.

978-1-5386-3693-0/18 $31.00 © 2018 IEEE

Combined Inference and Satisfiability Based Methods for Complete Signal Restoration in Post-Silicon Validation

Xiaobang Liu and Ranga Vemuri
University of Cincinnati, Cincinnati, OH 45220, USA
liu2xg@mail.uc.edu, ranga.vemuri@uc.edu

Abstract—**Forward inference and backward justification (FB) methods are commonly used for signal restoration in post-silicon validation and debug. We show that the FB method can miss a large number of restorable signal values. We propose a novel hybrid method, combining FB and satisfiability (SAT) checking based on time frames along with a signal prioritization heuristic, for efficient and accurate restoration of *all* signal values that can possibly be restored. Experimental results show that the proposed method can increase the signal restoration ratio by a factor of 18x. For large circuits where this method takes a long time and becomes impractical, we apply the time frames technique with SAT method employed locally in each frame to improve efficiency while compromising the restoration ratio. Using this method, experimental analysis shows up to 5x improvement in restoration ratio for large benchmarks.**

I. INTRODUCTION AND MOTIVATION

The process of detecting and localizing errors in integrated circuits after fabrication is referred to as post-silicon validation and debug. Limited visibility of internal signals poses a challenge to the use of at-speed functional and structural test methods. *Trace buffer* based run-time techniques have been proposed to enhance observability with relatively low hardware overhead. In these methods, a small set of pre-selected signals are tracked across a limited number of time steps and stored in an on-chip memory, named the trace buffer. Run-time data acquired in the trace buffer can be transferred off-chip and used to determine as many internal signal values as possible. This process is called *signal restoration* or *state restoration*. Restored signal values can be compared with those obtained using a simulator for error detection and localization. Clearly, the more the number of signals restored, the better the error detection [1][2][3].

The trace buffer stores the values of signals being traced over a predefined number time steps. *Width*, *w*, of the trace buffer is the number of signals being traced and the *depth*, *d*, of the trace buffer is the number of time steps (clock cycles) over which the signal values are stored. *Size* of the trace buffer is the product $w \cdot d$. The specific set of signals being traced and stored in the trace buffer are called *trace signals*. The quality of a state restoration algorithm is measured by the *Restoration Ratio* (R) which is defined as $(N_t + N_r)/N_t$, where N_t is the total number of signal instances traced over t time steps and N_r is the total number of signal instances restored by the restoration algorithm over the t time steps. t and N_t are usually same as d and $w \cdot d$ respectively. R is influenced not only by the restoration algorithm but also by the specific set of signals

traced and by the specific values of those signals recorded in the trace buffer. The set of signals to be traced is selected so as to maximize R. Various *trace signal selection algorithms* have been proposed based on this criterion. Quality of these algorithms is influenced by the quality of the state restoration algorithm they use.

A method called the forward propagation and backward justification (FB) method is commonly used for state restoration in signal selection algorithms. While being fast, the FB method cannot guarantee best restoration ratio due to its inability to infer non-trivial indirect implications. On the other hand, satisfiability (SAT) based methods for restoration can infer much larger instances of signal values than FB methods. It is possible that for two sets of trace signals, signal selection algorithms based on FB would choose one whereas signal selection algorithms based on SAT would suggest the other be the superior choice. For both choices, post-tracing restoration quality, and consequent error detection/localization quality, will be distinctly inferior with FB. While generally yielding much better restoration ratio than FB, SAT based methods are far less efficient.

In this paper we present a novel method, combining FB and SAT, to obtain maximum restoration ratio for a given set of traced signal values more efficiently than possible using the SAT method alone. In addition, we propose efficient methods to obtain better suboptimal restoration than possible using the FB method alone for large circuits over a large number of time steps.

The rest of this paper is organized as follows: Section II provides preliminary background and related research. Section III describes our restoration algorithms. Section IV presents results from experimental studies. Section V contains concluding remarks.

II. PRELIMINARIES AND RELATED RESEARCH

A. FB Method

Forward propagation is the process of uniquely determining the output value of a gate given some or all of its input values. Backward justification is the process of uniquely determining one or more input values of a gate given its output value and, possibly, some of the input values. FB method repeatedly applies the forward and backward implications over several passes across the logic circuit until no new signal values can be derived.

978-1-5386-3693-0/18 $31.00 © 2018 IEEE

Figure 1 provides an illustrative example. Flip-flop D is selected to be traced over 6 cycles ($w = 1$, $d = 6$, $N_t = 6$). Entries pointed by solid arrows in Figure 1(b) represent the states restored over the 6 cycles using FB method. $N_r = 6$ and $R = 12/6 = 2$.

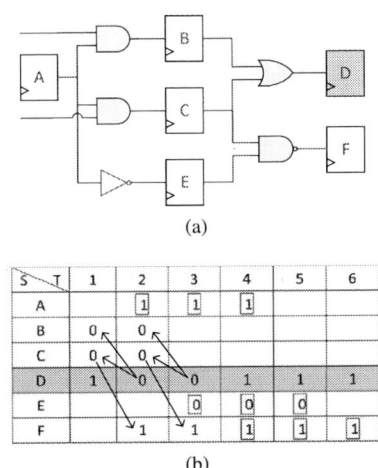

(a)

S\T	1	2	3	4	5	6
A		1	1	1		
B	0	0				
C	0	0				
D	1	0	0	1	1	1
E			0	0	0	
F		1	1	1	1	1

(b)

Fig. 1. Example to illustrate FB method

The signal restoration techniques used in most existing automated trace signal selection papers are based on FB method due to its efficiency. For example, in [4] [5] and [6], the signal restoration method is based on the FB restoration techniques. [7] and [8] have used a simulation approach to restore as many signal instances as possible exploiting a bit-parallel propagation technique to accelerate state restoration but the max width of the bit-parallel operations is limited to 64bits. [1] developed an event-driven simulator to conduct simulation in forward and backward direction. [9] proposed a fast state restoration algorithm by reducing the number of forward and backward restorations. [10] also implemented state restoration in forward/backward directions over multiple clock cycles and introduced a binary restore-once flag to gain faster reachability list computation to aid their signal selection algorithm. All of these approaches are inherently suboptimal in general as illustrated in Section I.

B. SAT Method

In this method, each gate is represented by a Boolean formula, in the conjunctive normal form (CNF), in terms of the input and output variables of the gate. A combinational logic circuit is represented by the conjunction of the CNF formulas of the constituent gates. A sequential circuit can be unrolled into a combinational iterative network over a user-defined number of time steps (N) as illustrated in Figure 2. Flip-flop outputs in slice i are same as the corresponding flip-flop inputs from slice i-1, for $2 \leq i \leq N$. Given a CNF formula F, the Boolean satisfiability (SAT) problem is to find a solution (assignment of binary values to variable) which would evaluate F to true. If such a solution (also called a *model*) exists, then F is said to be *satisfiable*; otherwise, it is *unsatisfiable*. The signal restoration problem can be rephrased

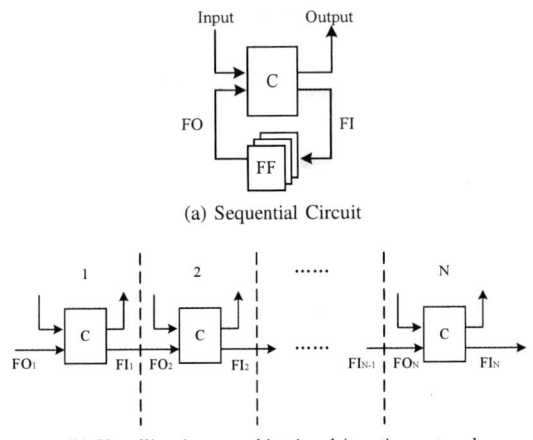

(a) Sequential Circuit

(b) Unrolling into combinational iterative network

Fig. 2. Unrolling of Sequential Circuits

as follows: Given a CNF representation and a set of fixed variable assignments, determine all the variable values which can be uniquely inferred. These variables assume the same values in *all* solutions to the satisfiability problem given the circuit (unrolled over N clock cycles) and the set of traced signal values from the trace buffer. However, SAT solvers usually provide only one solution making it necessary to make multiple calls as discussed below.

After unrolling the circuit in Figure 1.(a) over 6 cycles and using the SAT method, we can restore the additional values indicated by the boxed entries in Figure 1.(b). In this case, $N_r = 15$ and $R = 21/6 = 3.5$ which is significantly better than the FB results.

1) Probing and recording: Suppose we have a solution V of formula F. Let x be a variable of F and let $x = 0 (x = 1)$ in V. Define a new variable $l = \bar{x} (l = x)$. Consider a new formula $G = F \wedge \bar{l}$. If G is unsatisfiable then x can be inferred to be 0 (1) in all solutions of F. This is called *probing*. Conducting probing for each variable in F separately will detect all the variables whose values can be restored uniquely. During probing, if G is found to be satisfiable, then comparing the solutions of G and F yields a set of variables whose values differ in the two solutions. x would be one such variable but there may be others. Clearly, these variables cannot be restored uniquely and are not candidates for probing. This step called *recording* results in significant performance improvement by eliminating unnecessary SAT calls [2]. In the worst case $(n+1)$ calls to the SAT solver may be needed.

2) Core-based algorithm: This method focuses on taking advantage of the *unsat core* which is a set of clauses responsible for the unsatisfiability [11]. A subset of variables, called a *chunk*, is examined to identify if their values can be uniquely restored. Similar to probing, corresponding to the chunk $x_0, x_1, \cdots, x_{k-1}$, a probing formula $G = F \wedge \bar{l}_0 \wedge \cdots \wedge \bar{l}_{k-1}$ is formed where k is the chunk size. If G can be satisfied, then at least k variables cannot be uniquely restored. Otherwise, an unsat core is produced by the SAT solver. If the core contains a single literal then the value of corresponding variable is fixed, otherwise all literals in the core are removed from the chunk and the method reverts to other algorithms to check them later. This iterates on the shrunken chunk until it is empty, after which a new chunk will be inspected. The procedure repeats until all chunks are processed. This algorithm is effective due to its capability to handle multiple open variables in one SAT solver call, provided that the chunk size is not too large. We adopt it as part of our proposed method as it's one of the state-of-the-art approaches with superior performance.

3) Related Research: [12] introduced a SAT-based engine for multi-node implication to restore signals with a given set of signal assignments by using Binary Constraint Propagation (BCP). Implications are only obtained from the unit clauses in the simplified CNF formula, thus their method unable to identify all signals whose values can be restored in multi-time frames. Method discussed in [2] recovered values of non-traced signals by converting circuits to Boolean functions with SAT-based analysis. The paper studied several existing SAT-based techniques and then proposed some novel algorithms to recover values of non-traced signals. According to [2], the probing and recording method performs the best in the results.

III. PROPOSED ALGORITHM

Our approach to signal restoration is distinguished from the previous approaches in several aspects: *(i)* We perform SAT-based restoration after initial FB restoration is completed. While ensuring maximal restoration, this reduces the problem size for the solver which promotes efficiency. *(ii)* For large benchmarks for which SAT followed by FB still takes a long time, we propose a *time frame* method. While relatively efficient, this method cannot guarantee maximal restoration ratio. *(iii)* To ensure maximal restoration, we follow up the time frame method with global restoration using a combination of SAT and FB methods along with a signal ordering heuristic to reduce the problem size. We present the proposed method in several steps.

Let C be a sequential logic circuit. Let S be the set of signals in C. Let $M = |S|$. Let the signals be numbered $1, 2, \cdots, M$. Let $T \subseteq S$ be the set of state signals being traced. Let $w = |T|$ be the width of the trace buffer and let d be its depth. Let N be the number of clock cycles over which signal tracing and restoration are performed. Without loss of generality of the algorithms presented here, we assume $N = d$ in this paper. Let V be a $(M \times N)$ matrix consisting of signal instance values. Value of signal i at time t is given by $V(i,t)$. Entries in V can be one of the following: {*0, 1, non-unique, open*}. Entries '*0*' and '*1*' designate uniquely determined binary values (traced or restored). '*non-unique*' entry at $V(i,t)$ means that the SAT solver determined (during probing and recording) that the value of signal i at time t cannot be uniquely restored and doesn't warrant further examination in future calls to the solver. '*open*' implies that that signal instance is still open for further inspection. Before restoration begins, all entries in V are initialized to '*open*'. Then 0's and 1's are copied from the $(w \times d)$ signal instance values traced in the trace buffer into to V. As the restoration

process progresses, some of the other entries change from '*open*' to '*0*', '*1*', or '*non-unique*'. At the end, an optimal restoration algorithm will not leave any '*open*' instances in V whereas a suboptimal algorithm will leave some instances still '*open*'. $N_t = w \times d$. At the end, the number of known instances of signal values is given by $N_k = |\{(i,t)|(V(i,t) = 0 \ or \ V(i,t) = 1); 1 \le i \le M; 1 \le t \le N\}|$. Finally, $N_r = N_k - N_t$ and $R = N_k / N_t$.

A. FB and SAT Combined

Our first attempt at combining FB and SAT methods is shown as Algorithm 1. Given a circuit C, an initial signal instance value matrix V (with traced values from the trace buffer copied into it) and clock cycles N, the first step in the algorithm calls the FB method to restore as many signal instance values as possible (Section II part A). An updated V which contains all traced or inferred signal values up to that point is returned. The next step is to unroll the sequential logic circuit into a combinational form over N cycles (Figure 2). Step 3 invokes the SAT method, to restore as many remaining signal values as possible given the currently known signal instance values in V. This includes the construction of a CNF formula and using core based method which in turn uses probing and recording. In this process, some signal instances may be determined to be non-unique. Updated V is returned.

Algorithm 1 FB-SAT(C, V, N)
1: V ← FB-RESTORE(C, V, N)
2: C_N ← UNROLL(C, N)
3: V ← SAT-RESTORE(C_N, V)
4: **return** V

FB-SAT produces optimal R since it explicitly or implicitly probes every signal instance. The call to FB-RESTORE efficiently restores many signals thereby reducing the number of unknown variables for SAT-RESTORE. As a result, FB-SAT is more efficient than SAT-RESTORE alone and can be used for small to medium sized benchmarks. However, for larger benchmarks it becomes quite inefficient. Figure 1 has shown signals recovered by FB-RESTROE and additional signal values reconstructed by the followed SAT-RESTORE.

B. Time Frames

Our refinement introduces the idea of time frames to speed up the restoration process at the expense of restoration ratio. A *time frame* is a series of consecutive time steps (clock cycles). We divide the N time steps into a series of consecutive time frames. If f is the frame size (number of time steps in a frame), then the N steps are divided into $F = N/f$ frames. We will use SAT based restoration inside each frame and use FB based restoration across all the time steps. Algorithm 2 shows the method.

First two steps are same as in Algorithm 1. The next step organizes the unrolled iterative network into F frames. The *for* loop processes each frame as follows: Signal instances within the frame are restored using SAT-RESTORE. This

Algorithm 2 FRAMES-FB-SAT(C, V, N)

1: V ← FB-RESTORE(C, V, N)
2: C_N ← UNROLL(C, N)
3: $C_F[1...F]$ ← GENERATE-FRAMES(C_N, N, F)
4: **for** $r = 1$ to F **do**
5:      V ← SAT-RESTORE($C_F[r]$, V)
6:      V ← FB-RESTORE(C, V, N)
7: **end for**
8: **return** V

---

is followed by restoration across all time steps using FB-RESTORE. In this way, 'local' restoration in each frame using SAT is followed by 'global' restoration using FB. Since each frame is relatively small, SAT solving is efficient. However, restoration is suboptimal since each call to SAT-RESTORE uses only the fragment of unrolled circuit for that frame.

C. Reclaiming Complete Restoration

In order to ensure optimal restoration for large circuits, we may add, at the end of Algorithm 2, a call to SAT-RESTORE on the entire unrolled circuit over N cycles. This is shown as Algorithm 3 (Note that C_N is produced inside FRAMES-FB-SAT and can be used by SAT-RESTORE).

Algorithm 3 FRAMES-FB-SAT-SAT(C, V, N)

1: V ← FRAMES-FB-SAT(C, V, N)
2: V ← SAT-RESTORE(C_N, V)
3: **return** V

D. Prioritized Signal Restoration

We propose a heuristic, shown in Algorithm 4, to prioritize certain signal instances to be restored before the others. The first step is to call the frame based restoration shown in Algorithm 2. The heuristic depends on sampling signal data in some selected clock cycles called *sample cycles*, SC, after performing the frame based restoration. We determine all of the signal instances, OI, which are 'open' in the selected cycles (Step 2-3). Then, we attempt to restore these instances using the core based SAT method (Step 4). SAT-RESTORE-INSTANCES is similar to SAT-RESTORE except that the former focuses on restoring only the specified signal instances (while using the entire unrolled model C_N). SAT method will determine each open instance in OI to be either 0 or 1 or *non-unique*. At this point, all signal instances in the sampled cycles are resolved. $RI \subseteq OI$ are the signal instances successfully restored (0 or 1) by the SAT method. SAT-RESTORE-INSTANCES returns the updated V as well as RI.

We now perform a 'global' restoration using the FB method (Step 5). Now the *priority instances*, PI, are identified based on the idea that the successfully restored signals in SC are likely to have a good chance being restored in many of the remaining cycles as well. Open instances of these signals in the remaining cycles are collated as PI (Step 6). SAT method is invoked to attempt to restore these instances (Step 7). Next, FB method is used for one more pass of global restoration attempt across all

Algorithm 4 FRAMES-FB-SAT-PRIORITY(C, V, N)

1:   V ← FRAMES-FB-SAT(C, V, N)
2:   SC ← SELECT-SAMPLE-CYCLES(C, V, N)
3:   OI ← $\{(i,t)|V(i,t) = open; 1 \le i \le M; t \in SC\}$
4:   (V, RI) ← SAT-RESTORE-INSTANCES(C_N, V, OI)
5:   V ← FB-RESTORE(C, V, N)
6:   PI ← $\{(i,y)|(i,t) \in RI; V(i,y) = open; 1 \le y \le N; y \notin SC\}$
7:   (V, RI) ← SAT-RESTORE-INSTANCES(C_N, V, PI)
8:   V ← FB-RESTORE(C, V, N)
9:   NSC ← $\{t|1 \le t \le N; t \notin SC\}$
10: OI ← $\{(i,t)|V(i,t) = open; 1 \le i \le M; t \in NSC\}$
11: (V, RI) ← SAT-RESTORE-INSTANCES(C_N, V, OI)
12: **return** V

---

the cycles (Step 8). Finally, all of the 'open' signal instances in all non-sampled cycles NSC (i.e. all cycles not included in SC) are collated into OI and the SAT method is used to attempt to restore them (Steps 9-11). Note that all signal instances in the sampled cycles have already been resolved. Further, all signals whose instances appear in PI have all been probed across all cycles. None of these instances appear in the OI compiled in Step 10. At this point, since all open instances across all cycles have been explicitly or implicitly probed, V contains maximal restorations.

IV. EXPERIMENTAL RESULTS

A. Experimental setup

This section evaluates the performance of our algorithms against FB approach by comparing the restoration ratios attained for five publicly available ISCAS89 benchmark circuits which are commonly used in most related work published. Our algorithm is developed in C++ with Minisat2.2 API integrated to avoid frequent and costly CNF file read operations. The incremental solving feature of Minisat2.2 is exploited to speed up solving the same problem under different assumptions. Our experiments are run on an Intel quad-core mobile processor i7-6820HQ @2.7GHz with 16GB memory. The circuit statistics are shown in Table I.

TABLE I
CIRCUIT STATISTICS

Circuits	# of Flip-Flops	# of Gates	Control Signal [13]
s5378	179	2779	No control signal
s9234	211	5597	No control signal
s15850	534	9772	No control signal
s35932	1728	16065	RESET
s38584	1426	19253	g35

We perform Verilog simulation using ModelSim by applying random input vectors for each circuit to load trace buffer data, and to verify the correctness of our implementation by comparing our reconstructed data with the Verilog simulation data. It should be noted that there may be control signals in those circuits and they should be handled properly during

978-1-5386-3693-0/18 $31.00 © 2018 IEEE   419

TABLE II
COMPARISON OF RESTORATION RATIO AND TIME

BM	d	w	R_{fb}	R_{frm}	R_{opt}	T_{A1}	T_{A2}	T_{A3}	T_{A4}	A3 vs A1	A4 vs A3	A4 vs A1
s5378	64	8	4.049	9.014	9.014	5.01	2.05	6.58	5.8	-31.28%	11.85%	-15.72%
		16	5.336	5.865	5.87	2.27	1.17	3.54	3.16	-56.08%	10.73%	-39.33%
		32	3.157	3.628	3.644	1.18	0.78	1.48	1.72	-25.76%	-15.90%	-45.76%
	256	8	11.917	12.069	12.075	31.64	4.15	36.1	30.21	-14.08%	16.32%	4.54%
		16	5.835	6.406	6.43	43.88	4.89	38.96	34	11.22%	12.74%	22.53%
		32	2.508	3.749	3.754	30.91	3.34	23.27	20.18	24.72%	13.27%	34.71%
	512	8	10.793	11.25	11.26	262.31	12.3	265.44	197.45	-1.19%	25.61%	24.73%
		16	6.078	6.446	6.463	85.75	5.4	67.62	56.79	21.14%	16.01%	33.77%
		32	3.857	3.973	3.977	47.51	3.64	32.56	27.9	31.46%	14.32%	41.27%
	1024	8	4.222	10.596	10.598	988.16	19.56	861.22	671.01	12.85%	22.09%	32.10%
		16	4.293	6.561	6.568	632.62	16.63	533.61	430.44	15.65%	19.33%	31.96%
		32	4.138	4.367	4.384	126.54	6.18	94.43	80.02	25.38%	15.25%	36.76%
s9234	64	8	2.018	4.818	5.322	10.52	4.93	14.33	12.33	-36.24%	13.98%	-17.19%
		16	2.897	5.248	5.255	3.66	2.06	4.1	4	-11.91%	2.25%	-9.40%
		32	1.863	3.002	3.165	3.13	1.73	3.48	3.3	-11.13%	5.06%	-5.50%
	256	8	6.543	9.012	9.283	50.58	7.46	44.8	35.18	11.42%	21.48%	30.45%
		16	2.006	3.287	4.521	67.05	10.09	60.36	44.38	9.97%	26.47%	33.80%
		32	3.539	4.418	4.89	15.35	3.05	11.2	10.52	27.05%	6.11%	31.50%
	512	8	2.098	2.776	4.845	483.35	29.41	504.62	318.77	-4.40%	36.83%	34.05%
		16	2.007	4.836	4.954	271.14	16.46	160.86	107.1	40.67%	33.42%	60.50%
		32	3.001	4.186	4.575	75.26	6.51	40.6	37.45	46.05%	7.76%	50.23%
	1024	8	5.853	8.875	8.987	3641.98	33.49	3057.01	1665.46	16.06%	**45.52%**	54.27%
		16	3.989	4.647	5.604	746.5	23.55	841.22	514.06	-12.69%	38.89%	31.14%
		32	2.27	4.377	4.834	514.28	24.35	185.33	170.48	63.96%	8.01%	66.85%
s15850	64	8	1.07	18.506	19.91	25.72	13.01	29.18	26.76	-13.42%	8.29%	-4.01%
		16	3.06	15.567	15.839	21.69	9.56	20.53	20.26	5.33%	1.31%	6.57%
		32	4.899	8.749	9.083	16.66	8.14	18.78	17.7	-12.70%	5.73%	-6.24%
	256	8	1.926	19.456	22.116	420.66	48.44	306.38	231.61	27.17%	24.40%	44.94%
		16	3.552	16.23	19.259	341.1	37.1	193	170.88	43.42%	11.46%	49.90%
		32	2.527	10.651	11.151	273.37	26.47	112.02	108.43	59.02%	3.21%	60.34%
	512	8	10.708	23.294	24.708	1098.45	55.09	892.87	752.91	18.72%	15.68%	31.46%
		16	3.356	15.092	17.108	1394.24	82.39	700.3	655.63	49.77%	6.38%	52.98%
		32	7.659	12.171	12.638	601.36	33.16	287.19	285.6	52.24%	0.55%	52.51%
	1024	8	**1.464**	25.064	**28.51**	9468.26	175.25	4861.2	3910.75	48.66%	19.55%	58.70%
		16	2.495	20.035	22.4	6807.25	136.58	2219.74	2046.73	**67.39%**	7.79%	**69.93%**
		32	7.512	10.628	11.373	2665.4	70.81	1616.34	1494.2	39.36%	7.56%	43.94%
s35932	64	8	14.457	14.457	14.457	192.64	16.14	168.45	150.42	12.56%	10.70%	21.92%
		16	8.756	11.387	11.816	783.03	19.88	683.27	575.12	12.74%	15.83%	26.55%
		32	5.917	16.737	18.066	932.17	18.44	549.7	502.35	41.03%	8.61%	46.11%
s38584	64	8	1	6.264	8.402	671.71	124.47	580.83	381.36	13.53%	34.34%	43.23%
		16	1.384	4.968	5.282	448.68	79.28	435.89	292.95	2.85%	32.79%	34.71%
		32	1.162	5.656	6.572	346.75	77.43	318.18	210.08	8.24%	33.97%	39.41%

simulation. The application of constrained signal identification procedure introduced in [13] shows that the RESET signal in s35932 and g35 signal in s38584 are active-low reset. Therefore, these two signals are constrained to be value of 1 for normal operation.

For each benchmark, we have presented results for $w =$ 8, 16 and 32 and $d =$ 64, 256, 512 and 1024. After some experimentation, we have determined that frame size $f = 32$ works better. Sample cycles SC are selected to be the first 4 cycles plus the last four cycles plus the middle 8 cycles in every cycle in all experiments. This encourages sampling at the beginning, middle and end of tracing. For each w, trace

signal set is selected randomly and, of course, the same set is used for comparing different algorithms. We have performed experiments with several randomly selected trace signal sets and several simulation vector sequences and observed similar trends as reported here.

B. Results

Table II shows the experimental comparison of restoration ratios and run time. R_{fb}, R_{frm} and R_{opt} stand for restoration ratios obtained from FB-RESTORE, A2 and optimal algorithms(A1, A3, A4). T_{A1} stands for the runtime (in seconds) for Algorithm 1, similar meaning for T_{A2}, T_{A3} and T_{A4}. "A3 *vs A1*" means the run time improvement of A3 over A1, similar meaning for "*A4 vs A3*" and "*A4 vs A1*".

Generally SAT restores significantly more signals than FB from the result table. Notice that for several cases A2 produces near-maximal results, for example, circuit s15850 when d = 512, w = 32. Also, note that optimal algorithms A1, A3 and A4 are unable to complete execution even after several hours of CPU time for the two large benchmarks (*s35932* and *s38584*) when $d > 64$, thus results of the three algorithms not shown in the table. However, suboptimal algorithm A2 (SAT with frames) is able to restore up to 5x more signal instances compared to FB alone as shown in Table III for large benchmarks. For benchmark *s15850*, SAT is able to restore a large number of signal instances beyond FB. We have noticed up to 18x improvement in R when using FB&SAT compared with FB alone.

TABLE III
s35932 AND s38584 WHEN DEPTH>64

BM	d	w	R_{fb}	R_{frm}	T_{A2}
s35932	256	8	6.712	6.736	77.43
		16	10.055	10.433	97.6
		32	6.07	15.772	193.62
	512	8	9.619	9.622	145.13
		16	7.056	8.861	155.28
		32	5.18	11.21	940.84
	1024	8	14.496	14.564	287.28
		16	5.309	5.542	336.36
		32	5.584	18.805	891.1
s38584	256	8	1.249	4.14	441.74
		16	1.062	4.969	289.21
		32	1.314	5.117	241.55
	512	8	1.374	4.474	1069.6
		16	1.77	5.409	652.64
		32	1.218	6.078	441.59
	1024	8	1.532	7.25	1688.18
		16	1.187	6.239	1233.75
		32	**1.141**	**6.813**	814.53

Table II also shows the CPU time spent by Algorithms A1 to A4 for each benchmark circuit over different buffer depths and widths. The baseline FB-RESTORE method executes under 1 sec for all experiments thus not shown in the table. As expected A1 (basically the core-based algorithm) is the slowest in most cases and A2 is the fastest (but suboptimal). For scenarios where problem size is small(i.e. small *d* and *w*), the last 3 columns in a few cases exhibit negative percentages because the runtime is already very small. However, for cases where execution time becomes large, like s15850 when d = 1024, w = 8, the run-time reduction between A3 and A1, A4 and A3, or A4 and A1 are substantial. None of the optimal methods (A1, A3, A4) could be completed for *s35932* and *s38584* when $d > 64$. This underlines the need for further research in this area. When they complete for small and medium benchmarks, A4 is generally faster than A1 and A3 as expected. Finally, A3 is faster than A1 when the overhead due to framing is justified. This is the case with most experiments.

V. CONCLUSION

In this work, we have presented methods to effectively combine FB and SAT to achieve optimal and suboptimal signal restorations for post silicon validation based on trace buffers and restoration ratio metric. Our optimal algorithm can achieve maximal restoration for small and medium size benchmark circuits and our frame based heuristic can achieve much better suboptimal restoration compared with FB for larger benchmark circuits with good time efficiency compared with SAT. Thus, it is a compromise between pure FB based methods and pure SAT based methods. Further research in signal prioritization heuristics for SAT based restoration is needed to handle large benchmarks efficiently while maintaining maximal restoration. There is also research on new metric like the *assertion coverage* in [14] instead of using state restoration ratio, but it serves as an attempt to initiate a functional coverage based metric and not commonly recognized and widely used.

REFERENCES

[1] X. Liu and Q. Xu, "On signal selection for visibility enhancement in trace-based post-silicon validation," *IEEE Transactions on Computer-Aided Design of Integrated Circuits and Systems*, vol. 31, no. 8, pp. 1263–1274, Aug 2012.

[2] C. S. Zhu, G. Weissenbacher, D. Sethi, and S. Malik, "Sat-based techniques for determining backbones for post-silicon fault localisation," in *2011 IEEE International High Level Design Validation and Test Workshop*, Nov 2011, pp. 84–91.

[3] S. Mitra, S. A. Seshia, and N. Nicolici, "Post-silicon validation opportunities, challenges and recent advances," in *Design Automation Conference*, June 2010, pp. 12–17.

[4] S. Choudhary, A. M. Gharehbaghi, T. Matsumoto, and M. Fujita, "Trace signal selection methods for post silicon debugging," in *2015 IFIP/IEEE International Conference on Very Large Scale Integration (VLSI-SoC)*, Oct 2015, pp. 258–263.

[5] K. Rahmani, S. Ray, and P. Mishra, "Postsilicon trace signal selection using machine learning techniques," *IEEE Transactions on Very Large Scale Integration (VLSI) Systems*, vol. 25, no. 2, pp. 570–580, Feb 2017.

[6] K. Rahmani, P. Mishra, and S. Ray, "Efficient trace signal selection using augmentation and ilp techniques," in *Fifteenth International Symposium on Quality Electronic Design*, March 2014, pp. 148–155.

[7] D. Chatterjee, C. McCarter, and V. Bertacco, "Simulation-based signal selection for state restoration in silicon debug," in *2011 IEEE/ACM International Conference on Computer-Aided Design (ICCAD)*, Nov 2011, pp. 595–601.

[8] P. Komari and R. Vemuri, "A novel simulation based approach for trace signal selection in silicon debug," in *2016 IEEE 34th International Conference on Computer Design (ICCD)*, Oct 2016, pp. 193–200.

[9] S. BeigMohammadi and B. Alizadeh, "Combinational trace signal selection with improved state restoration for post-silicon debug," in *2016 Design, Automation Test in Europe Conference Exhibition (DATE)*, March 2016, pp. 1369–1374.

[10] M. Li and A. Davoodi, "A hybrid approach for fast and accurate trace signal selection for post-silicon debug," *IEEE Transactions on Computer-Aided Design of Integrated Circuits and Systems*, vol. 33, no. 7, pp. 1081–1094, July 2014.

[11] M. Janota, I. Lynce, and J. Marques-Silva, "Algorithms for computing backbones of propositional formulae," *AI Communications*, vol. 28, no. 2, pp. 161–177, 2015.

[12] S. Prabhakar and M. Hsiao, "Using non-trivial logic implications for trace buffer-based silicon debug," in *2009 Asian Test Symposium*, Nov 2009, pp. 131–136.

[13] H. F. Ko, "New algorithms and architectures for post-silicon validation," Ph.D. dissertation, McMaster University, 2009.

[14] S. Ma, D. Pal, R. Jiang, S. Ray, and S. Vasudevan, "Can't see the forest for the trees: State restoration's limitations in post-silicon trace signal selection," in *2015 IEEE/ACM International Conference on Computer-Aided Design (ICCAD)*, Nov 2015, pp. 1–8.

2018 31th International Conference on VLSI Design and 2018 17th International Conference on Embedded Systems

Emerging FETs for Low Power and High Speed Embedded Dynamic Random Access Memory

Md. Hasan Raza Ansari[1], Nupur Navlakha[1], Jyi-Tsong Lin[2], and Abhinav Kranti[1]

[1]Low Power Nanoelectronics Research Group, Discipline of Electrical Engineering,
Indian Institute of Technology Indore, Simrol, Indore 453552, India, Email: akranti@iiti.ac.in
[2]Department of Electrical Engineering, National Sun Yat-Sen University,
Kaohsiung 80424, Taiwan, R.O.C.

Abstract—**The work reports on the assessment of two emerging devices, namely Tunnel Field Effect Transistor (TFET) and Junctionless Transistor (JLT), for applicability as low power and high speed embedded Dynamic Random Access Memory (eDRAM) at 85 °C. The critical aspect of DRAM functionality being independent gate operation has been realized through twin/dual architecture in TFET and JLT. The first (front) gate primarily controls the read operation based on band-to-band tunneling (TFET) and drift-diffusion mechanism (JLT), whereas the second gate is responsible for charge storage in both the devices. The ability of TFET and JLT with write operation in short time (< 10 ns) and at low power shows their feasibility for embedded memory applications. TFET is benefited with a high retention time (~350 ms at 85 °C for gate length of 250 nm) while JLT requires further optimization for longer charge sustenance. The concept and insights into device operation highlight the advantages and limitations of devices for eDRAM.**

Keywords—**Dynamic memory, Retention time, Tunnel Field Effect Transistor, Junctionless Field Effect Transistor**

I. INTRODUCTION

Embedded memories have been evolving in integrated circuits that incorporate logic and memory on same chip for high performance computing and low power consumer electronics [1-9]. The performance of processor has slowed down in recent years, and therefore, the improvement in memory devices could enhance the overall system performance [10]. With technology advancement, Dynamic Random Access Memory (DRAM) has made significant development as embedded memory (eDRAMs) which are denser and power efficient than Static RAM (SRAM). Therefore, recent works are aimed at using eDRAM to replace SRAM for cheaper chips operating at lower voltage [3,7]. DRAM has been continuously evolving with technology scaling and is competitive in terms of performance and cost metric. Although computing segment has been shrinking, the advent of mobile, cloud computing, virtual reality, and Internet of Things (IoT) applications require the exploitation of innovative devices to deliver capacity, high performance, retention and low energy at reduced size [10-12].

The conventional DRAM with an access transistor (1T) and a capacitor (1C) for charge storage are compatible with logic devices. However, the several processing steps required for DRAM capacitors is a major issue [8-10]. This problem can be circumvented by use of single transistor that utilizes floating body effects to store the charges. Although 1T DRAMs were introduced to resolve the issue of capacitor scaling, they can be well-suited for embedded memory due to simple and cost-effective fabrication and most importantly, compatibility with CMOS devices and process [13-21]. An issue with 1T-1C DRAM is the scaling of transistor [6], where downscaling of gate length improves switching speed, functionality and density. It requires lowering of threshold voltage (V_{th}) along with the supply voltage (V_{dd}) to achieve a high on-current (I_{on}). However, reduction in V_{th} increases the off-current (I_{off}) and static power dissipation which can lower I_{on}/I_{off}, and thus, speed [22-23]. Therefore, the need to reduce V_{dd} without performance loss requires devices to replace conventional MOS for low power applications. Tunnel FET (TFET) [22-23] and Junctionless Transistor (JLT) [24] are two promising emerging energy efficient devices, and thus, the work investigates them for DRAM. Although, JLT [25-27] and TFET [28-31] devices have paved a way towards dynamic memory, further investigation is necessitated to improve the speed, power, and applicability for future embedded memory.

The device can be utilized as embedded memory if it can use low write time (~10 ns) and achieve high retention (~100 ms at 85 °C) at low voltages [8]. The investigation of JLT and TFET as capacitorless memory showcase their possibility as eDRAM due to full CMOS functionality and operation at low voltage. Other advantages include ease of fabrication due to same type of dopant in JL MOSFET and weak temperature dependency for TFETs along with better scaling perspective. The work presents insights into the operation of these devices, and their implementation as low power and high speed memory with longer charge sustenance for TFET and scope of further improvement in JLT for high retention time at 85 °C.

II. RESULTS AND DISCUSSION

A. Simulation and device description:

To analyze the performance, both devices were simulated by ATLAS software [32]. The simulation results of TFET are calibrated with the available experimental data [23] (Fig. 1(a)).

978-1-5386-3693-0/18 $31.00 © 2018 IEEE   422

The non-local model simultaneously with Klaassen model [32] for band-to-band tunneling (BTBT) has been used to capture the essential characteristics of TFET, while JL FET utilizes standard BTBT and impact ionization model. The other modules used are Band gap narrowing, Lombardi mobility, and SRH recombination model. The analysis was carried out using Fermi and Boltzmann statistics for TFET and JLT, respectively. The maximum value of carrier lifetime was 100 ns, along with temperature and doping dependent carrier lifetime for evaluation.

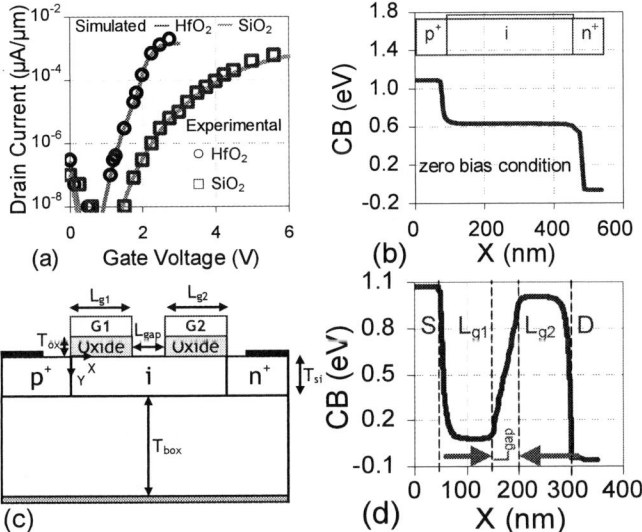

Fig. 1. (a) Comparison of drain current characteristics of our simulation for TFET with experimental data [23]. (b) Variation of Conduction Band (CB) for TFET along X, showing no well formation. (c) Schematic diagram of twin gate TFET with (d) its CB profile along X. The cutlines are at a cross-section of 1 nm below the front interface. V_d is the drain voltage.

The operation of dynamic memory is based on the creation of a dedicated volume for charge storage and sustenance, determined by respective device topology, its dimensions, physical effects, and applied voltages. The conventional MOS with n^+-p-n^+ structure uses p-type body for charge storage. Thus, a raised energy barrier between two regions (with same type of dopant) is utilized as dedicated storage region. TFET, with an intrinsic region between two differently doped p- and n-type regions, does not creates an energy barrier, as shown in Fig. 1(b). Thus, two front gates are utilized (Fig. 1(c)) to create storage region. The first front gate (G1) with workfunction of 4.15 eV is utilized for read mechanism based on BTBT. The low workfunction creates a virtual n-type region under G1. Further, the second gate (G2) with high workfunction of 5.25 eV creates an energy barrier between two n-type regions (G1 and drain). Thus, as demonstrated in Fig. 1(d) the creation of raised barrier (potential well) under G2 is governed by the position of gate, and also, by the workfunction. The twin gate TFET can be fabricated using the approaches as described in [33-35]. Such architectures have been previously utilized as Field Effect Diode (FED) to regulate the electrostatic discharge [35] and also, as dynamic memory [21], but with a different operating principle. A

device with different workfunctions can be fabricated as suggested by Tanaka [36] with several gate materials [37].

This is also evident from Fig. 2 where a JLT (Fig. 2(a)) is utilized as dynamic memory. JLT is a device with same type of dopants over the entire silicon film and therefore, a high workfunction gate can be utilized to form a barrier. Fig. 2(b) demonstrates the use of three different back gate workfunctions (φ_{m2}), where more raised barrier (profound well) is created with a high φ_{m2}. A higher workfunction gate can accumulate higher holes underneath it, and thus, forms a deep potential well which could maintain charges for longer duration and thus, attain a high retention time. In both the topologies, G1 is utilized for read mechanism while G2 is used for charge storage as well as sustenance. Therefore, dual gates are essential for the architectures. However, their position is based on the dopant type in source and drain, as described in this section. Fig. 2(c) shows the variation in transfer characteristic (I_d–V_{g1}) for n-type JLT for different φ_{m2}. The off-current is lower for high workfunction due to depletion of electrons from the silicon film. The workfunction optimization is essential [31] as it regulates the hole generation and recombination, and thus, aids to attain a high retention time. Fig. 2(c), also demonstrates JLT achieving an acceptable current at a low drain bias of 0.1 V, which highlights its applicability at low power.

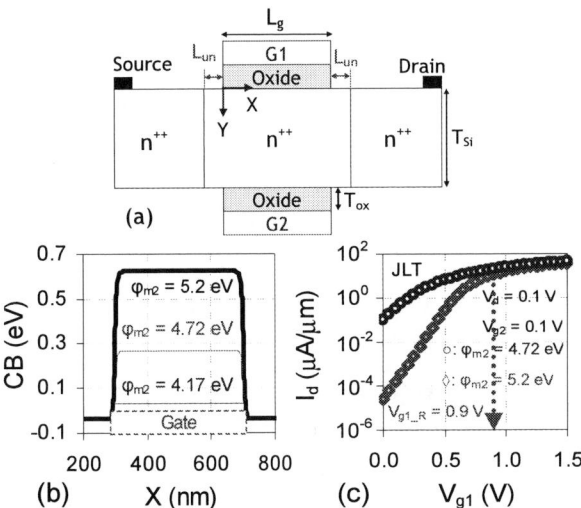

Fig. 2. (a) Schematic of Junctionless transistor (JLT) utilized as DRAM. (b) Variation in CB for different back gate workfunction (φ_{m2}) along X. The cutlines are taken at a cross-section of 1 nm above the back interface. (c) Drain current characteristics of JLT at a lower drain bias of 0.1 V. V_{g1} and V_{g2} are the bias applied at G1 and G2, respectively. V_{g1_R} is the bias at G1 during Read operation in JLT.

The device parameters of TFET and JLT are adopted to function efficiently as dynamic memory, mentioned in Table I. T_{si} for JLT and TFET are 10 nm and 20 nm, respectively. For JLT, a thicker film (> 10 nm) will require a higher workfunction (φ_{m2}) to deplete the electrons so as to create the potential well. With same workfunction and T_{si} > 10 nm, the potential well created will be shallower and would not avail

978-1-5386-3693-0/18 $31.00 © 2018 IEEE

holes to sustain for a longer duration and hence, degrade state '1', rapidly. On the other hand, TFET with a thinner T_{si} (< 20 nm) will result into enhanced BTBT during hold operation. It will generate holes and thus, degrade state '0' at a faster rate. Thus, device parameters are optimized to balance hole generation and recombination for high retention. Previous work on TFET based DRAM [31], has reported a minimum gate lengths, $L_{g1} = L_{g2} = 100$ nm to attain a Retention Time (RT) > 64 ms (target set by ITRS [3]). Therefore, our work utilizes same gate lengths to have a comparative analysis. RT > 64 ms, is not reported for JLT till date, and therefore, a longer gate length of 400 nm has been analyzed as DRAM in order understand the feasibility of JLT architecture for dynamic memory. An underlap is incorporated in JLT to reduce the hole generation in the storage region, associated with BTBT that degrades state '0'. Thus, device architectures for symmetric JLT with independent gate operation, and TFET with two front gates along with appropriate workfunction create a dedicated region for charge storage.

TABLE I. DEVICE PARAMETERS FOR TFET AND JLT BASED DRAMS

Parameters	TFET	JL FET
Gate length (L_g)[a]	250 nm	400 nm
S/D doping	10^{20} cm^{-3}	10^{20} cm^{-3}
Channel doping	10^{15} cm^{-3}	10^{19} cm^{-3}
Silicon film thickness (T_{Si})	20 nm	10 nm
Oxide thickness (T_{ox})	3 nm (HfO$_2$)	1 nm (SiO$_2$)
Underlap length (L_{un})	-	10 nm
Box thickness (T_{box})	145 nm	-
G1 Workfunction (φ_{m1})	4.15 eV	5.0 eV
G2 Workfunction (φ_{m2})	5.25 eV	5.2 eV

[a] For TFET: $L_g = L_{g1} + L_{g2} + L_{gap}$, where $L_{g1} = L_{g2} = 100$ nm and $L_{gap} = 50$ nm.

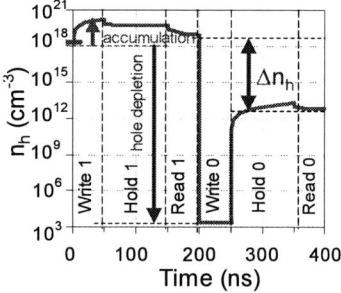

Fig. 3. Variation in hole concentration (n_h) in the potential well showing 1T-DRAM functionality with TFET in the sequence of operation.

B. Operating Mechanism:

Fig. 3 demonstrates DRAM operation for TFET in the sequence of Write, Hold and Read for state '1' and state '0', respectively. The functionality as DRAM can be summarized by hole distribution, where the accumulation of holes defines state '1', while their depletion represents state '0'. Thus, states are distinguished based on the presence and absence of holes in the storage region, as demonstrated in contour plots of Fig. 4(a) and (b) for JLT. The difference in hole concentration (Δn_h, Fig. 3) lead to variation in read currents, termed as Sense Margin (*SM*). Maintaining the generated holes defines

retention of state '1' which is disturbed by hole recombination, while hole generation disturbs state '0'. Thus, DRAM operation is based on regulating the hole generation and recombination, which is controlled by optimized set of bias values. The programming scheme for JLT and TFET DRAM is mentioned in Table II and Table III, respectively.

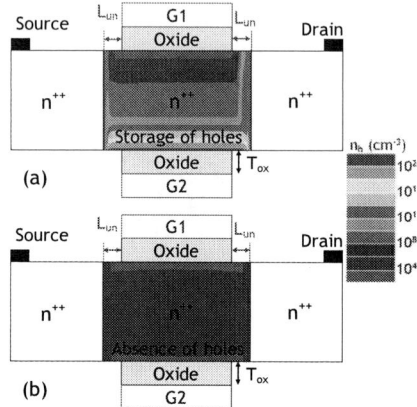

Fig. 4. Contour plot of hole concentration (n_h) for JLT in the silicon film after (a) Write '1' and (b) Write '0'.

TABLE II. PROGRAMMING SCHEME FOR JLT DRAM

Operation	V_{g1} (V)	V_{g2} (V)	V_s (V)	V_d (V)	Time
Write 1	1.0	-1.6	0.0	1.5	10 ns
Hold	0.0	-0.1	0.1	0.1	-
Read	0.9	0.1	0.0	0.1	40 ns
Write 0	1.5	1.5	0.0	0.0	10 ns

TABLE III. PROGRAMMING SCHEME FOR TFET DRAM

Operation	V_{g1} (V)	V_{g2} (V)	V_s (V)	V_d (V)	Time
Write 1	0.0	-1.0	0.0	1.0	5 ns
Hold	0.0	-0.2	0.0	0.0	----
Read	2.0	1.2	0.0	1.0	40 ns
Write 0	0.0	1.0	0.0	0.0	5 ns

Holes are generated and removed from the potential well through Write operation. The program or Write '1' operation in both the topologies is based on BTBT between G2 and drain [17]. This is performed by applying a negative bias at G2 and a positive bias at drain that tunnels the electrons from the region under G2 towards drain, thereby generating holes (Fig. 5(a)). The tunneling based write mechanism is power efficient and reliable as compared to write operation based on impact ionization [13]. The generated holes are accumulated at lower potential region which is the silicon film region under G2. The hole generation is influenced by programming time and voltage. Fig. 5(b) and (c) show the variation in sense margin with write time, where the sense margin saturates after write time of 10 ns and 5 ns for JLT and TFET, respectively. Fig. 5(d) shows *SM* as a function of back gate voltage during Write '1' for JLT. The maximum *SM* is attained at -1.6 V. As demonstrated in [13], although the hole concentration increases with bias and time, the difference between the states

978-1-5386-3693-0/18 $31.00 © 2018 IEEE

are determined by the read currents. Thus, the holes accumulated in potential well during Read are constant after a particular bias or time, and so is the *SM*. Therefore, the operation at low power and high speed (Table II and III) is useful for JLT and TFET devices as eDRAMs.

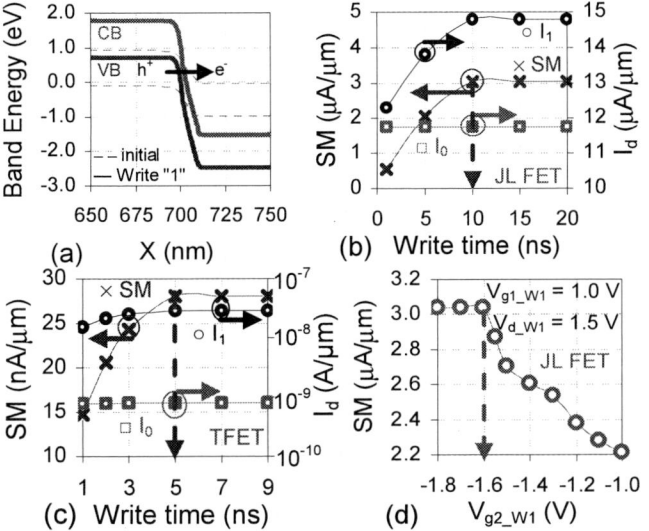

Fig. 5. (a) Energy band diagram of JLT during Write '1' extracted at the back surface of the silicon film. Variation of read currents and *SM* with write time for (b) JLT (c) TFET. (d) Dependence of *SM* as a function of back gate voltage during Write '1' for JLT.

Fig. 4(a) demonstrates the presence of excess holes at the back surface for JLT. The hole concentration (n_h) in the silicon film increases from its initial value of 2×10^{15} cm^{-3} to 3×10^{20} cm^{-3}. The same is observed for TFET in Fig. 3 where the initial n_h of $\sim 10^{18}$ cm^{-3} increases to $\sim 10^{20}$ cm^{-3}. Write '0' operation is performed through the forward bias mechanism with a positive bias at G2 that recombine holes at drain in TFET, and at drain and source in JLT. Fig. 4(b) demonstrates $n_h \sim 10^4$ cm^{-3} after Write '0' operation in JLT and $\sim 10^3$ cm^{-3} in storage region of TFET based DRAM (Fig. 3). The presence/absence of excess carriers is determined through the read current. This can be observed from Fig. 6(a) and (b), where higher electron current density in silicon film beneath G2 confirms higher current for state '1' as compared to state '0'. This is due to presence of excess holes that result into increased potential for state '1' as compared to state '0'. A higher effective potential at G2 results into lower barrier for electrons at G2 region (Fig. 6(c)) and thus, distinguishes the two currents (Fig. 6(d)). The difference between the read currents for state '1' and '0' is Sense Margin (*SM*) and the time required to reduce the maximum *SM* by 50% is estimated as Retention Time (*RT*).

The read operation in TFET is based on BTBT of electrons from the p^+ source region to the channel region, regulated by the first gate. A positive bias of 2 V at G1 lowers the band energy such that the valence band of source region lie adjacent to conduction band of channel region with a narrow barrier for

the electrons to tunnel. The electrons are further drifted towards the drain through a bias of 1.2 V and 1 V at G2 and drain, respectively. An inclined channel is formed in TFET during Read '1' as observed in (Fig. 6(a)). In JLT, the read current is based on drift/diffusion mechanism, where the states are distinguished through the current with bias of 0.9 V at front gate and 0.1 V at back. Similar to TFET, presence of excess holes result into higher current. The *SM* of JLT is ~ 3 μA/μm at 85°C at a lower drain bias of 0.1 V, which is advantageous as a low power application. In TFET, the read current is based on tunneling that limits I_{on} and thus, the sense margin is low (30 nA/μm) for TFET based dynamic memory, but the current ratio observed is higher as compared to JLT and other MOS based devices [14,36]. Thus, there exists a trade-off between *SM* and current ratio which needs to be addressed with optimization techniques.

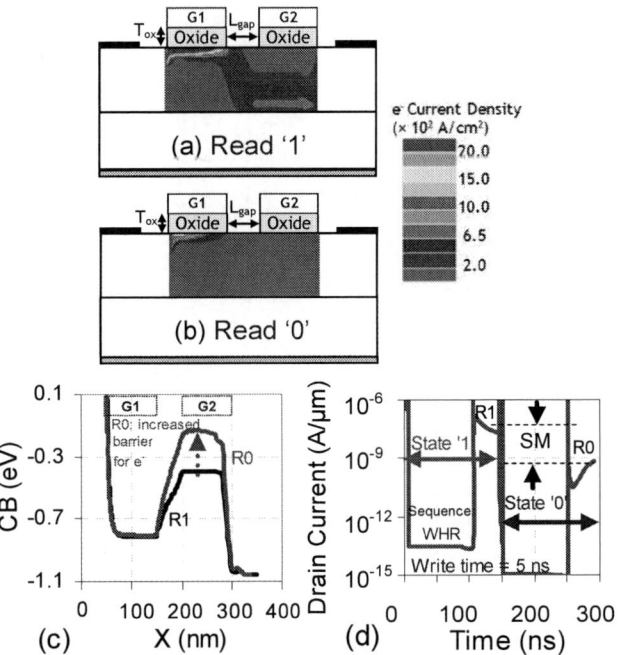

Fig. 6. Contour plots of TFET based DRAM illustrating electron current density in the channel for (a) Read '1' and (b) Read '0'. (c) Conduction band energy for Read operation, showing an increased barrier for electrons for Read '0' compared to '1'. (d) Transient currents in the sequence of operation (Write: W, Hold: H, Read: R) with *SM* of ~ 30 nA/μm and write time of 5 ns.

The maximum time between Write and Read where bit can be read correctly, is termed as the Hold time. A small voltage of -0.1 at G2 with 0.1 V at S/D for JLT, while -0.2 V at G2 for TFET, preserves the potential well that sustains the holes for a longer duration, thus, maintaining state '1'. Fig. 7 shows the contour plots highlighting the hole concentration in the storage region with hold time. To achieve high retention characteristics, an optimized bias is required to balance the hole generation/recombination. Greater negative bias at G2 during hold lead to hole generation, while more positive bias indicates hole recombination. The generation of holes in the potential well degrades state '0' while hole recombination

978-1-5386-3693-0/18 $31.00 © 2018 IEEE

degrades state '1'. The thermal generation and BTBT of electrons from region under G2 towards drain and also, source for JLT while G1 for TFET generates holes in the storage region. The hole generation can be reduced by regulating the lateral electric field associated with tunneling which is possible through an optimized underlap, utilized for JLT in this work. The thermal recombination and diffusion of holes from the storage region degrades state '1'. Thus, a negative bias at G2 for TFET as shown in Fig. 7, highlights the maintenance of state '1' with time but hole generation in the storage region, showcasing Hold '0' degradation.

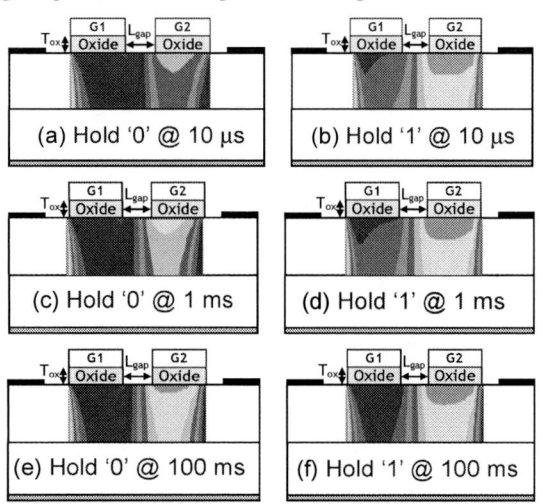

Fig. 7. Contour plots of TFET based DRAM illustrating n_h for hold times of (a)-(b) 10 µs, (c)-(d) 1 ms, and (e)-(f) 100 ms for Hold '0' and '1', respectively. The colour code indicated for n_h is same as given in Fig. 4.

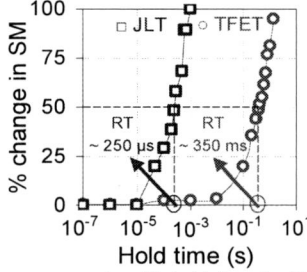

Fig. 8. Variation in sense margin with hold time for JLT and TFET DRAM showing retention time when maximum *SM* changes by 50%.

C. Performance Metrics and Comparative Analysis:

The maximum *RT* achieved for JLT is ~250 µs while for TFET is ~350 ms at 85 °C as shown in Fig. 8. The systematic analysis shows the improved performance in terms of writing speed and operation at low power with high retention characteristics of TFET, adequate for eDRAM. Fig. 9 shows the comparison of write time along with retention time as a function of storage region (L_s) for various devices (JLT, FED, TFET, and MOSFET). The previous work on TFET [30] based DRAM showed *RT* of 100 µs with L_s = 25 nm with write time of 25 ns, while in our work, TFET shows a *RT* of 350 ms with L_s = 100 nm at 85 °C. Table IV shows power consumption evaluated as in [14], for our optimized devices,

TFET and JLT. The evaluation of write power in Write '1' operation is essential as it is based on generation and results into higher current and thus, higher power consumption. Results for TFET DRAM, when compared to similar architectures [21, 39] show low power consumption due to tunneling based write mechanism. FED [21] and Z²FET [39] utilize forward bias current to store holes that yield higher current than that due to tunneling. However, these devices [21, 39] are benefited with a lower write time. Similarly, previous work on JLT [26] uses impact ionization which is faster than our work that utilizes BTBT for write operation, but consumes more power. Thus, trade-off between exists between write time and power during write operation. Therefore, the bias values in our work are adopted so as to consume low power while maintaining low write time (< 10 ns).

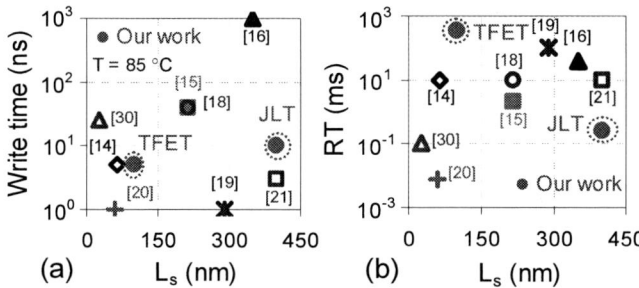

Fig. 9. Comparison of new emerging devices with published literature with varying (a) write time and (b) *RT* as a function storage region (L_s) at 85 °C.

TABLE IV. POWER CONSUMPTION DURING WRITE '1' OPERATION

Device architecture [Reference]	L_s (nm)	$|V_d|$ (V)	I_d (µA)	Power (µW)	Write time (ns)
FDSOI (n^+-p^+-n^+) [14]	65	1.1	314	345	5
DG finFET (n^+-p^+-n^+) [20]	60	2.0	400	800	1
FED (p^+-i-n^+) [21]	400	1.2	~100	120	4
JLT (n^+-n^+-n^+) [26]	10	2.0	900	1800	1
Z²-FET (p^+-i-n^+) [39]	400	1.3	500	650	1
Our work (JLT: n^+-n^+-n^+)	400	1.5	1.4	2.1	10
Our work (TFET: p^+-i-n^+)	100	1.0	0.3	0.3	5

TFET shows outstanding performance in terms of speed, retention and power, while JLT needs further exploitation for better retention. Although, the retention characteristics attained for JLT as DRAM is low, the operation with low power consumption and high speed along with a simple fabrication process that could be cost-effective, shows a path forward toward future embedded systems with additional optimization. These devices utilize innovative methodologies to deliver the capacity, retention, low energy at reduced size that can be suited for various emerging technologies with IoT, cloud computing and big data applications [11-12].

III. CONCLUSION

The work showcases potential benefits of Tunnel Field Effect Transistor (TFET) and Junctionless (JL) device for low power and high speed embedded memory applications. The performance indicates the possibility to replace conventional

978-1-5386-3693-0/18 $31.00 © 2018 IEEE

eDRAM that has issues related to complex fabrication of capacitor and leakage current. The device functionality as DRAM is based on the role of two gates of the respective architecture, where one gate is responsible for creation as well as maintenance of storage region, while the other gate for read operation. TFET shows a better capability as eDRAM while JLTs could be utilized but require additional optimization for longer retention. The operating principle and the analysis demonstrate the benefits and shortcoming of the device to be utilized as embedded memory and their further extension for IoT, cloud computing and big data applications.

ACKNOWLEDGMENT

This work is supported by Department of Science and Technology (DST), Government of India, through Global Innovation and Technology Alliance (GITA) under Grant no. GITA/DST/TWN/P-70/2015 and by National Science Council of Taiwan, R.O.C., under Contact 104-2923-E-110-001-MY3.

REFERENCES

[1] G. Hong, "Memory technology trend and future challenges," in *Proc. IEEE IEDM*, pp. 292–295, 2012.

[2] K. Kim, "Future memory technology: challenges and opportunities," in *International Symposium on VLSI Technology, Systems and Applications*, pp. 5-9, 2008.

[3] International Technology Roadmap for Semiconductors, 2015, www.itrs.net.

[4] C.H Kim and L. Chang, "Guest editors' introduction: Nanoscale Memories Pose Unique Challenges," *IEEE Design & Test of Computers*, vol. 28, no. 1, pp. 6-8, 2011.

[5] K. Itoh, "Embedded memories: Progress and a look into the future," *IEEE Design & Test of Computers*, vol. 28, no.1, pp. 10-13, 2011.

[6] S-K. Park, "Technology Scaling Challenge and Future Prospects of DRAM and NAND Flash Memory," 2015 *IEEE International Memory Workshop (IMW)*, pp.1-4, 2015.

[7] E. J. Marinissen, B. Prince, D. Keltel-Schulz, and Y. Zorian, "Challenges in embedded memory design and test," in *Proceedings of Design Automation and Test in Europe*, pp. 722–727, 2005.

[8] F. Gamiz, "Capacitor-less memory: Advances and challenges," in *Joint International EUROSOI Workshop and International Conference on Ultimate Integration on Silicon (EUROSOI-ULIS)*, pp. 68-71, 2016.

[9] S. Lurvi, J. Xu and A. Zaslavsky, "Future trends in microelectronics: Frontiers and Innovations,' *John Wiley & Sons*, pp. 59-70, 2013.

[10] S. H. Lee, "Technology Scaling Challenges and Opportunities of Memory Devices," *IEEE IEDM*, pp. 1-8, 2016.

[11] J. Ousterhout, *et al.*, "The case for RAMClouds: Scalable high-performance storage entirely in DRAM," *SIGOPS Operating Systems Review*, vol. 43, no. 4, pp. 92-105, 2009.

[12] H. Zhang, G. Chen, B. C. Ooi, K.-L. Tan, M. Zhang, "In-Memory big data management and processing: A survey," *IEEE Transactions on Knowledge and Data Engineering*, vol. 27, no. 7, pp. 1920-1949, 2015.

[13] G. Giusi, "Physical insights of body effect and charge degradation in floating-body DRAMs," *Solid-State Electron.*, vol. 95, pp. 1–7, 2014.

[14] S. Puget, *et al.*, "FDSOI Floating Body Cell eDRAM Using Gate-Induced Drain-Leakage (GIDL) Write Current for High Speed and Low Power Applications," *IEEE International Memory Workshop (IMW)*, pp. 1–2, 2009.

[15] K. R. A. Sasaki, *et al.*, "Improved retention times in UTBOX nMOSFETs for 1T-DRAM applications," *Solid-State. Electron.*, vol. 97, pp. 30–37, 2014.

[16] Q. Wu, *et al.*, "Experimental demonstration of the high-performance floating-body/gate DRAM cell for embedded memories," *IEEE Elect. Device Lett.*, vol. 33, no. 6, pp. 743–745, 2012.

[17] K. R. A. Sasaki, *et al.*, "Temperature influence on UTBOX 1T-DRAM using GIDL for writing operation," *2012 8th Int. Caribb. Conf. Devices, Circuits Syst. ICCDCS* 2012, pp. 6–9, 2012.

[18] T. Nicoletti, *et al.*, "The dependence of retention time on gate length in UTBOX FBRAM with different source/drain junction engineering," *IEEE Electron Device Lett.*, vol. 33, no. 7, pp. 940–942, 2012.

[19] C. W. Cao, *et al.*, "A Novel 1T-1D DRAM Cell for Embedded Application," *IEEE Trans. on Electron Devices*, vol. 59, no. 5, pp.1304-1310, 2012.

[20] J. Hou, Z. Shao and X. Miao, "A High Speed Low Power Capacitorless SOI-DRAM Cell Using Impact Ionization and GIDL Effect," *IEEE International Conference of Electron Devices and Solid-State Circuits (EDSSC)*, pp. 517- 520, 2009.

[21] A. Z. Badwan, Q. Li, and D. E. Ioannou, "On the nature of the memory mechanism of gated-thyristor dynamic-RAM cells," *IEEE J. Electron Devices Soc.*, vol. 3, no. 6, pp. 468–471, 2015

[22] H. Lu, and A. Seabaugh, "Tunnel field-effect transistors: State-of-the-art," *IEEE Journal of the Electron Devices Society*, vol. 2, no. 4, pp. 44-49, June 2014.

[23] J. Wan, C. L. Royer, A. Zaslavsky, and S. Cristoloveanu, "Tunneling FETs on SOI: suppression of ambipolar leakage, low-frequency noise behavior, and modeling," *Solid-State Electronics*, vol. 65-66, pp. 226-233, 2011.

[24] J.P. Colinge, *et al.* "Nanowire transistors without junctions," *Nature Nanotechnology*, vol. 5, pp. 225-229, 2010.

[25] R. Yu, *et al.*, "Impact ionization induced dynamic floating body effect in junctionless transistors," *Solid-State Electron.*, vol. 90, pp. 28–33, 2013.

[26] G Giusi and G. Iannaccone, "Junction Engineering of 1T-DRAMs," *IEEE Electron Device Lett.*, vol.34, pp. 408-410, 2013.

[27] M.S. Parihar, D. Ghosh, and A. Kranti, "Single transistor latch phenomenon in junctionless transistors," *Journal of Applied Physics.*, vol. 113, 184503, 2013.

[28] A. Biswas, *et al.*, "Investigation of tunnel field-effect transistors as a capacitor-less memory," *Applied Physics Letters*, vol. 104, no. 9, article 092108, 2014.

[29] A. Biswas, and A. M. Ionescu, "1T capacitor-less DRAM cell based on asymmetric tunnel FET design", *IEEE Journal of the Electron Devices Society*, vol. 3, no. 3, pp. 217-222, 2015.

[30] A. Biswas, and A. M. Ionescu, "Study of fin-tunnel FETs with doped pocket as capacitor-less 1T DRAM," in *Proc. IEEE SOI-3D-Subthreshold Microelectronics Technology Unified Conference*, 2014.

[31] N. Navlakha, J.-T. Lin, and A. Kranti, "Improving retention time in tunnel field effect transistor based dynamic memory by back gate engineering," *J. Appl. Phys.*, vol. 119, Jun. 2016, Art. no. 214501.

[32] ATLAS User's Manual, Silvaco, 2010.

[33] J.S. Park, "Dual Gate MOSFET Fabrication," US Patent 6168998 B1 2002

[34] Y. Jeon, M. Kim, D. Lim and S. Kim, " Steep subthreshold swing n- and p-channel operation of bendable feedback field-effect transistors with p^+-i-n^+ nanowires by dual-top gate voltage modulation," *Nano Lett.* Vol. 15, no. 8, pp. 4905–4913, 2015.

[35] A. Salman, S. G. Beebe, M. Emam, M. M. Pelella, and D. E. Ioannou, "Field effect diode: A novel device for ESD protection in deep sub-micron SOI technologies," in *Proc. IEEE IEDM*, pp. 109–112, 2006.

[36] L. M. Almeida, *et al.*, "Optimizing the front and back biases for the best sense margin and retention time in UTBOX FBRAM," *Solid-State Electronics*, vol. 90, pp. 149–154, 2013.

[37] T. Tanaka, K. Suzuki, H. Horie, and T. Sugii, "Ultrafast operation of V_{th}-adjusted p^+-n^+ double-gate SOI MOSFET's," *IEEE Electron Device Lett.*, vol. 15, no. 10, 386-388, 1994.

[38] T. Skotnicki, *et al.*, "Innovative Materials, Devices, and CMOS Technologies for Low-Power Mobile Multimedia," *IEEE Trans. Electron Devices*, vol. 55, no. 1, pp. 96-130, 2008.

[39] J. Wan, C. Le Royer, A. Zaslavsky, and S. Cristoloveanu, "A Compact Capacitor-Less High-Speed DRAM Using Field Effect-Controlled Charge Regeneration," *IEEE Elect. Device Lett.*, vol. 33, no. 2, pp. 179–181, 2012.

978-1-5386-3693-0/18 $31.00 © 2018 IEEE

2018 31th International Conference on VLSI Design and 2018 17th International Conference on Embedded Systems

Energy-Efficient Dynamic Data Encoding for Multi-Level STT-MRAM

Mohammad Gh. Alfailakawi, Imtiaz Ahmad, Sarah Elghandour

Computer Engineering Department

College of Engineering & Petroleum, Kuwait University

Email: {alfailakawi.m,imtiaz.ahmad}@ku.edu.kw

Abstract—**Spin-Torque-Transfer Magnetic RAM (STT-MRAM) is a promising candidate for next generation on-chip last level cache memory. Such technology offers non-volatility, excellent scalability, and CMOS process compatibility. Even though multi-level version of such memories offer more capacity than their single-level counterpart, it suffers from high write energy as well as performance overhead. These unwanted characteristics are due to Two-step Transition (TT) and Hard-Transition (HT). In this paper, we propose a dynamic resistance-to-logic state encoding to minimize energy in MLC STT-MRAM based caches. The proposed encoding/decoding scheme are presented algorithmically and at architectural level. Results on PARSEC benchmarks showed an average reduction of 55% in energy as compared to a recently proposed low power encoding approach.**

I. INTRODUCTION

Today's computing systems are becoming increasingly dependent on implementation technology of the devices used in the memory hierarchy. Currently, SRAM, DRAM, and flash are the dominant mainstream memory technologies serving as cache, main memory, and secondary storage, respectively [1]. SRAM and DRAM have played key roles during the evolution of modern computer systems. When the technology node of conventional memories is scaled down, DRAM and SRAM are all facing serious power consumption challenges due to increased leakage [2]. Moreover, multi-core processors demand large data transfers between cores hence requiring large capacity cache and main memory to achieve scalable performance. However, it is difficult to construct large caches that simultaneously achieve high performance and low power. For future caches, the use of SRAM is becoming less appealing due to its large power consumption and die area. Moreover, the performance of DRAMs is decreasing due to increased refresh penalty when large capacity devices are utilized. Therefore, in order to meet the growing demand for performance and low power, alternative memory technologies are being actively pursued [1], [3].

Resistance based emerging memory technologies are currently being investigated as possible replacement for next generation cache and main memory. One such candidate is Spin-Torque-Transfer Magnetic-RAM (STT-MRAM) [4]–[6]. Compared to SRAM, STT-MRAM advantages include non-volatility, CMOS compatibility, low power, and SRAM comparable performance. In addition, STT-MRAM is immune to high energy particle radiation hence an excellent candidate

for aerospace and aeronautical applications. As compared with DRAM, STT-MRAM requires no periodic refresh and has higher density. Thus, STT-MRAM is an attractive and promising candidate for the replacement of SRAM and DRAM in next generation memory systems [7]. Due to its advantages of high-density and low energy, STT-MRAM has attracted other applications such as energy efficient check-pointing [8], multi-port register file in GPUs [9], and video processing systems [10].

The key component to store information in STT-MRAM is the Magnetic Tunnel Junction (MTJ). An MTJ consists of two ferromagnetic layers separated by a dielectric layer. Magnetization direction of one ferromagnetic layer is fixed (called the reference layer) while that of the other one, called free layer, can be changed either by applying an external magnetic field or a spin polarized current. When magnetization direction of the free layer is in parallel with that of the reference layer, the MTJ has a low resistance state and represents logic value "1". When magnetization directions of the two ferromagnetic layers are anti-parallel, the MTJ has a high resistance state and denotes a logical "0".

A Single-Level Cell (SLC) stores one-bit of information using two resistance states [3], [5]. In the Multi-Level Cell (MLC) version of STT-MRAM, the free layer is partitioned into two magnetic domains with different magnetic properties commonly referred to as soft and hard domains. The soft domain can be switched using a small current while the hard one requires a much larger amount of current. Such configuration allows such cell to have the ability to store two bits of information in a single cell and thus enhance density [11], [12]. However, the MLC STT-MRAM cell suffers from reduced lifetime as it will be explained next.

This work is motivated by the key factor responsible for MLC STT-MRAM lifetime reduction, namely, the two-step state transitions [13]. Due to the fact that magnetization directions of the hard and soft domains cannot be flipped in opposite directions simultaneously, certain transitions must be performed using two write operations resulting in increased latency and energy consumption. For this reason, different data encoding schemes were proposed to reduce STT-MRAM write energy and latency [13]–[22]. The key idea in such algorithms is to increase energy-efficient states and decrease energy-inefficient ones when writing to the cache.

In this paper, we propose a dynamic data encoding tech-

978-1-5386-3693-0/18 $31.00 © 2018 IEEE

428

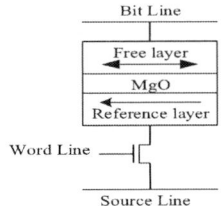

Fig. 1. SLC MTJ

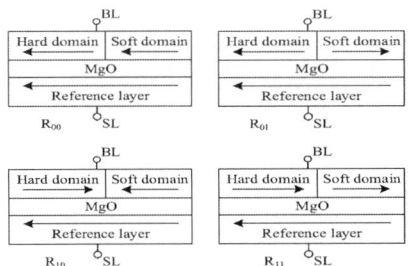

Fig. 2. MLC MTJ with Various Resistant States

nique that maps data to the most energy-efficient states at runtime. As compared with earlier approaches, we propose a cost-based encoding that is capable of reducing the overall cost of write operations. The cost model is used as an objective function that decides best encoding option possible. The proposed cost model is flexible and can be used to minimize either energy or latency. The encoding and decoding architectures are very simple but effective. The proposed encoding/decoding schemes are discussed and presented at both algorithmic and architectural levels. Results are compared with a recent technique to demonstrate the effectiveness and scalability of the proposed approach in reducing energy consumption.

The remainder of the paper is organized as follows. Section II gives a brief introduction to STT-MRAM and present related work. Details of the proposed algorithms and their implementation are discussed in Section III. Experimental results are presented in Section IV and conclusions are made in Section V.

II. BACKGROUND AND RELATED WORK

In this section, we first present general background of STT-MRAM followed by a brief description of related work that uses data encoding schemes to reduce energy consumption and enhance device lifetime.

A. SLC and MLC STT-MRAM

STT-MRAM cell uses magnetic tunnel junction (MTJ) to store binary information. The MTJ can be built with In-plane or Perpendicular Magnetic Anisotropy (IMA or PMA-MTJ) [23]. An IMA-MTJ is shown in Figure 1 [3]. The magnetization direction of one ferromagnetic layer is fixed and is referred to as reference or pinned layer. On the other hand, the other layer, called free layer, can be changed by injecting current through it. In SLC MTJ, when magnetization directions of the free and pinned layers are parallel to each other, the MTJ resistance is low representing a logic value "1". However, when magnetization directions are anti-parallel, the cell is in a high resistance state, thus stores a logic "1". In the MLC version, the free layer has two magnetic domains with different magnetic properties as shown in Figure 2. MLC MTJ can be stacked in series [11] or in parallel [12]. In this paper, we assume parallel MLC STT-MRAM. An MLC MTJ can store two-bit binary information since the two domains can have different magnetization directions. There are four resistance states defined by the combinations of the relative magnetization directions of the two domains as shown in Figure 2. The domain that requires large current to switch is referred to as hard-bit and represent most significant bit (MSB). Meanwhile, the other domain is called the soft-bit and stores the least significant bit (LSB).

Accessing the STT-MRAM cell is normally controlled using a single transistor, therefore current always passes through both domains during write/read operations. As mentioned earlier, the magnetization direction of the soft domain can be changed using a small current while the hard one requires a larger one to switch. Since the current required by the hard domain is always larger than that of the soft one, changing the hard domain always changes the magnetization of the soft one as well. Due to this fact, read and write operations of MLC STT-MRAM are relatively more complex compared to their SLC counterpart. Thus, transitions between different two-bit values cannot always be completed in a single step. For example, transition $R_{01} \rightarrow R_{10}$ where the bit on the left is hard bit will require two step write operation. The first step is to switch from state R_{01} to R_{11} using a large write current, and the second step will be to go from R_{11} to R_{10} using a small write current. In general, transition are commonly referred to as either Soft transition (ST) or Hard Transition (HT). In ST, only the soft domain is switched whereas in HT both the soft and hard domain are switched. Therefore, the energy of a write operation depends on the initial and final state of the cell as shown in Table I (for 2-bit MLC cell).

B. Related Work

Although STT-MRAM has many attractive features, it has some disadvantages that deny its adoption as a cache or main memory in today's memory systems. The write operation of MLC STT-MRAM is generally slower and more energy-consuming than the read operation. In the past decade, sev-

TABLE I
WRITE ENERGY FOR VARIOUS TRANSITIONS (IN PJ) [18]

From/To	R_{00}	R_{01}	R_{10}	R_{11}
R_{00}	0	0.045	0.185	0.120
R_{01}	0.021	0	0.194	0.128
R_{10}	0.144	0.189	0	0.001
R_{11}	0.164	0.209	0.065	0
Average	0.082	0.111	0.111	0.062

978-1-5386-3693-0/18 $31.00 © 2018 IEEE

eral encoding methods have been proposed to reduce energy consumption of the write operation and increase STT-MRAM lifetime [13]–[22]. Since read latency and energy of STT-MRAM are very low, replacing a write operation with a read-modify-write operation is an efficient way to reduce energy consumption. Yang et al. [14] were the first to propose a simple and effective Data-Comparison Write (DCW) technique to eliminate redundant bit write in NVMs with zero memory overhead. In DCW, input data word to be written is first compared to the already programmed word and a write is performed to bit positions with complementary values. Such practice leads to writing half of the bits within a word on average. However, the maximum number of bit updates can be as large as a word in the worst case.

In Chen et al. [15], the authors have shown that write energy can differ by as much as 27.5% for different mappings between resistance and logic states. They studied the impact of different logic-resistance mappings on power. The authors found that the optimal mapping between logic and resistance states to be 00→R11, 01→R10, 10→R01, 11→R00 for 2-bit MLC. However, this static encoding does not result in power saving for all applications as it application dependent.

In MLC STT-MRAM, hard bits are fast to read but slow to write while soft-bits are fast to write but slow to read. Jiang et al. [16] proposed a Line Pairing (LP) and Line Swapping (LS) approach that take advantage of such characteristics by organizing cache blocks in a single set into read-slow-write-fast soft-bit blocks and read-fast-write-slow hard-bit blocks. By promoting frequently written data into soft-bit blocks and frequently read ones to hard-bit blocks, hit latency is reduced resulting in improved overall performance. Similarly, a cell splitting mapping method was proposed by Bi et al. [17] where cache blocks are divided into fast and slow regions and data migration policies are used to allocate frequently-used data to fast regions.

The 2-bit logical values written to the cache vary with time and application. Since different resistance states have different write energies, Chen et al. [18] proposed a technique to map resistance states to logical values based on data-pairs frequency in application programs. The authors proposed two encoding schemes namely phase based and application based encoding. Since there are 4!=24 possible encoding for the four resistance states of 2-bit memory cell, such scheme requires 5-bit storage overhead per block to store encoding information. In addition, the proposed technique needs the assistance of the operating system to statically profile application data to decide the encoding in the different phases during application execution.

Due to the dependency of write energy on data values, Chi et al. [19] proposed a dynamic encoding technique to map most frequently appearing data patterns to the most-energy-efficient resistance states at run-time. Among the 24 mappings, the authors selected only 12 mappings which require only 4-bit of storage overhead. Unlike earlier approaches, this approach decide the encoding based on data values to be written, hence dynamic encoding encoding, instead of any static analysis of application data.

Hong et al. [20] proposed a three value MLC STT-MRAM cache in which the memory cell stores only three data values instead of four. By reducing the amount of information stored, higher read stability is achieved. A reliable write mechanism was also introduced to enhance writability. Their idea was to eliminate resistance state that overlap with adjacent states and causes read errors. To employ three-valued MLC STT-MRAM in caches, authors employed a simple binary-to-ternary coding technique, in which 32-bit data is divided into multiple 3-bit segments where each segment is mapped to 4-bit code words. Such ternary cache enhances the reliability while approaching the data density of 2-bit MLC MRAM. A similar technique where a three-valued STT-MRAM was proposed by Wen et al. [21] to mitigate read and write errors.

Magnetization directions of hard and soft domains of STT-MRAM cannot be flipped to two opposite directions simultaneously. Certain state transitions requires two step write operation. Such write operations result in more frequently updates to soft domain cell resulting in significant reduction in its lifetime. Based on this observation, Luo et al. [22] proposed a two-step transition minimization (TSTM) scheme that minimizes two-step write operations. In TSTM, certain cells are used as flags to indicate different encoding of the cells. The proposed technique divide a block of data into multiple 3-cell segments and every 2-bit are encoded using three bits. In such approach, states R_{00} and R_{11} have only one code (000 and 111) while R_{01} and R_{10} each have three different codes. The encoder and decoder design is simple and easy to implement, however requires a storage over head of almost 50%. For example, 512-bit were encoded as 768-bit in their approach which significantly reducing memory capacity.

III. PROPOSED ALGORITHM & ARCHITECTURE

In this section we present our data encoding and decoding algorithms. First, we formulate the cost functions used for the optimization of write operations followed by discussion on encoding/decoding algorithms and their implementation.

A. Encoding Options

Similar to previous approaches, the proposed technique explores different data encoding to find the one with the least cost. In this work, we extend the encoding options proposed in [24] resulting in the following available options:

1) No encoding used (W)
2) Invert all bits of W ($\overline{W}$)
3) Invert odd bits of W ($W_{\overline{odd}}$)
4) Invert even bits of W ($W_{\overline{even}}$)
5) Shift W (W_s)
6) Invert all bits of shifted W ($\overline{W_s}$)
7) Invert odd bits of shifted W ($W_s{}_{\overline{odd}}$)
8) Invert odd bits of shifted W ($W_s{}_{\overline{even}}$)

The proposed shift operation consists of reorganizing data word bit-pairs in such a way that bit pairs are created using one bit from the lower half while the other one from the upper one

Algorithm 1 Min-Cost Encoding
```
Shift, Odd, Even: Auxiliary coding bits
w: n-bit data word to be written
Shift=0, Odd=0, Even=0;
Compute cost according to all encoding options;
if cost(w even̄) is minimum then
    Invert even bits in w;
    Even=1;
else if cost(w odd̄) is minimum then
    Invert odd bits in w;
    Odd=1;
else if cost(w̄) is minimum then
    Invert all bits in w;
    Odd=Even=1;
else if cost(wₛ) is minimum then
    Shift=1;
else if cost(wₛ odd̄) is minimum then
    Invert odd bits in wₛ;
    Shift=Odd=1;
else if cost(wₛ even̄) is minimum then
    Invert even bits in wₛ;
    Shift=Even=1;
else if cost(w̄ₛ) is minimum then
    Invert all bits in wₛ;
    Odd=Even=Shift=1;
end if
```

as shown in Table II. The table shows an example of applying the various encoding options on W=11111001.

B. Cost Function

The proposed algorithm is based on DCW where the data word to be written is first compared to the existing one and only complementary bit positions are updated. Thus, the cost of writing a data word W under such scheme can be formulated as [24]:

$$Cost(W) = \alpha(T_{(00,01,10)\to11}) + \beta(T_{(00,01,11)\to10}) + \gamma(T_{(00,10,11)\to01}) + \delta(T_{(01,10,11)\to00}) \quad (1)$$

TABLE II
ENCODING EXAMPLE USING VARIOUS OPTIONS

Code	Type	Symbolic Representation	Example
000	W	$W_7\ W_6\ W_5\ W_4\ W_3\ W_2\ W_1\ W_0$	11 11 10 01
001	$\overline{W}$	$\overline{W_7}\ \overline{W_6}\ \overline{W_5}\ \overline{W_4}\ \overline{W_3}\ \overline{W_2}\ \overline{W_1}\ \overline{W_0}$	00 00 01 10
010	$W\ \overline{odd}$	$\overline{W_7}\ W_6\ \overline{W_5}W_4\ \overline{W_3}\ W_2\ \overline{W_1}\ W_0$	01 01 00 11
011	$W\ \overline{even}$	$W_7\ \overline{W_6}\ W_5\ \overline{W_4}\ W_3\ \overline{W_2}\ W_1\ \overline{W_0}$	10 10 11 00
100	W_s	$W_7\ W_3\ W_6\ W_2\ W_5\ W_1\ W_4\ W_0$	11 10 10 11
101	$\overline{W_s}$	$\overline{W_7}\ \overline{W_3}\ \overline{W_6}\ \overline{W_2}\ \overline{W_5}\ \overline{W_1}\ \overline{W_4}\ \overline{W_0}$	00 01 01 00
110	$W_s\ \overline{odd}$	$\overline{W_7}\ W_3\ \overline{W_6}\ W_2\ \overline{W_5}\ W_1\ \overline{W_4}\ W_0$	01 00 00 01
111	$W_s\ \overline{even}$	$W_7\ \overline{W_3}\ W_6\ \overline{W_2}\ W_5\ \overline{W_1}\ W_4\ \overline{W_0}$	10 11 11 10

where α, β, γ, and δ are scaling factors that determine the cost based on memory technology used and the objective function being optimized. By changing the values of these parameters, one can optimize for energy, performance, or both. For instance, to optimize for energy, write energy for each transition from Table I can be used for these parameters. This results in setting α to the average write energy of R_{11} (0.062), $\beta=\gamma$ to that for $R10/R_{01}$ (0.111), and δ to the energy of R_{00} (0.082).

Similar to equation 1, the cost for writing the inverted word ($\overline{W}$) can be found by inverting the corresponding new states in (1) as follows:

$$Cost(\overline{W}) = \alpha(T_{(00,01,10)\to00}) + \beta(T_{(00,01,11)\to01}) + \gamma(T_{(00,10,11)\to10}) + \delta(T_{(01,10,11)\to11}) \quad (2)$$

Note that in this case, the final states are the complement of those in (1) to signify the fact the all bits of W are now complemented. Below we present the cost functions for the remaining encoding options for W. The cost functions for W_s encoding options can be derived similarly but are not give here.

$$Cost(W\ \overline{odd}) = \alpha(T_{(00,01,10)\to01}) + \beta(T_{(00,01,11)\to00}) + \gamma(T_{(00,10,11)\to11}) + \delta(T_{(01,10,11)\to10}) \quad (3)$$

$$Cost(W\ \overline{even}) = \alpha(T_{(00,01,10)\to10}) + \beta(T_{(00,01,11)\to11}) + \gamma(T_{(00,10,11)\to00}) + \delta(T_{(01,10,11)\to01}) \quad (4)$$

To demonstrate the encoding options and cost calculation, consider the case when a new word W=00111001 to be written in instead of old data word X=00111001. Table III shows encoding code, name, transition costs, and overall cost respectively. The encoding cost was calculated using α=0.062, $\beta=\gamma$=0.111, and δ=0.082. It is apparent from the table that the best encoding option is W_s with a cost of 0.062.

C. Min-Cost Algorithm & Architecture

The proposed approach consists of an encoder and a decoder that implement algorithms 1 and 2 respectively. A block diagram for the encoder architecture is shown in Figure 3 whereas Figure 4 shows that of the decoder. The inputs to the encoder

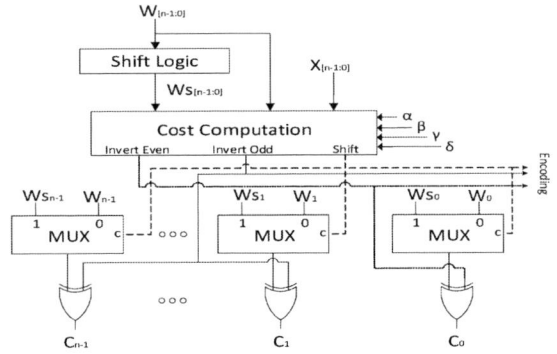

Fig. 3. Encoder Architecture

TABLE III
ENCODING & COST CALCULATION EXAMPLE

Encoding	Encoded Word	Transitions	Cost
W	11 10 11 01	$T_{(00,01,10)\to 11}=2$, $T_{(00,01,11)\to 10}=0$, $T_{(00,10,11)\to 01}=1$, $T_{(01,10,11)\to 00}=0$	0.235
$\overline{W}$	00 01 00 10	$T_{(00,01,10)\to 00}=0$, $T_{(00,01,11)\to 01}=1$, $T_{(00,10,11)\to 10}=1$, $T_{(01,10,11)\to 11}=1$	0.304
$W_{\overline{odd}}$	01 00 01 11	$T_{(00,01,10)\to 01}=0$, $T_{(00,01,11)\to 00}=0$, $T_{(00,10,11)\to 11}=1$, $T_{(01,10,11)\to 10}=1$	0.193
$W_{\overline{even}}$	10 11 10 00	$T_{(00,01,10)\to 10}=1$, $T_{(00,01,11)\to 11}=2$, $T_{(00,10,11)\to 00}=0$, $T_{(01,10,11)\to 01}=1$	0.366
W_s	11 11 10 01	$T_{(00,01,10)\to 11}=1$, $T_{(00,01,11)\to 10}=0$, $T_{(00,10,11)\to 01}=0$, $T_{(01,10,11)\to 00}=0$	0.062
$\overline{W_s}$	00 00 01 10	$T_{(00,01,10)\to 00}=0$, $T_{(00,01,11)\to 01}=1$, $T_{(00,10,11)\to 10}=1$, $T_{(01,10,11)\to 11}=1$	0.304
$W_{s\,\overline{odd}}$	01 01 00 11	$T_{(00,01,10)\to 01}=1$, $T_{(00,01,11)\to 00}=0$, $T_{(00,10,11)\to 11}=2$, $T_{(01,10,11)\to 10}=1$	0.366
$W_{s\,\overline{even}}$	10 10 11 00	$T_{(00,01,10)\to 10}=1$, $T_{(00,01,11)\to 11}=2$, $T_{(00,10,11)\to 00}=0$, $T_{(01,10,11)\to 01}=1$	0.366

Algorithm 2 Min-Cost Decoding

Odd, Even: Auxiliary coding bits;
w: n-bit data word to be written;
case {Shift,Odd,Even}:
  000: w;
  001: Invert even bits in w;
  010: Invert odd bits in w;
  011: Invert all bits in w;
  100: Shift w;
  101: Shift w then invert even bits;
  110: Shift w then invert odd bits;
  111: Shift w then invert all bits;
endcase

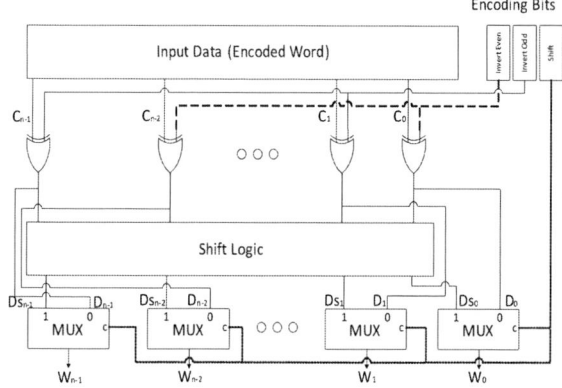

Fig. 4. Decoder Architecture

are old data word $X_{(n-1)}...X_1 X_0$ and the data to written denoted as $W_{(n-1)}...W_1 W_0$. A shifter unit is used to generate a shifted version of the data word to be written and is labeled as $W_{s_{(n-1)}}...W_{s_1} W_{s_0}$ in the figure. The "cost computation" module computes the cost for the various encoding options (i.e. using equations 1- 4) and generates appropriate control signals (InvertEven, InverOdd, Shift). These control signals are used to generate the encoded data word $C_{(n-1)}...C_1 C_0$. These control signals are also stored with the encoded data word to be used later in the decoding process. Note that the control signal "Shift" signifies the inversion to be applied to W_s instead of W. The encoding bits are utilized by the decoder to regenerate the original data word W as can be seen in Figure 4.

Compared to [19] which requires four bits to store encoding information, the proposed approach requires only three, thus reducing storage overhead by 25%. However, due to the dynamic nature and increased number of encoding options, the proposed approach incurs additional hardware overhead over the approach in [19]. Nevertheless, this increased hardware does not have serious performance impact since all computational modules operate in parallel.

IV. EXPERIMENTAL RESULTS

The proposed approach as well as that of [19] were implemented in C++ where 10 memory traces of PARSEC 2.1 benchmarks [25] were used to compare their performance.

Memory traces were extracted using Pin binary instrumentation tool [26] and were run on Gem5 simulator [27] in Full System mode (ALPHA ISA). A data word in the benchmarks consist of 48 bits with an average number of writes of approximately 9 million words.

Even though the approach proposed in [19] does not use DCW, in this work assumed that it does. Doing so will allow for fair and meaningful comparison between the proposed approach and [19]. The metric used to compare performance was the overall energy. Cost calculation for available encoding options was done using the values in Table I for α, β, γ, and δ as describe in Section III.

Figure 5 show energy reduction of the proposed algorithm as compared to [19]. On average, a 55% reduction in energy was achieved with a maximum of 64% and minimum of 24%. Note that the values given in the table are normalized to the approach proposed in [19]. The highest energy reduction was for animation benchmarks such as "facesim" and "fluidanimate". On the other hand, the least reduction was for Data Mining and Media processing benchmarks represented by benchmarks "freqmine" and "x264" respectively.

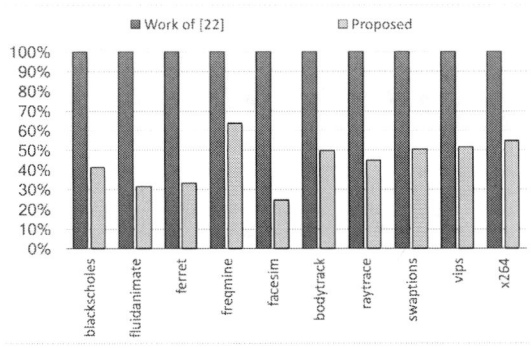

Fig. 5. Cost Reduction in Term of Power

V. CONCLUSION

STT-MRAM is becoming one of the most promising memory technologies for next generation memory systems due to its non-volatility, ultra-low power consumption and long endurance among other features. However, two-step state transitions of MLC STT-MRAM results in increased write energy and performance overhead. This work propose a dynamic data encoding technique that reduce write energy using using appropriate cost model that checks eight different encoding options to choose the one with least energy. Simulation results suggest that the proposed scheme reduces the average write energy per word by 55% on average compared with Dynamic Data-Resistance Encoding technique. Our future work is directed towards developing simpler encoding techniques for STT-MRAM [28].

REFERENCES

[1] L. Baldi, R. Bez, and G. Sandhu, "Emerging memories," *Solid-State Electronics*, vol. 102, pp. 2–11, 2014.

[2] http://public.itrs.net, "International technology roadmap for semiconductors (itrs)," *Semiconductor Industry Association*, 2016.

[3] S. Yu and P.-Y. Chen, "Emerging memory technologies: recent trends and prospects," *IEEE Solid-State Circuits Magazine*, vol. 8, no. 2, pp. 43–56, 2016.

[4] M. Hosomi, H. Yamagishi, T. Yamamoto, K. Bessho, Y. Higo, K. Yamane, H. Yamada, M. Shoji, H. Hachino, C. Fukumoto, *et al.*, "A novel nonvolatile memory with spin torque transfer magnetization switching: Spin-ram," in *Electron Devices Meeting, 2005. IEDM Technical Digest. IEEE International*, pp. 459–462, IEEE, 2005.

[5] J.-G. Zhu, "Magnetoresistive random access memory: The path to competitiveness and scalability," *Proceedings of the IEEE*, vol. 96, no. 11, pp. 1786–1798, 2008.

[6] M. H. Kryder and C. S. Kim, "After hard driveswhat comes next?," *IEEE Transactions on Magnetics*, vol. 45, no. 10, pp. 3406–3413, 2009.

[7] P. Chi, S. Li, Y. Cheng, Y. Lu, S. H. Kang, and Y. Xie, "Architecture design with stt-ram: Opportunities and challenges," in *Design Automation Conference (ASP-DAC), 2016 21st Asia and South Pacific*, pp. 109–114, IEEE, 2016.

[8] P. Chi, C. Xu, T. Zhang, X. Dong, and Y. Xie, "Using multi-level cell stt-ram for fast and energy-efficient local checkpointing," in *Computer-Aided Design (ICCAD), 2014 IEEE/ACM International Conference on*, pp. 301–308, IEEE, 2014.

[9] X. Liu, M. Mao, X. Bi, H. Li, and Y. Chen, "An efficient stt-ram-based register file in gpu architectures," in *Design Automation Conference (ASP-DAC), 2015 20th Asia and South Pacific*, pp. 490–495, IEEE, 2015.

[10] H. Zhao, H. Sun, Q. Yang, T. Min, and N. Zheng, "Exploring the use of volatile stt-ram for energy efficient video processing," in *Quality Electronic Design (ISQED), 2016 17th International Symposium on*, pp. 81–87, IEEE, 2016.

[11] T. Ishigaki, T. Kawahara, R. Takemura, K. Ono, K. Ito, H. Matsuoka, and H. Ohno, "A multi-level-cell spin-transfer torque memory with series-stacked magnetotunnel junctions," in *VLSI Technology (VLSIT), 2010 Symposium on*, pp. 47–48, IEEE, 2010.

[12] X. Lou, Z. Gao, D. V. Dimitrov, and M. X. Tang, "Demonstration of multilevel cell spin transfer switching in mgo magnetic tunnel junctions," *Applied Physics Letters*, vol. 93, no. 24, p. 242502, 2008.

[13] Y. Chen, X. Wang, W. Zhu, H. Li, Z. Sun, G. Sun, and Y. Xie, "Access scheme of multi-level cell spin-transfer torque random access memory and its optimization," in *Circuits and Systems (MWSCAS), 2010 53rd IEEE International Midwest Symposium on*, pp. 1109–1112, IEEE, 2010.

[14] B.-D. Yang, J.-E. Lee, J.-S. Kim, J. Cho, S.-Y. Lee, and B.-G. Yu, "A low power phase-change random access memory using a data-comparison write scheme," in *Circuits and Systems, 2007. ISCAS 2007. IEEE International Symposium on*, pp. 3014–3017, IEEE, 2007.

[15] Y. Chen, W.-F. Wong, H. Li, and C.-K. Koh, "Processor caches with multi-level spin-transfer torque ram cells," in *Proceedings of the 17th IEEE/ACM international symposium on Low-power electronics and design*, pp. 73–78, IEEE Press, 2011.

[16] J. Lei, Z. Bo, Z. Youtao, and Y. Jun, "Constructing large and fast multi-level cell sttmram based cache for embedded processors," in *Proc of the 49th ACM/EDAC/IEEE Design Automation Conf (DAC2012). Piscataway, NJ: IEEE*, vol. 20, pp. 907–912.

[17] X. Bi, M. Mao, D. Wang, and H. Li, "Unleashing the potential of mlc stt-ram caches," in *Computer-Aided Design (ICCAD), 2013 IEEE/ACM International Conference on*, pp. 429–436, IEEE, 2013.

[18] T. Chen, J. Meng, J. Ma, B. Quan, and J. Wu, "Phase based and application based dynamic encoding scheme for multi-level cell stt-ram," in *High Performance Computing and Communications & 2013 IEEE International Conference on Embedded and Ubiquitous Computing (HPCC_EUC), 2013 IEEE 10th International Conference on*, pp. 1–8, IEEE, 2013.

[19] P. Chi, C. Xu, X. Zhu, and Y. Xie, "Building energy-efficient multi-level cell stt-mram based cache through dynamic data-resistance encoding," in *Quality Electronic Design (ISQED), 2014 15th International Symposium on*, pp. 639–644, IEEE, 2014.

[20] S. Hong, J. Lee, and S. Kim, "Ternary cache: Three-valued mlc stt-ram caches," in *Computer Design (ICCD), 2014 32nd IEEE International Conference on*, pp. 83–89, IEEE, 2014.

[21] W. Wen, Y. Zhang, M. Mao, and Y. Chen, "State-restrict mlc stt-ram designs for high-reliable high-performance memory system," in *Proceedings of the 51st Annual Design Automation Conference*, pp. 1–6, ACM, 2014.

[22] H. Luo, J. Hu, L. Shi, C. J. Xue, and Q. Zhuge, "Two-step state transition minimization for lifetime and performance improvement on mlc stt-ram," in *Design Automation Conference (DAC), 2016 53nd ACM/EDAC/IEEE*, pp. 1–6, IEEE, 2016.

[23] X. Fong, Y. Kim, R. Venkatesan, S. H. Choday, A. Raghunathan, and K. Roy, "Spin-transfer torque memories: Devices, circuits, and systems," *Proceedings of the IEEE*, vol. 104, no. 7, pp. 1449–1488, 2016.

[24] I. Ahmad, A. Hamouda, and M. G. Alfailakawi, "Odd/even invert coding for phase change memory with thermal crosstalk," *Microprocessors and Microsystems*, vol. 49, pp. 150 – 163, 2017.

[25] C. Bienia, S. Kumar, J. P. Singh, and K. Li, "The parsec benchmark suite: Characterization and architectural implications," in *Proceedings of the 17th international conference on Parallel architectures and compilation techniques*, pp. 72–81, ACM, 2008.

[26] C.-K. Luk, R. Cohn, R. Muth, H. Patil, A. Klauser, G. Lowney, S. Wallace, V. J. Reddi, and K. Hazelwood, "Pin: building customized program analysis tools with dynamic instrumentation," in *Acm sigplan notices*, vol. 40, pp. 190–200, ACM, 2005.

[27] N. Binkert, B. Beckmann, G. Black, S. K. Reinhardt, A. Saidi, A. Basu, J. Hestness, D. R. Hower, T. Krishna, S. Sardashti, *et al.*, "The gem5 simulator," *ACM SIGARCH Computer Architecture News*, vol. 39, no. 2, pp. 1–7, 2011.

[28] I. Ahmad, M. Imdoukh, and M. Alfailakawi, "Extending multi-level stt-mram cell lifetime by minimizing two-step and hard state transitions in hot bits," *IET Computers & Digital Techniques*, August 2017.

978-1-5386-3693-0/18 $31.00 © 2018 IEEE

Switching-Time Dependent PUF Using STT-MRAM

Ashwani Kumar, Shubham Sahay, and Manan Suri

Department of Electrical Engineering
Indian Institute of Technology Delhi
New Delhi, India-110016
Email: manansuri@ee.iitd.ac.in

Abstract— **Physically Unclonable Function (PUF) has become an indispensable on-chip security primitive for hardware security. Area and power efficient PUFs are required to sustain the future security of IoT platforms. In particular, emerging magneto-resistive STT-MRAM based PUF circuits have gained much attention as an alternative to software/pure CMOS PUF owing to their area and power efficiency. In this work, we propose a novel PUF extraction architecture and methodology that exploits the probabilistic switching behavior of STT-MRAM devices. We exploit spatial (device to device) as well as temporal (cycle to cycle) switching stochasticity of STT-MRAM devices in an array of 1T-1MTJ cells. We performed all simulations using 90 nm CMOS technology node, and 32 nm (diameter) perpendicular anisotropic magnetic tunnel junction (PMA-MTJ). The proposed PUF exhibits 45.83 % inter-hamming distance and ~ 5.0 % intra-hamming distance without post processing of the extracted bit pattern.**

Keywords- IoT security, STT-MRAM, PUF, Hardware security

I. INTRODUCTION

The present digital era consists of an enormously large interconnected ecosystem of electronic devices. Cyber physical security has been a major issue concerning this vastly connected IoT era. Since non-volatile memory (NVM) constitutes an integral part of the IoT semiconductor landscape, it makes a lot of sense to exploit the NVM itself for realization of dedicated hardware security primitives such as physically unclonable functions (PUF). Hardware PUFs help in use cases such as- prevention of semiconductor counterfeiting, chip authentication/identification, and countering side-channel attacks [1]-[12]. PUFs generate a response when excited with a challenge. These challenge-response (C-R) pairs form a cryptographic key for the hardware. Ideally, a PUF should always generate the same response for a given challenge. This is measured as the intra-hamming distance; ideally close to zero. Furthermore, different PUF instances should generate different repose when excited with the same challenge. This property is characterized as the inter-hamming distance; ideally close to 50 %. Output response of an ideal PUF should be uniform; i.e. probability of occurrence of both 1 and 0 is 0.5. In literature, pure CMOS based PUFs have been realized using SRAM cells, consuming large area and power [8], [11]-[12]. PUFs based on emerging magneto-resistive memory devices

(STT-MRAM/MRAM) are shown to be more lucrative owing to their low power consumption and small area footprint. In literature, geometric variation of MRAM devices owing to device-to-device process variability has been exploited for PUF implementation [13]. MRAM with different geometrical shapes exhibits different orientation of the easy-axis (the metastable spin orientation). If such an array of MRAM is excited to the meta-stable state using external stimulus, due to the variability in the easy-axis, the final spin-state of the MRAM will be different generating a unique pattern [13]. Furthermore, a PUF exploiting the variation of the resistance in an array of STT-MRAM devices was also proposed [14], where the resistance of adjacent STT-MRAM devices in the array is compared and difference is used to generate response bits. Probabilistic switching behavior of STT-MRAM devices, in particular switching current has also been exploited for PUF application in literature [15]-[17]. In this work, we show how switching-time variability of STT-MRAM devices can be used as an effective tool for extracting PUF signatures from any array. We demonstrate a unique procedure to convert an input challenge to a unique PUF response from the STT-MRAM array. This paper is organized as follows: section II describes the fundamentals of STT-MRAM devices and their properties explored for extracting PUF. Section III describes our proposed approach for extracting the PUF signature. Analysis of the proposed methodology is presented in section IV, and section V concludes the work.

II. STT-MRAM: BASICS AND COMPACT MODEL

A. Basics of STT-MRAM

Spin transfer torque magneto-resistive RAMs (STT -RAM) devices exploit the tunnel magneto resistance (TMR) effect of a magnetic tunnel junction (MTJ) to switch between two different resistance states [18]. An MTJ is a layered stack as shown in Fig. 1, where a thin oxide barrier layer (like MgO) is sandwiched between a magnetic layer of fixed anisotropy and a free magnetic layer (generally CoFeB) [19]. When such stream of spin polarized electrons enters the free layer, it exchanges the spin with the electrons of the free layer and hence, alters the magnetic state of the free layer due to the spin-transfer torque (STT) phenomena [18]. If the spin orientation of both the free layer and the fixed layer are same, the device is said to be in parallel configuration while when the spin orientation is opposite, the device remains in anti-parallel configuration. The TMR ratio is defined as TMR = (RAP – RP) / RP where RP is

978-1-5386-3693-0/18 $31.00 © 2018 IEEE   434

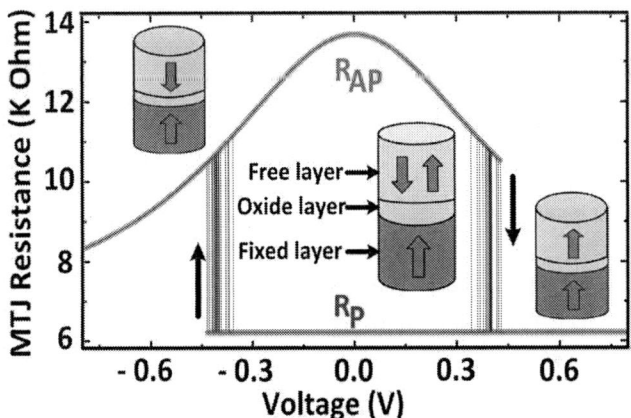

Figure 1. (Inset) Illustration of MTJ nanopillar structure containing thin films of CoFeB (free layer)/ MgO(oxide layer)/ CoFeB(fixed layer). Resistance-voltage switching characteristics for multiple STT-MRAM devices are obtained by performing the Monte-Carlo simulation using the compact model.

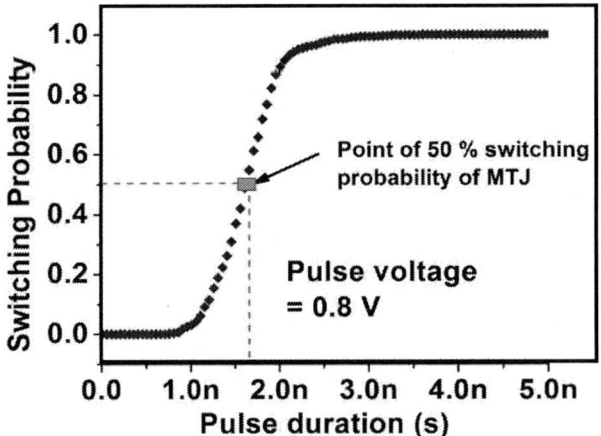

Figure 2. STT-MRAM switching probability (stohascitity) with respect to duration of applied programming pulse at a fixed programming voltage amplitude. Monte-Carlo simulation using the compact model is performed to imitate the programming of a single STT-MRAM device over 1000 cycles and to obtain the switching probability of STT-MRAM.

the resistance of the parallel state and RAP is the resistance of the anti-parallel state. A high TMR ratio means a broader resistance window and facilitates easy detection/reading of the resistance states [20].

Parallel configuration offers a low resistance and corresponds to the *set-state* while the anti-parallel configuration offers a high resistance and corresponds to the *reset-state* as also shown in Fig. 1. If the magnitude of the energy supplied by the applied pulse is too high, the magnetization of the free layer is bound to be switched due to STT. However, if the applied current magnitude is low, the magnetization of the free layer remains unperturbed. However, there exists a particular switching threshold for which the switching probability of the free layer is around 50 % as shown in Fig. 2. Therefore, STT-MRAM devices are known to exhibit probabilistic/stochastic *set-to-reset* or *reset-to-set* switching

behaviour upon excitation with weak programming pulse (i.e. pulse energy comparable to the threshold switching energy) [20]. Such probabilistic switching can be exploited by either using the voltage/current magnitude of the applied pulse or by using the pulse width for a constant applied voltage/current.

The probabilistic switching behaviour exhibits a spatial variation (i.e. device to device variability). Since the process variations lead to a difference in geometric variations and other physical parameters [20]-[21], the applied pulse current magnitude or pulse width for obtaining 50 % switching probability is different for different STT-MRAM devices even if they are fabricated using same process flow.

B. Compact Model of STT-MRAM

A compact model of STT-MRAM having a perpendicular magnetic anisotropy (PMA) is used in this work [22]. This model shows very good agreement with the device behaviour in all the current regimes (low to high). In addition to the intrinsic/temporal (cycle to cycle) and extrinsic/spatial (device to device) variations, this model helps to predict the mean switching time of the device. It also supports Monte-Carlo simulations over the intrinsic and extrinsic variation of STT-MRAM device which facilitates analysis and exploitation of the complete behaviour of the device for electronic circuit applications. Extrinsic and intrinsic variabilities are illustrated in Figs.1 and 2, respectively. We generated both Figs.1 and 2 by performing the Monte-Carlo simulations using the compact model. TABLE I lists the physical parameters of the MTJ used in simulations with their mean values and their corresponding deviations which are obtained from the literature [19]-[25]. A high TMR ratio of 120 % which has been reported for PMA-MTJ system (based on CoFeB/MgO/CoFeB material stack) is used in the simulations [19].

III. PROPOSED PUF EXTRACTION METHODOLOGY

A. Operating Principle of the PUF

We simulated a single STT-MRAM device for 1000 cycles using the compact model [22]. As shown in Fig. 3, the switching time from AP to P state of a single STT-MRAM device shows a broad distribution for 1000 cycles. The mean of the switching time of a single STT-MRAM device (T_m) is extracted from the distribution shown in Fig. 3.

TABLE I.   PHYSICAL PARAMETERS FOR MAGNETIC TUNNEL JUNCTION

Parameter	Parameter description	Value	Process variation
d	MTJ cross section diameter	32 nm	5 %
T_fl	Free layer thickness	1.3 nm	-
RA	Resistance Area product/coefficient	5 μm²	5 %
TMR	Tunnel magnetoresistance ratio	120 %	5 %
V_h	Voltage at half TMR	0.5 V	-
$\Delta E/K_B T$	Relative energy barrier	60	-

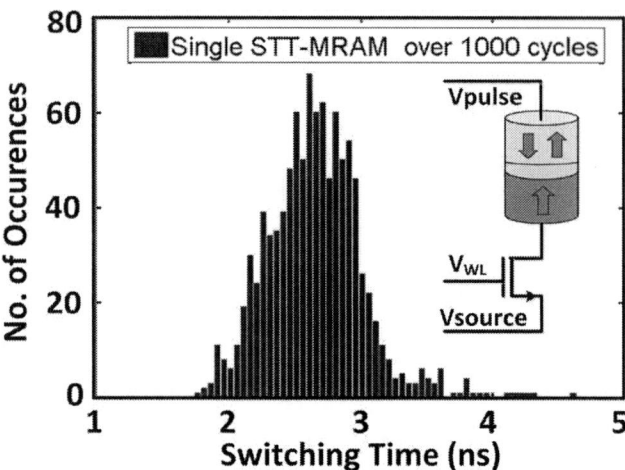

Figure 3. Switching time distribution of single STT-MRAM at Vpulse = 0.8 V & Vsource = 0 V with fully ON select transistor (V_{WL} = 1 V).

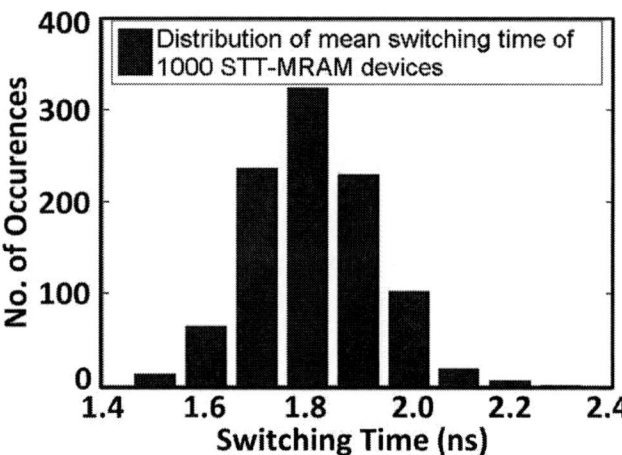

Figure 4. Distribution of the mean switching time of 1000 STT-MRAM devices. Y-axis represents the number of successfully switched devices corresponding to a particular programming pulse of a specific duration (X-axis).

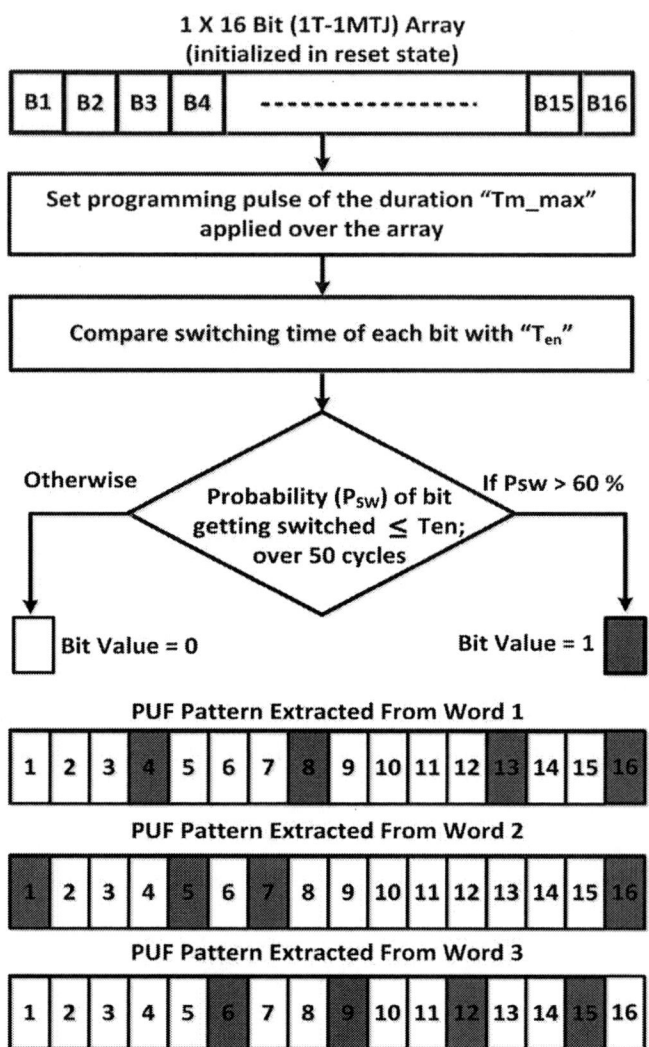

Figure 5. Illustration of the PUF extraction approach using an array (1 x 16) consists of 1T-1MTJ as unit cells. Also illustrates the obtained bit patterns in 3 different arrays for the generation of PUF signature.

Furthermore, a spatial device to device variability in switching time is also observed. Therefore, we have also simulated 1000 STT-MRAM devices using compact model to analyse distribution of the individual T_m. Fig. 4 shows the distribution of T_m for 1000 STT-MRAM devices obtained by performing Monte-Carlo simulations using standard deviation and mean values as mentioned in TABLE I. It can be observed from Fig. 4 that the distribution of T_m of 1000 STT-MRAM devices is narrower compared to the distribution of switching time of a single STT-MRAM device. The ensemble average (T_{en}) of T_m of 1000 STT-MRAM device is found to be 1.8 ns. The proposed methodology for PUF extraction from a 1x16 array of STT-MRAM devices is illustrated in Fig. 5 and explained further:

Step 1: The mean switching time (T_m at V_{pulse} = 0.8 V) of the individual STT-MRAM devices is evaluated for 1000 cycles similar to Fig 3. The ensemble average (T_{en} = 1.8 ns for this case) is extracted from the Monte-Carlo simulations for 1000 STT-MRAM devices.

Step 2: All STT-MRAM devices in the array are initialised to reset state using appropriate strong programming pulse.

Step 3: An input pulse (V_{pulse} = 0.8 V) with duration corresponding to the maximum T_m of the individual STT-MRAM (T_{m_max} = 2.3 ns in this case) is applied to switch the devices from reset to set.

Step 4: Switching time of the individual devices in the array is compared with T_{en}.

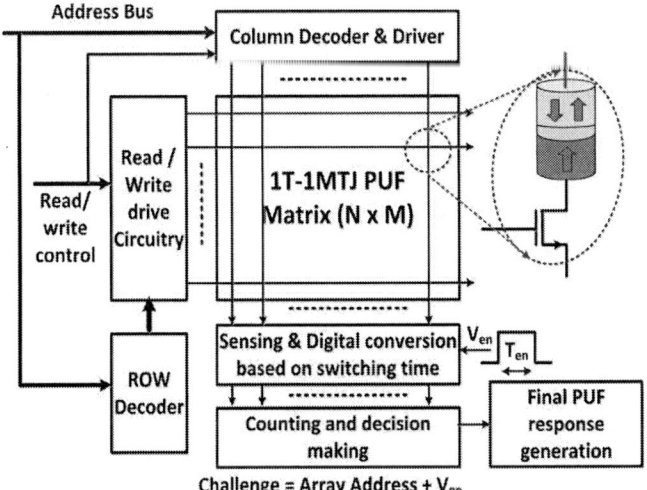

Challenge = Array Address + V_{en}

Figure 6. System level architecture for the STT-MRAM based PUF generation

Steps 2 – 4 are pre-processing steps and repeated multiple times to extract the switching probability distribution of the individual devices after step 4. For reliable operation of the proposed PUF extraction methodology, we repeated the aforementioned steps for 50 cycles and the probability of switching of individual STT-MRAM devices was extracted. The STT-MRAM devices were then segregated in two groups depending on the probability of switching. Devices with probability greater than 60 % are used to form the unique digital signature of the PUF. Such devices are assigned bit '1'. The remaining STT-MRAM devices are assigned bit '0'. The bit pattern forms the unique digital signature of the STT-MRAM based PUF. This pattern is obtained from the PUF only when the pulse width of the applied challenge pulse is T_{en}.

It may be noted that a probability greater than 50% can ensure the high reliability (explained in section IV) of the PUF as also mentioned in the PUF extraction procedure based on the resistance states of STT-MRAM [15]. In addition, a larger probability for extraction of bit "1" would lead to a smaller intra-hamming distance but at the same time it may affect the uniformity in the PUF response (explained in section IV). Therefore, choosing a probability becomes the function of a number of parameters i.e. number of cycles, total number of bits in the final response, and trade-off between reliability and uniformity. In this work, we present a case where bit "1" is assigned to only those STT-MRAM devices whose switching probability is greater than 60%. Such value of probability is considered to make a PUF more reliable with present given state of parameters (mentioned above). Moreover, the sacrificed uniformity can also be achieved during the final PUF response generation using post processing.

It may also be noted that although multiple programming and read cycles are required to obtain the PUF signature, the small ensemble average (1.8 ns) of switching time and the low power dissipation per switching event ($\sim 72~\mu W$) ensure that the delay and power penalty is significantly low while extracting the PUF signature.

B. System level implementation

Fig. 6 shows a system level overview of the STT-MRAM based PUF. To validate the proposed concept, we considered a 3x16 array. Each bit in the array corresponds to the 1T-1MTJ structure at the intersection of column and row lines (Fig. 6). As a pre-processing step, the T_m of the individual devices and the T_{en} of the array is extracted. The MTJ devices in the array are initialized to the reset state. The row decoder selects a particular row and a set programming pulse is applied. Simultaneously, a challenge pulse (V_{en}) with pulse width equal to the T_{en} is applied to the sensing circuitry block as shown in Fig. 6. Sensing circuitry evaluates the state of the individual MTJ which depends on their switching time. If the switching time of the MTJ > T_{en} the state of MTJ is treated as unperturbed (reset state) while if switching time of the MTJ is $\leq T_{en}$, the MTJ is treated as in set state. The sensing circuitry is followed by a digital conversion block which classifies the MTJ as bit '1' or bit '0' depending on its state. This digital output is then fed as the clock to the counter and based on the output of the counter, the probability of occurrence of bit '1' or '0' is calculated over multiple cycles (50 cycles in our case). The extracted probabilities enable the decision making block to segregate the MTJ bit cells in two categories. MTJ with switching probability greater than 60% are assigned bit '1' in the output response of PUF while others are assigned bit '0' and the final digital signature of the corresponding row is obtained. The aforementioned step is repeated for all the rows and their digital PUF signature is extracted. After this pre-processing step and extraction of the PUF signature of the individual rows, all the MTJs in the array are initialized to the reset state through the read/write circuitry.

Challenge for the proposed PUF is given as the address of the row and the applied pulse width. The row decoder selects a particular row and a programming pulse with a pulse width equal to the challenge is then applied to switch the state of the MTJ devices in that row to set state. Depending on the difference between the switching time of the individual MTJ devices and the applied challenge pulse, they are classified into bits and the response is generated. The decision making block matches the obtained response with the PUF response extracted during the pre-processing step by XOR operation and confirms whether the applied challenge is the cryptographic key or not. It may be noted that only when the applied pulse width is equal to the ensemble average (T_{en}), the correct response is obtained. Difficulty to predict exact value of T_{en}, makes the proposed PUF extremely reliable.

IV. PUF PERFORMANCE ANALYSIS

Performance of PUF mainly depends on parameters like uniqueness, reliability and uniformity. These parameters are quantified by calculating the values of inter-hamming distance and intra-hamming distance among the different responses of the PUF. The performance of such metrics also depends on number of PUF circuits to be evaluated, challenge to a PUF, the way of response extraction/generation and number of bits in a PUF's response. Here, we show the hamming distances on

978-1-5386-3693-0/18 $31.00 © 2018 IEEE

the PUF patterns shown in the Fig. 5 without any post processing using fuzzy logic.

Inter Hamming Distance: This is the measure of uniqueness of the PUF response. The average Inter Hamming distance of the response for the 3 different PUF patterns in Fig. 4 was obtained as 45.83 %.

Intra Hamming Distance: This is the measure of reliability of the PUF response. Average Intra-hamming distance of the proposed PUF was found to be ~ 5.0 % for 100 cycles.

Uniformity: This can be achieved by appropriately masking the bits which are constantly '0' in every PUF response like B2, B3, B10, B11, and B14 during final response generation. Since the digital signature of PUF can be modified by masking bits, we can have equal number of 1s and 0s in the output response. Therefore, the proposed PUF extraction methodology can be modified to ensure a uniform distribution of bits '1' and '0' in the output response.

V. Conclusion

In this work, we propose a PUF circuit based on the time dependent switching probability of STT-MRAM devices. We provide a methodology to extract PUF signature from a STT-MRAM device array. Using Monte-Carlo simulations with an experimentally calibrated compact model, we demonstrate that the proposed PUF circuit exhibits an inter-hamming distance 45.83 % close to ideal value and an intra-hamming distance 5.0 %. Fast, low-power and small area of STT-MRAM devices make it very comparable to existing PUF implementations in literature. The proposed PUF performance can be further improved by exploiting masking of individual bits.

References

[1] D. Lim et al., "Extracting secret keys from integrated circuits," IEEE Trans. Very Large Scale Integr. (VLSI) Syst., vol. 13, no. 10, pp. 1200–1205, Oct. 2005.

[2] A. Maiti, V. Gunreddy, and P. Schaumont, "A systematic method to evaluate and compare the performance of physical unclonable functions," in Embedded Systems Design with FPGAs. New York, NY, USA: Springer-Verlag, Nov. 2012, pp. 245–267.

[3] C. Herder et al., "Physical unclonable functions and applications: A tutorial," Proc. IEEE, vol. 102, no. 8, pp. 1126–1141, Aug. 2014.

[4] P. Tuyls, B. Škoric, S. Stallinga, A. H. M. Akkermans, and W. Ophey, "Information-theoretic security analysis of physical unclonable functions," in Proc. 9th Int. Conf. Financial Cryptography Data Security, 2005, pp. 141–155.

[5] T. Okamura, K. Minematsu, Y. Tsunoo, T. Iida, T. Kimura, and K. Nakamura, "DRAM PUF," in Proc. 29th Symp. Cryptography Inf. Security, Kanazawa, Japan, Jan./Feb. 2012.

[6] J. Guajardo, S. S. Kumar, G.-J. Schrijen, and P. Tuyls, "Physical unclonable functions and public-key crypto for FPGA IP protection," in Proc. Int. Conf. Field Program. Logic Appl., 2007, pp. 189–195.

[7] S. S. Kumar, J. Guajardo, R. Maes, G.-J. Schrijen, and P. Tuyls, "Extended abstract: The butterfly PUF protecting IP on every FPGA," in Proc. IEEE Int. Symp. Hardw.-Oriented Security Trust, Jun. 2008, pp. 67–70.

[8] B. Gassend, D. Clarke, M. van Dijk, and S. Devadas, "Silicon physical random functions," in Proc. 9th ACM Conf. Comput. Commun. Security, 2002, pp. 148–160.

[9] I. Verbauwhede and R. Maes, "Physically unclonable functions: Manufacturing variability as an unclonable device identifier," in Proc. GLSVLSI, 2011, pp. 455–460.

[10] M. O. Lehtonen, F. Michahelles, and E. Fleisch, "Trust and security in RFID-based product authentication systems," IEEE Syst. J., vol. 1, no.2, pp. 129–144, Dec. 2007.

[11] D. Puntin, S. Stanzione, and G. Iannaccone, "CMOS unclonable system for secure authentication based on device variability," Proc. Eur. Solid State Circuits Conf., pp. 130–133, Sep. 2008.

[12] R. Maes and I. Verbauwhede, "Physically unclonable functions: A study on the state of the art and future research directions," in Proc. Towards Hardware-Intrinsic Security Inf. Cryptography, 2010, pp. 3–37.

[13] J. Das, S. Kevin, R. Srinath, B. Drew, and B. Sanjukta, "MRAM PUF: A novel geometry based magnetic PUF with integrated CMOS." IEEE Transactions on Nanotechnology 14, no. 3, (2015): 436-443T.

[14] L. Zhang, X. Fong, C. H. Chang, Z. H. Kong, and K. Roy, "Feasibility study of emerging non-volatilememory based physical unclonable functions," IEEE 6th International Memory Workshop (IMW, 2014, pp. 1-4.

[15] Marukame, T. Tanamoto, and Y. Mitani, "Extracting physically unclonable function from spin transfer switching characteristics in magnetic tunnel junctions", IEEE Transactions on Magnetics, 2014 50(11), pp.1-4.

[16] Khaleghi, Soroush, Paolo Vinella, Soumya Banerjee, and Wenjing Rao. "An STT-MRAM based strong PUF." In *Nanoscale Architectures (NANOARCH), 2016 IEEE/ACM International Symposium on*, pp. 129-134. IEEE, 2016.

[17] Vatajelu, E.I., Natale, G.D., Barbareschi, M., Torres, L., Indaco, M. and Prinetto, P., 2016. STT-MRAM-based PUF architecture exploiting magnetic tunnel junction fabrication-induced variability. ACM Journal on Emerging Technologies in Computing Systems (JETC), 13(1), p.5.

[18] H. Meng, and J. P. Wang, "Spin transfer in nanomagnetic devices with perpendicular anisotropy," Applied physics letters, 2006, 88(17), p.172506.

[19] S. Ikeda, et. al., "A perpendicular-anisotropy CoFeB–MgO magnetic tunnel junction", IEEE Transactions on Electron Devices, 2010, 59(3), pp.819-826.

[20] W. Zhao, C. Chappert, V. Javerliac, and J. P. Noziere, "High speed, high stability and low power sensing amplifier for MTJ/CMOS hybrid logic circuits", IEEE Transactions on Magnetics, 2009, 45(10), pp.3784-3787.

[21] W. S. Zhao, et. al.,."Failure and reliability analysis of STT-MRAM. Microelectronics Reliability, 2012, 52(9), pp.1848-1852.

[22] A. F. Vincent, N. Locatelli, J. O. Klein, W. S. Zhao, S. Galdin-Retailleau, and D. Querlioz, "Analytical macrospin modeling of the stochastic switching time of spin-transfer torque devices," IEEE Transactions on Electron Devices, 2015, 62(1), pp.164-170.

[23] Y. Zhang, W. S. Zhao, Y. Lakys, J.O. Klein, J. V. Kim, D. Ravelosona, and C. Chappert, "Compact modeling of perpendicular-anisotropy CoFeB/MgO magnetic tunnel junctions," IEEE Transactions on Electron Devices, 2012, 59(3), pp.819-826.

[24] L.B. Faber, W. Zhao, J. O. Klein, T. Devolder, and C. Chappert, "April. Dynamic compact model of spin-transfer torque based magnetic tunnel junction (MTJ)," In Design & Technology of Integrated Systems in Nanoscal Era, 2009. DTIS'09. 4th International Conference on (pp. 130-135). IEEE

[25] H. Sato, E.C.I. Enobio, M. Yamanouchi, S. Ikeda, S. Fukami, S. Kanai, F. Matsukura, and H. Ohno, "Properties of magnetic tunnel junctions with a MgO/CoFeB/Ta/CoFeB/MgO recording structure down to junction diameter of 11 nm," 2014, Applied Physics Letters, 105(6), p.062403.

2018 31th International Conference on VLSI Design and 2018 17th International Conference on Embedded Systems

Floating Point Multiplication Mapping on ReRAM based In-Memory Computing Architecture

Tarun Vatwani, Arko Dutt, Debjyoti Bhattacharjee[‡] and Anupam Chattopadhyay

School of Computer Science and Engineering, Nanyang Technological University, Singapore 639798

Corresponding Email: [‡]debjyoti001@e.ntu.edu.sg

Abstract—**Low leakage power, high endurance and non-volatile storage capabilities have made memristive devices, such as Resistive RAM (ReRAM) popular. ReRAMs also offer in-memory computing capabilities by means of stateful logic operations. However, there are no standard libraries for floating point operations on ReRAM-based in-memory computing platforms. In this paper, we propose a mapping for one such mathematical function, namely multiplication for floating point numbers. We undertook a detailed study to derive a mapping with low memory footprint on ReVAMP architecture, which is an in-memory computing platform based on ReRAM crossbar arrays. For multiplication of two IEEE-754 compliant single (or double) precision numbers, the proposed mapping requires 1.944 Kb (or 8.904 Kb) of computation memory, 374.4 Kb (or 3457 Kb) of instruction memory while consuming an energy equivalent to 5.6 pJ (or 471 pJ).**

I. INTRODUCTION

Resistive Random Access Memories (ReRAMs) are one of the most promising emerging technologies for logic and storage applications. ReRAMs offer non-volatile, high endurance, high density and multi-state storage capabilities with fast access times, while at the same time allowing stateful logic operations [1]. ReRAMs have also been used for designing neuromorphic circuits [2], content addressable memories and in-memory computing platforms [3], [4]. Realization of large passive crossbar arrays can be achieved by means of a select device in series to a switch (1S1R) or a Complementary Resistive Switch (CRS) that prevents parasitic currents [5].

The adoption of a computation platform depends on the available tools and libraries for the platform. For example, BLAS libraries are available for scientific computing on GPG-PUs and CPUs [6]. Floating-point multiplication operation finds usage in multiple applications such as digital signal processing, data mining, etc. There has been existing works to map adder [7], integer multiplier [8] and binary matrix vector operations on ReRAMs [9]. In this work, we extend further in this direction to map floating point multiplication on ReVAMP, a ReRAM based in-memory computing platform [4]. ReVAMP harnesses bit-level parallelism inherent to ReRAM crossbar arrays. The key contributions of the current paper are :-

- We report the first in-memory implementation of IEEE-754 floating point multiplier on ReVAMP.
- A novel multiplier mapping for computing product of two binary numbers has been proposed.
- We studied the performance of the proposed implementation in terms of throughput and memory overhead.

The rest of the paper is organized as follows. In section II, a brief introduction to IEEE-754 floating point number representation is presented. In addition, the ReVAMP architecture is introduced. Section III presents the mapping for floating point multiplication on the ReVAMP architecture. In section IV, we present detailed experimental results and a summary of existing works. Section V presents a conclusion to the paper.

II. PRELIMINARIES

In this section, we present the IEEE-754 floating point representation of any real number. In addition, we also introduce the basics of ReVAMP – a ReRAM based in-memory computing architecture.

A. IEEE-754 Floating point representation

The IEEE format for a single precision (32-bit) number is depicted in Fig. 1. Any real number X is expressed as:

$$X = (-1)^{Sign} \times (1.Fraction)_2 \times 2^{(Exponent-127)}$$

1-bit	8-bit	23-bit
Sign	Exponent	Fraction

Fig. 1: IEEE 754 format for 32-bit floating point number.

A leading sign bit is used in the format with value '0' indicating the number is positive else it is negative. The fraction stores the values after decimal places of the number. It is to be noted that a 1 is always implied in the decimal place and is not directly used in the representation. The mantissa represents the implied 1 with it - $(1.Fraction)_2$. The Exponent is expressed in an excess$-B$ representation such that the exponent is always a positive number. If the Exponent consists of $e-$bits, then the bias B is $2^{e-1} - 1$.

B. ReVAMP Architecture

One of the recently proposed ReRAM based in-memory architecture is ReVAMP [4]. It supports word-serial execution of instructions by exhibiting bit-level parallelism. It uses two separate memories — data and computation memory (DCM) and instruction memory (IM). In-memory computation takes place in the DCM. The DCM is a ReRAM crossbar array which constitutes of multiple 1S1R ReRAM devices [10]. The DCM is accessed as w_D-bit wide words. Each 1S1R device has two-terminals, namely a wordline wl and bitline bl and an internal resistive state Z. A ReRAM device has an intrinsic property of implementing its next state as a Boolean Majority-three function M_3 with bitline input bl inverted i.e., the next state $Z_n = M_3(Z, wl, bl') = Z.wl + wl.bl' + bl'.Z$.

978-1-5386-3693-0/18 $31.00 © 2018 IEEE   439

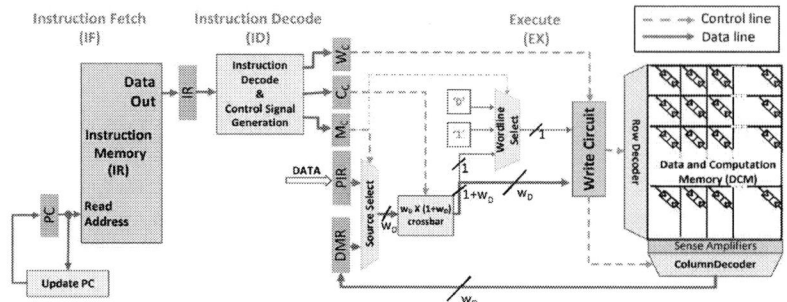

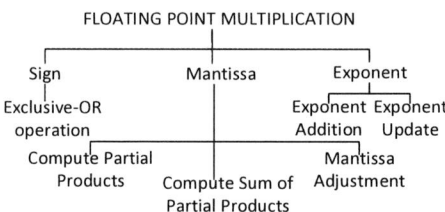

Fig. 2: ReVAMP Architecture [4].

The instruction set has two instructions, namely, 'Read' and 'Apply' whose format, as shown below.

> Read wl
> Apply wl s ws wb $(v$ $val_{w_D-1})$ ... $(v$ $val_0)$

A word wl is read out from DCM and stored in Data Memory Register (DMR) using the 'Read' instruction. This word available in the DMR can act as input for the following instructions. For computation, 'Apply' instruction is used. In 'Apply' instruction, wl specifies the word on which computation occurs, 1-bit flag s selects the input data source (either primary input register PIR or DMR), 2-bit flag ws selects the wordline input - 00 selects logic zero '0', 01 selects logic one '1', 10 is forbidden and 11 selects the bit specified by wb-address within the chosen data source for use as wordline input. Pairs (v, val) specify individual bitline inputs where $v=1$ indicates the input is valid else it is not used, and val specifies the address within the chosen data source from which the intended bit is used as bitline input.

III. IEEE-754 MULTIPLICATION ON ReVAMP

In this section, we present the mapping for multiplication of IEEE-754 floating-point numbers on ReVAMP, with major focus on low memory rather than highly-parallelized floating point operations. For generalization, let us consider that the n-bit floating point number in IEEE format is represented by e-bit exponent, $(k - 1)$-bit fraction with 1-bit leading sign-bit, such that $n = e + k$. The operations required to perform floating point multiplication is shown in Fig. 3. The individual operations are described below.

A. Computing Sign

For computing the sign bit of the result, we require to XOR the sign bits of the two inputs. The XOR of two one-bit operands a and b can be expressed as:

$$a \oplus b = a.b' + a'.b = a.b' + (a + b')'$$

The steps to realize XOR of 2-inputs (of 1-bit each) using ReRAM crossbar array is depicted in Fig. 4. It is implemented using a 3×1 ReRAM crossbar array, i.e., a crossbar with 3 wordlines and one bitline. We refer to the top wordline as wordline 1 and thereafter the next wordline as 2 and so one. Similarly, for referring to the bitline, we refer to the left-most bitline as 1, the next bitline on the right as 2 and

so on. In addition, we provide two basic formations using *Apply* instruction which will be used in the rest of the paper – applying a '1' as wordline input to a crossbar device will perform a Boolean *OR* operation of the inverted bitline input and crossbar state, while applying a '0' as wordline input to the device will perform a Boolean *AND* operation of the inverted bitline with crossbar state.

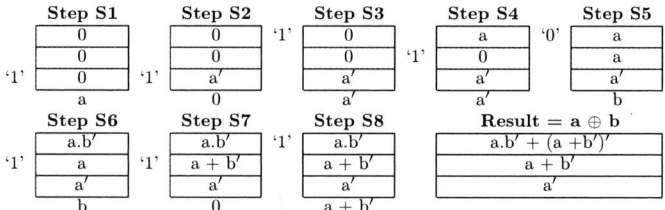

Fig. 4: 2-input (1-bit each) XOR Implementation Steps

Now, we present the steps to realize two input XOR.

Step S1: Input a is loaded into memory in inverted form using *Apply* instruction, by applying '1' to third wordline and input a to bitline. This is because $M_3(1, 0, a') = a'$.

Step S2: a' is read out using a *Read* instruction and is now available in the DMR.

Step S3-S4: a is computed and stored in memory locations corresponding to wordlines 1 and 2, by applying '1' to the wordlines respectively and a' to the bitline in subsequent steps, since $M_3(1, 0, a'') = a$.

Step	ReVAMP instruction
S1	Apply 3 0 01 0 1 1
S2	Read 3
S3	Apply 1 1 01 0 1 1

We formally introduce how these steps can be represented as ReVAMP instructions. In step S1, '1' is applied as wordline 3 input. So the opcode is *Apply* for this instruction, $wl = 3$, $ws = 01$ and the fifth bit after *Apply* is $wb = 0$ may be neglected. Since bitline 1 input is from primary input a, the second bit after *Apply* is $s = 0$. Only one (v,val) pair exists for this mapping since one bitline is used. The last two bits specify $v = 1$ and $val = 1$ respectively. In step S2, a *Read* instruction is used to read from memory corresponding to wordline 3. So for second instruction, opcode is *Read* and $wl = 3$. The read out data is stored in DMR, available for use in next cycles. Step S3 uses an *Apply* instruction with '1' as wordline 1 input and DMR data a' as bitline 1 input. So $wl = 1$, and $s = 1$ denoting DMR input; ws, wb and pairs (v,val) remain same as

978-1-5386-3693-0/18 $31.00 © 2018 IEEE   440

the instruction for step 1. Similarly, the rest of the steps can be represented in terms of ReVAMP instructions.

Step S5-S6: $a.b'$ and $a + b'$ are computed and stored in memory, by applying '0' and '1' to wordlines 1 and 2 respectively in subsequent steps and input b to bitline input, since $M_3(a, 0, b') = a.b'$ and $M_3(a, 1, b') = a + b'$.

Step S7-S8: $a + b'$ is read out from memory in step S7 and applied as bitline input along with '1' applied as wordline 1 input in step S8 to compute Result $= a \oplus b$.

The result of XOR is available in first wordline position of the DCM. Each Read or Apply instruction is effectively executed in a single cycle. Therefore, the computation of Sign bit requires 8 cycles in total.

B. Computing Exponent

It is evident that the exponents add up when the two numbers are multiplied. The exponent in the IEEE-754 format are biased, therefore the resultant exponent $E_R = E1 - B + E2$, where $B = 2^{e-1} - 1$ is the bias for a given precision. The computation of E_R can be treated as two subsequent addition operations. The exponent might have to be updated based on Mantissa MSB Carry or MMSBC. We will discuss about MMSBC later in mantissa adjustment phase of mantissa multiplication.

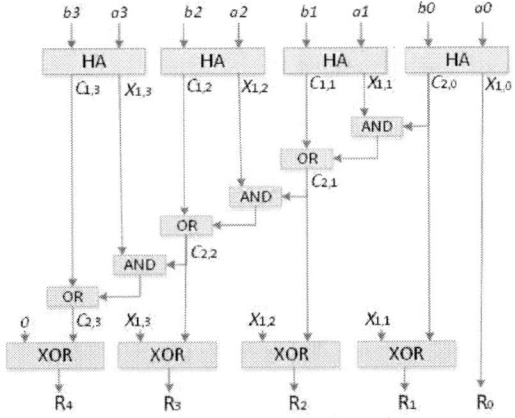

Fig. 5: 4-bit Adder Schematic used to compute exponent. HA - Half Adder, Result of Addition $R = (R_4 R_3 R_2 R_1 R_0)_2$.

We use a $3 \times (e+1)$ ReRAM crossbar to implement addition of two e-bit exponents. For demonstration, we consider adding 4-bit exponents E1 and E2 using the schematic shown in Fig. 5, which is mapped to a 3×5 ReRAM crossbar. The mapping steps are depicted in Fig. 6 and explained below.

Steps S1-S2: $4-$bit primary input a is loaded into memory in inverted form to wordlines 1, 2 as shown in Fig. 6.

Step S3: $a'.b'$ is computed by applying '0' to wordline input and $4-$bit primary input b via the bitlines.

Step S4: $a' + b'$ is computed by applying '1' to wordline input and $4-$bit primary input b to the bitlines.

Steps S5-S6: $a' + b'$ is read out and ORed with the first wordline to compute XNOR of a and b.

Steps S7-S9: XNOR of a and b is read out, all bits of first wordline are reset to '0' and XOR of a and b is computed

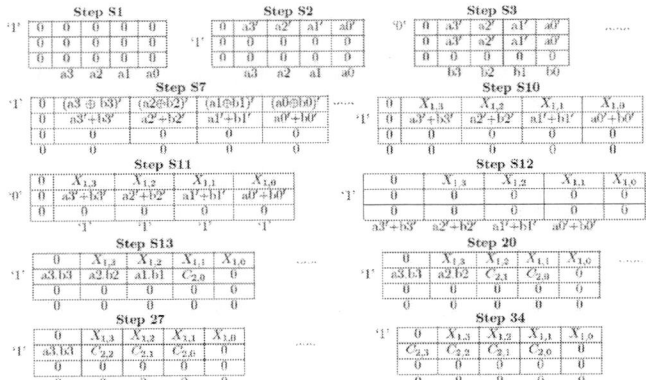

Fig. 6: 4-bit Adder Implementation Steps - Computation of $X_{1,i}$ and $C_{2,j}$.

and stored in first wordline of memory leaving the first bitline position. The output from step S9 represents X_1.

Steps S10-S11: $a' + b'$ are read out from memory locations corresponding to second wordline in step S10; the corresponding memory devices are then reset in step S11.

Step S12: $a.b$ is computed by applying '1' to second wordline input and the read-out data to bitline inputs. The output from this step represent C_1.

As of now, 4 half-adder circuits have been implemented in parallel to generate sum $X_{1,i}$ and carry-in $C_{1,i}$, represented by a set of two equations:

$$X_{1,i} = a_i \oplus b_i \qquad i \in \{0, 1, 2, 3\}$$
$$C_{1,i} = a_i \cdot b_i$$

The carry generated from each bit addition of two inputs needs to be propagated in a sequence of steps from the least significant bit position to the most significant bit position allowing addition of immediate left-adjacent input bits. The carry input for addition of immediate left-adjacent bits $C_{2,j}$ can be computed from

$$C_{2,j} = C_{1,j} + X_{1,j} \cdot C_{1,j-1} \qquad j \in \{1, 2, 3\}$$

Since addition of least significant bits of two inputs has no carry, we can say $C_{2,0} = C_{1,0}$. In steps S13-S19, it is shown how $C_{2,1}$ can be computed.

Steps S13: $C_{2,0}$ is read out using a *Read* instruction.

Steps S14: Using *Apply* instruction, $C_{2,0}'$ is computed by applying '1' as third wordline input and $C_{2,0}$ as third bitline input.

Steps S15: $X_{1,1}$ is read out from memory.

Steps S16-S17: $(X_{1,1}' + C_{2,0}')$ is computed by applying '1' as third wordline input and $X_{1,1}$ as third bitline input in step S16. $(X_{1,1}' + C_{2,0}')$ is read-out in step S17.

Steps S18-S19: $C_{1,1} + (X_{1,1}' + C_{2,0}')'$ is computed by applying '1' as second wordline input and $(X_{1,1}' + C_{2,0}')$ as third bitline input in step S18. This output is equivalent to $C_{2,1}$. ReRAM devices corresponding to third wordline input are then reset in step S19 for next carry generation steps.

Consequently, these 7 steps (S13-S19) are repeated to generate remaining $C_{2,j}$ ($C_{2,2}, C_{2,3}$ in this case) for adjacent left-bit sum computation and resetting devices. From Fig. 5, we can compute the final sum result $R = (R_4R_3R_2R_1R_0)_2$ as follows:

$$R_0 = X_{1,0}$$
$$R_j = X_{1,j} \oplus C_{2,j-1} \qquad j \in \{1,2,3\}$$
$$R_4 = 0 \oplus C_{2,3} = C_{2,3}$$

Three more instructions are required to read $X_{1,i}$ and $C_{2,i}$ ($C_{2,i}$ instead of $C_{2,j}$ since it includes $C_{2,0}$) from memory and reset devices corresponding to wordline 1 (except $X_{1,0}$). In total, $12 + (3 \times 7) + 3 + 9 = 45$ instructions are required in this case. The result will be available as five bits stored in DCM corresponding to first wordline. In general, an e-bit adder computation will require $24 + 7(e - 1)$ instructions. Between subsequent addition operations, 4 more instructions are required to read out the sum result and reset all devices corresponding to the three wordlines. So a total of $2[24 + 7(e - 1)] + 4$ instructions are required to compute the exponent (disregarding MMSBC update).

C. Computing Mantissa

The fractional number is $(k-1)$-bits long, which is normalized with a leading implied 1 to represent k-bits long mantissa. For demonstration of the computation, we consider a 4-bit mantissa ($k = 4$). Let the two operand mantissas be $M_a = a0 \bullet a1a2a3$ and $M_b = b0 \bullet b1b2b3$, where $a0 = a1 = 1$ and $\bullet$ represents decimal point. In Fig. 7, we show regular binary multiplication. In case of floating point representation, mantissa multiplication can take place similarly. However the resultant mantissa might need to be normalized depending on the value of $s0$. If $s0$ is 0, Mantissa will be $(1 \bullet s2s3s4)_2$ else it will be $(1 \bullet s1s2s3)_2$.

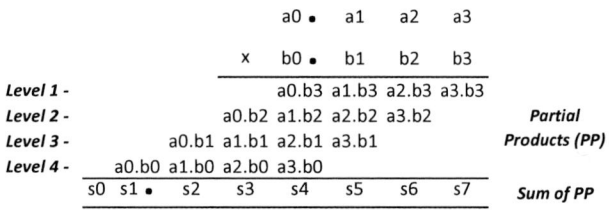

Fig. 7: 4-bit multiplication.

As evident from Fig. 7, three operations are needed to compute mantissa, namely, partial product calculation, sum of partial products and mantissa adjustment. All these operations are realized using $3(k + 1) \times k$ ReRAM crossbar array. We demonstrate a 4-bit mantissa multiplication implemented with a (15×4) crossbar array (with 15 wordlines and 4 bitlines) as follows.

1) Computing Partial Products: In the context of multiplication of two input mantissa $M_a = a0 \bullet a1a2a3$ and $M_b = b0 \bullet b1b2b3$, a partial product can be represented as:

$$p_{i,j} = ai \cdot bj \qquad i,j \in \{0,1,2,3\}$$

From Fig. 7, we can see that 16 partial products have to be computed for 4-bit multiplication. As evident from Fig. 7 with 4-bit multiplication, we will have 16 partial dot products (each level contains 4 partial dot products and there are 4 levels in

total, shown in Fig. 7). We have shown the computation of partial dot products for level 1 (with only wordline 1 from the set of 15 wordlines) in Fig. 8 and discuss the steps as follows.

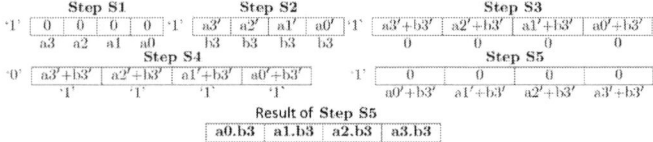

Fig. 8: Steps for 4-bit Partial Product Computation.

Step S1: Using *Apply* instruction, primary input bits from a are loaded in inverted form.

Step S2: Using *Apply* instruction, '1' is applied to the first wordline input and LSB (i.e $b3$) from primary input b is applied to all bitline inputs to compute first level partial products in inverted form, e.g. $(a0.b3)' = a0' + b3'$.

Step S3: Using *Read* instruction, data from devices corresponding to first wordline are read out and stored into DMR.

Step S4: Using *Apply* instruction, all bits of the first wordline are reset to zero.

Step S5: As shown in Fig 8, already read out first level partial products in inverted form are applied to the bitline inputs in reverse order and '1' is applied to the first wordline input to compute the required partial products and store them in memory locations corresponding to the first wordline.

The set of five steps are repeated in similar fashion with next set of bits ($b2$, $b1$, $b0$) from input b – three more times to compute and store all the remaining 12 partial dot products in memory locations corresponding to wordlines 2-4. This requires a total of $5 \times 4 = 20$ instruction cycles. In general with k-bit mantissa multiplication, it will require $5k$ cycles to compute all the partial products and store them in the devices corresponding to the first k-wordlines.

2) Computing Sum of Partial Products: The method to add the partial products of k-bit mantissa is explained here. Summation of multiple partial products can be performed by serial addition but that would lead to higher delay. Therefore, we use a fast adder construction that requires the carry to be propagated within same level only once in the final level. This adder construction can be mapped to the ReVAMP architecture using the steps illustrated as a flowchart in Fig. 10. We suppose $P_{i,j} = p_{3-i,3-j}$ and $R_m = s_{7-m}$. With this supposition, we can clearly relate how partial products shown in Fig. 7 are utilized in the adder implementation depicted in Fig. 9. For shifted partial product addition, there will be an additional carry generation level after the initial XOR and Carry generation steps and before final XOR operation. This will result in a $(k+1)$-bit XOR output and the MSB of 2-input sum result can be found in MSB of Carry generated from last second carry implementation step.

In Fig. 10, $XOR - 2$ and $CARRY - 2$ represents a 2-input k-bit XOR operation and 2-input k-bit carry generation respectively, together acts as a half adder. So, we have

$$XOR - 2(a,b) = a \oplus b$$
$$CARRY - 2(a,b) = a.b$$

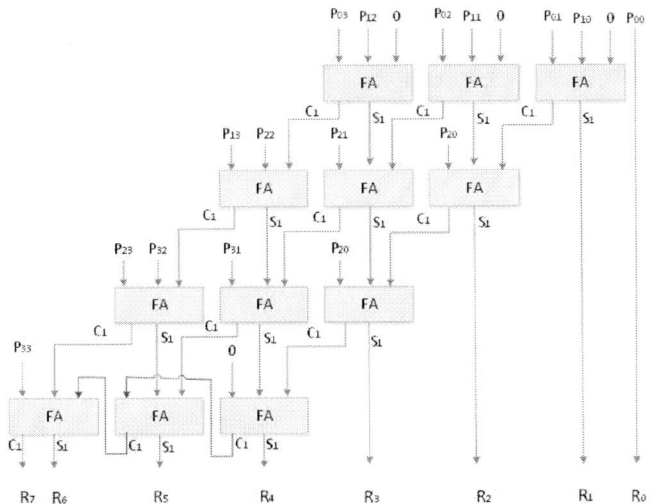

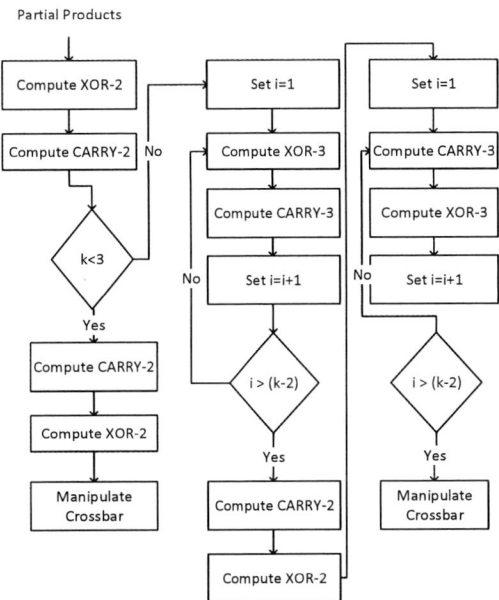

Fig. 9: 4-bit adder schematic to compute sum of partial products. FA represents a Full Adder and $\{S_1, C_1\}$ is $\{Sum, Carry\}$ generated by each FA. $P_{i,j} = p_{3-i,3-j}$ for $i,j \in \{0,1,2,3\}$ and $R_m = s_{7-m}$ for $m \in \{0,1,2,3,4,5,6,7\}$. Result $= (R_7 R_6 R_5 R_4 R_3 R_2 R_1 R_0)_2$.

Similarly, computing $XOR-3$ and $CARRY-3$ components yield a 3-input k-bit XOR output and 3-input k-bit Carry generation, acts as a full adder. Thus,

$$XOR - 3(a,b,c) = a \oplus b \oplus c$$
$$CARRY - 3(a,b,c) = a.b + b.c + c.a$$

Boolean AND, OR and XOR can be implemented by using a set of instructions similar to approach shown in Fig. 6. Since k-bit multiplication generates $2k$-bit output, the sum of products needs to be copied to two k-bit words. With Manipulate Crossbar operation, the resultant sum of products is copied to DCM devices corresponding to wordline position $3(k + 1) - 3$ and $3(k + 1) - 2$ with its significant k-bits in devices corresponding to wordline position $3(k + 1) - 3$.

The number of instructions required to compute all these operations can be determined as follows. Computation of XOR-2 and CARRY-2 operations together require 19 cycles. From Fig. 10, it is evident that two sets of $XOR - 2$, $CARRY - 2$ will be required irrespective of the value of k. Therefore, 38 instructions will be executed. Depending on the value of k, there will be iterations of $XOR - 3$ and $CARRY - 3$ operations. If $k > 2$ (signifying that fractional part is more than one bit), $XOR - 3$ and $CARRY - 3$ cycles will be executed. The $XOR - 3$ and $CARRY - 3$ operations together need 41 cycles for complete computation. From Fig. 10, it can be interpreted that two rounds of $(k-2)$ iterations of $XOR-3$, $CARRY - 3$ need to be implemented. So $2(k - 2) * 41$ instruction cycles will be required for this purpose. The 'Manipulate Crossbar' operation requires $5(k + 1)$ instruction cycles in total. Therefore, the number of instructions add up to $87k - 121$ for k-bit mantissa computation.

3) Mantissa Adjustment: The sum of the partial product will have $2k$ resultant bits available in the $3(k + 1) - 3$ and

Fig. 10: Flowchart to compute sum of partial products on ReVAMP.

$3(k + 1) - 2$ wordline memory. Now, there needs to be a set of instructions which can decide whether mantissa should be considered right from MSB (or MSB-1) bit-position of the result depending on the value of MSB being 1 (or 0). We call this as Mantissa MSB Carry or MMSBC since it is the carry output of last sum (from sum of products computation) operation and the MSB of the result of this sum. MMSBC may be referred as $s0$, depicted in Fig. 7. In regard to Fig. 7, if MMSBC = 1, Mantissa is selected as $(1 \bullet s1s2s3)_2$, and if MMSBC = 0, Mantissa is selected as $(1 \bullet s2s3s4)_2$. In this case, 15 instructions are required to adjust mantissa of the resultant product. This instruction count remains same even if the bit-length of the input is different.

D. Exponent Update

We need to incorporate addition of MMSBC value to sum of the exponents E_R computed in subsection III-B. After MMSBC is available, a sum operation similar to the method explained in subsection III-B is adopted to compute the new resultant exponent $E_R = E_R + MMSBC$. Therefore, an additional $24+7(e-1)+4$ cycles will be required to completely compute the Exponent. Moreover, an extra read instruction is needed to read MMSBC, totaling the number of instructions to $3[24 + 7(e - 1)] + 9$.

IV. Experimental Results and Analysis

The instruction sequence for multiplication of IEEE-754 compliant numbers has been developed using Matlab®. The correctness was verified by means of behavioural simulator for ReVAMP and also via device accurate simulations using Cadence Spectre® [10].

In order to compute latency of the proposed mapping, we assume a n-bit floating point number in IEEE-754 format that has a leading sign bit, e-bit exponent and $(k - 1)$-bit fraction

978-1-5386-3693-0/18 $31.00 © 2018 IEEE   443

and calculate the total number of instructions. The resultant sign computation needs 8 instructions, mantissa computation needs $92k - 106$ instructions and exponent computation requires $60 + 21e$ instructions. Each instruction requires one cycle effectively, except the first instruction which requires three cycles due to the 3-stage pipeline architecture. The number of instructions for 32-bit and 64-bit IEEE-754 floating point multiplication is reported in TABLE I.

TABLE I: Number of instructions for 32-bit and 64-bit IEEE FP multiplication on ReVAMP.

Operation		Number of Instructions	
		32-bit	64-bit
Sign		8	8
Mantissa	Partial Products	120	265
	Sum of Partial Products	1967	8306
	Adjustment	15	15
Exponent	Addition	150	192
	Update	78	99
Total		2338	8885

The DCM crossbar dimensions are (81×24) and (168×53) for 32-bit and 64-bit precision floating point multiplication, hence the corresponding crossbar size are $1.944\ Kb$ and $8.904\ Kb$ for respective precisions. The instruction size for the 32-bit and 64-bit are $160\ bits$ and $389\ bits$ respectively The overall instruction memory size is approximately $374.4\ Kb$ with 32-bit aligned access and $3457\ Kb$ with 64-bit aligned access.

As per ITRS report on emerging devices [11], the read/write cycle for ReRAM devices is projected to have a duration time of $1\ ns$ and a write cycle energy of $0.1\ fJ/bit$. The proposed mapping is estimated to achieve roughly 10×10^6 and 6×10^6 floating point operations per second (FLOPS) at the cost of $5.6\ pJ$ and $471\ pJ$ of energy consumption for 32-bit and 64-bit IEEE-754 multiplication. We should note that the current implementation is aimed at minimizing the number of devices used for multiplication and does not aim at parallelizing multiple floating point operations. In TABLE II, we report the throughput of the proposed floating point multiplication mapping with ReVAMP architecture.

TABLE II: Implementation Summary.

Implementation	Frequency (in GHz)		Throughput (in GFLOPS)	
	32-bit	64-bit	32-bit	64-bit
FPGA [12] (40nm Virtex-6)	0.25	0.25	0.010	0.005
GPU [12] (55nm)	1.3	1.3	0.002	0.001
CMOS ASIC [13] (250nm)	1.05	1.12	1.05	0.56
CMOS ASIC [13] (65nm)	1.05	1.12	1.05	0.56
CMOS ASIC [13] (90nm)	1.05	1.12	1.05	1.12
Using ReVAMP	1	1	0.010	0.006

It is to be noted that the throughput of one floating-point unit is considered for GPGPU and FPGA implementation and reported in TABLE II. A dual mode double precision floating point multiplier architecture based on ASIC implementation is proposed that can be configured to compute two single precision multiplications in parallel [13]. This concept was implemented to achieve an efficient resource sharing. A comparative analysis between GPU and FPGA implementations of matrix multiplications based on IEEE 754 floating point

formulation is presented in [12]. The results showed that GPUs were suitable for larger matrix multiplications, whereas FPGAs ensured higher throughput for multiplication of smaller matrices.

V. Conclusion

In this work, an efficient mapping of IEEE-754 floating point number multiplication on ReVAMP architecture has been proposed. The implementation uses a novel multiplication scheme for computing the resultant mantissa. The mapping is estimated to achieve $6 \times 10^{-3} GFLOPs$ throughput, which is comparable to the performance of FPGA based floating point multiplication unit. The implementation has an overall memory footprint($\approx 3831\ Kb$) with a low energy footprint ($\approx 471\ pJ$). We plan to extend the work in the direction of realizing floating point BLAS operations on the ReVAMP architecture using the proposed mapping.

References

[1] R. Waser, R. Dittmann, G. Staikov, and K. Szot, "Redox-based resistive switching memories–nanoionic mechanisms, prospects, and challenges," *Advanced Materials*, vol. 21, no. 25-26, pp. 2632–2663, 2009.

[2] S. H. Jo, T. Chang, I. Ebong, B. B. Bhadviya, P. Mazumder, and W. Lu, "Nanoscale memristor device as synapse in neuromorphic systems," *Nano letters*, vol. 10, no. 4, pp. 1297–1301, 2010.

[3] P.-E. Gaillardon, L. Amarú, A. Siemon, E. Linn, R. Waser, A. Chattopadhyay, and G. De Micheli, "The Programmable Logic-in-Memory (PLiM) Computer," *DATE*, 2016.

[4] D. Bhattacharjee, R. Devadoss, and A. Chattopadhyay, "ReVAMP: ReRAM based VLIW architecture for in-memory computing," in *Design, Automation Test in Europe Conference Exhibition (DATE), 2017*, Mar. 2017, pp. 782–787.

[5] E. Linn, R. Rosezin, C. Kügeler, and R. Waser, "Complementary resistive switches for passive nanocrossbar memories," *Nature materials*, vol. 9, no. 5, pp. 403–406, 2010.

[6] "BLAS (Basic Linear Algebra Subprograms)," http://www.netlib.org/ blas/, accessed: 2017-07-31.

[7] A. Siemon, S. Menzel, R. Waser, and E. Linn, "A complementary resistive switch-based crossbar array adder," *IEEE journal on emerging and selected topics in circuits and systems*, vol. 5, no. 1, pp. 64–74, 2015.

[8] D. Bhattacharjee, A. Siemon, E. Linn, and A. Chattopadhyay, "Efficient complementary resistive switch-based crossbar array booth multiplier," *Microelectronics Journal*, vol. 64, pp. 78–85, 2017.

[9] D. Bhattacharjee, F. Merchant, and A. Chattopadhyay, "Enabling in-memory computation of binary blas using reram crossbar arrays," in *Very Large Scale Integration (VLSI-SoC), 2016 IFIP/IEEE International Conference on.* IEEE, 2016, pp. 1–6.

[10] A. Siemon, S. Menzel, A. Marchewka, Y. Nishi, R. Waser, and E. Linn, "Simulation of TaOx-based complementary resistive switches by a physics-based memristive model," in *2014 IEEE International Symposium on Circuits and Systems (ISCAS)*, Jun. 2014, pp. 1420–1423.

[11] "Emerging Reseach Devices report, International Technology Roadmap for Semiconductors, 2013." [Online]. Available: https://www.semiconductors.org/clientuploads/Research_ Technology/ITRS/2013/2013ERD.pdf

[12] U. I. Minhas, S. Bayliss, and G. A. Constantinides, "GPU vs FPGA: A Comparative Analysis for Non-standard Precision," in *Reconfigurable Computing: Architectures, Tools, and Applications*, ser. Lecture Notes in Computer Science. Springer, Cham, Apr. 2014, pp. 298–305.

[13] M. K. Jaiswal and H. K. H. So, "Dual-mode double precision / two-parallel single precision floating point multiplier architecture," in *2015 IFIP/IEEE International Conference on Very Large Scale Integration (VLSI-SoC)*, Oct. 2015, pp. 213–218.

Parasitic Aware Automatic Analog CMOS Circuit Design Environment using ABC Algorithm

Subhash J Patel
Department of Electronics and Communication,
Indus University
Ahmedabad, India
Email: subhash.bvm@gmail.com

Rajesh A Thakker
Department of Electronics and Communication,
VGEC, Gujarat Technological University,
Ahmedabad, India
Email: rathakker2008@gmail.com

Abstract—In this work, we propose a novel concept of the parasitic-aware design automation for the high-performance analog CMOS circuit design. To achieve this, the concept of the automatic schematic-level circuit design is extended to layout-level using the configurable layouts. The configurable layouts allow consideration of exact parasitic from the beginning of the circuit design process. We designed two-stage operational amplifier at layout-level in $0.13\mu m$ CMOS technology using the proposed concept. The circuit was designed at layout-level with 0% average design error, satisfying all the specifications. The average design time was only 107.8 minutes. Further, we designed operational amplifier at schematic-level, where exact value of the layout-parasitics is not possible to consider, the design error was 0%. However, the post-layout simulation of this optimized schematic indicated the average design error of 6.46%. We also demonstrated the parasitic-aware automatic design of the operational amplifier considering process and temperature variations. The obtained results show the effectiveness of the proposed concept for designing high-performance analog circuits.

Index Terms—Parasitics, Layout, CMOS, Analog Circuits, Op-Amp, Optimization

I. INTRODUCTION

In order to design high-performance analog CMOS circuit in a time-efficient manner, many researchers have successfully applied evolutionary algorithms to design analog CMOS circuits [1]–[3]. However, their designs are confined to schematic-level optimization. Since the exact layout parasitics are not possible to consider in the schematic, the simulation results at schematic and layout-level are different, especially, for the frequency sensitive specifications. To overcome this problem and to extend the concept of algorithm based design of analog CMOS circuit from schematic-level to layout-

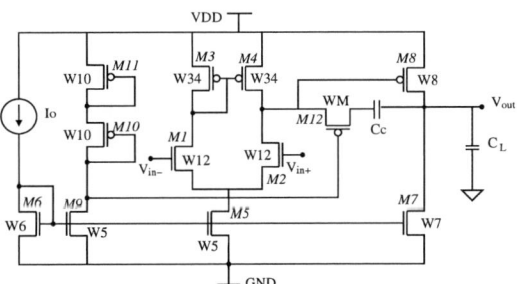

Fig. 2. Schematic of two stage CMOS op-amp

level, we present configurable (parameterizable) layouts. The configurable layouts enable consideration of the all kinds of parasitics from the beginning of the circuit design process and allow layout-level design automation. Based on the proposed concept, we presented the parasitic-aware automatic design of the two-stage operational amplifier (op-amp) at the layout level in $0.13\mu m$ CMOS technology using ABC algorithm.

II. ENVIRONMENT FOR PARASITIC AWARE OF DESIGN

Traditionally, the layouts of analog circuits are prepared using graphical interface. Once, the layout is prepared, the distance between various layout components and their dimensions become fixed. The changes in the layout require moderate to large efforts. Such layouts are not suitable for design automation. The configurable layouts proposed in this work are developed for the MAGIC VLSI Tool [4] and described using standard macros in the script file. In the script file, the distances between various layout components and their dimensions are described in terms of the parameters of layout, variables, and technology dependent parameters. By changing the values of layout parameters, the layout can be modified instantly. Further, the configurable layouts provide the possibility of exporting the layouts into other technologies. The conceptual diagram of the parasitic-aware automatic CMOS circuit design process is illustrated in Fig 1. In this work, we utilize the ABC algorithm [5]. However, any other evolutionary algorithm can also be used. The search space specifies the maximum and minimum values for the design parameters. Generally, the design parameters for the analog

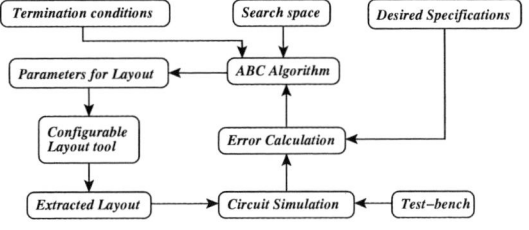

Fig. 1. Conceptual block diagram for parasitic aware automatic CMOS circuit design using ABC algorithm.

978-1-5386-3693-0/18 $31.00 © 2018 IEEE   445

CMOS circuits are length and width of various transistors. The well-defined search space avoids non-practical solutions. The ABC algorithm generates parameters for the configurable layout from search-space. The parasitics are extracted from the generated layout and circuit net-list with parasitics is simulated against pre-determined test-bench. Based on desired circuit specifications and simulation results, the design error is calculated using RMS error formula as suggested in [1]. This error is utilized by the algorithm to generate new solutions. This iterative process is automatic and does not require any intervention. The optimization process using the configurable layout enables parasitic-aware deisgn of analog circuit.

III. PARASITIC AWARE OF DESIGN OF TWO-STAGE OPERATIONAL AMPLIFIER

The circuit diagram of op-amp is shown in Fig. 2 [1]. The design parameters (variables), i.e. search space for the circuit are size of various transistors and value of Miller capacitor. The value of VDD is 1.2V and I_0 is set to $3\mu A$. The circuit is designed to drive a load of $0.05pF$.

TABLE I
AVERAGE OF SIMULATION RESULTS OVER 10 DESIGN RUNS.

Specifications		Schematic-level optimization		Parasitic aware optimization
		Schematic-level Simulation	Post-layout simulation	
Gain (dB)	≥ 80	81.66	81.66	81.77
PM (°)	≥ 60	66.52	52.71	62.72
UGB (MHz)	≥ 100	169.33	147.79	136.89
PSRR (dB)	≥ 75	80.33	80.33	79.09
CMRR (dB)	≥ 80	92.69	92.69	81.89
PC (μW)	≤ 30	28.25	28.25	28.65
RSR ($V/\mu S$)	≥ 35	53.98	54.61	37.82
FSR ($V/\mu S$)	≥ 35	35.58	31.34	37.03
Error (%)	$=0$	0.00	6.46	0.00
Time (Min)		16.7	-	107.8

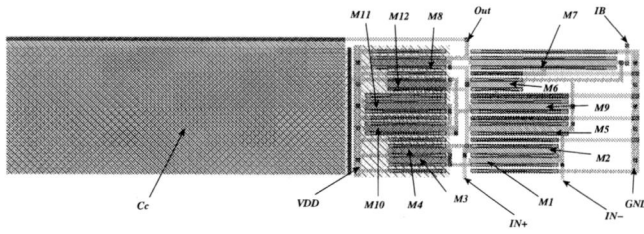

Fig. 3.   Layout of two stage CMOS op-amp.

In the first experiment, we designed the op-amp at schematic level 10 times using ABC algorithm where layout parasitics are not possible to consider. After each design run, the layout of the op-amp is generated based on the device dimensions obtained from the optimized schematic. The average of the obtained results with desired specifications is shown in Table I. The op-amp is designed at schematic-level successfully for all 10 design attempts with zero RMS error satisfying all design specifications. However, the post-layout simulation of

the layouts generated from the optimized schematics resulted in average RMS error of 6.46%. This is because, at schematic-level design, the layout parasitics were not considered. In the second experiment, the layout-level design of op-amp is carried out 10 times using the proposed approach of the parasitic-aware design. The obtained results (see Table I) indicated that the op-amp is designed all 10 times at layout-level with zero RMS error satisfying all desired specifications. The average parasitic-aware design time for the op-amp is found to be 107.8 minutes. In the Fig. 3, the layout generated in one of the parasitic-aware design runs is shown. In order to reduce the size of Miller capacitor, stacked capacitor built using multiple metal layers is used. In the third experiment, the op-amp is designed using the parasitic-aware design concept considering 10% process variation and temperature variation over range of $0°C$ to $70°C$. To consider process and temperature variations, five process corners TT, FF, FS, SF, and SS are simulated at three temperatures i.e. $0°C$, $25°C$ and $70°C$ during the parasitic-aware design process. The obtained results indicated that the op-amp is designed successfully satisfying all the specifications at all the process corner and over desired temperature range. The design time was 9.52 hrs. The minimum values of specifications we found in this experiment were 80.2dB (gain), 112MHz (UGB), 60.3° (phase margin), 75.5 dB (PSRR), 80.6 dB (CMRR), $39.8V/\mu S$ (RSR) and $35.3V/\mu S$ (FSR). The maximum power consumption was found to be $29.8\mu W$.

IV. CONCLUSION

We have proposed the concept of parasitic-aware automatic circuit design for analog and mixed signal CMOS circuits. The concept is implemented using MAGIC VLSI tool by developing configurable layouts and interfacing optimization algorithm with it. The proposed concept enables consideration of layout parasitic from the beginning of the design process. The parasitic-aware design of the two-stage op-amp is carried out in $0.13\mu m$ CMOS technology. The average layout-level design time is only 107.8 minutes. When the process and temperature variations are considered, the layout-level design time is found 9.52 hrs. The obtained results show the effectiveness of the proposed approach.

REFERENCES

[1] R. A. Thakker, M. S. Baghini, and M. B. Patil, "Automatic design of low-power low-voltage analog circuits using particle swarm optimization with re-initialization," *Journal of Low Power Electronics*, vol. 5, no. 3, pp. 291–302, 2009.

[2] G. Nicosia, S. Rinaudo, and E. Sciacca, "An evolutionary algorithm-based approach to robust analog circuit design using constrained multi-objective optimization," *Knowledge-Based Systems*, vol. 21, no. 3, pp. 175–183, 2008.

[3] B. P. De, R. Kar, D. Mandal, and S. Ghoshal, "An efficient design of CMOS comparator and folded cascode op-amp circuits using particle swarm optimization with an aging leader and challengers algorithm," *International Journal of Machine Learning and Cybernetics*, pp. 1–20.

[4] "Magic VLSI Layout Tool," http://opencircuitdesign.com/magic/, accessed: 2016-09-21.

[5] D. Karaboga and B. Basturk, "A powerful and efficient algorithm for numerical function optimization: artificial bee colony (abc) algorithm," *Journal of global optimization*, vol. 39, no. 3, pp. 459–471, 2007.

2018 31th International Conference on VLSI Design and 2018 17th International Conference on Embedded Systems

A Temperature Compensated Read Assist for Low Vmin and High Performance High Density 6T SRAM in FinFET Technology

[1]Vinay kumar , [1]Ravindra Kumar Shrivastava , [1]Madhav Mansukh Padaliya

[1]Synopsys India Pvt. Ltd. *vikumar@synopsys.com*

Abstract - A low Vmin, 6T-SRAM is realized in 7nm FinFET Technology using read and write assist methods. Read margin of the SRAM cell is recovered using a temperature compensated wordline lowering scheme. This temperature compensated Read Assist provides additional advantage that lowering on wordline is almost process independent that makes Read Assist very robust. This scheme makes design free from tuning after post silicon. Since Proposed Read Assist circuit lowers Wordline at high temperature while lowering at low temperature is very minimal, SRAM writability is not impacted by Read Assist Circuity at low temperature. At low voltage, SRAM performance is limited by Read cycle time. The proposed Read Assist scheme improves Read performance by 200%, which in-turn reflects the gain in operating frequency up to 100%. In the proposed Read assist implementation operating frequency is almost comparable to system when Read Assist is not enabled with added advantage of low voltage enablement.
Keywords— 7nm FINFET, Low Power, SRAM, High Density (HD), low Vmin, Write Assist (WA), Read Assist (RA), Write Margin (WM), High Performance (HP), Wordline Underdrive , Static Noise Margin (SNM) .

I. INTRODUCTION

Careful co-optimization between technology and design of memory assist circuits is required to deliver dense, low power memory operation at low voltages. High-density (HD) 6T SRAM cell is formed using single fins for each device in the bitcell which has 20-25% better bitcell area efficiency over High Performance (HP) SRAM bitcell in addition to leakage advantage. High Density SRAM bitcell has around 40-50% leakage advantage as compared to High Current bitcell.

With the process variations, the strength of PU transistor can be much stronger than the PG transistor. A stronger PU degrades the write margin significantly and results in severe write ability issue. In addition to writability issue, SNM study shows that it needs Read assist to enable SRAM operation beyond 0.65v. Word line lowering [1-4] is considered most prominent Read Assist technique in FinFET technology to enable low voltage operation. Read Assist further degrades writability for HD bitcell as shown in Figure 1.

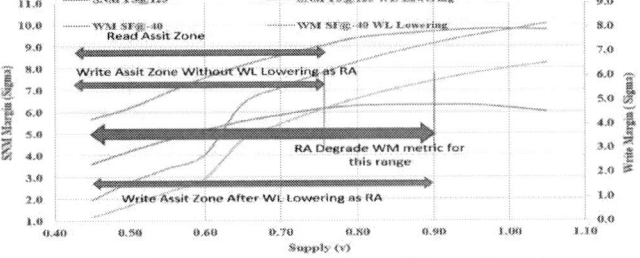

Figure 1. 7nm FinFET High Density SRAM SNM and Write Margin.

Wordline lowering RA scheme results in performance loss and reduces the efficiency of WA. This work presents a WL

lowering scheme that does not result in performance loss. The Scheme proposed here avoids under-drive at performance critical low temperature and provides required under-drive at high temperature to enhance SNM. This scheme minimizes under-drive at low temperatures where writability is major concern. This helps to avoid a penalty on the required negative bitline (used as WA) requirement.

II. Proposed Temperature Compensated Read Assist

To minimize the impact of Read Assist on Write Margin and Performance, temperature compensated RA scheme is implemented in 7nm FinFET technology so that WL dips only at higher temperatures to recover the SNM (WM and cell current should not see WL voltage drop at critical/low temperatures since both Write Margin and Read current degrades at Low temperature). In Conventional RA, Word line is lowered by PMOS whose gate is VSS as shown in Fig2(a), its impact on Word line Lowering across Process and temperature is shown in Figure 2(b).

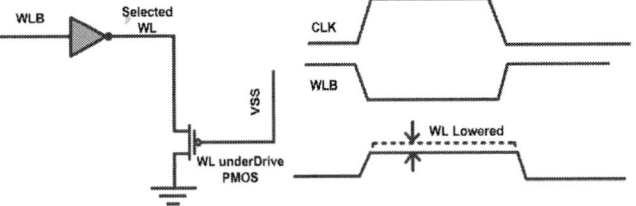

(a)WLUD Conventional Circuit.

VDDA(v)	WL Leve(%VDDA)				
	SS/-40	SF/-40	FS/-40	SF/125	FS/125
0.50	93%	93%	92%	92%	92%
0.55	93%	93%	93%	92%	93%
0.60	93%	93%	93%	92%	92%
0.65	93%	93%	94%	91%	92%

(b)WLUD across different PVT

Figure2 Conventional WLUD Read Assist.

To ensure that WL lowering is dominant only at high temperature, a temperature sensitive RA circuit is designed as shown in Figure 3.

In Proposed Read Assist the gate drive of PMOS "MRA" is used to lower the Wordline (Fig 3) which varies with temperature. When SRAM operating at Low temperature, PMOS "MCOM" threshold voltage is high and its drive strength is very weak. Due to the weak drive for "MCOMP" at low temperature, the signal "TEMP_COMP" settles at a value higher than VSS. In SMM worst condition when SRAM operates at high temperature, the drive for "MCOMP" improves and Signal "TEMP_COMP" value settles closer to VSS that provides strong gate drive to assist PMOS "MRA" and Wordline lowers sufficiently to improve SNM as shown in

978-1-5386-3693-0/18 $31.00 © 2018 IEEE   447

Fig 4. RA circuit in the proposed scheme is almost insensitive to memory periphery process variation to avoid post silicon issues/complexities when process mismatch occurs across SRAM and memory periphery devices. The sensitivity of proposed Wordline lowering scheme for node "TEMP_COMP" across different process and temperatures is shown in Figure4.

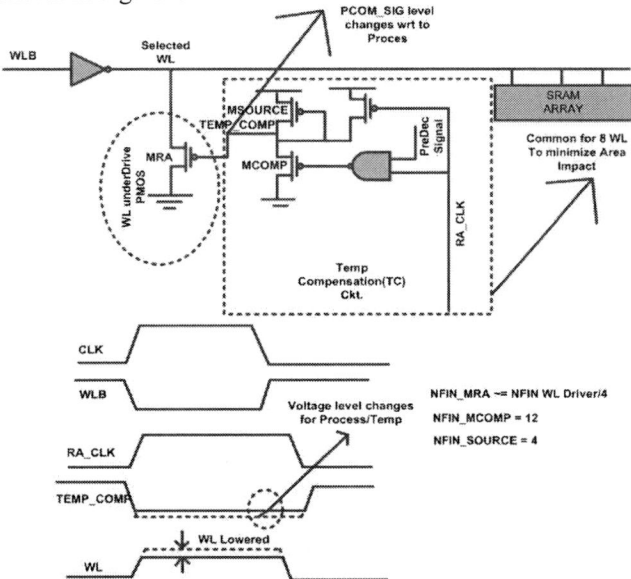

Figure 3 Proposed Temp Compensated Wordline Lowering Read Assist.

VDDA(v)	TEMP_COMP Signal (mv)			
	FS/125	FS/-40	SF/125	SF/-40
0.50	127	158	131	167
0.55	149	183	152	192
0.60	170	208	172	215
0.65	190	232	190	238
At high Temp, Pmos used to lower WL has better drive as compared to Low temperature				

Figure 4 voltage Level for signal "TEMP_COMP"

Simulation results show that the WL drop is almost nil in case of write margin (SF/-40C) and speed critical PVT (SS/-40) as shown in Table 1. The circuit is designed to ensure the required WL lowering at SNM critical PVT (FS/125). Compensation Block is designed to ensure least WL lowering at SF and SS conditions when temperature is low. The variation of WL lowering/drop across process and voltage is shown Table1 for 125C and -40C. This is observed that WL lowering is suppressed at lower temperatures.

	Wordline Under Drive Value as Read Assist(mv)			
	SF/-40	SF/125	FS/-40	FS/125
	Mean + 2 *sigma	Mean - 2*Sigma	Mean + 2 *sigma	Mean - 2*Sigma
VDDA(v)				
0.50	7	44	8	42
0.55	7	40	7	40
0.60	17	44	17	48
0.65	22	50	21	50

Table 1 Proposed WL Lowering RA Circuit Monte Carlo Analysis

Sensitivity for implemented WL lowering scheme with temperature is shown in Figure 5. SNM analysis for this bitcell shows that WL lowering requirement increases with temperature. Monte Carlo analysis performed on Read Assist circuitry, "Mean - 2*Sigma" value is considered to qualify SNM criteria while "Mean + 2*Sigma" is considered for writability and performance analysis to safeguard design against variation of proposed implementation.

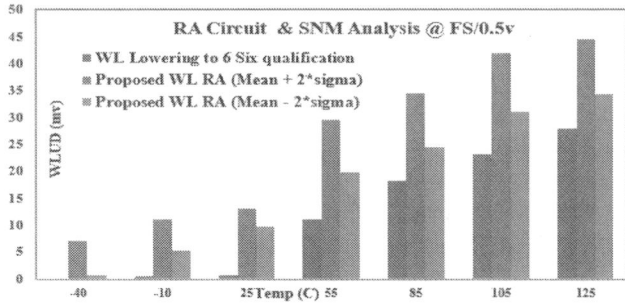

Figure 5 SNM sigma qualification with Temperature

III DISCUSSION AND CONCLUSION

Implementing temperature compensated WL lowering scheme, read stability at low voltage is ensured. Scheme uses ~40mV of WL lowering to achieve Vmin up to 0.5V. Stability of cell increases from 4 sigma to 6 sigma, ensuring 99% yield for a 256 Mb. The RA scheme is implemented with an area overhead of 2 percent. Vmin of the cell is 750mV without any read and write assist. In a conventional WL lowering scheme, underdrive at SS/-40C and SF/-40C is almost comparable to what is designed for FS/125C. This puts a performance penalty on memory. Considering similar degradation using conventional WL lowering scheme, cell current would have been reduced to one forth, resulting into severe penalty on operating frequency and access time. This performance loss is recovered using proposed temperature compensated WL lowering scheme. The improved read current values are shown in Figure 6. This improvement of read current (up to ~4x) results in the improvement of the access time for SRAM by more than 150%. In addition to access time, operating frequency is also improved up to 100% for low voltage range operation as reflected in Fig 6.

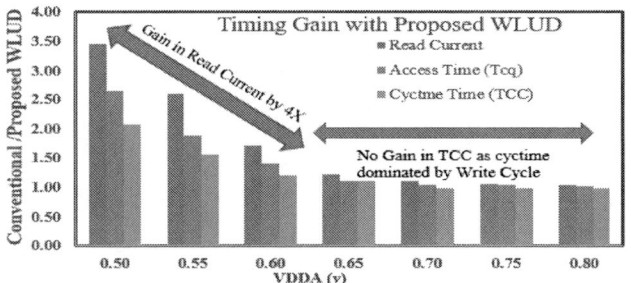

Figure 6 Timing Gain with Proposed Read Assist WL Lowering

References

[1] Jonathan Chang et al, "A 20nm 112Mb SRAM in High-κ Metal-Gate with Assist Circuitry for Low-Leakage and Low-VMIN Applications", ISSCC 2013, p.p. 316-317.

[2] Makoto Yabuuchi et al, "20nm High-Density Single-Port and Dual-Port SRAMs with Wordline-Voltage-Adjustment System for Read/Write Assists", ISSCC 2014, p.p. 234-235.

[3] Mudit Bhargawa et al, "Low VMIN 20nm Embedded SRAM with Multi-voltage Wordline Control based Read and Write Assist Techniques", Symposium on VLSI Circuits Digest of Technical Papers, 2014.

[4] Jonathan chang,A 7nm 256Mb SRAM in high-k metal-gate FinFET technology with write-assist circuitry for low-VMIN applications, ISSCC 2017

978-1-5386-3693-0/18 $31.00 © 2018 IEEE

FPGA Implementation of Power Management Algorithm for Wind Energy Storage System with Kalman Filter MPPT Technique

Vulisi Narendra Kumar
Dept. of Electrical Engineering
NIT Meghalaya
Shillong, India
narendrakumar.abc@gmail.com

Gayadhar Panda
Dept. of Electrical Engineering
NIT Meghalaya
Shillong, India
p_gayadhar@yahoo.com

Abstract— **The diminution of fossil fuels in future generations to come, imposes the use of non-conventional energy sources in the power sector. To avoid the stress and losses associated with the conventional converters, it is replaced by high-gain high-efficient converters in the selected wind energy storage systems (WESS). Also, an adaptive power management algorithm is developed to maintain the microgrid variables of the WESS system within the boundaries. The real-time checking of the proposed adaptive power management control algorithm will be done with the help of Virtex-7 FPGA kit and co-simulated using Xilinx system generator.**

I. INTRODUCTION

The wind generation associated with the energy storage systems like battery and supercapacitor bank is called as wind energy storage system (WESS) [1]. At present, the permanent magnet synchronous generators (PMSG) are widely used in wind generation. Other benefits of PMSG are its compactness, rugged construction and highly efficient operation. The maximum power from the wind turbine can be extracted by using Maximum Power Point Tracking (MPPT) technique. The Kalman filter (KF) based MPPT gives better results compared with existing MPPT techniques as described in [2].

The conventional DC-DC boost converters are associated with the high stress, high switching losses and low efficiency for high voltage applications. To eliminate this drawback, a high-gain, high-efficient converters are used by replacing with the conventional converters. The high-gain high-efficient Converter proposed in [3] uses a coupled inductor in an interleaved manner, an intermediate capacitor and a passive clamp circuit give better results in high voltage applications with reduced duty ratio along with improved efficiency. A bi-directional DC-DC converter for high voltage applications with two switches. An adaptive dynamic power management algorithm used for WESS is capable of maintaining the proper power sharing among the wind turbine, battery and supercapacitor in the microgrid. To check the real-time Implementation of the power management algorithm designed for the WESS, It is implemented in VIRTEX-7 FPGA kit through the Xilinx system generator.

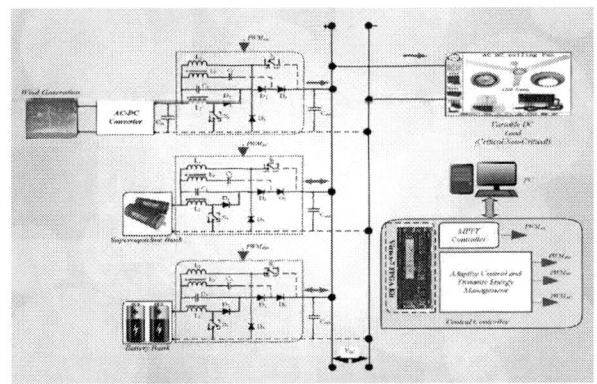

Fig 1. Wind energy storage system (WESS) under study.

The wind turbine accompanying with the energy storage systems connected to the varying high voltage DC loads through the high-gain high-efficiency converters as shown in Fig.1. The power balance equation in the microgrid is given by

$$P_{wn} = P_{load} + P_{sc} + P_{bat} \qquad (1)$$

Where P_{wn} is the instantaneous power generated by wind source; P_{load} is the power necessitated by DC loads; P_{sc} and P_{bat} are the amount of power shared by the supercapacitor and battery bank respectively.

II. HARDWARE-IN-LOOP ANALYSIS OF THE CONTROLLER

The main task of the controller algorithm is to operate the converters of the corresponding energy storing devices depending upon the grid variations. To perform the whole task a dual loop control is employed. The primary outer voltage controller loop containing with the PI control is involved to generate the reference current $(I_{ref}(s))$. The secondary inner current loop generates the pulses to the converters by comparing the reference current $(I_{ref}(s))$ with the converter input currents. The deviation in DC link voltage (V_{error}) is supplied to the voltage controller and the reference current generated is given by

978-1-5386-3693-0/18 $31.00 © 2018 IEEE

449

$$I_{ref}(s) = (k_p + (k_i/s)) \times V_{error} \qquad (2)$$

Where k_p is the proportional gain and k_i is the integral gain. The generated current reference(I_{ref}) is parted and delivered to the battery and supercapacitor as the reference current through a low-pass filter (LPF) having the cut-off frequency of 100 Hz. Xilinx system generator (XSG) performs the task of modelling, converting and programming the FPGA kit with the required control algorithm. Important sections in the Xilinx modelling of the controller is marked in Fig. 2.

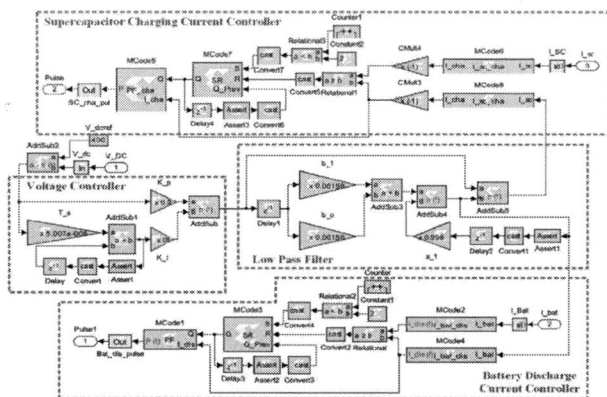

Fig 2. Xilinx modelling of the important sections of the controller

Fig 3. Experimental test bench for hardware-in-loop analysis

III. RESULTS AND DISCUSSION

The wind energy storage system (WESS) shown in Fig. 1 is modelled and synthesized by the support of MATLAB/SIMULINK. The DC Load of 400 W is turned on at 0.5 s and an additional load of 500 W is added at 1.5 s. Finally, a load of 200 W is applied at 3 s and removed at 5s. The wind speed is maintained at 12 m/s up to 6 s and it is reduced to 10 m/s between 6 s to 8 s. At last, a speed of 10 m/s is maintained from 8 s to 10 s. The power alterations of the wind power, battery bank, supercapacitor bank and DC load power are shown in Fig. 4. The voltage at the common bus bar and the load current are shown in Fig. 5. It shows that even though the load is changing in steps, the load voltage is maintained constant due to the controller action. The DC Voltage and the DC current after rectification process is shown in Fig. 6. Finally, it is supplied to the high-gain converter for boosting up to a high voltage of (400V). It is guaranteed that the voltage at the DC link bus bar is maintained constant irrespective of the load and wind speed variations.

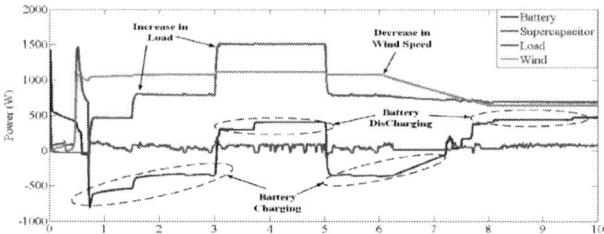

Fig 4. Power sharing among the microgrid participants.

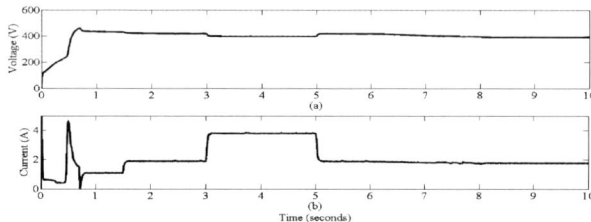

Fig 5. DC load parameters (a) Load Voltage (b) Load current

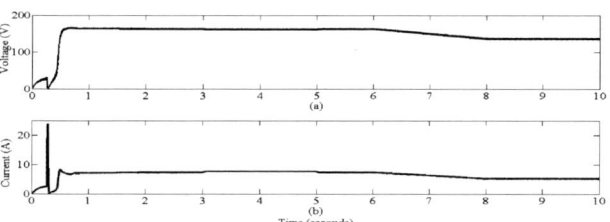

Fig 6. Boost converter input parameters (a) Voltage (b) Current

IV. CONCLUSION

The combination of wind turbine and ESS is chosen in this paper to form a stand-alone microgrid. The high-gain high-efficient converters are replaced by the conventional converters for an efficient microgrid operation. Finally, the controller performance is assured by doing the hardware-in-loop co-simulation by employing the ZYNQ ZC702 FPGA evaluation kit. The results showed that the microgrid can work effectively even in the dynamic changes caused by the wind speed and the DC loads.

V. ACKNOWLEDGEMENT

This work is supported by the REC Transmission Projects Company Limited Grant RECTPCL/CSR/2016-17/693.

REFERENCES :

[1] X. Zhao, Z. Yan and X. P. Zhang, "A Wind-Wave Farm System With Self-Energy Storage and Smoothed Power Output," in *IEEE Access*, vol. 4, no. , pp. 8634-8642, 2016.

[2] B. O. Kang and J. H. Park, " Kalman filter MPPT method for a solar inverter," in *IEEE Power and Energy Conference at Illinois*, Champaign, IL, 2011, pp. 1-5, 2011.

[3] Moumita Das and Vivek Agarwal, "Design and Analysis of a High-Efficiency DC–DC Converter With Soft Switching Capability for Renewable Energy Applications Requiring High Voltage Gain," in *IEEE Trans. Ind. Electron.* vol. 63, no. 5, may 2016.

978-1-5386-3693-0/18 $31.00 © 2018 IEEE

2018 31th International Conference on VLSI Design and 2018 17th International Conference on Embedded Systems

A Novel Tool for Synthesis by Direct Mapping of Asynchronous Circuits from Extended STG Specifications

Felipe Mendes, Tiago Curtinhas, Duarte L. Oliveira, Higor A. Delsoto, Lester A. Faria

Electronic Engineering Division
Technological Institute of Aeronautics - ITA
SJC – São Paulo, Brazil
felipe2855@gmail.com, {thiagohd, duarte, higordel, lester}@ita.br

Abstract— **In this paper we propose a novel tool called Dirmap to the synthesis, by direct mapping, of asynchronous controllers, described by the Extended Signal Transition Graph (XSTG) Specification. The XSTG specification combines the strengths of the XBM specifications and of the STG, while the direct mapping method presents the advantages of simplicity, requiring little computational effort and thus allowing synthesizing large specifications without any knowledge of the asynchronous logic.**

Keywords-asynchronous logic;STG specification; control cell

I. INTRODUCTION

Asynchronous circuits present several potential advantages over their synchronous counterparts: they tend to be faster, dissipate less power, do not present clock skew nor clock distribution problems and are more robust in respect to temperature variations and to electromagnetic interactions [1]. Concerning to this kind of circuits, two remarkable specification styles have been previously proposed to describe controllers, leading to an optimized synthesis: a) Signal Transition Graph *(STG)*, proposed by Chu [2], is a subclass of Petri nets; b) Extended Burst-Mode *(XBM*, proposed by Yun et al. [3]) is a kind of specification based on the State Transition Diagram. An important and promising area for research, focusing on the implementation of asynchronous controllers, is the different kinds of *heterogeneous systems* (controllers and/or synchronous processors), where the STG and XBM specifications show several limitations [4]. Due to limitations of the STG and XBM, different generalizations were proposed. A generalization was the Extended Signal Transition Graph Specification (XSTG) [4]. This specification incorporates both STG and XBM specifications.

We can classify in three different approaches to implement asynchronous controllers: logic synthesis [3], direct mapping [5] and de-synchronization [6]. One interesting style is direct mapping. The graph specification of a controller is translated to a circuit net-list. Each node/place of the graph corresponds to a memory element and the arcs between these elements represent the interconnections of the circuit. The advantage of the direct mapping method is that it allows the implementation of large specifications with just a little computational effort. It eliminates the steps of state minimization, state assignment and logic minimization.

Another advantage is that it does not require any knowledge of the asynchronous logic theory.

II. EXTENDED SIGNAL TRANSITION GRAPH (XSTG)

The Extended Signal Transition Graph (XSTG), proposed in [4], is a subclass of Petri nets, where each transition is sampled as a burst. The XSTG specification inherits all the properties of the STG and XBM specifications. Figure 1 shows the XSTG specification, where the input transition signals are a and b. The signal c is a level-sensitive decision signal with non-monotonic behavior. The output signals are x, y and z. For transitions with fan-in=1 and fan-out=1, the "bar" can be omitted. In Fig. 1, the transition (bar) $T1$ present fan-in=1 and fan-out=2. The node $P1$ presents the initial "TOKEM".

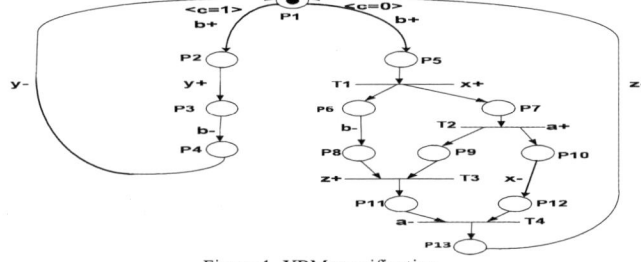

Figure 1. XBM specification.

III. AUTOMATIC SYNTHESIS BY DIRECT MAPPING

Figure 2 shows the flow of the proposed Dirmap tool (details see [7]), where an input file describes, textually, the XSTG specification. The tool verifies the consistency properties and persistent output of the STG, as well as the polarity property of XBM. Figure 3a shows the target architecture of the direct mapping used in the tool. Figure 3b shows the control cell for each node of XSTG.

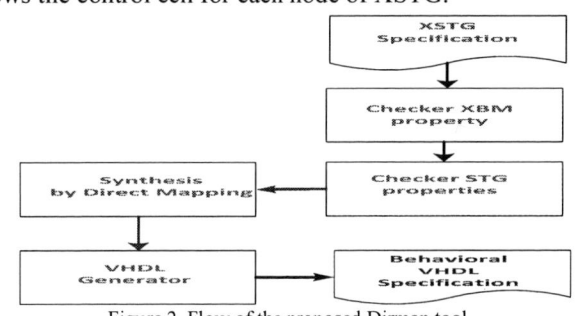

Figure 2. Flow of the proposed Dirmap tool

978-1-5386-3693-0/18 $31.00 © 2018 IEEE   451

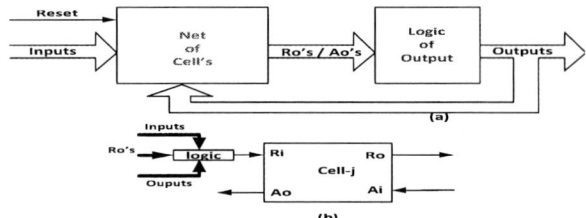

TABLE I. RESULTS OF DIRMAP TOOL

	Specification XSTG		Specifications STG / SG		Implementation VHDL (Dirmap Tool)	
	Places / Transitions	Primary In / Out	Places / Transitions	Number of states	Number of LUTs/FFs	Time of Processing
Biu-dma2fifo	16 / 18	7 / 9	26 / 51	67	12 / 18	6.2ms
Biu-fifo2dma	12 / 13	6 / 7	21 / 40	51	8 / 11	4.5ms
Ex-Des	15 / 16	7 / 8	41 / 71	18	12 / 17	5.7ms
Ex-ISQRT	43 / 46	20 / 23	99 / 167	780	37 / 49	17ms
I2C-bus [11]	19 / 22	2 / 2	30 / 56	52	17 / 27	6.8ms
Sbuf-send-pkt2	9 / 10	4 / 5	16 / 37	25	7 / 10	3.5ms
SCSI-Init-Send	19 / 21	9 / 11	41 / 78	121	13 / 21	7.5ms
SCSI-Targ-Send	25 / 28	11 / 14	38 / 73	152	19 / 29	10.3ms
Select2p	12 / 16	4 / 8	14 / 48	24	11 / 15	6.5ms
Selmerg2ph	20 / 24	8 / 12	22 / 63	51	15 / 24	8.7ms
Yun-Diffeq-Alu2	30 / 32	14 / 16	95 / 170	688	24 / 31	11.2ms

TABLE II. RESULTS OF DIRMAP x PETRIFY

	XSTG Specificatiion		Implementation VHDL		Petrify Tool
	Places / Transitions	In / Out	Number of LUTs	Processing Time (ms)	Processing Time (ms)
Alloc-outbound	21/22	4/5	12	3.7	30.0
Chu172	12/13	3/3	9	2.3	8.0
Master-read	40/28	6/8	31	8.2	600.0
RAM-read-sbuf	28/22	5/6	11	4.7	108.0
Sbuf-ram-write	29/24	5/7	15	5.2	114.0
Sender-done	9/8	2/2	6	1.9	8.0

Figure 3. Target architecture: a) structure; b) control cell

IV. CASE STUDY

In order to illustrate the use of Dirmap tool, an asynchronous version of *sbuf-send-pkt2* benchmark was designed by direct mapping. The description in XSTG used 9 nodes and 10 transitions, as shown in Fig. 4. The cell network is composed of 7 cells, as shown in Fig.5, where it needed an auxiliary cell. Figure 5 shows the cell network with the final connection.

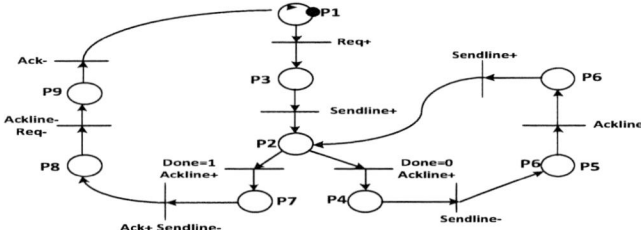

Fig. 4. XSTG specification: sbuf-send-pkt2.

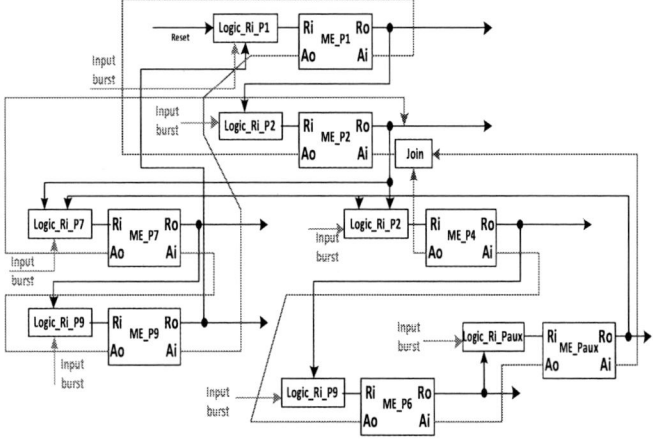

Fig. 5. Net of cells with final connection: sbuf-send-pkt2.

V. RESULTS

To illustrate the tool capabilities, it was applied in eleven different benchmarks, where ten were originally described in XBM, and one (*I2C-bus*) was described in XSTG. Table I shows these benchmarks, described in XSTG, and synthesized by the Dirmap tool. For eleven benchmarks, the Dirmap tool had an average processing time of 7.47ms to generate each behavioral VHDL. Table II shows six benchmarks originally described in the STG specification and shows the processing times by logic synthesis (using the Petrify tool [8]) and direct mapping.

VI. CONCLUSIONS

In this paper we propose a method, by direct mapping, for the XSTG specification. The XSTG specification combines the strengths of the asynchronous paradigm specifications, which are STG and XBM. The Dirmap tool was developed in PYTHON language, generating behavioral VHDL of XSTG.

REFERENCES

[1] C. J. Myers, "*Asynchronous Circuit Design*," Wiley & Sons, Inc., 2004, 2ª edition. .

[2] T. -A. Chu, "*Synthesis of Self-Timed VLSI Circuits from Graph-Theory Specifications*," Ph.D. thesis, June, 1987, Dept. of EECS, MIT.

[3] K. Y. Yun and D. L. Dill, "Automatic Synthesis of Extended *Burst-Mode* Circuits: Part I (Specification and Hazard-Free Implementation) and Part II (Automatic Synthesis)," *IEEE Trans. on CAD of Integrated Circuit and Systems*, Vol. 18:2, February, pp. 101-132, 1999.

[4] D. L. Oliveira and S. S. Sato, "FPGA Implementation of Bounded Wire Delay Asynchronous Controllers from Extended Signal Transition Graph," XV Workshop Iberchip, Buenos Aires, pp.300-304, 2009.

[5] D. Sokolov, A. Bystrov and A. Yakovlev, "Direct mapping of low-latency asynchronous controllers from STGs," *IEEE Trans. CAD of Integration Circuits and Systems*, vol. 26, no. 6, June 2007.

[6] R. Madsen, "*Desynchronization of Digital Circuits*," Master of Science, Technical University of Denmark, 2011.

[7] F. M. dos Santos, "A Tool for Synthesis by Direct Mapping of Asynchronous control Circuits from Extended STG specifications," work of undergraduate of Electronic Engineering, Technological Institute of Aeronautics, p.84, 2016.

[8] J. Cortadella, et al., "Petrify: A tool for manipulating concurrent specifications and synthesis of asynchronous controllers," *IEICE Trans. Inf. Syst.*, vol.E80-D, no. 3, pp.315-325, March 1997.

2018 31th International Conference on VLSI Design and 2018 17th International Conference on Embedded Systems

Optimized Concurrent Testing of Digital Microfluidic Biochips

Sourav Ghosh[1], Hafizur Rahaman[2], Chandan Giri[3]

[1]*Department of Information Technology, Academy of Technology, Adisaptagram, Hooghly, India*
[2,3]*Department of Information Technology, IIEST, Shibpur, India*
Email:[1]sourav.of.25@gmail.com, [2]rahaman_h@yahoo.co.in,[3]chandangiri@gmail.com,

Abstract—**In this work, we propose a testing technique for microfluidic biochip for minimization of test time including the constraint of droplet interference. Simulation results show that the proposed parallel droplet routing technique significantly reduces total test time compared to the earlier reported techniques.**

Keywords-**Biochip; electrowetting; trajectory path; test time**

I. INTRODUCTION

Microfluidic biochips are also referred to as lab-on-chips, which provide the precise control and manipulation of the nanoliter volume of biological fluids and chemical reagents. Microfluidic biochips arc mainly catcgorizcd into two types , namely (a) continuous flow based and (b) droplet based. Digital microfluidic biochips works with discrete droplets and it is easier to control each droplets independently. The basic cell of an EWOD (electrowetting-on-dielectric) based digital microfluidic biochip consists of two parallel glass plates. The droplets are dispensed from external reservoirs and then moved, splitted, merged and mixed on the basis of electrowetting actuation [1]. The concurrent and multiple bioassays analysis always incurred an operational complexity on biochip. In complex biochips, several types of manufacturing defects and physical defects [1] may arise which can lead to permanent or temporal failure of the biochips. As we use biochips in life critical situation, testing of biochip has received much attention in the last few years.

Various techniques for testing DMFBs have been reported in the literature. A graph model based on Hamiltonian path has been proposed in [5]. An Euler path based concurrent testing method has been proposed in [3] to minimize the total test time. The method maps a DMFB as an undirected graph and visits all the nodes of a graph. However, the test based on Euler path incurred a high testing time. An algorithm is presented in [6] to formulate the test time for any arbitrary layout. Multiple test planning techniques, e.g., parallel scan, peripheral scan, diagonal scan have been presented in [4], that finds out the physical defects. In this work, we present a concurrent testing procedure to test biochip cells independently. The method routes each droplet from source to sink using a predetermined path. The concurrent testing procedure using parallel droplet routes helps in minimizing the overall test time.

The remaining part of this work is organized as follows. Motivation and objective of the work have been discussed in Section II. Proposed method to solve the biochip testing is presented in Section III. Section IV presents the simulation results and comparison with other techniques. Finally, Section V draws the conclusion of the work.

II. MOTIVATION AND OBJECTIVE

Earlier methods on the testing of DMFBs are using a single test stimuli droplet. To reduce the overall test time, the biochip cells are partitioned and each partition is tested separately. For each partition, a separate test stimuli droplet is used to reduce the overall test time. The graph based model maps the DMFBs as an undirected graph. Here, the routing path is based on Euler or Hamiltonian path, which is either suffering from computational complexity or test time. Thus, the main objective of this method   (i) is to optimize the overall test time   (ii) using less number of test stimuli droplets and   (iii) total number of source and sink.

III. PROPOSED METHOD

Our proposed method divides a biochip into partitions based on the chip size. Here, each partition is tested concurrently using fixed number of droplets. The required number of droplets depends on the chip size. Now in this technique, we are dividing the $(M \times N)$ biochip into D number of partitions, where partition set is $P = \{p_{x \times y}^1, p_{x \times y}^2, \cdots, p_{x \times l}^D\}$, where $1 \leq x \leq M$, $1 \leq y \leq 3$ and $1 \leq l \leq y$.

The proposed algorithm is described with an example as shown in Figure 1. Here, D denotes the number of droplets, which is required for testing the entire biochip. The maximum size of a partition that is covered by a single droplet is $M \times 3$. Hence total number of discrete droplets is determined as $D = \left\lceil \frac{N}{3} \right\rceil$. For any partition, $p_{x \times y}^i$ is traversed by the ith droplet. For any $M \times N$ biochip if $N \% 3 = 0$, then the partitioned dimension of $p_{x \times y}^i$ must be equal to $p_{x \times y}^j$ $(i \neq j)$. We route the D droplets concurrently from source to sink under the observation of predefined neighbour constraints. In this problem, D droplets are transported from the input cell at 3 cycle intervals from $t = 0$. Generation of droplets at 3 cycles apart ensures that there will be no merging between two droplets. The movement of the droplets from source to sink is described by the alphabet set $\{l, d, r, u\}$. Here l, d, r, u denote a droplet movement to the *left, down, right* and *up* respectively with respect to a position. Droplet movement consists of three different types of trajectory paths which are given below.

i) 3 Columns trajectory path: It is feasible when we from a group of 3 adjacent columns to form a partition $\left(p_{x \times y}^i\right)$ with $x = M$ and $y = 3$. The trajectory is followed by the ith droplet described by the string $d^{i-1}r^{N-3i}d^{M-D}ru^{M-D}rd^{M-D}r^{3(i-1)}d^{D-i}$.

ii) 2 Columns trajectory path: It is feasible when we from a group of 2 adjacent columns to from a partition$\left(p_{x \times y}^i\right)$ with $x = M$ and $y = 2$. The trajectory is followed by the ith droplet described by the string $d^{D-1}r(dldr)^K(dl)^{M-(2K+D)}r^{N-(1+R)}$,where $K =$

978-1-5386-3693-0/18 $31.00 © 2018 IEEE   453

Table I
COMPARISON BETWEEN OUR METHOD AND EXISTING ALGORITHMS

N	M1:Vertical Strips Algorithm [2]		M2:Interleaved Rows Algorithm [2]		M3:Interleaved Zig-Zags Algorithm [2]		Our Method		Improvement of TT% from M1	Improvement of TT% from M2	Improvement of TT% from M3
	TT	ND	TT	ND	TT	ND	TT	ND			
8	35	3	37	8	29	4	30	3	14.29	18.92	-3.45
12	55	4	57	12	45	6	47	4	14.55	17.54	-4.44
16	75	6	77	16	61	8	62	6	17.33	19.48	-1.64
19	90	7	92	19	73	10	75	7	16.67	18.48	-2.74
21	100	7	102	21	81	11	86	7	14	15.69	6.17
22	105	8	107	22	85	11	88	8	16.19	17.76	-3.53
25	120	9	122	25	97	13	101	9	15.83	17.21	-4.12

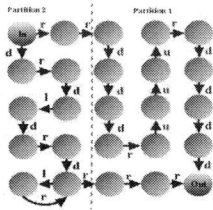

Figure 1. Droplet routing through 8×10 biochip by proposed method

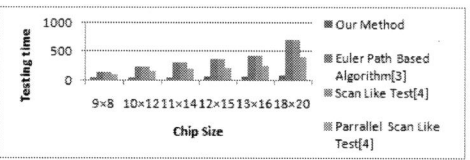

Figure 2. Comparing the testing time of different algorithms with our method

$$\left\lfloor \frac{M-D}{2} \right\rfloor \text{ and } R = \begin{cases} 1 & M = 2K + D \\ 0 & \text{Otherwise} \end{cases}$$

iii) 1 Column trajectory path: Here each column traverses by a single droplet. The trajectory followed by the ith droplet is described by the string $d^{M-1}r^{N-1}$.

It can be proved that for a given $M \times N$ size biochip where $M \geq 2$, 3 columns trajectory motion will take minimum time to traverse all the cells concurrently by $\lceil \frac{N}{3} \rceil$ droplets.

It can also be proved that, for any $(M \times N)$ Biochip where $M \geq 2$ and $N \geq 2$ the total testing time is calculated as,

$$T = \begin{cases} 3M + N + D - 8 & (N\%3 = 1 \text{ and } M \geq 2.5 \lceil \frac{N}{3} \rceil) \\ 3M + N + D - 8 & (N\%3 = 2 \text{ and } M \geq 2.5 \lceil \frac{N}{3} \rceil) + M\%2) \\ 3M + N + D - 5 & \text{Otherwise} \end{cases}$$

IV. SIMULATION RESULTS AND DISCUSSIONS

In this section, we have presented the experimental result and compared with the results of the technique presented in [2]. Experimental results are presented for the biochips, where the chip size is varying from (8×8) to (25×25). The comparison is made on the basis of two basic parameters. These are test time (TT) and number of droplets (ND). Table I show the simulation and comparison results of our proposed technique and the technique presented in [2]. Column 1 represents size of the Biochip chip size. From the Table I, it is observed that on an average reduction of 15.55%, 17.86% test time in our case compared to Vertical Strips Algorithm [2] and Interleaved Rows Algorithm [2] respectively. But the test time of Interleaved Zig-Zags Algorithm [2] is 1.96% lower in average case than our algorithm. Figure 2 presents the testing time of various algorithms applied on rectangular biochip and also comparing those testing times with our testing time. Hence, we can conclude that the proposed algorithm performs better than the techniques presented in [3], [4] due to the following reasons:- the proposed technique

(a) uses less number of droplets   (b) requires simple circuitry to handle less number of droplets and   (c) tests the chip fairly in less time.

V. CONCLUSIONS

In this work, we have presented a polynomial time $O(N)$ heuristic algorithm to minimize the test time of discrete droplet based biochip. Our testing approach tests the entire biochip using less number of droplets. But it would be better if we use this algorithm for on-line testing rather than off-line testing.

REFERENCES

[1] K. Chakrabarty and F. Su. *Digital Microfluidic Biochips: Synthesis, Testing, and Reconfiguration Techniques.* CRC Press, NY, 2007.

[2] Bogdan Pasaniuc, Robert Garifinkel, Ion Mandoiu , Alex Zelikovsky. Optimal testing of digital microfluidic biochip. *Informs Journal On Computing,* vol. 23 pp. 518-529 , 2011.

[3] Fei Su, William Hwang, Arindam Mukherjee and Krishnendu Chakrabarty. Defect-Oriented Testing and Diagnosis of Digital Microfluidics-Based Biochips. In *Proc. IEEE Int. Test Conf.,* pp. 487-496, 2005.

[4] T. Xu and K. Chakrabarty. Parallel scan-like test and multiple Defect diagnosis for digital microfluidic biochips. *IEEE Trans on Biomed Circuits Syst.,* vol. 1, no. 2, pp. 148-158, June 2007.

[5] F. Su, S. Ozev and K. Chakrabarty. Test planning and test Resource optimization for droplet-based microfluidic systems. *Journal of Electronics Testing: Theory and Application,* vol. 22, pp. 199-210, April 2006.

[6] Trung Anh Dinh, Shigeru Yamashita, Tsung-Yi Ho, K. Chakrabarty. A General Testing Method for Digital Microfluidic Biochips under Physical Constraints. *IEEE International Test Conference (ITC),* pp. 1-8, 6-8 Oct. 2015.

978-1-5386-3693-0/18 $31.00 © 2018 IEEE

2018 31th International Conference on VLSI Design and 2018 17th International Conference on Embedded Systems

Self-Powered IoT Device for Indoor Applications

Rolf Arne Kjellby, Thor Eirik Johnsrud, Svein Erik Loetveit,
Linga Reddy Cenkeramaddi, Mohamed Hamid and Baltasar Beferull-Lozano

Intelligent Signal Processing and Wireless Networks (WISENET) Lab.
Department of Information and Communication Technology
University of Agder, 4879, Grimstad, Norway

Abstract— This paper presents a proof of concept for self-powered Internet of Things (IoT) device, which is maintenance free and completely self-sustainable through energy harvesting. These IoT devices can be deployed in large scale and placed anywhere as long as they are in range of a gateway, and as long as there is sufficient light levels for the solar panel, such as indoor lights. A complete IoT device is designed, prototyped and tested. The IoT device can potentially last for more than 5 months (transmission interval of 30 seconds) on the coin cell battery (capacity of 120mAh) without any energy harvesting, sufficiently long for the dark seasons of the year. The sensor node contains ultra-low power sensors for temperature, humidity and light levels, with the possibility of adding several more sensors.

I. INTRODUCTION

Recent developments in semiconductor technology, integrated circuit (IC) technology and advancements in wireless technology enables the development of several applications related to wireless embedded systems and Internet of Things (IoT) devices. Wireless embedded systems such as wireless sensor networks (WSNs) has several applications, such as health care applications [1] and monitoring of industrial processes [2]. Internet of things (IoT) is essentially networking of smart embedded electronic devices which sense and exchange data without human intervention, with applications ranging from smart homes to industrial automation. It is expected that approximately 50 billion IoT devices will be connected by 2020. In majority of the WSNs and IoT applications, devices are powered by batteries with limited life time, ranging from hours to years. When these batteries are depleted, they need to be replaced or recharged. For devices in remote locations and harsh environments, it may not be possible to either recharge or replace the batteries. Energy harvesting from ambient sources is a viable solution to overcome these issues, especially for outdoor applications, however, it imposes many challenges in indoor applications, as the available ambient energy is drastically reduced. This paper presents a solution for an IoT device which is self-powered through energy harvesting from ambient sources for indoor applications.

This work was supported by the INFRASTRUCTURE ReRaNP grant 245699/F50 and the FRIPRO TOPPFORSK grant 250910/F20, from the Research Council of Norway.

The paper is organized as follows. Section II describes the brief overview of the system level design. Section III presents the energy harvesting and power management. Prototyping, testing, measurements results are described in section IV and section V concludes the paper.

II. SYSTEM LEVEL DESIGN

The design consists of sensor nodes communicating with a gateway node by using ultra low power (ULP) wireless communication as shown in Fig. 1. The sensor nodes consist of Arduino Pro Mini with nRF24L01 as 2.4GHz transceiver. An ESP8266, which also uses an nRF24L01, is used as a Wi-Fi gateway. A Raspberry Pi 3 receives data from the gateway through MQTT protocol, as well as storing and displaying the data.

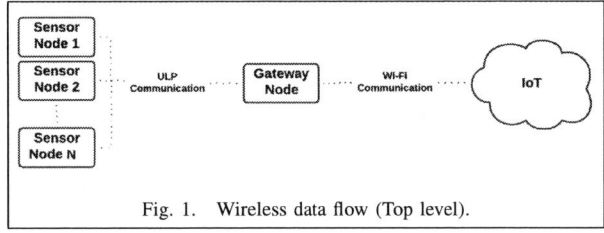

Fig. 1.   Wireless data flow (Top level).

III. ENERGY HARVESTING AND POWER MANAGEMENT

In order to power the wireless sensor network (WSN), an energy harvesting (EH) module is designed. Power available from different EH sources varies greatly with different environments. Therefore, Proof of Concept models are made to test how much energy is available in the indoor environment at University of Agder. These results will define how often the sensor node can measure and transmit the data to the IoT server. The most viable sources for EH for this project are Photovoltaic (light) and RF (electromagnetic energy).

A. *Photovoltaic energy harvesting*

Harvesting electrical energy from light is commonly done with Photovoltaic (PV) Cells. Most PV cells are designed to be most efficient in natural light, with the sun as the source. Sunlight has higher intensity and provides a wider spectrum than what is found in artificial lightning, as shown in Fig. 2. Fluorescent light, illustrated as blue, covers a very narrow spectrum, while LED light covers a broader spectrum, but peaks at a shorter wavelength.

978-1-5386-3693-0/18 $31.00 © 2018 IEEE

455

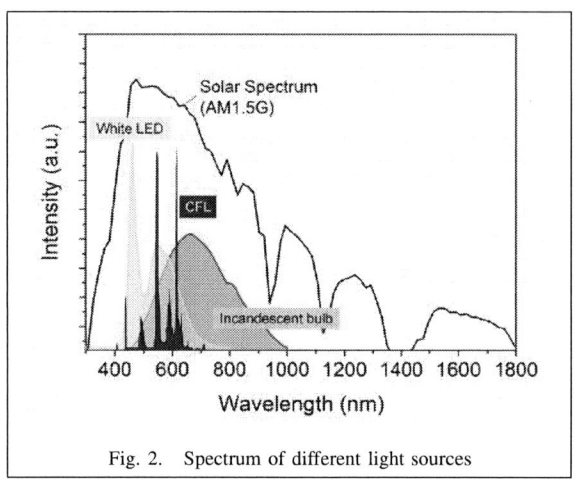

Fig. 2. Spectrum of different light sources

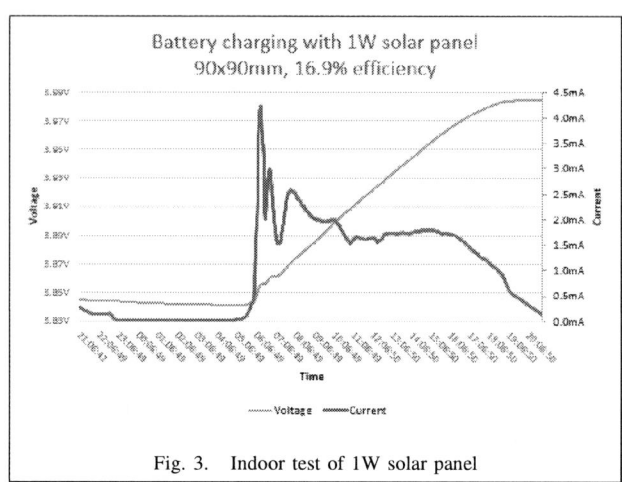

Fig. 3. Indoor test of 1W solar panel

B. Improving the output of PV cell

The output of a solar cell varies depending on light, load and temperature. To get the most of a PV-system, it is necessary to monitor output power and adjust load continuously through Maximum Power Point Tracking (MPPT). MPPT can be done with the open circuit voltage method, which can reach efficiency levels of more than 95% [3]. This involves measuring the open circuit voltage of the panel, and then regulating the load so that the output voltage of PV is set to a predefined percentage of an open circuit voltage.

A 90x90mm 1W epoxy-coated polycrystalline silicon cell panel is used for this prototype. Characterization of the panel at the Photovoltaics lab of University of Agder showed an efficiency of 16.9%. The maximum power point was measured at 80% of open circuit voltage.

C. RF energy harvesting

RF harvesting is the concept of collecting energy from electromagnetic signals in the air. Cell towers are continuously transmitting signals for mobile communication and TV-broadcasts. From measurements at University of Agder, it shows that harvested RF power is in range of 4-5nW indoors with a microstrip patch antenna designed for GSM frequencies. This is quite small compared to harvested solar energy, and is therefore not included in the prototype. It may however prove useful in other scenarios.

D. Power management

A 120mAh, 3.6V lithium ion coin cell battery is used for energy storage. BQ25570 is used for photovoltaic harvesting with MPPT, battery management and voltage conditioning [4]. It has a cold start voltage of 100mV, which is highly suitable for a solar panel in an indoor environment. During indoor testing the BQ25570, along with battery and the 1W panel, measurements showed an average current of 1.02mA, and an average power of 4.05mW, as shown in Fig. 3.

IV. Prototype Testing, Results and Discussion

Two sensors are included in the prototype. The HDC1010 which measures temperature and Relative Humidity (RH),

and TSL2561 which measures light levels [5][6].

A full-scale test of the WSN setup was conducted for 7 days in the WISENET Lab at University of Agder. The setup consists of the sensor node, the gateway and the IoT server. The functionality of the WSN was successfully verified throughout the test, with no major errors except for a few missed transmissions at 20 meter range through one drywall. The sensor node, with a transmission interval of 30 seconds, consumed $135.3\mu W$ on average, while the harvested energy through the 1W solar panel was 4.05mW average. The transmission included sensor readings of light levels, relative humidity and temperature, as well as node ID.

V. Conclusion

A prototype IoT device has been developed based on modular approach and tested successfully. The EH levels using low cost solar panels with relatively small sizes proved sufficient. The power consumption of the wireless sensor node was found to be many times lower than the amount of harvested energy on average with 30 seconds transmission intervals when using the nRF24l01 transceiver. The IoT device can potentially last for more than 5 months (transmission interval of 30 seconds) on the coin cell battery without any energy harvesting, sufficiently long for the dark seasons of the year.

References

[1] J. A Stankovic, Q. Cao, T. Doan, L. Fang, Z. He, R. Kiran, S. Lin, S. Son, R. Stoleru, and A. Wood, "Wireless sensor networks for in-home healthcare: Potential and challenges," 01 2005.

[2] F. Barac, M. Gidlund, and T. Zhang, "Scrutinizing bit- and symbol-errors of ieee 802.15.4 communication in industrial environments," *IEEE Transactions on Instrumentation and Measurement*, vol. 63, no. 7, pp. 1783–1794, July 2014.

[3] S. M. Ferdous, M. A. Mohammad, F. Nasrullah, A. M. Saleque, and A. Z. M. S. Muttalib, "Design and simulation of an open voltage algorithm based maximum power point tracker for battery charging pv system," pp. 908–911, Dec 2012.

[4] T. Instruments, "Bq25570 datasheet," http://www.ti.com/lit/ds/symlink/bq25570.pdf, accessed: 2017-05-02.

[5] ——, "Hdc1010 datasheet," http://www.ti.com/lit/ds/symlink/hdc1010.pdf, accessed: 2017-05-02.

[6] AMS, "Tsl2561 datasheet," http://ams.com/eng/content/download/250094/975485/142937, accessed: 2017-05-02.

978-1-5386-3693-0/18 $31.00 © 2018 IEEE

2018 31th International Conference on VLSI Design and 2018 17th International Conference on Embedded Systems

Fault Tolerance in Network on Chip using Bypass Path establishing Packets

Sharma Priya, Sukarn Agarwal and Hemangee K. Kapoor

Department of Computer Science and Engineering, IIT Guwahati, Assam, India-781039

Email: {sharma.priya, sukarn, hemangee}@iitg.ernet.in

Abstract—**Network-On-Chip (NOC) plays a vital role in on-chip communication infrastructure for Massively Parallel System On Chip environment. Large size NOC networks are prone to fault(s) on links and routers. This paper presents a fault tolerant routing algorithm by setting up bypass path around faulty nodes. Experimental evaluation shows significant reduction in latency and better reliability over an existing technique and the baseline network.**

Keywords—*Network-On-Chip, Reliability, Fault tolerance, Bypass path*

I. INTRODUCTION

The continuous reduction of transistor size enables us to incorporate many cores on the chip. With a large number of cores, the traditional bus architecture cannot handle the large communication demands of the applications. Network On Chip (NOC) is identified to solve these communication problems. NOC connects a large number of Processing Element (PE) in Massively Parallel System On Chip (MPSOC) environment. But as the size of the chip increased, with the incorporation of a vast number of PE, NOC became prone to the fault(s) in the network. These fault(s) take place due to: manufacturing defects, increase in traffic due to increased data demand, and, by aging of different components in the NOC. This paper presents a fault-tolerant routing algorithm.

With the fault(s) in the network, the reliability of the network to deliver the packet reduces. Previous fault tolerant turn based approaches such as Reconfig Route [1] provide the alternate path by avoiding some turns to avoid the contour and deadlock. But the major shortfall in these techniques is the large network latency to deliver the packet. Our turn-based fault-tolerant routing approach focuses to optimize the latency by establishing the bypass path around the faulty node. We compared our proposed technique with the Reconfig Route presented by Zhang et al. [1] and the baseline network. Ebrahimi et al. [2] provides a fault-free path by collecting information from the immediate neighbors. A VC based approach where the faults are tolerated in 3D mesh by XZXY and ZXY routing is presented in [3].

II. PROPOSED FAULT TOLERANT TECHNIQUE

The main idea of the proposed technique is to establish alternative paths between the immediate neighbors for a given faulty node. Each node in the network has a maximum of eight neighbors: four direct/immediate neighbor (located in the four directions: east, west, north, and south) and, four indirect neighbors (located in the four directions: NE, NW, SE and SW). The position of these neighbors with respect to node A is shown in the figure 1 (A). In case of a fault, the packet is transferred to one of the closest immediate neighbors

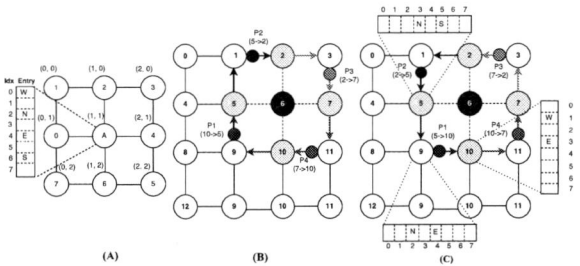

Fig. 1: (A): Position and co-ordinates of the neighbour node with the reconfiguration table. (B) & (C): Working Example of Alternate Path Establishment

of the faulty node from where the packet follows the fault-free path to the destination. The information to transfer the packet to one of the immediate neighbors is contained in the reconfiguration table stored with each router. The format of reconfiguration table is presented in the figure 1(A). Each entry of the reconfiguration table is the position of the immediate neighbor with respect to the current node.

A. Alternate Path Establishment

To establish an alternate/bypass route, a bypass packet P is generated from one immediate neighbor to another immediate neighbor in clockwise or anticlockwise direction. Note that these bypass paths are constructed only when the direct neighbor of the node detects faults. Each bypass packet is a standard data packet that maintains the stack which contains the information of the port through which the packet is transmitted. The establishment involves two basic operations:

Path Generation: Bypass packet generated from the source node traverse the network with the help of YX routing. During traversal, the packet stores the port number of node in its stack. The process continues until the packet reaches the destination. The working example is presented in figure 1(B).

Backtracking: Once the bypass packet reaches the destination, the packet starts to backtrack and follows the same path as the established path but in the reverse direction. In this case, the packet also updates the reconfiguration table in each node with the help of the stack. The modification in the reconfiguration table entries is according to the position of node concerning the given node as shown in figure 1(C). Finally, the bypass packet drops from the network when it completes the round trip.

B. Fault Tolerant Algorithm Description

Algorithm 1 shows the working approach of our proposed fault tolerant technique. With each packet P, we maintain two

978-1-5386-3693-0/18 $31.00 © 2018 IEEE   457

Algorithm 1 Fault Tolerant Algorithm

```
1:  Let c be the current node, d be the destination node and P be data packet.
2:  P.info: Information regarding the entry for next lookup.
3:  P.reconfBit: Distinguishes usual or alternate/bypass route.
4:  node_set: Set of neighbour nodes from where P is redirected in case of faults.
5:  rc_table: Reconfiguration table
6:  if P.reconfBit is not set then                         ▷ P follows Normal Path
7:      Port = XY(P)                                        ▷ XY routing algorithm
8:      if Port is faulty then
9:          BypassPathTraversal(Port)
10:     else
11:         P routed through the Port.                      ▷ P routed normally
12:     end if
13: else
14:     if c is immediate node then                        ▷ Direct neighbour of fault node
15:         P.reconfBit = 0                                 ▷ Reset redirection flag
16:         Port = XY(P)
17:         Go to line 8
18:     else                                                ▷ Indirect neighbour of faulty node
19:         Port = rc_table[P.info]                         ▷ find next redirected neighbour
20:         Packet is routed through the Port.
21:     end if
22: end if
23: function BYPASSPATHTRAVERSAL(Port)
24:     if Port is North then
25:         node_set[0/1] = 1/3                             ▷ 1 - North West 3 - North East
26:     else if Port is South then
27:         node_set[0/1] = 5/7                             ▷ 5 - South East 7 - South West
28:     else if Port is East then
29:         node_set[0/1] = 5/3                             ▷ 5 - South East 3 - North East
30:     else
31:         node_set[0/1] = 7/1                             ▷ 7 - South West 1 - North West
32:     end if
33:     if d.x != c.x then                                  ▷ Select appropiate neighbour for redirection
34:         if d.y > c.y then
35:             sel_option = node_set[0]
36:         else
37:             sel_option = node_set[1]
38:         end if
39:     else
40:         if c.x >= 1 then
41:             sel_option = node_set[0]
42:         else
43:             sel_option = node_set[1]
44:         end if
45:     end if
46:     if rc_table[sel_option] is empty then               ▷ handles multiple faults and faulty edge nodes
47:         Select other option from node_set as sel_option.
48:     end if
49:     P.info = getInfo(Port, sel_option)
50:     P.reconfBit = 1
51:     Port = rc_table[sel_option]
52:     Packet is routed through the Port
53: end function
```

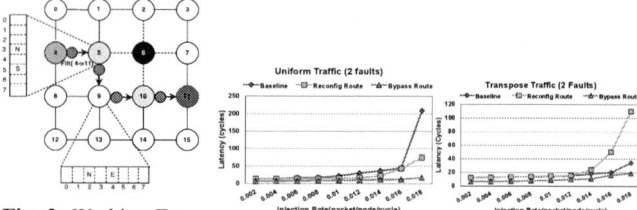

Fig. 2: Working Example of Proposed Fault Tolerant Algorithm

Fig. 3: Throughput analysis for 2 faults in uniform traffic

Fig. 4: Throughput analysis for 2 faults in transpose traffic

Traffic	Faults	Lat.Imp.		Eff. Imp.		Faults	Lat. Imp.		Eff. Imp.	
		B	R	B	R		B	R	B	R
Uniform	1	44.1%	42%	4.1%	-	2	38.6%	43.1%	17%	-
	3	56.4%	40.7%	24.8%	-	4	62.5%	37.4%	27%	-
	5	74.4%	44%	45.6%	-	6	90.2%	47.5%	60.6%	-
Transpose	1	37.6%	41.1%	4.1%	0.7%	2	36.2%	42.3%	5.2%	0.1%
	3	34.9%	43.6%	14.9%	2.1%	4	38.4%	43.4%	14.9%	3.1%
	5	35%	42.1%	20.5%	2.4%	6	39.6%	47%	56.2%	1.5%

TABLE I: Latency and efficiency improvement by our technique against the baseline (B) and the Reconfig Route (R) [1]

additional fields: One-bit Reconfiguration field: $reconfBit$ and the three-bit information field: $info$. We also maintain a set ($node\_set$) that stores the admissible neighbor nodes. The $node\_set$ is used for the redirection in case of a fault.

When the packet P comes to the node c for the routing, the $reconfBit$ of the P is examined. If not set, packet follows the usual path route (line 6 and 7). However, if the selected port is faulty, the set of direct neighbor nodes ($node\_set$) through which the packet can be redirected is assigned by the algorithm (line 24 to 32). Note that direction and number mentioned in the algorithm are the positions of the neighbor nodes with respect to current node as shown in figure 1(A). Later, based on the closeness with the destination, one node from the $node\_set$ is selected (line 33 to 45).

Once the node is picked from $node\_set$, the rc_table table entry corresponding to the node is examined. If it is empty, another node from the $node\_set$ is selected (line 46 to 48). Simultaneously, the $info$ field of the packet is updated with the index of the lookup entry in the rc_table for the next node. This value is provided by the getInfo function. Now, the $reconfBit$ of P is set to indicate that packet follows the alternate path. Finally, the packet is routed through the port provided by the rc_table (line 49 to 52).

In another case, when the $reconfBit$ is set, the possibility of the current node c being the immediate neighbor of the faulty node is verified. If it is, the packet follows the usual route and the $reconfBit$ of the packet is unset (line 14 to 17). Otherwise, the packet is routed through the port provided by the $info$ field of the packet (line 19 to 21). The working example is given in the figure 2.

III. Experimental Evaluation

A. Experimental Setup

We examine our proposed technique in cycle-accurate network-on-chip simulator Noxim [4]. All the simulations are performed in the 2D 8X8 mesh. We conducted an experiment on the uniform and transpose traffic.

B. Result and Analysis

Efficiency: Table I shows the improvement in the efficiency by the proposed technique. Note that our method maintains the same reliability against the existing technique. These developments are mainly due to the established alternate path where the data packet travels in case of fault(s).

Latency: Table I shows the improvement in the latency by the presented approach. This reduction in the latency compared to Reconfig Route is due to long distance traveled by the packet in Reconfig Route to avoid the northeast turn.

Saturation: Figures 3 and 4 present the networks saturation at a different injection rate. Our technique Bypass Route shows the network saturation at a prolonged rate as compared to baseline. This improvement is due to bypass path available for faulty nodes.

IV. Conclusion

We proposed a bypass route based fault-tolerant routing algorithm that supports many faults in the network topology. Our policy forms a bypass alternate path around the faulty node. When the packet encounters the faulty node, the alternate bypass route forwards the packet to another immediate neighbor node from where the packet follows the fault-free path according to XY routing algorithm.

References

[1] Z. Zhang et al., "A reconfigurable routing algorithm for a fault-tolerant 2d-mesh network-on-chip," in *DAC*, June 2008, pp. 441–446.

[2] M. Ebrahimi et al., "High performance fault-tolerant routing algorithm for noc-based many-core systems," in *PDP*, Feb 2013, pp. 462–469.

[3] S. Akbari et al., "Afra: A low cost high performance reliable routing for 3d mesh nocs," in *DATE*, March 2012, pp. 332–337.

[4] V. Catania et al., "Noxim: An open, extensible and cycle-accurate network on chip simulator," in *ASAP*, July 2015, pp. 162–163.

2018 31th International Conference on VLSI Design and 2018 17th International Conference on Embedded Systems

Impact of Variations on Synchronizer Performance: An Experimental Study

Joycee Mekie, Prashansa Mukim and Kimaya Kale

Department of Electrical Engineering, Indian Institute of Technology Gandhinagar, India

Abstract—Synchronizers play a crucial role in obtaining reliable operation in ASICs with multiple clock domains. In this paper, we study the impact of process variations and technology scaling on synchronizer parameter τ. For this we have carried out metastability measurements on FPGAs manufactured at different technology nodes, 90nm and 28nm. To capture the die-to-die variations we have used 4 FPGA boards for each technology node. For capturing within-die variations, we have implemented synchronizer circuits at 200 different locations for the 28nm FPGA and 50 locations for the 90nm chip and the metastability measurements are simultaneously carried out for all synchronizers. The same experiments are also done to capture propagation delay using ring oscillator setup. From the statistical data obtained, we show that as technology scales, the variations in τ are much more than that in FO4 delays. We also show that MTBF calculations based on average τ can be an underestimated value, and can lead to more failures than expected.

I. INTRODUCTION

Synchronizers play a key role in deciding the overall reliability of a multi-clocked SoC. Lack of proper synchronization can lead to failures due to metastability. Understanding synchronization process, and factors that affect synchronization, is an important area of study. To this end, study of power supply fluctuations, technology scaling, temperature effects, etc. have been reported. However, many of these are simulations-based or analytically derived. As technology scales, we know that the fluctuations in process parameters due to manufacturing are becoming worse, which degrades the overall performance. Moreover, complex SPICE models are inadequate to capture all the nuances related to manufacturing processes. Hence, experimental studies are required. In this paper, we report the experimental findings of effect of both process variations and technology scaling.

Process parameter fluctuations are becoming worse as technology scales. As reported in [1], a generation of performance may be lost due to systematic within-die fluctuations in process parameters. The authors have used critical path delays to experimentally and statistically derive the implications of within-die and die-to-die fluctuations in process parameters. Sedcole *et al* [2] report about ±3.54% variations in the delay of logic elements due to within-die variations. The effect of technology scaling on τ has been reported by Beer *et al* [3] based on experimental studies carried out on ASIC. The effect of voltage variations have also been studied analytically and the model for the same has been reported in [4]. We focus on study of effect of process and technology variations on τ.

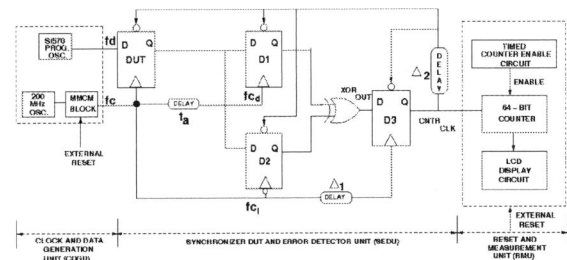

Fig. 1: Experimental setup for measuring τ

II. EXPERIMENT SETUP

The experimental setup used in this work is similar to that described in [5]. We have made use of state-of-the-art Kintex 7 KC-705 FPGA fabricated at 28nm and a versatile Spartan 3E FPGA manufactured at 90nm for our experiments. The experiments are carried out on multiple boards and at large number of locations on each FPGA to help us understand the effects of variations better. In our experimental set-up, all components are implemented on the FPGA chip, including the measurement circuit. The data clock (f_d) is generated externally to ensure that f_c and f_d are not related. Thus, the only off-chip component used is the data signal, making it a truly asynchronous input to the system. The key contributions of this work are the following:

- To study the effect of process variations on τ, we have collected experimental data from 4 boards per technology node and have placed flip-flop synchronizers [5] all over the FPGA. We have used the FPGA to the maximum extent possible, 52 locations for Spartan and 200 locations for Kintex boards. We have repeated each metastability experiment at least 10 times to check for consistency in the data obtained.

- We have completely automated the code-generation for large number of locations, so that synchronizers can be placed at exactly specified known locations with ease.

- For comparing τ and delay variations, we have placed both synchronizer and ring oscillator at exactly the same locations (LUTs and FFs) when carrying out the respective measurement. This allows us to compare the two under the effect of identical variations conditions (except voltage and temperature variations).

- From a detailed statistical analysis of the experimental results, we report that for a given process τ exhibits higher degree of variations, or degradation, than the delay

978-1-5386-3693-0/18 $31.00 © 2018 IEEE   459

parameter. Further, as technology scales, variations in τ are larger than that in delays.

While the results reported in this study are intuitive, it is for the first time, that this study is carried out experimentally on a large number of FPGAs with large number of locations on each FPGA to confirm the findings. Our work allows quantitative comparisons between the two parameters.

III. RESULTS AND DISCUSSIONS

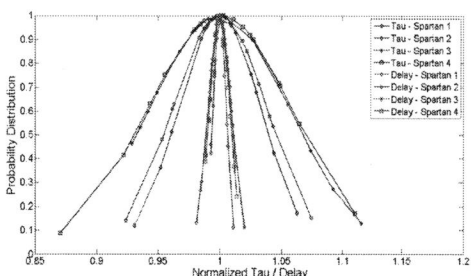

Fig. 2: Normalized probability distribution of τ and delay measured on Spartan 3E

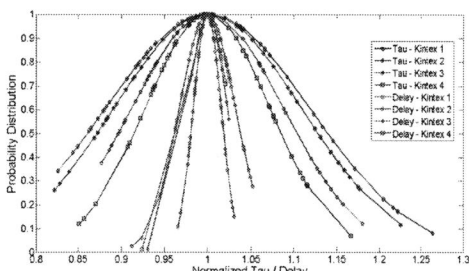

Fig. 3: Normalized probability distribution of τ and delay measured on Kintex 7

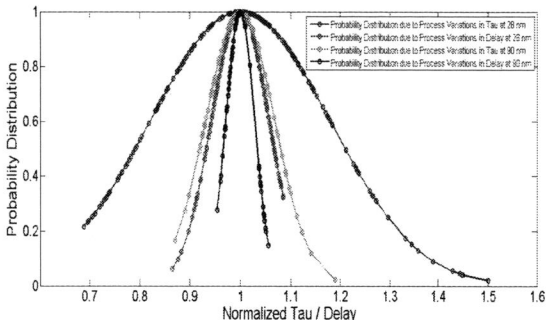

Fig. 4: Normalized probability distribution of τ and delay measured on Spartan 3E (90nm) and Kintex 7(28nm)

In this section we will discuss the effect of technology scaling and process variation on metastability parameter τ and delay, based on the experiments described in previous sections. Fig.2 and Fig.3 are the τ and delay distribution plots

for Spartan 3E and Kintex 7 FPGAs respectively, and Fig.4 is the combined graph that captures technology scaling. We observe the following: (a) the variations in τ are always more than that in delay (b) as technology scales, variations become worse for both delay and τ. However τ has larger variations compared to that of delay.

TABLE I: Comparison of MTBF values (for $t_a = 2500ps$)

MTBF	using average τ	using worst τ
Spartan 3E	5.6×10^5 yrs	Min: 4 days
Kintex 7	4.7×10^3 yrs	Min: 55 days

Conventionally, an average value τ of typical process corner (TT) analysis is used for projecting MTBF of synchronizer. However, it is clear from Table I that on a real hardware MTBF can be much worse or much better. Thus, it is critical for designers to measure τ before using a flip-flop as a synchronizer. All the experiments have been conducted at ambient temperatures of $24 - 26°C$. To specifically study the impact of temperature variations on τ, one set of our experiments was conducted at $16°C$. As expected, all τ values were found to be consistently 5% lower than the τ values obtained at $24 - 26°C$. However we have not specifically captured the effects of power fluctuations in this work.

IV. CONCLUSIONS AND FUTURE DIRECTIONS

In the paper, we have presented the results of metastability measurements done to study the effects of process variations and technology scaling on metastability parameter τ using FPGAs manufactured at 28nm and 90nm technology nodes. We observe that, as technology scales, the variations in both delay and τ increase, but the variation in τ is significantly more than that of delay. We also find that delay parameter scales with technology, whereas τ may not. We have automated the placement of multiple synchronizers on the FPGA. We show that MTBF predicted using an average (typical) τ will result in overestimation of MTBF.

V. ACKNOWLEDGEMENT

This work is supported by grant received from the Department of Science and Technology project SR/FTP/ETA-95/2010.

REFERENCES

[1] K. A. Bowman, S. G. Duvall, and J. D. Meindl, "Impact of die-to-die and within-die parameter fluctuations on the maximum clock frequency distribution for gigascale integration," in *IEEE Journal of Solid-State Circuits*, vol. 37, 2002, pp. 183–190.

[2] P. Sedcole and P. Y. K. Cheung, "Within-die delay variability in 90nm FPGAs and beyond," in *IEEE International Conference on Field Programmable Technology*, December 2006, pp. 97–104.

[3] S. Beer, R. Ginosar, M. Priel, R. Dobkin, and A. Kolodny, "The devolution of synchronizers," in *Proc. International Symposium on Advanced Research in Asynchronous Circuits and Systems*, 2010.

[4] S. Beer and R. Ginosar, "A model for supply voltage and temperature variation effects on synchronizer performance," in *IEEE Transactions on Very Large Scale Integration (VLSI) Systems*.

[5] T. Polzer and A. Steininger, "An approach for efficient metastability characterization of FPGAs through the designer," in *Proc. International Symposium on Advanced Research in Asynchronous Circuits and Systems*, 2013.

978-1-5386-3693-0/18 $31.00 © 2018 IEEE

2018 31th International Conference on VLSI Design and 2018 17th International Conference on Embedded Systems

TileNET: Scalable Architecture for High-throughput Ternary Convolution Neural Networks using FPGAs

Sahu Sai Vikram[*], Vibha Pant[†], Mihir Mody[‡] and Madhura Purnaprajna[†]

[*]Department of Electronics & Communication Engineering, Amrita University, Bengaluru, India
Email: saivikram6@gmail.com
[†]Department of Computer Science & Engineering, Amrita University, Bengaluru, India
Email: vibhapant89@gmail.com, Email: p_madhura@blr.amrita.edu
[‡]Texas Instruments, Bengaluru, India, Email: mihir@ti.com

Abstract—Convolution Neural Networks (CNNs) are becoming increasing popular in Advanced driver assistance systems (ADAS) and Autonomated driving (AD) for camera perception enabling multiple applications like object detection, lane detection and semantic segmentation. Ever increasing need for high resolution multiple cameras around car necessitates a huge-throughput in the order of about few 10's of TeraMACs per second (TMACS) along with high accuracy of detection. Existing implementations do not scale, with performance ranging only in the order of a few Giga operations per second. This paper, proposes a novel tiled architecture for CNNs that uses only ternarized weights, while input and output features are kept full precision resulting in minimal loss of accuracy. The proposed solution is implemented on Virtex-7 FPGA resulting in throughput of 13.76 TOPS. The post-implementation power simulation for AlexNet consumes 16 W, orders of magnitude lower than exist in GPUs.

I. INTRODUCTION

Methods to accelerate CNNs further have led to algorithmic explorations through trading-off accuracy to performance. The convolution layer in CNNs performs Multiply and Accumulate(MAC) operations is the most computationally intensive layer. The redundancy in the floating point operations that limit parallelism, is increasingly being replaced by quantization of weights and activation functions. This technique of quantization has led to binary (+1,-1) and ternary (0,+1,-1) representations, which leads simplification for MAC operations that compose a convolution layer. With fewer transistors required for MAC, higher number of MACs can now be accommodated in the same area. In this scenario, the input and output feature values kept at full precision. It has been proven that for only ternarized weights without quantizing input and output features, the loss in accuracy is within acceptable ranges [1]. This dimension of algorithmic acceleration is very beneficial for parallel and scalable implementation on FPGAs.

II. TILENET ARCHITECTURE

TileNET is a modular and scalable architecture. Tiling of input feature maps and the weight vectors are used in the convolution layers of CNN. A high-level representation of the TileNET accelerator template is shown in Figure 1.

A single compute tile accommodates the entire compute associated with a input tile size of 3x3. The input feature map is represented by an 8-bit fixed point vector [2]. As weights are ternarized, they are represented by values -1,

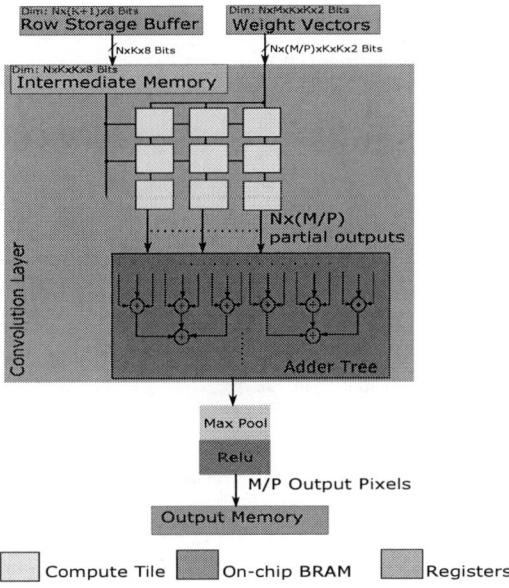

Figure 1. TileNET Accelerator Template

0, 1. In binary format, 2 bits (00, 01, 11 for 0, +1 and -1 respectively) are sufficient for ternary representation. For the ternarized weights, the multiplication operation is simply reduced to a multiplexing structure, as shown in Figure 2. This form of simplification for the multiplication reduces the resource requirement for a multiplier, which is simply replaced by a 4x1 multiplexer with a two's complement computation as shown in Figure 3. This reduction in resource requirement gives way to accommodating higher number of MAC units that can be realized in parallel. As shown in Figure 2, a single tile consists of 9 ternary multipliers followed by an adder tree. Multiple such tiles are instantiated in parallel for computing a part of a convolution layer. In any convolution layer, the input image size is considered to be of depth N with R rows and C Columns of pixel intensities. Due to resource constraints, all the output pixels cannot be computed in parallel for any convolution layer. So we introduce a parallelization factor P which determines the maximum number of pixels that can be generated at a time.

978-1-5386-3693-0/18 $31.00 © 2018 IEEE

461

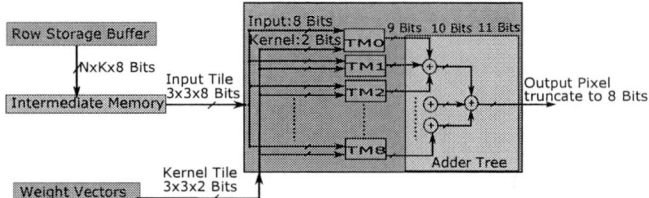

Figure 2. Tile

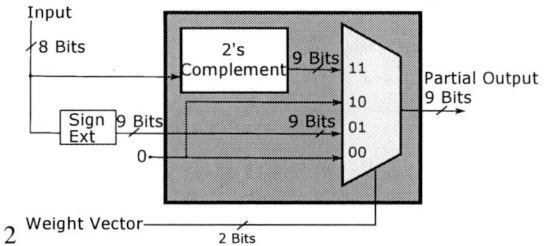

Figure 3. Ternary Multiplier

III. RESULTS AND ANALYSIS

A. Performance Estimation Model

The estimation is done for the maximum compute for AlexNet based on TileNET template considering the device as Virtex 7. The amount of parallelization for all the convolution layers in AlexNet is adjusted by varying P to achieve the maximum utilization of the device.

B. Implementation Results

Using the P values from the Estimation Model, all the convolutional Layers of AlexNet has been implemented on Virtex 7 FPGA device using Vivado design Suite resulting in maximum LUT Utilization of 89.63% with operational frequency of 391.69 MHz

C. Resource-Throughput Analysis

The main constraint to achieve the maximum throughput is the resources available on any device. As shown in Figure 4, for Virtex FPGA device which has the maximum resources from the given devices, the throughput achieved is also the maximum for TileNET implementation of AlexNet.

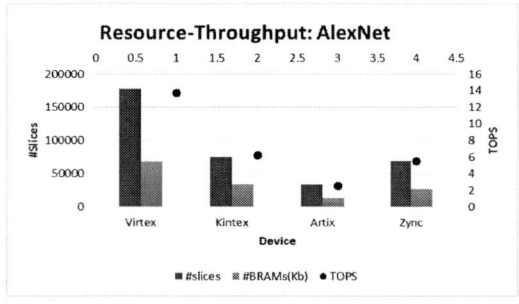

Figure 4. AlexNet:Resource - Throughput across devices

IV. RELATED WORK

As a recent rise of interest in accelerating machine learning,a range of custom architectural alternatives for CNNs such as CGRAs [3] and ASIC-based solutions [4], [5], [6], [7], [8] have also been explored.

The ternary scheme of quantization has been explored with this scheme of quantization applied to both inputs and weight vector representation [1]. In contrast, current implementations are limited in performance, while our approach to accelerating CNNs is performance scalable exploring parallelism through the use of Ternary Weighted Networks (TWN) [9]. TWNs are good enough to represent Full Precision Weighted Networks and are better than Binary Weighted Networks in the accuracy levels achieved across a range of networks.

V. CONCLUSION

This work proposes a novel modular and scalable architecture for achieving high throughput in CNNs. This parameterizable dataflow architecture can be scalable across different networks as well as devices. Estimated model is been developed and verified for the throughput analysis on different FPGA devices.

REFERENCES

[1] H. Alemdar, N. Caldwell, V. Leroy, A. Prost-Boucle, and F. Pétrot, "Ternary neural networks for resource-efficient ai applications," *CoRR*, vol. abs/1609.00222, 2016.

[2] P. Gysel, "Ristretto: Hardware-oriented approximation of convolutional neural networks," *CoRR*, vol. abs/1605.06402, 2016.

[3] M. Tanomoto, S. Takamaeda-Yamazaki, J. Yao, and Y. Nakashima, "A cgra-based approach for accelerating convolutional neural networks," in *2015 IEEE 9th International Symposium on Embedded Multicore/Many-core Systems-on-Chip*, Sept 2015, pp. 73–80.

[4] N. P. Jouppi, C. Young, and et al, "In-datacenter performance analysis of a tensor processing unit," *CoRR*, vol. abs/1704.04760, 2017.

[5] C. Farabet, B. Martini, P. Akselrod, S. Talay, Y. LeCun, and E. Culurciello, "Hardware accelerated convolutional neural networks for synthetic vision systems," in *Proceedings of 2010 IEEE International Symposium on Circuits and Systems*, May 2010, pp. 257–260.

[6] V. Gokhale, J. Jin, A. Dundar, B. Martini, and E. Culurciello, "A 240 g-ops/s mobile coprocessor for deep neural networks," in *Proceedings of the IEEE Conference on Computer Vision and Pattern Recognition Workshops*, 2014, pp. 682–687.

[7] K. Ovtcharov, O. Ruwase, J.-Y. Kim, J. Fowers, K. Strauss, and E. Chung, "Accelerating deep convolutional neural networks using specialized hardware," February 2015.

[8] D. Shin, J. Lee, J. Lee, and H. J. Yoo, "14.2 dnpu: An 8.1tops/w reconfigurable cnn-rnn processor for general-purpose deep neural networks," in *2017 IEEE International Solid-State Circuits Conference (ISSCC)*, Feb 2017, pp. 240–241.

[9] F. Li and B. Liu, "Ternary weight networks," *CoRR*, vol. abs/1605.04711, 2016.

978-1-5386-3693-0/18 $31.00 © 2018 IEEE

2018 31th International Conference on VLSI Design and 2018 17th International Conference on Embedded Systems

SHIRT (Self Healing Intelligent Real Time) Scheduling for Secure Embedded Task Processing

Krishnendu Guha[1], Sangeet Saha[2], Amlan Chakrabarti[3]

A. K. Choudhury School of Information Technology

University of Calcutta, Kolkata, India

kgchem_rs@caluniv.ac.in[1], ssakc_rs@caluniv.ac.in[2], acakcs@caluniv.ac.in[3]

Abstract—Scheduling in FPGAs are increasingly being employed in modern real-time embedded systems, which often impose strict timeliness constraints. Deploying such a schedule requires reconfiguration at various time instants with soft IPs procured from various vendors. However, performance degradation may be associated with a compromised soft IP from an untrustworthy vendor. The present work aims at tackling such a problem with a self aware approach. The security module checks the course of the proceedings at each intermittent point of a schedule and on detecting a malicious environment, heals the scenario and takes precautions to prevent similar malfunctions in future, without hampering the pre-determined schedule. Experimental validation shows effectiveness of our proposed methodology.

Keywords—Real-time Scheduling, Hardware Trojan, FPGA

I. INTRODUCTION

The property of quick reconfiguration of Field Programmable Gate Arrays (FPGAs) at runtime is being utilized in arenas ranging from avionics and automotive systems to nuclear reactors. Correctness of the system depends both on the logical results and the time of result generation, as output generation after deadline may be catastrophic. Soft IPs (Intellectual Properties) or bitstreams are basically HDL (Hardware Description Language) codes, which are procured from various third party IP (3PIP) vendors and stored in a resource pool [1]. Unreliable 3PIP vendors may implant Hardware Trojan Horses (HTHs) in the supplied IPs. HTHs in such soft IPs generally comprise a few additional lines of malicious codes or instructions which remain dormant during testing and get activated at runtime to disrupt normal functionality. Malfunctions may vary but in this work we focus only on delay inducing HTHs [2] which can jeopardize the scheduling and hinder real time tasks to complete their execution within the deadline.

A self aware module works autonomously based on the Observe-Decide-Act (ODA) paradigm [3]. However, no self aware methodologies exist which can detect scheduling strategies getting jeopardized and simultaneously heal the scenario to prevent catastrophic situations. In this work, we focus on "Deadline Partitioning Schedular for Fully Reconfigurable Systems (DPSFR)" [4], a real time scheduling technique for fully reconfigurable FPGA. We propose a self-healing, hybrid offline-online scheduling approach, coined as "Self Healing Intelligent Real Time (SHIRT) Scheduling". The main contributions are:

1. Analyzing the scenario where a pre-determined scheduling strategy could be jeopardized at runtime due to the effect of delay inducing Hardware Trojans.

2. Proposing a self aware security approach which possesses the ability to mitigate the malicious scenario at runtime.

II. ANALYSIS OF THREAT

A. Normal Scenario

We assume that five tasks T_1, T_2, ..., T_5, having weights ($\frac{e_i}{p_i}$) 14/60, 36/90, 42/60, 72/90 and 54/90, are running on a four tiled ($V_1, ..., V_4$) fully reconfigurable FPGA. We represent the real-time constraint as: "T_i must compute e_i number of instructions, within p_i time units". Assumed 1 instruction is completed in 1 time unit using standard FPGA clock frequency. Using DPSFR [4], the task shares to be executed within the time interval $ts_1 = 60$ time units are: $shr_1 = 14$; $shr_2 = 24$; $shr_3 = 42$ and $shr_4 = shr_5 = 36$ and we obtain the schedule as illustrated in Figure 1(a). Due to the inherent architectural constraint of synchronous reconfiguration of all tiles, slack times may remain within each time-frame (TF) for a tile V_i.

B. Threat Scenario

We consider the FPGA device to be trusted but take into account the trustworthiness of the soft IPs procured from various 3PIP vendors. An adversary may implant a HTH in the supplied soft IP or bitstream. Its presence cannot be detected due to its dormant nature during the testing period or its initial phases of operation. [5]. In the present context, objective of the adversary is to affect the scheduling strategy at runtime. Simple delay inducing HTHs is sufficient to hamper a schedule at runtime. Detecting them is challenging as they neither generate erroneous result, nor destroy the confidentiality of the system by leaking secret information. Existing mechanisms mostly focus on these two factors and dealing with delay threats is still in its infancy.

Considering the normal scenario, an IP allocated for execution of T_4 must complete 14 instructions in the first time-frame, assuming normal clock frequency, f. An HTH may prevent the execution of any further operation beyond 10 instructions and this will restrict T_4 to reach a desired output state. The threat scenario is depicted in Figure 1(b).

III. PROPOSED SECURITY METHODOLOGY

SHIRT comprises an offline and an online phase. In SHIRT-offline, aid of DPSFR [4] is sought to generate time slice wise offline schedule of all periodic tasks. SHIRT online comprises of three subphases, i.e. observe, decide and act, which facilitates our proposed methodology to be a self aware one. The mechanism is illustrated in Algorithm 1.

For the illustrated threat scenario, on detection of vulnerability for task T_4 during the full reconfiguration time of second time frame, the clock period of the clock is enhanced using the Dynamic Clock Management (DCM) for that portion of the FPGA which would be executing the soft IP for task T_4. The malfunctioning soft IP would also be replaced by another trusted soft IP for task T_4. Thus, T_4 completes 18 instructions instead of 14 in the second time-frame and aligns normality as shown in Figure 1(c).

978-1-5386-3693-0/18 $31.00 © 2018 IEEE

463

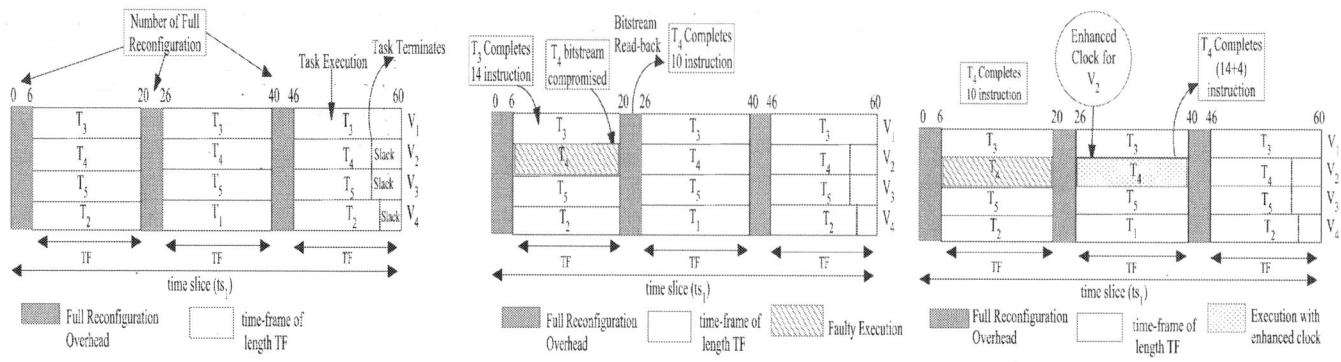

(a) Task Scheduling under DPSFR   (b) Task Scheduling under Attack   (c) Task Scheduling under Mitigation

Fig. 1: Task Scheduling under Different Scenarios

Algorithm 1: SHIRT Scheduling

Input: Given: n number of tasks and more than one soft IP for executing the same task; m number of tiles

Output: Detection and mitigation of vulnerabilty in schedule

Generate time slice wise fixed schedule of tasks in **Offline**

for *each time slice boundary,* **Online do**

     Select trusted bitstreams or soft IPs from the resource pool and set them for operation

     for *each end point of a time-frame* **do**

         /***Observe Phase***/

         Evaluate Detoriation Factor: $DF_k = I_i - I'_i$;

         for each tile/ task of that time-frame, where I_i and I'_i denotes the number of actual instructions and executed instructions respectively at any i^{th} time frame (having length TF_i).

         /***Decide and Act Phase***/

         if $DF_i == 0$ *for each task belongs to that time-frame* **then**

            execute next time-frame normaly;

         else

            Mark Soft IP as untrusted and removed from the resource pool *(for that particular task T_η for which $DF_\eta \neq 0$)*;

            Replace IP of T_η by trusted soft IP;

            Calculate $f_{enhanced} = \lceil TF_j/(DF_k + I_j) \rceil$;

            where TF_j: Length of the jth time-frame

            Execute T_η in the pre-fixed tile with $f_{enhanced}$ in next time-frame;

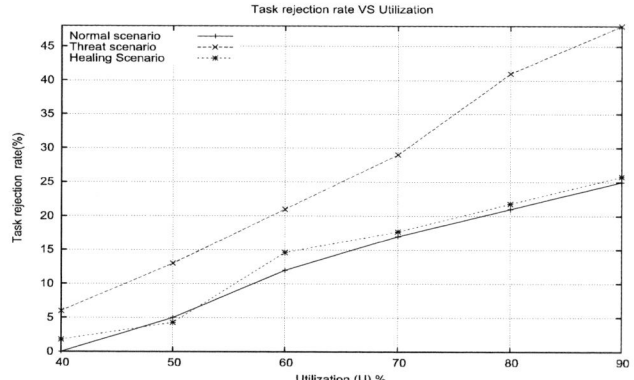

Fig. 2: TRR Vs. U; m = 4.

ever, as SHIRT healing mechanism operates, TRR becomes quantitatively comparable with the normal scenario.

V. CONCLUSION

Proposed SHIRT technique works on the ODA paradigm and not only self-heals in the attack scenario at runtime but also prevents such occurrence in future. Validation is performed via simulation over standard benchmarks as task sets. ∎

ACKNOWLEDGMENT

This work is supported in part by DST, Govt. of India, INSPIRE Fellowship No. 150916 granted to K.Guha, TCS Research Fellowship award granted to S.Saha and facilities of SMDP-C2SD Programme, University of Calcutta.

REFERENCES

[1] C. Liu, J. Rajendran, C. Yang, R. Karri, "Shielding heterogeneous mpsocs from untrustworthy 3PIPs through securitydriven task scheduling," *IEEE TETC*, vol. 2, no. 4, pp. 461-472.

[2] K. Guha, D. Saha, and A. Chakrabarti, "Self Aware SoC security to Counteract Delay Inducing Hardware Trojans at Runtime," in *VLSID 2017*, pp. 417-422.

[3] S. Sarma, N. Dutt, P. Gupta, N. Venkat., and A. Nicolau, "Cyberphysical-System-on-Chip (CPSoC): A self-aware mpsoc paradigm with cross-layer virtual sensing and actuation," in *DATE 2015*, pp. 625-628.

[4] S.Saha, A.Sarkar, A.Chakrabarti, "Scheduling dynamic hard real-time task sets on fully and partially reconfigurable platforms," *IEEE ESL*, vol.7, no.1, pp.23-26, 2015.

[5] S. Bhunia, M. S. Hsiao, M. Banga, S. Narasimhan, "Hardware Trojan Attacks: Threat analysis and Countermeasures," *Proc. of the IEEE*, vol. 102, no. 8, pp. 12291247, Aug 2014.

IV. EXPERIMENTS AND RESULTS

We target Virtex 4 as our FPGA platform and ITC99 benchmarks are used for constructing the hardware task sets. The attack scenario is generated by inserting some malicious codes in the HDL logic of some of the ITC 99 standard codes, and then synthesizing the bitstreams. The principal metric based on which simulation based evaluation has been carried out is *Task Rejection Rate (TRR)*, which is defined as the percentage of the total number of tasks rejected by the admission controller of the system over the entire schedule length (10000 time slots in our experiments). The avg. fault occurrence rate (λ) is assumed as 0.02.

As evident from the results of Figure 2, TRR increases in the attack scenario as the presence of vulnerabilities prevent a task from completing its required set of instructions. How-

2018 31th International Conference on VLSI Design and 2018 17th International Conference on Embedded Systems

Exploration of Loop Unroll Factors in High Level Synthesis

Preeti Ranjan Panda, Namita Sharma, Srikanth Kurra, Khushboo Anil Bhartia, and Neeraj Kumar Singh
Department of Computer Science and Engineering, Indian Institute of Technology Delhi, New Delhi, India

Abstract—The Loop Unrolling optimization can lead to significant performance improvements in High Level Synthesis (HLS), but can adversely affect controller and datapath delays. Unrolling causes additional operations to be scheduled, possibly leading to increased resource sharing, thereby increasing the MUX sizes and delay. We use an estimation based approach to compute the best unroll factor for loops in a behavioral specification by combining a prediction of controller and datapath delays with search space pruning methods. Our exploration strategy gives highly accurate unroll factor results significantly faster than an exhaustive synthesis procedure.

I. MOTIVATION

We take a closer look at the *Loop Unrolling* transformation in the context of HLS. In unrolling, copies of the loop body are merged in order to enable the discovery of parallelism across loop iterations. This can cause the synthesized controller to be more complex with a longer critical path. The datapath may have larger multiplexers at the function unit (FU) inputs because of additional operations leading to greater sharing. The HLS output is shown in Figure 1. If the impact is not carefully considered, the synthesized designs could be worse or the schedules could be incorrect because of longer critical paths.

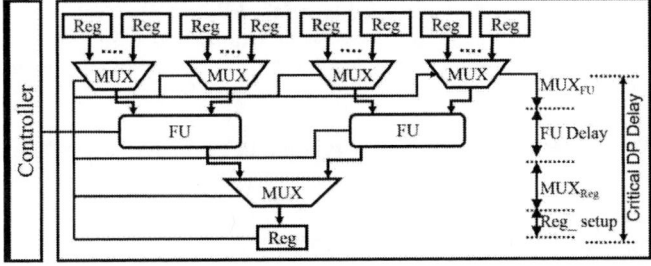

Figure 1: High level synthesis output

Kurra et al. [3] presented a technique for computing the best loop unroll factor based on controller delay estimation and a specified FSM delay bound. We extend this strategy by also accounting for datapath delay variation, and do not require any manual specification of a delay bound beyond the clock period. Finding the best loop unroll factor in HLS requires a fundamentally different analysis than in compilers [1], [2]. Loop unrolling creates a trailer loop when the unroll factor does not evenly divide the iteration count, which affects the overall latency negatively (Figure 2(a)). Figure 2(b) shows the variation of datapath delay with unroll factor for an example loop. The delay increases with increasing unroll

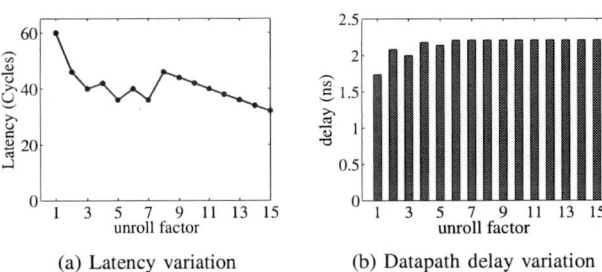

(a) Latency variation   (b) Datapath delay variation

Figure 2: Latency and datapath delay variation with unrolling

factors, but saturates at larger factors. The difference is due to the varying MUX delays; saturation occurs when paths are established through MUXes from all relevant registers and FUs providing data, to the MUXes at the inputs of the respective FUs.

II. DATAPATH DELAY ESTIMATION

Given an unroll factor, we can arrive at the critical datapath delay estimate by estimating the MUX widths at the FU and register inputs. Let the maximum resulting MUX width at the inputs of FUs of operation type t be represented by $FU\text{-}sharing_t(P)$ after running one pass of the FU resource sharing algorithm. If u_i is the unroll factor for loop l_i with iteration count b_i, and the number of type t operations in the original loop body is $LFU_t(l_i)$, then, the number of operations of type t in Program P, subject to unrolling vector U is given by [3]:

$$LUFU_t(P,U) = FU_t(P) + \sum_{i=1}^{n} (u_i - 1) \times LFU_t(l_i)$$
$$+ Even(u_i, b_i) \times LFU_t(l_i)$$

where $Even(u_i, b_i) = 0$ if $b_i \bmod u_i = 0$ and $Even(u_i, b_i) = 1$ otherwise.

The $FU\text{-}sharing_t(P)$ value increases with the unrolling factor only when all the resources of a particular operation type are exhausted. With resource sharing, the degree of multiplexing at the inputs of Type t FU (FU_t) increases by the sharing factor $\left\lceil \frac{LUFU_t(P,U)}{\#FU_t} \right\rceil$. Thus, the maximum degree of multiplexing at any FU input after unrolling is given by:

$$FU\text{-}sharing(P,U) = \max_t \left(FU\text{-}sharing_t(P) \times \left\lceil \frac{LUFU_t(P,U)}{\#FU_t} \right\rceil \right)$$

978-1-5386-3693-0/18 $31.00 © 2018 IEEE   465

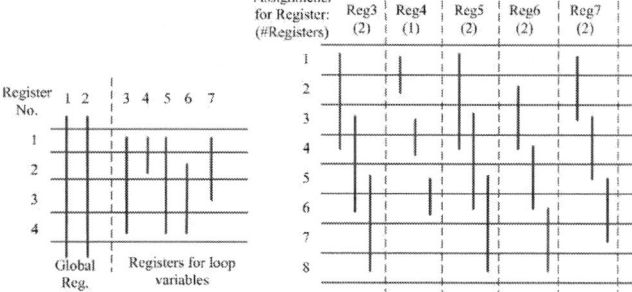

(a) Register allocation for rolled loop

(b) Maximum register requirement for each register assignment in (a)

Figure 3: Illustration for estimating maximum register count

However, the maximum MUX width at inputs of any FU is limited by the number of registers in the design. Figure 3(a) shows an example register allocation, with 2 global registers and 5 local registers. *Reg4* stores live variables for cycles 1–2, and is free during cycles 3–4. Figure 3(b) shows a possible register allocation for the unrolled loop in which new iterations are scheduled 2 cycles after the preceding iteration, corresponding to initiation interval II of 2. The total number of local registers for the kth unrolled loop is:

$$\#local\ registers_k = \sum_{r=1}^{N_k} \left\lceil \frac{L2_{k,r} - L1_{k,r} + 1}{II_k} \right\rceil$$

where N_k refers to the number of local registers in the rolled loop k, II_k is the initiation interval computed for this loop, and $L1_{k,r}$, $L2_{k,r}$, refer to the start and end of variable lifetimes mapped to physical register r in the rolled loop k.

The MUX width at the register inputs is determined by the number of paths from FU outputs to a specific register. We run one pass of the register allocation phase to determine the MUX widths with all loops rolled. Since the $FU{\rightarrow}Reg$ connections can be reused, we assume this to be the maximum MUX width. The final datapath delay is computed as the sum of the MUX delays at FU and registers, the FU delay, and the setup time. The critical path delay is estimated as the sum of the datapath and FSM delays [3]. When there are several loops, the individual unrolling decisions are not independent. In our exploration strategy we prioritize the loops using the ratio of latency reduction to critical path delay reduction. More details are given in [4].

III. Experimental Results

We implemented our exploration strategy on a behavioral synthesis framework that takes a C-style input and generates RTL-level VHDL datapath and FSM. These are then synthesized with Synopsys Design Compiler using a 90nm ASIC library. Figure 4 shows a comparison of the actual and estimated datapath and FSM delays at the explored

optimal unroll factors for two of the explored benchmarks. To account for the physical design effects in the delay estimation, an empirically derived scaling factor of 1.11 is used. Comparing the post-layout actual and estimated delays, we found an average deviation of 2.7% for FSM and 3.6% for datapath delays.

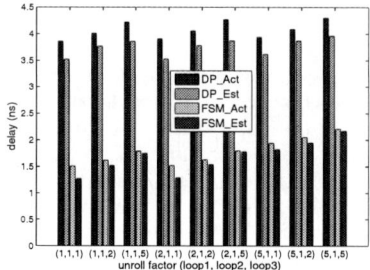

(a) Histogram

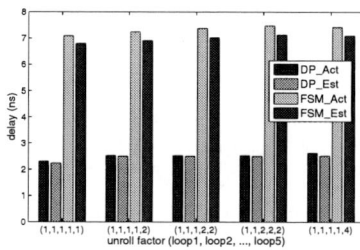

(b) FFT1K

Figure 4: Estimated and actual datapath and FSM delays

Our exploration strategy generated the selected unroll vectors in less than 0.1 seconds because we avoid explicitly unrolling and synthesizing the loops for every unroll factor candidate, and evaluate the effect of unrolling using simple estimations.

IV. Conclusions

We presented a strategy to predict the best loop unrolling factors during high level synthesis by estimating the effects on the datapath and FSM delays, without actually performing the unrolling transformation. Our experiments confirm both the accuracy of the delay predictions, as well as the quality of the unroll factors generated by our exploration.

References

[1] M. Bachir, S.-A.-A. Touati, F. Brault, D. Gregg, and A. Cohen, "Minimal unroll factor for code generation of software pipelining," *IJPP* 2013

[2] N. R. Miniskar, P. S. Gode, S. Kohli, and D. Yoo, "Function inlining and loop unrolling for loop acceleration in reconfigurable processors," *CASES* 2012

[3] S. Kurra, N. K. Singh, and P. R. Panda, "The impact of loop unrolling on controller delay in high level synthesis," *DATE* 2007.

[4] P. R. Panda, et al., "Loop unrolling in high-level synthesis – exploring unroll factors," Technical Report, IIT Delhi, 2017.

Securing Module-less Synthesis on Cyberphysical Digital Microfluidic Biochips from Malicious Intrusions

Sarit Chakraborty [1], Chandan Das [2], Susanta Chakraborty [3]

[1] Dept. of CST, IIEST-Shibpur, India, [2]Dept. of AEIE, BCREC-Durgapur, [3]Dept. of CST, IIEST-Shibpur, India
E-mail: [1]sarit.chak@ieee.org , [2]chandan.das@bcrec.ac.in, [3]sc@cs.iiests.ac.in

Abstract— **Digital Microfluidic biochips (DMFB's) lack the ability to recover from the errors incurred at assay runtime, which leads to erroneous assay results. Cyberphysical DMFB (CP-DMFB) can overcome such issues with the provision of real-time modification of its operations via interface sensors.**
In this work, we have modified the Module Less Synthesis (MLS) process for ensuring better security measures and demonstrated various attack scenarios for Modified-MLS (MMLS) method on a CP-DMFB. We have also proposed a checkpoint based novel attack detection method for MMLS technique. The effectiveness of our technique is checked with available benchmark assays and achieved results show much faster assay execution time and around 80% intrusion (error) detection rate on an average.

Keywords- Digital Microfluidics, Cyberphysical, Module-less, Synthesis, Checkpoint

I. INTRODUCTION

DMFB's have introduced a paradigm shift in healthcare technology. The biochemical assays are translated into a sequence of fluidic, thermal, and electronic operations e.g. dispensing, transport, mix/split, heating, incubation etc. This sequence is then implemented on-chip by the controller of the cyberphysical chip, which generates an actuation sequence for enabling various fluidic operations. The basic construction and a typical bioassay synthesis stages are shown in Fig. 1(a) and 1(b) respectively.

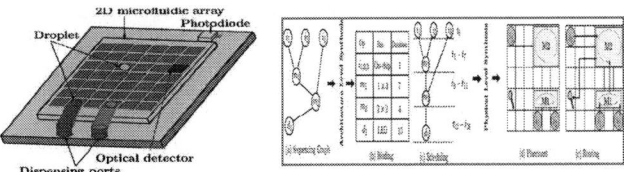

Fig. 1(a) Basic diagram of DMFB Fig. 1(b). Synthesis steps on DMFB

The assay execution is monitored through sensors or by overhead Charged Coupled Devices / CCD-cameras [1]. Different methodologies have been proposed for automating the DMFB design process as shown in Fig. 2. However, integration of cyberphysical paradigm to DMFBs can be potentially viable to malicious intrusions. Recent studies in this direction have identified several alarming backdoors in the various phases of a DMFB design flow that can be compromised by an attacker [3]. Intellectual property (IP) based CAD tools, which are usually, procured form Third Parties (3PIP) are utilized to generate actuation sequence for accomplishing the synthesis and thus may not be trusted [2]. Even during fabrication, the intrusion of malicious components (Trojans) in the foundries are quite common. A comprehensive survey of the various attack scenarios on a

CP-DFMB is illustrated in [2]. Different types of possible attacks like Denial of Service (DoS), Piracy attack, counterfeiting, reverse engineering on a DMFB etc. are discussed and threat model has been identified and microfluidic multiplexers based DMFB security approach is given in [3]. In [4], the entire supply-chain for a general purpose as well as custom DMFB design flow is considered.

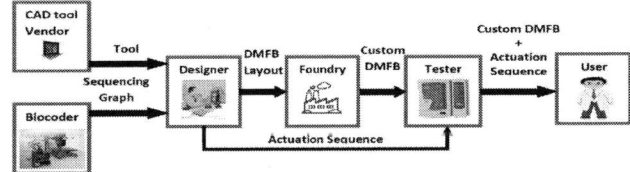

Fig. 2 An Automated Horizontal Design Flow of a Custom DMFB

In this work, we have modified the MLS process [5] for ensuring better security and introduced the concept of chain formation for homogeneous droplets. Various attack scenarios are demonstrated and checkpoint based attack detection mechanism is proposed for MMLS method.

II. PROBLEM FORMULATION

A. Motivation

Motivations for attacking a DMFB can be manifold. As the business opportunities increase, unscrupulous people or business organizations attacking DMFBs to gain illegal profits become a severe threat.

B. Routing Path Alteration Attack on MLS Process:

In MLS [5], the routing complexity increases many a fold as all mixing operations are of diffusion-based [6] mixing and are accomplished by finding various shift patterns on the chip. Thus, the possible chances of Routing Path Alteration Attack (RPAA) increase.

Scenario 1: Trojans implanted during fabrication of Biochip
Scenario 2: Trojans implanted in Routing files procured from Third Party Vendors (3PIP's)

CASE 1: As per MLS [5], a mixer droplet can take a straight run (0^0-Shift) for consecutive 3 time-steps. After that, a mandatory X-Shift has to take place. It may so happen that due to actuation sequence alteration the droplet takes straight run for more than three consecutive steps (Z_3) as shown below in Fig. 3a. We named such an attack scenario as 'Z-Shift Overrun attack".

CASE 2: For droplet 'M_j' (Fig. 3b), though space available for a straight run but M_j is unnecessarily diverted earlier and took an X-Shift (90^0-Shift). The actuation sequence has been altered in the middle of Z-Shift and such attach is considered as "Straight Run Violation attack" for MSL.

978-1-5386-3693-0/18 $31.00 © 2018 IEEE

467

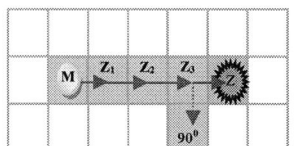
Fig. 3(a): Z-Shift Over-Run attack

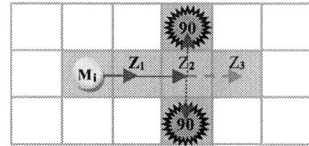
Fig. 3(b): Straight run violation attack

III. PROPOSED SECURITY METHOD

The MMLS technique ensures efficient droplet routing and better security measures by various proposed checkpoints.

A. Synthesis Process Using MMLS :

Here we have introduced the concept of chain formation by various shift-movements as shown in Fig. 4. We assume all the mixer droplets M_1, M_2 ...M_5 perform diffusion-based mixing and starts simultaneously on the 8x8 chip and out of these 5-mixer droplets M_1, M_2 and M_3 are able to form a chain on the chip shown in Red, Blue and Green color. Droplet M_4 and M_5 were not able to form the chain as there was not enough space on the chip to take two consecutive X-Shifts on 4th and 5th time step and hence they follow earlier MLS shift-movements.

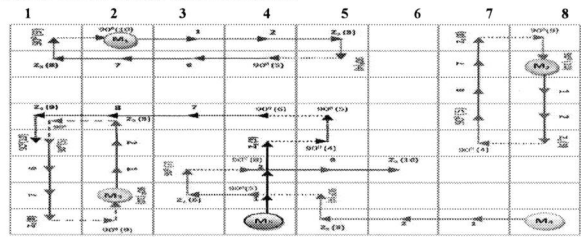

Fig. 4: Modified Module-Less Mixing Paths for M_1 to M_5 till 10th time step
*1-time step = 0.0625 Sec by considering CP DMFB working frequency f = 16Hz.

B. Checkpoint Based Attack Detection

To tackle security issues, checkpoint-based malicious droplet detection is proposed where CCD cameras are assumed to monitor the progress of an assay. The proposed arrangements of checkpoints for MMLS are as follows:

Type 1 Checkpoints: In MMLS, the mixing operations M_1, M_2, M_4, and M_5 (Fig. 4) are started at specific boundary cells of the chip and mixing-chain is formed for M_1,M_2 which reduces the mixing-space requirement as well as reduces routing complexity compared to earlier MLS process. The first type of checkpoint (static) is placed on the starting cell (Fig. 5a) from where each mixing operation is initiated and it will remain active until 100% mixing is completed. Thus, it ensures no other malicious droplet can come into the chain within that duration.

Type 2 Checkpoints: The second type of static checkpoint is placed whenever a mixer droplet takes a Z_3–shift and it has the provision to move straight (0^0-shift) further. From Fig. 4 it can be seen at (x, y) = (6, 1) coordinate position of the chip a checkpoint is placed. This checkpoint will be able to detect if any alteration happened and due to that the M_1 (RED) droplet mistakenly move to (6,1) position.

Type 3 Checkpoints: These are designated at 5th-time cycle of each mixing operations, which will detect the provision of probable chain formation as shown in Fig. 5(b) by detecting two consecutive X-shifts within the shift movements.

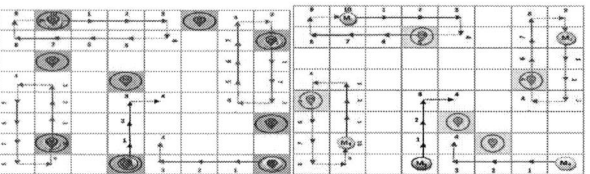
Fig. 5(a) Placement of Static Checkpoints at t = 0 & t = 10   **Fig. 5(b)** Placement of Type 3 Static Checkpoints at t = 5

IV. EXPERIMENTAL RESULTS

We simulate MMLS method on a 2.4 GHz Linux PC with 8GB RAM and randomly generate activation sequence alteration attacks. The completion time of PCR and IVD assay significantly improved compared to module-based synthesis method (Table 2). From Table 3, it is evident the arranged checkpoints are able to detect most of the errors for PCR, Example Problem, and other benchmark assays.

Table 2: Comparison of Assay Completion Time

Benchmarks	Chip Area	Bio-Assay Completion Time	
		Module Based Synthesis by Tabu Search Method [6]	Proposed MMLS Method
PCR	8X9	8.9	5.5
	7X8	13	6.25
IVD	8X9	12.5	8.54
	7X8	13.7	9.5

Table 3: Detection rate on various benchmarks

Test Benches (Chip- Size)	# of Mixing Operations (stage wise)	# of Chained Mixing Operations	# of Check points	% of error / intrusion Detection
PCR (8x9)	4 – 2 - 1	7	12	92.7
In_vitro I (16 x 16)	6	6	19	70.7
In_vitroII (14 x 14)	4 - 3 -2	9	12	71.0
Our Example (8x8)	5	3	18	85.5

V. CONCLUSION

The proposed MMLS method for cyberphysical-DMFB achieves faster assay execution. Chain formation facilitates to design the static checkpoints on specific cells (electrodes) of the chip, which ultimately helps to detect errors or Trojan intrusions if any. This work is first of its kind and further scope of improvement is there in terms of checkpoint placements to ensure full security of the chip in future.

REFERENCES

[1] Y. Luo, K. Chakrabarty, and T.-Y. Ho, "Error recovery in cyberphysical digital microfluidic biochips," IEEE Transactions on Computer-Aided Design of Integrated Circuits and Systems, vol. 32, no. 1, pp. 59–72, 2013.

[2] S. S. Ali et al., "Security assessment of cyberphysical digital microfluidic biochips," IEEE/ACM Trans. Comput. Biology Bioinform., vol. 13, no. 3, pp. 445–458, 2016.

[3] S. S. Ali, M. Ibrahim, O. Sinanoglu, K. Chakrabarty and R. Karri, "Microfluidic encryption of on-chip biochemical assays," *2016 IEEE Biomedical Circuits and Systems Conference (BioCAS)*, Shanghai, 2016, pp. 152-155

[4] S. S. Ali, M. Ibrahim, J. Rajendran, O. Sinanoglu and K. Chakrabarty, "Supply-Chain Security of Digital Microfluidic Biochips," in *Computer*, vol. 49, no. 8, pp. 36-43, Aug. 2016.

[5] S. Chakraborty and S. Chakraborty, "A Novel Approach towards Biochemical Synthesis on Cyberphysical Digital Microfluidic Biochip," *2017 30th International Conference on VLSI Design and 2017 16th International Conference on Embedded Systems (VLSID)*, Hyderabad, 2017, pp. 355-360

[6] P. Paik, V. K. Pamula, and R. B. Fair. *Rapid droplet mixers for digital microfluidic systems.*Lab on a Chip, 3:253–259, 2003.

978-1-5386-3693-0/18 $31.00 © 2018 IEEE

On the ESD Reliability issues in Carbon electronics: Graphene and Carbon Nano Tubes

Nagothu Karmel Kranthi, Abhishek Mishra, Adil Meersha And Mayank Shrivatava.

Abstract—**In this work, we present experimental investigations and new physical insights into the ESD behavior and failure of large area CVD graphene RF transistors and Multiwall carbon nanotubes. Unique two stage defect induced failure in graphene transistors is reported for the first time. Detailed study on the self-heating and its implication on the failure current and carrier transport in graphene FETs is addressed with the transient analysis in ESD time scales. A unique power law like behavior is also reported in Multi wall CNT's.**

Index Terms— **Graphene, Electrostatic Discharge, Multiwall Carbon Nano Tubes.**

INTRODUCTION

It's almost a decade since the carbon based materials like graphene and carbon nanotubes are being explored for electronic integrated circuit applications [1]. Graphene, the thinnest material reported till date, attributed to its superior electrical and thermal properties, is widely explored for RF transistor applications [2]. The other carbon allotrope, Carbon Nano Tubes (CNT's) with its extraordinary ability to carry current and immune to electro migration effects are potential candidates to replace copper interconnects in future applications. On the other hand, Electro Static Discharge (ESD) events are fundamental and major reliability threats to the IC. There were previous efforts to understand ESD behavior of graphene and CNT's, for example the very first demonstration on graphene was [3] reported using exfoliated material and then using Chemical Vapor deposition [4], more recently [5], a matured graphene technology was used for ESD explorations. The ESD physics of carbon nano tubes is also well explored [6-9]. In this work we are presenting unique failure mechanism in graphene transistors under ESD stress conditions. This work also addresses the self-heating effects in the both graphene and carbon Nano tubes that are critical under ESD like stress conditions.

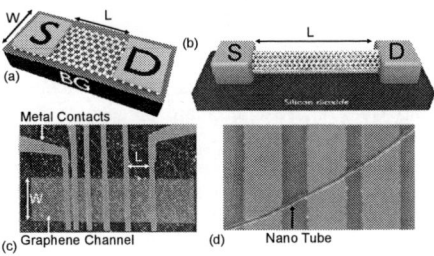

Fig.1 Atomistic view of the (a) back gated Graphene FET (b) Substrate supported Multiwall Carbon Nano Tubes. Here S&D are the source and drain metal pads. False color Scanning Electron Microscopic (SEM) image of (c) Graphene Transistors with different source to drain length (d) Carbon nanotubes.

All The authors are with Department of Electronic Systems Engineering, IISc Bangalore. This work was financially supported by Department of Science and Technology, Govt. of India, through the project grant number: SB/S3/EECE/063/2014

NEW INSIGHTS INTO THE ESD BEHAVIOUR OF GRAPHENE TRANSISTORS

Back gated GFETs (fig. 1(a) & (c)) and multiwall CNTs are stressed with commercial TLP testers. The technology details of this work can be found in [10].

1) Unique Defect Induced failure

Reported graphene failure modes are abrupt failure when channel is exposed to the ambient, gradual failure when graphene channel is encapsulated with the dielectric [8]. A unique failure signature is observed in this work along with the other reported types. Fig. 2 depicts the TLP IV characteristic of the back gated GFET with a channel length of 500 nm. The following observations can be made (a) The drain current increases linearly until the first collapse in the current (termed as first breakdown). (b) After first breakdown the drain current almost remain constant until the second or final breakdown /collapse in the current. (c) The two stage characteristic is also found in spot current data (Fig. 2(b)). The DC spot current after the first breakdown kept reducing with increase in the TLP stress. The reported two stage failure characteristics are unique. This behavior in graphene FET is attributed to the defect induced localized heating and consequent device degradation. If the graphene channel is prone to the severe defects, the localized heating near the defect causes the first failure. Increase in the TLP stress beyond this point will cause the graphene channel to degrade (no increase in the TLP current with increase in TLP voltage is observed). Increase in DC spot current during this stress range (after the first breakdown) confirms the device degradation near the defects induced failure.

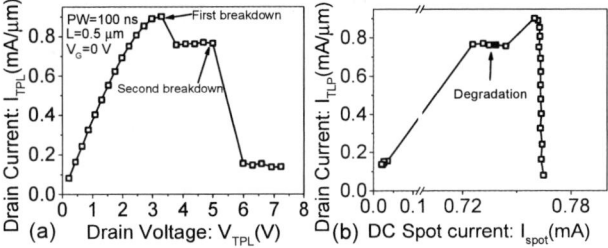

Fig. 2: (a) TLP I-V characteristics of short channel (L=0.5 µm) GFET device. (b) The corresponding DC spot measurement data. The device has shown unique two stage characteristic in both pulse I-V and spot data attributed to the defect induced localized heating and failure.

2) Self heating effect on failure current in graphene

Self-heating in graphene FET under ESD like stress conditions is not extensively discussed in the literature. In this work we use the transient voltage and current data for different stress durations with different power levels to understand the extent of self-heating and its implication on the device failure correlated with the carrier transport. Fig. 3(a) depicts the TLP IV characteristics of graphene transistors stressed with different pulse durations (25 ns-1500 ns). The following observations can be made. (a) Stressing with ultralow pulse widths (25 ns), the device current or failure current attain its maximum value, as the device self-heating is absent. (b) when

978-1-5386-3693-0/18 $31.00 © 2018 IEEE 469

increasing the pulse width, a saturating drain current characteristics are observed at sufficiently high fields (c) The onset of drain current saturation decreases with increasing pulse width. (d) The failure current scales down with increasing pulse width but the failure voltage almost remains constant. The observed drain current saturation in GFET at high enough fields, is attributed to the carrier scattering in the graphene channel with the substrate optical phonons. Though graphene has very large in plane thermal conductivity, the lower optical phonon energy of SiO_2 substrate (60 meV) dominates the optical phonon energy of graphene (180 meV). The substrate optical phonon modes cause the electron-phonon scattering in the graphene channel. As the pulse width increases electron phonon scattering increases, carrier mobility degrades

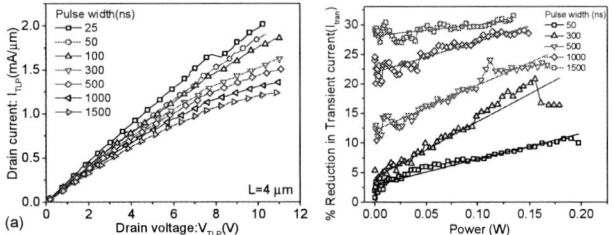

Fig.3: (a)TLP IV characteristics of back gated GFET stressed at different pulse duration ranging from 25 ns to 1500 ns (b) Percentage reduction in the transient drain current of plotted against the device power level.

Hence saturation in the drain current at sufficiently high fields, consequently reduction in the failure current. It's worth highlighting that device channel lengths used in this work are sufficiently large and are in the diffusive transport regime.

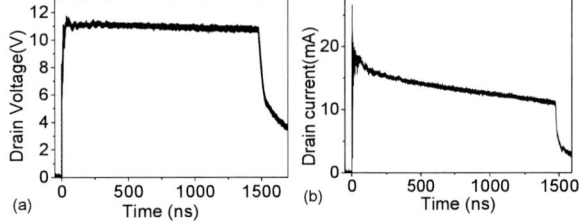

Fig.4. Transient (a) drain voltage and (b) corresponding drain current of graphene FET stressed under long pulse duration. The device channel length is 4 µm and width is 10 µm.

The severe substrate induced device self-heating can also be understand from the transient data presented in Fig. 4. At sufficiently high applied voltage, the transient current decreases with the time, a signature for self-heating. The degree of reduction in the transient current is a function of device power level and the stress duration. Fig.3(b) depicts the   percentage reduction in the drain current as a function of stress duration and device power level. Increasing the device power causes more device heating, reduction in the drain current for a fixed pulse width. For a fixed power level, increasing the pulse width causes more electron-phonon scattering, Increase in device self-heating.

MULTIWALL CARBON NANOTUBES

The investigations carried in this work are single multi wall tubes with different lengths. Single multi wall tubes can either be resting on the substrate or can be suspended between the two electrodes depending on the length. When stresses under ESD conditions, both the inner shells and outer shells contribute to the current in multiwall CNTs, however during the low bias DC measurements the conduction only happens through the outermost shell. The spot measurements keep track of the outer most shell resistance.

Fig.5 depicts the TLP IV characteristics and the corresponding DC spot data of the substrate supported MWCNT with tube length of 1 µm are stressed with different pulse widths. It is found that (a) Under ultra-low pulse width conditions (25 ns) the device doesn't experience the exponential raise in the current before the onset of device failure. (b) The failure current and voltage scales with the increasing pulse width (c) The power law like characteristics can be observed in MCNTS. The exponential raise in MCNTS is because of the phonon assisted increased conduction channels and the increased band to band tunneling in the individual shells [12]. The missing exponential raise in the CNT current when stress under ultralow pulse width (25 ns) conditions is attributed to reduced phonon population from hot contact to the cold contact and hence no additional conduction channels formed during the ESD discharge. Increasing pulse width has shown the exponential raise clearly before the onset of device failure. The scaling of failure current and failure voltage and as a result the observed power law behavior is attributed to the thermal failure nature of the Carbon Nanotubes. Finally, the Scanning Electron Microscopic (SEM) images, after the ESD induced damages are presented in fig.6.

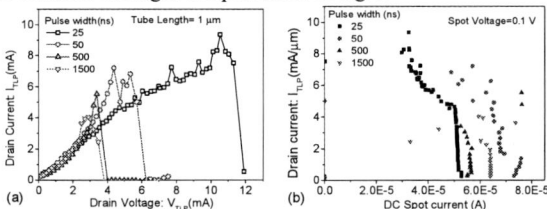

Fig.5 (a) The TLP IV characteristics of Multiwall Carbon nanotube with same tube length of 1 µm stressed with under different pulse durations. The observations are (i) The exponential increase in the tube current in the pre breakdown regime is absent under ultralow pulse widths. (ii) the failure current and failure voltage scales with increasing pulse duration and they follow power law like characteristics. (b) corresponding DC spot current data. The small variation in DC spot current for different devices is attributed to difference in the process induced charge surfactants in the outermost shell.

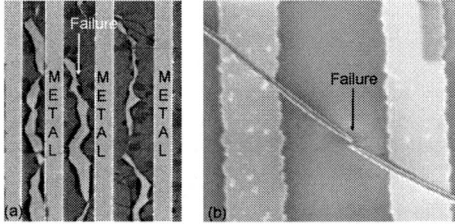

Fig .6. SEM image of (a) Graphene Channel with different channel length (b) CNT after the depicting the ESD induced damages.

CONCLUSION

A detailed physical insight into the ESD behavior of graphene transistors and CNT interconnects are presented. The unique two step failure in Graphene FETs, is because of the defect induced localized heating and further device degradation near the defects before the eventual failure. The self-heating in graphene devices studied through the transient analysis, found to reduce the drain current by almost 30%. The severe self-heating in graphene is attributed to the low energy substrate induced optical phonons scattering with the channel carriers. The reduction in the drain current because of self-heating is a function of pulse width and also the device power level. Finally, the investigations carried out on Multiwall CNTs have shown unique power law behavior which is attributed to the thermal failure nature of CNTs.

References: [1] Frank Schwierz, Nature Nano Technology, 2010. [2] G. Fiori, et.al, IEDM, 2012. [3] H. Li, et.al, TED,2014. [4] Q. Chen et al, TED,2016. [5] N. K. Kranthi, et.al, IRPS,2017. [6] M. Shrivastava, et.al, TDMR,2014. [7] M. Shrivastava, et.al, EOSESD,2014. [8] A. Mishra, et.al, IRPS,2016. [9] A. Mishra,et.al, EOSESD,2016 [10] Adil Meersha, et.al, IEDM, 2016.

978-1-5386-3693-0/18 $31.00 © 2018 IEEE

Author Index

Adimulam, Mahesh Kumar	19
Adusumalli, Ravi Kumar	325
Agarwal, Sukarn	457
Agarwal, Utkarsh	295
Ahamed, Shaik Rafi	283
Ahmad, Imtiaz	428
Ahmed, Alif	91
Aketi, Sai Aparna	250
Alam, Mahabubul	85
Alfailakawi, Mohammad Gh.	428
Ansari, Md. Hasan Raza	422
Azad, Chandrashekhar	295
Baghini, Maryam Shojaei	198, 220
Bahubalindruni, Pydi	319
Bajaj, Ronak	362
Banerjee, Sabyasachee	103
Bansal, Nitin	341, 353
Basu, Kanad	410
Beferull-Lozano, Baltasar	455
Bhaaskaran, V. S. Kanchana	149
Bhargava, Lava	79
Bhartia, Khushboo Anil	465
Bhasin, Shivam	155
Bhat, Sharath N.	13
Bhattacharjee, Debjyoti	439
Bhattacharya, Bhargab B.	103
Bhattacharya, Sarani	155
Bhattacharyya, Tarun Kanti	208
Bonizzoni, Edoardo	214
Cañedo, Janice	85
Cenkeramaddi, Linga Reddy	204, 455
Chakrabarti, Amlan	463
Chakrabarty, Krishnendu	121, 244
Chakraborty, Arpan	127
Chakraborty, Sarit	467
Chakraborty, Susanta	467
Chakravarty, Richa	139
Chang, Gregory	329
Chapagai, Kamal	319
Chatterjee, Baibhab	329
Chattopadhyay, Anupam	362, 398, 439
Chattopadhyay, Biman	13
Chauhan, Rajat	171, 347
Choudhury, Avishek	115
Chowdhury, Amrita Roy	386
Chunduri, Rama Mohan	37
Costa, Antonio Anastasio Bruto da	404
Cruz, Jonathan	91
Curtinhas, Tiago	451
Dalmia, Preyesh	289
Das, Abhijit	374
Das, Chandan	467
Das, Palash	380
Das, Shirshendu	31
Dasgupta, Pallab	37, 404
Datar, Mandar	368
Datta, Piyali	127
Datta, Pranoy	198
Deharia, Mukesh	177
Delsoto, Higor A.	451
Dey, Vishal	398
Dhar, Anindya Sundar	186
Dharade, Shriya	404
Dilip, Y.	55
Drechsler, Rolf	121
Dutt, Arko	439
Dutta, Ashudeb	335
Elghandour, Sarah	428
Elovici, Yuval	398
Farahmandi, Farimah	91
Faria, Lester A.	451
Fujita, Masahiro	410
Furth, Paul M.	232
G, SriHarsa Vardan	335
Gadiyar, Manasa	167
Garg, Bharat	73
Garimella, Annajirao	232
Ghosh, Narendra Nath	335
Ghosh, Sourav	453
Giri, Chandan	109, 453
Goel, Tarun	392
Gopinath, Anjali	325
Guha, Krishnendu	463
Guin, Ujjwal	85
Gupta, Hari Shanker	198
Gupta, Manish	133

Author Index

Gupta, Rahul	341	Kurra, Srikanth	465
Gupta, Shalabh	177	Lad, Kiran Kumar	167
Halliday, David M.	49	Lakhotia, Shivam	380
Hamid, Mohamed	455	Laxminidhi, Tonse	167, 347
Hanumolu, Pavan Kumar	226	Lenka, Prakash Kumar	335
Harkin, Jim	49	Lin, Jyi-Tsong	422
Hazra, Aritra	37	Liu, Junxiu	49
Ibrahim, Mohamed	121	Liu, Xiaobang	416
Jain, Nupur	61	Loetveit, Svein Erik	455
Jalasuthram, Maheedhar	97	M., Mohamed Asan Basiri	271
James, Diego	7	M., Prabhu Prasad B.	67
Jin, Shi	244	M.H., Vasantha	214
Jindal, Ankit	410	Mahapatra, Ipsita Biswas	295
Johnson, Anju P.	49	Mahapatra, Santanu	139
Johnsrud, Thor Eirik	455	Maheshwari, Pragya	177
Jose, John	374	Maity, Dilip Kumar	109
K., Sushmitha Din	167	Maity, Shovan	329
Kale, Kimaya	459	Majumder, Subhashis	103
Kapoor, Amit	19	Mandal, Sudipa	37, 404
Kapoor, Hemangee K.	31, 43, 380, 457	Mazumder, Pinaki	192
Karim, Shvan	49	McDaid, Liam	49
Karmakar, Abhijit	392	Meersha, Adil	469
Kawoosa, Mudasir S.	97	Mehendale, Mahesh	171
Keszocze, Oliver	121	Mehta, Ravi	13
Khan, Mohd. Tasleem	283	Mekie, Joycee	250, 459
Khan, Qadeer A.	226	Mendes, Felipe	451
Kim, Seong-Joong	226	Menezes, Vinod	171
Kjellby, Rolf Arne	455	Millard, Alan G.	49
Kranthi, Nagothu Karmel	469	Mishra, Abhishek	469
Kranti, Abhinav	133, 422	Mishra, Biswajit	61
Kudithipudi, Dhireesha	161	Mishra, Prabhat	91
Kulkarni, Uma Mukund	1	Mittal, Priyanka	301
Kumar, A. S. Kiran	198	Mittal, Rajesh K.	97
Kumar, Akash	307	Mody, Mihir	461
Kumar, Ashish	261	Moulik, Sanjay	43
Kumar, Ashwani	434	Mukhopadhyay, Debdeep	155
Kumar, Binod	410	Mukim, Prashansa	459
Kumar, Sudhir	266	Murty, Mahesh S.	277
Kumar, Vinay	447	Nambath, Nandakumar	177
Kumar, Vinay B. Y.	368	Nandy, S. K.	295
Kumar, Vulisi Narendra	449	Nath, Arijit	31
Kuncham, Sucheth S.	167	Navlakha, Nupur	422
Kunnath, Abishek T.	7	Nayak, Gopalkrishna	13

Author Index

Nishtha	319	Sankaranarayanan, Sowmya	347
Oliveira, Duarte L.	451	Sarkar, Arnab	43
P., Bhuvana B.	149	Sasikanth, Mannem Naga	208
P., Vijayanand	55	Saurabh, Sneh	301
Padaliya, Madhav Mansukh	447	Selvam, Ravikumar	143
Pal, Rajat Kumar	127	Sen, Shreyas	329
Palchaudhuri, Ayan	186	Sen, Subhajit	1
Panda, Gayadhar	449	Shah, Deval	368
Panda, Preeti Ranjan	465	Shah, Hemal	250
Pandey, Jai Gopal	392	Sharma, Dinesh K.	198
Pandey, Neeta	289	Sharma, G. K.	73
Panty, Vibha	461	Sharma, Madhvi	353
Parane, Khyamling	67	Sharma, Namita	465
Parashar, Abhinav	289	Shetty, Prabodh	380
Parekhji, Rubin A.	97	Shrestha, Rahul	277
Parikh, Chetan	1	Shrivastava, Mayank	469
Pasricha, Sudeep	238	Shrivatava, Ravindra kumar	447
Patel, Subhash J.	445	Shukla, Ankur	256
Patkar, Sachin B.	368	Shukla, Hemant	353
Priya, Sharma	457	Siddiqui, M. Sultan M.	266
Pudi, Vikramkumar	398	Sikdar, Biplab K.	115
Purnaprajna, Madhura	461	Singh, Adit	85
Purushothaman, A.	7	Singh, Mansi	61
Pyne, Sumanta	25	Singh, Neeraj Kumar	465
R., Manikandan R.	171	Singh, Saurabh Kumar	353
Rahaman, Hafizur	313, 453	Singhal, Vipul	171, 347
Ramakrishna, Vaikuntapu	79	Sinharay, Arindam	313
Ramanathan, Parameswaran	386	Sivanesan, Mayuran	362
Rana, Vikas	181	Sk, Noor Mahammad	271
Ranganathan, Rohit	325	Skjellum, Anthony	85
Rao, Rahul M.	256, 358	Sreekumar, Aswanth	347
Rathor, Vijaypal Singh	73	Srinivas, M.B.	19
Roy, Pranab	313	Srivastav, Sumit	266
Roy, Surajit Kumar	109	Srivastava, Abhishek	220
S., Arya	325	Suri, Manan	434
S., Dharshak B.	358	Surkanti, Punith R.	232
Sadhu, Pavan Kumar	177	Suthar, Manankumar	266
Saha, Dipankar	139	Talawar, Basavaraj	67
Saha, Sangeet	463	Thakker, Rajesh A.	445
Sahay, Shubham	434	Thesing, James	161
Sahoo, Bibhu Datta	7	Timmis, Jon	49
Sahoo, Siva Satyendra	307	Tomar, Akshi	289
Sahula, Vineet	79	Tyagi, Akhilesh	143

Author Index

Tyrrell, Andy M. 49

Umdekar, Alankar V. 31

Unni, Ravi Krishnan... 55

Vatwani, Tarun.. 439

Veeramachaneni, Sreehari.................................. 19

Veeravalli, Bharadwaj.. 307

Veerendranath, P. S. 214

Vemuri, Ranga.. 416

Vikas... 289

Vikram, Sahu Sai.. 461

Vinod, Kulkarni Chaitali..................................... 347

Visweswaran, G.S. 261

Wanjul, Dattatray Ramrao.................................. 266

Warnock, James D. 256

Wille, Robert.. 121

Y.B., Nithin Kumar.. 214

Zheng, Nan.. 192

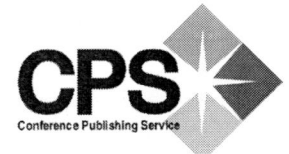

IEEE Computer Society
Technical & Conference
Activities Board

T&C Board Vice President
Hausi Müller
University of Victoria, Canada

IEEE Computer Society Staff
Evan Butterfield, *Director of Products and Services*
Patrick Kellenberger, *Manager, Conference Publishing Services*

IEEE Computer Society Publications

The world-renowned IEEE Computer Society publishes, promotes, and distributes a wide variety of authoritative computer science and engineering texts. These books are available from most retail outlets. Visit the CS Store at *http://www.computer.org/portal/site/store/index.jsp* for a list of products.

IEEE Computer Society *Conference Publishing Services* (CPS)

The IEEE Computer Society produces conference publications for more than 300 acclaimed international conferences each year in a variety of formats, including books, CD-ROMs, USB Drives, and on-line publications. For information about the IEEE Computer Society's *Conference Publishing Services* (CPS), please e-mail: cps@computer.org or telephone +1-714-821-8380. Fax +1-714-761-1784. Additional information about *Conference Publishing Services* (CPS) can be accessed from our web site at: *http://www.computer.org/cps*

Revised: 18 January 2012

CPS Online is our innovative online collaborative conference publishing system designed to speed the delivery of price quotations and provide conferences with real-time access to all of a project's publication materials during production, including the final papers. The ***CPS Online*** workspace gives a conference the opportunity to upload files through any Web browser, check status and scheduling on their project, make changes to the Table of Contents and Front Matter, approve editorial changes and proofs, and communicate with their CPS editor through discussion forums, chat tools, commenting tools and e-mail.

The following is the URL link to the ***CPS Online*** Publishing Inquiry Form:
http://www.computer.org/portal/web/cscps/quote

IEEE
445 Hoes Lane
Piscataway, NJ 08854-4141

ISBN 978-1-5386-3693-0